81 6509431 3
TELEPEN
DILEWYD O STOC
WITHDRAWN FROM STOCK
LLYFRGELL
LIBRARY
UNIVERSITY COLLEGE OF
SCHOOL OF
AF616119

FLAVINS AND FLAVOPROTEINS

DEVELOPMENTS IN BIOCHEMISTRY

Volume 1—Mechanisms of Oxidizing Enzymes, edited by Thomas P. Singer and Raul N. Ondarza, 1978

Volume 2—Electrophoresis '78, edited by Nicholas Catsimpoolas, 1978

Volume 3– Physical Aspects of Protein Interactions, edited by Nicholas Catsimpoolas, 1978

Volume 4—Chemistry and Biology of Pteridines, edited by Roy L. Kisliuk and Gene M. Brown, 1979

Volume 5—Cytochrome Oxidase, edited by Tsoo E. King, Yutaka Orii, Britton Chance and Kazuo Okunuki, 1979

Volume 6—Drug Action and Design: Mechanism-Based Enzyme Inhibitors, edited by Thomas I. Kalman, 1979

Volume 7—Electrofocus/78, edited by Herman Haglund, John G. Westerfeld and Jack T. Ball, Jr., 1979

Volume 8—The Regulation of Coagulation, edited by Kenneth G. Mann and Fletcher B. Taylor, Jr., 1980

Volume 9—Red Blood Cell and Lens Metabolism, edited by Satish K. Srivastava, 1980

Volume 10—Frontiers in Protein Chemistry, edited by Teh-Yung Liu, Gunji Mamiya and Kerry T. Yasunobu, 1980

Volume 11—Chemical and Biochemical Aspects of Superoxide and Superoxide Dismutase, edited by J.V. Bannister and H.A.O. Hill, 1980

Volume 12—Biological and Clinical Aspects of Superoxide and Superoxide Dismutase, edited by W.H. Bannister and J.V. Bannister, 1980

Volume 13—Biochemistry, Biophysics and Regulation of Cytochrome p-450, edited by Jan-Ake Gustafsson, Jan Carlstedt-Duke, Agneta Mode and Joseph Rafter, 1980

Volume 14—Calcium-Binding Proteins: Structure and Function, edited by Frank L. Siegel, Ernesto Carafoli, Robert H. Kretsinger, David H. MacLennan and Robert H. Wasserman, 1980

Volume 15—Gene Families of Collagen and Other Proteins, edited by Darwin J. Prockop and Pamela C. Champe, 1980

Volume 16—Biochemical and Medical Aspects of Tryptophan Metabolism, edited by Hayaishi, Ishimur, and Kido, 1980

Volume 17—Chemical Synthesis and Sequencing of Peptides and Proteins, edited by Teh-Yung Liu, Alan N. Schechter, Robert L. Heinrickson, and Peter G. Condliffe, 1981

Volume 18—Metabolism and Clinical Implications of Branched Chain Amino Acids, edited by Mackenzie Walser and John Williamson, 1981

Volume 19—Molecular Basis of Drug Action, edited by Thomas P. Singer and Raul N. Ondarza, 1981

Volume 20—Energy Coupling in Photosynthesis, edited by Bruce R. Selman and Susanne Selman-Reimer, 1981

Volume 21—Flavins and Flavoproteins, edited by Vincent Massey and Charles H. Williams, 1982

FLAVINS AND FLAVOPROTEINS

Proceedings of the Seventh International Symposium on Flavins and Flavoproteins

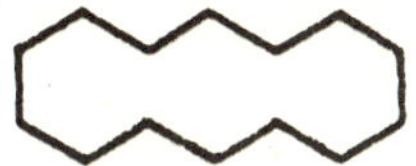

Ann Arbor, Michigan, June 21–26, 1981

Editors:
VINCENT MASSEY, Ph. D.
Department of Biological Chemistry
University of Michigan
Ann Arbor, Michigan

CHARLES H. WILLIAMS, JR, Ph.D.
Veterans Administration Medical Center
and Department of Biological Chemistry
University of Michigan
Ann Arbor, Michigan

ELSEVIER/NORTH-HOLLAND
NEW YORK • AMSTERDAM • OXFORD

©1982 by Elsevier North Holland, Inc.
All rights reserved.

Published by:
Elsevier North Holland, Inc.
52 Vanderbilt Avenue, New York, New York 10017

Sole Distributors outside the USA and Canada:
Elsevier Science Publishers B.V.
P.O. Box 211, 1000Ae
Amsterdam, The Netherlands

Library of Congress Cataloging in Publication Data

International Symposium on Flavins and Flavoproteins
(7th: 1981: Ann Arbor, Mich.)
Flavins and flavoproteins.

(Developments in biochemistry, ISSN 0165-1714; v. 21)
Bibliography: p.
Includes index.
1. Flavins—Congresses. 2. Flavoproteins—Congresses. I. Massey, Vincent. II. Williams, Charles H. (Charles Herbert), 1935- . III. Title. IV. Series.
[DNLM: 1. Flavins—Congresses. 2. Flavoproteins—Congresses. W1 DE997VG v.21 / QU 110 I62 1981f]
QP671.F5I57 1981 574.19'258 81-17280
ISBN 0-444-00672-9 AACR2

Manufactured in the United States of America

Contents

Acknowledgments

Organizing Committee:

V. Massey and C.H. Williams, Jr. (Ann Arbor) *Cochairmen*; H. Beinert (Madison); R. Bray (Brighton); H. Bright (Philadelphia); B. Curti (Milano); K. Decker (Freiburg); P. Hemmerich (Konstanz); H. Kamin (Durham); F. Labeyrie (Gif-sur-Yvette); T. Nakamura (Osaka); T. Singer (San Francisco); C. Veeger (Wageningen); C. Walsh (Cambridge, Mass.); K. Yagi (Nagoya); T. Yamano (Osaka).

Sponsors:

We wish to acknowledge with gratitude, financial support from:

The National Institutes of Health:
  Fogarty International Center
  National Institute of Mental Health
  National Institute of General Medical Sciences
  National Cancer Institute
The National Science Foundation
The University of Michigan
The Veterans Administration
Monsanto Co.
Merck & Co., Inc.
Burroughs Wellcome Co.
Syntex USA Inc.
Pfizer Inc.
Warner Lambert Co.

List of Participants

Robert Abeles
Waltham, USA

Brian Ackrell
San Francisco, USA

Samuel Adamson
Calgary, Canada

James Albarella
Elkhart, USA

Simon Albracht
Amsterdam, Netherlands

Robert Anderson
London, England

David Armstrong
Calgary, Canada

David Arscott
Ann Arbor, USA

June Ayling
San Antonio, USA

Adelbert Bacher
Munich, Germany

Steven Bailey
San Antonio, USA

Thomas Baldwin
Urbana, USA

David Ballou
Ann Arbor, USA

Michael Barber
Durham, USA

Chris Batie
Durham, USA

Joe Beckman
Milwaukee, USA

James Becvar
El Paso, USA

Helmut Beinert
Madison, USA

Michael Benecky
Princeton, USA

Shaun Black
Ann Arbor, USA

Robert Blake
Ann Arbor, USA

V. Boggaram
Stockholm, Sweden

Franco Bonomi
Milan, Italy

Robert Bray
Brighton, England

Thomas Bruice
Santa Barbara, USA

Christopher Bull
Ann Arbor, USA

Chantal Capiellere-Blandin
Gif-Sur-Yvette, France

Inger Carlberg
Stockholm, Sweden

Miles Choc
Ann Arbor, USA

J-D. Choi
Atlanta, USA

Jorge Churchich
Knoxville, USA

Al Claiborne
Ann Arbor, USA

Michael Clarke
Chestnut Hill, USA

Minor J. Coon
Ann Arbor, USA

Michael Coughlan
Galway, Ireland

Thomas Cromartie
Charlottesville, USA

Bruno Curti
Milan, Italy

Stanley Dagley
St. Paul, USA

Howard Dalton
Coventry, England

Colette Daubner
Ann Arbor, USA

Karl Decker
Freiburg, Germany

Arthur deKok
Wageningen, Netherlands

Daniel Dervartanian
Athens, USA

Kristina Detmer
Ann Arbor, USA

Mort Dolin
Ann Arbor, USA

John Dougherty
Urbana, USA

Jan Drenth
Groningen, Netherlands

Richard Dunham
Ann Arbor, USA

Dale Edmondson
Atlanta, USA

Dudley Eirich
Portland, USA

Paul Engel
Sheffield, England

James Fee
Ann Arbor, USA

Benjamin Feinberg
Milwaukee, USA

Gabrielle Fendrich
Waltham, USA

Helmut Fenner
W. Berlin, Germany

Jed Fisher
Minneapolis, USA

Paul Fitzpatrick
Ann Arbor, USA

Barbara Fox
Cambridge, USA

Lawrence Fox
Austin, USA

Peter Franck
W. Berlin, Germany

Verna Frasca
Ann Arbor, USA

Frank Frerman
Milwaukee, USA

Ted Gabig
Ann Arbor, USA

Robert Gennis
Urbana, USA

Sandro Ghisla
Konstanz, Germany

John Groves III
Ann Arbor, USA

Carole Hall
Atlanta, USA

Gordon Hamilton
University Park, USA

J. Woodland Hastings
Cambridge, USA

Cornelia Hentsch
W. Berlin, Germany

Jon Herriott
Seattle, USA

Russ Hille
Ann Arbor, USA

Jan Hofsteenge
New York, USA

Charles Holmes
Calgary, Canada

D.J. Hopper
Aberystwyth, Wales

Kihachiro Horiike
Shiga, Japan

Donald Hultquist
Ann Arbor, USA

Mazhar Husain
San Francisco, USA

Fred Jacobson
Cambridge, USA

Annette Jerchow
W. Berlin, Germany

Ken Jones
Ann Arbor, USA

Marilyn Jorns
Columbus, USA

Henry Kamin
Durham, USA

Pieter Kark
Los Angeles, USA

Andrew Karplus
Seattle, USA

Sabu Kasai
Osaka, Japan

Edna Kearney
San Francisco, USA

Albert Keston
New York, USA

Tokuji Kimura
Detroit, USA

Teizo Kitagawa
Osaka, Japan

Allen Krantz
Palo Alto, USA

Manfred Kurfurst
Konstanz, Germany

Francoise Labeyrie
Gif-sur-Yvette, France

David Lambeth
Atlanta, USA

Richard Lawton
Ann Arbor, USA

Florence Lederer
Gif-sur-Yvette, France

Jim Lee
Ann Arbor, USA

Gary Leisman
LaJolla, USA

Susan Lichtenberger
Yellow Springs, USA

John Likos
St. Louis, USA

Ylva Lindqvist
Uppsala, Sweden

Oksana Lockridge
Ann Arbor, USA

Robert Lorence
Ann Arbor, USA

Martha Ludwig
Ann Arbor, USA

Humphrey Mager
Delft, Netherlands

Bengt Mannervik
Stockholm, Sweden

Vincent Massey
Ann Arbor, USA

Mike Mather
Urbana, USA

R. Scott Mathews
St. Louis, USA

Kunio Matsui
Osaka, Japan

Rowena Matthews
Ann Arbor USA

Stephen Mayhew
Dublin, Ireland

Al Merrill
Atlanta, USA

Terrance Meyer
LaJolla, USA

Audrey Miller
Storrs, USA

Retsu Miura
Osaka, Japan

John Mizzer
Newark, USA

Fraser Morpeth
Ann Arbor, USA

David Morris
Elkhart, USA

Michael Morris
Ann Arbor, USA

Franz Müller
Wageningen, Netherlands

Donald McCormick
Atlanta, USA

James McFarland
Milwaukee, USA

William McIntire
San Francisco, USA

Martha McKean
Milwaukee, USA

Janos Nagy
Budapest, Hungary

Takao Nakamura
Osaka, Japan

Halina Neujahr
Stockholm, Sweden

Yasuzo Nishina
Okazaki, Japan

Takeshi Nishino
Yokohama, Japan

Mike O'Donnell
Ann Arbor, USA

Jack Omdahl
Dallas, USA

Dan Oprian
Ann Arbor, USA

Emil Pai
Heidelberg, Germany

Graham Palmer
Houston, USA

Michael Palmer
Edinburgh, Scotland

Ramesh Patel
Linden, USA

Harry Peck
Athens, USA

Julian Peterson
Dallas, USA

Justin Polowski
St. Paul, USA

Denis Pompon
Gif-sur-Yvette, France

Lawrence Poulsen
Austin, USA

Robert Presswood
Downers Grove, USA

James Profitt
Elkhart, USA

Stephen Rausch
Urbana, USA

Richard Rivlin
New York, USA

Severino Ronchi
Milan, Italy

Jonathan Ross
Ann Arbor, USA

James Salach
San Francisco, USA

Richard Sands
Ann Arbor, USA

Thor Santinga
Ann Arbor, USA

Dorothy Schafer
Ann Arbor, USA

Heiner Schirmer
Heidelberg, Germany

Jack Schmidt
Milwaukee, USA

Lawrence Schopfer
Ann Arbor, USA

Georg Schulz
Heidelberg, Germany

Jules Shafer
Ann Arbor, USA

Kiyoshi Shiga
Okazaki, Japan

Masateru Shin
Kobe, Japan

Seiji Shinkai
Nagasaki, Japan

Hirofumi Shoun
Tokyo, Japan

Lewis Siegel
Durham, USA

Richard Silverman
Evanston, USA

Royce Simondson
Tucson, USA

Thomas Singer
San Francisco, USA

Thomas Spiro
Princeton, USA

William Stallings
Ann Arbor, USA

Marian Stankovich
Minneapolis, USA

Daniel Steenkamp
San Francisco, USA

Kenneth Stevenson
Calgary, Canada

Richard Stewart
Ann Arbor, USA

V. Subramanian
Austin, USA

Kenzi Suzuki
Ishikawa, Japan

Richard Swenson
Ann Arbor, USA

Shigeki Takemori
Hiroshima, Japan

Lieselotte Tessendorf
W. Berlin, Germany

Colin Thorpe
Newark, USA

Hiromasa Tojo
Osaka, Japan

Gordon Tollin
Tucson, USA

R. Traber
Stuttgart, Germany

Daniel Trachewsky
Oklahoma City, USA

Haruhito Tsuge
Gifu, Japan

Shiao-Chun Tu
Houston, USA

Renate Untucht-Grau
Heidelberg, Germany

Cornelius Veeger
Wageningen, Netherlands

Ton Visser
Wageningen, Netherlands

Richard Von Korff
Midland, USA

Gerald Wagner
Urbana, USA

Li-ho Wang
Houston, USA

George Trainor
New York, USA

Diane Trainor
New York, USA

C.S. Tsai
Ottawa, Canada

Christopher Walsh
Cambridge, USA

Yoshitaka Watanabe
Osaka, Japan

Daniel Wellner
New York, USA

Alexandra Wenz
Konstanz, Germany

Albert Wessiak
Santa Barbara, USA

J.R. Wheeler
Edinburgh, Scotland

Carolyn Whitfield
Ann Arbor, USA

Charles Williams, Jr.
Ann Arbor, USA

Ralph Wolfe
Urbana, USA

Ron Woodard
Ann Arbor, USA

Kunio Yagi
Nagoya, Japan

Toshio Yamano
Osaka, Japan

Shigeko Yamazaki
Bethesda, USA

Tatsuro Yoshida
Ann Arbor, USA

Shinya Yoshikawa
Kobe, Japan

Toshitsuge Yubishui
Oita, Japan

Giuliana Zanetti
Milan, Italy

Albert Zeller
Evanston, USA

Miriam Ziegler
Urbana, USA

Preface

One of the characteristics of the Seventh Flavin Symposium was a major departure from the format of previous meetings. The field has grown so rapidly that we felt it was no longer possible, or even desirable, to have all participants present talks. Therefore we decided that there would be a limited number of lectures, held in the mornings, and that these should be scholarly reviews of particular topics, summarizing the most recent important developments, and discussing potentially exciting areas for future work. With ample time available, this idea succeeded very well, and for the first time in many years the audience felt sufficiently unconstrained by time to engage in meaningful and prolonged discussion.

The major experiment in format, and the one which caused us most apprehension in advance of the meeting, was that of integrated poster/discussion sessions. We decided that the main route for the transfer of new information would be by way of such sessions. Poster displays were held in the afternoons, being manned by an author for a nominal period of one hour, but as it turned out, usually for 2–3 hours. In order that the poster sessions would have optimal impact, evening sessions were held for further discussion of the posters of that particular day. To ensure that lively and informed discussion would result, each session was chaired by at least two discussion leaders. The authors of poster displays were asked to send one month in advance of the meeting an abstract and small photographic copies of graphs which they intended to use in their posters. This material was distributed promptly to the appropriate discussion leaders. This experiment was a great success, surpassing all our expectations. The success was, of course, in no small measure due to the expertise and the efforts of the discussion leaders. Our special thanks are due to them: Heiner Schirmer and Bruno Curti; Bob Abeles and Gordon Hamilton; Takao Nakamura, Woody Hastings, and Humphrey Mager; Dale Edmondson and Don McCormick; Rowena Matthews and Steve Mayhew; Graham Palmer and Lew Siegel. It is also a tribute to flavinologists, that when the formal evening discussion sessions finished at 10–10:30 pm, lively informal discussion continued for several more hours, aided by Canadian beer and Californian wines!

To organize the poster/discussion sessions, we had to divide the field of flavin research into several broad categories. In general, we have used the same categories and subdivisions in this publication. As in previous Symposia, certain areas of research stand out, either as coming to some fruition, or holding considerable promise for the future. This meeting was marked by a considerable amount of new information about flavoprotein structure, and was highlighted by three-dimensional models of the active site-regions of glutathione reductase, p-hydroxybenzoate hydroxylase, and flavodoxin, and progress reports on the structures of ferredoxin-NADP reductase and glycollate oxidase. The complete amino acid sequence of D-amino acid oxidase was also reported, along with beginning predictions of its secondary and tertiary structure, based on analogy and homology, and chemical modification studies.

The field of flavin chemistry and enzymology is a very vigorous and exciting one which has undergone a quantum jump in understanding and sophistication in the time since the first Symposium was held in Amsterdam a mere 16 years ago. In the intervening years, the enormous chemical reactivity of the flavin ring system has become recognized, and the great chemical versatility of flavins as catalysts has begun to be explored at a molecular level. New and important biological roles of flavins are beginning to be recognized, especially in many different types of detoxification processes, including those involving mercury, a large number of aromatic compounds, and a vast array of amines and sulfur-containing compounds. The last few years have also seen the development of many medical applications based on flavins, including their being targets of "mechanism-based" drugs, and their growing utility in a large number of sensitive clinical assays.

Another area of considerable interest to emerge recently is that of naturally-occurring flavins modified in the isoalloxazine ring system. Many of these are covalently linked to their specific protein through the flavin C(8α)-position or the C(6)-position; naturally-occurring flavins modified at the C(8)-position and C(6)-position, not covalently bound to protein, have also been discovered. In addition, a doubly modified flavin has been found as the F-420 cofactor of methane bacteria, and identified as a 5-deaza-8-hydroxyflavin derivative. This same modified chromophore has also been identified as the coenzyme involved in the photoreversion reaction leading to biological repair of thymine dimers back to the native DNA bases.

The above paragraphs give only an abbreviated account of the dynamic status of the field. We feel that the three-year interval between these Symposia is just right for the optimal advancement of the subject, being sufficiently long to accumulate significant new knowledge but not so long as to preclude all new advances coming under the scrutiny of most of the researchers in the field. But the main value of the Symposia is to gather together a group of scientists of diverse background, who share a common interest in the properties and functions of flavins. The backgrounds of the participants range from pure physics and mathematical chemistry, through X-ray crystallography to mechanistic organic chemistry, synthetic chemistry, inorganic and physical chemistry, protein chemistry, kinetics, microbiology and medical practice, to preparative, comparative, and mechanistic enzymology. This mixture of disciplines, meeting in a relaxed atmosphere where everyone can listen and participate, makes these Symposia of great value, and has been instrumental in promoting the rapid advances in the field. An important result, through publication of the proceedings, is to acquaint other scientists with the vigor and importance of the subject, and to attract them to experimental work in it. The Seventh Symposium was by far the largest so far held, with 212 participants. In order to keep this volume to a reasonable size, we have had to impose severe space limits. We wish to thank the participants for their cooperation and for providing their manuscripts at the time of the meeting.

We are indebted to our organizing committee, whose names appear in the front of this volume, and who contributed in various ways to the development of the meeting format, the selection of topics and speakers, and in particular, for help in securing funds from their local sources to provide travel support.

We are sorry that two of them, Harold Bright and Peter Hemmerich, were, at the last moment, unable to attend. The field as a whole owes much to their incisive experiments and intellect and their absence was sorely felt.

The cost of running such a meeting is very high, and we are indebted to our sponsors, who are listed in the front of the volume, for the financial support which made the Symposium possible. In particular, we wish to acknowledge the generous financial support and moral encouragement we received from Dr. George DeMuth, Associate Dean of the Medical School, and Dr. Charles Overberger, Vice President for Research, both of the University of Michigan. Their contributions, particularly in light of the financial difficulties facing the university at this time, are much appreciated.

Finally, we wish to acknowledge with gratitude the help and support we have obtained from colleagues here in Ann Arbor. Special thanks are due to Dona Rothwell and Susan Zuk for unstinting secretarial help, to Oksana Lockridge, who organized the T-shirt sales, to Larry Schopfer for many organizational details, to Al Claiborne, Dick Swenson, and Fraser Morpeth for taking care of the hospitality arrangements, to Martin Lutz, for manning the information desk during the meeting, and to our wives, Margot Massey and Angela Williams, who tolerated with good grace our long preoccupation with this meeting, who manned the registration desk, and who worked as editorial assistants in getting the contributions of this volume ready for the publisher.

Vincent Massey
Charles Williams, Jr.

Peter Hemmerich
1929–1981

Peter Hemmerich died on October 3rd, 1981, in his home in Konstanz, West Germany from a brain tumor. This is a fitting place to pay tribute to him, in a volume devoted to a field which he in no small measure helped to shape. In accord with the wishes of many of the participants of the Symposium, we dedicate this volume to him.

Peter Hemmerich was born on December 30th, 1929, in Frankfurt-am-Main, Germany, and so grew up in the turbulent years of Nazi Germany and World War II. His family was deeply opposed to Nazi ideology and anti-Semitism, and actively helped in providing shelter and hiding to Jewish friends. Toward the end of the war he entered Birklehof, a boy's school at Hinterzarten in the Black Forest, famous for its strict academic standards and humanistic philosophy. Here was born his fascination with science and his love of chemistry. Here also was developed his intense love of nature, which had a very great influence on his life.

With the shortage of places in German universities in the post-war years, he was not able to find a place as a student of chemistry, and so enrolled at the University of Freiburg in 1948 in the mathematics curriculum. With financial aid from an uncle he transferred next year to the University of Basel, where a few years later he entered the prestigious Institute of Inorganic Chemistry as a Ph.D. student of Hans Erlenmeyer. The latter was a leading authority on metal chelate chemistry and so it was natural that any research in his laboratory had to involve metal chelate studies. At around this time the first reports on riboflavin as a metal chelator began to appear, and so it was agreed that the young Hemmerich should work in this field. As a chemist, he soon became frustrated by the limited solubility of riboflavin and known isoalloxazine derivatives; thus he decided on the bold strategy of replacing the oxygen at the flavin 2 α-position by sulfur and of alkylating this, the N(3)-H and the C(4)-α position with various substituents. The resulting flavins, achieved by new and more efficient synthetic routes, were thus the product of the first original research of Peter Hemmerich. In the course of this brilliant research he established the structure of flavin, that we are now so used to seeing, as the 2,4 dicarbonyl tautomer A, rather than the iminohydroxyl tautomers B, which used to be seen frequently (and occasionally still are) in biochemistry texts.

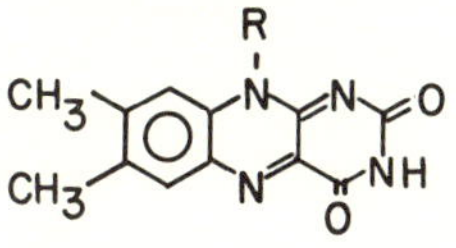

R
CH_3 N N O
CH_3 N NH
O
A

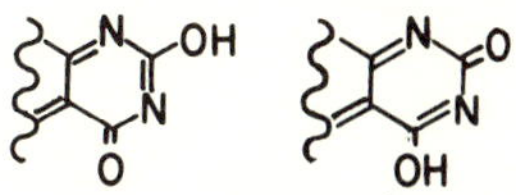

N OH N O
N N
O OH
B

C.P.C. ABERYSTWYTH LLYFRGELL

The Basel period of course also involved metal chelate studies. These came at the period when the involvement of metal centers in metalloflavoproteins was being uncovered, and provided the first contact of Peter Hemmerich with the biochemical community. His findings that oxidized flavin had only weak chelator properties, and that metal chelation was a property only of the flavin radical, had several important outcomes. It helped in the recognition that the metals were parts of discrete redox centers and not directly liganded to the flavin. It also led naturally to his very important studies on the properties of flavin radicals. His interests in these subjects resulted in close international collaborations with a number of biochemical colleagues, Helmut Beinert in Madison, Anders Ehrenberg in Stockholm, Cees Veeger in Wageningen, Tom Singer in San Francisco and Vince Massey in Ann Arbor to name a few. Several of these have spanned many years, from his first visits to the U.S.A. in 1964 and 1965, right up to his death. These collaborations were born out of his desire to relate his fundamental chemical studies to the biological roles of flavins, for he was among the first to realize the great chemical versatility of flavins, and that not every model chemical reaction necessarily pointed the way to the mechanism of a particular enzymic reaction. He was also quick to seize on correlations of properties of particular flavoproteins and the types of reactions which they catalyzed. Because of his impatience to ferret out the order which he believed must exist in nature, he sometimes made premature judgements and interpretations which proved misleading or confusing. However, even in those cases where it turned out that he was wrong, his catalytic influence was of great importance. His belief that different classes of flavoproteins must have different protein structural elements, but with common structural features within each class, led to the more rational classification scheme for flavoenzymes which he developed in his last years.

His scientific output was quite prolific. In the 25 years of his career, he published with students and colleagues 170 research papers and reviews. In addition to those already mentioned, other highlights of this research career include the realization of the ease of formation of covalent adducts of flavins and their possible participation in catalysis, fundamental investigations and practical applications of flavin photochemistry, the determination of properties of flavins in the reduced state, and participation in the elucidation of the structure of flavins covalently linked to proteins. He also had a keen interest in the structure and mode of action of copper and molybdenum sites in enzymes. He was among the first to point to the possibility of sulfur ligands to copper and the resulting ambiguity of the copper valency in such complexes, a property which remains a matter of debate even today.

His intense interest and love of flavin research have been passed on to several students and younger colleagues. Notable among these are Franz Müller and Sandro Ghisla, both of whom were speakers at the 7th Flavin Symposium.

In 1967 he was appointed a Professor of the new University of Konstanz, joining the Biology Faculty. He stayed there for the rest of his life, establishing a flourishing research group devoted to the study of flavins, as well as a group working with copper proteins. Konstanz quickly became a world-renowned center for flavin research, with a continuous succession of visiting scientists,

PART I:

Structure of Flavoproteins

Published 1982 by Elsevier North Holland, Inc.
Vincent Massey and Charles H. Williams, Editors
Flavins and Flavoproteins

CHAPTER 1

Structure and Catalytic Cycle of the Flavoenzyme Glutathione Reductase

E.F. Pai and G.E. Schulz

Max-Planck-Institut fuer medizinische Forschung, Jahnstr. 29, 6900 Heidelberg, West Germany

Introduction

Glutathione reductase (EC 1.6.4.2) is a ubiquitous FAD-containing enzyme. It catalyzes the reduction of oxidized glutathione (GSSG) at the expense of NADPH

$$GSSG + NADPH + H^{+} \rightleftharpoons 2\ GSH + NADP^{+}.$$

The enzyme's primary function is to maintain a low level of GSSG and a high level of GSH in the cell (1). Glutathione reductase is homologous with lipoamide dehydrogenase (2) and possibly also related to thioredoxin reductase (3). In our analysis, we used glutathione reductase from human erythrocytes (4–6), so that the structure elucidation may have direct medical implications. The amino acid sequence of glutathione reductase is known (7–11). The sequence has been fitted to an electron density map at 0.2 nm resolution, yielding the complete structure of the molecule (12).

X-Ray Diffraction Studies

Crystals of glutathione reductase were obtained already in the last step of the preparation, a dialysis against deionized water (7). The parameters of the resulting crystal form A are given in Table 1. However, crystal form-A was not suitable for an initial structure analysis because the crystals invariably deteriorated when soaked in heavy atom-containing solutions. This problem could be circumvented with crystal form-B, which was grown at 1.2 M $(NH_4)_2SO_4$. Crystal forms A and B resemble each other closely. For further analysis, we used crystal form B. Two other crystal forms of the enzyme, C and D, have been observed (Table 1). Form-D appears occasionally but not reproducibly under standard crystallization conditions. Form-C has been produced by prolonged soaking of form-B crystals with oxidized glutathione at high concentrations. Its additional asymmetry may indicate that substrate binding destroys the internal twofold symmetry of the molecule.

Glutathione reductase contains FAD as a prosthetic group. FAD can be removed by using high salt concentrations at pH 3.0 (13). The resulting apoenzyme has been crystallized (14). As shown in Table 1, the crystal parameters of the apoenzyme are very similar to those of forms-A and -B,

Table 1. Crystal Forms of Glutathione Reductase. The Closely-Related Enzyme Lipoamide Dehydrogenase from Torula Yeast Yielded Small Needle-Like Crystals.

Crystal form	Space group	Number of molecules per asymmetric unit	Unit cell parameters			
			a (nm)	b (nm)	c (nm)	γ (deg)
Human A	B2	1/2	11.94	8.47	6.37	58.3
Human B	B2	1/2	11.99	8.44	6.32	58.7
Human C	B2	1	12.23	17.67	5.79	54.4
Human D	$C222_1$	1/2	14.70	6.45	12.90	—
Human Apo	B2	1/2	11.48	8.51	6.37	57.5
E.coli	R32	1/2	11.07	11.07	15.54	—
yeast			crystal size about 100 μ·100 μ·300 μ			

except for a shortening of the a-axis by 0.5 nm, indicating that removal of FAD causes the molecule to shrink in this direction. In addition to the crystals of human glutathione reductase, we produced crystals of the enzyme from *E. coli* and yeast. Moreover, there exist crystals of the homologous enzyme (2) lipoamide dehydrogenase. The latter three crystal forms were obtained in collaboration with Dr. C.H. Williams, Ann Arbor, Michigan.

Glutathione reductase was analyzed by chemical and crystallographic methods in parallel. An interpretable electron density map was obtained at a time when only a couple of peptide sequences were known. Therefore we tried to facilitate chemical analysis by aligning peptides on the basis of the electron density map in a computerized manner (9,15). At a resolution of 0.2 nm, peptides as short as about 10 residues could be positioned (12). In the course of the chemical analysis, more and more peptide overlaps were established, diminishing the dependence on the electron density map. As of now, only 6 chemical overlaps are missing and these are unequivocally established by peptide alignment to the map (11). Furthermore, this fitting procedure revealed that the enzyme contains a flexible segment of 17 residues, which is not visible in the electron density map (10).

Domain Structure

Domains are defined as contiguous segments of the polypeptide chain having a high neighborhood correlation, that is, forming numerous internal and few external contacts (16). Already at 0.6 nm resolution, it turned out that glutathione reductase possesses a pronounced domain structure (17). As shown in Figure 1, its polypeptide chain can be subdivided into five consecutive sections: the flexible N-terminal segment, the FAD-domain, the NADPH-domain, the Central-domain, and the Interface-domain. The chain folds of the two dinucleotide binding domains resemble each other closely and they bind FAD and NADPH at equivalent positions. It has been shown that they have arisen by a gene duplication (18). A similar chain fold and FAD binding position is known from p-hydroxybenzoate hydroxylase (19). However, the chain fold differs from those of structurally known NAD-binding enzymes (20).

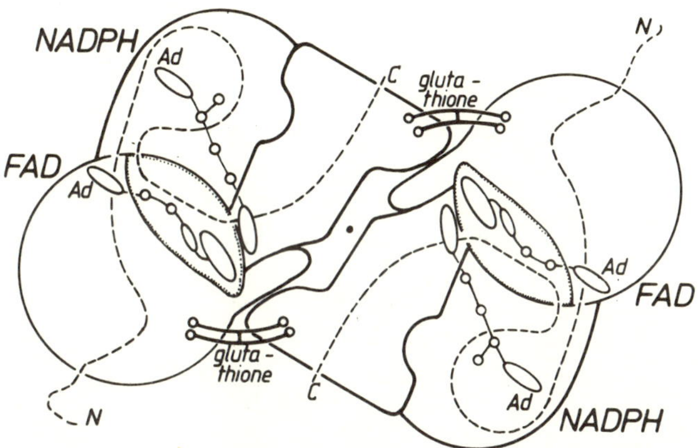

Figure 1. Domain structure of glutathione reductase. The polypeptide chain starts at the N-terminal flexible segment and runs through the FAD-domain, the NADPH-domain, the Central-domain, and the C-terminal Interface-domain. The symmetric arrangement of subunits and the binding positions of FAD, NADPH, and oxidized glutathione are indicated.

Binding of FAD

The prosthetic group FAD as bound to glutathione reductase is shown in the electron density map of Figure 2. This group could not be detected at 0.6 nm resolution, but it was clearly recognized at 0.3 nm, and later confirmed at 0.2 nm resolution (12). FAD is almost completely buried within the protein. In order to release this prosthetic group, the holoenzyme has to open its structure, which probably involves a relative movement of FAD- and NADPH-domains (Figure 1).

The flavin ring system is tightly attached to the Interface-domain of the other subunit via a hydrogen bond from its N3 position (Figure 3). Moreover, the N1 position of flavin is hydrogen bonded to the backbone of the Central-domain. The N5 position is in contact with the C_ϵ-methylene group of Lys-66 of the FAD-domain. It is presumed that such a bonding pattern prevents flavin from transferring single electrons, favoring the two electron transfer (21) actually observed in glutathione reductase.

Without NADPH being present, the side chain of Tyr-197 shields the NADPH-contacting re-face of the flavin from the solvent (22) as shown in Figure 2. Presumably, this lid protects the enzyme in its reduced state (probably a prevalent state of the enzyme *in vivo*) from losing reduction equivalents to molecular oxygen. The redox-active disulfide bridge of the protein is located at the opposite si-face of flavin so that position C4a of flavin is juxtaposed to one of the sulfurs. This geometry allows formation of the charge transfer complex between a thiolate anion and flavin, which had been postulated on the basis of absorption spectra in solution (23).

The ribitol part of FAD is hydrogen-bonded to backbone atoms and to the side chains of a threonine and a buried aspartate. The pyrophosphoryl group is squeezed between two strands of the polypeptide chain forming several hydro-

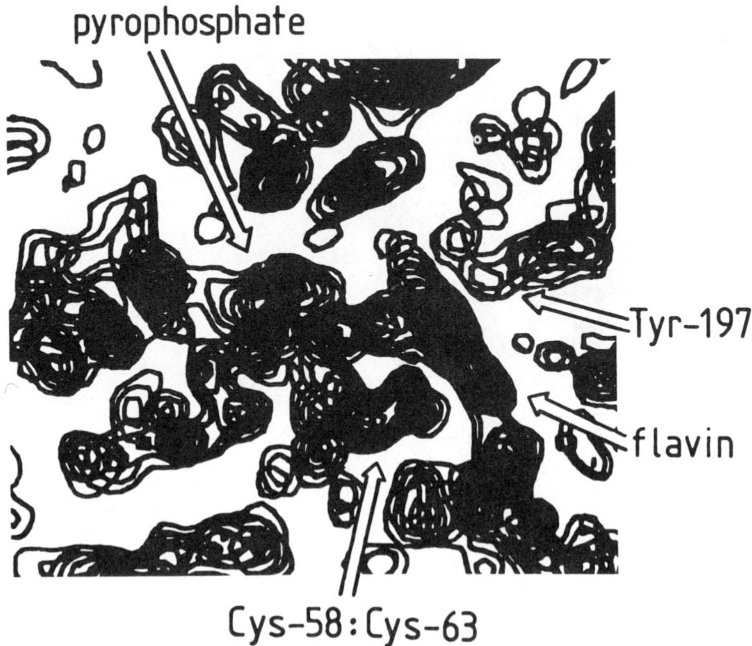

Figure 2. Electron density map at 0.2 nm resolution around the flavin binding site of glutathione reductase. The orientation is similar to that of the subunit on the left hand side of Figure 1. Indicated are the flavin ring, the pyrophosphate part of FAD, the redox-active disulfide bridge between CYS-58 and CYS-63, and the side chain of Tyr-197, which prevents access of solvent to the flavin.

gen bonds to the backbone. The ribose moiety of FAD is tightly attached to the carboxyl group of a glutamate via its 2′- and 3′-OH groups. Adenine is hydrogen-bonded to the side chains of a histidine and an asparagine, as well as to the backbones of further strands of the chain. FAD makes most of its contacts with the FAD-domain, fewer contacts with the Central-domain, and still fewer contacts with the NADPH- and both Interface-domains.

Binding of NADPH

Soaking experiments with $NADP^+$ using crystals of the oxidized enzyme showed no defined position of nicotinamide and its ribose, but precise binding of the 2′-phospho-5′-AMP moiety of the dinucleotide (22,24). Later studies using the half-reduced enzyme showed that under these conditions the nicotinamide moiety of $NADP^+$ and NADPH was bound firmly. Together, these findings may imply that in the oxidized enzyme $NADP^+$ is actively expelled from its binding site.

The bound nicotinamide stacks onto flavin with the S-proton at its C4 carbon atom close to position N5 of the flavin and to the ϵ-amino group of Lys-66 (Figure 3). This geometry explains the charge transfer interactions between nicotinamide and flavin found in spectroscopic experiments (1).

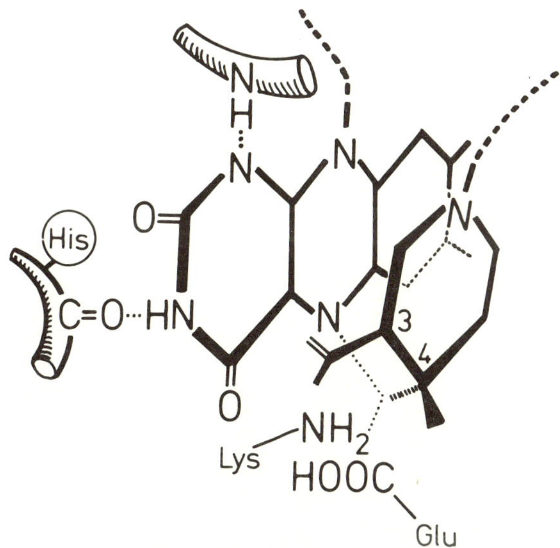

Figure 3. Sketch of the interaction between nicotinamide moiety of NADPH and flavin. The rings are stacked onto each other so that C3 of nicotinamide opposed C4a of flavin. There exists an internal salt bridge between Lys-66 and a glutamate of the NADPH-domain, which may contribute to catalysis. N1 of the flavin forms a hydrogen bond to the backbone of the Central-domain, whereas N3 of flavin is tightly hydrogen-bonded to the carbonyl group of His-467 of the Interface-domain of the other subunit.

Moreover, it allows the transfer of reduction equivalents as it was envisaged on the basis of solution studies. However, it does not differentiate between transfer of a hydride ion and other possible mechanisms of reduction (21,25,26).

The pyrophosphoryl group is squeezed between two strands of the polypeptide chain forming hydrogen bonds to their backbones. The 2′-phosphoryl group is attached to two arginines and to a histidine from the NADPH-domain. These interactions are the reason for the observed large contribution of the 2′-phosphoryl group to the overall binding energy. Adenine binds in a shallow hydrophobic depression at the protein surface. As shown in Figure 1, both NADPH and FAD bind in rather extended conformations to the protein. No base stacking between aromatic side chains and nucleotide ring systems is involved.

Further Enzyme-Ligand Complexes

A complete list of all enzyme-ligand complexes studies in crystals is given in Table 2. These investigations showed that the 2′-phosphoryl group is most essential for nucleotide binding. In the crystal (and most likely also in solution), AMP, ADP, ATP, ADP-ribose, NAD^+ were only loosely bound at their adenine moiety to the oxidized enzyme and to the carboxymethylated enzyme. Only the complex between NADH and the half-reduced enzyme was strong enough to fix the whole ligand. In contrast, all ligands containing the 2′-phosphoryl group were bound at their expected position.

Table 2. Enzyme-Ligand Complexes Investigated by X-Ray Analyses. The Ligands of the Oxidized Enzyme are Listed in Two Columns, the Second of Which Contains Ligands with a 2′-Phosphoryl Group.

Oxidized enzyme		**Half-reduced enzyme**	**Carboxymethylated enzyme**
AMP	2′-AMP	NADH	NAD^+
ADP	2′-phospho-5′-AMP	$NADP^+$	$NADP^+$
ATP	2′-phospho-5′-ADP	NADPH	
ADP-ribose	2′-phospho-5′-ADP-methylester	thio-$NADP^+$	
NAD^+	2′-phospho-ADP-ribose	3-acetylpyridine-ADP^+	
GSSG	$NADP^+$	3-aminopyridine-ADP^+	
Lipoic acid	3-acetylpyridine-ADP^+	GSH	
CoASSG	3-aminopyridine-ADP^+	lipoic acid	

When applied to the oxidized enzyme, $NADP^+$ and its analogs 3-acetyl-pyridine-ADP^+ and 3-aminopyridine-ADP^+ did bind only with their 2′-phospho-5′-AMP parts. The positive charge of the oxidized ring system seems to prevent binding of the pyridine moiety of the dinucleotide. However, these ligands and also the analog thio-$NADP^+$ formed tight complexes with the half-reduced enzyme. These results were corroborated by the observation that $NADP^+$ bound firmly to the carboxymethylated enzyme (Table 2). The ligand conformation in these complexes is very similar to the conformation of NADPH in the enzyme-substrate complex. Obviously, in its half-reduced state glutathione reductase can tolerate the positive charge.

As shown in Figure 2 in the oxidized enzyme the side chain of Tyr-197 shields flavin from the solvent. If the substrate NADPH is applied to the enzyme, this side chain swings back so that nicotinamide can stack onto flavin in order to deliver its reduction equivalents. In this induced-fit mechanism, the question arises how much of the NADPH molecule is necessary to make Tyr-197 move. The question was answered by binding a series of fragments of NADPH, namely 2′-AMP, 2′-phospho-5′-AMP, 2′-phospho-5′-ADP, 2′-phospho-5′-ADP-methylester, and 2′-phospho-5′-ADP-ribose to the oxidized enzyme (Table 2). Tyr-197 moved on binding the last compound of this series, showing that nicotinamide itself is not necessary for the induced fit.

As shown in Figure 1, the binding site of oxidized glutathione is quite different from the binding site of NADPH. It is located at the other side of the subunit and shared between both subunits. At this site in the oxidized enzyme, we could also bind the mixed disulfide between coenzyme-A and glutathione (CoASSG) as well as oxidized lipoic acid (27) (Table 2). Both compounds are poor substrates of glutathione reductase (13,28,29). On binding, the sulfurs of these compounds are displaced from the sulfur positions of the normal substrate oxidized glutathione, which seems to explain their slow reaction rates. Furthermore, we were able to stabilize the mixed disulfide between the half-reduced enzyme and glutathione. Here, glutathione-I interacted with Cys-58 (cf. Figure 4). In addition, the mixed disulfide between lipoic acid and the half-reduced enzyme could be produced and analyzed, showing that lipoic acid did bind to Cys-58, but was located in two equally occupied positions.

Catalytic Cycle

The steps of the catalytic reaction as revealed by biochemical (1) and X-ray analyses are given in Figure 4. Three true intermediates of the catalytic cycle have been trapped in the crystal: the complex between half-reduced enzyme and $NADP^+$, the half-reduced enzyme itself, and the mixed disulfide between protein and glutathione.

The cycle starts with NADPH binding to the oxidized enzyme. In this process, the 2′-phosphoryl group serves as an anchor (Table 2) and the ribose at the nicotinamide side induces the displacement of Tyr-197. Nicotinamide stacks onto the re-face of flavin (Figure 3). Two reduction equivalents are transferred to the flavin. The nature of these equivalents is still discussed (21,25,26). From the si-face of flavin two electrons flow via a C4a-adduct (30,31) to Cys-63 leading to the opening of the redox-active disulfide bridge. This is followed by the formation of a charge transfer complex between the thiolate anion of Cys-63 and reoxidized flavin (1). The negative charge of Cys-63 is assumed to be stabilized by an ion pair interaction (32) with the protonated His-467. This state is stable and corresponds to the half-reduced enzyme (Table 2). After oxidized glutathione is bound, Cys-58 attacks the sulfur of glutathione-I and forms a mixed disulfide, whereas glutathione-II takes a proton from His-467 and diffuses off. His-467 has now lost its charge, and cannot stabilize the charge transfer complex any longer. Thus, Cys-63 in a nucleophilic attack on the sulfur of Cys-58 restores the redox-active cysteine and releases glutathione-I.

Figure 4. Sketch of the catalytic cycle including biochemical results and intermediates which have been characterized by X-ray analyses. (1) Oxidized enzyme with the closed disulfide between Cys-58 and Cys-63. The base His-467 is unprotonated. The positions for noncovalent binding of NADPH, glutathione-I and -II are indicated. (2–4) Production of the stable half-reduced form of the enzyme. (5) Binding of oxidized glutathione (GSSG). (6) Disulfide interchange between Cys-58 and glutathione-I accompanied by the protonation of glutathione-II through His-467. (7–8) Release of glutathione-II and nucleophilic attack of the no-longer stabilized thiolate onto the mixed disulfide followed by the release of glutathione-I. Complexes 1, 3, 4, and 6 have been characterized by X-ray structure analyses at 0.3 nm resolution.

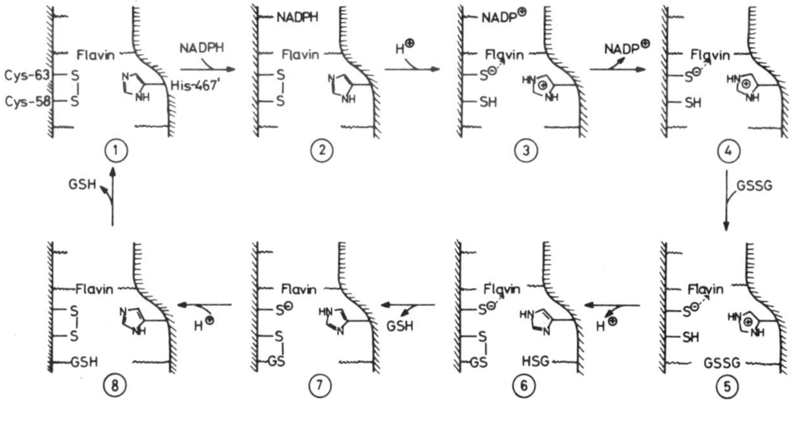

References

1. Williams, C.H., Jr. (1976) In *The Enzymes*. Boyer, P.D. (ed.) 3rd. Ed., Vol. 13, pp. 89–172. New York: Academic Press.
2. Arscott, L.D., Williams, C.H., Jr., and Schulz, G.E. (1982) Flavins and Flavoproteins, pp. 44–48.
3. Holmgren, A. (1980) *Experientia* Suppl 36:149–180.
4. Worthington, D.J. and Rosemeyer, M.A. (1974) *Eur J Biochem* 48:167–177.
5. Worthington, D.J. and Rosemeyer, M.A. (1975) *Eur J Biochem* 60:459–466.
6. Worthington, D.J. and Rosemeyer, M.A. (1976) *Eur J Biochem* 67:231–238.
7. Krohne-Ehrich, G., Schirmer, R. H., and Untucht-Grau, R. (1977) *Eur J Biochem* 80:65–71.
8. Schiltz, E., Blatterspiel, R., and Untucht-Grau, R. (1979) *Eur J Biochem* 102:269–278.
9. Untucht-Grau, R., Schulz, G.E., and Schirmer, R.H. (1979) *FEBS-Letters* 105:244–248.
10. Untucht-Grau, R., Schirmer, R.H., Schirmer, I., and Krauth-Siegel, R.L. (1981) *Eur J Biochem*, submitted.
11. Krauth-Siegel, R.L., Blatterspiel, R., Saleh, M., Schiltz, E., Schirmer, R.H., and Untucht-Grau, R. (1981) *Eur J Biochem*, submitted.
12. Thieme, R., Pai, E.F., Schirmer, R.H., and Schulz, G.E. (1981) *J Molec Biol*, submitted.
13. Icen, A. (1967) *Scand J Clin Lab Invest* Suppl 2, 96:1–67.
14. Fritsch, K.G., Pai, E.F., Schirmer, R.H., Schulz, G.E., and Untucht-Grau, R. (1979) *Z Physiol Chem* 360:261–262.
15. Schirmer, R.H., Pai, E.F., Schiltz, E., Schulz, G.E., and Untucht-Grau, R. (1980) In *Methods in Peptide and Protein Sequence Analysis*. Birr, C. (ed.), pp. 267–283. Amsterdam: Elsevier/North Holland Biomed. Press.
16. Schulz, G.E. and Schirmer, R.H. (1979) *Principles of Protein Structure*. New York: Springer Verlag.
17. Zappe, H., Krohne-Ehrich, G., and Schulz, G.E. (1977) *J Molec Biol* 113:141–152.
18. Schulz, G.E. (1980) *J Molec Biol* 138:335–347.
19. Wierenga, R.K., de Jong, R.J., Kalk, K.H., Hol, W.G.J., and Drenth, J. (1979) *J Molec Biol* 131:55–73.
20. Rossmann, M.G., Liljas, A., Brändén, C.-I., and Banaszak, L.J. (1975) In *The Enzymes*. Boyer, P.D. (ed.), Vol. 11, pp. 61–102. New York: Academic Press.
21. Hemmerich, P. (1976) In *Progress in the Chemistry of Organic Natural Products*. Herz, W., Grisebach, H., and Kirby, G. W. (eds.), Vol. 33, pp. 451–527. Wien: Springer Verlag.
22. Schulz, G.E., Schirmer, R.H., and Pai, E.F. (1980) In *Flavins and Flavoproteins*. Yagi, K. and Yamano, T. (eds.) pp. 557–567. Tokyo: Jap. Scient. Soc. Press.
23. Massey, V. and Ghisla, S. (1974) *Ann NY Acad Sci* 227:446–465.
24. Schulz, G.E., Schirmer, R.H., Sachsenheimer, W., and Pai, E.F. (1978) *Nature* 273:120–124.
25. Walsh, C. (1980) *Acc Chem Res* 13:148–155.
26. Blankenhorn, G. (1976) *Eur J Biochem* 67:67–80.
27. Pai, E.F. (1978) Dissertation, University of Heidelberg.
28. Carlberg, I. and Mannervik, B. (1975) *J. Biol Chem* 250:5475–5485.
29. Carlberg, I. and Mannervik, B. (1977) *Biochim Biophys Acta* 484:275–289.
30. Bruice, T.C. (1980) *Acc Chem Res* 13:256–262.
31. Loechler, E.L. and Hollocher, T.C. (1975) *J Amer Chem Soc* 97:3235–3237.
32. Arscott, L.D., Thorpe, C., and Williams, C.H., Jr. (1981) *J Biol Chem* 20:1513–1520.

Published 1982 by Elsevier North Holland, Inc.
Vincent Massey and Charles H. Williams, Editors
Flavins and Flavoproteins

CHAPTER 2

The Structure of p-Hydroxybenzoate Hydroxylase

R.K. Wierenga,* K.H. Kalk,* J.M. van der Laan,* J. Drenth,* J. Hofsteenge,** W.J. Weijer,** P.A. Jekel,** J.J. Beintema,** F. Müller,*** and W.J.H. van Berkel***

**Laboratory for Chemical Physics, University of Groningen; ***Biochemical Laboratory, University of Groningen; ****Biochemical Laboratory, University of Wageningen, The Netherlands*

Introduction

p-Hydroxybenzoate hydroxylase is a flavin monooxygenase (1,2). It catalyzes the following reaction:

$$HO-C_6H_4-CO_2^{\ominus} + NADPH + O_2 + H^{\oplus} \longrightarrow HO-C_6H_3(HO)-CO_2^{\ominus} + NADP^{\oplus} + H_2O.$$

The enzyme/substrate complex catalyzes more efficiently the reduction of FAD by NADPH than the free enzyme (3). After reduction, $NADP^+$ leaves the active site and then O_2 forms a hydroperoxide (4,15). Subsequently the oxygen O-O bond is split and one of the oxygen atoms is attached to the substrate as an hydroxyl group. The other oxygen atom leaves as a water molecule. At room temperature, the enzyme exists mainly as a dimer, although small amounts of higher oligomers also occur (6). The monomer consists of a single polypeptide chain which has 394 amino acid residues. Five of these are cysteines but the chain does not contain any disulfide bridge. One molecule of FAD is bound per monomer. The study of an enzyme is never complete without a structural framework in which the results of biochemical experiments can be integrated and which can be used for designing new experiments. In a combined effort from the Universities of Groningen and Wageningen, we determined the amino acid sequence and performed an X-ray crystallographic study for obtaining the three-dimensional structure. I would like to report on the results, part of which have been published before (5,7,8).

The Primary Structure

The enzyme was extracted from *Pseudomonas fluorescens* grown by Diosynth B.V., Oss, The Netherlands. For the isolation and purification, the procedure of Howell (9) as improved by Müller et al. (6), and van der Laan was used. The amino acid sequence was determined from peptides obtained by CNBr cleavage of carboxymethylated p-hydroxybenzoate hydroxylase. These fragments were sequenced by applying classical methods with automatic and manual Edman degradation. We shall not give technical details because they have been published elsewhere (7, 8). While the sequence results were produced, the electron density map was already available (5) and therefore any intermediate results of the sequence studies could be compared with the map. Also the connectivity of the CNBr fragments could be deduced from the electron density map. Not only was the map of indispensable value for the sequence work, but also the sequence provided us with a fully reliable interpretation of the map. Fortunately it turned out that the previous interpretation of the electron density map had been correct which confirms the high quality of the map. The complete sequence is presented in Figure 1.

The Three-Dimensional Structure

Details of the structure determination of the enzyme/substrate complex have been published (5). The 394 residue chain folds up to form a fairly elongated molecule with a length of 70 Å and a diameter of 45–50 Å. The molecule can be dissected into three domains: I, II, and III (Figure 2).

Domain I binds the AMP moiety of FAD. It contains three β-sheets, two small antiparallel sheets C and D, and a larger parallel sheet A, which has the same topology as the sheet in the familiar nucleotide binding domain in lactate dehydrogenase. It is sandwiched between some loops and sheet C on one side and helices on the other side. The $\beta\alpha\beta$ folding unit A1, H1, A2 binds the AMP part of FAD just as the nucleotides are bound in the dehydrogenases. The pyrophosphate group is located at the N-terminus of helix 1. This is a common feature of many phosphate binding proteins and is due to the overall electric dipole moment of an α-helix (10). This causes a positive potential near the N-terminus and a favorable binding site for negatively-charged groups.

The FAD is in an extended conformation with its isoalloxazine ring in a position where all three domains are in contact with each other. Next to it is the substrate-binding pocket. Part of its wall is formed by the antiparallel β-sheet B of domain II which is itself shielded from the solvent by helix 4.

Domain III is mainly helical. Helix 7 has a somewhat odd position outside the main body of the molecule. However this is not as odd as it seems because a crystallographic twofold axis is close to it. This means that in the crystal structure a second molecule has its helix 7 on the other side of the twofold axis touching helix 6 of the molecule in Figure 2, whereas helix 7 of this latter molecule is in contact with helix 6 of the other. Figure 3 illustrates that helix 7 makes a small angle with the twofold axis. In their C-terminal part, these helices come close and interact with each other and with helices 6 mainly through a cluster of aromatic residues. Near the N-terminus, the helices 7 are further apart. Monomer/monomer contact is here between helices 7 and 6, also through interaction of aromatic residues Trp 337, Tyr 344, and Phe 177 of helix 6 in one molecule with Phe 359, Ala 358, and Ile 363 of helix 7 in the other. The dimer interface is not completely hydrophobic because—besides

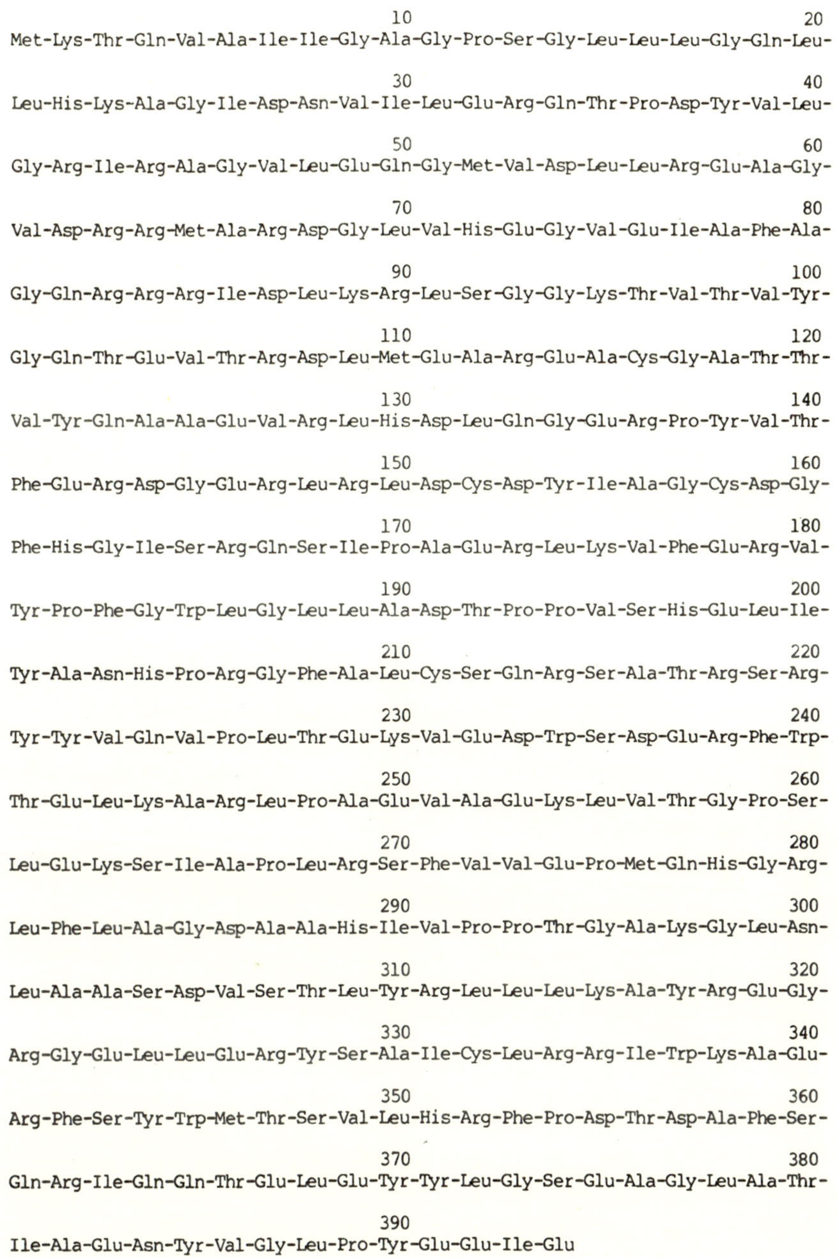

10 20
Met-Lys-Thr-Gln-Val-Ala-Ile-Ile-Gly-Ala-Gly-Pro-Ser-Gly-Leu-Leu-Leu-Gly-Gln-Leu-

30 40
Leu-His-Lys-Ala-Gly-Ile-Asp-Asn-Val-Ile-Leu-Glu-Arg-Gln-Thr-Pro-Asp-Tyr-Val-Leu-

50 60
Gly-Arg-Ile-Arg-Ala-Gly-Val-Leu-Glu-Gln-Gly-Met-Val-Asp-Leu-Leu-Arg-Glu-Ala-Gly-

70 80
Val-Asp-Arg-Arg-Met-Ala-Arg-Asp-Gly-Leu-Val-His-Glu-Gly-Val-Glu-Ile-Ala-Phe-Ala-

90 100
Gly-Gln-Arg-Arg-Arg-Ile-Asp-Leu-Lys-Arg-Leu-Ser-Gly-Gly-Lys-Thr-Val-Thr-Val-Tyr-

110 120
Gly-Gln-Thr-Glu-Val-Thr-Arg-Asp-Leu-Met-Glu-Ala-Arg-Glu-Ala-Cys-Gly-Ala-Thr-Thr-

130 140
Val-Tyr-Gln-Ala-Ala-Glu-Val-Arg-Leu-His-Asp-Leu-Gln-Gly-Glu-Arg-Pro-Tyr-Val-Thr-

150 160
Phe-Glu-Arg-Asp-Gly-Glu-Arg-Leu-Arg-Leu-Asp-Cys-Asp-Tyr-Ile-Ala-Gly-Cys-Asp-Gly-

170 180
Phe-His-Gly-Ile-Ser-Arg-Gln-Ser-Ile-Pro-Ala-Glu-Arg-Leu-Lys-Val-Phe-Glu-Arg-Val-

190 200
Tyr-Pro-Phe-Gly-Trp-Leu-Gly-Leu-Leu-Ala-Asp-Thr-Pro-Pro-Val-Ser-His-Glu-Leu-Ile-

210 220
Tyr-Ala-Asn-His-Pro-Arg-Gly-Phe-Ala-Leu-Cys-Ser-Gln-Arg-Ser-Ala-Thr-Arg-Ser-Arg-

230 240
Tyr-Tyr-Val-Gln-Val-Pro-Leu-Thr-Glu-Lys-Val-Glu-Asp-Trp-Ser-Asp-Glu-Arg-Phe-Trp-

250 260
Thr-Glu-Leu-Lys-Ala-Arg-Leu-Pro-Ala-Glu-Val-Ala-Glu-Lys-Leu-Val-Thr-Gly-Pro-Ser-

270 280
Leu-Glu-Lys-Ser-Ile-Ala-Pro-Leu-Arg-Ser-Phe-Val-Val-Glu-Pro-Met-Gln-His-Gly-Arg-

290 300
Leu-Phe-Leu-Ala-Gly-Asp-Ala-Ala-His-Ile-Val-Pro-Pro-Thr-Gly-Ala-Lys-Gly-Leu-Asn-

310 320
Leu-Ala-Ala-Ser-Asp-Val-Ser-Thr-Leu-Tyr-Arg-Leu-Leu-Leu-Lys-Ala-Tyr-Arg-Glu-Gly-

330 340
Arg-Gly-Glu-Leu-Leu-Glu-Arg-Tyr-Ser-Ala-Ile-Cys-Leu-Arg-Arg-Ile-Trp-Lys-Ala-Glu-

350 360
Arg-Phe-Ser-Tyr-Trp-Met-Thr-Ser-Val-Leu-His-Arg-Phe-Pro-Asp-Thr-Asp-Ala-Phe-Ser-

370 380
Gln-Arg-Ile-Gln-Gln-Thr-Glu-Leu-Glu-Tyr-Tyr-Leu-Gly-Ser-Glu-Ala-Gly-Leu-Ala-Thr-

390
Ile-Ala-Glu-Asn-Tyr-Val-Gly-Leu-Pro-Tyr-Glu-Glu-Ile-Glu

Figure 1. The amino acid sequence of p-hydroxybenzoate hydroxylase.

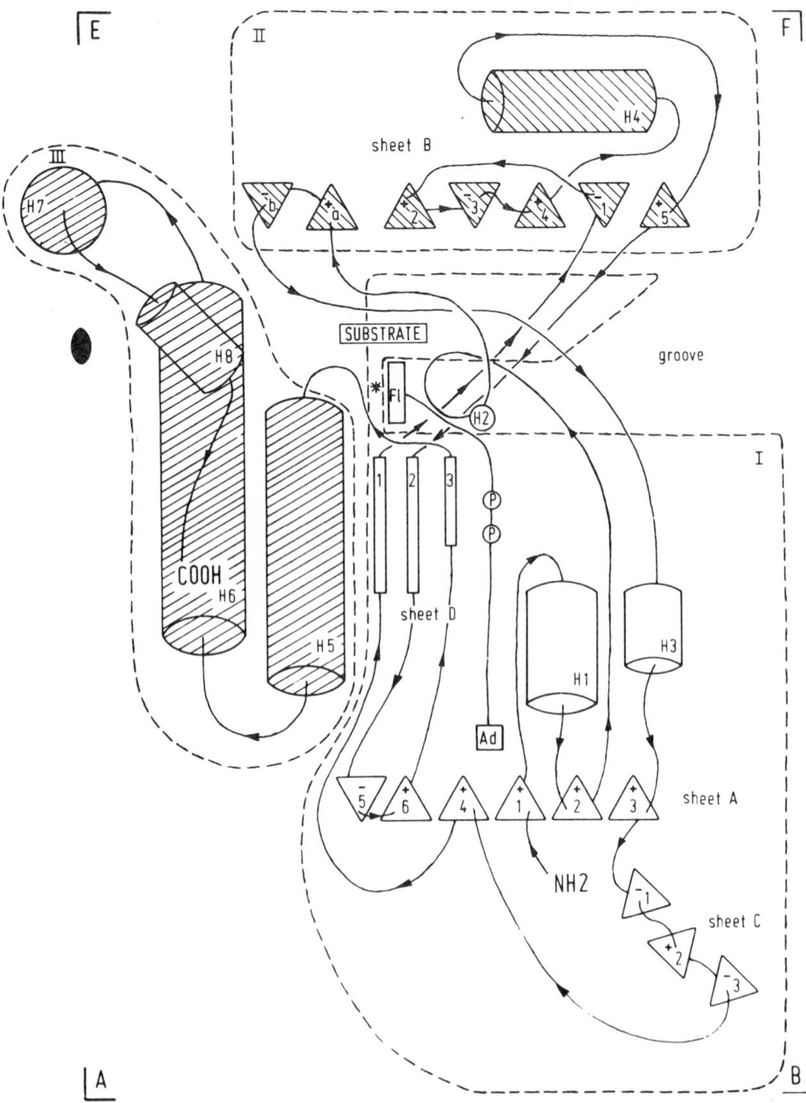

Figure 2. A schematic drawing of the secondary structure in p-hydroxybenzoate hydroxylase. The crystallographic twofold axis is indicated. The asterisk indicates the binding site for negatively-charged ions.

SOURCE: *Reproduced with permission from J Mol Biol (1979) 131:63.*

some hydrogen bonds—four salt bridges are present: Glu 274-Arg 362 and Glu 367-Arg 341. In addition to interactions between residues of helices 6 and 7, also the first turn of both helices 8 interact. If there was any influence of the dimer interface on the active site, it would most probably be through this helix 8, because Tyr 385 at its C-terminus is in the active site and could play a role in the catalytic mechanism, as we shall see later.

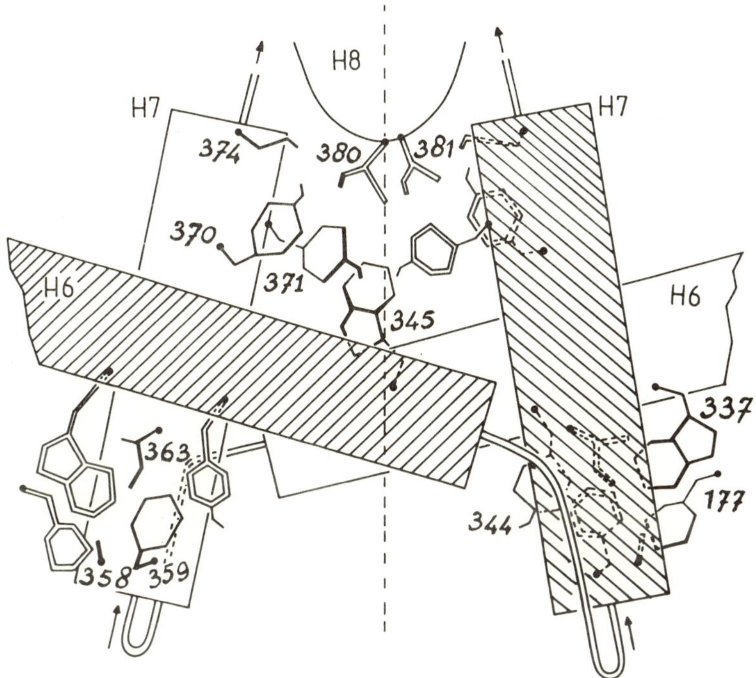

Figure 3. The dimer interface. The helices of one subunit are hatched and its amino acid side chains are drawn in open lines. Solid lines are used for the side chains of the related subunit. For the sake of clarity, related side chains are numbered once only and Trp 345, Thr 380, and Ile 381 of one subunit have been omitted. The broken line is the crystallographic diad. Only the hydrophobic clusters are explicitly shown.

The Active Site

Within the accuracy of the electron density map, the isoalloxazine ring is flat as one expects for the oxidized state. The electron density map suggests that its pyramidine ring is hydrogen-bonded to two pieces of the chain: 45–47 and 299–300 (Figure 4). Moreover these two pieces are connected by a salt bridge between Glu 49 and Lys 297. This network of interactions also incorporates the side chains of alanines 45 and 296. It is interesting that residues 299 and 300 are in the first turn of helix 5. Because of the helix electric dipole moment, this could polarize the isoalloxazine ring such that negative charge is attracted to the N(1)-O(2) side. The dimethylbenzene part is in contact with the solvent. N(5) has no interaction with enzyme or substrate atoms, at least not in the structure of the oxidized enzyme/substrate complex. The hydroxyl group of Tyr 222 is about 5 Å from N(5) but it could come closer if the substrate is removed.

The substrate is bound deep inside the protein. This is in agreement with the observation that a conformational change occurs on binding of a substrate molecule (9). The benzene ring is surrounded by hydrophobic residues and by the isoalloxazine ring. However both ends of the substrate make important

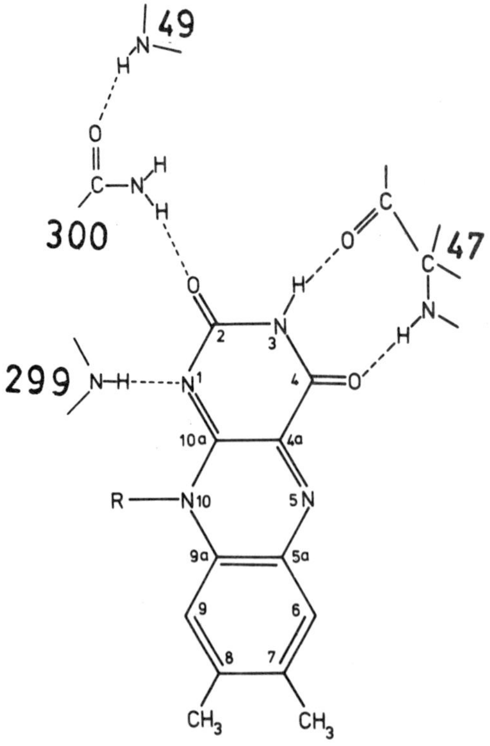

Figure 4. The hydrogen bond connections between the protein and the isoalloxazine ring system.

hydrophilic interactions (Figure 5). The carboxyl moiety binds to Arg 214 and the phenolic OH is hydrogen-bonded to two tyrosine hydroxyl groups: Tyr 385 and 201. The three phenolic hydroxyls form a hydrophylic cluster in a hydrophobic environment and therefore the hydrogen bonds will be fairly strong. In the enzyme without the substrate, this cluster is absent and the tyrosines will probably find other binding sites for their hydroxyl groups.

We have not yet determined the structure of the enzyme with NADPH incorporated. However from the structure of the enzyme/substrate complex, a reasonable suggestion can be made about the mode of binding of NADPH. In this complex, a cavity exists between the isoalloxazine ring system and the peptide loop from 293 to 300, where helix 5 begins. This cavity can bind negative ions, e.g., gold cyanide (11) and probably also Cl^- (12) and N_3^- (13). It is also likely that substrate inhibition at high concentration is caused by the binding of the substrate molecule in this cavity, and finally it is almost certainly the channel through which molecular oxygen reaches the flavin. We believe this to be sufficient support for accepting the cavity as binding site for the nicotinamide moiety of NADPH. Because the cavity is rather narrow the nicotinamide ring must be placed parallel to the isoalloxazine ring. The pyrophosphate can form salt bridges with arginines 166 and 269 and possibly

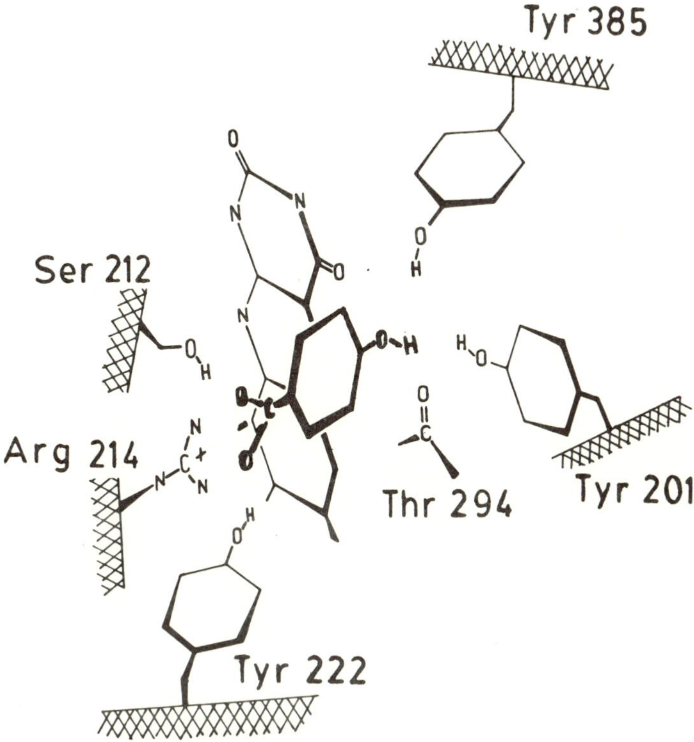

Figure 5. The position of the substrate in the active site.

with His 162. A one-turn α-helix dipole is available as binding site for the 2′-phosphate and a loop with hydrophobic side chains for the adenine. The cavity is rather narrow and the coenzyme will fill it up completely which is supported by the fact that $NADP^+$ has to dissociate away from the enzyme before O_2 can enter (3).

The Catalytic Mechanism

Before we can correlate the existing biochemical knowledge of the enzyme with the structural results, we should extend the structural work to other complexes of the enzyme, such as the free enzyme, the reduced form, and complexes with the coenzyme. However at this stage it is tempting to indulge in some speculation.

Binding of the substrate induces a conformational change and this is quite consistent with the structure of the complex. The substrate molecule is completely buried in the interior. Moreover three tyrosines and an arginine very likely change their position on substrate binding. The enzyme/substrate complex efficiently catalyzes the transfer of reducing equivalents from NADPH to FAD. This drastic change in catalytic properties compared with the substrate-free enzyme, is probably to some extent the result of a different position of Tyr

222. The hydroxyl group of this tyrosine could hydrogen-bond to N(5) in the absence but not in the presence of substrate.

In the reduction step intermediates with long wavelength absorption have been observed (3, 14). They are believed to be charge transfer complexes between flavin and pyridine nucleotide. Model building studies of the NADPH-binding suggest that the nicotinamide moiety can bind in a narrow cavity between the isoalloxazine ring and the loop Thr 294-Gly 295-Ala 296-Lys 297-Gly 298 which continues into helix 5. Molecular oxygen probably enters through this cavity and subsequently reacts with FAD to form the C(4)a)-OOH peroxide (4, 15). The distance between C(4a) and the C-atom of the substrate to be attacked is in this structure about 6 Å (Figure 5) (5).

The hydroxyl cluster formed by the substrate with Tyr 201 and Tyr 385 could function as a proton shuttle during the hydroxylation process. This hydrogen bonding network could also activate the substrate's benzene ring towards electrophilic attack (13).

Extension of the structural work and further integration with biochemical results will certainly lead to a reliable and detailed understanding of the catalytic mechanism.

ACKNOWLEDGMENTS
We thank Dr. Wim G.J. Hol for stimulating discussions. This work was carried out under the auspices of the Netherlands Foundation for Chemical Research (SON) with financial aid from the Netherlands Organization for the Advancement of Pure Research (Z.W.O.).

References

1. Hosokawa, K. and Stanier, R.Y. (1966) *J Biol Chem* 241:2453–2460.
2. Massey, V. and Hemmerich, P. (1980) *Biochem Soc Trans* 246–256.
3. Husain, M. and Massey, V. (1979) *J Biol Chem* 254:6657–6666.
4. Ghisla, S., Woodland Hastings, I., Favaudon, V., and Lhoste, J.-M. (1978) *Proc Natl Acad Sci USA* 75:5860–5863.
5. Wierenga, R.K., De Jong, R.J., Kalk, K.H., Hol, W.G.J., and Drenth, J. (1979) *J Mol Biol* 131:55–73.
6. Müller, F., Voordouw, G., Van Berkel, W.J.H., Steennis, P.J., Visser, S., and Van Rooyen, P.J. (1979) *Eur J Biochem* 101:235–244.
7. Hofsteenge, J., Vereijken, J.M., Weijer, W., Beintema, J.J., Wierenga, R.K., and Drenth, J. (1980) *Eur J Biochem* 113:141–150.
8. Vereijken, J.M., Hofsteenge, J., Bak, H.J., and Beintema, J.J. (1980) *Eur J Biochem* 112:151–157.
9. Howell, L.G., Spector, T., and Massey, V. (1972) *J Biol Chem* 247:4340–4350.
10 Hol, W.G.J., Van Duijnen, P.Th., and Berendsen, H.J.C. (1978) *Nature* 273:443–446.
11. Santema, J.S., Müller, F., Steennis, P.J., Jarbandhan, T., Drenth, J., and Wierenga, R.K. (1976) *Flavins and Flavoproteins, Proceedings of the 5th International Symposium*. Singer, T.P. (ed.), Chap. 15, pp. 155–160. Amsterdam: Elsevier.
12. Steennis, P.J., Cordes, M.M., Hilkens, J.G.H., and Müller, F. (1973) *FEBS Lett* 36:177–180.
13. Entsch, B., Ballou, D.P., and Massey, V. (1976) *J Biol Chem* 251:2550–2563.
14. Shoun, H., Beppu, T., and Arima, K. (1980) *Flavins and Flavoproteins, Proceedings of the 6th International Symposium*. Yagi, K. and Yamano, T. (eds.), pp. 117–124. Tokyo: Japan Scientific Societies Press.
15. Ghisla, S., Entsch, B., Massey, V., and Husain, M. (1977) *Eur J Biochem* 76:139–148.

Published 1982 by Elsevier North Holland, Inc.
Vincent Massey and Charles H. Williams, Editors
Flavins and Flavoproteins

CHAPTER 3

Comparisons of Flavodoxin Structures

Martha L. Ludwig,* Katherine A. Pattridge,*
Ward W. Smith,* Lyle H. Jensen,** and
Keith D. Watenpaugh**

**Biophysics Research Division and Department of Biological Chemistry, The University of Michigan, Ann Arbor, Michigan; **Department of Biological Structure, University of Washington, School of Medicine, Seattle, Washington*

Introduction

The three-dimensional structures of two "short-chain" flavodoxins, those from *D. vulgaris* (148 residues) and *Clostridium MP* (138 residues), have been determined by Watenpaugh, Sieker, and Jensen (1,2) and by Burnett et al. (3) and Smith et al. (4), respectively. Although interaction with each of these two proteins alters the redox potentials of FMN in a similar way, several structural features of the FMN binding site are dissimilar in these flavodoxins (5). The recent determination of the structure of a "long-chain" flavodoxin from the cyanobacterium, *Anacystis nidulans* (6), invites comparison with the flavodoxins from *Clostridium MP* and *D. vulgaris*. The oxidized/semiquinone potential of *A. nidulans* flavodoxin (−0.221V at pH 7) is nearly equal to that of free FMN. *D. vulgaris* and *A. nidulans* flavodoxins share the ability to bind riboflavin, whereas *Clostridium MP* flavodoxin has little or no affinity for the dephosphorylated prosthetic group.

The Structure of Flavodoxin from *Anacystis nidulans*

The folding of the polypeptide chain and the conformation of the FMN binding site were first deduced from isomorphous replacement maps at 2.5 Å resolution, with the aid of the chemical sequence of the first 55 and final 4 residues (6). To obtain the molecular model described here, we have extended the resolution of the native data to 2.0 Å and carried out two cycles of restrained group least squares refinement (7). The crystallographic R factor for this model, comprising 935 of the expected 1454 atoms (8), is 0.426 for data between 2.0 and 5.0 Å. In the absence of a complete sequence, the interpretation of several surface regions is still ambiguous. As a result, the residue numbering beyond position 55 is not unequivocally established.

Figure 1 is a schematic representation of the conformation of flavodoxin from *A. nidulans*. The five strands of the central parallel sheet are connected via several helical regions with the same topology found in flavodoxins from *Clostridium MP* (3,4) and *D. vulgaris* (1,2), but residues 121–142 form a small lobe which is not found in either of the shorter-chain structures. Superposition of the parallel sheets of *Clostridium MP* and *A. nidulans* flavodoxins (Cf.

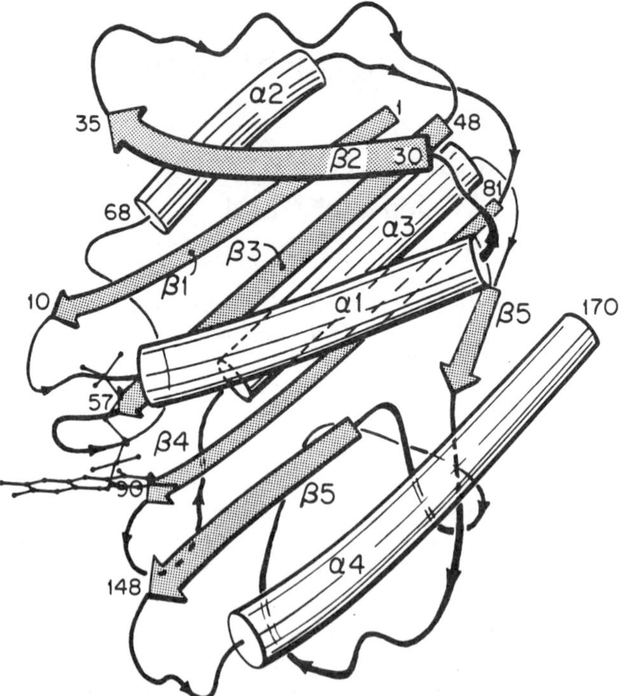

Figure 1. A drawing depicting the chain folding in flavodoxin from *Anacystis nidulans*. The central parallel sheet makes right-handed (9) connections via helices α_1 (above the sheet), and α_2 and α_3 (below). Residue numbers beyond 55 are based solely on the structural model and must be considered tentative. The lobe at the lower right, which interrupts strand β_5, accounts for 22 of the "extra" residues in this long-chain flavodoxin.

Figure 2) reveals the other regions where the chain of *A. nidulans* flavodoxin is lengthened, relative to the *Clostridium MP* chain: in the surface connection between residues 57 and 70, at the entry to the C-terminal helix, and beyond residue 90 (*A. nidulans* numbering).

The structure suggests that the major increment in the chain length could be the consequence of the insertion of a block of contiguous residues. Similar events have been postulated for the cytochrome c family (10), and the analogy implies that other long-chain flavodoxins contain insertions in the same vicinity as residues 121–142 of *A. nidulans* flavodoxin. However, flavin-binding domains of the larger flavoenzymes, glutathione reductase (11), and p-hydroxybenzoate hydroxylase (12), are not spliced into their longer polypeptide chains at a position corresponding to the middle of strand β_5 of flavodoxins. Instead, the polypeptide chains of these proteins leave the flavin-binding domain after forming the equivalent of strand β_3.

The conformation of the FMN binding site of *A. nidulans* flavodoxin is shown in Figure 3. Residues in the sequences 9–14, 55–62, and 89–100 provide the protein: prosthetic group contacts. The phosphate oxygens are positioned

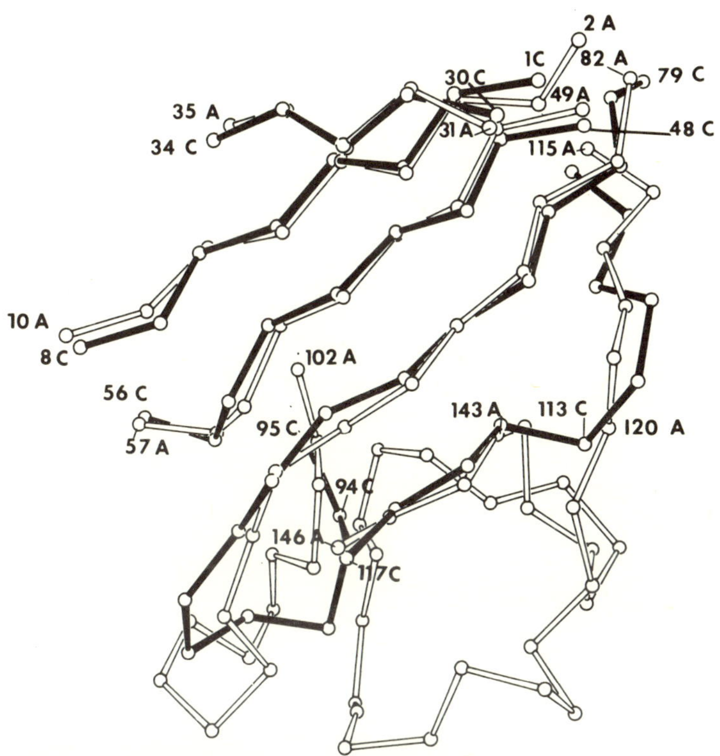

Figure 2. A superposition of the C_α atoms from portions of the polypeptide chains of flavodoxins from *Clostridium MP* (filled bonds, "C" labels) and *A. nidulans* (open bonds, "A" labels), comparing the middle of the molecules, and showing most of the residue insertions found in *A. nidulans* flavodoxin. The major insertion (residues 121–142 of *A. nidulans* flavodoxin) begins at a point where the interstrand hydrogen-bonding is disrupted in both *C. MP* and *D. vulgaris* flavodoxins by the presence of a small loop (residues 112–114 in clostridial flavodoxin). The two chains adopt very different conformations between 88C (90A) and 95C (102A) where, according to the present model, five additional residues appear in *A. nidulans* flavodoxin.

to hydrogen-bond to backbone NH groups in the sequence 10–13, and to the side chain hydroxyls of threonines 9, 11, and 14; the ring NH of Trp 57 may also interact with O I. In the electron density, the isoalloxazine ring lies between two strong planar features which can be fit by indole groups. The Trp ring at 57, "inside" the isoalloxazine, is tilted by about 35° relative to the flavin ring, but the "outer" Trp at 95 is nearly parallel to the isoalloxazine and stacked over its solvent side. Hydrogen bonds are probably formed between NH 91 and the flavin N(1) and/or O(2), between O 98 and N(3), and between NH 100 and O(2) (Figure 3). Interatomic distances between these donor-acceptor pairs are less than 3.5 Å in the current model.

The conformation of the chain reversal near residue 60 cannot be assigned unambiguously. For example, the present construction positions $C_\alpha 58$ near presumed side chain density but then places some of the other atoms of the

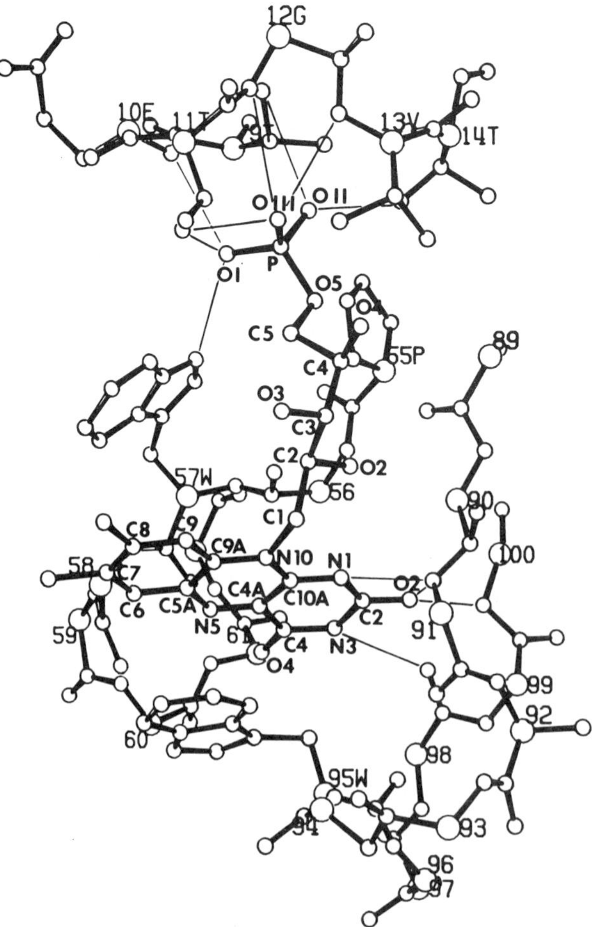

Figure 3. The FMN-binding site of flavodoxin from *A. nidulans*. Filled bonds distinguish the FMN from the protein; residue numbers and identities are indicated at C_α atoms. Except for Trp 57 and Trp 95, assigned from the electron densities, side chains have been removed or truncated if the chemical sequence is unknown. Possible hydrogen bonds involving phosphate oxygens or the flavin atoms N(1), O(2) and N(3) are indicated by fine lines which connect atoms with donor-acceptor distances less than 3.5 Å.

tight 3_{10} bend at 57–60 in weak density. Earlier studies (4,5) have implicated residues near position 60 in the stabilization of the semiquinone state via formation of NH(5): protein hydrogen bonds. The inference from the oxidized/semiquinone potential of *A. nidulans* flavodoxin is that formation of an NH(5): protein hydrogen bond in the semiquinone state must be almost precisely offset by compensating changes in other energies. Rearrangement of the conformation of Figure 3 would be required to bring a backbone oxygen within hydrogen-bonding distance of N(5); possible acceptors are 4.0 or more Å away from N(5) in this model of oxidized *A. nidulans* flavodoxin.

Comparisons of the FMN Binding Sites in *A. nidulans*, *D. vulgaris*, and *Clostridium MP* Flavodoxins

To detect homologies or differences in the conformations, we have superposed pairs of structures using coordinate transformations determined by least-squares fitting of equivalent atoms (13). Because the conformation of the central sheet is highly conserved, backbone atoms of the parallel strands were chosen to determine the coordinate transformations. The transformation of *Clostridium MP* coordinates was based on 132 atoms which matched with an rms discrepancy of 0.57 Å, and that of *D. vulgaris* coordinates, on 111 atoms which matched with an rms discrepancy of 0.67 Å. Coordinates for *D. vulgaris* flavodoxin were obtained by fitting the chemical sequence to an electron density map at 2.0 Å resolution, using the graphics system GRIP 75 at the University of North Carolina. The alignment of the flavin binding sites can be seen in Figures 4, 5, and 6.

It is immediately apparent that *A. nidulans* and *D. vulgaris* flavodoxins are more closely related than *A. nidulans* and *Clostridium MP* flavodoxins. In *A. nidulans* and *D. vulgaris* flavodoxins, the orientations of the isoalloxazine rings, of the inner hydrophobic residues (both Trp) and of the outer aromatics (Trp

Figure 4. A stereodrawing comparing the FMN-binding sites of flavodoxins from *D. vulgaris* (filled bonds, labelled N) and *A. nidulans* (open bonds, labelled CA). The drawing includes three sections of protein chain, one with three hydroxyamino acids which interact with the flavin phosphate (top), one near residue 60 (beneath the isoalloxazine ring), and one including residues 90–100 (right). Hydrogen bonds between FMN and protein are not drawn (Cf. text). Many of the side chains have been deleted or truncated for convenience; residues of *A. nidulans* flavodoxin are identified in Figure 3 and reference 6; the sequence of *D. vulgaris* flavodoxin appears in reference 15.

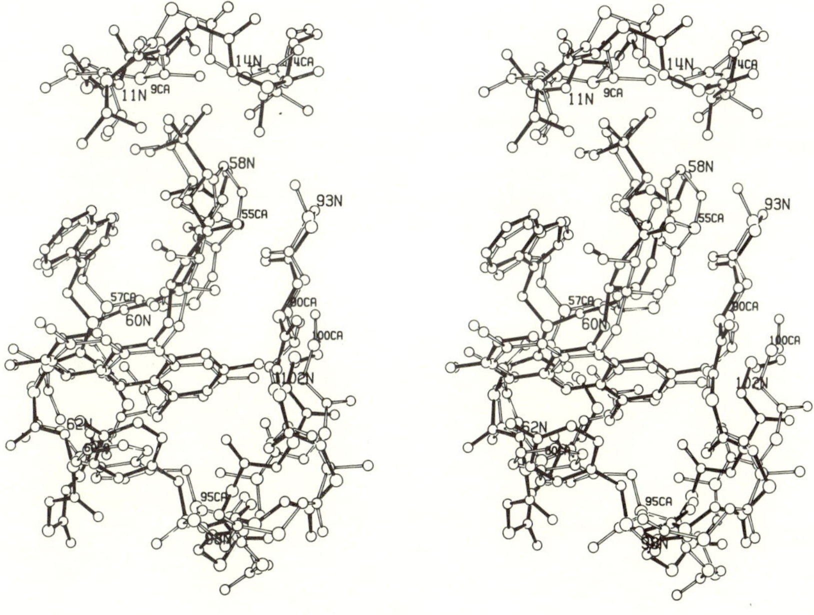

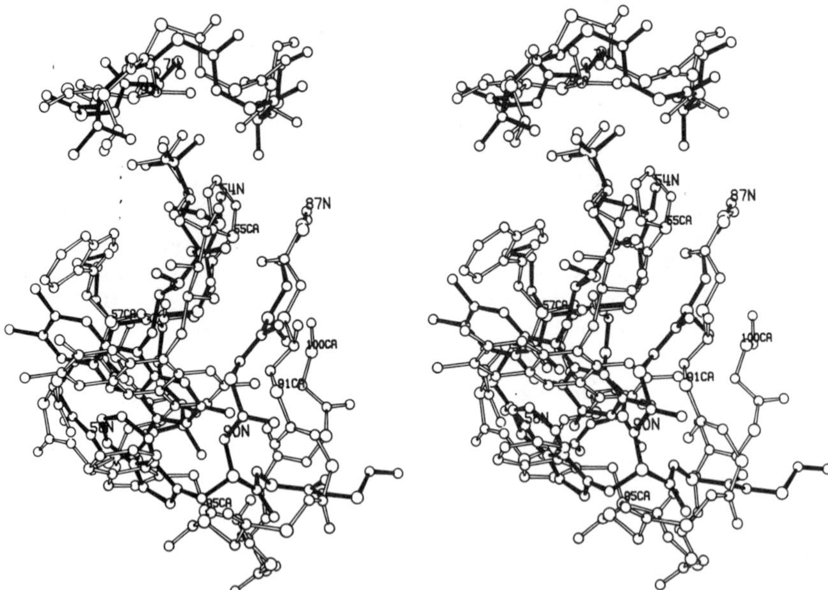

Figure 5. A stereodrawing comparing the-FMN binding sites of flavodoxins from *Clostridium MP* (filled bond, labelled N) and *A. nidulans* (open bonds, labelled CA). The orientation is the same as that of Figure 4, and the same portions of the structures are displayed. References 4 and 5 give the sequence of *Clostridium MP* flavodoxin.

or Tyr) are virtually identical. Interactions of the $-O-PO_3$ groups are analogous, although replacement of Ser 58 in *D. vulgaris* by Pro 55 in *A. nidulans* eliminates a hydrogen-bonding partner of O I; an apparent rotation of the side chain of Thr 11 may be an attempt to compensate for the loss of the serine interaction. The 89–100 segment of the *A. nidulans* model currently includes two more residues than the corresponding chain of *D. vulgaris* flavodoxin, but the two folds are nevertheless very similar in this region. The backbone NH to flavin N(1) or O(2), carbonyl O to N(3), and peptide NH to O(2) hydrogen bonds are closely reproduced in each molecule, and the C_α and C_β atoms of the outer aromatic occupy structurally equivalent positions. The largest variations within the FMN-binding site occur in the 60s region, where residue 59 of *A. nidulans* represents an insertion, while the *D. vulgaris* chain has extra residues at 63–64.

Alignment of the parallel sheets of *Clostridium MP* and *A. nidulans* flavodoxins does not superpose the two isoalloxazine rings. The flavin rings are actually inclined at an angle of about 30°, and the C_α and C_β of the outer aromatic residues, Trp 90 and Trp 95, are not spatially equivalent. The disposition of the outer aromatic rings is very different in Figure 5, partly because Trp 90 is not parallel to the isoalloxazine ring in *C. MP* flavodoxin. The C_α of the inner hydrophobic residues (Met 56 and Trp 57) coincide, but the ring of Trp 57 in *A. nidulans* flavodoxin assumes an orientation which would cause it to collide with the flavin ring of *Clostridium MP* flavodoxin.

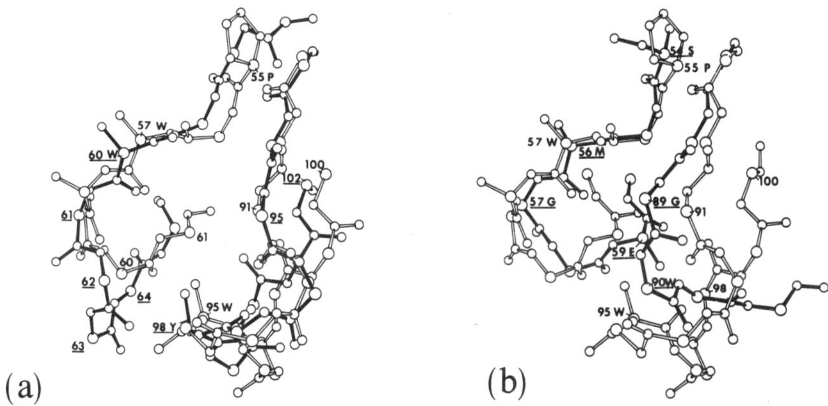

Figure 6. Drawings comparing the chain segments near residue 60 and near residues 90–100 in (a) *D. vulgaris* (filled bonds, C_α labels underlined) and *A. nidulans* (open bonds) flavodoxins and (b) *Clostridium MP* (filled bonds, C_α labels underlined) and *A. nidulans* (open bonds) flavodoxins. The view is rotated slightly from that of Figures 4 and 5.

The backbone NH to flavin N(1) interaction is similar in both *A. nidulans* and clostridial flavodoxins, but from residue 91 to residue 102 of *A. nidulans* flavodoxin the two chains fold quite differently (Figures 2 and 5). Following NH 91, the main chain of *Clostridium MP* flavodoxin makes no further interactions with the pyrimidine end of the flavin; instead, a carboxylate oxygen of Glu 59 forms a hydrogen bond at N(3). The chain reversal near residue 60 of *A. nidulans* flavodoxin does show considerable similarity to the corresponding region of clostridial flavodoxin, the principal discrepancy being the insertion of residue 58 in *A. nidulans* flavodoxin. Extension of the chemical sequence of *A. nidulans* flavodoxin, combined with continued refinement of the structure, should resolve ambiguities in the interpretation of the 60s segment, and define the relationship to clostridial flavodoxin. The phosphate-binding loop is the most constant portion of the active centers. Alignment of *Clostridium MP* and *A. nidulans* flavodoxins shows the atoms of residues 7–12 of clostridial flavodoxin in positions homologous to those of 9–14 of *A. nidulans*.

Discussion

None of the residues in the major insertion at 121–142, responsible for the classification of *A. nidulans* as a “long-chain” flavodoxin, makes immediate contact with the FMN. While it is conceivable that the inserted residues play some role in selective recognition of protein substrates, comparison of the structure of *D. vulgaris* and *A. nidulans* flavodoxins shows convincingly that the conformation of the FMN-binding site need not depend on the size classification. These representatives of “long”- and “short”-chain flavodoxins have clearly homologous active sites. In contrast, the distinctive spectra associated with the “rubrum” vs “pasteurianum” classifications, which correlate with the ability to bind riboflavin (14), must be related to structural variations in the FMN-binding site. The present comparisons suggest that the conformation at 90–100 (*A. nidulans* numbering) may be important for the interaction of flavodoxins with riboflavin. The similar structures found in *D. vulgaris* and *A. nidulans* flavodoxins produce good stacking interactions between the outer aromatic residues and the isoalloxazine rings, and provide a series of backbone: isoalloxazine hydrogen bonds.

Conservation of the phosphate-binding loop, residues 9–14 in *A. nidulans* flavodoxin, is strong enough to be evident in the primary structures, where a triplet of hydroxyaminoacids recurs, and is exhibited in the three-dimensional conformations. Residue substitutions may nevertheless alter the contribution of the phosphate: protein interaction to the overall K_a for binding of FMN. For example, the exchange of Pro 55 in *A. nidulans* for the serines found at 54 and 58 in clostridial and *D. vulgaris* flavodoxins affects the hydrogen-bonding pattern and the polarity of the phosphate site. Further, O I in *Clostridium MP* flavodoxin hydrogen bonds to a solvent molecule residing at a site which overlaps atoms of the "inner" Trp in *D. vulgaris* and *A. nidulans* flavodoxins.

Studies of the semiquinone forms of *D. vulgaris* (2) and *Clostridium MP* (4) flavodoxins have shown that the 60s region rearranges when an electron is added to the oxidized molecule. The present maps of oxidized *A. nidulans* flavodoxin indicate that the chain reversal near 60 cannot be superposed on the corresponding portion of either *Clostridium MP* or *D. vulgaris* flavodoxin. The structural variability in this region, and the dependence of the oxidized/semiquinone potential on species, raises the possibility that flavodoxins have developed several slightly different mechanisms for modulation of the oxidized/semiquinone potentials (4,5). Further analyses of the structures of the semiquinone forms will be necessary to test this notion.

ACKNOWLEDGMENTS
This research has been supported by NIH Grants GM 16429 (M.L.L.) and AM 3288 (L.H.J.).

References

1. Watenpaugh, K.D., Sieker, L.C., and Jensen, L.H. (1973) *Proc Nat Acad Sci USA* 70:3857–3860.
2. Watenpaugh, K.D., Sieker, L.C., and Jensen, L.H. (1976) In *Flavins and Flavoproteins*. Singer, T.P. (ed.) pp. 405–410. Amsterdam: Elsevier.
3. Burnett, R.M., Darling, G.D., Kendall, D.S., LeQuesne, M.E., Mayhew, S.G., Smith, W.W., and Ludwig, M.L. (1974) *J Biol Chem* 249:4383–4392.
4. Smith, W.W., Burnett, R.M., Darling, G.D., and Ludwig, M.L. (1977) *J Mol Biol* 117:195–225.
5. Mayhew, S.G. and Ludwig, M.L. (1975) In *The Enzymes*. Boyer, P.D. (ed.) 3rd ed., Vol. 12, pp. 57–118.
6. Smith, W.W., Pattridge, K.A., Ludwig, M.L., Petsko, G.A., Tsernoglou, D., Tanaka, M., and Yasunobu, K.T. (1981) *J Mol Biol*, submitted.
7. Hoard, L.H. and Nordman, C.E. (1979) *Acta Cryst* A35:1010–1015.
8. Smillie, R.M. and Entsch, B. (1971) *Meth Enz* 23:504–514.
9. Richardson, J.S. (1977) *Nature (London)* 268:495–500.
10. Dickerson, R.E. and Takano, T. (1978) In *Frontiers of Biological Energetics: Electrons to Tissues*. Dutton, P.L., Leigh, J.S., and Scarpa, A. (eds.) Vol. 1, pp. 101–108. New York: Academic Press.

11. Schulz, G.E., Schirmer, R.H., Sachsenheimer, A., and Pai, E.F. (1978) *Nature (London)* 273:120–129.

12. Wierenga, R.K., DeJong, R.J., Kalk, K.H., Hol, W.G.J., and Drenth, J. (1979) *J Mol Biol* 131:55–73.

13. Rao, S.T. and Rossmann, M.G. (1973) *J Mol Biol* 76:241–256.

14. D'Anna, J.A., Jr. and Tollin, G. (1972) *Biochemistry* 11:1073–1080.

15. Dubourdieu, M., Le Gall, J., and Fox, J.L. (1973) *Biochem Biophys Res Comm.* 52:1418–1425.

Published 1982 by Elsevier North Holland, Inc.
Vincent Massey and Charles H. Williams, Editors
Flavins and Flavoproteins

CHAPTER 4

The Structure of Ferredoxin-NADP Oxidoreductase: A Progress Report

P.A. Karplus and J.R. Herriott

Department of Biochemistry, University of Washington, Seattle, Washington

Introduction

Ferredoxin-NADP oxidoreductase is the enzyme responsible for catalyzing the transfer of a hydride from reduced ferredoxin to the pyridine nucleotide. Located on the chloroplast membrane, this enzyme participates in the final step in the conversion of light energy into one of the chemically usable forms, NADPH. We are engaged in the elucidation of the three-dimensional structure to provide a structural framework against which mechanistic models and hypotheses can be tested. It is our hope that structural information can be combined with chemical and spectroscopic data to provide greater understanding of how this flavoprotein functions.

Methods

Multiple isomorphous replacement was used to determine the phases to 3.7 A. To extend the resolution to 3.0 A, we have had to search for new and better derivatives. We have prepared a new derivative by changing the solvent system from ammonium sulfate to polyethylene glycol but this change has resulted in having one good derivative in each solvent system. The details of how we combined phase information from the two solvents will be published elsewhere. Since the phasing power of even our best derivatives was nearly exhausted at 3.0 A, we have used calculated phases to supplement those obtained by isomorphous replacement and to get approximate phases for reflections between 3.0 and 2.5 A. We have combined the phase information obtained from these various sources using the method of Hendrickson and Lattman (1) and weighting the individual contributions as described by Sim (2).

Results

An overall description of the molecule at 3.7 A resolution has been published (3). Then, as now, the electron density maps can be described as showing many sections of the chain clearly but the connections between these sections are more obscure. Alpha helices, for example, are easily recognized but often the density at either end fades out so that the path of the chain is difficult to follow. Because the contours representing low levels of electron density tend to wander, the course of the chain is sometimes ambiguous. The origin of these

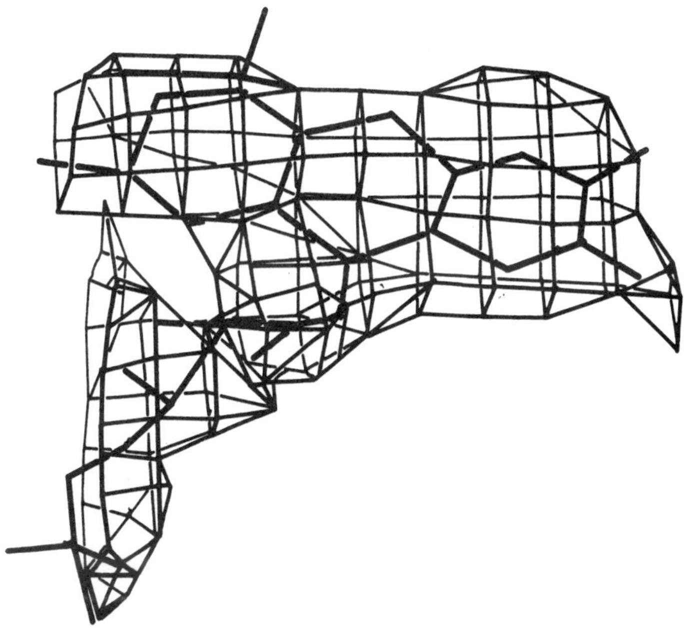

Figure 1. Redrawn from a photograph taken of the screen of a cathode ray tube attached to a computer graphics station. Scene shown is a superposition of cage contours of electron density and a stick figure of the isoalloxazine ring of FAD.

effects is unknown. Perhaps errors in the measured amplitudes, or the substantial errors in the phases are the cause. Perhaps those regions of the chain are disordered in the crystal. We are now engaged in a cyclic procedure of adjusting the model, calculating new phases, combining the new phases with those from isomorphous replacement, and then calculating new electron density and difference maps. The next cycle begins with making further adjustments to the model as indicated by these maps.

Concurrent with the X-ray crystallographic studies on FNR, we have begun a chemical determination of the sequence. The amino terminus of FNR is blocked. To generate peptides for sequencing, the protein was reduced and carboxymethylated and cleaved at methionines (CNBr), or additionally succinylated and cleaved at arginines (by trypsin). Sequence information obtained from peptides generated in this fashion sums up to 250 residues (out of a presumed 300). Some overlaps have been proven, but most remain to be shown.

Discussion

In contrast to some sections of the electron density map where the chain becomes hard to follow, the regions showing the coenzymes are quite clear. The

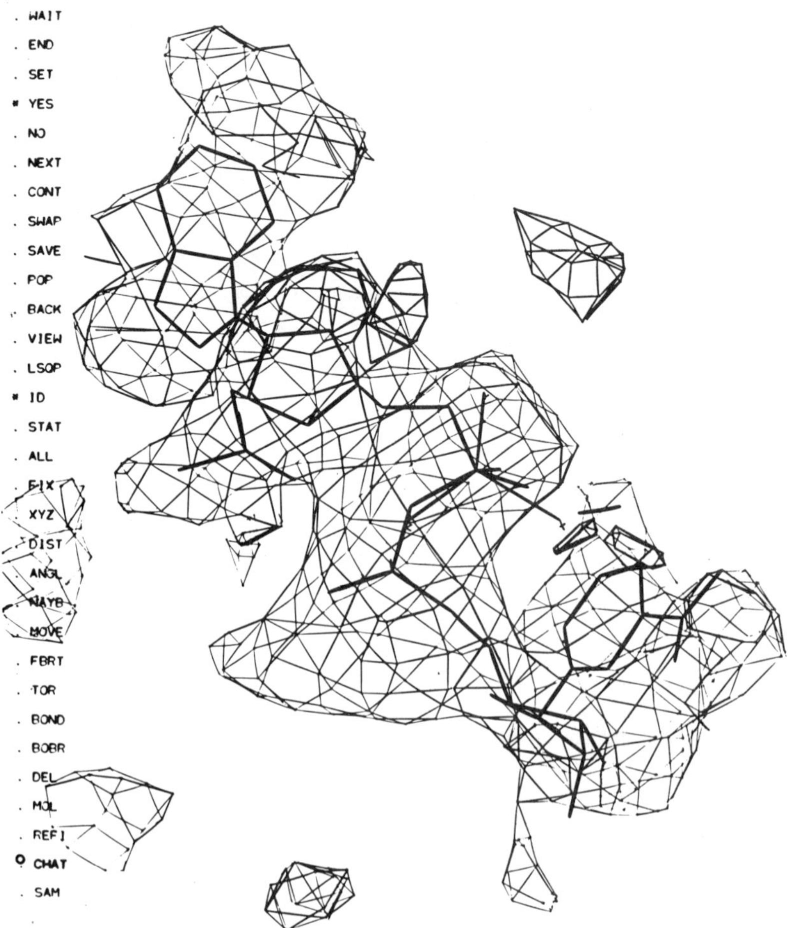

Figure 2. Fit of the NADPH to the difference electron density at 3.0 A resolution. Symbols along the left hand edge are commands to be used to manipulate the model with respect to the contours and are part of the "FRODO" system developed by T. Alwyn Jones.

FAD and NADPH are both in extended conformations and oriented so that the nicotinamide ring approaches to within six angstroms of the isoalloxazine ring. The fit of the isoalloxazine ring to its density is shown in Figure 1. The corresponding fit of the NADPH to the difference density is shown in Figure 2. In this view it appears that the nicotinamide could be adjusted to give a better fit to the density. This is indeed possible and makes the point that these results are preliminary and subject to minor adjustments. The orientation of the two-ring systems is shown in Figure 3. A schematic representation of the sequence results to date is shown in Figure 4.

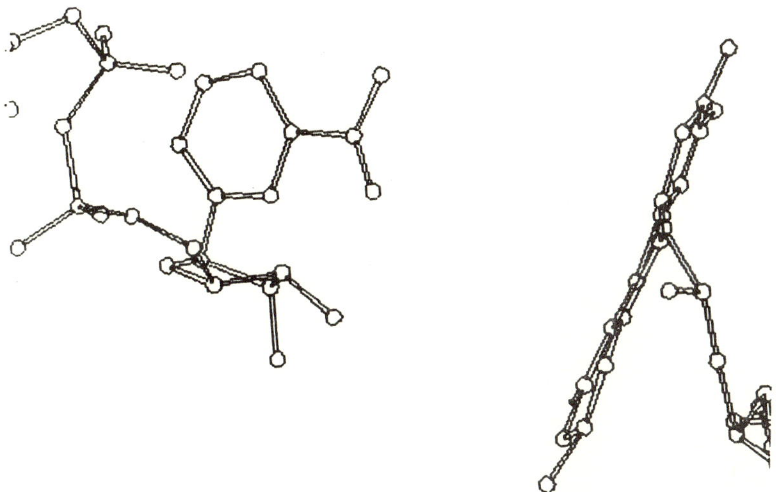

Figure 3. Drawing showing the relative orientation of the nicotinamide and isoalloxazine rings systems when the NADPH is bound to the crystal.

Figure 4. Schematic of the progress to date in determining the primary sequence of FNR. Parentheses indicate that the order of the enclosed peptides is as yet unknown.

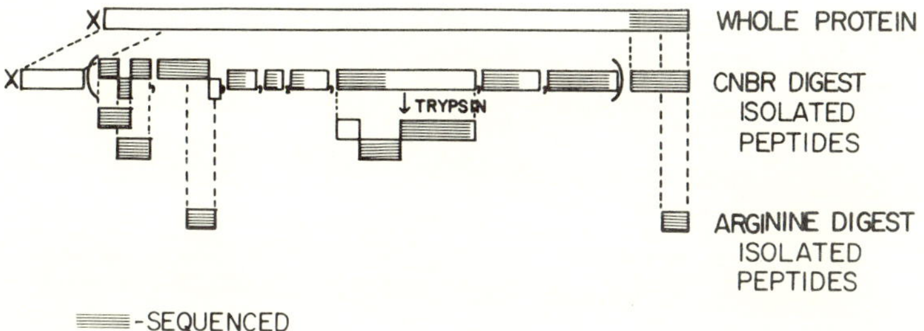

ACKNOWLEDGMENTS

We thank Albert J. Smith for his help. This work has been supported by The National Science Foundation Grant BMS75-07357. One of us (P.A.K.) has been supported during this time by NIH Predoctoral Training Grant GM-07270.

References

1. Hendrickson W.A. and Lattman, E.E. (1970) *Acta Cryst* B26:136–143.
2. Sim, Q.A. (1959) *Acta Cryst* 12:813.
3. Sheriff, S. and Herriott, J.R. (1981) *J Mol Biol* 145:441–451.

Published 1982 by Elsevier North Holland, Inc.
Vincent Massey and Charles H. Williams, Editors
Flavins and Flavoproteins

CHAPTER 5

Structural Studies on Glycolate Oxidase from Spinach

Ylva Lindqvist and Carl-Ivar Brändeń

Department of Chemistry and Molecular Biology, Swedish University of Agricultural Sciences, s-750 07 Uppsala, Sweden

Introduction

One step in the photorespiratory process of plants is the oxidation of glycolate to glyoxalate. This reaction is catalyzed by the FMN-containing enzyme glycolate oxidase (glycolate: oxygen oxidoreductase E.C.1.1.3.1.). We here report the structure determination to low resolution and an X-ray crystallographic study of the binding of a substrate analogue, thioglycolate. For a detailed account of this work see Lindqvist and Brändeń (1980) (3).

Materials and Methods

Glycolate oxidase was purified from spinach and crystallized as previously described (2). These crystals contain FMN. The spacegroup is I422 with cell parameters a=b=148.1 Å and c=134.9 Å. Derivatives were obtained by soaking crystals with $K_2Hg(CN)_4$ and $Pt(NH_3)_2Cl_2$. Crystals containing thioglycolate were obtained by cocrystallization. X-ray intensities to 5.5 Å resolution were measured on a Philips-Stoe, computer-controlled, 4-circle diffractometer in a cold room at 4°C. Phase angles were derived using data from the two heavy atom derivatives including anomalous dispersion effects.

Results

From the electron density map, it was deduced that the protein units are centered where the fourfold and the twofold axes intersect. Glycolate oxidase in our crystals are thus octameric molecules with 422 symmetry with a subunit molecular weight around 35,000. The dimension of each subunit is approximately 60 Å×45 Å×50 Å. A balsa model of the subunit is shown on Figure 1. The electron density showed ten regions which could be interpreted as helical regions. Eight of these formed a pattern which looked conspicuously like the characteristic barrel arrangement of eight consecutive α-helices found in the structure of triosephosphate isomerase, TIMase, (4) and in one domain of pyruvate kinase (5). We measured the coordinates of the two ends of each helical electron density, thus defining each helix as a vector. These vectors were then compared to the α-carbon atoms of the 8 consecutive helices of TIMase in

Figure 1. A balsa model of one subunit of glycolate oxidase showing the active site pocket and the position (dark circle) of thioglycolate. White regions correspond to electron density assigned to the barrel domain which here is partly hidden behind the darker regions of the second domain.

a Vector General 3400 display unit using the RING system of programs (1). This comparison showed excellent agreement between the diameters of the barrels and the twist of the helices. Figure 2 shows the electron density corresponding to these 8 helices from the glycolate oxidase map with the α-carbon atomic coordinates of the 8 helices of the TIMase barrel superimposed. We then plotted the remaining α-carbon atoms of TIMase into the glycolate oxidase electron density map. All β-strands were positioned reasonably well within density. One-third of the electron density could not be assigned to this barrel domain structure. This density is arranged in a second domain, which is outside one end of the barrel. It has not been possible to interpret this second domain in terms of a known folding pattern. The second domain is involved in the contact between subunits around one of the crystallographic axes. Four such dimers are in contact around the fourfold axis giving rise to the octameric molecules in the crystal lattice. The subunit interactions are illustrated in Figure 3. The roughly spherical octameric molecules have a diameter of approximately 100 Å. The centers of these molecules form a body-centered almost cubical lattice. By this arrangement, a stable three-dimensional network is obtained which contains large solvent channels, approximately 60 Å in diameter which pass through the crystal lattice.

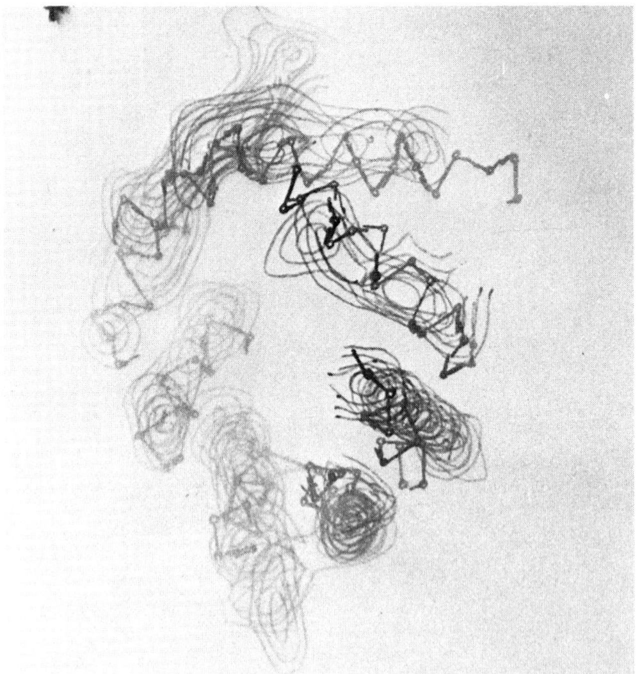

Figure 2. Part of the electron density map of glycolate oxidase which corresponds to the eight helices of the barrel domain. The positions of α-carbon atoms of the barrel helices in TIMase are superimposed.

Figure 3. Schematic diagram of the subunit contacts in the octameric glycolate oxidase molecule. (Left) viewed along the fourfold axis. (Right) viewed perpendicular to the fourfold axis.

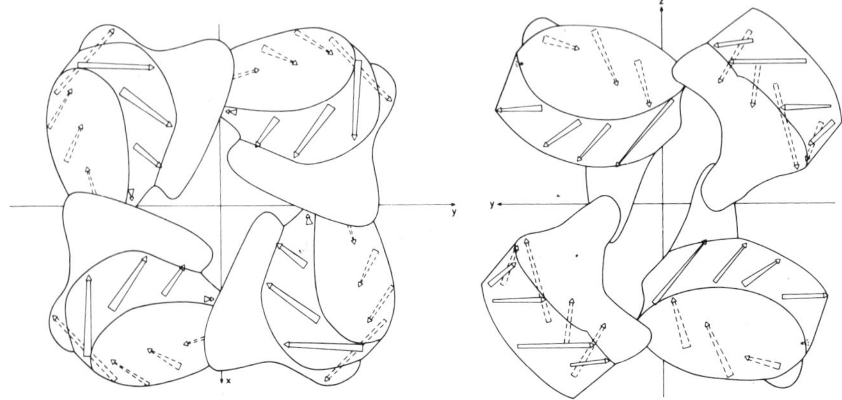

A difference Fourier map between native X-ray data and data from a crystal containing thioglycolate was computed. This map showed only one main peak located in the interface region between the two domains of the subunit at one end of the barrel close to its axis. The arrangement of the domains is such that this site is at the bottom of a deep cleft almost in the middle of the subunit (see Figure 1).

References

1. Jones, T.A. (1978) *J Appl Cryst* 11:268–272.
2. Lindqvist, Y. and Brändeń, C.-I. (1979) *J Biol Chem* 254:7403–7404.
3. Lindqvist, Y. and Brändeń, C.-I. (1980) *J Mol Biol* 143:201–211.
4. Phillips, D.C., Sternberg, M.J.E., Thornton, J.M., and Wilson, I.A. (1978) *J Mol Biol* 119:329–351.
5. Stuart, D.I., Levine, M., Muirhead, H., and Stammers, D.K. (1979) *J Mol Biol* 134:109–142.

Published 1982 by Elsevier North Holland, Inc.
Vincent Massey and Charles H. Williams, Editors
Flavins and Flavoproteins

CHAPTER 6

Preliminary X-Ray Diffraction Studies of Crystals of Lactate Oxidase from *Mycobacterium smegmatis*

William C. Stallings, Katherine A. Pattridge, Martha L. Ludwig, Vincent Massey, and Yee Soon Choong

Biophysics Research Division and Department of Biochemistry, The University of Michigan, Ann Arbor, Michigan

Single crystals (Figure 1) of lactate oxidase have been grown from ammonium sulfate at pH 6.0 in the presence of 0.2 M phosphate. The crystals are square parallelepipeds of maximally observed dimensions: 1.0 mm×1.0 mm×0.4 mm; they form gradually after the appearance of small rhombs resembling those reported by Sullivan et al. (1) and are probably identical to the square plates noted by these authors. We presume from the phosphate concentration that the enzyme has been crystallized as the phosphate-inhibited complex (2). The diffraction patterns (Figure 2) extend to greater than 3.0 Å resolution and the crystals are generally stable in the X-ray beam. Approximate lattice constants are: $a=146, b=146, c=273$ Å. Evidence of 4/mmm diffraction symmetry indicates a tetragonal space group; a nonprimitive tetragonal lattice with the crystallographic a and b axes increased by a factor of $\sqrt{2}$ is not ruled out.

Figure 1. Crystals of lactate oxidase from *Mycobacterium smegmatis*.

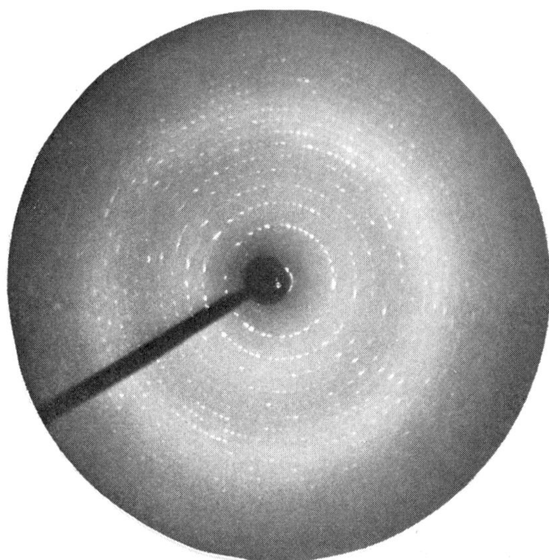

Figure 2. Still photograph: beam approximately along the *c* axis.

The molecule is an octamer composed of apparently identical 43,600 MW subunits (1). Calculations of volume: dalton ratio suggest that the unique portion of the unit cell houses more than one of the subunits, and could contain the entire octamer. We expect that the noncrystallographic symmetry can be exploited in the determination of this relatively large structure.

ACKNOWLEDGMENTS
This work was supported by NIH Grants GM 16429 and GM 11106.

References

1. Sullivan, P.A., Choong, Y.S., Schreurs, W.J., Cutfield, J.F., and Shepherd, M.G. (1977) *Biochem J* 165:375–383.
2. Lockridge, O., Massey, V., and Sullivan, P.A. (1972) *J Biol Chem* 247:8097–8106.

Published 1982 by Elsevier North Holland, Inc.
Vincent Massey and Charles H. Williams, Editors
Flavins and Flavoproteins

CHAPTER 7

The Complete Primary Structure of Glutathione Reductase from Human Erythrocytes

Renate Untucht-Grau, Rosi Blatterspiel,*
R. Luise Krauth-Siegel, Mansoor Saleh,
Emile Schiltz,* R. Heiner Schirmer, and
Brigitte Wittmann-Liebold**

*Max-Planck-Institut für Medizinische Forschung, Heidelberg; *Physiologisch-Chemisches Institut der Universität Würzburg; **Max-Planck-Institut für Molekulare Genetik, Berlin, Germany*

Summary

The FAD-free apoenzyme consists of two identical polypeptide chains. The amino acid sequence of one chain was determined to be:

1 Ala-Cys-Arg-Gln-Glu-Pro-Gln-Pro-Gln-Gly-Pro-Pro-Pro-Ala-Ala-Gly-Ala-Val-Ala-Ser-Tyr-Asp-Tyr-Leu-Val-
26 Ile-Gly-Gly-Gly-Ser-Gly-Gly-Leu-Ala-Ser-Ala-Arg-Arg-Ala-Ala-Glu-Leu-Gly-Ala-Arg-Ala-Ala-Val-Val-Glu-
51 Ser-His-Lys-Leu-Gly-Gly-Thr-Cys-Val-Asn-Val-Gly-Cys-Val-Pro-Lys-Lys-Val-Met-Trp-Asn-Thr-Ala-Val-His-
76 Ser-Glu-Phe-Met-His-Asp-His-Ala-Asp-Tyr-Gly-Phe-Pro-Ser-Cys-Glu-Gly-Lys-Phe-Asn-Trp-Arg-Val-Ile-Lys-
101 Glu-Lys-Arg-Asp-Ala-Tyr-Val-Ser-Arg-Leu-Asn-Ala-Ile-Tyr-Gln-Asn-Asn-Leu-Thr-Lys-Ser-His-Ile-Glu-Ile-
126 Ile-Arg-Gly-His-Ala-Ala-Phe-Thr-Ser-Asp-Pro-Lys-Pro-Thr-Ile-Glu-Val-Ser-Gly-Lys-Lys-Tyr-Thr-Ala-Pro-
151 His-Ile-Leu-Ile-Ala-Thr-Gly-Gly-Met-Pro-Ser-Thr-Pro-His-Glu-Ser-Gln-Ile-Pro-Gly-Ala-Ser-Leu-Gly-Ile-
176 Thr-Ser-Asp-Gly-Phe-Phe-Gln-Leu-Glu-Glu-Leu-Pro-Gly-Arg-Ser-Val-Ile-Val-Gly-Ala-Gly-Tyr-Ile-Ala-Val-
201 Glu-Met-Ala-Gly-Ile-Leu-Ser-Ala-Leu-Gly-Ser-Lys-Thr-Ser-Leu-Met-Ile-Arg-His-Asp-Lys-Val-Leu-Arg-Ser-
226 Phe-Asp-Ser-Met-Ile-Ser-Thr-Asn-Cys-Thr-Glu-Glu-Leu-Glu-Asn-Ala-Gly-Val-Glu-Val-Leu-Lys-Phe-Ser-Gln-
251 Val-Lys-Glu-Val-Lys-Lys-Thr-Leu-Ser-Gly-Leu-Glu-Val-Ser-Met-Val-Thr-Ala-Val-Pro-Gly-Arg-Leu-Pro-Val-
276 Met-Thr-Met-Ile-Pro-Asp-Val-Asp-Cys-Leu-Leu-Trp-Ala-Ile-Gly-Arg-Val-Pro-Asn-Thr-Lys-Asp-Leu-Ser-Leu-
301 Asn-Lys-Leu-Gly-Ile-Gln-Thr-Asp-Asp-Lys-Gly-His-Ile-Ile-Val-Asp-Glu-Phe-Gln-Asn-Thr-Asn-Val-Lys-Gly-
326 Ile-Tyr-Ala-Val-Gly-Asp-Val-Cys-Gly-Lys-Ala-Leu-Leu-Thr-Pro-Val-Ala-Ile-Ala-Ala-Gly-Arg-Lys-Leu-Ala-
351 His-Arg-Leu-Phe-Glu-Tyr-Lys-Glu-Asp-Ser-Lys-Leu-Asp-Tyr-Asn-Asn-Ile-Pro-Thr-Val-Val-Phe-Ser-His-Pro-
376 Pro-Ile-Gly-Thr-Val-Gly-Leu-Thr-Glu-Asp-Glu-Ala-Ile-His-Lys-Tyr-Gly-Ile-Glu-Asn-Val-Lys-Thr-Tyr-Ser-
401 Thr-Ser-Phe-Thr-Pro-Met-Tyr-His-Ala-Val-Thr-Lys-Arg-Lys-Thr-Lys-Cys-Val-Met-Lys-Met-Val-Cys-Ala-Asn-
426 Lys-Glu-Glu-Lys-Val-Val-Gly-Ile-His-Met-Gln-Gly-Leu-Gly-Cys-Asp-Glu-Met-Leu-Gln-Gly-Phe-Ala-Val-Ala-
451 Val-Lys-Met-Gly-Ala-Thr-Lys-Ala-Asp-Phe-Asp-Asn-Thr-Val-Ala-Ile-His-Pro-Thr-Ser-Ser-Glu-Glu-Leu-Val-
476 Thr-Leu-Arg

Ala-1 is N-acetylated. Cys-58 and Cys-63 (formerly designated Cys-41 and Cys-46) constitute the catalytic redox-active disulfide; Cys-90 forms a disulfide bridge with its counterpart of the other chain. We recommend using the position numbers of the chemically determined primary structure also in other studies on glutathione reductase.

Introduction

Glutathione disulfide reductase (1) isolated from human erythrocytes (2) catalyzes the reaction

$$NADPH + GSSG + H^+ \rightleftharpoons 2\ GSH + NADP \text{ (EC 1.6.4.2.)}$$

The crystal structures of several enzyme-substrate-complexes are known at high resolution so that the pathway of the reduction equivalents through the protein's interior can be traced (3,4). The sequence information which is necessary for more detailed studies on glutathione reductase is given here.

Experimental Procedures

15 CNBr-produced fragments of reduced and carboxymethylated glutathione reductase were isolated and sequenced using established procedures (5–7). The phenylthiohydantoin derivatives of amino acids delivered by the solid phase sequenator were analyzed by high performance liquid chromatography (8).

Many sequenced peptides were ordered by computerized assignment (9) to the map-sequence, that is to a complete sequence deduced from the electron density map at 3Å (or 2Å) resolution (3,10). In retrospect, it was found that the map-sequence is correct in approximately one-third of all residue positions and that it contains 3% single insertions or single deletions (5–7). This quality of the map-sequence was sufficient for our purpose when the following criteria were met. The sequenced peptide had to be longer than 20 residues when it was assigned to an electron density map of 3Å resolution, and longer than 10 residues when it was assigned to a map of 2Å resolution (10). Both the best and the second best assigned positions were considered. The quality of fit for the best position had to be at least 4.0′ standard deviation units above the mean of the frequency distribution of all possible assignments. The second best position was scrutinized according to the rules of jigsaw puzzling: it was demonstrated that there was another piece fitting better to this position than the sequence under consideration. For illustration, let us take the CNBr-produced fragment V2 whose sequence was determined as Ile-Arg-His-Asp-Lys-Val-Leu-Arg-Ser-Phe-Asp-Ser-Hse. The sequence was assigned by the computer to position 197–209 of the map-sequence (=position 217–229 in the new numbering system). The quality of fit for this best position was 6.0 standard deviation units. The second best assignment—with a quality of fit of 3.8σ—corresponds to position 374–386. To this region, however, another sequenced peptide was assigned with a quality of fit of 10.3σ. In conclusion, the best position of V2, namely 197—209, is unambiguous. Most computerized assignments were confirmed by sequences of overlapping peptides and vice versa (5–7).

Results and Discussion

Data on the primary structure of human glutathione reductase are summarized in Figure 1 and in Table 1. A few aspects are to be detailed here.

Glutathione Reductase as a Monomeric Protein

The two polypeptide chains of glutathione reductase are cross-linked by a disulfide bond located at the twofold molecular axis (6,10). As discussed by Klotz et al. (11) and Vickers (12), glutathione reductase must be regarded as

Table 1. Some Properties of Human Glutathione Reductase.

Holoenzyme (Mr 105 000) = 2 identical chains connected by a disulfide bond + 2 FAD-molecules as dissociable prosthetic groups.

Amino acid composition (478 residues per chain): Cys_{10}, Asp_{21}, Asn_{17}, Thr_{31}, Ser_{31}, Glu_{29}, Gln_{11}, Pro_{24}, Gly_{43}, Ala_{42}, Val_{44}, Met_{15}, Ile_{29}, Leu_{34}, Tyr_{13}, Phe_{14}, Lys_{34}, His_{16}, Arg_{17}, Trp_{3}.

CNBr-produced fragments

Name	R1	R2	R3	T1	V1	V2	S1	V3/V4	Q1	V5
No. of residues	69	10	80	43	14	13	36	13	128	13
N-terminus	A1	W70	H80	P160	A203	I217	I230	V266	I279	Y407
Name	V6	V7	V8	V9	U1					
No. of residues	2	14	8	10	25					
N-terminus	K420	V422	Q436	L444	G454					

Domains: FAD-domain (residues 1–157, segment 1–17 being flexible); NADPH-domain (158–293); central domain (294–364); interface domain (365–478).

Conspicuous residues: N-acetyl-Ala-1 = N-terminus. Cys-2 = reactive thiol located in the flexible extension. Gly-27 = first residue of the pyrophosphate-binding loop, reference point for aligning flavoproteins. Cys-58 – Cys-63 = redox active dithiol; when viewing from the S-atom of Cys-63, flavin has the textbook orientation with the xylene moiety to the left. Cys-90 – Cys-90′ = interchain disulfide. Tyr-197 = lid protecting the flavin at the nicotinamide site. Arg-218, His-219, Arg-224 = residues binding to the 2′-phospho group of NADPH. His-467 = catalytic base. Arg-478 = C-terminus.

monomer since its constituent chains are not in an association-dissociation equilibrium which is characteristic for an oligomeric structure. It remains to be seen, however, whether the redox state of the interchain disulfide bond is subject to biochemical control in situ. Interchain disulfide bonds—which might be there in order to guarantee the specificity of a binding site made up from residues of two chains—have been observed in immunoglobulins as well as in other proteins (13). Glutathione reductase fits into this category since both polypeptide chains contribute to the two sites where GSSG is bound and reduced. This arrangement is most clearly demonstrated by the catalytic ensemble Cys-58—Cys-63 (of one chain) and His-467′ (of the other chain).

The N-Terminal Extension of the FAD-Domain

Almost 20 N-terminal residues of glutathione reductase are not visible in the crystal structure (3,6) and therefore assumed to be flexible in the native enzyme. However, in view of the 5 prolyl residues located here, it cannot be excluded that this segment participates in a defined structure within cells. The acetyl group at the N-terminus and the accumulation of undigestible residues may serve to protect this lose end against proteolysis in situ. Cys-2 is a reactive thiol (14) at the tip of the extension; the aggregation properties (15) of the enzyme in the absence of protecting thiols, for instance, could be due to disulfide bridges involving Cys-2. Other flavoenzymes which are probably related to glutathione reductase appear not to have the long extension. The N-terminal sequences of the FAD-domains of several proteins can be aligned

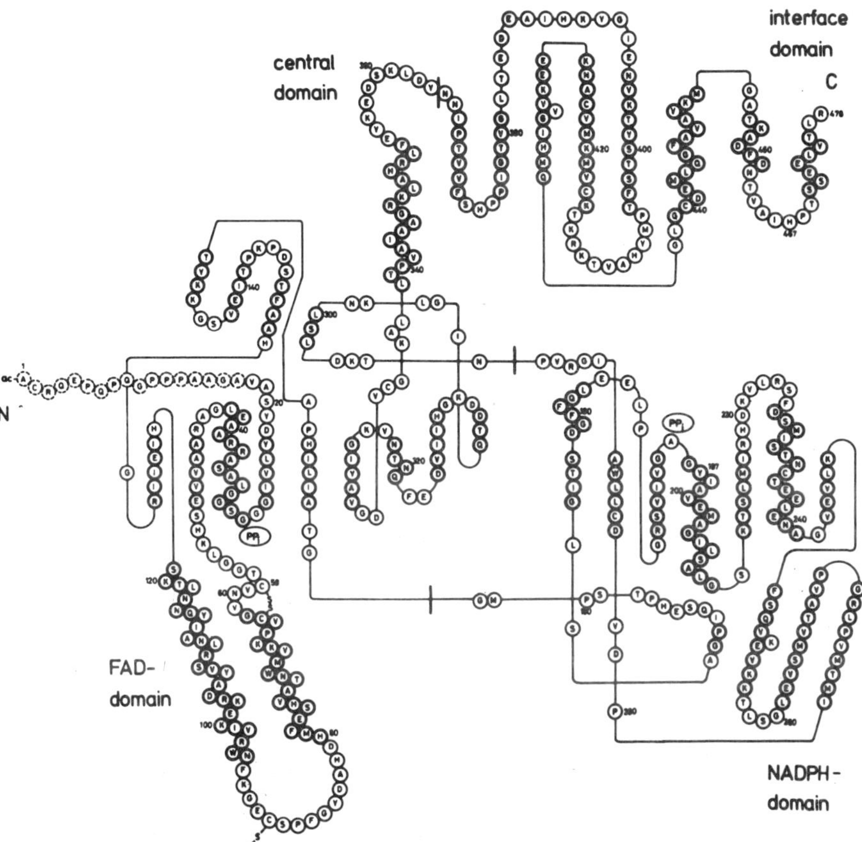

Figure 1. The chain fold of human glutathione reductase in diagrammatic representation. The one-letter code for amino acid residues (26) is used. PP_i indicates the binding positions of the pyrophosphoryl moieties of FAD and NADPH, respectively. S-S, not encircled, represents a disulfide bond. The domain borders along the chain are as in Table 1. Residues which are not visible in the electron density map (3, 10) are encircled with dashed lines, residues involved in helical structures or in β-sheets with heavier lines. The representation of sheet strand residues in straight lines perpendicular to the strands indicates that adjacent residues are hydrogen-bonded; an example is the line I123-A46-Y23-H151-G325-N322-I314-Q306. The twisted parallel-stranded sheets of the Rossmann-folds in the FAD- and in the NADPH-domain are depicted in the paper plane; the β-meanders (residues 129–148 and 248–279) must be imagined below this plane. The Greek key motif (27) of the interface domain contains two palindromic sequences in adjacent sheet strands, namely T-Y-S-T401-S-F-T and C-V-M-K420-M-V-C. The numbers are the residue positions in the chemically determined sequence. In previous papers, a numbering system based on the crystal structure was used: Tyr-21, Cys-58, Cys-63, Lys-66, Tyr-197, and His-467, for instance, were formerly called "res. 4," "Cys-41," "Cys-46," "Lys-49," "res. 177," and "His-450," respectively.

using as reference-sequences the first sheet strand of the Rossmann-fold and the following pyrophosphoryl-binding loop (residues 23 through 31 in Figure 1). On the basis of this alignment, glutathione reductase has an extension of 14 residues when compared with lipoamide dehydrogenase (16), of 18 residues when compared with p-hydroxybenzoate hydroxylase (17) and of 20 residues when compared with D-amino acid oxidase (18).

The Structurally-Known Relatives of the NADPH-Domain: One NAD-Domain and Several FAD-Domains

The FAD-domain and the NADPH-domain of glutathione reductase possess homologous three-dimensional structures (19). A computerized comparison of the two corresponding sequences (segment 1–157 vs segment 158–293) revealed practically no similarity (6), which indicates that the phylogenetic relationship between the 2 domains is very distant. In contrast, each of the 4 domains of NADPH-dependent glutathione reductase is closely related to its counterpart in NAD-dependent lipoamide dehydrogenase, the sequence homology being as high as 40% (16).

The similarity between the two pyridine nucleotide domains indicates that NAD- and NADP-dependent domains can belong to the same protein family; in our case this family even includes the FAD-domains of several proteins as distant relatives (4, 16, 17, 20).

Other known NAD(P)-dependent domains possess different chain folds and therefore belong to different protein families. It will be interesting to study NADH-dependent glutathione reductase from *Chromatium vinosum* (21) in order to test the statement (22) that the discrimination between NADPH and NAD may concern only structural details, e.g., the chemical nature of a few side chains (Table 1), and not the overall architecture of a protein. This in turn would suggest that catabolic NAD-dependent pathways and anabolic NADPH-dependent pathways are based on related protein structures.

Outlook

1. As shown by Williams et al. (16), the primary structure of glutathione reductase can serve as the interface when assigning peptide sequences of lipoamide dehydrogenase to the geometric structure of glutathione reductase.
2. The sequence data will be helpful when interpreting a variety of chemical modifications of glutathione reductase with respect to specific functions of specific residues (23).
3. The interaction of FAD-(analogues) and apoenzyme can now be investigated more systematically (S. Ghisla and R. L. Krauth-Siegel, unpublished). A fragment containing the 135 N-terminal residues of glutathione reductase (and thus more than four-fifths of the FAD-domain) may be of particular use in these studies (6). The fragment results from cleaving the acid-labile bond between Asp-135 and Pro-136 (Figure 1).
4. Glutathione reductase is a target of the antitumor drug BCNU (24) and of other compounds of medical interest (25). In light of the enzyme's atomic

structure, it should be possible to tailor these drugs so that undesirable side effects are minimized.

ACKNOWLEDGMENTS
Frau Irene König kindly helped with the manuscript. Our project was supported by the Deutsche Forschungsgemeinschaft (Schiltz 145/4 and Schirmer 102/5).

References

1. Williams, C.H., Jr. (1976) In *The Enzymes*. Boyer, P.D. (ed.). Vol. 13, 3rd ed. pp. 89–173. New York, San Francisco, and London: Academic Press.
2. Worthington, D.J. and Rosemeyer, M.A. (1976) *Eur J Biochem* 67:231–238 and earlier papers. Krohne-Ehrich, G., Schirmer, R.H., and Untucht-Grau, R. (1977) *Eur J Biochem* 80:65–71.
3. Schulz, G.E., Schirmer, R.H., Sachsenheimer, W., and Pai, E.F. (1978) *Nature* 273:120–124.
4. Pai, E.F. and Schulz, G.E. (1981) This volume.
5. Schiltz, E., Blatterspiel, R., and Untucht-Grau, R. (1979) *Eur J Biochem* 102:269–278.
6. Untucht-Grau, R., Schirmer, R.H., Schirmer, I., and Krauth-Siegel, R.L. (1981) *Eur J Biochem*, 120:407–419.
7. Krauth-Siegel, R.L. Blatterspiel, R., Saleh, M., Schiltz, E., Schirmer, R.H., and Untucht-Grau, R. (1981) *Eur J Biochem*, in press.
8. Frank, R. (1979) PhD thesis, Universität Heidelberg.
9. Untucht-Grau, R., Schirmer, R.H., and Schulz, G.E. (1979) *FEBS Lett* 105:244–248.
10. Thieme, R., Pai, E.F., Schirmer, R.H., and Schulz, G.E. (1981) *J Mol Biol*, 152:763–782.
11. Klotz, I.M., Darnall, D.W., and Langerman, N.R. (1975) In *The Proteins* H. Neurath (ed.), 3rd ed., Vol. I, pp. 293–441. New York: Academic Press.
12. Vickers, T.G. (1981) *Trends Biochem Sci* 6: No. 5, p. V.
13. Schulz, G.E. and Schirmer, R.H. (1979) *Principles of Protein Structure*. New York: Springer Verlag.
14. Untucht-Grau, R. (1981) PhD thesis, Heidelberg.
15. Worthington, D.J. and Rosemeyer, M.A. (1975) *Eur J Biochem* 60:459–466.
16. Arscott, L.D., Williams, C.H., Jr, and Schulz, G.E. (1981) This volume.
17. Weijer, W.J., Hofsteenge, J., Vereijken, J.M. Jekel, P.A., Bak, H.J., Beintema, J.J., Wierrenga, R.K., and Drenth, J. (1981) This volume.
18. Ronchi, S., Galliano, M., Minchiotti, L., Curti, B., Swenson, R.P., Williams, C.H., Jr., and Massey, V. (1981) This volume.
19. Schulz, G.E. (1980) *J Mol Biol* 138:335–347.
20. Karplus, P.A. and Herriott, J.R. (1981) This volume.
21. Chung, Y.C. and Hulbert, R.E. (1975) *J Bacteriol* 123:203–211.
22. Matthews, D.A., Alden, R.A., Bolin, J.T., Filman, D.J., Freer, S.T., Hamlin, R., Hol, W.G.J., Kisluk, R.L., Pastore, E.J., Plante, L.T., Xuong, Nguyenhuu, and Kraut, J. (1978) *J Biol Chem* 253:6946–6954.
23. Boggaram, V. and Mannervik, M. (1981) This volume.
24. Frischer, H. and Ahmad, T. (1977) *J Lab Clin Med* 89:1080–1091.
25. Krauth-Siegel, R.L., Pai, E.F., Schirmer, R.H., Schulz, G.E., and Untucht-Grau, R. (1980) *J Clin Chem Clin Biochem* 18:724–725.
26. Dayhoff, M.O. (1978) *Atlas of Protein Sequence and Structure*. Vol. 5, suppl. 3. Washington, D.C.: Natl. Biomed. Res. Foundation.
27. Richardson, J.S. (1977) *Nature* 268:495–500.

Published 1982 by Elsevier North Holland, Inc.
Vincent Massey and Charles H. Williams, Editors
Flavins and Flavoproteins

CHAPTER 8

Pig Heart Lipoamide Dehydrogenase and Human Erythrocyte Glutathione Reductase—Homology in Each of the Three Domains

L. David Arscott, Charles H. Williams, Jr., and Georg E. Schulz*

*Veterans Administration Medical Center and Department of Biological Chemistry, University of Michigan, Ann Arbor; *Max-Planck-Institute für Medizinische Forschung, Heidelberg*

Lipoamide dehydrogenase and glutathione reductase have both mechanistic and structural properties in common. Both flavoproteins contain a cystine residue which undergoes reduction-oxidation during catalysis. In catalysis, both cycle between the oxidized and two-electron reduced (EH_2) forms. We have previously reported on the high degree of amino acid sequence homology around this cystine (1).

Based on chemical modification studies on the 2-electron reduced enzymes, the thiol nearer the amino terminus of each protein interacts with the respective substrate in thiol-disulfide interchange while the other thiol interacts with the FAD (2–4). The same roles have been assigned based on X-ray diffraction analysis and chemical sequencing of glutathione reductase (5–7).

The structural and mechanistic equivalence of the nascent thiols at the EH_2 level in these enzymes, together with the extensive homology in the region of the cystine residues, suggests that homology may exist at other points between the lipoamide dehydrogenase and glutathione reductase molecules. Therefore, each tryptic peptide from lipoamide dehydrogenase thus far sequenced has been tested for homology with the sequence of human erythrocyte glutathione reductase which has recently been completed (8,9).

Each of the 20 peptides thus far sequenced has been tested for homology using a computer program which incorporates the empirical amino acid exchange frequencies of Dayhoff (10). Using this program, two very different sequences which nevertheless contain many frequently encountered changes can still be recognized as related (11–13). Extensive homology has now been found between nine tryptic peptides of pig heart lipoamide dehydrogenase and the sequence of human erythrocyte glutathione reductase (Table 1). The average homology is 40%. Peptides "a," "b," "c," "d," "f," and "g" of lipoamide dehydrogenase are homologous with segments of the two parts of the FAD domain of glutathione reductase, peptide "e" with the NADPH

Table 1. Comparison of the Sequences of Tryptic Peptides from Pig Heart Lipoamide Dehydrogenase with Homologous Regions of Human Erythrocyte Glutathione Reductase.

Peptide	Sequences	Score distance from mean (Standard deviation units)	Identity (%)
a	A D Q P I D A D V T V I G S G P G G Y V A A I K	5.1	38
	A G A V A S Y D Y L V I G G G S G G L A S A R R		
b	A A Q L G F K	4.3	57
	A A E L G A R		
c	T V C I E K N E T L G G T C L N V G C I P S K	7.6	52
	A A V V E S H K - L G G T C V N V G C V P K K		
d	L L N N G H A * H * A H * K	5.0	29
	I I R - G H A A F T S D P K		
e	S E E Q L K E E G I E - Y K	4.1	36
	C T E E L E N A G V E V L K		
f	P F T Q N L G L E E L G I E L R	5.1	44
	P N T K D L S L N K L G I Q T D		
g	I P N I A A I G D V V A G P	3.7	43
	V K G I Y A V G D V C - G K		
h	C V P S V I Y T H P E V A W V G K	6.1	35
	N I P T V V F S H P P I G T V G L		
i	V L G A H I I G P G A G E M I N E A A L A L E Y G A S C E D I A R	6.1	36
	V V G I H M Q G L G C D E M L Q G F A V A V K M G A T K A D F D N		

The lipoamide dehydrogenase sequence is on the top in each comparison. An * indicates that the residue at that position could not be identified in the lipoamide dehydrogenase sequence; for the purpose of the computer comparison, an alanine residue has been inserted. A dash indicates a deletion from that sequence in order to maximize homology.

domain, and peptides "h" and "i" with the interface domain. Thus, the homology extends through the molecule (Figure 1). The 11 peptides not placed with certainty are all short, 11 residues or less. Short peptides even with 40% identity cannot be placed with certainty unless they fit exactly between two other peptides. However their positions can presumably be determined as soon as the adjacent peptides are located.

One of the tryptic peptides ("a") placed in the sequence of glutathione reductase is the amino terminal peptide of lipoamide dehydrogenase. Together with the adjacent peptide "b," it constitutes part of the binding region of the adenine of FAD. It is homologous not only with glutathione reductase (9) but also with the functionally analogous regions of thioredoxin reductase (14), D-amino acid oxidase (15), and p-hydroxybenzoate hydroxylase (16) as well as the lactate dehydrogenase family (17) (Table 2). Homology in this region was first noted between p-hydroxybenzoate hydroxylase and D-amino acid oxidase (16). This region is highly conserved in these enzymes which represent three of the major classes of flavoproteins: transhydrogenases (class 1), oxidases (class 2), and oxygenases (class 3)(18). Peptide "c" is adjacent to peptide "b" and contains the redox-active disulfide, Cys-45 and Cys-50 (19).

We feel that these findings offer strong evidence for an evolutionary relationship between lipoamide dehydrogenase and glutathione reductase which have previously been shown to be mechanistically very similar. It should be noted that in the classical pair of homologous differentiated proteins, lysozyme and lactalbumin, the proportion of identical amino acid residues is also approximately 40% (20). Since glutathione reductase presumably diverged from lipoamide dehydrogenase during the oxygen buildup period (21) while lysozyme and lactalbumin diverged only with the emergence of mammals, the former pair have diverged much more slowly.

While the mechanism of lipoamide dehydrogenase is essentially preserved in glutathione reductase, both substrate specificities have changed. One important difference between the two mechanisms is that whereas in two-electron reduced lipoamide dehydrogenase the presence of NAD^+ changes the oxidation-reduction potential of the FAD relative to the dithiol so that electrons pass to the FAD, no such change is observed in glutathione reductase (4,22). This evolution allows glutathione reductase to function physiologically in the opposite chemical direction. The structural basis of the difference is not known and

Figure 1. Homology between lipoamide dehydrogenase and glutathione reductase. The thick bar represents the amino acid sequence of human erythrocyte glutathione reductase with the boundaries of the amino terminal extension (N), part 1 of the FAD domain (FAD-1), the NADPH domain (NADPH), part 2 of the FAD domain (FAD-2), and the interface domain (INTERFACE), indicated by vertical dashed lines. The small bars above the glutathione reductase sequence represent the tryptic peptides from pig heart lipoamide dehydrogenase listed in Table 1. They are aligned with the glutathione reductase sequence.

a b c d e f g h i
LIPOAMIDE DEHYDROGENASE
N FAD-1 NADPH FAD-2 INTERFACE
GLUTATHIONE REDUCTASE

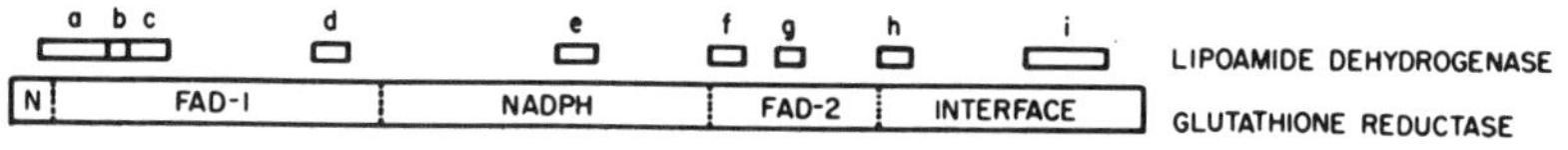

Table 2. AMP Binding Site Homology.

LipDH	-	I	D	A	D	V	T	V	I	G	S	G	P	G	G	Y	V	A	A	I	K	A	A	Q	L	G	-	-	F	K	T	V	C	I	E	K
GR	-	A	S	Y	D	Y	L	V	I	G	G	G	S	G	G	L	A	S	A	R	R	A	A	E	L	G	-	-	A	R	A	A	V	V	E	S
TRR	-	K	H	S	K	L	L	I	L	G	S	G	P	A	G																					
DAAO				M	R	V	V	V	I	G	A	G	V	I	G	L	S	T	A	L	C	I	H	E	R	Y	-	H	S	V	L	Q	P	L	D	V
PHBH		M	K	T	Q	V	A	I	I	G	A	G	P	S	G	L	L	L	G	Q	L	L	H	K	A	G	-	-	I	N	D	V	I	L	E	R
LDH	-	Y	N	N	K	I	T	V	V	G	V	G	A	V	G	M	A	C	A	I	S	I	L	M	K	D	L	A	D	E	V	A	L	V	D	V
								*	*	*		*			*	*			*																*	

LipDH = lipoamide dehydrogenase; GR = glutathione reductase (9); TRR = thioredoxin reductase (14); DAAO = D-amino acid oxidase (15); PHBH = p-hydroxybenzoate hydroxylase (16); LDH = lactate dehydrogenase (17). LipDH, GR, TRR, and LDH have 4, 18, 3, and 19 additional residues respectively between the sequences shown and their aminotermini. *Highly conserved positions.

it is hoped that relevant information will emerge as the amino acid sequence of pig heart lipoamide dehydrogenase is fitted to that of human erythrocyte glutathione reductase.

ACKNOWLEDGMENTS
We thank Drs. Schirmer, Krauth-Siegel, Saleh, and Untucht-Grau for making available to us sequence data on glutathione reductase prior to publication. This research has been supported by the Medical Research Service of the Veterans Administration and in part by Grant GM-21444 from the National Institute of General Medical Sciences, Public Health Service.

References

1. Williams, C.H., Jr. (1976) In *The Enzymes*. P.D. Boyer (ed.) New York: Academic Press, Vol. 13, pp. 89–173.
2. Williams, C.H., Jr., Thorpe, C., and Arscott, L.D. (1978) In *Mechanisms of Oxidizing Enzymes*. T.P. Singer and R.N. Ondarza (eds.) New York: Elsevier/North-Holland, pp. 2–16.
3. Thorpe, C. and Williams, C.H., Jr. (1981) *Biochem* 20:1507–1513.
4. Arscott, L.D., Thorpe, C., and Williams, C.H., Jr. (1981) *Biochem* 20:1513–1520.
5. Pai, E.F., Schirmer, R.H., and Schulz, G.E. (1978) In *Mechanisms of Oxidizing Enzymes*. T.P. Singer and R.N. Ondarza (eds.) New York: Elsevier/North-Holland, pp. 17–22.
6. Krohne-Ehrich, G., Schirmer, R.H., and Untucht-Grau, R. (1977) *Eur J Biochem* 80:65–71.
7. Schulz, G.E., Schirmer, R.H., Sachsenheimer, W., and Pai, E.F. (1978) *Nature* 273:120–124.
8. Schlitz, E., Blatterspiel, R., and Untucht-Grau, R. (1979) *Eur J Biochem* 102:269–278.
9. Untucht-Grau, R. et al., this volume.
10. Dayhoff, M.O. (1978) *Atlas of Protein Sequence and Structure*. Vol. 5, suppl. 3, p. 354, Washington, D.C.: Natl. Biomed. Res. Foundation.
11. Untucht-Grau, R., Schulz, G.E., and Schirmer, R.H. (1979) *FEBS Lett* 105:244–248.
12. Schirmer, R.H., Pai, E., Schiltz, E., Schulz, G.E., and Untucht-Grau, R. (1980) In *Methods of Peptide and Protein Sequence Analysis*. C. Birr (ed.) Amsterdam: Elsevier/North-Holland Biochemical Press, pp. 267–283.
13. Schulz, G.E. and Schirmer, R.H. (1979) *Principles of Protein Structure*, New York: Springer-Verlag, ch. 9.
14. O'Donnell, M.E. and Williams, C.H., Jr., unpublished results.
15. Ronchi, S. et al., this volume.
16. Hofsteenge, J., Vereijken, J.M., Weijer, W.J., Beintema, J.J., Wierenga, R.K., and Drenth, J. (1980) *Eur J Biochem* 113:141–150.
17. Eventoff, W., Rossmann, M.G., Taylor, S.S., Torff, H.J., Meijer, H., Keil, W., and Kiltz, H.H. (1977) *Proc Nat Acad Sci USA* 74:2677–2681.
18. Massey, V. and Hemmerich, P. (1980) *Biochem Soc Trans* 8:246–257.
19. Jones, E.T. and Williams, C.H., Jr. (1975) *J Biol Chem* 250:3779–3784.
20. Brew, K., Castellino, F.J., Vanaman, T.C., and Hill, R.L. (1970) *J Biol Chem* 245:4570–4582.
21. Fahey, R.C. (1977) *Adv Expt Med Bio* 86A:1–30.
22. Matthews, R.G., Ballou, D.P., and Williams, C.H., Jr. (1979) *J Biol Chem* 254:4974–4981.

Published 1982 by Elsevier North Holland, Inc.
Vincent Massey and Charles H. Williams, Editors
Flavins and Flavoproteins

CHAPTER 9

Role of Arginine Residues in the Catalytic Function of Glutathione Reductase

Vijayakumar Boggaram and Bengt Mannervik

Department of Biochemistry, Arrhenius Laboratory, University of Stockholm, S-106 91 Stockholm, Sweden

Functional Groups in Glutathione Reductase

The importance of FAD (1) and a redox-active cystine disulfide (2) for the catalytic function of glutathione reductase has long been recognized, but knowledge of additional groups has been lacking until recently. It was proposed that the ε-amino group of a lysine residue ("Lys-49" in the human enzyme [3]) was involved as a base in catalysis, since lysine was present in corresponding positions on the C-terminal side of the redox-active disulfide of both glutathione reductase and lipoamide dehydrogenase (4). Indeed, this amino group is positioned near the isoalloxazine ring of FAD in the interior of the glutathione reductase molecule (3) but its possible role in catalysis is unclear. In a search for amino-group blocking reagents, it was found that 2,4,6-trinitrobenzenesulfonate interacts strongly with the dithiol in the G-site of reduced enzyme (5,6). Various additional reagents capable of modifying amino groups have been tested but none, so far, has given an effect that could be assigned to reaction with an essential lysine residue of glutathione reductase. An essential histidine residue was found by chemical modification and by protection experiments. Its location was ascribed to the G-site (6,7). This assignment has support from the identification by X-ray diffraction analysis of a histidine residue near the redox-active disulfide (3,8).

So far, no functional groups had been identified in the N-site of glutathione reductase. It was assumed that arginine residues might be involved in binding of the nucleotide substrate and modification of guanidino groups was consequently attempted (9).

Modification of Arginine Residues in Glutathione Reductase

2,3-Butanedione, 1,2-cyclohexandione, and phenylglyoxal were all found to inactivate irreversibly glutathione reductase from human erythrocytes. The activities inhibited involved GSSG reduction, TNBS-dependent O_2 reduction (10) as well as the transhydrogenase reaction (cf. [6]).

The requirement of borate for inactivation with 2,3-butanedione and the regain of activity of 2,3-butanedione-modified enzyme upon dialysis against phosphate buffer or upon treatment with hydroxylamine indicated reaction with arginine residues. Amino acid analysis of enzyme modified with 2,3-

butanedione demonstrated loss of approximately 5 arginine residues per enzyme molecule (i.e., 2–3 per subunit). Spectrophotometric analysis of glutathione reductase inactivated with phenylglyoxal indicated modification of 2 arginine residues per enzyme molecule.

The kinetics of the inactivation were more complex than those of a first order process (Figure 1), indicating that more than one class of residues in the enzyme was modified. The time-progress curve of inactivation could be described as the sum of two exponentials:

$$A_t = A_1 \cdot \exp(-k_1 t) + A_2 \cdot \exp(-k_2 t),$$

where A_t is remaining activity at time t, A_1 and A_2 are constants, and k_1 and k_2 are apparent rate constants for two inactivation processes.

Figure 1. Inactivation of GSSG-reducing activity of glutathione reductase (1 μM) by 20 mM 2,3-butanedione in 50 mM sodium borate buffer (pH 8.3) at 30°C in the absence (○) and presence of 4 mM NADP$^+$ (●), 4 mM NADPH (□), or 4 mM 2′,5′-ADP (△); control (×).

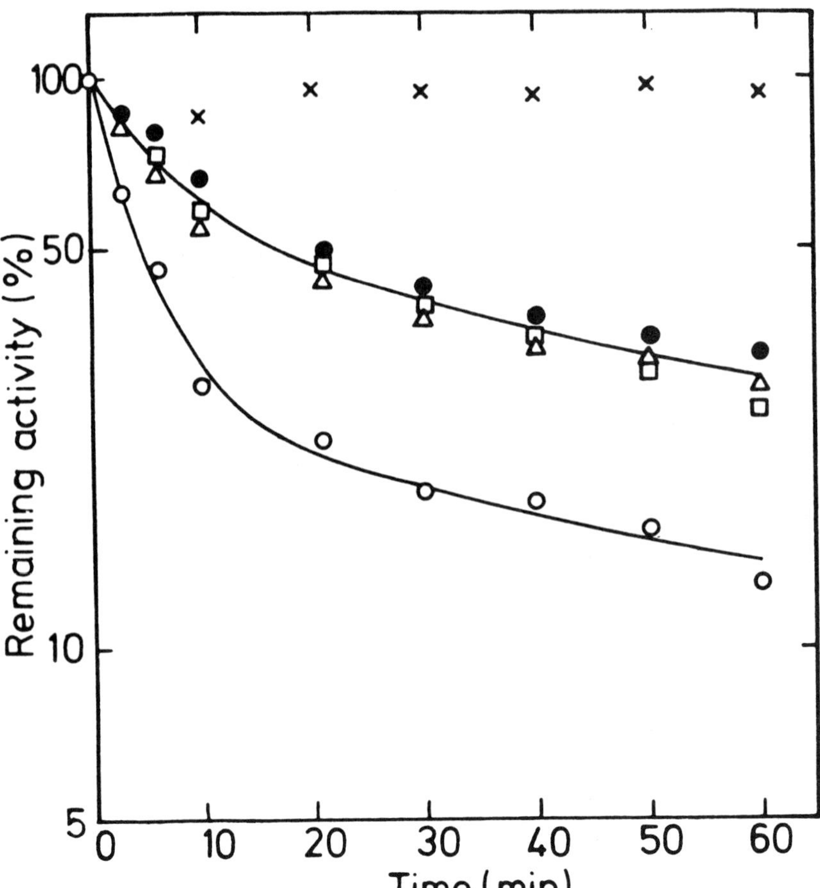

Protection of Glutathione Reductase Against Inactivation by Dicarbonyl Reagents

NADPH, $NADP^+$, and other adenosine 2′-phosphate nucleotides partially protected glutathione reductase against inactivation by dicarbonyl reagents. Figure 1 shows that NADPH, $NADP^+$, and 2′,5′-ADP lower the rate of the initial rapid phase of the inactivation, whereas the rate of the second slower phase is unchanged. GSSG did not provide any significant protection. These findings indicate the presence of an arginine residue in the N-site interacting with the 2′-phosphate of the nucleotides. Such an interaction may explain the high affinity of the enzyme for NADPH in comparison with the low affinity for NADH.

The activity of glutathione reductase in Figure 1 was measured as GSSG reduction. When the transhydrogenase activity was monitored, the protection involved not only the rapid phase but also the slower phase of inactivation. In the latter case, the slow phase appeared to be completely eliminated in the presence of protectors. This finding implies that arginine residues play a role, not only in binding nucleotides, but also in the catalytic events leading to reduction of GSSG.

Kinetic Studies of Glutathione Reductase Modified with 2,3-Butanedione

In order to simplify the steady-state kinetics of glutathione reductase and to obtain some measure of the "affinity" of the enzyme for nucleotides, the kinetics of the transhydrogenase reaction were studied with the concentrations in fixed ratio (cf. [6]). NADPH and thionicotinamide adenine dinucleotide phosphate were used as substrates for the comparison of native and 2,3-butanedione modified enzyme. It was found that both the apparent V_{max} and the apparent affinity ($1/K_{m,app}$) were decreased significantly. These findings support the above conclusions about roles of arginine residues both in binding and in the catalytic events per se.

The kinetics of reduction of glutathione reductase by NADH were measured at 530 nm by stopped-flow spectrophotometry. Enzyme modified with 2,3-butanedione was reduced severalfold less rapidly than native enzyme. Reduction of the enzyme with NADH has been reported to reflect the binding of the nucleotide (11) and the results thus lend further support to the conclusion about the importance of arginine residues for binding of nucleotide substrates.

Effect of Modification of Glutathione Reductase by 2,3-Butanedione on Affinity for 2′,5′-ADP-Sepharose

Glutathione reductase is bound effectively to 2′,5′-ADP-Sepharose (12,13). After modification with 2,3-butanedione, a large fraction of an amount of enzyme that is normally adsorbed on the affinity matrix was not adsorbed. The adsorbed enzyme had approximately the same specific activity as the enzyme not bound to the matrix. Accordingly, these findings also show loss of binding affinity for nucleotides after modification with a dicarbonyl reagent.

Conclusion

The results of the modification studies suggest that an arginine residue in the N-site of glutathione reductase interacts with the 2′-phosphate of the adenosine moiety of NADP(H). Further, an additional role of arginine residues in the catalytic events leading to GSSG reduction is indicated.

ACKNOWLEDGMENTS
This work was supported by Grants (to B.M.) from the Swedish Natural Science Research Council.

References

1. Asnis, R.E. (1955) *J Biol Chem* 213:77–85.
2. Massey, V. and Williams, C.H., Jr. (1965) *J Biol Chem* 240:4470–4480.
3. Schulz, G.E., Schirmer, R.H., Sachsenheimer, W., and Pai, E.F. (1978) *Nature* 273:120–124.
4. Williams, C.H., Jr. (1976) *The Enzymes*, 3rd ed. 13:89–173.
5. Carlberg, I. and Mannervik, B. (1979) *FEBS Lett* 98:263–266.
6. Mannervik, B., Boggaram, V., Carlberg, I., and Larson, K. (1980) In *Flavins and Flavoproteins* Yagi, K. and Yamano, T. (eds.), pp. 173–187, Tokyo: Japan Scientific Societies Press.
7. Boggaram, V. and Mannervik, B. (1978) *Biochem Biophys Res Commun* 83:558–564.
8. Untucht-Grau, R., Schulz, G.E., and Schirmer, R.H. (1979) *FEBS Lett* 105:244–248.
9. Boggaram, V. and Mannervik, B. (1979) *Acta Chem Scand* B33:593–594.
10. Carlberg, I. and Mannervik, B. (1980) *FEBS Lett* 115:265–268.
11. Bulger, J.E. and Brandt, K.G. (1971) *J Biol Chem* 246:5570–5577.
12. Brodelius, P., Larsson, P.-O., and Mosbach, K. (1974) *Eur J Biochem* 47:81–89.
13. Mannervik, B., Jacobsson, K., and Boggaram, V. (1976) *FEBS Lett* 66:221–224.

Published 1982 by Elsevier North Holland, Inc.
Vincent Massey and Charles H. Williams, Editors
Flavins and Flavoproteins

CHAPTER 10

Reaction of 2, 4, 6-Trinitrobenzenesulfonate with the Active Site of Glutathione Reductase

Inger Carlberg and Bengt Mannervik

Department of Biochemistry, Arrhenius Laboratory, University of Stockholm, S-106 91 Stockholm, Sweden

Inhibition of Glutathione Reductase by TNBS

In a search for reagents that might modify chemical groups essential for the catalytic function of glutathione reductase, 2,4,6-trinitrobenzenesulfonate (TNBS) was found to be a potent inhibitor of the enzyme from the seven biological species investigated (1,2). The reduction of GSSG by NADPH catalyzed by the enzyme was inhibited by 60% at pH 7 in the presence of 0.05 μM TNBS. Unexpectedly, the inhibition was reversible and did not increase upon preincubation of the enzyme with TNBS. The inhibition increased with pH and demonstrated an apparent pK_a of 6.9. The effect of TNBS was assigned to the active site of glutathione reductase by the following findings. The charge transfer absorption band at 530 nm, characteristic of the two-electron-reduced enzyme, was eliminated and the spectrum changed to that of oxidized enzyme upon treatment of the reduced enzyme with an amount of TNBS equimolar to the enzyme-bound FAD. The inhibition was generalized competitive (3) with GSSG and generalized uncompetitive (3) with NADPH. This inhibition pattern as well as the lack of effect on the transhydrogenase activity of glutathione reductase located the inhibitory action of TNBS to the GSSG-binding site (G-site [2]) of the enzyme. Also the protection of the dithiol of reduced glutathione reductase against alkylation demonstrated the involvement of the G-site.

TNBS-Dependent NADPH Oxidase Activity

High concentrations of TNBS (1 mM) were found to elicit oxidase activity of glutathione reductase (4). This NADPH-dependent oxidase activity amounted to 15–20% of the GSSG-reducing activity and was dependent on the presence of FAD in the enzyme. The activity was proportional to the concentration of holoenzyme. At high concentrations of glutathione reductase (5 μM FAD) also, low concentrations of TNBS (1 μM) gave rise to oxygen reduction (5). The stoichiometry between O_2 consumed and NADPH added was approximately 0.9 : 1 at the end point and the ratio of the velocities of oxygen reduction and NADPH oxidation was 0.8 : 1 under steady-state conditions. Upon addition of catalase, both ratios changed to approximately 0.5 : 1, demonstrating that the final product of oxygen obtained in the absence of catalase was H_2O_2. TNBS acted catalytically in the presence of O_2; a small amount of TNBS effected the oxidation of a large excess of NADPH. Anaerobically, however, the amount of TNBS limited the amount of NADPH oxidized.

Other substances such as thiol-blocking reagents and nitro compounds other than TNBS were essentially without effect when tested as promotors of oxidase activity of glutathione reductase (5). Approximately 2% of the oxidase activity effected by 1 mM TNBS was obtained with 2,5-dinitrobenzoate and 4-chloro-7-nitrobenzo-2-oxa-1,3-diazole used at the same concentration. It should be noted that the latter reagent as well as TNBS have been demonstrated to elicit oxidase activity also in lipoamide dehydrogenase (4).

Stimulatory Effect of $NADP^+$ on Oxidase Activity

The TNBS-dependent oxidase activity of glutathione reductase increased with the TNBS concentration. Half-saturation was obtained at approximately 0.5 mM TNBS. The activity was largely independent of the NADPH concentration in the range of 2–100 μM, but decreased somewhat with increasing NADPH concentration (Figure 1). However, $NADP^+$ in concentrations ≤ 0.3 mM exerted a marked stimulatory effect on the oxidase activity (Figure 1). At TNBS concentrations >0.5 mM, the oxidase activity decreased when $NADP^+$ concentrations exceeded 0.6 mM (at 0.1 mM NADPH) (not shown). A similar inhibition by excess of $NADP^+$ (at concentrations >0.3 mM) was demonstrated for NADPH concentrations <0.05 mM (using 0.5 mM TNBS). Evidently the oxidized and reduced forms of the pyridine nucleotide had a cooperative stimulatory effect. Both the inhibition exerted by excess of NADPH and the stimulatory effect of $NADP^+$ were more pronounced in the presence than in the absence of the anions phosphate or chloride. The thio analog of $NADP^+$, thionicotinamide adenine dinucleotide phosphate, also stimulated the oxidase activity, but 3-aminopyridine adenine dinucleotide phosphate and 2′,5′-ADP were without effect. NADH could serve as donor substrate in place of NADPH, and in this case both NAD^+ and $NADP^+$ enhanced the oxidase activity. A role of NAD^+ as an activator in the disulfide-reductase activity of lipoamide dehydrogenase reaction has been discussed (6).

TNBS-Dependent Reduction of Cytochrome *c* Catalyzed by Glutathione Reductase

Cytochrome *c* was found to serve as an acceptor of electrons from NADPH in the presence of TNBS and glutathione reductase (Figure 1). Approximately 2 moles of cytochrome *c* were reduced per mole of NADPH. Addition of superoxide dismutase had only a minor effect on the rate of cytochrome *c* reduction. This finding indicates that superoxide anion is not an obligatory intermediate in the reduction of cytochrome *c*. Likewise, it was found that both the rate of the reduction of cytochrome *c* and the rate of oxidation of NADPH

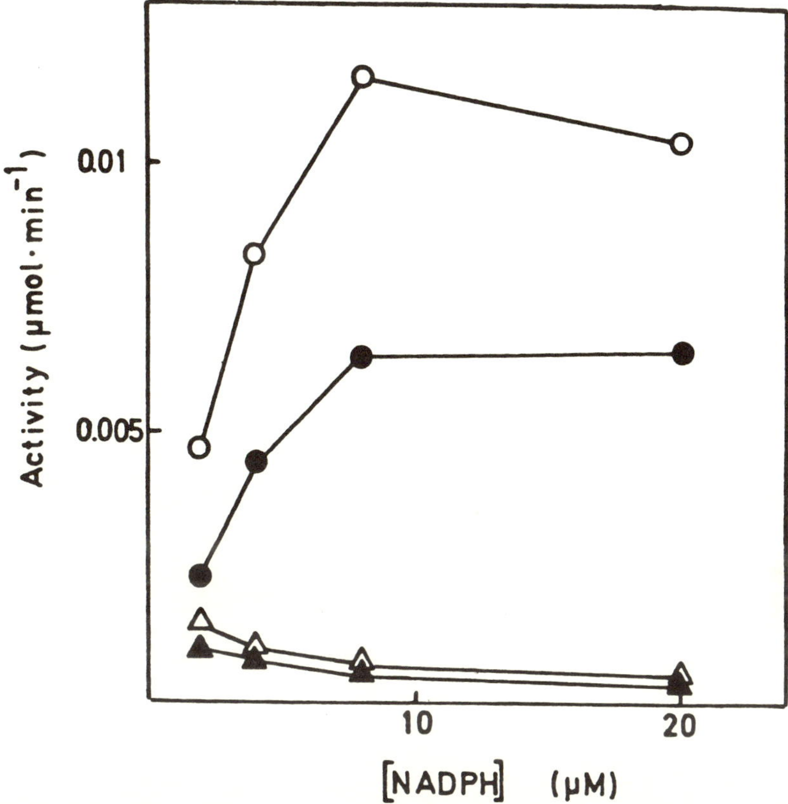

Figure 1. TNBS-dependent reduction of cytochrome *c* by glutathione reductase and the effect of addition of $NADP^+$. The reaction system contained at 30°C: 0.1 M sodium phosphate (pH 7.4), 1 mM EDTA, 15 μM cytochrome *c*, 0.1 mM TNBS, 0.5 μg human glutathione reductase, and varying concentrations of NADPH. Cytochrome *c* reduction (open symbols) was measured at 550 nm and NADPH oxidation (filled symbols) at 340 nm in the absence (△) and the presence (○) of 0.3 mM $NADP^+$.

were essentially unchanged upon removal of oxygen from the reaction medium. These results indicate that electrons are accepted directly from reduced enzyme or from TNBS.

Figure 1 shows that the concentration dependence of NADPH and the effect of addition of $NADP^+$ are essentially the same for cytochrome *c* reduction as for the oxidase activity.

Conclusion

TNBS interacts specifically with the dithiol in the G-site of reduced glutathione reductase. This interaction inhibits GSSG reduction and makes possible reduction of O_2 and cytochrome *c*. Since FAD is buried in the interior of the protein molecule (7), it appears most likely that TNBS serves as a carrier of electrons from the reduced enzyme to oxygen and cytochrome *c*. A nitroradical is proposed as an intermediate to account for one-electron transfer to cytochrome *c*.

The stimulatory effect of $NADP^+$ on the above enzymatic activities strongly indicates cooperativity between two (or more) pyridine-nucleotide binding sites. Such a cooperative effect in glutathione reductase has not been demonstrated before even if the kinetics of the reverse reaction, in which $NADP^+$ oxidizes GSH, are consistent with cooperative binding of $NADP^+$ and NADPH (8).

ACKNOWLEDGMENTS
This work was supported by Grants (to B.M.) from the Swedish Natural Science Research Council.

References

1. Carlberg, I. and Mannervik, B. (1979) *FEBS Lett* 98:263–266.
2. Mannervik, B., Boggaram, V., Carlberg, I., and Larson, K. (1980) In *Flavins and Flavoproteins*. Yagi, K. and Yamano, T. (eds.), pp. 173–187, Tokyo: Japan Scientific Societies Press.
3. Mannervik, B. (1978) *FEBS Lett* 93:225–227.
4. Carlberg, I. and Mannervik, B. (1980) *Acta Chem Scand* B 34:144–146.
5. Carlberg, I. and Mannervik, B. (1980) *FEBS Lett* 115:265–268.
6. Williams, C.H., Jr. (1976) *The Enzymes*, 3rd ed. 13:89–173.
7. Schulz, G.E., Schirmer, R.H., Sachsenheimer, W., and Pai, E.F. (1978) *Nature* 273:120–124.
8. Mannervik, B. (1976) In *Flavins and Flavoproteins*. Singer, T.P. (ed.), pp. 485–491, Amsterdam: Elsevier.

Published 1982 by Elsevier North Holland, Inc.
Vincent Massey and Charles H. Williams, Editors
Flavins and Flavoproteins

CHAPTER 11

Evidence for an Essential Histidine Residue at the Active-Site of Lipoamide Dehydrogenase from the Pyruvate Dehydrogenase Multienzyme Complex of *Escherichia coli*

S. Robert Adamson and Kenneth J. Stevenson

Division of Biochemistry, Department of Chemistry, University of Calgary, Calgary, T2N 1N4, Alberta, Canada

Introduction

Trivalent arsenicals (R-AsO, R-$AsCl_2$) interact covalently with vicinal thiols to form cyclic dithioarsinites of differing ring size (1). The interactions can be efficiently decomposed by the addition of 2,3-dithiopropanol (BAL) which forms a stable 5-membered ring complex with the arsenical (1–3). Recently, the interactions of mono- and bifunctional organoarsenoxides with pyruvate dehydrogenase multienzyme complex (PD complex) from *E. coli* have been investigated (2,3). The bifunctional reagent p-[(bromoacetyl)-amino] phenyl arsenoxide ($BrCH_2CONHPhAsO$) (2,4), in the presence of substrates, irreversibly inhibited the overall PD complex activity and the activity of the lipoamide dehydrogenase (E3) component. This paper provides evidence for the alkylation of a histidine residue at or near the active-site of the lipoamide dehydrogenase component of the PD complex by $BrCH_2[^{14}C]CONHPhAsO$.

Experimental Procedures

Lipoamide dehydrogenase, alkylated by the bifunctional reagent as a component of the PD complex, was purified by first degrading the lipoate acetyltransferase (E2) component with trypsin in 10 mM phosphate buffer pH 7. The trypsin was inactivated with lima bean proteinase inhibitor. The solution was made 8 M in urea and applied to hydroxylapatite. Step-wise elution with phosphate buffer (pH 7) of increasing molarity and containing 8 M urea led to the recovery of pure E3 in 500 mM buffer.

Results and Discussion

The inactivation of PD complex in the presence of substrates by $BrCH_2[^{14}C]CONHPhAsO$ is shown in Figure 1. The activity of the PD

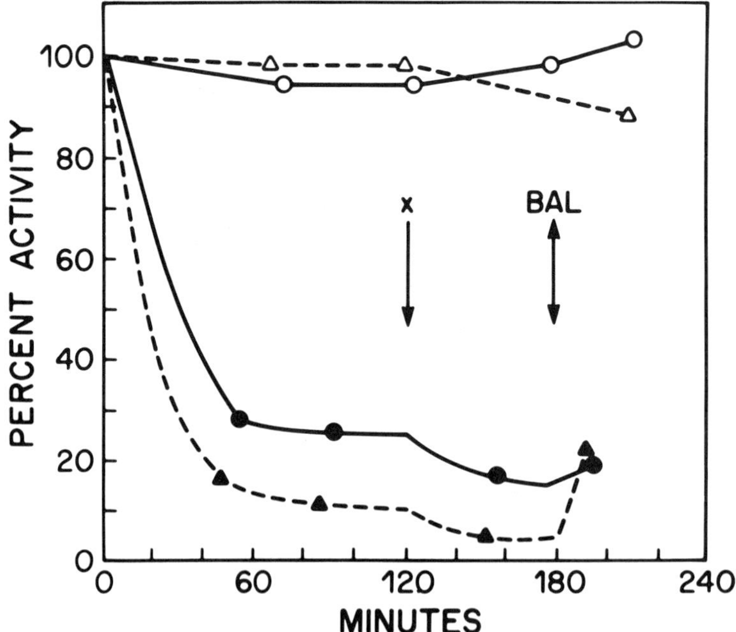

Figure 1. Inhibition of PD complex by $BrCH_2[^{14}C]CONHPhAsO$ in the presence of 0.5 mM thiamine pyrophosphate, 2 mM pyruvate, 0.13 mM coenzyme A, and 10 mM Mg^{2+}. Two mL of PD complex (10 mg/mL) in 20 mM potassium phosphate buffer, pH 7, containing 2 mM EDTA and 0.02% sodium azide was bubbled with N_2 and the bifunctional reagent was added from a 10 mM ethanol solution to a final concentration of 0.1 mM. A further addition of substrates was introduced at "X." The control sample did not contain substrates. The solutions were kept under N_2 throughout. BAL (2,3-dithiopropanol) was added from a 10 mM ethanol solution to give a 10-fold excess over reagent. (▲) PD complex activity of the reaction sample; (●) E3 activity of the reaction sample; (△) PD complex activity of the control; (○) E3 activity of the control.

complex and E3 was progressively lost and, after treatment with BAL, the PD complex activity increased only to the level of residual E3 activity. The PD complex in the absence of substrates (which served as the control) did not lose activity when incubated with the bifunctional reagent. Sodium dodecyl sulfate polyacrylamide gel electrophoresis (3) conducted on the inhibited, BAL-treated, PD complex indicated that the $[^{14}C]$-label resided predominantly in E3 (89%) with some present in E2 (11%) but none in E1. The PD complex control incorporated only traces of radioactivity. The mode of inactivation of E3 likely occurred by initial "anchoring" of the bifunctional reagent (via its $-AsO$ group) to a reduced lipoyl group of E2 (generated by substrates) followed by the delivery of the $BrCH_2[^{14}C]CO$-moeity into the active-site of E3 where an irreversible alkylation ensued (2).

Acid hydrolysis of the alkylated E3 and of the control E3 (both isolated from their respective PD complexes) produced radiolabelled carboxymethyl amino acids that could be readily identified (Figure 2A) and quantitated

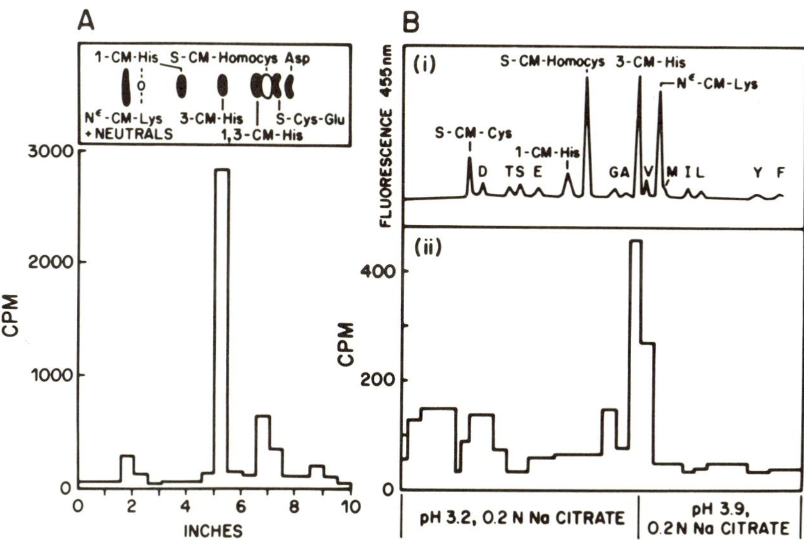

Figure 2. Identification of radiolabelled carboxymethylated amino acids in the acid hydrolysate of purified E3 inhibited with $BrCH_2[^{14}C]CONHPhAsO$ while part of the PD complex. (A) High voltage electrophoresis was conducted at pH 6.5 for 1 hr and 3 kV. N^{ϵ}-CM-Lys is N^{ϵ}-carboxymethyl lysine; 1-CM-His is N^1-carboxymethyl histidine; 3-CM-His is N^3-carboxymethyl histidine; 1,3-CM-His is N^1,N^3-dicarboxymethyl histidine; S-CM-Cys is S-carboxymethyl cysteine, S-CM-Homocys is S-carboxymethyl homocysteine; 0 is the origin. The ninhydrin-stained ionograms were cut into half-inch strips and the radioactivity was determined by scintillation counting. (B) Amino acid analysis was conducted using a modified Beckman 121 analyzer with a programmed in-line fraction collector. (i) The elution position of a number of carboxymethyl amino acids and standard amino acids (single letter code). (ii) The occurrence of radioactivity determined by liquid scintillation counting of each peak.

Table 1. Quantitation of N^3-Carboxymethyl Histidine in the Acid Hydrolysate of Purified E3 Inhibited by $BrCH_2[^{14}C]CONHPhAsO$ when in the PD Complex.

	Number of N^3-carboxymethyl histidine residues per mole of E3	
Sample	**(a) Specific activity 7,300 dpm/nmol**	**(b) Specific activity 8,165 dpm/nmol**
1	0.64	0.57
2	0.87	0.78
Mean ± SD	0.75 ± 0.16	0.68 ± 0.15

Samples were duplicates from the inhibition profile shown in Figure 1. The specific activity of $BrCH_2[^{14}C]CONHPhAsO$ was determined by (a) thin layer chromatography of known amounts of reagent ($\epsilon_{ethanol}^{262\ nm} = 1.72 \times 10^4\ L\ mol^{-1}\ cm^{-1}$ (4) and scintillation counting of sections of the plate and (b) synthesis of S-[^{14}C]-carboxymethyl cysteine from cysteine and $BrCH_2[^{14}C]CONHPhAsO$ followed by acid hydrolysis. The specific activity of S-[^{14}C]-carboxymethyl cysteine was determined by amino acid analysis. The amino acid composition of E3 was in excellent agreement with published values. The compound error of analysis was about ±15%.

(Figure 2B). The inhibited sample contained N^3-carboxymethyl histidine and a radiolabelled residue tentatively identified as S-carboxymethyl cysteine. These residues were not present in significant amounts in the control. The remainder of the trace radiolabelled derivatives seen in the inhibited sample were also present in the control. The quantitation of N^3-carboxymethyl histidine in the alkylated E3 is presented in Table 1. The recovery of S-[^{14}C]-carboxymethyl cysteine was determined from the amino acid analyses at ~0.1 residues per mole of E3. The total alkylation of about 0.8 moles of combined histidine and cysteine residues is in good agreement with the irreversible loss of biological activity (81%) (cf. Figure 1).

In lipoamide dehydrogenase, the participation of a base, essential for catalysis and residing close to the redox disulfide, has been suggested (5,6). This may be a histidine residue (7–9). The present study, using a radiolabelled bifunctional arsenoxide to inactivate the PD complex, provides chemical evidence that a histidine residue is located at or near the active-site of lipoamide dehydrogenase.

References

1. Webb, J.L. (1966) In *Enzyme and Metabolic Inhibitors.* Vol. 3, p. 595, New York: Academic Press.
2. Stevenson, K.J., Hale, G., and Perham, R.N. (1978) *Biochemistry* 17:2198.
3. Adamson, S.R. and Stevenson, K.J. (1981) *Biochemistry*, in press.
4. Robinson, J.A. (1980) M.Sc. Thesis, University of Calgary, Calgary, Alberta, Canada.
5. Matthews, R.G. and Williams, C.H., Jr. (1976) *J Biol Chem* 251:3956.
6. Matthews, R.G., Ballou, D.P., Thorpe, C., and Williams, C.H., Jr. (1977) *J Biol Chem* 252:3199.
7. Schulz, G.E., Schirmer, R.H., Sachsenheimer, W., and Pai, E.F. (1978) *Nature (London)* 273:120.
8. Boggaram, V. and Mannervik, B. (1978) *Biochem Biophys Res Comm* 83:558.
9. Untucht-Grau, R., Schulz, G.E., and Schirmer, R.H. (1979) *FEBS Lett* 105:244.

Published 1982 by Elsevier North Holland, Inc.
Vincent Massey and Charles H. Williams, Editors
Flavins and Flavoproteins

CHAPTER 12

Modification of Lipoamide Dehydrogenase and the Pyruvate Dehydrogenase Complex of *Azotobacter vinelandii* with Maleimides

A. de Kok, A.J.W.G. Visser, and A.C. de Graaf-Hess

Dept. of Biochemistry, Agricultural University, Wageningen, The Netherlands

The pyruvate dehydrogenase complex from *A. vinelandii* is composed of 2–3 pyruvate dehydrogenase dimers, one transacetylase trimer to which 9 lipoyl groups are covalently attached, and one lipoamide dehydrogenase dimer (1). We are interested to know how the complex channels the many reducing equivalents on the lipoyl-SH groups via the two redox-active S-S bridges of lipoamide dehydrogenase to NAD^+ and vice versa. One approach to this problem is to study the reaction of the SH groups, generated by the addition of reduced pyridine nucleotides, with maleimides and to use fluorescent and ESR-active maleimide analogues as probes for their environment.

Lipoamide dehydrogenase from *A. vinelandii* contains 3 half-cysteines per subunit of Mr 56.000, as determined by amino acid analysis and other methods. Thus apart from the redox-active S-S bridge, the enzyme contains one SH group. This group is not accessible to alkylating reagents in the oxidized enzyme, but can be modified in the NADH-reduced enzyme. Thus, upon anaerobic incubation with 1 mM NADH and 1 mM N-ethyl [2,3-^{14}C]maleimide, a maximum incorporation of 2.92 moles of NEM[1] per FAD is obtained after 2 hr at 0°C. The activity decreases in a much more rapid pseudo-first-order reaction (Figure 1A). However, complete specificity is not obtained and 1.2–1.3 moles of NEM/FAD are bound upon extrapolation to ±100% inactivation. The enzyme is protected against inactivation by excess NAD^+ in which case only the pre-existent SH group is modified. The enzyme can also be reduced slowly by NADPH to yield the uncomplexed 540 nm compound. NADPH cannot transfer its reducing equivalents to lipoamide nor can $NADP^+$ function as an electron acceptor in the complex. In the presence of NADPH, 1.0 SH/FAD is titrated at 100% inactivation (Figure 1B). With 1 mM NEM, 2.04 moles of NEM/FAD are incorporated. The reactivity of the essential SH group is 20 times lower when the enzyme is bound to the complex (Figure 1A). Thus it is possible to titrate the lipoyl-SH groups with NEM without noticeable inactivation of lipoamide dehydrogenase. Such a titration (Figure 1B), shows that all lipoyl-SH groups can be modified, but modification of only one lipoyl-SH/FAD results in a large decrease in complex activity.

[1]Abbreviations: NEM=N-ethylmaleimide; ANM=N-(1-anilinonaphtyl-4)maleimide.

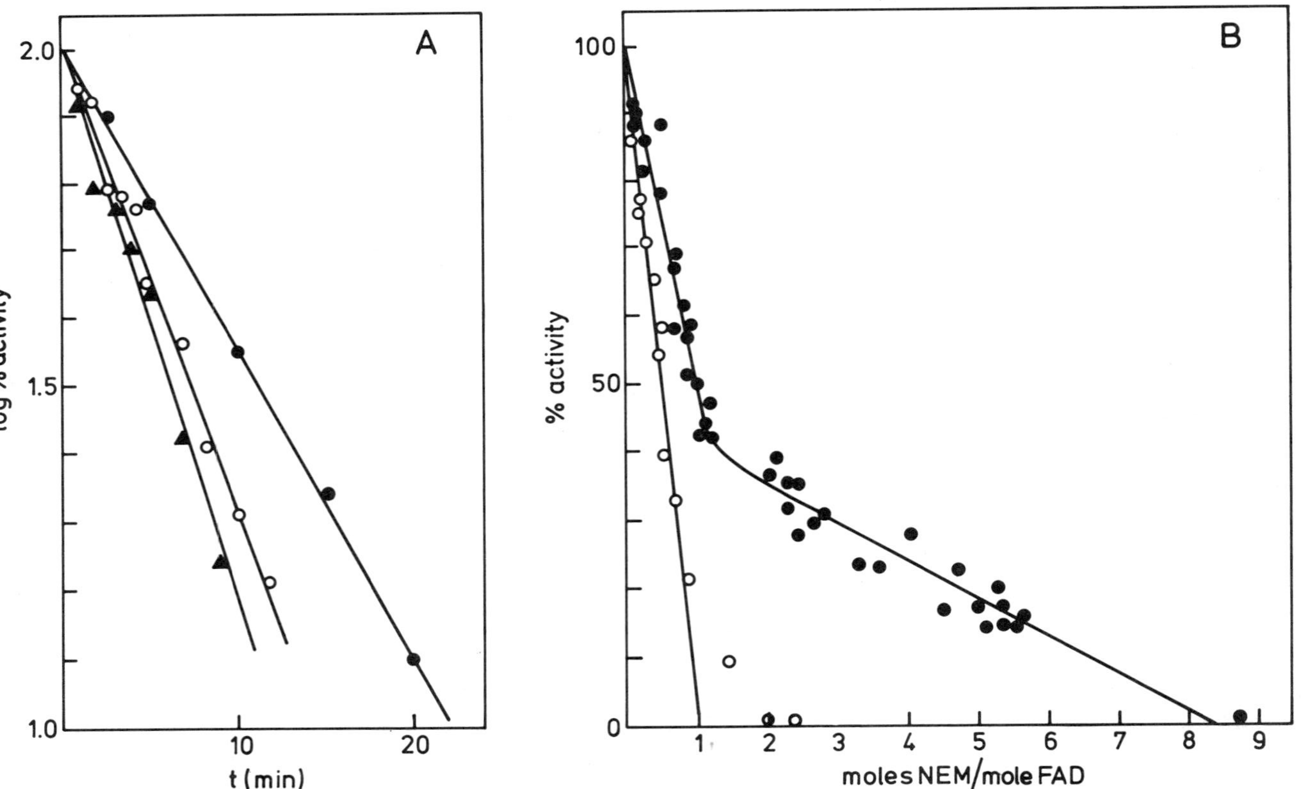

Figure 1. (A) Reaction of lipoamide dehydrogenase and the pyruvate dehydrogenase complex of *A. vinelandii* with NEM in the presence of NADH. 2 μM enzyme (FAD) was incubated at 0°C anaerobically with 1 mM NADH in 0.1 M phosphate—1 mM EDTA pH 7.0. Reactions were started by the addition of NEM and stopped in assay buffer, containing 5 mM β-mercaptoethanol. (▲▲▲) free enzyme, lipoamide dehydrogenase activity, 0.1 mM NEM; (●●●) complex, lipoamide dehydrogenase activity, 1 mM NEM; (○○○) complex, "overall" activity, 40 μM NEM. (B) Incorporation of N-ethyl[2,3-^{14}C]maleimide in free lipoamide dehydrogenase and the lipoyl groups of the complex. (○○○) free lipoamide dehydrogenase, 10 μM (FAD), reductant 5 mM NADPH; (○○○) complex pretreated with NEM, 17 μM (FAD), aerobic conditions, reductant 1 mM NADH.

One explanation (2) is that S-S interchange equilibria exist among the many lipoyl groups and the active site SH groups in lipoamide dehydrogenase:

In this way, both the protection of the thiol groups which participate in interchange such as the active site thiolate of lipoamide dehydrogenase and the enhanced reactivity of one lipoyl-SH/FAD can be understood.

After reoxidation and gel filtration of alkylated lipoamide dehydrogenase, the 455 nm band is shifted to the blue with loss of resolution (Figure 2A), indicating an increase in the polarity of the FAD environment. The fluorescence properties change accordingly. The average fluorescence lifetime decreases from 2.4 to 1.7 ns, which is mainly due to an increase in the contribution of the short component to 76% of the 2-component decay (1). This seems to exclude the previous conclusion that the two FAD sites have different environments (3) and to strengthen those explanations which are based on excited state processes (1). Increased accessibility to solvent is also clearly demonstrated by the 300% increase in quenching by I^-. An increase in the mobility of the FAD is indicated by the 40% decrease in the short component of the rotation correlation time and by the decrease in the steady state anisotropy (Figure 2B). A further indication that the compact structure of the oxidized enzyme changes to a more open structure upon reductive cleavage of the S-S bridge is obtained by limited proteolysis. The oxidized enzyme is stable to proteolysis by 2% trypsin during at least 24 hr at 0°C but loses 50% of its lipoate activity in 10 min at a 1% trypsin level in the presence of NADH or after alkylation and reoxidation of the flavin. The DCIP activity remains above 90%. Preliminary experiments indicate that the loss in lipoate activity is accompanied by the loss of a 3000 D peptide. This indicates that the conformational change affects the lipoyl entrance site and may be related to the function of this enzyme in multienzyme catalysis.

Short incubation with the spinlabel 4-maleimide 2,2,6,6 tetramethyl piperidine 1-oxyl after reduction with either NADH or NADPH yields an immobile signal with a rotational correlation time of the same order as that of the enzyme, 30 ns (1). Upon longer incubation, the spectrum of an extremely mobile species appears in addition, $\tau_c = .2\text{–}1$ ns. This indicates that the two redox-active cysteines are in a completely different environment, at least after alkylation. The high mobility, which indicates a position on the outside on a flexible portion of the polypeptide chain (like the lipoyl-group which has a comparable mobility), seems in contrast with the low reactivity but this could be visualized if this SH group is in interaction with another group in the protein prior to alkylation.

The fluorescent probe ANM was bound to either of the redox-active SH groups. The enhanced reactivity and position of the emission maximum (Figure 2A) indicates an apolar environment for the most reactive thiol. Only when bound to this thiol, is energy transfer to the flavin observed. The limiting

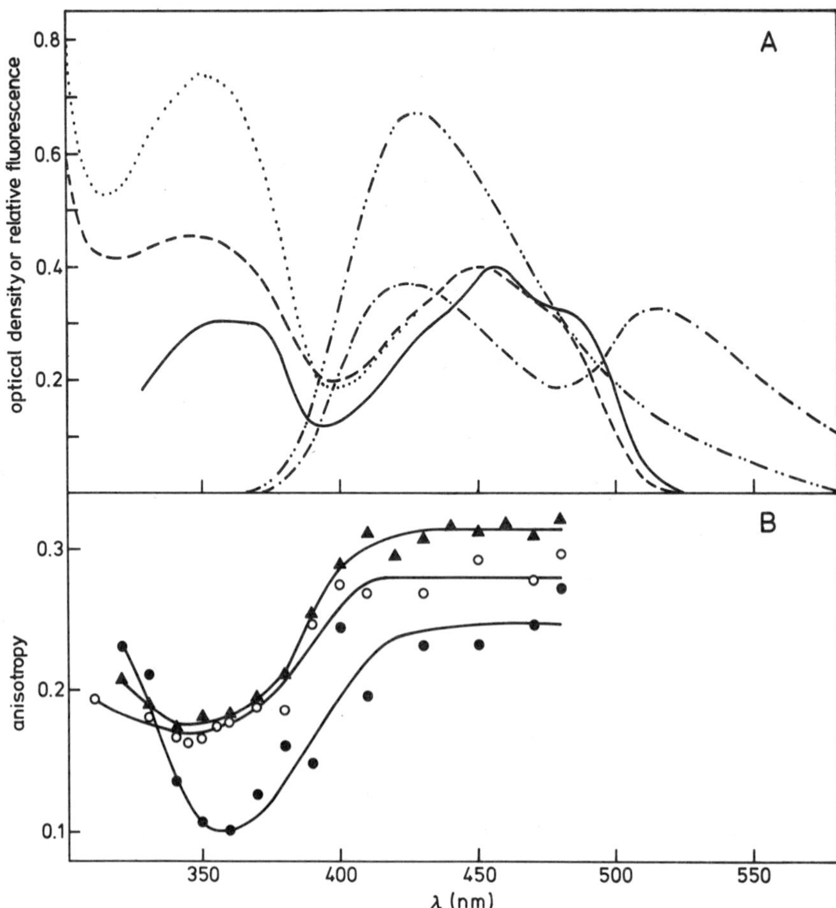

Figure 2. Absorption, fluorescence, and anisotropy excitation spectra of alkylated lipoamide dehydrogenase. Samples in 0.1 M phosphate buffer—1 mM EDTA, pH 7.0, temp. 20°C. (A) Absorption spectrum of unmodified enzyme. (---) Absorption spectrum of NEM-treated enzyme; (···) Absorption spectrum of ANM-treated enzyme; (–·–) Emission spectrum (λ_{ex} 350 nm) of ANM-treated enzyme; (–··–) Emission spectrum of ANM-treated enzyme after dithionite reduction. (B) Anisotropy excitation spectra (λ_{em}>550 nm) of unmodified (▲▲▲), NEM-treated (○○○), and ANM-treated (●●●) lipoamide dehydrogenase.

anisotropy of 0.25 for both donor and acceptor indicates rather rigid binding. Assuming complete rigid binding, an upper limit of 3.0 nm was calculated for the distance between donor and acceptor (4, 5).

ACKNOWLEDGMENTS

We thank Dr. S. Visser for the determination of the cysteine content, Drs. F. Müller and W.H. Scouten for the synthesis of the maleimides. The investigations were partly subsidized by the Netherlands Foundation for Chemical Research (SON) with financial aid from the Netherlands Organization for the Advancement of Pure Research (ZWO).

References

1. Bosma, H.J., de Graaf-Hess, A.C., de Kok, A., Veeger, C., Visser, A.J.W.G., and Voordouw, G. (1981) Intern. Thiamine Conference, New York, in press.
2. de Kok, A., Kornfeld, S., Benziman, M., and Milner, Y. (1980) *Eur J Biochem* 106:49–58.
3. Wahl, P., Auchet, J.C., Visser, A.J.W.G., and Veeger, C. (1975) *Eur J Biochem* 50:413–418.
4. Förster, T.H. (1965) In *Modern Quantum Chemistry*, Istanbul lectures, part III, New York: Academic Press, pp. 93–137.
5. Dale, R.E. and Eisinger, J. (1974) *Biopolymers* 13:1573–1605.

Published 1982 by Elsevier North Holland, Inc.
Vincent Massey and Charles H. Williams, Editors
Flavins and Flavoproteins

CHAPTER 13

The Complete Amino Acid Sequence of D-Amino Acid Oxidase From Pig Kidney

Severino Ronchi,* Lorenzo Minchiott,** Monica Galliano,** Bruno Curti,*** Richard P. Swenson,[†] Charles H. Williams, Jr.,[†] and Vincent Massey[†]

**Institute of Veterinary Physiology and Biochemistry, University of Milano, Milano, Italy;* ***Institute of Biological Chemistry, Faculty of Medicine, University of Pavia, Pavia, Italy;* ****Institute of General Physiology and Biochemistry, Faculty of Science, University of Milano, Milano, Italy;* [†]*Veterans Administration Medical Center and Department of Biological Chemistry, University of Michigan, Ann Arbor, Michigan*

In recent years, it has become apparent that flavin enzymes can act in a wide variety of enzymatic reactions, transferring either one or two electrons between chemically diverse donor and acceptor molecules. This multiplicity of function has led to a classification of the flavoproteins based on donor and acceptor specificities (1). This classification, which rests upon a large body of kinetic, physical, and chemical properties, implies the unique roles that the protein moieties play in the modulation of the catalytic processes of the flavin cofactor. As a consequence, it is clear that the knowledge of the primary and three-dimensional structure of at least a few representative flavoproteins is of utmost importance. For the most part, however, relatively little structural work has been done on the protein moiety. Of the 5 major functional classes of flavoproteins, plus subdivisions (1), representatives of only 3 have had their three-dimensional structures deduced, and either full or partial primary structures determined: glutathione reductase [Class 1(b), carbon-sulfur transhydrogenase (2, and references therein)], p-hydroxybenzoate hydroxylase [Class 3, dehydrogenase/oxygenase (3)], and flavodoxin [Class 5, pure electron transferase (4)].

Within the dehydrogenase/oxidase group (Class 2), D-amino acid oxidase (DAAO) (EC 1.4.3.3) represents a model system; its properties have been thoroughly investigated, and a catalytic mechanism has been envisioned based mainly on spectroscopic evidence (5,6). A determination of the primary structure of the protein is clearly needed in order to gain a better insight into the physical-chemical properties of this enzyme, and to allow comparisons with the other flavoprotein structures already described.

In this paper, we report the complete amino acid sequence of D-amino acid oxidase from pig kidney together with some of the main features of the primary structure, and the localization of the amino acid residues in or near the active center of the enzyme, some of which may be involved in catalysis.

Primary Structure of the Enzyme

The primary structure of D-amino acid oxidase has been determined through the use of chemical and enzymatic cleavages of the S-carboxymethylated protein. Preliminary characterization of fragments derived from the cyanogen bromide hydrolysis of the protein and their alignment in the protein has been described (7). The amino acid sequences of individual regions in the molecule were determined by isolation and analysis of all tryptic peptides, which accounted for all lysine and arginine residues determined by amino acid analysis. These peptides accounted for the 347 amino acid residues observed in the while protein. The complete primary structure of the enzyme was deduced from the alignment of these tryptic peptides by fragments obtained from the chymotryptic and *Staphylococcal aureus* protease digests of the whole protein and the cyanogen bromide fragments, CNBr-1 and CNBr-2, as well as from the cyanogen bromide fragments themselves. All alignments were established by peptide overlaps of two or more residues. The complete primary structure is shown in Figure 1. The sequences of most of the smaller peptides (less than 15 residues) were determined by the manual dansyl-Edman procedure (8), whereas the larger peptides were sequenced using the automated Edman-Begg sequenator (7,9). It should be emphasized that the sequence surrounding each residue was determined at least twice with complete agreement, and in most cases, the amino acid sequences were determined in separate laboratories using different degradation and identification procedures. For these reasons, we are quite confident of the accuracy of this structure.

The amino acid composition as determined from the sequence is in very good agreement with the published amino acid analysis of the protein (10), (see Table 1). The values observed for most amino acids are within the accuracy of the analysis method; however, a few exceptions require further comment. The low values of isoleucine by amino acid analysis could be related to the presence of 4 Ile-Ile and 5 Val-Ile dipeptidyl sequences in the protein which tend to be hydrolyzed very slowly. Recent analyses gave an average value of 8.5 ± 0.1 histidine residues per monomer, in good agreement with the sequence results. Tryptophan has now been analyzed after methanesulfonic acid hydrolysis in the presence of tryptamine (11) yielding an average content of 9.1 ± 0.1 residues per monomer, again, in good agreement with the sequence data.

Five cysteine residues were observed; two are contiguous in the sequence. As demonstrated previously, no disulfide groups are present (10). The amide content, as reported in Table 1, is lower than the published value (10). A theoretical pI was calculated from the sequence by the method described by Tanford (12). A value of 7.24 was determined, which is in good agreement with that reported for the apoenzyme by Yagi and Ohishi (13). The distribution of acidic and basic residues throughout the protein is represented schematically in Figure 2. Three regions (39–98, 177–198, and 222–248) show a notable lack of charged residues, and appear to be quite hydrophobic. The molecular weight of

Met-Arg-Val-Val-Val-Ile-Gly-Ala-Gly-Val-Ile-Gly-Leu-Ser-Thr- 15
Ala-Leu-Cys-Ile-His-Glu-Arg-Tyr-His-Ser-Val-Leu-Gln-Pro-Leu- 30
Asp-Val-Lys-Val-Tyr-Ala-Asp-Arg-Phe-Thr-Pro-Phe-Thr-Thr-Thr- 45
Asp-Val-Ala-Ala-Gly-Leu-Trp-Gln-Pro-Tyr-Thr-Ser-Glu-Pro-Ser- 60
Asn-Pro-Gln-Glu-Ala-Asn-Trp-Asn-Gln-Gln-Thr-Phe-Asn-Tyr-Leu- 75
Leu-Ser-His-Ile-Gly-Ser-Pro-Asn-Ala-Ala-Asn-Met-Gly-Leu-Thr- 90
Pro-Val-Gly-Ser-Tyr-Asn-Leu-Phe-Arg-Glu-Ala-Val-Pro-Asp-Pro- 105
Tyr-Trp-Lys-Asp-Met-Val-Leu-Gly-Phe-Arg-Lys-Leu-Thr-Pro-Arg- 120
Glu-Leu-Asp-Met-Phe-Pro-Asp-Tyr-Arg-Tyr-Gly-Trp-Phe-Asn-Thr- 135
Ser-Leu-Ile-Leu-Glu-Gly-Arg-Lys-Tyr-Leu-Gln-Trp-Leu-Thr-Glu- 150
Arg-Leu-Thr-Glu-Arg-Gly-Val-Lys-Phe-Phe-Leu-Arg-Lys-Val-Glu- 165
Ser-Phe-Glu-Glu-Val-Ala-Arg-Gly-Gly-Ala-Asp-Val-Ile-Ile-Asn- 180
Cys-Thr-Gly-Val-Trp-Ala-Gly-Val-Leu-Gln-Pro-Asp-Pro-Leu-Leu- 195
Gln-Pro-Gly-Arg-Gly-Gln-Ile-Ile-Lys-Val-Asp-Ala-Pro-Trp-Leu- 210
Lys-Asn-Phe-Ile-Ile-Thr-His-Asp-Leu-Glu-Arg-Gly-Ile-Tyr-Asn- 225
Ser-Pro-Tyr-Ile-Ile-Pro-Gly-Leu-Gln-Ala-Val-Thr-Leu-Gly-Gly- 240
Thr-Phe-Gln-Val-Gly-Asn-Trp-Asn-Glu-Ile-Asn-Asn-Ile-Gln-Asp- 255
His-Asn-Thr-Ile-Trp-Glu-Gly-Cys-Cys-Arg-Leu-Glu-Pro-Thr-Leu- 270
Lys-Asp-Ala-Lys-Ile-Val-Gly-Glu-Tyr-Thr-Gly-Phe-Arg-Pro-Val- 285
Arg-Pro-Gln-Val-Arg-Leu-Glu-Arg-Glu-Gln-Leu-Arg-Phe-Gly-Ser- 300
Ser-Asn-Thr-Glu-Val-Ile-His-Asn-Tyr-Gly-His-Gly-Gly-Tyr-Gly- 315
Leu-Thr-Ile-His-Trp-Gly-Cys-Ala-Leu-Glu-Val-Ala-Lys-Leu-Phe- 330
Gly-Lys-Val-Leu-Glu-Glu-Arg-Asn-Leu-Leu-Thr-Met-Pro-Pro-Ser- 345
His-Leu

Figure 1. The complete primary structure of D-amino acid oxidase.

the protein can be calculated from the sequence and was determined to be 39,336. This value compares very well with the value of 39,6000 calculated from amino acid analysis and the FAD content, and 38,000 as determined by SDS polyacrylamide electrophoresis (10).

Table 1. Comparison of the Amino Acid Composition of D-Amino Acid Oxidase by Amino Acid Analysis and Sequence Analysis.

Amino acid	Amino acid analysis[a]	Sequence analysis
CM-cysteine	5.0	5
Aspartic acid	32.2	32[b]
Threonine	22.6	22
Serine	13.6	13
Glutamic acid	36.9	36[c]
Proline	23.4	22
Glycine	33.2	32
Alanine	17.9	17
Valine	24.8	26
Methionine	5.0	5
Isoleucine	16.8	20
Leucine	36.2	36
Tyrosine	13.9	14
Phenylalanine	14.9	15
Histidine[d]	7.5	9
Lysine	11.9	12
Arginine	21.2	21
Tryptophan[e]	8.0	10
Total	345	347
Amides	38	34

[a]Values given as residues per DAAO monomer (see reference 10).

[b]13 as aspartate; 29 as asparagine.

[c]21 as glutamate; 15 as glutamine.

[d]Recent analyses yield 8.5 ± 0.1 residues rather than 7.5 as reported in ref. 10.

[e]9.1 residues were determined after hydrolysis in methanesulfonic acid in the presence of tryptamine (11).

Localization of Chemically Modified Amino Acid Residues Within the Primary Structure

A number of investigators have identified, by chemical modification, several amino acid residues which may be in or near the active center of D-amino acid oxidase. Some of these residues may participate in the catalytic mechanism of the enzyme. In many cases, the peptide containing the modified residue has been isolated and characterized; therefore, it is now possible to localize its exact position within the primary structure. The results obtained for five different modifying reagents are shown in Table 2.

The enzyme becomes modified during oxidation of D-propargylglycine, an acetylenic compound that has been reported to act as a suicide inhibitor (14). Marcotte and Walsh (15) have found that the modified enzyme is not inactivated but is converted into a mixture of catalytic species during covalent alkylation of the protein. The modified enzyme preparation is demonstrably heterogeneous. Two sites of alkylation have been determined (16, 17), and can now be localized to histidine-307 and tyrosine-228. The modified enzyme displays residual activity which is insensitive to further inactivation (15);

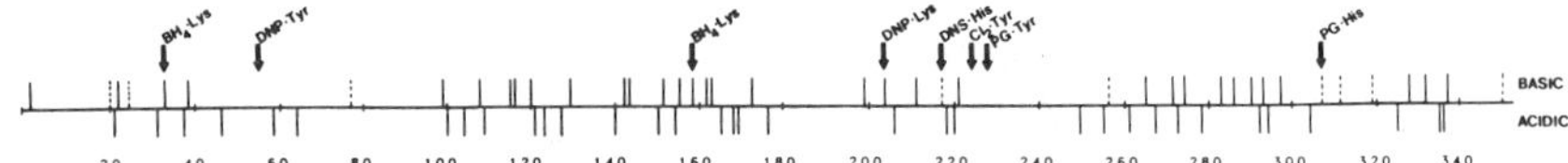

Figure 2. Schematic representation of the distribution of acidic and basic amino acid residues in the sequence of D-amino acid oxidase. Also included are the locations of the amino acid residues that have been chemically modified by a variety of reagents. The reagent and modification are represented as follows: PG=D-propargylglycine; Cl_2-=N-chloro-D-leucine; DNP-=dinitrofluorobenzene; DNS-=dansyl chloride; and $NaBH_4$-=sodium borohydride reduction in the presence of D-alanine.

therefore, it is doubtful whether the sites of modification are truly in the active center of the enzyme. However, the structure of this center may be altered as demonstrated by the change in substrate specificity and in the rate determining step (15).

Rudie et al. (18) have reported the partial inactivation of DAAO by the affinity reagent, N-chloro-D-leucine. This reagent chlorinates a specific tyrosine residue (19), now localized as tyrosine-224. This modification alters the kinetic parameters of the enzyme such that the V_{max} for D-alanine oxidation is reduced to approximately 30% that of the native enzyme, and the K_m increases 5-fold. This tyrosine may be involved in the mechanism of flavin reduction because its modification results in a 2000-fold decrease in the rate of reduction by D-alanine.

Nishino et al. (20) have shown that dinitrofluorobenzene acts as an active site-directed reagent, completely inactivating the enzyme during the incorporation of two dinitrophenyl groups. Inactivation and the incorporation of approximately one dinitrophenyl group were prevented by inclusion of the competitive inhibitor, benzoate. The specific active site-directed modification involved predominately tryrosine residue(s). Swenson et al. (9) have demonstrated that dinitrofluorobenzene modifies either a specific tyrosine residue (tyrosine-55) or a lysine residue (lysine-204) in a mutually exclusive manner. The tyrosine was the major site of dinitrophenylation; however, the degree of

Table 2. Location of the Chemically Modified Amino Acids in D-Amino Acid Oxidase.

Reagent[a]	Enzyme inactivation	Protection by benzoate	Amino acid modified	Sequence position
Fluorodinitrobenzene	Complete	Complete	Tyr	55
		Complete	Lys	204
Dansyl-Cl	Complete	Partial	His	217
N-chloro-D-Leu	Partial	Partial	Tyr	224
Alanine + $NaBH_4$	None	Partial	Lys	33
			Lys	158
Propargylglycine	Partial	Not done	His	307
			Tyr	228

[a]See text for references.

lysine modification observed was significant. Both modifications are required to account for the levels of active site-directed reagent covalently incorporated, and for the degree of inactivation observed. These observations argue strongly for both these residues being in the active center of the enzyme. The specific roles that these residues may play in catalysis is not known, although the involvement of the tyrosine in substrate binding has been postulated (20).

Nishino et al. (20) have also modified a histidine residue in the substrate-binding site of the protein using dansyl chloride. The modification is quite specific, and is largely prevented by benzoate. A peptide containing the labile dansyl-histidine adduct has been isolated using high performance liquid chromatography (21). Sequence analysis indicated that histidine-217 was the major site of modification. It has been proposed that a monoprotonated base abstracts the α-hydrogen from the substrate during the initial steps of catalysis (22). The dansylated histidine may represent this base (20).

Lysine residues have been modified by sodium borohydride reduction of the enzyme in the presence of D-alanine (23); however this does not lead to loss of activity (24). Mizon et al. (25) have extended this work by isolating two peptides containing modified lysine residues and report the following sequences: Gly-Val-Lys-Phe and Asx-Val-Val-Lys-Tyr. While the first lysine is identifiable as lysine-158 in the primary structure, we are not able to localize the second peptide. Tentatively, we have assigned the lysine in the later peptide to position 33, in the middle of the sequence: -Asp-Val-Lys-Val-Tyr-.

It is worthwhile to note that Lys-204, His-217, Tyr-224, and Tyr-228 are all clustered closely in the primary structure of DAAO. These residues as well as His-307 are all located in the second half of the molecule. Two of the modified residues, Lys-33 and Tyr-55, are located near the region, at the amino terminus, found to be homologous to the AMP-binding domain of other nucleotide dependent enzymes (see following paper). It is reasonable to speculate that the protein may be divided into a coenzyme binding domain and a catalytic domain, analogous to other flavoproteins.

ACKNOWLEDGMENTS

This research has been supported by Grants from the Italian Research Council (B.C. and S.R.) and the National Institute of General Medical Sciences, Public Health Service, GM 21444 (to C.W) and GM 11106 (to V.M.) and by the Medical Research Service of the Veterans Administration.

References

1. Massey, V. and Hemmerich, P. (1980) *Biochem Soc Trans* 8:246–257.
2. Schiltz, E., Blatterspiel, R., and Untucht-Grau, R. (1979) *Eur J Biochem* 102:269–278.
3. Hofsteenge, J., Vereijken, J.M., Weijer, W.J., Beintema, J.J., Wierenga, R.K., and Drenth, J. (1980) *Eur J Biochem* 113:141–150.
4. Mayhew, S.G. and Ludwig, M.L. (1975) *The Enzymes*. 3rd ed. 12:57–118.
5. Bright H.J. and Porter, D.J.T. (1975) *The Enzymes*. 3rd ed. 12:421–505.
6. Porter, D.J.T., Voet, J.G., and Bright, H.J. (1977) *J Biol Chem* 252:4464–4473.
7. Ronchi, S., Minchiotti, L., Curti, B., Zapponi, M.C., and Bridgen, J. (1976) *Biochim Biophys Acta* 634–643.
8. Gray, W.R. (1972) *Methods Enzymol* 25:333–344.

9. Swenson, R.P., Williams, Jr., C.H., and Massey, V. (1981) *J Biol Chem*, submitted.
10. Curti, B., Ronchi, S., Branzoli, U., Ferri, G., and Williams, Jr., C.H. (1973) *Biochim Biophys Acta* 327:266–273.
11. Simpson, R.J., Neuberger, M.R., and Liu, T.Y. (1976) *J Biol Chem* 251:1936–1940.
12. Tanford, C. (1961) In *Physical Chemistry of Macromolecules*. New York: Wiley and Sons, Inc., pp. 548–586.
13. Yagi, K. and Ohishi, N. (1972) *J Biochem (Tokyo)* 71: 993–998.
14. Horiike, K., Nishima, Y., Miyake, Y., and Yamano, T. (1975) *J Biochem (Tokyo)* 78:57–63.
15. Marcotte, R. and Walsh, C.T. (1978) *Biochemistry* 17:2864–2868.
16. Ronchi, S., Minchiotti, L., Galliano, M., Curti, B., and Ghisla, S. 1980) In *Flavins and Flavoproteins*. (Yagi, K. and Yamano, T. (eds.) Tokyo: Japan Scientific Societies Press, pp. 511–515.
17. Ronchi, S., Minchiotti, L., Galliano, M., Curti, B., and Ghisla, S., unpublished results.
18. Rudie, N.G., Porter, D.J.T., and Bright, H.J. (1980) *J Biol Chem* 255:498–508.
19. Ronchi, S., Galliano, M., Minchiotti, L., Curti, B., Rudie, N.G., Porter, D.J.T., and Bright, H.J. (1980) *J Biol Chem* 255:6044–6046.
20. Nishino, T., Massey, V., and Williams, Jr., C.H. (1980) *J Biol Chem* 255:3610–3616.
21. Swenson, R.P., Williams, Jr., C.H., and Massey, V., unpublished results.
22. Walsh, C.T., Krodel, E., Massey, V., and Abeles, R.H. (1973) *J Biol Chem* 248:1946–1955.
23. Coffey, D.S., Heims, A.H., and Hellerman, L. (1965) *J Biol Chem* 240:4058–4064.
24. Massey, V., Curti, B., Muller, F., and Mayhew, S.G. (1968) *J Biol Chem* 243:1329–1332.
25. Mizon, J., Han, K., Dautrevaux, M., and Biserte, G. (1972) *Biochim Biophys Acta* 276:63–69.

Published 1982 by Elsevier North Holland, Inc.
Vincent Massey and Charles H. Williams, Editors
Flavins and Flavoproteins

CHAPTER 14

D-Amino Acid Oxidase: Predictions on the Secondary Structure and Comparisons with Other Flavoproteins

Richard P. Swenson,* Charles H. Williams, Jr.,*
Vincent Massey,* Severino Ronchi,**
Lorenzo Minchiotti,† Monica Galliano,†
Martino Bolognesi,‡ and Bruno Curti††

**Veterans Administration Medical Center and Department of Biological Chemistry, University of Michigan, Ann Arbor, Michigan; **Institute of Veterinary Physiology and Biochemistry and ††Institute of General Physiology and Biochemistry University of Milano, Milano; †Institute of Biological Chemistry, Faculty of Medicine; and ‡Institute of Crystallography, Faculty of Science, University of Pavia, Pavia, Italy*

The primary structure of D-amino acid oxidase (DAAO) from pig kidney has been determined by our groups and is reported in the preceding paper (1). Several physical-chemical studies have been carried out on this protein including an effort to produce crystals suitable for detailed X-ray crystallographic investigation. As a result, two different crystalline forms containing high molecular weight oligomers of the enzyme have been obtained (2). Unfortunately, neither form has been amenable to high resolution analysis. In the framework of a more thorough structural investigation, we report in this paper an analysis of the amino acid sequence data to predict, using the methods developed by Chou and Fasman (3), the secondary structure of DAAO, as well as to determine if any similarities exist with other flavoproteins whose three-dimensional structures are already known. After a preliminary secondary structure was determined, it was "refined" by boundary analysis, i.e., incorporating in the predicted structure the information available on residues most frequently found at the amino- and carboxyl-termini of secondary structure elements. The results of these predictions are summarized schematically in Figure 1. Approximately 30% of the residues are predicted to adopt β-structure, while 23% are in helical structures. These predictions compare quite well with the circular dichroism data of Ohama et al. (32% β-structure, 17% α-helix) (4). The DAAO monomer appears to be composed of 16 β-strands (average length, 6.4 residues) and 10 α-helices (average length, 7.9 residues). Sixteen β-turns seem to be involved in the building of the tertiary structure. If one were to depict $\beta\alpha\beta$ supersecondary structural elements within the molecule, only four

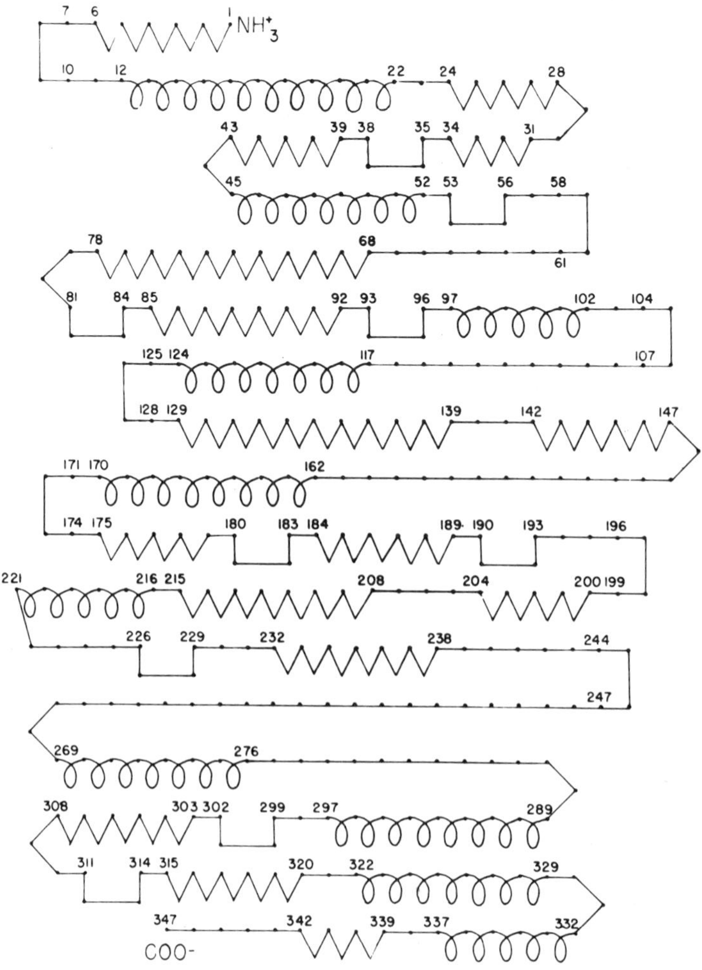

Figure 1. Schematic diagram of the predicted secondary structure of D-amino acid oxidase. Residues are represented in helical (ʊ), β-sheet (V), β-turn (⊔), and random coil (●—●).

such units seem to be present, none of which are strictly contiguous to the other. Thus no $\beta\alpha\beta\alpha\beta$ unit (Rossmann fold) can be built. On the other hand, β-meanders ($\beta\beta\beta$ units) can be localized at at least two structural elements such as $\beta_2\beta_3\beta_4$ and $\beta_9\beta_{10}\beta_{11}$ or $\beta_{10}\beta_{11}\beta_{12}$. Additionally, the number of sulfur-aromatic interactions likely to occur in the enzyme was calculated on the basis of the amino acid composition (5). These interactions seem to play a role in the stabilization of the tertiary structure and occur between sulfur atoms and aromatic residues. Fourteen such interactions are possible within each DAAO monomer.

The secondary structural predictions described above provide limited information for comparison with the tertiary structures that have been reported for the other flavoproteins. In a parallel study, however, we considered the possible structural and evolutionary relatedness of DAAO to the other flavoproteins based on their amino acid sequences. At the time of preparation of this paper, the only flavoprotein primary structures available to us were: pig kidney DAAO (entire structure, 347 residues) (1); several flavodoxins—however, comparison to only that of *Clostridium MP* was made—(entire structure, 138 residues) (6); p-hydroxybenzoate hydroxylase (PHBH) from *Pseudomonas fluorescens* (residues 1–138, 35% of the structure) (7); and glutathione reductase (GR) from human erythrocytes [residues 18–51, (8), and fragment "Q," a 128-residue region which includes part of the FAD-binding domain (9)]. For this preliminary analysis, the method described by Dickerson (10) was used to calculate the minimum base change (m.b.c.) per codon averaged over the entire length of the segments being compared. From the results of the m.b.c. analysis, it can be concluded that no significant homologies exist between DAAO and the flavodoxin from *Clostidium MP*. No indications of any extended internal homologies within the DAAO sequence itself were observed. However, homologies in certain regions of DAAO to both GR and PHBH were apparent. Significant homology exists at the amino terminal regions of these molecules as summarized in Fig. 2a. Residues 3 to 31 of DAAO were found to be similar to the region between residues 23 to 50 of GR (28% homology) and residues 5 to 32 of PHBH (31% homology) (7). If one allows for interchange of amino acid residues with aliphatic side chains and interchange of aspartate and glutamate, an even more impressive homology is apparent, 38% with GR and 41% with PHBH. M.b.c. parameters of 1.07 and 1.00 were determined for these regions during the comparison of DAAO to GR and PHBH, respectively, and are consistent with these apparent homologies. These values are significantly less than the value of 1.6, which has been determined for random sequences showing no apparent homology (10). The tertiary structures of GR and PHBH as determined by X-ray crystallography indicate that these regions form a $\beta\alpha\beta$ fold which interacts with portions of the adenosyl portion of the FAD molecule. Examination of the secondary structure predictions for DAAO (Figure 1) also indicates that the region between residues 2 and 28 has the potential to form such a structure. The greatest sequence homology exists near the carboxyl-terminal ends of the β-strands and the amino-terminal end of the α-helix of this fold. It is this portion of the structure that interacts with the ribose and pyrophosphate portions of the cofactor (7). Particularly striking is the invariant -Gly-X-Gly-Y-Z-Gly- sequence found in this region of these flavoproteins as well as in the nucleotide-binding domains of a number of dehydrogenases (7). As discussed by Hofsteenge et al. (7), these glycine residues have important structural roles. The first interacts closely with the ribose of the adenosyl moiety, while the other two are involved in the bend connecting the first β-strand and the α-helix of the fold. Substitution of a side group on the α-carbon at these positions would interfere greatly with these interactions. An acidic residue is invariably found following the second β-strand in these flavoproteins and in

b

DAAO Q T F N Y L L S H L G S P N A A N M G L T P V S G Y N L F R

PHBH Q T P D Y V L G R I R A G V L E Q G M V D L L R E A G V D R

Figure 2. Panel a: Sequence homology between the amino terminals of D-amino acid oxidase (DAAO), (residues 3 to 31); glutathione reductase, (residues 23 to 50); and p-hydroxybenzoate hydroxylase (PHBH), (residues 5 to 32). Panel b: Additional sequence homology between DAAO, (residues 70 to 99) and PHBH, (residues 34 to 63).

many pyridine nucleotide-dependent dehydrogenases. The side chain carboxyl group is hydrogen-bonded to the 2′-OH of the ribose of the adenosyl group (7). The overall length (29 to 30 residues) of this fold is very similar in these proteins.

Additional homologies between DAAO and the published sequence data available for GR and PHBH were searched for. No homology with fragment "Q" of GR was found even when comparing only the portion which forms part of the FAD domain. However, during the comparison of DAAO to the available sequence of PHBH, an additional region of homology was observed—residues 70 to 99 in DAAO as compared to residues 34 to 63 in PHBH (Figure 2b). By allowing for interchange of amino acid side chains with similar chemical structures and properties, an apparent homology of 37% was observed. A m.b.c. of 1.13 was calculated for these regions. Interestingly, portions of this region in PHBH were found to be hydrogen-bonded to the pyrimidine subnucleus of the isoalloxazine ring of FAD (7). If one joins the two regions in DAAO (3 to 31 and 70 to 99) displaying homology with the amino terminal portion of PHBH (5 to 63) by excluding a loop of polypeptide (32 to 68), the resultant region of apparent homology is striking. Interestingly, the nonhomologous loop contains two amino acid residues (Lys-33 and Tyr-55) which have been chemically modified, one of which (the tyrosine) is in the active center of the enzyme (1).

Within the limits of the prediction methods applied, we believe that structural similarities exist between these three flavoproteins. X-ray crystallographic analyses have revealed topographical similarities between the FAD-binding domains of GR and PHBH. It would seem reasonable to speculate, based on these preliminary sequence homology comparisons and the secondary structural predictions, that DAAO may also adopt a similar structure in this domain. As the complete sequences of GR and PHBH become available in the literature, a more thorough analysis of structural homologies will be made. More detailed comparisons await the elucidation of the three-dimensional structure of DAAO which is presently under investigation.

ACKNOWLEDGMENTS
This work has been supported by the Medical Research Service of the Veterans Administration and by Grants from the Italian Research Council to B.C., S.R. and M.B. and by Grants GM-21444 to C.W. and GM-11106 to V.M. from the National Institute of General Medical Sciences, Public Health Service.

References

1. Ronchi, S. et al., preceding paper in this volume.
2. Bolognesi, M., Ungaretti, L., Curti, B., and Ronchi, S. (1978) *J Biol Chem* 253:7513–7514.
3. Chou, P.Y. and Fasman, G.D. (1979) *Adv Enzymol* 47:45–148.
4. Ohama, H., Sugiura, N., and Yagi, K. (1974) *J Biochem* (*Tokyo*) 76:1–6.
5. Morgan R.S. and McAdon, J.M. (1980) *Int J Pept Prot Res* 15:177–180.
6. Mayhew, S.G. and Ludwig, M.L. (1975) *The Enzymes* 3rd ed., 12:57–118.
7. Hofsteenge, J., Vereijken, J.M., Weijer, W.J., Beintema, J.J., Wierenga, R.K., and Drenth, J. (1980) *Eur J Biochem* 113:141–150.
8. Untucht-Grau, R. et al., this volume.
9. Schiltz, E., Blatterspiel, R., and Untucht-Grau, R. (1979) *Eur J Biochem* 102:269–278.
10. Dickerson, R.E. (1971) *J Mol Biol* 57:1–15.

Published 1982 by Elsevier North Holland, Inc.
Vincent Massey and Charles H. Williams, Editors
Flavins and Flavoproteins

CHAPTER 15

Multiple Conformational States of L-Amino Acid Oxidase

Daniel Wellner and Louise A. Lichtenberg

Department of Biochemistry, Cornell University Medical College, New York, New York

Introduction

L-Amino acid oxidase is an unusual protein in that it is capable of existing in several stable or metastable conformations. These conformational states can be distinguished from each other by a number of physical and biological properties, including enzymatic activity, absorption spectrum, optical rotatory dispersion, circular dichroism, interaction with specific antibodies, and interconvertibility. The inactivation of L-amino acid oxidase by warming at slightly alkaline pH was first reported and studied in detail by Kearney and Singer (1–3). They found that the process was completely reversible. Their results, as well as those of another study of this phenomenon by Wellner (4), are consistent with a model in which the enzyme undergoes a limited conformational change involving the portion of the molecule to which the FAD is bound. In this report, the native, enzymatically active conformation is referred to as State I, while the form obtained by warming the enzyme to 37° at pH 7.5 is designated as State II. State II is inactive, but readily converted to State I at pH 5 and 37°. Another conformation of the enzyme, State V, was first reported by Wellner and Hayes (5). Its spectrum differs markedly from that of States I or II, and it is not reactivated at pH 5 and 37°. It is obtained from State II by incubation at pH 5 and 0° in the presence of p-chloromercuribenzoate. State IV is an intermediate between States II and V (see Results). A model for the interconversion of these conformational states was discussed at a previous International Symposium on Flavins and Flavoproteins (6). The reversible formation of another inactive form of the enzyme (State III), obtained by freezing, was first reported by Singer and Kearney (7) and subsequently studied in several laboratories (8–10). It can be distinguished from the other forms by immunochemical and spectral properties.

This report shows that various properties of the enzyme, such as affinity for the antibody, catalytic activity, spectrum, and ability to be reactivated, change at different rates in the course of the conformational interconversions. It is concluded that the conformational changes consist of discrete, sequential steps involving different parts of the protein molecule.

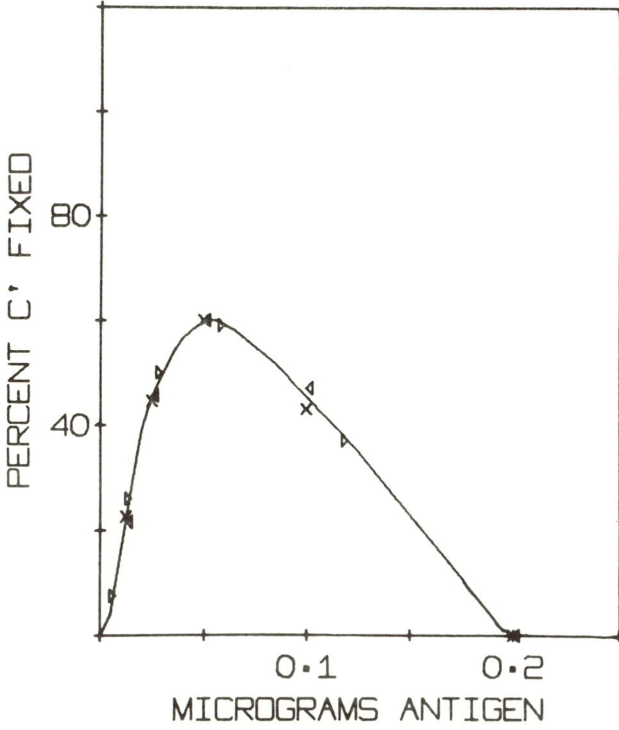

Figure 1. Complement fixation curves of L-amino acid oxidase before and after removal of sialic acid. (×) native enzyme; (▷) enzyme incubated for 120 hr 23°, pH 5, in the presence of neuraminidase; (◁) enzyme incubated as above in the absence of neuraminidase. Complement fixation was measured as described under Materials and Methods.

Materials and Methods

Crystalline L-amino acid oxidase was prepared from *Crotalus adamanteus* venom as previously described (11). Antibodies were obtained in New Zealand white rabbits by injections of native enzyme. Antibodies from three different rabbits gave similar results. Enzyme assays were carried out by a sensitive fluorometric method, using L-phenylalanine as substrate (12). The reaction of the enzyme with antibody was measured quantitatively by the complement fixation procedure described by Wasserman and Levine (13). The relative number of antibody binding sites on different forms of the enzyme was estimated from the position of the peak along the ordinate (see Figure 1) while the relative affinity of the enzyme for the antibody was calculated from the height of the peak.

Results

Figure 1 shows that there is no change in the complement fixation curve after incubation of L-amino acid oxidase with neuraminidase under conditions which have been shown to remove the sialic acid. Earlier studies showed that

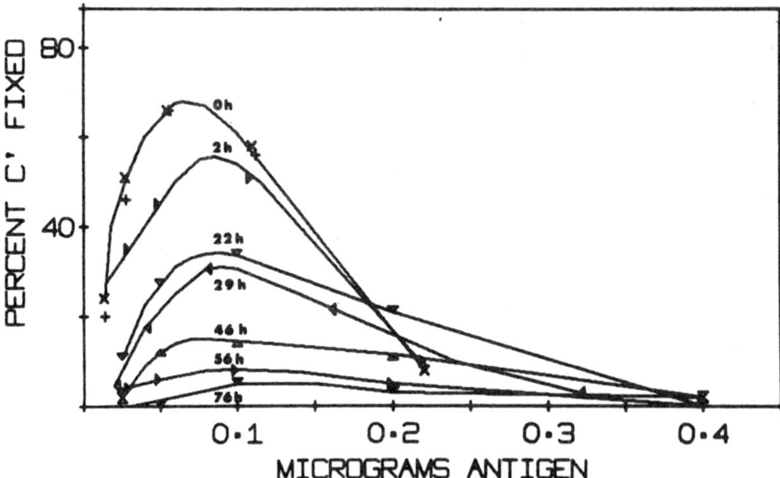

Figure 2. Changes in the complement fixation curve of L-amino acid oxidase during the transition from State II to State V. Enzyme in State II was reacted with p-chloromercuribenzoate and incubated at pH 5, 0°, for the indicated times. The reaction was stopped by adjusting the pH to 7.3 with Tris-HCl buffer. Complement fixation was measured as in Figure 1. The curve at 0 time was identical to the one for the native enzyme.

Figure 3. Kinetics of formation of States IV and V from State II. Incubation conditions as for Figure 2. Curve II: Enzymatic activity recovered following incubation at 37° for 1 hr (this represents the amount of enzyme in State II). Curve V: Formation of State V as determined by the decrease in A_{387} (assumed to be 100% at 24 hr). Curve IV: Formation of State IV, calculated by subtracting States II and V from the total enzyme. Curve Ab: antibody binding sites recovered following incubation at 37° for 1 hr.

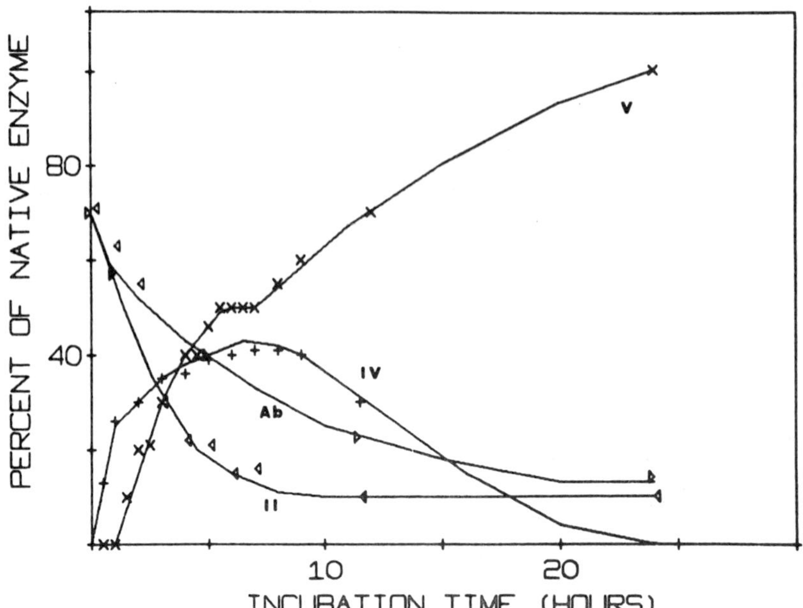

there was no inhibition of the antibody reaction by the free sugars which are constituents of the carbohydrate moiety of the enzyme (4). Thus, the antibody does not appear to be directed at the carbohydrate portion of the molecule. Similarly, it is not directed at the active site, since (a) the antigen-antibody complex is fully active, and (b) a conformational change involving the active site (State I to State II) does not affect the complement fixation curve (4).

Figure 2 shows that the transition from State II to State V alters the antibody binding site of the enzyme. This results in almost complete loss of antibody affinity, or alternatively, in the formation of an antigen-antibody complex which lacks complement-fixing properties.

Figure 3 shows that the loss of ability to be reactivated (curve II) is more rapid than the appearance of State V, as determined spectrophotometrically. This demonstrates the intermediate formation of State IV, with a spectrum similar to that of State II. Thus, the conformational change resulting in loss of ability to be reactivated precedes that causing the change in flavin absorption. This conclusion is supported by the finding that the spectra recorded during the transition from State II to State V are characterized by isosbestic points at 370, 415, and 462 nm (D. Wellner and M.B. Hayes, unpublished results). Figure 3 also shows that the loss of ability to recover antibody binding sites is slower than the loss of ability to be reactivated.

Figure 4. Kinetics of interconversion of States I and III. The native enzyme (State I) was kept frozen at −20° in 0.1 M sodium acetate buffer, pH 5, for 7 days. The enzyme, now in State III, was reactivated by warming the solution to 37° for 1 hr. Curve a: Number of antibody binding sites; curve b: antibody affinity; curve c: enzymatic activity.

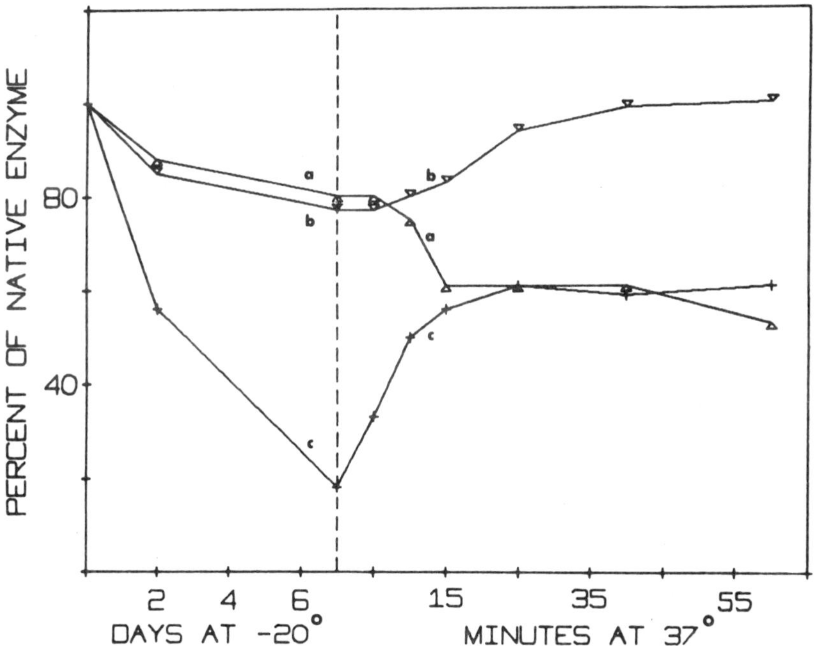

Figure 4 shows that, on being converted to State III, the enzyme loses both antibody affinity and binding sites. The lag in return of antibody affinity upon reactivation indicates a sequential reformation of the active site and antibody binding site. The incomplete recovery of activity and the loss of some antibody binding sites upon reactivation are indicative of some protein denaturation under the conditions used.

Discussion

Unlike most other proteins, L-amino acid oxidase does not exist in a unique conformation representing the free energy minimum dictated by its amino acid sequence. It appears to be capable of undergoing slow shifts to new stable or metastable conformational states in response to changes in its environment. Some of these states may represent local, not global minima in the potential energy surface. These findings may be of significance in relation to the mechanism of protein folding and the role of FAD in this process.

ACKNOWLEDGMENT
This work was supported by Grant No. AM-12068 from the National Institutes of Health, U.S. Public Health Service.

References

1. Kearney, E.B. and Singer, T.P. (1951) *Arch Biochem Biophys* 33:377–396.
2. Kearney, E.B. and Singer, T.P. (1951) *Arch Biochem Biophys* 33:397–413.
3. Kearney, E.B. and Singer, T.P. (1951) *Arch Biocem Biophys* 33:414–426.
4. Wellner, D. (1966) *Biochemistry* 5:1585–1591.
5. Wellner, D. and Hayes, M.B. (1968) *Ann NY Acad Sci* 151:118–132.
6. Wellner, D. (1971) In *Flavins and Flavoproteins*. Kamin, H. (ed.) Baltimore: University Park Press, pp. 322–326.
7. Singer, T.P. and Kearney, E.B. (1950) *Arch Biochem Biophys* 29:190–209.
8. Curti, B., Massey, V., and Zmudka, M. (1968) *J Biol Chem* 243:2306–2314.
9. Zimmerman, S.E., Brown, R.K., Curti, B., and Massey, V. (1971) *Biochim Biophys Acta* 229:260–270.
10. Coles, C.J., Edmondson, D.E., and Singer, T.P. (1977) *J Biol Chem* 252:8035–8039.
11. Wellner, D. and Meister, A. (1960) *J Biol Chem* 235:2013–2018.
12. Lichtenberg, L.A. and Wellner, D. (1968) *Anal Biochem* 26:313–319.
13. Wasserman, E. and Levine, L. (1961) *J Immunol* 87:290–295.

Published 1982 by Elsevier North Holland, Inc.
Vincent Massey and Charles H. Williams, Editors
Flavins and Flavoproteins

CHAPTER 16

Artificial Flavins as Active Site Probes of Flavoproteins

Vincent Massey and Peter Hemmerich

Department of Biological Chemistry, University of Michigan, Ann Arbor, Michigan and Fakultät Biologie, Universität Konstanz, D-7750 Konstanz, West Germany

The last two Flavin Symposia have witnessed a growing awareness of the utility of employing artificial flavins either as probes of reaction mechanism or of the protein environment immediately around the flavin prosthetic group. In the last couple of years, new and versatile probes have become available with the recognition that the chloro substituent of 8-chloroflavins is readily displaced by nucleophiles such as thiolates and sodium sulfide (1,2). These reactions form the basis of methods to study the accessibility to solvent of the flavin 8-position in various flavoproteins. Additional probes of the same position come from 8-mercaptoflavins, which have been shown to react with alkylating agents such as iodoacetic acid and iodoacetamide, and with methyl methane thiolsulfonate (3). Similar reactivity with 2-thioflavins permits the use of these flavins as probes of the solvent accessibility of the pyrimidine subnucleus of the flavin (4)

The Native Flavin as Active Site Probe

Before detailing the uses of artificial flavins, it is well to remember that the natural flavin associated with a particular protein itself conveys a lot of information concerning its immediate environment. Thus, the absorption spectra of flavoproteins differ very significantly one from the other, and often from that of the free coenzyme. Studies with model flavins show that the absorption spectrum is very dependent on the polarity of the solvent, as illustrated in Figure 1. Accordingly it is generally assumed that flavoproteins showing a distinct resolution of the 450 nm absorption band are ones where the flavin is surrounded by a nonpolar region of the protein. In similar fashion, if on binding a substrate or inhibitor the spectrum changes from unresolved to resolved, it is commonly assumed that ligand binding must be accompanied by a protein conformational change giving a more nonpolar environment to the flavin. However, it is not at all clear that these are the only factors influencing the absorption spectrum; there are frequent examples among flavoproteins where distinct resolution of the 450 nm band is observed, consistent with a nonpolar environment, but rarely is the near-UV band influenced in the same way as that shown in Figure 1.

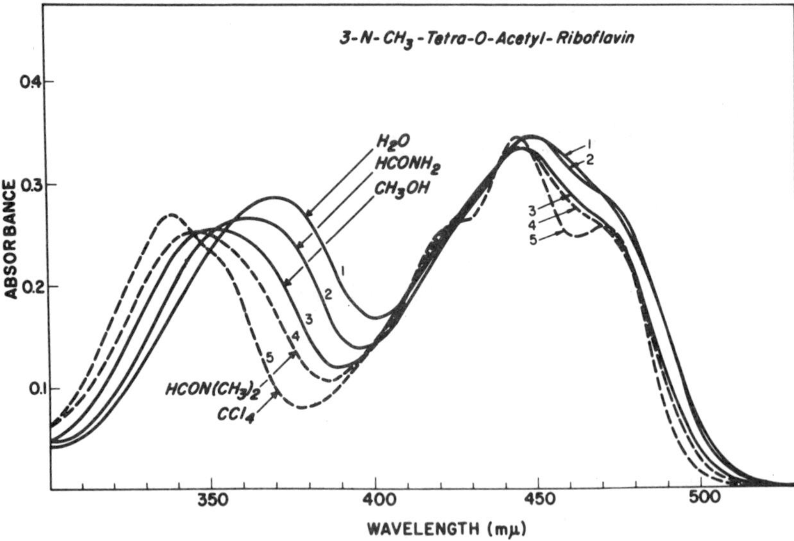

Figure 1. Influence of solvent polarity on the spectrum of 3-methyl-tetraacetyl riboflavin.

Another distinctive feature of flavins which varies greatly from one flavoprotein to another is that of fluorescence. In many cases, the oxidized flavin fluorescence is completely quenched, in some cases partially quenched, and in a few cases even enhanced. On the basis of model studies showing quenching of flavin fluorescence on interaction with phenols and indoles, it is generally assumed that nonfluorescence in a flavoprotein implies interaction between flavin and protein tyrosine or tryptophan residues. Again, it is not at all clear that this is the only explanation. The structure of the active site of p-hydroxybenzoate hydroxylase, reported by Drenth et al. at this meeting (5), shows a cluster of tyrosine residues around the flavin, yet this enzyme has appreciable fluorescence, even in the presence of p-hydroxybenzoate (6). The almost complete quenching of fluorescence of FMN in flavodoxin appears to be nicely accounted for by a single tryptophan residue lying over and roughly parallel to the isoalloxazine ring (7). However, again there may be some question as to whether this overlap is the only cause of the fluorescence quenching. The reduced flavin of flavodoxin has significant fluorescence (8) and yet there are only minor changes in protein structure accompanying reduction.

Another indicator of active site geometry is the ability to form charge transfer complexes. Nearly all of the pyridine nucleotide-linked flavoenzymes exhibit charge transfer complexes, either between the oxidized flavin and NAD(P)H or between reduced flavin and $NAD(P)^+$, or both (9). While the mechanistic significance of these charge transfer interactions remains a matter of conjecture, their existence argues at least that the planar pyridine nucleotide must be able to position itself over part of the flavin in order to obtain efficient π electron orbital overlap. This conclusion is nicely borne out by the X-ray crystallographic studies, reported at this meeting, on glutathione reductase (10).

The ability of a given flavoprotein to stabilize either the blue neutral flavin radical or the red anion radical must also depend on particular protein-flavin interactions, given the fact that in free solution the equilibrium:

$$FL_{ox} + FL_{red}H_2 \rightleftharpoons 2FLH^{\bullet}$$

lies considerably in favor of dismutation. In this connection, it is important to distinguish between true thermodynamic stabilization of the radical as opposed to kinetic stabilization. In a flavoprotein where 1-electron reduction has been obtained by chemical or photochemical means, there may be severe kinetic barriers against two flavin radicals coming in contact, particularly if they are negatively charged. Many of the flavoprotein radicals reported in the literature appear to be kinetically stabilized, disproportionating to oxidized and reduced flavin forms if sufficient time is allowed. If, on the other hand, true thermodynamic stabilization is obtained, then it has to be attained by protein-flavin interactions. As will be detailed in a later section, the use of artificial flavins has been invaluable in elucidating the causes of such stabilization, as well as striking structure-function correlations observed within different classes of flavoproteins (11)

Modified Flavins as Active Site Probes and Probes of Mechanism

General Considerations

With most flavoproteins it is possible to dissociate the flavin from the apoprotein, and reconstitute the holoenzyme by adding back either FAD or FMN, whatever nature originally provided. The specificity relies mainly on the nature of the flavin N(10)-side chain; most flavoenzymes will accept into their active sites chemically modified ring structures, provided that the modified flavin is at the appropriate FMN or FAD level. It has been an act of faith that such artificial flavins are bound in the same way as the native flavin with respect to flavin-protein geometry. This assumption appears to be borne out by the experimental results so far accumulated; clearly one must be wary, however, when very bulky substituents are introduced into the isoalloxazine ring. The principle of flavin replacements was recognized many years ago, and extensive use of the technique has been made by McCormick, Veeger, Tollin, Edmondson, and their coworkers (12–14). These early studies were hampered by the difficulty of converting the modified flavins from the riboflavin level to the appropriate FMN or FAD level. These difficulties have now been largely overcome with the discovery of the flavokinase/FAD synthetase system of *Brevibacterium ammoniagenes*, which has been found to convert almost any artificial flavin presented to it, from the riboflavin to the FAD-level (15). The FMN-level modified flavin is most conveniently obtained from the FAD form by hydrolysis with *Naja naja* venom.

Use of a Series of Flavins of Different Redox Potential

Depending on the nature and postion of the substituent, ring-modified flavins are avilable with E'_0 values ranging from approximately −370 mV (1,5-

dideazaflavin) to −30 mV (8-phenyl sulphonyl flavin). (For useful tables of redox potentials see references 1, 15–17). The first reported use of a series of artificial flavins of different redox potentials to provide information not available with the native protein was with Old Yellow Enzyme. This protein had been isolated in the native state also as a green form, due to complex formation with p-hydroxybenzaldehyde (18). It was found that similar long wavelength bands could be formed with a wide variety of aromatic and heteroaromatic compounds, whose one common feature was an ionizable hydroxy group. The obvious conclusion was that the spectral properties were due to charge transfer interactions between phenolate anions as donors and the oxidized flavin as acceptor. Substantive evidence for this conclusion was obtained by replacement of the native FMN by a series of artificial flavins of different potential, and finding that the position of the long wavelength band changed in the fashion predicted for such a charge transfer interaction (17). A similar series of flavins at the FMN level has been used as replacement of the native FMN of the FMN- and FAD-containing enzyme, NADPH cytochrome P-450 reductase. These studies provided further support to the conclusion that the air-stable semiquinone of this enzyme is due to the FMN component, and that this is the flavin of higher redox potential (19). Results of mechanistic importance were also obtained, indicating that the FMN and FAD have separate roles in catalysis, with FAD being the flavin reacting with NADPH, and $FMNH_2$ as the flavin reacting with cytochrome P-450 (20).

Another recent example of use of a series of flavins of different redox potentials is with the related enzyme NADPH-adrenodoxin reductase (16). In this study, it was shown that the rate of reduction of the enzyme by NADPH varied in a linear logarithmic fashion with the redox potential of the bound flavins, and that this step is rate-limiting in the catalytic turnover with ferricyanide as acceptor. In the case of adrenodoxin as acceptor, a definite maximum in the steady state turnover was obtained with flavin potentials between −220 and −160 mV. The conclusion was drawn that the apoprotein of adrenodoxin reductase tailors the midpoint potential of its bound FAD in order to balance the activation energies of the reductive and oxidative half reactions.

Yet another example of the utility of this general approach is given by the results with a series of FAD analogs of different potential introduced as replacements of the native FAD of xanthine oxidase. This has yielded direct proof of the important concept formulated by Olson et al. (21) that in this multiredox center enzyme, the observed time course of reduction of the molybdenum, flavin and iron-sulfur centers is dictated solely by the number of reducing equivalents which have entered the protein (a total uptake of 6-electron equivalents is possible) and the different redox potentials of the various centers, with rapid equilibration of reducing equivalents between the centers. Dr. Russell Hille has studied the reduction properties of enzymes where the FAD has been replaced by one of higher potential (8-chloro-FAD) and two of lower potential (6-hydroxy-FAD and 8-mercapto-FAD). In all cases, the reduction behavior, both qualitatively and quantitatively, is that predicted on the basis of the Olson rapid equilibrium model (22).

5-Deazaflavins as Mechanism Probes

The previous two symposium volumes have devoted a great deal of attention to 5-deazaflavins, as have several review articles (23–25). Hence, we will not discuss the results with this flavin except to say that from various substrates there appears to be direct transfer of a "hydride" equivalent to the C(5)-position of 5-deazaflavin. This has been taken as evidence for similar transfer to native flavin. However, caution should be exercised in this interpretation, owing to the fact that 5-deazaflavin appears to be as good or better a nicotinamide analogue as it is a flavin analogue.

1-Deazaflavins as Mechanism Probes

1-Deazaflavins appear to be much more promising flavin analogues, retaining most of the chemical reactivities of normal flavins, including O_2-reactivity and the stability of the semiquinone form (26). This flavin has been incorporated into a number of enzymes, with very interesting results. 1-Deaza-FAD-reconstituted glucose oxidase has a catalytic turnover approximately 10% that of native enzyme, presumably because of the lower redox potential (E'_0 of −280 mV compared to −208 mV for FAD). However, 1-deaza-FAD-D-amino acid oxidase has the same turnover number as native enzyme (26), providing good supporting evidence for the previous conclusion that the rate limiting step in catalysis of this enzyme is the physical release of product (27,28).

Particularly informative results have come from the use of this flavin with monooxygenases. The 1-deazaflavin form of cyclohexanone monooxygenase is fully competent in carrying out the normal monooxygenation reaction of this enzyme (29). However, when introduced into the apoproteins of a number of aromatic hydroxylases, viz. orcinol hydroxylase (29), melilotate hydroxylase (30), and p-hydroxybenzoate hydroxylase (31), substrate-stimulated NAD(P)H oxidation is observed, but no hydroxylation. In the case of p-hydroxybenzoate hydroxylase, the molecular events have been dissected with the aid of stopped flow spectrophotometry. The reduced enzyme-substrate complex reacts rapidly with oxygen to form a C(4a)-1-deazaflavin hydroperoxide, in an analogous fashion to native enzyme. However, at this point the analogy stops. With native enzyme, the C(4a)-hydroperoxide undergoes a series of steps resulting in hydroxylation of the substrate, with distinctive oxygenated flavin intermediates preceding the final reformation of oxidized flavin. With the 1-deazaflavin hydroperoxide, however, these steps do not occur, and the compound breaks down to H_2O_2 and oxidized 1-deazaflavin enzyme. These results indicate that interactions between the protein and N(1)-C(2=0) region of the flavin may be important in controlling the fate of the flavin C(4a)-hydroperoxide. Whatever the explanation, the results clearly indicate different routes of oxygen activation in the electrophilic oxygen insertion reactions catalyzed by the aromatic hydroxylases and the nucleophilic attack in the cyclohexanone monooxygenase reaction. It would seem unlikely that hydroxylation is accomplished directly from the flavin C(4a) hydroperoxide, or a C(4a)-N(5) ring opened form, since such reactions should be equally feasible with 1-deazaflavin as with normal flavin.

Use of Flavin Analogues to Identify Flavin Covalent Intermediates

The absorption spectra of flavin N(5)- and C(4a)-adducts are sufficiently similar to each other and to those of reduced flavins, that the ascription of the position of attachment of a covalent intermediate can be very difficult, especially if it is transitory in nature. This situation is helped somewhat by the recognition of the fluorescence properties of dihydroflavins and flavin adducts (8), but this property is not uniformly shown by all flavoenzymes. Fortunately, an alternative is available through the different spectral properties of the artificial flavins. In this respect, 2-thioflavin is particularly useful since its N(5)- and C(4a) adducts have quite different absorption spectra. Use of this property has been made to identify unequivocally the glycolyl adduct of lactate oxidase as the N(5)-adduct (32). Iso-FMN or Iso-FAD (6,7-dimethyl-8 nor-flavin) has also proved useful in a similar way. The product of the suicide reaction of α-hydroxybutynoate with D-lactate dehydrogenase has been identified as a C(6)-N(5) cyclic addition product (33); as expected this addition cannot occur with the Iso-FAD-enzyme (34).

Chemically Reactive Flavins as Probes of Active Site Topography

The facile nucleophilic displacement of the chloro- (and bromo-) substituent from 8-chloro- (and 8-bromo-) flavins by thiolates and sulfide and the accompanying substantial spectral changes (1,2) provide very sensitive methods for testing whether the benzene subnucleus of enzyme-bound flavin is exposed to solvent.

λ_{max} 445, 360
ε ~ 11 $mM^{-1} cm^{-1}$

$RS^{\ominus}$ → $Cl^{\ominus}$
λ_{max} ~ 480
ε ~ 25 $mM^{-1} cm^{-1}$

$S^{\ominus}$ → $Cl^{\ominus}$
λ_{max} ~ 520
ε ~ 30 $mM^{-1} cm^{-1}$

Many 8-chloroflavin-substituted flavoproteins have been tested in this way, and the results are summarized in Table 1 and in an accompanying paper in this volume (3). While reaction rates of sulfide or thiophenol with 8-chloroflavoproteins as fast or faster than those of the free 8-chloroflavins clearly indicate exposure of the flavin 8-position to the solvent, low rates of reaction are more difficult to interpret, since they may be due to geometric constraints in forming the tetrahedral intermediate which presumably is involved in these reactions. This difficulty in interpretation can be overcome by making use of the further chemical reactivity of 8-mercaptoflavins, which react readily with alkylating agents such as iodoacetic acid and iodoacetamide, and with methylmethanethiolsulfonate.

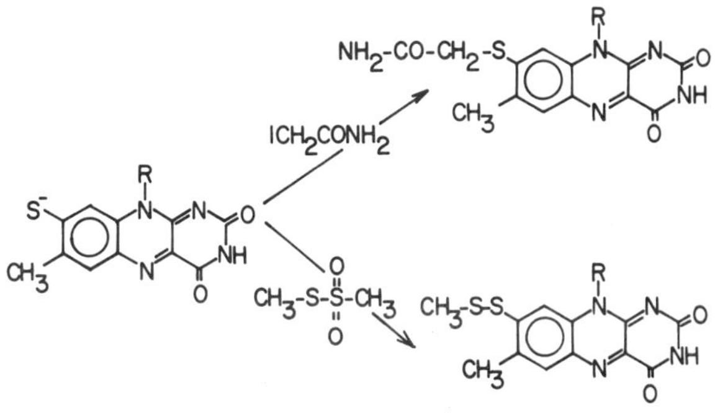

Both methods, reactivity of 8-chloroflavins and reactivity of 8-mercaptoflavins, have been used to determine whether the flavin 8-position is exposed or buried in the protein matrix, in the examples listed in Table 1.

In similar fashion, the chemical reactivity of 2-thioflavins can be used to probe the solvent accessibility of the flavin pyrimidine ring. In this case the reaction rates with iodoacetic acid and iodoacetamide are too slow to be of practical use, but methylmethanethiolsulfonate reacts sufficiently rapidly to permit its reactivity to be used as a measure of the protein topography in the vicinity of the flavin 2-position. This work is discussed in more detail in another article in this volume (4) and the results are summarized in Table 1.

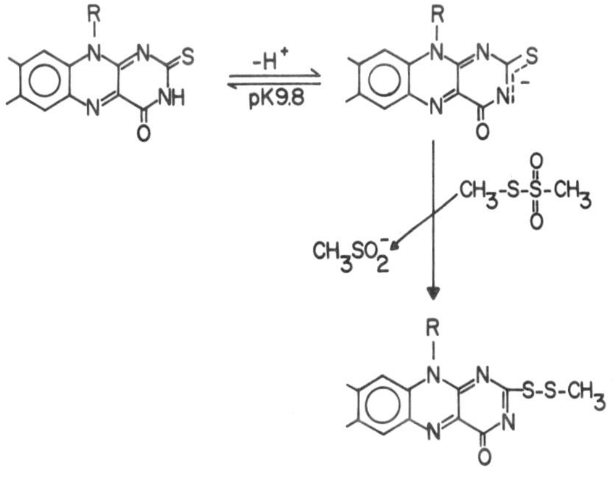

Environmentally Sensitive Flavins as Active Site Probes

The principle underlying this technique was first formulated by Ghisla and colleagues for the naturally occurring 6-hydroxy- and 8-hydroxyflavins, which in their anionic states exist mainly in the benzoquinoid mesomeric form, with the negative charge localized in the N(1)-C(2α) region of the flavin (35,36). The pKs for ionization are 4.8 for 8-hydroxyflavin (36) and 7.1 for 6-hydroxyflavin (35), and in both cases there are substantial spectral shifts on ionization. Thus, in both cases, a positively-charged protein residue in the vicinity of the flavin N(1)-position would stabilize the benzoquinoid mesomer and lower the pK of the protein-bound flavin. On the other hand, the

Table 1. Correlations of Structure and Function Among Flavoproteins.

Flavoprotein	Class	Radical	Topography		8-Mercapto-form	6-Hydroxy-form
			8-Position	2-Position		
Class 1						
D-lactate dehydrogenase	C-N transhydrogenase	Blue(unstable)	Exposed	—	Smooth benzoquinoid	pK 6.4
Pig Kidney Acyl CoA dehydrogenase	C-N transhydrogenase	Blue(-subst)	Buried	Exposed	Resolved benzoquinoid	Neutral (pK 8.7)
		Red(+subst)	Buried	—	Resolved benzoquinoid	Benzoquinoid (pK < 7)
M. elsdenii ETF	N-N transhydrogenase	Red(unstable)	Buried? Covalent with 8-ClFAD		Partly resolved benzoquinoid (unstable)	Benzoquinoid (pK < 7)
Class 2						
D-amino acid oxidase	Dehydrogenase/oxidase	Red(stable)	Exposed	Buried	Resolved benzoquinoid	Benzoquinoid (pK < 5)
L-lactate oxidase	Dehydrogenase/oxidase	Red(stable)	Buried	Buried	V. resolved benzoquinoid	Benzoquinoid (pK < 5)
Glucose oxidase	Dehydrogenase/oxidase	Red/Blue pK 7.3	Buried	Exposed?	Resolved benzoquinoid	Benzoquinoid (pK 5.6)
Putrescine oxidase	Dehydrogenase/oxidase	Red(stable)	Buried	—	Resolved benzoquinoid	—
Class 3						
Meliotate hydroxylase	Dehydrogenase/oxygenase	None(−subst)	Exposed	—	Smooth benzoquinoid	—
		None(+subst)	Exposed	—	Thiolate	—
P-hydroxybenzoate hydroxylase	Dehydrogenase/oxygenase	None/Red(−subst)	Exposed	Exposed	Thiolate	pK 6.9
		Blue(+subst)	Exposed	Buried	Smooth benzoquinoid	
Luciferase	Oxygenase	Blue	Exposed	—	Smooth benzoquinoid	

Class 4						
Cytochrome P-450 reductase	Dehydrogenase/electron transferase	Blue(stable)	Exposed?	—	Thiolate (λ max 550)	—
Putidaredoxin reductase	Dehydrogenase/electron transferase	Blue	Exposed?	—	Smooth benzoquinoid (λ max 580)	—
Adrenodoxin reductase	Dehydrogenase/electron transferase	None(blue)	—	—	Smooth benzoquinoid (λ max 570)	—
Class 5						
Flavodoxin	Pure electron transferase	Blue(stable)	Exposed	Buried	Thiolate	Neutral (pK $\sim$ 9)
Miscellaneous						
Old yellow enzyme	Function unknown	Red(unstable)	Exposed	Buried	Smooth benzoquinoid	Benzoquinoid (pK 5.7)
Riboflavin binding protein of egg white	No known redox function	Blue(stable)	Buried	Exposed	Neutral (pK 9)	Neutral (pK > 9.6)
Oxynitrilase	Non-redox enzyme	Red	Buried	—	Partly resolved benzoquinoid	
Pyruvate oxidase (*E. coli*)	Also contains thiamin pyrophos (dehydrogenase/ electron transferase)	Red	Buried	Buried	Resolved benzoquinoid	
Xanthine oxidase	Also contains Molybdenum and iron-sulfur centers (electron transferase)	Blue	Exposed		Smooth benzoquinoid	pK $\sim$ 6.9

The original references to the type of radical, the classification, and the specral forms of the 8-mercapto flavoproteins and 6-hydroxy flavoproteins are given in references 2 and 11. The information on topography of the 8-position and the 2-position is presented in references 3 and 4. Information on other enzymes is as follows: pig kidney acyl CoA dehydrogenase (40). Putidaredoxin reductase (Wagner, G., Gunsalus, I.C., and Massey, V., unpublished); *E. coli* pyruvate oxidase (41); xanthine oxidase (22); D-lactate dehydrogenase (Morpeth, F. and Massey, V., unpublished); Old Yellow Enzyme (Schopfer, L.M. and Massey, V., unpublished).

benzoquinoid form could also be favored by negatively-charged residues in the vicinity of either the flavin 6-position or 8-position, but in these instances the pK of the protein-bound form would actually be raised. Hence a study of the ionization behavior of both forms bound to a particular protein can yield valuable information about the environment of the flavin.

To these two examples can now be added an even more versatile and sensitive "active site indicator," 8-mercaptoflavin (1,2) whose various tautomeric and mesomeric forms are shown below:

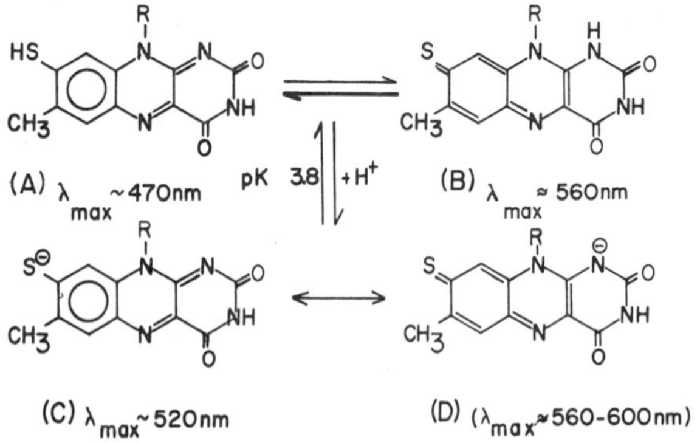

(A) λ_{max} ~470nm   pK 3.8 +H$^+$ (B) λ_{max} ≈ 560nm

(C) λ_{max} ~520nm   (D) (λ_{max} ≈ 560-600nm)

In this case the pK is very low, pH 3.8, and in free solution the preferred form of the anion is the 8-thiolate mesomer, rather than the N(1)-ionized benzoquinoid form (2). The two forms are dramatically different, the thiolate being deep red in color, with λ max in the 520–550 nm region, while the benzoquinoid form is bright purple-blue in color, with λ max in the region 570–610 nm (Figure 2). In this case it is only when there is a positive charge in the protein located near the flavin N(1)-position, or a negative charge by position-8, that the blue benzoquinoid form would be stabilized. Thus, a combined study of the spectral properties of the 8-mercapto enzyme and the pKs of the 6-hydroxyenzyme and 8-hydroxyenzyme should allow definite conclusions to be drawn. As summarized in Table 1, and detailed in previous studies (2,11), it is clear that with all the enzymes studied of the dehydrogenase-oxidase class, the benzoquinoid forms of 8-mercaptoflavin and 6-hydroxyflavin are stabilized, indicating the existence of a protein positive charge interacting with positon N(1) of the flavin. These same enzymes all stabilize the red anionic semiquinone as well as the anionic form of the fully reduced native flavin, and with few exceptions all form N(5)-sulfite adducts with the native enzyme (37). As discussed in more detail elsewhere (2,11), we take these results as providing strong evidence for a common type of flavin-protein interaction among all the enzymes of this class, viz. a positively-charged protein residue in the vicinity of the N(1)-C(2α) locus of the flavin, and that this interaction with the reduced enzyme in some way directs its high reactivity with O_2. The same interaction would serve to promote sulfite attack at the N(5)-position and stabilize the N(5)-sulfite adduct, except in those cases, such as putrescine oxidase, where formation of a sulfite adduct appears to be prevented because of a negatively-

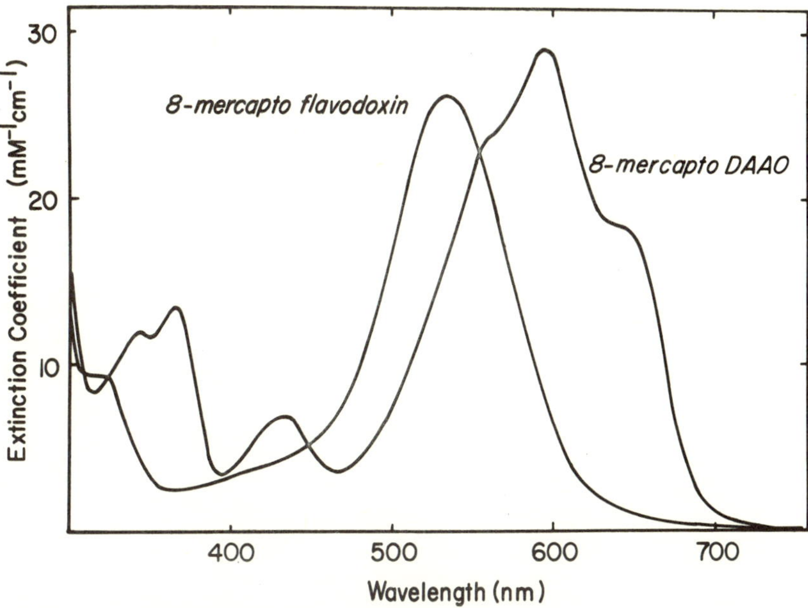

Figure 2. Spectra of 8-mercaptoflavoproteins in the 8-thiolate form (flavodoxin) and in the benzoquinoid form (D-amino acid oxidase).

charged group of the protein in the vicinity of the flavin N(5)-position (38). This negatively-charged group is probably involved in binding the positively-charged substrate.

The spectrum of the benzoquinoid form of 8-mercaptoflavin is typified by three visible-near UV absorption bands, whereas that of the 8-thiolate form has a single intense visible band centered around 520–550 nm (1,2). The spectrum of the benzoquinoid form also appears to be quite sensitive to the polarity of the environment. Figure 3 shows the spectrum of an N(10)-N(1) bridged 8-mercaptoflavin in water, acetonitrile, and benzene. It is seen that in the nonpolar solvents the long wavelength band becomes markedly resolved, in much the same way as is observed with native flavin (Figure 1). Similar effects are found with various 8-mercapto flavoproteins, as illustrated in Figure 4. These results would indicate that the chromophore in 8-mercapto Old Yellow Enzyme is located in a hydrophilic region, an interesting observation in view of the evidence that the 8-position is exposed to solvent. In contrast, the 8-mercaptoflavin form of lactate oxidase would appear to be in a hydrophobic environment, a result in keeping with the observation that both the benzene and pyrimidine subnuclei appear to be buried in the protein matrix, and not exposed to solvent. D-amino acid oxidase also appears to offer a largely hydrophobic environment to 8-mercaptoflavin, even though the 8-position is exposed to solvent (Table 1). However, on binding of benzoate, which appears to cause a conformational change which buries the benzene subnucleus, the

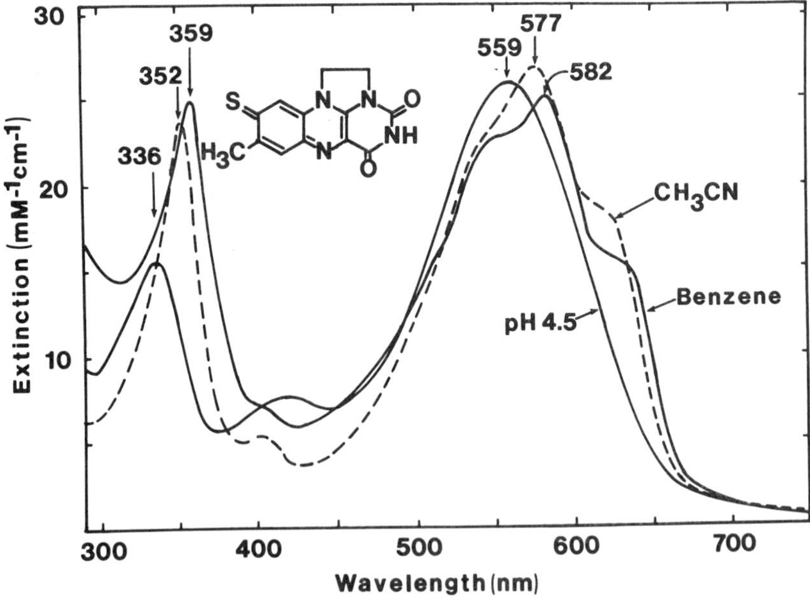

Figure 3. Effect of solvent on the spectrum of 1, 10-ethano-8-mercaptoflavin (from ref. 2).

spectrum becomes even more resolved, indicating a more hydrophobic environment of the flavin.

An inspection of Table 1 reveals that most of the enzymes which have the flavin 8-position exposed to solvent exhibit either an 8-thiolate spectrum or that of the unresolved benzoquinoid form. On the other hand, all of the enzymes with a buried 8-position exhibit the resolved 8-mercapto benzoquinoid spectrum.

Table 1 also reveals some other tantalizing correlations. For instance, all of the enzymes which react with pyridine nucleotides appear to have the flavin 8-position exposed to solvent. Does this imply that all pyridine nucleotide-linked flavoproteins are structurally related, with a pyridine nucleotide-binding site juxtaposed to the benzene subnucleus of the flavin, as appears to be the case with glutathione reductase (10)?

Finally, it should be emphasized that the conclusions regarding flavin active site topography obtained from chemical modification studies are consistent with those presently available from X-ray crystallographic studies of flavodoxins, glutathione reductase, and p-hydroxybenzoate hydroxylase, which are the subject of other articles in this volume. The results with egg white riboflavin-binding proteins are also in agreement with the conclusions of McCormick and coworkers, based on the binding affinity of various substituted flavins to the apoprotein (39). These agreements give good support to the validity of the assumption mentioned earlier that the artificial flavins are bound in substantially the same geometry as the native flavin.

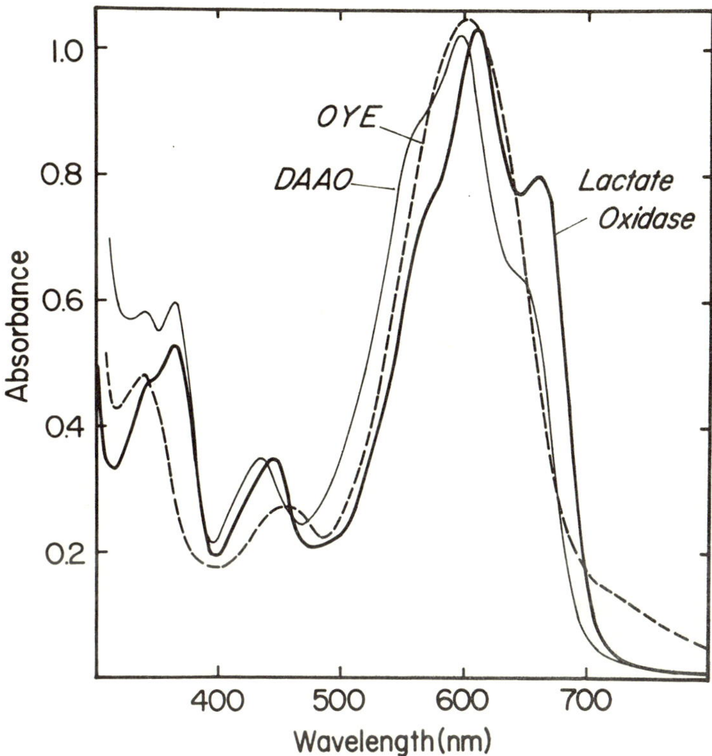

Figure 4. Absorption spectra of some typical 8-mercaptoflavoproteins in the presumed benzoquinoid form. OYE=Old Yellow Enzyme; DAAO=D-amino acid oxidase.

References

1. Moore, E.G., Ghisla, S., and Massey, V. (1979) *J Biol Chem* 254:8173–8178.
2. Massey, V., Ghisla, S., and Moore, E.G. (1979) *J Biol Chem* 254:9640–9650.
3. Schopfer, L.M., Massey, V., and Claiborne, A. (1982), this volume.
4. Claiborne, A., Massey, V., and Schopfer, L.M. (1982), this volume.
5. Weijer, W.J., Hofsteenge, J., Vereijken, J.M., Jekel, P.A., Bak, H.J., Beintema, J.J., Wierenga, R.K., and Drenth, J. (1982), this volume.
6. Howell, L.G., Spector, T., and Massey, V. (1972) *J Biol Chem* 247:4340–4350.
7. Smith, W.W., Burnett, R.M., Darling, G.D., and Ludwig, M.L. (1977) *J Mol Biol* 117:195–225.
8. Ghisla, S., Massey, V., Lhoste, J-M., and Mayhew, S.G. (1974) *Biochemistry* 13:589–597.
9. Massey, V. and Ghisla, S. (1974) *Ann NY Acad Sci* 227:446–465.
10. Pai, E.F. and Schulz, G. (1982), this volume.
11. Massey, V. and Hemmerich, P. (1980) *Biochemical Society Transactions* 8:246–257.
12. McCormick, D.B., Chassy, B.M., and Tsibris, J.C.M. (1964) *Biochim Biophys Acta* 89:447–452.
13. Visser, J. and Veeger, C. (1970) *Biochim Biophs Acta* 206:224–241.
14. Edmondson, D.E. and Tollin, G. (1971) *Biochemistry* 10:113–124, 124–132, 133–145.
15. Walsh, C., Fisher, J., Spencer, R., Graham, D.W., Ashton, W.T., Brown, J.E., Brown, R.D., and Rogers, E.F. (1978) *Biochemistry* 17:1942–1951.

16. Light, D.R. and Walsh, C. (1980) *J Biol Chem* 255:4264–4277.
17. Abramovitz, A.S. and Massey, V. (1976) *J Biol Chem* 251:5327–5336.
18. Matthews, R.G., Massey, V., and Sweeley, C.C. (1975) *J Biol Chem* 250:9294–9298.
19. Vermilion, J.L., Massey, V., and Coon, M.J. (1980) In *Flavins and Flavoproteins*. K. Yagi and T. Yamano (eds.) Baltimore:University Park Press, pp. 693–702.
20. Vermilion, J.L., Ballou, D.P., Massey, V., and Coon, M.J. (1981) *J Biol Chem* 256:266–277.
21. Olson, J.S., Ballou, D.P., Palmer, G., and Massey, V. (1974) *J Biol Chem* 249:4363–4382.
22. Hille, R., Fee, J.A., and Massey, V. (1981) *J Biol Chem* in press.
23. Hemmerich, P., Massey, V., and Fenner, H. (1977) *FEBS Letters* 84:5–21.
24. Walsh, C. (1980) *Accounts of Chemical Research* 13:148–155.
25. Bruice, T.C. (1980) *Accounts of Chemical Research* 13:256–262.
26. Spencer, R., Fisher, J., and Walsh, C. (1977) *Biochemistry* 16:3586–3594, 3594–3602.
27. Massey, V. and Gibson, Q.H. (1964) *Fed Proc* 23:18–29.
28. Porter, D.J.T., Voet, J.G., and Bright, H.J. (1977) *J Biol Chem* 252:4464–4473.
29. Walsh, C., Jacobson, F. and Ryerson, C.C. (1980) *Biomimetic Chemistry*, *Advances in Chemistry Series*, American Chemical Society, 119–138.
30. Schopfer, L.M. and Massey, V., unpublished.
31. Entsch, B., Husain, M., Ballou, D.P., Massey, V., and Walsh, C. (1980) *J. Biol. Chem.* 255:1420–1429.
32. Ghisla, S. and Massey, V. (1980) *J Biol Chem* 255:5688–5696.
33. Ghisla, S., Olson, S.T., Massey, V., and Lhoste, J-M. (1979) *Biochemistry* 18:4733–4742.
34. Olson, S.T. (1979) PhD Thesis, University of Michigan.
35. Mayhew, S.G., Whitfield, C.D., Ghisla, S., and Jorns, M. (1974) *Eur J Biochem* 44:579–591.
36. Ghisla, S. and Mayhew, S.G. (1976) *Eur J Biochem* 63:373–390.
37. Massey, V., Müller, F., Feldberg, R., Schuman, M., Sullivan, P.A., Howell, L.G., Mayhew, S.G., Matthews, R.G., and Foust, G.P. (1969) *J Biol Chem* 244:3999–4006.
38. Swain, W.F. and DeSa, R.J. (1976) *Biochim Biophys Acta* 429:331–341.
39. Choi, J-D and McCormick, D.B. (1980) *Archiv Biochem Biophys* 204:41–51.
40. Thorpe, C. and Massey, V. (1982), this volume.
41. Mather, M., Schopfer, L., Massey, V., and Gennis, R.B. (1982), this volume.

Published 1982 by Elsevier North Holland, Inc.
Vincent Massey and Charles H. Williams, Editors
Flavins and Flavoproteins

CHAPTER 17

2-Thioflavins as Active Site Probes of Flavoproteins

Al Claiborne, Vincent Massey, and
Lawrence M. Schopfer

Department of Biological Chemistry, The University of Michigan, Ann Arbor, Michigan

Introduction

Recent studies of flavoproteins reconstituted with modified flavins have provided quite useful information concerning the interactions within the flavin-binding site between the coenzyme and the respective apoprotein (1–6). We have reconstituted a number of flavoproteins with the appropriate 2-thioflavin (sulfur replacing oxygen substituent at position 2). The ionization behavior of this modified coenzyme as well as its reactivity toward the thiol reagent methyl methanethiosulfonate (MMTS) provide sensitive probes of the flavin environments within these proteins.

Methods

Spectral pK_a determinations were performed with a Cary 219 double-beam spectrophotometer thermostatted at 4°. MMTS reactions were monitored either with a Cary 118 spectrophotometer or with a stopped-flow instrument. Rates were measured at 520–530 nm, at 4°, after mixing of samples with MMTS.

Results and Discussion

Scheme 1 depicts the ionization of 2-thioflavin; the pK_a determined for the N(3) proton of 2-thio riboflavin (Figure 1) is 9.8. Methyl methanethiosulfonate (MMTS) reacts with 2-thioflavin to yield the corresponding 2-$SSCH_3$ flavin disulfide; the accompanying spectral changes are shown in Figure 2. The close agreement between the 2-thioflavin pK_a values obtained by spectral titration and by the kinetic method suggests the participation of the 2-thioflavin anion in the MMTS reaction.

Similar spectral pK_a determinations for several 2-thio flavoproteins have been performed as well; the intrinsic pK_a values thus obtained can be correlated with the limiting rates of MMTS reactivity for the respective bound 2-thioflavins. Figure 3 represents the Bronsted analysis of those 2-thio flavoproteins examined. We conclude that the flavin pyrimidine subnuclei in 2-thio p-hydroxybenzoate hydroxylase (PHBH) and 2-thio riboflavin-binding

Scheme 1.

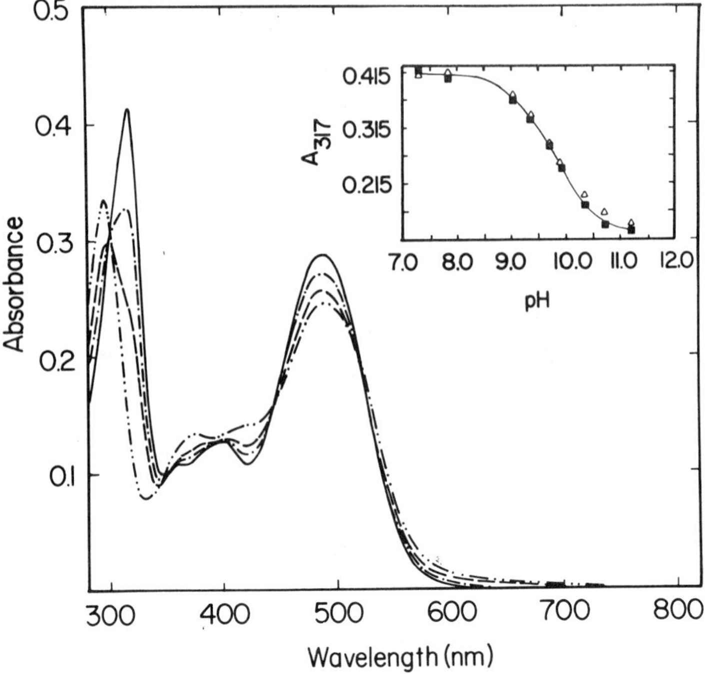

Figure 1.

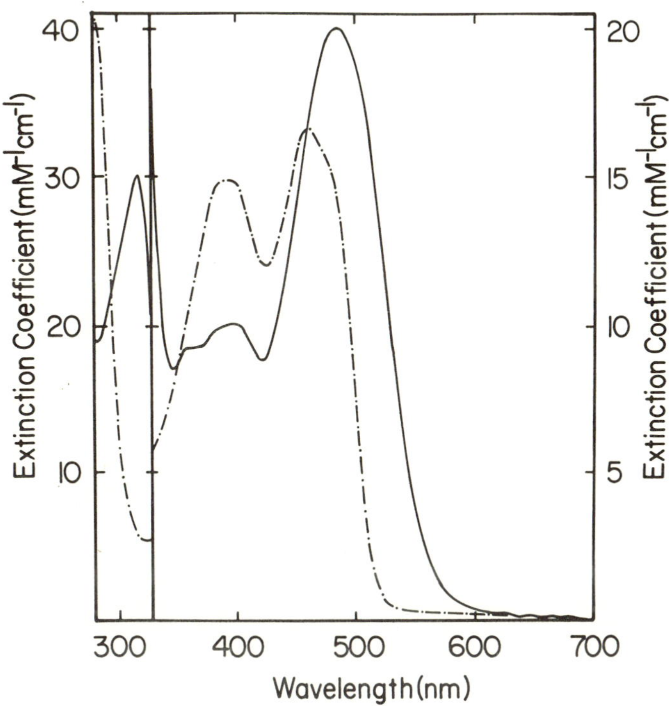

Figure 2. Reaction of 2-thio riboflavin with MMTS. 2-Thio riboflavin (—) was reacted with 1.76 mM MMTS in 0.1 M potassium phosphate, pH 7.0, plus 0.6 mM EDTA at 4°. The product 2-$SSCH_3$ riboflavin spectrum (–·–) was taken 41 minutes after MMTS addition. Flavin concentration was 4.9 μM.

protein are accessible to solvent; in 2-thio flavodoxin and the 2-thio PHBH/p-hydroxybenzoate complex, however, there is considerable burial of the flavin pyrimidine subnucleus. 2-Thio Old Yellow Enzyme, 2-thio lactate oxidase, and 2-thio D-amino acid oxidase react only poorly (if at all) with MMTS; the flavin pyrimidine subnuclei in these cases as well must be buried within their respective protein matrices.

The observed shift in the accessibility of the flavin pyrimidine subnucleus in 2-thio PHBH upon substrate-binding is consistent with interpretations from crystallographic studies of the native PHBH/p-hydroxybenzoate complex (7-9), and may well underlie other documented changes in the catalytic properties of the enzyme upon binding of the oxygenatable substrate (10). In addition, previous high-resolution analysis of the three-dimensional structure of

---

Figure 1. pH titration of 2-thio riboflavin. 2-Thio riboflavin (14.4 μM; in 50 mM potassium phosphate, plus 0.6 mM EDTA) was titrated with solid Na_2CO_3 and 5 M NaOH. Intermediate spectra are shown at pH 7.27 (—); pH 9.36 (–·–); pH 9.90 (– –); and pH 11.16 (–··–). The inset shows the resulting titration curve; titration data (■) and points calculated for a pK_a of 9.75 (△) are given. Temperature was 4°.

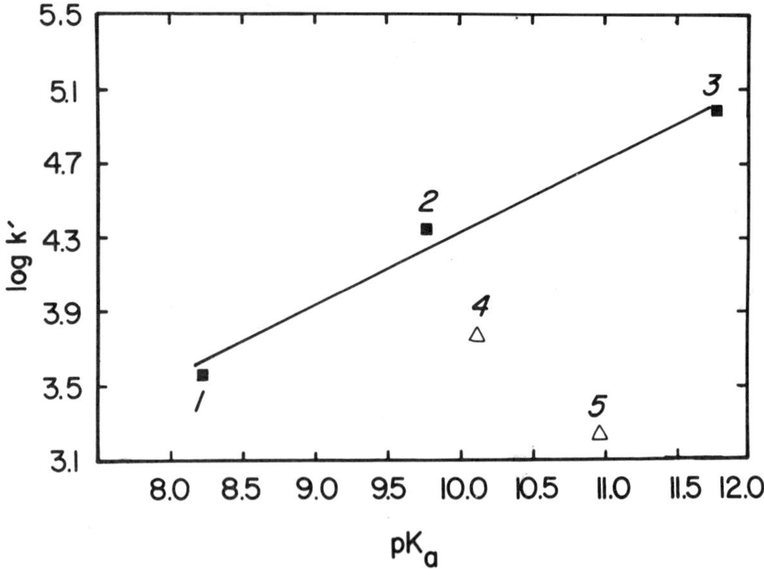

Figure 3. Bronsted correlation for 2-thio flavoproteins:MMTS reactivities. All protein pK_a values (except for *4*) were measured in the presence of excess apoprotein at 4°. k′ represents the intrinsic second-order rate constant (M^{-1} min^{-1}), after correction to 100% anionic species present. *1* = 2-thio PHBH; *2* = free 2-thio riboflavin; *3* = 2-thio riboflavin-binding protein; *4* = 2-thio PHBH/p-hydroxybenzoate complex; and *5* = 2-thio flavodoxin.

flavodoxin from *Clostridium MP* indicates that the flavin pyrimidine subnucleus is inaccessible to solvent (11); our findings with the reconstituted 2-thio flavodoxin from *Megasphera elsdenii* suggest a similar orientation of the flavin within this functionally related protein.

ACKNOWLEDGMENTS
We thank Dr. Peter Hemmerich, University of Konstanz, West Germany, for his gift of 2-thio riboflavin and Dr. Jules Shafer, University of Michigan, for methyl methanethiosulfonate. This work was supported by NIH Grant No. GM 11106 from the U.S. Public Health Service.

References

1. Massey, V. and Hemmerich, P. (1980) *Biochem Soc Trans* 8; 246–257.
2. Hemmerich, P. and Massey, V. (1979) In *3rd International Symposium on Oxidases*. King, T.E., Mason, H.S., and Morrison, M. (eds.), Baltimore: University Park Press, in press.
3. Massey, V., Ghisla, S., and Moore, E.G. (1979) *J Biol Chem* 254:9640–9650.
4. Mayhew, S.G., Whitfield, C.D., Ghisla, S., and Schuman-Jorns, M. (1974) *Eur J Biochem* 44:579–591.
5. Ghisla, S. and Mayhew, S.G. (1976) *Eur J Biochem* 63:373–390.
6. Schopfer, L.M., Massey, V., and Claiborne, A. (1981) *J Biol Chem*, in press.

7. Wierenga, R.K., de Jong, R.J., Kalk, K.H., Hol, W.G.J., and Drenth, J. (1979) *J Mol Biol* 131:55–73.
8. Hofsteenge, J., Vereijken, J.M., Weijer, W.J., Beintema, J.J., Wierenga, R.K., and Drenth, J. (1980) *Eur J Biochem* 113:141–150.
9. Hofsteenge, J. (1981) PhD dissertation, University of Groningen, The Netherlands.
10. Howell, L.G., Spector, T., and Massey, V. (1972) *J Biol Chem* 247:4340–4350.
11. Burnett, R.M., Darling, G.D., Kendall, D.S., LeQuesne, M.E., Mayhew, S.G., Smith, W.W., and Ludwig, M.L. (1974) *J Biol Chem* 249:4383–4392.

Published 1982 by Elsevier North Holland, Inc.
Vincent Massey and Charles H. Williams, Editors
Flavins and Flavoproteins

CHAPTER 18

Active Site Probes of Flavoproteins: Determination of the Solvent Accessibility of the Flavin 8-Position for a Series of Flavoproteins

Lawrence M. Schopfer, Vincent Massey, and Al Claiborne

Department of Biological Chemistry, The University of Michigan, Ann Arbor, Michigan

The availability of chemically modified flavins at the FAD and FMN levels, has opened many new and intriguing possibilities for probing flavin enzyme mechanism (1–3). We have recently exploited the chemical reactivity of 8-chloro and 8-mercapto flavins as a means for assessing the solvent accessibility of the 8-position of protein-bound flavins (4).

Our experimental approach capitalized on the chemical reactivity of the 8-position chloro group toward thiols (sulfide and thiophenol) and the 8-position mercapto group toward thiol reagents (iodoacetamide and iodoacetic acid). We measured the rates of reaction for these reagents with both protein-bound and free flavins, and took the rate of the flavoprotein reaction as a measure of the accessibility of the 8-position to solvent.

Progress of the reaction was conveniently followed spectrophotometrically, using the large differences in the visible absorbance spectra of these various flavins (4).

It was normally found that the protein-bound flavin reacted either significantly faster than the free flavin or very much slower. The former condition could best be accounted for if the flavin was reacting while still protein-bound, thus the 8-position was accessible to solvent. The latter condition could represent either a slow reaction between the protein-bound flavin and reagent or dissociation of the flavin from the protein followed by reaction with the reagent free in solution. We have found that the slow reactions invariably resulted from flavin dissociation. The accumulated results from this study are shown in Table 1.

Free flavins. The reaction of free 8-mercaptoflavins with both iodoacetamide and iodoacetic acid were simple, second order. Reaction of 8-chloroflavin with thiophenol was also second order. However, reaction of 8-chloroflavin with sodium sulfide was complex (4,5). Due to its complexity, sodium sulfide was used with only a few proteins. The rates reported for sulfide reflect the major (approximately logarithmic) middle phase of the reaction and are given

Table 1. Reaction Rates.[a]

Enzymes	Reaction of the 8-Cl-form with		Reaction of the 8-mercapto-form with	
	Na_2S[b]	Thiophenol[c]	Iodoacetamide[d]	Iodoacetic Acid[d]
FAD	0.034 min^{-1}	1.84 min^{-1}	0.46 min^{-1}	0.10 min^{-1}
FMN	0.064 min^{-1}[e]	1.84 min^{-1}	0.47 min^{-1}	0.10 min^{-1}
			0.096 min^{-1}[f]	
L-lactate oxidase		$t_{1/2} > 1$ day	$t_{1/2} \sim 1$ wk	ND[g]
D-amino acid oxidase	0.08 min^{-1}	2.65 min^{-1}	$t_{1/2} \sim 6$ hr	ND[g]
Glucose oxidase	$t_{1/2} > 1$ day	No reaction[h]	$t_{1/2} \sim 12$ hr	ND[g]
Putrescine oxidase		No reaction[h]	$t_{1/2} \sim$ wk	ND[g]
Melilotate hydroxylase	1.2 min^{-1}	11.5 min^{-1}	0.21 min^{-1}	0.10 min^{-1}
p-OH-benzoate hydroxylase	Equal to 8-Cl-FAD[i]	2.24 min^{-1}	1.2 min^{-1}	0.14 min^{-1}
Old Yellow Enzyme	Very fast[j]	20.8 min^{-1}	0.80 min^{-1}	2.0 min^{-1}
Flavodoxin		0.0039 min^{-1}	0.49 min^{-1}	0.04 min^{-1}
Luciferase	0.12 min^{-1}[e]	No reaction[k]	0.003 min^{-1}[f]	ND[g]
Xanthine oxidase		No reaction[h]	>3.5 min^{-1}	$t_{1/2} \sim 3$ hr
D-lactate dehydrogenase		0.008 min^{-1}	1.27 min^{-1}	0.016 min^{-1}
Riboflavin binding protein	No reaction	0.0035 min^{-1}	ND[g]	ND[g]

[a] All reactions were performed in 0.1 M KPi buffer, pH 7.3 at 25°C unless otherwise indicated. Results are reported as pseudo-first-order rates using a fixed concentration of reagent for the sake of comparison. Most reactions were second order in reagent concentration, see reference 4.

[b] Reactions were performed with 12.5 mM sulfide at pH 7.8.

[c] Reactions were performed with 575 μM thiophenol.

[d] Reactions were performed with 12.5 mM iodoacetamide or iodoacetic acid.

[e] Reactions were performed with 12.5 mM sulfide at 4°C, pH 8.1.

[f] Reactions were performed at 4°C.

[g] ND indicates not determined.

[h] Reaction was monitored for 12 hr.

[i] Reaction carried out at pH 8.5, 25°C; the k_{obs} was the same for 8-Cl-FAD p-OH-benzoate hydroxylase and 8-Cl-FAD ($k_{obs} = 0.26$ min^{-1} at 5 mM sulfide).

[j] Reaction carried out at pH 8.5, 20°C, and was complete within 3 min with 10 mM sulfide.

[k] Reaction followed for 5 hr at 4°C.

as a qualitative measure of the relative reactivity of sulfide with some 8-chloroflavins and flavoproteins.

L-lactate oxidase, glucose oxidase and purtrescine oxidase. The 8-chloro form of these three flavoenzymes was insensitive to thiophenol. The 8-mercapto form was essentially untouched by iodoacetamide. These findings suggest that the 8-position of the flavin on these enzymes is inaccessible to solvent.

D-amino acid oxidase. Unlike with the oxidases above, the 8-position of D-amino acid oxidase was found to be quite reactive. 8-Chloro-FAD D-amino acid oxidase reacted with both sulfide and thiophenol at rates faster than those found for free 8-chloro-FAD. Thus the 8-position must be available to solvent.

Old Yellow Enzyme and p-hydroxybenzoate hydroxylase. Both of these enzymes were more reactive than the corresponding free flavin (whether in the 8-chloro or 8-mercapto form), arguing that the 8-position of the protein-bound flavin must be accessible to solvent for both.

Melilotate hydroxylase. This enzyme reaction was faster than free flavin for the 8-chloro form and equivalent to free flavin for the 8-mercapto form. Hence the 8-position of the enzyme-bound flavin must be exposed. The dependence of the rate on reagent concentration was, in all cases, nonlinear. For this reason, the reactions involving melilotate hydroxylase were compared to those of the free flavins under fixed sets of conditions.

Luciferase, xanthine oxidase, and D-lactate dehydrogenase. These three enzymes were similar in that each reacted rapidly with only one of the four reagents tried. 8-Chloro-FMN luciferase reacted quite rapidly with sulfide. We thus tentatively concluded that the 8-position of luciferase-bound flavin is exposed to solvent.

With xanthine oxidase and D-lactate dehydrogenase, only the 8-mercapto form was found to react rapidly, and that species reacted only with the neutral iodoacetamide reagent. In neither case was there a linear dependence of apparent rate on iodoacetamide concentration. But the pseudo-first-order rate for the proteins was faster than that for free 8-mercapto-FAD, suggesting that the 8-position for bound flavin on both of these proteins may be exposed to solvent. It is important to recall however that iodoacetamide is readily reactive toward amino acid side chains. Thus it is possible that these proteins may have been modified and that our results reflect the accessibility of the flavin on a modified, not the native, protein.

Flavodoxin. The flavin on this protein was expected to have an exposed 8-position from X-ray crystallographic studies (6). We were therefore surprised to find that 8-chloro-FMN flavodoxin was poorly reactive toward thiophenol (and sulfide). However, 8-mercapto-FMN flavodoxin was quite reactive toward the thiol reagents. Controls indicated that the thiol reagents were not reacting with the protein. It is possible that the lack of reaction of the 8-chloro-FMN flavodoxin reflects an unfavorable geometry for formation of the tetrahedral intermediate expected in the nucleophilic displacement of the chlorine residue.

Riboflavin-binding protein. The 8-chloro form of this protein was found to be unreactive toward either sulfide or thiophenol. We have therefore concluded, in agreement with earlier workers (7,8), that the 8-position of flavin bound to this protein is unavailable to solvent. The 8-mercapto form of this protein is protonated at neutral pH and hence unreactive toward thiol reagents.

Effect of ligand binding on the accessibility of the 8-position to solvent. We have examined 4 enzyme-ligand complexes for changes in the accessibility of the flavin 8-position upon complexation: melilotate/melilotate hydroxylase, p-hydroxybenzoate/hydroxybenzoate hydroxylase, benzoate/D-amino acid oxidase, and NADPH/Old Yellow Enzyme (8-mercapto form only). The latter two complexes caused a dramatic decrease in the rate of reaction at the 8-position, suggesting that a major decrease in the accessibility of the 8-position occurs on complexation.

The hydroxylase complexes showed only minor changes in reactivity, suggesting that the 8-position remains available to solvent in these cases.

ACKNOWLEDGMENT
This work was supported by a Grant from the U.S. Public Health Service, GM-11106.

References

1. Moore, E.G., Cardemil, E., and Massey V. (1978) *J Biol Chem* 253:6413–6422.
2. Massey, V., Ghisla, S., and Moore, E.G. (1979) *J Biol Chem* 254:9640–9650.
3. Massey, V. and Hemmerich, P. (1980) *Biochem Soc Trans* 8:246–257.
4. Schopfer, L.M., Massey, V., and Claiborne, A. (1981) *J Biol Chem*, in press.
5. Moore, E.G., Ghisla, S., and Massey, V. (1979) *J Biol Chem* 254:8173–8178.
6. Burnett, R.M., Darling, G.D., Kendall, D.S., LeQuesne, M.E., Mayhew, S.G., Smith, W.W., and Ludwig, M.L. (1974) *J Biol Chem* 249:4383–4392.
7. Becvar, J.E. (1973) PhD Dissertation, University of Michigan, Ann Arbor.
8. Choi, J. and McCormick, D.B. (1980) *Arch Biochem Biophys* 204:41–51.

Published 1982 by Elsevier North Holland, Inc.
Vincent Massey and Charles H. Williams, Editors
Flavins and Flavoproteins

CHAPTER 19

Reconstitution of Liver NADH Cytochrome b_5 Reductase and of *Desulfovibrio vulgaris* Flavodoxin with 1-Deazaflavin

Denis Pompon, Bernard Guiard, and Florence Lederer

Centre de Génétique Moléculaire, CNRS - 91190 Gif-sur-Yvette, France

1- and 5-deazaflavins have been used in recent years as probes for flavoenzyme mechanisms. We published in 1978 the results of reconstitution with these analogs of flavocytochrome b_2 from baker's yeast, a typical dehydrogenase-electron transferase (1). When these experiments were carried out, no studies had been published concerning reconstitution of electron transferases and dehydrogenases-electron transferases, apart from the reconstitution of flavodoxin with 5-deaza FMN (2). Comparison of our results with those obtained for lactate oxidase reconstituted with 5-deaza FMN (33) led us to conclude that the N(5)-C(4) protein environment was highly similar in those two proteins which are expected to use the same dehydrogenation mechanism. On the other hand, no binding of 1-deaza FMN to flavocytochrome b_2 could be detected, in contrast with results obtained with oxidases (4). We thus suggested that for flavoproteins which are reoxidized by O_2 and by monoelectronic transfer, the difference might reside in the N(1)-O(2α) area environment (1,5).

In order to test this suggestion, we used 1-deaza FMN for reconstitution studies with flavodoxin from *Desulfovibrio gigas* and with hog NADH cytochrome b_5 reductase. The choice of the latter enzyme was dictated by the known homology between cytochrome b_5 and the heme-binding domain of flavocytochrome b_2 (6) which suggest that binding features important for flavin to heme transfer should be identical in both proteins.

Reconstitution of Flavodoxin with 1-Deaza FMN

Apoflavodoxin was prepared as described in (7). 1-deaza FMN (kindly given by Dr. C. Walsh) showed high affinity for the protein, both in the oxidized and the reduced state (0.1 μM deazaflavodoxin exchanged in one hr 10% of its cofactor in the presence of a 35-fold FMN excess). Oxidized reconstituted flavodoxin could be slowly reduced by excess dithionite in anaerobiosis. At pH 7, the final spectrum was that of a blue type semiquinone with fine structure (Figure 1); at pH 9, it was the fully reduced spectrum (Figure 2). Abrupt lowering of pH from 9 to 7 elicited the immediate appearance of the semiquinone spectrum (Figure 2); reciprocally full reduction was achieved by raising the pH. The isosbestic wavelength and the apparent absorbance index at the isosbestic point and at 780 nm varied between pH 6.4 and 7. This suggested that at those pH's the reduction mixtures also contained some totally reduced flavin in rapid equilibrium with the semiquinone. These results can all be explained by the known variation of dithionite redox potential as a function of pH (8).

Reoxidation of the reconstituted flavodoxin mixture after O_2 introduction showed rapid reoxidation at pH 9 with transitory appearance of long wavelength absorbance; slow reoxidation at pH 7: the maximum at 780 nm initially increased then slowly decreased and was still incomplete after 24 hr.

These results are qualitatively similar to those previously described for native *D. vulgaris* flavodoxin (9).

Reconstitution of NADH-Cytochrome b_5 reductase with 1-Deaza FMN

Enzyme and apoenzyme were prepared following (10). Reconstitutions were carried out with 1-deaza FMN [the FMN enzyme shows as much as two-thirds the activity of the native enzyme; $K_D^{FMN} = 1\text{–}2 \times 10^{-8}$ M, $K_D^{FAD} < 10^{-9}$ M (11)].

Binding

The spectrum of bound oxidized deaza FMN shows a 20 nm red shift and resolved structure compared to the free cofactor (Figure 3). The dissociation

Figure 1. Anaerobic reduction of 1-deazaflavodoxin by dithionite as a function of time at pH 7; 20 mM phosphate buffer, 0.5 mg/ml dithionite, 20°C. There is a 5-min interval between each curve except for the last one, which was recorded after a total reaction time of 80 min.

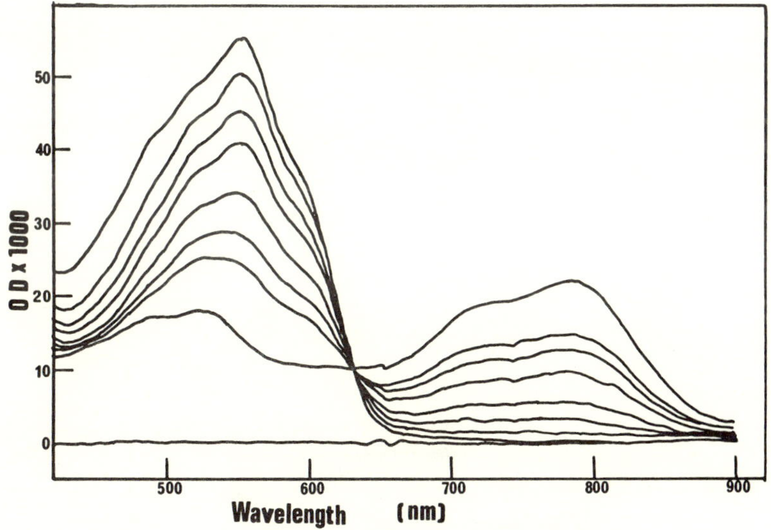

constant could be measured kinetically in competition studies with FMN in the normal enzymatic assay, because the activity of the deazaenzyme was very low. Affinity of FMN and FAD were determined directly in the same fashion. The following values were thus obtained:

$$\text{FAD: } K_D = 5 \pm 3 \times 10^{-10}\ \text{M}$$

$$\text{FMN: } K_D = 8 \pm 1 \times 10^{-8}\ \text{M}$$

$$\text{1-dFMN: } K_D = 4 \pm 0.8 \times 10^{-8}\ \text{M}$$

Catalytic Competence

The deaza FMN-reductase was reduced by NADH (Figure 3), although never completely, even in anaerobiosis. This incomplete reduction is not understood at present; we know at least that it does not arise from an equilibrium between the couples $NADH/NAD^+$ and deaza FMN/deaza $FMNH_2$. In the presence of O_2, long wavelength absorbance appeared with a maximum around 850 nm. In order to provide support for the idea that this was due to semiquinone formation, the production of superoxide anion in the system was studied by looking at the reduction of nitrotetrazolium blue (NTB) in the presence and absence of superoxide dismutase (SOD). The following relative rates of NTB reduction were observed:

	−SOD	+SOD
native enzyme+NADH+NTB	100%	4%
1-deaza FMN enzyme+NADH+NTB	320%	18%

Figure 2. Anaerobic reduction of 1-deazaflavodoxin by dithionite as a function of time at pH 9; 0.1 M borate buffer, 1 mM EDTA. Otherwise conditions are the same as for Figure 1. Dashed line: After reduction was complete , 10% (v/v) 1 M H_2KPO_4 was added in anaerobiosis, so as to bring the pH back to 6.5.

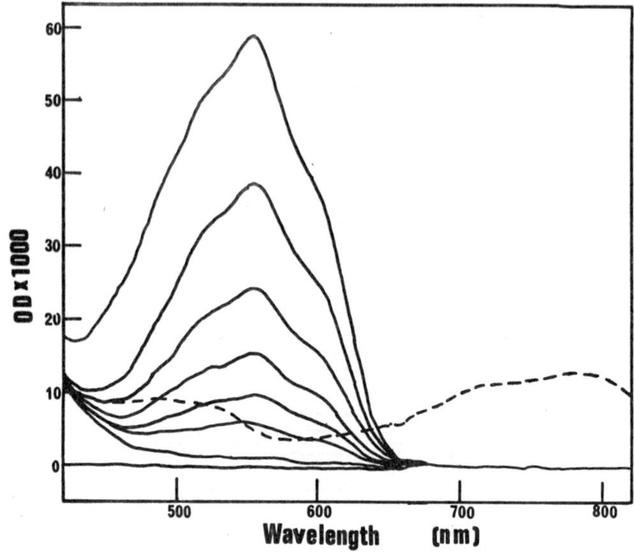

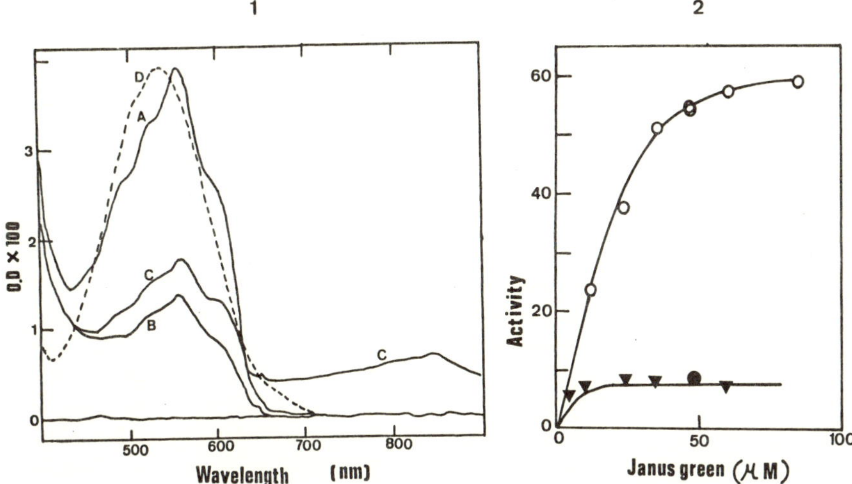

Figure 3. (1) Spectral properties of cytochrome b_5 reductase reconstituted with 1-deaza-FMN. A=reconstituted enzyme; B=same after addition of NADH (0.05 mg/ml) in anaerobiosis; C=reduced enzyme 2 min after O_2 admission; after 5 min the spectrum was back to A'; D=free deaza FMN. Conditions: 20 mM Tris-HCl buffer pH 8.5, 1 mM EDTA. (2) Kinetics of oxidation of NADH (○) or NADD (▼,●) in the presence of Janus green. Conditions: 0.1 M phosphate buffer, pH 7, 30°C. NADD was chemically (●) or enzymatically (▼) prepared.

The overall rate of transfer from NADH to ferricyanide was found to be on the order of 1% that of FMN enzyme, a value in the same range as that arising from residual holoenzyme. Therefore we resorted to Janus green as an acceptor. That dye ($E'_0 = -230$ mV at pH 6) did not act as an electron acceptor for the native enzyme [$E_m = -160$ mV (12)] and reacted only weakly with NADH; its use was spectrally compatible with absorbance measurements at 340 nm. The rate that was thus measured amounted to 0.7% the rate of transfer between NADH and ferricyanide by native enzyme (Figure 3).

Deuterium Isotope Effect

With enzymatically synthesized NADD, the specific activity was slowed down (Figure 3). At dye concentrations above 70 μM, an isotope effect of 7 was thus determined [compared to a value of 4 found for native enzyme (13), and of about 5 for FMN-enzyme].

Conclusion

While this work was in progress, results describing the successful reconstitution of flavodoxin from *Peptostreptococcus elsdenii* and adrenodoxin reductase with 1-deazaflavin were published (14, 15). Thus it appears that the absence of detectable binding of 1-deazaflavin to flavocytochrome b_2 is the exception rather than the rule among electron transferases and flavoproteins in general.

Hemmerich proposed that "le$^-$- and 2e$^-$-transfer can be switched on and off in flavoproteins by formation of strongly directed hydrogen bridges between apoprotein and N(5)- or N/O(1/2 α)- sites of the flavin, respectively" (16). This idea was elaborated upon in subsequent papers (17–19). If one considers 1-deazaflavin as a probe of the protein environment in the N(1)-O(2α) area, results published thus far (4, 14, 15, 20), including the present ones, show that indeed binding in that area is critical in many cases, but not exactly as predicted (16). Whereas for 5-deazaflavins binding is generally good and the chemical reactivity of the bound cofactor is the one expected from the chemistry of the free cofactor, in the case of 1-deazaflavin the situation is completely different [cf. for example results with hydroxylases and oxygenases (4, 14, 20)]. These observations point to the importance of the N(1)-O(2α) environment but they do not appear to be correlated with the flavoprotein function. NADH cytochrome b_5 reductase is the best model that can exist for flavocytochrome b_2 as far as electron output is concerned (see Introduction). The fact that the 1-deazaflavoreducase is catalytically competent, while the cytochrome shows no detectable affinity for the analog implies binding differences in the N(1)-O(2α) area which can hardly be trivial. Thus the present results do not provide support for Hemmerich's hypothesis.

References

1. Lederer, F. and Pompon, D. (1979) In *Flavins and Flavoproteins*. Yagi, K. (ed.) Tokyo: Japan Scientific Societies Press, pp. 609–615.
2. Edmondson, D., Barman, B., and Tollin, G. (1972) *Biochemistry* 11:1133–1138.
3. Averill, B., Schunbrunn, A., Abeles, R., Weinstock, L., Cheng, G., Fisher, J., Spencer, R., and Walsh, C. (1975) *J Biol Chem* 250:1603–1615.
4. Spencer, R., Fisher, J., and Walsh, C. (1977) *Biochemistry* 16:3594–3602.
5. Pompon, D. and Lederer, F. (1979) *Eur J Biochem* 96:571–579.
6. Guiard, B. and Lederer, F. (1979) *J Mol Biol* 135:639–650.
7. Hinkson, J.W. (1968) *Biochemistry* 7:2666–2672.
8. Mayhew, S.G. (1978) *Eur J Biochem* 85:535–547.
9. Dubourdieu, M., Le Gall, J., and Favaudon, V. (1975) *Biochim Biophys Acta* 376:519–532.
10. Pompon, D., Guiard, B., and Lederer, F. (1980) *Eur J Biochem* 110:565–570.
11. Strittmatter, P. (1961) *J Biol Chem* 236:2329–2335.
12. Iyanagi, T. (1977) *Biochemistry* 16:2725–2730.
13. Strittmatter, P. (1962) *J Biol Chem* 237:3250–3254.
14. Entsch, B., Husain, M., Ballou, D.P., Massey, V., and Walsh, C. (1980) *J Biol Chem* 255:1420–1429.
15. Light, D.R. and Walsh, C. (1980) *J Biol Chem* 255:4264–4277.
16. Hemmerich, P. (1976) *Fortschr Chem Org Naturst* 33:451–527.
17. Massey, V. and Hemmerich, P. (1980) *Biochem Soc Transac* 8:246–257.
18. Hemmerich, P. and Massey, V. (1979) In *3rd International Symposium on Oxidases*. King, T.E., Masson, H.S., and Morrison, M. (eds.) Baltimore: University Park Press.
19. Massey, V., Ghisla, S., and Moore, E.G. (1979) *J Biol Chem* 254:9640–9650.
20. Walsh, C., Ryerson, C., and Jacobson, F. (1979) In *Biomimetic Chemistry*. Dolphin, D. (ed.) Washington, D.C.: American Chemical Society.

Published 1982 by Elsevier North Holland, Inc.
Vincent Massey and Charles H. Williams, Editors
Flavins and Flavoproteins

CHAPTER 20

Physical and Chemical Studies on the FMN and Non-Flavin Phosphate Residues in *Azotobacter* Flavodoxin

Dale E. Edmondson[†] and Thomas L. James*

[†]*Department of Biochemistry, Emory University, Atlanta, Georgia; and *Department of Pharmaceutical Chemistry, University of California, San Francisco, California*

Introduction

Over the past several years, we have been developing approaches to obtain structural information on flavoproteins not elucidated by X-ray crystallography. One approach has been application of ^{31}P NMR to monitor the environment of phosphorus groups in flavin coenzymes, i.e., the phosphate monoester in FMN and the pyrophosphate linkage in FAD. In addition to any phosphorus in the coenzyme, it was found that two well-known flavoproteins [*Azotobacter* flavodoxin (1) and glucose oxidase (2)] each contain a protein-bound disubstituted phosphorus residue. These two cases represent the first demonstration of such residues in proteins. We present here both chemical and physical (NMR) studies on the flavin and non-flavin phosphates of *Azotobacter* flavodoxin.

Experimental Procedures

Azotobacter flavodoxin was isolated from cells grown under nitrogen-fixing conditions following a modified procedure of Hinkson and Bulen (3) as described previously (1). Samples were prepared for NMR spectral experiments by lyophilization, dissolution in 90% $^{2}H_2O$, and passage of the resultant solution through a small Chelex column. pH values were uncorrected for the deuterium isotope effect at the glass electrode. Total phosphorus concentrations were determined by the method of Bartlett (4). β-elimination experiments and analysis of the resulting α-ketoacids arising on acid hydrolysis were performed following the procedure described by Plantner and Carlson (5). ^{31}P NMR spectra were obtained in 12 mm NMR tubes at 40.5 MHz using a Varian XL-100-15 spectrometer equipped with a Nicolet Fourier transform accessory and computer. Quadrature phase detection was used and field-locking employed the deuterium resonance of D_2O in the sample. Spectra of anaerobic samples were measured using special NMR tubes (Wilmad 514-A-SJ) with an atmosphere under O_2-free argon. Chemical shifts were determined relative to an external standard of 85% H_3PO_4.

Results and Discussions

^{31}P NMR Spectral Studies

Before performing ^{31}P NMR experiments on *Azotobacter* flavodoxin, we analyzed the preparation for phosphorus content. Our expectation was the FMN phosphorus as the only constituent; however, it was important to test whether any phosphate impurity still remained with the protein inasmuch as the flavoprotein was isolated in phosphate buffer and subsequently exchanged for Tris-acetate buffer on Sephadex chromatography. The analytical data showed the presence of three phosphorus groups per mole of protein rather than one and suggested the presence of two non-FMN phosphorus residues associated with the flavodoxin. An identical analysis was found if the isolation procedure was carried out in Tris buffer or if a modified isolation procedure (6) was used.

Treatment of flavodoxin with 5% trichloroacetic acid resulted in liberation of two phosphorus residues from the protein. One residue can be assigned to that of the FMN moiety while the other is due to an acid-labile form of a phosphorus linked to the protein. Analysis of the apoprotein revealed one mole of phosphorus per mole of protein showing that one of the three phosphorus residues is covalently bound to the protein in an acid-stable linkage. Analysis of a flavodoxin sample which had been stored at $-15°$ for several months and then subjected to Sephadex chromatography showed the presence of only two phosphorus residues per mole of protein; one was due to the acid dissociable FMN while the other was due to the acid-stable covalent phosphorus bound to the protein. These data showed that the acid-labile non-FMN phosphorus becomes labilized on storage of purified protein.

As expected from the chemical analysis, 3 resonances are observed in the ^{31}P NMR spectrum of freshly prepared native flavodoxin (Figure 1) and occur at chemical shift values of -5.6, -0.8, and -0.2 ppm, respectively. ^{31}P NMR spectral data on "aged" flavodoxin samples (spectra not shown) show two resonances at -5.6 and -0.8 ppm, so the resonance at -0.2 ppm can be assigned to the acid-labile non-FMN phosphorus residue. Assignment of the other two resonances was easily accomplished by comparing the spectrum of the apoprotein (Figure 1) with that of the native sample. Only one resonance, at -0.9 ppm. was observed for the apoprotein corresponding to that of the acid-stable covalent phosphorus residue.

From these data, the unequivocal assignment of each of the resonances to their respective phosphorus residues can be made. The FMN resonance of -5.6 ppm observed with *Azotobacter* flavodoxin is downfield from the respective resonances observed with the flavodoxins of *D. vulgaris* or *D. gigas* (4.99 ppm) (7) or from the FMN resonance of *M. elsdenii* flavodoxin (-4.7 ppm) (8). No significant chemical shift change was observed in the pH range of 5.5 to 9.5 for any of the 3 resonances of *Azotobacter* flavodoxin, which shows the 3 phosphorus residues do not ionize in this pH range.

The resonance frequency of the covalent phosphorus residue in the apoprotein was also independent of pH in the presence of 8 M urea or when samples of the protein were degraded to component peptides by trypsin/chymotrypsin digestion. These treatments would diminish any protein

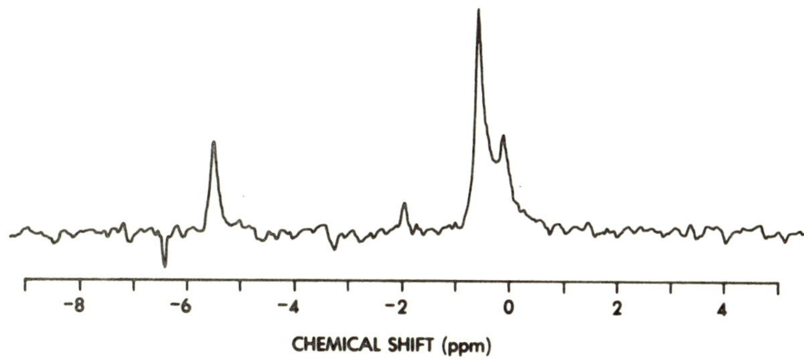

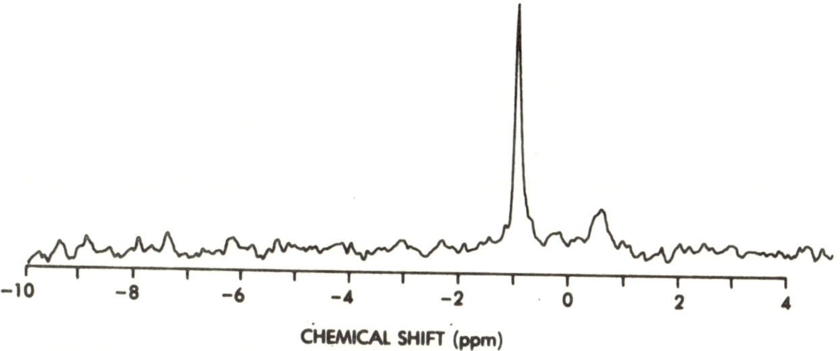

Figure 1. (Top) 40.5 MHz ^{31}P NMR spectrum (proton-decoupled) of native *Azotobacter* flavodoxin (1.5 mM) in 0.1 M Tris-acetate, pH 8.0. The spectra were obtained from 14,000 transients with a pulse repetition time of 0.34 sec and exponential multiplication equivalent to 2 H_z line broadening. (Bottom) ^{31}P NMR spectrum (proton-decoupled) of *Azotobacter* apoflavodoxin (1.15 mM) in 8 M urea - 0.1 M Tris-acetate, pH 8.0. The spectrum was obtained under conditions of the top spectrum except that 11,000 transients were acquired.

structural influence on the possible ionization of a monoester and thus the NMR data conclusively show that the covalent phosphorus residue in *Azotobacter* flavodoxin is disubstituted.

Chemical Studies on the Non-FMN Phosphate Residues

With demonstration of a covalent, disubstituted phosphorus residue in the protein, it was important to decide whether the linkage was between two amino acids, was the result of adenylation of the protein, or perhaps was due to a tightly-bound acid nonextractable phospholipid. The flavodoxin (either native or apoprotein) gave a negative orcinol reaction showing the absence of sugars or nucleotides. Attempts to release bound phosphorus by extraction with various organic solvents were unsuccessful thus showing the absence of any phospholipid. The NMR parameters (chemical shift of -0.8 ppm and a proton-phosphorus coupling of ~ 10 Hz) were consistent with the properties of a phosphodiester linkage between two hydroxyl amino acids (either serine and/or threonine). To provide evidence for such a linkage, approaches used in glycoprotein chemistry (5) were adopted. Alkaline treatment of the apoprotein should lead to elimination of the phosphate with formation of the dehydro form of the appropriate hydroxylamino acid. The results in Table 1 show that upon acid hydrolysis of the alkali-treated protein, approximately one mole of pyruvate[1] and one mole of α-ketobutyrate are formed per mole of phosphorus released. The pyruvate arises from serine and the α-ketobutyrate arises from threonine. These data show that the covalent phosphorus residue in *Azotobacter* flavodoxin is a phosphodiester linkage between a serine and a threonine residue. Previous sequence studies of *Azotobacter* flavodoxin (10) failed to detect the presence of any phosphorylated peptides in the protein and it remains for future work to locate its position in the amino acid sequence.

Attempts to determine the structure of the acid-labile non-FMN phosphorus residue have been unsuccessful to date. It has been found to be dissociated from the protein on storage, acid-precipitation, or on treatment of the protein with 1 M hydroxylamine and is of low molecular weight. These properties suggest this residue to be in an acyl phosphate linkage with a carboxyl group on the protein. It is not simply inorganic phosphate, as acid digestion is required before it will form the "Mo-blue" complex. The "R" group on the phosphate is not a peptide since it fails to give a positive ninhydrin reaction on acid hydrolysis.

Effect of Mn^{++} on ^{31}P NMR Spectral Properties

Addition of a paramagnetic ion, such as Mn^{++}, has been used effectively to distinguish solvent-accessible from inaccessible phospholipids in ^{31}P NMR studies on membrane systems (11). The electron spin relaxation time of hydrated Mn^{++} is long enough ($\sim 10^{-9} - 10^{-10}$ sec) that dipolar coupling between Mn^{++} and phosphorus results in ^{31}P line broadening (16). This approach delineates whether flavin coenzyme phosphate residues or protein-bound phosphate residues are near the protein surface or "buried" within the protein. In the case of *Cl. MP* and *D. vulgaris* flavodoxin (12,13), the FMN phosphate is "buried" and not accessible to solvent. Previous studies on glucose oxidase (2) have shown the resonance arising from the pyrophosphate linkage of bound FAD to be insensitive to paramagnetic Mn^{++} in solution. The spectra in Figure 2 show the FMN phosphate resonance at -5.6 ppm to be unaffected by Mn^{++} concentrations up to 0.1 mM. Unpublished data in our laboratories also show the same behavior for the FMN phosphate of *M. elsdenii* flavodoxin. It can be concluded that, like flavodoxins from the *Cl. MP* and *D. vulgaris*, the FMN side chains in flavodoxins from *A. vinelandii* and *M. elsdenii* are extended into the protein and are not in contact with the solvent.

In contrast, extensive broadening is observed for the covalently-bound phosphodiester resonance at -0.8 ppm (Figure 1). The phosphodiester linkage, therefore, is near the surface of the protein (<10 A°) to interact with

[1]The stoichiometry of pyruvate to phosphate of greater than one is due to the well-known β-elimination of cysteine [there is one cysteine residue present (9)] to form dehydroalanine. Alkylation of this single thiol residue with iodocetate decreases the extent of this interfering reaction (Table 1).

Table 1. Pyruvate and α-Ketobutyrate Formed After Alkaline Treatment and Subsequent Acid Hydrolysis of Apoflavodoxin.

Sample	Moles pyruvate/mole P_i	Moles α-ketobutyrate/mole P_i
Apoflavodoxin	1.49[a]	1.04[b]
Apoflavodoxin after carboxymethylation	1.26[a]	1.07[b]

[a]Determined from the amount of NADH oxidized on the addition of 1 unit of lactate dehydrogenase.

[b]Determined from the amount of NADH oxidized on the addition of 50 units of lactate dehydorgenase.

Mn^{++} ions in solution. A similar behavior is observed in "aged" or hydroxylamine-treated protein samples in which the postulated acyl phosphate residue is missing; the FMN resonance is unaffected by Mn^{++} but the phosphodiester resonance is extensively broadened.

The effect of Mn^{++} on the ^{31}P NMR properties of the proposed acyl phosphate linkage does not follow the simple concept of either being exposed to the surface or "buried" within the protein milieu. At 10 μM Mn^{++} a peak at −5.3 ppm begins to appear and seems to be completely formed in the

Figure 2. Effect of Mn^{++} on the proton-decoupled ^{31}NMR spectrum of *Azotobacter* flavodoxin (1 mM) in 0.1 M Tris-acetate, pH 8.0. (A) no Mn^{++}, (B) in the presence of 10 μM $MnCl_2$, (C) in the presence of 0.1 mM $MnCl_2$. All spectra were obtained with a pulse repetition time of 0.34 sec and exponential line broadening of 3 H_z. The number of transients acquired were 14,800 (A), 20,000 (B), and 20,000 (C).

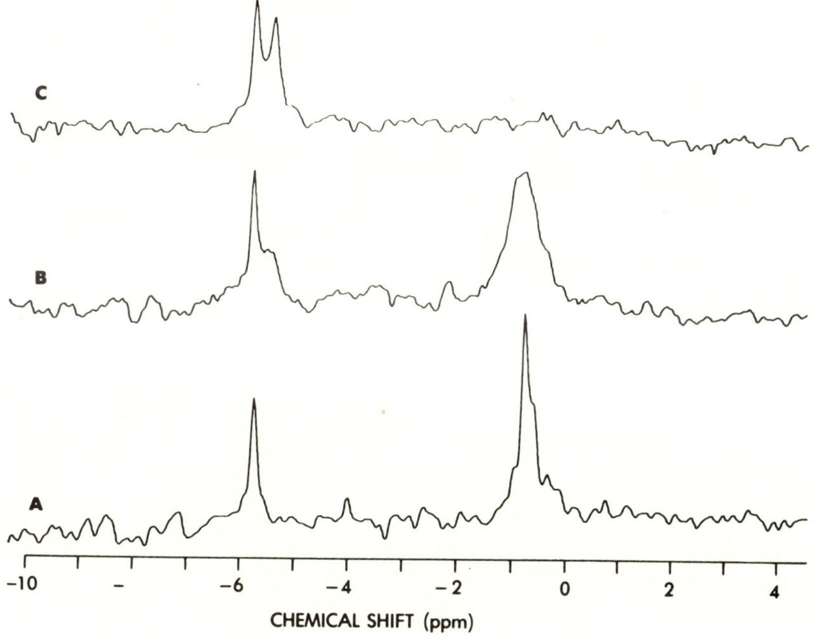

presence of 0.1 mM Mn^{++} (Figure 3). The data show an unusual situation where less than stoichiometric quantities of Mn^{++} (the protein concentration is 1 mM) interact with the protein to cause a change in chemical shift rather than a broadening of the resonance. Although further work is required for a definitive explanation of this behavior, a reasonable rationale is the following. For the electron spin relaxation time of Mn^{++}, any contact shifts can be disregarded. Metal ions have long been known to catalyze hydrolysis of acyl phosphate (anhydride) linkages (14) to their constitutive carboxylate and phosphate forms. The NMR data suggest that rather than being released from the protein (where its resonance would be so extensively broadened as to be unobservable), the phosphate component of the acyl phosphate remains bound to the protein in such a manner as to be shielded from interaction with the Mn^{++}. The chemical shift change (from −0.2 to −5.3 ppm) could arise from ionization to a dianionic form, from interaction with protein groups (i.e., a different environment) or a combination of these processes. If this explanation

Figure 3. Influence of FMN reduction on the ^{31}P NMR spectrum (proton-decoupled) of *Azotobacter* flavodoxin (1 mM) in 0.1 M Tris-acetate, pH 8.0 (A) oxidized flavodoxin, (B) semiquinone form obtained by reduction with 0.5 mM dithionite, (C) hydroquinone form obtained on reduction with excess (10 mM) dithionite. All spectra were obtained as in Figure 3 with 14,800 transients (A), 23,000 transients (B), and 20,000 transients (C).

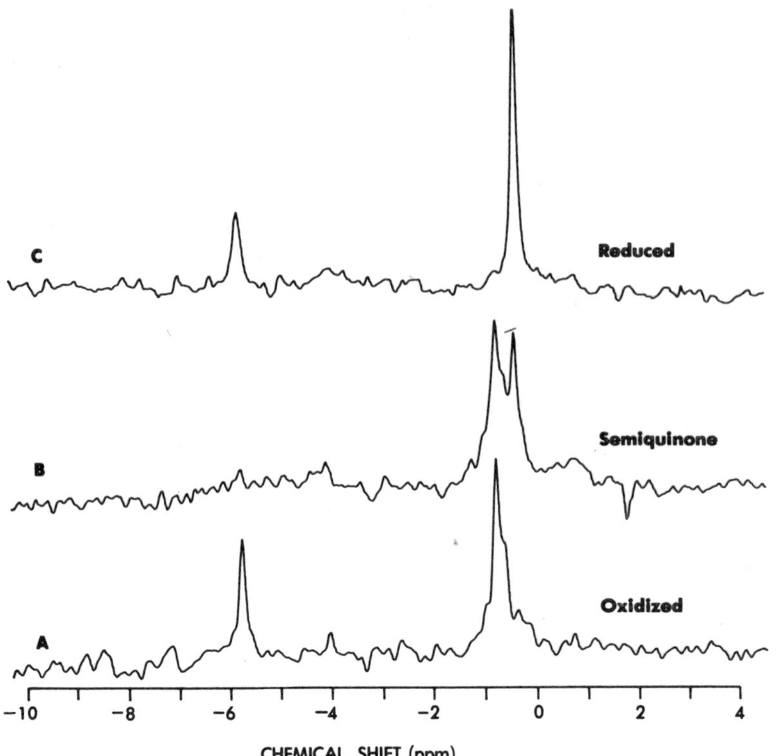

is correct, it would imply that the proposed acyl phosphate linkage is near the surface of the protein for interaction with Mn^{++} to occur.

Effect of FMN Reduction on ^{31}P NMR Spectral Properties

In addition to the studies on the oxidized form, it was of interest to determine the effect of FMN reduction on the phosphorus resonances. The lifetime of the free electron in the flavin semiquinone is sufficient (10^{-8} sec) (15) to induce measurable dipolar line broadening (~ 5Hz) if the phosphorus is <10 A° from the flavin ring system according to calculations using the Solomon-Bloombergen equation (16). Precise distance calculations from such data are not possible due to the distribution of spin density over the benzenoid and pyrazine portions of the flavin ring system (17). Reduction of the FMN to its semiquinone form results in extensive line broadening of the FMN phosphate resonance as observed for the flavodoxins of *D. vulgaris*, *D. gigas* (8), and of *M. elsdenii* (8). The two upfield peaks due to the non-FMN phosphorus residues become more resolved upon semiquinone formation with no observable line broadening (Figure 3). Therefore, these residues are distant from the flavin ring. However, 1-electron reduction of the FMN results in alterations in the environment of the postulated acylphosphate linkage, slightly altering its resonance frequency. Upon reduction of the FMN to its hydroquinone form, the FMN resonance reappears with a chemical shift of −5.4 ppm (Figure 3). The upfield shift (0.2 ppm) suggests the 5′-phosphate is in a slightly different environment compared with the oxidized form. The non-FMN phosphate resonances become magnetically equivalent with a single sharp peak at −0.4 ppm (Figure 3). The difference in chemical shifts of the non-FMN phosphate with the different FMN redox forms may result from limited protein conformational changes. ^{31}P Chemical shifts are quite sensitive to environment; so only small environmental changes need occur to effect the observed differences. Similar conclusions have also been reached from ^{1}H NMR studies of *Desulfovibrio* flavodoxins (7).

ACKNOWLEDGMENTS

This work was supported by NSF Grant PCM-81-00770 (D.E.E.) and NIH Grants HL-16251 (D.E.E.) and GM-25018 (T.L.J.) TLJ also gratefully acknowledges receipt of a Career Development Award (AM 00291) from the National Institutes of Health. Portions of this work were carried out at the Molecular Biology Division, Veterans Administration, Medical Center, San Francisco, California. The NMR facility used in this work was supported by Grant RR00892-01 A1 from the Division of Research Resources, National Institutes of Health, to the UCSF Magnetic Resonance Laboratory.

References

1. Edmondson, D.E. and James, T.L. (1979) *Proc. Natl Acad Sci* USA 76:3786–3789.
2. James, T.L., Edmondson, D.E., and Husain, M. (1981) *Biochemistry* 20:617–621.
3. Hinkson, J.W. and Bulen, W.A. (1967) *J. Biol Chem* 242:3345–3351.
4. Bartlett, G.A. (1959) *J Biol Chem* 234:466–468.
5. Plantner, J.J. and Carlson, D.M. (1972) In *Methods in Enzymology*. V. Ginsburg (ed.) Vol. 27. New York: Academic Press. pp. 46–48.
6. Yoch, D.E. and Arnon, D.I. (1972) *J Biol Chem* 247:4514–4520.

7. Favaudon, V., LeGall, J., and Lhoste, J.M. (1980) In *Flavins and Flavoproteins.* K. Yagi and T. Yamano (eds.) Tokyo: Japan Scientific Societies Press, pp. 373–386.
8. Müller, F., van Schagen, G.G., and R. Kaptein (1980) In *Methods in Enzymology.* D.B. McCormick and L.D. Wright (eds.) Vol. 66. New York: Academic Press, pp. 385–416.
9. Edmondson, D.E. and Tollin, G. (1977) *Biochemistry* 10:124–132.
10. Tanaka, M., Hanin, M., Yasunobu, K.T., and Yoch, D.C. (1977) *Biochemistry* 16:3525–3537.
11. Michaelson, D.M., Horwitz, A.F., and Klein, M.P. (1974) *Biochemistry* 13: 2605–2612.
12. Burnett, R.M., Darling, G.D., Kendall, D.S., LeQuesne, M.E., Mayhew, S.G., Smith, W.W., and Ludwig, M.L. (1974) *J Biol Chem* 249:4383–4392.
13. Watenpaugh, K.D., Sieker, L.C., Jensen, L.H., LeGall, J., and Dubourdien, M. (1972) *Proc Natl Acad Sci USA* 69:3185–3188.
14. Koshland, D.E. (1951) *J Am Chem Soc* 73:4103.
15. Palmer, G., Müller, F., and Massey, V. (1972) In *Flavins and Flavoproteins.* H. Kamin (ed.) Baltimore: University Park Press, pp. 123–140.
16. James, T.L. (1975) *Nuclear Magnetic Resonance in Biochemistry: Principles and Applications.* New York: Academic Press, pp. 46–48, 177–181.
17. Müller, F., Hemmerich, P., and Ehrenberg, A. (1971) In *Flavin and Flavoproteins.* H. Kamin (ed.) Baltimore: University Park Press, pp. 107–122.

PART II:

Reaction Mechanisms

Published 1982 by Elsevier North Holland, Inc.
Vincent Massey and Charles H. Williams, Editors
Flavins and Flavoproteins

CHAPTER 21

Scope of Chemical Redox Transformations Catalyzed by Flavoenzymes

Christopher Walsh

Departments of Chemistry and Biology, Massachusetts Institute of Technology, Cambridge, Massachusetts

Flavoenzymes catalyze a large number of diverse redox interconversions among substrates, reflecting the chemical versatility of the tricyclic isoalloxazine ring system of the flavin coenzyme as an electron conduit between substrates of various structures. Perhaps the single key feature of flavins, and one long recognized, is the ability to act as a two-electron/one electron switching center, for example, placing the isoalloxazine system at redox crossroads in membrane respiratory chains between two electron donors and obligate one-electron acceptors.

There have been several attempts to collect the kinds of flavoenzyme-mediated redox transformations into categorizations which reveal similarity. Thus, Massey and Hemmerich (21) have elsewhere noted 5 classes based on the nature of the (flavin) reductive half reaction and of the subsequent (dihydro-flavin) oxidative half reaction which occurs in any flavoenzyme catalytic cycle. This categorization suggested: transhydrogenases

$$S_1H_2 + E-FL_{ox} \rightleftharpoons S_1 + FlH_{2red} \qquad \text{(reductive half reaction)}$$

$$S_2 + E-FLH_{2red} \rightleftharpoons S_2H_2 + E-FL_{ox} \qquad \text{(oxidative half reaction)}$$

(C→C, C→N, S→N, N→N), dehydrogenases/oxidases, dehydrogenases/oxygenases, dehydrogenases/electron transferases, pure electron transferases. This classification scheme is quite a useful one.

The focus in this paper is a somewhat different one, dealing as much with the substrate functional group undergoing oxidation or reduction as with the flavin coenzyme.

Reductive Half Reactions

Table 1 indicates substrate functional groups which are oxidized, in net two-electron processes, during flavoenzyme catalysis. These substrates are in general soluble, low molecular weight metabolites with equilibria favoring oxidation in many cases. There seems little question about the mode of electron transfer for dihydronicotinamide oxidation or for dithiol oxidation to disulfide, highlighting the C_{4a}, N_5 region as the major port of entry of electrons into oxidized flavins. Chemical expectation for the olefin-forming class is abstraction of an acidic proton α to a carbonyl followed by β-hydride transfer

Table 1. Oxidizable Substrate Groups.

Reduced substrate	Product oxidized (substrate)	Proposed mechanisms of electron transfer
1. NADH, NADPH	NAD, NADP	H^- to and from N^5 of flavin
2. RSH + RSH	RSSR	covalent adduct at C_{4a} of flavin
3. $-\overset{H}{\underset{H}{C}}-C-\underset{O}{\overset{}{C}}-$ (H on first C; O double bonded to third C)	$-C=C-\underset{O}{C}-$	α-proton abstraction, β-hydride transfer to N^5 of flavin
4. $-\overset{H}{\underset{X}{C}}-$	$-\underset{X}{C}-$ (C=X)	proton abstraction, covalent $C-N^5$ adduct
5. $SO_3^=$ + AMP	$AMP-SO_4^=$	$N^5-SO3^=$ adduct
6. 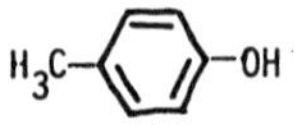 $H_3C-C_6H_4-OH$	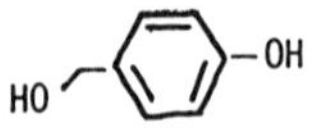 $HO-CH_2-C_6H_4-OH$	H^- transfer to N^5, H_2O addition to quinone methide

to N^5. Good evidence derives from the HF elimination on 2,2-difluorosuccinate by succinate dehydrogenase (30). Glutaryl CoA decarboxylase apparently uses a similar mechanism to desaturate the substrate transiently and set up a low-energy decarboxylation route (13). (Scheme 1)

A recent suggestion was made that conversion of dihydro orotate to orotate in trypanosomatid parasites involved a pterin-dependent hydroxylation at C_5 followed by dehydration, rather than by a flavoprotein desaturase (15). (Scheme 2) This claim turns out to be incorrect. We have now purified the responsible enzyme from *Crithidia fasiculata* and find it is an FMN-containing oxidase (Pascal, R., unpublished results).

The dehydrogenation of alcohols, amines, hemiacetals, and aldehydes to their corresponding aldehydes or ketones, lactones, and acids may or may not all proceed by processes involving C-H cleavage to a carbanionic species and a proton, depending on the structural possibilities for delocalization and stabilization of carbanionic intermediates. Homolytic cleavage in enzymes such as amine oxidase or glucose oxidase has not yet been definitively ruled out (4).

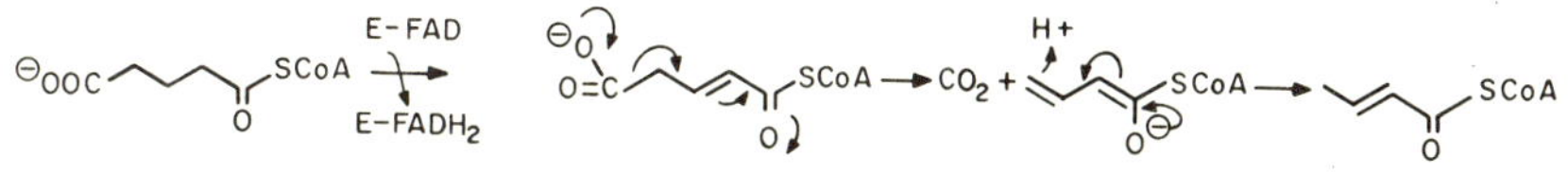

Scheme 1.

When carbanionic species are formed, there is evidence in the case of glycollate oxidase that collapse to an N^5 adduct is a chemically and kinetically competent pathway (12). (Scheme 3)

The conversion of cresol to the hydroxymethyl product is not an oxygenase reaction but rather apparently involves hydride ejection (to N^5 of flavin presumably) with quinone methide formation and capture of this electrophilic species by water (17). (Scheme 4) The substrate quinone methide chemistry is analogous to quinone methide chemistry that is relevant in covalent joining of coenzyme to apoprotein amino acid side chain nucleophiles at C_{8a} as illustrated by recent model studies (21). (Scheme 5)

The versatility of quinone methide exocyclic methylene as a base in reaction with a proton and as an electrophile in reaction with a nucleophile is well illustrated in these two cases. In analogy is the likely route of covalent modification of electrophilic flavin coenzyme at C_6 by an enzymic cysteinyl thiolate anion in trimethylamine dehydrogenase biosynthesis (29). The initial adduct is again trapped out by reoxidation of the dihydro state, pointing out that these covalent joining reactions are categorizable as redox processes. (Scheme 6)

Dihydroflavin Oxidative Half Reactions

Table 2. Substrate Groups Reduced in Dihydroflavin Oxidative Half Reaction.

Oxidized substrate	Reduced product	Mode of electron transfer	Examples
1. Quinones	Hydroquinones	$2e^-$ and/or $1e^-$	(a) *E. coli* D-alanine dehydrogenase (b) Vitamin K reductase
2. FMN	$FMNH_2$	$2e^-$ and/or $1e^-$	(a) Electron transfer flavoproteins
3. Heme-Fe^{III}	Heme-Fe^{II}	$1e^-$	(a) NADH cytochrome b_5 reductase (b) Yeast cytochrome b_2 (L-lactate dehydrogenase)
4. Fe/S-$cluster^{n-}$	Fe/S-$cluster^{n+1(-)}$	$1e^-$	(a) Adrenodoxin reductase and analogous reductases
5. Mo^{VI}	Mo^{IV}	$2e^-$	(a) Xanthine oxidase (b) Aldehyde xoidase
6. Hg^{II}	Hg^{0}	$2e^-$ and/or $1e^-$	(a) NADPH mercuric reductase
7. O_2	O_2^-, H_2O_2, or H_2O	1, 2, or $4e^-$	

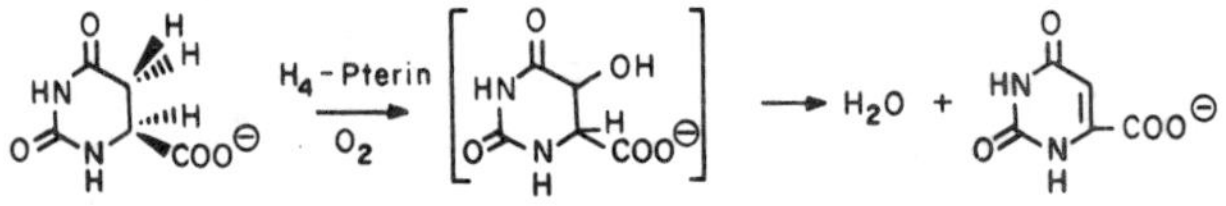

Scheme 2.

Analysis of the common reoxidants of dihydroflavin enzymes (Table 2) points out that many of the physiological substrates which thereby undergo reductions are facultative or obligate one-electron acceptors. Not listed in Table 2 are the functional groups in Table 1, although all such processes are necessarily reversible and most likely involve a single two-electron redox step. Many flavoenzymes undergoing reoxidation by iron systems or by quinones are membrane-associated where they perform their step down redox transforming mode.

The reduction of inorganic mercuric ions to elemental mercury is one of the key steps accounting for the molecular basis of bacterial resistance to organomercurial compounds (28). The mer system found on drug-resistant transposons codes for a permease, an organomercury lyase, and flavoprotein mercuric reductase, and at full induction 4–6% of bacterial cell protein is the reductase which we have recently purified from Tn501-infected pseudomonads (Fox, B., unpublished experiments).

$$\text{R-Hg} \rightarrow \text{Hg}^{II} + \text{RH} \qquad \text{organomercury lyase}$$

$$\text{NADPH} + \text{Hg}^{II} \rightarrow \text{NADP} + \text{Hg}^{\circ} \qquad \text{mercuric reductase}$$

The Hg° produced is volatile and leaves the immediate microenvironment of the bacterium. The mechanism of mercury reduction is as yet unclear because the reductase also has a redox active disulfide group and shows many of the properties of glutathione reductase and dihydro lipoamide dehydrogenase, well-known enzymes with four-electron redox capacity. It is clear that on substrate reduction of the enzyme and subsequent denaturation two additional cysteinyl-SH groups are titratable by dithiobisnitrobenzoate compared to oxidized, denatured enzyme. A possible mechanism for reduction of a mercuric

Scheme 3.

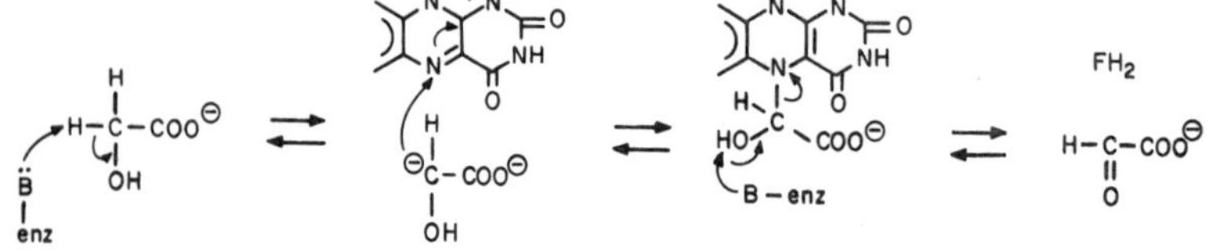

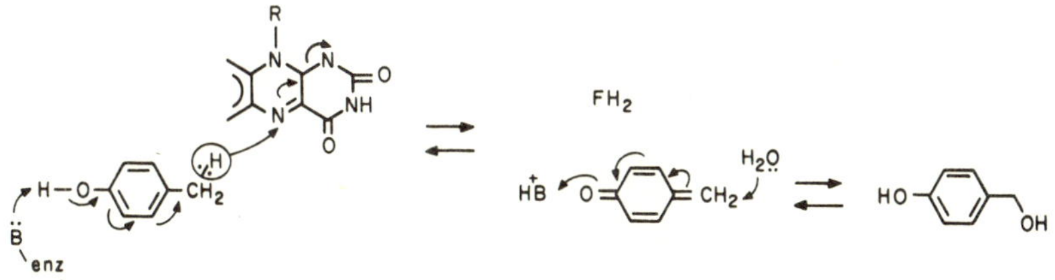

Scheme 4.

dithiolate complex, the presentable substrate form, is indicated below in analogy to the dithiol-disulfide interchange step of glutathione reductase. (Scheme 7) The possibility of a direct hydride transfer from N^5 of dihydro-FAD to Hg^{II} is also not ruled out. Reconstitution studies with 5-deazaFAD, which is a hydride transfer agent when reduced but very sluggish towards C_{4a} thiolate addition, may shed light on this pathway. The pure mercuric reductase also reduces gold ^{III} and silver ^{I} ions apparently to the elemental forms of the metals.

Scope of Substrates Oxygenated

When molecular oxygen is the reducible substrate for dihydroflavin reoxidation, reactions involving net reduction of O_2 by one-electron, two-electron, and four electron are known (10,20), quite possibly all arising from an initial common reaction pathway involving rapid one-electron transfer to superoxide and flavin semiquinone (32,33). Rapid recombination would lead to the now well-known flavin 4_a-hydroperoxide, the key intermediate. With flavoprotein oxidases, no leakage of $O_2^{-\cdot}$ is ever detected, leaving open the possibility of a direct adduct formation without discrete $O_2^{-\cdot}$, semiquinone involvement (16,21). (Scheme 8)

Scheme 5.

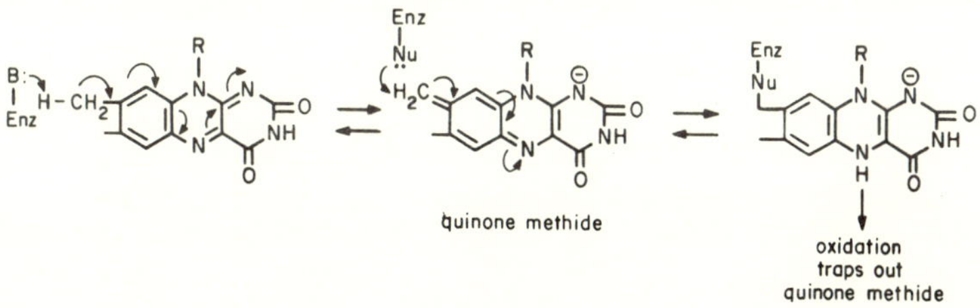

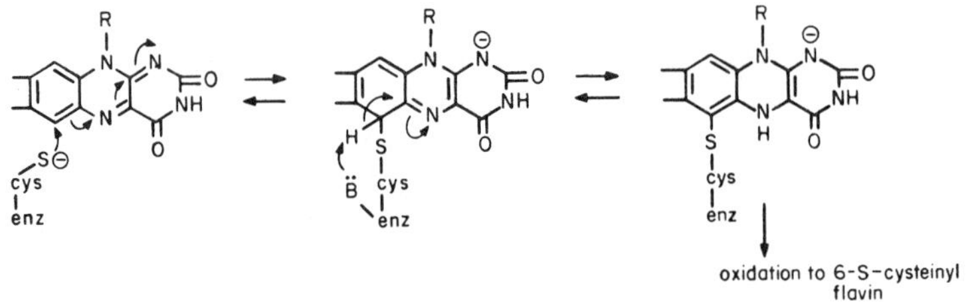

oxidation to 6-S-cysteinyl flavin

Scheme 6.

Superoxide is clearly the physiological product in at least one case, the respiratory burst enzyme in activated leukocyte membranes involved as a first-line defense in killing white-cell ingested microbes (2). The stoichiometry of a partially purified FAD-enzyme from activated human neutrophils is indicated with quantitative flux to released $O_2^{-\cdot}$ (19).

$$NADPH + O_2 \rightarrow NADP + 2O_2^{-\cdot}$$

The partially purified enzyme has heme as well and it is unclear if FAD or heme iron is the actual O_2-reductant, although precedent in xanthine oxidase reveals FAD as the redox site of $O_2^{-\cdot}$ yield in that enzyme (22).

In other oxidases and oxygenases, the fate of the 4a-flavin hydroperoxide intermediate is distinct, in one case breaking down intramolecularly to H_2O_2 and oxidized flavin, in the other case undergoing O-O bond fission and cosubstrate oxygenation.

Table 3 lists the four types of cosubstrate groups known to be oxygenated by flavoenzymes, all with a generalized stoichiometry as indicated. The phenolic hydroxylations are the most extensively studied and, thanks

$$\underset{(NADH)}{NADPH} + RH + O_2 \rightarrow \underset{(NAD)}{NADP} + R{-}OH + H_2O$$

largely to the efforts of Massey, Ballou, and their colleagues (8,9,11), one pictures the flavin 4a-OOH as donor of the distal oxygen of the bound flavin 4a-OOH as an electrophilic oxygen to a phenolic carbanion equivalent to generate the ring-opened variant of the flavin 4a-OOH, a facile dehydration away from oxidized flavin, and the hydroxycyclohexadienone tautomer of the catecholic product. (Scheme 9) Reactions 2, 3, 4 in Table 3 are catalyzed by the microsomal FAD-linked S,N-oxygenase, the only mammalian flavoprotein

Scheme 7.

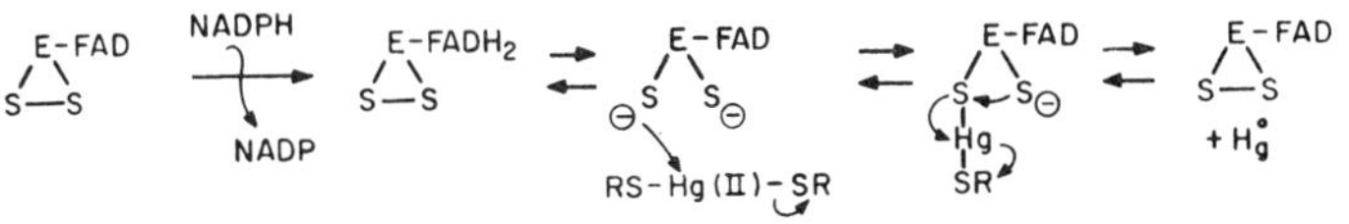

E-FAD (S—S) —NADPH → NADP→ E-FADH2 (S—S) ⇌ E-FAD ($S^{\ominus}$ $S^{\ominus}$) + RS-Hg(II)-SR ⇌ E-FAD (S—Hg—SR, $S^{\ominus}$) ⇌ E-FAD (S—S) + Hg^{0}

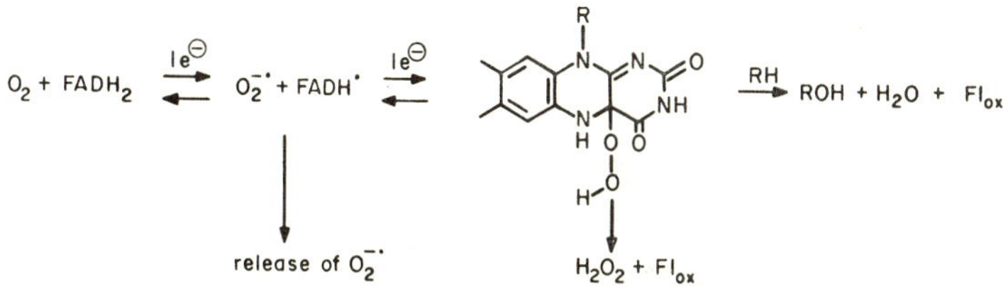

$$O_2 + FADH_2 \underset{}{\overset{1e^{\ominus}}{\rightleftharpoons}} O_2^{-\cdot} + FADH^{\cdot} \overset{1e^{\ominus}}{\rightleftharpoons} \text{[4a-hydroperoxyflavin]} \xrightarrow{RH} ROH + H_2O + Fl_{ox}$$

$O_2^{-\cdot} + FADH^{\cdot} \downarrow$ release of $O_2^{-\cdot}$

[4a-hydroperoxyflavin] $\downarrow$ $H_2O_2 + Fl_{ox}$

Scheme 8.

Table 3. Substrate Functional Groups Oxygenated by Flavoproteins.

1. 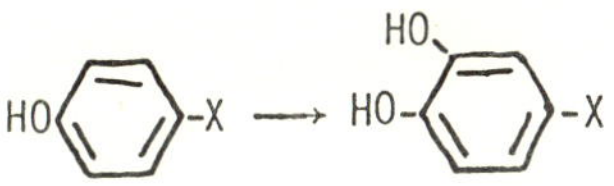 HO-C_6H_4-X ⟶ (HO)$_2C_6H_3$-X

2a) R-S ⟶ R-S-OH (unstable)

b) RSR' ⟶ RSR'

3a) R_3N ⟶ R_3NO

b) R_2NH ⟶ R_2N-OH

4.  RC(=O)R' ⟶ RC(=O)OR'

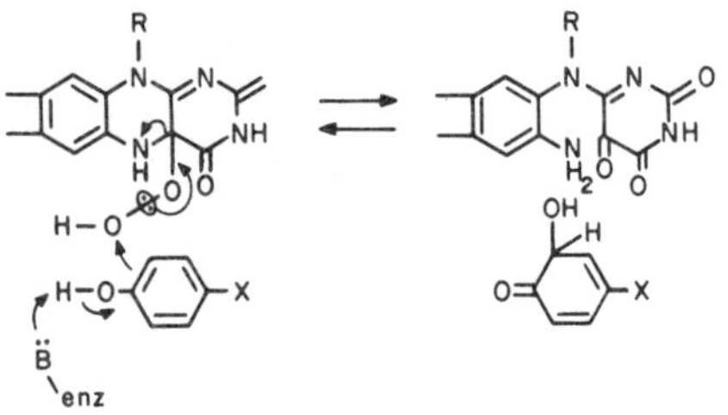

Scheme 9.

oxygenase (35), and there is good evidence that this enzyme sits around in the steady state as the Enz-FAD-4a-OOH as an "oxygen gun" equivalent, able to deliver electrophilic oxygen rapidly to any bound nucleophilic cosubstrate (3,23). We have recently analyzed the stereochemical outcome at sulfur during oxygenation of ethyl tolylsulfide by this enzyme and find a 95:5 preference for generation of the S-sulfoxide isomer (Light, D.; Waxman, D., unpublished experiments). (Scheme 10)

Category 4 of Table 3 has recently attracted our attention. For the general structure where R′=H, the transformation is that catalyzed by bacterial luciferase. For R′≠H, the sequence is an oxygen insertion into a carbon-carbon bond as a ketone is converted to a lactone. We have recently examined the cycloyhexanone oxygenase of *Acinetobacter* (14) and validated a flavin 4a-OOH species in this ring-expanding lactonization sequence (25). (Scheme 11) We (25) and also Schwab (27) have found that 2,2,6 6-D_4-cyclohexanone gives tetradeutero-ϵ-caprolactone with no loss of deuterium, strong evidence against an enolization-epoxidation sequence and for a Baeyer-Villiger type process as illustrated (26). (Scheme 12) In such a path, the distal peroxide oxygen of the flavin 4a-OOH must act as a *nucleophilic equivalent*, a functional reversal of polarity from the other flavoprotein monooxygenase classes (32,33). These facts argue that the flavin 4a-OOH, or a species derived readily therefrom, [e.g., the 10a-OOH in Hemmerich and Massey's suggestions (20)], is an ambident oxygenating reagent capable, on specific substrate demand, of delivering the distal oxygen as electrophile or nucleophile. Indeed, we have now

Scheme 10.

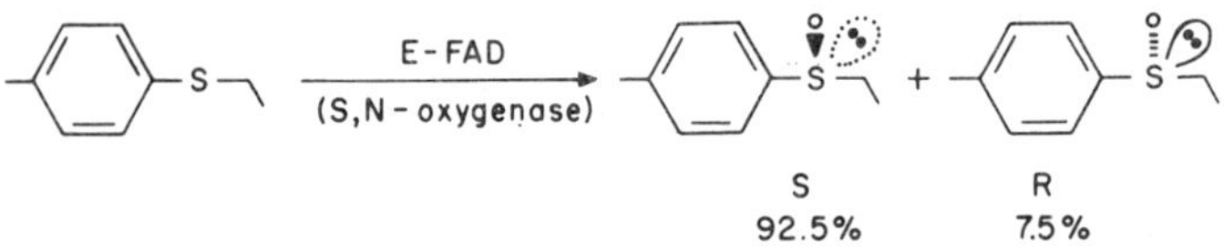

E-FAD (S,N-oxygenase)

S 92.5%

R 7.5%

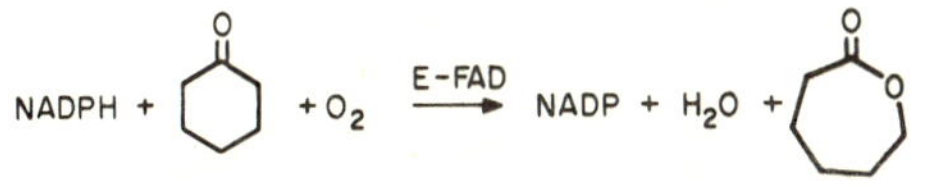

NADPH + + O_2 $\xrightarrow{\text{E-FAD}}$ NADP + H_2O +

Scheme 11.

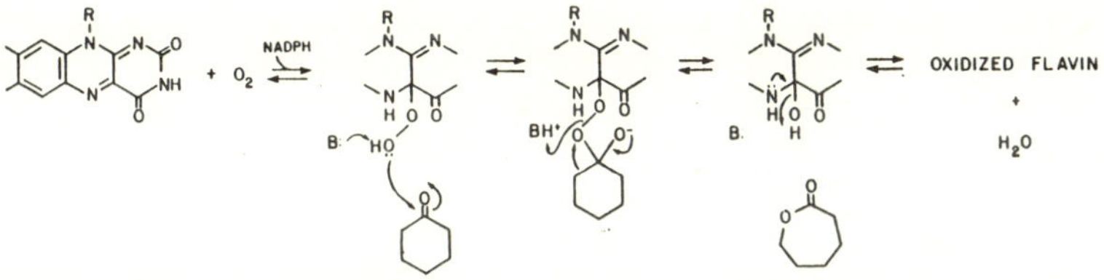

NADPH + O_2 ⇌ ⇌ ⇌ ⇌ OXIDIZED FLAVIN + H_2O

Scheme 12.

Scheme 13.

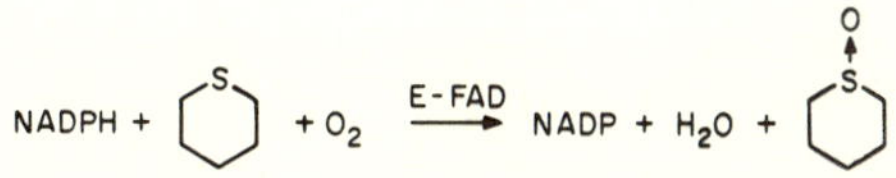

NADPH + + O_2 $\xrightarrow{\text{E-FAD}}$ NADP + H_2O +

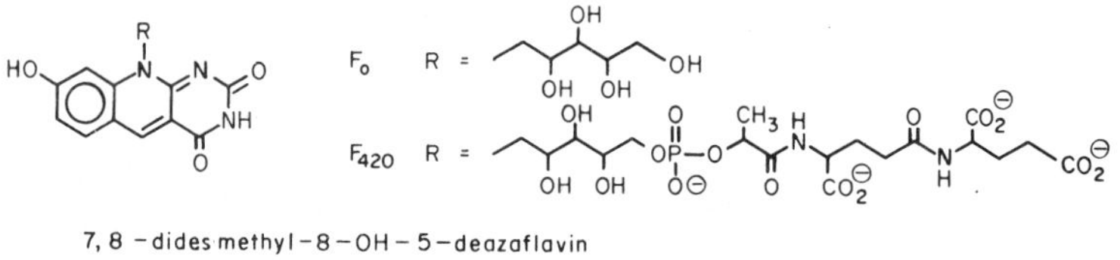

7, 8 -didesmethyl-8-OH-5-deazaflavin

Scheme 14.

established that the *Acine-tobacter* cyclohexanone monooxygenase will also oxygenate cyclohexyl sulfide, thiane, to thiane-1-oxide, demonstrating ambident reactivity for a single flavoenzyme monooxygenase (26). (Scheme 13) At this point, it may be interesting to see whether luciferase will S-oxygenate long chain alkyl sulfides with or without light production.

Other Reactions

We (31,33) and Massey's group (21) have recently used synthetic flavin analogs modified at key loci in the isoalloxazine nucleus to change thermodynamic and kinetic parameters in catalysis and dissect out specific steps in flavoenzyme mechanisms. With some significant effort invested in synthetic 5-carba-5-deazaflavins, we were delighted to observe Ralph Wolfe's discovery (5) that the methanogenic bacterial factor 420 is a 7-desmethyl-8-hydroxy-5-deazaflavin, (Scheme 14) a natural flavin analog, at quite a different redox crossroads. We have begun a systematic study of its redox properties (1) and those of the riboflavin level, partial hydrolysis product, F_o, with an NADP-oxidoreductase and the hydrogenase from methanogens (34). In particular, in collaboration with Orme-Johnson's group at M.I.T., we have now purified a three subunit FAD and Fe/S-containing hydrogenase which uses F_o or F_{420} as

Scheme 15.

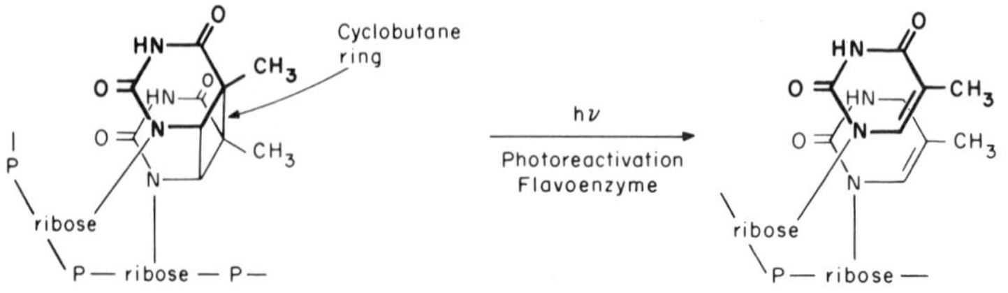

Cyclobutane ring

$h\nu$

Photoreactivation Flavoenzyme

cis-syn-intrachain thymine dimer

the immediate electron acceptor, raising mechanistic questions about modes of obligate two-electron transfer from reduced Fe/S clusters.

As a final example of the scope of chemistry available to flavins (in yeast, *E. coli*) and the Factor 420 natural flavin analogs (in *Streptomyces griseus*), we note the existence of photoreactivation enzymes which use the dihydro forms of these coenzymes to repair pyrimidine dimers in DNA (6,7,8). This is a directed flavin (5-deazaflavin) photochemistry flux where the enzymes catalyze photo reversion, with cleavage of 2 C-C bonds in the cyclobutyl ring, back to native DNA bases as illustrated for reversion of thymine dimers. (Scheme 15) Whether these are concerted, stepwise, or radical processes for DNA repair remains to be elucidated. Analyses of these enzymic processes should illuminate both the dark and light chemistry of flavins in biology.

ACKNOWLEDGMENTS
I thank my coworkers at M.I.T. who were responsible for much of the conception and all of the execution of research from my laboratories noted in this paper: C. Ryerson, F. Jacobson, B. Fox, D. Light, P. Olsiewski, K. Haldar, R. Pascal, and our colleagues and collaborators W. Orme-Johnson, L. Daniels, and L. Tsai (NIH), S. Yamazaki (NIH), A. Tauber (Harvard). Supported in part by National Institutes of Health Grant No. GM21643.

References

1. Ashton, W., Brown, R., Jacobson, F., and Walsh, C. (1979) *J Am Chem Soc* 101: 4419.
2. Babior, B. (1978) *New Eng J Med* 298:659, 721.
3. Beatty, N. and Ballou, D. (1980) *J Biol Chem* 255:3817.
4. Bruice, T. (1980) *Accts Chem Res* 13:256.
5. Eirich, D., Vogels, G., and Wolfe, R. (1978) *Biochemistry* 17:4583.
6. Eker, A. and Fichtinger-Schepman, A. (1975) *Biochem Biophys Acta* 378:54–63.
7. Eker, A. (1980) *Photochemistry and Photobiology* 32:593.
8. Entsch, B., Ballou, D., and Massey, V. (1976a) *J Biol Chem* 251:2550.
9. Entsch, B., Ballou, D., Husain, M., and Massey, V. (1976b) *J Biol Chem* 251:7367.
10. Flashner, M. and Massey, V. (1974) In *Molecular Mechanisms of Oxygen Activation.* O. Hayaishi (ed.), New York: Academic Press, p. 245.
11. Ghisla, S., Entsch, B., Massey, V., and Husain, M. (1977) *Eur J Biochem* 76:139.
12. Ghisla, S. and Massey, V. (1980) *J Biol Chem* 255:5688.
13. Gomes, B., Fendrich, G., and Abeles, R. (1981) *Biochemistry* 20:1481.
14. Griffin, M. and Trudgill, P. (1972) *Biochem J* 129:505.
15. Gutteridge, W., Dave, D., and Richards, W. (1979) *Biochem Biophys Acta* 582:390.
16. Hemmerich, P. (1976) In *Progress in Natural Product Chemistry* Vol. 33. New York: Springer-Verlag, p. 451.
17. Hopper, D. and Taylor D. (1977) *Biochem J* 167:155.
18. Iwatsaki, N., Joe, C., and Werbin, H. (1980) *Biochemistry* 19:1172.
19. Light, D., Walsh, C., O'Callaghan, A., Goetzl, E., and Tauber A. (1981) *Biochemistry* 20:1468.
20. Massey, V. and Hemmerich, P. (1975) In *The Enzymes.* P. Boyer (ed.), Vol. 12, New York: Academic Press, p. 191.
21. Massey, V. and Hemmerich, P. (1980) *Biochemical Society Transactions* 8:246

22. Massey, V. (1973) In *Iron-Sulfur Proteins*. W. Lovenberg (ed.), Vol. 12, New York: Academic Press, p. 301.
23. Poulsen, L. and Ziegler, D. (1979) *J Biol Chem* 254:6449.
24. Rastetter, W. and Frost, J. (1981) *Biochemistry,* submitted.
25. Ryerson, C. (1981) PhD Dissertation, M.I.T. Chemistry Department.
26. Ryerson, C., Ballou, D., and Walsh, C. (1981) *Biochemistry*, submitted.
27. Schwab, J. (1981) *J Am Chem Soc* 103:1876.
28. Silver, S. and Kencherf, T. (1981) In *Enzymatic Basis of Detoxification*. W. Jakoby (ed.), Vol. II. New York: Academic Press.
29. SteenKamp, D., Kenney, W., and Singer, T. (1978) *J Biol Chem* 253:2812.
30. Tober, C., Nicholls, P., and Brodie, J. (1970) *Archivs Biochem Biophys* 138:506.
31. Walsh, C. (1978) *Ann Rev Biochem* 45:891.
32. Walsh, C. (1979) *Enzymatic Reaction Mechanisms*. San Francisco: W.H. Freeman.
33. Walsh, C. (1980) *Accts Chem Res* 13:148.
34. Yamazaki, S., Tsai, L., Stadtman, T., Jacobson, F., and Walsh, C. (1980) *J Biol Chem* 255:9025.
35. Ziegler, D. (1981) In *Enzymatic Basis of Detoxification.* W. Jakoby (ed.), Vol. I. New York: Academic Press.

Published 1982 by Elsevier North Holland, Inc.
Vincent Massey and Charles H. Williams, Editors
Flavins and Flavoproteins

CHAPTER 22

Dehydrogenation Mechanism in Flavoprotein Catalysis

Sandro Ghisla

Fakultät für Biologie, Universität Konstanz, D-7750 Konstanz, Germany

Introduction

The term "dehydrogenation" was originally used to describe a chemical process involving the abstraction of molecular hydrogen from an organic compound. Following Wieland's suggestion, it now has a broader use and means the rupture of (at least) one C—H bond, with concomitant transfer of its 2 electrons to a suitable acceptor:

$$-\overset{|}{\underset{\underset{H}{|}}{C}}-X-H+\mathrm{Acc(ox)}\rightarrow\ \rangle C=X+\mathrm{Acc(red)}H_2$$

This type of reaction should be differentiated from the oxygenase reaction, in which the substrate electrons are transferred directly to oxygen. The limiting chemical step of the dehydrogenation, i.e., the rupture of the kinetically stable substrate C—H bond, is catalyzed by enzymes which have either pyridine nucleotide, pyridoxal-phosphate, flavin-, or pteridine as coenzymes.

The mechanism of the biological dehydrogenation step is undoubtedly a cardinal question and has been addressed since the discovery of redox enzymes. Michaelis' discovery of flavin radicals prompted the proposal that biological redox processes mandatorily proceed through single-electron transfer steps (1). This type of mechanism was revived in the early sixties by Beinert's discovery of stable flavoprotein radicals (2). Cornforth, however, was the first to make a concrete mechanistic suggestion proposing in 1959 that the oxidation of Acyl-CoA substrates by Acyl-CoA dehydrogenase is initiated by abstraction of the proton, and that it goes to completion by subsequent transfer of radical entities (3). In 1964, Hemmerich independently proposed a carbanion-initiated mechanism for the oxidation of "activated" substrates (4). It was not until the late 60's, however, that new inputs and ideas were put forward which constituted the stimulus for the progress that led to our present knowledge. This experimental work will be discussed below.

Definition of the Problem

Detailed mechanisms for the dehydrogenation reaction can be formulated as in Scheme 1.

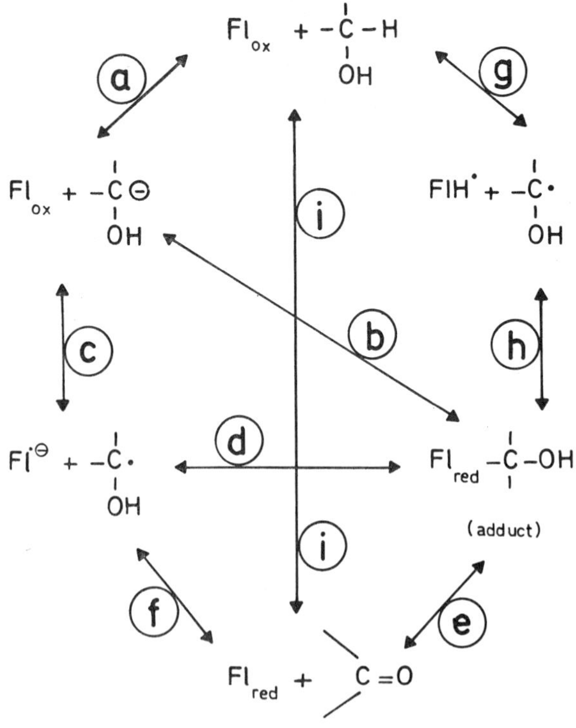

Scheme 1. Alternative mechanisms for the transfer of redox equivalents between a "C—H" substrate and oxidized Flavin (Fl_{ox}). Further alternatives such as those initiated by $1e^-$ abstraction from the -OH group, and different modes of radical transfer subsequent to step g have been omitted for the sake of clarity.

They have been discussed elsewhere *in extenso* (5–11), and shall be summarized as "hydride mechanism" (step i), "radical mechanism" (g and following ones) or "carbanion mechanism" (step a). A dehydrogenation reaction initiated by step a can proceed via a covalent intermediate and its fragmentation (steps b+e) or by transfer of a single electron (c) to form a radical pair. The latter can then either collapse to a covalent adduct (d) or transfer a further radical to form the products (f). As discussed elsewhere, an experimental differentiation between steps b and c+d might be impossible, since such reactions are very fast (10, 11). The relevant differentiation between sequences c+f and b+e might, however, be possible as discussed below.

Chemical Work

The recognition of the flavin positions 4a and 5 as the most chemically reactive functions that are also likely to be involved in catalysis, and the investigation of the different reduced flavin isomers, including their mode of formation, is the principal merit of extensive work by Hemmerich's group (see reference 2 for a review). Hamilton's proposal (12) that flavin catalysis is initiated by addition of the -X function of a R_2-CHXH substrate to the isoalloxazine 4a position is based on poor experimental evidence and was later demonstrated to be untenable for theoretical reasons as well (7). However, this proposal did have the merit of stirring up discussion and prompted fruitful experiments conceived to prove it wrong. On the other hand, work initiated by Weatherby and Carr (13), then completed and extended by Bruice's group (7, 14), and also results by Rynd and Gibian (15) provided the first evidence that carbanion species can chemically reduce flavin models. An extensive and comprehensive work by Bruice et al. on the interaction of "C—H acids" with flavin models was reviewed recently (7, 10). The most relevant conclusion from this work is that deprotonation, i.e., step a, Scheme 1, is chemically feasible in those instances where the electron pair of the carbanion can be delocalized and precedes Fl_{ox} reduction. With respect to the alternative modes of redox transfer from carbanion to Fl_{ox}, Bruice concludes that it might proceed via radical intermediates [c+f, Scheme 1; (10)]. While this deduction is not being questioned for a free chemical system, where specifically modified substrate and coenzyme models have to be used, within an enzyme active site different reaction profiles can be preferred, as will be discussed below. Also, in organic systems, transfer occurring via covalent adducts has been reported (16). Experiments involving addition of sulfite and phosphines (17, 18), or photosubstrates (6), and theoretical calculations (19) concur in predicting that the N(5) position is the one involved in transfer of redox equivalents:

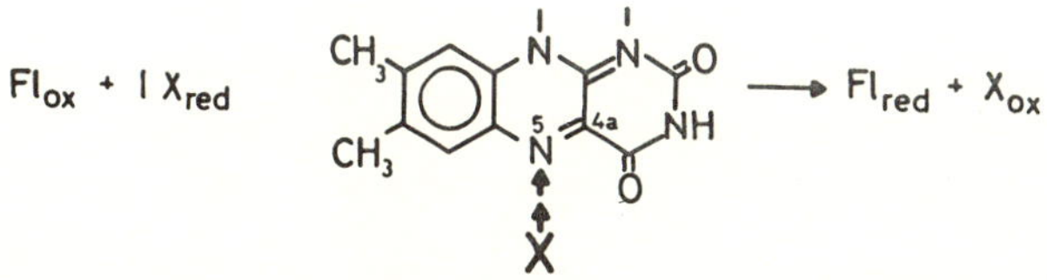

$$Fl_{ox} + |X_{red} \longrightarrow Fl_{red} + X_{ox}$$

Enzymatic Work

In the early 70's, a new chapter of flavin enzymology was initiated by a series of papers dealing particularly with molecular mechanisms: Tober et al. (20) demonstrated that succinate dehydrogenase catalyzes the elimination of F^- from 2,2-difluorosuccinate, a reaction likely to be initiated by abstraction of the relatively acidic substrate α-proton. Shortly thereafter, in 1971, Miyake et al. (21) reported at an "ISOX" conference that D-amino-acid oxidase processes catalytically the substrate along β-Cl-alanine. In the discussion of these results, the formation of pyruvate as a product was mentioned; however, the crucial deduction that this might occur via abstraction of the α-H as a proton, and Cl^- elimination to form the enamine was not put forward. The merit of recognizing such a mechanism thus goes to Walsh et al. (22). In their elegant work, elimination of Cl^- from β-Cl-alanine was demonstrated to be a process indeed catalyzed by D-amino acid oxidase. This work received the title "Evidence for Removal of Substrate Hydrogen as a Proton" and was interpreted as direct evidence for a carbanion mechanism (cf. step a, Scheme 1). While it is not my intention to diminish the importance of this work, it should be stated, as has also been discussed elsewhere (11), and in particular by Bright (5), that these results are compatible with a carbanion mechanism but do not prove it. This is unfortunately taken for granted in some uncritical review articles and discussions which have recently appeared (6,23). Later work by Abeles, Walsh, as well as by Massey's group, confirmed the ambiguity of such an interpretation (24). Yagi, in his review of D-amino acid oxidase reaction mechanism presents and discusses evidence in agreement with these conclusions (25).

Of similar importance is the work by Bright's group involving the use of nitroalkane artificial substrates (26). These molecules are very efficient in the catalytic reduction of D-amino acid oxidase and glucose oxidase and in the presence of the CN^- as a trapping agent progressively and irreversibly inactivate the enzyme. Catalysis was thus proposed to proceed via covalent flavin N(5) adducts (5,26) as shown:

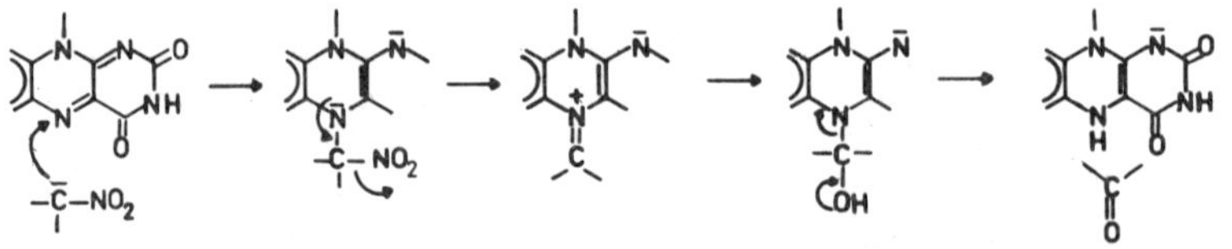

Bruice, however, points out (7,10) that nitroalkane anions are oxidized by a radical mechanism and that the formation of covalent adducts can be unique to the specific reactions with these enzymes.

Similar evidence for the involvement of position N(5) in catalysis comes from the "suicide" inactivation of flavin oxidases by acetylenic substrates, where in all cases formation of covalent adducts to N(5) was observed (see reference 27 for a review). This is also consistent with, but not direct proof for a carbanion mechanism.

Direct evidence for a carbanion intermediate and for covalent catalysis was obtained recently for the reaction of L-lactate oxidase and glycollate (11). From this alternate substrate, two glycollyl flavin N(5) adducts of different stability were formed and characterized. They are derived from chiral abstraction of either the Re or Si protons of the prochiral substrate via two parallel pathways (Scheme 2):

In the case of the stable adduct, which is derived from abstraction of the Si proton, it was shown that it arises directly from oxidized enzyme and glycollate and that it cannot be formed by alternate pathways ("Si path"). This is clearly incompatible with a hydride mechanism (step i, Scheme 1) and is evidence that a carbanion must initiate the reactions (see also Scheme 1, step a, and then either steps b or c). The companion, labile adduct is catalytically competent and occurs during abstraction of the substrate Re proton ("Re path"), i.e., it decays to the reduced enzyme glyoxylate complex which in turn reacts with oxygen to form the final products. Although in the second case a direct formation of the adduct from oxidized enzyme and glycollate cannot be obtained, it was concluded from the analogies between the "Si and Re

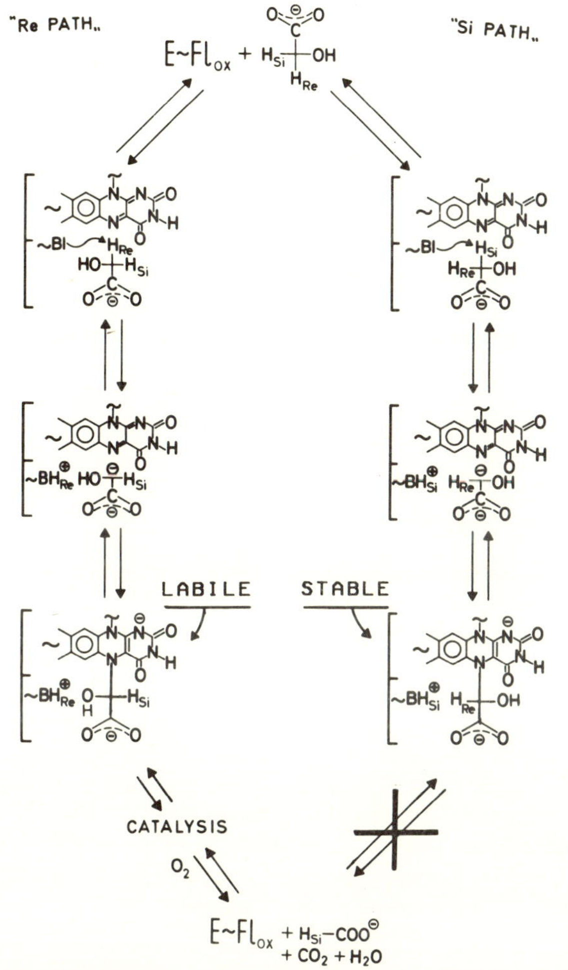

Scheme 2. This scheme represents the reactions of the prochiral substrate glycollate with L-lactate oxidase from *Mycobacterium smegmatis*. The left side ("Re path") represents the catalytic process proceeding via a labile flavin N(5) glycollyl adduct. The right side ("Si path") shows the dead end reaction leading to the stable diastereomeric adduct.

pathways" that the catalytic process proceeds by the same mechanism. The mechanism of Scheme 2 is in agreement with the work and proposals of Lederer's group concerning the flavocytochrome b_2 (28).

Evidence in favor of a carbanion mechanism was also obtained recently with acyl-CoA dehydrogenases and glutaryl-CoA dehydrogenase by Gomes et al. (29). With the first, enzyme inactivation by acetylenic inhibitors does not involve flavin reduction and is proposed to proceed via the corresponding allene intermediate:

$$-C{\equiv}C-CH(H\cdots B\sim)-CO- \rightarrow -C(H)=C=CH-CO- \rightarrow -C(H)=C(B)-CH_2-CO-$$

Similarly, Wenz et al. (30) have shown that 3,4-pentadienoyl-CoA is tautomerized to the 2.4 analogue by general acyl CoA dehydrogenase, possibly over a flavin covalent intermediate:

$$\left[\begin{array}{l} Fl_{ox} \quad HB \\ H_2C{=}C{=}CH{-}CH{-}CO{-} \end{array}\right] \rightarrow \left[\begin{array}{l} \text{Fl(N}^{\ominus}\text{)} \quad HB^{\oplus} \\ HC{=}C(HC){-}CH{-}CO{-} \\ ? \end{array}\right] \rightarrow \left[\begin{array}{l} Fl_{ox} \quad B \\ H_2C{=}CH{-}HC{=}CH{-}CO{-} \end{array}\right]$$

Is a Carbanion Transient Energetically Feasible?

Several authors have questioned the feasibility of a carbanion transient state from thermodynamic considerations, in particular by suggesting that the developing negative charge cannot be stabilized by delocalization (23). These claims can be disputed (10), and although they cannot be discussed in detail here, it should suffice to point out that nature has created several enzyme types that work by the carbanion mechanism (31). The pK of a weak acid such as C—H of lactate can be estimated to be around 25–30 (32). With the enzyme L-lactate oxidase, binding of transition state analogs such as dianionic oxalate cause the pK of an enzyme base to be increased by 7 pK units (33). Clearly, at the enzyme active center and with a properly fitting substrate, the microscopic pK of a C—H bond in question could be lowered by at least the same amount, thus bringing it into the range accessible for catalysis (33).

The Deazaflavin Dilemma

5-Deazaflavin have been used as coenzyme analogs of most classes of flavin enzymes (6,8,34). Four sets of experiments are directly pertinent to the question of flavoprotein dehydrogenation: (a) The reaction of enzyme-bound 5-deazaflavins with reduced pyridine nucleotides leads in incorporation of the 4-H of the latter into position C(5) of the coenzyme analogue (35); (b) The same is the case with the α-H of lactate or of amino acids (35,36); (c) No elimination of halide from β-halogenated substrates is observed with 5-deaza enzymes (35); (d) No covalent adducts were observed in the cases where reduction occurred (35–37) and no inactivation (and covalent adduct formation) was observed with α-hydroxybutynoic acid (37). Also, no reduction or adduct formation was observed with glycollate and 5-deaza FMN lactate oxidase (35).

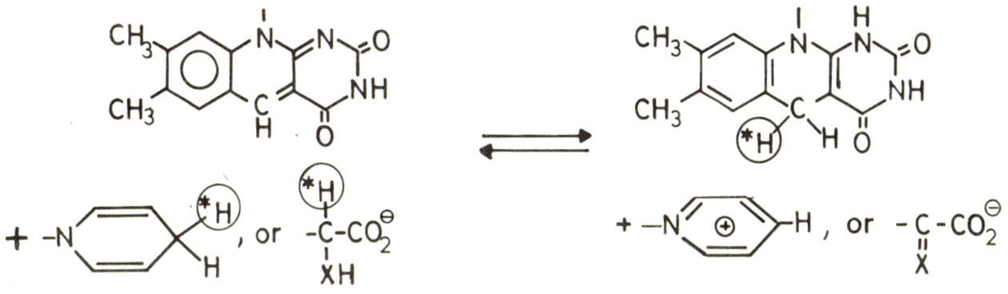

These results clearly can be taken as apparent evidence for a hydride, and against a carbanion mechanism, and contrast sharply with the procarbanion evidence discussed above. As has also been suggested by Fisher et al. and by Pompon and Lederer (35, 37), there are escapes from the dilemma: according to Hemmerich (6) deazaflavins are pyridine nucleotide rather than flavin analogues and might thus work by a hydride mechanism. Walsh's group has provided some evidence that covalent adducts of deazaflavins might not be as stable as one would expect (35). The transfer of redox equivalents from a carbanion to deazaflavins might be unique for the latter and proceed by an alternate mechanism involving sequential transfer of electrons and protons (35). Thus, in spite of the wealth of experiments carried out with deazaflavins, uncertainty remains. On the other hand, at least one point emerges: namely, that the differences between this and the native coenzyme are such that direct mechanistic deductions cannot be extrapolated from one case to the other (6).

Comparison of Substrate Reactivities Within Different Types of Enzymes

If one assumes that pyridine nucleotide-dependent enzymes work by a hydride and pyridoxal phosphate ones by a carbanion mechanism, then the reactivity of flavin enzymes with the same (type of) substrates or pseudo-substrates can be used as a suitable mode of comparison (Scheme 3).

Thus pyridoxal phosphate enzymes catalyze eliminations and can be inactivated by acetylenic or comparable inhibitors (31, 39), while the same types of reactions have never been observed within pyridine-nucleotide enzymes. This places flavin enzymes unequivocally in the same "activation" category as pyridoxal phosphate-dependent ones, a strong argument in favor of a carbanion mechanism for both.

Conclusions

An objective interpretation of the results discussed above leads to the conclusion that a direct and definitive proof of a carbanion mechanism does not as yet exist. The sum of the evidence is, however, most clearly in its favor. With respect to the further course of the reaction, a mechanistic differentiation reduces to the problem of definition of the sequence of steps such as those defined in Scheme 1. A kinetic approach is bound to be of very difficult

COMPARISON OF SUBSTRATE REACTIVITIES WITH ENZYMES CATALYZING RUPTURE OF THE C-H BOND

SUBSTRATE TYPE	PYRIDOXAL	FLAVIN	NICOTINAMIDE
$Y-CH_2-C(H)(XH)-$ Y = leaving group X = OH, NH_2	$Y-CH_2-C^{\ominus}(^{+}NH)-$ ↓ $Y^{\ominus}$ + $H_2C=C(XH)-$ (elimination)	$Y-CH_2-C^{\ominus}(XH)-$	$Y-CH_2-C(H)(XH)-$ ↓ $Y-CH_2-C(=X)-$ (oxidation)
$H-C≡C-C(H)(X)-$ X = OH, NH_2	$H-C≡C-C^{\ominus}(XH)-$ ↓ $H_2C=C=C(XH)-$ (inactivation)	$H-C≡C-C^{\ominus}(XH)-$	$H-C≡C-C(H)(XH)-$ ↓ $H-C≡C-C(=X)-$ (oxidation)

Scheme 3. Comparison of reactions catalyzed by different classes of enzymes with the same type of substrate.

experimental verification for the reasons outlined above and elsewhere (11). On the other hand, a thermodynamic approach comparing the reaction profiles of the reaction paths involving covalent intermediates (Scheme 1, b, e), with those proceeding via radical transfer (c, f) might give a clue: such a comparison was made by Williams and Bruice for the chemical system, and suggested a radical transfer sequence (32). In the case of flavin-dependent dehydrogenase reactions, where reliable data exist, e.g., with glucose oxidase, it appears that radical stabilization results from kinetic as well as from thermodynamic factors (40). In the cases of methanol oxidase (41, 42), and bacterial luciferase (43), an extremely strong radical stabilization was observed; with these enzymes, however, a catalytic role for these radicals is excluded. In other cases, e.g., lactate oxidase (44), and acyl CoA dehydrogenase (this volume), a thermodynamic stabilization of the radical is at least possible. However, no clear-cut example of flavin radical involvement in catalysis has ever been reported for this class of enzymes. On the other hand, a clear case of covalent intermediate stabilization has been documented with L-lactate oxidase, where the catalytically competent N(5) covalent adduct is in a ~1 : 1 (rapid) equilibrium with the products (45). Thus, even if at present the experimental evidence is in favor of covalent catalysis, I feel that a generalization cannot yet be made, radical transfer being a reasonable alternative.

ACKNOWLEDGMENTS
I wish to express my thanks to Dr. V. Massey and Dr. P. Hemmerich for many valuable discussions, and to the Deutsche Forschungsgemeinschaft (DFG Project Gh 2/3) for financial support.

References

1. Michaelis, L. and Schwarzenbach, G.J. (1938) *J Biol Chem* 123:527–542.
2. Beinert, H. (1960) In *The Enzymes*. Boyer, P.D., Lardy, H., and Myrbäck, K. (eds.), Vol. 2, 2nd ed., New York: Academic Press, pp. 339–416.
3. Cornforth, J.W. (1959) *J Lipid Res* 1:3–28.
4. Hemmerich, P. (1964) In *Mechanism Enzymatischer Reaktionen*. Mosbach Symposium. Berlin: Springer-Verlag, pp. 183–209.
5. Bright, H.J. and Porter, D.J.T. (1975) In *The Enzymes*. Boyer, P.D. (ed.) 3rd ed., Vol. 12. New York: Academic Press, pp. 421–505.
6. Hemmerich, P. (1976) In *Progress in Natural Product Chemistry*. Grisebach, H. (ed.) Vol. 33, New York: Springer-Verlag, pp. 451–526.
7. Bruice, T.C. (1975) In *Progress in Bioorganic Chemistry*. Kaiser, E.T. and Kezdy, F.J. (eds.) Vol. 4, New York: Wiley Interscience, pp. 1–87.
8. Walsh, C.T. (1978) *Ann Rev Biochem* 47:881–931.
9. Ghisla, S. and Massey, V. (1978) In *Mechanism of Oxidizing Enzymes*. Singer, T.P. and Ondarza, R.N. (eds.) Amsterdam: Elsevier-North Holland, pp. 55–88.
10. Bruice, C.T. (1980) *Accts Chem Research* 13:256–262.
11. Ghisla, S. and Massey, V. (1980) *J Biol Chem* 255:5688–5696.
12. Hamilton, G.A. (1971) *Progr Bioorg Chem* 1:83–159.
13. Weatherby, G.D. and Carr, D.O. (1970) *Biochemistry* 9:351–354.
14. Main, L., Kasparek, G.J., and Bruice, T.C. (1972) *Biochemistry* 11:3991–3999.
15. Rynd, J.A. and Gibian, M.J. (1970) *Biochem Biophys Res Commun* 41:1097–1103.
16. Szwarc, M. (1972) *Accts Chem Research* 5:169–176.
17. Müller, F. and Massey, V. (1971) In *Methods in Enzymology*, Vol. 18B. McCormick, D. and Wright, L.D. (eds.) New York: Academic Press, p. 468 ff.
18. Müller, F. (1972) *Zeitschrift Naturforsch* 27b:1023–1025.
19. Platenkamp, R.J. (1981) PhD Thesis, Univ. of Leiden, pp. 100–122 and Lit. cited therein.
20. Tober, C., Nicholls, P., and Brodie, J. (1970) *Arch Biochem Biophys* 138:506–514.
21. Miyake, Y., Abe, T., and Yamano, T. (1973) In *Oxidases and Related Redox Systems*. King, T.E., Mason, H.S., and Morrison, M. (eds.) Baltimore, London, Tokyo: University Park Press, pp. 209–215.
22. Walsh, C.T., Schonbrunn, A., and Abeles, R. (1971) *J Biol Chem* 246:6855–6866.
23. Visser, C.M. (1980) *Naturwiss* 67:549–555.
24. Massey, V., Ghisla, S., Ballou, D.P., Walsh, C.T., Cheung, Y.T., and Abeles, R.H. (1976) In *Flavins and Flavoproteins*. Singer, T.P. (ed.) Amsterdam, Oxford, New York: Elsevier, pp. 199–212.
25. Yagi, K. (1980) In *New Horizons in Biological Chemistry*. Koike, M., Nagatsu, T., Okuda, J., and Ozawa, T. (eds.) Tokyo: Japan Scientific Societies Press, pp. 257–270.
26. Porter, D.J.T., Voet, J.B., and Bright, H.J. (1973) *J Biol Chem* 248:4400–4416.
27. Ghisla, S., Wenz, A., and Thorpe, C. (1980) In *Enzyme Inhibitors*. Brodbeck, U. (ed.) Verlag Chemie, pp. 44–60.

28. Pompon, D., Iwatsubo, M., and Lederer, F. (1980) *Eur J Biochem* 104:479–488.
29. Gomes, B., Fendrich, G., and Abeles, R. (1981) *Biochemistry* 20:2481–2490.
30. Wenz, A., Ghisla, S., and Thorpe, C., this volume.
31. Walsh, C.T. (1979) *Enzymatic Reaction Mechanisms*. San Francisco: W.H. Freeman.
32. Williams, R.F. and Bruice, T.C. (1976) *J Amer Chem Soc* 98:7752–7768.
33. Ghisla, S. and Massey, V. (1977) *J Biol Chem* 252:6729–6735.
34. Walsh, C. (1980) *Accts Chem Research* 13:148–155.
35. Fisher, J., Spencer, R., and Walsh, C. (1976) *Biochemistry* 15:1054–1064.
36. Averill, B.A., Schonbrunn, A., Abeles, R., Weinstock, L.T., Cheng, C.C., Fisher, J., Spencer, R., and Walsh, C. (1975) *J Biol Chem* 250:1603–1605.
37. Pompon, D. and Lederer, F. (1979) *Eur J Biochem* 96:571–579.
38. Massey, V., unpublished.
39. Jung, M.J. (1980) In *Enzyme Inhibitors*. Brodbeck, U. (ed.) Verlag Chemie, pp. 86–95.
40. Stankovich, M.T., Schopfer, M.L., and Massey, V. (1978) *J Biol Chem* 253:4971–4979.
41. Geissler, J. and Hemmerich, P. (1981) *FEBS Letters*, in press.
42. Mincey, T., Tayrien, G., Mildvan, A.S., and Abeles, R.H. (1980) *Proc Natl Acad Sci USA* 77:7099–7101.
43. Kurfuerst, M., Ghisla, S., Presswood, R., and Hastings, W., this volume.
44. Choong, Y.S. and Massey, V. (1980) *J Biol Chem* 255:8672–8677.
45. Massey, V., Ghisla, S., and Kieschke, K. (1980) *J Biol Chem* 255:2796–2806.

PART II A:

Carbon-Sulfur Transhydrogenases

Published 1982 by Elsevier North Holland, Inc.
Vincent Massey and Charles H. Williams, Editors
Flavins and Flavoproteins

CHAPTER 23

Mechanisms of GSSG and Oxygen Reduction Catalyzed by Glutathione Reductase

Bengt Mannervik, Vijayakumar Boggaram, and Inger Carlberg

Department of Biochemistry, Arrhenius Laboratory, University of Stockholm, S-106 91 Stockholm, Sweden

Functional Differentiation Between the Substrate Binding Sites of Glutathione Reductase

Kinetic studies of glutathione reductase showed that the two substrates glutathione disulfide (GSSG) and NADPH (or $NADP^+$) could be bound simultaneously to the enzyme, and it was concluded that different binding sites exist for donor (NADPH) and acceptor (GSSG) substrates (1,2). These sites were established as topographically discrete by use of X-ray diffraction analysis (3,4) and were named N-site (for nucleotides) and G-site (for glutathione) (5). The finding that glutathione reductase catalyzes transhydrogenase reactions involving pyridine nucleotides (5–7) demonstrated that the N-site has catalytic activity per se. Reduction of GSSG as well as the TNBS-dependent O_2 (8,9) and cytochrome c (9) reductions involve both N-site and G-site. So far no enzymatic activity involving only the G-site has been found. Consequently, a reagent interacting with the N-site is expected to affect all activities, whereas a reagent interacting with the G-site should not interfere with the transhydrogenase activity. By use of these rationales the action of various inhibitors have been located.

Reagents Modifying the N-Site

Several dicarbonyl reagents, known to modify arginine residues, inhibit the GSSG and O_2 reductions as well as the transhydrogenase activity of glutathione reductase (10,11). These findings and the protection obtained by nucleotides demonstrate that the N-site is involved. Antibodies to glutathione reductase were also found to inhibit the three above activities and their effect was counteracted by nucleotides (12). Determinants for binding of these antibodies are apparently located at or near the N-site.

Reagents Modifying the G-Site

Ethoxyformic anhydride modifies histidine residues and inactivates GSSG reduction but has only a minor effect on the transhydrogenase activity (5,13).

Thus, an essential histidine residue in the G-site was indicated. Increased fluorescence of FAD upon modification suggested that the imidazole was close to the isoalloxazine ring. Indeed, structural studies demonstrated a histidine in such a position (14). Diazonium-1-H-tetrazole, another histidine-modifying reagent, caused a similar inactivation of the enzyme in oxidized as well as in two-electron-reduced form.

Reduced glutathione reductase has a redox-active dithiol in the G-site which reacts with thiol-blocking reagents (15). The "distal" sulfur [i.e., that of "Cys-41" in the human enzyme (4)] is the most reactive of the two sulfur atoms (16). Alkylation of the dithiol with, e.g., iodoacetamide causes loss of GSSG-reducing as well as O_2 reducing activities. A prerequisite for modification is reduction of the enzyme to the dithiol form. 2,4,6-Trinitrobenzenesulfonate (TNBS) interferes with alkylation by reacting with a thiol group (probably the distal) and/or by reoxidizing the enzyme (17). On the basis of this effect and additional criteria, TNBS has been classified as a reagent reacting with the G-site (9,17).

Cross-Linking of the Two Subunits by a Reaction in the G-Site

The X-ray diffraction analysis indicated that a histidine residue in one subunit was close to the isoalloxazine ring of FAD and the redox-active disulfide/dithiol of the other subunit (4). If this histidine residue were the essential residue (13), it might be possible to simultaneously inactivate the GSSG reduction and cross-link the two subunits by a bifunctional reagent. The cross-linking would be expected to involve a thiol group of a subunit of the reduced enzyme and the histidine of the other subunit. Indeed, 1,5-difluoro-2,4-dinitrobenzene was found to inactivate reduced enzyme, but not oxidized enzyme at 0°C (18). Both the GSSG and O_2 reductions were inhibited. Upon polyacrylamide gel electrophoresis in the presence of small amounts of 2-mercaptoethanol, a major fraction of the modified enzyme migrated corresponding to $M_r = 100{,}000$, whereas native enzyme was completely resolved into subunits of $M_r = 50{,}000$. High 2-mercaptoethanol concentrations caused the large component of modified enzyme to dissociate into subunits of $M_r = 50{,}000$ as expected for a derivative obtained by arylation of thiol and imidazole groups. The transhydrogenase activity was largely unaffected by the bifunctional reagent. Further support for the interpretation that the two subunits were cross-linked in the G-site was obtained by the finding that TNBS protected the enzyme against the action of 1,5-difluoro-2,4-dinitrobenzene (18). Figure 1 shows a diagram of the proposed linkage of the two subunits by the reagent.

The above findings support the conclusion that both subunits of glutathione reductase contribute essential residues to the G-site (4).

Sequence of Events in the Electron Transport Catalyzed by Glutathione Reductase

Glutathione reductase may accept two electrons per FAD from NADPH. This process is facilitated by arginine residues, probably binding the 2′-phosphate of the adenosine moiety of NADPH. The electrons can be transferred to an

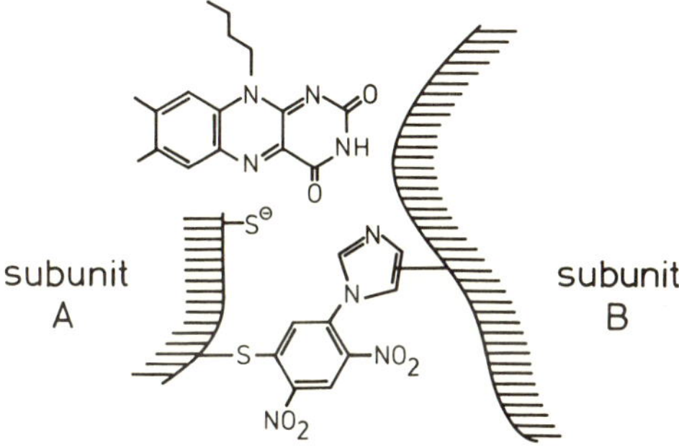

Figure 1. Proposed cross-linking of subunits A and B by 1,5-difluoro-2,4-dinitrobenzene in the G-site of glutathione reductase.

oxidized pyridine nucleotide in a transhydrogenase reaction (5–7) or to a dye in a diaphorase reaction (7). Alternatively the electrons are taken up by the cystine disulfide in the G-site. The resulting dithiol reacts with GSSG to yield 2 GSH as products (5, 15). In the catalytic events in the G-site, histidine is proposed to serve as an acid-base catalyst. All the above partial reactions are believed to be heterolytic processes in which the electrons are transferred pair-wise.

TNBS intercepts the flow of electrons to GSSG by interacting with the dithiol in the G-site of the reduced enzyme. In the presence of TNBS one-electron transfer takes place as evidenced by the reduction of cytochrome c. Cytochrome c reductions takes place both aerobically and anaerobically, indicating electron transfer from reduced enzyme via a nitroradical of TNBS. It seems probable that also the TNBS-dependent O_2 reduction is effected by one-electron transfer via TNBS. The superoxide anion thus formed dismutes to H_2O_2 and O_2.

ACKNOWLEDGMENT
This work was supported by Grants (to B.M.) from the Swedish Natural Science Research Council.

References

1. Mannervik, B. (1973) *Biochem Biophys Res Commun* 53:1151–1158.
2. Mannervik, B. (1976) In *Flavins and Flavoproteins*. Singer, T.P. (ed.), Amsterdam: Elsevier, pp. 485–490.
3. Zappe, H.A., Krohne-Ehrich, G., and Schulz, G.E. (1977) *J Mol Biol* 113:141–152.
4. Schulz, G.E., Schirmer, R.H., Sachsenheimer, W., and Pai, E.F. (1978) *Nature* 273:120–124.
5. Mannervik, B., Boggaram, V., Carlberg, I., and Larson, K. (1980) In *Flavins and Flavoproteins*. Yagi, K. and Yamano, T. (eds.), Tokyo: Japan Scientific Societies Press, pp. 173–187.
6. Moroff, G., Ochs, R.S., and Brandt, K.G. (1976) *Arch Biochem Biophys* 173:42–49.
7. Boggaram, V., Larson, K., and Mannervik, B. (1978) *Biochim Biophys Acta* 527:337–347.
8. Carlberg, I. and Mannervik, B. (1980) *FEBS Lett* 115:265–268.
9. Carlberg, I. and Mannervik, B. (1981) This volume.
10. Boggaram, V. and Mannervik, B. (1979) *Acta Chem Scand* B33:593–594.

11. Boggaram, V. and Mannervik, B. (1981) This volume.
12. Carlberg, I., Altmejd, B., and Mannervik, B. (1981) *Biochim Biophys Acta*, in press.
13. Boggaram, V. and Mannervik, B. (1978) *Biochem Biophys Res Commun* 83:558–564.
14. Untucht-Grau, R., Schulz, G.E., and Schirmer, R.H. (1979) *FEBS Lett* 105:244–248.
15. Massey, V. and Williams, C.H., Jr. (1965) *J Biol Chem* 240:4470–4480.
16. Arscott, L.D., Thorpe, C., and Williams, C.H., Jr. (1981) *Biochemistry* 20:1513–1520.
17. Carlberg, I. and Mannervik, B. (1979) *FEBS Lett* 98:263–266.
18. Boggaram, V. and Mannervik, B. (1980) *Acta Chem Scand* B34:147–148.

Published 1982 by Elsevier North Holland, Inc.
Vincent Massey and Charles H. Williams, Editors
Flavins and Flavoproteins

CHAPTER 24

Multifunctionality of Lipoamide Dehydrogenase

C.S. Tsai, D.M. Templeton, A.J. Wand, and J.R.P. Godin

Department of Chemistry and Institute of Biochemistry, Carleton University, Ottawa K1S 5B6, Canada

Summary

ABSTRACT

The role of apoenzyme on NADH-linked reductase, transhydrogenase, electron transferase, and diaphorase activities of pig heart lipoamide dehydrogenase was investigated. Six isozymes have been separated and each exhibits multifunctional activities. The dimeric structure is required for the reductase and transhydrogenase activities while the monomeric structure favors the diaphorase activity. Changes in conformations and/or the flavin environment are manifested in the reciprocal effect on the reductase versus diaphorase activities. Chemical modifications allow correlations between the critical amino acid residues and changes in multifunctional activities. The obligatory function of the active site disulfide is assisted by the histidine residue in the reductase reaction while carboxyl group(s) suppress the electron transferase activity.

Lipoamide dehydrogenase (EC 1.6.4.3) from pig heart is a multifunctional enzyme which catalyzes NADH-linked lipoate reductase (RDase), nicotinamide nucleotide transhydrogenase (THase), inorganic electron transferase (ETase), and quinone diaphorase (DPase) reactions (1,2). In view of the fact that free flavins also facilitate these reactions (3,4), lipoamide dehydrogenase represents a favorable case for studying the role of apoenzyme in specification and regulation of enzyme multifunctionality.

Six isozymes have been separated by isoelectric focusing (5). Table 1 lists kinetic parameters of multifunctional reactions catalyzed by two cationic and two anionic isozymes. In spite of a partial denaturation of the isolated isozymes, it can be inferred from these results that the multifunctional activities are molecular properties of every isozyme. The contribution of subunit structure to the multifunctional behavior of lipoamide dehydrogenase has been studied by the use of chemically trapped monomeric and dimeric enzymes (6). Chemical monomerization greatly enhances the DPase activity while the dimeric structure, which suppresses the DPase activity, is essential to the RDase and THase activities.

The polarity of the flavin environment is amenable to absorption spectroscopic studies (7). A spectral transition from the bound FAD to the free flavin

Table 1. Kinetic Parameters of Isozymes.

	Isozymes			
	I	II	V	VI
Isoelect. point	5.82	6.05	7.26	7.65
Yield (%)	6.8	11.2	17.2	9.6
RDase: V/E_t		2900	1600	
K_a		360	110	
K_b		600	610	
THase V/E_t	355	300	435	700
K_a	8.5	42	59	8.0
K_b	105	30	28	16
ETase V/E_t	3500		780	4050
K_a	65		530	23
K_b	370		100	180
K_{ia}	62		120	150
DPase V/E_t	2275	2100	1900	2200
K_a	225	560	342	200
K_b	575	330	190	150
K_{ia}	22	32	46	18

Isozymes are separated by isoelectric focusing (5) and steady-state kinetic studies were carried out as described (2). Non-Michaelian kinetic behaviors were observed for all isozymes. Parameters (V/E_t in min^{-1}, K_a, K_b, and K_{ia} in μM) were evaluated from asymptotic regions. Partial denaturation was evidenced by a reduced RDase activity and an enhanced DPase activity.

is indicative of an increased polar environment. An increased FAD mobility is detectable by facile flavin removal during dialysis at neutral pH and a shortened fluorescence lifetime (Table 2). These changes characterize the enhanced DPase activity of the monomeric enzyme, isolated isozymes, and some modified enzymes. An alteration in the flavin environment is accompanied by conformational change(s). Various structural perturbations, except complete denaturation, fail to inactivate DPase.

Table 2. Fluorescence Decay Parameters.

	τ_1(ns)	τ_2(ns)	A_1/A_2
Native enzyme	4.04	0.92	1.02
Isozyme I	3.92	0.81	0.88
Isozyme II	3.48	0.46	1.15
E_{rcM}	3.10		
Free FAD	2.84		

Fluorescence decay measurements were performed under the guidance of Dr. D.M. Rayer (8). The exponential decay is fitted to

$$F(t) = \sum_{i=1}^{n} A_i e^{-t/\tau_i}$$

where F(t) is the fluorescence intensity at time t. τ_i is the lifetime of the ith decay component and the pre-exponential factor, A_i is an expression of the amount of ith component present. E_{rcM} is reductive carboxylated monomer (6).

Table 3. Variations in Multifunctional Activities by Chemical Modifications.

Enzyme derivative	Effect	% Activity of control			
		RDase	THase	ETase	DPase
E_{As}	-s-s-	4	44	77	160
E_{Cd}	-s-s-	2	16	110	273
E_{ma}	-s-s-	0	139	148	188
$E_{h\nu,RB}$	His	28	64	106	129
$E_{h\nu,MB}$	His, Met, Tyr	3	10	65	107
$E_{NH_3^+}$	COOH	62	86	680	169
$E_{SO_3^-}$	COOH	68	95	105	136

Chelations with arsenite (E_{As}) (9) and cadmium (E_{Cd}) (10) were performed as described. Monoalkylation (E_{ma}) follows the literature procedure (11). Photooxidations (12) were sensitized by rose bengal ($E_{h\nu,RB}$) and methylene blue ($E_{h\nu,RB}$). Carbodiimide-facilitated modifications of carboxyl groups (13) were carried out by the use of ethylene diamine which introduces $-CH_2CH_2NH_3^+$ (E_{NH_3+}) and taurine which introduced $-CH_2CH_2SO_3^-$ ($E_{SO_3^-}$).

Table 3 shows that multifunctional activities of lipoamide dehydrogenase can be modulated independently by specific chemical modifications. Blocking of the active site disulfide eliminates the RDase activity. The effect on ETase is marginal while the DPase activity is enhanced. A parallel trend, though less profound, is observed for rose bengal-sensitized photooxidation of histidines (12). Methylene blue sensitization which destroys methionine, phenylalanine, tyrosines, and histidines results in an overall inactivation. Thus, the active site disulfide is obligatory to the RDase activity but not imperative to the ETase and DPase activities. The histidine residue is required for RDase presumably to stabilize the thiolate ion of the reduced enzyme intermediate. The presence of carboxyl groups suppresses the ETase activity. A charge effect is implicated. These observations are summarized in Figure 1 which expresses perturbation

Figure 1. Multifunctionality scale expressing effects of apoenzyme structures on multifunctional activities of lipoamide dehydrogenase.

R Dase THase ETase DPase
Subunit Structure : Dimer — Monomer
FAD Environment : Buried Rigid Nonpolar — Exposed Mobile Polar
Conformation : Sensitive — Insensitive
Disulfide : Essential — Nonessential
Histidine : Required — Indifference
Carboxylic Suppression

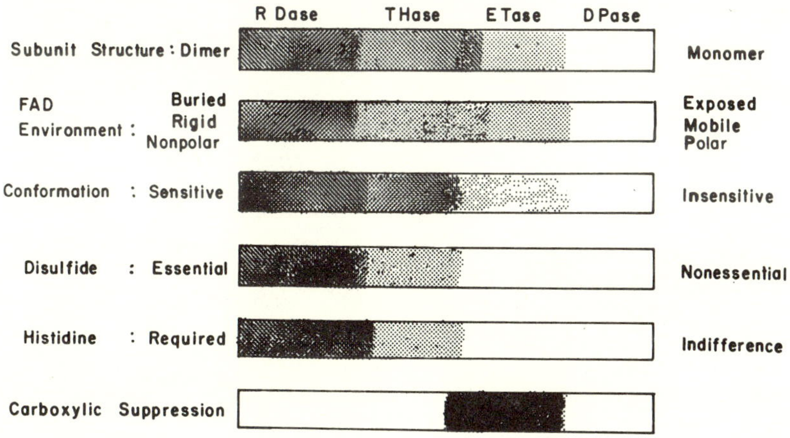

effects of the apoenzyme structure on the multifunctional activities of lipoamide dehydrogenase.

Dixon (14) noted, with few exceptions, that the reductions carried out by flavoenzymes occurred only with substrates which would also be reduced by free reduced flavins. Immobilization of flavin to amphipathic polymers greatly facilitates its reduction by NADH (15). Therefore, the apoenzyme of lipoamide dehydrogenase may be considered as a natural means to immobilize FAD for reduction. The reactivity of the reduced FAD is subsequently regulated by generating the RDase activity and suppressing the ETase and DPase activities through the coordinated structural effects (Figure 1) of the apoenzyme.

ACKNOWLEDGMENT

This work was supported by a grant from Natural Sciences and Engineering Research Council of Canada.

References

1. Massey, V. (1963) In *The Enzymes*, 2nd Ed. Boyer, P.D., Lardy, H., and Myrbäck, K. (eds.), Vol. 7, New York: Academic Press, pp. 275–306.
2. Tsai, C.S. (1980) *Int J Biochem* 11:407–413.
3. Walsh, C. (1980) *Acc Chem Res* 13:148–155.
4. Bruice, T. (1980) *Acc Chem Res* 13:256–262.
5. McManus, I.R. and Cohen, M.L. (1974) In *Isozymes*. Markert, C.L. (ed.) Vol. 1. New York: Academic Press, pp. 621–636.
6. Tsai, C.S., Templeton, D.M., and Wand, A.J. (1981) *Arch Biochem Biophys* 206:77–86.
7. Williams, C.H., Jr. (1976) In *The Enzymes*, 3rd Ed. Boyer, P.D. (ed.) Vol. 13. New York: Academic Press, pp. 89–173.
8. Rayer, D.M., McKinnon, A.E., Szabo, A.G., and Hackett, P.A. (1976) *Can J Chem* 54:3246–3259.
9. Massey, V. and Palmer, G. (1962) *J Biol Chem* 237:2347–2358.
10. Stein, A.M. and Stein, J.H. (1971) *J Biol Chem* 246:670–678.
11. Thorpe, C. and Williams, C.H., Jr. (1976) *J Biol Chem* 251:3553–3557.
12. Templeton, D.M. and Tsai, C.S. (1979) *Biochem Biophys Res Comm* 90:1085–1090.
13. Hoare, D.G. and Koshland, D.E. Jr. (1967) *J Biol Chem* 242:2447–2453.
14. Dixon, M. (1971) *Biochim Biophys Acta* 226:259–268.
15. Shinkai, S., Ando, R., and Kunitake, T. (1978) *Biopolymers* 17:2757–2760.

Published 1982 by Elsevier North Holland, Inc.
Vincent Massey and Charles H. Williams, Editors
Flavins and Flavoproteins

CHAPTER 25

Proton Stoichiometry of the Reduction of FAD and the Disulfide of Thioredoxin Reductase, Oxidation Reduction Potentials as a Function of pH, FAD Replacement by 1-Deaza-FAD

Michael E. O'Donnell and Charles H. Williams, Jr.

Veterans Administration Medical Center and Department of Biological Chemistry, University of Michigan, Ann Arbor, Michigan

Thioredoxin reductase, TRR, catalyzes the transfer of electrons between NADPH and thioredoxin, a small protein containing a redox-active disulfide (1,2). TRR contains FAD and a redox-active disulfide (1–3). It is thought that the electrons flow sequentially from NADPH to the FAD, to the disulfide, and from the dithiol on TRR to the disulfide on thioredoxin. This last reaction is a thiol-disulfide interchange known to be initiated via attack by a thiol anion (4). Thus, a knowledge of the protonic stoichiometry of the reduction of the FAD and the disulfide is important in defining the mechanism of TRR. This can be derived from a study of the pH dependence of the oxidation reduction potentials of the FAD and the disulfide. The oxidation reduction potentials have been determined by measuring equilibria of the enzyme species with $NADH/NAD^+$ or $APADH/APAD^+$. The midpoint potentials, E_m, of the disulfide and the FAD are within 40 mv at all pH values. Thus, four microforms are in equilibrium with pyridine nucleotide (Figure 1). Knowledge of the concentrations of one of the microforms, oxidized and reduced FAD, together with the total electrons added, allows the linked equilibria to be solved and the four microscopic midpoint potentials to be calculated.

The FAD in TRR can readily be replaced by 1-deaza-FAD which has a more negative oxidation reduction potential. The spectral changes observed on reduction of 1-deaza-FAD TRR indicate the partial formation of a thiolate C-4a adduct. The pH dependence of 1-deaza-FAD adduct formation combined with the proton stoichiometry results suggests the presence of a base at the active site of TRR.

Measurement of Oxidation Reduction Potentials

Establishment of Equilibrium

Since it is essential to the calculation of oxidation reduction potentials from equilibrium measurements that the system under study be at equilibrium,

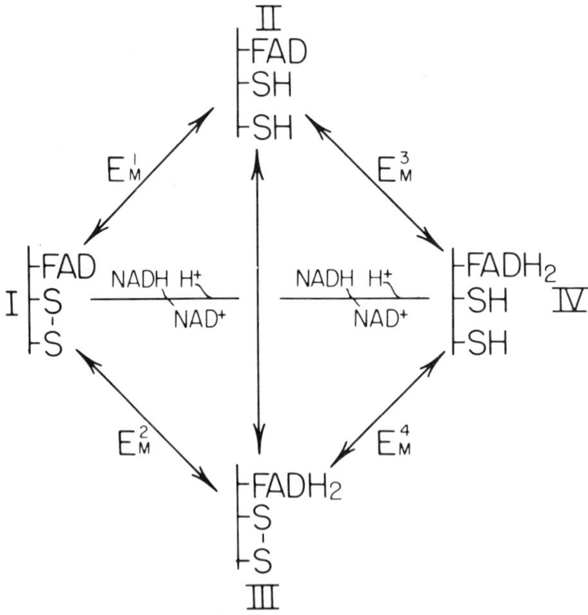

Figure 1. Distribution of enzyme forms in equilibrium with NAD^+ and NADH.

two-electron transhydrogenation among enzyme species was measured and was found to take place with a half-life of about 8 minutes at pH 6.0 under the conditions used in this study. Thus, changes in absorbance during a titration were allowed to stabilize before recording data to ensure that the criterion of equilibrium was fulfilled. A small amount of TRR neutral semiquinone is formed through 1-electron comproportionation (5). The rate of semiquinone formation is too slow to collect equilibrium data. Thus, corrections to the data have been made for semiquinone; these amount to not more than 7%.

The accuracy of midpoint potentials determined from equilibrium measurements is dependent on the ability to measure concentrations of unliganded species. Two lines of evidence suggest that the pyridine nucleotides used do not form a complex with TRR. A long wavelength charge transfer band associated with complex formation between $NADP^+$ and reduced TRR (1) is not observed between reduced TRR and either NAD^+ or $APAD^+$ under the conditions employed. Also, the equilibria are not dependent on the concentration of protein. Titrations of TRR with pyridine nucleotide at different concentrations of enzyme are shown in Figure 3. E_m values agree to within 2 mv.

Measurement of Equilibria

The oxidation reduction potentials of the four couples of TRR were determined by equilibrium measurements of enzyme with $NADH/NAD^+$ or

APADH/APAD$^+$. These measurements require a knowledge of equilibrium concentrations of NAD$^+$, NADH, and each enzyme form. Measurement of NADH at equilibrium takes advantage of the isosbestic at 347 nm for reduction of TRR. The concentration of NAD$^+$ at equilibrium is obtained by difference.

$$[\mathrm{NADH}]_{eq} = \left(A^{347}_{obs} - A^{347}_{initial}\right)/\varepsilon^{347}_{NADH}$$

$$[\mathrm{NAD^+}]_{eq} = [\mathrm{NADH}]_{initial} - [\mathrm{NADH}]_{eq}$$

Two macroscopic and one microscopic quantities are required to solve the linked equilibria of Figure 1. Two macroscopic quantities, the amounts of reduced FAD and dithiol, can be measured:

$$[\mathrm{FAD}]_{red} = \left(A^{455}_{initial} - A^{455}_{obs}\right)/\left(\varepsilon^{455}_{FAD_{ox}} - \varepsilon^{455}_{FAD_{red}}\right)$$

$$[\mathrm{SH}]_2 = e^- \,\mathrm{EQUIVALENTS} - [\mathrm{FAD}]_{red}$$

where $\quad e^- \,\mathrm{EQUIVALENTS} = [\mathrm{NADH}]_{initial} - [\mathrm{NADH}]_{eq}$

These data are obtained at different levels of reduction in titrations with NADH as shown in Figure 2 for pH 6.0 and 8.15.

The concentration of one microform is required to solve for the concentrations of the other three. This can be achieved at one level of reduction in a separate experiment. Phenyl mercuric acetate (PMA) forms a tight complex with the dithiol of microforms II and IV (1). In this type of experiment, 1.0 equivalent of dithionite is added to an anaerobic solution of TRR. After reduction, addition of a five-fold excess (enzyme FAD) of PMA rapidly displaces the equilibrium between microforms II and III toward form II. The equilibrium concentration of form III before addition of PMA is obtained from the rapid jump in absorbance at 455 nm observed upon addition of PMA. The equilibrium concentrations of the other three microforms at this one level of reduction are calculated.

$$[\mathrm{III}] = [\mathrm{FAD_{ox}}]_{after\ PMA} - [\mathrm{FAD_{ox}}]_{before\ PMA}$$

$$[\mathrm{IV}] = [\mathrm{ENZYME}]_{total} - [\mathrm{FAD_{ox}}]_{after\ PMA}$$

$$[\mathrm{II}] = e^- \,\mathrm{EQUIVALENTS} - [\mathrm{III}] - 2[\mathrm{IV}]$$

$$[\mathrm{I}] = [\mathrm{ENZYME}]_{total} - [\mathrm{II}] - [\mathrm{III}] - [\mathrm{IV}]$$

Sodium dithionite was used as the reductant in PMA experiments since it was of concern that pyridine nucleotide could catalyze 2-electron transfer between forms I and IV to yield forms II and III, resulting in an anomalously large absorbance change at 455 nm upon addition of PMA. Since sodium dithionite was the reductant, the equilibrium between enzyme and reductant could not be measured. However, the results of a PMA experiment and a titration experiment can be combined to calculate the equilibria of Figure 1. The ratios shown below are obtained from a PMA experiment.

$$k_1/k_2 = [\mathrm{II}]/[\mathrm{III}]$$

$$k_1/k_4 = [\mathrm{II}][\mathrm{III}]/[\mathrm{I}][\mathrm{IV}]$$

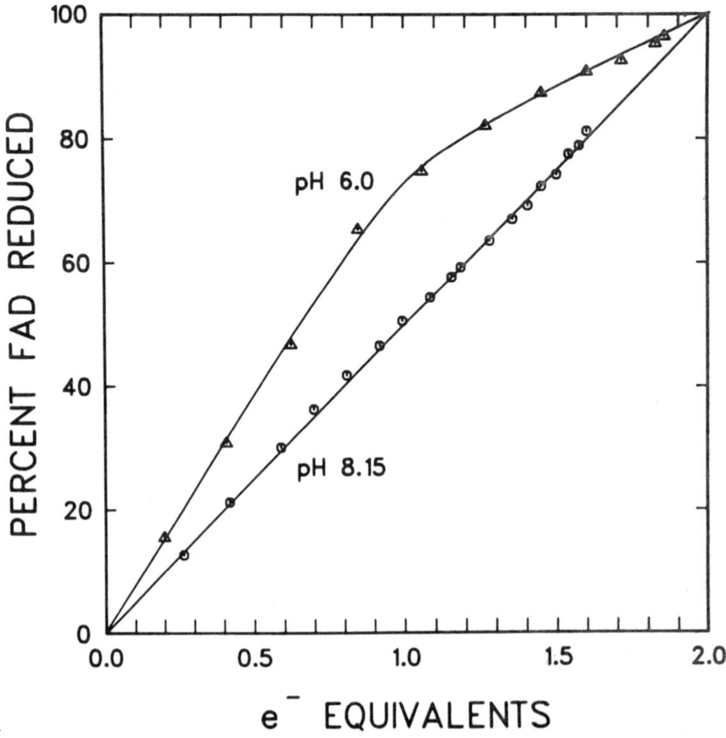

Figure 2. Relationship of reduced FAD to the number of equivalents of NADH taken up by thioredoxin reductase at pH 6.0 and at pH 8.15. Enzyme was 17.7 μM in 0.1 M phosphate buffer, 0.3 mM EDTA, 12°. The solid lines are calculated from the ratio of k_1/k_2 and k_1/k_4 derived from PMA addition experiments performed under the same conditions as titrations with pyridine nucleotide. pH 6.0: $k_1/k_2=0.273$, $k_1/k_4=6.04$; pH 8.15: $k_1/k_2=0.99$, $k_1/k_4=3.73$.

These ratios characterize the linked equilibria of Figure 1. These ratios, in combination with the $[\text{ENZYME}]_{\text{total}}$ and the e^- EQUIVALENTS allow calculation of the four microscopic oxidation reduction potentials at each point in a titration. This procedure was carried out at pH values spanning the range from 5.5 to 8.15. Data for the two FAD couples, E_m^2 and E_m^3, at pH values 6.0, 7.0 and 8.15 are plotted in Figure 3.

The E_m vs pH for the four couples is shown in Figure 4. The E_m^1 and E_m^2 couples have slopes of −0.057 V and −0.060 V respectively from pH 5.5 to 7.2. This indicates a 2-proton stoichiometry for reduction of both the FAD and the disulfide in this pH range. Above pH 7.2, the slopes of both couples increase to −0.083 V. This indicates a 3-proton stoichiometry for both the FAD and the disulfide above pH 7.2. This suggests the presence of a base in oxidized enzyme (ca pK 7.2) whose ionization is perturbed by reduction of the FAD or the disulfide (6). The data-point at pH 8.15 for the E_m^1 and E_m^2 couples

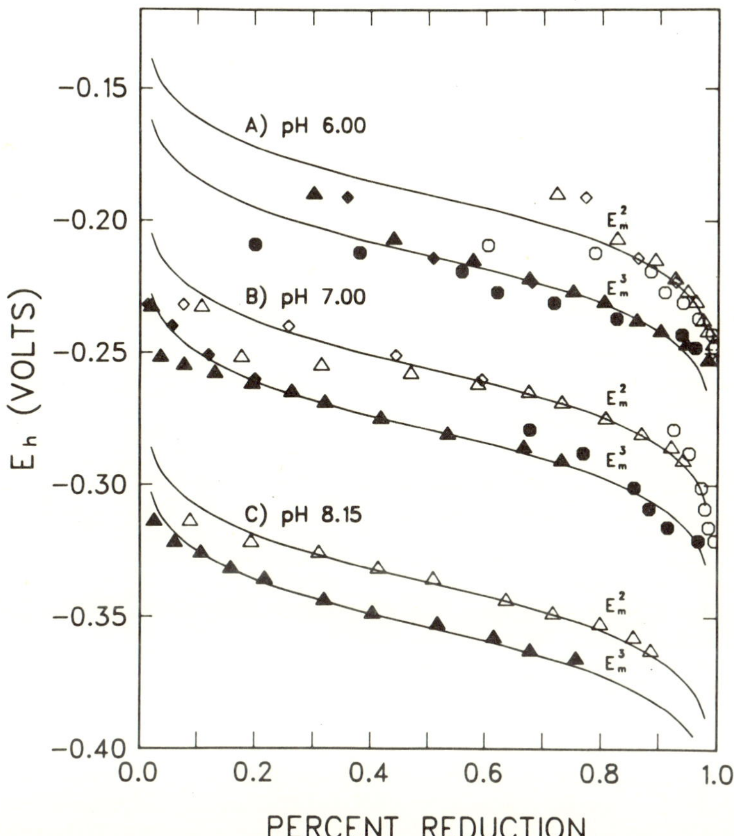

Figure 3. Nernst curves of the two FAD couples, E_m^2 (open symbols) and E_m^3 (closed symbols) at three different values of pH. The different symbols represent separate titration experiments—the details are specified. pH 6.0: (○), 6.76 μM enzyme and 624 μM added NAD^+ titrated with NADH; (△), 16.6 μM enzyme and 628 μM added NAD^+ titrated with NADH; (◇), 44.9 μM enzyme titrated with APADH. pH 7.0: (○), 17.7 μM enzyme titrated with NADH; (△), 17.2 μM enzyme and 478 μM added NAD^+ titrated with NADH; (◇), 43.7 μM enzyme titrated with APADH. pH 8.15: (△), 17.7 μM enzyme titrated with NADH. The solid lines represent theoretical curves calculated for 2-electron reduction assuming the E_m calculated from the data points.

is above the line for a −0.083 V slope. If this data-point is taken seriously, the break to a shallower slope marks an ionization on 2-electron reduced enzyme (6). The E_m^3 and E_m^4 couples have slopes of −0.065 V and −0.057 V respectively from pH 5.5 to 8.15. This indicates a 2-proton stoichiometry for these couples throughout the pH range studied.

1-Deaza-FAD Thioredoxin Reductase

Apo-TRR was prepared by resolving the FAD in 5.0 M guanidine hydrochloride and adsorption of FAD by addition of activated charcoal. Charcoal was

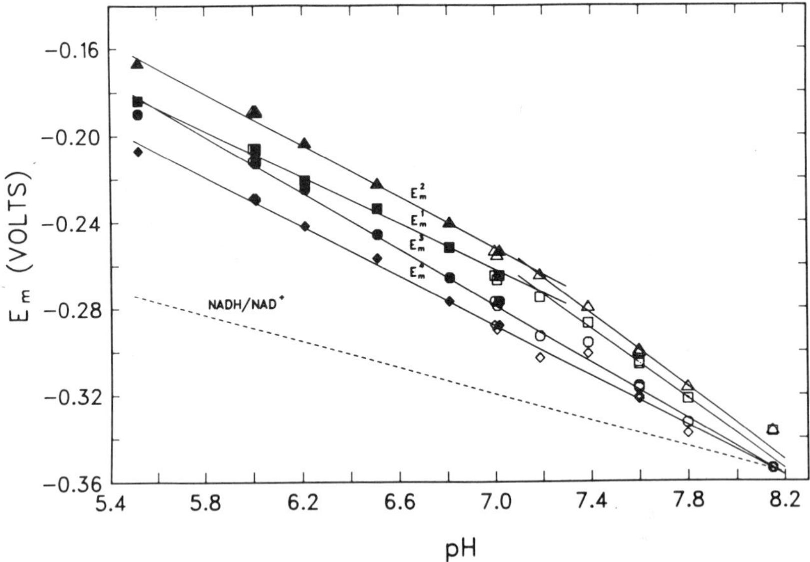

Figure 4. Effect of pH on the midpoint potentials of thioredoxin reductase. (□), E^1_m; (△), E^2_m; (○), E^3_m; (◇), E^4_m. Open symbols, data obtained from reduction by NADH; closed symbols, data obtained from reduction by APADH. The dashed line is the curve relating E_m to pH for the $NADH/NAD^+$ couple.

removed by centrifugation and the supernatant dialyzed to remove guanidine hydrochloride. The apo-TRR contained 0.7% residual activity determined by the DTNB coupled assay (2) and was 87% reconstitutable with FAD. Reconstitutable protein had 99% of the activity of native enzyme. Reconstitution of apo-TRR with 1-deaza-FAD was 90% complete assuming $\varepsilon_{552} = 6.8$ mM^{-1} cm^{-1}(7) and had 6.8% of the activity of native enzyme. Binding of 1-deaza-FAD to apoenzyme is accompanied by spectral changes which are of the same nature as those of FAD bound to TRR (i.e., vibronic resolution of the main peak).

The E_m of 1-deaza-FAD bound to apo-TRR is 0.07 V more negative than that of the disulfide at pH 7.6. Thus, the reduction of 1-deaza-FAD TRR is expected to be a 2-step process, reduction of the disulfide followed by reduction of the 1-deaza-FAD. An anaerobic dithionite titration of 1-deaza-FAD is shown in Figure 5. The reduction is composed of two stages. The first stage shows only a small loss of absorbance at 552 nm and an isosbestic at 475 nm. The second stage shows nearly complete bleaching of the 552 nm peak and an isosbestic at 426 nm. The inset to Figure 5 shows that each stage is a 2-electron reduction. The experiment shown in Figure 6 shows the effect of pH on the spectrum of 2-electron reduced 1-deaza-FAD TRR. The pH was obtained indirectly in a model experiment. The spectral changes are of the same nature as the changes associated with reduction by the first 2 electrons. This behavior implies that 2-electron reduced enzyme is a pH-dependent equilibrium mixture of two species. The dashed spectrum in Figure 6 is a

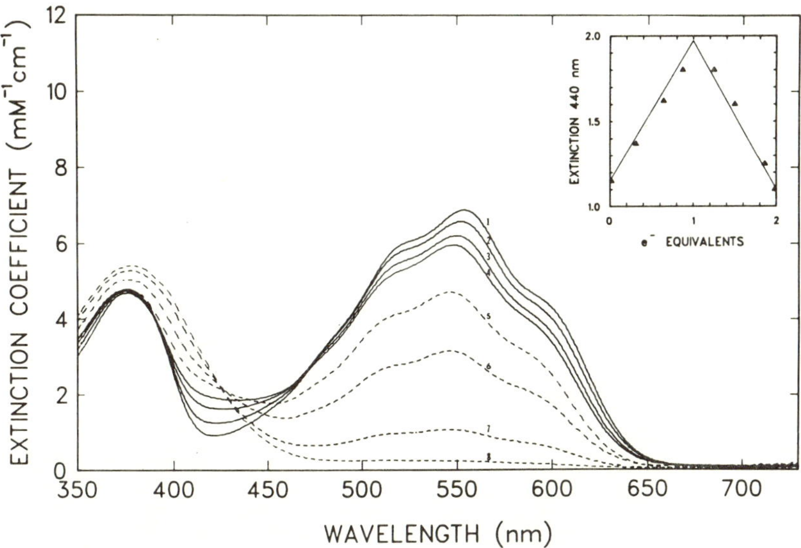

Figure 5. Anaerobic reduction of 1-deaza-FAD thioredoxin reductase with sodium dithionite. (1) 23.6 μM enzyme in 0.1 M phosphate buffer, 0.3 mM EDTA, pH 7.6, 12°. Methyl viologen, 0.47 μM was added to facilitate reduction. Spectra were recorded after the addition of dithionite: (2) 0.30, (3) 0.62, (4) 0.81, (5) 1.25, (6) 1.50, (7) 1.86, (8) 2.00 mol/mol enzyme FAD. All spectra have been corrected for semiquinone. The solid spectra indicate reduction up to 1.0 equivalent. The dashed spectra indicate reduction by more than 1.0 equivalent. Inset: Relationship of extinction at 440 nm to equivalents of dithionite added.

computer extrapolation of the spectral changes, assuming that the modified species has an $\varepsilon_{552} = 0.6$ mM^{-1} cm^{-1}(8). This spectrum is similar to C-4a adducts of 1-deaza-FAD. The equilibrium mixture of 2-electron reduced 1-deaza-FAD TRR is shown below:

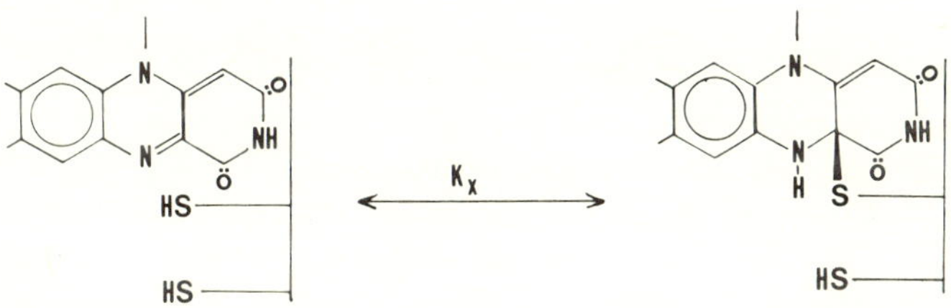

The dependence of putative C-4a adduct formation on pH is shown in the inset to Figure 6. The equilibrium does not predict a dependence on pH. The value of K_x may be influenced by the ionization behavior of a nearby base.

FAD Fluorescence of Thioredoxin Reductase

The fluorescence of the FAD of TRR was measured at different values of pH (Figure 7). The dependence of fluorescence on pH suggests the presence of a

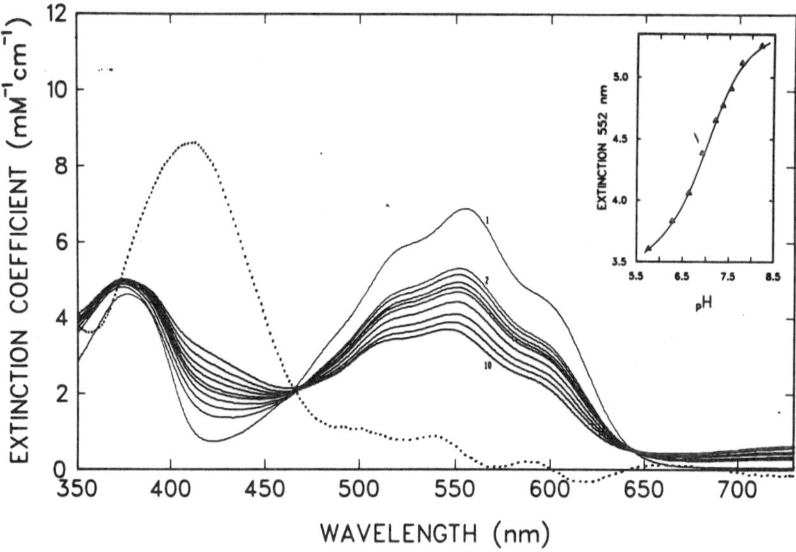

Figure 6. Effect of pH on partially reduced 1-deaza-FAD thioredoxin reductase. (1) 26.2 μM enzyme in 0.01 M phosphate buffer, 0.3 mM EDTA, pH 8.22, 12°, (2) 1.26 equivalents of sodium dithionite. The solution was then titrated anaerobically with 0.5 M acetic acid pH 3.0. Spectra were recorded at the following values of pH; curves 3 through 10: (3) pH 7.72, (4) pH 7.55, (5) pH 7.37, (6) pH 7.21, (7) pH 6.92, (8) pH 6.64, (9) pH 6.28, (10) pH 5.77. The dotted spectrum is an estimate of the modified 1-deaza-FAD species assuming $\varepsilon_{552} = 0.6$ mM^{-1} cm^{-1} and after correcting for semiquinone. Inset: Effect of pH on the absorbance at 552 nm.

base near the FAD with a pK similar to the pK associated with 1-deaza-FAD adduct stabilization and the pK of the base on the oxidized enzyme as determined by the E_m vs pH.

Discussion

The slopes of E^1_m and E^2_m appear different so that although the couples have equal midpoint potentials at pH 8.15, E^1_m is 18 mv more negative than E^2_m at pH 6.0 (Figures 2 and 4). A possible explanation for this result is that the reduction of either the FAD or the disulfide in oxidized enzyme increases the pK of a base in TRR. The principle of microscopic reversibility implies that protonation of the putative base on oxidized enzyme will increase the E_m of the FAD and the disulfide. If the interaction between the base and the FAD is stronger than the interaction between the base and the disulfide, the redox potential of the FAD couple is expected to increase relative to the disulfide couple as the base becomes more fully protonated. This may explain the difference in slope of the E^1_m and E^2_m couples. Thus, the difference in slopes of E^1_m and E^2_m, the pH dependence of 1-deaza-FAD C-4a adduct formation (Figure 6), and the pH dependence of FAD fluorescence (Figure 7), may all result from the presence of a single base in the active site. The indicated pK of this putative base in oxidized enzyme is 7.2 and is raised to approximately 8.0

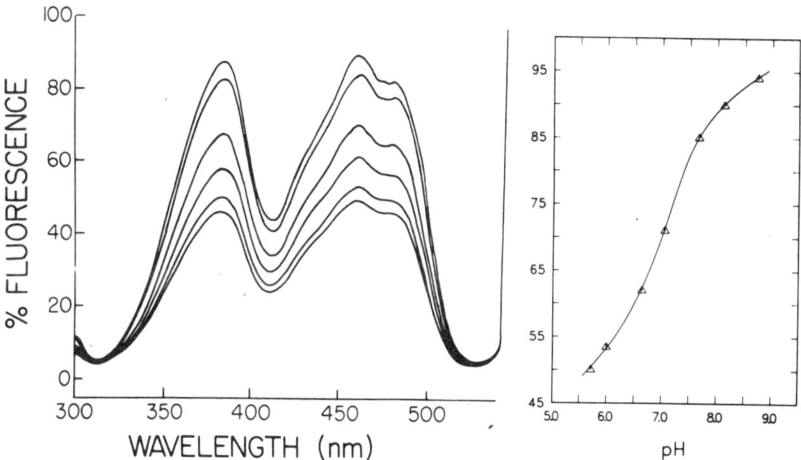

Figure 7. Effect of pH on the fluorescence excitation spectra of oxidized thioredoxin reductase. Enzyme was 8.16 μM in 0.055 M phosphate buffer, 0.15 mM EDTA, 12°.

in two-electron reduced enzyme. A base with these properties would be a good candidate for thermodynamic regulation of catalysis.

The E_m vs pH data (Figure 4) show that the disulfide takes up 2-protons upon reduction. This does not exclude formation of a thiol anion. Pig heart lipoamide dehydrogenase is another flavoprotein which contains a redox active disulfide (9). Two-electron reduction of lipoamide dehydrogenase results in the formation of a thiol anion (ca pK 5) stabilized by ion-pair interaction with a protonated base (10). Thus, protonation of the base is concomitant with formation of the thiol anion. The E_m vs pH for the disulfide of lipoamide dehydrogenase shows a 2-proton stoichiometry over the range of pH from 5.5 to 7.6 (11).

Reduction of 1-deaza-FAD TRR yielded spectral evidence suggestive of a C-4a covalent adduct presumably involving one of the active site thiols (Figure 6). Model studies have suggested that electron transfer between thiolate and the FAD occurs at the C-4a position of the isoalloxazine ring (12–14). Evidence for the FAD C-4a species has been presented in specifically modified lipoamide dehydrogenase (15).

ACKNOWLEDGMENTS
This work has been supported by the Medical Research Service of the Veterans Administration and in part by Grant GM-21444 from the National Institute of General Medical Sciences, Public Health Service. We are indebted to Dr. Colin Thorpe for suggestions pertaining to the preparation of apo-TRR and anaerobic changes of pH and to Dr. Vincent Massey for supplying the 1-deaza-FAD and suggesting its use in an anaerobic reduction of 1-deaza-FAD TRR.

References

1. Zanetti, G. and Williams, C.H., Jr. (1967) *J Biol Chem* 242:5232–5236.
2. Moore, E.C., Reichard, P., and Thelander, L. (1964) *J Biol Chem* 239:3445–3452.

3. Ronchi, S. and Williams, C.H., Jr. (1972) *J Biol Chem* 247:2083–2086.
4. Semenow-Garwood, D. and Garwood, D.C. (1972) *J Org Chem* 37:3804–3810.
5. Zanetti, G., Williams, C.H., Jr., and Massey, V. (1968) *J Biol Chem* 243:4013–4019.
6. Clark, W.M. (1960) *Oxidation-Reduction Potentials of Organic Systems.* Baltimore: The Williams and Wilkins Co.
7. Spencer, R., Fisher, J., Walsh, C. (1977) *Biochemistry* 16:3586–3593.
8. Entsch, B., Husain, M., Ballou, D.P., Massey, V., and Walsh, C. (1980) *J Biol Chem* 255:1420–1429.
9. Massey, V. and Veeger, C. (1961) *Biochim Biophys Acta* 48:33–47.
10. Matthews, R.G., Ballou, D.P., Thorpe, C., and Williams, C.H., Jr. (1977) *J Biol Chem* 252:3199–3207.
11. Matthews, R.G. and Williams, C.H., Jr. (1976) *J Biol Chem* 251:3956–3964.
12. Gascoigne, I.M. and Radda, G.K. (1967) *Biochim Biophys Acta* 131:498–507.
13. Loechler, E.L. and Hollocher, T.C. (1975) *J Am Chem Soc* 97:3235–3237.
14. Yokoe, I. and Bruice, T.C. (1975) *J Am Chem Soc* 97:450–451.
15. Thorpe, C. and Williams, C.H., Jr. (1976) *J Biol Chem* 251:7726–7728.

PART II B:

Carbon-Carbon Transhydrogenases

Published 1982 by Elsevier North Holland, Inc.
Vincent Massey and Charles H. Williams, Editors
Flavins and Flavoproteins

CHAPTER 26

Purification and Properties of Methylenetetrahydrofolate Reductase from Pig Liver

S. Colette Daubner and Rowena G. Matthews

Department of Biological Chemistry and Biophysics Research Division, The University of Michigan, Ann Arbor, Michigan

Methylenetetrahydrofolate reductase from pig liver is a flavoprotein which catalyzes the following physiological reaction:

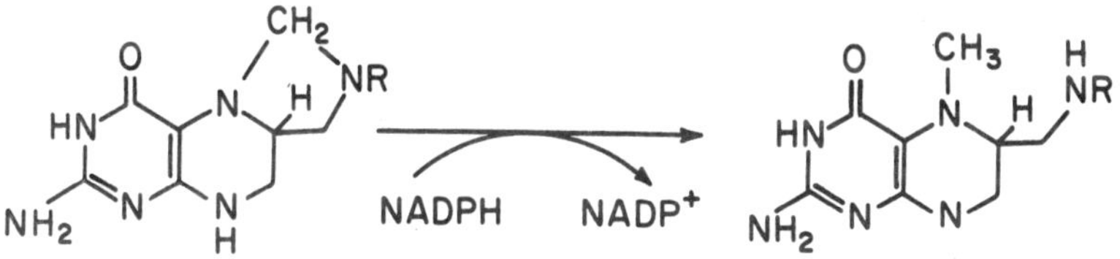

The side chain on CH_2-H_4folate[1] designated by R consists of p-aminobenzoyl-(glutamate)$_n$ and may contain up to 7 glutamyl residues linked by peptide bonds involving the γ-carboxylates of the glutamyl residues. Donaldson and Keresztesy (1) first identified the enzyme and Kutzbach and Stokstad (2) established the identity of the physiological substrates and products and identified AdoMet as an allosteric inhibitor of the enzyme. Since methylenetetrahydrofolate reductase catalyzes an effectively irreversible physiological reaction (3) which commits one-carbon units to the pathways of AdoMet-dependent biological methylations, inhibition of the enzyme by AdoMet can be considered as a typical example of feedback inhibition. Dihydrofolate and its polyglutamyl analogues are also inhibitors of the enzymatic activity (4) and are competitive with respect to CH_2-H_4folate and uncompetitive with respect to NADPH. H_2PteGlu$_1$ (dihydrofolate) is the most potent inhibitor of the monoglutamyl folate derivatives studied, and has a K_i of 6.5 μM. The K_i for H_2PteGlu$_n$ inhibition decreases as the polyglutamate chain length increases, reaching a minimum value of 0.013 μM for H_2PteGlu$_6$ (5). Since methylenetetrahydrofolate reductase is a possible control point in the distribution of one carbon units to the pathways of purine biosynthesis, thymidylate synthesis and AdoMet-dependent methylations (Figure 1), inhibition of the enzyme by

[1]Abbreviations used are: CH_2-H_4folate = methylenetetrahydrofolate; AdoMet = S-adenosylmethionine; H_2PteGlu$_n$ = dihydropteroylpolyglutamate with n glutamyl residues; CH_3-H_4folate = methyltetrahydrofolate; deazaflavin = 5-carba-5-deaza-3,10-dimethylisoalloxazine, SDS = sodium dodecyl sulfate.

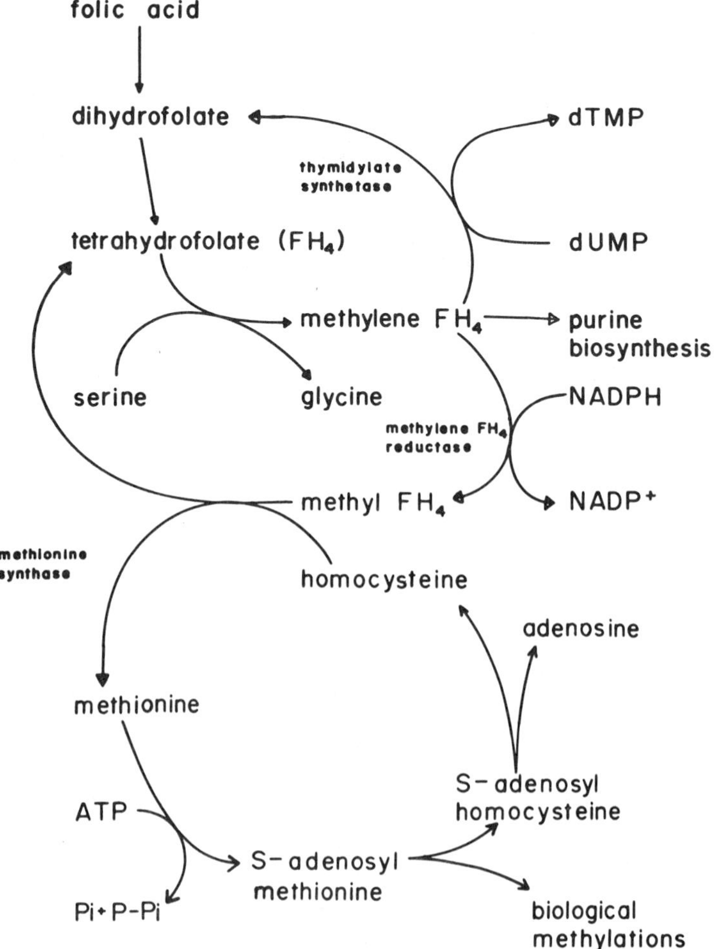

Figure 1. Pathways of mammalian folate metabolism.

$H_2PteGlu_n$ derivatives may be physiologically important. Cellular conditions which lead to a rise in steady-state levels of $H_2PteGlu_n$ and a depletion of CH_2-$H_4PteGlu_n$, e.g., enhanced rates of thymidylate biosynthesis as a cell initiates replication, may also lead to inhibition of methylenetetrahydrofolate reductase, thus sparing CH_2-H_4folate for purine and pyrimidine biosynthesis.

Studies on the inhibition of methylenetetrahydrofolate reductase have also provided insights into the enzyme's catalytic mechanism. Mechanistic work on the reductase has centered on examining the various possibilities shown in Scheme 1. The first step shows a ring-opening as N_{10} of CH_2-H_4folate is protonated and the bond between N_{10} and the methylene carbon is broken. This step results in the formation of a 5-iminium cation (6,7). The 5-iminium cation can now react by one of the pathways indicated (A, B, or C). Pathway A

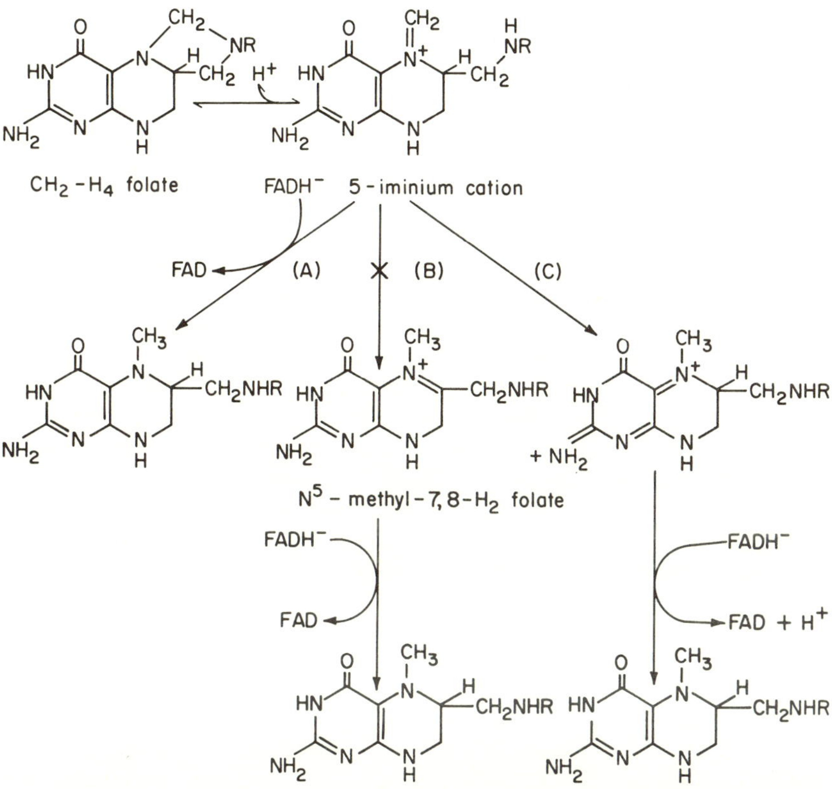

Scheme 1.

involves direct reduction of the exocyclic double bond of the iminium cation with no involvement of the electrons in the tetrahydropterin ring system. The second pathway (B) involves a shift of the double bond from the N_5-exocyclic position to the 5,6-endocycylic position, followed by reduction of the endocyclic double bond. This possibility was considered unlikely after experiments showed that tritium label at C6 was neither released during the course of catalysis, nor transferred to the nascent 5-methyl substituent, and that no isotopic discrimination against tritiated substrate was associated with formation of product (4). Pathway C is most probably the mechanism by which enzyme-catalyzed reduction of CH_2-H_4folate occurs. Consistent with this mechanism is the observation that methylenetetrahydrofolate reductase can reduce quinonoid dihydropterin derivatives which lack a 5-substituent, but which are otherwise analogous in structure to the intermediate depicted in pathway C, and that the rates of reduction of such quinonoid dihydropterins are comparable to the rate of reduction of CH_2-H_4folate (8). In the presence of a suitable electron acceptor, such as menadione, rapid oxidation of CH_3-H_4folate is catalyzed by the reductase. Such facile oxidation is readily understood if it proceeds according to pathway C. Oxidation of the tetrahydropterin ring by the enzyme-bound flavin leads to the introduction of α, β unsaturation adjacent to the 5-methyl substituent of CH_3-H_4folate, facilitating removal of a methyl group hydrogen as a proton and subsequent resonance stabilization of the ylide, which is a canonical form of the 5-iminium cation. Such a mechanism also exploits the chemical similarities between dihydroflavins and tetrahydrofolates, both of which are subject to facile oxidations. In this case, the oxidoreduction of the tetrahydrofolate ring system (while not apparent from the overall reaction stoichiometry) serves to stabilize intermediates in the reduction of the exocyclic methylene group.

We hope that our development of a scheme to purify pig liver methylenetetrahydrofolate reductase will allow further elucidation of the catalytic mechanism and mode of regulation of the enzyme. Our method of purification provides us with homogeneous enzyme, which is both active and stable, although only modest quantities are obtained due to the scarcity of the enzyme in mammalian tissues. We estimate that an adult pig liver weighing 1.5 kg contains 3.8 mg of enzyme, and the isolation of homogeneous enzyme requires a 32,000-fold purification.

Our isolation procedure is described in Table 1. We have listed total units of activity as determined by both NADPH-menadione oxidoreductase assay (column B) and by CH_3-H_4folate-menadione oxidoreductase assay (column A). The latter assay utilizes [methyl-^{14}C]-H_4folate and is suitable for use in very crude samples, while the spectrophotometric assay is less specific and is only used after purification of the enzyme on DEAE-52. We report both activities to document their copurification after this stage in the procedure; note that their ratio does not change with further purification. We believe there to be no contaminating NADPH-menadione oxidoreductases present after the elution from DEAE-52.

Table 1. Purification of Methylenetetrahydrofolate Reductase from Pig Liver.

Step	Activity units (A)	Activity units (B)	Ratio A/B	Protein mg	Specific activity (assay A)	Yield
Supernatant from pH 5.9 homogenization and centrifugation	146	ND	ND	239,000	0.00061	100%
Batch chromatography on DEAE-52	145	621	0.23	12,900	0.0112	99%
Chromatography on Amicon Matrex Gel Blue A	72.1	314	0.23	217	0.332	49%
Chromatography on DEAE-Sephadex A-50	34.3	137	0.25	2.6	13.2	23%
Chromatography on Sephacryl S-300	20.6	81	0.25	1,06	19.4	14%

Activity A: CH_3-H_4folate-menadione oxidoreductase activity, units are μmoles CH_3-H_4folate oxidized per minute. Activity B: NADPH-menadione oxidoreductase activity, units are μmoles NADPH oxidized per minute. The reaction is monitored at 343 nm where menadione and menadiol are isosbestic. ND = not determined.

The addition of 10% glycerol to the isolation buffers used after the DEAE-52 step was found to be essential for the maintenance of stable pure enzyme and for the attainment of satisfactory yields at intermediate steps. The addition of detergents such as Triton-X100 or Brij 58 did not have as powerful a stabilizing effect.

Protein concentrations were determined by using the dye-binding assay mix marketed by Bio-Rad Laboratories. Assays of enzyme samples were compared with standard curves based on reaction of the protein assay mixture with bovine serum albumin. Pure enzyme samples subjected to amino acid analysis were found to contain 76.5% of the protein concentration determined by the Bio-Rad assay, so this factor was used to correct the dye-binding assay values for enzyme samples purified through the DEAE-Sephadex and Sephacryl chromatography steps.

Our enzyme preparation was shown to be pure using several criteria. Only one band was seen after electrophoresis of samples of the purified enzyme on polyacrylamide gels containing SDS (9). During the final chromatography on Sephacryl S-300, the eluting fractions were analyzed for activity and protein content. Constant specific activities were observed for the peak fractions. Finally, amino acid analyses gave a molecular weight per flavin which agreed well with the subunit molecular weight obtained by electrophoresis on polyacrylamide gels. When purified enzyme was electrophoresed in the presence of SDS on a 10% polyacrylamide gel, an apparent subunit molecular weight of 77,300 was found (Figure 2). The molecular weight per flavin obtained by amino acid analyses was 74,500. This value does not include contributions to the molecular weight per flavin from tryptophanyl and cysteinyl residues, which were not determined. The agreement between the subunit molecular weight and the molecular weight per flavin suggests to us not only that the enzyme has a subunit molecular weight of about 78,000 (including FAD), but also that each subunit of the native enzyme, as isolated, contains bound FAD.

When methylenetetrahydrofolate reductase from any stage of purification is chromatographed on a Sephacryl S-300 column which has been calibrated with standard proteins, an apparent holoenzyme molecular weight of 210,000 is obtained (Figure 3). This value is not a clear multiple of 78,000 and we are not sure whether the native enzyme exists as a trimer or an ellipsoidal dimer. We have shown that the enzyme does not contain bound carbohydrate since the subunit does not stain for glycoprotein following electrophoresis in the presence of sodium dodecyl sulfate (10, 11).

The visible absorption spectrum of the purified enzyme appears in Figure 4. The enzyme is clearly a flavoprotein, with peaks at 450 and 380 nm, a shoulder at 475 nm and troughs at 407 and 330 nm. Reduced enzyme has also been prepared by photoreduction in the presence of deazaflavin and EDTA (not shown). The isosbestic point for the reduction is at 340 nm and no semiquinone intermediate is seen in the course of the reduction.

The flavin cofactor of methylenetetrahydrofolate reductase was shown to be FAD by the method of Wassink and Mayhew (12). The extinction coefficient of the enzyme-bound FAD at 450 nm was 12,100 M^{-1} cm^{-1}, as determined by the method of Thorpe et al. (13).

Kinetic parameters were determined for the physiological reaction, NADPH-linked reduction of CH_2-H_4folate (4). Double reciprocal plots of

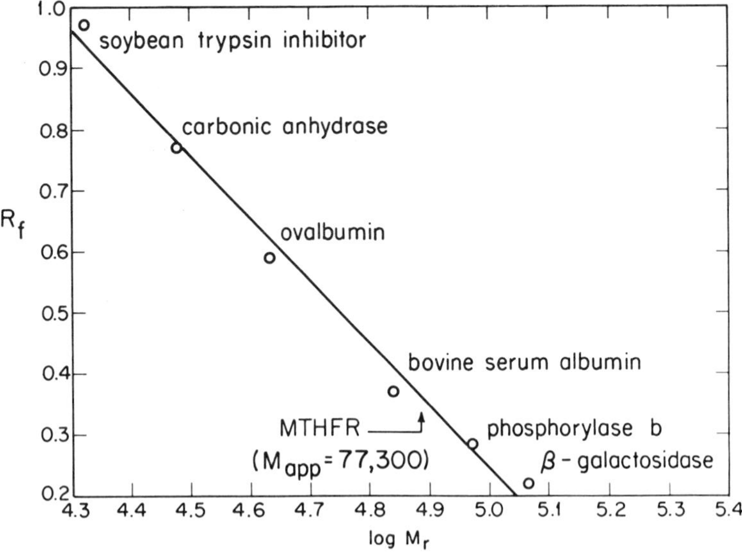

Figure 2. Determination of the subunit molecular weight of methylenetetrahydrofolate reductase by polyacrylamide gel electrophoresis in the presence of sodium dodecyl sulfate. A sample of purified enzyme was dialyzed to remove salt. Enzyme and standard were denatured with 1% sodium dodecyl sulfate, 1% 2-mercaptoethanol, by heating at 70° for 20 minutes, before separation on a 10% polyacrylamide slab gel. After electrophoresis, gels were stained for protein with Coomassie Blue. The migration distance for each standard named above relative to the migration of bromophenol (R_f) was plotted versus the log of the subunit molecular weight. The R_f for the single band of methylenetetrahydrofolate reductase was compared to the standard curve.

velocity versus varied CH_2-H_4folate at different NADPH concentrations were parallel and the secondary plot of intercepts versus reciprocal NADPH concentrations was linear. The kinetic parameters obtained were: V_{max}, 1600 mol NADPH oxidized per min per mol enzyme-bound FAD; K_{NADPH}, 16 μM; and $K_{CH_2\text{-}H_4folate}$, 19 μM. The substrate used for these kinetic studies was enzymatically prepared, chromatographically purified (6-R)-CH_2-H_4folate (4), the active isomer for methylenetetrahydrofolate reductase (14). Parallel lines in the primary plot are consistent with a ping-pong mechanism for methylenetetrahydrofolate reductase. This mechanism is also supported by other findings. When enzyme purged of oxygen is reacted with deoxygenated NADPH in the absence of CH_2-H_4folate, full reduction of the flavin is observed on addition of one mol of NADPH per mol of flavin. Reduced enzyme, prepared by photoreduction in the presence of deazaflavin and EDTA, is rapidly reoxidized by (6-R)-CH_2-H_4folate, and the extent of reoxidation is nearly stoichiometric with the amount of CH_2-H_4folate added. Figure 4 shows the reaction of oxidized enzyme with (6-S)-CH_3-H_4folate under anaerobic conditions. After the addition of 25 equivalents of CH_3-H_4folate the enzyme was about 73% reduced. These experiments demonstrate that the pyridine nucleotide substrate can react with the enzyme-bound flavin in the absence of the folate substrate and *vice versa*.

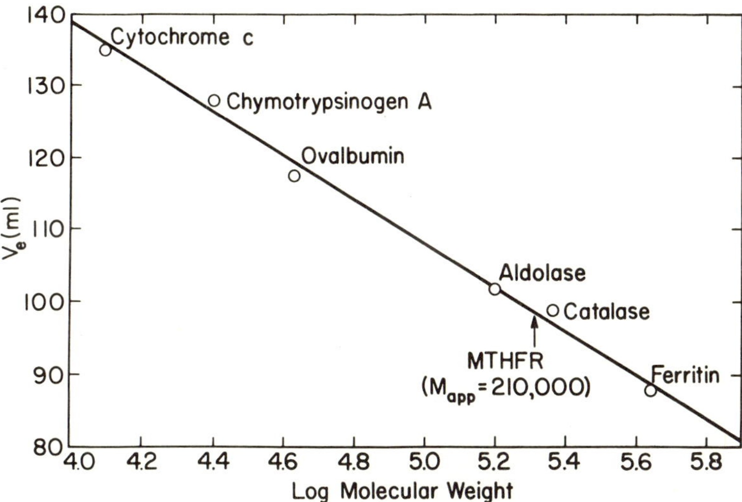

Figure 3. Chromatography of methylenetetrahydrofolate reductase on a calibrated Sephacryl S-300 column. Each standard protein named above was applied to a Sephacryl S-300 column and the elution volume (V_e) determined. The calibration curve of V_e versus the log M_r (apparent molecular weight based on Stoke's radius) was plotted. The V_e for purified methylenetetrahydrofolate reductase was determined and compared to the standard curve.

Figure 4. Reduction of methylenetetrahydrofolate reductase by CH_3-H_4folate. Enzyme (4.47 nmol FAD) was titrated with aliquots of 1.01 mM (6-S)-CH_3-H_4folate under anaerobic conditions. Curve 1: oxidized enzyme; Curves 2-12: after addition of specified amounts of CH_3-H_4folate—curve 2: 12.1 nmol; curve 3: 14.1 nmol; curve 4: 16.2 nmol; curve 5: 18.2 nmol; curve 6: 22.2 nmol; curve 7: 26.3 nmol; curve 8: 32.3 nmol; curve 9: 40.4 nmol; curve 10: 48.5 nmol; curve 11: 64.6 nmol, and curve 12: 88.9 nmol. Addition of the first 10.1 nmol CH_3-H_4folate resulted in reoxidation by residual oxygen. The inset shows A_{450} changes, corrected for dilution, resulting from the successive additions of CH_3-H_4folate.

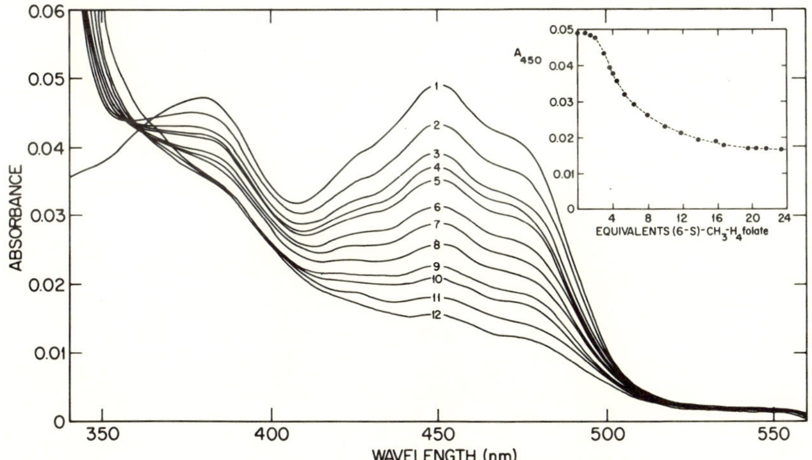

Turnover numbers were determined for the NADPH-menadione oxidoreductase assay (5900 per minute under standard assay conditions), and for the CH_3-H_4folate-menadione oxidoreductase assay (1500 per minute under standard assay conditions).

In summary, we have developed a method for the isolation of homogeneous methylenetetrahydrofolate reductase which can be used to obtain sufficient protein for studies involving substrate levels of enzyme and we have reported initial characterizations of the enzyme's physical and catalytic properties.

ACKNOWLEDGMENTS

This work has been supported in part by Grant GM 24908 from the Institute of General Medical Sciences, National Institutes of Health, by Training Grant GM 00187 (SCD) from the Institute of General Medical Sciences, and by an Established Investigatorship of the American Heart Association (RGM).

References

1. Donaldson, K.O. and Keresztesy, J.C. (1959) *J Biol Chem* 234:3235–3240.
2. Kutzbach, C. and Stokstad, E.L.R. (1971) *Biochim Biophys Acta* 250:459–477.
3. Katzen, H.M. and Buchanan, J.M. (1965) *J Biol Chem* 240:825–835.
4. Matthews, R.G. and Haywood, B.J. (1979) *Biochemistry* 18:4845–4851.
5. Matthews, R.G. and Baugh, C.M. (1980) *Biochemistry* 19:2040–2045.
6. Kallen, R.G. and Jencks, W.P. (1966) *J Biol Chem* 241:5851–5861.
7. Benkovic, S.J. and Bullard, W.P. (1973) *Prog Bioorg Chem* 2:133–175.
8. Matthews, R.G. and Kaufman, S. (1980) *J Biol Chem* 255:6014–6017.
9. Laemmli, U.K. (1970) *Nature* 227:680–685.
10. Zacharius, R.M., Zell, T.E., Morrison, J.H., and Woodlock, J.J. (1969) *Anal Biochem* 30:148–152.
11. Racusen, D. (1979) *Anal Biochem* 99:474–476.
12. Wassink, J.H. and Mayhew, S.G. (1975) *Anal Biochem* 68:609–616.
13. Thorpe, C., Matthews, R.G., and Williams, C.H., Jr. (1979) *Biochemistry* 18:331–337.
14. Fontecilla-Camps, J.C., Bugg, C.E., Temple, C., Jr., Rose, J.D., Montgomery, J.A., and Kisliuk, R.L. (1979) *J Am Chem Soc* 101:6114–6115.

PART II C:

Dehydrogenase – Oxidase Flavoproteins

Published 1982 by Elsevier North Holland, Inc.
Vincent Massey and Charles H. Williams, Editors
Flavins and Flavoproteins

CHAPTER 27

Thermodynamic Studies on the Self-Associating System of D-Amino Acid Oxidase

Hiromasa Tojo, Kihachiro Horiike,** Kiyoshi Shiga,* Yasuzo Nishina,* Retsu Miura, Hiroshi Watari,* and Toshio Yamano

*Department of Biochemistry, Osaka University Medical School, Osaka Japan; *National Institute for Physiological Sciences, Okazaki, Japan, and **Department of Biochemistry, Shiga University of Medical Science, Shiga, Japan*

Summary

The self-association of subunits of hog kidney D-amino acid oxidase was investigated by low-angle laser light scattering at 25°C, and pH 8.3. The stoichiometry of the association reaction was found to be very complex. However, it is likely that the class of indefinite self-associations approximately gives a description of the observed system.

Introduction

The self-association system of hog kidney D-amino acid oxidase (DAO) has been extensively studied by various methods (1, and references cited therein). However, there are some discrepancies with regard to the stoichiometry of the system obtained from the measurement of molecular weight (2–4). The reproducibility of molecular weight determinations performed previously was not always good (2,3). In the present study, the concentration dependence of the apparent weight-average molecular weight of DAO was determined by low-angle laser light scattering method, and was analyzed according to several equilibrium models.

Materials and Methods

Hog kidney DAO (5,6) and its apoenzyme (7) were purified as described previously. The enzyme concentrations were determined by the $A^{1\%}_{1cm} = 18.1$ at 280 nm for the apoenzyme and the $A^{1\%}_{1cm} = 2.75$ at 455 nm for the holoenzyme, obtained in this study by dry weight measurements. Light scattering measurements were performed with a KMX-6 low-angular laser light scattering photometer (Chromatix). The 6-7 degree annulus was used. The average scattering angle observed was 4.86 degrees. Refractive index increments were determined to be 0.186±0.002 ml/g at 632.8 nm by a KMX-16 differential refractometer (Chromatix) at 25.0±0.07°C. The densities of the solvents and enzyme solutions were measured with a digital precision density meter DMA 602-60 (Anton Parr) at 25.0±0.01°C. Sedimentation experiments were performed with a Hitachi 282 analytical ultracentrifuge using photoelectric scanner optics. Enzyme solutions were prepared so as to satisfy the requirements of the thermodynamic theory of multicomponent solutions (8).

Results and Discussion

Monomer Molecular Weight

The monomer molecular weight of DAO was determined from low- and high-speed sedimentation equilibrium experiments in 5.9 M guanidine-HCl solution. In the three-component system, the effect of preferential interactions between solvent components and protein was critical. The apparent partial specific volume, ϕ_2', at constant chemical potential of each diffusible component was determined to be 0.723±0.005 ml/g. Then, the molecular weight of the subunit was calculated to be 38000±1200.

Light Scattering Studies

The conventional plots of Kc/R_θ vs c are shown in Figure 1. The ordinate corresponds to the reciprocal of the apparent weight-average molecular weight (M_w^a). Both in the holoenzyme and apoenzyme, the value of $1/M_w^a$ decreases monotonically upon increasing the enzyme concentration, and the extrapolated values of M_w^a to zero concentration were practically identical to the monomer molecular weight. In particular, since the apoenzyme appreciably dissociates into its monomer at low concentrations, the extrapolation procedure is very reliable. The data from different experiments (with different enzyme preparations) are included in Figure 1. Each point reproducibly lies on the single curve with statistical scatter. These findings indicate that the oxidase is in a reversibly associating system.

Analysis of Self-Associating System

Henn and Ackers (4) proposed a monomer-dimer model for the apoenzyme in the low concentration range (<0.9 mg/ml). This model was tested based on Equation (27.1) (9),

$$1/M_w^a = \left[1+(1+4kc)^{-1/2}\right]/2M_1 + Bc \qquad (27.1)$$

where M_1 is the monomer molecular weight, k the association constant based on c (mg/ml)-scale, and B the second virial coefficient. The values of k and B for apoenzyme were calculated to be 1.7 ml/mg, -1.6×10^{-6} mole·l/g^2, respectively. This large negative virial coefficient is unreasonable and the regenerated $1/M_w^a$ vs c plot gave a very poor fit to the data. Another monomer-n-mer (n>2) model did not fit the data, which was judged by the graphical analysis of Stafford (10). In the holoenzyme also, no monomer-n-mer model did fit the data.

We considered the following reaction model, $2\ M_1 \rightleftharpoons M_2$, $M_2+M_1 \rightleftharpoons M_3$, —, $M_i+M_1 \rightleftharpoons M_{i+1}$, —. Assuming that each step has the same molar association constant K_0, this model (isodesmic indefinite self-association) was analyzed

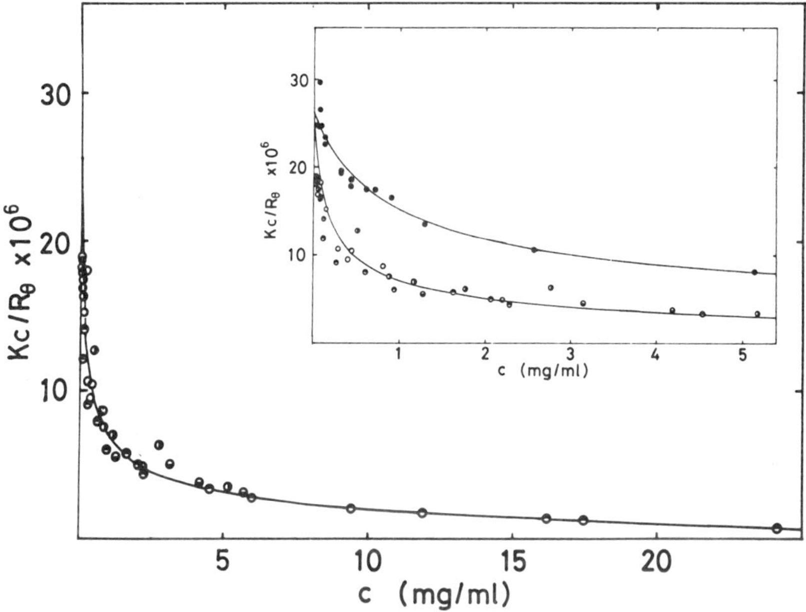

Figure 1. Light scattering data for D-amino acid oxidase in 0.1 M sodium pyrophosphate (pH 8.3) at 25°C. All the experiments for the holoenzyme were performed in the presence of FAD (50 μM.). Samples were: apoenzyme (●); holoenzyme dialyzed against the buffer in the presence of FAD (11 μM) (○); holoenzyme eluted from a Sephadex G-50 column just before measurement (◑,◒,◓). Different symbols represent different experiments. The solid lines are theoretical curves calculated on the basis of indefinite self-association model.

by the methods of both Van Holde and Rossetti (11, Equation (27.2) and Stafford (10).

$$R_a = M_w^a / M_1, k_0 = K_0 / M_1$$

$$k_0 = \left[R_a^2/(1 - BM_1R_ac)^2 - 1\right]/4c \qquad (27.2)$$

In this case, the association constant, k_0, defined by weight concentration (mg/ml) was used. The values of k_0 and B were determined to be 3.0 ml/mg and -3.2×10^{-8} mole$\cdot$1/g^2 for holoenzyme, respectively and 0.51 ml/mg and 4.6×10^{-8} mole$\cdot$1/g^2 for apoenzyme, respectively. The plot of $1/M_w^a$ vs c calculated from k_0 and B fairly fit the data (Figure 1). It is needless to say that this is not the only model that fit the data, since the holoenzyme exists predominantly in a high polymerized form so that it is very difficult to discriminate one model from others (e.g., discrete or attenuated equilibrium constant model for indefinite self-association) (12).

References

1. Tojo, H., Horiike, K., Shiga, K., Nishina, Y., Watari, H., and Yamano, T. (1980) *FEBS Lett* 114:4.
2. Charlwood, P.A., Palmer, G., and Bennett, R. (1961) *Biochim Biophys Acta* 50:17.

3. Antonini, E., Brunori, M., Bruzzesi, M.R., Chiancone, E., and Massey, V. (1966) *J Biol Chem* 241:2358.
4. Henn, S.W. and Ackers, G.K. (1969) *Biochemistry* 8:3829.
5. Curti, B., Ronchi, S., Branzoli, U., Ferri, G., and Williams, C.H., Jr. (1973) *Biochim Biophys Acta* 327:266.
6. Horiike, K., Shiga, K., Isomoto, A., and Yamano, T. (1976) *J Biochem* 80:1073.
7. Massey, V. and Curti, B. (1966) *J Biol Chem* 241:3417.
8. Casassa, E.F. and Eisenberg, H. (1964) *Adv Protein Chem* 19:287.
9. Adams, E.T., Jr. and Fujita, H. (1962) *Ultracentrifugal Analysis in Theory and Experiment.* Williams, J.W. (ed.) New York: Academic Press, p. 119.
10. Stafford, W.F., III (1980) *Biophys J* 29:149.
11. Van Holde, K.E. and Rossetti, G.P. (1967) *Biochemistry* 6:2189.
12. Adams, E.T., Jr., Wan, P.J., and Crawford, E.F. (1978) *Methods in Enzymology.* Hirs, C.H.W. and Timasheff, S.N. (eds.) New York: Academic Press, p. 69.

Published 1982 by Elsevier North Holland, Inc.
Vincent Massey and Charles H. Williams, Editors
Flavins and Flavoproteins

CHAPTER 28

Resonance Raman Study on the Reaction Intermediates of D-Amino Acid Oxidase

Retsu Miura, Yasuzo Nishina,* Kiyoshi Shiga,* Hiromasa Tojo, Hiroshi Watari,* Yoshihiro Miyake,** and Toshio Yamano

*Department of Biochemistry, Osaka University Medical School, Kita-ku, Osaka 530; *National Institute for Physiological Sciences, Myodaiji, Okazaki, Aichi 444; and **Department of Biochemistry, National Cardiovascular Center Research Institute, Suita, Osaka 565, Japan*

Summary

Resonance Raman (RR) spectra of the reaction intermediates of D-amino acid oxidase were obtained. The intermediates studied are the one in the reaction with β-cyano-D-alanine and the other in the reverse reaction with chloropyruvate and ammonium. Their RR spectra excited at 488.0 or 514.5 nm are those of oxidized flavin. The RR spectra excited at 632.8 nm exhibited resonance enhancement for the lines at 1585, 1550, and 1350 cm^{-1}, while in the latter intermediate a line unassignable to flavin was observed at 1657 cm^{-1} which shifted to 1653 cm^{-1} when ^{15}N-ammonium was used.

Introduction

We have previously observed stable intermediary charge transfer complexes in the reaction of D-amino acid oxidase (DAO) with β-cyano-D-alanine (D-BCNA) and in the reverse reaction of DAO with chloropyruvate and ammonium, and assigned the charge transfer complexes of oxidized flavin with α-amino-β-cyanoacrylate and with α-imino-β-chloropropionate to the former and the latter intermediates, respectively (1-3). Unambiguous identification of these intermediates and studies of their charge transfer interaction will inevitably be valuable for the complete understanding of DAO catalysis. In the present study, RR spectra of the intermediates mentioned above were measured and analyzed in terms of the charge transfer interaction between flavin and the substrates modified.

Materials and Methods

Porcine kidney DAO was purified according to the method reported previously (4,5). RR spectra were measured with a JASCO R-800 Raman spectrometer (Japan Spectroscopic Co.) and an argon laser (Spectraphysics model 164) or a

He-Ne laser (Kinmon Electronics, model KLG-103). The wave number calibration of the spectrometer was carried out with indene (6). All the sample solutions contained about 3% (w/w) of ammonium sulfate as an internal standard, whose Raman line at 981 cm^{-1} is marked SO_4^{2-} in the figures. All the experiments were performed in 0.1 M sodium pyrophosphate buffer, pH 8.3.

Results

RR Spectra of the Intermediate in the Reaction with D-BCNA

Aerobic incubation of DAO (0.72 mM) with D-BCNA (1.50 mM) produced the stable reaction intermediate with a charge transfer band, which is identical to that reported previously (2). The RR spectra of the intermediate in three regions are shown in Figure 1. The spectra excited at 488.0 (Figure 1a) and 514.5 nm (Figure 1b) are similar to each other and typical of oxidized flavin. The RR spectrum excited at 632.8 nm within the region of the charge transfer band is shown in Figure 1c, where only a few lines are resonance enhanced.

Figure 1. RR spectra of the reaction intermediate in the aerobic reaction of DAO (0.72 mM) with D-BCNA (1.50 mM). The excitations were at 488.0 (a), 514.5 (b), and 632.8 nm (c).

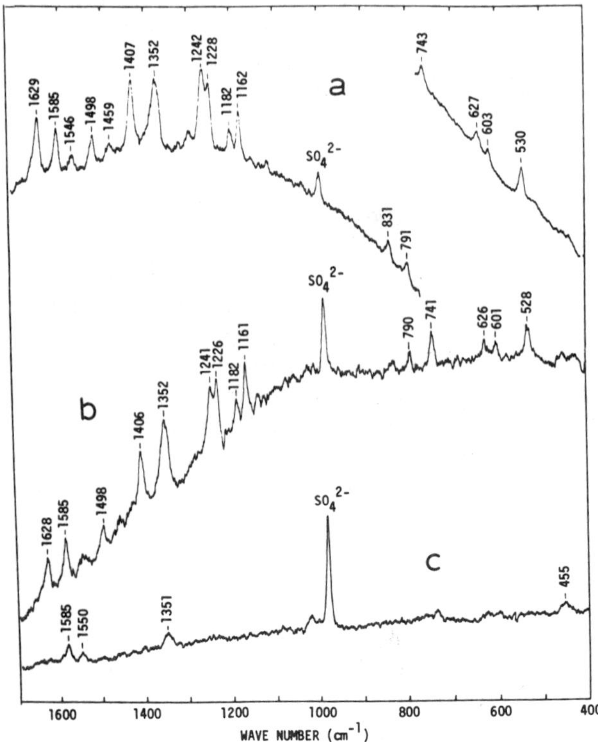

RR Spectra of the Intermediate in the Reverse Reaction with Chloropyruvate and Ammonia

The intermediate in the reverse reaction was produced by incubation of DAO (0.92 mM) with chloropyruvate (4.6 mM) and ammonium sulfate (1.5 M). Ammonium sulfate added in this case served not only as an internal standard but also as a substrate. The absorption spectrum of the intermediate was identical to that reported previously (1,3). The RR spectrum excited at 514.5 nm is shown in Figure 2a, which is similar to Figures 1a and 1b and characteristic of oxidized flavin. The RR spectrum excited at 632.8 nm within the region of the charge transfer band is shown in Figure 2b, which is similar to Figure 1c except that two additional lines at 1403 and 1657 cm^{-1} were observed. The latter line at 1657 cm^{-1} is unassignable to flavin vibrational mode and shifted to 1653 cm^{-1} (Figures 2c and 3) when ^{15}N-ammonium sulfate was used. This indicates that this line originates from a substrate

Figure 2. RR spectra of the reaction intermediate in the reverse reaction of DAO with chloropyruvate and ammonium (curves a, b, and c). Curve b is the control spectrum without substrates. The excitations were at 514.5 (a) and 632.8 nm (b, c, and d). Final concentrations of DAO were 0.92 mM (a and b), 0.74 mM (c), and 0.93 mM (d). Final concentrations of the substrates were: a, chloropyruvate (4.6 mM), ammonium sulfate (0.15 M); b, chloropyruvate (5.0 mM), ^{15}N-ammonium sulfate (0.16 M).

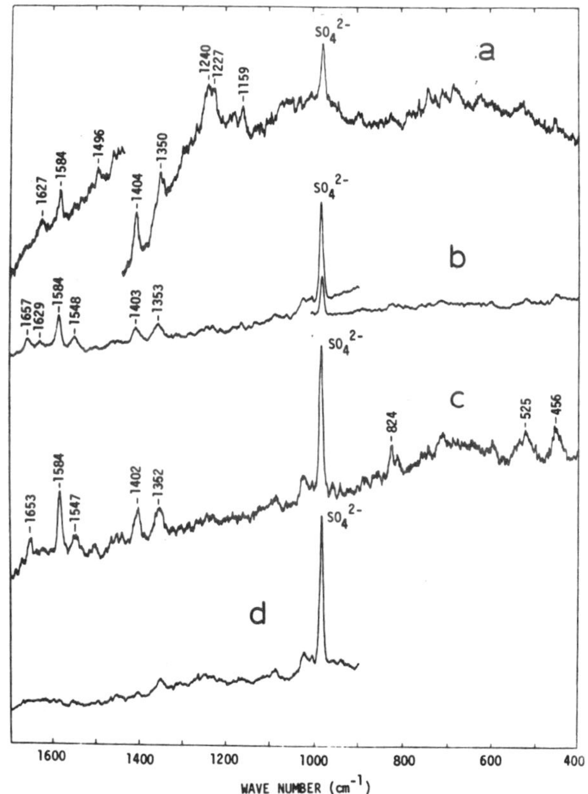

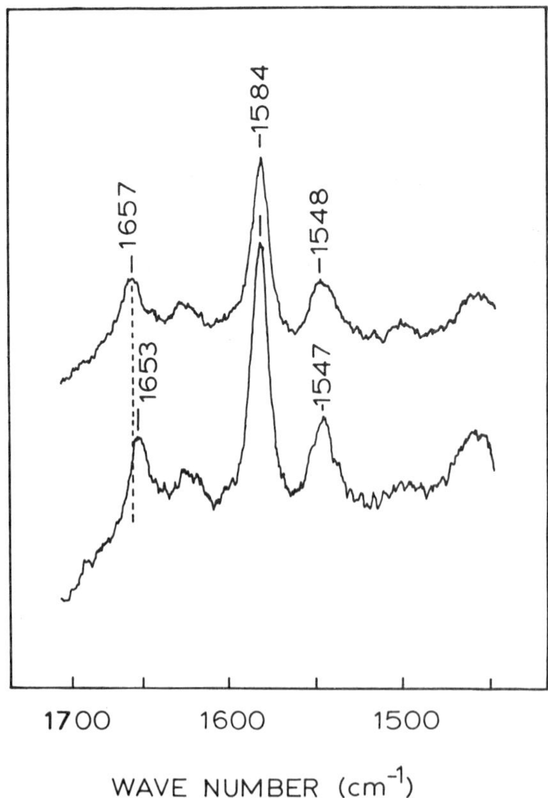

Figure 3. Effect of isotopic substitution of ammonium to ^{15}N-ammonium on the RR spectra of the reaction intermediate in the reverse reaction of DAO with chloropyruvate and ammonium. Final concentrations were: upper, DAO (0.92 mM), chloropyruvate (4.6 mM), ammonium sulfate (0.15 M); lower, DAO (0.74 mM), chloropyruvate (5.0 mM), ^{15}N-ammonium sulfate (0.16 M). Each spectrum was an average of four scans.

intermediate rather than from flavin and can be assigned to C=N stretching of α-imino-β-chloropropionate, the postulated counterpart in the charge transfer complex with oxidized flavin (1).

Discussion

The RR spectra of the intermediates excited in the region of the first π-π^* band of flavin (Figures 1a, b, and 2a) are those of oxidized flavin indicating that flavin is in the oxidized state in these intermediates. This is in agreement with our previous deduction that oxidized flavin is involved as an electron acceptor in these charge transfer complexes. The Raman line around 1240 cm^{-1} is peculiar in that it is shifted considerably to lower frequency compared to flavins free or bound to other proteins. This line was reported to be sensitive to N(3)-H···protein interaction (7,8). It may be that the flavin-protein interaction in these intermediates is distinct from that in other flavoproteins. The resonance enhancement observed in the RR spectra excited in the region of the charge transfer band (Figures 1c and 2b) offer important suggestions. The resonance enhancement in the 1585 cm^{-1} line in both spectra suggests that the charge transfer interaction involves the C(4a)-N(5) region of oxidized flavin since this line is associated with the displacement of C(4a) and N(5) atoms (8). The resonance enhancement in the line at 1657 cm^{-1} in Figure 2b and its isotopic frequency shift to 1653 cm^{-1} (Figures 2c and 3) upon usage of ^{15}N-ammonium not only substantiates the involvement of α-imino-β-chloropropioninate as the counterpart of the charge transfer complex, but also the participation of the imino group as the electron-donating site.

References

1. Miyake, Y., Matsushita, S., Miura, R., and Yamano, T. (1980) *Flavins and Flavoproteins.* Yagi, K. and Yamano, T. (eds.) Tokyo: Japan Scientific Societies Press, p. 93.
2. Miura, R., Shiga, K., Miyake, Y., Watari, H., and Yamano, T. (1980) *J Biochem* 87:1469.
3. Matsushita, S., Miura, R., Yamano, T., and Miyake, Y. (1980) *J Biochem* 88:121.
4. Shiga, K., Nishina, Y., Horiike, K., Tojo, H., Watari, H., and Yamano, T. (1980) *Medical J Osaka Univ* 30:71.
5. Curti, B., Ronchi, S., Branzoli, U., Ferri, G., and Williams, C.H., Jr. (1973) *Biochem Biophys Acta* 327:266.
6. Hendra, P.J. and Loader, E.J. (1968) *Chem Ind (London)* 718.
7. Nishina, Y., Kitagawa, T., Shiga, K., Horiike, K., Matsumura, Y., Watari, H., and Yamano, T. (1978) *J Biochem* 84:925.
8. Kitagawa, T., Nishina, Y., Kyogoku, Y., Yamano, T., Ohishi, N., Takai-Suzuki, A., and Yagi, K. (1979) *Biochemistry* 18:1804.

Published 1982 by Elsevier North Holland, Inc.
Vincent Massey and Charles H. Williams, Editors
Flavins and Flavoproteins

CHAPTER 29

Utilization of D- and L-Thiazolidine-2-Carboxylic Acids as Substrates for D-Amino Acid Oxidase

Paul F. Fitzpatrick and Vincent Massey

Department of Biological Chemistry, University of Michigan Ann Arbor, Michigan

Recently, Hamilton et al. proposed that the true substrate for hog kidney D-amino acid oxidase was thiazolidine-2-carboxylic acid, formed from cysteamine and glyoxylate as shown in Scheme 1 (1). We have examined the properties of such a cysteamine-glyoxylate mixture as a substrate for D-amino acid oxidase. When we determined the stoichiometry of the oxidation of racemic thiazolidine-2-carboxylic acid at 19°C, the results shown in Figure 1 were obtained. The racemic mixture was completely oxidized, suggesting that both the D- and L-isomers are substrates for D-amino acid oxidase.

The straightforward way to determine if this were indeed the case would have been to synthesize D- and L-thiazolidine-2-carboxylic acid and test each separately as a substrate. We were able to synthesize thiazolidine-2-carboxylic acid ethyl ester and resolve the D- and L-isomers (2). However, upon hydrolysis the free amino acids racemized quite rapidly. We measured the rate of this racemization by generating free D- or L-thiazolidine-2-carboxylic acid from the ester in the polarimeter with pig liver esterase. Direct measurement of the rate of racemization showed that both isomers lost optical activity at 20°C, pH 8.5, with a half-time of 3.3 minutes. At 4°C, the half-time was 52 minutes. The effect of lowering the temperature on the stoichiometry is shown in Figure 2. At 2°C, one-half of the thiazolidine-2-carboxylic acid was rapidly consumed, followed by a very slow oxidation of the remainder.

We next determined whether this slow phase could be completely accounted for by racemization of L-thiazolidine-2-carboxylic acid to D-thiazolidine-2-carboxylic acid, by comparing the rate of oxidation of the L-isomer with the rate of isomerization as a function of temperature. At temperatures above 15°C, the rate of oxidation was measured as turnover with an oxygen electrode.

Scheme 1. Formation of thiazolidine-2-carboxylic acid from cysteamine and glyoxylate.

$$NH_2\text{-}CH_2\text{-}CH_2\text{-}SH + CHO\text{-}COO^- \rightleftharpoons H_2C(SH)\text{-}H_2C\text{-}N{=}CH\text{-}COO^- \rightleftharpoons H_2C\text{-}S\text{-}CHCOO^-\text{-}NH\text{-}CH_2 \text{ (ring)}$$

thiazolidine-2-carboxylic acid

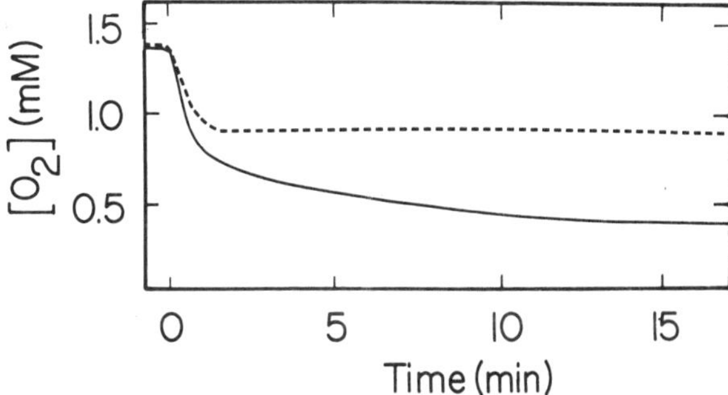

Figure 1. Stoichiometry of oxygen consumption during oxidation of thiazolidine-2-carboxylic acid at 19°C in 20 mM sodium pyrophosphate, pH 8.5. A 3 ml solution of 1 mM D,L-proline (---) or 1 mM D,L-thiazolidine-2-carboxylic acid (—) was equilibrated with oxygen in a Yellow Springs instrument equipped with a Clarke oxygen electrode. D-amino acid oxidase was added at t_0 to a final concentration of 5.6 μM.

As shown in Figure 3, at 15°C, use of D-thiazolidine-2-carboxylic acid resulted in rapid consumption of some 80–90% of the substrate; the slow second phase was due to isomerization of some of the D-thiazolidine-2-carboxylic acid during the experiment. The same experiment with L-thiazolidine-2-carboxylic acid gave a small burst, due to the presence of some D-isomer, followed by logarithmic consumption of the remainder. The first-order rate constant from such a trace was used for comparison with the measured isomerization rate. For temperatures below 15°C, the rate of oxidation was measured as the rate of reduction of D-amino acid oxidase. The

Figure 2. Stoichiometry of oxygen consumption during oxidation of thiazolidine-2-carboxylic acid at 2°C. (–··–), 0.17 mM D-proline; (—), 0.17 mM D,L-thiazolidine-2-carboxylic acid, in air-equilibrated buffer.

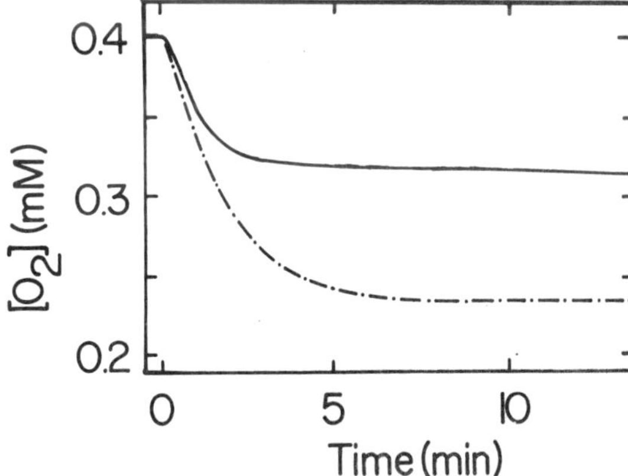

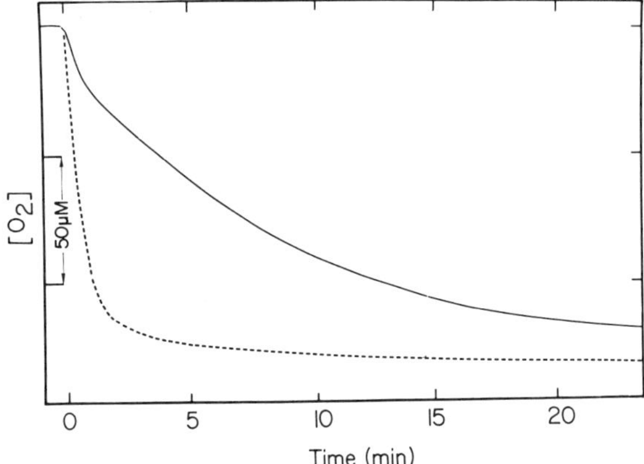

Figure 3. Consumption of (---) D- or (—) L-thiazolidine-2-carboxylic acid by D-amino acid oxidase at 15°C. The amino acid was generated by incubating the appropriate ethyl ester (5 mM) with pig liver esterase (0.2 mg/ml) for 5 minutes at 4°C. Eighty μl of this was then injected into the oxygen electrode chamber which contained 10 μM FAD, 1.6 μM D-amino acid oxidase in 3 ml 0.1 M sodium pyrophosphate, pH 8.5, equilibrated with 50% O_2 at 15°C.

reduction, followed at 455 nm, was slow with the L-isomer and too rapid to observe with the D-isomer, even at 2°C (Figure 4).

An Arrhenius plot of the rate of oxidation of L-thiazolidine-2-carboxylic acid by D-amino acid oxidase and the rate of isomerization as a function of temperature is shown in Figure 5. The excellent agreement between both types of measurements demonstrates that the oxidation of L-thiazolidine-2-carboxylic

Figure 4. Anaerobic reduction of 5.7 μM D-amino acid oxidase by 50 μM (---) D- or (—) L-thiazolidine-2-carboxylic acid at 2°C. The appropriate ethyl ester from the sidearm of an anaerobic cuvette was mixed with enzyme and 0.1 mg/ml pig liver esterase in 0.1 M sodium pyrophosphate, pH 8.5.

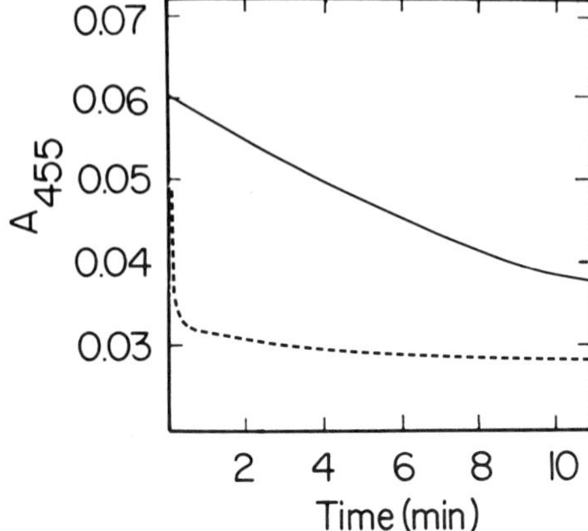

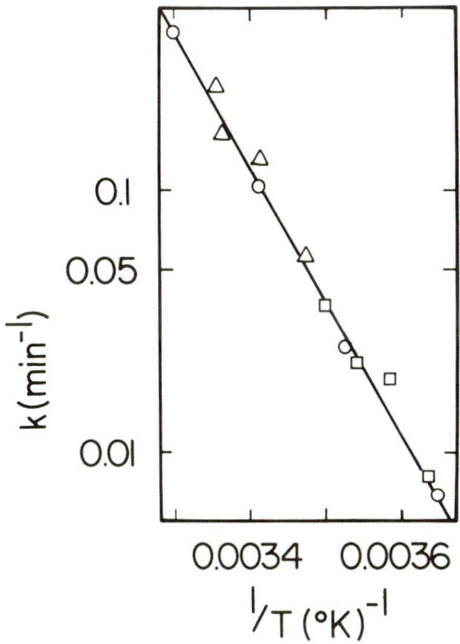

Figure 5. Arrhenius plot of the rate of isomerization of thiazolidine-2-carboxylic acid and the rate of consumption of L-thiazolidine-2-carboxylic acid by D-amino acid oxidase. (□) rate of reduction of D-amino acid oxidase on addition of L-thiazolidine-2-carboxylic acid, determined as an initial rate. (△) rate of consumption of L-thiazolidine-2-carboxylic acid by D-amino acid oxidase measured with the oxygen electrode, determined as a first-order rate. (○) directly measured rate of isomerization of L-thiazolidine-2-carboxylic in 0.1 M sodium pyrophosphate, pH 8.5.

acid by D-amino acid oxidase can be completely explained by prior isomerization to D-thiazolidine-2-carboxylic acid.

ACKNOWLEDGMENTS
This work was supported by Grant GM 11106 from the United States Public Health Service and by a Rackham Predoctoral Fellowship (P.F.).

References

1. Hamilton, G.A., Buckthal, D.J., Mortensen, R.M., and Zerby, K.W. (1979) *Proc Natl Acad Sci USA* 76:2625–2629.
2. Johnson, R.L., Smissman, E.E., and Plolnikoff, N.P. (1978) *J Med Chem* 21:165–169.

Published 1982 by Elsevier North Holland, Inc.
Vincent Massey and Charles H. Williams, Editors
Flavins and Flavoproteins

CHAPTER 30

A New Material for Flavoprotein Researchs—Purification and Properties of Hog Kidney Quasi D-Amino Acid Oxidase

Kiyoshi Shiga,* Yasuzo Nishina,* Kihachiro Horiike,** Hiromasa Tojo,*** Toshio Yamano,*** and Hiroshi Watari*

**National Institute for Physiological Sciences; **Shiga University of Medical Science; ***Osaka University Medical School, Japan*

Introduction

Hog kidney D-amino acid oxidase (DAO) is one of the most extensively studied flavoproteins and the purification method is well established (1–3). During the course of the purification of DAO, another type of DAO, referred to as quasi D-amino acid oxidase (Q-DAO) (4), is obtained, which seems to be an isozyme of DAO. In this report, the purification method of Q-DAO together with that of DAO is reported, which is rather simpler than previous ones (2–4). Some properties of Q-DAO are also described.

Experimental Procedure

To carry out the amino acid analyses using a Hitachi 855 amino acid analyzer, holoenzyme-benzoate complexes of Q-DAO and DAO were hydrolyzed with 6N HCl in a sealed evacuated tube at 110°C. The concentration of these enzymes were determined spectrophotometrically, assuming $\varepsilon = 11{,}300$ $M^{-1}cm^{-1}$ at 463 nm for both enzymes (3,4). A content of cystine and cysteine was detected as cysteic acid by performic acid oxidation (5). Cysteine content was also detected by spectrophotometric titration with DTNB in the presence of 8 M urea (6). Tryptophan content was determined by the MCD method (7) using apoenzymes whose concentrations were determined by Lowry's method (8), taking bovine serum albumin as a standard and assuming the values of 38,200 for DAO and 37,500 for Q-DAO as molecular weights. The COOH-terminal amino acid was determined by carboxypeptidase Y digestion (9). The NH_2-terminal amino acid was determined by the method of Edman, using a JEOL sequence analyzer, JAS-47K (10).

Results and Discussion

Purification of Q-DAO and DAO

All following procedures were performed below 10°C, unless otherwise specified.

The cortex of hog kidney was homogenized with 0.017 M sodium pyrophosphate buffer, pH 8.3 containing 3 g/liter sodium benzoate (this buffer is designated as A). The pH was adjusted to 5.2 by acetic acid and the temperature was maintained at 48°C for 5 minutes. After cooling, the supernatant was obtained by centrifugation. After the addition of ammonium sulfate (200 g/liter), the precipitate was obtained by centrifugation and dissolved in A. The temperature was maintained at 59°C, pH 5.2, for 3 minutes. After cooling, the supernatant was obtained by centrifugation, and then ammonium sulfate (200 g/liter) was added to obtain the precipitate by centrifugation. This precipitate contains DAO, Q-DAO, and impurity. This was dissolved in and dialyzed against 0.02 M phosphate buffer, pH 6.7 (or 6.8) containing 3 g/liter sodium benzoate. After centrifugation, the supernatant was applied on a hydroxyapatite column preequilibrated with the same buffer. The column was washed with the same buffer and then DAO moved as a yellow band followed by another yellow band (Q-DAO). These two yellow bands were collected separately and ammonium sulfate (250 g/liter) was added to obtain the precipitates of Q-DAO and DAO (at this step, these were still not pure). It is noted that at pH 6.3, Q-DAO is strongly adsorbed on a hydroxyapatite column and therefore eluted at too dilute a concentration.

The precipitates were separately applied on a DEAE-Sepharose (CL-6B) column as described previously (4) [the heat treatment before this step used in the previous paper (4) was not necessary; DEAE-Sepharose was used instead of DEAE-Sephadex (3,4)]. The enzyme preparations obtained after chromatography were precipitated by the addition of ammonium sulfate (300/liter). Thus obtained Q-DAO was pure, as checked by SDS-polyacrylamide gel and polyacrylamide gel electrophoresis. DAO obtained was again treated by a hydroxyapatite column under the conditions of 0.02 M phosphate buffer pH 6.3, containing sodium benzoate (3 g/liter). Eluted DAO was precipitated by the addition of ammonium sulfate (200/liter). DAO thus obtained was pure.

Although the steps mentioned above usually give satisfactory results, when some impurity is contaminating, chromatography on a Sephadex G-100 column is effective for both enzymes.

The amount of Q-DAO obtained was about half the amount of DAO, and L-alanine was not a substrate for Q-DAO. Also, the following properties were reported previously (4). The absorption (250–600 nm) and CD (visible region) spectra were the same in both enzymes. The kinetic parameters for D-alanine and D-proline and the binding constants for benzoate were the same in both enzymes. The species of flavin of Q-DAO was FAD. Q-DAO moved very slightly faster than DAO on SDS-polyacrylamide gel electrophoresis, suggesting that the molecular weight of the former (37,500) is a little smaller than the latter (38,200).

Hydroxyapatite and DEAE-Sepharose Column Chromatographies

DAO was eluted faster than Q-DAO on a hydroxyapatite column chromatography as shown in Figure 1. The ratio of the peak elution volumes of Q-DAO and DAO was about 2–3. However, the peak elution volumes of these enzymes were not much different on a DEAE-Sepharose column chromatography (Figure 1). Therefore, to separate Q-DAO and DAO, the chromatography on a hydroxyapatite column was effective.

Amino Acid Compositions and Terminal Amino Acids

The amino acid compositions of the enzymes were only slightly different (Table 1). While 5 moles of methionine and 22 moles of proline per mole of DAO were determined which were consistent with the previous reports (3, 11), 4 moles of methionine and 20 moles of proline per mole of Q-DAO were observed. The NH_2- and COOH-terminal residues of DAO were methionine and leucine, respectively (3). The same results were obtained in Q-DAO.

Q-DAO is not a hydrolytic product derived from DAO by cleavage near the NH_2- and/or COOH-terminus for the following reasons: (a) About 80 amino acid residues exist between the NH_2-terminal methionine and the next methionine (12). The differences of molecular weights and amino acid compositions between the two enzymes are too little to consider Q-DAO as the hydrolytic product of DAO by the cleavage near the NH_2-terminus. (b) The COOH-terminal sequence of DAO is -Ser-His-Leu (12). Both enzymes have the same number of serine and histidine (Table 1), and also the same COOH-terminal leucine. Therefore, Q-DAO is not a product of DAO hydrolyzed near the COOH-terminus. Therefore, it is suggested that the two enzymes have different primary structures.

Figure 1. Chromatography on a hydroxyapatite column (a) and a DEAE-Sepharose (b) column. The abscissa shows elution volume and the ordinate the absorbance at 463 nm. DAO and Q-DAO were applied on the column, independently. (a) The elution buffer was 0.02 M phosphate buffer, pH 6.7, containing 3 g/liter sodium benzoate at 25°C (●, DAO; ×, Q-DAO). The column size was 2×14 cm. (b) The elution buffer was 10 mM Tris-HCl buffer, pH 8.0, containing 60 mM $CaCl_2$ and 200 mg/liter sodium benzoate (---●---; ---×---, Q-DAO) or, containing 125 mM KCl and 29 mg/liter sodium benzoate (—●—, DAO; —×—, Q-DAO) at 5°C. The column size was 1.8×11 cm.

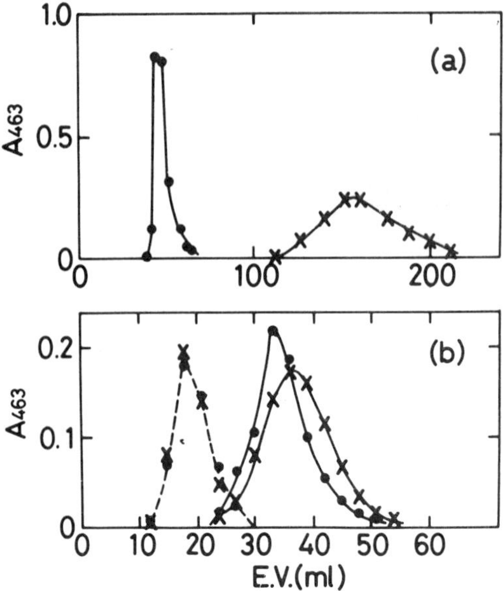

Table 1. Amino Acid Compositions of Q-DAO and DAO.

		Hydrolysis (hours)	Amino Acid																	
			Lys	His	Arg	Trp	Asp	Thr	Ser	Glu	Pro	Gly	Ala	Cys	Val	Met	Ile	Leu	Tyr	Phe
Residues per mole of FAD	Q-DAO	24	11.6	7.7	21.6		31.9	20.5	11.5	36.1	20.3	33.5	17.8	4.6	23.0	4.2	15.3	32.9	13.9	15.3
		24	11.6	7.7	21.8		32.2	20.8	11.7	36.5	19.8	33.6	17.9		23.0	4.0	15.5	33.0	14.0	15.3
		48	11.7	8.0	21.9		32.1	20.2	10.7	36.6	20.1	33.9	18.0	4.9	24.9	4.0	17.7	33.4	14.1	15.3
		48	11.6	8.0	22.1		32.1	20.2	10.7	36.6	19.4	33.9	18.0		24.9	4.0	17.8	33.4	14.1	15.4
		72	10.9	8.0	22.2		30.1	18.2	9.1	34.6	18.7	32.3	17.0		24.8	4.0	18.8	32.9	14.1	15.1
		Average or extrapolated	11.5	7.9	21.9	6.8	31.7	22.1	13.2	36.1	19.7	33.4	17.7	4.8 5.3*	24.9	4.0	18.1	33.1	14.0	15.3
		** ±	0.4	0.2	0.3	0.6	1.1	0.0	0.0	1.1	0.8	0.8	0.5	1.9 0.9	0.1	0.1	1.5	0.3	0.1	0.1
	DAO	24	11.5	7.7	21.9		31.7	21.3	11.7	36.3	22.7	35.0	17.9	4.6	23.6	5.0	16.0	34.5	13.6	15.2
		24	11.5	7.7	21.8		31.9	21.6	11.9	36.6	21.9	35.2	18.0		23.8	5.1	16.0	34.5	13.6	15.2
		48	11.3	7.7	21.1		30.9	20.2	10.5	35.6	21.6	34.3	17.6	5.1	24.1	4.9	17.4	33.8	13.5	14.7
		48	11.3	7.7	21.3		31.2	20.3	10.6	35.9	22.2	34.6	17.6		24.3	4.9	17.5	34.1	13.6	14.8
		72	11.1	7.7	21.1		31.4	19.9	9.9	35.5	22.2	34.3	17.3		24.6	4.9	17.8	33.5	13.5	15.0
		Average or extrapolated	11.3	7.7	21.4	7.2	31.4	22.2	12.8	36.0	22.1	34.7	17.7	4.9 5.1*	24.3	5.0	17.6	34.1	13.6	15.0
		** ±	0.2	0.0	0.5	0.1	0.5	0.0	0.0	0.6	0.5	0.5	0.3	3.1 1.2	0.6	0.1	0.5	0.5	0.1	0.3

Values of serine and threonine were extrapolated to 0 time and those of valine and isoleucine were the average at 48 and 72 hours of hydrolysis time.

*The values were determined by spectrophotometric titration with DTNB.

**Confidence interval of 95%.

References

1. Kubo, H., Yamano, T., Iwatsubo, M., Soyama, T., Shiga, T., and Isomoto, A. (1960) *Bull Soc Chim Biol* 42:569.
2. Brumby, P.E. and Massey, V. (1968) *Biochem Prep* 12:29.
3. Curti, B., Ronchi, S., Branzoli, U., Ferri, G., and Williams, C.H. Jr., (1973) *Biochim Biophys Acta* 327:266.
4. Shiga, K., Nishina, Y., Horiike, K., Tojo, H., Watari, H., and Yamano, T. (1980) *Medical J of Osaka Univ* 30:71.
5. Moore, S. (1963) *J Biol Chem* 238:235.
6. Ellman, G.L. (1959) *Arch Biochem Biophys* 82:70.
7. Holmquist, B. and Vallee, B.L. (1973) *Biochemistry* 12:4412.
8. Lowry, O.H., Rosenbrough, N.J., Farr, A.L., and Randall, R.J. (1951) *J Biol Chem* 193:265.
9. Hayasi, R., Moore, S., and Stein, W.H. (1973) *J Biol Chem* 248:2296.
10. Edman, P. and Begg, G. (1967) *Eur J Biochem* 1:80.
11. Tu, S.C., Edelstein, S.J., and McCormick, D.B. (1973) *Arch Biochem Biophys* 159:889.
12. Ronchi, S., Minchiotti, L., Curti, B., Zapponi, M.C., and Bridgen, J. (1976) *Biochim Biophys Acta* 427:634.

Published 1982 by Elsevier North Holland, Inc.
Vincent Massey and Charles H. Williams, Editors
Flavins and Flavoproteins

CHAPTER 31

Suicide Inactivation of the Flavoenzyme Alcohol Oxidase by Cyclopropanone

Thomas H. Cromartie

Department of Chemistry, University of Virginia, Charlottesville, Virginia

Alcohol oxidase (EC 1.1.3.13) can be isolated from a variety of yeasts grown with methanol as the sole source of carbon (1,2) as a soluble, octameric enzyme which has FAD as the only cofactor and which oxidizes small primary alcohols while reducing O_2 to H_2O_2 (3,4). It is remarkable in that some of the FADs in the native enzymes from *Hansenula polymorpha* and from *Candida boidinii* are stable to air as the anionic flavin semiquinone (5,6). The physiological and mechanistic significance of this observation is not yet clear. In addition, it has been found that the enzyme from *C. boidinii* is inactivated by alcohol substrates having a β-acetylenic functionality in a manner different from other flavoenzymes (7) and undergoes substantial conformational changes on binding products and competitive inhibitors (8).

Like the enzyme from *H. polymorpha* (5), alcohol oxidase from *C. boidinii* is irreversibly inactivated by reaction with cyclopropanol. In the course of studying this inactivation, it was found that cyclopropanone (as the hydrate) was an irreversible inactivator of alcohol oxidase. Cyclopropanone hydrate is known to be a potent, but reversible, inhibitor of several enzymes, including aldehyde dehydrogenase, which contain essential sulfhydryl residues (9). The inhibition was found to result from formation of a stable thiohemiketal between cyclopropanone and the essential sulfhydryl group. The inhibition could be completely reversed by dilution of the inhibited enzyme. Although alcohol oxidase has eight sulfhydryl groups per subunit, these are unreactive toward electrophiles other than metals (8). Therefore, a study of the inactivation of alcohol oxidase by cyclopropanone was begun.

Incubation of alcohol oxidase with cyclopropanone hydrate causes a loss of enzymatic activity which shows pseudo-first-order behavior (Figure 1). The relationship between the observed rate of inactivation and various concentrations of cyclopropanone hydrate employed was found to be hyperbolic, suggesting that a reversible complex of enzyme and inactivator is formed prior to inactivation. From a double reciprocal plot of these data, an apparent K_I of 2.0 mM and a V_{max}^{inact} of 0.14 min^{-1} were determined. Since the rate of formation of cyclopropanone from tetrahedral adduct precursors is known to be quite slow (10), the rate of inactivation of alcohol oxidase suggests that the hydrate of cyclopropanone rather than the ketone itself is the actual inactivating species. Since alcohol oxidase has been found to oxidize aldehydes, such as formaldehyde (11) and trifluoroacetaldehyde (12), which exist primarily as

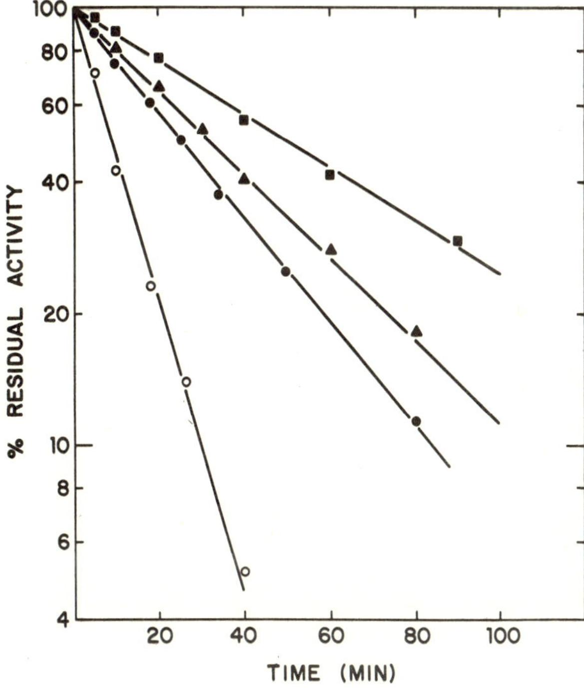

Figure 1. Rate of inactivation of alcohol oxidase at 25°C in 50 mM sodium phosphate at pH 7.5 by (●) 1 mM or (○) 5 mM cyclopropanone hydrate. The effect of the presence of (■) 12 mM methanol or (▲) 10 mM glutathione on the inactivation by 1 mM cyclopropanone hydrate is also shown.

hydrates, the binding of cyclopropanone hydrate to the enzyme is not unprecedented. The inactivation proceeds essentially to completion and cannot be reversed by dilution of the inactivated enzyme. Dialysis or Sephadex gel chromatography, even the presence of thiol compounds and of alcohol substrates, do not lead to recovery of enzymatic activity. Therefore, the inactivation is irreversible in contrast to the inhibition of other enzymes by cyclopropanone hydrate (9). In the presence of the substrate methanol, the rate of inactivation of the enzyme under aerobic conditions is reduced (Figure 1), which suggests the inactivation is an active-site-directed process. The presence of glutathione also decreases the rate of inactivation, probably due to formation of an unreactive thiohemiketal with cyclopropanone since it does not change the final extent or irreversibility of the inactivation. As expected, the inactivation can be completely prevented by destruction of cyclopropanone hydrate by heating at 90°C or by standing in acid prior to addition of the enzyme. If the enzyme is reduced under anaerobic conditions by either substrate methanol or by dithionite, subsequent addition of cyclopropanone hydrate does not cause any inactivation until the enzyme is reoxidized by the admission of air. Thus the inactivation requires oxidized FAD.

Inactivation of alcohol oxidase by cyclopropanone hydrate is exactly paralleled by a loss of the 460 nm absorbance of oxidized FAD. Spectroscopic studies on alcohol oxidase are complicated by the presence of flavin semiquinones which have appreciable absorbance in the 300–450 nm region but which do not appear to be involved in enzymatic catalysis (5,6). By spectroscopic estimation (5), the alcohol oxidase used in these studies has four FAD and four anionic flavin semiquinones per enzyme octamer. The loss of oxidized flavin on inactivation by cyclopropanone hydrate is best demonstrated by difference spectroscopy (Figure 2). The modified spectrum, indicative of the loss of oxidized FAD, is stable to air at room temperature for at least 18 hr. It is not further changed by the addition of substrate alcohols or of dithionite under aerobic or anaerobic conditions, which demonstrates that all of the oxidized FAD has been lost. Denaturation of native alcohol oxidase liberates all of the flavin as oxidized FAD, but denaturation of cyclopropanone hydrate-inactivated enzyme with urea, heat, ethanol, or guanidinium hydrochloride does not liberate oxidized FAD from the modified fraction of the enzyme flavin. Therefore, the inactivation of alcohol oxidase must involve a stable, covalent modification of the catalytic FAD which has the characteristics of a "suicide" inactivation.

After denaturation of the modified enzyme with urea, the absorbance spectrum is not significantly altered by standing at pH 1.5 for 8 hr or at pH 12

Figure 2. Effect of inactivation of alcohol oxidase by cyclopropanone hydrate on the flavin spectrum. Line A: spectrum of identical samples of alcohol oxidase (2.5 mg/ml) in sample and reference cuvettes. Line B: following complete inactivation of enzyme in the reference cuvette with 5 mM cyclopropanone hydrate. Line C: following denaturation of native (sample cuvette) and inactivated (reference cuvette) enzyme by treatment with 4 M urea.

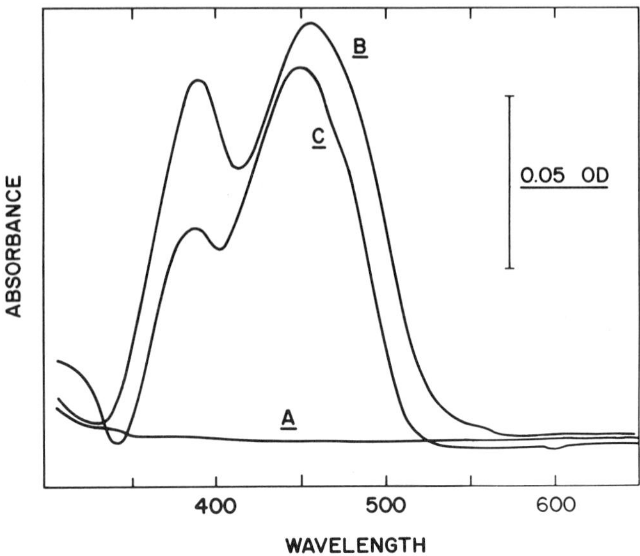

for 8 hr. Since the modified FAD is not returned to FAD by these treatments, the modification must involve changes which cannot be reversed by simple acid-base treatment. Denaturation of the modified enzyme with ethanol and removal of the precipitated protein gives a solution with absorption maxima at 380 and at 450 with a 2:1 ratio. TLC of this solution shows FAD as the only fluorescent component. Further studies to elucidate the structure of the modified flavin are underway. Knowledge of the structure of the adduct is required before the mechanism of the inactivation can be determined. However, the modification of the oxidized FAD is similar to a chemical reduction, and so the cyclopropanone hydrate is probably oxidized on reaction. Oxidation of the hydrate could occur by ring opening and perhaps via radical pathways, which are known for the cyclopropanone (13).

References

1. Sahm, H. (1977) *Adv Biochem Eng* 6:77–103.
2. Cardemil, E. (1978) *Com Biochem Physiol* 60B:1–7.
3. Fujii, T. and Tonomura (1972) *Agr Biol Chem* 36:2297–2306.
4. Sahm, H. and Wagner, F. (1973) *Eur J Biochem* 36:350–356.
5. Mincey, T., Tayrien, G., Mildvan, A.S., and Abeles, R.H. (1980) *Proc Natl Acad Sci* 77:7099–7101.
6. Geissler, J. and Hemmerich, P. (1981) *FEBS Lett* 126:152–156.
7. Nichols, G.S. and Cromartie, T.H. (1980) *Biochem Biophys Res Comm* 97:216–221.
8. Cromartie, T.H. (1981) *Biochemistry*, in press.
9. Wiseman, J.S. and Abeles, R.H. (1979) *Biochemistry* 18:427–435.
10. Wasserman, H.H. and Clagett, D.C. (1966) *J Am Chem Soc* 88:5368–5369.
11. Sahm, H. (1975) *Arch Microbiol* 105:179–181.
12. Cromartie, T., unpublished observations.
13. Wasserman, H.H., Clark, G.M., and Turley, P.C. (1974) *Topics Curr Chem* 47:73–156.

Published 1982 by Elsevier North Holland, Inc.
Vincent Massey and Charles H. Williams, Editors
Flavins and Flavoproteins

CHAPTER 32

Microbial Oxidation of Methanol: Properties of Crystallized Alcohol Oxidase from *Pichia* Sp.

Ramesh N. Patel, C.T. Hou, A.I. Laskin,
and P. Derelanko

Exxon Research and Engineering Company, Corporate Research Science Laboratory, P.O. Box 45, Linden, New Jersey

Materials and Methods

Enzyme Assay

Alcohol oxidase activity was assayed polarographically with a Clark oxygen electrode cell. The reaction mixture (total volume of 3.0 ml) contained: 50 mM potassium phosphate buffer, pH 7.0, 2.8 ml; 100 μmoles substrate, and extract. The incubation temperature was 30°C. The reaction was started by injecting substrate into the reaction chamber and the rate of oxygen consumption was measured. The specific activity was expressed as μmoles of oxygen consumed/min/mg of protein.

Assay of Aldehydes

The aldehydes formed by the oxidation of primary alcohols (C1-C4) by the purified alcohol oxidase were assayed by gas chromatography. The reaction mixture at 30°C contained (total 1 ml volume) 50 μmoles sodium phosphate buffer, pH 9.0; 20 μmoles substrate and the purified alcohol oxidase (50 μgm). A stainless steel column (12 ft by 1/8 inch) packed with 10% Carbowax 20 M on 80/100 Chromosorb W (Perkin Elmer Corporation, Norwalk, Connecticut) was used. The column temperature was maintained isothermally at 120°C, and carrier gas flow was 35 ml of helium per min. The aldehydes were identified by retention time comparisons and co-chromatography with authentic standards.

Assay of Formate

Formate was determined with the NAD-dependent formate dehydrogenase of *M. capsulatus* (1). Formate estimation was performed immediately after formaldehyde oxidations were completed, in a polarographic assay system with purified enzyme. After oxygen consumption ceased, reaction mixtures were supplemented with NAD and the soluble fraction of *M. capsulatus* and the formate dehydrogenase assay was carried out as described previously (1). Suitable controls in the absence of soluble fraction and assays with standard formate solution were carried out.

Preparation of Apo-Alcohol Oxidase

Alcohol oxidase (50 mg) in 10 ml was dialyzed for 24 hours against two changes of 500 ml 3 M KI, 0.15 M Tris-HCl, pH 8.5, containing 5×10^{-4}M EDTA. The enzyme was transferred for 4 hours to 0.5 M KI containing Tris-EDTA, and finally for 24 hours to 0.5 M KCI buffer for dialysis.

Results

The procedure for the purification of alcohol oxidase from a soluble fraction of methanol-grown yeast, *Pichia* sp. is summarized in Table 1. The crystalline enzyme preparations migrated as a single protein band when subjected to electrophoresis on polyacrylamide gel.

Substrate Specificity and Product Formation

The purified enzyme catalyzed the oxidation of primary alcohols (C_1 to C_6), 2-propan-1-ol, 2-chloroethanol, 3-chloro-1-propanol, 4-chloro-1-butanol, 2-mercaptoethanol, 2-methoxyethanol, 2-methyl-1-butanol, and isobutanol. The affinity of the enzyme toward primary alcohols decreased with increasing the length of the alkyl chain. Secondary alcohols (2-propanol, 2-butanol, 3-pentanol), diols (1, 2-butanediol, 2, 3-butanediol), aromatic alcohols, (2-phenoxyethanol, phenethylalcohol, benzoate), cyclohexanol, tert. butanol, 2-aminoethanol and 2-amino-1-propanol were not oxidized (Table 2). The purified alcohol oxidase from *Pichia* sp. also catalyzed the oxidation of formaldehyde. The oxidation of formaldehyde by alcohol oxidase is due to the hydration (99%) of formaldehyde in aqueous solution. The K_m values for methanol and formaldehyde were 0.5 mM and 3.5 mM, respectively.

Aldehydes were detected as the product of primary alcohol (C_1-C_4) oxidation. The rates of production of aldehydes by the purified enzyme preparations are shown in Figure 1. Formate was detected as the product of formaldehyde oxidation. The stoichiometry of substrate utilized (propanol or formaldehyde), oxygen consumed, and product formed (propionaldehyde or formate) was found to be approximately 1 : 1 : 1.

Table 1. Purification of Alcohol Oxidase from *Pichia* Sp.

Step	**Total protein (mg)**	**Total units**	**Specific[a] activity units/mg protein**
Crude extracts	1435	574	0.40
Protamine sulfate treatment	1200	504	0.42
Ammonium sulfate fractionation (50–80% fraction)	200	400	2.0
DEAE-celulose chromatography	60	288	4.8
Bio-Gel chromatography	40	264	6.6
Crystallization	25	165	6.6

[a]Alcohol oxidase activity was assayed polarographically as described in the Materials and Methods.

Table 2. Substrate Specificity of Alcohol Oxidase from *Pichia* Sp.

Substrate	Rate of oxidation[a] (%)
Methanol	100
Ethanol	92
1-Propanol	74
1-Butanol	52
1-Pentanol	30
1-Hexanol	4
2-Chlorethanol	70
3-Chloro-1-propanol	22
4-Chloro-1-butanol	11
2-Mercaptoethanol	25
2-Methoxyethanol	15
2-Aminoethanol	0
2-Amino-1-propanol	0
2-Methyl-1-butanol	22
2-Phenoxyethanol	0
2-Propanol	0
2-Butanol	0
3-Pentanol	0
1, 2-Butanediol	0
2, 3-Butanediol	0
Isobutanol	2
Tert. Butanol	0
Cyclohexanol	0
Phenethylalcohol	0
Formaldehyde	15
Acetaldehyde	0

[a] A 100% rate of oxidation = 6.6 μmoles of oxygen consumed per min per mg of protein by the purified enzyme.

Effect of Inhibitors

Various alcohols not oxidized by the alcohol oxidase were tested as potential inhibitors of enzyme activity. The oxidation of methanol was inhibited by 2-aminoethanol. Enzyme activity was slightly inhibited by hydroxylamine, thiosemicarbazide, and imidazole among various metal-chelating compounds tested. Sulfhydryl agents (p-hydroxymercuribenzoate, 5-5′-dithiobis-2 nitrobenzoate, iodoacetate, and mercuric chloride), hydrogen peroxide and cupric sulfate inhibited the enzyme activity.

Molecular Weight, Subunit Size, and Proesthetic Group

The molecular weight of the alcohol oxidase was determined by gel filtration to be 300,000. The subunit size determined by sodium dodecyl sulfate gel electrophoresis was 76,000. The enzyme appears to be a tetramer composed of four identical subunits.

The purified enzyme from *Pichia* sp. has absorption peaks in the visible region at 380 nm and 460 nm, and at 280 nm in the ultraviolet region. The addition of methanol to the enzyme resulted in a decrease in absorbance at 380

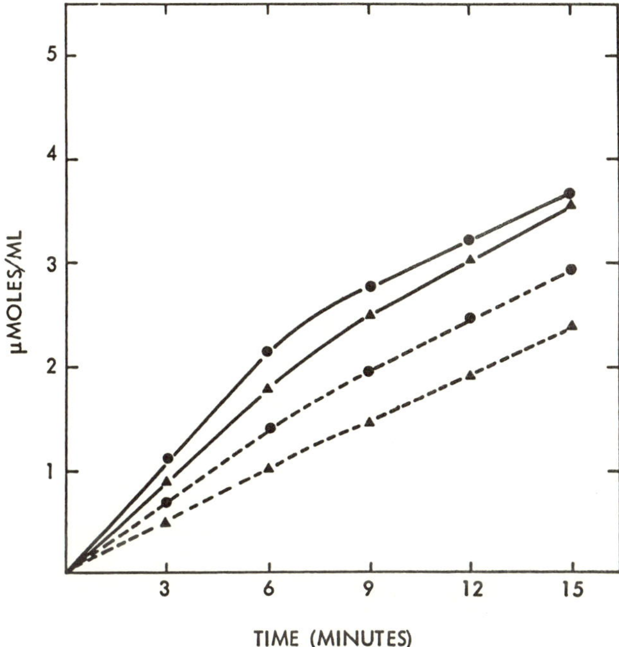

Figure 1. Production of aldehydes by the purified alcohol oxidase from *Pichia* sp. Reaction mixture in a total of 1.0 ml contained: 50 mM phosphate buffer pH 9.0, 0.9 ml; substrate, 20 μmoles; and the purified enzyme, 50 μgm. Aldehydes were identified and estimated by gas chromatography as described in the Materials and Methods. ●—●—● formaldehyde; ▲—▲—▲ acetaldehyde; ●--●--● propionaldehyde; ▲--▲--▲ butyaldehyde.

nm and 460 nm (Figure 2). Apoalcohol oxidase was prepared by treatment of holoenzyme with potassium iodide and following the disappearance of absorption peaks at 380 nm, 460 nm with simultaneous loss of enzyme activity. The fluorescence spectra of holoenzyme and apoenzyme were examined.

The excitation spectrum (Em 525 nm) of holoalcohol oxidase is similar to that of FAD and cofactor extracted from the holoenzyme, with peaks at 375 nm, 452 nm, and 466 nm. The emission spectrum (Ex 375 nm) of holoalcohol oxidase has a strong emission peak at 525 nm. The apoalcohol oxidase has much weaker emission signals and excitation peaks (Em 525 nm) than those of holoalcohol oxidase. The supernatant obtained after boiling the purified enzyme at 100°C for 5 min showed the typical absorption and fluorescence spectrum of a flavin, which was further identified as FAD by paper and thin layer chromatography.

Discussion and Summary

Oxidation of primary alcohols (C_1 to C_5) in methanol-utilizing yeasts has been shown to be catalyzed by alcohol oxidase (2-5). Alcohol oxidase has been

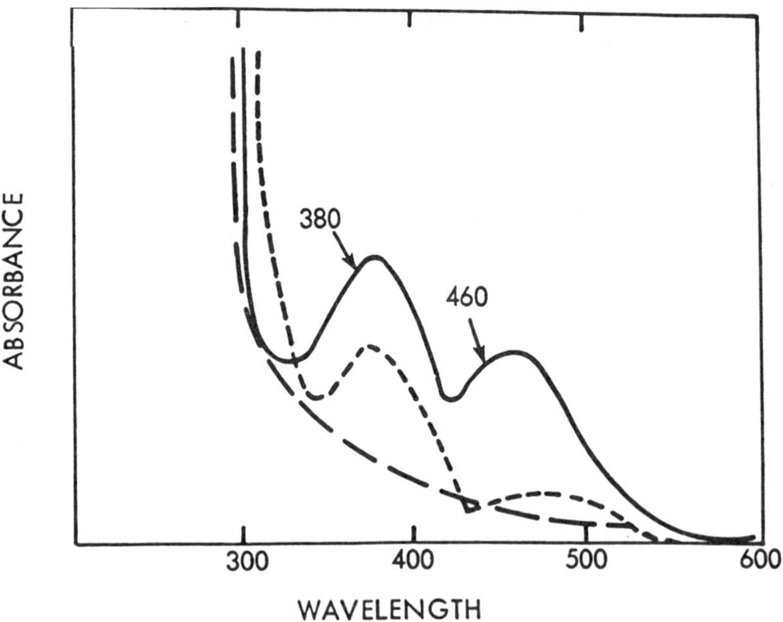

Figure 2. Absorption spectrum of alcohol oxidase from *Pichia* sp. The concentration of holo- and apoenzyme was 1.0 mg/ml. (—) Holoalcohol oxidase; (------) Holoalcohol oxidase + 10 μmoles methanol; (– – –) Apoalcohol oxidase.

purified from *Kloeckera* sp. (6, 7), *Hansenula polymorpha* (6), and *Candida* sp. (2, 4).

The purified alcohol oxidase from *Pichia* sp. is similar in many respects to the alcohol oxidase from other yeasts (2, 4, 6, 7). It catalyzed the oxidation of short-chain primary alcohols and formaldehyde. The subunit size of the enzyme is similar (75,000) and contains FAD as a prosthetic group.

The alcohol oxidase from *Pichia* sp. differs from alcohol oxidases from other yeasts in its molecular weight. The purified enzyme has a molecular weight of 300,000 and consists of four identical subunits. The alcohol oxidase from other yeasts are octamers with a molecular weight of 600,000–670,000 (5-7). The only exception is the alcohol oxidase from *Candida* sp. (2) with a molecular weight of 210,000. The purified alcohol oxidase from *Pichia* sp. has absorption maxima at 380 and 460 nm and is similar in absorption spectrum to the purified alcohol oxidase from *Candida* sp. (2). However, the shoulder at 395 nm as reported for alcohol oxidase from *Hansenula polymorpha* (6) and *Kloeckera* sp. (7) was not observed.

There has been no previous report on the fluorescence spectra and preparation of apoalcohol oxidase from yeasts. The excitation spectrum (Em 525 nm) of holoalcohol oxidase from *Pichia* sp. is similar to that of FAD and cofactor extracted from the holoenzyme with peaks at 375 nm, 452 nm, and 466 nm.

Alcohol oxidase and catalase have been shown to localize in the microbodies known as peroxisomes (8–10) in methanol-grown yeasts. The hydrogen peroxide formed by the alcohol oxidase is immediately metabolized by catalase.

References

1. Patel, R.N. and Hoare, D.S., (1971) *J Bacteriol* 107:187.
2. Fugii, T. and Tonomura, K. (1972) *Agric Biol Chem* 36:2297.
3. Tani, Y., Miya, T., Nishikawa, H., and Ogata, K. (1972) *Agric Biol Chem* 36:68.
4. Sahm, H. and Wagner, F. (1973) *Eur J Biochem* 36:250.
5. van Dijken, J.P., Oho, R., and Harder, W. (1976) *Arch Microbiol* 11:137.
6. Tani, Y., Miya, T., and Ogata, K. (1972) *Agric Biol Chem* 36:76.
7. Kato, N., Omori, Y., Tani, Y., and Ogata, K. (1976) *Eur J Biochem* 64:341.
8. Fukui, S., Kawamoto, S., Yasurhara, S., Tanaka, T., Osumi, M., and Imaizumi, F. (1975) *Eur J Biochem* 59:651.
9. Roggenkamp, R., Sahm, H., Hinkelman, W., and Wagner, F. (1975) *Eur J Biochem* 59:231.
10. van Dijken, J.P., Veenhuis, M., Kreger-van Rij, N.J.W., and Harder, W. (1975) *Arch Microbiol* 102:41.

Published 1982 by Elsevier North Holland, Inc.
Vincent Massey and Charles H. Williams, Editors
Flavins and Flavoproteins

CHAPTER 33

Fluorescent Substrates and Probes of the Catalytic Binding Site of Pyridoxine-5-P Oxidase

Christine Wu, Kathryn Jones, and Jorge E. Churchich

Department of Biochemistry, University of Tennessee, Knoxville, Tennessee

Introduction

The formation of pyridoxal-5-P from pyridoxine-5-P is catalyzed by pyridoxine-5-P oxidase, an enzyme which has been detected in various tissues (1). Although it is known that pyridoxine-5-P is the preferred substrate of the oxidase isolated from brain tissues (2), the enzyme also catalyzes the oxidation of secondary amines.

Despite these studies which indicate the important role played by the oxidase in the formation of pyridoxal-5-P, little is known about the mechanism of action of this enzyme.

In our laboratory, we have isolated and characterized pyridoxine-5-P oxidase from pig brain (2). The aims of the present investigation are 2-fold. Firstly, to obtain information about the mechanism of action of this enzyme and secondly, to investigate the amino acid residues critically connected with enzyme activity.

Results and Discussion

Substrate Specificity

In an attempt to define the specificity of the reaction catalyzed by the oxidase, the oxidation of the substrates pyridoxine-5-P and pyridoxamine-5-P was studied over the pH range 7.5–9.5. At pH 8.0, the oxidation of pyridoxine-5-P as catalyzed by the enzyme proceeds with a specific activity of 90 units/mg at 25°C, whereas the oxidation of pyridoxamine-5-P is characterized by a specific activity of 9 units/mg of protein.

The introduction of the aromatic groups m-aminobenzoic, p-amino benzoic, and aniline into the structure of P-pyridoxyl, yielded derivatives which are substrates of the enzyme. As shown by the results included in Table 1, the V_{max} values obtained for P-pyridoxyl-m-aminobenzoate and P-pyridoxyl-p-aminobenzoate are slightly larger than the V_{max} for the natural substrate pyridoxine-5-P. In addition, the affinity of those synthetic substrates for the binding site is at least ten times greater than the affinity of the natural substrate (Table 1). Thus, the attachment of meta and para amino benzoates to the structure of P-pyridoxyl has a dramatic effect on both stabilization of the transition state and affinity of the substrate for the catalytic site.

Table 1. Catalytic Properties of Pyridoxine-5-P Oxidase.

Sample	V_{max}	K_M (μM)
p-pyridoxamine	0.10	100
p-pyridoxine	0.75	15
p-pyridoxyl-aniline	0.23	1
p-pyridoxyl-m-aminobenzoate	0.90	1
p-pyridoxyl-p-aminobenzoate	1.00	2

Assays conducted at pH 8.

It should be noted that P-pyridoxyl-m-aminobenzoate and P-pyridoxyl-p-aminobenzoate are better substrates than P-pyridoxyl-aniline, indicating that the introduction of electron-withdrawing substituents tends to stabilize carbanionic species formed as intermediates during the catalytic step. A similar mechanism has been postulated for D-amino acid oxidase (3).

Lysyl Residues Implicated in Catalysis

In view of the preceding results, it was thought of interest to investigate whether lysyl residues can act as proton acceptors during the catalytic step leading to the formation of carbanionic species. Pyridoxine-5-P oxidase is inactivated by o-phthalaldehyde, a specific reagent for lysyl residues in proteins (4). A substantial loss of catalytic activity (80%) was observed upon preincubation of the enzyme (4 μM) with o-phthalaldehyde (100 μM) at 25°C for 50 minutes (Figure 1). The time course of inactivation was significantly changed when the synthetic substrate P-pyridoxyl-p-aminobenzoate was added to the enzyme prior to the reaction with the modifying reagent. Judging from the results included in Figure 2, it is evident that P-pyridoxyl-p-aminobenzoate afforded complete protection against the inactivating effect of o-phthalaldehyde.

The pH-rate profile for the reaction of the oxidase with o-phthalaldehyde leading to loss of catalytic activity is shown in Figure 2.

If the reaction of the lysyl groups with the reagent can be described by the reaction scheme:

$$\begin{array}{c} OP + E-NH_2 \xrightarrow{K_o} E-N=OP \\ \downarrow\uparrow K \\ E-NH_3^+ \end{array}$$

then the observed rate constant of inactivation (K_{obs}) is related to the dissociation constant (K) and the proton concentration (H^+) by means of Equation (33.1)

$$K_{obs} = \frac{K_o(OP)_o}{1+(H^+)/K} \qquad (33.1)$$

where K_o is the second order rate constant and $(OP)_o$ the concentration of o-phthalaldehyde.

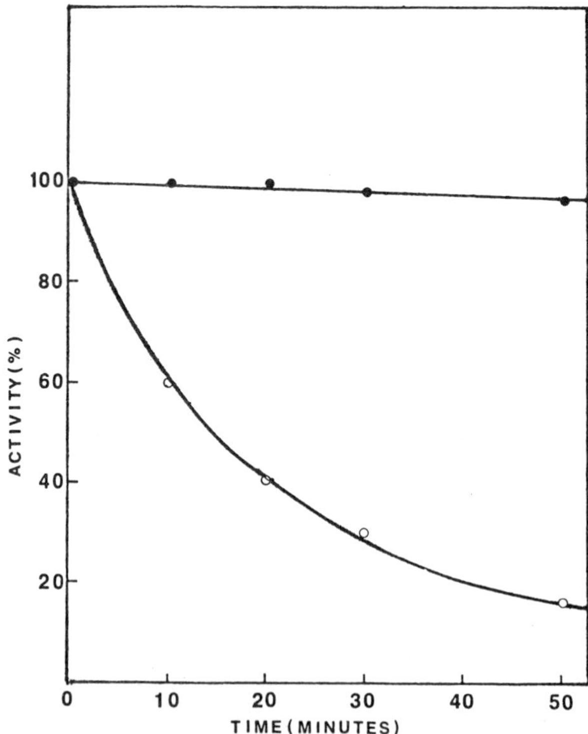

Figure 1. Inactivation of p-pyridoxine oxidase (4 μM) incubated with o-phthaldehyde (100 μM) in the absence (○) and presence (●) of the synthetic substrate p-pyridoxyl-p-aminobenzoate (170 μM).

The experimental values of K_{obs} fit well a theoretical curve calculated for pK=8.25 and $K_o = 1.7 \times 10^3\ M^{-1}\ min^{-1}$.

The reaction of o-phthalaldehyde with the enzyme can be followed by absorption and fluorescence spectroscopy. The reaction proceeds to completion at a readily measurable rate; and the spectroscopic changes facilitate determination of the number of lysyl residues modified in the course of the reaction. Figure 3 shows the kinetic results obtained when the enzyme (12 μM) was allowed to react with o-phthalaldehyde (120 μM) at pH 8.25 in 0.1 M pyrophosphate buffer (pH 8.25). The profile of the reactivity curve is biphasic, exhibiting an initial rapid increase in absorbance after the reaction of approximately two lysyl residues per dimer. Analysis of the kinetic data yields a K_1 for the first step 0.2 min^{-1}, whereas that for the slow step (K_2) is 0.03 min^{-1}.

The observed rate constant of inactivation ($K_{obs} = 0.14\ min^{-1}$) is comparable in magnitude to the rate constant corresponding to the fast reaction of the lysyl residues of the enzyme ($K_1 = 0.2\ min^{-1}$). Hence, the observed inactivation of pyridoxine-5-P oxidase by o-phthalaldehyde can be related to the modification of two lysyl residues per dimer. The possibility exists that those

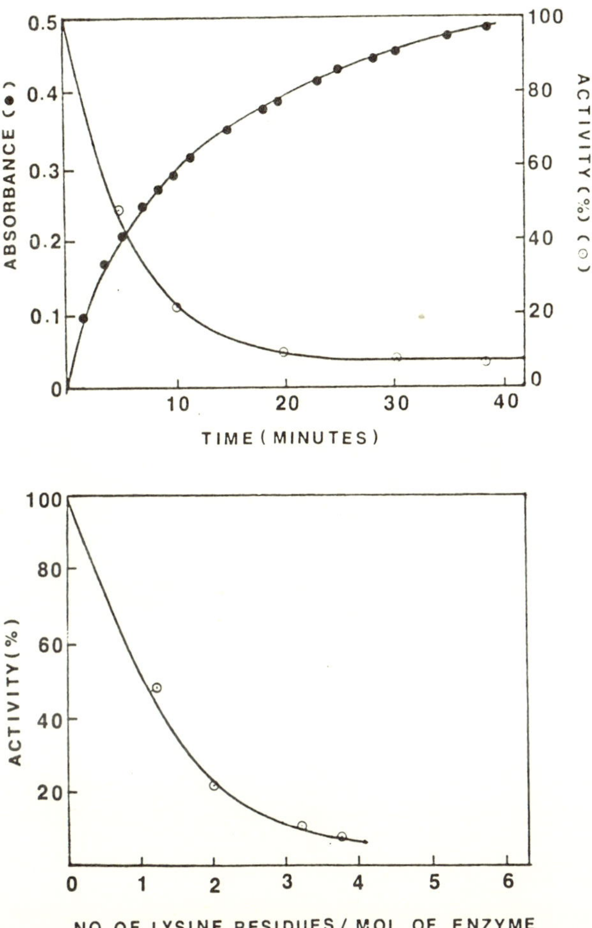

Figure 2. Reaction of p-pyridoxine oxidase (12 μM) with o-phthaldehyde (120 μM) at pH 8.25. Upper: Change in absorbance at 335 nm (●) due to the reaction of lysyl residues with o-phthaldehyde and change in enzymatic activity (⊙) as a function of incubation time. Bottom: Change in enzymatic activity as a function of number of lysyl residues modified.

lysyl residues are the bases acting as proton acceptors during the formation of carbanionic intermediates.

Binding of The Substrate to The Apoprotein

Table 2 shows the spectroscopic properties of the substrate P-pyridoxyl-aniline, P-pyridoxyl-m-aminobenzoate, and P-pyridoxyl-p-aminobenzoate.

When the apoenzyme (10 μM) was mixed with P-pyridoxyl-m-aminobenzoate (10 μM) at pH 8, and the emission spectrum immediately recorded, it was observed that the fluorescence quantum yield of the substrate is enhanced when compared to free P-pyridoxyl-m-aminobenzoate (Figure 4). The change

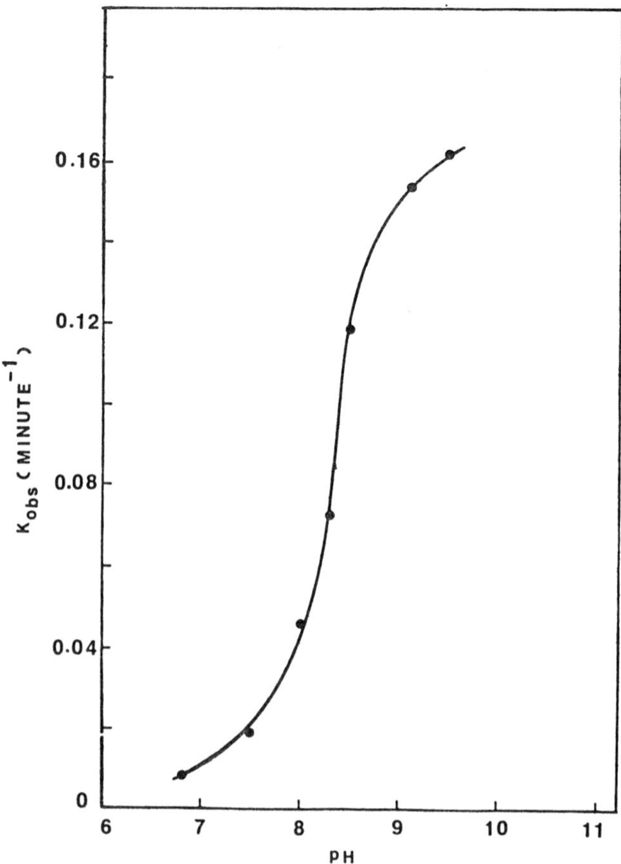

Figure 3. Dependence of the inactivation rate constant on pH. p-Pyridoxine oxidase (4 μM) was incubated with o-phthaldehyde (100 μM) at the indicated pH values. Pseudo-first-order rate constants for the inactivation (K_{obs}) were calculated from the slopes of semilogarithmic plots of activity against incubation time. The solid line is drawn assuming a single critical group which has a $pK_a \approx 8.25$ and a bimolecular rate constant $K_o = 1.7 \times 10^3$ min^{-1} M^{-1}.

Table 2. Spectroscopic Properties of the Substrates.

Sample	λ ABS (nm)	λ FL (nm)	τ
p-pyridoxyl-aniline	325	395	2 nanoseconds
p-pyridoxyl-m-aminobenzoate	320	410	6.5 nanoseconds
p-pyridoxyl-o-aminobenzoate	320	400	8 nanoseconds
p-pyridoxyl-p-aminobenzoate	320, 285	—	—

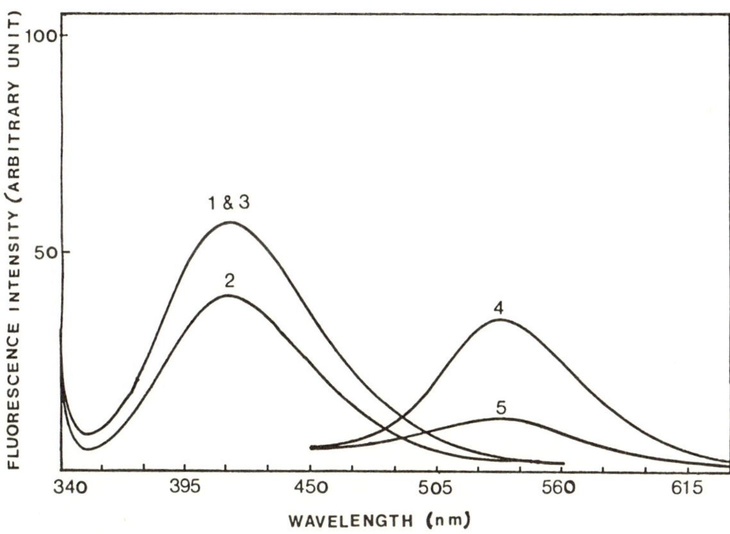

Figure 4. (Curves 1, 2, 3): Fluorescence spectra of m-carboxyphenyl-pyridoxamine phosphate (10 μM) in the presence (1) and absence (2) of apoPNP oxidase (10 μM) in 0.1 M phosphate buffer (pH 8). Curve 3 is the fluorescence spectrum of CPP_p in the presence of apoPNP oxidase and 3-carboxymethyl FMN (10 μM). Excitation wavelength 315 nm. (Curves 4, 5): Fluorescence spectra of 3-carboxymethyl FMN (10 μM) in the absence (4) and presence (5) of apoPNP oxidase (10 μM) in 0.1 M phosphate buffer (pH 8). Excitation wavelength 380 nm.

in fluorescence intensity is not due to the formation of m-aminobenzoate since the fluorescence signal recorded over the wavelength range 340–500 nm remains essentially the same after preincubation of the mixture for 10 minutes at 25°C. In addition, the sample of apoenzyme did not show any residual activity in the presence of the natural substrate pyridoxine-5-P. It follows from these considerations that the fluorescence enhancement detected upon mixing the apoprotein with P-pyridoxyl-m-aminobenzoate can be attributed to binding of this substrate to the protein. It implies that the substrate recognizes its binding site even in the absence of the cofactor FMN. It is important to note that the addition of 3-carboxymethyl-FMN to the mixture containing apoprotein and synthetic substrate is accompanied by a decrease in the fluorescence of the FMN analogue without any effect on the magnitude of the fluorescence emitted by P-pyridoxyl-m-aminobenzoate as indicated by the results included in Figure 4.

References

1. Kazarinoff, M.N. and McCormick D. (1975) *J Biol Chem* 250:3436–3442.
2. Kwok, F. and Churchich, J.E. (1980) *J Biol Chem* 255:882–887.
3. Neims, A., Deluca, H. and Hellerman, L. (1966) *Biochemistry* 5:203.
4. Blaner, W.S. and Churchich, J.E. (1979) *J Biol Chem* 254:1974–1978.

Published 1982 by Elsevier North Holland, Inc.
Vincent Massey and Charles H. Williams, Editors
Flavins and Flavoproteins

CHAPTER 34

Steady-State Kinetic Properties of Pyridoxamine (Pyridoxine) 5′-Phosphate Oxidase from Rabbit Liver

J. D. Choi, M.D. Davis, D.M. Bowers-Komro, D.E. Edmondson, and D.B. McCormick

Department of Biochemistry, Emory University, Atlanta, Georgia

Introduction

An FMN-dependent oxidase (E.C.1.4.3.5) is involved in conversion of pyridoxamine 5′-phosphate (PMP) and pyridoxine 5′-phosphate (PNP) to pyridoxal 5′-phosphate (PLP) (1). The oxidase can also catalyze the oxidation of 4′-(phosphopyridoxyl) secondary amines with O_2 to form H_2O_2, PLP, and primary amines (2, 3). Product inhibition of this oxidase has been postulated to exert control in vitamin B_6 metabolism (4). The oxidase contains one FMN per 54,000 molecular weight monomer and is composed of two noncovalently linked subunits with blocked N-termini (5, 6). Chemical modification studies have shown that one cysteinyl sulfhydryl is critical for catalytic function (6), at least one tryptophanyl residue is involved in the binding of coenzyme (7), an arginyl residue contributes to binding interaction with the 5′-phosphate of substrate and product (8), and one histidyl residue is essential for substrate binding and probably functions as a base during hydrogen abstraction (9). In this paper, we describe steady-state kinetic studies of the oxidase using an artificial as well as natural substrate.

Experimental Procedures

Enzyme from rabbit liver was purified and assayed as described previously (10). N-(5′-Phospho-4′-pyridoxyl)-N′-(1-naphthyl)ethylenediamine (NPP-NED) was synthesized by the method of DePecol and McCormick (3); the 4′-dideutero compound was prepared by NaB^2H_4 reduction of the Schiff base formed from condensing N-(1-naphthyl)ethylenediamine with [4′-^{2}H]PLP. The latter was obtained by oxidation of [4′-^{2}H]PMP with MnO_2-celite according to published procedure (11). [4′-^{2}H]PMP was synthesized by similar borodeuteride reduction of PLP in the presence of NH_4OH in D_2O.

Results and Discussion

Steady-state kinetic studies of the oxidase were initiated using the NPP-NED substrate, since the reaction can be followed by continuously monitoring the increase in fluorescence due to release of NED (3). From analysis of double-reciprocal plots of the kinetic data obtained at various substrate concentrations, two features of the reaction become apparent (Figure 1A). First, parallel lines are observed; this is consistent with the kinetic behavior of other flavoprotein oxidases (12). Second, at relatively low substrate concentrations, there is an upward curvature indicative of substrate inhibition (Figure 1A, inset). This deviation from linearity is dependent on the oxygen concentration, being completely eliminated by ~1 mM O_2. Replotting the apparent V_{max} and K_m values obtained (Table 1) allows the true values of V_{max} (22 min^{-1}) and K_m for both NPP-NED (3.2 μM) and oxygen (235 μM) to be determined graphically (Figure 1B).

To investigate steps that may be rate-limiting, kinetic isotope effects were examined with both NPP-NED and PMP. Dideuteration of the methylene group (which becomes oxidized to the aldehyde) results in small but significant changes in the Michaelis constants in both cases. For the reaction with NPP-NED, the V_{max} is decreased 2-fold while the K_m (NPP-NED) is increased by a factor of 2.5 (Table 1). In the analogous study using PMP, the V_{max} decreases only slightly (from 5 min^{-1} to 4.6 min^{-1}) while the k_m (PMP) increases from 2.2 to 3.0 μM (Figure 2). Finally, aerobic and anaerobic stopped-flow results (not shown) indicate the reduction of pyridoxamine phosphate oxidase with both substrates is significantly faster than reoxidation. The alleviation of substrate inhibition by oxygen during the steady-state catalysis of NPP-NED by PMP oxidase can be rationalized within a scheme in which the reoxidation of free enzyme is competitive with complex formation of reduced enzyme with substrate. Thus, inhibitory effect of high substrate levels probably reflects formation of a nonproductive complex with EH_2. A similar interpretation has been previously suggested for the reaction of L-amino acid oxidase with phenylalanine and oxygen (12, 13). The lack of substrate inhibition in the analogous reaction with PMP, even in the presence of high concentration of substrate [~200 times K_m(PMP)] (Figure 2), is consistent with our previous observation (14).

The deuterium effect observed with PMP displaying a small decrease (~10%) in V_{max} and a larger increase (~50%) in K_m(PMP), is consistent with the suggestion of previous workers that the rate-limiting step in catalysis is product

Table 1. Kinetic Constants for Oxidation of NPP-NED and [4′, 4′-^{2}H]NPP-NED by Pyridoxamine Phosphate Oxidase at Various Oxygen Concentrations.

O_2 (mM)	Substrate	K_m(μM)	V_{max}(min^{-1})
0.12	NPP-NED	1.2	7.1
	[4′, 4′-^{2}H]NPP-NED	3.0	3.7
0.14	NPP-NED	1.2	8.1
0.25	NPP-NED	1.7	11.9
	[4′, 4′-^{2}H]NPP-NED	4.8	5.9
0.66	NPP-NED	2.4	16.4
1.10	NPP-NED	2.6	18.5

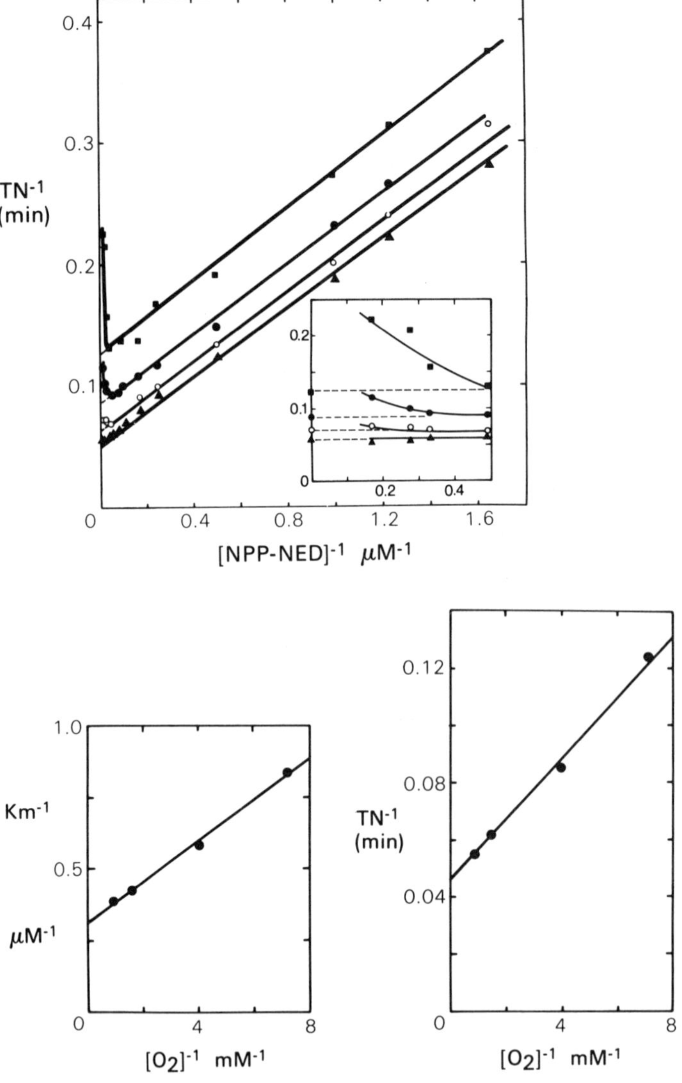

Figure 1. (Top) Double-reciprocal plot for the oxidation NPP-NED by the oxidase. Assays were performed in 0.2 M Tris-HCl, pH 8, 25°, as a function of substrate concentration at various oxygen concentrations: (■) 0.14 mM; (●) 0.25 mM; (○) 0.66 mM; (▲) 1.10 mM. Initial rates are plotted as mol min^{-1} mol of $enzyme^{-1}$. (Bottom) Secondary plots of V_{max} and K_m from the data (top) against reciprocals of oxygen concentration.

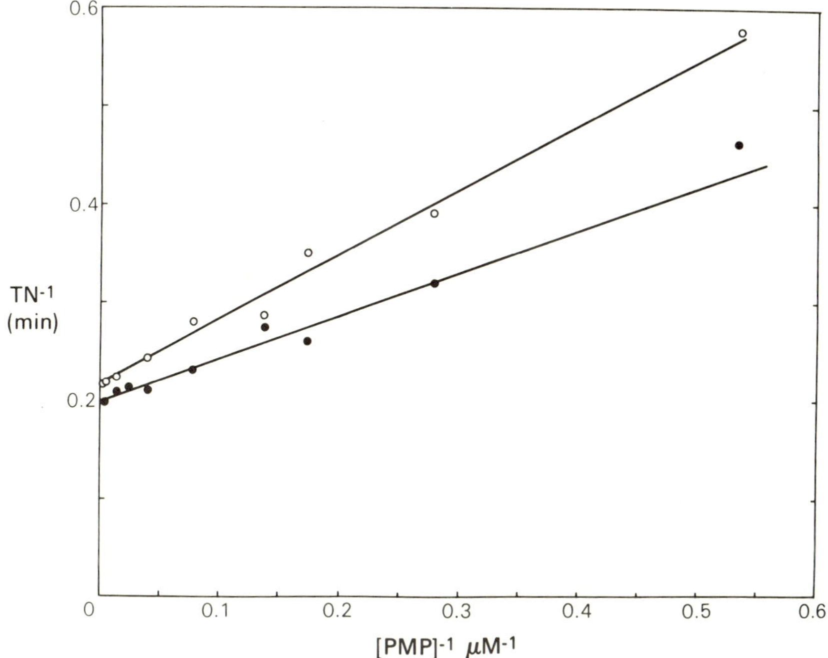

Figure 2. Deuterium isotope effect on rate for oxidation of PMP by the oxidase. The rates (at 25°) are plotted in the double-reciprocal manner at 0.25 mM oxygen: (●) PMP; (○) [4′,4′-^{2}H]PMP.

release (14). A similar kinetic isotope behavior has been observed with xanthine oxidase (15, 16). In the analogous reaction with NPP-NED, the deuterium isotope effect on V_{max} (~2) is slightly larger than that with PMP. In this case, probably the rate of reduction is somewhat closer to the rate of product release.

ACKNOWLEDGMENTS

This work was supported by N.I.H. Grant AM-26746 (to D.B.M.) and N.S.F. Grant PCM-81-00770 (to D.E.E.).

References

1. McCormick, D.B. and Merrill, A.H., Jr. (1980) In *Vitamin B_6 Metabolism and Role in Growth*. Tyfiates, G.P. (ed.) Westport, Connecticut: Food and Nutrition Press, pp. 1–26.
2. Kazarinoff, M.N. and McCormick, D.B. (1975) *Biochem Biophys Res Commun* 56:440–446.
3. DePecol, M.E. and McCormick, D.B. (1980) *Anal Biochem* 101:435–441.
4. Merrill, A.H., Jr., Horiike, K., and McCormick, D.B. (1978) *Biochem Biophys Res Commun* 83:984–990.
5. Kazarinoff, M.N. and McCormick, D.B. (1975) *J Biol Chem* 250:3436–3442.
6. McCormick, D.B., Kazarinoff, M.N., and Tsuge, H. (1976) In *Flavins and Flavoproteins*. Singer, T.P. (ed.) Amsterdam: Elsevier, pp. 708–719.

7. McCormick, D.B. (1977) *Photochem Photobiol* 26:169–182.
8. Choi, J.D. and McCormick, D.B. *Biochemistry*, in press.
9. Horiike, K., Tsuge, H., and McCormick, D.B. (1979) *J Biol Chem* 254:6638–6643.
10. Merrill, A.H., Kazarinoff, M.N., Tsuge, H., Horiike, K., and McCormick, D.B. (1979) *Methods Enzymol* 62:563–574.
11. Peterson, E.A. and Sober, H.A. (1954) *J Am Chem Soc* 76:169–175.
12. Bright, H.J. and Porter, D.J. (1975) In *The Enzymes*. Boyer, P.D. (ed.). Third Edition Volume XII Port B 421–505, New York: Academic Press.
13. Desa, R.J. and Gibson, Q.H. (1966) *Fed Proc* 25:649.
14. Horiike, K., Merrill, A.H., Jr., and McCormick, D.B. (1979) *Arch Biochem Biophy* 195:325–335.
15. Edmondson, D., Ballou, D., Van Heuvelen, A., Palmer, G., and Massey, V. (1973) *J Biol Chem* 248:6135–6144.
16. Olson, J.S., Ballou, D.P., Palmer, G., and Massey, V. (1974) *J Biol Chem* 249:4363–4382.

Published 1982 by Elsevier North Holland, Inc.
Vincent Massey and Charles H. Williams, Editors
Flavins and Flavoproteins

CHAPTER 35

A Mechanism for Mitochondrial Monoamine Oxidase-Catalyzed Amine Oxidation

Richard B. Silverman, Stephen J. Hoffman, and William B. Catus, III

Department of Chemistry, Northwestern University, Evanston, Illinois

Mitochondrial monoamine oxidase (MAO, EC 1.4.3.4) is a flavin adenine dinucleotide-containing enzyme responsible for the oxidative deamination of physiologically-important monoamines. Although some mechanistic features of this enzyme-catalyzed reaction are known, the most crucial aspect of the mechanism, namely, the mechanism for the oxidation of the amine to produce an iminium ion, has not been addressed (1). Here we describe enzymatic and electrochemical experiments directed toward the elucidation of this mechanism.

The mechanism for electrochemical (2), chemical (3), and photochemical (4) amine oxidations is well established and appears to be the same under all of these conditions; the electrochemical mechanism is shown below.

$$RCH_2\ddot{N}R'_2 \xrightarrow[\text{slow}]{-e^-} RCH_2\overset{\cdot+}{N}R'_2 \xrightarrow[\text{fast}]{-H^+} R\dot{C}H\ddot{N}R'_2$$

$$R\dot{C}H\ddot{N}R'_2 \xrightarrow[\text{fast}]{-e^-} RCH{=}\overset{+}{N}R'_2 \xrightarrow{H_2O} RCHO + R'_2NH$$

It seems reasonable that a mechanism which is common to electrochemical, chemical oxidant, and photochemical oxidations should be operative as well in an enzyme-catalyzed oxidation. This mechanism, as applied to MAO-catalyzed reactions, is shown below. According to this two one-electron transfer mechanism, a nitrogen nonbonded electron is transferred to the flavin to give an amine radical cation intermediate (**1**). This is plausible since the unprotonated (free base) form of the amine has been implicated as the reactive substrate form (5). The protons adjacent to amine radical cations are quite acidic (2). Proton removal generates the carbon radical which can be oxidized further by two pathways. Pathway a involves radical combination followed by two-electron transfer; pathway b is the transfer of a second electron from the radical to the flavin semiquinone. Both pathways would give the iminium ion and the reduced flavin observed as products of the enzyme reaction.

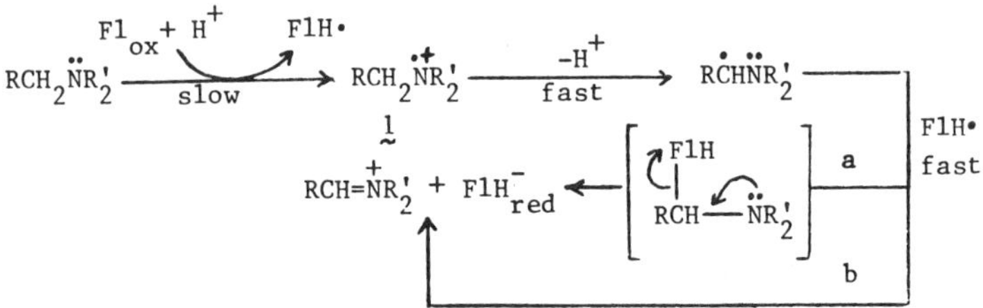

We previously reported (6) that N-cyclopropylbenzylamine (N-CBA) was a mechanism-based inactivator of MAO. Since this type of inactivator depends upon the mechanism of the enzyme, information about the mechanism of MAO should be obtainable by studying the mechanism of inactivation of MAO by N-CBA. Because of the increased acidity of *all* protons adjacent to radical cations, one-electron oxidation of a secondary amine could lead to removal of a proton from *either* α-carbon atom. Therefore, either carbon could be oxidized. As shown below, oxidation of the secondary amine, N-CBA, could give initially the nitrogen radical cation **2**. Removal of the benzyl methylene proton (pathway c) would produce benzaldehyde and cyclopropylamine, whereas removal of the cyclopropyl carbon proton (pathway d) would lead to enzyme inactivation or to benzylamine and some derivative of cyclopropanone if hydrolysis occurred. It is known that electrochemical (2) and photochemical (4) oxidations of nonsymmetrical amines produce more than one product derivable from a common nitrogen radical cation intermediate.

-e$^-$ d -H$^+$ -e$^-$ 2 c -H$^+$ -e$^-$ PhCHO H$_2$O

Enzyme inactivation or hydrolysis to PhCH$_2$NH$_2$

In our studies on the inactivation of MAO by N-CBA (6), we found that every time a proton is released from the cyclopropyl carbon of N-CBA, the enzyme is inactivated. The amount of enzyme inactivated, then, should be a measure of the amount of cyclopropyl carbon oxidized. The approach taken for the study of the mechanism of MAO-catalyzed amine oxidations is an indirect one. The ratio of benzyl methylene carbon oxidation to that of the cyclopropyl carbon which is observed during MAO-catalyzed oxidation is compared with the same oxidation ratio observed during electrochemical oxidation. Since the electrochemical amine oxidation mechanism has been established (2), a correlation in these two systems may suggest intermediates and mechanisms.

Pig liver MAO (7) was treated with [phenyl-^{14}C]-, [cyclopropyl-1-^{3}H]-, and [benzyl-7-^{3}H]N-CBA and the ratio of benzyl methylene to cyclopropyl carbon oxidation determined. In the case of [phenyl-^{14}C]N-CBA, the ratio was given by the amount of [phenyl-^{14}C] benzaldehyde (or benzoic acid as a result of air oxidation) produced (Dowex 50) to the amount of enzyme inactivated. This ratio, 0.6–0.8 to 1, indicates that N-CBA can be partitioned between oxidation of the two α-carbon atoms. When [cyclopropyl-1-^{3}H]N-CBA was used, the ratio was derived from the amount of [1-^{3}H]cyclopropylamine to $^{3}H_2O$ (which is the same as the amount of enzyme inactivated) isolated by ion-exchange chromatography. This ratio was found to be 2.3 to 1; the increase in the ratio can be attributed to the higher activation energy required to break the C-^{3}H bond of the cyclopropyl group relative to the C-H bond of the benzyl methylene carbon. When the benzyl methylene carbon contains a tritium, the oxidation ratio, as measured by the ratio of tritiated non-amines to inactivated enzyme, shifts in favor of cyclopropyl carbon oxidation (0.3–0.4 to 1). These results can be explained in terms of an amine radical cation intermediate which can partition between removal of the benzyl methylene proton or the cyclopropyl proton. When only protons are affixed to these carbons, the lower energy pathway is removal of the cyclopropyl proton, but when a tritium is attached to the cyclopropyl group, there is a shift in activation energy favoring removal of the benzyl methylene proton. A tritium placed on the benzyl methylene carbon results in a larger preference for cyclopropyl proton removal than when protium is at both positions. In other words, there is a tritium isotope effect on the *partitioning* of proton removal. This is consistent with a radical cation intermediate. However, since this is not the rate-determining step in the overall mechanism (5), this partitioning isotope effect is not reflected in the overall rate of the reaction.

Although the results are consistent within themselves, it was necessary to relate them to a system in which the mechanism of oxidation was known, i.e., the electrochemical amine oxidation. Controlled potential electrolysis of [phenyl-^{14}C]N-CBA produced non-amine radioactivity (as a result of benzyl methylene oxidation) and [phenyl-^{14}C]-benzylamine (as a result of cyclopropyl oxidation) in the ratio of 0.2 to 1. As in the case of the MAO-catalyzed oxidation of [phenyl-^{14}C]N-CBA, oxidation of the "abnormal carbon" (the cyclopropyl carbon instead of the benzyl methylene carbon) was predominant. When N-[cyclopropyl-1-^{3}H]CBA was electrochemically oxidized, this ratio increased to 0.6 to 1. Considering the very different conditions for the enzyme-catalyzed (buffer, pH 9.0) and electrochemical (acetonitrile) reactions, the oxidation ratios are quite similar. The two important comparisons are, first, that in both cases the carbon not present in the normal substrates of MAO (i.e., the cyclopropyl carbon) is oxidized preferentially. Second, when a tritium is placed at the cyclopropyl carbon, the oxidation ratio increases by a factor of about 3 in both cases. These results suggest that the known mechanism of electrochemical amine oxidation may be related to the mechanism of MAO-catalyzed amine oxidation. Recently, Krantz et al. (8) also suggested a radical cation mechanism based on their studies of reactions of tertiary amines with photogenerated triplet 3-methyllumiflavin.

Mechanistic studies of electrochemical (2), chemical (3), and photochemical (4) amine oxidations have indicated that the transfer of the electron from the nitrogen is the slow step in the oxidation process followed by a fast proton removal and transfer of a second electron. If this is so enzymatically, it may be unlikely that the radicals will be observable by esr spectroscopy using the substrate; a pseudosubstrate may have more chance for success.

It should be noted that the Hammett relationship found for substituted benzylamines (9) as MAO substrates is perfectly consistent with a radical cation mechanism as well as a carbanion mechanism. However, because of the very high pKa value for the alpha protons in the monoamine free base form, it is highly unlikely that the substrates are oxidized via a carbanionic mechanism.

References

1. Silverman, R.B., Hoffman, S.J., and Catus, W.B., III (1980) *J Am Chem Soc* 102:7126.
2. Mann, C.K. and Barnes, K.K. (1970) *Electrochemical Reactions in Nonaqueous Systems*. New York: Marcel Dekker, Chap. 9.
3. (a) Lindsay Smith, J.R. and Mead, L.A.V. (1973) *J Chem Soc. Perkin Trans* 2:206. (b) Smith, P.A.S., and Loeppky, R.N. (1967) *J Am Chem Soc* 89:1147.
4. (a) Cohen, S.G., Parola, A., and Parsons, G.H. (1973) *Chem Rev* 73:141. (b) Lewis, F.D. and Ho, T. (1980) *J Am Chem Soc* 102:1751.
5. McEwen, C.M., Jr., Sasaki, G., and Jones, D.C. (1969) *Biochemistry* 8:3952.
6. Silverman, R.B. and Hoffman, S.J. (1980) *J Am Chem Soc* 102:884.
7. Oreland, L. (1971) *Arch Biochem Biophys* 146:410.
8. Krantz, A., Kokel, B., Claesson, A.,, and Sahlberg, C. (1980) *J Org Chem* 45:4245.
9. (a) Zeller, E.A., Hsu, M., Li, P.K., Ohlsson, J., and SubbaRao, K. (1971) *Chimia* 25:403. (b) Hellerman, L., Chuang, H.Y.K., and DeLuca, D.C. (1972) *Adv Biochem Psychopharm* 5:327.

Published 1982 by Elsevier North Holland, Inc.
Vincent Massey and Charles H. Williams, Editors
Flavins and Flavoproteins

CHAPTER 36

Stopped-Flow and Steady-State Kinetic Studies on the Reaction Mechanism of Bovine Liver Monoamine Oxidase

Mazhar Husain, Dale E. Edmondson,[1] and Thomas P. Singer

Department of Biochemistry and Biophysics, University of California, San Francisco, California and Molecular Biology Division, Veterans Administration Medical Center, San Francisco, California

The mitochondrial monoamine oxidase (MAO) (EC 1.4.3.4), a ubiquitous flavoprotein, catalyzes the oxidative deamination of various monoamines to form the corresponding aldehyde, hydrogen peroxide and ammonia. The enzyme consists of two subunits of

$$RCH_2NH_2 + O_2 + H_2O \rightarrow RCHO + H_2O_2 + NH_3$$

approximately equal molecular weight (~60,000), only one of which bears an FAD, covalently attached via 8 α-methylene in a thioether linkage to cysteine. Until now, kinetic approaches used to elucidate the mechanism of this interesting and important enzyme included only steady-state methods. Parallel-line double reciprocal plots were obtained for enzyme from different sources using a variety of amine substrates (1–4). These data were consistently interpreted as indicating that MAO operates by a substitution (ping-pong) mechanism. Studies with other flavoprotein oxidases have shown, however, that steady-state data are often insufficient to establish a ping-pong mechanism, unless supplemented by pre-steady-state data (5,6). Study of the two half-reactions by rapid reaction techniques has become possible only recently because of the availability of a procedure for preparing substantial quantities of the enzyme in homogeneous form (7).

The present paper summarizes the results and interpretations of stopped-flow measurements of the two half-reactions, as well as steady-state experiments using benzylamine (BA), α,α-dideutero benzylamine, and β-phenylethylamine (PEA) as substrates. Because of the limited space available, some of the pertinent data will be presented elsewhere in this volume (8).

Methods

MAO was purified from bovine liver (7). [α,α-^{2}H]-Benzylamine was synthesized by reducing benzonitrile with $LiAlD_4$ (9). All kinetic experiments were

[1]Present address: Department of Biochemistry, Emory University, School of Medicine, Atlanta, Georgia.

performed at 25° in 50 mM HEPES buffer pH 7.5, containing 0.5% (w/v) Triton X-100. Initial rates for the oxidation of BA (both α,α-^{1}H and α,α-^{2}H) were measured spectrophotometrically (10). With PEA as substrate, initial rates were measured with an oxygen electrode. Rates are expressed as moles min^{-1} mole of $enzyme^{-1}$. Rapid reaction kinetic experiments were performed using an Aminco-Morrow stopped-flow spectrophotometer interfaced with a Nova 2/4 minicomputer.

Results and Discussion

Initial rate measurements were made both by conventional methods using catalytic amounts of the enzyme and by stopped-flow (11). The two methods showed good agreement. When initial rates were plotted in double reciprocal form with amines and O_2 as the variable substrates, linear parallel line patterns resulted for all the three substrates used. Estimates of Michaelis-Menten parameters for BA from steady-state were: V_{max} (∞ BA and ∞O_2)~800 min^{-1}, $K_m^{BA}(\infty O_2)=0.55$ mM, and $K_m^{oxygen}(\infty BA)=0.28$ mM. Substitution of deuterium for hydrogen at the α-carbon in BA resulted in a kinetic isotope effect of ~7 on V_{max}, with little or no effect on the K_m values. This large kinetic isotope effect is a strong indication that the dehydrogenation of the substrate is fully rate-limiting in the oxidation of BA.

Steady-state data with PEA as substrate gave a value of 1250 min^{-1} for V_{max} and 0.18 mM for $K_m^{PEA}(\infty O_2)$. These values are only moderately different from those for BA. However, K_m^{oxygen} in the presence of PEA is increased dramatically (10-fold), so that the secondary plot of intercepts vs reciprocal of O_2 concentration appeared to be nearly a second-order kinetic plot.

The addition of substrates to MAO in anaerobiosis resulted in a rapid bleaching of the flavin. Examination of the reaction at different wavelengths showed that the flavin is reduced directly to its hydroquinone form, without the transient formation of semiquinone or charge transfer-type intermediates. The rate of reduction of the flavin chromophore followed pseudo-first-order kinetics. The data in Figure 1 for BA and its dideutero analogue show saturation kinetics, as illustrated by the following scheme:

$$E_{ox} + S \underset{k_2}{\overset{k_1}{\rightleftharpoons}} E_{ox}S \overset{k_3}{\rightarrow} E_rP$$

where E_{ox} and E_r are the oxidized and reduced forms of the enzyme, respectively, K_S the dissociation constant for the enzyme substrate complex, and k_3, the limiting value for the rate of reduction of flavin by a given substrate, can be calculated from double reciprocal plots, such as those in Figure 1 (12) and are summarized in Table 1. The k_3 values for BA and its deuterated analogue agree well with the V_{max} values, substantiating the conclusion reached from the steady-state data that reduction of the flavin by substrate is the rate-limiting step. In contrast, k_3 value for PEA is ~30 times higher than the V_{max}, eliminating the possibility that flavin reduction by the substrate is rate-limiting.

The oxidative half-reaction was studied by stopped-flow by mixing an anaerobic solution of the reduced enzyme with a buffer containing a known concentration of dissolved O_2, and following the rate of appearance of

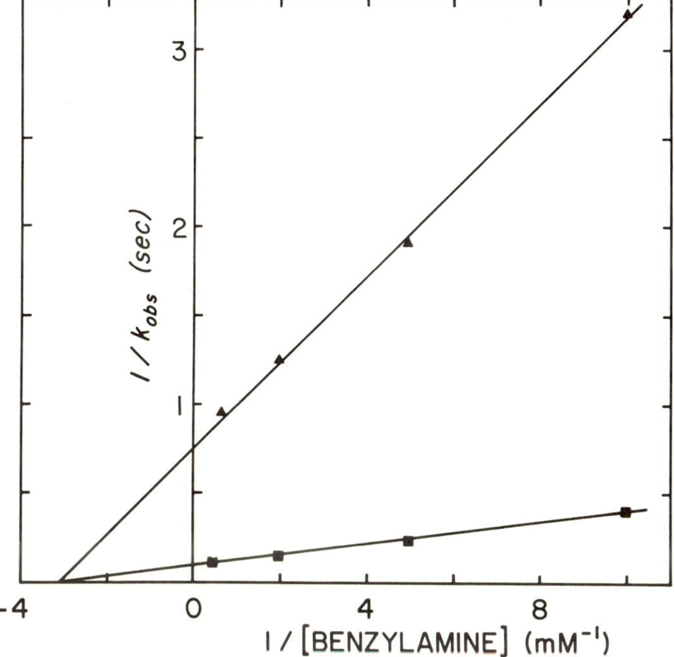

Figure 1. Deuterium isotope effect on the apparent first-order rate constant for the reduction of MAO by benzylamine. The reactions were carried out anaerobically by mixing in the stopped-flow, equal volumes of the enzyme (~15 μM) and buffer solutions containing different amounts of BA (■) or [α, α-^{2}H]-BA (▲). The first-order rate constants were calculated from the absorbance changes at 450 nm.

oxidized flavin band at 450 nm. When the first-order rate constants for the reoxidation of BA-reduced enzyme were plotted against O_2 concentration, a straight line passing through origin was obtained. Evidently, a second-order reaction is involved (3.6×10^5 $M^{-1}min^{-1}$). The dithionite-reduced enzyme gave the same results. The observed rate at any given O_2 concentration is much less than the turnover number for BA. Clearly, then, what we are measuring here cannot be part of the catalytic cycle and most likely represents the reaction of *free* reduced enzyme with O_2, whereas in the turnover studies the species reacting with O_2 is presumably the reduced enzyme-product complex.

Table 1. Kinetic Constants Calculated from Reductive Half-Reaction.

	Substrate		
Constant	**Benzylamine**	**[α, α-^{2}H]-benzylamine**	**β-Phenylethylamine**
k_3 (min^{-1})[a]	700	80	34,300
$K_S = k_2/k_1$ (mM)[a]	0.36	0.33	4.5

[a]Calculated from the slope and intercept values of the double reciprocal plots such as those shown in Figure 1, from the following relationship: k_3 = 1/intercept; K_S = slope/intercept (12).

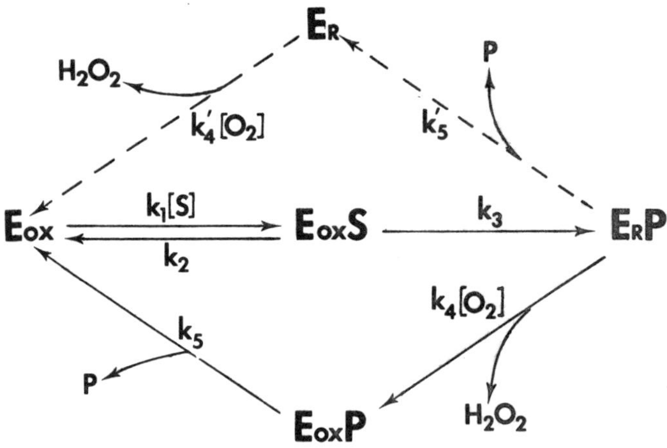

Scheme 1.

Oi et al. have reported that benzaldehyde is released under anaerobic conditions when MAO is reduced by BA in static experiments (2). In contrast to these observations with BA, when the rate of oxidation of PEA is compared with the rate of oxidation of the free, reduced enzyme by O_2, the two are in satisfactory agreement. These data, together with those from turnover and reductive half-reaction studies, strongly suggest that in the oxidation of PEA the enzyme-O_2 half-reaction is rate-determining.

In conclusion, the data presented above suggest that the mechanistic path followed by MAO, i.e., upper or lower loop in Scheme 1, will depend on the nature of the substrate, as proposed for other flavoprotein oxidases by Bright and Porter (6). Thus, in the oxidation of BA the lower loop is followed rather than a binary complex (ping-pong) mechanism suggested by several authors (1–4). In contrast, with PEA, the upper loop—the binary complex mechanism—appears to be operative.

ACKNOWLEDGMENTS

This investigation was supported by the Veterans Administration, by the National Institutes of Health (Program Project HL-16251), and by the National Science Foundation (Grants PCM 78-05382 and PCM 80-01913).

References

1. Tipton, K.F. (1968) *Eur J Biochem* 5:316–320.
2. Oi, S., Shimada, K., Inamasu, M., and Yasunobu, K.T. (1970) *Arch Biochem Biophys* 139:28–37.
3. Fowler, C.J. and Oreland, L. (1979) In *Monoamine Oxidase: Structure, Function, and Altered Functions.* Singer, T.P., Von Korff, R.W., and Murphy, D.L. (eds.) New York: Academic Press, pp. 145–151.
4. Roth, J.A. (1979) In *Monoamine Oxidase: Structure, Function, and Altered Functions.* Singer, T.P., Von Korff, R.W., and Murphy, D.L. (eds.) New York: Academic Press, pp. 153–168.

5. Palmer, G. and Massey, V. (1968) In *Biological Oxidations*. Singer, T.P. (ed.) New York: J. Wiley & Sons, pp. 263–300.
6. Bright, H.J. and Porter, D.J.T. (1975) *The Enzymes* 12:421–505.
7. Salach, J.I. (1979) *Arch Biochem Biophys* 192:128–137.
8. Singer, T.P. and Husain, M., this volume.
9. Amundsen, L.H. and Nelson, L.S. (1951) *J Am Chem Soc* 73:242–244.
10. Tabor, C.W., Tabor, H., and Rosenthal, S.M. (1960) *J Biol Chem* 208:645–661.
11. Gibson, Q.H., Swoboda, B.E.P., and Massey, V. (1964) *J Biol Chem* 239:3927–3934.
12. Strickland, S., Palmer, G., and Massey, V. (1975) *J Biol Chem* 250:4048–4052.

Published 1982 by Elsevier North Holland, Inc.
Vincent Massey and Charles H. Williams, Editors
Flavins and Flavoproteins

CHAPTER 37

The Effect of Superoxide Dismutase on Phenazine Methosulfate Inhibition of Monoamine Oxidase

R.W. Von Korff and A.R. Wolfe

Michigan Molecular Institute, Midland, Michigan

In 1977, we reported (1) a reversible inhibition of monoamine oxidase (MAO) by phenazine methosulfate (PMS). Subsequently we observed (2) the formation of 5-methylphenazine-3-one (MPO) when μM concentration of MAO reacted with PMS (10^{-4} to 10^{-3}M) in the absence of other substrates. Visible spectra for PMS, and MPO are shown in Figure 1. In 1962, Rajagopalan and Handler (3) reported that liver aldehyde dehydrogenase oxidized PMS to MPO (Figure 2, curve 1). We have observed that xanthine oxidase also forms MPO from PMS in the absence of other substrates (Figure 1).

Although aldehyde oxidase appears to give a complete conversion of PMS to MPO, this does not seem to be true for xanthine oxidase. In another experiment (not shown), although the 516 nm maximum reached 90% of its expected value, the 368 nm peak was still 28% higher than the expected value.

All data for MAO were obtained with beef liver enzyme isolated by the method of Salach (4). One unit is equal to 1 μmole of product formed/min or 1 μatom of oxygen consumed/min with catalase present. Superoxide dismutase(SOD) was obtained from the Sigma Chemical Company. SOD units are McCord and Fridovich units as reported by Sigma. We synthesized MPO by the procedure of Kehrmann and Cherpillod (5) and found that it did not inhibit MAO, nor did pyocyanine, formed by the action of light of PMS (see Figure 3).

As part of our investigations on the mode of action of reversible inhibitors of MAO, we have been studying the mechanism by which PMS inhibits MAO. Since (a) superoxide ($O_2^{\cdot-}$) is formed during the reactions of aldehyde oxidase and of xanthine oxidase (6) and during the oxidation of PMSH (7); (b) PMS is partially converted to MPO by the action of KO_2 (Figure 2); and (c) NADH plus PMS reduces nitro blue tetrazolium in a reaction inhibited by SOD (7), we tested the effect of SOD on the inhibitory action of PMS on MAO. SOD protected against PMS inhibition in all assay procedures tested (ΔO by polarographic assay, Figure 4 and Table 3; ΔA_{249nm} using benzylamine as substrate, Table 2 and radiometric assays using [^{14}C] tyramine (Table 3).

The data of Figure 5 show that some protection of MAO against inhibition by PMS is obtained with as little as 0.5–5.0 units of SOD. However, complete protection against the action of PMS is not obtained with as much as 300–600 units of SOD (Tables 1 and 2).

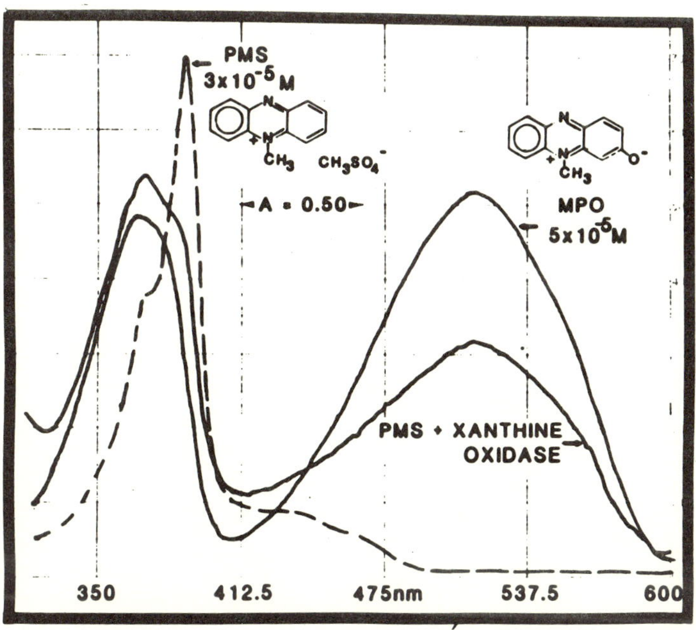

Figure 1. Visible spectra of PMS, MPO, and of the reaction products of PMS (5×10^{-5}M) and xanthine oxidase.

Figure 2. Visible spectra of (curve 1) product of aldehyde oxidase action on PMS (6.5×10^{-5}M) and (curve 2) PMS (6.7×10^{-5}M) treated with a suspension of KO_2 in dimethylsulfoxide.

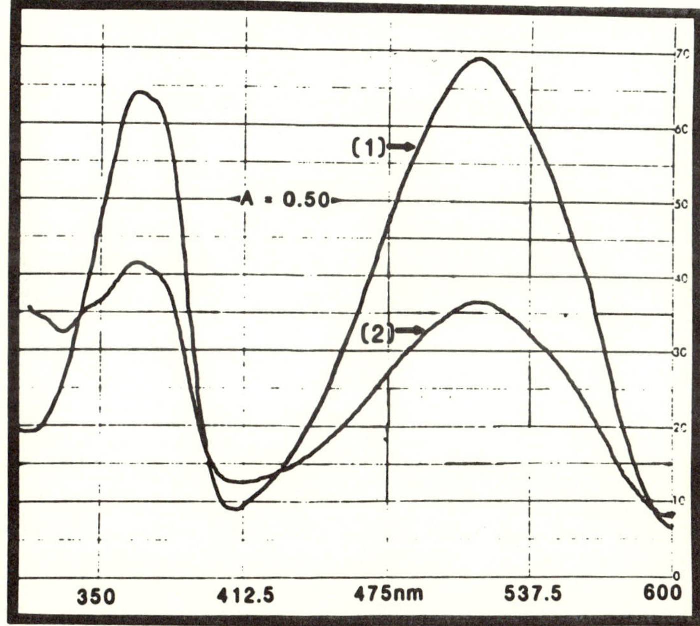

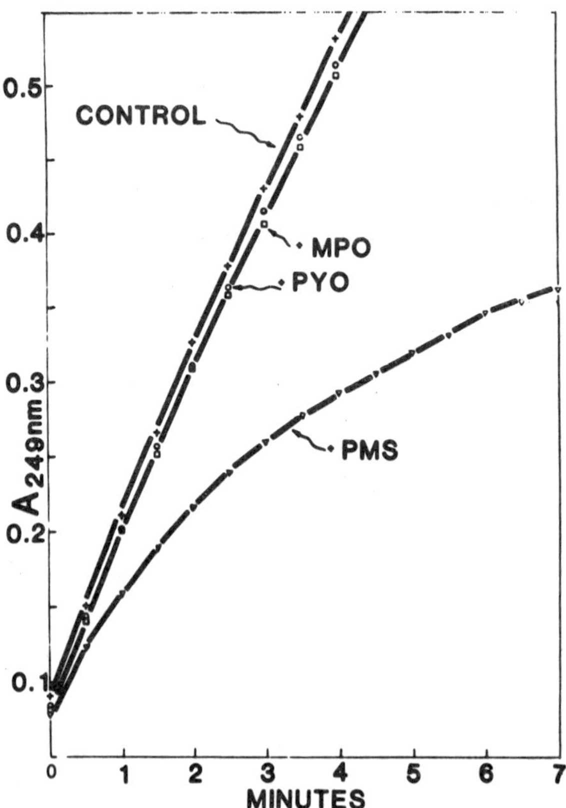

Figure 3. Inhibition of MAO by PMS and the lack of an inhibitory effect of MPO or pyocyanine (PYO) on MAO activity. Benzylamine (3.3 mM) was the substrate.

It was observed by one of us (A.W.) that preincubation of MAO with PMS in the cell compartment of the Cary spectrophotometer for two minutes produced less inhibition than when the incubation was performed with the usual diffuse daylight and fluorescent ceiling lights of the laboratory. Further examination showed that even less inhibition was observed for samples preincubated in the dark (Table 4). In addition, samples preincubated with PMS in the light show an increase in the rate during assay in contrast to those preincubated with PMS in the dark. The latter may decrease or yield linear rates. Over the range of PMS concentrations studied, there appears to be a greater dependence on the PMS concentration for the dark reaction than for samples incubated in the light. The nature of the light activation is not known; intermediates in the light-induced oxidation of PMS to pyocyanine may be involved, although the final oxidation product, pyocyanine is not an MAO inhibitor. Currently, we are examining the effect of PMS on A-type MAO activity to determine if any enzyme transformation occurs, such as that reported by Gorkin (8), to result from lipid peroxidation processes.

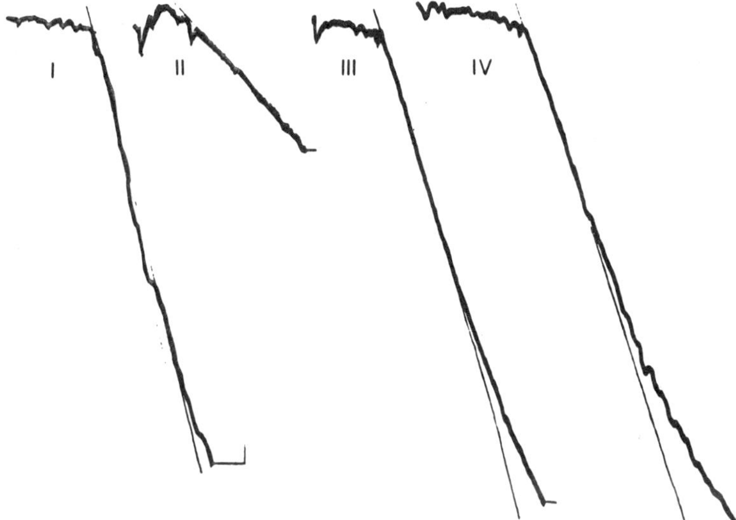

Figure 4. Polarographic ΔO traces showing the effect of SOD on PMS inhibition of MAO. All samples contained catalase, 640 U; Pi, 10 mM, pH 7.4; and MAO, 0.083 U. Samples II–IV were preincubated for 2 min at 30° with 9.4×10^{-6}M PMS. In addition, sample III contained 150 U SOD and sample IV contained 300 U SOD during the incubation period. Reactions were started with benzylamine (3.3 mM).

Table 1. Protection by SOD Against MAO Inhibition by PMS. (Calculated from Data of Figure 4. For Details see Legend for Figure 4.)

PMS (10^{-6}M)	SOD (U)	ACTIVITY (U/ml)	INHIBITION (%)
0	0	1.66	0
9.4	0	0.45	73
9.4	150	1.25	24
9.4	300	1.34	19

Table 2. Protection of MAO by SOD Against Inhibition by PMS (Spectrophotometric A_{249nm} Assays). Benzylamine (3.3 mM) Was the Substrate.

ENZYME	PMS (M)	SOD (U)	MAO (U/ml)	INHIBITION (%)
144-89-22	0	0	4.21	0
	0	84	4.35	0
	1.6×10^{-5}	0	0.45	89
	1.6×10^{-5}	84	3.10	26

Table 3. Protection against PMS Inhibition of MAO by SOD (Radiometric Assays). Samples Were Incubated for 5 minutes at 30°C, Then Terminated with 50 μl 6N HCl and 0.25 ml Passed Through amberlite CG-50 (pH 6.5), Washed with Water and Counted with a Beckman Model LS-9000 Scintillation Counter. The MAO Had a Specific Activity of 3.20 U/mg as measured by ΔA_{249} with Benzylamine (3.3 mM) in Phosphate Buffer (10 mM, pH 7.5).

EXPT	ENZYME	^{14}C TYR (μM)	PMS (M)	SOD (U)	MAO (U/ml)	INHIBITION (%)
1	144-94-26S	136	0	0	0.59	0
	"	"	5×10^{-5}	0	0.03	95
		"	"	240	0.17	71
2	151-83-43	136	0	0	1.15	0
		"	1×10^{-5}	0	0.44	62
	"	"	"	240	0.92	20
			"	600	0.96	16
	"	"	5×10^{-5}	0	0.03	97
	"	"	"	240	0.36	69

Table 4. Stimulation of PMS Inhibition of MAO by Room Light. Benzylamine (3.3 mM) Was the Substrate.

PMS ($M \times 10^{-5}$)	ΔA_{249} SPECTROPHOTOMETRIC		ΔO POLAROGRAPHIC	
	PREINCUBATION CONDITIONS			
	2' LIGHT	2' DARK	2' LIGHT	2' DARK
	INHIBITION (%)			
1	87	20	–	–
2	94	68	–	–
3	–	–	70	54
5	95	86	–	–

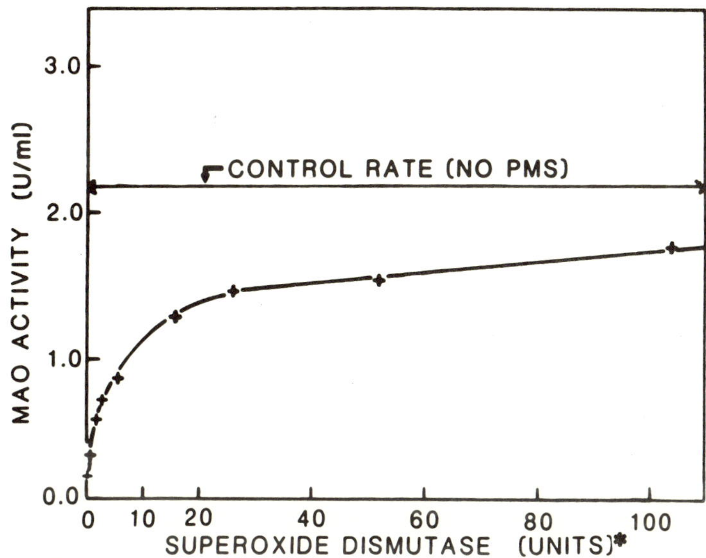

Figure 5. Protection of MAO against inhibition by PMS as a function of superoxide dismutase activity. MAO was preincubated for 2 min on the benchtop with PMS (10^{-5}M) or with PMS plus SOD. The activity was followed by measuring ΔA_{249nm} using benzylamine (3.3 mM) as substrate.

ACKNOWLEDGMENTS

The authors acknowledge with thanks the extensive help of Mr. Richard Hackbarth with laboratory work and in preparing the figures and tables. Supported in part by USPHS Grant No. MH33481.

References

1. Von Korff, R.W. (1977) *Biokhimiya* Consultants Bureau English Ed. 42:302–308.
2. Von Korff, R.W. and Salach, J.I. (1978), unpublished.
3. Rajagopalan, K.V. and Handler, P. (1962) *Biochem Biophys Res Commun* 8:43–47.
4. Salach, J.I. (1979) *Arch Biochem Biophys* 192:128–137.
5. Kehrmann, F. and Cherpillod, F. (1924) *Helv Chim Acta* 7:973–980.
6. McCord, J.M., Beauchamp, S., Goscin, S., Misra, H.P., and Fridovich, I. (1973) In *Oxidases and Related Redox Systems*. T.S. King, H.S. Mason, and M. Morrison (eds.) University Park Press, Vol. 1, pp. 52–53.
7. Nishikimi, M., Appaji, Rao, N., and Yagi, K. (1972) *Biochim Biophys Res Commun* 46:849–854.
8. Gorkin, V.Z. (1979) In *Monoamine Oxidase, Structure, Function,* and *Altered Functions*. T.P. Singer, R.W. Von Korff, and D.L. Murphy (eds.) New York: Academic Press, pp. 347–350.

Published 1982 by Elsevier North Holland, Inc.
Vincent Massey and Charles H. Williams, Editors
Flavins and Flavoproteins

CHAPTER 38

Photochemical Reactions of a Model Flavin with Inhibitors of MAO: Relevance to the Enzyme-Induced Reaction

Allen Krantz,[1]* Bruno Kokel,* Alf Claesson,† and Christer Sahlberg†

**The Departments of Chemistry and Pharmacology, State University of New York at Stony Brook, Stony Brook, New York; †Uppsala University, Biomedical Center, Uppsala, Sweden*

Krantz (1,2) has previously reported that N-2,3-butadienyl-N-benzylmethylamine (**1**, R′=$CH_2\phi$) forms a 1:1 covalent adduct with mitochondrial monoamine oxidase [MAO, (E.C. 1.4.3.4.)], which is not a flavocyanine but has the character of a reduced (alkylated) flavin. In contrast, the isomeric N-2-butynyl-N-benzylmethylamine (**2**, R=$CH_2\phi$), when treated with MAO under identical conditions, forms a flavocyanine adduct analogous to that observed for pargyline with MAO (1,3).

In connection with the elucidation of the structure of the adduct between **1** (R′=$CH_2\phi$ or CH_3) and MAO, we have irradiated **1** (R′=$CH_2\phi$ or CH_3) in the presence of 3-methyllumiflavin (3-MLF). Surprisingly, the allenic amine **1** (R′=CH_3 or $CH_2\phi$) reacted with 3-MLF to produce flavocyanine **3**, in a photoreaction which was isosbestic (isosbestic points with **1**, R′=CH_3, are at 411, 280, and 253 nm). As judged from the electronic spectra of products, the reaction to give flavocyanine adducts selectively is not confined to **1** but occurs with other 3° allenic amines

4, $R_1=CH_3$, $R_2=R_3=H$, $R_4=CH_2\phi$;

$$R_{1\text{-}3}=H,\ R_4=\begin{array}{l}\diagdown\\ CHCH_2\phi,\\ |\\ CH_3\end{array}$$

and with the only secondary amine (**4**, $R_{1\text{-}4}=H$) that was tested (but not with primary amines). With two geminal substituents on the allene (**4**, $R_1=R_2=CH_3$, $R_4=CH_3$, $R_3=H$, or **4**, $R_1=R_2=H$, $R_3=R_4=CH_3$), the product spectrum is that of a reduced flavin. Of additional interest is the observation that the photoreaction of 3-MLF with β, γ-acetylenic amine **2**, which leads to

[1]Present address: Syntex Research, 3401 Hillview Avenue, Palo Alto, California.

both the C_{4a}-N_5 adduct **5** and flavocyanine **3** (and minor products), gives a more complex reaction mixture than its allenic amine isomer **1** ($R'=CH_3$ or $CH_2\phi$).

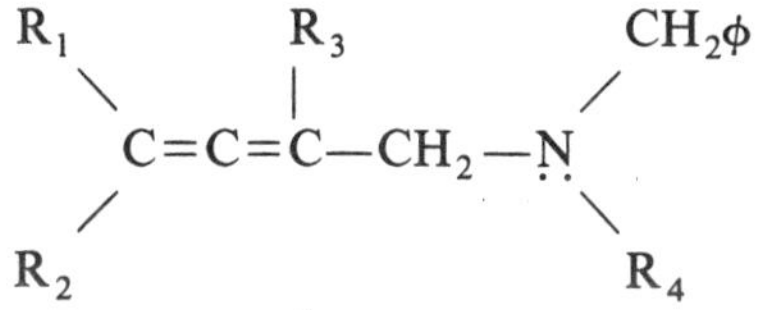

$$\begin{array}{c} R_1 \quad\quad R_3 \quad\quad\quad CH_2\phi \\ \diagdown \quad\quad | \quad\quad\quad\quad \diagup \\ C{=}C{=}C{-}CH_2{-}\ddot{N} \\ \diagup \quad\quad\quad\quad\quad\quad \diagdown \\ R_2 \quad\quad\quad\quad\quad\quad\quad R_4 \end{array}$$

4

Although the mechanisms of both the above enzymatic and photoreactions of amines with flavins are matters of continuing research and intense controversy, a reasonable working hypothesis can be formulated which rationalizes the formation of the known products. The primary assumption is that the enzymatic inhibition and the photoreaction resemble the turnover of substrate by enzyme, in that they are all redox reactions leading to imines and reduced flavin. Hence both the $\underline{C_{4a}}-\underline{N_5}$ and flavocyanine products can be accounted for by assuming a competition between the nucleophilic C_{4a} and N_5 atoms of the reduced flavin, for the strongly electrophilic acetylenic or (allenic) immonium ion (Scheme 1).

An entirely parallel path for the allenic amine involving the allenic immonium ion ($>C{=}C{=}CH{-}C{=}N^{+}<$) may be written.

The photoredox reaction is in accord with the well-known roles of (excited) flavins (4) and amines (5–7) as one-electron acceptors and donors, respectively.

Weller (5) has uncovered two different mechanisms for electron transfer reactions between excited molecules and amines. One path involves intermediate formation of an ion pair, and is characteristic of exothermic reactions in polar media.

$$A^* + D \rightarrow A_s^- \cdots D_s^+ \qquad \text{Type I}$$

solvated ion pair

The other leads to the formation of an excited molecular complex $(A^- D^+)^*$;

$$A^* + D \rightarrow (A^- D^+)^* \qquad \text{Type II}$$

The latter represents an excited charge-transfer complex in which the electrons of the complex partners are mutually coupled by exchange. Reactions occurring in nonpolar media often conform to the type II path, rather than producing discrete ions in an endothermic process.

Like the examples of Weller, we suggest that flavin photoreactions with amines in aqueous media, and the corresponding enzymatic reaction, may represent two extremes involving electron transfer. The photoreaction proceeds essentially via a Type I Weller mechanism (with full electron transfer between triplet flavin and amine) mediated by proton transfer from solvent. In aprotic polar solvents, the exothermic reaction can be formulated as in Equation (38.1).

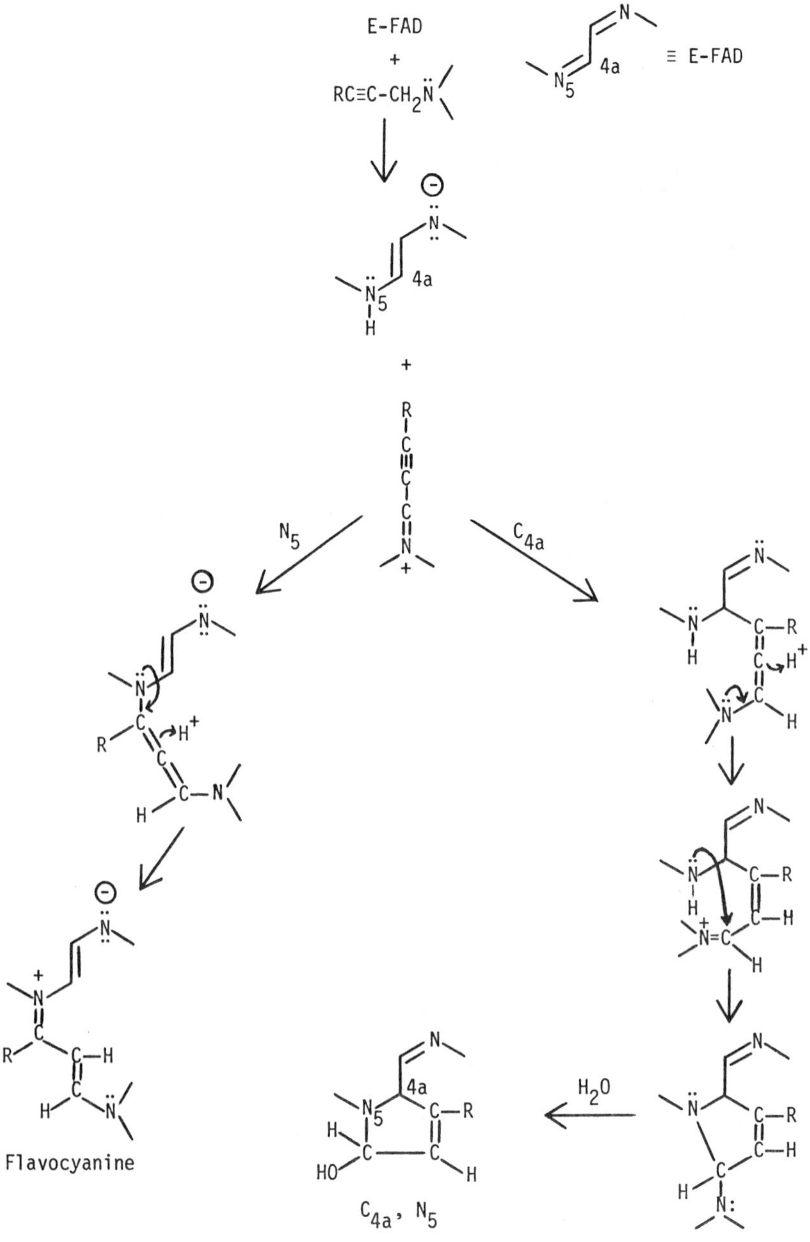

E-FAD + RC≡C-CH$_2$N̈< → ... ≡ E-FAD

N_5 ; C_{4a} ; H_2O

Flavocyanine

C_{4a}, N_5

Scheme 1.

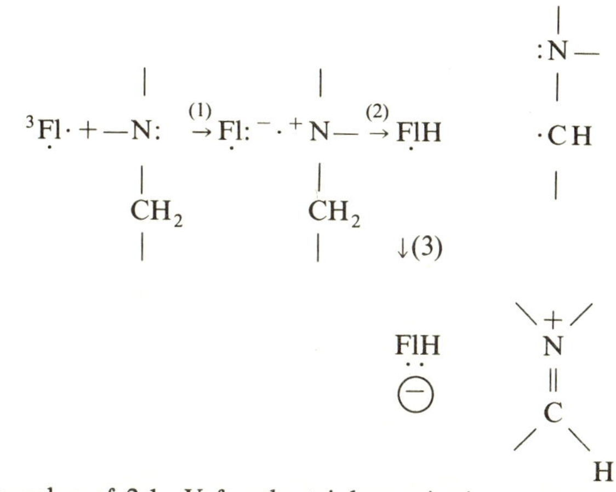

$$^{3}\dot{Fl}\cdot + -\underset{\substack{|\\CH_2\\|}}{\overset{|}{N}}: \xrightarrow{(1)} \dot{Fl}:^{-}\cdot\ {}^{+}\underset{\substack{|\\CH_2\\|}}{\overset{|}{N}}- \xrightarrow{(2)} \dot{Fl}H \quad \cdot CH \text{ (with } :N- \text{ above)}$$

$$\downarrow (3)$$

$$\ddot{Fl}H^{\ominus} \quad \overset{\diagdown + \diagup}{N} = \underset{\diagup \quad \diagdown H}{C} \tag{38.1}$$

(Assuming a value of 2.1 eV for the triplet excitation energy of 3-MLF (8), −0.24 eV for the reduction potential of 3-MLF (9), and a range of 0.73–1.23 eV for 3° amines (10), electron transfer to triplet 3-MLF is exothermic by 0.6 to 1.1 eV.)

Structure 1, 2, 3, 4, 5

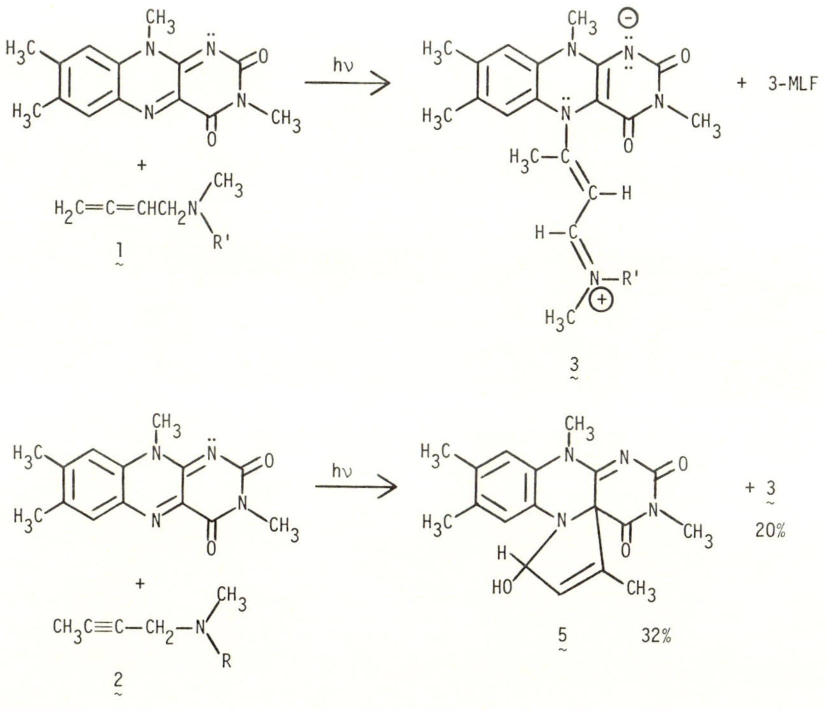

Subsequent steps involving proton transfer within the radical pair (Step 2) and electron transfer (Step 3) would produce the reduced flavin-imine (redox) couple.

The enzymatic reaction probably involves molecular complex formation in which charge-transfer interactions between amine and flavin play a crucial role in stabilizing intermediates and activated complexes that would otherwise be kinetically inaccessible.

We speculate that the formation of distinct enzymatic and photoadducts from **1** with flavins may be traced to the forces that determine binding at the active site of the enzyme (and thus site-specific attack on the flavin prosthetic group), which are certain to be different from those forces that determine the structure of molecular or encounter complexes generated from amines and photo-excited flavins.

ACKNOWLEDGMENTS

This work was supported by Grant No. 1 R01 NS13220 from the National Institutes of Health to A. Krantz, Biomedical Research Support Grant No. 2S07RR057-3607, and a stipend from the donors of the Petroleum Research Fund, administered by the American Chemical Society, to B. Kokel, for which we express our deep gratitude. Support (A.C.) from IF:s Stiftelse för farmaceutisk forskning, Apotekare C. D. Carlssons stiftelse, and Lennanders fond is also gratefully acknowledged.

References

1. Krantz, A. and Lipkowitz, G. (1977) *J Am Chem Soc* 99:4156.
2. Krantz, A., Kokel, B., Sachdeva, Y.P., Salach, J., Claesson, A., and Sahlberg, C. (1979) *Drug Action and Design: Mechanism-Based Enzyme Inhibitors.* Kalman, T. (ed.), New York: Elsevier North Holland, p. 145.
3. Gärtner, B., Hemmerich, P., and Zeller, E.A. (1976) *Eur J Biochem* 63:211.
4. Novak, M., Miller, A., Bruice, T.C., and Tollin, G. (1980) *J Am Chem Soc* 102:1465, and references therein.
5. Knibbe, H., Rehm, D., and Weller, A. (1968) *Ber Bunsenges Phys Chem* 72:257.
6. Cohen, S.G., Parola, A., and Parsons, G.H., Jr. (1973) *Chem Rev* 73:141.
7. Inbar, S., Linschitz, H., and Cohen, S. (1981) *J Am Chem Soc* 103:1048.
8. Sun, M., Moore, T.A., and Song, P.-S. (1972) *J Am Chem Soc* 94:1730.
9. Bruice, T.C. (1976) *Progress in Bio-Organic Chemistry.* Kaiser, E.T. and Kezdy, F.J. (eds.) New York: John Wiley and Sons, Inc., p. 1.
10. Lindsay Smith, J.R. and Mastheder, D. (1976) *J Chem Soc Perkin Trans II*:47.

Published 1982 by Elsevier North Holland, Inc.
Vincent Massey and Charles H. Williams, Editors
Flavins and Flavoproteins

CHAPTER 39

Iron in Monoamine Oxidase: Effect of Cytochrome Impurities on Spectra of the Enzyme and Its Adducts with Suicide Inhibitors

J.I. Salach and W. Weyler

Veterans Administration Medical Center, San Francisco, California, and Department of Biochemistry and Biophysics, University of California, San Franciso, California

Despite the availability of almost pure preparations of monoamine oxidase for a number of years (1–5), it is surprising that the question of its iron content is not resolved. The reason for this is that an iron-containing component appears to be loosely bound to the enzyme. The absorption spectrum of most preparations contains a distinct peak at 410–412 nm long suspected to be due to a cytochrome impurity (3, 6). It is not bleached on reduction of the enzyme by substrate and the maximum appears shifted on reduction with dithionite.

The simple expedient of centrifuging the enzyme prepared from beef liver in a discontinuous sucrose gradient succeeded in removing almost all of the cytochrome impurity from this enzyme preparation.

Materials and Methods

The enzyme from beef liver was prepared by the method of Salach (5), applied to the top of a gradient comprised of successive layers of 65, 55, 50, 45, and 35% sucrose in 0.1 M TEA, pH 8.0. and centrifuged for 17 hr at $107{,}000 \times g_{max}$ in a Beckman SW28 rotor. Following fractionation and pooling of the enzyme from the gradients, purity was verified by slab gel electrophoresis, flavins were determined as described (5), and iron content of trichloroacetic acid extracts or nitric/perchloric acid digests was measured by the method of Brumby and Massey (7). Spectra were recorded either with a Cary 219 or a Cary 14 with a computer interface.

Results and Discussion

The absorption spectrum of the oxidized and substrate-reduced enzyme obtained from the sucrose gradient is shown in Figure 1.

The cytochrome impurity is absent in this preparation and anaerobic reduction by excess substrate gives a typical reduced flavoprotein spectrum.

The iron content of preparations of this purity is shown in Table 1. Nonheme iron was less than 5 mol % of the enzyme flavin in these prepara-

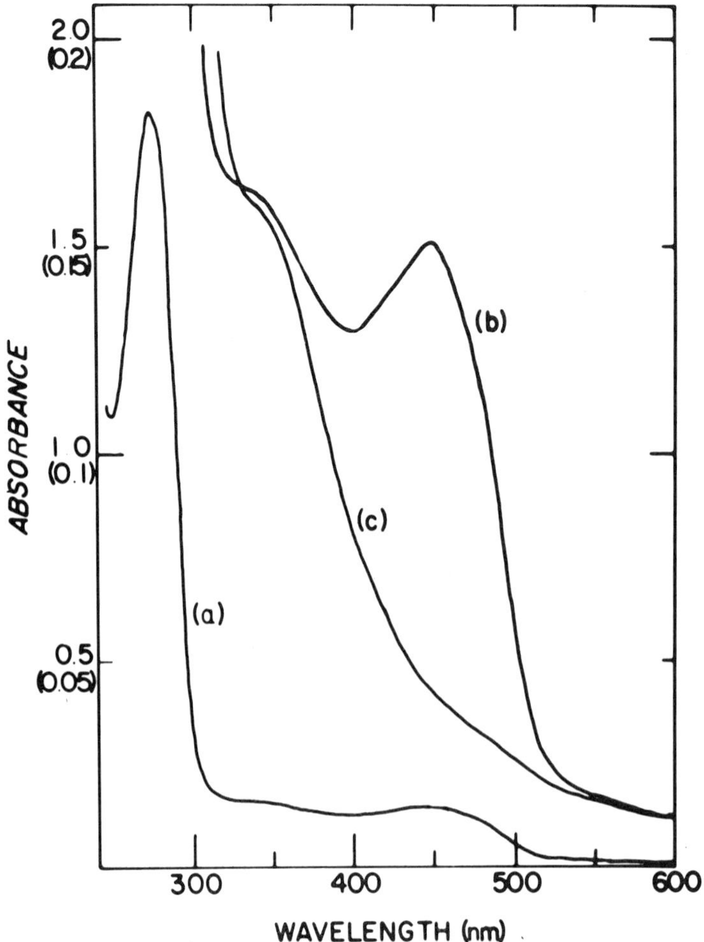

Figure 1. Spectrum of monoamine oxidase after sucrose gradient centrifugation. Enzyme, specific activity 5.4, was in 45% sucrose, 0.1 M triethanolamine, pH 8.0. Spectra were recorded: (a) Absorbance scale: 2.0; the 275 nm : 448 nm ratio was 12.2. (b) Same, absorbance scale: 0.2. (c) Same after reduction with excess benzylamine added anaerobically.

tions; total iron, however, was substantially higher, reflecting the fact that some iron-containing impurity still remains. This was not more than 25 mol % of the enzyme flavin. The spectra shown in Figure 1 were taken with the lower band material in which total iron was approximately 10 mol % of the enzyme flavin.

Unlike reduction by substrate, anaerobic titration with dithionite resulted not only in the bleaching of the 445 nm absorbance of the flavin, but in the appearance of new maxima at 360 and 410–412 nm and a broad absorbance from 440–530 nm. These changes are shown in Figure 2 and are very similar to those described by Massey and Palmer (8) for the red flavin semiquinone produced by photoreduction with EDTA in other flavoenzymes.

Table 1. Iron Content of Monoamine Oxidase Preparations Following Sucrose Gradient Centrifugation.

Sample	FAD (nmols)	Nonheme iron (nmols)	Ratio: NHI/FAD	Total iron (nmols)	Ratio: total iron/FAD
Preparation I					
Upper	63.9	2.8	0.044	15.2	0.238
Lower	58.3	2.7	0.046	5.4	0.093
Preparation II					
Upper	60.3	2.8	0.046	14.0	0.232
Lower	62.0	1.8	0.029	6.4	0.103

Flavins and iron were determined as described in Materials and Methods. All determinations were duplicates. Values shown are the average of the flavin and iron content of the samples.

Figure 2. Titration of monoamine oxidase with sodium dithionite. Enzyme, specific activity 5.4, 24 nmols, was in 50 mM NaPi, pH 7.5 containing sufficient Triton X-100 to make 0.04% (w/v) final concentration. The spectrum was recorded, the sample was made anaerobic, sodium dithionite was added, mixed, and the spectrum was immediately recorded. No addition, (—); 10.8 nmols dithionite (– –); 28.8 nmols dithionite (· · ·); 76 nmols dithionite (– · – ·).

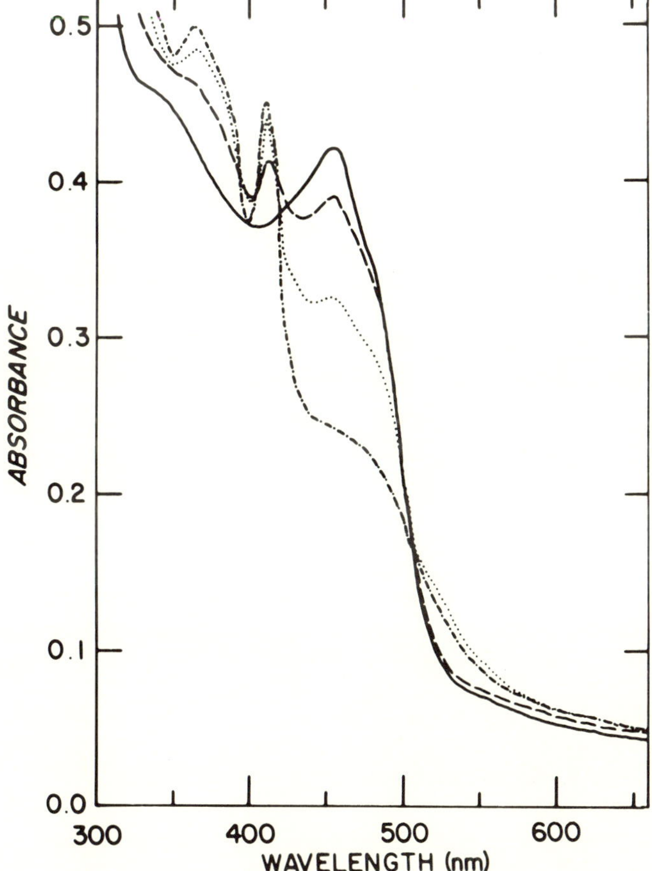

After standing overnight with excess dithionite, this semiquinone absorbance decreased considerably, indicating that further reduction was occurring, though very slowly and incompletely, to the dihydroflavin. This is seen in Figure 3, solid line. When air was admitted to the sample, the semiquinone absorbances reintensified initially, then disappeared as the flavin became fully reoxidized. Sixty minutes after admitting air, the enzyme had recovered 80% of its initial catalytic activity.

Since the 410–412 maximum of the semiquinone underlies the Soret band of the cytochrome impurity of less pure preparations, the formation of the semiquinone by dithionite reduction had not been recognized, although Igaue et al. (9) had observed it in light/EDTA-reduced preparations.

Figure 3. Reoxidation of enzyme following sodium dithionite reduction. The enzyme from the experiment shown in Figure 2 was allowed to stand overnight at 23°C, the spectrum was recorded (—) and the cuvette opened to admit air. Spectra were taken after 30 min (– –) and 60 min (· · ·) exposure to air.

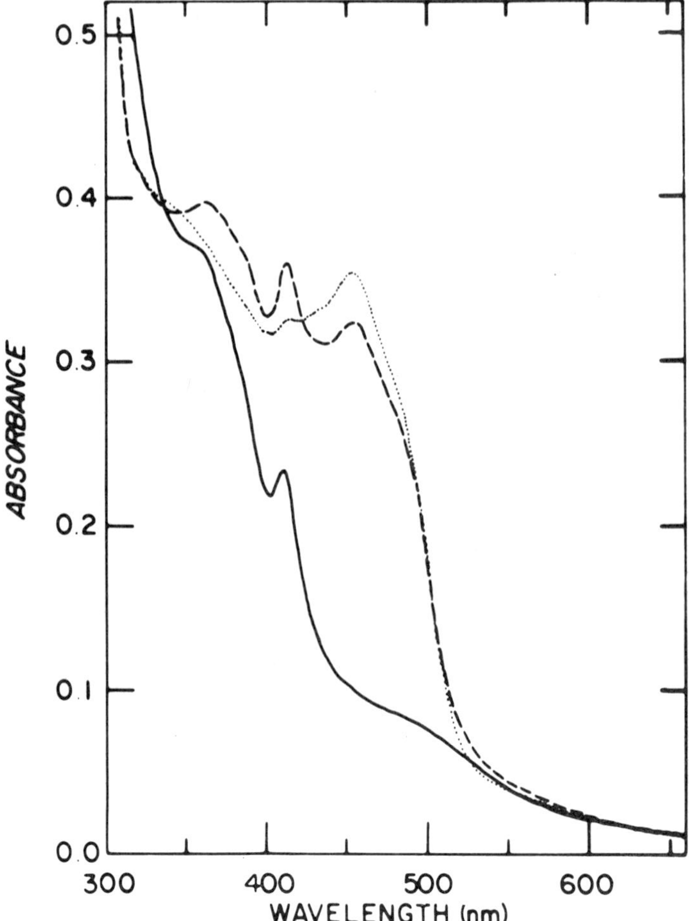

Removal of the cytochromes from this preparation of monoamine oxidase had no significant effect on the spectrum of the adduct formed with the suicide inhibitor *trans*-2-phenylcyclopropylamine (tranylcypromine).

Comparison of the spectra obtained following phenylhydrazine inhibition of enzyme preparations both before and after a sucrose gradient purification, Figures 4 and 5, shows a substantial change as a consequence of cytochrome removal.

The difference spectra of the less pure preparation, Figure 4 (11), showed changes in the Soret region unrelated to the extent or the time course of enzyme inhibition. These were due to slow reactions between the cytochrome and phenylhydrazine, since on removing most of the cytochrome these changes largely disappeared, Figure 5. In addition, the spectrum of the isolated peptide adduct, Figure 6, does not resemble that of the inhibited enzyme, Figure 5, indicating that the adduct initially formed with the enzyme is probably altered in the course of proteolysis and purification of the flavin peptide.

Figure 4. Spectrum of monoamine oxidase preparation containing cytochromes before and after phenylhydrazine inhibition (11). Enzyme, 21 nmols in 45 mM HEPES, pH 7.0 containing 2 mM EDTA and 1% (w/v) Triton X-100 was inhibited at ambient temperature with 250 nmols of phenylhydrazine. Spectra were taken before (—) and 15 min (×—×); 65 min (·—·); 124 min (---); and 20 hr (lower—) later. The inset shows the corresponding difference spectra. The enzyme was 95% inhibited after 5 min.

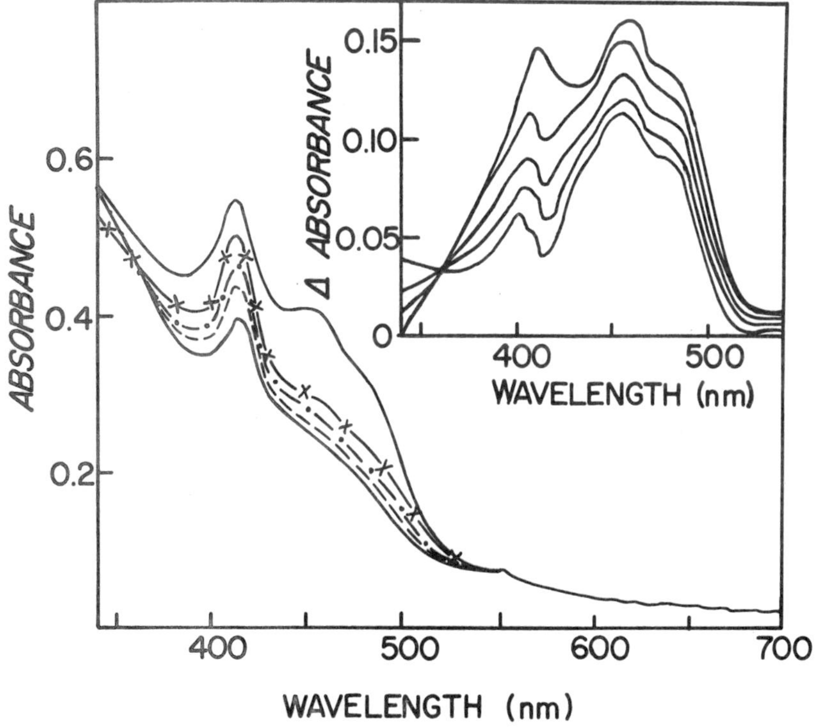

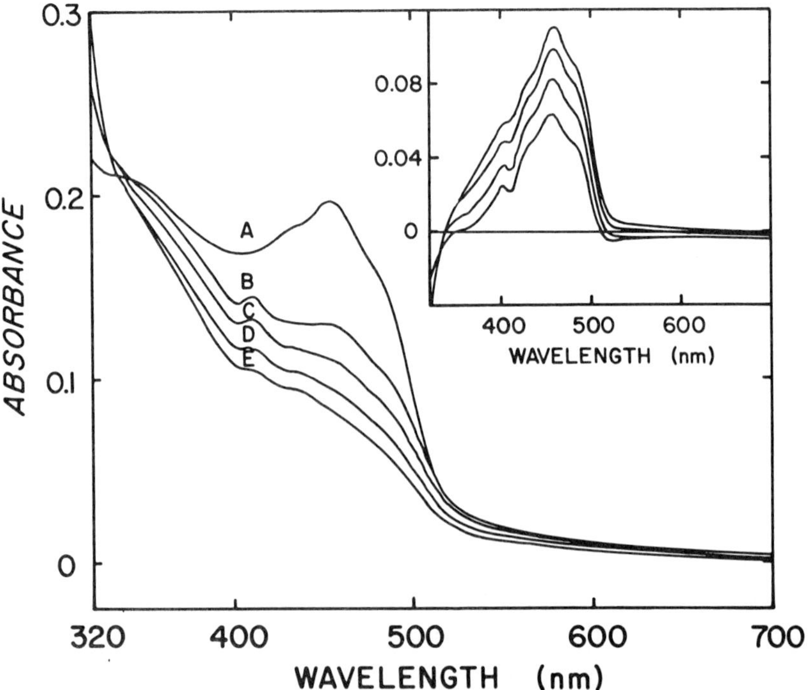

Figure 5. Spectrum of cytochrome-free enzyme before and after phenylhydrazine inhibition. The enzyme, 13 nmols, in the same buffer system as given in Figure 4 was inhibited with 250 nmols of phenylhydrazine. A is before the addition, B through E are 1.8, 6.25, 14.5, and 28 min after adding phenylhydrazine, respectively. Difference spectra are shown in the inset.

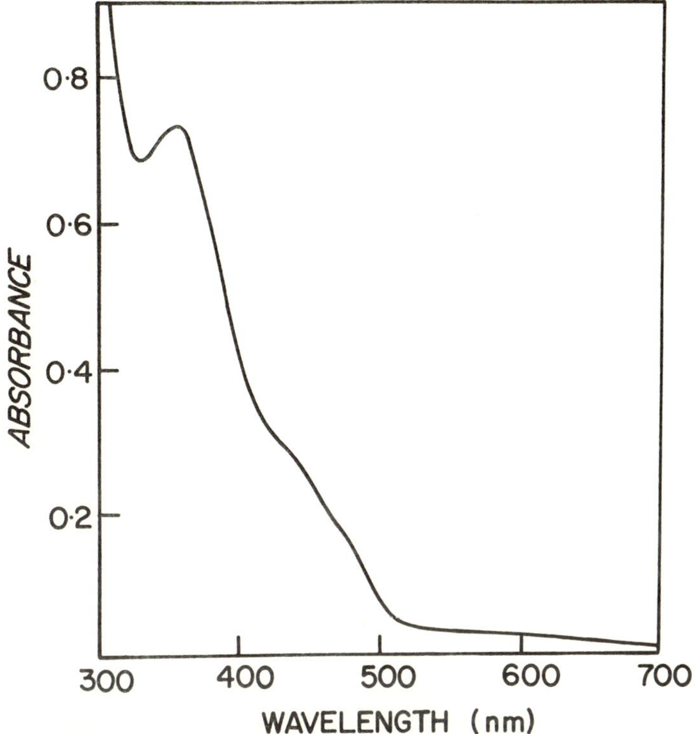

Figure 6. Spectrum of the flavin peptide isolated from proteolytic digests of phenylhydrazine inhibited monoamine oxidase at neutral pH.

ACKNOWLEDGMENTS

This work was supported by the Veterans Administration and the National Institutes of Health Grant #NS-14222.

References

1. Minamiura, N. and Yasunobu, K.T. (1978) *Arch Biochem Biophys* 189:481–489.
2. Youdim, M.B.H. (1976) In *Flavins and Flavoproteins*. T.P. Singer (ed.) Amsterdam: Elsevier, pp. 593–604.
3. Oreland, L. (1971) *Arch Biochem Biophys* 146:410–421.
4. Chuang, H.Y.K., Patek, D.R., and Hellerman, L. (1974) *J Biol Chem* 249:2381–2384.
5. Salach, J.I. (1979) *Arch Biochem Biophys* 192:128–137.
6. Erwin, V.G. and Hellerman, L. (1967) *J Biol Chem* 242:4230–4238.
7. Brumby, P.E. and Massey, V. (1967) *Methods Enzymol* 10:463–474.
8. Massey, V. and Palmer, G. (1966) *Biochemistry* 5:3181–3189.
9. Igaue, I., Gomes, B., and Yasunobu, K.T. (1967) *Biochem Biophys Res Communs* 29:562–570.
10. Maycock, A.L., Abeles, R.H., Salach, J.I., and Singer, T.P. (1976) *Biochemistry* 15:114–125.
11. Kenney, W.C., Nagy, J., Salach, J.I., and Singer, T.P. (1979) In *Monoamine Oxidase: Structure, Function, and Altered Functions*. T.P. Singer, R.W. Von Korff, and D.L. Murphy (eds.) New York: Academic Press, pp. 25–37.

Published 1982 by Elsevier North Holland, Inc.
Vincent Massey and Charles H. Williams, Editors
Flavins and Flavoproteins

CHAPTER 40

Mechanistic Aspects of Mitomycin c Activation by Flavoprotein Transhydrogenases

Jed F. Fisher and Roger A. Olsen

Department of Chemistry, University of Minnesota

Mitomycin c is an antitumor antibiotic that is used for the palliative treatment of adenocarcinomas (1–3). On the basis of model system studies, mitomycin c is believed to require metabolic activation as a necessary component to its *in vivo* action (4). This activation process involves initial reduction of its quinone functional group to the hydroquinone. Under aerobic conditions, the hydroquinone reduces molecular oxygen; under anaerobic conditions, it rearranges to electrophiles that alkylate macromolecules (Figure 1). Although the ability of cell extracts to accomplish this activation has been recognized for some time (5,6), a specific enzyme catalyst has not been identified previously. Such a catalyst is necessary to the study of the chemistry of the reductive activation process. To this end, we have found that several flavin-dependent transhydrogenases possess this catalytic ability.

The initial mechanistic studies have used NADPH as a reducing agent and homogeneous yeast Old Yellow Enzyme (7) as the catalyst. Incubation of mitomycin c with these reagents under anaerobic conditions leads to a progressive consumption of the antibiotic, as monitored by liquid chromatography and by a pronounced color change from blue to purple. This color change (at 550 nm) permits the observation of the kinetic course of the reaction. The kinetics are complex; one observes an initial lag phase, followed by a zero order phase that relaxes into a slower, curved phase as the reaction approaches completion (Figure 2). The lag phase is not abolished by rigorous exclusion of oxygen (maintained by a glucose and glucose oxidase scrubbing system); its origins are not known. Oxygen-initiated radical pathways are excluded, as the overall kinetics are unaffected by superoxide dismutase. The zero order velocity is strictly proportional to enzyme concentration and corresponds to a turnover of 0.2 s^{-1} (0.10 M Tris pH 8.0, 0.5 mM EDTA buffer at 30°C). Complete retention of the catalytic ability of the enzyme (assayed with O_2 as an electron acceptor) is observed following mitomycin c activation. Yet, isolation of this Old Yellow Enzyme shows chromophore incorporation at approximately 370 nm and 560 nm (Figure 3). Increased absorption at these wavelengths is consistent with the presence of mitosenes (from reductively activated mitomycin), although a less equivocal identification is precluded by the strong dependence of the mitosene absorption spectrum on its environment. A substantial portion of this increased absorption is retained with the

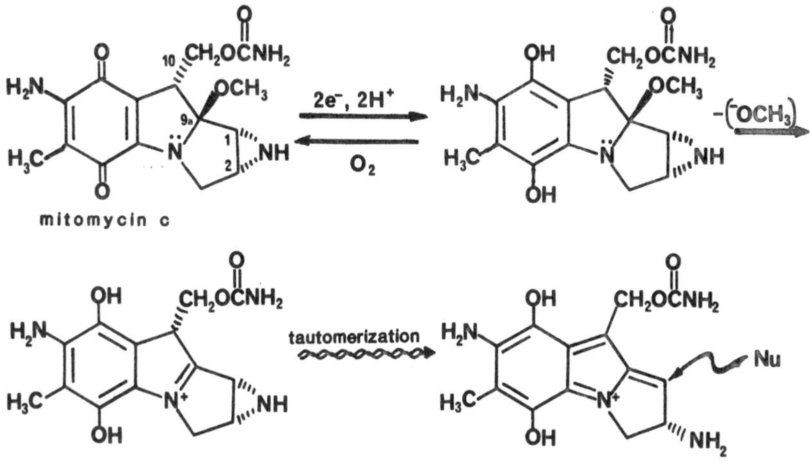

Figure 1. This figure summarizes the events that are believed to occur subsequent to reduction of mitomycin c to the hydroquinone. Loss of the methoxide from C-9a, tautomerization to the eneamine, and internally assisted opening of the aziridine provide an electrophilic intermediate capable of nucleophile addition at C-1. Further reaction leading to carbamate elimination and nucleophile addition to C-10 is possible.

Figure 2. Progress curves at 550 nm at three different Old Yellow Enzyme concentrations, for mitomycin c reductive activation, are illustrated. The reaction is monitored under anaerobic conditions in a 5.00 mm path-length Thunberg cell, in a total volume of 2.0 mL 0.10 M Tris pH 8.0, 0.5 mM EDTA buffer. The reagent concentrations are mitomycin c, 1.50 mM; NADPH, 4.5 mM, and Old Yellow Enzyme catalyst (1.49 μM, 2.98 μM, and 5.96 μM).

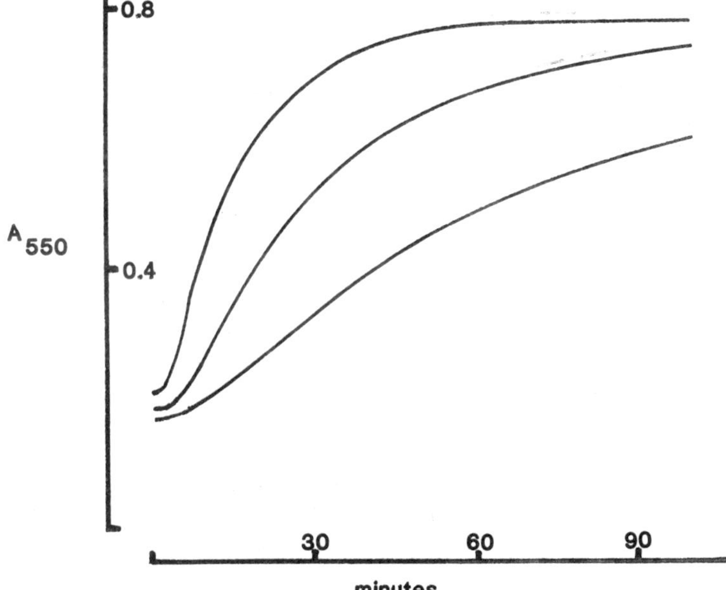

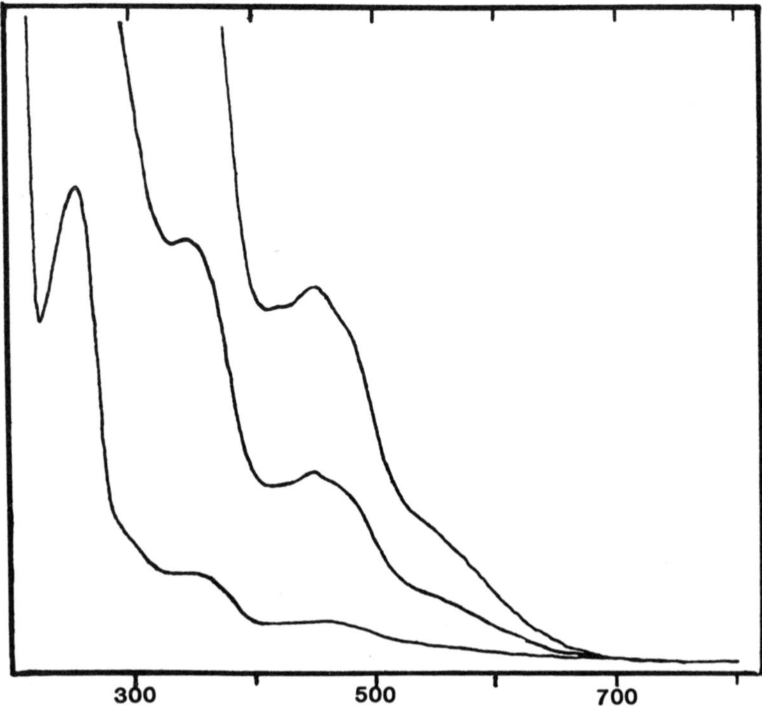

Figure 3. The absorption spectrum of the Old Yellow Enzyme catalyst, isolated and purified following its catalysis of mitomycin c reductive activation, is shown at three different full scale settings. The spectrum shows the following λ_{max} (nm) and approximate extinction coefficients ($M^{-1} cm^{-1}$): 274, 130,000; 368, 23,000; 470, 10,250; 560 sh, 3900.

protein, as the FMN coenzyme is lost, during purification under denaturing conditions. These observations are consistent with random alkylation of exterior amino acid nucleophiles of the enzyme by the mitomycin-derived electrophiles. This implies that the rate at which the hydroquinone is released from the active site exceeds the rate for its conversion to electrophiles. Additional support for the rearrangement processes occurring outside of the active site is obtained from two experiments. In the presence of thiolate nucleophiles, sulfur trapping dominates (8) and the product distribution is independent of reducing agent (dithionite or enzyme). Secondly, inclusion of a second enzyme (yeast alcohol dehydrogenase) leads to a decrease in its specific activity as mitomycin c reductive activation ensues.

Other flavoenzymes that catalyze mitomycin c reductive alkylation are (in decreasing order of efficiency) human erythrocyte cytochrome b_5 reductase, pig heart lipoamide dehydrogenase, and yeast glutathione reductase. The kinetic course for these three enzymes is similar to that for Old Yellow Enzyme. Cytochrome b_5 reductase is the most efficient catalyst (zero order velocity of approximately 10 s^{-1}). This enzyme is of interest due to its poor use of O_2 as an electron acceptor; this permits the study of mitomycin c reduction under aerobic conditions. In the presence of NADH, mitomycin c, and O_2, the cytochrome b_5 reductase catalyzes oxygen reduction only, presumably via the hydroquinone as an electron carrier. Only after complete oxygen depletion do the spectral changes characteristic of mitomycin reductive activation occur.

Several conclusions are suggested: (1) Flavoenzymes with liberal tolerances for the electron acceptor are likely candidates for catalysis of these events. These include enzymes (such as cytochrome b_5 reductase) with relatively positive redox potentials. (2) Release of the hydroquinone into solution is generally faster than its rearrangements to electrophiles. Thus, alkylation of the protein occurs outside of the active site. This does not disrupt normal turnover but may well disrupt essential protein-protein contacts. (3) If the specific alkylation of a particular macromolecule is an essential aspect of the therapeutic activity provided by mitomycin c, it is likely that the mitomycin is activated in a mitomycin-macromolecule complex rather than as the free antibiotic. Alone, mitomycin will be efficient only in oxygen reduction. Under anaerobic conditions, the electrophiles derived from the uncomplexed mitomycin hydroquinone are neither potent nor specific in their reactivity.

ACKNOWLEDGMENTS

We acknowledge the generous support of this research by the following organizations: Bristol Laboratories, for a gift of mitomycin c; the American Cancer Society, for initial support; Du Pont, The University of Minnesota, The Research Corporation, and the Petroleum Research Fund (administered by the American Chemical Society), for grants for equipment; and the National Institute of General Medical Sciences, for sustaining support.

References

1. Crooke, S.T. (1979) *Mitomycin c.* Carter, S.K. and Crooke, S.T. (eds.) New York: Academic Press, pp. 1–4.
2. Franck, R.W. (1979) *Prog Chem Org Nat Products* 38:1–45.
3. Remers, W.A. (1980) *Anticancer Agents Based on Natural Product Models.* Douros, J. (ed.) New York: Academic Press, pp. 131–145.
4. Lown, J.W. (1979) *Mitomycin c.* Carter, S.K. and Crooke, S.T. (eds.) New York: Academic Press, pp. 5–26.
5. Schwartz, H.S., Sodergren, J., and Philips, F. (1963) *Science* 142:1181–1182.
6. Iyer, V. and Szybalski, W. (1964) *Science* 145:55–58.
7. Abramovitz, A. and Massey, V. (1976) *J Biol Chem* 251:5327–5336.
8. Hornemann, U., Keller, P.J., and Kozlowski, J.F. (1979) *J Am Chem Soc* 101:7121–7124.

Published 1982 by Elsevier North Holland, Inc.
Vincent Massey and Charles H. Williams, Editors
Flavins and Flavoproteins

CHAPTER 41

Are Adducts of Thiols with Glyoxylate the Physiological Substrates for Mammalian L-α-Hydroxy Acid Oxidase?

Gordon A. Hamilton and Edward J. Brush

Department of Chemistry, The Pennsylvania State University, University Park, Pennsylvania

Summary

L-α-hydroxy acid oxidase is a flavoprotein found in the peroxisomes of many plants and animals. However, despite its wide distribution in animal tissues its actual physiological substrate is unknown. Recently, we reported (6) that adducts formed from the reaction of glyoxylate with various amines (expecially cysteamine) are good substrates for the related peroxisomal enzyme, D-amino acid oxidase, and we presented evidence that the cysteamine reaction is probably occurring physiologically. In a similar study, we now have found that rat kidney L-α-hydroxy acid oxidase rapidly catalyzes O_2 uptake when both glyoxylate and one of a number of thiols are present. Reactive thiols in this system include ethanethiol, 1-propanethiol 2-mercaptoethanol, N-acetylcysteamine, propane-1, 3-dithiol, coenzyme A and its metabolites, and dihydrolipoic acid which, based on inhibition studies, appears to have a specific binding site on the enzyme. Presumably the substrate is a thiol-glyoxylate adduct, since both thiol and glyoxylate must be present in order to obtain a rapid enzyme-catalyzed reaction. Moreover, the products of the enzymic reactions are hydrogen peroxide and oxalyl thioesters, further confirming that thiol-glyoxylate adducts are the substrates. Several of the adducts have lower K_ms (some, an order of magnitude lower) than α-hydroxybutyrate, which has long been considered a good substrate for the enzyme, and they all exhibit ca the same V_{max} as α-hydroxybutyrate. This high reactivity strongly implies that a thiol-glyoxylate adduct may be the physiological substrate for this enzyme. A possible function for this reaction in metabolic control, mediated either by the oxalyl thioester or by oxalate, is briefly considered.

Introduction

Peroxisomes and microperoxisomes are small organelles (ca 0.1 to 1 μm diameter) which characteristically contain catalase and several oxidative enzymes, including the flavoprotein oxidases, L-α-hydroxy acid oxidase and D-amino acid oxidase (1–3). Although these organelles are present in virtually all animal cells (4), their metabolic function has remained unknown. However, there is now increasing evidence that they may be involved in controlling metabolism (2). The specific function of the flavoprotein oxidases has remained even more obscure because, until recently, no reasonable suggestion concerning the physiological substrates for these enzymes had been made. Their traditional good substrates namely, D-amino acids for D-amino acid oxidase, and L-α-hydroxy acids with four or more carbons (5) for at least one type of L-α-hydroxy acid oxidase, are not present to any significant extent physiologically.

Recently, we have been considering the possibility that perhaps the true physiologic substrates for these enzymes are not simple α-amino or α-hydroxy acids, but instead are adducts formed in situ from the combination of two or more other molecules. From chemical considerations, as well as from the known specificity of the enzymes, it became evident that adducts of glyoxylate with various nucleophiles might be effective substrates. Following up on these rationalizations we experimentally demonstrated (6) that adducts of glyoxylate with various amines are substrates for D-amino acid oxidase, and we presented arguments that the most likely physiological substrate for this enzyme is the cysteamine-glyoxylate adduct, thiazolidine-2-carboxylate. Because an adduct of any nucleophile with glyoxylate would be an α-hydroxy acid [Equation (41.1)], more recently we have been investigating whether any such unstable intermediates

$$\mathrm{RXH + O{=}CH{-}COO^- \rightleftharpoons RX{-}\underset{\displaystyle OH}{\underset{|}{CH}}{-}COO^-}$$

$$X = O,S,NH \qquad (41.1)$$

might be substrates for rat kidney L-α-hydroxy acid oxidase. Some results of this investigation are summarized here.

Experimental Procedure

Rat kidney L-α-hydroxy acid oxidase was purified by the method of Cromartie and Walsh (5,7); with D,L-α-hydroxybutyrate as a substrate, it had activity comparable to that reported for the homogeneous enzyme. Pantetheine was prepared by hydrogenation of pantetheine (8), and phosphopantetheine as described by Michelson (9).

For all kinetic data reported here the enzyme activity was determined by following oxygen consumption using an O_2 electrode. The reactions were run at 25°C with an air atmosphere ($[O_2]=0.25$ mM) unless specified otherwise. Each reaction solution (3 ml) contained 0.1 M sodium phosphate buffer, pH 7.5, 2 to 3 mM EDTA, 11 μg/ml crystalline catalase, ca 60 μg/ml L-α-hydroxy acid oxidase, and glyoxylate and the various nucleophiles at the given concentrations. Initial rates of enzyme catalyzed O_2 uptake were corrected for the small nonenzymic autooxidation rate of the thiol (usually less than 5% the enzymic rate), and were multiplied by a factor of two because catalase is present. Enzyme activity is given in katals/kg; one katal is equal to one mole of substrate converted per sec.

In product studies, the chromatographic analysis of oxalyl monohydroxamate, formed by treating reaction mixtures with hydroxylamine, followed closely the procedure of Quayle (10).

Results

In the absence of any added nucleophile, glyoxylate (presumably its hydrate) is a poor substrate for the enzyme, but if its concentration is kept at 1 mM or less, the rate due to glyoxylate alone is negligible (less that 2% of the rate with 25 mM α-hydroxybutyrate). With 1 mM glyoxylate and 10 mM nucleophile present, it was found in preliminary experiments that simple alcohols and amines do not lead to any additional enzyme-catalyzed O_2 uptake, but various thiols cause a rapid enzyme-catalyzed reaction. As the results in Table 1 indicate, several thiol-glyoxylate adducts are at least as good substrates for hydroxy acid oxidase as α-hydroxybutyrate, which is one of the better known substrates for the enzyme (5). However, not all thiols are reactive in this system. At 10 mM thiol and 1 mM glyoxylate, 2-mercaptopropionate, mercaptoacetate, L-cysteine, o-mercaptobenzoate and 3-mercaptopicolinate cause no appreciable enzyme-catalyzed O_2 uptake (less than 2% that given with

Table 1. Effects of Thiol Structure and Thiol and Glyoxylate Concentrations on the Rates of Reactions Catalyzed by L-α-Hydroxy Acid Oxidase.

Thiol	Glyoxylate conc. (mM)	Relative rate[a] at the following thiol concentrations (mM)		
		0.1	1.0	10.0
2-Mercaptoethanol	0.1	6	12	18
	1.0	11	47	93
Coenzyme A	0.1	0	2	3
	1.0	4	9	24
Dephospho CoA	0.1	0	5	—
	1.0	6	20	—
Phosphopantetheine	0.1	2	4	4
	1.0	8	27	13
Pantetheine	0.1	3	9	9
	1.0	12	47	51
N-acetylcysteamine	0.1	7	18	20
	1.0	24	62	80
Propane-1, 3-dithiol	0.1	45	75	79
	1.0	73	97	98
Dihydrolipoate	0.1	28	37	13
	1.0	68	108	55
Ethanethiol	1.0	—	—	110
1-Propanethiol	1.0	—	—	99

[a]The numbers given are the rates of enzyme catalyzed O_2 uptake relative to that obtained with 25 mM D, L-α-hydroxybutyrate which was set at 100. The actual rate with 25 mM hydroxybutyrate varies from 3 to 6 mkat/kg depending on the enzyme preparation. For comparison, the relative rate with 1 mM α-hydroxybutyrate is 42 and with 0.1 mM is 13.

25 mM α-hydroxybutyrate), and the rate with cysteamine and glutathione is just barely detectable (4 and 3% respectively). The high reactivities of dihydrolipoate and its analog, propane-1,3-dithiol, are particularly noteworthy; even at 10 μM dihydrolipoate and 100 μM glyoxylate, enzyme-catalyzed O_2 uptake is readily detectable. There appears to be a specific binding site for dihydrolipoate since the results in Table 1 indicate it is an inhibitor of the dihydrolipoate-glyoxylate reaction. Using D-L-α-hydroxybutyrate as substrate, the inhibition by dihydrolipoate was found to be competitive with substrate, and a K_i of 0.13 mM was obtained.

Some apparent K_ms and V_{max}s obtained using an air atmosphere are given in Table 2. The kinetic constants should be those for the thiol-glyoxylate adducts because sufficient thiol was used to convert essentially all the glyoxylate to adduct (11). An interesting feature of these results is that V_{max} appears to be about the same for all reactants. This suggests that the slow step is reoxidation of a reduced enzyme by O_2, and this was found to be the case. With saturating amounts of substrate present, the K_m for O_2 was found to be 0.3 mM when α-hydroxybutyrate is the substrate, and 0.46 mM when the propane-1,3-dithiol-glyoxylate adduct is the substrate. With saturating amounts of both O_2 and substrate, V_{max} is 10 and 13 mkat/kg for the same two substrates respectively.

When thiol-glyoxylate adducts are substrates, the expected reaction catalyzed by hydroxy acid oxidase is that shown in eq. 2. That H_2O_2 is a product is evident from the observation that a burst of O_2 is produced when

$$RS\!-\!\underset{\underset{OH}{|}}{CH}\!-\!COO^- + O_2 \rightarrow RS\!-\!\underset{\underset{O}{\|}}{C}\!-\!COO^- + H_2O_2 \qquad (41.2)$$

catalase is added to an assay mixture not containing catalase but which had been allowed to react for a brief time. Also, the rate of O_2 uptake increases as expected (12) when ethanol is present in a complete reaction mixture containing catalase. Initial evidence that oxalyl thioesters are products is the observation of a new absorption centered at 260 nm (13). The addition of hydroxylamine to such reaction mixtures leads to the disappearance of the 260 nm absorption and the generation of a new absorption at 240 nm, identical to

Table 2. Apparent Kinetic Constants for Some L-α-Hydroxy Acid Oxidase Catalyzed Reactions.[a]

Substrate	K_m (mM)	V_{max} (mkat/kg)
D, L-α-hydroxybutyrate	1.2	5.5
2-mercaptoethanol + glyoxylate	0.75	ca 5
N-acetylcysteamine + glyoxylate	0.40	5.8
Pantetheine + glyoxylate	0.70	3.4
Coenzyme A + glyoxylate	2.2	ca 4
Propane-1, 3-dithiol + glyoxylate	0.03	4.8

[a]Unless otherwise specified the values for the thiol-glyoxylate reactions were obtained using an air atmosphere, various concentrations of glyoxylate (0.01 to 1 mM) and thiol always in excess (3 to 50 mM).

that given by authentic oxalyl monohydroxamate. Furthermore, the oxalyl monohydroxamate formed from enzymic reaction mixtures treated with hydroxylamine was shown to have the same R_Fs in two different solvent systems as authentic material.

Discussion

The results reported here show that various thiol-glyoxylate adducts are very effective substrates for rat kidney L-α-hydroxy acid oxidase, and that the products of the enzymic reactions are oxalyl thioesters. The maximum rates observed with some of the thiol-glyoxylate adducts are similar to those obtained with the best-known substrates, and the apparent K_ms are considerably lower than those reported for the best-known substrates (5). Such results imply that a thiol-glyoxylate adduct may be the physiological substrate for this enzyme.

One cannot definitely conclude from the present research what specific thiol is likely to be the most important substrate physiologically, but, in conjunction with literature data on the probable concentrations of glyoxylate (6) and the reactive thiols (14–16), the more likely possibilities would appear to be coenzyme A, phosphopantetheine, and dihydrolipoate. In many respects, the reactivity of the latter is the most interesting. Not only are 1,3-dithiols especially reactive, but also the enzyme has a specific binding site for dihydrolipoate as evidenced by the observed inhibition by this compound. Although the free lipoic acid content of animal tissues is probably less than 1 μM (16), the intriguing possibility still remains that dihydrolipoate, or possibly a derivative of dihydrolipoate, is a natural substrate for the enzyme. It is, of course, entirely possible that some other thiol not tested in this research is the true physiological substrate with glyoxylate. Essentially any cysteine-containing peptide is a potential substrate, although the low reactivity of glutathione might suggest that peptides containing crysteine will be unreactive.

What might be the physiological role of this enzymic reaction? As indicated earlier, it seems possible that the peroxisomal oxidases, including L-α-hydroxy acid oxidase and D-amino acid oxidase (6), are involved in the control of metabolism. Since oxalyl thioesters are very susceptible to nucleophilic attack, one possibility is that they could modify the activity of proteins or nucleic acids by transferring the oxalyl group to an amino, hydroxy or mercapto group on the surface of the biopolymer. Another possibility is that oxalate, which will eventually be formed from the hydrolysis of the oxalyl thioesters, is the important species. Oxalate is usually considered to be a useless end product of animal metabolism which can be harmful under some physiological conditions that lead to calcium oxalate precipitation (17,18). However, since the conversion of oxalate to other products is a biochemically trivial process which several other biological systems are able to do, it can well be argued that oxalate must be of critical importance to animal metabolism, or otherwise animals would have evolved enzymes to remove it in order to avoid the harmful effects. Pyruvate kinase, pyruvate carboxylase, lactate dehydrogenase, and malic enzyme are a few important enzymes in animal metabolism (19,20) which are known to be markedly inhibited by oxalate when its concentration is in the 10 to 100 μM range, a concentration range which is certainly present in animal systems (19). The suggestion is made here that part of the control on these enzymes (and thus also on glycolysis, gluconeogenesis, etc.) is due to oxalate, whose concentration is partially controlled by the hydroxy acid oxidase reaction (and the D-amino acid oxidase reaction as well since its suspected physiological product (6) also would give oxalate on hydrolysis). Consistent with this suggestion and the results of the present research is the observation that lipoic acid has profound effects on carbohydrate metabolism in animals, including the lowering of pyruvate carboxylase activity (21–24).

ACKNOWLEDGMENTS
This research was supported by research Grants from the National Institute of Arthritis, Metabolism, and Digestive Diseases (AM-13448) and the National Institute of General Medical Sciences (GM-27083), Public Health Service.

References

1. De Duve, C. and Bauduin, P. (1966) *Physiol Rev* 46:323–357.
2. Masters, C. and Holmes, R. (1977) *Physiol Rev* 57:816–882.
3. Hruban, Z. and Rechcigl, M. (1969) *Microbodies and Related Particles.* New York: Academic Press.
4. Novikoff, A.B., Novikoff, P.M., Davis, C., and Quintana, N. (1973) *J Histochem Cytochem* 21:737–755.
5. Cromartie, T.H. and Walsh, C.T. (1975) *Biochemistry* 14:2588–2596.
6. Hamilton, G.A., Buckthal, D.J., Mortensen, R.M., and Zerby, K.W. (1979) *Proc Natl Acad Sci USA* 76:2625–2629.
7. Meyer, S.E. and Cromartie, T.H. (1980) *Biochemistry* 19:1874–1881.
8. Whittle, E.L., Moore, J.A., Stipete, R.W., Peterson, F.E., McGlohon, V.M., Bird, O.D., Brown, G.M., and Snell, E.E. (1953) *J Am Chem Soc* 75:1694–1700.
9. Michelson, A.W. (1964) *Biochim Biophys Acta* 93:71–77.
10. Quayle, J.R. (1963) *Biochem J* 87:368–373.
11. Kanchuger, M.S., Leong, P.M., and Byers, L.D. (1979) *Biochemistry* 18:4373–4379.
12. Keilin, D. and Hartree, E.F. (1945) *Biochem J* 39:293–301.
13. Koch, J. and Jaenicke, L. (1962) *Ann* 652:129–139.
14. Brown, G.M. (1959) *J Biol Chem* 234:379–382.
15. Nakamura, T., Kusunoki, T., Soyama, K., and Kuwagata, M. (1969) *Bitamin* 40:412–415.
16. Kawashima, S. (1964) *Bitamin* 29:285–288.
17. Liao, L.L. and Richardson, K.E. (1972) *Arch Biochem Biophys* 153:438–448.
18. Smith, L.H. and Williams, H.E. (1974) In *Heritable Disorders of Amino Acid Metabolism.* Nyhan (ed.) New York: Wiley, p. 343–358.
19. Hogkinson, A. (1977) *Oxalic Acid in Biology and Medicine.* New York: Academic Press.
20. Anderson, D.B., Ferguson, S.M.F., and Lardy, H.A. (1971) *FEBS Letters* 14:283–284.
21. Singh, H.P.P. and Bowman, R.H. (1970) *Biochem Biophys Res Commun* 41:555–561.
22. Hangaard, N. and Hangaard, E.S. (1970) *Biochim Biophys Acta* 222: 583–586.
23. Harris, J.B., Alonso, D., Park, O.H., Cornfield, D., and Chacin, J. (1975) *Am J Physiol* 228:964–971.
24. Chacin, J., Park, O.H., Harris, J.B., and Alonso, D. (1976) *Am J Physiol* 231:209–215.

PART II D:

Chemical Models

Published 1982 by Elsevier North Holland, Inc.
Vincent Massey and Charles H. Williams, Editors
Flavins and Flavoproteins

CHAPTER 42

Flavins as Transient Carbanion Traps

Seiji Shinkai

Department of Industrial Chemistry, Faculty of Engineering, Nagasaki University, Nagasaki 852, Japan

It has been proposed that some flavin-dependent enzymes such as amino acid oxidase and lactate oxidase employ a carbanion intermediate during the oxidation by flavin of bound substrates (1,2). This proposition is also supported by model investigations in nonenzymatic systems (3). Application of this concept "flavin oxidation of carbanions" to organic chemistry should be interesting, since carbanion intermediates are proposed for a number of organic reactions. We here demonstrate the trapping by flavin of transient carbanion intermediates, diverting the reactions via the carbanions to the oxidation reactions.

It is well-known that cyanide ion catalysis of benzoin condensation proceeds via a carbanion intermediate, $Ph\bar{C}(CN)(OH)$. Franzen and Fikentscher (4) also found that the decarboxylation of α-keto acids yields similar carbanion intermediates and finally gives benzoin-type condensation products. We found that in the presence of flavins (e.g., 3-methyltetra-O-acetylriboflavin: MeFl), these reactions are easily diverted to the oxidation affording the corresponding carboxylic acids (5–7). For example, the main product in the reaction of KCN and 4-chlorobenzaldehyde was, of course, 4,4′-dichlorobenzoin (72% yield), whereas the addition of 3-methyl-10-butylisoalloxazine (20 mol% of 4-chlorobenzaldehyde) strikingly suppressed the formation of 4,4′-dichlorobenzoin (8.6%) and simultaneously enhanced the yield of 4-chlorobenzoic acid (22%). Importantly, the addition of CTAB (hexadecyltrimethylammonium bromide, cationic micelle) above the critical micelle concentration suppressed the benzoin condensation almost completely (4,4′-dichlorobenzoin 0.8%, 4-chlorobenzoic acid 74.2%). Thus, the combination of cyanide ion and cationic micelle remarkably facilitates the flavin oxidation of aldehydes and α-keto acids. These results are summarized in the following schemes.

Scheme 1.

benzoin
condensation
↑ ArCHO

$$ArCHO \underset{k_{-1}}{\overset{k_1}{\rightleftharpoons}} Ar{-}C(H)(CN){-}O^- \underset{k_{-2}}{\overset{k_2}{\rightleftharpoons}} Ar{-}\bar{C}(CN){-}OH$$

CN^- (returns to ArCHO)

$Ar\bar{C}(CN)OH \xrightarrow{Fl \to FlH^-} ArCOCN \xrightarrow{H_2O} ArCOO^- + CN^-$

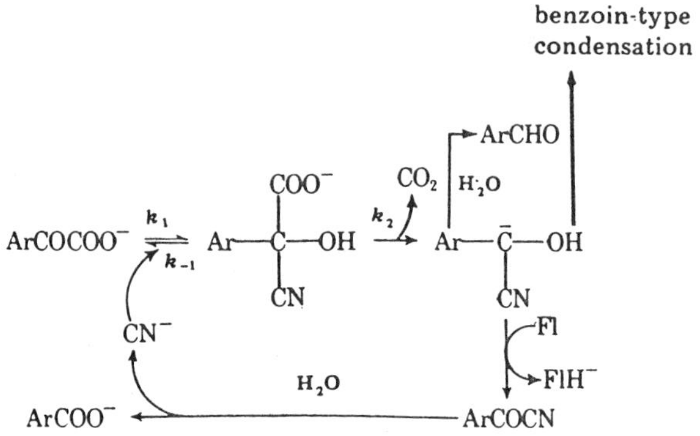

Scheme 2.

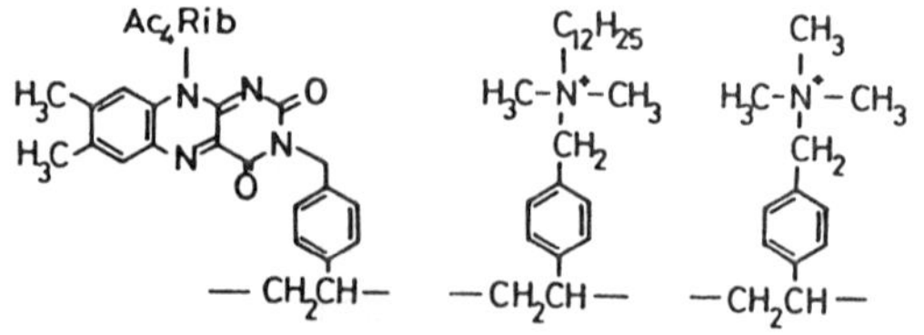

Scheme 3.

The role of the cationic micelle can be replaced by the flavin immobilized in cationic polyelectrolytes (8). This result suggests that the cationic micellar surface plays an important catalytic role to enhance the trapping efficiency of micelle-bound flavin.

Thiazolium ion, the catalytic moiety of thiamine pyrophosphate, catalyzes acyloin condensation in a manner similar to cyanide ion (9). We found that the reaction sequence of acyloin condensation of aldehydes which is catalyzed by N-hexadecylthiazolium bromide (HxdT) in the CTAB micelle (10) was readily diverted by added MeFl to the oxidation reaction to afford carboxylic acids (11–13). Similarly, the micellized thiazolium ion plus MeFl system efficiently catalyzed the decarboxylative oxidation of aliphatic α-keto acids but scarcely catalyzed that of aromatic α-keto acids. Thiamine and thiamine pyrophosphate also served as catalysts but the activities were much lower than that of HxdT. On the basis of kinetic examination and product analysis, we proposed that the reactions involve oxidative trapping by MeFl of the intermediate (active aldehyde) formed by the rate-limiting deprotonation or decarboxylation from the HxdT-substrate adducts. The reaction schemes are illustrated at right. It is worthwhile mentioning that the reactions serve as a relevant model system

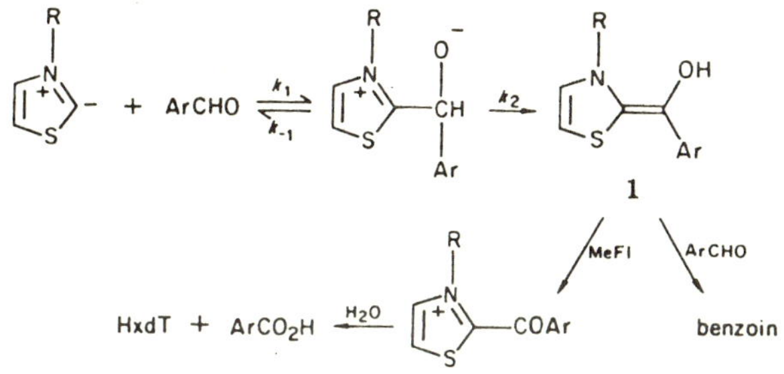

Scheme 4.

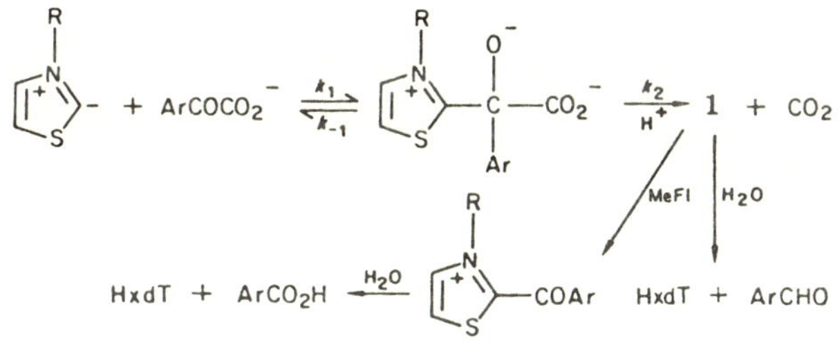

Scheme 5.

for pyruvate dehydrogenase which requires FAD and thiamine pyrophosphate as cofactors and catalyzes the conversion of pyruvic acid to acetic acid.

The similar reaction was catalyzed by a thiazolium ion immobilized in cationic polymers (14). The reactivity of the thiazolium ion is comparable with that of HxdT in the CTAB micelle.

In the foregoing systems, the flavin oxidation is always competing with the conventional benzoin (or acyloin) condensation. In order to apply the flavin trapping to synthetic purposes or to enzyme model investigations, it would be desirable to suppress one of the competing reactions completely. We thus synthesized a biscoenzyme Fl-T which has within a molecule both flavin and thiazolium ion (15). When the concentration of MeFl was lowered to 1 mM and those of 4-chlorobenzaldehyde and HxdT were enhanced to 100 mM and 10 mM, respectively, the trapping efficiency was only 1.6 (=28/17). On the other hand, the condensation is suppressed almost completely in the presence of Fl-T, the trapping efficiency being always greater than 115. In particular, the condensation products were not detected at all in the presence of 5 mM of Fl-T. The remarkably high trapping efficiency should be attributed to the high proximity term of the intramolecular isoalloxazine.

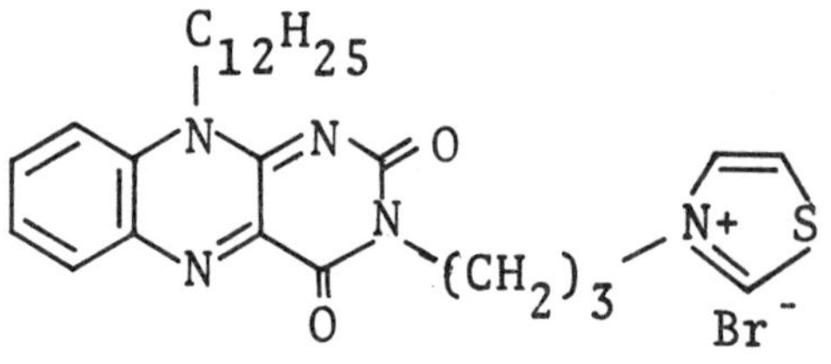

Scheme 6.

The flavin trapping also becomes a useful strategy to assess the reaction mechanism. The glyoxalase enzyme catalyzes the conversion of glyoxals to corresponding α-hydroxy acids with the aid of glutathione (GSH) as cofactor. Two mechanisms have been proposed for the 1,2-hydrogen shift step: the 1,2-hydride shift mechanism and 1,2-proton shift mechanism via the enediol. We found that the rearrangement of the hemithiol acetal to the α-hydroxythiol ester is almost completely inhibited on the addition of MeFl due to the occurrence of the flavin-trapping of the enediol intermediate (16,17). This result suggests that the enzymic catalysis by glyoxalase I occurs via the enediol intermediate.

In conclusion, the flavin trapping is a useful method to divert the reactions involving carbanion intermediates to their oxidation and to assess reaction mechanisms via carbanion intermediates.

References

1. Walsh, C. (1980) *Acc Chem Res* 13:148–155.
2. Bruice, T.C. (1980) *Acc Chem Res* 13:256–262.
3. Shinkai, S., Kunitake, T., and Bruice, T.C. (1974) *J Am Chem Soc* 96:7140–7141.
4. Franzen, V. and Fikentscher, L. (1958) *Justus Liebigs Ann Chem* 613:1–6.
5. Shinkai, S., Ide, T., and Manabe, O. (1978) *Chem Lett* 583–586.
6. Shinkai, S., Yamashita, T., and Manabe, O. (1979) *J Chem Soc Chem Commun* 301–302.
7. Shinkai, S., Yamashita, T., Kusano, Y., Ide, T., and Manabe, O. (1980) *J Am Chem Soc* 102:2335–2340.
8. Shinkai, S., Kusano, Y., and Manabe, O. (1980) *Makromol Chem* 181:1791–1798.
9. Breslow, R. (1957) *J Am Chem Soc* 79:1762–1763.
10. Tagaki, W. and Hara, H. (1973) *J Chem Soc Chem Commun* 891.
11. Shinkai, S., Yamashita, T., Kusano, Y., and Manabe, O. (1980) *Tetrahedron Lett* 21:2543–2546.
12. Shinkai, S., Yamashita, T., Kusano, Y., and Manabe, O. (1980) *J Org Chem* 45:4947–4952.
13. Yano, Y., Hoshino, Y., and Tagaki, W. (1980) *Chem Lett* 749–752.
14. Shinkai, S., Hara, Y., and Manabe, O., to be submitted.
15. Shinkai, S., Yamashita, T., and Manabe, O. (1981) *Chem Lett* in press.
16. Shinkai, S., Yamashita, T., Kusano, Y., and Manabe, O. (1979) *Chem Lett* 1323–1326.
17. Shinkai, S., Yamashita, T., Kusano, Y., and Manabe, O. (1981) *J Am Chem Soc* 103:2070–2074.

Published 1982 by Elsevier North Holland, Inc.
Vincent Massey and Charles H. Williams, Editors
Flavins and Flavoproteins

CHAPTER 43

Studies on the Active Centers of Flavoproteins: Reaction of Peroxides with 5-Deazaflavin

M.S. Jorns, A. Pokora, and D. Vargo

Department of Chemistry, The Ohio State University, Columbus, Ohio

FAD bound to oxynitrilase exhibits properties similar to the flavoprotein oxidases and is important as a structural component of the active site. FAD can be replaced by deazaFAD·X, a derivative unreactive in oxidation-reduction reactions, without affecting catalytic activity, in contrast to the slow turnover observed with 5-deazaFAD (1). Reaction of 5-deazaFAD bound to oxynitrilase with either H_2O_2 or m-chloroperoxybenzoate proceeds via a rapidly-formed intermediate (I) (Figure 1, curve 2), followed by a slower, peroxide-independent conversion of intermediate I to deazaFAD·X enzyme (curves 2–5). The same intermediate is formed with peroxides and 5-deazaflavin bound to glucose oxidase (Figure 1, curve 2), glycolate oxidase (Figure 2, curve 2) or D-amino acid oxidase, but no reaction occurs with flavodoxin or the riboflavin-binding protein. Differences in reactivity with peroxides can be correlated with the susceptibility of the bound coenzyme towards nucleophilic attack (e.g., sulfite, cyanide) at position 5 (2). Intermediate I (λ_{max} ~340 nm) has been identified as 4a,5-epoxy-5-deazaflavin based on the similarity of its properties with those observed for model epoxide compounds (Scheme 1). These derivatives are nonfluorescent, decompose in neutral solution with a loss of absorption in the visible region (Figure 3) and react with iodide to regenerate the original parent compound (2,3).

Conversion of intermediate I to deazaflavin·X is observed with glucose oxidase (Figure 1, curves 2–5) and glycolate oxidase (Figure 2, curves 2–6), similar to oxynitrilase, except a second intermediate (II, λ_{max} = 355, 375 nm) is detected with glycolate oxidase. The protein moiety facilitates the conversion which can be inhibited by compounds which bind near the coenzyme site [e.g., benzoate plus oxynitrilase intermediate I (2)] or prevented by denaturation of intermediate I (Figure 3). Phosphodiesterase converts deazaFAD·X to deazaFMN·X. The original side chain at position 10 is not modified since deazaFMN·X(λ_{max} = 340, 356 nm) can be tightly bound to apoflavodoxin (K_d = 9.3 nM). Deazaflavin·X emits an intense purple fluorescence (emission λ_{max} = 383 nm) and exhibits a single pKa in acid solution. Modification of the original pyrimidine ring is suggested by the absence of a pKa attributable to ionization at N(3)H. Deazaflavin·X is not formed with D-amino acid oxidase where intermediate I decomposes similar to that observed upon denaturation

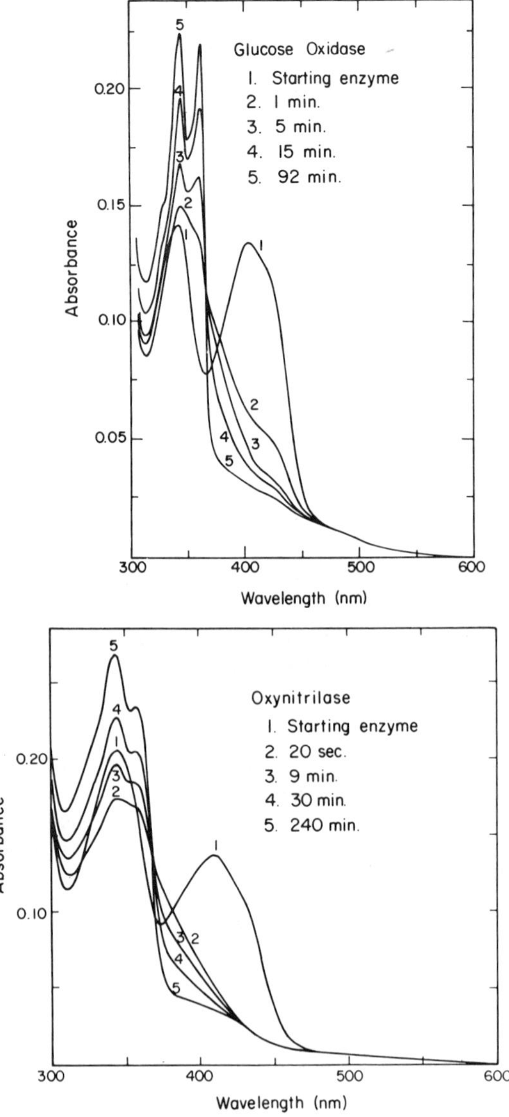

Figure 1. Reaction of peroxides with enzyme-bound 5-deazaFAD at 25°C. Conditions, glucose oxidase (10.2 μM) in 0.1 M sodium phosphate pH 5.7 containing 0.3 M NaCl and 0.3 mM EDTA plus 100 mM H_2O_2; oxynitrilase (14.9 μM) in 0.1 M sodium phosphate pH 7.5 plus 16.3 μM m-chloroperoxybenzoate.

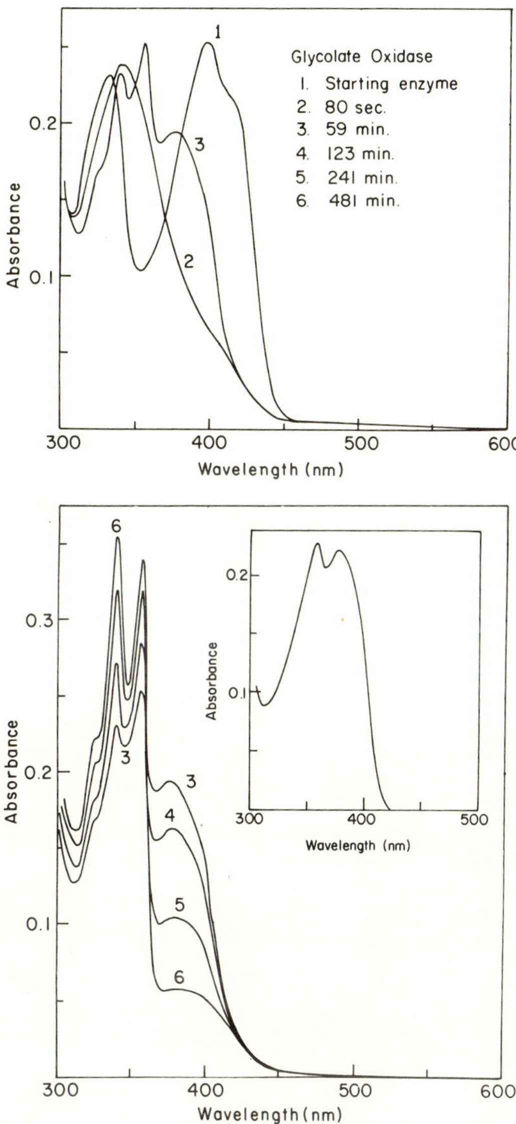

Figure 2. Reaction of 5-deazaFMN bound to glycolate oxidase (21.0 μM) with 3.0 mM H_2O_2 at 15°C in 0.1 M sodium phosphate pH 7.0 containing 0.3 mM EDTA. The inset shows the spectrum of intermediate II (17.5 μM), calculated (2) by correcting curve 3 for contributions from intermediate I, deazaFMN·X enzyme, and unreactive 5-deazaFMN enzyme (16.7%).

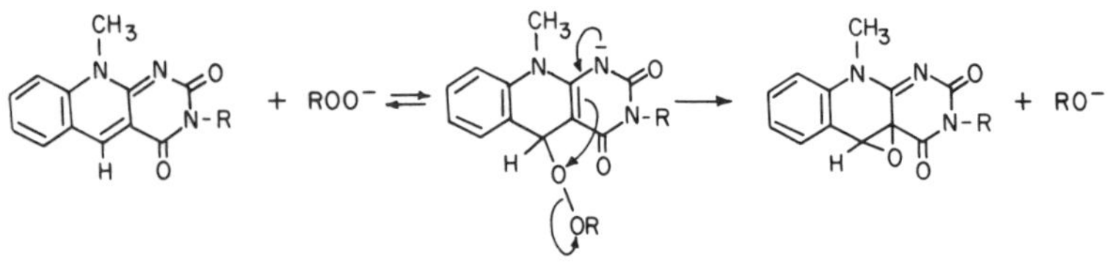

Scheme 1.

of glycolate oxidase intermediate I or with the model epoxide compounds at neutral pH (Figure 3). In contrast, an analogous reaction, accompanied by the release of CO_2 and RNH_2, is observed with the model epoxide compounds at alkaline pH (Scheme 2, R=H, CH_3). The purple fluorescent product (4-methyl-oxazolo[4,5-b]quinolin-2(4H)-one, compound C) is stable in neutral solution (λ_{max} =330, 345 nm), does not react with iodide and exhibits a single pKa (pKa~0.2), similar to deazaflavin·X. The structure of compound C is consistent with mass spectral (M^+ =200), ^{1}H-NMR[$CDCl_3$: δ 4.15(s, 3H, N(4)CH_3), 7.41(s, 1H, C(9)H), 7.47–7.80(m, 4H, ArH), no exchangeable protons], ^{13}C-NMR and infrared data (Figure 4).

Immediate conversion of the epoxide to intermediate A (λ_{max} =290 nm, Figure 5, curve 1) is observed at pH 11.1 where intermediate B is barely detectable owing to a rate-determining conversion of intermediate A to intermediate B. The rate observed for the latter step is pH-independent (7.1×10^{-2}

Figure 3. (A) Effect of denaturation on glycolate oxidase intermediate I. Curves 2–5 were recorded 0.2, 30, 60, and 300 min, respectively, after adding sodium dodecyl sulfate (2%) to intermediate I, formed by reacting the initial enzyme (curve 1) with 1.0 mM H_2O_2 for 3 min at 25°C in 0.1 M sodium phosphate pH 7.0, 0.3 mM EDTA. Inset: Curve I is the difference spectrum obtained by correcting curve 2 for unreactive 5-deazaFMN enzyme. Curve II is the spectrum of 4a,5-epoxy-5-deazariboflavin. (B) The decomposition of the latter (curves 1–5) under the same conditions was monitored 0.2, 45, 90, 180, and 600 min, respectively, after dissolving the solid.

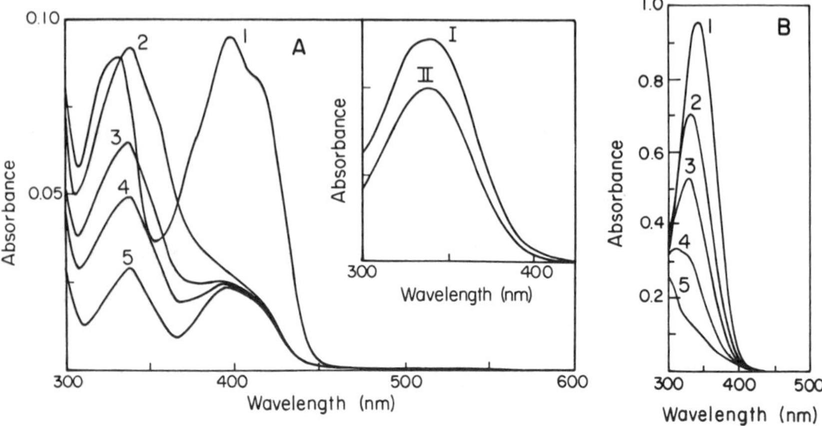

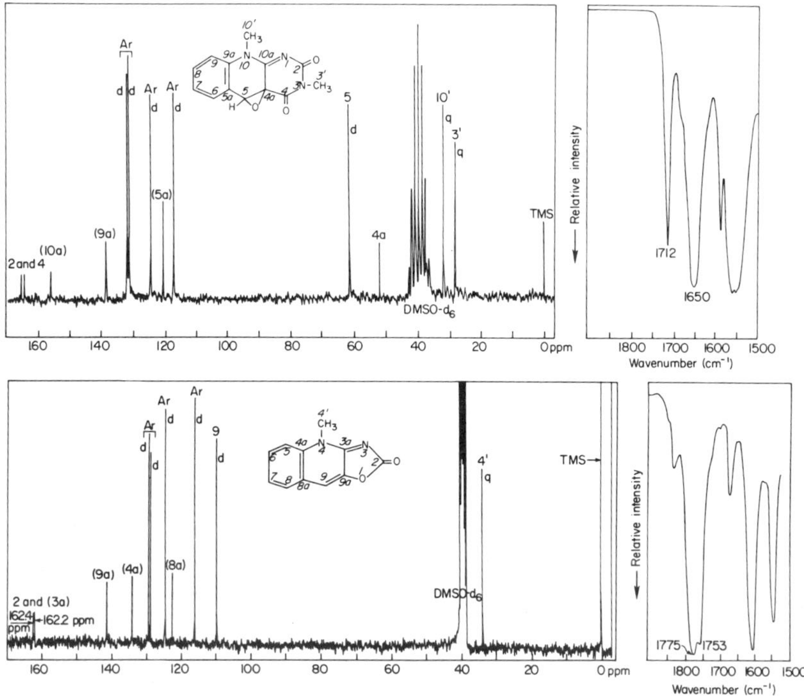

Figure 4. Wide-band ^{1}H decoupled ^{13}C-NMR spectra (in DMSO-d_6) and infrared spectra (KBr pellet) of 4a,5-epoxy-3,10-dimethyl-5-deazaisoalloxazine and compound C. (NMR) Unless otherwise indicated, singlets were observed in off-resonance ^{1}H decoupled experiments. Peak assignments shown in parentheses are tentative. (IR) Carbonyl group assignments are indicated. The carbonyl group in compound C appears as a doublet due to Fermi resonance, characteristically observed with unsaturated γ-lactones (4).

min^{-1},25°C, pH>7) whereas the rate of conversion of intermediate B to compound C decreases with increasing pH. Intermediate B (λ_{max}=346, 357 nm) is readily observed when the last step is rate-determining (pH>12) and can also be formed by reacting compound C with excess RNH_2 at lower pH values (Figure 5). Intermediate B exhibits properties similar to intermediate II in the glycolate oxidase reaction. Intermediate II is stabilized by the enzyme since denaturation results in immediate formation of deazaflavin·X, similar to

Scheme 2.

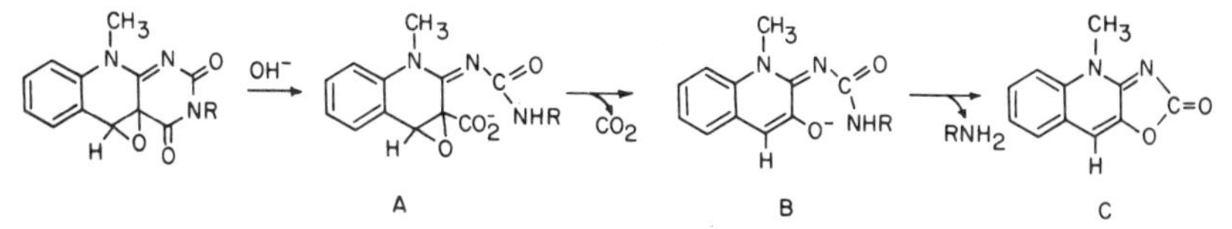

OH^- → A → (CO_2) → B → (RNH_2) → C

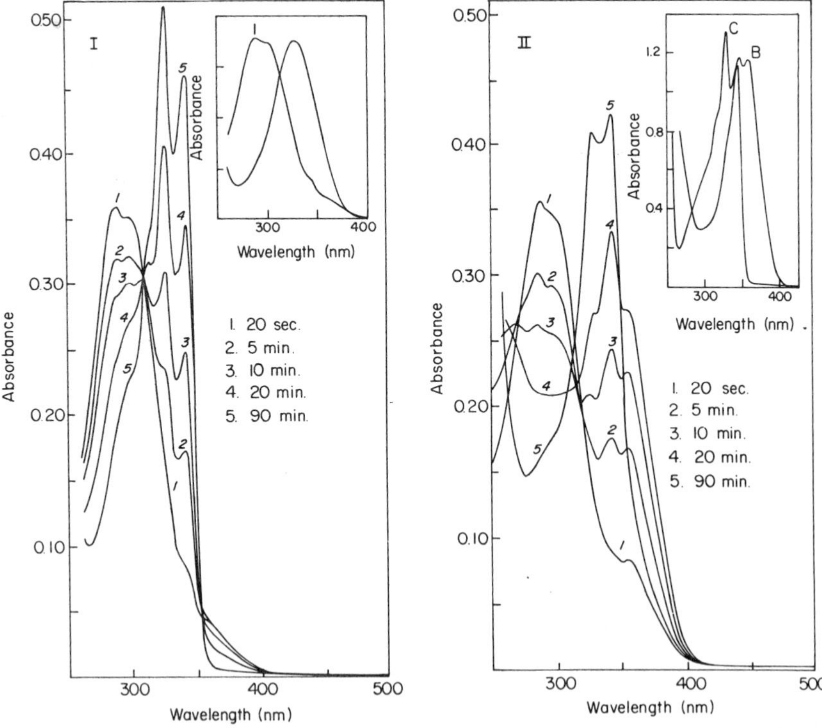

Figure 5. Reaction of 4a,5-epoxy-3,10-dimethyl-5-deazaisoalloxazine (22.6 μM) at pH 11.1 (I) or pH 12.2 (II). Conditions, 0.1 M sodium phosphate, 25°C. (I) The spectrum of intermediate A (curve 1) is compared in the inset with the spectrum of the epoxide (22.6 μM) obtained at pH 7.0. (II) The reaction is not complete in 90 min but the subsequent spectral course (not shown) is complicated by the instability of the product at pH 12.2. The inset shows the conversion of compound C (4.5×10^{-5} M) to intermediate B observed upon addition of methylamine (4.84 M) at pH 11.1.

that observed with intermediate B at neutral pH, and in contrast to the slower conversion ($k = 5.8 \times 10^{-3}$ min^{-1}, 15°C) observed with intact enzyme. Alkaline conditions are required only for conversion of the epoxide to intermediate A. This suggests that the protein moiety facilitates an analogous step in the formation of deazaflavin·X with enzyme-bound epoxide at neutral pH. The results show that the enzyme environment strongly influences the reaction of peroxides with 5-deazaflavin which may prove useful as an active site probe with different flavoproteins.

References

1. Jorns, M.S. (1979) *J Biol Chem* 254:12145–12152.
2. Vargo, D., Pokora, A., Wang, S., and Jorns, M.S. (1981) *J Biol Chem* 256:6027–6033.
3. Vargo, D. and Jorns, M.S. (1979) *J Am Chem Soc* 101:7623–7626.
4. Silverstein, R.M., Bassler, G.C., and Morrill, T.C. (1974) *Spectrometric Identification of Organic Compounds*. New York: John Wiley & Sons, pp. 77–78, 149.

PART III:

Oxygen Activation by Flavins and Flavoproteins

PART III A:

Model Reactions

Published 1982 by Elsevier North Holland, Inc.
Vincent Massey and Charles H. Williams, Editors
Flavins and Flavoproteins

CHAPTER 44

A Progress Report on Studies of the Activation of Molecular Oxygen by Dihydroflavins

Thomas C. Bruice

Department of Chemistry, University of California, Santa Barbara, Santa Barbara, California

The object of this contribution is to provide an overview of our contributions to the understanding of the means by which 1,5-dihydroflavins and molecular oxygen react to form 4a-hydroperoxyflavins and how mono- and dioxygen transfer from the 4a-hydroperoxyflavins to substrates occur.

It is well-known that oxidation of 1,5-dihydroflavin by oxygen is an autocatalytic process which involves the formation of flavin radical in the initiation and chain propagating steps (1). The direct reaction of oxygen with 1,5-dihydroflavin may only be important in the radical initiation step [Equation (44.1a, b)].

$$1,5\text{-FlH}_2 + O_2 \rightarrow \text{Fl}_{ox} + H_2O_2 \qquad (44.1a)$$

$$\text{Fl}_{ox} + 1,5\text{-FlH}_2 \rightarrow 2\text{FlH}\cdot \qquad (44.1b)$$

For this reason, little information has been obtained about the mechanism of the direct reaction of $^3O_2$ with 1,5-dihydroflavin. This is so because once a trace of flavin radical is formed, the ensuing autocatalytic process may (depending on pH) involve only the oxidation of flavin radical and not the reaction of $^3O_2$ with 1,5-FlH_2.

There are two possible modes for the $2e^-$ reduction of oxygen by 1,5-dihydroflavin. The first involves $2e^-$ transfer from 1,5-dihydroflavin to $^3O_2$. This process, and we shall not concern ourselves with hypothetical mechanisms, requires that $^3O_2$ undergo a spin inversion in order to react with singlet 1,5-dihydroflavin or that 1,5-dihydroflavin be promoted to its triplet state in order to react with the triplet $^3O_2$. The second mechanism involves two consecutive $1e^-$ transfers from 1,5-FlH_2 to $^3O_2$ and consequently involves a radical pair intermediate [Equation (44.2)] (1e).

$$1,5\text{-FlH}^- + O_2 \rightarrow \text{FlH}\cdot O_2^{\dot{-}} \rightarrow \text{Fl}_{ox} + HO_2^- \qquad (44.2)$$

In order to understand the mechanism of Equation (44.1a), we chose to study the reaction of oxygen with 5-ethyl-1,10-ethano-1,5-dihydrolumiflavin (**I**). The 1,5-dihydrolumiflavin **I** can transfer one electron to $^3O_2$ providing **II** and $O_2^{\dot{-}}$ but further oxidation of **II** to provide the dication **III** and H_2O_2 is not favored. Since **III** does not form, its reaction with **I** to yield **II** is not feasible so that the autocatalytic oxidation of **I** by oxygen cannot occur. (Scheme 1)

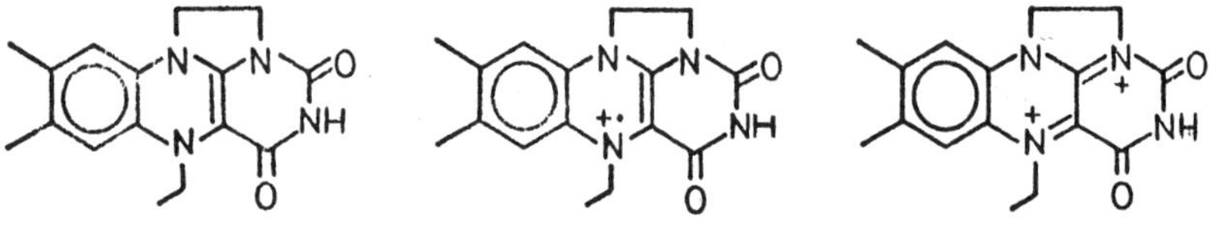

I II Not likely III

Scheme 1.

Dr. Gert Eberlein (2) has found that the reaction of $\mathbf{I} + O_2$ in H_2O is first order (to completion) in both components. The products of the reaction are **II** (100% yield) and H_2O_2 (50% yield). His findings firmly establish the mechanism of Equation (44.3). One may approximate the second-order rate constants for the reaction (Scheme 2)

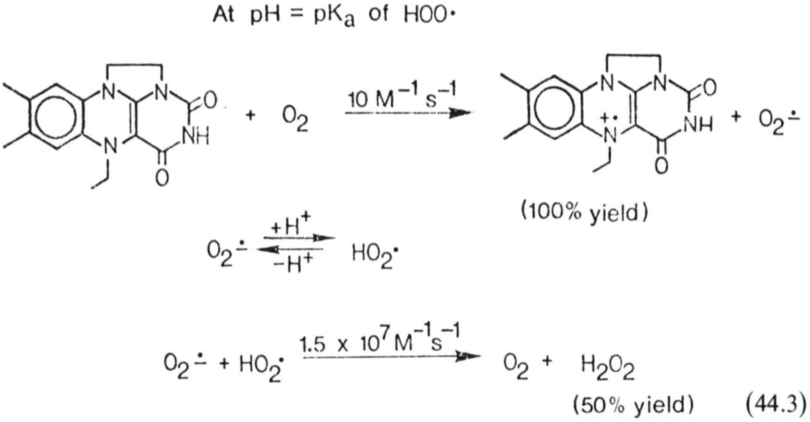

At pH = pK_a of HOO·

$$\mathbf{I} + O_2 \xrightarrow{10\ M^{-1}\ s^{-1}} \mathbf{II}\ (100\%\ \text{yield}) + O_2^{\cdot-}$$

$$O_2^{\cdot-} \underset{-H^+}{\overset{+H^+}{\rightleftharpoons}} HO_2^{\cdot}$$

$$O_2^{\cdot-} + HO_2^{\cdot} \xrightarrow{1.5 \times 10^7\ M^{-1}s^{-1}} O_2 + H_2O_2\ (50\%\ \text{yield}) \qquad (44.3)$$

Scheme 2.

of a series of 1,5-dihydroflavins with oxygen from the tangential slope to the initial portion of the autocatalytic time course. These rate constants may then be compared to the electrochemical potential for $1e^-$ donation by the dihydroflavins. It is found that the more negative the potential, the greater the rate constant for oxygen reduction by dihydroflavins. When the potential for $1e^-$ donation by **I** and the second-order rate constant for oxygen reduction by **I** are compared to the like constants for other flavins, it is found that they form a common series. These results strongly suggest that the reaction of **I** with O_2 occurs by a mechanism shared by other 1,5-dihydroflavins. Thus, the reaction of 1,5-dihydroflavins with $^3O_2$ involves rate limiting $1e^-$ transfer to yield $O_2^{\cdot-}$ and flavin radical [Equation (44.4)].

$$\mathrm{Fl_{red}} + {}^3\mathrm{O_2} \xrightarrow{\text{slow}} \left[\dot{\mathrm{Fl}}_{\mathrm{sem}}\mathrm{O_2}^{\dot{-}}\right]$$

$$\left[\dot{\mathrm{Fl}}_{\mathrm{sem}}\mathrm{O_2}^{\dot{-}}\right] \xrightarrow{\text{fast}} \text{4a-FlOOH} \qquad (44.4)$$

$$\left[\dot{\mathrm{Fl}}_{\mathrm{sem}}\mathrm{O_2}^{\dot{-}}\right] \xrightarrow{\text{fast}} \mathrm{Fl_{ox}} + \mathrm{H_2O_2} \leftarrow \text{4a-FlOOH}$$

I believe that it is safe to say that it is now generally accepted that a 4a-hydroperoxyflavin is the species formed at the active site of flavoenzyme mixed-function oxidases when enzyme-bound reduced flavin cofactor reacts with ${}^3O_2$ (3). We find that the 4a-hydroperoxyflavin 4a-FlEtOOH is formed on reaction of: (a) ${}^3O_2$ with FlEtH; (b) $O_2^{\dot{-}}$ with FlEt·; or (c) H_2O_2 with $Fl_{ox}^{+}Et$ [Equation (44.5)] (1e,4). The role of the N^5-ethyl group in the model reactions of Equation (44.5) is to prevent the dissociation of hydrogen peroxide as would occur with an unsubstituted flavin [Equation (44.6)].

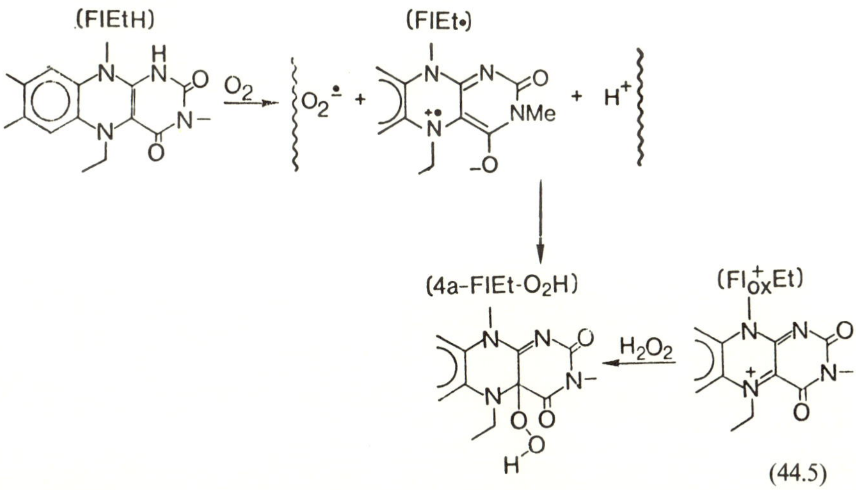

(FlEtH) $\xrightarrow{O_2}$ { $O_2^{\dot{-}}$ + (FlEt•) + H^+ } → (4a-FlEt-O_2H) $\xleftarrow{H_2O_2}$ ($Fl_{ox}^{+}Et$)

(44.5)

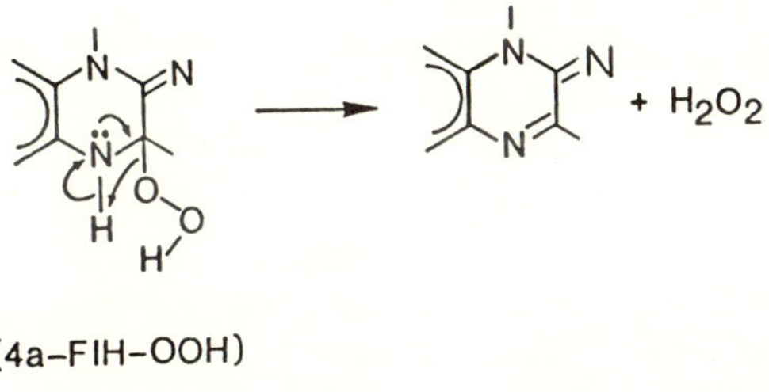

(4a-FlH-OOH) → Fl_{ox} + H_2O_2

(44.6)

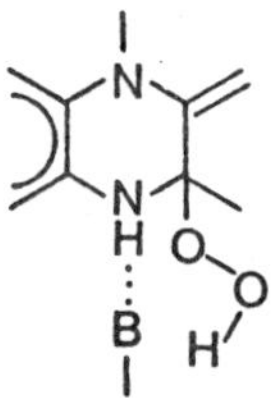

It is known that bacterial luciferase and other flavoenzyme mixed-function oxidases will not bind 1,5-dihydroflavin cofactor alkylated at N^5 (5). This complete lack of bulk tolerance on replacing the $-H$ at N^5 by even $-CH_3$ establishes that the proton on N^5 is closely associated with a functional group at the active site. This functional group may well be a base which, by forming a strong H-bond, slows the rate of elimination of hydrogen peroxide from 4a-FlH-OOH *viz.*

Thus, the N^5-ethyl substituent in 4a-FlEtOOH performs the same task as a basic functional group at the active site of the enzyme in stabilizing 4a-FlH-OOH. In any event, 4a-FlH-OOH is the monooxygen donor, or a precursor to the monooxygen donor, in flavoenzyme mixed-function oxidase reactions.

4a-FlEtOOH reacts with hydroxylamines [Equation (44.7)] tertiary amines [Equation (44.8)], secondary amines [Equation (44.9)] (6) and alkylsulfides [Equation (44.10)] (5a, 6), to produce the same

$$\text{4a-FlEtO}_2\text{H} + \text{PhCH}_2\text{N(OH)CH}_3 \longrightarrow \text{4a-FlEtOH} + \text{PhCH}_2\text{N}^+(\text{OH})(\text{CH}_3)\rightarrow\text{O} \xrightarrow{-H_2O} \text{PhCH}{=}\text{N}(\text{CH}_3)\rightarrow\text{O} + \text{PhCH}_2\text{N}(=\text{CH}_2)\rightarrow\text{O} \quad (44.7)$$

ENZYMATIC 3 : 1
MODEL REACTION 2.2 : 1

$$\text{4a-FlEtOOH} + \text{PhCH}_2\text{N(CH}_3)_2 \rightarrow \text{4a-FlEtOH} + \text{PhCH}_2{}^+\text{N}(\text{CH}_3)_2\text{—O}^- \quad (44.8)$$

$$\text{4a-FlEtOOH} + \text{PhCH}_2\text{NH(CH}_3) \rightarrow \text{4a-FlEtOH} + \text{PhCH}_2\text{—N(CH}_3)\text{—OH} \quad (44.9)$$

$$\text{4a-FlEtOOH} + \text{S}\langle\text{ring}\rangle\text{O} \rightarrow \text{4a-FlEtOH} + \text{OS}\langle\text{ring}\rangle\text{O} \quad (44.10)$$

products as does the flavoenzyme microsomal mixed-function oxidase (7). The reactions of [Equation (44.7−44.10)] are first-order in [4a-FlEtOOH] and first order in [substrate]. The monooxygenated substrate and 4a-hydroxyflavin

(4a-FlEtOH) are both formed in 100% yield. There is no evidence for the formation of any intermediate in these reactions. Separate studies have established that the N^5-blocked 4a-hydroperoxy flavin does not rearrange to form either the 9a- or 10a- isomer (8). The common mechanism of Equation (44.11) suffices. All evidence supports the very same

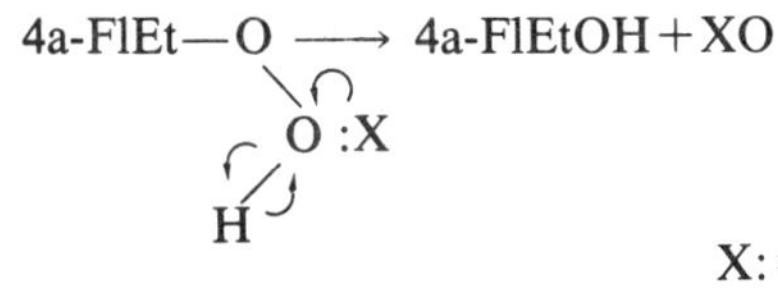

$$\begin{array}{l} \text{4a-FlEt—O} \longrightarrow \text{4a-FlEtOH} + \text{XO} \\ \quad\quad \text{O :X} \\ \quad \text{H} \end{array}$$

$$\text{X:} = \text{—N:—S}^- \qquad (44.11)$$

mechanism for the enzymatic reaction (7,9). The relative second-order rate constants for the oxidation of $I^- \rightarrow I_2$ and the N and S oxygenation of amines and an alkyl sulfide by different kinds of hydroperoxides are compared in Table 1 (10).

Table 1. Comparison of the Second-Order Rate Constants [Relative to Those for 2] For the Hydroperoxide Oxidation of I^- (k_I), S-Oxidation of Thioxane (k_S), and N-Oxidation of N, N-Dimethylbenzyl Amine (k_N).

Substrate	k_I	k_S	k_N
1 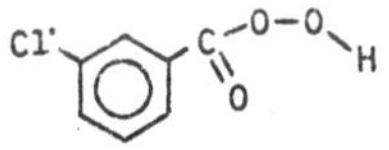	7.9×10^2		
2 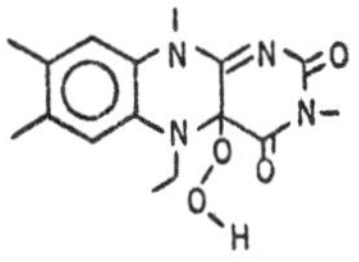	1.0 ($6.0\ M^{-1}s^{-1}$)	1.0 ($0.12\ M^{-1}s^{-1}$)	1.0 ($0.12\ M^{-1}s^{-1}$)
3 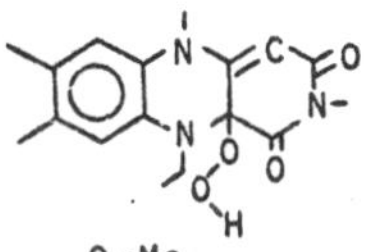	3×10^{-1}	5.7×10^{-2}	8.6×10^{-2}
4 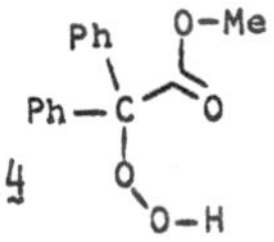	5.5×10^{-2}	3.2×10^{-3}	1.1×10^{-2}
5 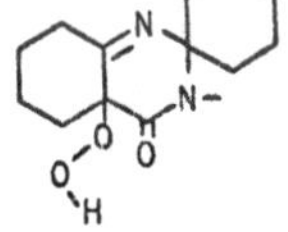	9.0×10^{-3}	9.2×10^{-4}	1.2×10^{-4}
6 H_2O_2	1.0×10^{-3}	1.4×10^{-4}	2.8×10^{-5}
7 CH_3-C(CH_3)(CH_3)-O-OH	1.0×10^{-3}	5.8×10^{-6}	$> 1 \times 10^{-6}$

Examination of the table establishes that 4a-FlEtOOH is not as good an oxygen donor as the peracid m-$ClC_6H_4CO_3H$, but that it is a much better oxygen donor than other hydroperoxides which we have investigated.

The hydroxylation of electron-rich aromatic ring structures by flavoenzyme mixed-function oxidases is not yet understood. Hamilton proposed that 4a-FlH-OOH is converted into a carbonyl oxide **VII** or vinylogous ozonide **VIII** and that these species serve as monooxygen donors (11). The flavin derived species **IX**, formed on transfer of an oxygen atom from **VII** or **VIII** to substrate, was then proposed to yield 4a-FlHOH which then dissociated HO^- to yield Fl_{ox} [Equation (44.12)]. A carbonyl oxide type intermediate has also been proposed for the biopterin

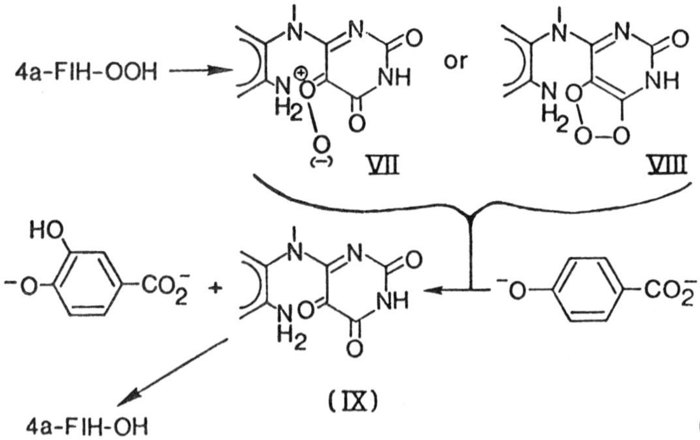

(44.12)

requiring phenylalanine hydroxylase. In this instance, oxygen transfer to phenylalanine is proposed to yield **X** as the immediate product (12). The structural similarity between **IX** and **X** is obvious. Entsch, Ballou and Massey (13) have proposed a mechanism for p-hydroxybenzoate hydroxylase which also requires the intermediacy of **IX**. In these studies, it was observed that when certain alternate substrates were employed with p-hydroxybenzoate hydroxylase, the enzyme-bound 4a-FlHOOH provided a strongly absorbing species (λ_{max} 395–420; $\varepsilon = 15{,}000$) which converted to 4a-FlHOH. The mechanism of Equation (44.13) was proposed.

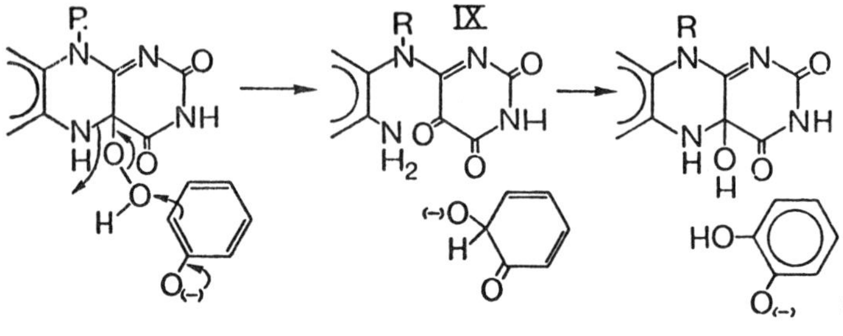

(44.13)

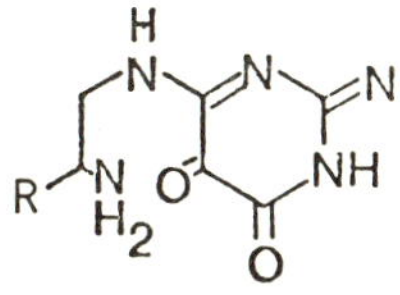

(X)

The aforementioned investigation is of considerable importance since it represents the single study to date in which an intermediate has been observed on conversion of 4a-FlH-OOH→4a-FlH-OH with accompanying mono-oxygenation of substrate. The necessity of obtaining the spectrum of authentic **IX** for comparison to that of the observed enzyme intermediate is obvious. An important step in this direction is the synthesis and characterization of **XI** by Dr. Albert Wessiak in the author's laboratory (14). From comparison of the

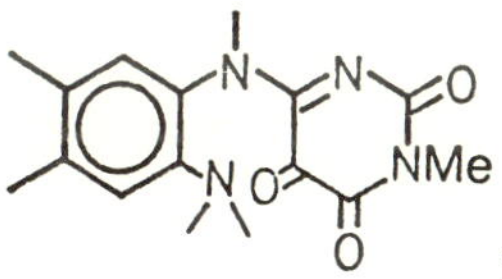

XI

λ_{max} = 342 nm

ε = 7,120

SOLVENT CH_3CN

From comparison of the spectrum of the enzyme intermediate to the spectrum of **XI**, one may conclude that the species formed in the enzymatic reaction does not exhibit the spectral characteristics of the 6-[R]-pyrimidine-2,4,5-trione **XI**. Thus, there is no basis for the assignment of structure **IX** to this intermediate. The structure **XI** has been established by elemental analysis, ir, ^{1}H-nmr and ^{13}C-nmr. It is possible, of course, that the strongly absorbing intermediate observed by Entsch et al. (13) represents a quinoid isomer of **IX** (i.e., **IX′**) which perhaps could only be stabilized at the active site. The synthesis of suitable analogues to structures **IX′** is being explored in our laboratory. It should be noted that the spectrum of **XI** is very similar to that of a 4a-flavin adduct. This feature may becloud spectral identification of intermediates in enzymatic reactions.

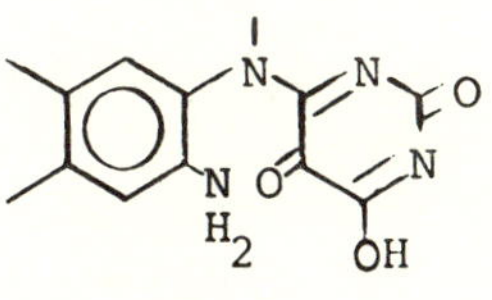

or

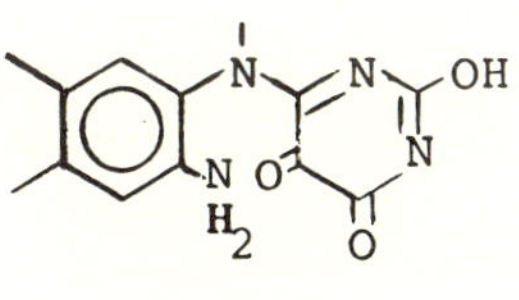

IX′

Another class of flavoenzyme mixed-function oxidases are the bacterial luciferases (15). The evidence for the involvement of an enzyme-bound 4a-FlHOOH species in the oxidation of the aldehyde substrate is irrefutable (3c, 16). However, the mechanism of the enzymatic reaction is not well understood. It is believed that a mixed peroxide of general structure **XII** is formed at the active site and that a step-wise decomposition provides product plus a photon [Equation (44.14)] (17).

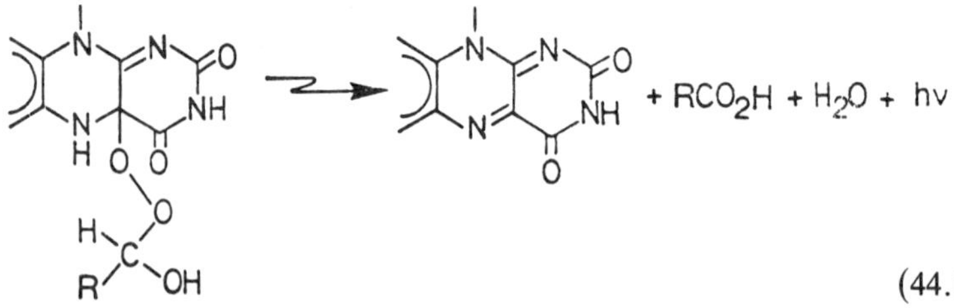

XII $\rightarrow$ + RCO_2H + H_2O + $h\nu$ (44.14)

There are known to be two types of bacterial luciferases. One (e.g., *Beneckea harveyi*) derives its photon emission from an excited flavin species formed on decomposition of **XII**. The other (*Photobacterium*) luciferase does not produce light unless in the presence of a protein which contains the fluorescer 6,7-dimethyl-8-(1′-D-ribityl) lumazine (**XIII**) (18). It has been shown that an excited but nonfluorescent species is produced at the active site of the photobacterium enzyme and that light is only seen when energy can be transferred to a fluorescent compound.

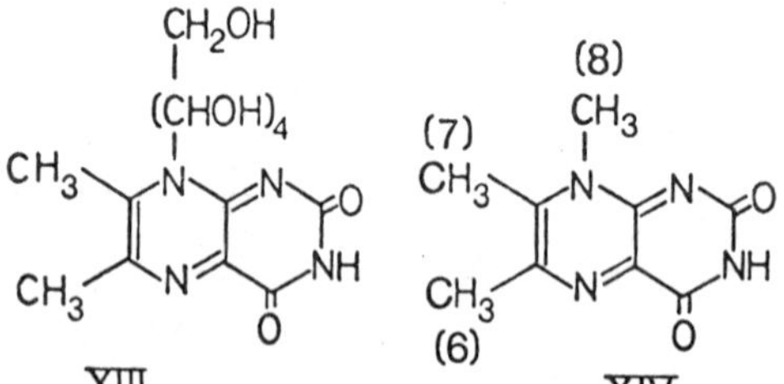

XIII XIV

We have observed, in model systems the chemiluminescent reactions of [Equation (44.15)] (19), wherein two excited species are produced. One, excited flavin ($\{Fl_{ox}\}^{\ddagger}$), emits a photon ($\Phi = 4 \times 10^{-4}$) while the other ($\{X\}^{\ddagger}$) is nonfluorescent and the observation of chemiluminescence requires the presence of a fluorescer. When the fluorescer is Rhodamine-B, the quantum yield (Φ) = 0.01 which is comparable to that for the *Photobacterium* system (Φ = 0.03). Also, the lumazine **XIV** works

$$\{Fl_{ox}\}^* \longrightarrow Fl_{ox} + h\nu \quad \text{(a)}$$

$$\{X\}^* \xrightarrow{[\text{fluorescer}] \times 10^9\ s^{-1}} X + \{\text{fluorescer}\}^* \quad \text{(b)}$$

$$\{X\}^* \xrightarrow{10^5\ s^{-1}} X + \Delta$$

(44.15)

well as a fluorescer in the model system just as protein-bound **XIII** serves as the natural fluorescer for the *Photobacterium* enzyme. The excited species $\{Fl_{ox}\}^{\ddagger}$ and $\{X\}^{\ddagger}$ are both formed on cleavage of the peroxide and the C-H(D) bonds, of the mixed peroxide, but the chemical nature of $\{X\}^{\ddagger}$ is unknown.

The oxygen-dihydroflavin species involved in the few known flavodioxygenase enzyme reactions is unknown. We feel that it is likely that 4a-FlH-OOH is formed at the active site, much as in the flavoenzyme mixed-function oxidases. This opinion is based on our finding that 4a-FlEtOO$^-$ serves as well as a dioxygen transfer agent to ambient nucleophilic substrates (20). Interestingly, oxygen transfer from 4a-FlEtOO$^-$ results in the regeneration of 1,5-dihydro-5-ethyl flavin anion (FlEt$^-$). Since FlEt$^-$ reacts with molecular oxygen to provide 4a-FlEtOO$^-$ there is obtained a catalysis, by FlEt$^-$, of the peroxidation of substrate by molecular oxygen. As an example:

$$\text{4a-FlEtO}_2^- + \text{(phenolate)} \longrightarrow \text{FlEt}^- + \text{(peroxide anion)} \quad (O_2)$$

>95% >95%

(44.16)

The formation of a peroxide anion from the ambident nucleophile may result in C-C bond scission. Dr. Muto, in this laboratory, has established the reactions of Equation (44.17) and Equation (44.18) (20b,c). The C-C bond scissions result from a Criegee-like

$$\text{4a-FlEtO}_2^- + \text{(O}^-\text{, OEt phenanthrene)} \longrightarrow \text{FlEt}^- + {}^-O_2C\text{-biphenyl-}CO_2Et$$

(44.17)

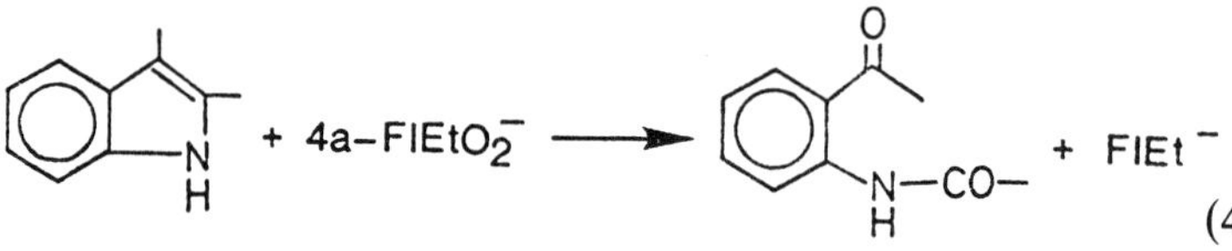

$$\text{(indole)} + \text{4a-FlEtO}_2^- \longrightarrow \text{(o-acylanilide: O, N–CO–, H)} + \text{FlEt}^- \qquad (44.18)$$

rearrangement of the initially-formed hydroperoxide anion [for example, Equation (44.19)] (21).

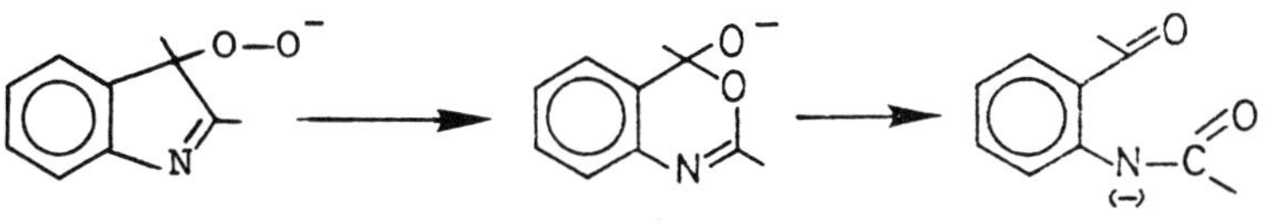

$$\text{(indolenine 3-hydroperoxide anion, O–O}^-\text{)} \longrightarrow \text{(cyclic peroxide, O}^-\text{, O, N)} \longrightarrow \text{(O, N–C=O, (–))} \qquad (44.19)$$

The transfer of the dioxygen moiety from 4a-FlEtO$_2^-$ to substrate takes place by the trapping of a species (**XV**) formed from 4a-FlEtO$_2^-$ in an endothermic reaction [Equation (44.20)]. The constant k_1 is independent of the nature of the substrate:

$$\text{4a-FlEtO}_2^- \underset{k_2}{\overset{k_1}{\rightleftharpoons}} \text{XV} \xrightarrow{k_3[\text{substrate}^\ominus]} \text{FlEt}^- + \text{Substrate-O}_2^-$$

Substrate	$k_1(s^{-1})$	$k_2/k_3(M)$
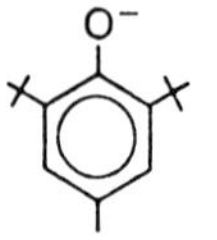 O⁻ (phenolate)	0.36	2.2 x 10^{-4}
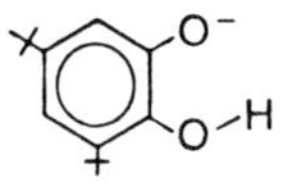 O⁻, O–H (phenolate)	0.37	2.8 x 10^{-4}
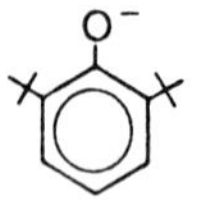 O⁻ (phenolate)	0.39	6 x 10^{-4}
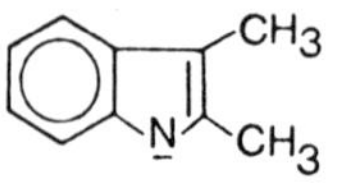 CH$_3$, N, CH$_3$ (indole)	0.33	8.3 x 10^{-3}
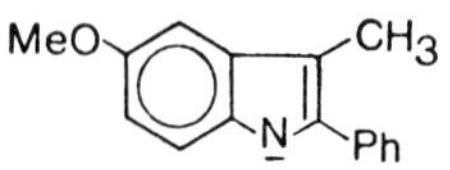 MeO, CH$_3$, N, Ph (indole)	0.37	1.3 x 10^{-2}

(44.20)

The intermediate **XV** must represent a species in which the peroxide O-O bond is intact. Choices for solvent-separated flavin and oxygen species include **XVI, XVII**, or **XVIII** which can be imagined to be formed upon complete heterolytic or homolytic dissociation of 4a-FlEtO$_2^-$.

$FlEt\cdot + O_2^{\dot{-}}$ $FlEt^- + {}^3O_2$ $FlEt^- + {}^1O_2$

XVI XVII XVIII

If dioxygen transfer from 4a-$FlEtO_2^-$ were to involve the intermediacy of solvent separated $FlEt\cdot + O_2^{\dot{-}}$, then the mechanism of oxygen transfer would be as shown in eq. (44.21) (20c). The homolytic dissociation mechanism of eq. (44.21a) is the

a. $4a\text{-}FlEtO_2^- \underset{fast}{\overset{0.33\ s^{-1}}{\rightleftharpoons}} FlEt\cdot + O_2^{\dot{-}}$

b. FlEt· + [2,6-di-t-butyl-4-methylphenolate] $\xrightarrow{fast}$ $FlEt^-$ + [2,6-di-t-butyl-4-methylphenoxy radical]

c. [2,6-di-t-butyl-4-methylphenoxy radical] + $O_2^{\dot{-}}$ $\xrightarrow{fast}$ [4-methyl-4-peroxy-2,6-di-t-butylcyclohexadienone anion, O–$O^{(-)}$]

(44.21)

microscopic reverse of the coupling of FlEt· with $O_2^{\dot{-}}$ to yield 4a-$FlEtOO^-$ (4c). All substrates which have been investigated and which accept the dioxygen moiety from 4a-$FlEtO_2^-$, have been shown to 1e$^-$ reduce FlEt· to yield substrate radical and $FlEt^-$ [Equation (44.21b)]. Furthermore, the rate constants for these 1e$^-$ reductions were found to be great enough to allow the kinetic competency of this step in the dioxygen transfer reaction [Equation (44.22)]. The mechanism of Equation (44.21) would appear to suffer, however, from reports that $O_2^{\dot{-}}$ does not couple with the radical of 2,6-di-*t*-butyl-4-methylphenol [as required in Equation (44.21c)] (22).

$$A^{(-)} + FlEt(\cdot) \xrightarrow{k_r\ M^{-1}s^{-1}} A(\cdot) + FlEt^{(-)}$$

$A^{(-)}$ = [2,6-di-t-butylphenolate], [10-methyl-9-phenanthrolate (CH_3)], [10-ethoxy-9-phenanthrolate (OEt)], [t-butyl hydroquinone monoanion (OH)], [3-methylindolide (N(−))], [5-methoxy-3-methyl-2-phenylindolide (MeO, Ph, N(−))]

k_r =	>10^6	10	1 x 10^3	2 x 10^4	2 x 10^2	2.6 x 10^2

(44.22)

The mechanism of Equation (44.23) would also appear to be unreasonable on the basis that the determined second-order rate constants for the reactions of substrate

$$4a\text{-}FlEtO_2^- \rightleftharpoons FlEt^- + {}^3O_2$$

$$ {}^3O_2 + \text{(2,6-di-}t\text{-butyl-4-methylphenolate)} \longrightarrow \text{(4-peroxy-2,6-di-}t\text{-butyl-4-methylcyclohexadienone, } {}^-O_2\text{)} \qquad (44.23)$$

anions with $^3O_2$ are too small to allow the kinetic competency of Equation (44.23) (20). The involvement of $^1O_2$ can be ruled out by the finding that $^1O_2$ traps as 2,3-dimethylbutene and 2,5-dimethylfuran do not serve as substrates [Equation (44.24)] (20b).

$$4a\text{-}FlEtO_2^- \rightleftharpoons FlEt^- + {}^1O_2$$

$$ {}^1O_2 + \text{(2,5-dimethylfuran)} \not\longrightarrow \text{(endoperoxide, O-O)} \qquad (44.24)$$

Since **XV** does not apparently represent solvent-separated flavin and oxygen species, [Equations (44.21), (44.23), and (44.24)] it must represent a species in which the oxygen has not departed from the flavin moiety. **XV** might represent a charge transfer complex $^3O_2$ and reduced flavin or an isomer of 4a-$FlEtO_2^-$ formed by back addition of the peroxy anion on the flavin moiety.

ACKNOWLEDGMENTS
The work described herein has been supported by grants from the National Institutes of Health and the National Science Foundation.

References

1. (a) Gibson, Q.H. and Hastings, J.W. *Biochem J* 83:368 (1962); (b) Massey, V., Palmer, G., and Ballou, D (1971) In *Flavins and Flavoproteins*. H. Kamin (ed.) Baltimore, Maryland: University Park Press, p. 349; (c) Massey, V., Palmer, G., and Ballou, D.P. (1973) In *Oxidases and Related Redox Systems*. Vol. I. J.E. King, H.S. Mason, and M. Morrison (eds.) Baltimore, Maryland: University Park Press, p. 25; (d) Kemal, C., Chan, T.W. and Bruice, T.C. (1977) *J Am Chem Soc* 99:7272; (e) Favaudon, V. (1977) *Eur J Biochem* 78:293.
2. Eberlein, G. and Bruice, T.C., work in progress.
3. (a) Spector, T. and Massey, V. (1972) *J Biol Chem* 247:5632; (b) Entsch, B., Ballou, D.P., and Massey, V. (1976) *ibid*. 251:2550 (c) Strickland, S. and Massey, V. (1973) *ibid*. 248:2953; (d) Hastings, J.W., Balny, C., Le Peuch, C., and Douzou, P. (1973) *Proc Natl Acad Sci USA* 70:3468 (e) Poulsen, L.L. and Ziegler, D.M. (1979) *J Biol Chem* 254:6449.
4. (a) Kemal, C. and Bruice, T.C. (1976) *Natl Acad Sci USA* 73:995; (b) Iwata, M., Bruice, T.C., Carrell, H.L., and Glusker, J. (1980) *J Am Chem Soc* 102:5036; (c) Nanni, F.J., Sawyer, D.T., Ball, S.S., and Bruice, T.C. (1981) *J Am Chem Soc* 103:.

5. (a) Kemal, C., Chan, T.W., and Bruice, T.C. (1977) *Proc Natl Acad USA* 74:405; (b) Ghisla, S., Entsch, B., Massey, V., and Husain, M. (1977) *Eur J Biochem* 76:139.
6. Ball, S. and Bruice, T.C. (1980) *J Am Chem Soc* 102:6498.
7. Poulsen, L.L., Kadlubar, F.F., and Ziegler, D.M. (1974) *Arch Biochem Biophys* 164:774; Ziegler, D.M. and Mitchell C.H. (1972) *Arch Biochem Biophys* 150:116; Hajjar, N.P. and Hogson, E. (1980) *Science* 209:1134.
8. Miller, A. and Bruice, T.C. (1979) *J Chem Soc Chem Comm* 896; Bruice, T.C. and Miller, A. (1980) *J Chem Soc Chem Comm* 693.
9. Beaty, N.B. and Ballou, D.P. (1980) *J Biol Chem* 255:3817; Ballou, D.P. *J Biol Chem*, in Press.
10. Bruice, T.C., Ball, S., Noar, B., and Vankataram, U., work In Progress.
11. Hamilton, G.A. (1971) *Prog in Bioorganic Chemistry* 1:83.
12. Bailey, S.W. and Ayling, J. (1980) *J Biol Chem* 255:7774.
13. Entsch, B., Ballou, D.P., and Massey, V. (1976) *J Biol Chem* 251:2550.
14. Wessiak, A. and Bruice, T.C., work in progress.
15. Hastings, J.W. (1978) *CRC Crit Rev Biochem* 163.
16. Ghisla, S., Hastings, J.W., Favaudon, V., and Lhoste, J.-M. (1978) *Proc Natl Acad Sci USA* 75:5860.
17. Presswood, R.P. and Hastings, J.W. (1978) *Biochem Biophys Res Comm* 82:990; *ibid*, (1979) 30:93; Shannon, P., Presswood, R.B., Spencer, R., Becvar, J.E., Hastings, J.W., and Walsh, C. (1978) In *Mech. of Oxidizing Enzymes* T.P. SInger and R.N. Ondarza (eds.) New York: Elsevier North Holland, p. 69.
18. Small, E.E., Koka, P., and Lee, J. (1980) *J Biol Chem* 255:8804; Irwin, R.M., Visser, A.J.W.G., Lee, J., and Carriera, L.A. (1980) *Biochemistry* 19:4639; Visser, A.J.W.G. and Lee, J. (1980) *Biochemistry* 19:4366.
19. Shepherd, P.T. and Bruice, T.C. (1980) *J Am Chem Soc* 102:7774.
20. (a) Kemal, C. and Bruice, T.C. (1979) *J Am Chem Soc* 101:1635; (b) Muto, S. and Bruice, T.C. (1980) *J Am Chem Soc* 102:4472; (c) Muto, S. and Bruice, T.C. (1980) *J Am Chem Soc* 102:7559.
21. Muto, S. and Bruice, T.C. (1980) *J Am Chem Soc* 102:7379.
22. Nishinaga, A., Itahara, T., Tomita, H., Nishizawa, K., and Matsuura, T. (1978) *Photochem Photobiol* 28:687; Sawyer, D., private communication.

Published 1982 by Elsevier North Holland, Inc.
Vincent Massey and Charles H. Williams, Editors
Flavins and Flavoproteins

CHAPTER 45

Flavin-Oxygen Complex Formed on the Reaction of Superoxide Ions With Flavosemiquinone Radicals

Robert F. Anderson

Cancer Research Campaign Gray Laboratory, Mount Vernon Hospital, Northwood, Middlesex HA6 2RN, United Kingdom

Summary

Superoxide ions react with flavosemiquinone radicals to form an intermediate whose spectral characteristics (λ_{max}385 nm, ε=9000 dm^3 mol^{-1} cm^{-1}) are consistent with the known 5-HF1-4a-OOH species. Below pH 6.5, the intermediate decays at a rate independent of pH ($k=2.6\times10^2$ s^{-1}) to give restoration of oxidized flavin. This reaction is also catalyzed by hydroxyl ions ($k=7.0\times10^8$ dm^3 mol^{-1} s^{-1}). The entropies of activation of these two reactions are -90.4 and 23.5 J K^{-1} mol^{-1} respectively.

Introduction

Although kinetic investigations on the reaction of dihydroflavin $FlH_2(FlH^-)$ with oxygen have provided evidence for a reduced flavin-oxygen complex (HFlOOH) being formed (1,2) it has not been seen for free flavin and its existence has been questioned (3). Recently, we reported that $O_2^{\cdot}$ reacts with flavosemiquinones to form an intermediate (4) which absorbs in the region attributed to a reduced flavin-oxygen complex. The spectrum of this complex has been partially observed by stopped-flow experiments with 3-methyl lumiflavin (5).

Superoxide ions ($O_2^{\cdot}$) have been shown to be produced during the oxidation of $FlH_2(FlH^-)$ by molecular oxygen in basic solution (6), but as no hydroxylations by $O_2^{\cdot}$ have been observed, it is suspected that a reduced flavin-oxygen complex is the hydroxylating species of flavomonooxygenases (7).

By generating flavosemiquinone and $O_2^{\cdot}$ simultaneously by pulse radiolysis, we have now obtained the full spectrum of this species, HFlOOH, together with kinetic data on its decay to fully oxidized flavin using riboflavin.

Methods

Flavosemiquinone radicals and $O_2^{\cdot}$ ions were produced by pulse radiolysis and detected by kinetic spectrophotometry. Details of the pulse radiolysis equipment used have been published (8). Radiation doses were determined by irradiating aerated KSCN (10 mmol dm^{-3}) assuming the $(SCN)_2^{\cdot-}$ radical produced had G(radicals per 100 eV)=2.9 and $\varepsilon=7100\ dm^3\ mol^{-1}\ cm^{-1}$ at 480 nm (9). In aerated solutions containing formate ions G(radical)=6.2 was assumed. Transient spectra were recorded on a Datalab DL 905 transient recorder and kinetic parameters determined using a Wang 2200 computer system (10).

Results and Discussion

Spectral Properties of the HFlOOH Species

Superoxide ions were produced by pulse radiolysis (60 Gy in 2 μs) of aerated solutions containing formate ions (0.01 mol dm^{-3}).

$$H_2O \rightsquigarrow \cdot OH + H + e_{aq}^- + H_2O_2 + H_2 + H_3O^+ \quad (45.1)$$

$$\cdot OH(H\cdot) + HCO_2^- \rightarrow CO_2^{\cdot-} + H_2O(H_2) \quad (45.2)$$

$$e_{aq}^-(CO_2^{\cdot-}) + O_2 \rightarrow O_2^{\cdot-}(+CO_2) \quad (45.3)$$

When riboflavin (80 μmol dm^{-3}) is also present, further electron transfer reactions take place

$$e_{aq}^-(CO_2^{\cdot-}) + Fl \rightarrow Fl^{\cdot-}(+CO_2) \quad (45.4)$$

$$HFl\cdot \rightleftharpoons Fl^{\cdot-} + H^+ (pKa=8.5[11,12]). \quad (45.5)$$

$$Fl^{\cdot-} + O_2 \rightarrow Fl + O_2^{\cdot-} \quad (45.6)$$

Under the above conditions, the only radicals present 20 μs after the pulse are the HFl· and $O_2^{\cdot-}$ species. The concentration of the HFl· species produced can be deduced by comparing its absorption spectrum with that produced in deaerated solution (11, 12). It is found that following a 60 Gy electron pulse, 8% of the total radical yield of $G=6.2$ produced HFl·([HFl·]=3.1 μmol dm^{-3}). As HFl· reacts only slowly with oxygen ($k<4\times10^4\ mol^{-1}\ dm^3\ s^{-1}$ [13]), the observed fast exponential decay of its absorption at 540 nm is ascribed to its reaction with the large excess of $O_2^{\cdot-}$ present ($[O_2^{\cdot-}]=35$ μmol dm^{-3}) to yield a peroxide intermediate which at neutral pH probably quickly protonates (see below):

$$HFl\cdot + O_2^{\cdot-} \rightarrow HFlOO^- \quad (45.7)$$

$$HFlOO^- + H^+ \rightarrow HFlOOH \quad (45.8)$$

The rate constant for the reaction of riboflavinsemiquinone radicals and $O_2^{\cdot-}$ is $7.2\times10^8\ dm^3\ mol^{-1}\ s^{-1}$ (4) and the spectrum 200 μs after the pulse is shown in Figure 1. Correction was made for the bleaching of the ground-state absorption of riboflavin assuming G(HFlOOH)=G(FlH·)=0.50 [8% of total G(radical)].

The HFlOOH species has λ_{max} at 385 nm with $\varepsilon=9000\ dm^3\ mol^{-1}\ cm^{-1}$. These spectral characteristics are similar to: (a) the labile enzyme-bound oxygen complex reported to be formed between oxygen and flavoprotein hydroxylases (14): (b) the similar complex with bacterial luceriferase (15): and (c) the 4a-pseudobases of 5-alkylflavoquinonium salts described by Walker

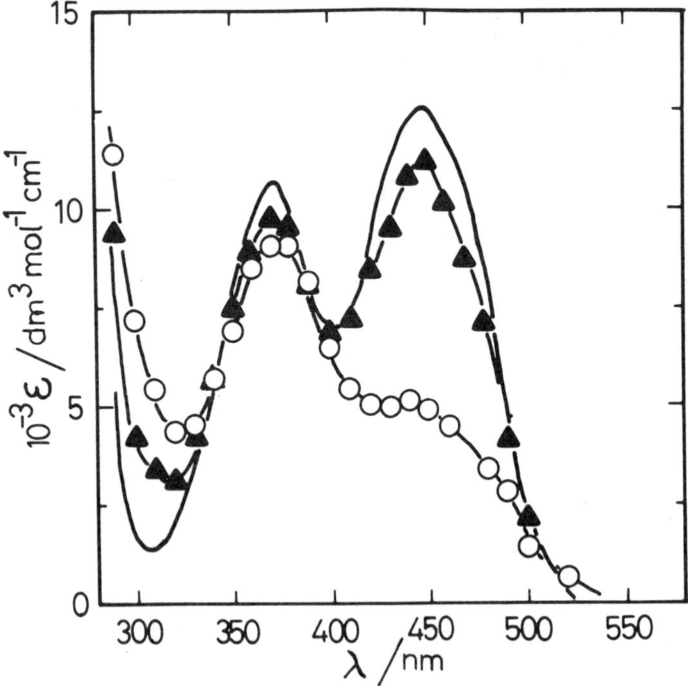

Figure 1. Spectra of transients formed following pulse radiolysis of aerated solutions of riboflavin (80 μmol dm^{-3}); (○) spectrum at pH 7.0~200 μs after the pulse; (▲) spectrum at pH 7.35~8 ms after the pulse, corrected for 50% restoration of oxidized flavin by reaction [Equation (45.13)] at this pH.

et al. (16). We suggest that reaction [Equation (45.7)] results in the formation of the 5-HFl-4a-OOH species. It is probably identical to the proposed complex (1) initially formed on the reaction of oxygen with reduced flavin:

$$FlH_2(FlH^-) + O_2 \rightarrow HFlOOH(^-FlOOH)$$

Decay of the HFlOOH Species

The decay of the HFlOOH species beyond 200 μs has been studied over a wide pH range. Below pH 6.5, HFlOOH decayed to give oxidized flavin at a rate independent of $[H^+]$. Above pH 6.5, the rate of restoration increased with increasing $[OH^-]$ (Figure 2). It is likely that a deprotonation occurs, probably on N3, followed by a fast breakdown.

$$HFlOOH \rightarrow Fl + HO_2^-(H_2O_2) + H^+ \quad (45.9)$$

$$HFlOOH + OH^- \rightarrow {}^-FlOOH + H_2O \quad (45.10)$$

$$^-FlOOH \rightarrow Fl + HO_2^-(H_2O_2) \quad (45.11)$$

At room temperature (298K)$k_9 = 2.6 \times 10^2$ s^{-1} and $k_{10} = 7.0 \times 10^8$ dm^3 mol^{-1} s^{-1}.

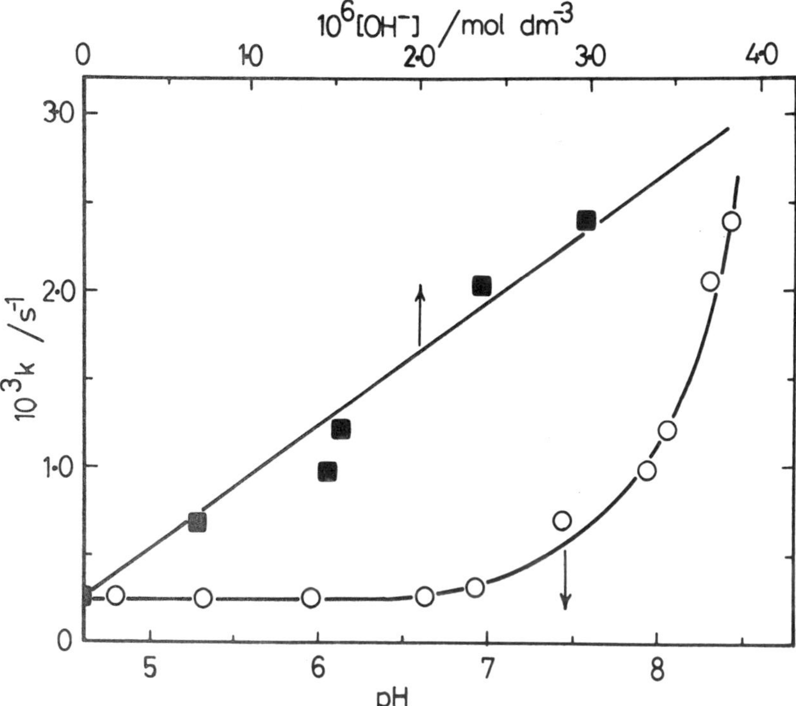

Figure 2. Influence of (○) pH and (■) [OH$^-$] on the decay of the HFlOOH species measured at 480 nm.

However at high pH, incomplete restoration of the oxidized flavin was seen on this short timescale (Figure 1). This is probably due to a small amount of fully reduced flavin, HFl^- (5% of radical yield) being produced during the pulse

$$CO_2^{\cdot -} + Fl^{\cdot -} + H^+ \rightarrow HFl^- + CO_2 \tag{45.12}$$

Further information on the above proposed reactions [Equations (45.9) and (45.10)] can be obtained from Arrhenius plots (Figure 3). The temperature dependence of the pH independent reaction [Equation (45.9)] was studied at pH 5.2. The measured decay rates for this reaction at various temperatures were subtracted from the observed increased rates at pH 7.0 to obtain rate constants for the hydroxyl ion dependent reaction [Equation (45.10)] at the same temperature. The Arrhenius energies E_{act} and the entropy of activation are given in Table 1.

The large loss in entropy of reaction [Equation (45.9)] is consistent with the release and solvation of a proton. The small positive change in entropy of reaction [Equation (45.10)] is also consistent with a transfer of charge to a larger molecule.

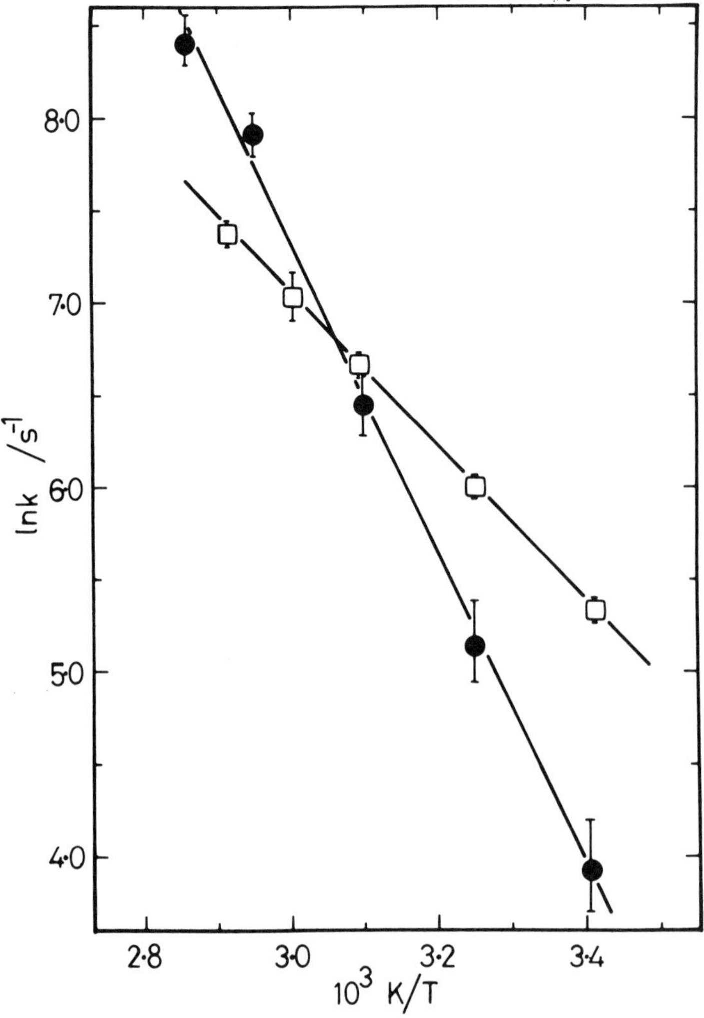

Figure 3. Arrhenius plots for the decay of the HFlOOH species (□) pH 5.2 and (●) pH 7.0 (corrected for pH independent decay) measured at 480 nm.

Table 1. Activation Energies and Entropy Changes for the Transition Complexes.

Reaction	E_{act}/k J mol^{-1}	$\Delta S^{\ddagger}$/J K^{-1} mol^{-1}
(45.9)	34.7 ± 0.5	-90.4 ± 1.5
(45.10)	71.8 ± 2.8	23.5 ± 8.8

These results support the contention (17) that below pH 8 the reduced flavin oxygen complex decays to give oxidized flavin and hydrogen peroxide without flavosemiquinone being formed as an intermediate. The greatly improved time resolution facilitated by pulse-radiolytic production of the flavosemiquinone and O_2^- species compared to stopped-flow rapid-mixing techniques has enabled observations of these reactions with free flavins to be made for the first time.

ACKNOWLEDGMENT
This work is financially supported by the Cancer Research Campaign.

References

1. Massey, V., Palmer, G., and Ballou, D. (1973) In *Oxidases and Related Redox Systems*. King, K.E., Mason, H.S., and Morrison, M. (eds.), Baltimore: University Park press, vol. 1, pp. 25–49.
2. Favaudon, V. (1977) *Eur J Biochem* 78:293–307.
3. Bruice, T.C. (1980) *Acc Chem Res* 13:256–262.
4. Anderson, R.F. (1981) In *Oxygen and Oxy-Radicals in Chemistry and Biology*. Rodgers, M.A.J. and Powers, E.L. (eds.), New York: Academic Press, in press.
5. Massey, V., Palmer, G., and Ballou, D. (1971) In *Flavins and Flavoproteins*. Kamin, H. (ed.) Baltimore: University Park Press, pp. 349–361.
6. Ballou, D., Palmer, G., and Massey, V. (1969) *Biochem Biophys Res Commun* 36:898–904.
7. Massey, V. (1979) In *Biochemical and Clinical Aspects of Oxygen*. Caughey, W.S. (ed.) New York: Academic Press, pp. 477–490.
8. Adams, G.E., Boag, J.W., and Michael, B.D. (1965) *Trans Faraday Soc* 61:492–505.
9. Adams, G.E., Boag, J.W., and Michael, B.D. (1965) *Trans Faraday Soc* 61:1674–1680.
10. Sehmi, D.S. (1978) *Measurement and Control* 11:142–146.
11. Land, E.J. and Swallow, A.J. (1969) *Biochemistry* 8:2117–2125.
12. Anderson, R.F. (1976) *Ber Bunsenges Physik Chem* 80:969–972.
13. Vaish, S.P. and Tollin, G. (1971) *Bionergetics* 2:61–72.
14. Entsch, B., Massey, V., and Ballou, D.P. (1974) *Biochem Biophys Res Commun* 57:1018–1025.
15. Hastings, J.W., Balny, C., Le Peuch, C., and Douzou, P. (1973) *Proc Natl Acad Sci USA* 70:3468–3472.
16. Walker, W.H., Hemmerich, P., and Massey, V. (1967) *Helv Chim Acta* 50:2269–2279.
17. Hemmerich, P. and Wessiak, A. (1979) In *Biochemical and Clinical Aspects of Oxygen*. Caughey, W.S. (ed.) New York: Academic Press, pp. 491–511.

Published 1982 by Elsevier North Holland, Inc.
Vincent Massey and Charles H. Williams, Editors
Flavins and Flavoproteins

CHAPTER 46

Ring-Opening Reactions in Flavin Oxygen Transfer Models

H.I.X. Mager and R. Addink

Biochemical and Biophysical Laboratory of the University of Technology, 67 Julianalaan, 2628 BC DELFT, The Netherlands

Some Types of Bond Scissions and New Bond Formations

The mechanisms of flavoenzymatic monooxygenations are still unknown. So far, oxygen transfer in flavin model systems has proven to occur by radical mechanisms not requiring ring-opened intermediates (1–4). The actual oxygen transfer may be *followed* by ring-opening reactions (cf **4** → **5**, Scheme 1, page 285). On the other hand, alternative oxene mechanisms (5,6) are supposed to take place *essentially* by ring-opened transients (**2** → **3**; **2** → **5**; **9** → **8**^{b}). Further studies on these subjects are of fundamental importance.

The autoxidative transformation of a N^1-blocked dihydroflavin **1** (R=R′=R″= −CH_3) into the C^{4a}-spirohydantoin **5** was the first example (7) of a rearrangement indicating a 1,10^a-bond scission either of the 10^a-pseudobase (**4**→**5**) or, directly, of the 10^a-hydroperoxide (**2**→**5**), *without the intermediacy of the 10^a-pseudobase* (6).

In the early studies on the formation of the spirohydantoin, the occurrence of a blue intermediate was already observed. We established that 10^a-adducts like **2** are not blue (8–12). In 1978, we were the first to propose a 10,10^a-ring-opened structure **3** for the blue transient (8,9). This was supported by the formation of benzimidazole **6** as the product of a competitive rearrangement. The absorption spectrum (11) of **3** (X= −OH) or its nonquinonoid isomer is similar to the spectra of products of failed ring closures in flavin syntheses (13–15).

In the N^5-alkyl-alloxazine and flavin series (4,12), both 1,10^a- and 4,4^a-bond scissions can take place giving the C^{4a}- and C^{10a}-spirohydantoins **8**^{a} and **8**^{b}, respectively (Scheme 2). Mild experimental conditions were found (12) under which the 4^a-pseudobase (5-RFl-4^a-OH) remained stable, while the 4^a-hydroperoxide **9** rearranged directly into the C^{10a}-spirohydantoin **8**^{b} justifying the postulate on the intermediacy of a C^{10a}-spirohydantoincarbonyloxide (6,12).

The 1,10^a- and 4,4^a-bond scissions resulting in the two types of spirohydantoins are not likely to be biochemically relevant, since spirohydantoins are not easily reconverted into the original flavins (16). In contrast, if a 4,4^a-bond scission has finally led to a hexahydroimidazoquinoxaline **11** (Scheme 3), repair of the original flavin model was found to be possible (10). In fact, the

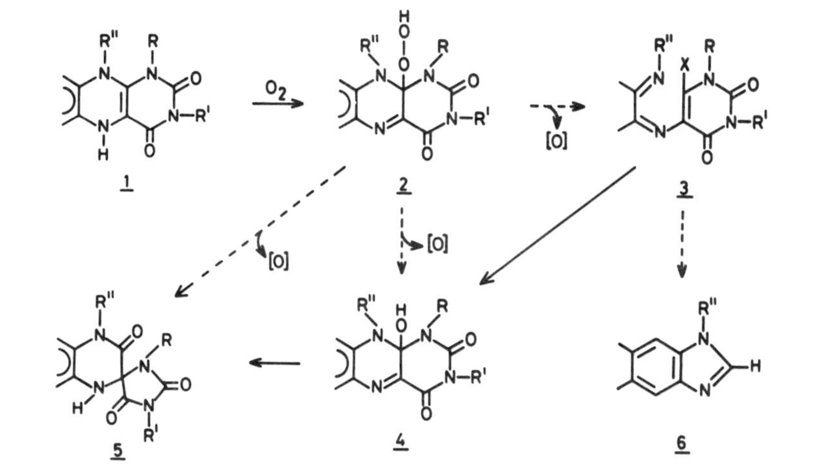

Scheme 1. Rearrangements indicating a $1,10^a$- or $10,10^a$-bond scission.

first observation on a scission of the original $10,10^a$-bond was then made. The ring-opened derivative **12**

$$\left(R''' = -\overset{\overset{\displaystyle O}{\|}}{C} - CH_3 \right)$$

was isolated and characterized and converted into the benzimidazole **13**.

Scheme 2. Rearrangements indicating a $1,10^a$-, $4,4^a$- or $4^a,5$-bond scission.

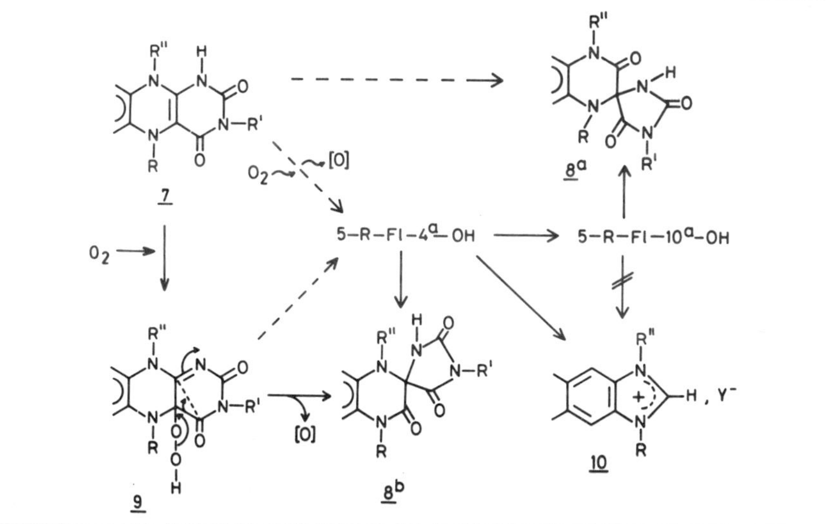

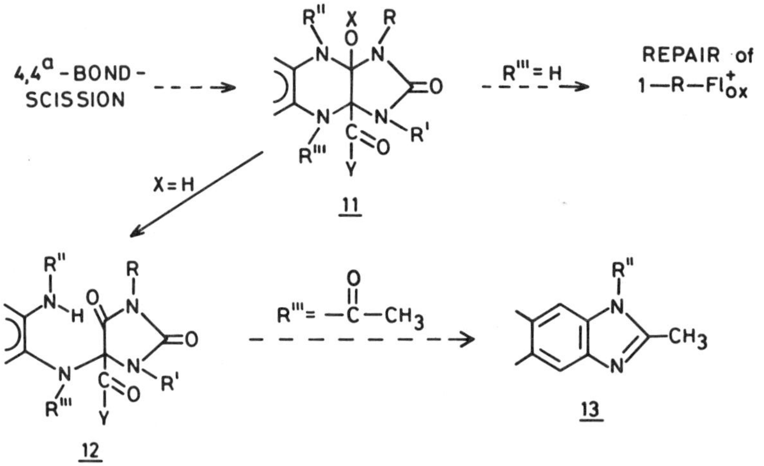

Scheme 3. Rearrangements following the transformation into hexahydroimidazo-quinoxalines.

The formations of the spirohydantoins $\mathbf{8}^a$ and $\mathbf{8}^b$ can be displaced by a rearrangement leading to the benzimidazolinium salts **10**.

Rearrangements into benzimidazoles are of diagnostic value in distinguishing a flavin ring opening either by a $10,10^a$- or a $4^a,5$-bond scission. A differentiation can be made by subjecting ^{13}C-enriched flavin derivatives to a benzimidazole transformation (17). The formation of **10** results from a $4^a,5$-ring opening and not from a $10,10^a$-bond scission, which was a possibility on account of the occurrence of 5-RFl-10^a-OH.

It has already been emphasized (5,6) that in particular the 10, 10^a- and $4^a,5$-bond scissions could lead to ring-opened transients of biochemical relevance.

Detailed studies on the $10,10^a$-bond scission have now been facilitated by the finding (8,11) that the blue colored intermediates **3** do not only arise by oxidation of dihydroflavins, but also in some spontaneous conversions of flavinium salts. These are easily observed on replacing the usual perchlorate anion by one derived from an aliphatic or aromatic carboxylic acid. In the present paper, this will be exemplified by the spontaneous behavior of 1,3,10-trimethylalloxazinium trifluoroacetate and the trichloroacetate in solution which is in remarkable contrast with the stability of the salts in the crystalline state.

Interactions of the N^1-Blocked Flavinium Cation Model 15 With Acid Anions

Starting from the $4^a,10^a$-ethylenedioxy adduct **14** (Scheme 4) alloxazinium salts can be prepared in situ or isolated like the trifluoroacetate, the trichloroacetate, and some substituted benzoates.

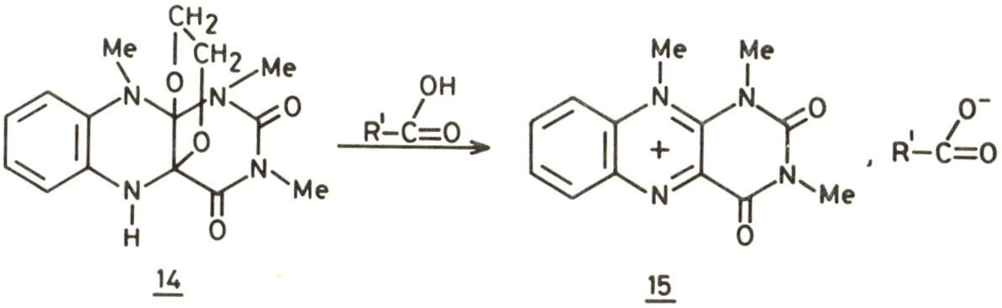

Scheme 4 (chemical structures): 14, 15, reagent R'–C(=O)OH, product R'–C(=O)O⁻

Scheme 4. Preparation of 1,3,10-trimethylalloxazinium carboxylates.

The factors influencing the rate and the degree of the conversions of the alloxazinium salts into the blue $10,10^a$-ring-opened intermediates **3** have now appeared to be connected with: (a) the nature of the anion; (b) the nature of the solvent; (c) the presence of small amounts of water and alcohols in the solvent; (d) the presence of molecular oxygen; and (e) the concentration of the alloxazinium salts.

Solutions of the trifluoroacetate in acetonitrile are reasonably stable, although the reagent-grade solvent may contain up to 0.05% of water. They are very useful as stock solutions for analytical work.

The absorption spectrum in acetonitrile (λ_{max} = 365 nm) is identical with the one of the alloxazinium perchlorate. On dissolving these salts in water, no differences were observed in the conversion to the spirohydantoin **5**, particularly with respect to the appearance of the 300 nm peak, to the disappearance of the 370 nm peak of the cation, and to the fact that no other intermediates including the 10^a-pseudobase **4** could then be detected (Figure 1^a) (11). However, characteristic differences were observed between the perchlorate and the trifluoroacetate on diluting the respective acetonitrile stock solutions with solvents like benzene, toluene, xylenes, chloroform, ethyl acetate, dioxane, etc. The dilutions of the perchlorate only showed the presence of the cation or its conversion into the spirohydantoin **5** (R″ = –H; pathway A, Scheme 5), while the dilutions of the trifluoroacetate could rapidly change to give striking blue colorations (curves a_1 and b_1, Figure 1b), depending on some other conditions also. *Apparently, there are several pathways in the conversion of flavinium salts (Scheme 5). The solvent determines the conversion of the ionic bond into a covalent one at a bridge carbon atom, for example with* $R''O^-$ *(*$R'' = H$*). A covalent bond formation with* $R'COO^-$ *is supposed to give transient esters acting as key intermediates in other pathways.*

The rate and the degree of the blue coloration is stimulated by the presence of small amounts of water in the organic solvents. In benzene dried over sodium (curve b_2, Figure 1^b), $10,10^a$-ring opening decreased by 40% as compared with the coloration in water-saturated benzene (curve b_1). (The use of a dried solvent did not mean that water had been absolutely excluded. The final dilutions contained 0.4% of acetonitrile by which water might have been introduced up to a content of 2 ppm).

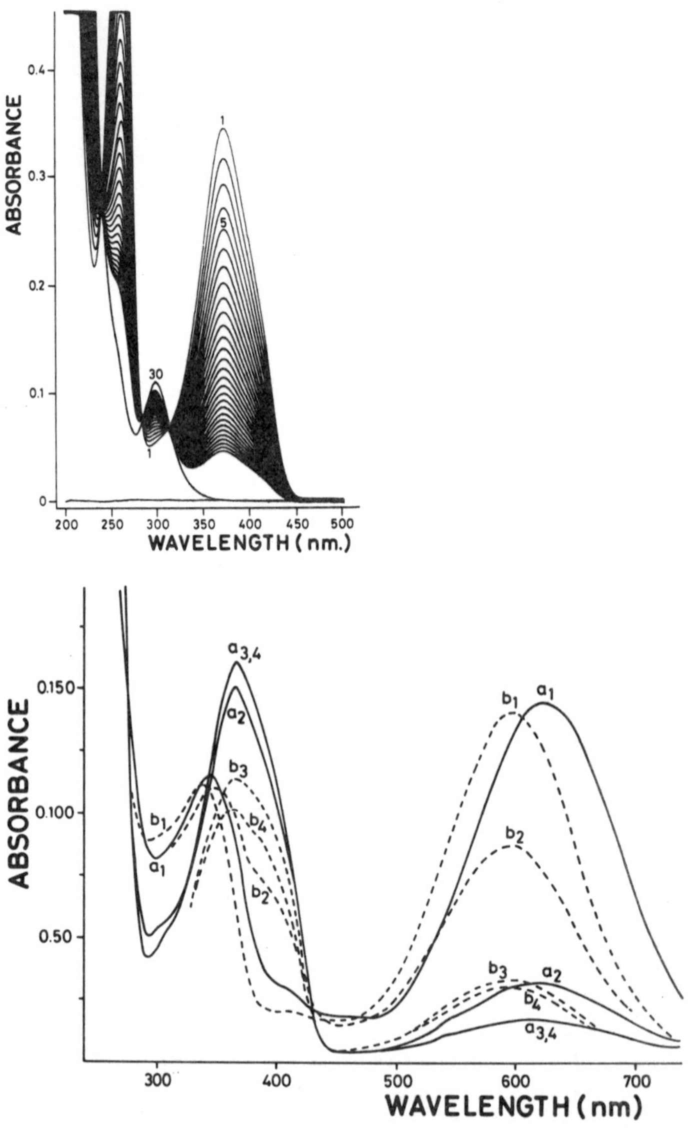

Figure 1. (Top) Repetitive scan of 15, CF_3COO^- in H_2O (2.56×10^{-5} M) at 22°; $\Delta_t = 230$ s; λ_{max} (370 nm) of curve 1 was taken at t=269 s; isobestic points at 242, 285, and 314 nm. (Bottom) Spectra recorded at optimal blue coloration of 1.86×10^{-5} M solns of 15, CF_3COO^- at 25° in: (a_1) H_2O-satd $CHCl_3$/air; (a_2) $CaCl_2/K_2CO_3$-dried $CHCl_3$/air; (a_3) H_2O-satd $CHCl_3/N_2$; (a_4) $CaCl_2/K_2CO_3$-dried $CHCl_3/N_2$; (b_1) H_2O-satd benzene/air; (b_2) sodium-dried benzene/air; (b_3) H_2O-satd benzene/nitrogen; (b_4) sodium-dried benzene/N_2. (Procedure: MeCN-stock solns of the salt were diluted with $CHCl_3$ or benzene in the ratio of 1:250).

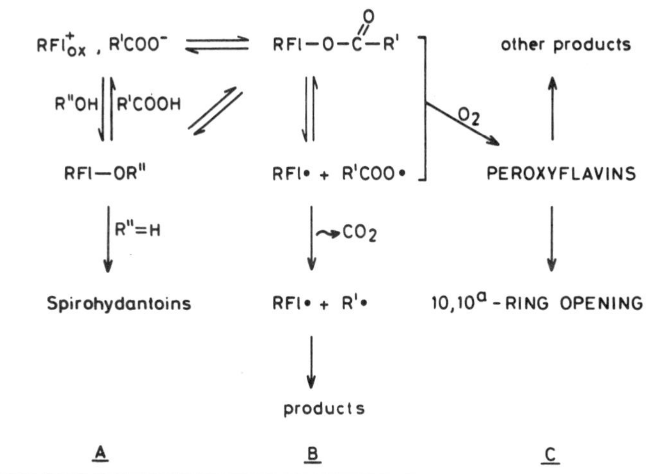

Scheme 5. Competitive pathways in the conversion of flavinium salts.

Compared with the coloration in water-saturated chloroform (curve a_1), a more spectacular decrease of 77% was obtained in chloroform dried over $CaCl_2/K_2CO_3$ (curve a_2), while the time to reach the optimal coloration (t_m) increased from about 8 to 20 minutes (curves 1 and 7 Figure 2). Curves 2–6 show the effects of the stepwise decrease of the water content of the chloroform.

The presence of 0.05–0.1% of ethanol or methanol in the dried solutions may slightly increase the blue coloration. However, such interpretations should be given very carefully, since a concomitant introduction of traces of water will have a similar effect. This is illustrated by curve 8: the increase of the relative absorbance was caused by less than 1.5 ppm of water, concomitantly introduced on raising the final content of the acetonitrile from 0.4 to 0.7%.

In water-saturated chloroform containing 0.1% of ethanol (curve 9) the blue coloration is slightly stimulated, while the t_m is increased to 11.5 minutes. Curves 10–12 were obtained on further increasing the content of ethanol to 0.2, 0.3, and 1%, respectively. The formation of 10^a-ethoxy adduct transients appeared from the increased absorbance in the 410 nm region and from the increased fluorescence at 529 nm (18).

It is concluded that water can play an important part in the $10,10^a$-ring-opening mechanism. Alcohols may have a similar but smaller effect, but their main role is in the formation of alkoxy adducts (RFl-OR", Scheme 5) acting as buffer intermediates for the flavinium carboxylates. The adduct ***14*** *can also behave as such a storage compound when the salts are prepared in situ.*

The salts are on the level of a fully oxidized flavin. Therefore, the requirement for molecular oxygen to obtain a $10,10^a$-ring opening (Figure 1b) was quite unexpected and an observation not made in the preliminary studies (11). The optimal absorbance at 624 nm in H_2O-saturated chloroform under nitrogen (curve a_3) was inhibited by 88% as compared with the optimal absorbance

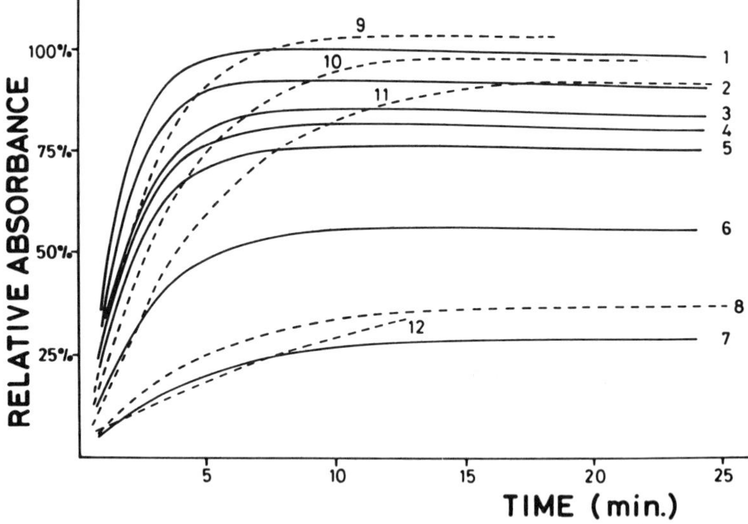

Figure 2. Relative E_{624} vs time of 1.86×10^{-5} M solns of 15, CF_3COO^-/air at 25° in: (1) H_2O-satd $CHCl_3$; (7) $CaCl_2/K_2CO_3$-dried $CHCl_3$; (2–6) in mixtures of H_2O-satd $CHCl_3$ and $CaCl_2/K_2CO_3$-dried $CHCl_3$ in the ratios of: 5 : 1 (2); 2 : 1 (3); 1 : 1 (4); 1 : 2 (5); 1 : 5 (6); (8) in $CaCl_2/K_2CO_3$-dried $CHCl_3$/MeCN (0.7% instead of 0.4%); 9–12) in H_2O-satd $CHCl_3$/EtOH (0.1% (9); 0.2% (10); 0.3% (11); 1% (12)). (The final dilutions except (8) contained 0.4% of MeCN).

of a similar dilution made in the presence of air (curve a_1). The spectrum a_4 given by an anaerobic reaction solution in dried chloroform was identical with a_3 showing that water was not an important factor any more. Anaerobic dilutions in H_2O-saturated and dried benzene gave the spectra b_3 and b_4, respectively, with relative absorbances of 23% and 22%.

The requirement for oxygen proves that in some way the alloxazinium salts have rapidly reacted to give reduced flavin species, subsequently converted to peroxyflavins, liable to $10,10^a$-ring opening (pathway C, Scheme 5; cf **1**→**2**→**3**, Scheme 1). The view that the interaction of a flavinium cation and a carboxylate anion in the dark can result in an electron transfer is supported by experiments showing that transients can be withdrawn from $10,10^a$-ring opening in competitive reactions indicated in Scheme 5 as the radical producing and consuming pathway B (cf section III).

A competitive reaction is revealed by a fluorescence with an emission maximum at 445 nm and excitation maxima at 324 nm and 382 nm in chloroform and at 321 nm and 380 nm in benzene. The product responsible for this was isolated and identified as 1,3-dimethylalloxazine resulting from a spontaneous N^{10}-dealkylation. The product was also determined by the absorbance spectra of the completely bleached reaction mixture obtained after standing for some days at room temperature.

On suppressing the $10,10^a$-ring opening (cf Fig. 1b) either by excluding water or oxygen, N^{10}-dealkylation increased: (a_1) $CHCl_3/H_2O/O_2$: 22%; (a_2)

$CHCl_3/O_2$: 48%; (a_3) $CHCl_3/H_2O/N_2$: 42%; (a_4) $CHCl_3/N_2$: 37%; (b_1) $C_6H_6/H_2O/O_2$: 2%; (b_2) C_6H_6/O_2: 6%; (b_3) $C_6H_6/H_2O/N_2$: 16%; (b_4) C_6H_6/N_2: 11%.

For every solvent the concentration of the alloxazinium salts giving a maximal blue coloration has to be determined. For the trifluoroacetate, it proved to be in the order of 1.5 to 2×10^{-5} M in chloroform, while twice this concentration is admissable in benzene. This is connected with the fact that in benzene less competitive reactions occur by pathway B (Scheme 5).

Depending on the concentration of the salt and the nature of the anion, dealkylation/decarboxylation (pathway B) may even become the main process at room temperature, resulting in a rapid and quantitative evolution of carbon dioxide. This was shown by the trichloroacetate which unlike the trifluoroacetate is absolutely unstable in acetonitrile solution.

Interactions of the N^5-Blocked Flavinium Cation **17** With Acid Anions

In pathway B, we propose that carboxylate radicals are formed which then lose carbon dioxide by homolytic fission. This is in contrast with the more general view (19) that decarboxylations proceed by an ionic mechanism. Unambiguous evidence for the intermediacy of the *counterparts* of the carboxylate radicals was obtained in the N^5=alkylflavin series. The expectation that on account of the higher stability, accumulation of 5-alkylsemiquinones could be effected, proved to be correct (Figure 3).

5-Ethyl-4^a-methoxy-3-methyl-4^a,5-dihydrolumiflavin **16** (Scheme 6) proved to be a suitable starting compound. The conversion to a flavinium salt in benzene solution is exemplified by curve b (Figure 3), representing the immediate effect of β-naphtalenesulfonic acid. The addition of trichloroacetic acid to a solution of **16** in benzene causes a rapid accumulation of the flavin radical (curve c). A slower but not less efficient accumulation of the radical was effected by an aromatic carboxylic acid like salicylic acid (curve d). The strength and the concentration of the acid may also influence the wavelength of the maxima and the absorbance. However, no such influences are shown by curve d, from which the radical accumulation was calculated to be 74%.

The electron transfer partially led to N^5-dealkylation. The anaerobic production of 3-methyllumiflavinium was 14% (c) and 11% (d), respectively. In the same order of magnitude, 3-methyldihydrolumiflavin occurred as another component of reaction mixture d.

On adding sodium nitrite and dilute hydrochloric acid, an immediate recovery of the 5-ethyl-3-methyllumiflavinium cation was accomplished in yields of 65% (c) and 58% (d).

It is noticed that the electron transfer does not exclusively result from a *carboxylic acid anion* /flavinium cation interaction. The solution showing curve b gradually changed within 15 hours to give the radical cation. Rapid oxidation gave a recovery of the flavinium cation in a yield of about 70%. A very slow radical formation was observed even in flavinium perchlorate solutions in the dark.

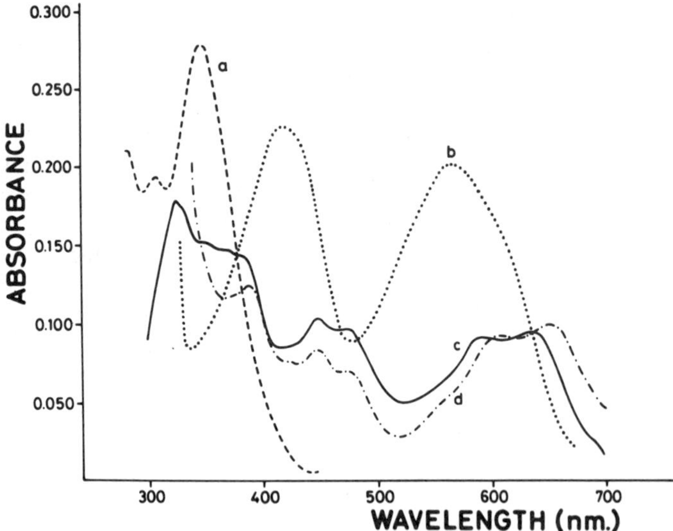

Figure 3. Spectra of 2.78×10^{-5} M solutions of **16** at 25° in: (a) sodium-dried benzene/air; (b) sodium-dried benzene/β-naphtalenesulfonic acid (3.2×10^{-4} M)/air; (c) sodium-dried benzene/trichloroacetic acid (4.8×10^{-4} M)/N_2/$t_m = 4$ min; (d) sodium-dried benzene/salicylic acid (4.5×10^{-4} M)/nitrogen/$t_m = 3$ hr; (Content of MeCN: 0.4% (a, c, d) and 0.8% (b)).

Concluding Remarks

New oxygenation model systems have been developed. They are based on the fact, so far unknown, that an electron transfer can take place in flavinium salt solutions, resulting merely from a flavinium cation/acid anion interaction. Flavin esters probably act as key intermediates.

The requirement for molecular oxygen to obtain a $10, 10^a$-ring opening was established. Consequently, on suppressing the oxygen activation and other factors which determine the $10, 10^a$-ring opening, radicals are accumulated or increasingly consumed in competitive dealkylation/decarboxylation reactions. The fate of the decarboxylated fragments has not yet been elucidated.

Scheme 6. Preparation of 5-ethyl-3-methyllumiflavinium carboxylates.

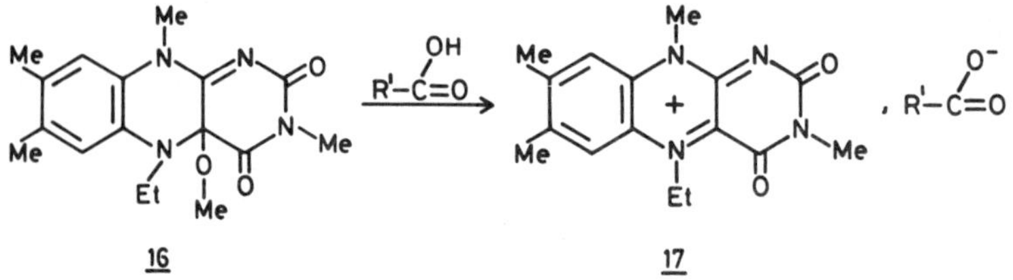

The possibility of a relation between these results and certain oxygenases having carboxylic acids as substrates will be studied.

Concerning the bacterial bioluminescence, we wonder whether the excited species could be a C^{4a}-flavin ester, arisen from a rearrangement of the transient hydroxyalkylperoxyflavin. Formation of the semiquinone (20) as a product of a side path would be in agreement with our results.

References and Notes

1. Mager, H.I.X. and Berends, W. (1972) *Rec Trav Chim* 91:611–629.
2. Mager, H.I.X. and Berends, W. (1974) *Tetrahedron* 30:917–927.
3. Mager, H.I.X. (1976) In Flavins and Flavoproteins. T.P. Singer (ed.), pp. 23–37, Amsterdam: Elsevier.
4. Mager, H.I.X. and Berends, W. (1976) *Tetrahedron* 32:2303–2312.
5. Hamilton, G.A. (1974) In *Progress in Bioorganic Chemistry*. E.T. Kaiser and F. J. Kézdy (eds.), vol. 1, pp. 134–142; In *Molecular Mechanisms of Oxygen Activation*. O. Hayaishi (ed.), pp. 405–451, New York: Academic Press (1974).
6. Mager, H.I.X. In *Oxidases and Related Redox Systems*. T.E. King, H.S. Mason, and M. Morrison (eds.) Oxford: Pergamon Press (1981).
7. Mager, H.I.X. and Berends, W. (1966) *Biochim Biophys Acta* 118:440–441.
8. Mager, H.I.X. In *Flavins and Flavoproteins*. K. Yagi and T. Yamono (eds.), pp. 455–468, Japan Scientific Societies Press (1980).
9. Mager, H.I.X. (1979) *Tetrahedron Letters* 2423–2426.
10. Mager, H.I.X. (1977) *Tetrahedron* 33:981–989.
11. Mager, H.I.X. and Addink, R. (1979) *Tetrahedron Letters* 3545–3548.
12. Mager, H.I.X. (1979) *Tetrahedron Letters* 3549–3552.
13. Tishler, M., Wellman, J.W., and Ladenburg, K. (1945) *J Am Chem Soc* 67:2165–2181.
14. Clark-Lewis, J.W. and Moody, K. (1970) *Aust J Chem* 23:1229–1248, 1249–1273; **24**:2593–2610 (1971).
15. Trout, G.E., Wessiak, A., and Hemmerich, P. (1980) *FEBS Letters* 120:171–174.
16. *Note*: A contrary statement was made by Yoneda, F., Sakuma, Y., and Shinozuka, K. (1977) *J Chem Soc Chem Comm* 175–177.
17. Mager, H.I.X., to be published.
18. *Note*: In preliminary experiments (11) on the $10,10^a$-ring opening the intermediacy of 10^a-adducts was already revealed by the time course fluorescence curves. It has now appeared that a difference between chloroform and benzene solutions had then been caused by the additional presence of the 10^a-ethoxy adduct arisen from the ethanol (0.2–0.3%) still present in the chloroform.
19. Clark, L.W. (1969) In *The Chemistry of Carboxylic Acids and Esters*. S. Patai (ed.), pp. 589–622, London: Interscience-Publishers.
20. Hastings, J.W. et al. In *Second Int. Symp. on Bio- and Chemiluminescence*, La Jolla (1980), in press.

Published 1982 by Elsevier North Holland, Inc.
Vincent Massey and Charles H. Williams, Editors
Flavins and Flavoproteins

CHAPTER 47

Phenylalanine Hydroxylase: A Model for Dihydroflavin-Dependent Monooxygenases?

June E. Ayling and Steven W. Bailey

College of Medicine, University of South Alabama, Mobile, Alabama

Phenylalanine hydroxylase is a tetrahydrobiopterin (6-[dihydroxypropyl]-2-amino-5,6,7,8-tetrahydro-4-pteridinone)-dependent enzyme which catalyzes the oxidation of phenylalanine to tyrosine. The structural similarity of dihydroflavin to tetrahydrobiopterin has prompted speculation that the two classes of monooxygenases which utilize these cofactors may function by related intermediates. The study of phenylalanine hydroxylase is facilitated, in comparison to flavin monooxygenases, by the free dissociation of cofactor from enzyme, by modest cofactor autooxidation rates, and an order of magnitude slower turnover.

Two pyrimidine active-sites probes have been designed which have provided the first strong evidence for the formation of a cofactor-oxygen adduct in pteridine-dependent enzymes. 2,5,6-Triamino-4-pyrimidinone (TP), equivalent to tetrahydropterin minus C-6 and C-7, and its 5-N benzyl derivative (BDP), allow enzymatic tyrosine production at maximum rates several hundred times slower than those observed with tetrahydrobiopterin (1). These analogs cannot be used catalytically since, as a result of one turnover, the 5-amine substituent is cleaved, yielding ammonia or benzylamine, respectively, and a quinoid pyrimidine carrying a keto at position 5. In the presence of a reductant (e.g., mercaptoethanol), this latter product can be recovered as 2,6-diamino-4,5-dihydroxypyrimidine (divicine) (2). Despite the fact that both TP and BDP stimulate somewhat uncoupled reactions (TP=1.8, BDP=1.3 cofactor molecules consumed per molecule of tyrosine produced), all of the enzymatically oxidized pyrimidine is converted to divicine (3).

Origin of the Divicine 5-Substituent

Enzymatic phenylalanine hydroxylations with the above pyrimidine cofactors were performed in the presence of $^{18}O_2$. The isotropic purity of the molecular oxygen in the reaction mixtures (94±1%) was monitored by isolation (4) and mass spectrometry of the tyrosine formed. Synthetically-labeled ^{18}O-divicine (obtained by $H_2^{18}O$ hydrolysis of quinoid TP in MeOH/trifluoracetic acid, followed by catalytic reduction) was found to be quite susceptible to exchange with the medium. Following deproteinization, the enzymatically-produced divicine was, therefore, quickly derivatized with t-butyl-dimethylsilyl chloride, so that it could be purified by HPLC prior to analysis. Mass spectra of derivatized divicine, isolated from hydroxylation reactions initiated with either TP or BDP, revealed 86% and 88%, respectively, of the maximum possible label. However, synthetically-labelled ^{18}O-divicine, carried through identical parallel reaction, derivatization, and isolation procedures, showed $12 \pm 1\%$ loss of label. Thus, when corrected for this control, the divicine obtained from TP and BDP is initially $98 \pm 5\%$, and $100 \pm 3\%$ labelled with oxygen derived from $^{18}O_2$ (5) (Figure 1).

Uncoupled Enzymatic Reactions

To further investigate the uncoupled pathway of cofactor oxidation, phenylalanine was replaced with o-methylphenylalanine. This substrate analog greatly emphasizes an unproductive pathway of cofactor oxidation (cofactor oxidized/substrate hydroxylated is higher than 20). The majority of the divicine recovered from these highly uncoupled reactions still contained label from molecular oxygen (75% from TP, 90% from BDP). These figures must be considered minimum values as they have not been corrected for the effect of a low level (less than 1%) of divicine in the TP and BDP. The rates simulated by o-methylphenylalanine are significantly lower than with the natural substrate, and not all cofactor is converted to product during the 15-min reaction period. An effect of the impurity is quite probable in the case of TP, where o-methylphenylalanine induced a shift of cofactor K_m to an unmeasurably high value, resulting in a very low percentage of enzymatic consumption of pyrimidine.

Position of ^{18}O-Label; Evidence that Peramide is not an Intermediate

The rapid loss of label from underivatized enzymatically-generated divicine in the presence of catalytic amounts to an oxidant, e.g., oxygen, is circumstantial

Figure 1. Reaction of phenylalanine hydroxylase at 27° and pH 7.4, utilizing ^{18}O-TP as cofactor in the presence of $^{18}O_2$. R=H; (●) oxygen-18.

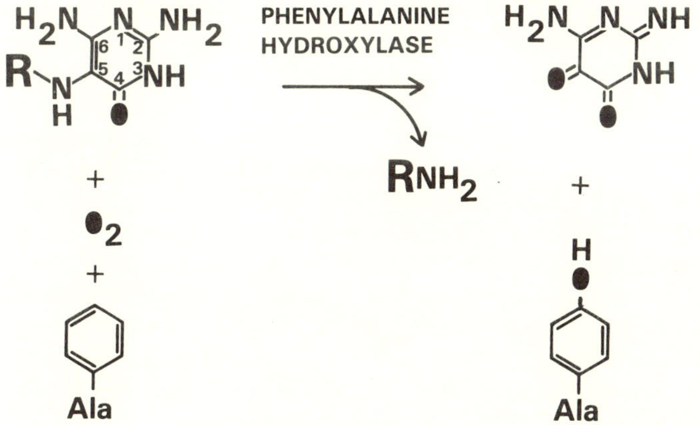

evidence that the label is located at carbon 5. This exchange was shown to be promoted by oxidation to quinoid divicine, followed by hydration, and subsequent disproportionation with the remaining unoxidized divicine. The position of the label in the phenylalanine hydroxylase pyrimidine product was unambiguously assigned through the use of synthetically-labelled ^{18}O-TP. This material was prepared by the action of Na^{18}OH upon 4-chloro-2,6-diamino-5-nitropyrimidine, followed by dithionite reduction. The mass spectrum of the t-butyldimethylsilyl divicine obtained via phenylalanine hydroxylation in the presence of ^{18}O-TP and $^{18}O_2$ revealed two atoms of label. Molecular oxygen is thus linked to carbon-5 of the pyrimidine cofactor (analogous to C-4a in a pteridine) (5) (Figure 1). A corollary of this observation is the elimination of a peramide as a possible intermediate in pteridine-dependent monooxygenases (6).

Only three of the molecular oxygen/cofactor adducts which have been proposed as intermediates in enzymatic hydroxylation are compatible with these results: (a) C-4a hydroperoxide, (b) C-4a carbonyl oxide (or its cyclic trioxide tautomer) (7), and (c) C-4a–C-8a perepoxide (which must open exclusively to the C-4a carbinolamine) (6). Flavoproteins have been shown to interact covalently with molecular oxygen by the shift of the C-4a ^{13}C nuclear magnetic resonance upon adding O_2 to bacterial luciferase/reduced FMN complex at low temperatures (8). Evidence for a similar occurrence in a flavin-dependent monooxygenase is the agreement between the spectrum of a short-lived early intermediate of p-hydroxybenzoate hydroxylase (9) and that of synthetic 5-ethyl-3-methyl-4a-hydroperoxylumiflavin (10). Thus both external dihydroflavin and tetrahydrobiopterin-dependent monooxygenases appear to utilize a covalent oxygen adduct of the same bridgehead carbon, even though the latter have been shown to be iron-containing enzymes (11,12), and the former are metal free.

The most significant difference in the two classes of monooxygenase cofactors may be the properties of their 5-nitrogens in the transition state. Unbound tetrahydrobiopterin and dihydroflavin have pKs of 5.0 and less than zero, respectively (13). Upon formation of a C-4a-oxygen adduct, the flavin N-5 will still remain fairly acidic as a result of the conjugation with the xylene ring. The same nitrogen in pteridine will, however, become quite basic. Moderated somewhat by the inductive effect of the hydroperoxide group, it would not be unreasonable to expect a pK near neutral. Thus, protonation of the 5-nitrogen of a pteridine-4a-oxygen adduct is more likely than that of flavin.

It has been suggested that the aromatic amino acid hydroxylases may take advantage of the pteridine 2-amino group (which allows an extended guanidinium resonance) to create the electrophilic oxenoid reagent necessary for the oxidation of the nearly unactivated phenylalanine and tryptophan rings (7). The pK for this transition is, however, likely to be more acidic than that of N-5, although it is possible that a pteridine hydroperoxide intermediate could be made dicationic by the enzyme. Furthermore, N-5 protonation of tetrahydropteridine-4a-hydroperoxide is a prerequisite to the ring opened carbonyl oxide proposed by Hamilton (7). The cleavage of the amine substituent from the pyrimidine cofactor, although consistent with this hypothesis, might also be caused by preferential loss of amine from the carbinolamine product of the direct utilization of the C-5 (C-4a in a pteridine) hydroperoxide.

The study of tetrahydrobiopterin-dependent enzymes via pyrimidine cofactor analogs may allow, by their drastic alteration of the rate-limiting step in the reaction, the elucidation of many of the remaining questions concerning activation of molecular oxygen by enzyme-bound cofactors.

ACKNOWLEDGMENTS
This investigation was supported by NIH Grants CA-20708 and GM-27880 and by a Grant, AX-706, from the Robert A. Welch Foundation.

References

1. Bailey, S.W. and Ayling, J.E. (1978) *Biochem Biophys Res Comm* 85:1614–1621.
2. Bailey, S.W. and Ayling, J.E. (1979) In *Chemistry and Biology of Pteridines*. Kisliuk, R.L. and Brown, G.M. (eds.) Elsevier/North Holland, pp. 171–176.
3. Bailey, S.W. and Ayling, J.E. (1980) *J Biol Chem* 255:7774–7781.
4. Bailey, S.W. and Ayling, J.E. (1980) *Anal Biochem* 107:156–164.
5. Bailey, S.W., Weintraub, S., Hamilton, S.M., and Ayling, J.E. *J Biol Chem* in press.
6. Dmitrienko, G.I., Snieckus, V., and Viswanatha, T. (1977) *Bioorganic Chemistry* 6:421–429.
7. Hamilton, G.A. (1974) In *Molecular Mechanisms of Oxygen Activation*, Hayaishi, O. (ed.) New York: Academic Press, pp. 405–451.
8. Lhoste, J-M., Favaudon, V., Ghisla, S., and Hastings, J.W. (1980) In *Flavins and Flavoproteins*, Yagi, K. and Yamano, T. (eds.) University Park Press, pp. 131–138.
9. Entsch, B., Ballou, D.P., and Massey, V. (1976) *J Biol Chem* 251:2550–2563.
10. Kemal, C. and Bruice, T.C. (1976) *Proc Nat Acad Sci USA* 73:995–999.
11. Fisher, D.B., Kirkwood, R., and Kaufman, S. (1972) *J Biol Chem* 247:5161–5167.
12. Gotschall, D., Lazarus, R., Dietrich, R., Henrie, R., and Benkovic, S.J. (1981) *Fed Proc* 40:1652.
13. Hemmerich, P. and Wood, H.C.S. (1965) *Angew Chem (Int)* 4:671–687.

PART III B:

Flavoprotein Monooxygenases

Published 1982 by Elsevier North Holland, Inc.
Vincent Massey and Charles H. Williams, Editors
Flavins and Flavoproteins

CHAPTER 48

Flavoprotein Monooxygenases

David P. Ballou

Department of Biological Chemistry, University of Michigan, Ann Arbor, Michigan

Flavoproteins are extremely versatile catalytic molecules as has been well documented in the preceding papers. The monooxygenases catalyze reactions in which molecular oxygen is reduced by four electrons. In these reactions, an organic substrate is oxidized and one atom of the oxygen molecule is added to it, hence the term monooxygenase. In addition, the second atom of oxygen is reduced to form water at the expense of either a reduced pyridine nucleotide, the substrate itself, or another organic molecule. This paper will focus on those flavoprotein monooxygenases which utilize pyridine nucleotides.

Reduction Properties

The reactions of the flavoprotein monooxygenases can be discussed in terms of two separate half-reactions; the reductive step in which the flavin is reduced by two electrons from a pyridine nucleotide, and an oxidative step whereby oxygen and substrate interact with the reduced flavin to bring about the oxygenation. The reaction with the reduced pyridine nucleotides in all cases so far studied, involves the transfer of only the 4R-H from the nicotinamide to the flavin (1). This maintenance of 4R-stereochemistry in nicotinamide oxidation implies that there are some close similarities in the active site geometries of these enzymes. In those cases which have been studied by rapid reaction techniques (2–4), the reduction is accompanied by a 5–10-fold primary kinetic deuterium isotope effect. In some cases (3,5–8), direct demonstration of charge transfer interactions for both the NAD(P)H-FAD and the NAD(P)-FADH complexes (i.e., the complexes preceding and following the reduction) has been reported. Thus, the reduction mechanism for the oxygenases is very similar to that for the pyridine nucleotide-linked flavoprotein dehydrogenases (9).

In all cases of the phenolic hydroxylases, it is found that the substrate stimulates enormously the rate of reduction (10); enhancement factors of more than 10^5 have been reported (5,7,8). It is known that the reduced forms of these enzymes react readily with oxygen. Hence substrate stimulation provides an elegant means of control, whereby reduction in the absence of substrate is extremely slow, and wasteful utilization of high energy reducing equivalents by an NAD(P)H oxidase activity is avoided. It has been found for many of these enzymes that not only are substrates effective in this role, but other substrate-like molecules can also enhance the reductive step; in the latter case, pyridine nucleotide oxidase activity occurs, but there is little or no oxygen incorporated into the effector (10). Presumably, these nonsubstrate effector molecules are normally not available to the cell.

C.P.C. ABERYSTWYTH LLYFRGELL

Some of the phenolic hydroxylases bind only the phenolate form of the substrate (6,8), others bind only the phenolic forms (6,11), while some effectors have no phenolic group at all. Thus the phenolic group alone cannot be the primary determining factor in the effector role. The minimum requirement seems to be that the effector be an aromatic compound of similar size to the substrate. Other common features of effectors are not as yet clear. Binding of effector does not necessarily improve the binding of pyridine nucleotide (8), and random order of addition frequently occurs (6,7). The redox potential of the flavin in melilotate hydroxylase is increased upon binding the effector, but the magnitude of this increase can only account for part of the overall rate enhancement (Dr. L. Schopfer, personal communication). Perhaps the phenyl ring somehow induces a better geometry at the active site so that the charge-transfer interactions which are seen between the pyridine nucleotide and the flavin and which seem to correlate with reduction efficacy, can be improved.

The effector role of the substrate is not a universal phenomenon among the flavoprotein monooxygenases. The microsomal FAD-containing monooxygenase, cyclohexanone monooxygenase, and luciferase to show no such substrate control over the reductive step. Rather, the control occurs in the oxidative step, as will be detailed below. It should be noted that these enzymes are not phenolic hydroxylases, so that an aromatic ring, which may be important in the effector role, cannot participate.

Reaction with Oxygen

All flavoprotein monooxygenases which have so far been investigated with rapid reaction techniques form C(4a)-hydroperoxyflavin intermediates (12–19). In the absence of substrate, the nonphenolic hydroxylases (microsomal FAD-containing and cyclohexanone monooxygenases and luciferase) form relatively stable C(4a)-hydroperoxyflavin intermediates. In contrast to the phenolic hydroxylases where in the absence of substrate the reductive step severely inhibits the pyridine nucleotide oxidase activity, it is the stability of these C(4a)-hydroperoxyflavin intermediates which prevents these enzymes from wastefully acting as pyridine nucleotide oxidases. Thus, in the absence of organic substrates (since the intracellular space is normally a reducing environment), these oxygenases exist mainly as their potentially reactive hydroperoxides. To paraphrase Henry Kamin, "these enzymes go around cocked, just looking for a substrate to attack."

A remarkable feature of the reactions of the reduced forms of these flavoproteins with oxygen is the degree of rate enhancement over that of model systems. It is found with free flavins that the rate is 50 $M^{-1}s^{-1}$ at 25° C (20), whereas with the monooxygenases the rates range from $1–9 \times 10^5$ $M^{-1}s^{-1}$ at 4° C, a factor of greater than 10^4. Recently, Carol Ryerson, Chris Walsh, and I have found with cyclohexanone monooxygenase that the rate of formation of the C(4a)-hydroperoxyflavin is too rapid to observe in the stopped flow instrument, even at 30 μM O_2 (18). This implies a rate constant in excess of 3×10^7 $M^{-1}s^{-1}$ at 4°C and thus the rate of this reaction may approach diffusion controlled limits. Realizing that this overall process requires a singlet reduced flavin to reach with triplet O_2, a spin forbidden process, it is unlikely that such a reaction involves a simple 2-electron transfer. Kemal et al. (20) have postulated [See Equation (48.1)] that the reduced flavin transfers an electron to the oxygen to form a radical pair, a step which is not spin forbidden, and one which will be greatly aided by electron delocalization on the flavin.

$$\mathrm{FlH}^- + \mathrm{O}_2 \rightarrow \mathrm{FlH}\cdot \uparrow\uparrow \mathrm{O}_2^{\cdot -} \rightarrow \mathrm{FlH}\cdot \uparrow\downarrow \mathrm{O}_2^{\cdot -} \rightarrow \mathrm{FlH{-}OOH} \qquad (48.1)$$

This radical pair, after spin inversion, collapses to the hydroperoxy flavin (presumably at a rate faster than radical dissociation). Free energy calculations, coupled with kinetic studies (20) support their hypothesis. Dr. Anderson has shown in a pulse radiolysis experiment (see this volume) in which he produced FlH· and $O_2^{\cdot -}$, that these two radicals combine to form the C(4a)-hydroperoxyflavin at a rate of about $5 \times 10^8 M^{-1}s^{-1}$. This will be a minimum rate for the collapse of the two radicals mentioned above, since in Anderson's experiment the radicals are solvated and probably would not react as fast as the radical-complex proposed in this scheme. Thus one can say that this radical-pair mechanism is probably correct.

The evidence that the intermediates formed by reaction of the reduced forms with oxygen actually are C(4a)-hydroperoxyflavins is quite convincing and comes from comparing their spectral forms with those of models (21), and from ^{13}C-NMR studies of the intermediate obtained with luciferase (22). Spectra typical of C(4a)-hydroperoxyflavin species are shown in Figure 1.

Oxygen Reactions in the Presence of Substrates

The general kinetic schemes describing the flavoprotein monooxygenases fit quite well into two classifications (Scheme 1 and Scheme 2). In the first, which is characteristic of the phenolic hydroxylases, substrate and NAD(P)H add randomly (or preferentially) and reduce the enzyme. Before oxygen adds, NAD(P) must be released, whereupon oxygenated intermediates are formed which participate in the oxygenation of the organic substrate. Finally, product (SOH) and water are released, probably in that order, although this point has not been carefully analyzed. In the second scheme, which is characteristic of both cyclohexanone and the microsomal FAD-containing monooxygenases (16, 17), the pyridine nucleotide is bound tightly throughout the catalysis and is released last. The bound NAD(P), in fact, is required for stabilizing the C(4a)-hydroperoxyflavin intermediate in the absence of substrate. The mechanism of this stabilization is completely unknown at this point, although model studies (21) would suggest that whatever the details are, it involves stabilization of the N(5) proton.

Several of these monooxygenases have been investigated by rapid kinetic techniques (6, 12–16, 18). When the reduced forms of the enzymes in the presence of substrate are reacted with oxygen, two, and in some cases three, intermediates are observed (13). Figure 1 shows spectra obtained with para-hydroxybenzoate hydroxylase in the presence of the substrate, 2,4-dihydroxybenzoate. To fully resolve these intermediates, monovalent anions are frequently used, since they differentially retard various steps in the reactions (12, 13, 23); the chemical mechanism does not appear to be altered. The

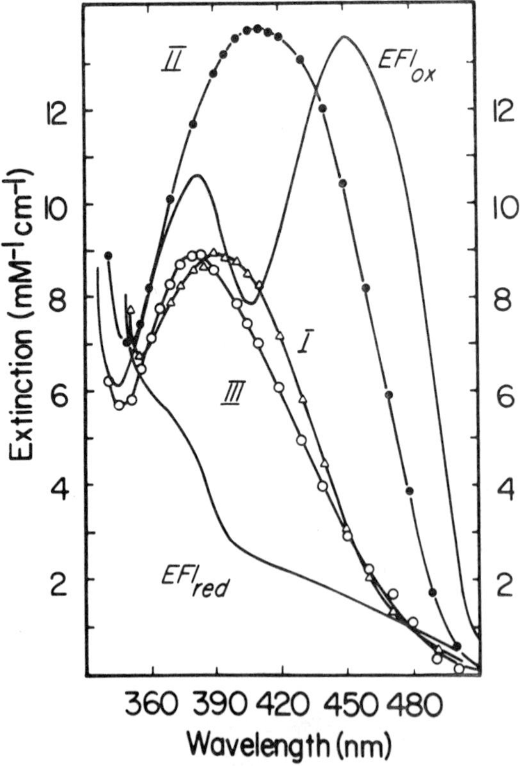

Figure 1. Spectra of intermediates I, II, and III as calculated from the data in reference 13. Reduced enzyme in complex with 2,4-dihydroxybenzoate was reacted with 0.6 nM oxygen.

spectra of those species labeled I and III are extremely similar to those of the model C(4a) species and have been identified as the C(4a)-hydroperoxyflavin and the C(4a)-hydroxyflavin (pseudobase) species (13). Intermediate II has a very high absorbance in the 370–440 nm region and has never been satisfactorily accounted by chemical models. The kinetics of its formation have been shown to be identical to the kinetics of product formation, as determined by

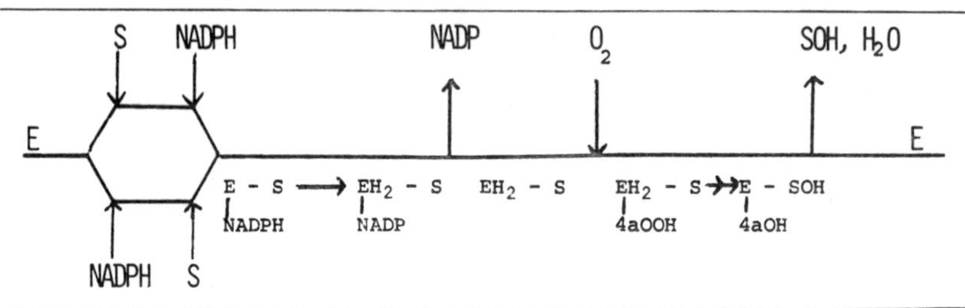

SCHEME I

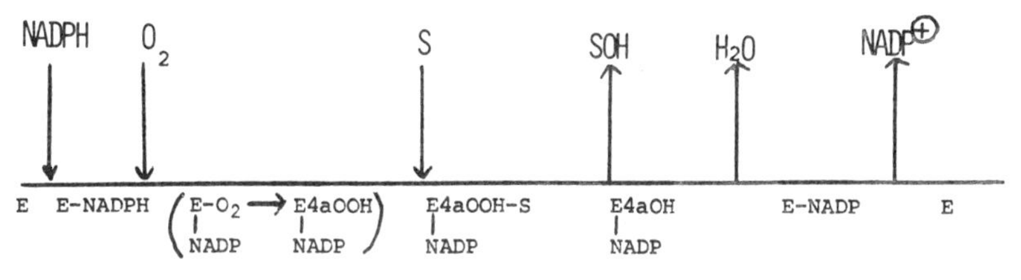

SCHEME II

parallel stopped flow and rapid quench techniques (24). Thus one must consider this species as very likely relevant in the overall scheme. With this enzyme, intermediate II has been seen with 2,4-dihydroxybenzoate, para-aminobenzoate, and paramercaptobenzoate as substrate, while other known substrates do not show this species. Presumably, this is due to the decay of II being more rapid than its formation, so that it is kinetically invisible. It should be pointed out, however, that parahydroxybenzoate hydroxylase is the only monooxygenase (or for that matter, model flavin system) with which intermediate II has been observed. Perhaps the inability to form intermediate II will account for the fact that no model system has ever shown competence in aromatic hydroxylation. Therefore the structure of this intermediate, which may be the crux of the aromatic hydroxylation mechanism, is the most important unsolved problem in this area of flavin research. Intermediate III has also been seen with melilotate hydroxylase (23), microsomal FAD-containing monooxygenase (16), cyclohexanone monooxygenase (tentative), and luciferase (tentative). Other systems have not been investigated in detail for intermediates.

The microsomal enzyme is somewhat unique in that it is the only mammalian enzyme of this type which has been studied, and furthermore, there is such a wide variety of substrates which can be oxygenated. Nearly any hydrophobic secondary or tertiary amine, or N-oxide, mercaptan, thiourea, thioether, or even disulfide will serve as substrate (25).

With this enzyme, the turnover numbers using any of the numerous substrates vary no more than about 3-fold (25). It is believed that this is due to the rate-limiting step being the breakdown of the pseudobase. This was supported by rapid reaction studies in which Intermediate I was reacted with various substrates (16) (this species is very stable at ice temperatures). With dimethylaniline, a second C(4a) species with a slightly blue shifted spectrum could readily be observed to form, which then decayed at the rate-limiting step of catalysis. When cysteamine was used, a different initial rate was observed, but the final rate-limiting decay was again that of a C(4a) species decaying back to oxidized enzyme. The decay of intermediate III, the pseudobase, is usually one of the primary rate-limiting steps in monooxygenase reactions. Thus the difference in reactivity expected with various substrates does not occur at the rate-determining step of catalysis, and is therefore not seen to any extent.

The reaction of cyclohexanone monooxygenase also involves C(4a)-hydroperoxyflavin intermediate(s) (18). The decay of the first intermediate follows very complex kinetics and has not been fully resolved. This enzyme demonstrates the versatility of the C(4a)-hydroperoxyflavin species, since this Baeyer-Villiger oxygenation surely involves nucleophilic attack on the substrate (as probably does the luciferase reaction), in contrast to the electrophilic attack involved with the other monooxygenases. Curiously, this enzyme is perfectly capable of catalyzing electrophilic attack as well, as has been demonstrated by its oxygenation of thiane (26).

Replacement With 1-Deaza-FAD

Several of the flavoprotein monooxygenases have had the normal FAD replaced with the 1-deaza analogue (27). The phenolic hydroxylases are thereby converted to oxidases; the C(4a)-hydroperoxyflavin intermediate is formed, but simply breaks down unproductively to form hydrogen peroxide and oxidized flavin (28). However, both cyclohexanone monooxygenase (26) and luciferase (29) have nearly normal activity with this analogue. This seems to indicate that the N(1)-position is very important for inducing electrophilicity, but not for nucleophilicity. Further investigations along this line may prove to be very informative. A good test would be to try the 1-deazaflavin forms of these enzymes as catalysts for the oxygenation of sulfides.

Mechanistic Hypotheses

Scheme 3 shows a general mechanism which explains many of the known facts about the oxygen half reaction of the phenolic hydroxylases. The reduced enzyme in complex with substrate reacts rapidly with O_2 to produce intermediate I, the C(4a)-hydroperoxyflavin. The nucleophilic substrate then attacks the electrophilic C(4a)-hydroperoxyflavin with the concomitant rearrangement of the X group (usually a phenol), and possibly with the aid of an enzyme base, B. This step results in the oxygenation and leads to a cyclohexadienone structure, which in complex with the pseudobase or perhaps a transitory ring-opened flavin (13), may comprise intermediate II. In most cases, the rearomatiziation occurs very quickly so that intermediate III, the free pseudobase, is seen directly in kinetic experiments. The pseudobase simply releases water to form oxidized flavin as oxygenated product is released.

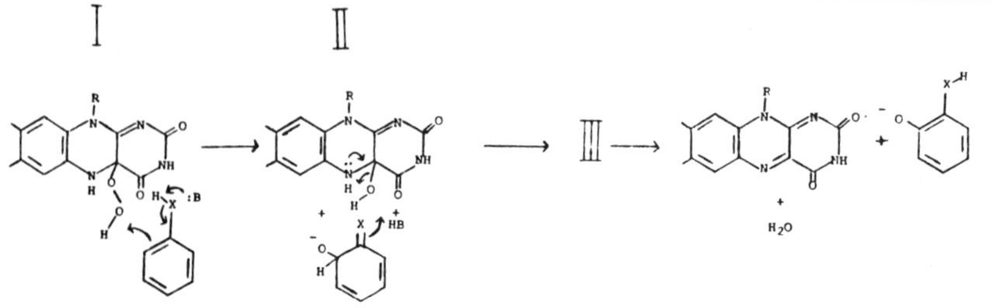

SCHEME III

GENERAL MECHANISM FOR PHENOL HYDROXYLASES

Scheme 4 shows the mechanism as it applies to several of the known substrates of parahydroxybenzoate hydroxylase. The center three substrates all cause formation of observable intermediate II. When the tetrafluoro-substrate is oxygenated, fluoride is released and a highly reactive orthoquinone is the first observable product (30). During normal catalysis, this product is rapidly reduced by NADPH so that two equivalents of NADPH are used. In this case, the orthoquinone is seen to form simultaneously with the appearance of III. The amino-substrate is a direct analogy to anthranilate hydroxylase (31) which hydrolyzes the Schiff base analogue (II) to release NH_3 and form 3-carboxycatechol. However, with parahydroxybenzoate hydroxylase apparently very little hydrolysis occurs (13). It would be interesting to check this more carefully.

Finally, this mechanism can be applied to the only known dioxygenation reaction catalyzed by a flavoprotein, 2-methyl-3-hydroxypyridine-5-carboxylate oxygenase (32) (Scheme 5). After insertion of oxygen, either water or the pseudobase itself attacks the 3-position to form the gem diol, which simply breaks down to form the observed product. This would account for the incorporation of ^{18}O at essentially 100% in position 2 and only about 15% in position 3. Thus this reaction may actually be a monooxygenation in mechanism, but a fortuitous dioxygenase reaction. Other properties of the mechanism of this enzyme (33) are very similar to the phenolic hydroxylases.

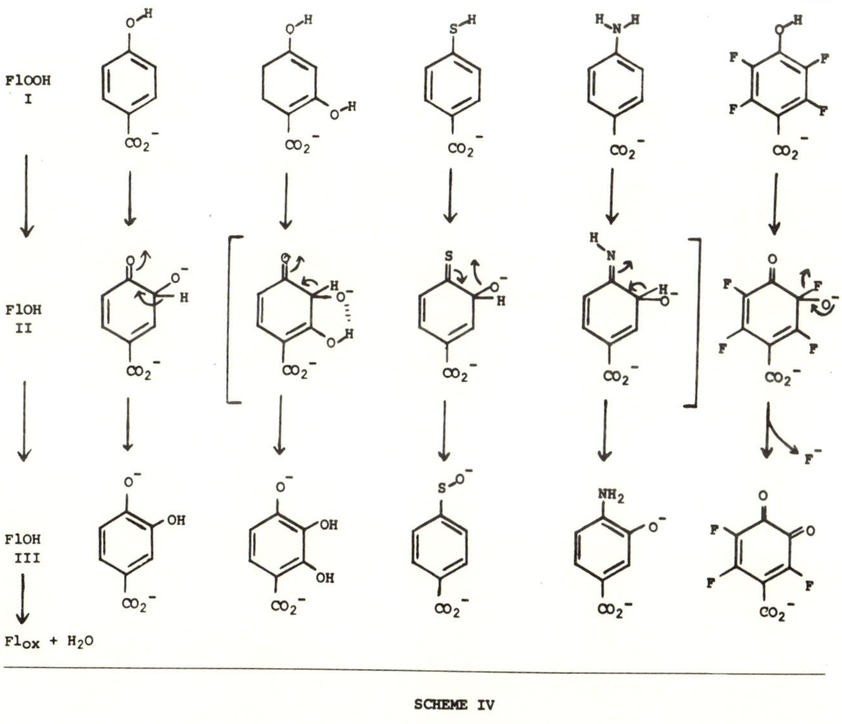

SCHEME IV

REACTIONS OF P-HYDROXYBENZOATE HYDROXYLASE WITH VARIOUS SUBSTRATES

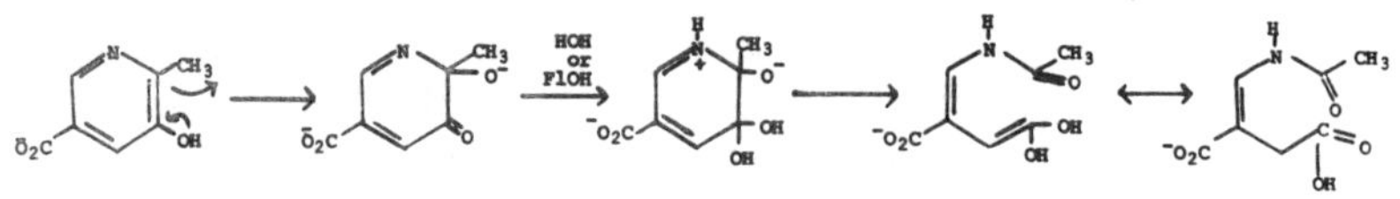

SCHEME V

"DIOXYGENATION" OF 2-METHYL-3-HYDROXYPYRIDINE-5-CARBOXYLATE

The main difficulty with this overall mechanistic proposal is that intermediate II is not easily accounted. This mechanism requires that the cyclohexadienone-like species somehow interacts with the pseudobase (or some other flavin form) to give rise to the spectrum of intermediate II, and that certain substrates are able to provide kinetic stabilization of this species. In addition, the mechanism as applied to mercaptobenzoate is probably not correct, since one would expect direct oxygenation at the sulfur atom. If, indeed, oxygenation does occur at the sulfur atom, this species could not make the same contribution to the spectrum as would a cyclohexadienone, yet the spectra of all the observed intermediate II are substantially similar. If some other form of the flavin is actually II, it could be a ring opened form as proposed earlier (13), although the model of Bruice (34) suggests that it is not. Studies with the model incorporated into the enzyme are required for verification.

The microsomal FAD-containing monooxygenase probably proceeds by a mechanism very similar to that proposed for the model systems (35) involving direct nucleophilic displacement of the distal oxygen from the C(4a)-hydroperoxyflavin to form the pseudobase. Even the substrate specificity is similar for the model and the enzyme system.

Cyclohexanone monooxygenase and luciferase can be envisioned to proceed by nucleophilic attack of the C(4a)-hydroperoxyflavin on the substrate. Scheme 6 shows a proposed luciferase mechanism. Shown is the step where the deuterium isotope effect (Deffect) would be predicted (36) and possible points of emission of light. The kinetics of this system have not been completely

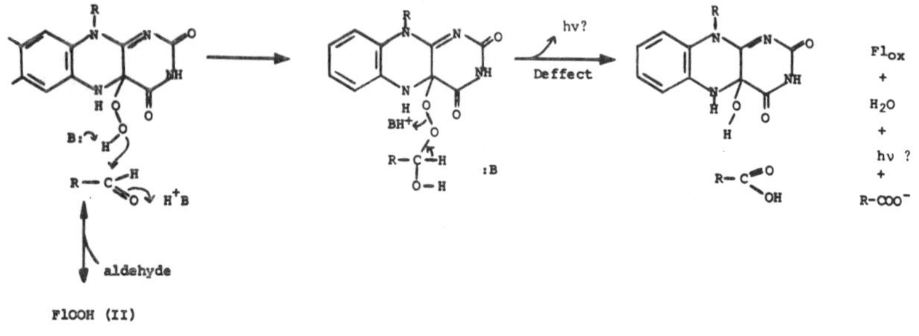

SCHEME VI

PROPOSED LUCIFERASE MECHANISM

worked out so that its verification remains to be done. The cyclohexanone mechanism is thought to be very similar to this with an alkyl group migrating from the mixed peroxide, rather than a hydrogen atom. This enzyme has been described in this volume by Dr. Walsh (37).

References

1. You, K., Arnold, L.J., and Kaplan, N.O. (1977) *Arch Biochem Biophys* 180:550–554.
2. Ryerson, C.C., Ballou, D.P., and Walsh, C. (1981) *J Biol Chem* 256, in press.
3. Schopfer, L.M. and Massey, V. (1979) *J Biol Chem* 254:10634–10643.
4. Beaty, N.B. and Ballou, D.P. (1981) *J Biol Chem* 256:4611–4618.
5. Howell, L.G., Spector, T., and Massey, V. (1972) *J Biol Chem* 247:4340–4350.
6. Entsch, B., Ballou, D.P., Husain, M., and Massey, V. (1976) *J Biol Chem* 251:7367–7379.
7. Husain, M. and Massey, V. (1979) *J Biol Chem* 254:6157–6666.
8. Shoun, H., Higashi, W., Beppu, T., Nakamura, S., Hiromi, K., and Arima, K. (1979) *J Biol Chem* 254:10944–10951.
9. Matthews, R.G., Ballou, D.P., and Williams, C.H., Jr. (1979) *J Biol Chem* 254:4974–4981.
10. Massey, V. and Hemmerich, P. (1975) *The Enzymes*, 3rd ed. 12:191–252.
11. Schopfer, L.M., personal communication.
12. Detmer, K., Massey, V., Ballou, D.P., and Neujahr, H.Y., this volume.
13. Entsch, B., Ballou, D.P., and Massey, V. (1976) *J Biol Chem* 251:2550–2563.
14. Presswood, R.P. and Kamin, H. (1976) In *Flavins and Flavoproteins*. T.P. Singer (ed.) N.Y.: Elsevier, pp. 145–154.
15. Strickland, S. and Massey, V. (1973) *J Biol Chem* 248:2953–2962.
16. Beaty, N.B. and Ballou, D.P. (1981) *J Biol Chem* 256:4619–4625.
17. Poulsen, L.L. and Ziegler, D.M. (1979) *J Biol Chem* 254:6449–6455.
18. Ryerson, C.C., Ballou, D.P., and Walsh, C., in preparation.
19. Hastings, J.W., Balny, C., LePeuch, C., and Douzou, P. (1973) *Proc Natl Acad Sci USA* 70:3468–3472.
20. Kemal, C., Chan, T.W., and Bruice, T.C. (1977) *J Am Chem Soc* 99:7272–7286.
21. Kemal, C. and Bruice, T.C. (1977) *Proc Natl Acad Sci USA* 74:405–409.
22. Ghisla, S., Hastings, J.W., Favaudon, V., and Lhoste, J.M. (1978) *Proc Natl Acad Sci USA* 75:5860–5863.
23. Schopfer, L.M. and Massey, V. (1980) *J Biol Chem* 255:5355–5363.
24. Entsch, B., Ballou, D.P., and Massey, V. (1974) *Biochem Biophys Res Commun* 57:1018–1025.
25. Ziegler, D.M. (1980) In *Enzymatic Basis of Detoxification. Vol. 1*. New York: Academic Press, pp. 201–227.
26. Walsh, C., Jacobson, F., and Ryerson, C.C. (1980) In *Biomimetic Chemistry, Advances in Chemistry Series*. D. Dolphin, C. McKenna, Y. Murakami, and A. Tabushi (eds.) Amer. Chem. Soc., pp. 119–138.
27. Massey, V. and Hemmerich, P. (1981), this volume.
28. Entsch, B., Husain, M., Ballou, D.P., Massey, V., and Walsh, C. (1980) *J Biol Chem* 255:1420–1429.
29. Hastings, W., Ghisla, S., and Kurfurst, M. (1981), this volume.
30. Husain, M., Entsch, B., Ballou, D.P., Massey, V., and Chapman, P. (1980) *J Biol Chem* 255:4189–4197.
31. Powlowski, J. and Dagley, S. (1981), this volume.
32. Sparrow, L.G., Ho, P.P.K., Sundarum, T.K., Zach, D., Nyns, E.J., and Snell, E.E. (1969) *J Biol Chem* 244:2590–2600.

33. Kishore, G.M. and Snell, E.E. (1981) *J Biol Chem* 256:4228–4233.
34. Bruice, T.C. (1981), this volume.
35. Ball, S. and Bruice, T.C. (1980) *J Am Chem Soc* 102:6498–6503.
36. Presswood, R.P., Shannon, P., Spencer, R., Walsh, C., Becvar, J.E., Tu, S.C., and Hastings, J.W. (1980) In *Flavins and Flavoproteins*. K. Yagi and T. Yamano (eds.) Baltimore: University Park Press, pp. 155–160.
37. Walsh, C. (1981), this volume.

Published 1982 by Elsevier North Holland, Inc.
Vincent Massey and Charles H. Williams, Editors
Flavins and Flavoproteins

CHAPTER 49

The Role of Flavoproteins in Aromatic Catabolism

Stanley Dagley

Department of Biochemistry, College of Biological Sciences, University of Minnesota, St. Paul, Minnesota

The object of this survey is to identify some of the physiological functions of flavoprotein enzymes that are of prime importance for the operation of the earth's carbon cycle. Such terrestrial cycles are analogous to the metabolic cycles of individual cells: they reflect the dynamic states of the various systems in which they occur. The evolution of living organisms, from simple forms to those of high complexity, required the establishment of a continuous flow of energy through the biosphere; this was promoted by accumulation of dioxygen, providing an atmosphere in which carbon dioxide is the most stable compound of carbon. Higher organisms were able to evolve in the energy gradient set up between photosynthesized biochemicals and carbon dioxide. A carbon cycle is a prerequisite for evolution, for within the cycle organisms can reproduce and undergo selection: they die, and dead material is recycled, largely through the reactions of microbial catabolism.

This perspective of evolution emphasizes two processes: harnessing of solar energy by higher plants and degradation of photosynthesized biochemicals by microorganisms. These processes are complementary. For plants to have made the most of sunlight as they evolved, they needed structural material to enable them to stand erect, namely, the three-dimensional biopolymer, lignin, composed of substituted benzene rings combined through linkages particularly resistant to biochemical attack. After cellulose, with which it is usually associated, lignin is the most abundant biopolymer in nature; but unlike proteins, carbohydrates, and nucleic acids, it is not extensively attacked by hydrolytic enzymes. Details of the biodegradation of lignin itself remain obscure, but it is generally believed to be initiated by fungal oxygenases; however, the catabolism of aromatic acids, arising from partial degradation of lignin and from many other natural sources, is carried on mainly by bacteria and this process has been extensively investigated. The benzene nucleus is one of the most frequently encountered units of chemical structure, and bacteria elaborate a wide range of oxygenases that are used for hydroxylation and ring-cleavage.

In view of the vast turnover of aromatic compounds in the biosphere, it is not surprising that separate, well-worn channels of catabolism have been evolved for those compounds most commonly encountered. These are shown in Figure 1. Before a benzene nucleus can be opened by a ring-fission dioxygenase, it must usually carry two hydroxyl groups situated either on adjacent carbons, as in catechol or protocatechuic acid, or placed across the ring as in gentisic or homogentisic acid. Of course, the original bacterial growth substrate may frequently possess a more complex structure, and the ring-fission substrates of Figure 1 will themselves have been formed by prior degradative reactions. Details of the reaction sequences that follow ring-fission and lead to metabolites of the Krebs cycle are given in various reviews (1–3). For the present discussion, it is sufficient to observe that these pathways are kept separate by the operation of two factors: substrate specificity and specificity of enzyme derepression. As regards the first, one example must suffice: the structures of gentisic and homogentisic acids differ only by one methylene group, but the dioxygenase of the first of these acids does not attack the second; and the subsequent enzymes for one pathway do not significantly metabolize the homologous substrates for the other. As for enzyme derepression, it is generally found that if one particular bacterial strain oxidizes, say, 2,5-dihydroxybenzoate (gentisate) after growth with benzoate, it does not oxidize other feasible ring-fission substrates such as 3,4-dihydroxybenzoate (protocatechuate), or catechol until it is grown on another appropriate substrate. Conversely, another strain might derepress, and use, either catechol or

Figure 1. Summary of some of the central reaction sequences of bacterial aromatic catabolism. The products, acetyl-CoA, succinyl-CoA, and fumarate enter the Krebs cycle directly; pyruvate, succinic semialdehyde, acetaldehyde, and acetoacetate are readily converted into metabolites of the cycle by well established reactions.

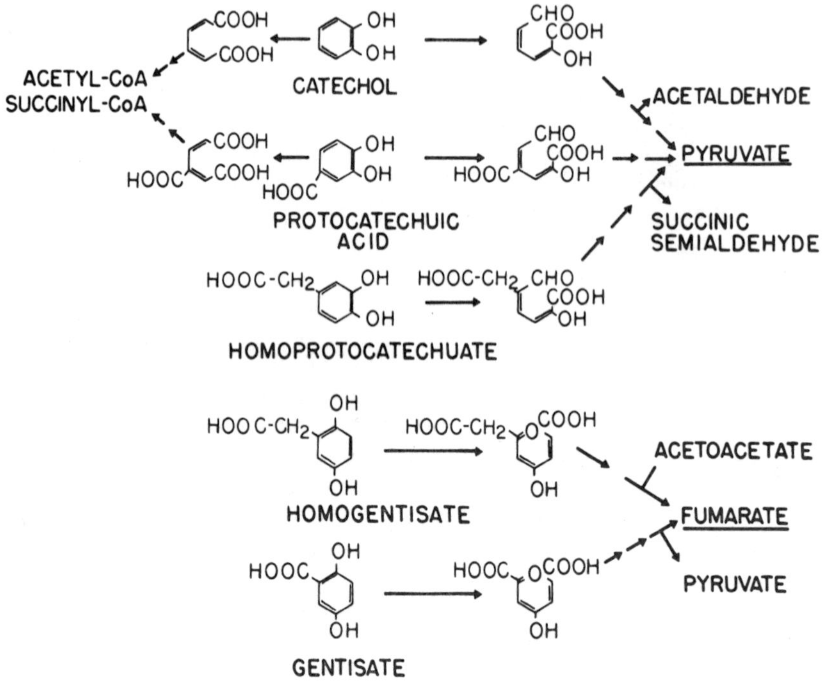

protocatechuate enzymes for metabolizing benzoate; but gentisate would not be oxidized. This behavior is the basis of Simultaneous Adaptation, an experimental technique by which pathways of aromatic catabolism have often been elucidated from measurements of oxygen consumption by cell suspensions (4).

When an aromatic acid carries one hydroxyl group, a second must be introduced by a hydroxylase and, as might be predicted from the necessity to select only one of several feasible pathways, these bacterial flavoprotein monooxygenases exhibit narrow substrate specificity. Thus, 4-hydroxybenzoate 3-hydroxylase (5) and 3-hydroxybenzoate 4-hydroxylase (6) are separate, specific enzymes that give the same product, protocatechuic acid, from their respective substrates. This restricted substrate specificity sometimes gives the impression that one of the pathways of Figure 1 is "chosen" in preference to another. Thus, *Pseudomonas acidovorans* can be induced to elaborate all of the enzymes involved in oxidizing 3,4-dihydroxyphenylacetate (homoprotocatechuate) to completion (7); but when grown with 4-hydroxyphenylacetate, the organism does not make use of them by hydroxylating the nucleus at C-3 as might be expected. Instead, a flavin monooxygenase hydroxylates at C-1, displacing the side chain to C-2 to give 2,5-dihydroxyphenylacetate (8); this compound is then oxidized by a set of enzymes entirely different from those used for homoprotocatechuate. Several other examples might be given to show that one function of an aromatic flavoprotein hydroxylase is to direct catabolism of its substrate into one, and only one, of the reaction sequences shown in Figure 1.

Recent work (9) shows that specific hydroxylases can so direct the course of catabolism that methanol is released when methoxylated aromatic compounds are degraded by soil bacteria. This methanol then serves as nutrient for other bacterial strains. Lignins, and aromatic acids derived from them, contain methoxyl-substituted benzene rings. Enzymic demethylation gives a phenol (10):

$$\text{Aryl}\cdot OCH_3 \rightarrow [\text{Aryl}\cdot O\cdot CH_2OH] \rightarrow \text{Aryl}\cdot OH + HCHO \qquad (49.1)$$

The formaldehyde released is oxidized to carbon dioxide under aerobic conditions; but when air becomes depleted, reduction to methanol might then occur. However, methanol is certainly formed from methoxyl groups by another route when air is readily available (9). Thus, when *P. putida* oxidizes 3,4,5-trimethoxybenzoic acid, demethoxylation invariably takes place at C-4 according to Equation (49-1) and formaldehyde is oxidized to carbon dioxide. One, but not both, of the remaining methoxyl groups is also oxidized, so that the ultimate demethoxylation product is 3-methoxy-4,5-dihydroxybenzoic acid. The enzymes responsible are NADH-dependent monooxygenases, and Bernhardt et al. (11) found, for a similar reaction, that the system involved both a flavoprotein and an iron-sulfur protein. When two of the three methyl groups have been removed, a dioxygenase cleaves the nucleus of 3-methoxy-4,5-dihydroxybenzoate between carbons 3 and 4, so that the remaining methoxyl group is now part of a methyl ester grouping, from which methanol is readily released by enzymic hydrolysis. This work has been extended in my laboratory, using various bacterial isolates from soil, to include substituted cinnamic and phenylacetic acids; and since methanol was found in these experiments wherever its release could be predicted from the pattern of methoxyl substitution, there is little doubt that this process makes a significant contribution to supplying soil methylotrophs with nutrient.

Striking examples of how catabolic pathways are made available by the use of hydroxylases are provided by studies using the soil yeast *Trichosporon cutaneum* (12). When degrading aromatic acids, this eukaryote is as versatile as any pseudomonad, and yet it lacks ring-fission dioxygenases for three of the central metabolites of Figure 1, namely, protocatechuic, homoprotocatechuic, and gentisic acids; moreover it suffers the additional handicap of not being able to elaborate any metafission dioxygenases. Some pseudomonads have orthofission reactions at their disposal for one substrate and metafission reactions for another. As shown in the summary of its aromatic catabolism presented in Figure 2, *T. cutaneum* uses hydroxylases to compensate for dioxygenase deficiencies. Thus, 4-hydroxybenzoate (XI) is hydroxylated to give homoprotocatechuate (XII) a common bacterial ring-fission substrate for which, however, this organism possesses no dioxygenase. A third hydroxyl is now inserted (reaction **8**) to give a substrate which is cleaved by the same dioxygenase that attacks homogentisate (XV). As in bacteria, salicylate (VIII) is oxidatively decarboxylated to give catechol (IX); however, salicylate hydroxylase from *T. cutaneum* also attacks gentisate (VII), catalyzing reaction **4** to give a ring-fission substrate, hydroxyquinol (IV) which again carries three hydroxyl groups, in contrast to the two which suffice for bacterial substrates. Another contrast is seen in the three consecutive hydroxylations used to convert benzoate (I) into IV. Many soil bacteria possess a single enzyme system that oxidizes benzoate, with loss of CO_2, to give the ring-fission substrate catechol. Although most of the hydroxylases catalyzing reactions of Figure 2 have been obtained cell-free and shown to be NAD(P)H-dependent, it is not known whether they are flavoproteins; however, it is probable that this is, in fact, the case, since Neujahr and Gaal (13) showed that phenol hydroxylase from *T. cutaneum* contains FAD, as does anthranilate 3-hydroxylase which we have purified from this organism. It may be mentioned that mechanisms of enzyme derepression in *T. cutaneum* are far less substrate-specific than those in bacteria: thus growth with benzoate strongly induces enzymes for catabolizing 2- and 3-hydroxybenzoates, catechol and gentisate, none of which lie upon the direct pathway for degrading benzoate. This renders the time-honored bacterial technique of Simultaneous Adaptation useless for investigating *T. cutaneum* (12).

Although the catabolic versatility of *T. cutaneum* is increased by its capacity to add additional hydroxylations to oxidative pathways, this is achieved at the expense of energy that would otherwise be devoted to synthesizing cellular material. Each hydroxylation uses one molecule of NADH which would otherwise be oxidized at the electron transport chain and so provide for the regeneration of ATP to be used in metabolism; hydroxylations fritter away metabolic energy as heat. This has been confirmed by direct microcalorimetric measurements (14). Experimental data were handled as follows: heat evolution ($-H_r$) was measured, using a cell suspension, and then expressed as a fraction (H_r/H_c) of the total energy release expected for the substrate, calculated from heat of combustion data, and assuming no assimilation. With this procedure,

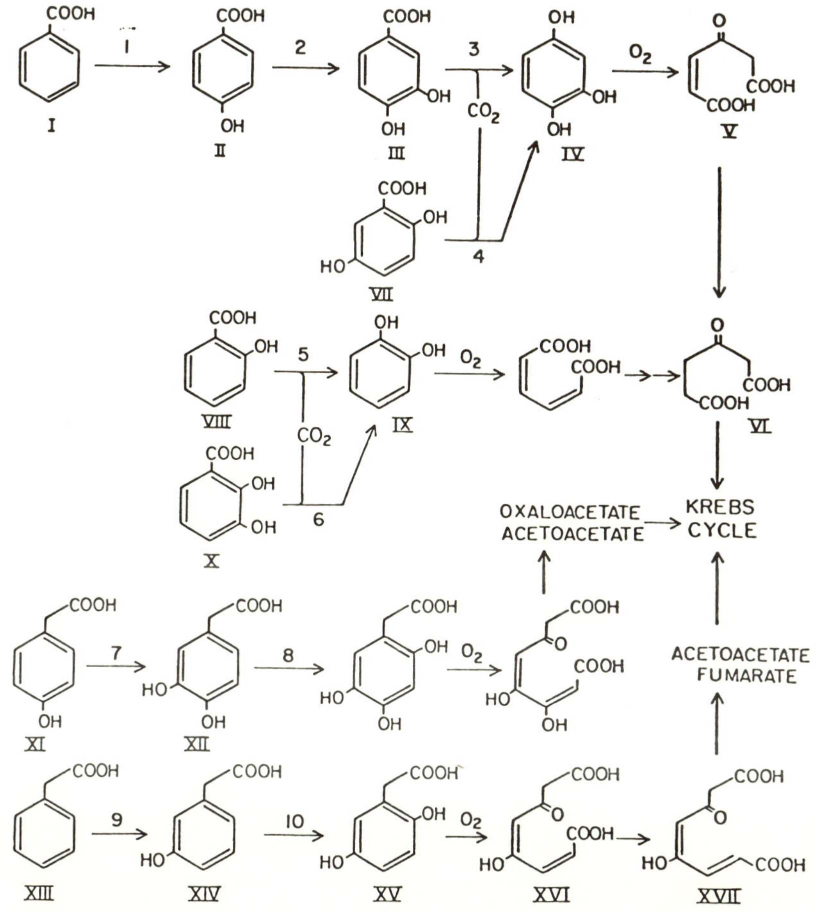

Figure 2. Summary of reactions used by the soil yeast, *Trichosporon cutaneum*, to degrade aromatic acids (from reference 12). All reactions shown by arrows are catalyzed by soluble enzymes, except **1** and **9** which were demonstrated for intact cells using ^{14}C-substrates. Reactions **2** and **7** are catalyzed by hydroxylases that require NADPH specifically; hydroxylases for reactions **3**, **4**, **5**, **8**, and **10** require NADH, replaced less effectively by NADPH. Reaction **6** is a nonoxidative decarboxylation; reactions **4** and **5** are catalyzed by one and the same hydroxylase.

H_r/H_c is proportional to the heat evolved, and $1\text{-}H_r/H_c$ to the energy assimilated by the cells. To take two examples from Figure 2 of how successive hydroxylations on a catabolic pathway can be "counted," ratios for benzoate (I), 4-hydroxybenzoate (11), and protocatechuate III were respectively 0.641, 0.530, and 0.376; and for phenylacetate (XIII), 3-hydroxyphenylacetate (XIV), and homogentisate (XV), these were 0.566, 0.417, and 0.233 (14).

The vast majority of hydroxylases that function in microbial catabolism are flavoproteins; but by contrast, those that are found in liver and serve for

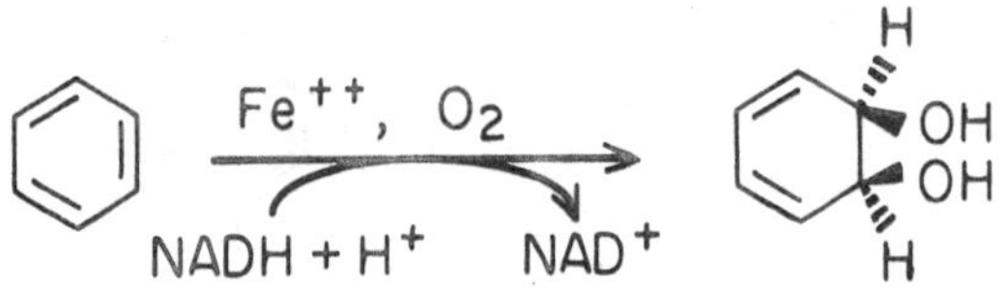

Fe^{++}, O_2 — NADH + H$^+$ → NAD$^+$

Figure 3. Enzymatic oxidation of benzene to *cis*-benzeneglycol.

detoxication are in the cytochrome P-450 category. It is true that the hydroxylase that initiates camphor degradation by *Pseudomonas putida* belongs to the second type; but this occurrence appears to be as exceptional as it was fortunate, having regard to the fruitful research that stemmed from its discovery (15). Other fully reduced ring systems, however, are attacked by flavoprotein-activated oxygen (16, 17); and when oxygen is incorporated from water into the methyl group of p-cresol (18), dehydrogenation by an enzyme containing tyrosine-bound FAD (19) precedes the hydration step. The use of flavoproteins in catabolism and cytochrome P-450 in detoxication has been beautifully demonstrated by David T. Gibson and his coworkers for microorganisms. Some species use benzene, toluene, and other aromatic hydrocarbons for growth, forming first a *cis*-dihydrodiol and then a catechol, the nucleus of which is cleaved by a dioxygenase. For benzene, Axcell and Geary (20) showed an enzyme system used by *Pseudomonas putida* to catalyze the first of these steps contains a flavoprotein and two nonheme iron proteins (Figure 3). A similar system operates for toluene, and the flavoprotein component has been purified by Subramanian et al. (21). By contrast, Cerniglia and Gibson (22) showed that the filamentous fungus *Cunninghamella elegans* forms several hydroxylation products from benzo[a]pyrene; these are not further metabolized and can be regarded as detoxication products. More recent work by this group (private communication) has shown that enzymes for conjugating the hydroxylation products with sulfate or glucuronate are elaborated by this fungus. The organism uses cytochrome P-450 for converting naphthalene to 1-naphthol and *trans*-1,2-dihydroxy-1,2-dihydronaphthalene (23). In its metabolism of aromatic hydrocarbons, we can regard *C. elegans* as behaving like a minute, filamentous liver; whereas *P. putida*, on the other hand, uses flavoprotein oxygenases to prepare these compounds for degradation and assimilation as nutrients.

ACKNOWLEDGMENTS
My research is supported by Grants from the N.I.H. (ES AI 00678 and R01 AM 21981).

References

1. Dagley, S. (1977) *Surv Prog Chem* 8:121–170.
2. Dagley, S. (1978) In *The Bacteria*. Ornston, L.N. and Sokatch, J.R. (eds.) Vol. 6, New York: Academic Press. pp. 305–388.
3. Dagley, S. (1978) *Q Rev Biophys* 11:577–602.
4. Dagley, S. and Chapman, P.J. (1971) *Methods Microbiol* 6A:217–268.
5. Howell, L.G., Spector, T., and Massey, V. (1972) *J Biol Chem* 277:4340–4350.

6. Michalover, J.L. and Ribbons, D.W. (1973) *Bioch Biophys Res Commun* 55:888–896.
7. Sparnins, V.L. and Dagley, S. (1975) *J Bacteriol* 124:1374–1381.
8. Hareland, W.A., Crawford, R.L., Chapman, P.J., and Dagley, S. (1975) *J Bacteriol* 121:272–285.
9. Donnelly, M.I. and Dagley, S. (1980) *J Bacteriol* 142:916–924.
10. Ribbons, D.W. (1970) *FEBS Lett* 8:101–104.
11. Bernhardt, F.-H., Erdin, N., Staudinger, H., and Ullrich, V. (1973) *Eur J Biochem* 35:126–134.
12. Anderson, J.J. and Dagley, S. (1980) *J Bacteriol* 141:534–543.
13. Neujahr, H.Y. and Gaal, A. (1973) *Eur J Biochem* 35:386–400.
14. Anderson, J.J. and Dagley, S. (1980) *J Bacteriol* 143:525–528.
15. Yu, C.-A. and Gunsalus, I.C. (1970) *Biochem Biophys Res Commun* 40:1431–1436.
16. Griffin, M. and Trudgill, P.W. (1976) *Eur J Biochem* 63:199–209.
17. Donoghue, N.A., Norris, D.B., and Trudgill, P.W. (1976) *Eur J Biochem* 63:175–192.
18. Hopper, D.J. (1976) *Biochem Biophys Res Commun* 69:462–468.
19. McIntire, W., Edmondson, D.E., Singer, T.P., and Hopper, D.J. (1980) *J Biol Chem* 255:6553–6555.
20. Axcell, B.C. and Geary, P.J. (1975) *Biochem J* 146:173–183.
21. Subramanian, V., Liu, T.-N., Yeh, W.-K., Narro, M. and Gibson, D.T. (1981) *J Biol Chem* 256:2723–2730.
22. Cerniglia, C.E. and Gibson, D.T. (1979) *J Biol Chem* 254:12174–12180.
23. Cerniglia, G.E. and Gibson, D.T. (1978) *Arch Biochem Biophys* 186:121–127.

Published 1982 by Elsevier North Holland, Inc.
Vincent Massey and Charles H. Williams, Editors
Flavins and Flavoproteins

CHAPTER 50

Chemical Modification of p-Hydroxybenzoate Hydroxylase from *Pseudomonas fluorescens* with Diethylpyrocarbonate

R. Wijnands and F. Müller

Department of Biochemistry, Agricultural University, Wageningen, The Netherlands

Introduction

The inducible enzyme p-hydroxybenzoate hydroxylase belongs to the class of external flavoprotein monooxygenases. A purification procedure for this enzyme from *P. fluorescens* has been published (1), which has been modified recently (2).

Molecular properties, three-dimensional structure, and amino acid sequence are being studied by three Dutch groups in close cooperation. Results on the three-dimensional structure and amino acid sequence have been published (3) (cf also Drenth et al., this volume). We have devoted ourselves to the study of chemical modification of amino acid residues important for the catalytic activity of the enzyme.

Results and Discussion

Diethylpyrocarbonate (DEP) was used for chemical modification. Carbethoxylation of histidine is specific at pH 4–7 (increase of absorbance at 240 nm), at a higher pH also tyrosine reacts (decrease of absorbance at 278 nm and 240 nm). The reactions are reversible. The removal of the carbethoxy group from both of the modified amino acids can be accelerated by hydroxylamine, but O-carbethoxy-tyrosine is less reactive towards hydroxylamine than N-carbethoxy-histidine.

Ethoxy formylation of p-hydroxybenzoate hydroxylase (pHBH) in the pH range 5 to 9 results in a complete loss of activity when the DEP to pHBH ratio is about 20 or higher. At a DEP to pHBH ratio of 10, there is a quick decrease of activity to a residual value of 75% at pH<6, while at higher pH a second reaction can be distinguished resulting in a lower residual activity depending on the pH. At higher DEP to pHBH ratios, the first reaction is quickly camouflaged by the second reaction as can be seen in Figure 1. A pseudo-first-order rate constant for loss of activity at pH 6.15 can be determined from the slope of ln (residual activity) vs time but an initial faster reaction also occurs.

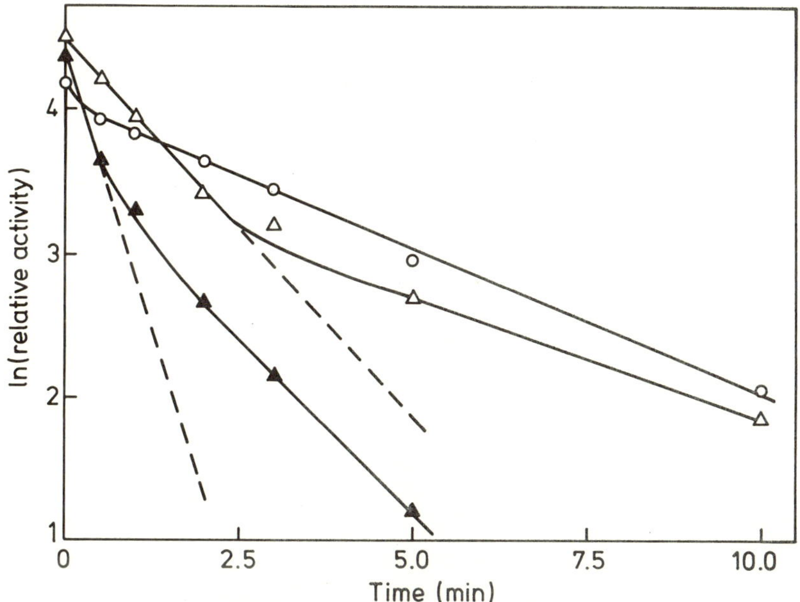

Figure 1. ln (natural logarithm) of relative activity of pHBH as a function of time after adding 0.4 mM DEP to 20 μM pHBH at 0°C. Aliquots of 4μl were withdrawn from the incubation mixture and added to 1 ml of activity assay mixture containing 0.1 M Tris/H_2SO_4, pH 8, 0.15 mM NADPH, and 0.15 pOHB. (○) pH 6.15, (△) pH 6.5; (▲) pH 7.55.

At higher pH, a third reaction can be distinguished from the second one (Figure 1). Eventually, even a fourth reaction seems to occur (after 95% of activity is lost). The pH-dependence of the second-order rate constants for loss of activity is shown in Figure 2A. These values were calculated from slopes of curves of activity measurements as a function of DEP concentration. The curve was calculated using two ionizing groups of $pK'_a=6.3$ and a single ionizing group of $pK'_a=7.2$. At pH>7.5, the influence of a fourth ionizing group, not included in our calculations, is obvious.

When the carbethoxylation experiments were done in the presence of NADPH at pH 6, some protection against inactivation was observed. Benzoate gave no protection at pH 6, but it did at pH 8.

Carbethoxylation yields also an increase of the flavin fluorescence. In Figure 2B the pseudo-first-order rate constant of this increase is shown as a function of pH. The theoretical curve was calculated using a single ionizing group of $pK'_a=7.2$ and two ionizing groups of $pK'_a=8.4$. When these values are compared with those found with the inactivation studies, the similarities are obvious.

The dissociation constants of the enzyme-NADPH and -p-hydroxybenzoate complex were determined by fluorometric titration (Table 1). It must be emphasized that the values calculated for the enzyme-NADPH complex should

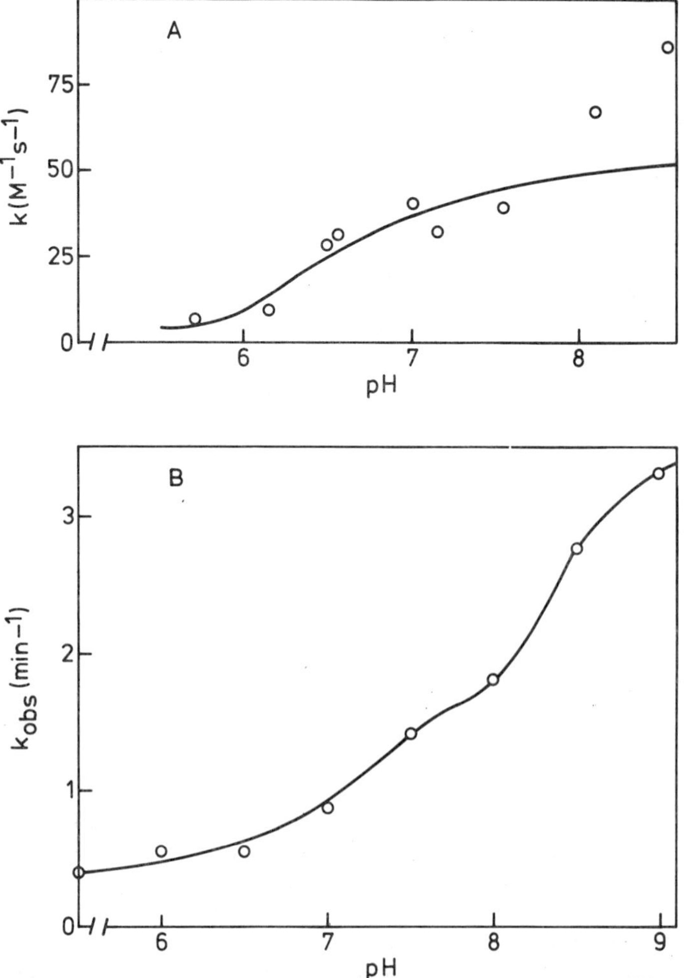

Figure 2. (A) pH dependence of the second-order rate constant of inactivation at 0°C. (B) pH dependence of the pseudo-first-order rate constant of the increase of flavin fluorescence at 25°C. The enzyme concentration was 20 μM and the DEP concentration was 0.5 mM, excitation at 450 nm, emission at 525 nm.

not be compared with those determined kinetically (1) because of interface of the fluorescence of NADPH. However for the purpose of comparison they are quite satisfactory.

The results strongly indicate that a dissociating group with a $pK'_a=6.3$ is involved in the binding of NADPH. Modification with DEP leads to a decrease of NADPH binding and therefore also to a decrease in enzymatic activity. Because of the higher specificity of DEP at pH 6 the pK'_a value of 6.3 is most probably due to a histidine residue.

Table 1. Dissociation Constants of Enzyme Complexes Determined at Various pH Values.

			K in mM	
			Enzyme modified at	
Substrate	**pH**	**Native enzyme**	**pH 6**	**pH 7–8**
NADPH	6	0.49	5.0	
NADPH	8	0.70	3.0	
pOHB	6	5.8×10^{-2}	7.4×10^{-2}	
pOHB	8	3.2×10^{-2}		∞

On the other hand, a dissociating group with a $pK'_a = 7.2$ is involved in the binding of the substrate. Modification of this group leads to a decrease of substrate binding. This is accompanied by an increase in flavin fluorescence indicating the proximity of the isoalloxazine ring system to this group, in contrast with the histidine residue mentioned. As modification of this group occurs at $pH > 6.5$, where DEP is not as specific as at pH 6, the modified amino acid residue is probably not histidine.

Modification at pH 6 is accompanied by an increase of absorbance at 240 nm. This is just what is to be expected when histidine is modified. However modification at pH 8 shows a smaller increase at 240 nm and a decrease at 278 nm, which indicates that at higher pH values modification of a tyrosine residue takes place. Several kinetic experiments to compare rates of chemical modification, inactivation, and increase of absorbance are still to be done but at pH 6 a pseudo-first-order constant of chemical modification monitored at 240 nm was found to be in very good agreement with the constant found for inactivation under the same conditions.

The original activity, fluorescene quantum yield, and substrate binding capacity could be restored by treatment of the modified enzyme with hydroxylamine. Further studies are in progress.

Conclusions

Our results strongly indicate that a histidine residue with a $pK'_a = 6.3$ is involved in the binding of NADPH and a tyrosine residue with a $pK'_a = 7.2$ in the binding of p-hydroxybenzoate. A possible candidate for the histidine residue is His-162 which could form a hydrogen bridge with one of the oxygen atoms of the pyrophosphate of NADPH. For the tyrosine residue, there are several possible candidates: Tyr-202, Tyr-223, and Tyr-386 (cf Drenth et al., this volume).

ACKNOWLEDGMENTS

We thank Mr. M.M. Bouwmans for the preparation of the figures and Mrs. J.C. Toppenberg-Fang for typing the manuscript. This study was carried out under the auspices of the Netherlands Foundation for Chemical Research (SON) with financial aid from the Netherlands Organization for the Advancement of Pure Research (ZWO).

References

1. Howell, L.G., Spector, T., and Massey, V. (1972) *J Biol Chem* 247:4340–4350.
2. Müller, F., Voordouw, G., Van Berkel, W.J.H., Steennis, P.J., Visser, S., and Van Rooijen, P. (1979) *Eur J Biochem* 101:235–244.
3. Hofsteenge, J., Vereijken, J.M., Weijer, W.J., Beintema, J.J., Wierenga, R.K., and Drenth, J. (1980) *Eur J Biochem* 113:141–150.

Published 1982 by Elsevier North Holland, Inc.
Vincent Massey and Charles H. Williams, Editors
Flavins and Flavoproteins

CHAPTER 51

An Essential Histidine Residue at the NADPH-Binding Site of p-Hydroxybenzoate Hydroxylase

Hirofumi Shoun and Teruhiko Beppu

Department of Agricultural Chemistry, The University of Tokyo, Bunkyo-ku, Tokyo 113, Japan

p-Hydroxybenzoate hydroxylase (EC 1.14.13.2) is an NADPH-dependent flavoprotein hydroxylase and catalyzes the conversion of p-hydroxybenzoate to 3, 4-dihydroxybenzoate. We have been studying the enzyme from *Pseudomonas desmolytica* and have revealed some part of the reaction mechanism involving the protein moiety of the enzyme (1–3). We have shown that arginine, tyrosine, and histidine residues might be implicated in the substrate-binding site of the enzyme. Involvement of a histidine residue in the binding of NADPH was also implied from pH dependence of the K_d for the enzyme·NADPH complex in the absence of substrate (2). To confirm the involvement of histidine residue(s) in the catalytic mechanism, chemical modification of the enzyme with diethylpyrocarbonate was employed.

Modification of p-Hydroxybenzoate Hydroxylase with Diethylpyrocarbonate

p-Hydroxybenzoate hydroxylase was inactivated by treatment with diethylpyrocarbonate. The inactivation seemed to stabilize at the remaining activity of 5–10% (pH 6.2) when concentration of the reagent was low. The inactivation accompanied modification of 3 moles of histidine and 1 mol of tyrosine residues (per mol of enzyme).

Effects of pH and Ligands on the Modification

Among various ligands, NADPH was noticeable in protecting the enzyme from the inactivation. The apparent inactivation rate (k_{app}) was obtained at various pH values in the presence and absence of saturating amounts of some ligands, at 25° (Figure 1A) or 10°C (Figure 1B). The rate increased with increasing pH value, particularly on the alkaline side. p-Hydroxybenzoate and benzoate lowered the rate slightly in the acidic pH region, and markedly in the alkaline region. Chloride ion (an inhibitor competitive with NADPH) showed no effect.

The pH profile in the presence of p-hydroxybenzoate (or benzoate) suggested an ionizing group with a pK of ca 6.7 (25°) or 7.2 (10°C) participating

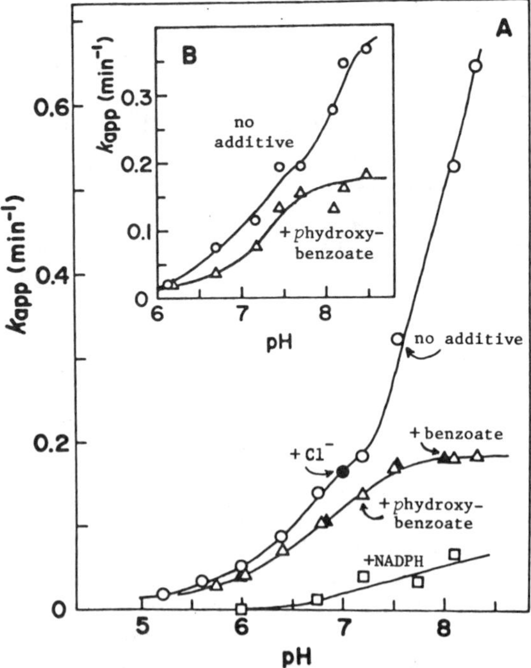

Figure 1. pH dependence of the apparent inactivation rate, at 25° (A) or 10°C (B). p-Hydroxybenzoate hydroxylase, 10 μM; diethylpyrocarbonate, 0.1 mM (A) or 0.2 mM (B); ligands, with saturating amount; in 80 mM potassium phosphate buffer.

in the inactivation. NADPH possibly inhibits modification of the ionizing group. The remarkable increase of the rate above pH 7 (without ligand) might be due to modification of some group(s) at the substrate site, since it was repressed by p-hydroxybenzoate (or benzoate). All the amino acid residues suggested in our previous study to be present at the substrate site could possibly react with the reagent. Deviation of the pK value between 25° and 10°C is reasonable, assuming the ionizing group as a histidine residue.

Reversal of the Modification

The inactivated enzyme (modified at pH 6.2) was treated with a lower concentration (20 mM) of hydroxylamine, at pH 7. The treatment regenerated all of the modified histidine as well as most of the lost activity, but not the modified tyrosine.

The modification of histidine cannot be followed when NADPH is present, because of its strong absorbance in the UV region. To examine effects of NADPH on the modification, two modified enzyme species were obtained by modification in the presence of NADPH or benzoate, followed by gel filtration (Sephadex G-25), and used in the reversal experiment. Figure 2 shows the time course of the reactivation and regeneration of histidine during the hydroxyl-

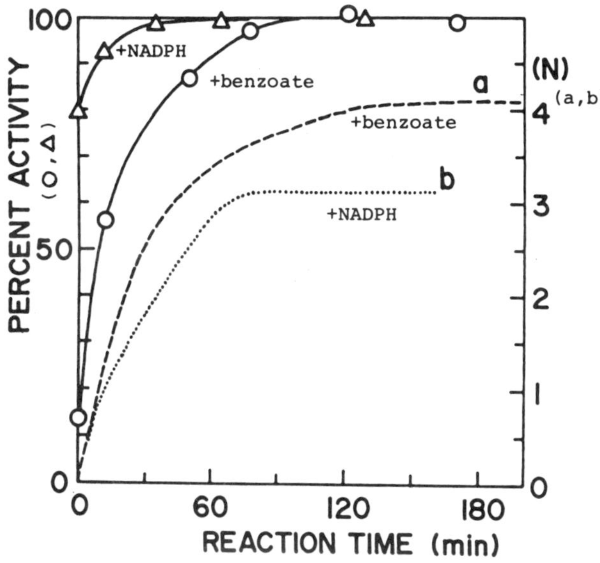

Figure 2. Reversal of the diethylpyrocarbonate-modification of p-hydroxybenzoate hydroxylase with hydroxylamine. The enzyme was treated with diethylpyrocarbonate in the presence of 4 mM NADPH or 1.8 mM benzoate, at pH 6.1, 25°C, followed by gel filtration (Sephadex G-25, pH 7). The modified enzyme sample (2.4 μM) was treated with 20 mM hydroxylamine (50 mM potassium phosphate, pH 7, 25°C), and the process of reactivation and recovery of histidine was followed. The remaining activity after the modification treatment corresponds to the plots at reaction time of 0 min on the figure, for each enzyme species. (N) regeneration of histidine residues (moles per mol of bound FAD) which was determined from decrease in absorbance at 242 nm. +NADPH (or benzoate) means that it was added during the diethylpyrocarbonate-treatment.

amine-treatment of the modified enzyme species. The result indicated that NADPH decreases the amount of modified histidine just by 1 mol per mol of enzyme, since benzoate exerted only slight effects on modification of histidine as well as on the inactivation in the acidic pH region (pH 6.1). The result also showed that the marked difference between the two enzyme species in the extent of the remaining activity before the reversal treatment might correspond to the difference in the extent of the modified histidine residues.

The possibility of involvement of sulfhydryl groups in the diethylpyrocarbonate-caused inactivation of the enzyme was excluded by the results of double modification of the enzyme with diethylpyrocarbonate and p-mercuribenzoate, and its reversal with hydroxylamine and dithiothreitol.

Properties of the Modified Enzyme Species

The inactivated enzyme (remaining activity, 8.5%; modified in the presence of p-hydroxybenzoate at pH 7) was shown to retain the substrate-binding ability almost completely, while the K_d for NADPH of the enzyme species (at pH 6) increased 14-fold. The K_m for NADPH in the overall reaction (at pH 8) increased 34-fold in the modification, but the apparent V_{max} did not change.

All the results above are consistent with the presence of a histidine residue at the NADPH site of p-hydroxybenzoate hydroxylase. Ionization of the histidine residue seems to be reflected on the pH profile of the K_d for the enzyme·NADPH complex without effector or in the presence of benzoate, but not on that in the presence of p-hydroxybenzoate (2). It should be a most interesting problem to clarify how the shift of the optimal pH for the NADPH-binding occurs in the presence of p-hydroxybenzoate (2). It would appear that the histidine residue at the NADPH site plays some significant role in the mechanism (shift of the optimal pH).

References

1. Shoun, H., Beppu, T., and Arima, K. (1979) *J Biol Chem* 254:899–904.
2. Shoun, H., Higashi, N., Beppu, T., Nakamura, S., Hiromi, K., and Arima, K. (1979) *J Biol Chem* 254:10944–10951.
3. Shoun, H., Beppu, T., and Arima, K. (1980) *J Biol Chem* 255:9319–9324.

Published 1982 by Elsevier North Holland, Inc.
Vincent Massey and Charles H. Williams, Editors
Flavins and Flavoproteins

CHAPTER 52

Displacement of FAD from Phenol Hydroxylase by Monovalent Anions and the Effector Function of Phenol

Halina Y. Neujahr

Department of Biochemistry, The Royal Institute of Technology, S-100 44 Stockholm, Sweden

Introduction

The activity of phenol hydroxylase EC [1.14.13.7] is inhibited by chloride and other monovalent anions (1,2). Figure 1 shows this inhibition with nine such anions, all but azide supplied as potassium salts. Controls with sodium salts give identical results. The inhibition is most severe with cyanide and azide, followed—in decreasing order—by fluoride, thiocyanate, chloride, nitrate, iodide, bromide, and acetate. The inhibitory effects increase with increasing concentrations of phenol as if the anion interfered with oxygen uptake. This report shows that such phenomena reflect the weakening of FAD-attachment and, ultimately, removal of the prosthetic group from the enzyme. It also demonstrates that a few equivalents of phenol per enzyme-bound FAD protect the enzyme against this effect.

Results

The spectrum of free and phenol-complexed enzyme (Figure 2A) shows characteristic shoulders on either side of the maximum at 443 nm. The shoulder around 465 nm is especially pronounced. This shoulder is smoothed out in the presence of 1–2 equivalents of phenol per enzyme-bound FAD, the shoulder and the peak at 443 nm being shifted upwards and towards shorter wavelengths.

Figure 2B–E shows how the spectrum of phenol hydroxylase is affected by potassium thiocyanate. Also in this case, there is a smoothing out of the shoulder(s), but the peak at 443 nm is shifted downwards and, together with the smoothed-out shoulders, towards longer wavelengths, instead. At 0.25 M–1.0 M potassium thiocyanate (Figure 2D–E), the spectrum of phenol hydroxylase is very similar to that of free FAD (cf Figure 2F).

Similar effects on the spectrum of phenol hydroxylase are observed with cyanide, azide, chloride, iodide, and nitrate (Figure 3A–E). There is also an effect with acetate (Figure 3F), although it is much less pronounced than with the other anions in Figure 3.

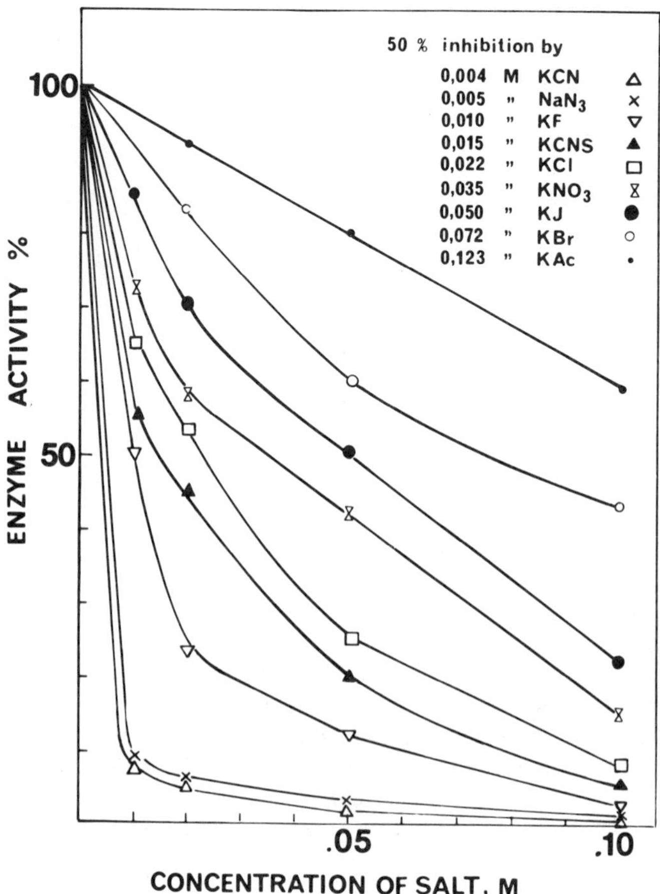

Figure 1. Inhibition of phenol hydroxylase by monovalent anions. The uninhibited activity was 3.2 enzyme units/mg protein (100%). Enzyme assay at 25° by recording $-\Delta A_{340}$ in 3 ml 0.05 M potassium phosphate pH 7.5 containing 0.15 mM NADPH, 100 μg purified enzyme protein and 0.17 mM phenol added 1 min after the indicated salts. Selected controls, incubated with salts for longer times, gave identical results.

The spectral effects of bromide and fluoride are shown in Figure 4A–B. Both anions cause some haziness and, after 5–10 min at 25°, extensive precipitation. The effect of bromide (Figure 4A) is similar to that of the other monovalent anions. It changes the spectrum of phenol hydroxylase to that of free FAD. The effect of fluoride (Figure 4B), however, is different. There is neither a shift of the peak at 443 nm nor any smoothing-out of the shoulder around 465 nm. Rather, this shoulder seems to become more accentuated. Higher concentrations of fluoride caused prompt precipitation.

Conclusive evidence that monovalent anions (except fluoride) remove FAD from phenol hydroxylase was obtained in experiments where the enzyme was subjected to gel permeation chromatography in their presence (Figure 5A–D).

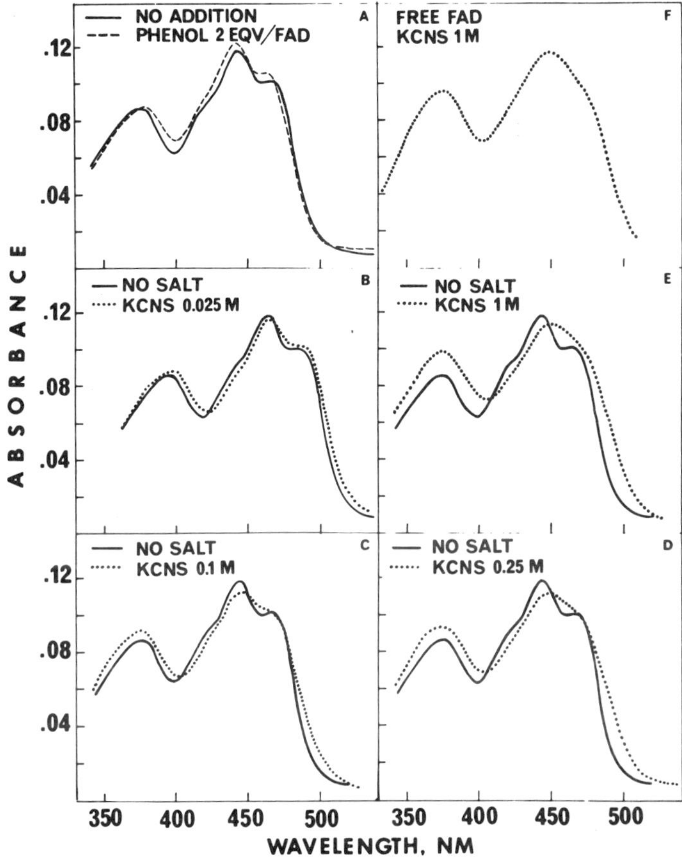

Figure 2. Absorption spectra of phenol hydroxylase and free FAD in the presence of KCNS. The spectra of the enzyme in the absence of salt, free and complexed with phenol (A) are shown for comparison. All spectra in 0.05 M HEPES pH 7.5 25°. Potassium phosphate 0.05 pH 7.5 gave identical results.

Appropriate controls, that have been subjected to nearly identical conditions of gel chromatography in the absence of salt, are included (solid lines). As seen in Figure 5, 1 M sodium azide (A) or 1 M potassium cyanide (B) removes FAD from the enzyme. In the case of sodium azide, a nearly complete protection against the removal is obtained with 1 mM phenol. No protection by phenol is observed in the case of 1 M potassium cyanide. Also 0.25 M potassium thiocyanate removes FAD from the enzyme (Figure 5C). Considerable, but far from complete, protection is obtained with 1 mM phenol. Least efficient of the four anions in Figure 5 is chloride (Figure 5D). Under the conditions of the experiment (5–10 min exposure to the salt) only about one-fourth of the enzyme's FAD is removed by 1 M KCl. Phenol, 1 mM, offers complete protection. Selected controls with only 2 equivalents phenol per enzyme-bound FAD indicated similar results.

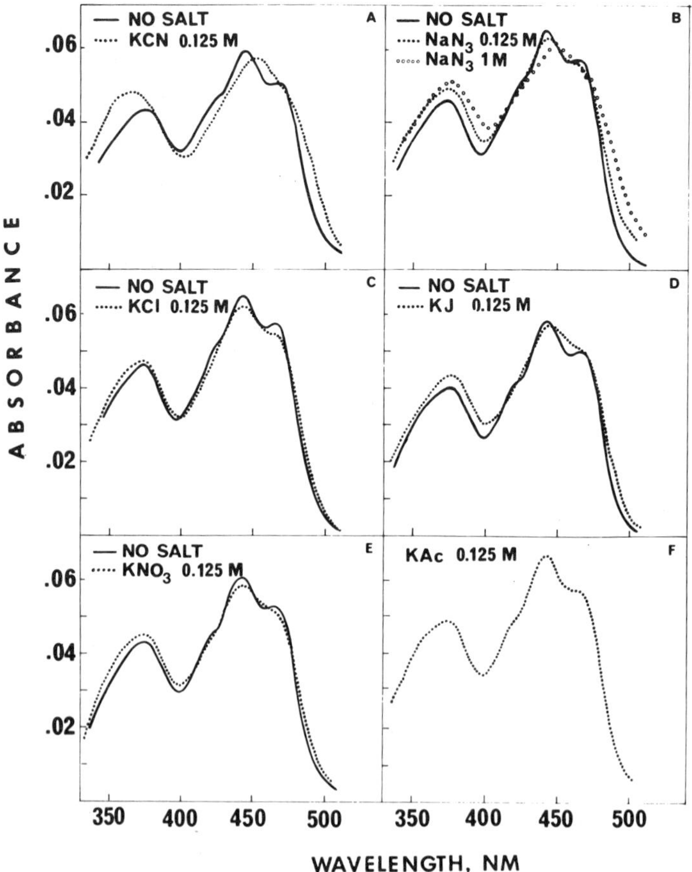

Figure 3. Absorption spectra of phenol hydroxylase in the presence of various monovalent anions; 0.05 M potassium phosphate pH 7.5 25°.

Discussion

The occurrence of shoulders on both sides of the absorption maximum at 443 nm indicates that FAD in phenol hydroxylase is anchored in a predominantly hydrophobic environment. Analogous features have been observed in the spectra of free flavins in low polarity solvents (7). Hydrophobic interactions are associated with an an ordered structure of water. Most anions that were tested here are classified as "chaotropic ions." Hydrated forms of such ions carry large positive entropies, hence they are thought to destroy the ordered structure of water. This leads to weakening of the hydrophobic bond (8). Fluoride stands apart in this respect. Its hydrated form carries negative entropy and it is thus an "antichaotropic ion" (8). This fits into the pattern of ion effects observed here. In contrast to the other ions which more or less efficiently remove FAD from the enzyme, fluoride seems to bring about a closer attachment of the prosthetic group. This is indicated by the deepening of

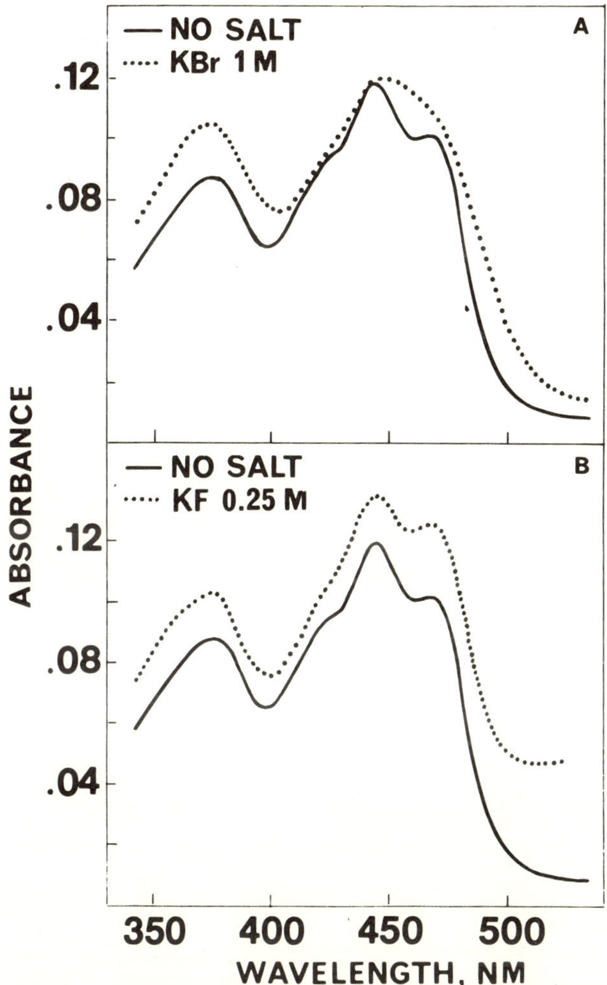

Figure 4. The effect of KBr and KF on the absorption spectra of phenol hydroxylase in 0.05 M HEPES pH 7.5 25°. Both salts cause some haziness, KF causes precipitation after 5–6 min incubation.

the shoulder around 465 nm (Figure 4B). On the other hand, fluoride is one of the anions which inhibit the activity most (Figure 1). It also readily precipitates the enzyme.

Phenol, in concentrations of a few equivalents per enzyme-bound FAD, counteracts the FAD displacing effect of the monovalent anions. Thus, one manifestation of the effector function of phenol is a conformational change which seems to bring the FAD-binding group(s) deeper inside the catalytic center. This may protect them from the monovalent anions in the solvent. Another aspect of the effector function of phenol is the earlier observation that it makes the enzyme's SH groups more "buried" as indicated by their de-

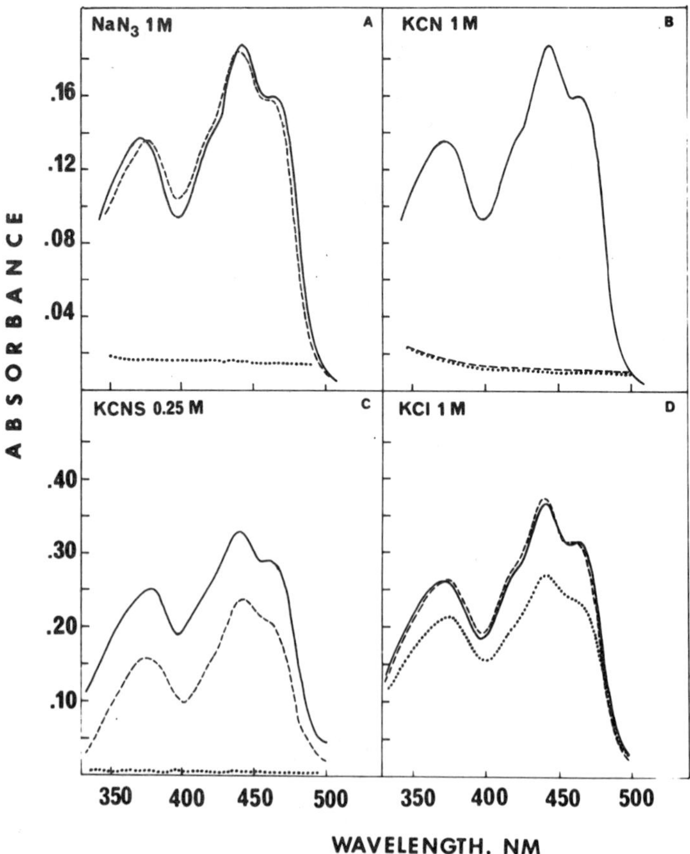

Figure 5. Removal of FAD from phenol hydroxylase by monovalent anions and the effector function of phenol. Spectra of protein eluates from Sephadex G-25 columns. (A–B) 1.23 mg/ml; (C) 2.23 mg/ml; (D) 2.38 mg/ml; 0.05 M potassium phosphate pH 7.5 25°. (—) controls without salt; (· · ·) salt; (---) salt plus 1 mM phenol.

creased reactivity towards SH-reagents. This occurs concomitantly with a firmer attachment of FAD (3). A third manifestation is that phenol enhances the enzyme's reactivity towards NADPH (4) by promoting the exposure of a reactive lysyl residue (5). The three effects of phenol, so far observed, actualize the question whether the binding site for phenol in its capacity of effector is the same or different from its binding site in its capacity of substrate. Spectrophotometric titration of the enzyme with phenol gives two isobestic points, indicating that there may be two binding sites for phenol (K. Detmer and V. Massey, personal communication). Evidence of the identity or disparity of the two sites has yet to be obtained.

The interference of monovalent anions with proper attachment of FAD to the oxidized enzyme correlates with the recent observation that one of the short-lived transients in the catalytic cycle accumulates in the presence of chloride or azide (Detmer et al., this volume). This transient is similar to intermediate III, also observed to accumulate during the reaction of p-hydroxybenzoate hydroxylase in the presence of azide (6). Judging by this effect of azide or chloride, the inhibiting effect of monovalent anions could be the result of their interaction with some positively charged group(s) on the protein, essential to proper attachment of FAD. The two types of effects, the "chaotropic" and the "electrostatic," are, of course not mutually exclusive. A combination of both may underly the inhibiting effect of monovalent anions.

References

1. Neujahr, H.Y. and Gaal, A. (1973) *Eur J Biochem* 35:386–400.
2. Neujahr, H.Y. and Gaal, A. (1973) *Proc 9th Intern Congr Biochemistry* Abstr. No 7a7, Stockholm.
3. Neujahr, H.Y. and Gaal, A. (1975) *Eur J Biochem* 58:351–357.
4. Neujahr, H.Y. and Kjellén, K.G. (1978) *J Biol Chem* 253:8835–8841.
5. Neujahr, H.Y. and Kjellén, K.G. (1980) *Biochemistry* 19:4967–4972.
6. Entsch, B., Ballou, D.P., and Massey, V. (1976) *JBC* 251:2550–2563.
7. Koziol, J. (1971) In *Methods Enzymology 18*. D.B. McCormick and L.D. Wright (eds), p. 257–261.
8. Hatefi, Y. and Hanstein, W.G. (1969) *Proc Natl Acad Sci US* 62:1129.

Published 1982 by Elsevier North Holland, Inc.
Vincent Massey and Charles H. Williams, Editors
Flavins and Flavoproteins

CHAPTER 53

Steady State and Rapid Reaction Studies on Phenol Hydroxylase

Kristina Detmer,* Vincent Massey,* David P. Ballou,* and Halina Y. Neujahr‡

**Department of Biological Chemistry, University of Michigan, Ann Arbor, Michigan; and*
‡Royal Institute of Technology, Stockholm, Sweden

Phenol hydroxylase from the soil yeast, *Trichosporon cutaneum*, is an NADPH-dependent flavoprotein monooxygenase which catalyzes the hydroxylation of a variety of phenols by molecular oxygen to form catechols:

$$\text{phenol (C}_6\text{H}_5\text{OH)} + \text{NADPH} + H^+ + O_2 \longrightarrow \text{catechol} + \text{NADP}^+ + H_2O$$

Substrate specificity is quite broad; fluorophenols, m- and p-chlorophenols, cresol, and dihydroxybenzenes are all substrates for phenol hydroxylase (1), but the carboxylated phenols, p-hydroxybenzoate and salicylate, are not substrates (2). This paper presents the results of kinetic studies by steady state and rapid reaction techniques designed to provide information about the mechanism by which the enzyme inserts an oxygen atom into the aromatic ring.

Steady State Kinetics

All steady state experiments were performed with resorcinol (1,3-dihydroxybenzene) as substrate since substrate inhibition occurs with phenol, even at low concentrations (1).

When double reciprocal plots are made of initial rates as a function of resorcinol concentration at several fixed NADPH concentrations, lines are obtained which intersect at a point on the left hand side of the velocity axis (Figure 1). The secondary plots of the slopes and intercepts versus reciprocal NADPH concentration are linear. The intersecting lines in the primary plot indicate that NADPH and resorcinol bind to enzyme species which are "reversibly connected" (3), in this case, to the oxidized form of the enzyme.

Parallel lines are obtained from double reciprocal plots of initial rate data when resorcinol and oxygen concentrations are varied at a constant NADPH concentration and when NADPH and oxygen concentrations are varied at a

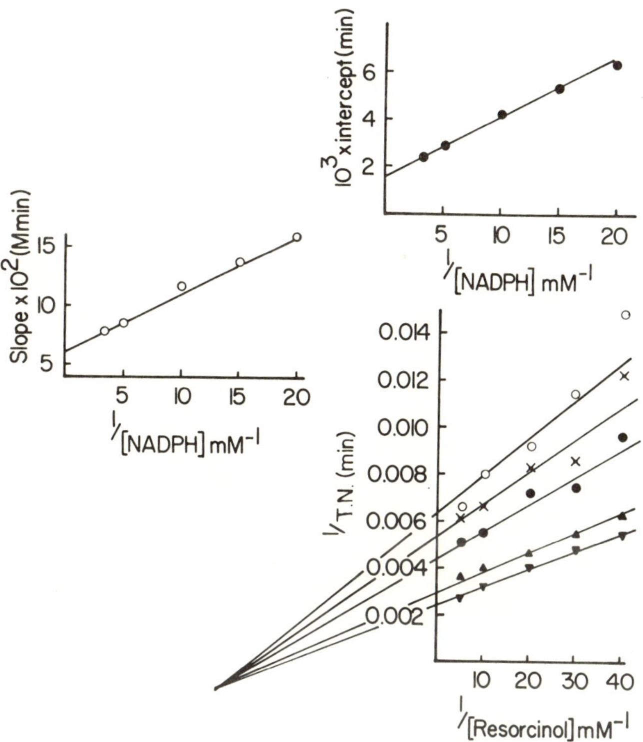

Figure 1. Steady state analysis of phenol hydroxylase at varying resorcinol and NADPH concentrations. Reaction rates were determined by measuring oxygen consumption using a Clark Yellow Springs oxygen electrode. Conditions: 0.05 M KPi, pH 7.6, 0.256 mM O_2, 25°C. Lower right: Lineweaver-Burk plots of primary data. Left: Secondary plot of slopes versus reciprocal NADPH concentration. Upper right: Secondary plot of velocity axis intercepts versus reciprocal NADPH concentration. These plots combined with the data from steady state plots with oxygen as variable substrate yield the following constants: $V_{max} = 840\ min^{-1}$, $K_{resorcinol} = 50\ \mu M$, $K_{NADPH} = 210\ \mu M$, $K_{O_2} = 66\ \mu M$.

constant concentration of resorcinol. A mechanism consistent with these results is shown below according to the notation of Cleland (4):

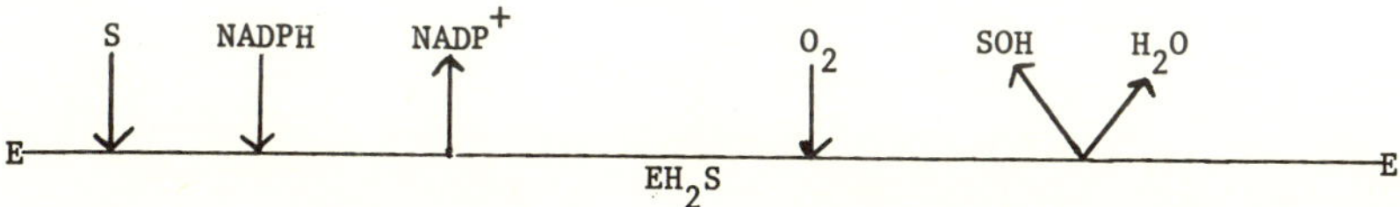

According to this mechanism, a complex is formed by enzyme, resorcinol, and NADPH which leads to reduction of the flavin and the subsequent release of $NADP^+$. The data are consistent with either random or ordered addition of the first two substrates. Oxygen then adds to the reduced enzyme-resorcinol complex to generate a second ternary complex. Enzyme-bound oxygen is activated and reacts with the resorcinol. The reaction of oxygen with the reduced enzyme-substrate complex is essentially irreversible. Such a mechanism has also been proposed for salicylate hydroxylase (5), melilotate hydroxylase (6), and p-hydroxybenzoate hydroxylase (7).

Rapid Reaction Studies

When the reduced enzyme-phenol complex was reacted with oxygen, three kinetic phases were detected, indicating the occurrence of two consecutive intermediates on the path from reduced to oxidized flavin. In the presence of inhibitory monovalent anions, such as Cl^- or N_3^-, the rates of formation and decay of the intermediates were selectively affected. The rate of decay of the second observed intermediate was decreased to the point that repetitive spectra could be recorded with a Cary spectrophotometer as the oxidized enzyme was being reformed (Figure 2).

Figure 2. Reoxidation of phenol hydroxylase in the presence of 0.25 M KCl. Phenol hydroxylase (30 μM) in 50 mM KPi, pH 7.6, containing 0.25 M KCl, 8.3 mM EDTA, 0.5 mM phenol, and 0.9 μM N(5)-deazaflavin was made anaerobic and photoreduced (11). A spectrum of the reduced enzyme was taken before transfer to an anaerobic stopped flow chamber. Temperature throughout the experiment was 4°C. The enzyme was mixed with phosphate buffer containing 0.25 M KCl and 0.5 mM phenol which had been equilibrated with 100% oxygen. A spectrum was taken immediately (2 sec) with the stopped-flow spectrophotometer (12) and the enzyme in the mixing chamber was transferred to a cold spectrophotometer cell.

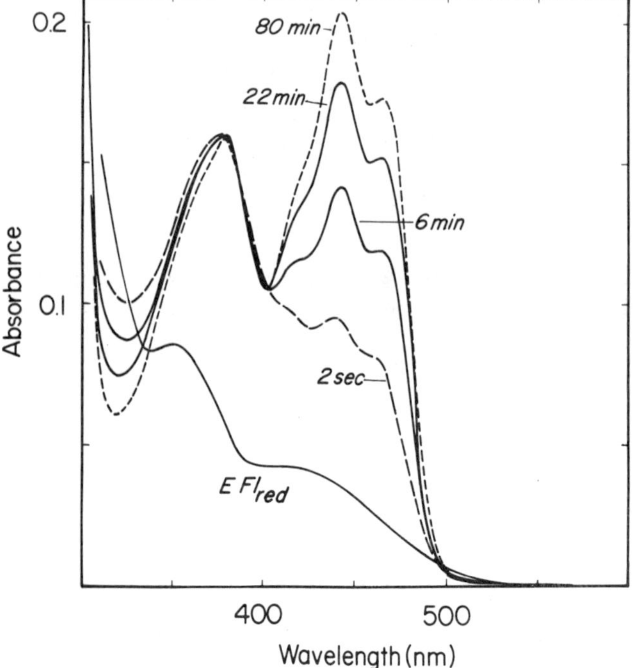

The spectra of the two transient intermediates were calculated as described previously (8). The spectra (Figure 3) are characteristic of C(4a)-substituted flavins (9, 10). The first appears to be a C(4a)-hydroperoxyflavin (λ_{max}, 390 nm), and the second, a C(4a)-hydroxyflavin (λ_{max}, 360 nm), results which are similar to those obtained previously with melilotate hydroxylase (6) and p-hydroxybenzoate hydroxylase (7, 8). With the latter enzyme, a sequence of three intermediates was observed in the oxygen half-reaction (8), the C(4a)-hydroperoxyflavin being identified as intermediate I, and the C(4a)-hydroxyflavin as intermediate III. To maintain a consistent nomenclature, the second observed intermediate with phenol hydroxylase has been labeled III, even though intermediate II has not yet been detected.

Further evidence that the second intermediate is a C(4a)-hydroxyflavin comes from the following experiment. When a complex of reduced enzyme and phenol is mixed with oxygenated buffer in the presence of Cl^-, and the reaction is quenched 2–5 sec later by squirting into acid, catechol can be extracted from the reaction mixture in quantities stoichiometric with that of the initial reduced enzyme-phenol complex. This suggests strongly that oxygen transfer takes place prior to or concommitant with the formation of this

Figure 3. Calculated spectra of the transient intermediates observed on mixing a complex of reduced phenol hydroxylase and phenol with oxygenated buffer in the stopped flow spectrophotometer. Reaction mixtures contained 15 μM enzyme, 0.5 mM phenol, 0.25 M KCl, 4 mM EDTA, 0.24 mM oxygen, and 50 mM KPi, pH 7.6 at 4°C. The absorbance changes on mixing were followed at a variety of wavelengths. Intermediate I (---) was calculated from the fast phase of the reaction. Intermediate III (—) was calculated from the slow phase.

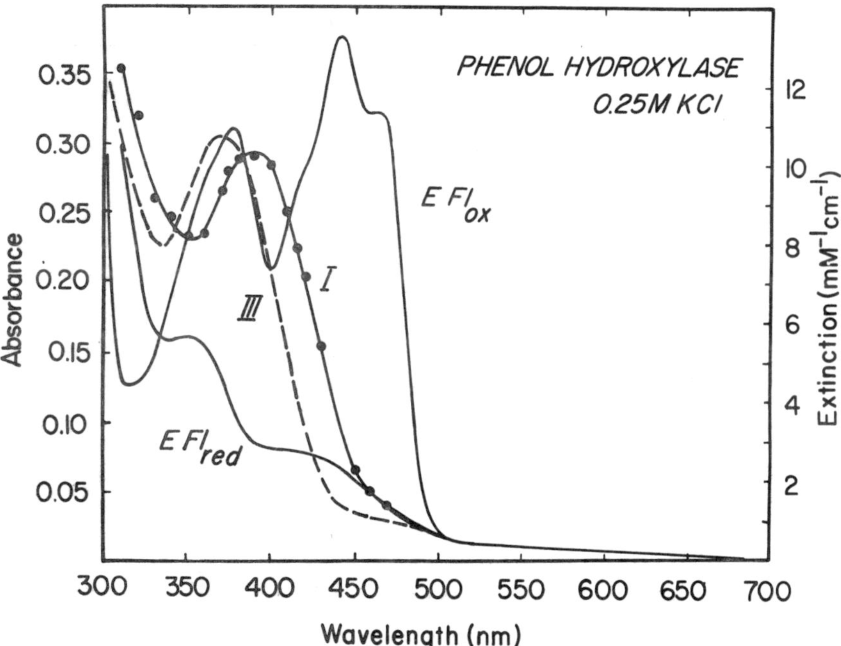

"trapped" intermediate. This fact coupled with the spectral characteristics (Figure 3) implies that the second intermediate is the C(4a)-hydroxyflavin. Moreover, the inhibition by monovalent anions is due to the stabilization of the C(4a)-hydroxyflavin, whose conversion to the oxidized flavoenzyme and water is slowed dramatically.

References

1. Neujahr, H.Y. and Gaal, A. (1973) *Eur J Biochem* 35:386–400.
2. Neujahr, H.Y. (1976) In *Flavins and Flavoproteins.* T.P. Singer (ed.) Amsterdam: Elsevier, pp. 161–168.
3. Cleland, W.W. in *The Enzymes.* Boyer, P. (ed.) 3rd ed. Vol. 2, New York: Academic Press, p. 15.
4. Cleland, W.W. (1963) *Biochem Biophys Acta* 67:173–187.
5. White-Stevens, R.H., Kamin, H., and Gibson, Q.H. (1972) *J Biol Chem* 247:2371–2381.
6. Strickland S. and Massey, V. (1973) *J Biol Chem* 248:2953–2962.
7. Spector, T. and Massey, V. (1972) *J Biol Chem* 247:5632–5636.
8. Entsch, B., Ballou, D.P., and Massey, V. (1976) *J Biol Chem* 251:2550–2563.
9. Ghisla, S., Entsch, B., Massey, V., and Husain, M. (1977) *Eur J Biochem* 76:139–145.
10. Kemal, C. and Bruice, T.C. (1976) *Proc Nat Acad Sci USA* 73:995–999.
11. Massey, V. and Hemmerich, P. (1978) *Biochemistry* 17:9–17.
12. Beaty, N.B. and Ballou, D.P. *J Biol Chem* 251:4611–4618.

Published 1982 by Elsevier North Holland, Inc.
Vincent Massey and Charles H. Williams, Editors
Flavins and Flavoproteins

CHAPTER 54

Anthranilate Hydroxylase (Deaminating) from *Trichosporon cutaneum*

Justin Powlowski and Stanley Dagley

Department of Biochemistry, College of Biological Sciences, University of Minnesota, St. Paul, Minnesota

Microorganisms degrade L-tryptophan to anthranilic acid; this is first oxidized to catechol and then converted into the Krebs cycle intermediates, succinate and acetyl CoA, by reactions of the β-ketoadipate pathway (1). Two routes for the conversion of anthranilate into catechol have been described. A bacterial system, consisting of two protein fractions (2), oxidizes anthranilate directly to catechol with liberation of ammonia and simultaneous incorporation of both of the atoms of $^{18}O_2$ (3). This enzyme has been designated anthranilate 1,2-dioxygenase (deaminating, decarboxylating; EC 1.14.12.1). The second pathway has been investigated for the eukaryotes *Aspergillus niger* (4), *Claviceps paspali* (5), and *Trichosporon cutaneum* (6). It involves two separate enzymatic steps. The first enzyme of the pair has been listed as anthranilate 2,3-dioxygenase [EC 1.14.12.2], but we shall refer to it as anthranilate hydroxylase (deaminating) since Anderson and Dagley (6) showed that a partially purified preparation from *T. cutaneum* converts anthranilate into 2,3-dihydroxybenzoate with incorporation of ^{18}O either from $H_2{}^{18}O$ or from $^{18}O_2$; one mol of NADPH is consumed in the reaction (see Figure 1). The second enzyme of this eukaryotic sequence, namely 2,3-dihydroxybenzoate decarboxylase, has been purified and characterized (6).

Anthranilate hydroxylase (deaminating) was purified from 225 g, wet weight, of *T. cutaneum* grown at the expense of salicylic acid, a substrate which strongly derepresses synthesis of this enzyme (6). Procedures for aerobic growth of the yeast, preparation of cell extract (810 ml) in phosphate buffer, pH 7.4, and spectrophotometric assay of the hydroxylase by monitoring NADPH utilization at 340 nm, were as previously described (6). The purification is summarized in Table 1; 0.1 mM EDTA was present throughout in order to stabilize the enzyme.

A fraction precipitating at 50–70% saturation with ammonium sulfate contained the enzyme; this was applied to a column of DEAE-cellulose and eluted with a linear gradient of KCl. Fractions collecting at 0.12 M KCl were passed first through a column of phenyl Sepharose CL-4B, which was equilibrated with 10% saturated ammonium sulfate, and then applied to a column of Ultrogel AcA34 and eluted with 0.05 M phosphate buffer, pH 7.4. The purified enzyme, which gave a single protein band when examined by polyacrylamide gel electrophoresis, was yellow, showed maximum light absorbance at about 380 and 450 nm, and was bleached by dithionite. Enzymatic activity

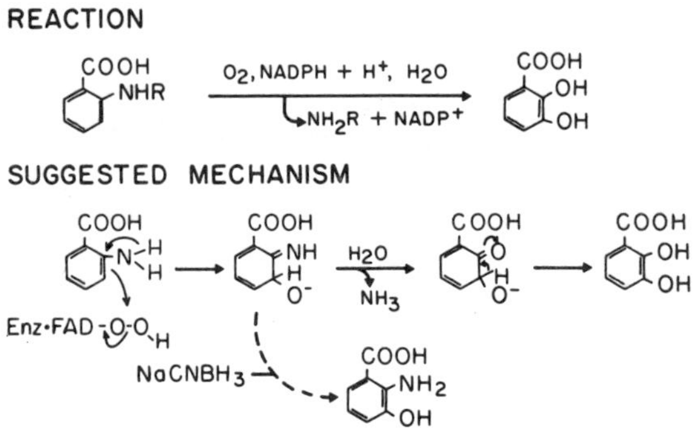

REACTION

SUGGESTED MECHANISM

Figure 1.. Anthranilate hydroxylase (deaminating). The reaction is formulated at the top. Below: suggested mechanism, showing the reduction of an intermediate by sodium cyanoborohydride, giving 3-hydroxyanthranilate (broken line).

was lost upon prolonged dialysis against KBr, and was partially restored when FAD was added. After incubation for 1.5 hr with 3 M KBr, and precipitation with saturated ammonium sulfate, a yellow supernatant solution was obtained from which FAD was extracted and identified by thin layer and paper chromatography.

Gel filtration of the enzyme through a column of Sephacryl S-200 gave an apparent molecular weight of 94,000; a value of 50,000 was obtained for the single subunits indicated by sodium dodecyl sulfate-polyacrylamide gel electrophoresis. From spectrophotometric measurements, it appeared that 2 mol of FAD were present per mol of enzyme. The pH optimum of the hydroxylase was 7.7, and from a variety of simple aromatic acids tested that carried amino, hydroxyl and other substituents, the only substrates for the enzyme were anthranilic and *N*-methylanthranilic acids. Salicylic, 3-hydroxyanthranilic and benzoic acids were nonsubstrate effectors, catalyzing uptake of oxygen and production of hydrogen peroxide.

A mechanism for the reaction is suggested in Figure 1, by which one atom of oxygen is incorporated into 2,3-dihydroxybenzoate when an imine intermediate is hydrolyzed. This suggestion is supported by the observation that at pH 7, relatively high concentrations of the hydroxylase catalyze the oxidation

Table 1. Purification of Anthranilate Hydroxylase (Deaminating).

Fraction	Volume (ml)	Protein (mg)	Sp Act[a] (U/mg)	Yield %	Purification (fold)
Crude extract	810	8900	0.18	100	1
50–75% $(NH_4)_2SO_4$, pH 5.6	192	1900	0.63	75	3.5
DEAE-cellulose	21	150	3.0	28	17.0
Phenyl Sepharose	11	68	4.2	18	23.0
Ultrogel AcA34	15	60	4.8	18	30.0

[a]Micromoles of NADPH consumed per minute per mg.

of anthranilic acid to give a mixture of 2,3-dihydroxybenzoic and 3-hydroxyanthranilic acids in the presence of sodium cyanoborohydride, a mild reagent that reduces imines preferentially under these conditions (7). Further experiments to study incorporation of ^{18}O into 3-hydroxyanthranilate are in progress.

ACKNOWLEDGMENTS
This research was supported by Grants from the N.I.H. (ES AI 00678 and R01 AM 21981).

References

1. Stanier, R.Y. and Ornston, L.N. (1973) *Adv Microb Physiol* 9:89–151.
2. Kobayashi, S. and Hayaishi, O. (1970) *Methods Enzymol* 17A:505–510.
3. Kobayashi, S., Kuno, S., Itada, N., Hayaishi, O., Kozuka, S., and Oae, S. (1964) *Biochem Biophys Res Comm* 16:556–561.
4. Subba Rao, P.V., Moore, K., and Towers, G.H.N. (1967) *Biochem Biophys Res Commun* 28:1008–1012
5. Gröger, D., Erge, D., and Floss, H.G. (1965) *Z Naturforsch* 206:856–858.
6. Anderson, J.J. and Dagley, S. (1981) *J Bacteriol* 146:291–297.
7. Borch, R.F., Bernstein, M.D., and Durst, H.D. (1971) *J Amer Chem Soc* 93:2897–2904.

Published 1982 by Elsevier North Holland, Inc.
Vincent Massey and Charles H. Williams, Editors
Flavins and Flavoproteins

CHAPTER 55

Functional Consequence of Modifying an Arginyl Residue of Salicylate Hydroxylase

Kenzi Suzuki and Masayuki Katagiri

Department of Chemistry, Faculty of Science, Kanazawa University, Kanazawa 920, Japan

Introduction

Salicylate hydroxylase (salicylate, NADH : oxygen oxidoreductase [1-hydroxylating, decarboxylating] EC 1. 14. 13. 1), from *Pseudomonas putida* is known to bind salicylate to form an enzyme-substrate complex, which, by the addition of NADH, is further converted to the two-electron reduced intermediate capable of supporting the hydroxylation of the bound salicylate with O_2. As reported previously (1), we found treatment of the enzyme protein with glyoxal resulted in modification of arginine residues, revealing changes in the catalyzed monooxygenase characteristics of the native enzyme to those of a "salicylate-dependent NADH dehydrogenase." From a plot of log (half-time of the conversion) against log (glyoxal) as previously used by Levy et al. (2), a straight line was obtained with a slope equal to 0.75. This suggests that the change in the catalytic nature is due to the modification of one of the 20 arginyl residues of the enzyme molecule. We report here the separation of the single-arginine modified enzyme and its characterization.

Preparation of the Modified Enzyme

Preparation of the bulk of the modified enzyme was performed by treating 50 μM salicylate hydroxylase with 120 mM glyoxal at 25° for 6 hr in 160 ml of 0.3 M potassium phosphate buffer, pH 6.9. The reaction mixture was applied to a DEAE-Sephadex A-50 column (2.7×70 cm), which had been equilibrated with 50 mM Tris-Cl buffer, pH 7.5. After washing with a linear gradient of NaCl from 70 mM to 140 mM (500 ml each) in the same buffer, elution was further continued with the same buffer containing 140 mM NaCl. The combined fractions eluted between 140 ml and 270 ml of the latter buffer were concentrated by adding solid ammonium sulfate to 70% saturation and dialyzed against 30 mM Tris-Cl buffer, pH 7.0. The concentrated material was rechromatographed under the same conditions. The preparation showed a single disc band on polyacrylamide gel electrophoresis at pH 8.3.

Properties of the Modified Enzyme

Analytical properties of the modified enzyme are similar to those of the native enzyme with the exception that only one arginyl residue is lost in the modified enzyme. The spectral properties of the modified enzyme thus prepared were also similar to those of the native enzyme in exhibiting a marked shoulder around 480 nm upon addition of salicylate. From the data of the spectral titration curve, the dissociation constant of salicylate for the modified enzyme was calculated to be 4 μM (for the native enzyme, 5.8 μM).

Under the standard assay conditions for the salicylate hydroxylase reaction (5), as monitored by the rate of catechol formation by using the metapyrocatechase method (6), the modified enzyme revealed only 26% of the monooxygenase activity. It seems unlikely that the remaining monooxygenase activity due to the presence of unreacted native enzyme, since preparations which had been modified up to 3 out of 20 arginine residues per molecule, followed by separation through a DEAE-Sephadex column, revealed a similar activity ratio of catechol formation per NADH consumption. At higher FAD concentrations, the modified enzyme catalyzed efficient oxidation of NADH which is also dependent on the presence of salicylate. Figure 1 illustrates the dependence of the two NADH oxidation activities with varying FAD concentrations. Since the overall rate of NADH oxidation under these conditions may be assumed to be the sum of the rates of the monooxygenase- and dehydrogenase-dependent reactions, the "dehydrogenase" activity is represented, after correction, by subtracting the monooxygenase activity, as measured by the rate of catechol formation, from the total NADH oxidase activity.

Figure 1. Effect of FAD on the monooxygenase activity and the dehydrogenase activity of the native (A) and the glyoxal-modified (B) salicylate hydroxylase. The assay mixture for the measurement of the total NADH oxidation activity contained 100 μM salicylate, 40 μM NADH, and varying amounts of FAD in 1 ml of 30 mM potassium phosphate buffer, pH 7.0. The activity was followed at 20° by the rate of change in absorbancy at 340 nm for the first 2 min. For the measurement of the monooxygenase activity (-○-), the assay conditions were the same with the exception that the medium contained 0.1 unit of metapyrocatechase (6) and absorbance at 400 nm was followed. The NADH dehydrogenase activity (-●-) is based on the calculated values by subtracting the monooxygenase activity from the total NADH oxidation activity.

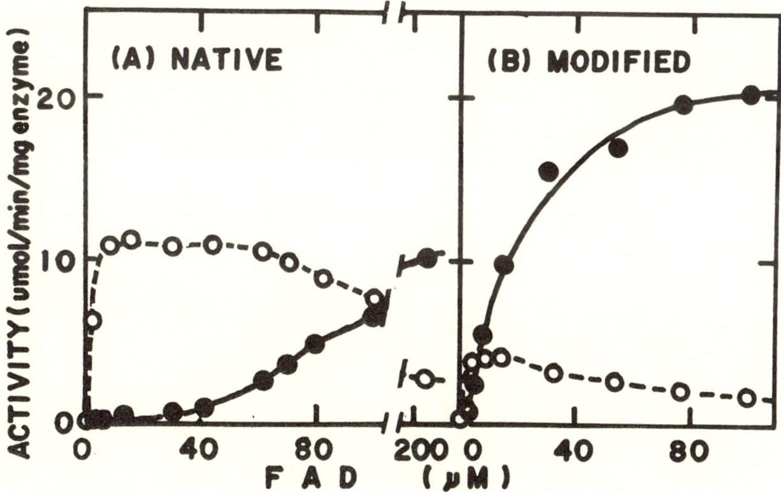

Table 1. Effect of Various Electron Acceptors on the Dehydrogenase Activity of the Modified Enzyme.

Acceptor	K_m μM	V_{max} min^{-1}
FAD	20	970
FMN	31	850
Riboflavin	15	880
DCIP	0.9	370
$K_3Fe(CN)_6$	40	1,260

Experiments were carried out under conditions similar to those described for Figure 1, with the exception that FAD is replaced by the indicated electron acceptors. When DCIP or ferricyanide was the acceptor, the rate of the reaction was measured by the decrease in absorbancy at 600 nm and 420 nm, respectively.

As is evident from the plot, increasing concentrations of FAD affected either the native or the modified enzyme, producing an increase in the rate of the dehydrogenase reaction. It is of considerable interest that, in the dehydrogenase reaction, the apparent K_m value of the electron acceptor FAD for the modified enzyme is an order of magnitude smaller than that for the native enzyme. In addition to the higher affinity for the acceptor FAD, the modified enzyme utilized compounds such as FMN, riboflavin, DCIP, and ferricyanide as active electron acceptors (Table 1). This is in contrast to the already reported fact that no such compounds replaced FAD in the dehydrogenase reaction catalyzed by the native enzyme under anaerobic conditions (7). It is also noted that changing the partial pressure of O_2 did not result in a change in either the K_m value of FAD for the modified enzyme or its catalyzed rate of NADH oxidation, indicating that O_2 is not an obligatory electron acceptor.

Figure 2. Proposed mechanism for the relationship between salicylate hydroxylase-catalyzed monooxygenase (a), oxidase (b), and dehydrogenase (c) reactions. EF: salicylate hydroxylase; S: salicylate or its analog; P: catechol + CO_2; and A: electron acceptor.

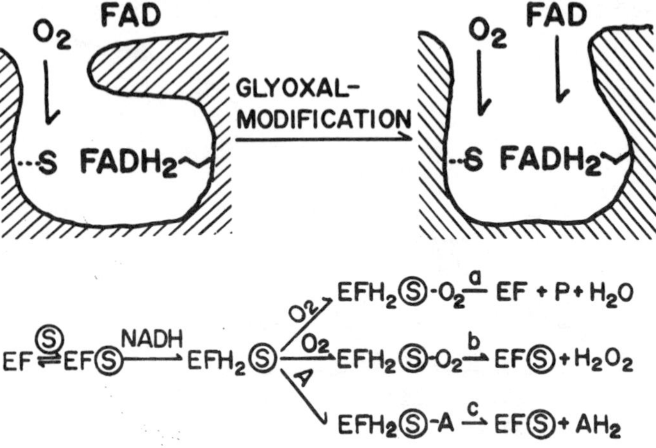

Conclusions

The results presented in this paper suggest that the dehydrogenase reaction which is catalyzed by the modified enzyme differs in mechanism from the salicylate analog-induced NADH oxidation, since the latter reaction is known to be an "NADH oxidase" reaction (8). The results also suggest that the changes induced by the modification in catalytic properties of the enzyme are mainly caused by the conformational change at the level of the reduced form of the enzyme-salicylate complex, increasing the reactivity of the complex toward the exogenous electron acceptor. Figure 2 is a schematic description of the proposed mechanism. Stopped flow experiments revealed that the modification had little effect on the rates of either formation of the enzyme-salicylate complex or its reduction by NADH.

ACKNOWLEDGMENTS
We thank Messrs. K. Ohnishi, M. Kitamura, and S. Toyoda for technical assistance. This work was supported in part by Grants from the Ministry of Education, Science, and Culture, Japan, and from the Yamada Science Foundation.

References

1. Takemori, S., Suzuki, K., and Katagiri, M. (1976) *Flavins and Flavoproteins*. Singer, T.P. (ed.), Amsterdam: Elsevier Scientific Publishing Company, p. 178–183.
2. Levy, H.M., Leber, P.D., and Lyan, F.M. (1963) *J Biol Chem* 238:3654–3659.
3. Messineo, L. (1966) *Arch Biochem Biophys* 117:534–540.
4. Yamada, S. and Itano, H.A. (1966) *Biochim Biophys Acta* 130:538–540.
5. Yamamoto, S., Katagiri, M., Maeno, H., and Hayaishi, O. (1965) *J Biol Chem* 240:3408–3413.
6. Takemori, S., Komiyama, T., and Katagiri, M. (1971) *Eur J Biochem* 23:178–184.
7. Katagiri, M., Maeno, H., Yamamoto, S., Hayaishi, O., Kitao, T., and Oae, S. (1965) *J Biol Chem* 240:3414–3417.
8. White-Stevens, R.H. and Kamin, H. (1972) *J Biol Chem* 247:2358–2370.
9. Takemori, S., Nakamura, M., Suzuki, K., Katagiri, M., and Nakamura, T. (1972) *Biochim Biophys Acta* 284:382–393.

Published 1982 by Elsevier North Holland, Inc.
Vincent Massey and Charles H. Williams, Editors
Flavins and Flavoproteins

CHAPTER 56

On the Mechanism of Salicylate Hydroxylase: Studies Using Deuterated Substrates

Lee-Ho Wang, Riyad Y. Hamzah, and Shiao-Chun Tu

Department of Biophysical Sciences, University of Houston, Houston, Texas

Salicylate hydroxylase is an FAD-dependent monooxygenase which utilizes NAD(P)H as an external reductant for the decarboxylative hydroxylation of salicylate to produce catechol. In the present study, deuterium-incorporated NADH and salicylate have been synthesized and used for investigating the reaction mechanism of this hydroxylase from cells of *Pseudomonas cepacia* (1) previously isolated from soil organisms and maintained as *P.* sp. ATCC 29351 (2).

(4S)-[4-^{2}H]NADH (B-NADD) was synthesized by the lipoamide dehydrogenase-catalyzed reduction of NAD^+ in ^2H_2O containing $(NH_4)_2CO_3$ (0.05 M), lipoamide, and dithiothreitol, and then isolated as barium salt precipitate (3). (4R)-[4-^{2}H]NADH (A-NADD) was obtained by the method of You et al. (4). This involves the oxidation of B-NADD by the A-stereospecific yeast alcohol dehydrogenase in the presence of acetaldehyde, purification of [4-^{2}H]NAD^+ by Dowex-1 column chromatography, and reduction of [4-^{2}H]NAD^+ to A-NADD as that described for B-NADD except that the reaction was carried out in H_2O.

Both A- and B-NADD were tested for their deuterium isotope effects on the salicylate hydroxylase-catalyzed reaction. Results of such studies using saturating salicylate as a cosubstrate are shown in Figure 1. No significant effects on either K_m or k_{cat} were found with B-NADD. A-NADD was found to exhibit no significant effect on the K_m but a primary kinetic isotope effect of 1.6 for k_H/k_D was detected. This latter effect remains constant in the pH range of 5 to 9. When 2,4-dihydroxybenzoate, 2,3-dihydroxybenzoate, p-aminosalicylate, and benzoate were used in place of salicylate, the deuterium isotope effects of A-NADD on k_{cat} were found to be 1.8, 1.9, 2.5, and 4.0, respectively (Table 1). These findings confirm the earlier report that salicylate hydroxylase is A-stereospecific for NADH (4). Furthermore, it is demonstrated that the step of reduction of holoenzyme-substrate (or pseudosubstrate) by NADH is at least partially rate-limiting. In this connection, the importance of the step of reoxidation in the regulation of the overall reaction rate has also been indicated for presumably the same hydroxylase species (2).

For the enzymatic conversion of salicylate to catechol, the site of hydroxylation could be either at position 1 or 3 of the salicylate ring. The determination

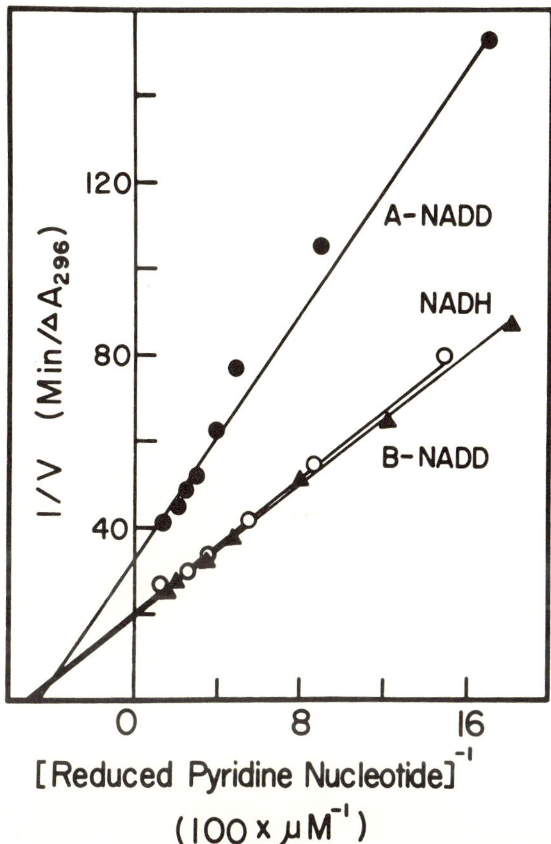

Figure 1. Lineweaver-Burk plots of salicylate hydroxylase activity versus reduced proteo/deutero pyridine nucleotide concentration. The reaction was carried out in 1 ml 0.02 M phosphate, pH 7.6, containing 0.13 mM salicylate, 0.9 μg salicylate hydroxylase and NADH, A-, or B-NADD at the designated concentration.

of the site for substrate hydroxylation is important for an understanding of the reaction mechanism. Studies concerning such a determination have been carried out using synthetic [3,5-^{2}H]salicylate as a substrate for the *P. cepacia* hydroxylase (5). A mixture of equal amounts of salicylic acid and sodium salicylate was heated at 100°C in ^2H_2O. Once a week, the solvent was removed under reduced pressure and replaced with fresh ^2H_2O. The hydrogen/deuterium exchange occurred specifically at positions 3 and 5 (Figure 2A and B) and reached ~97% completion after 40 days. The [3,5-^{2}H]salicylate sample so obtained and NADH were used as substrates for the hydroxylase and the product catechol was isolated as described previously (6). Since the enzymatic reaction was carried out in H_2O, the product catechol from [3,5-^{2}H]salicylate would retain both deuterium atoms if the site of hydroxylation was the same as the position of decarboxylation but only the deuterium at position 5 would remain if position 3 was hydroxylated. The NMR spectrum of the isolated

Table 1. Deuterium Isotope Effects of A- and B-NADD on Salicylate Hydroxylase Activity.

	k_H/k_D	
Substrate	**A-NADD**	**B-NADD**
Salicylate	1.6	0.9
2, 4-Dihydroxybenzoate	1.8	1.1
2, 3-Dihydroxybenzoate	1.9	1.1
p-Aminosalicylate	2.5	1.1
Benzoate	4.0	

catechol showed two singlets for the protons at positions 4 and 6; the ratio of total protons for the isolated catechol over that for the standard, based on the integrated signals normalized to the same sample concentration and signal scale, was found to be 1.9:4 (Figure 2C and D). The retention of both deuterium atoms by the reaction product catechol was also demonstrated by mass spectroscopy analysis (5). These findings indicate unequivocally that salicylate is decarboxylated and hydroxylated at the same ring position.

Figure 2. NMR spectra of [2,5-^{2}H]salicylate and enzymatic reaction product derived from it, in comparison with those of salicylate and catechol standards. The spectra of a salicylate standard (A) and the deuterated salicylate (B), both at 200 mg·ml^{-1}, were determined at 60 MHz. The spectra of a catechol standard (C; 10 mg·ml^{-1}) and catechol produced enzymatically from the deuterated salicylate (D; 5 mg·ml^{-1}) were recorded at 80 MHz. Identical vertical scales are used for A and B, whereas one intensity unit for C is 5.38-fold larger than that for D. All spectra were determined in ^2H_2O.

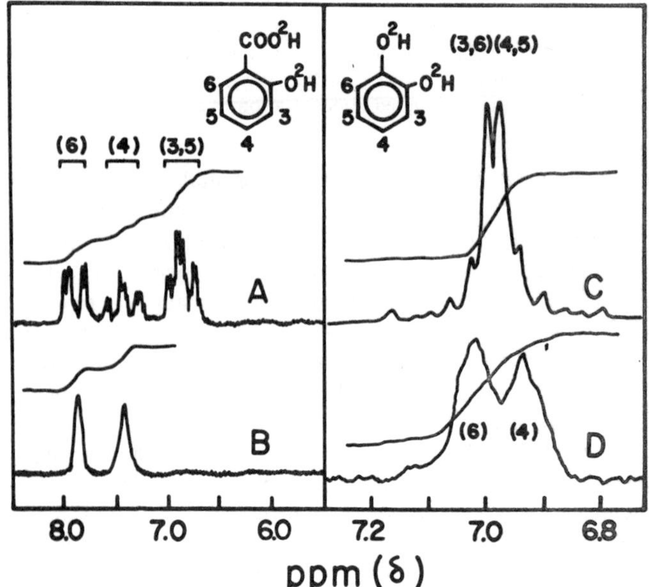

$[3,5\text{-}^2H]$Salicylate was also tested at two saturating levels for kinetic isotope effect. The k_H/k_D values were found to be 1.00 ± 0.02, indicating that there is no detectable kinetic isotope effect of the deuterated salicylate. Such a finding is consistent with the conclusion that the hydroxylation does not occur at position 3 of salicylate.

Previously, pyrogallol has been detected as a reaction product from 2,3- or 2,6-dihydroxybenzoate using *P. putida* salicylate hydroxylase, suggesting that the sites for decarboxylation and hydroxylation were the same (6). However, the expected product hydroxyhydroquinone from 2,4- or 2,5-dihydroxybenzoate could not be detected in the same study. It has also been reported that both 2,3-dihydroxybenzoate and 3-methyl salicylate exhibit partial substrate activities with the hydroxylase obtained from yet another cell strain (*P.* sp. ATCC 29352) (7), suggesting that position 1 was the hydroxylation site. However, the reaction products of these salicylate analogs were not identified, thus providing no direct evidence for decarboxylation and/or hydroxylation. The approach of using $[3,5\text{-}^2H]$salicylate demonstrated here for the *P. cepacia* enzyme should be quite applicable to the two other salicylate hydroxylase species mentioned above.

ACKNOWLEDGMENTS
This work was supported by Robert A. Welch Foundation Grant E-738 and NIH Grant GM 25953.

References

1. Tu, S.-C., Romero, F.A., and Wang, L.-H. (1981) *Arch Biochem Biophys*, in press.
2. Presswood, R.P. and Kamin, H. (1976) In *Flavins and Flavoproteins*. Singer, T.P. (ed.) Amsterdam: Elsevier, pp. 145–154.
3. Oppenheimer, N.J., Arnold, L.J., and Kaplan, N.O. (1971) *Proc Natl ACad Sci USA* 68:3200.
4. You, K.-S., Arnold, L.J., Jr., and Kaplan, N.O. (1977) *Arch Biochem Biophys* 180:550.
5. Hamzah, R.Y. and Tu, S.-C. (1981) *J Biol Chem*, in press.
6. Yamamoto, S., Katagiri, M., Maeno, H., and Hayaishi, O. (1965) *J Biol Chem* 240:3408.
7. White-Stevens, R.H. and Kamin, H. (1972) *J Biol Chem* 247:2358.

PART III C:

Luciferase

Published 1982 by Elsevier North Holland, Inc.
Vincent Massey and Charles H. Williams, Editors
Flavins and Flavoproteins

CHAPTER 57

Bioluminescence Emission of Bacterial Luciferase with FMN Analogs, in Particular with 1-Deaza-FMN

Manfred Kurfuerst, Sandro Ghisla, and J. Woodland Hastings

Fakultaet fuer Biologie der Universitaet, Postfach 5560, D-7750 Konstanz, BRD, and The Biological Laboratories, Harvard University, 16 Divinity Avenue, Cambridge, Massachusetts

Introduction

Bacterial luciferase catalyzes the concomitant oxidation of reduced FMN, and long-chain aldehyde with molecular oxygen, via two sequential and experimentally separable steps with the emission of light (λ_{max}, 490 nm):

$$FMNH_2 + O_2 \rightarrow FMNH{-}OOH$$

$$FMNH{-}OOH + RCHO \rightarrow FMN + RCOOH + H_2O + \text{light}$$

The structure of the metastable luciferase flavin peroxide, which was isolated by low temperature chromatography (10), has been established by ^{13}C-NMR spectroscopy as that of a C(4a) derivative of 4a,5-dihydro FMN (5). The subsequent reaction involves the activation of the flavin peroxide oxygen for the oxidation of the aldehyde, and the population of the emitting chromophore excited state. For this crucial process, a variety of mechanisms have been put forward (1,14,15,19,23). In some schemes, it is proposed that the peroxide oxygen is activated via rearrangement to a (C-10a) adduct and opening of the pyrimidine or pyrazine rings. In contrast to this, our groups (12), as well as Shepherd and Bruice (20), assume that oxidation can proceed from the 4a-peroxide without structural modifications of the flavin rings.

A second crucial question pertains to the chemical nature of the emitter. Can it be 1-HFMN$^+$ as proposed by Eley et al. (2)? Is it an enzyme-bound reduced or oxidized flavin, as concluded by Matheson et al. (16)? Or is it a 4a-substituted FMN, as suggested by us (12)?

In order to address these questions, we have extended previous studies (11,18) involving the use of modified FMN derivatives in the bioluminescent reaction. In particular it was our hope that 1-deaza-FMN, as suggested by Walsh (personal communication), might yield an unequivocal answer to some of the questions put forward above. This flavin analog clearly cannot be protonated at position N(1) in the same pH range as normal FMN. In addition, reversible ring opening reactions are very unlikely in this case, due to the different chemistry to be expected upon replacement of the N(1) nitrogen with carbon.

Results and Discussion

That major modifications of the isoalloxazine chromophore do not lead to complete loss of its ability to activate oxygen, and to emit light in the bacterial luciferase reaction, is known from earlier studies (18). With two such flavin isomers, the color (emission spectrum) of the bioluminescence produced differed considerably from that obtained with unaltered $FMNH_2$. With one of these (iso-FMN), the emission was shifted to the blue, while with the other (2-thio-FMN) the color was red shifted (Table 1). The striking fact was that these emissions were not at all correlated with the fluorescence emissions of the oxidized forms of these flavins. And, of course, the bioluminescence emission with FMN (λ_{max}, 490 nm) also does not correspond to its fluorescence emission (λ_{max}, 525 nm). The isolated luciferase FMN peroxide intermediate was subsequently shown to exhibit fluorescence whose spectrum did correspond to the bioluminescence emission spectrum (9). The emitter was thus postulated to be a hydroxy, 4a,5-dihydroflavin (12). This postulate has now received strong support from the observation that the specific luciferase-bound peroxides of the above two modified flavins exhibit fluorescence that also closely matches the corresponding bioluminescence (11). This correlation of bioluminescence with the fluorescence of the isolated luciferase-bound peroxides of FMN and modified flavins is also better than that found for the emission of the different oxidized flavin cations (Table 1), a species earlier suggested as the emitter (2). Our postulate contrasts with the recent report of Matheson et al. (16) who suggest that "the bioluminescent emitter is probably a reduced flavin species for the iso-$FMNH_2$ and $FMNH_2$ reactions, and an oxidized flavin species in the 2-thio-$FMNH_2$ reaction." The possibility that an oxidized flavin is the emitter in one case, and a reduced form in others appears less reasonable than an hypothesis that involves a similar mechanism and a similar excited state with different flavins as substrates.

We have recently been able to demonstrate that reduced 1-deaza-FMN will also act as a substrate for luciferase and result in appreciable light emission,

Table 1.

Flavin structure	Bioluminescence emission	Fluorescence of the peroxyflavin	Fluorescence in water	Fluorescence of the flavin cation
FMN	490[a]	490[b]	525[a]	480–493[c]
Iso-FMN	472[a]	465–470[b]	543[a]	530[c]
2-thio-FMN	534[a]	540–550[b]	non. fl.[a]	510–530[c]
1-deaza-FMN	485	[e]	non. fl.	[e]
4-thio-FMN	490	[e]	non. fl.	[e]
8-O-CH_3-FMN	500	[e]	500[d]	[e]
2-NH_2-FMN	490	[e]	530	[e]

[a]Mitchell and Hastings, 1969 (18).

[b]Hastings et al., 1981 (11).

[c]Eley et al., 1970 (2).

[d]Ghisla and Mayhew, 1976 (7).

[e]Not measured.

peaking at about 485 nm (Figure 1). The bioluminescence intensity found with this analog at saturating concentrations was approximately 15% that found with normal FMN. In the case of this molecule, we can exclude its protonated form as the emitter, since the λ_{max} of this emission is at a higher energy than the lowest energy transition of 1-deaza-FMN at $H_o = -1$ (Figure 1). More generally, it seems most improbable that a protein (e.g., luciferase) could generate a microenvironment having a pH<0, and/or generate a different species at the pH. Figure 1 also shows the absorption spectrum of a flavin 4a-peroxide bound to p-hydroxybenzoate hydroxylase, this being closely similar in its λ_{max} and spectral shape to the 4a-peroxide of normal p-hydroxybenzoate hydroxylase (3). It also matches the absorption spectrum of the luciferase 4a-peroxide obtained with normal FMN (10). These coincidences indicate that also in the case of 1-deaza-FMN a 4a,5-dihydro flavin chromophore can serve as the emitter. That 1-deaza-FMN was previously reported not

Figure 1. Bioluminescence emission of bacterial luciferase (*Beneckea harveyi*) generated with 1-deaza-FMN. 1-deaza-FMN was prepared from 1-deaza-riboflavin by the method of Spencer et al. (21), and was used in this experiment at a final concentration of 5×10^{-5} M. It was dissolved in 0.35 M phosphate buffer pH 7.0, reduced at 4° with a small excess 0.2 M dithionite in the same buffer (full reduction is indicated by disappearance of the 1-deazaflavin purple color), and in the presence of 7.5×10^{-6} M luciferase, final concentration, [prepared and assayed according to Hastings et al. (8)]. The light emission was initiated by addition of 1.0 ml of a 0.1% solution of dodecanal in the same buffer. Curve (—) represents the spectrum of the emitted light recorded with a Perkin Elmer MPF-3 fluorimeter. This spectrum is corrected for a minor time dependence in the emission intensity, but not for photomultiplier response. Curve (–·–·) shows the absorption spectrum of the 4a-peroxide of 1-deaza-FAD bound to p-hydroxybenzoate hydroxylase (3). Curve (···) represents the absorption spectrum of 1-deaza-FMN at $H_0 = -1$.

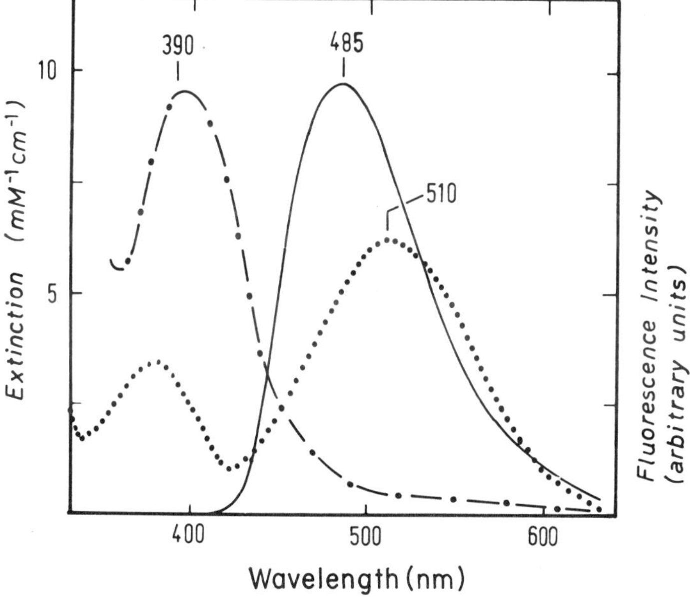

to function in the bacterial bioluminescence reaction (3,22), might be due to the fact that reduced 1-deaza-FMN binds to luciferase with a K_d (4×10^{-5} M, Figure 2), which is 50-fold higher than that found with normal FMN (17).

Scheme 1 represents a pathway proposed for the catalytic reaction of bacterial luciferase using 1-deaza-FMN. This sequence should logically apply also with normal FMN, and is analogous to that originally proposed by Hastings and Nealson (12). It should be pointed out that the 4a,5-dihydroflavin chromophore, although not fluorescent in solution, can be fluorescent when bound to proteins (6). In particular, 4a-hydroxy-5-ethyl-4a,5-dihydro-FMN has been shown to be fluorescent when bound to luciferase (4).

Conclusions

The results discussed above, and in particular the bioluminescence emission found with 1-deaza-FMN, clearly refute rearrangement reactions involving migration of the peroxide group, ring openings and closure at the position C(10a), and 1-protonated flavins as the emitter. The coincidence of the *in vitro* emission with that of the fluorescence of the luciferase-bound 4a-OR flavins also argues against alternative proposals (16) in which the emitter is the oxidized or reduced form of the flavin. The proposal put forward in Scheme 1 is in agreement with the findings of Bruice's group (see Chapter 44 in this

Figure 2. Dependence of bioluminescence emission intensity upon the concentration of 1-deaza-FMN. The assays were carried out as described in the legend of Figure 1, except at 20°, with 1.3×10^{-6} M luciferase, and at the different 1-deaza-FMN concentrations shown. The points represent the average of three measurements of the maximal light intensity obtained upon addition of decanal to the cuvette.

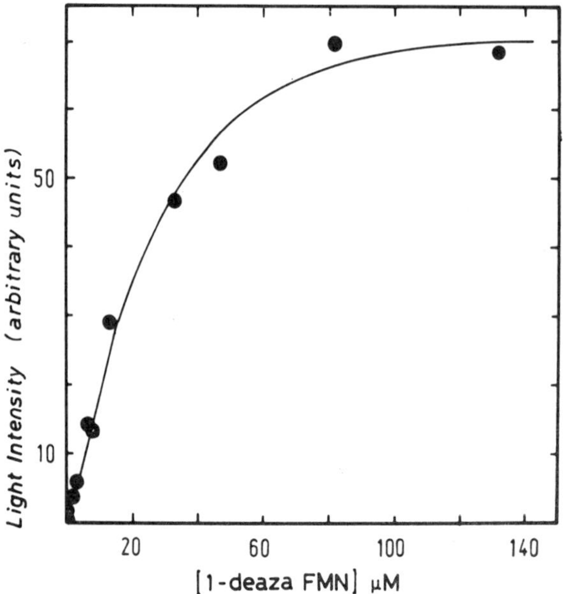

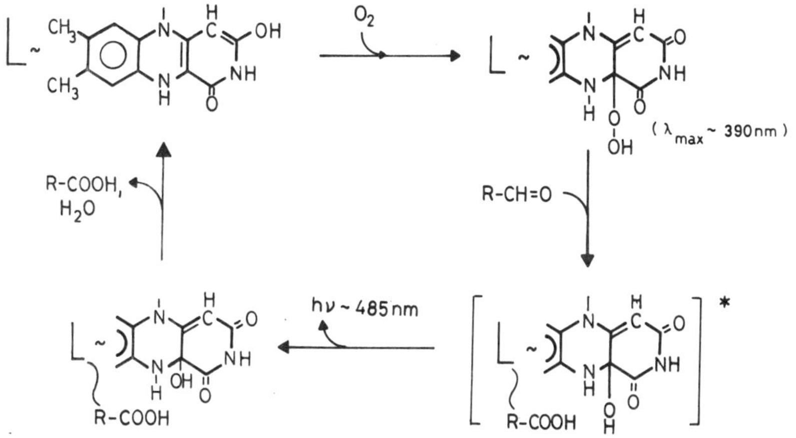

Scheme 1. Pathway proposed for the bioluminescence reaction obtained with bacterial luciferase (L), 1-deaza-FMN, long chain aldehyde (R-CH=0), and molecular oxygen.

volume), and suggests that the primary product to be expected from reaction of the 4a-peroxide is the 4a-hydroxy flavin shown, and that it can serve as the primary emitter. Our results cannot yield information as to whether this flavin chromophore is the primary excited state or merely a serendipitous acceptor, which might thus compete with a different chromophore in the *in vivo* reaction, as implied by Koka and Lee (13). Nevertheless, our proposal is consistent with the experimental facts presently available.

ACKNOWLEDGMENTS

We thank Dr. P. Hemmerich for fruitful discussions, and for the gift of 2-thio and 4-thioflavins; Dr. C. Walsh for 1-deaza-FMN; Dr. V. Massey, for a gift of 1-deaza-riboflavin; the Deutsche Forschungsgemeinschaft for support to SG (Project Gh 2/3), and NSF for support to J.W.H. (PCM 80-19504). J.W.H. was an awardee of the Alexander Von Humboldt Foundation.

References

1. Eberhard, A. and Hastings, J.W. (1972) *Biochem Biophys Res Commun* 47:348–353.
2. Eley, M., Lee, J., Lhoste, J.M., Lee, C.Y., Cormier, M.J., and Hemmerich, P. (1970) *Biochemistry* 9:2902.
3. Entsch, B., Husain, M., Ballou, D.P., Massey, V., and Walsh, C. (1980) *J Biol Chem* 255:1420–1429.
4. Ghisla, S., Entsch, B., Massey, V., and Husain, M. (1977) *Eur J Biochem* 76:139–148.
5. Ghisla, S., Hastings, J.W., Favaudon, V., and Lhoste, J.M. (1978) *Proc Nat Acad Sci* 75:5860–5863.
6. Ghisla, S., Massey, V., Lhoste, J.M., and Mayhew, S.G. (1974) *Biochemistry* 13:589–597.
7. Ghisla, S. and Mayhew, S. (1976) *Eur J Biochem* 63:373–390.
8. Hastings, J.W., and Baldwin, T.O., and Nicoli, M.Z. (1978) In *Methods in Enzymology*. M. DeLuca (ed.) New York: Academic Press, pp. 135–152.
9. Hastings, J.W. and Balny, C. (1975) *J Biol Chem* 250:7288–7292.

10. Hastings, J.W., Balny, C., Le Peuch, C., and Douzou, P. (1973) *Proc Nat Acad Sci* 70:3468–3472.
11. Hastings, J.W., Ghisla, S., Kurfürst, M., and Hemmerich, P. (1981) In *Proceedings of the Second International Congress of Chemiluminescence and Bioluminescence.* M. DeLuca and W.D. McElroy (eds.) New York: Academic Press, pp. 97–102.
12. Hastings, J.W. and Nealson, K.H. (1977) *Ann Rev Microbiol* 31:549–595.
13. Koka, P. and Lee, J. (1979) *Proc Nat Acad Sci* 76:3068–3072.
14. Kosower, E.M. (1980) *Biochem Biophys Res Commun* 92:356–364.
15. Mager, H.J. and Addink, R. (1979) *Tetrahedron Letters* 37:3545–3548.
16. Matheson, J.B.C., Lee, J., and Müller, F. (1981) *Proc Nat Acad Sci* 78:948–952.
17. Meighen, E.A. and Hastings, J.W. (1971) *J Biol Chem* 246:7666–7674.
18. Mitchell, G. and Hastings, J.W. (1969) *J Biol Chem* 244:2572–2576.
19. McCapra, F. and Hysert, D.W. (1973) *Biochem Biophys Res Commun* 52:298.
20. Shepherd, P.T. and Bruice, T.C. (1980) *J Am Chem Soc* 102:7774–7776.
21. Spencer, R., Fisher, J., and Walsh, C. (1976) *Biochemistry* 16:1043–1053.
22. Walsh, C., Fisher, J., Jacobson, F., Spencer, R., Ashton, W., and Brown, R. (1980) In *Flavins and Flavoproteins.* K. Yagi and T. Yamano (eds.) Japan Scientific Soc. Press, pp. 13–21.
23. Wessiak, A., Trout, G.E., and Hemmerich, P. (1980) *Tetrahedron Letters* 21:739–742.

Published 1982 by Elsevier North Holland, Inc.
Vincent Massey and Charles H. Williams, Editors
Flavins and Flavoproteins

CHAPTER 58

The Bacterial Luciferase Neutral Flavin Radical: Identification and Catalytic Inactivity

Manfred Kurfuerst, Sandro Ghisla, Robert Presswood, and J. Woodland Hastings

Fakultaet fuer Biologie der Universitaet, D-7750 Konstanz, BRD; and The Biological Laboratories, Harvard University, 16 Divinity Avenue, Cambridge, Massachusetts

Introduction

The first step in the bioluminescent reaction catalyzed by bacterial luciferase involves the reaction of reduced flavin with molecular oxygen to form an oxygen-containing luciferase-bound 4a-peroxy flavin intermediate (2,4):

$$E—FMNH_2 + O_2 \rightarrow E—FMNH—OOH$$

When originally isolated and characterized, the peroxy intermediate was shown to exhibit absorption peaking at 373 nm, a shoulder at around 450 nm, tailing off at 500 nm, with no absorption above 520 nm.

In later work however, it was reported that an appreciable absorption in the 500–700 nm region occurred in such preparations, both in the presence and absence of long chain aldehyde (9,10). This observation prompted a series of speculations concerning the role of this new species and the mechanism of the luciferase reaction, (5,6,11,12). We were interested in the nature of this blue (red absorbing) intermediate, particularly with respect to whether or not it had some role in the sequence of processes involving oxygen activation, and population of the emitting excited state:

$$E—FMNH—OOH + RCHO \rightarrow (\ )^* \rightarrow h\nu + E—FMN_{ox} + RCOOH + H_2O$$

As described below, we have studied its formation and properties; the results show unambiguously that the species is a luciferase semiquinone neutral flavin radical and that it has no evident role in the bioluminescent reaction.

Results and Discussion

When the luciferase peroxide intermediate is prepared by reacting enzyme with FMN reduced photochemically in the presence of EDTA, followed by aeration directly in the cuvette, material with a spectrum such as that shown on Figure 1 is obtained. It is evident that a considerable absorption at >500 nm is present, which apparently corresponds to the red absorption reported previ-

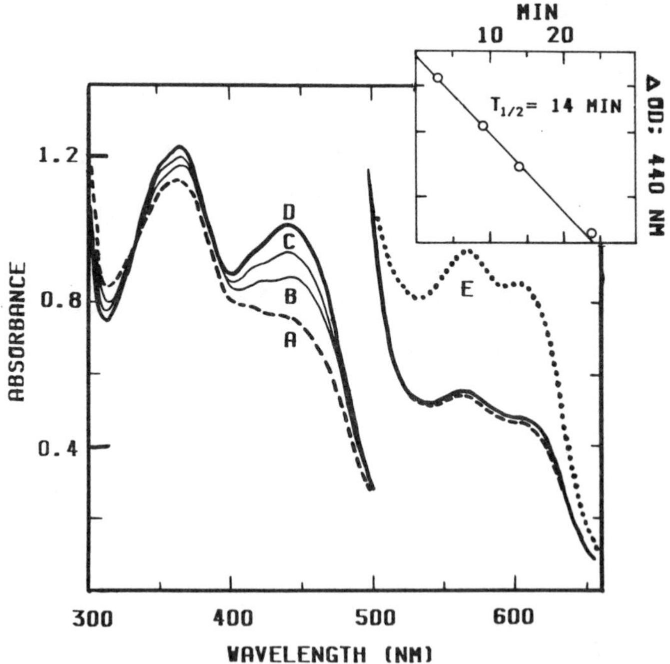

Figure 1. Absorption spectra of the products of the reaction of the luciferase-$FMNH_2$ complex with oxygen, and changes with time. *Beneckea harveyi*, (mutant strain M-17) luciferase (8×10^{-5} M), FMN(10^{-4} M), and EDTA (10^{-3} M), in phosphate buffer (0.2 M, pH 7) was made anaerobic by degassing, and then irradiated for 3 min to achieve full reduction of the flavin, as judged by the disappearance of the FMN_{ox} fluorescence. Oxygen was then admitted, and the first spectrum was recorded after 1 min (trace A). The absorption of the flavin peroxide peaks at about 370 nm, and its decay to oxidized FMN over the next 25 min (inset) was monitored by the increase at 440 nm (traces B, C, and D). The absorption of the blue species in the 500–700 nm range (scale at right, one division =0.05) was already present at the time trace A was recorded, and remained essentially constant during the decay of the peroxide. Repetition of the reduction-oxidation procedure as described resulted in an even greater amount of the blue species (trace E). Inset: ordinate, O.D. at 440 nm, plotted on a log scale; abscissa, time.

ously. In addition, there is absorption in the 300 to 500 nm range, this being identified with the luciferase flavin peroxide species ($_{max}$ =370 nm) previously reported. At 2°, the absorption in the 440 nm region increased ($t_{1/2}$ for increase to final value, 14 min), indicative of the decay of the peroxy flavin to form oxidized FMN (1). During this period, the absorption at 600 nm did not decay; in fact it increased slightly, so that after 26 min the OD at 560 nm was about 0.13. Subsequent to this, the photoreduction-oxidation cycle was repeated; this resulted in the formation of even more radical (Figure 1, trace E).

This result clearly differs from previous reports in two respects. In the earlier reports, there was no absorption in the 500 to 700 nm range, and the peroxide intermediate had a longer life time (40 to 50 min) at 2° (1,4). But such a behavior was indeed duplicated in the course of the present experiment by loading a dithionite (or photo-) reduced FMN-luciferase complex directly on a Sephadex G-25 column, and allowing oxidation to occur during separation (3). In this way, the luciferase-bound reduced flavin encountered oxygen only after its separation from small molecules, notably oxidized flavins, and no blue species was generated. When a similar preparation was allowed to react with molecular oxygen in the test tube for approximately 3 min prior to application to the column, a preparation was obtained with an absorption similar to that shown in Figure 1.

Similar amounts of the blue species were obtained with preparations reduced by dithionite, followed by aeration. When a sample prepared by this method was aged at 2° (thereby allowing the peroxide to decay), and then chromatographed on a short Sephadex G-25 column, a preparation was obtained containing essentially homogeneous blue species. Its spectrum (Figure 2) has maxima at 610, 568, 470, 384, and 336 nm, and is very similar to the spectra of neutral (blue) flavin radicals in apolar solvent (8), and in particular to the spectrum of the radical of *Azotobacter vinelandii* flavodoxin (7). From this and

Figure 2. Absorption spectrum of the luciferase neutral (blue) radical and its decay product, enzyme-FMN_{ox}. The radical was prepared as follows: FMN, 2×10^{-4} M, in 0.5 ml 0.4 M phosphate buffer, pH 7 was reduced at 0° in the presence of 8×10^{-5} M luciferase with a small excess of dithionite in the same buffer. The mixture was then oxygenated, and the absorbance at 600 nm was measured. The formation of some additional absorbance at 600 nm occurred over the next 20 minutes, at which time the sample was chromatographed on a Sephadex G-25 column (void volume 4.5 ml) equilibrated with the same buffer. The protein fraction was collected, and its spectrum recorded immediately (---). (–·–·) shows the spectrum of the same sample at the end of the decay process, after approximately 12 hr. The inset shows the time course of this decay 11°.

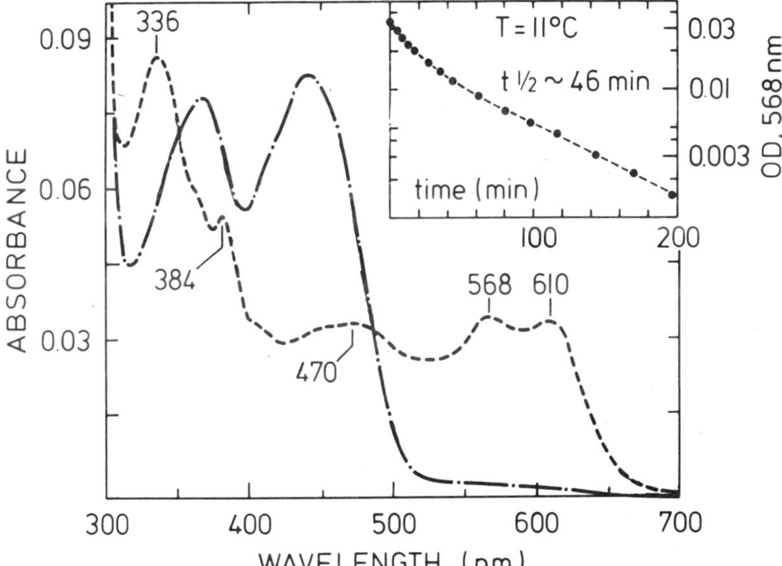

the results described below it was deduced that the blue species is a luciferase semiquinone radical. The radical nature of the blue species is confirmed by the finding of an essentially quantitative ESR radical signal, which decreases concomitantly with the disappearance of the 600 nm absorption. An identical blue radical can be generated by titration of the FMN_{ox}-luciferase complex with dithionite, by conproportionation of equimolar concentrations of FMN_{ox}, and FMN_{red} in the presence of luciferase, and also, presumably, as in preparations of Figures 1 and 2, by 1-e^- oxidation of the FMN_{red}-luciferase complex (Scheme 1).

The radical species is metastable, and decays isobestically to yield FMN_{ox} (Figure 2), the rate of the process being biphasic (Figure 2, inset), and highly temperature-dependent (Arrhenius activation energy, 40 kcal/Mole). Addition of decanal (a substrate in the luminescence reaction, reacting with the peroxide) to the radical generates no light emission. On the contrary, the stability of the radical is increased approximately threefold; the activation energy of the decay, however, remains the same. The high activation energy suggests that a large conformational change in the luciferase is involved in the decay of the flavin radical. The luciferase resulting from this decay possesses full activity.

Scheme 1. Proposed modes of formation and decay of the luciferase neutral (blue) radical. The reaction of the luciferase-$FMNH_2$ complex with oxygen to form the peroxide, and the reaction of the latter with long-chain aldehyde represent the catalytic pathway. $1e^-$ oxidation of the luciferase-reduced FMN complex (by, e.g., dithionite, EDTA-hν, or by excess FMN_{ox}) can form the radical (upper structure). It can similarly be formed by $1e^-$ reduction of the luciferase-FMN_{ox} complex. A further way of generation can consist of the enzyme binding preferentially $HFMN^{\cdot}$, which is formed by conproportionation of the oxidized and reduced FMN species. Direct interconversion of the luciferase peroxide and the radical does not appear to occur, at least not on a rapid time scale relevant for catalysis.

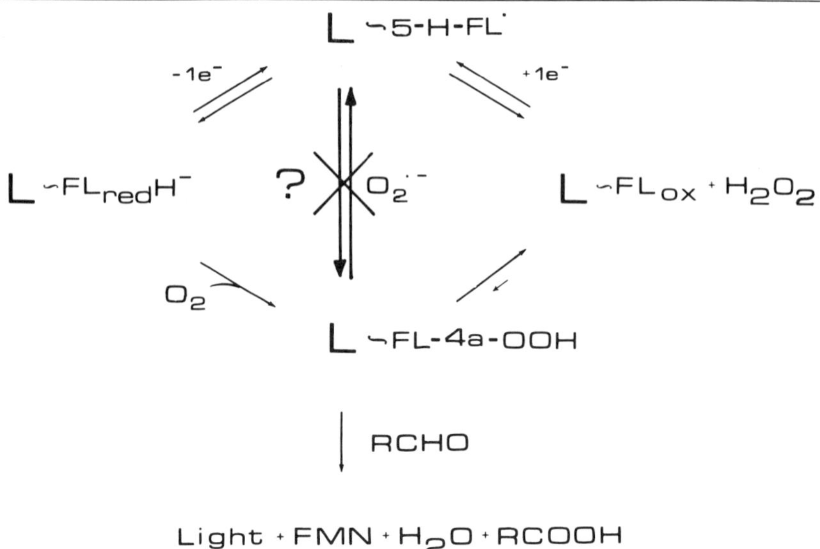

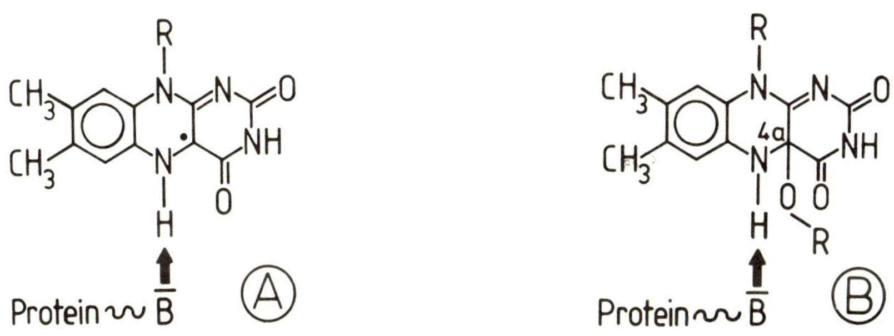

Scheme 2. Proposed mode of stabilization of the luciferase neutral radical, and luciferase-flavin 4a-intermediates. Structure (B) shows the hydrogen bridge which serves in the stabilization (i.e., prevention of elimination of R-O-H) of the peroxide, and hydroxy intermediates. The same interaction causes a stabilization of a neutral (blue) radical (A).

No luciferase radical is formed upon the decay of pure peroxyflavin, the latter prepared as described above. It is thus apparent that the blue neutral luciferase radical is not being formed from the peroxy species, and is not in a (relevantly rapid) equilibrium with it (Scheme 1). In fact, if such an equilibrium did exist, superoxide dismutase would effect it, this in contrast to the experimental findings, which show that superoxide dismutase or catalase have no effect on the decay rates of either the peroxide or the radical.

Conclusions

The mode of formation and decay of the luciferase blue radical, as well as its unusual stability, precluded any role of this species in the light-emitting catalysis. Its high stability however cannot be fortuitous. A comparison of the structure of such a radical [Scheme 2, structure (A)], with that of the transient species occurring during catalysis, i.e., the 4a-peroxide, and the 4a-hydroxy-FMN derivatives [Scheme 2, Structure (B)], indicate that in both cases the same mechanism of stabilization may be operative. This must consist of a strong interaction (hydrogen bridge) between the flavin (5) nitrogen, and a protein function. Thus, the occurrence of the blue radical is interpreted as reflecting a key property of the luciferase active center.

ACKNOWLEDGMENTS
We thank Dr. P. Hemmerich for fruitful discussions, and the Deutsche Forschungsgemeinschaft for support to S.G. (Project Gh 2/3), and NSF for support to J.W.H. (PCM 80-19504). J.W.H. was an awardee of the Alexander von Humboldt Foundation.

References

1. Becvar, J.E., Tu, S.C., and Hastings, J.W. (1978) *Biochemistry* 17:1807–1812.
2. Ghisla, S., Hastings, J.W., Favaudon, V., and Lhoste, J.M. (1978) *Proc Nat Acad Sci* 75:5860–5863.

3. Hastings, J.W., Presswood, R., Ghisla, S., Kurfuerst, M., and Hemmerich, P. (1981) In *Proceedings of the Second International Congress on Chemiluminescence and Bioluminescence.* M. DeLuca and W.D. McElroy (eds.) New York: Academic Press, pp. 403–408.
4. Hastings, J.W., Balny, C., Le Peuch, C., and Douzou, P. (1973) *Proc Natl Acad Sci* 70:995–999.
5. Kosower, E.M. (1980) *Biochem Biophys Res Commun* 92:356–364.
6. Mager, H.J.D. and Addink, R. (1979) *Tetrahedron Letters* 37: 3545–3548.
7. Massey, V. and Palmer, G. (1966) *Biochemistry* 5:3181–3189.
8. Mueller, F., Bruestlein, M., Hemmerich, P., Massey, V., and Walker, W.H. (1972) *Eur J Biochem* 25:573–580.
9. Presswood, R. and Hastings, J.W. (1978) *Biochem Biophys Res Commun* 82:990–996.
10. Presswood, R. and Hastings, J.W. (1979) *Photochem Photobiol* 30:93–99.
11. Shannon, P., Presswood, R.P., Spencer, R., Becvar, J.E., Hastings, J.W., and Walsh, C.T. (1978) In *Mechanisms of Oxidizing Enzymes.* T.P. Singer and R. Ondarza (eds.) Elsevier, North Holland, pp. 69–78.
12. Wessiak, A., Trout, G.E., and Hemmerich, P. (1980) *Tetrahedron Letters* 21:739–742.

Published 1982 by Elsevier North Holland, Inc.
Vincent Massey and Charles H. Williams, Editors
Flavins and Flavoproteins

CHAPTER 59

Properties of Oxygenated Bacterial Luciferase Intermediates Formed With Different Flavins

Shiao-Chun Tu

Department of Biophysical Sciences, University of Houston, Houston, Texas

The bacterial bioluminescent monooxygenation of long-chain aliphatic aldehyde and $FMNH_2$ involves a long-lived luciferase-oxygenated flavin species as a key intermediate. This intermediate (designated II),[1] formed by the reaction of enzyme-$FMNH_2$ with O_2, has been isolated and characterized at $-20°C$ in 50% ethylene glycol-phosphate buffer (1–4). Based on spectroscopic properties, the intermediate II was proposed to be a luciferase-4a-hydroperoxyFMN species (1, 2, 4). II has also been isolated at 0°C in neutral phosphate buffer and found to exhibit full bioluminescence activity upon reacting anaerobically with decanal (5). Since II decayed appreciably at 0°C, detailed chemical and spectral studies could not be satisfactorily carried out at this temperature.

We have found that II can be markedly stabilized at 0°C by long-chain aliphatic alcohols, carboxylic acids, and methyl esters of the acids (Table 1). II was prepared at 23°C in the absence or presence of saturating (20–200 μM) nonaldehyde aliphatic compounds (6). Aliquots (50 μl) withdrawn were each added to 1 ml aerobic solution containing 0.26 mM decanal to initiate the bioluminescence reaction. The stability of the intermediate was determined by following the rate of decrease in bioluminescence potential. None of the nonaldehyde aliphatic compounds tested reacted with II to emit light. The association of these compounds with II is apparently reversible; all complexes exhibited full bioluminescence light yields when reacted with decanal. Aliphatic alcohols were generally more effective than acids and acid methyl esters in stabilizing II. Since 4a-hydroperoxyflavins are highly labile in aqueous solution (7), the stabilizing effects of these aliphatic compounds are likely attributable to an increased hydrophobicity at the luciferase active site that resulted from the binding. The binding of substrates and nonsubstrate effectors to flavohydroxylases has been known to enhance the rate of reduction of bound flavin and, in some cases, the subsequent oxygenation of reduced flavin (8). Our

[1]Abbreviations: II = oxygenated luciferase intermediate formed with $FMNH_2$; CP = ω-carboxypentylflavin; DA = 2′,3′-diacetylFMN; RF = riboflavin; CPH_2, DAH_2, and RFH_2 = reduced CP, DA, and RF, respectively; CPII, DAII, and RFII = oxygenated luciferase intermediates formed with CPH_2, DAH_2, and RFH_2, respectively.

Table 1. Decay Rates of Aliphatic Compound-II Complexes[a]

Aliphatic compound	K (min^{-1})
None	4.33
$CH_3(CH_2)_7OH$	0.12
$CH_3(CH_2)_9OH$	0.21
$CH_3(CH_2)_{11}OH$	0.04
$CH_3(CH_2)_{13}OH$	0.09
$CH_3(CH_2)_8COOH$	1.73
$CH_3(CH_2)_{10}COOH$	1.73
$CH_3(CH_2)_{12}COOH$	0.26
$CH_3(CH_2)_8COOCH_3$	0.14
$CH_3(CH_2)_{10}COOCH_3$	0.63
$CH_3(CH_2)_{12}COOCH_3$	1.28

[a]Determined at 23°C in 0.1 M phosphate, pH 7.

finding with luciferase (6) provides the first demonstration that a nonsubstrate effector can stabilize the oxygenated flavin intermediate.

The decay of oxygenated luciferase intermediates formed with several reduced flavins could all be effectively retarded by dodecanol (Table 2). Taking advantage of this effect, oxygenated flavin intermediates were prepared at 0°C in 0.1 M phosphate, pH 7, containing saturating dodecanol (6, 9). Samples were applied to a Sephadex G-25 column and eluted at 0°C with the dodecanol-containing buffer. Following this procedure, II, CPII, and DAII have been isolated (~10–20 μM) as dodecanol complexes. Very little RFII-dodecanol was obtained by the same method but RFII was similarly isolated (~3 μM) at −22°C in 40% ethylene glycol-0.1 M pH* 7. The absorption spectra of these four isolated intermediates are essentially the same as that of II at −20°C (1).

The yields of quantum, H_2O_2, and oxidized flavin of these intermediates have been determined (Table 3). About equal molar quantities of oxidized flavin and H_2O_2 were recovered, with very little light emission, from II-, CPII-, and DAII-dodecanol as decay products in the absence of aldehyde. With

Table 2. Effect of Dodecanol on the Decay Rate of Oxygenated Luciferase Intermediates Formed with Different Flavins[a]

	k($10^3 \times min^{-1}$)	
Flavin	**− Dodecanol**	**+ Dodecanol**
FMN	63	2
ω-Carboxypentylflavin	347	9
3-CarboxymethylFMN	50	5
2-ThioFMN	50	2
Riboflavin	41	17
2′, 3′-DiacetylFMN	28	5

[a]Determined at 0°C in 0.05 M phosphate, pH 7, containing zero or saturating levels of dodecanol.

Table 3. Yields of Products of *In Vitro* Bacterial Bioluminescence Reaction in the Absence and Presence of Decanol.

Starting reactant[a]	Decanol	H_2O_2/flavin	Quantum yield
II-dodecanol	−	0.97	0.008
	+	0.17	0.15
CPII-dodecanol	−	0.99	0.001
	+	0.16	0.077
DAII-dodecanol	−	1.18	0.0002
	+	0.59	0.021
RFII	+	—	0.029
$FMNH_2$	+		0.16
CPH_2	+		0.03
DAH_2	+		0.0003
RFH_2	+		0.0006

[a]Aliquots (0.5 ml) of II-, CPII-, and DAII-dodecanol were each added at 23°C to 0.7 ml 0.1 M phosphate, pH 7, containing zero or 0.6 μmol decanol for measurements of quantum output. Subsequently, the flavin and H_2O_2 contents were quantitated (9). A l-ml RFII sample was allowed to decay and the flavin content was determined. Aliquots (50μl) of RFII were also each injected into decanol-containing buffer at 23°C for quantum yield measurements. The H_2O_2 yield was not determined due to interference by ethylene glycol. The quantum yields of all intermediates are based on the flavin contents. The light yields were also determined at 23°C by injecting 1 ml of a decanol-containing buffer to 1 ml solution containing luciferase and different amounts of reduced flavin. The quantum yields shown are values determined by extrapolation to infinite flavin concentration and based on the limiting luciferase content. One luciferase molecule binds one reduced flavin (10).

decanal added in excess, the light yields were enhanced to different levels. Correspondingly, the production of H_2O_2 was reduced substantially for II- and CPII-dodecanol but only partially for DAII-dodecanol. The quantum yields determined at saturating levels of CPH_2, DAH_2, and RFH_2 are substantially lower than those of the corresponding oxygenated intermediates (Table 3). However, no significant difference in the light output was observed between $FMNH_2$ and intermediate II as the starting reactants. It appears that the coupling of luciferase monooxygenation and oxidation activities is subject to controls by flavins at the stage of bound reduced flavin reacting with O_2 and that of oxygenated intermediate reacting with aldehyde.

ACKNOWLEDGMENTS
This work was supported by Robert A. Welch Foundation Grant E-738 and NIH Grant GM25953.

References

1. Hastings, J.W., Balny, C., LePeuch, C., and Douzou, P. (1973) *Proc Natl Acad Sci USA* 70:3468.
2. Balny, C. and Hastings, J.W. (1975) *Biochemistry* 14:4719.
3. Hastings, J.W. and Balny, C. (1975) *J Biol Chem* 250:7288.
4. Ghisla, S., Hastings, J.W., Favaudon, V., and Lhoste, J.-M. (1978) Proc Natl Acad Sci USA 75:5860.
5. Becvar, J.E., Tu, S.-C., and Hastings, J.W. (1978) *Biochemistry* 17:1807.

6. Tu, S.-C. (1979) *Biochemistry* 18:5940.
7. Kemal, C. and Bruice, T.C. (1976) *Proc Natl Acad Sci USA* 73:995.
8. Massey, V. and Hemmerich, P. (1975) *Enzymes*, 3rd Ed. 12:191.
9. Tu, S.-C. (1981) *Proceedings of the Second International Congress of Chemiluminescence and Bioluminescence*. DeLuca, M. and McElroy, W.E. (eds.) New York: Academic Press, pp. 161–167.
10. Becvar, J.E. and Hastings, J.W. (1975) *Proc Natl Acad Sci USA* 72:3374.

Published 1982 by Elsevier North Holland, Inc.
Vincent Massey and Charles H. Williams, Editors
Flavins and Flavoproteins

CHAPTER 60

Interaction of Bacterial Luciferase with Long-Chain Fatty Acid[1]

Takao Nakamura[2]

Department of Biology, Faculty of Science, Osaka University, Toyonaka, Osaka 560, Japan

Summary

Long-chain fatty acid is one of the reaction products and a competitive inhibitor in the bioluminescent reaction of bacterial luciferase. The affinity and mode of interaction of luciferase with saturated fatty acids of various chain length (C_{10}–C_{26}) were investigated by kinetic and spectroscopic methods including the measurement of EPR spectrum of stearic acid spin-label bound to luciferase. The value of $-\log K_i$ (K_i:inhibitor constant) of the fatty acids increased with an increase of the chain length and approached a plateau value of 5.4 at the carbon number of C_{24}. It was suggested that the binding of the alkyl chain of stearic acid to luciferase occurs along the entire length of the chain, and the binding affinity of the methylene group in stearic acid is maximal at positions 16-C and 17-C of the alkyl chain.

Introduction

The bioluminescent reaction of bacterial luciferase requires $FMNH_2$, O_2, and a long-chain aliphatic aldehyde as substrates, and the affinity ($-\log K_m$ (Michaelis constant of the bioluminescent reaction)) of the aldehyde for luciferase increases with an increase of the chain length (1,2). It was shown (3,4) that luciferase was inactivated by chemical modification of a single sulfhydryl group in a hydrophobic region of the enzyme molecule, and the sulfhydryl group was more effectively protected against modification by aldehydes of longer chain length. It was also indicated by studies on ^{13}C NMR spectra of dodecanol in luciferase (5) that the aldehyde was bound to the enzyme molecule along the entire length of the alkyl chain. All these results suggest a hydrophobic binding site in the luciferase molecule, which provides a better fit for aldehydes of longer alkyl chain. However, in previous work, the affinity of the aldehyde for luciferase has not been determined for those of chain length of more than 16. One reason for this is that the aldehydes of

[1]Abbreviations: St*(12,3): 2-(3-carboxypropyl)-4,4-dimethyl-2-tridecyl-3-oxazolidinyloxyl; St*(5,10): 2-(10-carboxydecyl)-2-hexyl-4,4-dimethyl-3-oxazolidinyloxyl; St*(1,14): 2-(14-carboxytetradecyl)-2-ethyl-4,4-dimethyl-3-oxazolidinyloxyl; S: order parameter.

[2]Present address: Department of Molecular Physiology, National Cardiovascular Center Research Institute, 5-125 Fujishiro-dai, Suita, Osaka 565, Japan.

longer chain length rapidly polymerize and become unreactive with luciferase. In the luciferase reaction, aldehyde is oxidized to the corresponding fatty acid (6,7). Long-chain fatty acids inhibit the bioluminescent reaction competitively with aldehyde by reversible binding to the aldehyde-site in luciferase (8). In the present investigation, we studied the mode of interaction of long-chain (C_{10}–C_{26}), saturated fatty acids including spin-labelled stearic acid with the aldehyde-binding site in luciferase by kinetic and spectroscopic methods. All the spectrophotometric, EPR, and bioluminescence measurements were conducted at 25°C in 0.05 M K-phosphate buffer, pH 7.0. A luciferase from *Photobacterium phosphoreum* (9) was used for the experiments.

Results

Inhibitory Effect of Long-Chain Fatty Acids in the Luciferase Reaction

The bioluminescent reaction of luciferase was carried out in the absence and presence of various concentrations of saturated fatty acids of various chain length, and the I_o value (peak luminescence intensity) measured from a Dixon plot against fatty acid concentration. As has already been reported for decanoic, dodecanoic, and tridecanoic acids (8), a competitive type inhibition of the bioluminescent reaction was demonstrated with stearic acid and all other fatty acids (C_{10}–C_{26}) tested. K_i was determined in the Dixon plot for each fatty acid and pK_i values ($-\log K_i$) calculated are summarized in Figure 1. Taking K_i as the dissociation constant of the luciferase-fatty acid complex, the increment of Gibbs energy change (ΔG) for binding of fatty acid with luciferase, divided by the increment of carbon number of the fatty acid, was calculated from the adjoining pK_i values in the figure and presented in the same figure. Spin-labelled stearic acid also inhibited the luciferase reaction, and the values of K_i obtained were 10 μM for St*(12,3), 12 μM for St* (5,10), and 15 μM for St*(1,14), respectively.

Spectrophotometric Characteristics of the Luciferase-Stearic Acid Complex

UV fluorescence emission and absorption spectra of a luciferase solution (5.06 μM) were examined in the absence and presence of 98 μM stearic acid. A 10 mM ethanolic solution of stearic acid was dissolved in 0.05 M K-phosphate buffer, pH 7.0, to give a final concentration of 500 μM which is just under the critical micelle concentration of K-stearate at 25°C (10), and used for spectrophotometric measurements. Since K_i for stearic acid is 10 μM, the enzyme was calculated to be 90% in stearate-bound state under this experimental condition. No change in the fluorescence emission spectrum in the range of 300–400 nm (excited at 288 nm) or absorption spectrum in the range of 240–340 nm was detected in the recorded spectra on addition of stearic acid. CD spectra of a luciferase solution (1.22 μM) were also recorded in the absence and presence of 100 μM stearic acid with which the enzyme was calculated to be 90% in stearate-bound state. Only negligibly small change in the spectrum in the range of 185–260 nm was observed on binding of stearic acid to the enzyme. From the recorded spectrum, molecular ellipticity of luciferase at 222 nm was

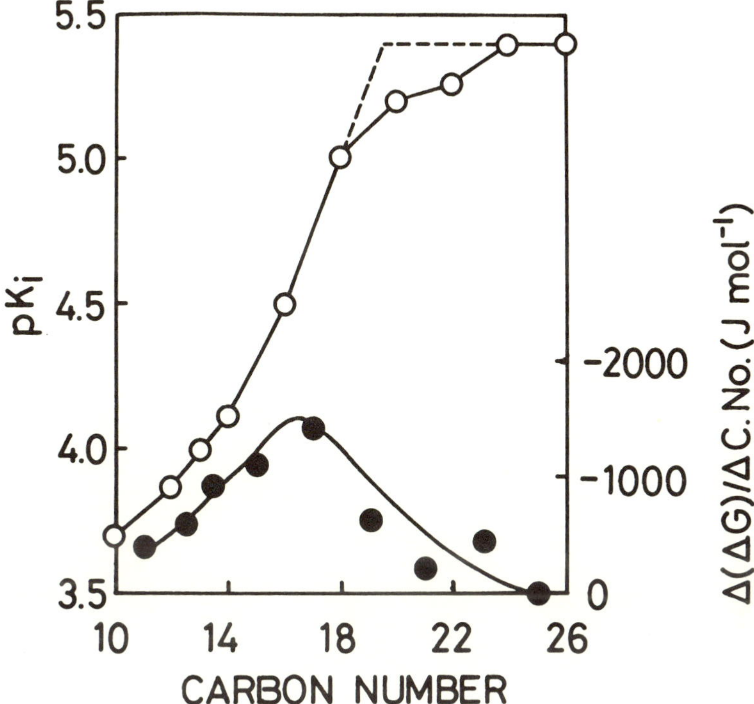

Figure 1. pK_i values of saturated fatty acids in the luciferase reaction. pK_i values were determined in Dixon plots and plotted against carbon number of the fatty acid (○). The increment of Gibbs energy change for binding of fatty acid with luciferase divided by the increment of carbon number, $\Delta(\Delta G)/\Delta C.No.$, calculated from adjoining pK_i values in this figure, is also plotted (●).

determined as -10.9×10^3 degrees cm^2 $decimole^{-1}$ on the basis of published amino acid composition of luciferase (11). The α-helix content of luciferase was calculated to be 31%.

EPR Studies

The EPR spectrum of a solution of a stearic acid spin-label and luciferase was examined. Since the binding of stearic acid with luciferase is reversible, the EPR spectrum of stearic acid spin-label could not be recorded at the 100% luciferase-bound state. Experimental conditions were chosen so that as much as possible of the spin-label was in the enzyme-bound state, using the highest possible concentration of luciferase in the enzyme-stearic acid mixture. EPR spectra of stearic acid spin-label were recorded in a mixture of 58 μM of St*(12,3), 38 μM of St*(5,10), or 25 μM of St*(1,14) with 125 μM of luciferase. In these conditions, 88% of stearic acid spin-label was calculated to be in the luciferase-bound state. EPR absorption bands due to luciferase-bound and free stearic acid spin-label were clearly distinguished in the recorded spectrum (Figure 2). A 12% portion of the intensity of a control spectrum,

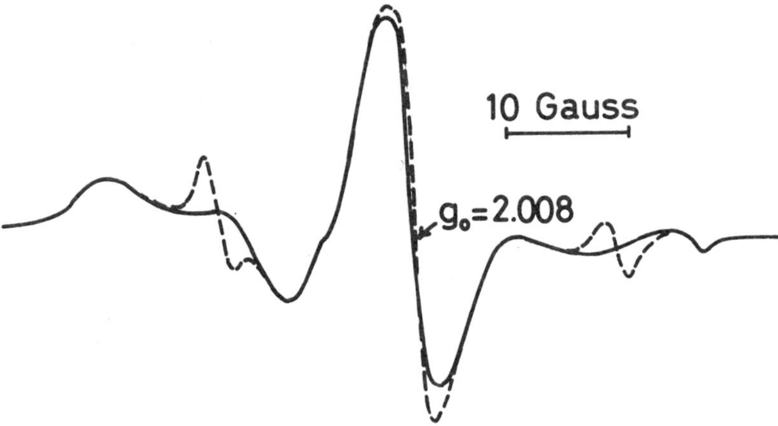

Figure 2. Derivative curve of EPR spectrum of stearic acid spin-label St*(12,3) bound to luciferase. The EPR spectrum (---) was recorded with a sample containing 58 μM St*(12,3) and 125 μM luciferase, total concentrations. In this sample, the stearic acid spin-label was 88% in luciferase-bound and 12% in free states. A 12% portion of the intensity of a control spectrum recorded with the same concentration of the spin-label but without luciferase, was graphically subtracted and the resultant difference spectrum is shown (—). Modulation frequency, 100 kHz, modulation amplitude, 2.5 Gauss. In the difference spectrum, apparent hyperfine splitting $T'_{\parallel}$ and $T'_{\perp}$ were estimated as 24.4 and 9.5 Gauss, respectively.

which was recorded under identical instrumental conditions with the same concentration of stearic acid spin-label but without luciferase, was graphically subtracted from the obtained spectrum, and $T'_{\parallel}$ and $T'_{\perp}$ (apparent hyperfine splittings) were estimated in the resultant difference spectrum (solid curve in Fig. 2). Approximate values of the order parameter S (12) were calculated from $T'_{\parallel}$ and $T'_{\perp}$ using Gaffney's equation (13). Results obtained are summarized in Table 1. Slightly larger S values were observed when the spin-label was at the 12-C [St*(5,10)] or at 16-C [St*(1,14)] position than when it was at position 5-C [St*(12,3)].

Discussion

In the present work, the affinity (pK_i) of long-chain, saturated fatty acids with luciferase was determined by a kinetic method. Results as shown in Figure 1 indicate that the affinity increases with an increase of the chain length, most markedly from C_{14} to C_{18} acids, and approached a plateau value of 5.4 with C_{24} and C_{26} acids. Total length of the binding site for the alkyl chain of the fatty acid in luciferase may be graphically estimated as the intercept of the dotted lines in Figure 1, and the intercept corresponded to about 23 Å. Assuming that the carboxyl group of the long-chain fatty acid is bound in luciferase at the same site as that for binding of the carbonyl group of the substrate aldehyde, this figure estimates the total length of the hydrophobic binding site for the aldehyde in luciferase.

Table 1. Order Parameter S of Stearic Acid Spin-Label in Luciferase.

Spin-label	S
St*(12, 3)	0.53
St*(5, 10)	0.63
St*(1, 14)	0.61

When $FMNH_2$ was bound to luciferase, marked changes in UV fluorescence and absorption spectra of the enzyme were observed (14, 15). In contrast to this, bound stearic acid did not cause an appreciable change in the UV fluorescence, absorption, or CD spectrum of luciferase, suggesting that the microenvironment of aromatic amino acid residues in luciferase or backbone structure of luciferase do not change on binding of stearic acid. The increment of ΔG for binding of fatty acid with luciferase, divided by the increment of the carbon number may represent the individual contribution of the methylene group for the binding. As shown in Figure 1, maximal contribution ($-1{,}400$ J $mole^{-1}$) was observed by those which correspond to 16-C and 17-C of the fatty acid. This is in accordance with the observations (1) that stearic acid which has the spin-label group in the alkyl chain at farther positions from the carboxylic group, showed less affinity to luciferase, e.g., $K_i = 10$ μM for St*(12,3), 12 μM for St*(5,10), and 15 μM for St*(1,14), due possibly to a stronger steric hindrance caused by the attached spin-label group when the methylene group was bound to a subsite of better fit, and (2) larger S values were observed when the spin-label was at farther positions from the carboxylic group (Table 1). The latter observation is in contrast to the finding (12) that when incorporated with the membrane of nerve axon, the S value of the spin-label in stearic acid became smaller as the position of the spin-label in the alkyl chain moved toward the methyl terminal, since the local structure of the membrane at the position of the methyl terminal of the incorporated stearic acid is supposed to be more fluid than at the membrane surface.

ACKNOWLEDGMENTS

This study was supported in part by Grant-in-Aid for Special Project Research No. 311808 and Grant-in-Aid for Scientific Research No. 444075 from the Ministry of Education, Science, and Culture of Japan.

References

1. Hastings, J.W., Spudich, J., and Malnic, G. (1963) *J Biol Chem* 238:3100–3105.
2. Watanabe, T. and Nakamura, T. (1972) *J Biochem* 72:647–653.
3. Nicoli, M.Z., Meighen, E.A., and Hastings, J.W. (1974) *J Biol Chem* 249:2385–2392.
4. Nicoli, M.Z. and Hastings, J.W. (1974) *J Biol Chem* 249:2393–2396.
5. Viswanathan, T.S., Campling, M.R., and Cushley, R.J. (1979) *Biochemistry* 18:298–304.
6. McCapra, F. and Hysert, D.W. (1973) *Biochem Biophys Res Comm* 52:298–304.
7. Dunn, D.K., Michaliszyn, G.A., Bogacki, I.G., and Meighen, E.G. (1973) *Biochemistry* 12:4911–4918.
8. Yoshida, K. and Nakamura, T. (1974) *J Biochem* 76:985–990.

9. Watanabe, T. and Nakamura, T. (1976) *J Biochem* 79:489–495.
10. Klevens, H.B. (1953) *J Am Oil Chem Soc* 30:74–80.
11. Yoshida, K. and Nakamura, T. (1973) *J Biochem* 74:915–922.
12. Hubbel, W.L. and McConnell, H.M. (1971) *J Am Chem Soc* 93:314–326.
13. Gaffney, B.J. (1976) In *Spin-Labelling*. Berliner, J.J. (ed.) New York: Academic Press, pp. 567–571.
14. Watanabe, T., Tomita, G., and Nakamura, T. (1974) *J Biochem* 75:1249–1255.
15. Watanabe, T., Yoshida, K., Takahashi, M., Tomita, G., and Nakamura, T. (1976) In *Flavins and Flavoproteins*. Singer, T.P. (ed.) Amsterdam: Elsevier Scientific Publishing Co., pp. 62–76.

Published 1982 by Elsevier North Holland, Inc.
Vincent Massey and Charles H. Williams, Editors
Flavins and Flavoproteins

CHAPTER 61

Structural Analysis of Fragments Resulting From Limited Proteolysis of Bacterial Luciferase

Steven K. Rausch, John J. Dougherty, Jr., and Thomas O. Baldwin

Department of Biochemistry and Biophysics, Texas A&M University, College Station, Texas

Bacterial luciferase is exceedingly sensitive to proteases. Any of several proteolytic activities inactivate luciferase by hydrolysis of bonds within a discrete region of the α subunit thought to comprise a portion of the active center (1–3). The active center of luciferase is also thought to contain a highly reactive cysteinyl residue (4,5). The experiments reported here were undertaken to determine the location of the protease sensitive region relative to the primary structure of the α subunit and to determine which fragment(s) resulting from limited proteolysis contain the reactive thiol.

The fragment location of the reactive thiol was determined by fluorographic detection of tritium label attached to the reactive thiol by reaction with N-ethylmaleimide (Figure 1). The fluorographs as well as the stained gels show striking similarities between the patterns generated by trypsin and chymotrypsin, despite the distinctly different specificities of these proteases. At early times, the tritium migrates from α (ca 42000 daltons) into a series of fragments designated γ (ca 28000 daltons). At later times, label can also be seen in the fragment designated δ_L (ca 10000 daltons). The percentage of total label detected in γ reached a maximum after a relatively short time, and then decreased slowly, while the label in δ_L accumulated slowly but steadily.

The fragments designated γ, δ_H, and δ_L (from chymotrypsin) were purified by chromatography on Sephadex G-100 in 8 M urea-containing buffer; fragments γ and δ_H required additional purification by preparative (1.3 cm thick) SDS gel electrophoresis (10). The fragments were subjected to automated Edman degradation; the amino terminal sequences together with intact α subunit are shown in Figure 2. The γ fragment has the same amino-terminal sequence as intact α, demonstrating that γ is derived from the amino-terminal 2/3 of the α subunit. The δ_L fragment contains the reactive thiol, as does γ, and therefore must contain residues in common with γ. The amino terminal 26 residues of δ_L were distinct from the amino terminal sequences of α and γ. The sequence surrounding the reactive thiol (phe-gly-ile-cys-arg; 11) was not found in either γ or δ_L. The δ_H fragment (ca 17000 daltons) does not contain the reactive thiol. Its amino-terminal sequence is distinct from both γ and δ_L.

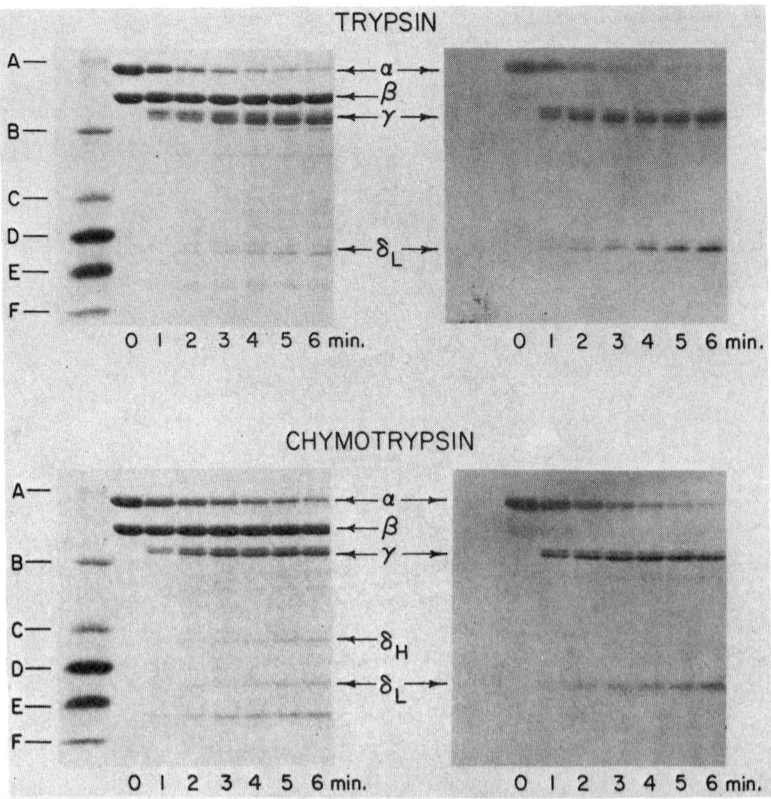

Figure 1. Treatment of ^{3}H-NEM-labelled luciferase with trypsin and chymotrypsin. Luciferase at 3.0 mg/ml, purified from *Vibrio harveyi* according to Gunsalus-Miguel et al. (6) with an extra chromatography step over aminohexyl-Sepharose and labelled with ^{3}H-N-ehtylmaleimide (1.1 Ci/mmol) to an extent of 0.52 mol NEM per mol luciferase was treated with either trypsin (top) or chymotrypsin (bottom) at concentrations of 25 ug/ml. Aliquots were removed at 60-sec intervals and diluted into SDS gel sample buffer at 100°C to stop proteolysis. 7-μg samples of each time point were run on 15% polyacrylamide gels (7), then stained in 1% Fast Green FCF in 25% isopropanol, 10% acetic acid, and de-stained. Lane 1 contains molecular weight standards: (A) ovalbumin (MW 43000); (B) α-chymotrypsinogen (25700); (C) β-lactoglobulin (18400); (D) lysozyme (14300); (E) bovine trypsin inhibitor (6200); (F) insulin (3000). The stained gels appear on the left. These were fluorographed according to the methods of Bonner and Laskey (8) and Laskey and Mills (9). The fluorograph corresponding to its original gel appears on the right.

Figure 3 shows the subunit location of the protease-sensitive sites, and therefore the origin of γ, δ_H, and δ_L, and the location of the reactive thiol as suggested by these data. The δ_L fragment, or a portion of it, must derive from γ, since both contain the reactive thiol. Since the γ fragments appear first, it is clear that the sites of initial attack of peptide bond hydrolysis are about 2/3 the distance from the amino- to the carboxyl-terminus. The δ_L fragment, which appears after the γ fragment, must come from hydrolysis of bonds within a second exposed region about 100 residues from the carboxyl-terminus of γ.

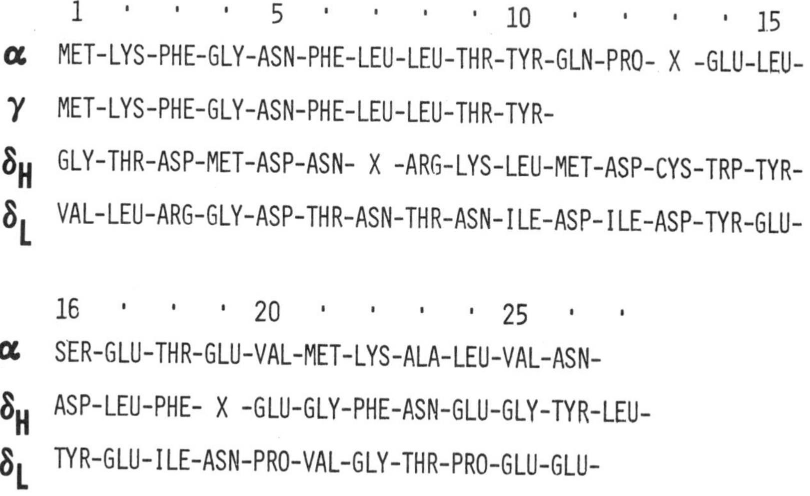

Figure 2. Amino-terminal amino acid sequences of α subunit and fragments γ, δ_H, and δ_L (chymotrypsin). The α sequence is taken from ref. 12. Anilinothiazolinone derivatives from a Beckman Model 890C Sequenator (run with the "Protein Quadrol" program # 122974) were converted to the phenylthiohydantoins by treatment with 1 N HCl for 10 min at 80°C. Following ethyl acetate extraction, the samples were spotted onto fluorescent silica gel plates and chromatographed in the systems of Jeppsson and Sjoquist (13) or in chloroform : isopropanol : xylene : propionic acid (30 : 5 : 2 : 1). Identification was made by comparing spot mobilities with standards under UV irradiation. Whenever possible, redundant analyses were performed on a gas chromatograph (CFC blend, 155° for 16 min, then 8°/min to a final temperature of 270° held for 8 min; carrier flow 60 ml/min).

Figure 3. Model of relative locations of protease-sensitive sites and the reactive thiol in the primary structure of the α subunit. See text for details.

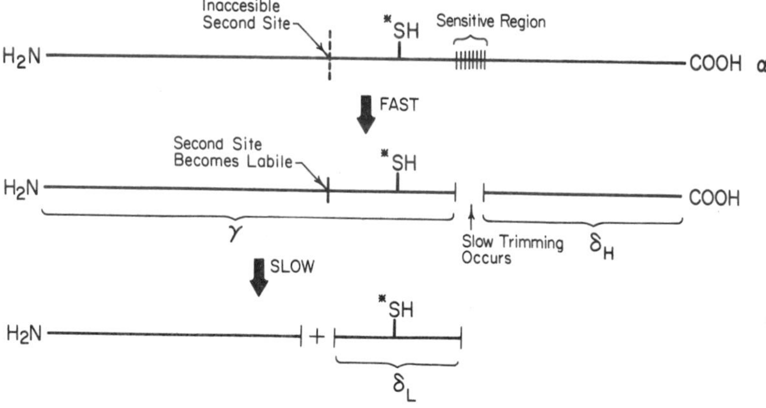

This second clip occurs only after the clip that generates γ has been made. The apparent molecular weight of δ_H, its lack of the reactive thiol, and its amino-terminal sequence suggest that it is derived from the remaining 1/3 of the α subunit liberated with the concomitant formation of γ. This model predicts a fragment without the reactive thiol and having the same amino-terminal sequence as both α and γ generated at the same time as δ_L. There are additional fragments visible in Figure 1 that have not been analyzed. The model also predicts that δ_H should have the same carboxyl-terminal sequence as α; this has not yet been determined.

ACKNOWLEDGMENT
This research was supported by NSF PCM 79-25335.

References

1. Baldwin, T.O., Hastings, J.W., and Riley, P.L. (1978) *J Biol Chem* 253:5551–5554.
2. Holzman, T.F. and Baldwin, T.O. (1980) *Proc Natl Acad Sci USA* 77:6363–6367.
3. Holzman, T.F., Riley, P.L., and Baldwin, T.O. (1980) *Arch Biochem Biophys* 205:554–563.
4. Meighen, E.A., Nicoli, M.Z., and Hastings, J.W. (1971) *Biochemistry* 10:4062–4068.
5. Meighen, E.A., Nicoli, M.Z., and Hastings, J.W. (1971) *Biochemistry* 10:4069–4073.
6. Gunsalus-Miguel, A., Meighen, E.A., Nicoli, M.Z., Nealson, K.H., and Hastings, J.W. (1972) *J Biol Chem* 247:398–404.
7. Laemmli, U.K. (1970) *Nature* 227:680–685.
8. Bonner, W.M. and Laskey, R.A. (1974) *Eur J Biochem* 46:83–88.
9. Laskey, R.A. and Mills, A.D. (1975) *Eur J Biochem* 56:335–341.
10. Hanaoka, F., Shaw, J.L., and Mueller, G.C. (1979) *Anal Biochem* 99:170–174.
11. Nicoli, M.Z. (1972) PhD Thesis, Harvard University.
12. Baldwin, T.O., Ziegler, M.M., and Powers, D.A. (1979) *Proc Natl Acad Sci USA* 76:4887–4889.
13. Jeppsson, J.-O. and Sjoquist, J. (1967) *Anal Biochem* 18:264–269w.

Published 1982 by Elsevier North Holland, Inc.
Vincent Massey and Charles H. Williams, Editors
Flavins and Flavoproteins

CHAPTER 62

Effect of Limited Proteolysis on the Microenvironment Surrounding the Essential Cysteine of Bacterial Luciferase

John J. Dougherty Jr., Steven K. Rausch, and Thomas O. Baldwin

Department of Biochemistry and Biophysics, Texas A&M University, College Station, Texas

The active center of bacterial luciferase resides primarily, if not exclusively, on the α subunit (1–3); the role of the β subunit is unknown, but it is required for bioluminescence activity (4,5). Modification of a specific cysteinyl residue, thought to be located in or near the active center, by any of a variety of sulfhydryl specific reagents renders the enzyme inactive (6–8). Hydrolysis by proteases of a peptide bond(s) within a discrete region of the α subunit, thought to comprise an integral component of the active center, likewise causes inactivation of the enzyme without gross disruption of the quaternary structure of the enzyme (9–13). If these interpretations of the available data are correct, then one would predict that the environment of the reactive thiol should be dramatically altered by the limited action of proteolytic enzymes. To test this hypothesis, we have attached spectroscopic probes to the reactive thiol and analyzed the alterations of the spectroscopic properties of the probes due to limited proteolysis and due to denaturation by urea.

The reaction of bacterial luciferase with N-[p-2(benzoxazolyl)phenyl]-maleimide (NBPM) and the fluorescence properties of the labelled enzyme have been described (14). We have measured an apparent second-order rate constant of 2.53×10^5 M^{-1} min^{-1} for this reaction at 25°C, pH 7.0 (data not shown), in agreement with the rate constants reported for other hydrophobic, N-substituted maleimides (7). The polarization of fluorescence of NBPS-luciferase (Table 1) indicates that the probe is highly immobilized (i.e., not free to rotate) when covalently bound to the protein, in agreement with the observations of nitroxide spin-labelled luciferase (15).

Following limited proteolysis, both the polarization of fluorescence and the fluorescence lifetime of the probe have decreased (Table 1). These observations indicate a decrease in the rotational relaxation time from 58.8 ns to 6.8 ns (trypsin-treated) or 7.8 ns (chymotrypsin-treated). These results demonstrate that the limited action of proteases causes a dramatic increase in the rotational freedom of the NBPS probe, even though it remains attached to the high molecular weight complex.

Table 1. The Effects of Limited Proteolysis and Urea Denaturation On the Fluorescence Polarization and the Fluorescence Lifetime of NBPS-Luciferase.

	P	τ [ns]	ρ [ns][b]
NBPS-LUCIFERASE	0.330 (100.0)[a]	1.41 (100.0)	58.8
TRYPSIN TREATED	0.244 (13.2)	1.12 (25.6)	6.8
CHYMOTRYPSIN TREATED	0.241 (3.2)	1.34 (14.4)	7.8
6 M UREA	0.234 (0)	—	—

Limited proteolysis was carried out on purified luciferase (18) that had been labelled with ca 0.9 moles NBPM/mole of enzyme (Eastman; dissolved in methyl cellosolve) in 50 mM PO_4 (pH 7.0, 25°C) (10). Protein concentrations were then adjusted so that the sample had an $A_{308\ nm} = 0.3$. Fluorescence polarization measurements were performed on a photon counting apparatus (19) and fluorescence lifetimes were obtained using a cross correlation phase fluorimeter (20) with an excitation wavelength of 310 nm, an excitation modulation frequency of 30 MHz, and Corning 0–52 filters at the emission slits (25°).

[a]Parentheses refer to percent activity. The 100% activity reported here corresponds to the ca 10% activity remaining after treatment with N-[p-2(benzoxazolyl)phenyl]maleimide.

[b]The rotational relaxation time, ρ, was calculated from the expression, $(1/P-1/3) = (1/P_0-1/3) \times (1 + 3\tau/\rho)$, by assuming a limiting polarization (P_0) of 0.351 (14).

Figure 1. Protein samples were prepared as described in Table 1. Fluorescence emission spectra were recorded on an instrument of standard design equipped with bipolar averaging (21) and a digital integrator (22). Spectra were obtained using an excitation wavelength of 310 nm and a Corning 0-52 filter at the emission slit (25°C).
[a]The 100% activity reported here corresponds to the ca 10% activity remaining after treatment with the fluorescent maleimide.

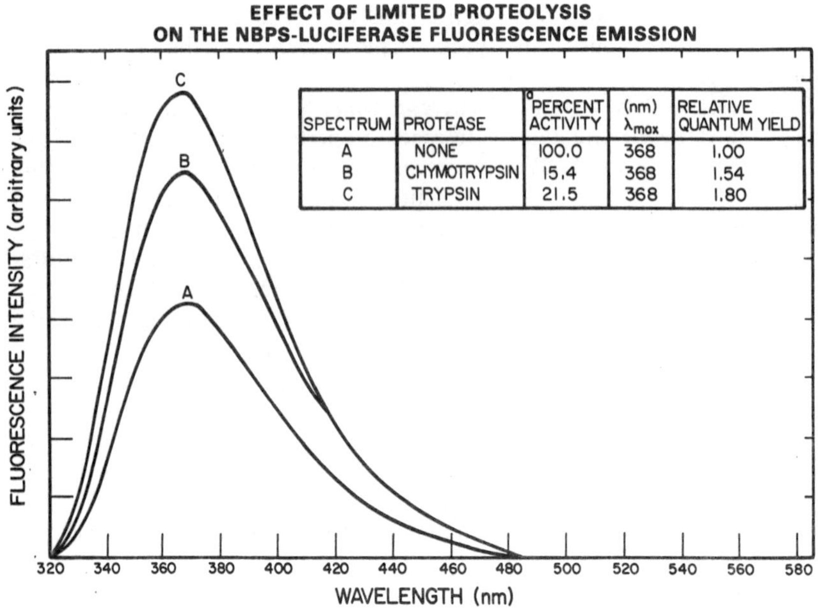

SPECTRUM	PROTEASE	PERCENT ACTIVITY[a]	(nm) λ_{max}	RELATIVE QUANTUM YIELD
A	NONE	100.0	368	1.00
B	CHYMOTRYPSIN	15.4	368	1.54
C	TRYPSIN	21.5	368	1.80

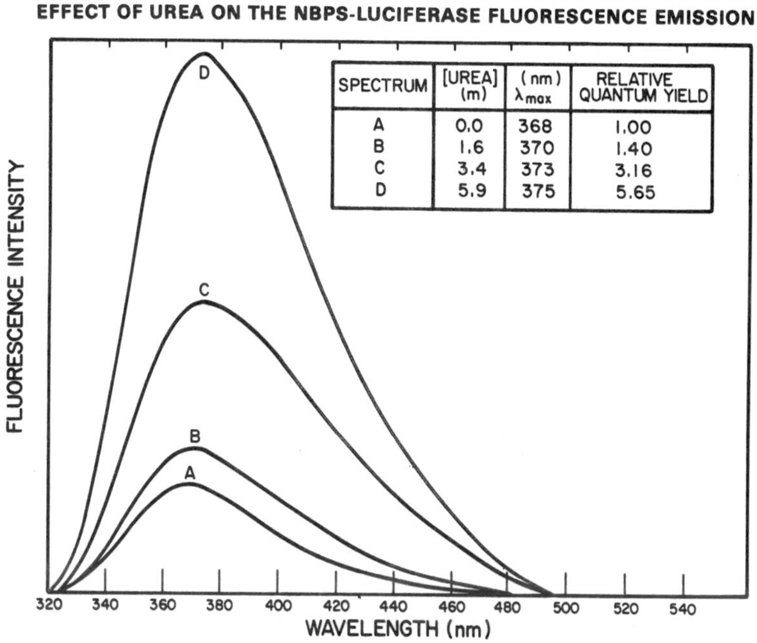

EFFECT OF UREA ON THE NBPS-LUCIFERASE FLUORESCENCE EMISSION

SPECTRUM	[UREA] (m)	(nm) λ_{max}	RELATIVE QUANTUM YIELD
A	0.0	368	1.00
B	1.6	370	1.40
C	3.4	373	3.16
D	5.9	375	5.65

FLUORESCENCE INTENSITY

WAVELENGTH (nm)

Figure 2. Purified luciferase (18) was labelled with ca 0.9 moles NBPM/mole of enzyme and concentrations were adjusted as described in Table 1. The samples were brought to the indicated urea concentrations by the addition of aliquots of an 8 M urea stock solution. Each sample was allowed to incubate for 10 min at 25°C before the emission spectra were recorded as described in Figure 1.

The fluorescence quantum yield of NBPS-luciferase was dramatically increased by limited proteolysis (Figure 1) and by urea denaturation (Figure 2). The increase in the fluorescence quantum yield following limited proteolysis was not accompanied by a change in the emission maximum, while in urea, there was an 8 nm red-shift in the emission maximum. Comparison of these results with the model compound studies reported by Kanaoka et al. (16) suggests that the spectral alterations following limited proteolysis are due to protein conformational changes and not merely to exposure of the fluorescent probe to bulk water.

These studies show that following limited proteolysis the microenvironment of the probe was sufficiently expanded to allow rotational freedom, but the solvent polarity of this region was not significantly altered. The results reported here show that the reactive thiol is close to the protease sensitive site(s) on the α subunit. The fragments resulting from limited proteolysis have been described (17). The reactive thiol resides in the γ fragments (ca 28,000 daltons), which comprise the amino-terminus of the α subunit, and appears upon continued proteolysis in the δ_L (ca 10,000 daltons). These results suggest that the reactive thiol resides close to and on the amino side of the protease-sensitive region that gives rise to the γ fragments.

ACKNOWLEDGMENT

This work was supported by a Grant from the National Science Foundation #PCM 79-25335.

References

1. Meighen, E.A., Nicoli, M.Z., and Hastings, J.W. (1971) *Biochemistry* 10:4062–4068.
2. Meighen, E.A., Nicoli, M.Z., and Hastings, J.W. (1971) *Biochemistry* 10:4069–4073.
3. Cline, T.W. and Hastings, J.W. (1972) *Biochemistry* 11:3359–3370.
4. Meighen, E.A. and Bartlet, I. (1980) *J Biol Chem* 255:11181–11187.
5. Welches, W.R. and Baldwin, T.O. (1981) *Biochemistry* 20:512–517.
6. Nicoli, M.Z., Meighen, E.A., and Hastings, J.W. (1974) *J Biol Chem* 249:2385–2392.
7. Nicoli, M.Z. and Hastings, J.W. (1974) *J Biol Chem* 249:2393–2396.
8. Ziegler, M.M. and Baldwin, T.O. (1981) In *Proceedings of the Second International Congress of Chemiluminescence and Bioluminescence* M. DeLuca and W.D. McElroy (eds.) New York: Academic Press, 155–160.
9. Baldwin, T.O., Hastings, J.W. and Riley, P.L. (1978) *J Biol Chem* 253:5551–5554.
10. Holzman, T.F. and Baldwin, T.O. (1980) *Biochem Biophys Res Comm* 94:1199–1206.
11. Holzman, T.F., Riley, P.L., and Baldwin, T.O. (1980) *Arch Biochem Biophys* 205:554–563.
12. Holzman, T.F. and Baldwin, T.O. (1980) *Proc Natl Acad Sci* USA 77:6363–6367.
13. Ruby, E.G. and Hastings, J.W. (1979) *Curr Microbiol* 3:157–159.
14. Tu, S.C., Wu, C.W., and Hastings, J.W. (1978) *Biochemistry* 17:987–993.
15. Merritt, M.V. and Baldwin, T.O. (1980) *Arch Biochem Biophys* 202:499–506.
16. Kanaoka, Y., Machida, M., Kokubun, H. and Sekine, T. (1968) *Chem Pharm Bull Tokyo* 16:1747–1753.
17. Rausch, S.K., Dougherty, J.J. Jr., and Baldwin, T.O., this volume.
18. Hastings, J.W., Baldwin, T.O., and Nicoli, M.Z. (1978) *Methods Enzymol* 57:135–152.
19. Jameson, D.M. and Weber, G. (1978) *Rev Sci Instrum* 49:510–514.
20. Spencer, R.D. and Weber, G. (1969) *Ann NY Acad Sci* 158:361–376.
21. Wehrly, J.A., Williams, J.F., Jameson, D.M., and Kolb, D.A. (1976) *Anal Chem* 48:1424–1426.
22. Jameson, D.M., Williams, J.F., and Wehrly, J.A. (1977) *Anal Biochem* 79:623–626.

Published 1982 by Elsevier North Holland, Inc.
Vincent Massey and Charles H. Williams, Editors
Flavins and Flavoproteins

CHAPTER 63

Characterization of a Yellow Fluorescent Protein from *Vibrio* (*Photobacterium*) *fischeri*

Gary Leisman and Kenneth H. Nealson

Marine Biology Research Division, Scripps Institution of Oceanography, La Jolla, California

Introduction

Almost thirty years after the components for light emission by luminous bacteria were elucidated (1), the identity of the emitter is still not resolved. Using pure luciferase in a flavin-initiated ($FMNH_2$) assay, the emitter was proposed to be an excited state of some flavin molecule but not FMN itself (2,3). Currently, it is thought that the emitter *in vitro* is a 4a-substituted flavin, possibly the 4a-hydroxy species (4). However, other workers recently proposed that the emitting complex *in vitro* is a dihydroflavin bound to luciferase (5).

It has long been known that the spectrum of light emitted by bacteria *in vivo* is slightly different from that obtained with luciferase *in vitro* (6), suggesting that the emitter *in vivo* might be different from that identified *in vitro*. Lee and coworkers have isolated, from *Photobacterium phosphoreum*, a lumazine protein with a molecular weight of 20,000 that blue shifts the emission maximum by 14–16 nm when added to luciferase *in vitro* (7–10). The chromophore of this protein has been identified as 6,7-dimethyl-ribityl-lumazine (11). The hypothesis is that the lumazine protein sensitizes the bioluminescent reaction in whole cells and thus the lumazine protein is the emitter *in vivo*. This report describes the properties of a yellow fluorescent protein (YFP) isolated from a yellow emitting strain (Y-1) of *Vibrio* (*Photobacterium*) *fischeri* that participates by energy transfer in the production of yellow light by this strain.

Materials and Methods

Strain Y-1 was grown in sea water complete medium at 18° in 2.8 liter flasks that contained 1 liter of medium. After growth to late log phase, the cells were harvested and stored as a pellet at −70°. Purification procedures for luciferase and YFP were carried out at 0–4°. Frozen cells were thawed and lysed by osmotic shock in 10 mM EDTA, 1 mM DTT, pH 7.0. After 10 minutes, the cell debris was removed by centrifugation and discarded. The fraction precipitating between 40 and 70% ammonium sulfate was dissolved in 10 mM $KNaHPO_4$, 1 mM DTT, pH 7.0 and applied to a column of Sephadex G-100 (1.9×35 cm) and eluted with the same buffer. Corrected fluorescence spectra were recorded on a Perkin Elmer MPF 44A spectrofluorimeter equipped with a Hammamatsu R777 photomultiplier tube (now called the R928 with an s-20 red-sensitive spectral response), a thermostated cell holder (at 15°), a differential corrected spectra unit (DCSU-1) and a Hitachi 057 XY recorder. Bioluminescence spectra were recorded with the xenon lamp turned off and are uncorrected. The coupled (NADH-initiated) assay that was used produced light that was constant for at least the several minutes required to record the spectrum. The yellow/blue or Y/B ratio was obtained by dividing the intensity of bioluminescence at 535 nm by the intensity at 485 nm. The ratio is a measure of the relative amount of energy transfer occurring.

Results

The protein YFP isolated from strain Y-1 has a molecular weight of approximately 20,000 based on gel filtration. YFP has fluorescence excitation maxima of 270, 380, and 452 nm for emission at 540 nm and appears to be a flavoprotein.

Figure 1 shows the effect of adding YFP to a luciferase assay. Luciferase has a bioluminescence emission with a maximum of 485 nm and a Y/B ratio of

Figure 1. Bioluminescence assays of luciferase and YFP. Lower trace (YFP): 90 μl YFP fraction from Sephadex G-100+10 μl phosphate buffer, pH 7.0+substrates (decanol −0.01% sonicated suspension, NADH and FMN). Middle trace (Luciferase): 90 μl phosphate buffer+10 μl luciferase fraction from Sephadex G-100+substrates. Upper trace (YFP+Luciferase): 90 μl YFP fraction+10 μl luciferase fraction+substrates. The final volume in each assay was 115 μl, the final NADH concentration was 500 in μM and the final FMN concentration was 6 μM.

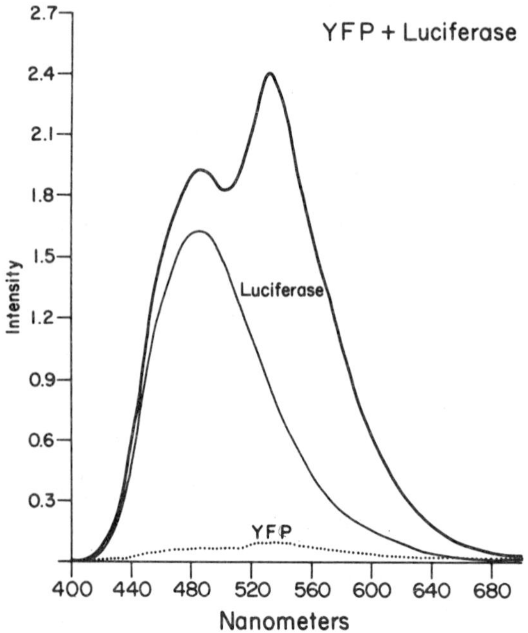

0.46. Upon adding YFP to the assay mixture with the same amount of luciferase and the same final volume, there is a 50 nm red shift in the emission maximum to 535 nm with a shoulder at 485 nm; the Y/B ratio is 2.0.

The color of bioluminescence is greatly affected by the concentration of the extracts (Figure 2). The standard assay with both YFP and luciferase present has a Y/B ratio of 2.0; on 20-fold dilution with buffer, the color changes to blue-green (maximum at 480 nm with a Y/B ratio of 0.66). When the same solution is reconcentrated to the original volume using a Millipore Immersible CX filter (nominal molecular weight cutoff 10,000) and NADH and decanol are added back, the original yellow color and Y/B ratio are observed.

Discussion

Strain Y-1, *Vibrio* (*Photobacterium*) *fischeri* is a natural isolate that produces yellow bioluminescence *in vivo* (12). YFP is apparently required for the yellow light emission since purified luciferase from strain Y-1 emits a blue-green bioluminescence unless YFP is added to the assay. The interaction between YFP and luciferase is temperature-sensitive both *in vitro* (unpublished observa-

Figure 2. Effect of dilution. (A) 270 μl crude extract (25,000×g supernatant)+ substrates (decanol, FMN and NADH). After the first scan (Original), the solution was diluted 20-fold with phosphate buffer (Diluted 1:20). This solution was then concentrated using a Millipore Immersible CX filter with a 50 ml syringe and a one-way valve (Millipore) as the vacuum source (Reconcentrated). (B) The spectrum of light emitted by the diluted solution was recorded on a more sensitive scale to show the blue maximum.

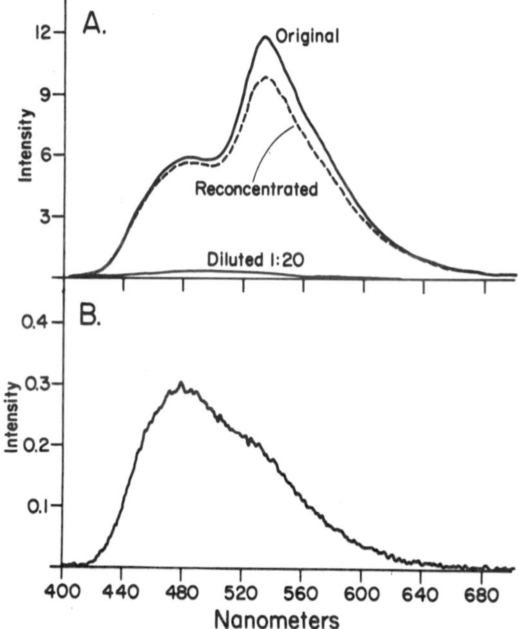

tions) and *in vivo* (13); pH and aldehyde chain length also influence the energy transfer from YFP to luciferase.

Thus, in whole cells and cell-free extracts, depending on the conditions of growth and assay, the emitter can be either YFP or an excited intermediate of luciferase itself. With a 50 nm separation between the emission maxima of the two emitters, the distinction is unambiguous. It seems likely, considering these data and those of Lee and coworkers, that many of the luminous bacteria may have secondary emitters *in vivo*.

ACKNOWLEDGMENTS
We thank Drs. Thomas Baldwin and Frederick I. Tsuji for many useful discussions. This work was supported in part by NSF PCM80-06991.

References

1. Cormier, M.H. and Strehler, B.L. (1953) *J Am Chem Soc* 75:4864–4865.
2. Hastings, J.W. and Nealson, K.H. (1977) *Ann Rev Microbiol* 31:549–595.
3. Baldwin, T.O., Nicoli, M.Z., Becvar, J.E., and Hastings, J.W. (1975) *J Biol Chem* 250:2763–2768.
4. Hastings, J.W., Ghisla, S., Kurfürst, M., and Hemmerich P. (1981) In *Bioluminescence and Chemiluminescence: Basic Chemistry and Analytical Applications*. DeLuca, M.A. and McElroy, W.D. (eds.) pp. 97–102, New York: Academic Press.
5. Matheson, I.B.C., Lee, J., and Müller, F. (1981) *Proc Natl Acad Sci USA* 78:948–952.
6. Seliger, H.J. and Morton, R.A. (1968) In *Photophysiology*. Giese, A.C. (ed.), Vol. 4, pp. 253–314, New York: Academic Press.
7. Gast, R. and Lee, J. (1976) *Proc Int Congr Photobiol Rome*, Abstr, p. 324.
8. Gast, R., Neering, I.R., and Lee, J. (1978) *Biochem Biophys Res Commun* 80:14–21.
9. Gast, R. and Lee, J. (1978) *Proc Natl Acad Sci USA* 75:833–837.
10. Small, E.D., Koka, P., and Lee, J. (1980) *J Biol Chem* 255:8804–8810.
11. Koka, P. and Lee, J. (1979) *Proc Natl Acad Sci USA* 76:3068–3072.
12. Ruby, E.G. and Nealson, K.H. (1977) *Science* 196:432–434.
13. Leisman, G. and Nealson, K.H. (1979) *Am Soc Photobiol* Abstr, 7th Ann. Meeting, p. 70.

PART IV:

Biomedical Aspects of Flavins and Flavoproteins

Published 1982 by Elsevier North Holland, Inc.
Vincent Massey and Charles H. Williams, Editors
Flavins and Flavoproteins

CHAPTER 64

Monoamine Oxidase and Its Reaction with Suicide Substrates

Thomas P. Singer and Mazhar Husain

Molecular Biology Division, Veterans Administration Medical Center, San Francisco, California and Department of Biochemistry and Biophysics, University of California, San Francisco, California

Introduction

Interest in the biochemistry and pharmacology of the flavoprotein monoamine oxidase (MAO) is so wide-spread that for several years now at least one, often two or more international symposia have been held annually on the subject. One reason for this may be that this is one of the few systems where basic research can be directly related, and is clearly relevant, to clinical medicine. The vast literature on the subject makes it, of course, impossible to offer in this limited space more than a brief overview of those aspects of recent research which are apt to interest fellow workers in the flavoenzyme field.

For biochemists who have not been following the literature on MAO actively, one of the most confusing questions must be the difference between the A and B forms of MAO. The literature is replete with claims and hypotheses that the two are isoenzymes, that they are the same protein with different lipids attached, that they are different active sites on the same enzyme, or that they are different proteins coded for by separate genes. Perhaps we might start by clarifying the status of this problem, since much of what follows concerns the different behavior of the two forms.

The most satisfactory definition remains that the acetylenic amine, clorgyline, inhibits the A form at very low concentrations, but blocks MAO B only at much higher concentrations (1), while low concentrations of deprenyl inhibit the B form but block MAO A only at much higher levels (2) (Figure 1). The distinction is even more poignantly shown by studying the binding of "A-specific" and "B-specific" inhibitors (3). The B-selective acetylenic amine, pargyline, binds stoichiometrically to the B enzyme and in so doing quantitatively inactivates it. The same is true of deprenyl. In contrast, while clorgyline is also bound to the B form, the inhibitor must be added in large excess, for only a small fraction is bound to the enzyme but what binding occurs is nonstoichiometric, so that many mols of clorgyline may become bound to the protein with relatively little loss of activity. The converse is true when comparing the binding of clorgyline and pargyline to the A form of MAO (3).

The difference between the A and B forms is also reflected in their substrate specificities. There are typical B substrates, such as benzylamine and phenylethylamine, and typical A substrates, such as 5-hydroxytryptamine. The im-

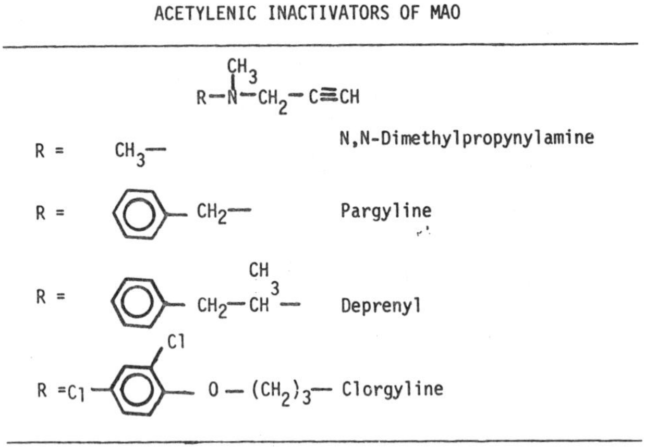

ACETYLENIC INACTIVATORS OF MAO

$R-N(CH_3)-CH_2-C\equiv CH$

R =		
R =	CH_3-	N,N-Dimethylpropynylamine
R =	$C_6H_5-CH_2-$	Pargyline
R =	$C_6H_5-CH_2-CH(CH_3)-$	Deprenyl
R =	$Cl_2C_6H_3-O-(CH_2)_3-$	Clorgyline

Figure 1. Structures of some acetylenic inhibitors of MAO.

portant thing to remember is that this is a quantitative, rather than qualitative, distinction, for virtually all A substrates are slowly oxidized by MAO-B, and B substrates by MAO-A. Thus, there is no rational basis for estimating the relative contents of the A and B enzymes, respectively, in a given tissue from the rates of oxidation of these two classes of substrates, although this fallacious shortcut to knowledge is still used by some workers.

Different mammalian tissues may contain only the A enzyme, only the B form, or both. Thus, human placenta contains the A form but little or no B (4), beef liver only the B form (5), while rat liver and brain tissue from many species contain both forms. The two forms of MAO are thought to have different physiological functions (6). The A form is specialized for the oxidation of hydroxylated amines, such as norepinephrine and serotonin, and, hence, plays an important role in the sympathetic nervous system, while the B form is considered to function primarily in regulating the level of pressor amines, such as phenylethylamine, in the circulation and in tissues, a function shared with the A enzyme to a certain extent.

Since the structural basis of the differences in the properties of the A and B forms of the enzyme is one of the most important problems in this field, the current status of the evidence will be summarized below. First, however, it is necessary to survey the molecular properties of the enzyme.

Molecular Properties of Monoamine Oxidase

Among the two forms of MAO, the B type has been available in nearly homogeneous form from kidney (7) and liver (8,9) for some time. Highly purified preparations of the A form from human placenta have only recently been obtained (10) but the details have not yet appeared.

MAO-B has a pronounced tendency to aggregate and to bind lipids tenaciously. As a result, its molecular weight cannot be easily measured by physical methods. Reliable values have been based on cysteinyl flavin content and [^{14}C]-pargyline binding, another expression of the flavin content, with correction for trace impurities. It is now generally agreed that the molecular weight is 110,000 to 120,000 (7–9). After dissociation of the quarternary structure a single subunit with molecular weight of 52,000 to 57,000 has been found by a variety of methods (9, 11). Thus, MAO-B is made up of two subunits with very similar molecular weights, only one of which contains the flavin site. Interestingly, Pintar et al. (12) reported that MAO-A from the same cell line has essentially identical charge and molecular weight as the B form and the respective subunits migrate also identically in denaturing gels.

8 α-S-Cysteinyl FAD is the prosthetic group of both A and B forms of the enzyme (13, 14). In fact, the peptide sequence at the flavin site of the A enzyme from human placenta is the same as that of the B form from bovine liver (15).

An important question which has recently been resolved concerns the presence of metals in MAO. At one time, copper was stated to be present in the liver enzyme and, although this claim was withdrawn and disproven many years ago, periodic reports in the literature still appear claiming the presence and catalytic function of copper in the enzyme (16). A more substantial case has existed for Fe, since MAO activity declines substantially in iron deficiency (17) and the presence of 1 to 2 g atoms of Fe per mol has been reported in highly purified preparations from many laboratories (18, 19), although the fact that labile S was absent and no EPR signal characteristic of Fe-S clusters was ever detected, suggested that it might be an impurity.

Since the question as to what role, if any, Fe might play in MAO has continued occupying the interest of many investigators (20), we tried to resolve the matter once and for all. Clearly, if the iron is associated with a protein impurity, the latter would be tenaciously bound to MAO, since it accompanies the enzyme through a variety of purification procedures. In an analogous situation several years ago, Cremona and Kearney (21) succeeded in separating tightly associated FAD-containing impurities from NADH dehydrogenase by centrifugation in a sucrose gradient at alkaline pH. The same treatment worked with MAO from beef liver: the resulting preparation contained <0.1 g atom of Fe/mol of flavin and the Fe was recovered in a hemoprotein fraction (11, 12). The resulting preparation also gave the correct molecular weight (110,000), in harmony with the findings of others (9).

Besides flavin, the active site of MAO-B is known to contain an -SH group which is essential for activity (23). The action of -SH directed inhibitors on MAO has been studied by many investigators over the years (for a review cf Yasunobu et al. [24]). The number of thiol groups reacted depends on the reagent used and even with the same reagent (e.g., NEM, iodoacetate, DTNB), different laboratories find very different titers for the same type of enzyme preparation. The reaction with all -SH reagents is sluggish and the "essential" -SH groups (those presumably located at the catalytic site), in our hands, do not react preferentially with any of the reagents tested. Attempts in our laboratory to label the -SH group(s) at the substrate binding site by first treating the enzyme with cold alkylating agents in the presence of substrates, then, following gel exclusion to remove the protective substrates and cold

NEM, incubating the enzyme with [^{14}C]-NEM, have failed, largely because of the slow reaction of the enzyme with alkylating agents. To our knowledge, the only practical way of reacting the substrate site thiol selectively is the use of certain suicide inhibitors, such as *trans*-phenylcyclopropylamine or arylhydrazines (cf below).

There have been periodic speculations in the literature on the possible role of lipids in the action of MAO and in the differences between the A and B forms. One of the few direct statements is a report that Triton X-100 plus diphosphatidylglycerol potentiates the action of rat liver fractions on a 5-hydroxytryptamine and tryamine (25) but the data did not provide an unambiguous demonstration of lipid dependence. Very recently, Yagi's laboratory (26,27) claimed that lipids are required for the activity of MAO on certain substrates and that they are also the basis for the differences between the A and B forms. Since these conclusions are contrary to current thinking in most other laboratories, they must be examined in some detail.

The experimental material used by these workers is beef heart mitochondria, an unfortunate choice, since the content of the enzyme is very low, whereas that of lipids quite high. Moreover, this choice precludes comparison with the data of other workers using conventional sources. The heart enzyme, purified by a series of chromatographic steps, was stated to be homogeneous in acrylamide electrophoresis, although its specific activity on benzylamine (0.089 μmol/min/mg) (26) was <2% of that of preparations isolated from beef liver and kidney (7–9). Moreover, the subunit molecular weight was 28,000, in contrast to the accepted value of 52,000 to 57,000. We think that such a difference in subunit structure and activity of an enzyme in different organs of the same species is most unlikely.

Naoi and Yagi delipidated their preparation with sodium perchlorate or with 1% sodium dodecyl sulfate and DTT. Such treatment was stated to reduce the phospholipid content from 238 to 140 mols/mol enzyme and result in over 90% loss of activity on benzylamine (and somewhat lower inactivation on other substrates). Sonication with a phospholipid mixture restored more than the original activity (26). In another paper, digestion with phospholipase A_2 was stated to abolish all activity on A substrates without any effect on B substrates, while phospholipase D selectively destroyed activity on B substrates (27). The latter experiments are at variance with the fact that extensive digestion of placental mitochondria with phospholipase A_2 has little or no effect on the activity of MAO-A (15). As regards "delipidation" with perchlorate or SDS, the possibility that the harsh treatment caused some reversible conformation change, responsible for the loss of activity, cannot be excluded. It is noteworthy that the pure beef liver enzyme contains only ~4 mols of lipid P per mol and that this can be reduced to <2 mols/mol by passage through hydroxylapatite, with loss of activity (28). These findings may be contrasted with the presence of 140 mols of lipid P in Yagi's delipidated sample (27). We submit, therefore, that more convincing data are needed before we can conclude that lipids are required for the activity of MAO on any of its substrates, although it is quite possible that the lipid environment stabilizes the enzyme.

Suicide Inhibitors of Monoamine Oxidase

Several reviews have appeared on the action of suicide inhibitors on MAO (22, 28, 29). Therefore, the biochemical mechanisms involved will be only briefly summarized. Acetylenic inhibitors of MAO, which include the compounds of greatest pharmacological interest (Figure 1), act by forming a stoichiometric adduct, a flavocyanine with N(5) of the flavin, which is not dissociable even on denaturation. The compound was originally characterized after isolation from the B form of MAO (29) but the same structure is formed on reaction of clorgyline with MAO-A (14), as well as photochemically, using model flavins (30).

Suicide inhibition by acetylenic compounds is usually thought to involve the intermediate formation of the reactive allene. Interestingly, with MAO this does not seem to occur, since Krantz et al. (31) have shown that while the reaction of N-2-butynyl-N-benzylmethylamine with MAO-B yields the expected flavocyanine at N(5) of the flavin, the tautomeric allene, N-2,3-butadienyl-N-benzylmethylamine yields a different adduct on inactivation, probably the cyclic N(5)-C(4a) compound. Also, in this instance, the photochemical model reaction takes the opposite course from the enzymatic one.

As is true of acetylenic amines, the inactivation of MAO by arylcyclopropylamines, as exemplified by 2-phenylcyclopropylamine (*tranylcypromine*, PCPA), has been extensively studied over the years and is also of considerable pharmacological interest. The structure of the adduct formed with MAO-B from liver has been studied in Silverman's (32, 33) and our laboratory (34, 35). Silverman and Hoffman (33) initially thought that N-cyclopropyl arylalkylamines form an adduct with N(5) of the flavin, which would account for the bleaching of the 450 nm band during inactivation (Figure 2). We concluded

Figure 2. Spectral changes during the inactivation of beef liver MAO by *trans*-phenylcyclopropylamine. Upper curve: active enzyme; lower curve: fully inactivated enzyme; inset: difference spectrum.

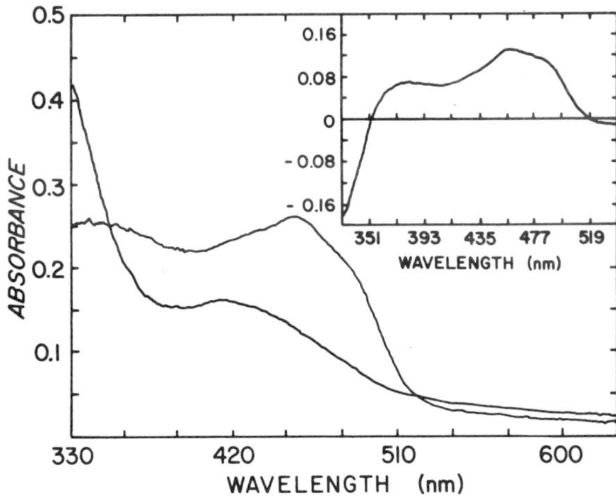

that the -SH group at the substrate site, not the flavin, is the target, based on the following evidence. We have reported (34,35) that the difference spectrum obtained on inactivation of beef liver MAO is not typical of flavin reduction, as would be expected if alkylation of N(5) had occurred, but shows a pronounced shoulder in the near u.v. Since in our published studies a preparation containing traces of a heme protein was used, we show here the same experiment with an Fe-free preparation. The perturbation in the near u.v. is quite evident. Second, when radioactive PCPA was used to inactivate the enzyme, the radioactive inhibitor was released even on denaturation at neutral pH and anaerobosis. These conditions would not be expected to hydrolyze the N(5)-alkyl linkage, but dissociation of a thiohemiketal or thioaminoketal (Figure 3) would be expected, since stabilization by the tertiary structure of the protein is removed. Most importantly, when the rates of return of the 450 nm absorbance and of the release of the radioactive inhibitor were compared, the former occurred far faster than the latter. We rationalized this by suggesting that the initial bleaching is due to reduction of the flavin during oxidation of the PCPA to the imine, which then combines with the nucleophilic -SH group and in the resulting adduct the access of O_2 to the flavin is blocked. Denaturation permits rapid reoxidation of the flavin, because the protein is unfolded, but release of the inhibitor can occur more gradually. This hypothesis has also been adopted by Silverman and Hoffman (33) in a recent paper.

Inactivation of MAO by arylhydrazines is even more complex. Using [^{14}C]-phenylhydrazine, Kenney et al. (36) concluded that, following dehydrogenation by the enzyme to the highly reactive phenyldiazene, about 50% of inhibitor attacks C(4a) of the flavin, while the other 50% goes to the active site -SH group to form a thioether. The assignment of C(4a) as the target is provisional and is based, in part, on the similarity of the absorption spectra at neutral and acid pHs with those of synthetic 4a-phenyl FMN, in part on the reaction of the related enzyme, trimethylamine dehydrogenase, which forms

Figure 3. Proposed structures of the active site of MAO inactivated with PCPA.

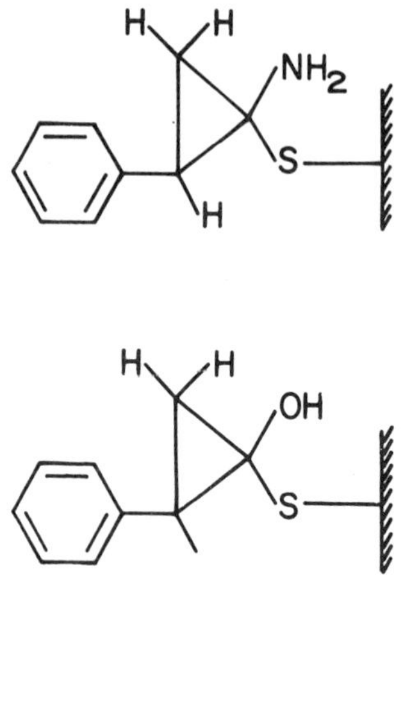

H H NH2 S H

H H OH S

only the C(4a) adduct on inactivation by phenylhydrazine (37). It is emphasized, however, that the reaction of MAO with phenylhydrazine is much more complex. For instance, the absorbance changes in the visible continue long after inactivation is complete and show evidence of major perturbation of the flavin, supporting the notion that at least part of the phenyldiazene attacks a protein group. So far, however, a pure nonflavin peptide containing the [^{14}C]-phenyl group has not been isolated, perhaps because of the instability of the adduct.

We have also studied the inactivation of MAO-A from human placenta by [^{14}C]-phenylhydrazine. The reaction is a typical suicide inactivation, involving biphasic loss of activity, protection by the substrate (kynuramine), and irreversibility. Approximately 1 mol of inhibitor becomes bound to the enzyme on complete inactivation. In this instance, however, only ~20% of the radioactivity was recovered in the flavin peptide fraction, showing that the majority of the phenyldiazene reacted with a protein group, probably the -SH. The application of this reaction to defining the differences in the peptide sequences of MAO-A and B is discussed below.

A Critical Look at the A-B Differences

The origin of the notions that the A and B forms of MAO represent the same protein but with different lipids attached and that the two forms may be interconvertible probably goes back to the experiments of Ekstedt and Oreland (38) who found that the treatment of rat liver preparations with methyl ethyl ketone results in extraction of the B but complete loss of the A activity. Oreland emphasized (39), however, that this cannot be taken as evidence of the hypothesis cited above, since the solvent merely inactivated the A enzyme, rather than converting it to the B form. Similarly, the report of Houslay and Tipton (40) that treatment of rat liver preparations with sodium perchlorate abolishes the biphasic response to clorgyline when tested on A-B substrates is more likely due to structural damage to one or both proteins than to lipid removal, all the more since no effect on the rates of oxidation of either type substrate was noted.

The experiments of Naoi and Yagi (26,27), already discussed, imply different lipid requirements for the activity of the two forms. As already discussed, however, the data on phospholipase A_2 squarely contradict findings in other laboratories, while those based on the use of phospholipase D and chaotropic agents can be readily interpreted in other ways, without assuming that lipids are required for the activity of either enzyme or that they distinguish the two forms.

Studies in the laboratories of Breakefield and ours have been designed on the assumption that MAO-A and B are two closely related but different proteins. The former group worked with various types of cell cultures containing both forms, particularly rat liver hepatoma. The mitochondrial fraction was labelled with [^{3}H]-pargyline, with and without prior treatment with a low concentration of clorgyline, an A-selective inhibitor. The rationale was that in a sample containing both forms in the presence of clorgyline only the B would be labelled and in its absence both A and B. It should be noted that pargyline will inactivate and combine with both forms, but only in the case of the B enzyme is inactivation brought about by 1 : 1 combination with N(5) of the flavin; in the case of the A type, high levels of pargyline evoke combination of several mols of the inhibitor without major inactivation. Following *limited* digestion with various proteolytic enzymes, 2 large (>10,000 mw) peptides were obtained in the absence of clorgyline and only 1 in its presence (41, 12). These experiments strongly suggest that the peptides which bind pargyline are different and electrophoretically separable in the A and B forms.

In our laboratory, it was shown that the flavin pentapeptides from beef liver MAO-B and human placental MAO-A, obtained by *extensive* proteolysis, have identical sequences (15). This is not unexpected, since in order to account for the differences in binding A vs B substrates and selective suicide inhibitors, differences should be sought at the substrate binding site. Nor is there a contradiction between these results and those of Cawthon et al. just quoted (41), since they worked with a much larger polypeptide, which might have included the substrate site region. It is also possible that in their studies the predominant binding of pargyline to the A form was not to the flavin, but, in any event, their data clearly suggest that A and B from the same cell are different proteins. Final proof will have to come from sequencing the substrate sites from the two forms of MAO. To this end, we are engaged in attempts to label the substrate site -SH specifically in MAO-A and B. Since conventional -SH reagents do not react preferentially with this group (cf above), we have turned to suicide inhibitors, which generate the reactive electrophile while attached to the active site. PCPA is an obvious choice, but it dissociates on denaturation. However, arylhydrazines appear promising, provided that we can find one which labels largely or even exclusively the -SH, without significant reaction with the flavin. Figure 4 compares the rates of inactivation of both forms by arylhydrazines with increasing lengths of alkyl group. It is seen that with both enzymes, benzylhydrazine appears to be the most effective suicide inhibitor. Determination of the distribution of the benzyl group between the -SH and the flavin will require synthesis of radioactive benzylhydrazine. This would permit isolation and sequencing of relatively short peptides originating from the substrate binding region.

Before leaving this subject, it may be well to emphasize the importance of clearly defining the structural differences between the two forms. Suicide inhibitors of MAO-B which react only slightly with the A form have already proven to be highly useful in clinical practice, e.g., in the treatment of Parkinsonism. The recent report of Knoll (42) that deprenyl reverses some of the effects of senescence stresses the additional potential of this class of drugs. Elucidation of the complete structure of the substrate sites of the A and B form would represent a giant step forward in the rational design of still more potent and specific inhibitors of MAO-B, one for which the biochemist would be able to take just credit.

Mechanism of Action

Studies of the reaction mechanism of MAO have so far relied on steady-state kinetic experiments. From the parallel lines obtained in double reciprocal

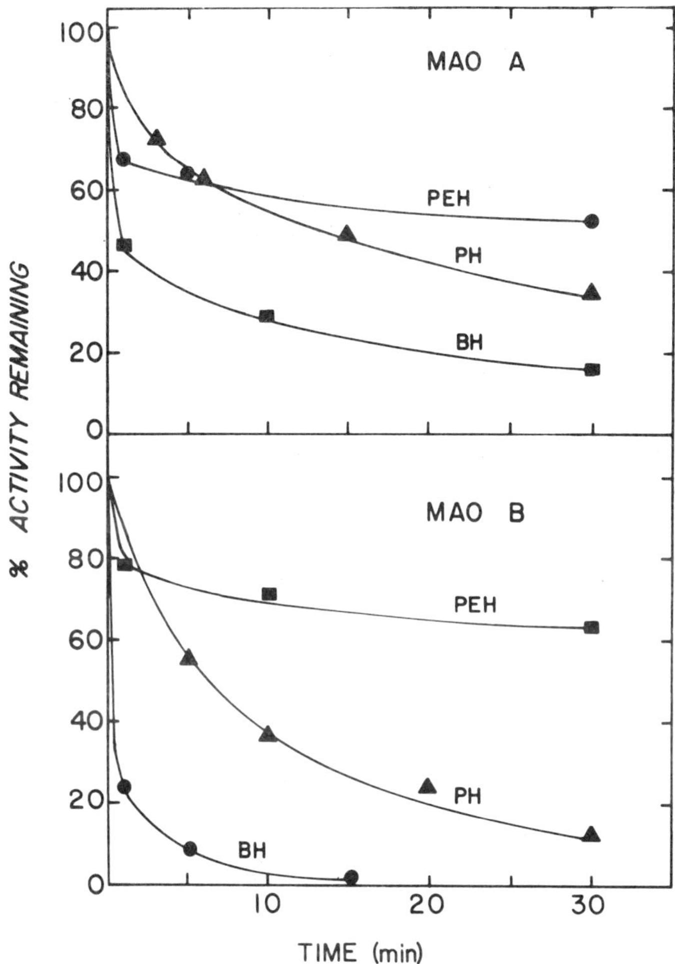

Figure 4. Inactivation of monoamine oxidases by benzyl- (BH), phenyl- (PH), and phenylethyl (PEH) hydrazines. MAO-A (1.5 μM) and MAO-B (4.4 μM) were incubated at 30° with a 13- and 11-fold molar excess of the inhibitors, respectively, in 50 mM KP_i, pH 7.2. The enzymes were assayed spectrophotometrically, using benzylamine (MAO-B) and kynuramine (MAO-A) as substrates.

plots, it was concluded that MAO-B operates by a ping-pong mechanism (43,44). From work on other flavoenzymes, it is clear that such data must be supplemented with pre-steady state experiments. We have recently concluded a mechanistic study on the pure enzyme from beef liver. A full report will appear elsewhere (28). The present summary and the brief paper elsewhere in this volume (45) can only touch on the main conclusions.

Using the highly purified enzyme from beef liver, we confirmed that steady state kinetic analysis yields parallel lines in Lineweaver-Burk plots for a series of substrates (Figure 5). Table 1 shows that dideuterated benzylamine gives

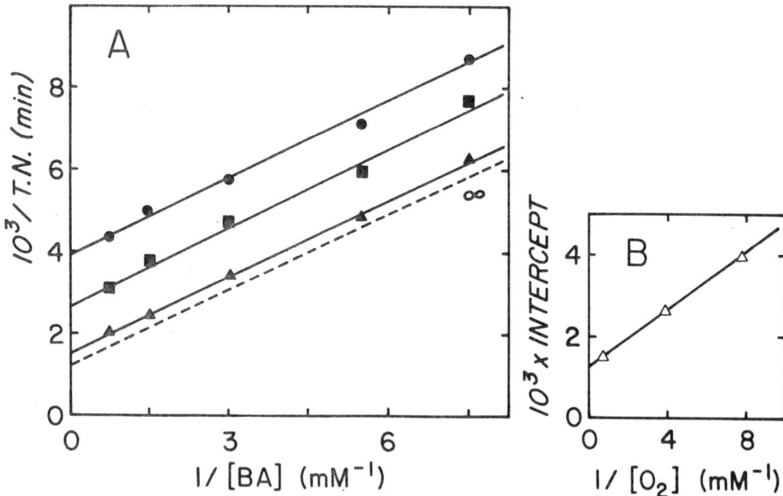

Figure 5. (A) Double reciprocal plot of initial rate data for the oxidation of benzylamine (BA) by beef liver monoamine oxidase. Spectrophotometric assays at 25° as a function of benzylamine concentration at the following fixed O_2: (●) 0.13 mM; (■) 0.26 mM; and (▲) 1.3 mM. Abscissa, moles of substrate oxidized/min/mole of enzyme. (B) Secondary plot of intercepts from the data in A against the reciprocal of the O_2.

about the same K_m value as benzylamine but the V_{max} is 6.7 times higher for the hydrogen containing form. This large kinetic isotope effect clearly shows that the dehydrogenation step is rate-limiting in the oxidation of this substrate. This is not true of phenylethylamine, however, since in this case the K_m for the substrate is only moderately lower than for benzylamine, but the K_m for O_2 is 10 times higher when phenylethylamine is oxidized. In consequence, in air the turnover rate for benzylamine (400/min) is 3.5 times higher than for phenyl-

Table 1. Steady-State Kinetic Parameters of Momamine Oxidase with Various Substrates.

Kinetic parameters	Benzylamine	[α, α-^{2}H]-Benzylamine	β-Phenylethylamine
K_m^S(mM)[a]	0.51 0.58[b]	0.48 0.46[b]	0.19
$K_m^{O_2}$(mM)[a]	0.28 0.27[b]	0.18 0.21[b]	2.8
V_{max}(min^{-1})[c]	800 740[b]	120 115[b]	1250

[a]The Michaelis constant is calculated by dividing ϕ_S by ϕ_O.

[b]Calculated from stopped-flow monitored turnover data.

[c]Maximal velocity at infinite concentrations of amine and oxygen calculated from the relationship: $V_{max} = 1/\phi_O$.

ethylamine, but at infinite O_2 concentration V_{max} for the latter is 50% higher than for benzylamine.

Anaerobic stopped-flow experiments on the reductive half-reaction gave a limiting value (k_3) of 740 at 25° for benzylamine, in agreement with the TN calculated from steady state experiments. The same studies confirmed the large kinetic isotope effect with this substrate. In contrast, the k_3 for phenylethylamine calculated from anaerobic bleaching was 30 times higher than the TN in steady state assays, again ruling out the possibility that dehydrogenation of the substrate is rate-limiting with this substrate. Figure 6 is a stopped-flow monitored turnover experiment, confirming these conclusions. With benzylamine and its dideutero form, most of the enzyme is in the oxidized form in the steady state, because reduction of the flavin is the slow step, while with phenylethylamine and kynuramine most of the enzyme is in the reduced form in the steady state, because reoxidation of the flavin is the slow step.

These results emphasize that conventional kinetic assays do not give an accurate picture of the substrate specificity of this enzyme. Another important conclusion, discussed in the paper of Husain et al. (45), is that stopped-flow measurements of the oxidative half-reaction showed that, depending on the substrate, the prevailing kinetic mechanism is different. With phenylethylamine the rate of reoxidation of the reduced enzyme by $[O_2]$ agreed with the turnover

Figure 6. Variation in the steady state concentration of the oxidized enzyme during stopped-flow monitored turnover experiments with different substrates. The enzyme (final concentration ~7 μm) was mixed in the stopped-flow with equal volumes of solutions containing excess of substrate and the decrease in absorbance at 450 nm recorded for the time indicated. Symbols: (—), 10 mM [α,α-^{2}H] benzylamine; (---), 10 mM benzylamine; (···), 10 mM kynuramine; and (-·-·-), 20 mM β-phenylethylamine.

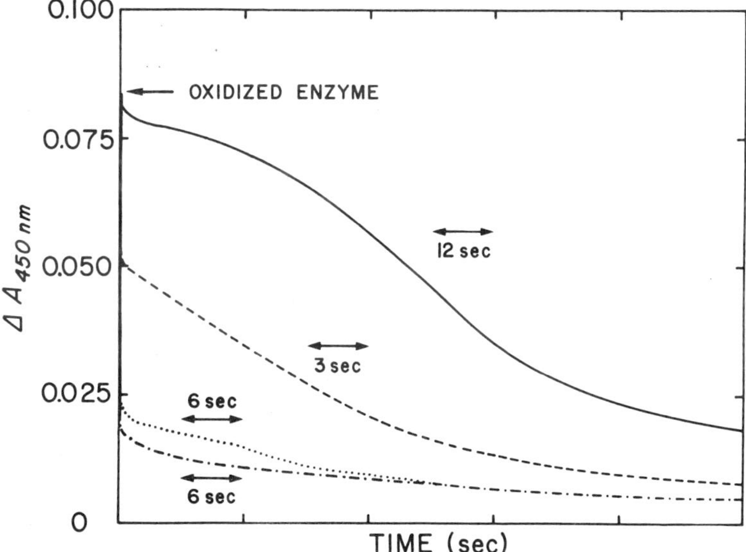

number from steady state assays, showing that the reoxidation step is rate-limiting and that a ping-pong mechanism describes the events. In contrast, with benzylamine at any given [O_2] the rate of reoxidation of substrate- or dithionite-reduced MAO was much less than the TN in steady state assays. This suggests that in normal catalysis the product does not dissociate from the reduced enzyme prior to reaction with O_2. Thus, the rate of reoxidation of the reduced enzyme is a function of the type of imine still attached, ruling out a ping-pong mechanism.

ACKNOWLEDGMENTS
This investigation was supported by the Veterans Administration, by the National Institutes of Health (Program Project HL-16251), and by the National Science Foundation (PCM 78-23716).

References

1. Johnston, J.P. (1968) *Biochem Pharmacol* 17:1285–1297.
2. Knoll, J. and Magyar, K. (1972) *Advan Biochem Psychopharmacol* 5:393–408.
3. Singer, T.P. (1979) In *Monoamine Oxidase: Structure, Function, and Altered Functions*. Singer, T.P., Von Korff, R.W., and Murphy, D.L. (eds.) New York: Academic Press, pp. 7–24.
4. Youdim, M.B.H. and Sandler, M. (1967) *Biochem J* 105:43P.
5. Salach, J.I. (1979) *Meth Enzymol* 53:495–501.
6. Zeller, A.E. (1981) In *Monoamine Oxidase: Basic and Clinical Frontiers*. Kamijo, K. (ed.), Excepta Medica, in press.
7. Chuang, H.Y.K., Patek, D.R., and Hellerman, L. (1974) *J Biol Chem* 249:2381–2384.
8. Salach, J.I. (1979) *Arch Biochem Biophys* 192:128–137.
9. Minamiura, N. and Yasunobu, K.T. (1978) *Arch Biochem Biophys* 189:481–489.
10. Zeller, E.A. (1981) In *Function and Regulation of Monoamine Enzymes: Basic and Clinical Aspects*. Usdin, E., Weiner, N., and Youdim, M.B.H. (eds.) Reading, Massachusetts: Addison-Wesley Publishing Co., in press.
11. Weyler, W. and Salach, J.I. (1981) *Arch Biochem Biophys*, submitted.
12. Pintar, J.E., Cawthon, R.M., Castro Costa, M.R., and Breakefield, X.O. (1979) In *Monoamine Oxidase: Structure, Function, and Altered Functions*. Singer, T.P., Von Korff, R.W., and Murphy, D.L. (eds.) New York: Academic Press, pp. 185–196.
13. Walker, W.H., Kearney, E.B., Seng, R.L., and Singer, T.P. (1971) *Eur J Biochem* 24:328–331.
14. Salach, J.I. and Detmer, K. (1979) In *Monoamine Oxidase: Structure, Function, and Altered Functions*. Singer, T.P., Von Korff, R.W., and Murphy, D.L. (eds.) New York: Academic Press, pp. 121–128.
15. Nagy, J. and Salach, J.I. (1981) *Arch Biochem Biophys* 208:388–394.
16. Carper, W.R., Stoddard, D.D., and Martin, D.F. (1974) *Biochem Biophys Acta* 334:287–296.
17. Symes, A.L., Sourkes, T.L., Youdim, M.B.H., Gregoriadis, G., and Birnbaum, H. (1969) *Canadian J Biochem* 47:999–1002.
18. Oreland, L. (1971) *Arch Biochem Biophys* 146:410–421.
19. Youdim, M.B.H. (1976) In *Flavins and Flavoproteins*. Singer, T.P. (ed.) Amsterdam: Elsevier, pp. 593–604.
20. Usdin, E. (1979) In *Monoamine Oxidase: Structure, Function, and Altered Functions*. Singer, T.P., Von Korff, R.W., and Murphy, D.L. (eds.) New York: Academic Press, pp. 539–552.
21. Cremona, T. and Kearney, E.B. (1964) *J Biol Chem* 239:2328–2334.
22. Singer, T.P. and Salach, J.I. (1981) In *Monoamine Oxidase Inhibitors: The State of the Art*. Youdim, M.B.H. and Paykel, E.S. (eds.) New York: John Wiley and Sons Ltd., pp. 17–29.
23. Barron, E.S.G. and Singer T.P. (1945) *J Biol Chem* 157:241–253.

24. Yasunobu, K.T., Watanabe, K., and Zeidan, H. (1979) In *Monoamine Oxidase: Structure, Function, and Altered Functions*. Singer, T.P., Von Korff, R.W., and Murphy, D.L. (eds.) New York: Academic Press, pp. 251–263.

25. Kandaswami, C. and D'Iorio, A. (1978) *Arch Biochem Biophys* 190:847–849.

26. Naoi, M. and Yagi, K. (1980) *Arch Biochem Biophys* 205:18–26.

27. Naoi, M. and Yagi, K. (1980) *Biochem Intern* 1:371–376.

28. Husain, M., Edmondson, D.E., and Singer T.P. (1981) In *Function and Regulation of Monoamine Enzymes: Basic and Clinical Aspects*. Usdin, E., Weiner, N., and Youdim, M.B.H. (eds.) Reading, Massachusetts: Addison-Wesley Publishing Co., in press.

29. Maycock, A.L., Abeles, R.H., Salach, J.I., and Singer T.P. (1976) *Biochemistry* 15:114–125.

30. Gärtner, B. and Hemmerich, P. (1975) *Angew Chem Int Ed Engl* 14:110–111.

31. Krantz, A., Kokel, B., Sachdeva, Y.P., Salach, J.I., Detmer, K., Claesson, A., and Sahlberg, C. (1979) In *Monoamine Oxidase: Structure, Function, and Altered Functions*. Singer, T.P., Von Korff, R.W., and Murphy, D.L. (eds.) New York: Academic Press, pp. 51–70.

32. Silverman, R.B. and Hoffman, S.J. (1979) In *Monoamine Oxidase: Structure, Function, and Altered Functions*. Singer, T.P., Von Korff, R.W., and Murphy, D.L. (eds.) New York: Academic Press, pp. 71–79.

33. Silverman, R.B., Hoffman, S.J., and Catus, W.B. (1980) *J Am Chem Soc* 102:884–886.

34. Paech, C., Salach, J.I., and Singer, T.P. (1979) In *Monoamine Oxidase: Structure, Function, and Altered Functions*. Singer, T.P., Von Korff, R.W., and Murphy, D.L. (eds.) New York: Academic Press, pp. 39–50.

35. Paech, C., Salach, J.I., and Singer, T.P. (1980) *J Biol Chem* 255:2700–2704.

36. Kenney, W.C., Nagy, J., Salach, J.I., and Singer, T.P. (1979) In *Monoamine Oxidase: Structure, Function, and Altered Functions*. Singer, T.P., Von Korff, R.W., and Murphy, D.L. (eds.) New York: Academic Press, pp. 25–37.

37. Nagy, J., Kenney, W.C., and Singer, T.P. (1979) *J Biol Chem* 254:2684–2688.

38. Ekstedt, B. and Oreland, L. (1975) *Biochem Pharmacol* 25:119–124.

39. Oreland, L. (1976) In *Monoamine Oxidase and its Inhibition*. Amsterdam: Elsevier, p. 17.

40. Houslay, M.D. and Tipton, K.F. (1973) *Biochem J* 135:173–186.

41. Cawthon, R.M., and Pintar, J.E., Haselfine, S.P., and Breakefield, X.O. (1981) *J Neurochem*, in press.

42. Knoll, J. (1981) In *Monoamine Oxidase Inhibitors—The State of the Art*. Youdim, M.B.H. and Paykel, E.S. (eds.) New York: John Wiley and Sons Ltd., pp. 45–61.

43. Tipton, K.F. (1968) *Eur J Biochem* 5:316–320.

44. Oi, S., Shimada, K., Inamasu, M., and Yasunobu, K.T. (1970) *Arch Biochem Biophys* 139:28–37.

45. Husain, M., Edmondson, D.E., and Singer, T.P., this volume.

Published 1982 by Elsevier North Holland, Inc.
Vincent Massey and Charles H. Williams, Editors
Flavins and Flavoproteins

CHAPTER 65

Biomedical Aspects of Flavoproteins

Kunio Yagi, Nobuko Ohishi, and Hiroshi Ohkawa

Institute of Biochemistry, Faculty of Medicine, University of Nagoya, Nagoya 466, Japan

The biological significance of riboflavin is mainly due to its coenzyme action in flavoproteins after conversion to FMN or FAD. Accordingly, medical phenomena provoked by ariboflavinosis are ascribed to the decreased activity of flavoproteins.

Recently, increased lipid peroxidation has been considered as an important cause for many degenerative diseases such as atherosclerosis and retinopathy. It must be emphasized, in the field of flavin research, that lipid peroxides are increased in ariboflavinosis. Since it is well known that flavoproteins intervene in the processes both for the formation and decomposition of lipid peroxides, it seems that the balance of formation and decomposition of lipid peroxides should be understood by the changes in activities of these flavoproteins. The present paper reviews the problems relating to the above mechanism. In connection with this problem, nonenzymatic effect of flavins on lipid peroxidation might not be ruled out, and the recently-obtained data are presented.

Metabolism of Flavins

The general feature of the absorption, metabolism, and excretion of flavins in higher animals is well elucidated as schematically indicated in Figure 1 (1,2), even though some of the detailed mechanism is still unknown. The site of absorption of riboflavin is concluded to be the upper gastrointestinal tract (3,4), but the detailed mechanism of absorption is unclear at present. Many factors such as food, bile acid, and drugs affect the rate of absorption of the vitamin. In plasma, a large part of riboflavin occurs in bound form; it combines with riboflavin-binding protein (5). As to the excretion of flavin into urine, it was hitherto believed that the excreted form is only riboflavin. However, our recent result (6) obtained by using high performance liquid chromatography indicates that some other metabolites are excreted into human urine. Some of these metabolites seem to arise from the covalently-bound flavins.

Lesions due to riboflavin deficiency are also well documented, and it is accepted that these lesions could be provoked by metabolic disorders caused by the decrease in the activities of flavoproteins due to the decreased level of the coenzymes, FMN and FAD. Accordingly, so-called "ariboflavinosis" can be provoked by a series of causes. Deficiency of flavins in the diet, decreased synthesis of flavins by intestinal flora, and decreased absorption through the intestine result in ariboflavinosis. Any inhibition of the synthesis of FMN and

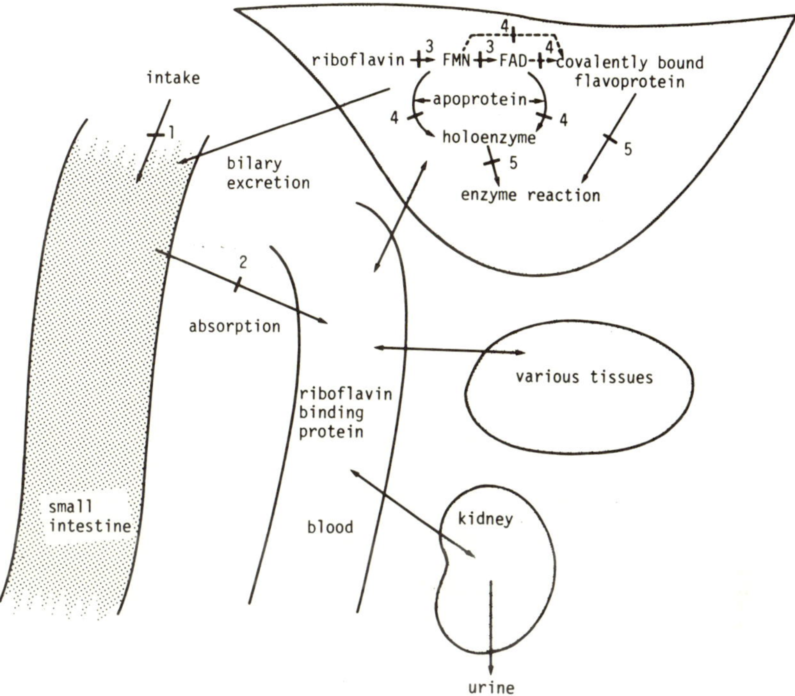

Figure 1. Metabolism of flavins in mammals. Numbered solid bars indicate the causes of ariboflavinosis: (1) decrease in intake, (2) inhibition of absorption, (3) inhibition of the conversion of riboflavin to the coenzymes, (4) inhibition of the formation of holoenzymes, and (5) inhibition of enzyme reaction.

FAD from riboflavin results in a decreased level of flavin coenzymes. Moreover, inhibition of some flavoproteins might cause lesions equal to those in riboflavin deficiency. Some antibiotics affect the reaction of flavoproteins (7).

Some hereditary diseases are due to the lack of some flavoproteins. An example is chronic granulomatous disease, in which a membrane-bound flavoenzyme, NADPH oxidase, is lacking in neutrophils (8, 9).

Response of Flavoproteins to Riboflavin Deficiency

Data are well documented as to the decrease in flavoproteins in riboflavin deficiency. The decrease in rats fed on a riboflavin-deficient diet is found not only in dissociable flavoproteins such as glycollate oxidase (10), D-amino oxidase (10), glycine oxidase (10), xanthine oxidase (11), L-glycerophosphate dehydrogenase (11), NADPH cytochrome c reductase (12), and glutathione reductase (13), but also in covalently-bound flavoproteins, such as succinate dehydrogenase (11), monoamine oxidase (14), and L-gulonolactone oxidase (15). The decrease in covalently-bound flavoprotein is not surprising, since it was found that administered riboflavin is incorporated into the covalently-bound flavin after the incorporation into FAD has reached the maximum level (16, 17). These reduced enzyme activities are restored by the administration of riboflavin, and this recovery in some flavoenzymes has been demonstrated within a few hours after realimentation (10).

Glutathione reductase, one of the flavoenzymes in the erythrocytes, is stimulated by FAD. The stimulation is more marked in riboflavin deficient rats than in those administered with riboflavin (13). Another flavoenzyme in the erythrocytes, NADH-methemoglobin reductase, is also influenced by the riboflavin-deficient status. Recently, the activity of glutathione reductase in hemolysates has been shown to reflect the level of flavin intake, and its measurement is useful to monitor the nutritional condition of flavins.

Lipid Peroxidation in Riboflavin Deficiency

In riboflavin-deficiency, lipid metabolism is fairly affected (18–21). At late stages of the deficiency, a decrease in the serum and liver concentration of polyunsaturated fatty acids was observed. These results led us to approach the problem of "lipid peroxidation" from the standpoint of the action of flavin.

A reliable micromethod for the assay of lipid peroxides in serum was devised by one of us (22), and it was applied to the specimens of riboflavin-deficient rats produced by feeding on a riboflavin-deficient diet (23). As shown in Table 1, the plasma lipid peroxide level tends to increase in riboflavin-deficient rats compared with riboflavin-sufficient rats.

From the data on lipid peroxidation in organs or tissues, it is reasonably accepted that increased lipid peroxidation always provokes some damage in the primary site of the peroxidation, especially in the membranes, and in addition the lipid peroxides formed are transferred via blood to intact organs or tissues where secondary damages are provoked. When linoleic acid hydroperoxide was injected into a rabbit through the ear vein, the $t_{1/2}$ of the hydroperoxide in the blood stream was found to be approximately 50 min, and a significant increase in lipid peroxide level was found in the intima of the aorta (24). Electron microscopic observation of the endothelial surface of the thoracic aorta after injection revealed the exposure of subendothelial fibrous

Table 1. Preventive Effect of Riboflavin on Levels of Liver Flavin and Blood Lipid Peroxides Upon Administration of Oxidized Corn Oil to Rats.

Administration		Total flavins in liver		Lipid peroxides in blood plasma	
Riboflavin	**Oxidized corn oil**	**(μg/g)**		**(nmol/ml)**	
+	−	31.6 ± 2.26		1.80 ± 0.10	
+	+	29.5 ± 1.33		1.84 ± 0.21	] b (vs. 5.99)
−	−	15.0 ± 0.61	] a	2.53 ± 0.63	] b
−	+	12.5 ± 0.85	] a	5.99 ± 0.52	] b

Mean ±SE (n = 5); $^a p < 0.05$; $^b p < 0.01$.

Daily administration: riboflavin, 20 μg; oxidized corn oil, 100 nmol in terms of malondialdehyde. Breeding: 22 days. Total flavins and lipid peroxides are assayed by lumiflavin fluorescence method (31) and thiobarbituric and method (22) respectively.

tissue and the adhesion of platelets with pseudopodia (24). These morphological changes provoked by a high level of lipid peroxides in the blood are regarded as features of the initial event of atherogenesis.

In these respects, the increase in lipid peroxide level in the blood in ariboflavinosis mentioned above should be considered to be deleterious to our body in provoking degenerative disorders. As clinical features of riboflavin deficiency, the changes in the skin and mucous membranes, such as angular stomatitis, cheilosis, seborrheic dermatitis, and scrotal dermatitis, are generally observed. Probably, these disorders relate to some extent to the increase in lipid peroxidation in riboflavin deficiency.

Roles of Flavoproteins in the Formation and Decomposition of Lipid Peroxides

It has been demonstrated that many flavoproteins take part in the processes of formation and decomposition of lipid peroxides *in vivo*, the major part of which is illustrated in Figure 2. Microsomal peroxidation with NADPH-cytochrome P-450 reductase was extensively studied. It has been reported that various flavoproteins produce an active oxygen, superoxide anion (O_2^-), during their oxidoreduction (9,25–29). When O_2^- or other active oxygen species derived from O_2^- can react with polyunsaturated fatty acids, the formation of lipid peroxides is provoked.

The *in vivo* decomposition of lipid hydroperoxides is mainly catalyzed by glutathione peroxidase to form the corresponding hydroxy derivatives. This reaction is coupled with the reduction of oxidized glutathione by the NADPH requiring flavoprotein, glutathione reductase, in the cytosol.

From these relations, it is considered that the increase in lipid peroxides in riboflavin deficiency is mainly ascribed to the decrease in the decomposition of lipid hydroperoxides, due to the reduced activity of glutathione reductase.

Figure 2. Role of flavoproteins in formation and decomposition of lipid peroxides.

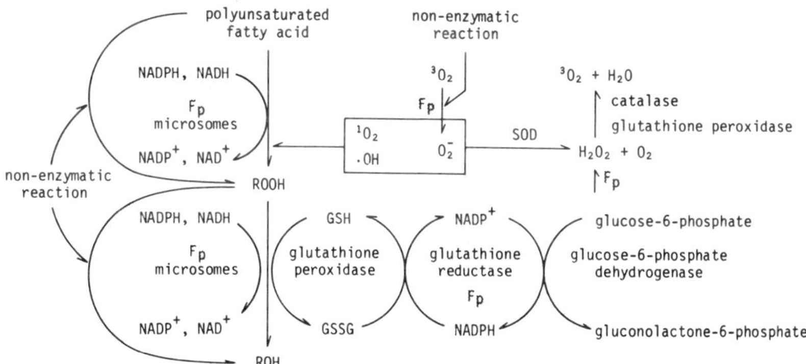

Preventive Effect of Riboflavin on the Toxicity of Lipid Peroxides

From the view points mentioned above, the effect of riboflavin on chronic intoxication of lipid peroxides was examined (23,30). As shown in Figure 3, the body weight gain of rats administered with riboflavin-deficient diet containing oxidized corn oil (D), was markedly reduced. On the other hand, the daily administration of riboflavin-sufficient diet with oxidized corn oil (C) brought about fairly good body weight gain. The administration of the oxidized corn oil tends to decrease the flavin level in the liver and the degree of decrease was marked for the rats administered with riboflavin-deficient diet (Table 1). Elevation of the lipid peroxide level of blood plasma in riboflavin deficiency was strengthened by daily administration of the oxidized corn oil (Table 1). It should be noted that this elevation was suppressed by simultaneous administration of riboflavin.

The results in Table 2 clearly demonstrate the preventive effects of riboflavin on the elevation of lipid peroxide levels in some tissues and on the acceleration

Figure 3. Effects of administration of lipid peroxides and riboflavin on body weight gain of rats. The mean values of body weight gain (n=5) were plotted. Administration of riboflavin: A and C, 20 μg/day/rat; B and D, riboflavin-deficient. Administration of lipid peroxides: (A and B) fresh corn oil; (C and D) oxidized corn oil (lipid peroxides content, 100 nmol in terms of malondialdehyde/day/rat).

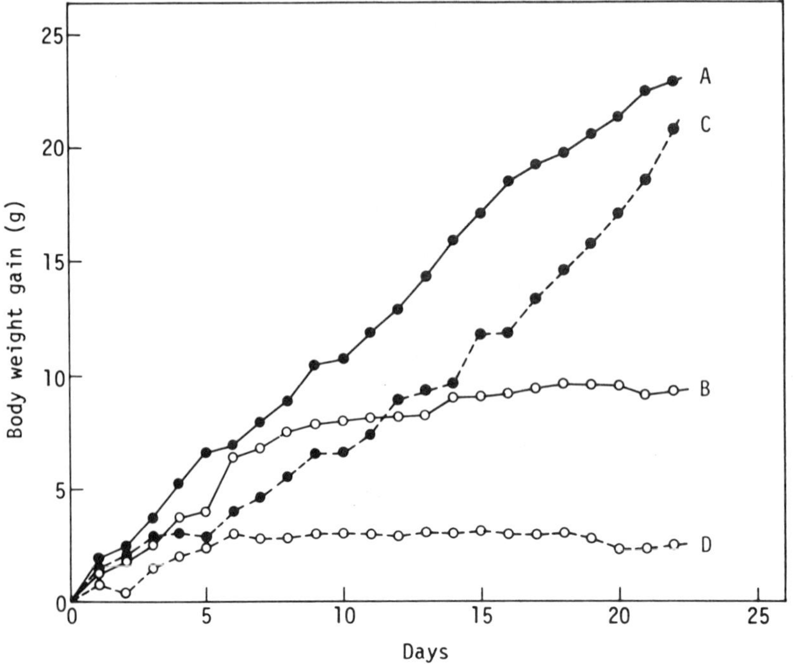

Table 2. Preventive Effect of Riboflavin on Lipid Peroxides Levels and Platelet Aggregation Upon Administration of Linoleic Acid Hydroperoxide to Rats.

Daily administration		Lipid peroxides (nmol/mg protein)			
Riboflavin (μg)	Linoleic acid hydroperoxide (nmol)	Retina	Aorta	Liver	Platelet aggregation (%/min)
5	—	0.76 ± 0.097	1.42 ± 0.141	1.13 ± 0.076	20.1 ± 1.01 ⌉ [a]
5	100	1.40 ± 0.347	1.53 ± 0.138 ⌉ [b]	1.34 ± 0.063	26.5 ± 2.90 ⌋
100[c]	100	0.70 ± 0.226	0.88 ± 0.082 ⌋	1.54 ± 0.100	21.0 ± 2.81

Mean ± SE (n = 6–12); [a]$p < 0.05$; [b]$p < 0.01$; [c]administered in the form of riboflavin 2′, 3′, 4′, 5′-tetrabutyrate. Breeding: 3 months. Linoleic acid hydroperoxide and lipid peroxides were assayed by thiobarbituric acid method (32) and expressed in terms of malondialdehyde. Platelet rich plasma contained 5×10^8 platelet/ml, and platelet aggregation was carried out with 4.5×10^{-6}M ADP.

Figure 4. Changes in absorption spectrum of riboflavin tetrabutyrate during peroxidation of linoleic acid with lipoxygenase. Reaction mixture consisted of riboflavin tetrabutyrate (200 μM), linoleic acid (2 mM), and soybean lipoxygenase (5×10^4 U/ml). Top curve: initial; botttom curve: after 45 min. Inset: the absorption spectrum of the purified product.

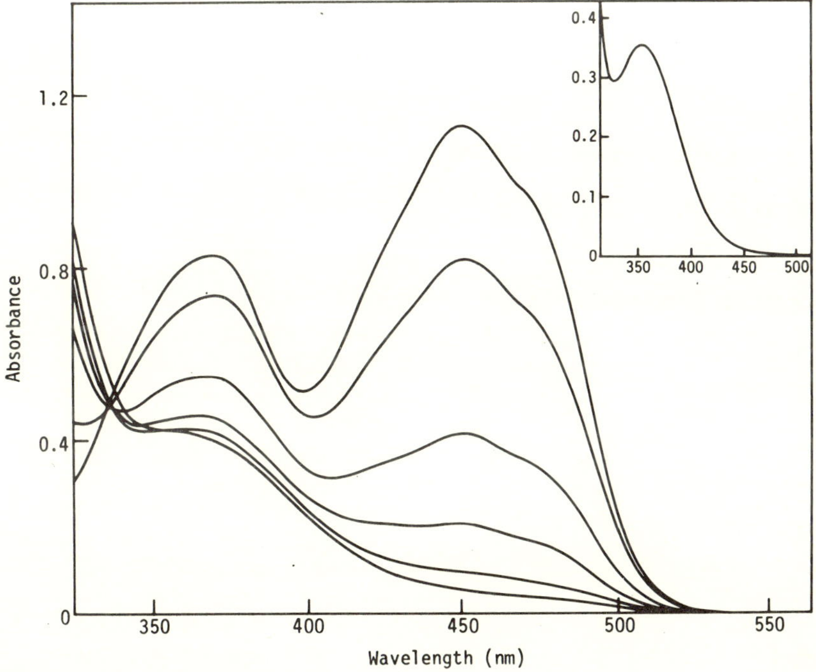

of platelet aggregation caused by the administration of linoleic acid hydroperoxide to rats. The preventive effect of riboflavin on some deleterious action of lipid peroxides should be noted from the biomedical point of view.

Nonenzymatic Effects of Flavin on Lipid Peroxidation

To examine the possibility of nonenzymatic effects of flavin on lipid peroxidation *in vivo*, the behavior of riboflavin 2′,3′,4′,5′-tetrabutyrate, a fat-soluble derivative of riboflavin (33), in the peroxidation of linoleic acid with lipoxygenase, was followed (34). The absorption spectrum of riboflavin tetrabutyrate gradually changed to that like the reduced form (Figure 4), and a new derivative of flavin was detected on high performance thin layer chromatogram. The absorption spectrum of the purified product is shown in the inset of Figure 4, and this product is considered to be an adduct form. In the presence of riboflavin tetrabutyrate, the rate of formation of linoleic acid hydroperoxide was decreased and the amount of the hydroperoxide formed was suppressed to some extent.

These findings suggest the occurrence of nonenzymatic inhibition of lipid peroxidation by flavin *in vivo*. If this is the case, riboflavin tetrabutrate, a fat-soluble riboflavin derivative, is useful, because lipid peroxidation generally occurs in lipid-rich biomembranes.

References

1. Okuda, J. and Yagi, K. (1979) In *Biochemical Aspects of Nutrition*. Yagi, K. (ed.), Tokyo: Japan Scientific Societies Press, pp. 155–160.
2. Jusko, W.J. and Levy, G. (1975) In *Riboflavin*. Rivlin, R.S. (ed.) New York: Plenum Press, pp. 99–152.
3. Campbell, J.A. and Morrison, A.B. (1963) *Amer J Clin Nutr* 12:162–169.
4. Levy, G. and Jusko, W.J. (1966) *J Pharm Sci* 55:285–289.
5. Merrill, A.H., Jr., and McCormick, D.B. (1980) *Fed Proc* 39:1698.
6. Ohkawa, H., Ohishi, N. and Yagi, K. (1981) *Biochem Int* 2:447–453.
7. Yagi, K., Yamamoto, Y., and Kobayashi, M. (1968) *J Vitaminol* 14:271–277.
8. Baehner, R.L., Nathan, D.G., and Karnovsky, M.L. (1970) *J Clin Invest* 49:865–870.
9. Babior, B.M. and Peters, W.A. (1981) *J. Biol Chem* 256:2321–2323.
10. Burch, H.B., Lowry, O.H., Padilla, A.M., and Combs, A.M. (1956) *J Biol Chem* 223:29–45.
11. Zaman, Z. and Verwilghen, R.L. (1975) *Biochem Biophys Res Commun* 67:1192–1198.
12. Patel, J.M. and Pawar, S.S. (1974) *Biochem Pharmac* 23:1467–1477.
13. Glatzle, D., Weber, F., and Wiss, O. (1968) *Experientia* 24:1122.
14. Sourkes, T.L. (1979) In *Monoamine Oxidase: Structure, Function, and Altered Functions*. Singer, T.P., Von Korff, R.W., and Murphy, D.L. (eds.), New York: Academic Press, pp. 291–307.
15. Kiuchi, K., Nishikimi, M., and Yagi, K. (1980) *Biochim Biophys Acta* 630:330–337.
16. Yagi, K., Nakagawa, Y., Suzuki, O., and Ohishi, N. (1977) *J Biochem* 79:841–843.
17. Muttart, C., Chaudhuri, R., Pinto, J., and Rivlin, R.S. (1977) *Amer J Physiol* 233:E397–E401.
18. Sugioka, G., Porta, E.A., Corey, P.N., and Hartroft, W.S. (1969) *Amer J Path* 54:1–19.

19. Mookerjea, S. and Hawkins, W.W. (1960) *Brit J Nutr* 14:231–238.
20. Koyanagi, T. and Oikawa, K. (1965) *Tohoku J Exp Med* 86:19–22.
21. Taniguchi, M., Yamamoto, T., and Nakamura, M. (1978) *J Nutr Sci Vitaminol* 24: 363–381.
22. Yagi, K. (1976) *Biochem Med* 15:212–216.
23. Saito, Y., Ohishi, N., and Yagi, K. (1981) *J Nutr Sci Vitaminol* 27:17–21.
24. Yagi, K., Ohkawa, H., Ohishi, N., Yamashita, M., and Nakashima, T. (1981) *J Appl Biochem*, 3:58–65.
25. Rajagopalan, K.V. and Handler, P. (1964) *J Biol Chem* 239:2022–2026.
26. Aleman, V. and Handler, P. (1967) *J Biol Chem* 242:4087–4096.
27. Ballou, D., Palmer, G., and Massey, V. (1969) *Biochem Biophys Res Commun* 36:898–904.
28. Massey, V., Strickland, S., Mayhew, S.G., Howell, L.G., Engel, P.C., Matthews, R.G., Schuman, M., and Sullivan, P.A. (1969) *Biochem Biophys Res Commun* 36:891–897.
29. McCord, J.M. and Fridovich, I. (1970) *J Biol Chem* 245:1374–1377.
30. Yagi, K., Ohishi, N., Ohkawa, H., Matsuoka, S., Asai, J., and Iijima, S., unpublished results.
31. Yagi, K. (1962) In *Methods of Biochemical Analysis*. Glick, D. (ed.) New York: Interscience, pp. 319–356.
32. Ohkawa, H., Ohishi, N., and Yagi, K. (1979) *Anal Biochem* 95:351–358.
33. Yagi, K., Okuda, J., Dmitrovskii, A.A., Honda, R., and Matsubara, T. (1961) *J Vitaminol* 7:276–280.
34. Ohishi, N., Ohkawa, H., and Yagi, K. (1981) *Biochem Int*, 3:213–216.

Published 1982 by Elsevier North Holland, Inc.
Vincent Massey and Charles H. Williams, Editors
Flavins and Flavoproteins

CHAPTER 66

Riboflavin Analogues as Antihypertensive Drugs

Daniel Trachewsky

Departments of Medicine and of Biochemistry and Molecular Biology, University of Oklahoma Health Sciences Center, Oklahoma City, Oklahoma

Introduction

Mineralocorticoid hypertension induced by chronic administration of deoxycorticosterone acetate (DOCA) to rats in conjunction with saline-loading simulates, in some respects, the salt-dependent form of essential hypertension in humans (1,2). Abnormalities associated with salt metabolism have been observed in many forms of experimental and clinical hypertension. Expansion of body fluid volumes by renal retention of Na^+ is known to raise blood pressure in susceptible individuals. It is commonly held that this is the principal factor in the pathogenesis of hypertension caused by mineralocorticoid excess (3). It has been reported that at two months of age, blood pressure was definitely elevated in the spontaneously hypertensive rat (SHR), a genetic model of hypertension, and was associated with relatively high basal levels of plasma aldosterone (4). The functional renal abnormality that is responsible for the rise of blood pressure in the Dahl salt sensitive (S) rat, another genetic model of hypertension, appears to be a difficulty in its ability to eliminate Na^+ and not a mineralocorticoid dysfunction (5).

The present investigation was designed to examine whether riboflavin analogues are effective antihypertensive drugs in the above three animal models of human hypertension and at dosages which would not present any problem of chronic toxicity. These studies are based on our previous observation that aldosterone enhances the biosynthesis of renal FMN and FAD, and that the riboflavin analogues 7,8-dimethyl-10-formylmethyl isoalloxazine (FMI) and 7,8-dimethyl-10-(2′-hydroxyethyl) isoalloxazine (HEI) competitively inhibit the conversion of riboflavin to FMN and reabsorption of Na^+ in the kidney of adrenalectomized rats (6).

Results and Conclusions

When 1.6 mg of FMI or HEI were administered simultaneously with 3.0 mg of DOCA twice weekly for 9 weeks, the tail systolic blood pressure (SBP) of unanesthetized rats rose only to 136 ± 5 mm Hg (SEM), compared to 165 ± 5 mm Hg during DOCA therapy alone ($p < 0.0005$, Table 1, Figure 1) (7). The saline polydipsia that occurs with DOCA administration was not abated by simultaneous treatment with FMI and HEI (Table 1); consequently the reduc-

Table 1. Effect of FMI Administration on the Mean Systolic BP and Mean Na^+ Intake of DOCA-Salt Hypertensive Rats During Treatment, Weeks 5 Through 8.

			Significance*			Significance*	
Group	Rats (no.)	SBP (mm Hg)†	DOCA vs control (p)	DOCA vs DOCA+FMI (p)	Na^+ Intake (mEq/24 hrs/rat)†	Control vs DOCA (p)	Control vs DOCA−FMI (p)
Control	14	129 ± 4	< 0.0005		13.63 ± 1.19		
DOCA	21	163 ± 5			18.91 ± 0.80	< 0.01	
DOCA + 1.2 mg FMI	8	147 ± 5		< 0.01	18.97 ± 1.67		< 0.01
DOCA vs 1.6 mg FMI	8	136 ± 5		< 0.0005	18.74 ± 1.26		< 0.025

*Significance determined by t test for unpaired data.
†Mean ± 1 SEM for Weeks 5 through 8.

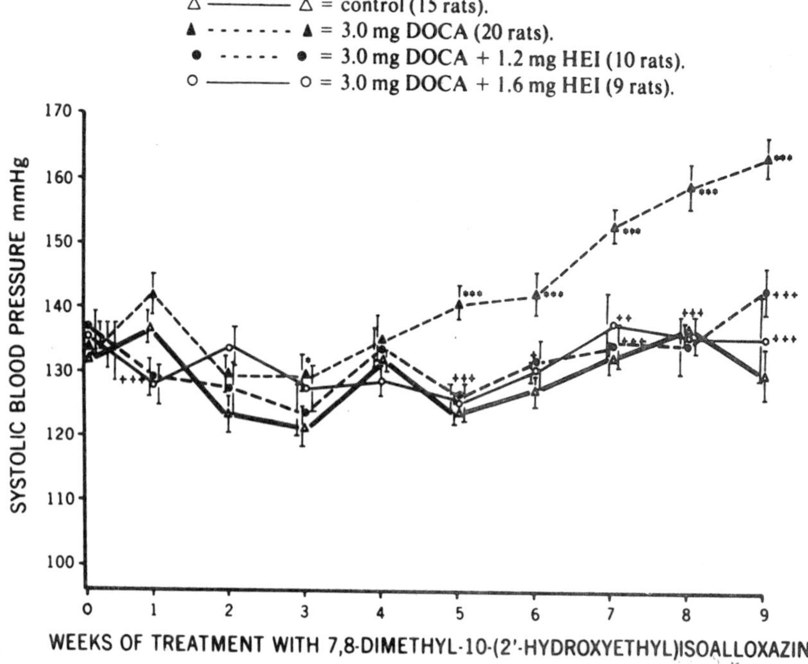

Figure 1. Effect of treatment with HEI on the systolic BPs of right adrenalectomized-nephrectomized rats made hypertensive by DOCA and salt. Each point on the figure represents the mean ±SEM. *$p<0.05$; **$p<0.01$; ***$p<0.0005$; DOCA-treated group vs control group: $^{+}p<0.05$; $^{++}p<0.01$; $^{+++}p<0.0005$; DOCA+1.2 mg HEI or DOCA+1.6 mg HEI group vs DOCA-treated group.

tion of SBP by FMI and HEI is not due to a reduced fluid exchange. DOCA-treated rats were found to have a 24% increase in iliopsoas muscle Na^{+} concentration (186±4 vs 150±4 μEq/g dry wt, $p<0.0005$, Table 2) and a 0.8% increase in the water content of the muscle (77.04±0.15% vs 76.44±0.25%, $p<0.05$), suggesting a positive Na^{+} balance. Administration of FMI and HEI blunted the ability of DOCA to increase muscle Na^{+} concentration (164±7 vs 186±4 μEq/g dry wt, $p<0.025$, Table 2) and water content (76.01±0.41% vs 77.04±0.15%, $p<0.01$). In contrast to the DOCA-treated animals, HEI administration for 12 weeks at 0.5 mg/100 g body wt twice weekly did not lower the SBP of the SHR (217±7 vs 212±3 mm Hg, ns); HEI treatment did not significantly alter the iliopsoas muscle Na^{+} concentration (178±12 vs 163±26 μEq/g dry wt, ns).

In subsequent studies, the more potent riboflavin antagonist 7,8-dimethyl-10-(3-chlorobenzyl) isoalloxazine (CBI) was employed. Administration of 0.5 mg CBI/100 g body wt twice weekly for 7 weeks lowered the SBP of the unanesthetized SHR from 188±7 mm Hg to 148±2 mm Hg ($p<0.0001$, Figure 2). As in the experiments employing FMI and HEI, the rats tolerated CBI well and showed no signs of riboflavin deficiency. CBI was more effective than hydrochlorothiazide (HCTZ) which at 2.2±0.3 mg/100 g body wt daily

Table 2. Effects of 8 Weeks of FMI and 9 Weeks of HEI Administration, Respectively, On the Na^+ and K^+ Contents of Iliopsoas Muscle of DOCA-Salt Hypertensive Rats.

			Significance*			Significance*	
Group	Rats (no.)	Na^+ (μEq/g dry weight)†	DOCA vs control (p)	DOCA vs DOCA + analog (p)	K^+ (μEq/g dry weight)†	DOCA vs control (p)	DOCA vs DOCA + analog (p)
Control	14	117 ± 2	< 0.0005		311 ± 10	< 0.0025	
DOCA	21	149 ± 3			273 ± 7		
DOCA + 1.2 mg FMI	8	138 ± 5		< 0.05	268 ± 10		ns
DOCA + 1.6 mg FMI	8	132 ± 3		< 0.025	274 ± 15		ns
Control	15	184 ± 7	< 0.0005		465 ± 8	< 0.0005	
DOCA	20	223 ± 5			404 ± 6		
DOCA + 1.2 mg HEI	10	216 ± 8		ns	406 ± 13		ns
DOCA + 1.6 mg HEI	9	196 ± 11		< 0.01	401 ± 11		ns

Significance determined by Student's *t* test for unpaired data.
†Mean ± 1 SEM.

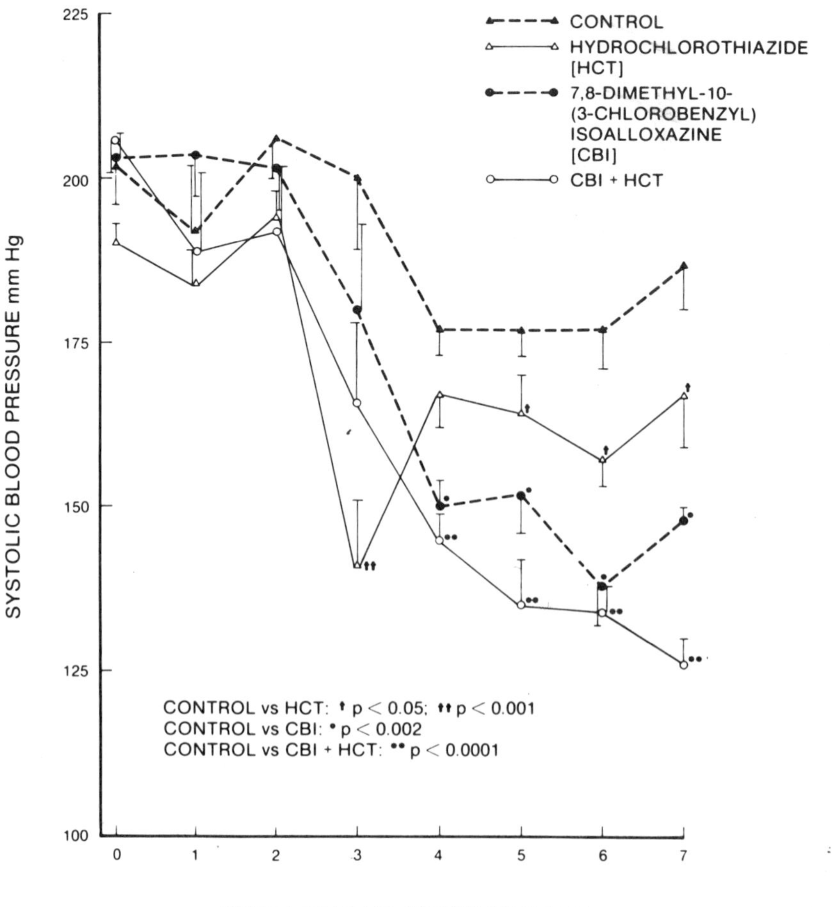

Figure 2. Effect of administration of CBI and HCT on the SBP of the SHR. Each point in the figure represents the mean ± SEM with 7 rats in each group. Each rat received 0.5 mg CBI/100 g body wt twice weekly for seven weeks. Drinking water contained 200 mg of HCT/liter and each rat averaged 2.2 ± 0.3 mg of HCT daily/100 g body wt. The SBP of the unanesthetized animals was measured ten times per week by a tail cuff plethysmographic method for 5 weeks prior to therapy to condition the rats.

only lowered the SBP by 21 mm Hg to 167 ± 5 mm Hg ($p<0.0001$, Figure 2). The simultaneous administration of CBI and HCTZ resulted in an additive reduction of the SBP from 188 ± 7 mm Hg to 126 ± 4 mm Hg ($p<0.0001$, Figure 2). CBI treatment gave a 36% decrease in iliopsoas muscle Na^+ concentration (117 ± 18 vs 182 ± 11 μEq/g dry wt, $p<0.05$, Table 3) and an 11% decrease in the water content of the muscle (65.15 ± 3.45% vs 73.20 ± 1.04%, $p<0.025$, Table 3), suggesting that the untreated SHR is retaining Na^+ (4).

Table 3. Effect of CBI and HCTZ Administration on Several Paramters in the SHR.

Group	Systolic Blood Pressure mmHg	Control vs Treated p†	Plasma Aldosterone ng%	Control vs Treated p†	Iliopsoas Muscle					
					Na^+ μEq per g dry wt	Control vs Treated p†	K^+ μEq per g dry wt	Control vs Treated p†	H_2O %	Control vs Treated p†
Control(7)	188±7*		12.40±3.66*		182±11*		311±14*		73.20±1.04*	
HCTZ(7)	167±8	<0.04	47.44±8.32	<0.01	147±15	<0.05	253±21	<0.025	65.60±2.74	<0.01
CBI(7)	147±2	<0.0001	20.20±5.94	ns	117±18	<0.005	243±35	<0.05	65.15±3.45	<0.025
HCTZ+CBI(7)	126±4	<0.0001	23.30±4.35	ns	130±7	<0.0025	331±6	ns	71.58±0.63	ns

*Mean ± SEM

†Significance determined by Student's t test for unpaired data.

Numbers in parentheses refer to number of rats in each group.

Plasma aldosterone was measured. The HCTZ-treated SHR exhibited elevated levels of plasma aldosterone when compared with controls (47.44±8.32 vs 12.40±3.66 ng%, $p<0.01$, Table 3) (8). CBI administration had no significant effect on this parameter (20.20±5.94 vs 12.40±3.66 ng%, Table 3, ns). There was a profound effect on the secondary hyperaldosteronism generated by HCTZ when CBI was administered together with the diuretic (23.30±4.35 vs 47.44±8.32 ng%, $p<0.05$, Table 3). These findings on plasma aldosterone were also reflected in the iliopsoas muscle K^+ concentration (Table 3). This discovery opens the way for consideration of the riboflavin analogues as dual-mechanism diuretic agents which may be K^+-sparing. In contrast to the SHR, the administration of CBI for 8 weeks at 0.5 mg/100 g body wt twice weekly to the Dahl S rats did not reduce their mean SBP (205±5 vs 200±4 mm Hg, ns). These animals became hypertensive over a period of weeks on a chow containing 8% NaCl; CBI treatment did not significantly decrease their iliopsoas muscle Na^+ concentration (225±32 vs 178±21 μEq/g dry wt, ns) or water content (75.64±0.73% vs 76.04±0.29%, ns).

These results are consistent with the hypothesis that the antihypertensive actions of the riboflavin analogues are closely associated with their ability to modify the effects of mineralocorticoids on Na^+ balance. These data suggest that riboflavin analogues may function as a novel class of antihypertensive drugs and with no apparent side effects.

ACKNOWLEDGMENTS

The author thanks Dawn Conway and Marla Stover for technical assistance. This investigation would not have been possible without generous donations of FMI and HEI from Dr. Paul W. O'Connell of the Upjohn Company, Kalamazoo, Michigan and CBI from Dr. Donald W. Graham of Merck and Company, Rahway, New Jersey. This study was supported by U.S. Public Health Service Grant HL-19279 from the National Heart, Lung, and Blood Institute and a Grant from the H.E. and Carolyn Bailey Research Funds.

References

1. Green, D.M. (1953) *Ann Int med* 39:333.
2. Stanton, H.C. and White, J.B., Jr. (1965) *Arch Int Pharmacodyn Ther* 154:351.
3. Dustan, H.P., Bravo, E.L., and Tarazi, R.C. (1974) *Am J Cardiol* 31:606.
4. Sowers, J., Tuck, M., Asp, N.D., and Sollars, E. (1981) *Endocrinology* 108:1216.
5. Tobian, L., Lange, J., Iwai, J., Hiller, K., Johnson, M.A., and Goosens, P. (1979) *Hypertension* 1:316.
6. Trachewsky, D. (1978) *J Clin Invest* 62:1325.
7. Trachewsky, D. (1981) *Hypertension* 3:75.
8. Kaplan, N.M. (1978) In *Clinical Hypertension*, 2nd edition. Baltimore: Williams and Wilkins Co., p. 261.

Published 1982 by Elsevier North Holland, Inc.
Vincent Massey and Charles H. Williams, Editors
Flavins and Flavoproteins

CHAPTER 67

Regulation of Flavin Metabolism by Thyroid Hormones and Psychotropic Drugs

Richard S. Rivlin, John Pinto, Yee Ping Huang, and Nicholas Pelliccione

Department of Medicine, Memorial Sloan-Kettering Cancer Center and New York Hospital-Cornell Medical Center, New York, New York

Our laboratory has explored the hormonal, nutritional, and pharmacological regulation of flavin metabolism. Thyroid hormones act at several sites in the riboflavin to FAD pathway. Overall, hypothyroidism decreases and hyperthyroidism increases the rate of conversion of riboflavin into FMN and FAD in experimental animals (1). Hepatic activity of flavokinase, which converts riboflavin to FMN, is increased by physiological doses of thyroxine; thyroid hormones also increase the activity of FAD pyrophosphorylase, which converts FMN to FAD (2). In hypothyroidism, flavokinase activity is half that of controls, and hepatic concentrations of FMN and FAD are reduced as a consequence (3). Degradation of FAD and FMN to riboflavin is less subject to hormonal regulation.

To determine whether thyroid hormone-induced changes in activities of enzymes metabolizing riboflavin as measured *in vitro* are likely to be physiologically significant, the rate of formation of FMN and FAD from radioactive riboflavin was determined *in vivo*. These results clearly showed that under a wide variety of conditions, changes in activities of riboflavin-metabolizing enzymes are reflected both in direction and magnitude by changes in actual rates of flavin formation (4).

Effects of nutritional factors have been investigated by inducing dietary riboflavin deficiency. Highly selective effects upon riboflavin-metabolizing enzymes were observed with depressed flavokinase activity, elevated FAD pyrophosphorylase activity, and unaltered FMN phosphatase activity (5). These findings may explain why riboflavin deficiency depletes tissue FMN concentrations to a greater extent than those of FAD.

Recent studies on effects of psychotropic drugs on flavin metabolism derived their impetus from recognition of the remarkably similar structures of riboflavin, phenothiazine derivatives, and tricyclic antidepressants. We determined that drugs of these two types markedly impair FAD formation.

Groups of rats received i.p. injections of chlorpromazine, imiparamine, or amitriptyline, 2–25 mg/kg body weight, for varying periods of time. Controls were age- and sex-matched, and injected with a similar volume of isotonic saline. 1 hour prior to sacrifice, each rat received a single s.c. injection of (^{14}C)riboflavin, 25 μCi/kg body weight. Rats were sacrificed by decapitation and tissues promptly removed. Formation of (^{14}C)FAD was determined by reverse isotope dilution and anion exchange column chromatography with DEAE-Sephadex A-25 (6). This procedure for extraction of flavins yields 98% of total radioactivity in liver.

Some of the data obtained by studies of this kind are shown in Table 1. In this experiment, rats were treated with each of three drugs twice daily for three days. (^{14}C)FAD formation was diminished by approximately one-third in all four organs studied, liver, cerebrum, cerebellum, and heart.

At this high concentration, chlorpromazine, imipramine, and amitriptyline are equally effective in depressing (^{14}C)FAD formation (7,8). Further investigations revealed that chlorpromazine inhibits flavin synthesis more rapidly and more effectively at lower doses than either imipramine or amitriptyline. Each of these drugs inhibits hepatic flavokinase *in vitro*. Psychoactive drugs structurally unrelated to riboflavin, including phenytoin, haloperidol, phenobarbital, diazepam, and chlordiazepoxide, are ineffective in inhibiting riboflavin metabolism.

In further studies, rats received chlorpromazine (2 mg/kg body weight) for a 3- and 7-week period. This dose was selected because it is comparable on a

Table 1. Inhibition of Incorporation of [^{14}C]Riboflavin into [^{14}C]FAD One Hour After a Subcutaneous Injection[a] in Liver, Cerebrum, Cerebellum, and Heart of Adult Male Rats Treated with Chlorpromazine, Imipramine or Amitriptyline (25 mg/kg Body Weight I.P.) Twice Daily for Three Days.

Organ	Treatment group	[^{14}C]FAD (dpm/100 mg)	Significance of difference from control
Liver	Control	12,769 ± 787[b]	
	Chlorpromazine[c]	9,994 ± 499	$p < 0.02$
	Imipramine	10,389 ± 632	$p < 0.05$
	Amitriptyline	10,022 ± 683	$p < 0.02$
Cerebrum	Control	774 ± 37	
	Chlorpromazine	499 ± 43	$p < 0.001$
	Imipramine	503 ± 24	$p < 0.001$
	Amitriptyline	510 ± 32	$p < 0.001$
Cerebellum	Control	941 ± 26	
	Chlorpromazine	608 ± 36	$p < 0.001$
	Imipramine	676 ± 38	$p < 0.001$
	Amitriptyline	698 ± 52	$p < 0.001$
Heart	Control	4,932 ± 362	
	Chlorpromazine	3,227 ± 214	$p < 0.01$
	Imipramine	3,295 ± 128	$p < 0.01$
	Amitriptyline	3,340 ± 172	$p < 0.01$

[a]25 μCi/kg body wt.

[b]Data are shown as mean ± SEM with 7–9 animals per group.

[c]Groups of animals received i.p. injections of each drug; controls were age- and sex-matched and were injected with a similar volume of isotonic saline.

(*Source*: *Adapted from Pinto et al.* [*8*].)

Table 2. Erythrocyte Glutathione Reductase Activity and Urinary Riboflavin Excretion in Adult Male Rats Treated with Chlorpromazine, 2.0 mg/kg Body Weight, by Daily I.P. Injection for 3 to 7-Week Periods, and in Age- and Sex-Matched Pair-Fed Control Rats Treated with Saline.[a]

Study period (weeks)	Treatment group	Erythrocyte glutathione reductase Activity[b]: Without FAD (μmoles NADPH oxidized/gHb/min)	Erythrocyte glutathione reductase Activity[b]: FAD added (μmole NADPH oxidized/gHb/min)	Erythrocyte glutathione reductase Activity[b]: Activity coefficient	Urinary riboflavin excretion[b] μg/mg creatinine
3	Control	7.06 ± 0.92	7.88 ± 1.03	1.13 ± 0.02	3.92 ± 0.35
	Chlorpromazine	9.73 ± 1.00	12.24 ± 1.07	1.28 ± 0.04	9.13 ± 0.48
	Significance of difference	N.S.	$p < 0.02$	$p < 0.005$	$p < 0.001$
7	Control	5.49 ± 0.26	5.48 ± 0.26	1.00 ± 0.03	3.80 ± 0.35
	Chlorpromazine	7.03 ± 0.33	9.25 ± 0.53	1.33 ± 0.08	8.57 ± 0.97
	Significance of Difference	$p < 0.005$	$p < 0.001$	$p < 0.005$	$p < 0.001$

[a]Data are shown as mean $\pm$ SEM with 8–13 animals per group.

[b]Urinary riboflavin excretions were obtained on the last day of the treatment period; erythrocyte glutathione reductase activities were measured in samples obtained at sacrifice.

(*Source*: *Adapted from Pinto et al.* [*8*].)

weight basis to that used to treat patients with psychiatric disorders. All experimental and age- and sex-matched controls were housed in metabolic cages, permitting collection of urine samples. Rats ate Purina Chow determined to contain 24 ppm riboflavin, which represents 30 times the Recommended Dietary Allowance for this vitamin. The activity of erythrocyte glutathione reductase (9, 10), urinary riboflavin excretion (11) and tissue FMN and FAD concentrations (2) were determined.

Results of some of these experiments are shown in Table 2. A highly significant difference was noted between control and chlorpromazine-treated groups in the activity coefficient of erythrocyte glutathione reductase activity after both 3 and 7 weeks of treatment. This finding is similar to that observed with dietary riboflavin deficiency (12). It is essential to recognize that this index of riboflavin deficiency was present in drug-treated animals even while receiving a diet abundant in riboflavin.

There was also a marked increase in urinary riboflavin excretion in drug-treated animals. Subsequent experiments revealed that increased riboflavin excretion occurs within hours of treatment with chlorpromazine. As would be expected following chronic drug treatment and continued urinary losses of the vitamin, hepatic concentrations of FMN and FAD are reduced below normal levels (8).

This line of investigation has revealed, in summary, that chlorpromazine has four significant effects upon riboflavin metabolism in the rat as a whole:

1. Conversion of riboflavin to FAD is impaired, probably as a result of inhibition of flavokinase. Chlorpromazine inhibits both basal synthesis and thyroid hormone-induced increases in FAD synthesis.
2. The activity coefficient of erythrocyte glutathione reductase, an FAD-containing enzyme widely utilized, both in experimental animals and in man as an index of riboflavin deficiency, is elevated. This finding is compatible with a deficiency state.
3. The urinary excretion of riboflavin is more than twice that of age- and sex-matched pair-fed control rats.
4. Hepatic concentrations of FMN and FAD are significantly lower than those of pair-fed littermates, despite consumption of a diet abundant in riboflavin.

Thus, certain psychotropic drugs interfere with flavin metabolism, inhibiting conversion of riboflavin to FMN and FAD, promoting excretion and impairing overall utilization. Nutritional status, hormones, and certain drugs all interact to regulate flavin metabolism.

ACKNOWLEDGMENTS

Supported by NIH Grants 1 P30 CA-29502, AM-15265, and CA-12126, and by Grants from the Guttman Foundation, Hoffmann-La Roche, Inc., National Dairy Council, National Live Stock and Meat Board, U.S. Brewers Association, a Future Leaders Award, the Nutrition Foundation (J.P.) and a Postdoctoral Training Grant, #5T32 HL 07379-04 (N.P.).

References

1. Rivlin, R.S., Fazekas, A.G., Huang, Y.P., and Chaudhuri, R. (1976) In *Flavins and Flavoproteins. Fifth International Symposium*. T.P. Singer (ed.) Amsterdam: Elsevier, pp. 747–753.
2. Rivlin, R.S., Menendez, C., and Langdon, R.G. (1968) *Endocrinology* 83: 461–469.
3. Rivlin, R.S. (1970) *N Engl J Med* 283:463–472.
4. Fazekas, A.G., Pinto, J., Huang, Y.P., and Rivlin, R.S. (1978) *Endocrinology* 102: 641–648.
5. Fass, S. and Rivlin, R.S. (1969) *Am J Physiol* 217:988–991.
6. Fazekas, A.G. In *Riboflavin*. (1975) R.S. Rivlin (Ed.) New York: Plenum Press, pp. 81–98.
7. Pinto, J., Wolinsky, M., and Rivlin, R.S. (1979) *Biochem Pharmacol* 28:597–600.
8. Pinto, J., Huang, Y.P., and Rivlin, R.S. (1981) *J Clin Invest* 67:1500–1506.
9. Beutler, E. (1969) *J Clin Invest* 48:1956–1966.
10. Fieldings, H.A. and Langley, P.E. (1958) *Am J Clin Pathol* 30:528–529.
11. King, T.E., Howard, R.L., Wilson, D.F., and Li, J.C.R. (1962) *J Biol Chem* 237:2941–2946.
12. Bamji, M.S. (1969) *Clin Chim Acta* 26:263–269.

Published 1982 by Elsevier North Holland, Inc.
Vincent Massey and Charles H. Williams, Editors
Flavins and Flavoproteins

CHAPTER 68

Lipoamide Dehydrogenase and Inherited Ataxias

R.A. Pieter Kark, David M. Becker, and Susan L. Perlman

Department of Neurology and the Reed Neurological, Jerry Lewis Neuromuscular and Mental Retardation Research Centers, UCLA School of Medicine, Los Angeles, California

Defects of pyruvate metabolism are associated with several recessively-inherited brain diseases (1). For example, severe defects of enzymes of the pyruvate dehydrogenase complex (PDHC) (1-3) (including lipoamide dehydrogenase) of pyruvate carboxylase and of Krebs' cycle enzymes (2) occur in infants with mental and motor retardation, seizures, and lactic acidosis progressing to early death. Moderate defects of PDHC are associated with progressive neurological defects, especially ataxia (incoordination and poor balance), beginning at age 1 to 3 years. The severity gets worse when blood levels of pyruvate and lactate are elevated (1,4).

Mild abnormalities of pyruvate metabolism have been associated with recessively-inherited ataxias presenting between 3 and 20 years of age and progressing over 1 to 4 decades. Sensation, reflexes, strength, the heart, and the bones are also involved. The syndromes include Friedreich's syndrome (Friedreich's ataxia plus clinical variants) and less well-defined illnesses known as ragged-red ataxias, Kearns-Sayre syndrome, or progressive external ophthalmoplegia-plus with ataxia (5–7). The biochemical abnormalities are found in some families but not all and include elevated levels of pyruvate, lactate, or alanine (6,8–10), low rates of oxidation of [2-^{14}C]- or [U-^{14}C]-pyruvate by tissue slices or whole cells (8,11), decreased activity of PDHC (sometimes with decreased activity of the ketoglutarate dehydrogenase complex, (KGDHC) (10,12–14), and in some families, decreased activity of lipoamide dehydrogenase (15,16).

There are several problems with the latter findings. The enzyme abnormalities are only partial, some 30% to 50% of residual enzyme activity remaining. Some workers do not believe that partial defects can be pathogenic (but see 17,18). Because of the residual activity, it is necessary to pay careful attention to many technical details and to do replicate assays on each of 4 to 8 separate samples per subject in order to demonstrate the abnormalities (8,12,14–16,19). When the original methods have been followed, the results have been reproduced (13), but when there have been changes in critical points, or entirely other assays have been used, assays not equivalent to the original ones (8,13,19–21), the results have not always been reproduced either by ourselves or by others (13,22–25). Finally, the defects only occur in patients from certain families with Friedreich's or ragged-red ataxia. Some feel that the same biochemical defect must underlie every case (10,25,26). An alternate view is based on the fact that patients' clinical findings differ from family to family (26–28). It is not known whether the clinical differences reflect different primary mutations or whether genetic or enviromental factors can modify the effect of a primary mutation. There are many examples in medicine of a clinical syndrome's being related to each of several biochemical defects (17,18). There are also defects of a single catalytic step—even a single mutation—that present as several clinical syndromes (17,18).

With these caveats, we shall review evidence for an association between partial abnormalities of lipoamide dehydrogenase and recessively-inherited ataxias.

In 1969, partial defects of pyruvate decarboxylase were found in two families with childhood ataxia (29,30). A family with ragged-red ataxia had abnormalities of pyruvate metabolism compatible with defective oxidation of HADH (6). This and other data led to the speculation that defects of pyruvate metabolism might be associated with inherited ataxias or inherited peripheral neuropathies in some families, perhaps especially when, as in Friedreich's ataxia, the two syndromes overlapped (Figure 1).

Figure 1. Ven diagram of early hypothesis. The latter was that some families with inherited disorders producing both ataxia and peripheral neuropathies might have inherited disorders of pyruvate metabolism.

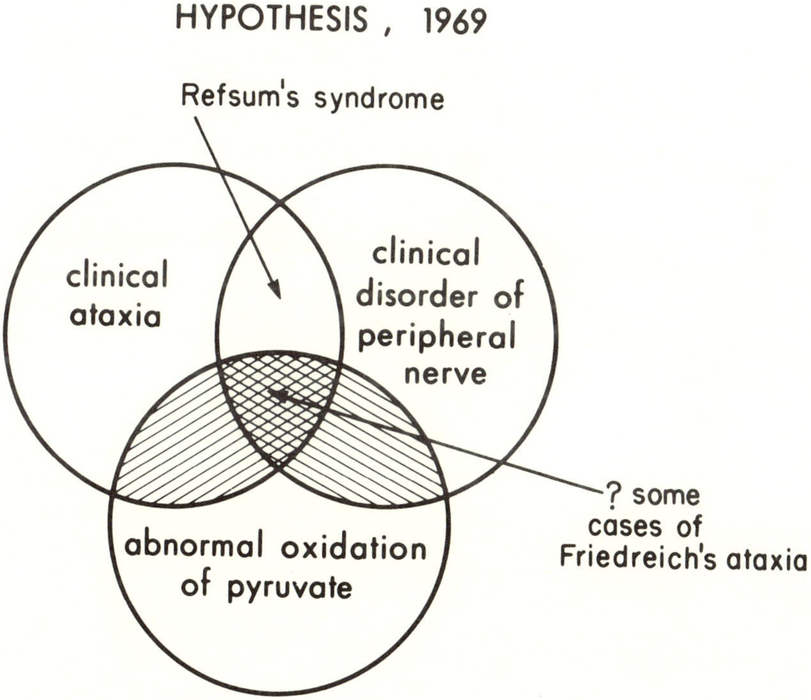

Sliced muscle and intact cultured fibroblasts from about a third of a group of patients with inherited ataxias oxidized pyruvate at rates more than 2 standard deviations below the control mean, as did muscle from a few patients with inherited neuropathies but none with myopathies (8). Other substrates of the Krebs' cycle were oxidized normally by the Friedreich's patients. Affected siblings had similar rates of oxidation, normal or low.

Fibroblast cultures were monitored for contamination by mycoplasma because the latter have up to 100 times the specific activity of PDHC of fibroblasts (31). Homogenates of fibroblasts derived from Friedreich's patients had activities of PDHC and KGDHC 50% or more below the mean value for normals and disease-controls (12). The activities of pyruvate decarboxylase and of cytochrome c oxidase were normal (12). Lipoamide dehydrogenase, present both in PDHC and KGDHC (32–35), was assayed by a modification of the method of Sakurai et al. (32). The activity was decreased in homogenized cells from two Friedreich's families (21). The mitochondrial content of cell samples from the same subject varied considerably even when replicate samples were taken at the same time. The resulting variation in the specific activity of lipoamide dehydrogenase could be decreased by expressing the activity as a ratio to a mitochondrial marker, cytochrome c oxidase. This we have since done routinely.

To ascertain the prevalence of these abnormalities and their correlation with clinical syndromes, lipoamide dehydrogenase was assayed in platelet-enriched fractions (19) of 20 ml to 100 ml fresh blood from 19 normal controls and 26 patients with various inherited ataxias, including 17 patients with Friedreich's syndrome (27). Eleven Friedreich's patients and one with ragged-red ataxia had activities of lipoamide dehydrogenase between 35% and 48% of the control mean. Affected siblings again had similar values. The population of patients with Friedreich's syndrome differed from the normal population by the χ^2-test ($p<0.001$) (Figure 2). There were, in retrospect, several consistent clinical differences between the Friedreich's patients with low enzyme activity and those with normal activity. However, we have since found that these clinical points do not always predict the enzyme's activity.

Could an hypothesis as simple as that illustrated in Figure 3 account for the results? "A" represents the normal gene for the enzyme (or for any protein giving posttranslational changes in the enzyme) and "a," a mutation. The hypothesis is that the mutation is inherited in the same pattern as the clinical disease; allelic compounds (36) are ignored. The normal products are shown as straight lines and the abnormal ones as curved lines. The human enzyme has the same relative molecular weight as pig heart enzyme (Becker, Knipprath, and Kark, unpublished) and is shown as homodimers and a possible heterodimer. Alterations could be due to a change in conformation, to a posttranslational change in structure, or to a primary change in the amino acid sequence.

Several findings are consistent with this hypothesis but none distinguishes between the possible explanations for an alteration in the enzyme's shape. K_m values for lipoamide, NADH, or both were estimated in 8 of the 11 Friedreich's patients with low activity. Three were normal while 5, from 4 families, had elevated values (15). The abnormalities, plus rapid heat inactivation (16) have since been found in all affected siblings in these families. Some abnormalities

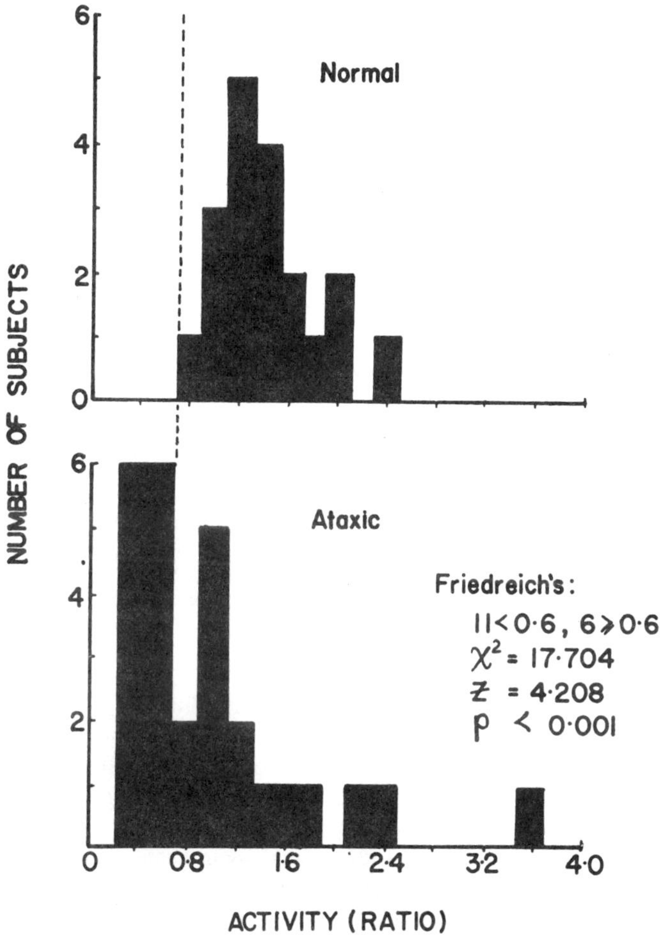

Figure 2. Distribution of activity of lipoamide dehydrogenase in population of normal controls vs patients with inherited ataxias. Activities were expressed as a ratio to that of the mitochondrial marker, cytochrome c oxidase. Determinations were in platelet-enriched fractions of blood. (Data recalculated from ref. 27.)

were found in one seemingly normal sibling 2 years before he became symptomatic (15). Unfortunately, we have not so far been able to get permission to study the other normal siblings.

The parents from 3 families could be studied with improved assay conditions (15). The affected children again had low enzyme activity, averaging 48% of the new normal mean. The parents' enzymes' activities averaged 66% of the normal mean, intermediate between that and the mean for the affected children. This is a conventional criterion for the biochemical heterozygous state in human disease. In addition, the parents' enzymes had abnormal K_m values that we believe are compatible with the postulated alteration in enzyme shape (cf Figure 3).

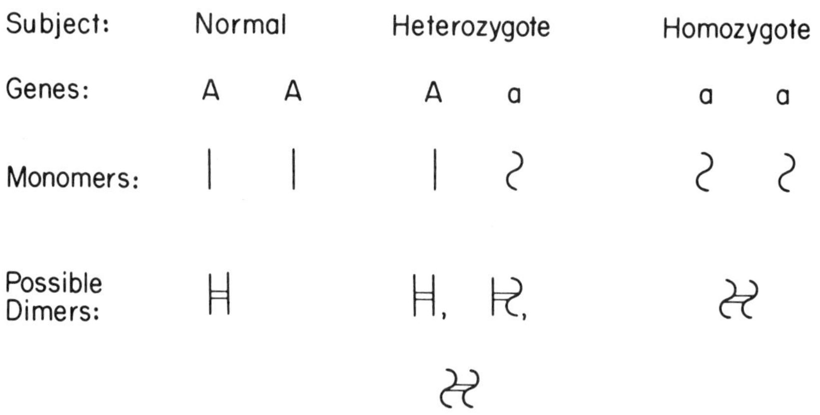

Subject:	Normal	Heterozygote	Homozygote
Genes:	A A	A a	a a
Monomers:			
Possible Dimers:			

Figure 3. Genetic hypothesis of abnormal lipoamide dehydrogenase values in families with ataxia. The lower case "a" represents a hypothetical mutation of a gene coding for the enzyme or for some other protein that produces post-translational changes in the enzyme. The straight lines represent the normal monomers, the curved lines the monomers with altered shape and the pairs of lines, hypothetical homo- and hetero-dimers. Note that one cannot predict the likely kinetic values of the mixtures of dimers for the heterozygote, but that an abnormal value or a heterozygote could be indicative of some abnormality of shape within the mixture.

The data are consistent with the motion that abnormalities of lipoamide dehydrogenase are in some way associated with the ataxia in these families. The abnormalities may serve as a genetic marker, a biochemical predisposition, or even a clue to the primary biochemical defect. However, the results have all been obtained in simple homogenates of tissue. Lipoamide dehydrogenase is presumably present largely as a part of multiprotein dehydrogenase complexes. As the complexes can impose reversible changes in the conformation of the enzyme (32–35), the findings do not distinguish between a conformational change and a direct alteration in the enzyme's structure. Answers to these questions, and more readily reproducible assays than are now available, must await studies of isolated, highly purified human lipoamide dehydrogenase.

ACKNOWLEDGMENTS

Supported in part by Grants from NIH, NS-15245, HD-04612 and RR-865; by Grants and a fellowship to S.L.P. From the Muscular Dystrophy Association, a Grant from the Easter Seal Research Foundation, and by donations from the Marjorie L. Johnson Fund, the Friedreich's Ataxia Group in American and the National Ataxia Foundation.

References

1. Blass, J.P. (1979) *Neurology (New York)* 29:280–286.
2. Robinson, B.H., Taylor, J., and Sherwood, W.G. (1980) *Pediatr Res* 14:956–962.
3. DeVivo, D.C., Haymond, M.W., Obert, K.A., Nelson, J.S., and Pagliara, A.S. (1979) *Ann Neurol* 6:483–494.
4. Falk, R.E., Cederbaum, S.D., Blass, J.P., Gibson, G.E., Kark, R.A.P., and Carrel, R.E. (1976) *Pediatrics* 58:713–721.

5. Berenberg, R.A., Pellock, J.M., Dimauro, S., Schotland, D.L., Bonilla, E., Eastwood, A., Hays, A., Vicale, C.T., Behrens, M., Chutorian, A., and Rowland, L.P. (1977) *Ann Neurol* 1:37–54.
6. Kark, R.A.P., Weinbach, E.C., Blass, J.P., and Engel, W.K. (1973) In Kakulas, B.A. (ed.) *Clinical Studies in Myology* (ICS 295). Amsterdam: Excerpta Med., pp. 98–107.
7. Roberts, N.K., Perloff, J.K., and Kark, R.A.P. (1979) *Amer J Cardiol* 44:1396–1400.
8. Kark, R.A.P., Blass, J.P., and Engel, W.K. (1974) *Neurology (Minneap.)* 24:964–971.
9. Dunn, H.G. and Dolman, C.L. (1969) *Neurology (Minneap.)* 19:536–549.
10. Barbeau, A., Butterworth, R.F., Ngo, T., Breton, G., Melancon, S., Shapcott, D., Geoffroy, G., and Lemieux, B. (1976) *Canad J Neurol Sci* 3:379–388.
11. Darsee, J.R., Miklozek, C.L., Hopkins, L.C. Jr., Heymsfield, S.B., Wenger, N.K., and Nutter, D.O. (1981) *Clin Res* 29:429A.
12. Blass, J.P., Kark, R.A.P., and Menon, N.K. (1976) *New Eng J Med* 295:62–67.
13. Bertagnolio, B., Uziel, G., Bottachi, E., Crenna, G., D'Angelo, A., and DiDonato, S. (1980) *Canad J Neurol Sci* 7:409–412.
14. Kark, R.A.P. and Rodriguez-Budelli, M. (1979) *Neurology (New York)* 29:126–131.
15. Kark, R.A.P., Rodriguez-Budelli, M., Perlman, S., Gulley, W.F., and Torok, K. (1980) *Neurology (New York)* 30:502–508.
16. Kark, R.A.P., Budelli, M.M.R., Becker, D.M., Weiner, L.P., and Forsythe, A.B. (1981) *Neurology (New York)* 31:199–202.
17. Rosenberg, R.N. (1980) *Trends in Neuro Sciences* 3:144–148.
18. Kark, R.A.P. and Becker, D.M. (1981) *Muscle and Nerve* 4:31–40.
19. Blass, J.P., Cederbaum, S.D., and Kark, R.A.P. (1977) *Clin Chem Acta* 75:21–30.
20. Blass, J.P., Hinman, L., and Soricelli, A. (1981) *Neurology (New York)*, in press.
21. Rodriguez-Budelli, M. and Kark, R.A.P. (1978) *Neurology (New York)* 28:1283–1286.
22. Stumpf, D.A. and Parks, J.K. (1978) *Ann Neurol* 4:366–368.
23. Constantopoulos, G., Chang, C.S.C., and Barranger, J.A. 91980) *Ann Neurol* 8:636–639.
24. Evans, O.B. (1981) *Ann Neurol* 9:93–94.
25. Barbeau, A., Melancon, S., Butterworth, R.F., Filla, A., Izumi, K., and Ngo, T.T. (1978) In *Advances in Neurology*, Vol. 21. Kark, R.A.P., Rosenberg, R.N., and Schut, L.J. (eds.) chap. 14, New York: Raven Press, pp. 203–217.
26. Barbeau, A. (1979) *Canad J Neurol Sci* 6:145–319.
27. Kark, R.A.P. and Rodriguez-Budelli, M.M. (1979) *Neurology (New York)* 29:1006–1013.
28. Skre, H. (1976) *Hereditary Ataxias and Their Associated Traits* (Thesis) pp. 1–288, University of Bergen and Oslo.
29. Blass, J.P., Avigan, J., and Uhlendorf, B.W. (1970) *J Clin Invest* 49:423–432.
30. Blass, J.P., Kark, R.A.P., and Engel, W.K. (1971) *Arch Neurol* 25:449–460.
31. Farrell, D.F. (1977) In *Research to Practice in Mental Retardation. Vol. III. Biomedical Aspects.* Mittler, P. (ed.) Baltimore: Univ. Park Press, pp. 147–155.
32. Sakurai, Y., Fukuyoshi, Y., Hamada, M., Hayakawa, T., and Koike, M. (1970) *J Biol Chem* 245:4453–4462.
33. Kenney, W.C., Zakim, D., Hogue, P.K., and Singer, T.P. (1972) *Europ J Biochem* 28:253–260.
34. McManus, I.R. and Cohn, M.L. (1974) In *Isozymes: Molecular Structure.* Markert, C.L. (ed.) Nw York: Acad. Press, pp. 621–636.
35. Koike, M. and Koike, R. (1976) *Advances in Biophysics* 9:187–227.
36. O'Brien, J.S. (1978) *Am J Hum Genet* 30:672–675.

Published 1982 by Elsevier North Holland, Inc.
Vincent Massey and Charles H. Williams, Editors
Flavins and Flavoproteins

CHAPTER 69

Monoamine Oxidase Type A from Human Placental Mitochondria: Isolation Procedure and Properties

Daniela H. Gürne and E. Albert Zeller

Department of Pathology and Laboratory Medicine, Evanston Hospital, Evanston, Illinois

In 1948, human placenta was found to be a rich source of monoamine oxidase [MAO; amine: oxygen oxidoreductase (deaminating) (flavin-containing) 1.4.3.4] (1). Since the same organ also exhibits a very high diamine oxidase activity (2, 3), it must play an important role in amine metabolism. Later, placental MAO was shown to belong almost exclusively to the A-type (4). Since it had been purified only partially before (4–6), we undertook its isolation so that we could study the properties of a typical MAO-A in a near-pure state and to compare the results with those reported for MAO-B or for the enzyme present in "intact" placental mitochondria.

Isolation of MAO-A

Mitochondria (900 mg, 32 units) prepared from two fresh placentae, were washed with 90 ml of 0.5% Triton X-100 in 50 mM potassium phosphate buffer pH 7.6, and MAO was extracted several times with 1% Triton (protein concentration 5 mg/ml). The extracts were centrifuged at 46,000 g for 30 minutes. Those with highest activities were combined and the enzyme precipitated with ammonium sulfate, 35% saturation at 4°. After centrifugation at 46,000 g, the upper layer was collected, diluted with potassium phosphate buffer (8 ml, 120 mg, 12 units) and applied to a Sepharose 6B column (2.5×100 cm), equilibrated with 50 mM potassium phosphate, containing 1 mM of DL-dithiothreitol (DTT) and 0.05% Triton. The fractions with the highest activity were concentrated in an Amicon cell to 15 ml (45 mg, 10 units). The concentrate was further fractionated on DEAE-Sepharose (1.5×17 cm), equilibrated with 20 mM potassium phosphate in 20% glycerol and 1 mM DTT, pH 7.4. The enzyme was eluted with 20/400 mM potassium phosphate gradient, containing 20% glycerol, 0.8% of n-octyl-glucoside, and 1 mM DTT. We dialyzed the fractions containing most of the activity with several changes of 20 mM phosphate buffer pH 7.4 at 4°, (6.9 mg, 7 units, 23 ml). This preparation is stable for several weeks at 4° with only a small activity loss. The oxidase was next concentrated with an Amicon Diaflo cell and purified further with Bio-Gel HTP, equilibrated with 20 mM potassium phosphate, 20% glycerol, and 1 mM DTT. We eluted the enzyme with 140 mM potassium phosphate, 20% glycerol, and 1 mM DTT. Thus, for the first time, no detergent was required for eluting MAO. The total yield from mitochondria varied from 8–20%. As judged by the sodium dodecyl sulfate (SDS) electrophoresis performed by the method of Neville (7) and Maizel (8) in different systems, with respect to pH and gel concentrations, one major band was stained in all cases by Coomassie Brilliant Blue (Figure 1).

The enzymic activity of MAO was routinely assayed by measuring the rate of oxidation of M-methyl-β-phenylethylamine by a modified Köchli and von Wartburg method. One unit of activity is defined as the amount of enzyme which oxidizes one μmole of substrate per minute at 25°.

Determination of Molecular Weight of Subunits

For MW determinations of subunits we carried out the SDS electrophoresis by the method of Maizel (8) for which known standards were run simultaneously with the enzyme samples. The proteins used as standards included phosphorylase b, albumin, ovalbumin, carbonic anhydrase, trypsin inhibitor, and lactalbumin. After staining, the gels were scanned with Varian gel scanner attached to Varian 219 spectrophotometer. The subunit MW was calculated to be 63,000.

Absorption Spectrum

The absorption spectrum of the purified MAO-A (Figure 2) clearly resembles that of purified MAO-B from beef liver and other sources (9–11). It has been suggested that the 410 nm peak belongs to a heme-like impurity (10, 11) which can be removed from beef liver enzyme without loss of activity. So far, we have failed to extract any heme with acetone/HCl from our preparation or to change the spectrum by alkaline sucrose gradient centrifugation (according to a personal communication from T.P. Singer et al.). However, after treatment of

Figure 1. Polyacrylamide gel electrophoresis (SDS) scan of placental MAO (10 μg after DEAE) in tris-boric acid system pH 8.3. Sample was incubated with 1% SDS and 20 mM DTT at 95° for 5 minutes.

SDS GEL ELECTROPHORESIS SCAN

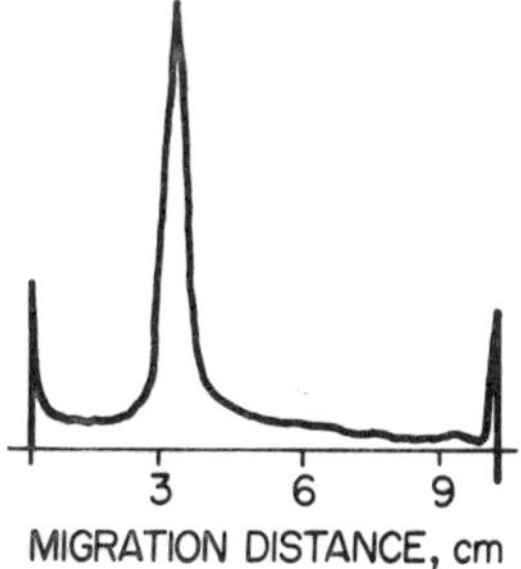

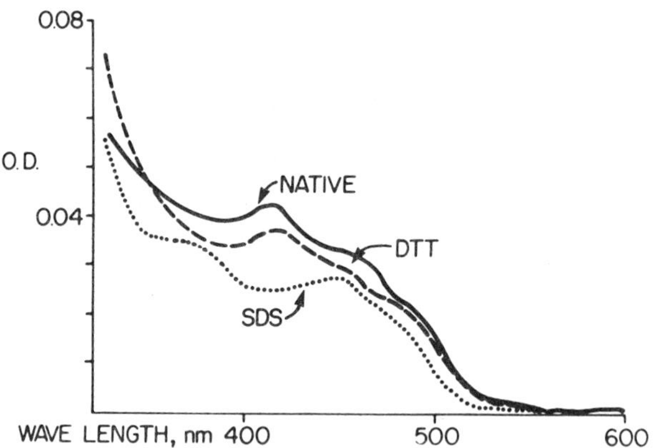

Figure 2. Absorption spectrum change of purified MAO. Protein concentration 0.3 mg/ml. Native MAO (—); Incubated with 40 mM of DTT (---); Incubated with 40 mM DTT and 28 mM SDS (···).

the enzyme with 0.8% of SDS and in the presence of DTT, the peak at 410 nm disappeared (Figure 3), accompanied by a complete loss of MAO activity.

When the purified enzyme interacts with clorgyline, a decrease of absorbance in the range of 440 to 500 nm takes place, with a concomitant increase of the maximum at 410 nm, the latter disappearing on treatment with SDS and DTT. This trend became already apparent when the first cyanoflavin was produced by the photochemical interaction of flavins with acetylenic inhibitors (12) and when it was shown that the increase of the 410 nm peak exactly paralleled the degree of inhibition by purified beef liver MAO caused by Su 11-723 (13).

Flavin Content

The purified enzyme retains its remarkable sensitivity towards clorgyline, ($pI_{50} = 9$). Assuming binding of one mole clorgyline to one mole flavin, the number of active flavins per one subunit of MAO (63,000) was found to be 0.65. A similar value, 0.62, was computed from the 450 nm peak, in the spectrum of native MAO (Figure 2) and in the trypsin/chymotrypsin digest, prepared according to Salach (14), taking the value of 11,000 as the molar extinction coefficient (15). No flavin was extracted by 2% perchloric acid or 5–10% TCA, suggesting covalent binding of flavin to protein, as in MAO-B (16).

Turnover Number and pH Optimum

Based on estimates of the enzyme concentration obtained by titration with clorgyline, the turnover number reached approximately 100 per minute for serotonin at 25° and in air, a rather low reaction rate.

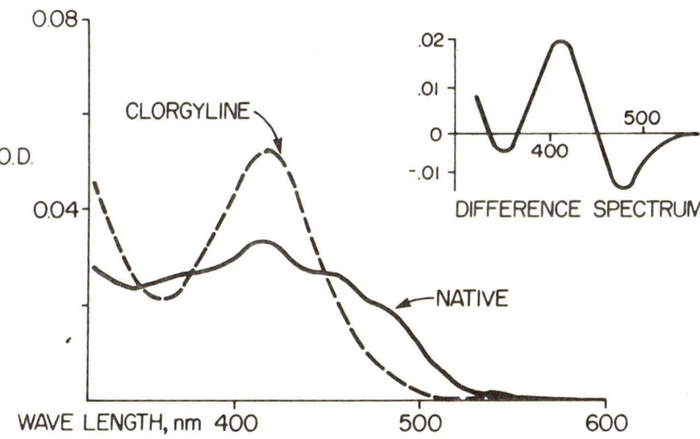

Figure 3. Flavin spectrum change after inactivation of purified MAO (0.3 mg/ml) by 8 μM of clorgyline. Inset: Difference spectrum for the same conditions.

The V_{max} for placental MAO is not constant for a given substrate at various pH values tested, in contrary to findings for MAO-B from porcine brain (17) and human platelets (18). With tyramine and phenylethylamine, it reaches its optimum around pH 9. The rate measurements were carried out with the Clark electrode.

Specificity Pattern of Purified Placental MAO

The purified enzyme displayed a substrate specifity typical for MAO-A (Table 1): Serotonin (5HT), tyramine, and N-methyl-PEA exhibit the highest reaction rates. PEA which is often considered to be a type B-specific substrate, is degraded almost half as fast as compared with 5HT! It was found in this laboratory that, in contrast to MAO-B, the shorter aliphatic amines are only slightly attacked by MAO-A, if at all, while the longer n-alkylamines are increasingly better oxidized by the type A. Here, n-hexylamine undergoes degradation in the presence of the purified enzyme with 73% of the rate of 5HT. Thus, we found another nonhydroxylated amine which is fairly well attacked by both MAO types. It shares this reactivity with N-methyl-PEA (19, 20). The para-halogenated PEAs offer a striking difference in their response to A- and B-activity: while with MAO-B preparations the logarithm of V_{max} obtained for the halogenated compounds follow a Hammett pattern with a positive slope, we find that the pure (and mitochondrial) MAO-A is increasingly less capable of degrading the Cl, Br-, and I-derivative. This phenomenon may be explained in part by the occurrence of nonproductive binding of substrate: large groups in paraposition of the PEA molecule may not fit well into the active site of MAO-A and force the substrate amine into nonproductive orientations. The phenomenon of nonproductive binding has already been associated with MAO in 1965 (21, 22).

Table 1. Substrate Pattern of Purified and Mitochondrial MAO from Placenta.

	Mitochrondria				Purified MAO			
Substrate	**Km**	**$\widetilde{Km}$**	**Vm**	**Rr**	**Km**	**Km**	**Vm**	**Rr**
Serotonin	125	0.030	0.66	100%	145	0.36	1.0	100%
Tyramine	45	0.04	0.33	50	75	0.06	0.70	70
N-Ch_3-PEA	86	0.17	0.27	41	101	0.20	0.92	92
n-Hexylamine					194	0.21	0.73	73
Phenylethyamine	100	0.40	0.18	27	188	0.75	0.49	49
p-F-PEA	100	1.56	0.33	50	170	2.60	1.10	110
p-Cl-PEA	21	0.10	0.15	23	16	0.08	0.22	22
p-Br-PEA	13	0.04	0.07	11	17	0.05	0.11	11
p-I-PEA	16	0.05	0.01	2	45	0.02	0.02	2

Km and $\widetilde{Km}$: μM; Vm: μmole/min/mg protein at 25°; Rr: relative rate. $\widetilde{Km}$: computed for nonprotonated amine concentration.

ACKNOWLEDGMENT
Thus study was generously supported by Shriners Hospital for Crippled Children, Chicago Unit.

References

1. Luschinsky, H.L. and Singher, H.O. (1948) *Arch Biochem Biophys* 19:95.
2. Zeller, E.A., Schär, B., and Staehlin, S. (1939) *Helv Chim Acta* 22:837.
3. Danforth, D.N. (1939) *Proc Soc Exp Biol Med* 40:319.
4. Bathina, H.B., Huprikar, S.V., and Zeller, E.A. (1975) 34:293.
5. Egashira, T. (1976) *Jpn J Pharmacol* 26:493.
6. Youdim, M.B.H. and Sandler, M. (1967) *Biochem J* 105:43P.
7. Neville, D.M. (1971) *J Biol Chem* 246:6328.
8. Maizel, J.V., Jr. (1971) In *Methods in Virology*. K. Maramorosch and H. Koprowski (eds.) Vol. 5, New York: Academic Press, p. 179.
9. Singer, T.P. (1979) In *Monoamine Oxidase: Its Structure, Function, and Altered Functions*. T.P. Singer, R.W. Von Korff, and D.L. Murphy (eds.) New York: Academic Press, pp. 7–24.
10. Oreland, L. (1971) *Arch Biochem Biophys* 146:410.
11. Chuang, H.Y.K., Patek, D.R., and Hellerman, L. (1974) *J Biol Chem* 249:2381.
12. Zeller, E.A., Gärtner, B., and Hemmerich, P. (1972) *Z Naturforsch* 27b:1052.
13. Zeller, E.A. and Hsu, M. (1973) In *Frontiers in Catecholamine Research*. E. Usdin and S. Snyder (eds.) New York: Pergamon Press Inc., pp. 153–155.
14. Salach, J.I. (1979) *Arch Biochem Biophys* 192:128.
15. Ghisla, S. (1980) *Methods of Enzymology* 66:360.
16. Kearney, E.B., Salach, I.J., Walker, W.H., Seng, R.L., Kenney, W., Zeszotek, E., and Singer, T.P. (1971) *Eur J Biochem* 24:321 and references cited therein.
17. Williams, C.H. (1973) *Biochem Pharmacol* 23:615.
18. Gürne, D.H. and Zeller, E.A., unpublished results.
19. Zeller, E.A. and Arora, K.L. (1979) In *Catecholamines: Basic and Clinical Frontiers*. E. Usdin, I. Kopin, and J. Barchas (eds.) New York: Pergamon Press Inc., pp. 17–22.
20. Zeller, E.A., Arora, K.L., Gürne, D.H., and Huprikar, S.B. (1979) In *Monoamine Oxidase: Its Structure, Function, and Altered Functions*. T.P. Singer, R.W. Von Korff, and D.L. Murphy (eds.) New York: Academic Press, pp. 7–24.
21. Zeller, E.A. (1963) *Ann NY Acad Sci* 107:811.
22. Zeller, E.A. (1963) *Biochem* Z 339:13.

PART V:

Modified Flavins, Natural, and Synthetic

Published 1982 by Elsevier North Holland, Inc.
Vincent Massey and Charles H. Williams, Editors
Flavins and Flavoproteins

CHAPTER 70

An Unusual Flavin, Coenzyme F_{420}

L. Dudley Eirich,* Godfried D. Vogels,** and Ralph S. Wolfe[†]

**Portland State University, Portland, Oregon; **University of Nijmegen, Nijmegen, The Netherlands; [†]University of Illinois, Urbana, Illinois*

In 1967, Paul Cheeseman noted that aerobically harvested cells of *Methanobacterium bryantii* exhibited a blue-green fluorescence under UV light. He also observed that the fluorescence disappeared as hydrogen was bubbled through a suspension of these cells. The organism had been mass cultured on hydrogen and carbon dioxide as substrates; so he correctly assumed that the fluorescent compound was reduced by electrons provided via hydrogenase. The blue-green fluorescence was found to be associated with a yellow compound that had a strong absorption at 420 nm; so the compound was referred to as Factor 420 (F_{420}) (1). Cheeseman found that cell suspensions or cell extracts in which F_{420} had been reduced by hydrogenase could be stored at $-20°$ for long periods with full retention of enzymic activity, a finding that greatly benefited our research program on the biochemistry of methanogenesis.

F_{420} was discovered first in methanogens simply because these organisms synthesize large amounts of it (2). The amount of F_{420} found in the following selected organisms is expressed in mg per kilogram of wet cells: *M. bryantii*, 160; *Methanobacterium thermoautotrophicum*, 324; *Methanobacterium formicicum*, 206; *Methanospirillum hungatei*, 319; *Methanogenium marisnigri*, 120; *Methanobrevibacter ruminantium*, strain M1, 6; *Methanosarcina barkeri*, 16. Edwards and McBride (3) first pointed out that the fluorescence of methanogens formed the basis of a powerful technique for the selective recognition of colonies of methanogens, when isolating organisms from nature. McBride (personal communication, 1975) also was the first to use an epifluorescence UV microscope to detect individual cells of methanogens in wet amounts of sediments, sludge, and rumen fluid; he also noted the bright fluorescence of certain rumen protozoa and suggested that methanogens might be involved. This technique was reported by Mink and Dugan (4). The procedure was significantly improved by use of filters, surprisingly good results being obtained with water immersion objectives and filter systems which match the excitation and emission wavelengths of either F_{420} or a group of blue fluorescent compounds (5). The latter group consists of (a) methanopterin (6), a coenzyme probably involved in the binding and reduction of CO_2, (b) the reduced carboxylated form of methanopterin, formerly known as YFC (7), and (c) F_{342}, previously described by Gunsalus and Wolfe (8) and probably a precursor in the synthesis of methanopterin (6). For excitation at 420 nm Leitz Ploemopak filter system D (3 mm BG 3+KP-425) can be used with barrier filter K-460, whereas for excitation of the blue fluorescent compounds at 350 nm filter system A (2×2 UG 1) with barrier filter K-430 yields optimal results.

This technique allows the specific recognition of the cells of methananogenic bacteria even in such very complex ecosystems. In this way, an ectosymbiosis of rumen ciliates of the family *Ophryoscolecidae* (order *Entodiniomorphida*) with methanogenic bacteria was found (9, 10) (Figure 1). The somatic interaction may reflect a metabolic interaction in which efficient interspecies hydrogen transfer benefits both partners. The interaction of ciliates with methanogenic bacteria is strongly influenced by the feeding of the ruminant; during starvation the association is high, but soon after the intake of food the frequency of association decreases. Preliminary results (C.K. Stumm and G.D. Vogels, unpublished data) indicate that methanogenic bacteria in the rumen tend to accumulate rapidly at places where the source of hydrogen is most strong, i.e., the rumen fluid in the well-fed animal or the surface of ciliates in the starving animal.

In studying the reduction of NADP from hydrogen or formate by extracts of *M. ruminantium*, Tzeng, in the laboratory of M.P. Bryant, resolved the enzyme system (by use of a gel diffusion column) for an unknown factor that was required for the formation of NADPH. When F_{420} was added to the resolved extract (Figure 2A), NADP was reduced with hydrogen as the electron donor, the apparent K_m for F_{420} being 5×10^{-6} M; NAD was not reduced by this system (11). F_{420} also was found to be the natural coenzyme that accepted electrons in the formic dehydrogenase reaction of *M. ruminantium* (Figure 2B), an apparent K_m of 8.3×10^{-4} M being found. In these extracts, chemically reduced F_{420} could donate electrons for the formation of hydrogen or for the reduction of NADP by an oxidoreductase (12). Later a similar formate system was found in *M. hungatei* (13) and recent studies have been reported on an F_{420}-NADP oxidoreductase in *Methanococcus vannielii* (16, 17). Recently F_{420} has been found to be the electron acceptor for pyruvate dehydrogenase and for α-ketoglutarate dehydrogenase of *M. thermoautotrophicum* (14, 15).

Figure 1. Ectosymbiosis of a methanogen on the surface of a rumen ciliate. (A) Phase-contrast photomicrograph of methanogenic cells. (B) Photomicrograph of the same area taken with an epifluorescence microscope showing fluorescence of the methanogenic cells. Photomicrogaphs by C. Stumm.

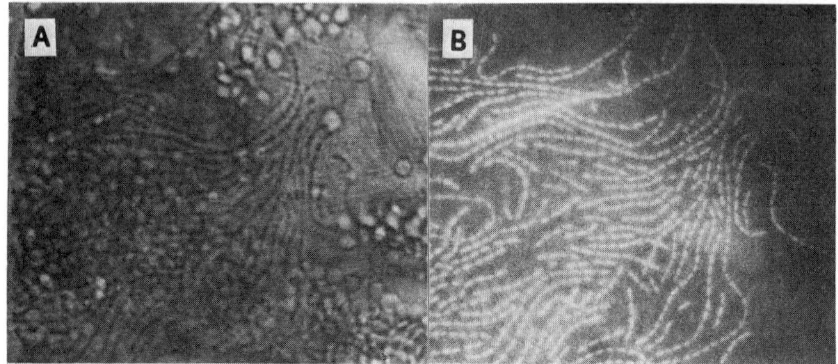

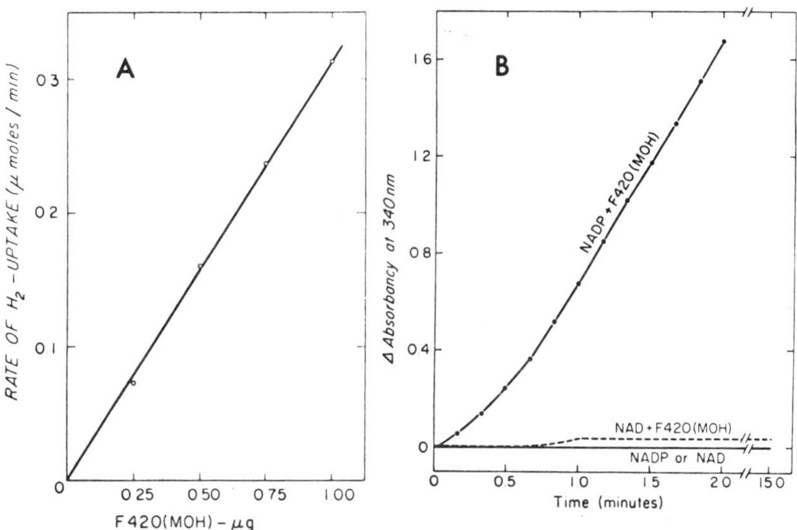

Figure 2. F_{420} dependent reactions in extracts of *Methanobrevibacter ruminantium*. (A) Dependence of hydrogen uptake and reduction of NADP on addition of F_{420}. (B) F_{420}-Dependent NADP reduction from formate dehydrogenase.

SOURCE: *From J. Bacteriology, with permission.*

Following the work of Cheeseman et al. (1), Eirich studied the structure of F_{420} (Ph.D. Dissertation, University of Illinois, 1978). The structure of the coenzyme is presented in Figure 3 (18, 19). UV-visible spectra of F_{420} at 4 pH values are presented in Figure 4. Hydrolysis of F_{420} at 110° produced a series of hydrolytic fragments that were named P, N-2, N-1, F^+, FO, FA, and SAC on the basis of electrophoretic mobility. When F_{420} was treated with periodate, a chromophore, PA with cationic properties, and ALD, a strongly anionic aldehyde were produced. The position of each fragment in the side chain was established by use of various techniques (18). The structure of the chromophore was established by use of spectrophotometric techniques (UV-visible, ^{1}H, and ^{13}C NMR) as well as quantitative analysis. The analytical data of F_{420} and its fragments were compared to that of known flavin derivatives, 8-hydroxy-FMN, FMN, and 5-deazariboflavin (18).

The proposed structure of the chromophore of F_{420} was supported in studies by Walsh et al. (20) and confirmed by Ashton et al. (21) who synthesized 7,8-didemethyl-8-hydroxy-5-deazariboflavin and 8-demethyl-8-hydroxy-5-deazariboflavin; the former compound was shown to be identical with FO and possessed similar properties when tested in enzymic systems of *M. thermoautotrophicum*.

The position of the hydroxy group on the 5-deazaisoalloxazine ring was substantiated (Pol et al., 22) by comparison of the properties of F_{420} with 7- as well as 8-hydroxy-10-methyl-5-deazaisoalloxazine. The physicochemical properties (absorption spectra of oxidized and reduced forms, pK_a values of the hydroxy group, proton NMR spectra, and redox potential) of the 8-hydroxy

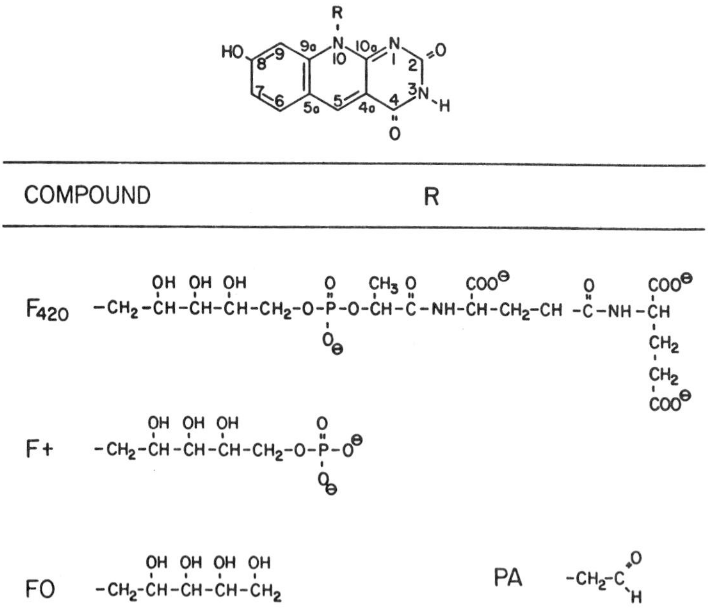

COMPOUND	R
F_{420}	$-CH_2-CH(OH)-CH(OH)-CH(OH)-CH_2-O-P(=O)(O^{\ominus})-O-CH(CH_3)-C(=O)-NH-CH(COO^{\ominus})-CH_2-CH_2-C(=O)-NH-CH(COO^{\ominus})-CH_2-CH_2-COO^{\ominus}$
F+	$-CH_2-CH(OH)-CH(OH)-CH(OH)-CH_2-O-P(=O)(O^{\ominus})-O^{\ominus}$
FO	$-CH_2-CH(OH)-CH(OH)-CH(OH)-CH_2(OH)$
PA	$-CH_2-CHO$

Figure 3. Structures of F_{420} and fragments, F^+, FO, and PA.
SOURCE: *From J. Bacteriology*, with permission.

derivative, in contrast to those of the 7-hydroxy compound were very consistent with those of F_{420}. Moreover, only the 8-hydroxy form could replace F_{420} in the NADP-linked hydrogenase assay which was found to be specific for F_{420} and some closely related derivatives (11, 19).

Reduction of the chromophore of F_{420} was found to occur at positions 1 and 5 (18). The stoichiometry of the reduction of F_{420} with hydrogen revealed that 13.6 μmol of hydrogen was used for the reduction of 13.6 μmol of F_{420}; thus, each molecule of F_{420} received 2 electrons. No ESR signal to indicate the presence of a semiquinone was obtained as a solution of F_{420} was tested during stages of partial reduction at pH 8. It appears that F_{420} may participate only in 2-electron transfer reactions (20). The hydrolytic fragments F^+ and FO were tested for their ability to accept electrons from hydrogenase and to participate in the hydrogenase-F_{420}-NADP oxidoreductase reaction (19). The apparent K_m for each compound in the hydrogenase reaction was: F_{420}, 25 μM; F^+, 100 μM; FO, 100 μM. For NADP reduction, the apparent K_m for each compound was: F_{420}, 27 μM; F^+, 110 μM; D FO, 44, μM. Fragments PA and FA were inactive in these enzymic systems.

When the electrophoretic mobility of F_{420} from each of several methanogens was compared to F_{420} from *M. bryantii*, all were similar except that from *Methanosarcina barkeri* which possessed a more negative charge at pH 4.4 (19).

In this organism, five forms of F_{420} have been found that differ only in the number of glutamate moieties (1 to 5) per molecule of F_{420} (A. Pol, unpublished results).

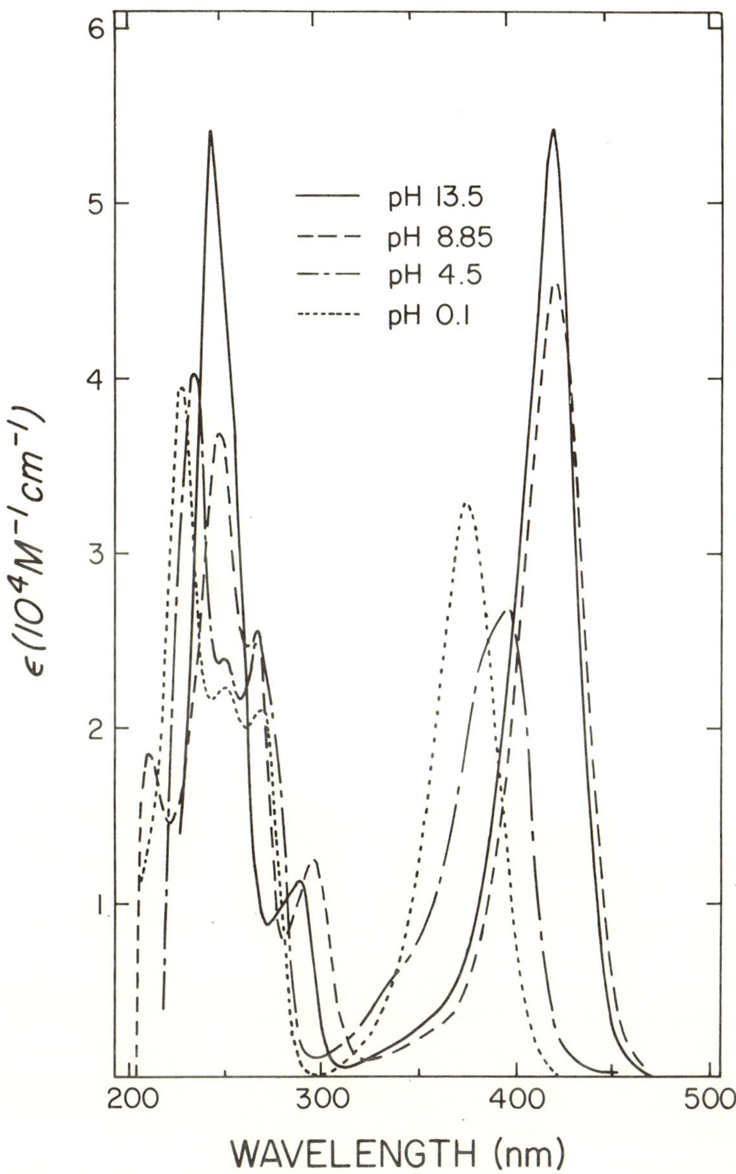

Figure 4. UV-visible absorption spectra of coenzyme F_{420} obtained at the indicated pH values.

SOURCE: *From Biochemistry, with permission.*

F_{420} in Nonmethanogens

Recently (23), 8-hydroxy-5-deazaflavin derivatives (called SF_{420}) were isolated from cells of *Streptomyces griseus*. The chromophoric part of SF_{420} was identified on the basis of absorption spectra at various pH-values, proton NMR data, and pK_a-values, which were all very similar to those observed for F_{420}. Moreover, the changes observed on reduction and reoxidation confirmed the proposed structure of the chromophore. On acid hydrolysis of SF_{420}, two fluorescent compounds were formed, which were identical to F^+ and FO, products also obtained from F_{420} (18). Since these compounds contain a ribitylphosphate and a ribityl group, respectively, SF_{420} must contain a side chain similar to F_{420}. The presence of a lactyl moiety in SF_{420} was evident from the data obtained in the proton NMR spectrum. SF_{420} differed from F_{420} isolated from *M. bryantii* since the former compound yielded two additional ninhydrin-positive substances as compared to L-glutamic acid and γ-L-glutamyl-glutamic acid which were formed from both compounds. The additional ninhydrin-positive substances were converted to L-glutamate on prolonged hydrolysis. It is therefore very tempting to propose that SF_{420} is a mixture of F_{420} and two derivatives containing one and two additional glutamyl residues. SF_{420} was tested in both the hydrogenase and the NADP-linked oxidoreductase systems of *M. bryantii* and was equally as active as F_{420} in both enzyme systems (23).

The amount of SF_{420} was estimated to be 8.1 μmol/kg wet cells of *S. griseus*; this value lies in the same range as the contents of F_{420} measured for *Methanosarcina barkeri* and *Methanobrevibacter ruminantium*, but in other methanogenic bacteria the levels of F_{420} were from 20 to 40 times higher.

The function of SF_{420} in *S. griseus* is quite different (24,25) from the electron carrier function demonstrated for F_{420} in methanogenic bacteria (11–14,19). In *S. griseus*, a photoreactivating enzyme (EC 4.1.99.3) mediates the splitting of pyrimidine dimers in UV-irradiated DNA in the presence of visible or near-UV light; the light absorbed provides the energy needed for the splitting of the cyclobutane ring in the dimer. The photoreactivating enzyme contains an intrinsic chromophoric cofactor, which in its bound form exhibits a characteristic absorption spectrum with a maximum at 445 nm and a shoulder at 425 nm (24). The cofactor was isolated from the highly purified enzyme and was identified as a 7,8-didemethyl-8-hydroxy-5-deazaflavin derivative (25). Native photoreactivating enzyme spontaneously loses its chromophore and simultaneously its biological activity. Both losses are strongly prevented by the presence of UV-irradiated DNA, the substrate of the enzyme. These results strongly indicate that the photoreactivating enzyme contains an intrinisic and biologically active cofactor which is structurally and by origin related, if not identical, to SF_{420}. Moreover, SF_{420} itself was able to sensitize the monomerization of isolated thymine dimers with the help of visible light.

The occurrence of 8-hydroxy-5-deazaflavins in photoreactivating enzymes is probably not limited to *S. griseus*. The chromophore was also demonstrated in the cyanobacterium *Anacystis nidulans* (A.P.M. Eker, unpublished observations) and, as can be judged from the action spectra, 8-hydroxy-5-deazaflavins could be a good candidate for the cofactor of the photoreactivating enzymes of this organism, another cyanobacterium *Agmenellum quadruplicatum*, the green alga *Platymonas subcordiformis*, the moss *Bryum pseudotriquetrum*, and the midge *Smittia* (25).

References

1. Cheeseman, P., Toms-Wood, A., and Wolfe, S. (1972) *J Bacteriol* 112:527–531.
2. Eirich, L.D., Vogels, G.D., and Wolfe, R.S. (1979) *J Bacteriol* 140:20–27.
3. Edwards, T. and McBride, B.C. (1975) *Appl Microbiol* 29:540–545.
4. Mink, R.W. and Dugan, P.R. (1977) *Appl Environ Microbiol* 33:713–717.
5. Doddema, H.J. and Vogels, G.D. (1978) *Appl Environ Microbiol* 36:752–754.
6. Keltjens, J.T. and Vogels, G.D. (1981) *Proc. 3rd Int. Sympos. Microbial Growth on C_1-Compounds*. London: Heyden and Sons, Ltd.
7. Daniels, L. and Zeikus, J.G. (1978) *J Bacteriol* 136:75–84.
8. Gunsalus, R.P. and Wolfe, R.S. (1978) *FEMS Microbiol Letters* 3: 191–193.
9. Vogels, G.D. and Stumm, C. (1980) *Antonie van Leeuwenhoek* 46:108.
10. Vogels, G.D., Hoppe, W.F. and Stumm, C.K. (1980) *Appl. Environ Microbiol* 40:608–612.
11. Tzeng, S.F., Wolfe, R.S., and Bryant, M.P. (1975) *J Bacteriol* 121:184–191.
12. Tzeng, S.F., Bryant, M.P., and Wolfe, R.S. (1975) *J Bacteriol* 121:192–196.
13. Ferry, J.G. and Wolfe, R.S. (1977) *Appl Environ Microbiol* 34:371–376.
14. Zeikus, J.G., Fuchs, G., Kenealy, W., and Thauer, R.K. (1977) *J Bacteriol* 132:604–613.
15. Fuchs, G., Stupperich, E., and Thauer, R.K. (1978) *Arch Microbiol* 119:215–218.
16. Jones, J.B. and Stadtman, T.C. (1980) *J Biol Chem* 255:1049–1053.
17. Yamazaki, S. and Tsai, L. (1980) *J Biol Chem* 255:6462–6465.
18. Eirich, L.D., Vogels, G.D., and Wolfe, R.S. (1978) *Biochemistry* 17:4583–4593.
19. Eirich, L.D., Vogels, G.D., and Wolfe, R.S. (1979) *J Bacteriol* 140:20–27.
20. Walsh, C., Jacobson, F., and Ryerson, C.C. (1980) *Biometric Chemistry, Advances in Chemistry Series* 191:119–138.
21. Ashton, W.T., Brown, R.D., Jacobson, F., and Walsh, C. (1979) *J Am Chem Soc* 101:4419–4420.
22. Pol, A., van der Drift, C., Vogels, G.D., Cuppen, T.J.H.M., and Laarhoven, W.H. (1980) *Biochem Biophys Res Commun* 98:255–260.
23. Eker, A.P.M., Pol, A., van der Meyden, P., and Vogels, G.D. (1980) *FEMS Microbiol Letters* 8:161–165.
24. Eker, A.P.M. (1980) *Photochem Photobiol* 32:593–600.
25. Eker, A.P.M., Dekker, R.H., and Berends, W. (1980) *Photochem Photobiol* 33:65–72.

Published 1982 by Elsevier North Holland, Inc.
Vincent Massey and Charles H. Williams, Editors
Flavins and Flavoproteins

CHAPTER 71

8-Hydroxy-5-Deazaflavin Cofactor Analogues: Relative Activity as Substrates for 8-Hydroxy-5-Deazaflavin-Dependent NADP$^+$ Reductase

Shigeko Yamazaki

The Laboratory of Biochemistry, National Heart, Lung, and Blood Institute, National Institutes of Health, Bethesda, Maryland

The 8-hydroxy-5-deazaflavin cofactor (8-OH-5dF1) (1) which is abundant in methane-producing bacteria, has been shown to serve as electron carrier in certain coupled enzyme systems that utilize either molecular hydrogen or formate as electron donor (2–4). In the formate-NADP$^+$ oxidoreductase system of *Methanococcus vannielii*, which consists of formate dehydrogenase and 8-hydroxy-5-deazaflavin-dependent NADP$^+$ reductase (5-deazaflavin-NADP$^+$ reductase), 8-OH-5dF1, **1**, is specifically required as cofactor (4). Homogeneous preparations of 5-deazaflavin-NADP$^+$ reductase catalyze a reversible direct hydride transfer reaction [Equation (71.1)] which is the first example of the obligatory participation of a natural

$$\underline{1} + \text{NADPH} \rightleftharpoons \text{(5-H,H reduced form)} + \text{NADP}^+$$

$$\text{R} = \text{CH}_2(\text{CHOH})_3\text{CH}_2\text{O}-\underset{\text{O}^\ominus}{\overset{\text{O}}{\overset{\|}{\text{P}}}}-\text{O}\underset{\text{CH}_3}{\text{CH}}\text{CONH}\underset{\text{COO}^\ominus}{\text{CH}}(\text{CH}_2)_3\text{CONH}\underset{\text{COO}^\ominus}{\text{CH}}(\text{CH}_2)_2\text{COO}^\ominus \quad (71.1)$$

5-deazaflavin in a flavin pyridine nucleotide redox reaction (5,6). In order to probe the nature of the interaction between the cofactor and the 5-deazaflavin-NADP$^+$ reductase, the enzymic reduction of a number of analogues of 8-OH-5dF1 was examined. To correlate the effects of structural variations of the analogues with enzymic activity, the K_m and k_{cat} values for reduction of these compounds by the enzyme were determined (Table 1).

Table 1. Reduction of 5-Deazaflavin Analogues.[a]

No.	Compound	pK_a (8-OH)	5-Deazaflavin-NADP$^+$ Reductase[b] Km (μM)	k_{cat} (min^{-1})	k_{cat}/Km ($min^{-1} \cdot \mu M^{-1}$)	$NaBH_4$[c] Rate of Reaction (nmol/min)	$NaBH_3CN$[d] Rate of Reaction (nmol/min)
1	HO, R, N, N, O, NH, O	6.3(1)	2.8(5)	442(5)	158	nd	nd
2	HO, R′, N, N, O, NH, O	6.1(8)	3.1	174	56	0.04	4.4
3	HO, CH_3, R′, N, N, O, NH, O	6.1	3.2	10.8	3.4	0.02	5.0
4	HO, C_2H_5, N, N, O, NH, O	6.1	4.2	53	12.6	0.02	5.9
5	HO, C_2H_5, N, N, O, N, CH_3, O	6.1	1.1	0.6	0.6	0.16	6.3
6	CH_3O, C_2H_5, N, N, O, NH, O		—	—	—	1.11	1.8

7	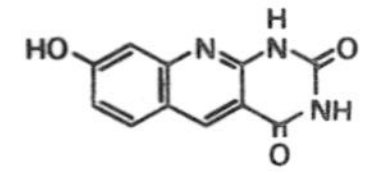	7.2	35.9	1090	30.4	<0.01	2.6
8	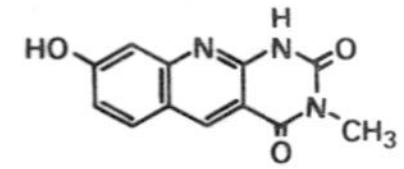	7.2	4.0	0.1	0.02	<0.01	3.0
9	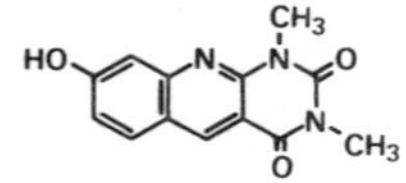	7.2	—	—	—	<0.01	0.5
10	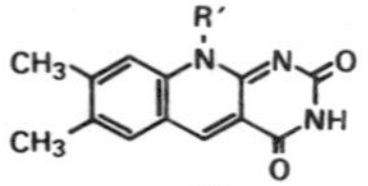		39.9	124	3.1	2.40	12.4
11	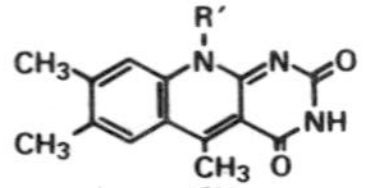		—	—	—	0.04	0.1
12	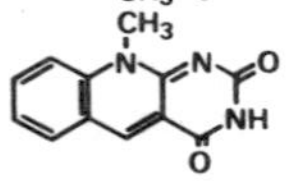		27.4	3990	146	7.66	31.3
13	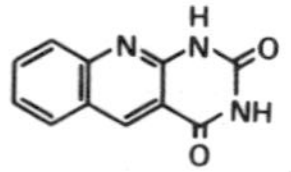		—	—	—	<0.01	<0.01

[a]R, see equation I; R′, ribityl; nd, not determined; 1 and 5-deazaflavin-$NADP^+$ reductase were isolated from *M. vannielii* (5).

[b]For the sake of comparison, the fluorometric (5) or spectroscopic assay of the enzyme was performed at 20°C and pH 7 although the optimal pH for the reduction of some of these compounds may be different. The K_m and k_{cat} values were obtained from Lineweaver-Burk plots (7) and the k_{cat} values were estimated as the number of moles of substrates reduced per min per mole of enzyme.

[c]The compound (30 nmoles) in 0.1 N NaOH was reduced at 20°C by the addition of 10 μl of 10% (w/v) $NaBH_4$.

[d]The compound (30 nmoles) in 1% acetic acid was reduced at 20°C by the addition of 10 μl of 10% (w/v) $NaBH_3CN$.

Effects of Modification of the Chromophore

The relative effectiveness of compounds **4**, 7, and **12** as substrates of the enzyme are of special interest. Compound **4** possesses the essentials of the chromophoric group of the natural cofactor, **1**, except the N-10 side chain is replaced by an ethyl group; thus the interaction of **4** with the enzyme may be similar to that of **1**. The pH profile of **4** follows the same pattern as that observed for the natural cofactor (5). From the effect of pH on enzymic activity with the natural cofactor it was concluded earlier (5) that the reacting species is the neutral form of **1**. This interpretation can be applied equally well to the case of **4**. In contrast, **12** showed a broad pH profile with an optimal pH at 7. Since **12** is virtually a neutral molecule in the pH range of 4 to 10, the effect of pH on the rate of its reduction must be ascribed to the ionization state of the enzyme. The ability of the enzyme to reduce **7** at an appreciable rate is somewhat surprising because this compound, having no substituent at N-10, is a different heteroaromatic system than that of **1** or **4**. Furthermore **13** is not reduced by the enzyme. Among the several possible tautomeric forms for **7**, **7′** is especially interesting because it is derived from replacement of a N-10 alkyl group of **1** with a hydrogen.

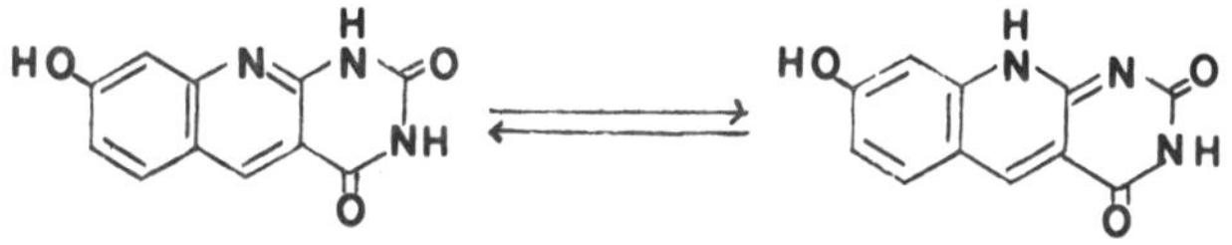

The site of reduction of **1** and related compounds has been shown to be the 1,5-conjugated system which is present in **7′** but not in **7**. Although the electronic absorption spectrum of **7** shows that **7** is the predominant tautomeric state at neutral pH, in the presence of enzyme, it is possible that the 1,5-conjugated tautomer **7′** is present in sufficient concentration for the enzymic reaction. This view is supported by the observation that the 1,3-dimethyl derivative, **9**, is not reduced by the enzyme. Since both riboflavin and 1,5-dideazariboflavin are enzymically inert, the above results suggest that the structural requirement of the chromophoric group as an enzyme substrate is 10-alkyl or 8-hydroxy 2,4-dioxopyrimido[4,5-b]quinoline.

Effects of Varying Ring Substituents

Position 8. Although the natural cofactor, **1**, possesses a hydroxy group at C-8, the 8-hydroxy function is not an absolute requirement for enzymic activity. However, when the 8-hydroxy group is methylated as in **6**, the resulting analogue is not reduced enzymically even though it is reduced at about the same rate as **4** by $NaBH_3CN$. The addition of a 10-fold excess of **6** did not affect the rate of reduction of a half saturating level of **4**, which indicates that methylation of the 8-hydroxy group significantly decreases the affinity of the compound for the enzyme.

Position 7. Substitution at position 7 produces a profound effect on reactivity in the enzymic reaction as indicated by the large difference in the k_{cat} values of **2** and **3**. However, the rates of reduction by $NaBH_3CN$ for these compounds are comparable.

Position 5. In view of the importance of position 5 as a site of the reduction, it was especially interesting to compare **10** and **11**. Both were reduced by either $NaBH_4$ or $NaBH_3CN$ but the rate of reduction of **11** was much slower than that of **10**. The enzyme, however, reduced only **10**. Compound **11** was shown to inhibit the enzymic reduction of **10**. In control experiments, the chemical oxidation of dihydro-5-deazariboflavin by **11** was not observed. These results show that the lack of enzymic activity for **11** is not due to its lack of affinity for the enzyme but probably is a result of the steric and inductive effect of the methyl group at the 5 position which decreases the rate of reduction.

Position 3. The presence of an N-methyl (**5** and **8**) instead of NH (**4** and **7**) at position 3 results in marked changes in the kinetic constants for reduction of these analogs by the enzyme. With both pairs, the presence of the methyl group resulted in drastic reductions of k_{cat} values. In the riboflavin series, it was shown that substitution at the N-3 position has very little effect on the binding energy (10). Thus the effects produced by the methyl group at the N-3 position of the 5-deazaflavins will be interesting to examine in detail.

ACKNOWLEDGMENTS
The author would like to thank Dr. Thressa C. Stadtman and Dr. L. Tsai for their encouragement and advice in the course of this work, and Dr. Wei-Mei Ching for helpful discussions. Part of this work was submitted to *Biochemistry* for publication.

References

1. Eirich, L.D., Vogels, G.D., and Wolfe, R.S. (1978) *Biochemistry* 17:4583–4593.
2. Tzeng, S.F., Wolfe, R.S., and Bryant, M.P. (1975) *J Bacteriol* 121:154–191.
3. Tzeng, S.F., Bryant, M.P., and Wolfe, R.S. (1975) *J Bacteriol* 121:192–196.
4. Jones, J.B. and Stadtman, T.C. (1980) *J Biol Chem* 255:1049–1053.
5. Yamazaki, S. and Tsai, L. (1980) *J Biol Chem* 255:6462–6465.
6. Yamazaki, S., Tsai, L., Stadtman, T.C., Jacobson, F.S., and Walsh, C. (1980) *J Biol Chem* 255:9025–9027.
7. Lineweaver, H. and Burk, D. (1934) *J Am Chem Soc* 56:658–666.
8. Eirich, L.D., Vogels, G.D., and Wolfe, R.S. (1979) *J Bacteriol* 140:20–27.
9. Becvar, J. (1973) PhD Thesis, University of Michigan.

Published 1982 by Elsevier North Holland, Inc.
Vincent Massey and Charles H. Williams, Editors
Flavins and Flavoproteins

CHAPTER 72

Raman Spectra of 8-Mercapto-Flavin, Protein-Bound and Free

Lawrence M. Schopfer* and Michael D. Morris†

Departments of Biological Chemistry and Chemistry,† The University of Michigan, Ann Arbor, Michigan*

Raman spectroscopy offers the intriguing potential for examining the effect of specific protein interactions on the electronic distribution of the isoalloxazine ring in protein-bound flavins. The last few years have seen much progress toward the characterization of the Raman spectrum of flavins (1,2). In the course of these studies, a number of 8-position modified flavins have been examined. It has been found that when groups more electron-donating than methyl are substituted into the 8 position, the Raman spectra are markedly perturbed (3–5). These results have been interpreted as a shift in the electronic distribution reflecting a greater contribution from a quinonoid form (Figure 1).

We have examined the Raman spectra of 8-mercapto-flavin, both free in solution and bound to the apoproteins of L-lactate oxidase, glucose oxidase, and Old Yellow Enzyme. These proteins were previously found to stabilize the quinonoid form of 8-mercapto-flavin (6). Thus their Raman spectra would be expected to provide a model for the quinonoid electronic distribution.

The Raman spectra of protein-bound 8-mercapto-flavin is very different from that of riboflavin (Figure 2). Though there is a significant amount of protein-dependent variation in band positions, the quinonoid spectra show the following trends: the riboflavin bands at 1501 cm^{-1}, 1462 cm^{-1}, 1407 cm^{-1}, 1250 cm^{-1}, and 1178 cm^{-1} are missing; bands at 1281 cm^{-1}, 1228 cm^{-1}, 1161 cm^{-1}, and 1072 cm^{-1} appear to be essentially unshifted; bands at 1631 cm^{-1}, 1582 cm^{-1}, 1547 cm^{-1}, and 1354 cm^{-1} are shifted to lower frequencies; and new bands appear at about 1110 cm^{-1}, 1090 cm^{-1}, 1040 cm^{-1}, and 1000 cm^{-1}.

Figure 1. The structure of the isoalloxazine portion of flavin.

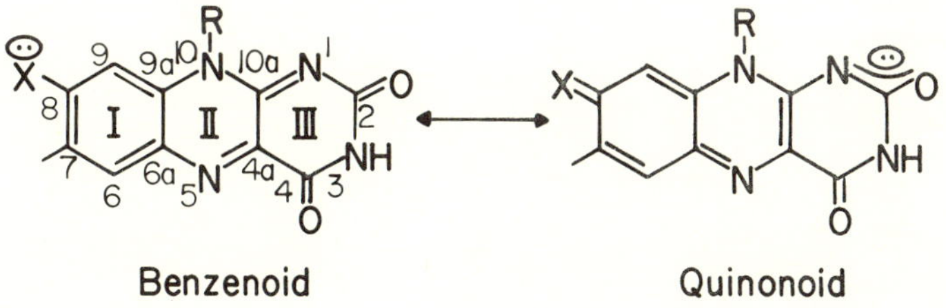

Though band assignments for the quinonoid flavin must be tentative at this time, we have chosen to equate the three highest wave number bands of the quinonoid form (e.g., 1614 cm^{-1}, 1536 cm^{-1}, and 1504 cm^{-1} for 8-mercapto-FMN L-lactate oxidase) with the bands at 1631 cm^{-1}, 1582 cm^{-1}, and 1547 cm^{-1} of riboflavin. This assignment is based primarily upon the observation that the Raman spectra of flavins containing electron-donating substituents at position 8 such as hydroxyl (4), amino (4), methoxy (5), methylamino (5), dimethylamino (data not shown), or thiomethyl (3) have their three highest wave number peaks at positions between those found for riboflavin and

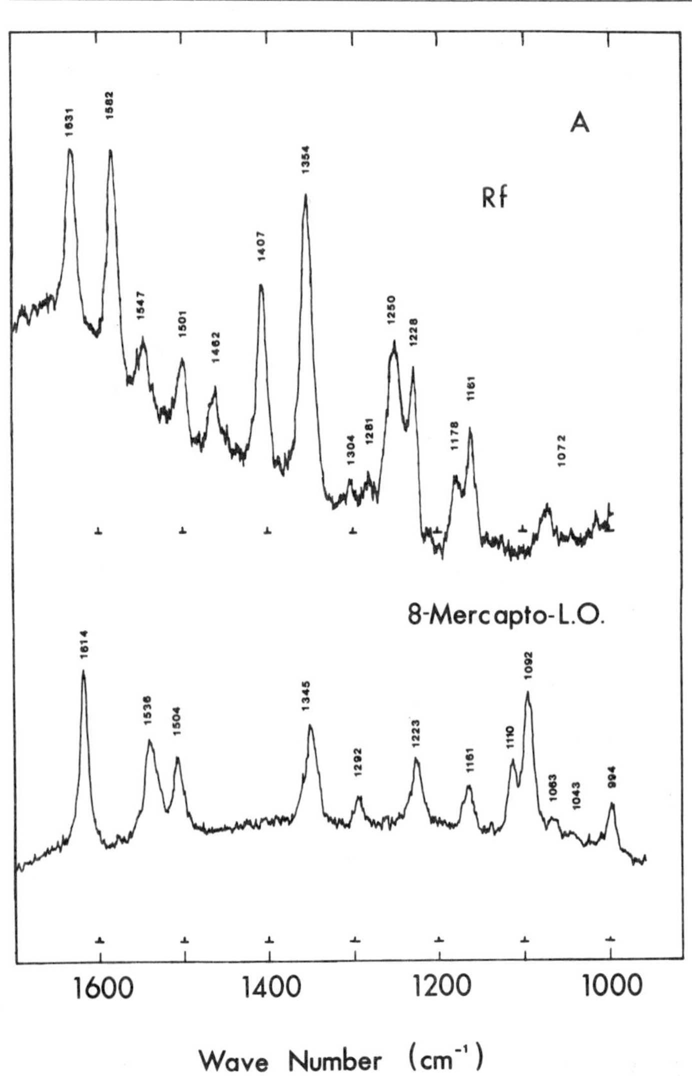

quinonoid flavin. Thus these flavins would appear to have electron distributions which are composed of significant contributions from both benzenoid and quinonoid forms. An additional example of this situation is seen with 8-mercapto-riboflavin free in solution (Figure 3). Under these conditions, the Raman spectrum of 8-mercapto-flavin has bands at 1620 cm^{-1}, 1548 cm^{-1}, 1507 cm^{-1}, and a broad band stretching from about 1360 cm^{-1} to 1320 cm^{-1}.

The quinonoid band at about 1340 cm^{-1} has been correlated with the riboflavin 1354 cm^{-1} band for the same reasons as outlined above. Those bands which appear to remain fixed in the spectra of both the quinonoid and

Figure 2. Raman spectra of protein-bound 8-mercapto-flavin. (Panel A) Riboflavin (Rf), data taken from reference 3 shown here for comparison. 8-Mercapto-FMN L-lactate oxidase (8-Mercapto L.O.), excited at 632.8 nm with a helium-neon laser (about 40 mW power). The enzyme (0.2 mM at neutral pH, ambient temperature) was flushed with oxygen-free nitrogen and sealed into a capillary tube for irradiation. Under these conditions, the sample was completely stable in the laser. (Panel B) 8-Mercapto-FAD glucose oxidase (8-Mercapto-G.O.) conditions were the same as for lactate oxidase, however the enzyme was much less stable in the laser. 8-Mercapto-FMN Old Yellow Enzyme (8-Mercapto-O.Y.E.) conditions were the same as for lactate oxidase, and this enzyme was also less stable in the laser.

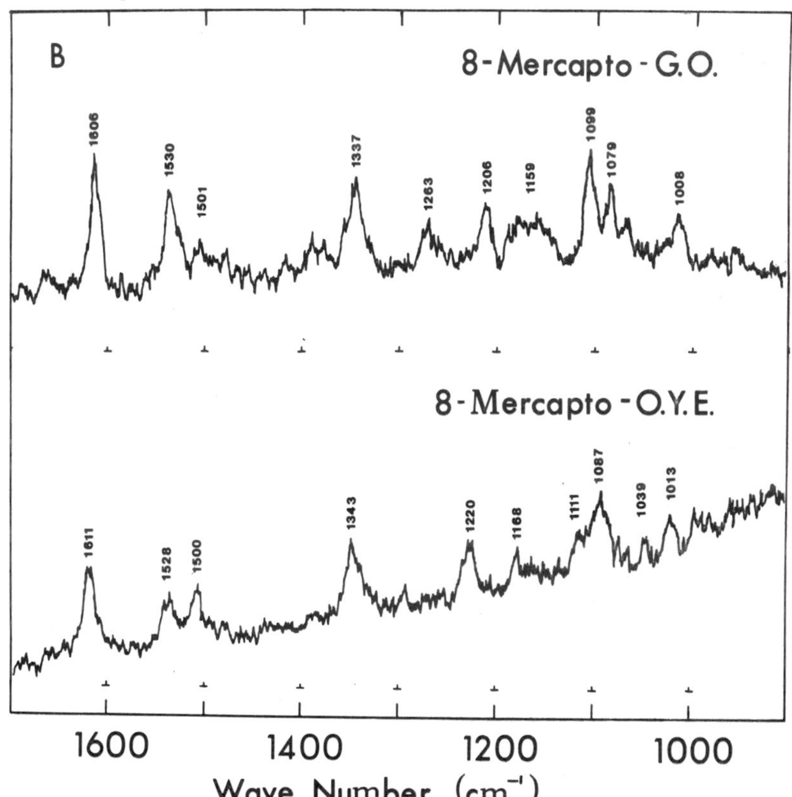

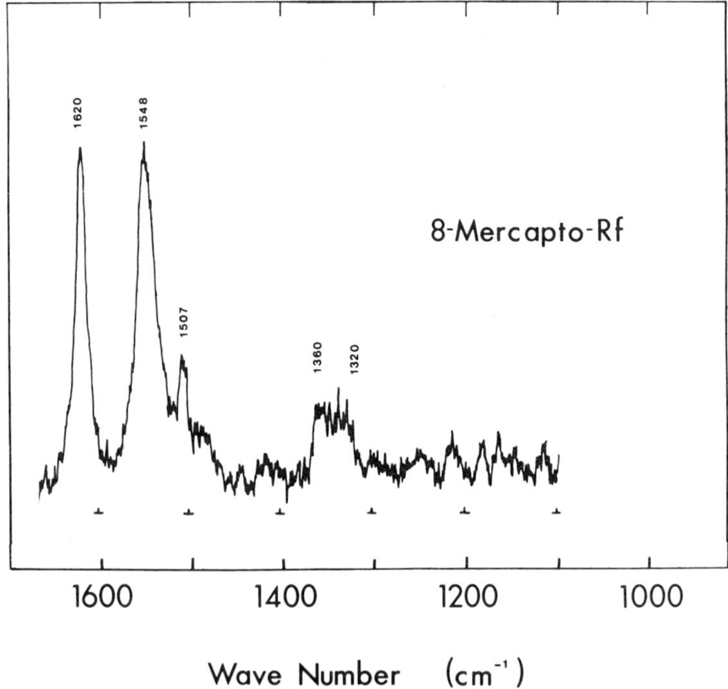

Figure 3. Raman spectrum of 8-mercapto-riboflavin. A solution of 8-mercapto-riboflavin (about 0.4 mM in 0.1 M NaPPi buffer pH 8.4, containing 3 M potassium iodide) was flushed with oxygen-free nitrogen and sealed into a capillary tube for irradiation at 514.5 nm with an argon laser (about 50 mW power). Under these conditions, the sample was relatively stable in the laser.

the benzenoid forms are also equated. The new bands have not yet been assigned.

ACKNOWLEDGMENTS

This work was supported by a Grant from the U.S. Public Health Service, GM-11106, and the National Science Foundation, CHE 79-15185.

References

1. Kitagawa, T., Nishina, Y., Kyogoku, Y., Yamano, T., Ohishi, A., and Yagi, K. (1979) *Biochemistry* 18:1804–1808.
2. Dutta, P.K., Nestor, J., and Spiro, T.G. (1978) *Biochem Biophys Res Commun* 83:209–216.
3. Schopfer, L.M. and Morris, M.D. (1980) *Biochemistry* 21:4932–4935.
4. Dutta, P.K., Spencer, R., Walsh, C., and Spiro, T.G. (1980) *Biochim Biophys Acta* 623:77–83.
5. Nishina, Y., Shiga, K., Horiike, K., Tojo, H., Kasai, S., Yanase, K., Matsui, K., Watari, H., and Yamano, T. (1980) *J Biochem (Tokyo)* 88:403–409.
6. Massey, V., Ghisla, S., and Moore, E.G. (1979) *J Biol Chem* 254:9640–9650.

Published 1982 by Elsevier North Holland, Inc.
Vincent Massey and Charles H. Williams, Editors
Flavins and Flavoproteins

CHAPTER 73

Reactions of 5-Thia-5-Deazaflavins

Helmut Fenner and Lieselotte Tessendorf

Fachbereich Pharmazie, Freie Universtität Berlin, D-1000 Berlin 33 Dahlem, Germany

In addition to 5-deazaflavin, which is a flavin analog lacking 1 e^--transfer, we have studied a modification of the flavin nucleus which would remove the $2e^-$-transfer activity in favor of 1 e^--transfer (1–4). By replacing the N-5 atom in reduced flavocoenzymes by sulfur, 5-thiadihydroflavins are obtained. We have reported the structures and properties of the fully-reduced and half-reduced thiaflavin species (3,4), which are isoelectronic and isochromic with flavin in the corresponding state of oxidoreduction and protonation.

Scheme 1.

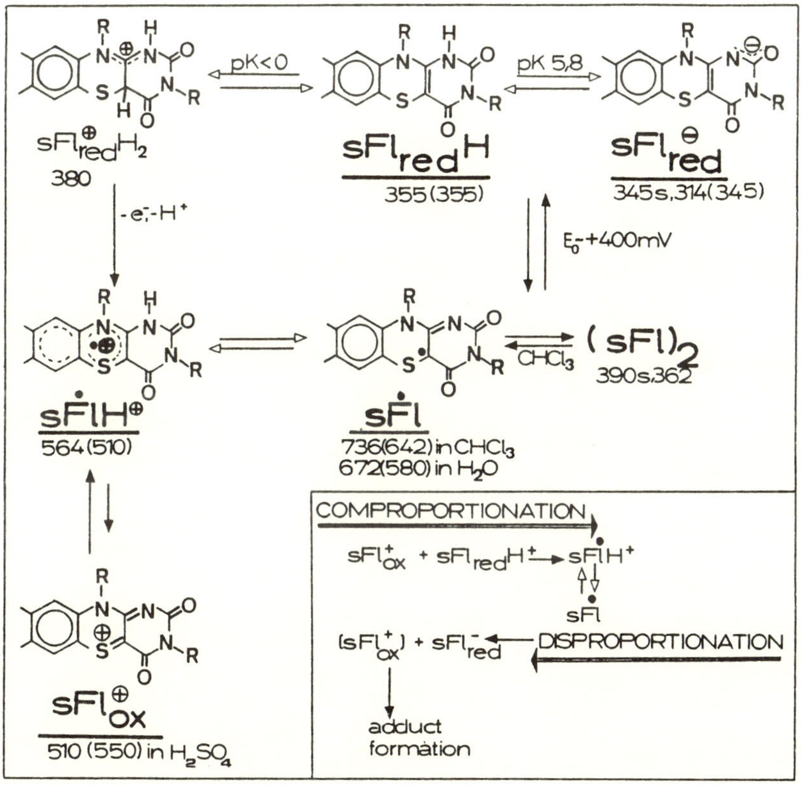

1o-Alkylpyrimido [5,4-b]benzothiazines (5-thiaflavins) can be prepared by different methods yielding a mixture of 6,7- and 7,8-dimethyl-10-alkylpyrimidobenzothiazines (5). By cyclization of anilinouracils with $SOCl_2$ the 4a-chloropyrimidobenzothiazines are formed, which yield 5-thiaflavins upon reduction with hydrazine. Starting from alkyl-anilinouracils only the 7,8-dimethyl derivates are obtained; in the case of ribitylanilinouracils the main product is the 7,8-dimethyl isomer, which can be purified by repeated recrystallization. As outlined in scheme 1, the sFl_{ox}^{+} species is not stable under aqueous conditions; it can be obtained in pure form only in concentrated H_2SO_4 solution. Studies on the reactivity of the fully-oxidized thiaflavin species with nucleophiles therefore have to start from other sFl_{ox}^{+} sources. We have studied the adduct formation of sFl_{ox}^{+} by means of two procedures, which in most cases yield the identical products: $s\dot{F}l$, obtained by deprotonation of $s\dot{F}l\text{-}1\text{-}H^{+}$, disproportionates in the presence of nucleophiles, forming one part of $sFl_{red}H$ and one part of the 4a- or S-5 adduct. Among the nucleophiles studied were different alcohols, mercaptans, phenols, primary and secondary amines and carbanions. The same adduct formation occurred with sFl_{ox}-4a-Cl. In all cases, the thus-obtained adducts could be isolated, except the mercaptans studied, which are oxidized by $sFl_{ox}{}^{+}$. This reaction was found to proceed to completion giving $sFl_{red}H$ and disulfide in quantitative yield. Some $sFl_{ox}{}^{+}$ reactions yielding either C-4a or S-5 adducts are outlined in scheme 2. [All adducts given were identified by UV/VIS, MS and ^{1}H-NMR spectra, their

Scheme 2.

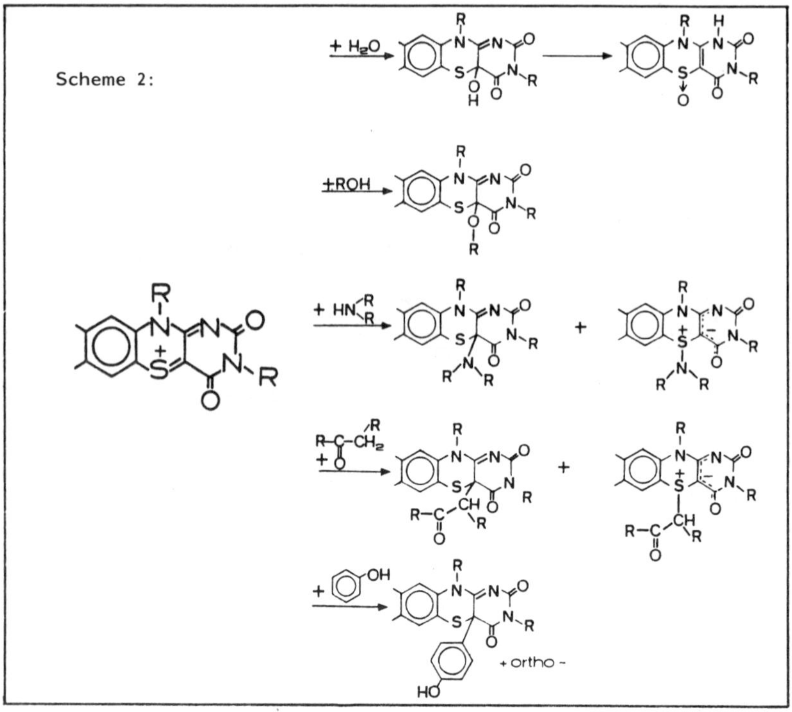

Scheme 2:

+ H_2O

+ROH

+ HN(R)R

R-C(=O)-CH$_2$-R

+ (phenol)OH

+ ortho -

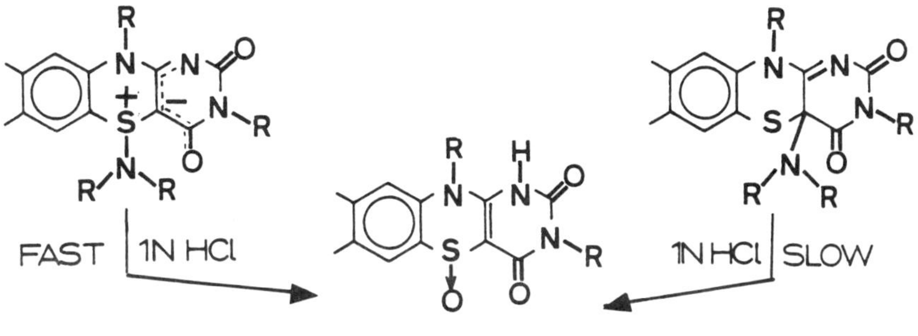

Scheme 3.

spectral data were compared with known C-4a and S-5 derivatives of thiaflavin (3,4)]

S-5 addition of primary and secondary amines yields sulfilimines which are similar to adducts known from phenothiazines (7). In contrast to the C-4a type, the S-5 adducts are converted under acidic conditions into the sulfoxide in a very fast reaction; in the case of sFl$_{ox}$-4a-NR$_2$ this hydrolysis is slow.

Spectral data of some adducts are given in Table 1.

Table 1.

	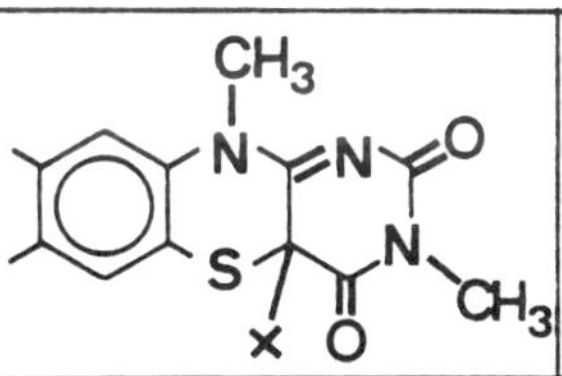 C-4a adduct (X at C-4a)	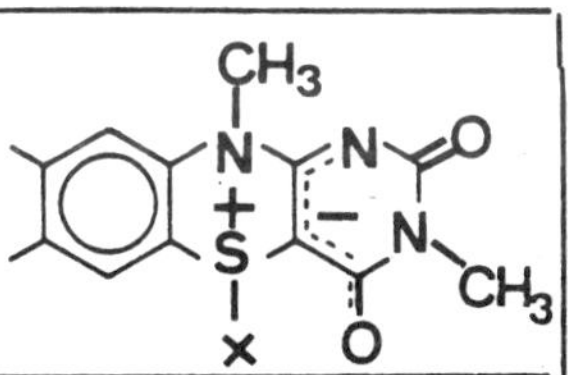 S-5 adduct (X at S-5)
X = $CH_2CO-C_6H_4-OCH_3$ 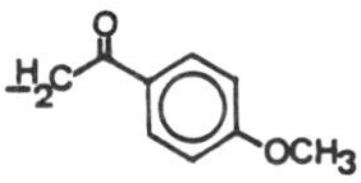	CH_3OH: 340(7000) 310(9800) 278(28800)	CH_3OH: 277(26200)
	pH 2: 355(6600) 295(22600) 279(25200)	pH 2: 355(3000) 298(19100)
X = $-N(H)CH_3$	CH_3OH: 335(6800) 312(8000) 273(9700)	CH_3OH: 310(3400) 264(38500)
	6N HCl: 347(6900) 333(6400) 274(7300)	
X = $-N(CH_3)_2$	CH_3OH: 340(7200) 313(7900) 276(5600)	CH_3OH: 313(8000) 278(39400)

Conclusion

Studies on the reactivity of fully-oxidized thiaflavins indicate the formation of two types of adduct formation (C-4a vs S-5). Because of the high redox potential of these "mutilated flavins" further studies on modified thiaflavins with a lower potential are necessary to demonstrate the relevance of adduct formation.

References

1. Fenner, H. *Arzneim.-Forsch. (Drug Res)* 20:1815 (1970).
2. Hemmerich, P., Massey, V., and Fenner, H. *FEBS Lett* 84:5 (1977).
3. Fenner, H., Grauert, R.W., and Hemmerich, P. *Liebigs Ann Chem* (1978) 193.
4. Fenner, H., Grauert, R. W., Hemmerich, P., Michel, H., and Massey, V. *Eur J Biochem* 95:183 (1979).
5. Janda, M., and Hemmerich, P. *Angew Chem*, 88:475 (1976).
6. Fenner, H., Grauert, R.W., and Tessendorf, L. *Arch Pharm*, in press.
7. Bandlish, B.K., Mani, S.R., and Shine, H.J. *J Org Chem* 42:1538 (1977).

Published 1982 by Elsevier North Holland, Inc.
Vincent Massey and Charles H. Williams, Editors
Flavins and Flavoproteins

CHAPTER 74

The Reactivity of Flavin and 5-Deazaflavin in Photoredox Reactions

Rainer Traber,* Horst E.A. Kramer,* and Peter Hemmerich[†]

**Institut für Physikalische Chemie der Universität Stuttgart, Pfaffenwaldring 55, D-7000 Stuttgart 80, Federal Republic of Germany; †Fakultät für Biologie der Universität Konstanz, Postfach 5560, D-7750 Konstanz 1, Federal Republic of Germany*

The flavin chromophore exhibits three mechanistically different main activities, namely (de)hydrogenation, which is "$2e^-$ only," O_2 activation ($2e^-$ or $1e^-$) and thirdly electron transfer ($1e^-$ by definition). $1e^-$ transfer as well as O_2 activation are at the same time chemical activities inherent in the isolated flavin chromophore. On the other hand, for a chemical verification of flavin dependent (de)hydrogenation one is referred to flavin photochemistry. We have studied stoichiometry of flavin (Fl_{ox}) and 5-deazaflavin ($5dFl_{ox}$) photoreduction as depending on the nature of the photosubstrate. Unlike flavin-dependent enzymatic (de)hydrogenation, we find here a competition between $1e^-$ and $2e^-$ transfer (Figure 1), where the latter mechanism is similar to the carbanion transfer observed with flavin-dependent dehydrogenases, and, therefore, serves us as chemical model reaction.

The $1e^-$ Transfer Mechanism ①

As shown by Rehm and Weller (1), the reaction free enthalpy of the $1e^-$ transfer process in the excited singlet state may be calculated from the oxidation and reduction potentials in the ground state, the excitation energy $E_{o,o}(A^{z*})$ and the Coulomb interaction:

$$\Delta G = FE_D^{ox} - FE_{A^z}^{red} - E_{o,o}(A^{z*}) + (z-1)\frac{Fe_o}{4\pi\varepsilon\varepsilon_o a} \quad (74.1)$$

Considering endergonic electron transfer processes only, a linear relationship is obtained between the overall reaction rate constant k_q of the reaction scheme in Equation (74.2) (1) and the change of the free enthalpy ΔG_{23} of the electron transfer process in the encounter complex $A^* \cdots D$.

$$A^* + D \underset{k_{21}}{\overset{k_{12}}{\rightleftharpoons}} A^* \cdots D \underset{k_{32}}{\overset{k_{23}}{\rightleftharpoons}} (\dot{A}^- \cdots \dot{D}^+) \begin{matrix} \xrightarrow{k_{34}} \dot{A}^- + \dot{D}^+ \\ \xrightarrow[k_{30}]{} A + D \end{matrix} \quad (74.2)$$

This is derived from the Arrhenius equation, since for endergonic reactions $\Delta G_{23}^{\neq}$ is approximated to be ΔG_{23}:

$$\lg k_q = C - (1/(2.3RT))\Delta G_{23} \quad (74.3)$$

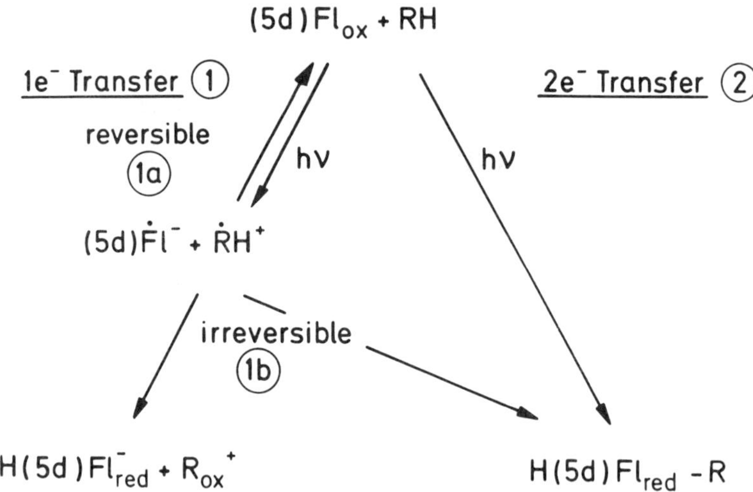

Figure 1. Pathways of flavin and 5-deazaflavin photoreduction.

Applying Equation (74.1) the following relation holds:

$$\lg k_q = C' - (F/(2.3RT))\, E_D^{ox} \tag{74.4}$$

From this equation, a straight line with a slope of $-17.0\ V^{-1}$ is expected when $\lg k_q$ is plotted versus E_D^{ox}. The position of this line on the scale of oxidation potentials of the electron donors defines the reactivity of the electron-accepting dye.

For the photoexcited states of flavin and 5-deazaflavin, the position of the straight lines was determined in methanolic solution, using a set of methyl and methoxy-substituted benzenes and naphthalenes of known oxidation potential (2). The result is shown in Figure 2 and summarized as follows:

1. The protonated flavin triplet is more reactive than the neutral flavin triplet. The difference, obtained experimentally, agrees well with the value which is calculated according to the Michaelis cycle from the difference of the pK values of flavosemiquinone and flavin triplet (2).
2. Applying Equation (74.1), the difference in reactivity of photoexcited singlet and triplet states is expected to be due only to the difference in excitation energy. Experimentally, however, an additional difference of 0.13 eV is found, exceeding the singlet-triplet energy splitting of flavin (2). The effect can be explained by the influence of the multiplicity of the reacting state on the competition between the kinetic rates of deactivation (k_{30}) and electron back transfer (k_{32}), if the solvent shared radical pair ($\dot{A}^- \cdots \dot{D}^+$) in Equation (74.2) is replaced by an exciplex for endergonic $1e^-$ transfer reactions (3).
3. The "reactivity lines" of the corresponding singlet and triplet states of Fl_{ox} and $5dFl_{ox}$ lie rather close together. Thus, the higher excitation energy of $5dFl_{ox}$ compensates the bad $1e^-$ acceptor quality of $5dFl_{ox}$ in the ground state (4).

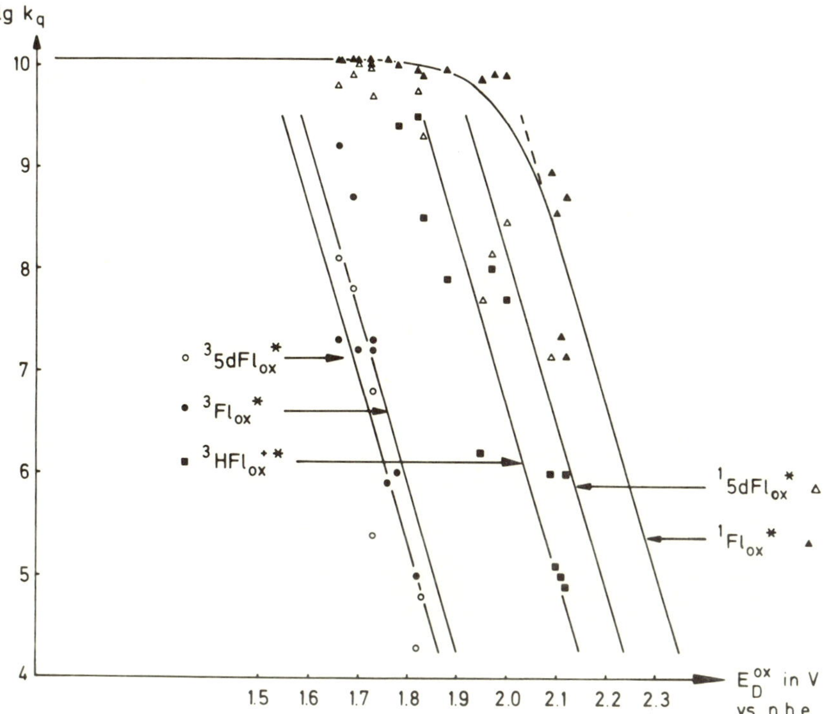

Figure 2. The "reactivity lines" of excited states of Fl_{ox} and $5dFl_{ox}$ in $1e^-$ transfer reactions.

The Reversible (1a) and Irreversible (1b) $1e^-$ Transfer

From energetic considerations, it follows that the photoinduced charge separation should be reversible in the dark. Indeed this is found in the photocomproportionation reaction of flavin in presence of 1,5-dihydroflavin (5), or with substrates which form stable radicals like phenol (6), nitrite (7), or 1,4-diazabicyclo-[2,2,2] octan (8).

If the substrate radical, however, can undergo competing processes like deprotonation or decarboxylation, irreversible bleaching of Fl_{ox} and $5dFl_{ox}$ occurs. Since by these reactions the redox properties of the substrate radicals are changed, a further transfer of an electron equivalent to an oxidizing molecule (radical combination inclusive) is possible, instead of back donation. Examples of such substrates are N-allylthiourea, (ethylenedinitrilo)tetraacetate (EDTA) and nitrilotriacetate (unpublished results).

The $2e^-$ Transfer Mechanism (2)

A second mechanism, yielding reduced forms of flavin, is the nucleophilic addition of a substrate to the quinoide system of Fl_{ox} (9). This mechanism also occurs in the photoexcited triplet state of Fl_{ox}, as shown by flash photolysis with borohydride (5) (10) and cyanide (7) as nucleophiles. With the latter substrate, a C-C bond formation can be photoinduced in position 6 and 9 of Fl_{ox}, which is not found in the ground state (7). Analogously, the photochemical 5-deazaflavin dimer formation was revealed to occur by addition of 1,5-dihydro-5-deazaflavin to the photoexcited triplet state of $5dFl_{ox}$ and not by 5dFl-radical combination (8). In contrast to this, the sulfite ion, which is a nucleophile in flavin ground state chemistry (11), does not add, but donates single electrons to the photoexcited flavin triplet state, as revealed by flash photolysis experiments (unpublished results). This shows that the competition between the reduction modes is altered in favor of the $1e^-$ transfer mechanism upon light excitation. From this it follows that substrates undergoing $2e^-$ transfer are soft Lewis bases, which are characterized by a high $1e^-$ oxidation potential.

ACKNOWLEDGMENTS
The financial support of the Deutsche Forschungsgemeinschaft and Fonds der Chemischen Industrie is gratefully acknowledged.

References

1. Rehm, D. and Weller, A. (1969) *Ber Bunsenges Physik Chem* 73:834–839.
2. Traber, R., Vogelmann, E., Schreiner, S., Werner, T., and Kramer, H.E.A. (1981) *Photochem Photobiol* 33:41–48.
3. Vogelmann, E., Rauscher, W., Traber, R., and Kramer, H.E.A. (1981) *Z Phys Chem NF* 124:13–22.
4. Blankenhorn, G. (1977) *Pyridine Nucleotide-Dependent Dehydrogenases*. Sund, H. (ed.) Berlin: Walter de Gruyter, pp. 185–205.
5. Hemmerich, P., Knappe, W.-R., Kramer, H.E.A., and Traber, R. (1980) *Eur J Biochem* 104:511–520.
6. Vaish, S.P. and Tollin, G. (1970) *Bioenergetics* 1:181–192.
7. Traber, R., Kramer, H.E.A., Knappe, W.-R., and Hemmerich, P. (1981) *Photochem Photobiol* 33:807–814.
8. Goldberg, M., Pecht, I., Kramer, H.E.A., Traber, R., and Hemmerich, P. (1981) *Biochim Biophys Acta* 673:570–593.
9. Hemmerich, P. (1976) *Prog Chem Org Nat Prod* 33:451–527.
10. Traber, R., Werner, T., Schreiner, S., Kramer, H.E.A., Knappe, W.-R., and Hemmerich, P. (1980) *Flavins and Flavoproteins, Proceedings of the 6th International Symposium*. Yagi, K. and Yamano, T. (eds.) Tokyo: Jap. Sci. Soc. Press and Baltimore, Maryland: University Park Press, pp. 431–442.
11. Müller, F. and Massey, V. (1969) *J Biol Chem* 244:4007–4016.

Published 1982 by Elsevier North Holland, Inc.
Vincent Massey and Charles H. Williams, Editors
Flavins and Flavoproteins

CHAPTER 75

Structure of P-Flavin from *P. phosphoreum*

Sabu Kasai, Kunio Mastui, and Takao Nakamura*

*Research Institute for Atomic Energy, Osaka City University, Osaka and *National Cardiovascular Center, Research Institute, Suita, Japan*

P-flavin (PF) is a flavin compound which was isolated from bacterial luciferase and characterized by K. Matsuda and T. Nakamura (1). In this communication, purification, properties and structure of PF are reported and the relationship between PF and light-emitting reaction is discussed (Figure 1).

Purification of PF

PF was extracted from whole cells of *P. phosphoreum* with 80% ethanol and the solution was evaporated. The aqueous residue was reextracted with n-butanol and the organic phase was evaporated to dryness. The residue was applied to a large-scale droplet countercurrent chromatograph. The sample was chromatographed by ascending method using a solvent system of n-butanol-n-propanol-water (2 : 1 : 3) and the peak of PF was quite separated from other flavins. PF fraction was purified further by a column chromatography of Biogel P-2 and crystallized in 60% ethanol.

Properties and Structure of PF

Molecular formula of PF. The molecular weight of PF was measured by mass spectrometry. A solution of PF was acidified with hydrochloric acid and extracted with n-butanol and the solution was charged on the emitter of field desorption equipment. As shown in Figure 2A, the molecular ion peak of PF hydrochloride was found at m/e 739 $(M+5)^+$, and another peak at m/e 761 corresponded to $(M+4+Na)^+$. Qualitative elemental analysis of PF was achieved by using X-ray fluorescence analysis, and calcium and phosphate were found in the crystal. Phosphate was determined by Fiske and Subbarow's method on the sample of PF (719 μg, 0.908 μmoles) oxidized with nitric acid and 0.890 μmoles of phosphoric acid were found. Carbon, hydrogen, and nitrogen were determined by automatic gas chromatographic methods. The results are as follows.

Calcd. for $C_{31}H_{46}N_4O_{12}PCa_{1/2}.3H_2O$:	C, 47.02;	H, 6.57;	N, 7.08%
Found:	47.33	7.00	7.10

pKa values and electronic and fluorescence spectra of PF. pKa values were measured by spectrophotometric titration and were 11.0 and 0.4. These values are comparable with those of riboflavin. Electronic spectra of the anion,

Figure 1. Structure of P-flavin.

neutral, and cation species of PF are shown in Figure 3A and spectra of neutral species of PF in buffer solution and organic solvent (n-butanol) are compared in Figure 3B. In buffer solution, the spectrum of PF is similar to that of isoflavin; the second peak was higher than the first and shifted to longer wavelength. But in butanol the height of the second peak was diminished and a new peak was observed in the shorter wavelength region. λ_{max}'s of fluorescence emisson spectra were 525 nm (pH 7.0) and 517 nm (chloroform).

Redox properties of PF. PF was reduced to fully-reduced form by dithionite and the semiquinone form was not observed. The redox potential of PF was measured to be −0.47 V (vs SCE, 25°C, pH 7.0) by the polarographic method. This value is also comparable with that of riboflavin (−0.45 V, same conditions). PF was decomposed during reduction by sodium borohydride.

PMR spectrum of PF. 400 MHz PMR spectrum of PF in DMSO d-6 is shown in Figure 4. Four acidic protons were observed: 11.31 and 11.28 ppm

Figure 2. Mass spectra of P-flavin (A), and periodate oxidation product of P-flavin (B).

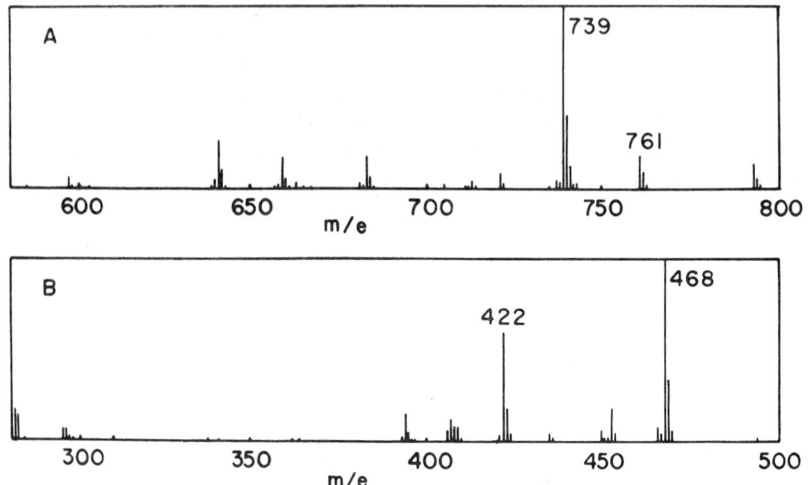

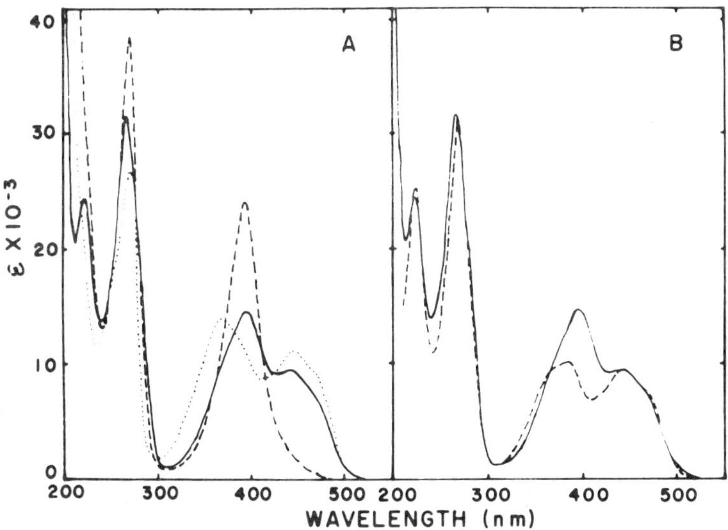

Figure 3. Electronic spectra of P-flavin. (A) (—) at pH 7.0, neutral species; (···) in 0.1 N NaOH, anion; (---) in 6 N HCl, cation. (B) (—) in buffer solution (at pH 7.0); (---) in n-butanol.

(P-OH), 9.90 ppm (broad, 3-NH), and about 7.7 ppm (broad, COOH). The phenyl proton signals were found at 8.57 and 7.89 ppm. Two very big peaks at about 3.5 and 2.5 ppm corresponded to the protons of water and DMSO respectively. Two peaks of methyl groups attached to the isoalloxazine ring were found at 2.41 and 2.33 ppm, but one of these peaks was diminished to

Figure 4. 400 MHz PMR spectrum of P-flavin in DMSO d-6. δ values are expressed in ppm from tetramethylsilane.

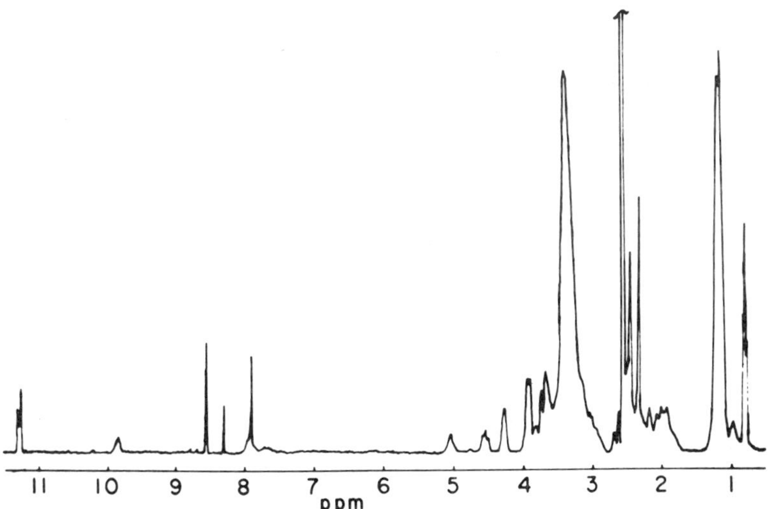

2/3 height. It is noteworthy that this peak was also a singlet. The peak at 1.18 ppm corresponded to protons of alkyl methylene groups and at 0.84 ppm a terminal methyl signal was found.

Mass spectrum of periodate oxidation product of PF (PFOX). PF was oxidized by periodate and extracted with chloroform and the sample was charged on the mass spectrometer and ionized by electron impact (EI) method. The very high base peak at m/e 468 in Figure 2B was analyzed to be $C_{26}H_{36}N_4O_4$ or $C_{28}H_{38}N_1O_5$.

Structure of PF. As shown above, pKa values and redox potential were not remarkably different from those of riboflavin, suggesting that the isoalloxazine ring of PF is not highly modified. The molecule of PF ($C_{31}H_{47}N_4O_{12}P$) is larger than that of FMN ($C_{17}H_{21}N_4O_9P$) by $C_{14}H_{26}O_3$. The PMR spectrum of PF showed that a fatty chain is attached to FMN and the attaching position is the 7 or 8 methyl group (probably 8), as one of these signals was diminished. As this smaller peak was also a singlet, the fatty chain might attach to the methylene group by an ether linkage. EI mass spectra of PFOX showed that the base peak is heavier than the periodate oxidation product of FMN by $C_{12}H_{25}O$. The extraordinary height of the base peak showed that fragmentation occurred at the branched position of the fatty chain and it was calculated that a CH_3COOH group was reduced by this fragmentation. Such a fragmentation is well-known as McLafferty rearrangement. From this evidence, the structure of PF was elucidated as Figure 1. The position attaching the fatty chain (7 or 8) and the absolute configuration of β-carbon of the fatty acid are not certain yet.

Relationship Between PF and Light-Emitting Reaction

We have a working hypothesis that PF is the product of the light-emitting reaction in the bacterial cell, from the following evidence: (1) The amount of light emitted from the bacteria is proportional to the produced amount of PF in the bacterial cell. (2) *Cis-α, β*-unsaturated tetradecenyl aldehyde reacts with $FMNH_2$ and O_2 in the presence of bacterial luciferase, emitting light with a very long decay time.

ACKNOWLEDGMENTS

This work was supported in part by a Grant-in-aid for encouragement of young scientists to S.K. (No. 479112) and Grant-in-aid for special project research to T.N. (No. 311808) from the Ministry of Education, Science, and Culture of Japan.

The authors thank the following persons or companies: Dr. T. Tokuyama, Dr. T. Kinoshita, Dr. T. Fukumura, Dr. A. Ichimura, Mr. J. Gohda, Dr. Y. Kimura, Toyo Jozo Co. Ltd., Institute for Fermentation, Otsuka Pharmaceutical Co. Ltd., and Hitachi Ltd.

Reference

1. Mastuda, K. and Nakamura, T. (1972) *J Biochem* 72:951–955.

PART VI:

Covalently-Bound Flavins

Published 1982 by Elsevier North Holland, Inc.
Vincent Massey and Charles H. Williams, Editors
Flavins and Flavoproteins

CHAPTER 76

Biosynthesis of Covalent Flavoproteins

Karl Decker

Biochemisches Institut, Albert-Ludwigs-Universität, Hermann-Herder-Str. 7, D-7800 Freiburg im Breisgau, Federal Republic of Germany

Flavoproteins (FPs)[1] have been known since 1935, when H. Theorell (1) elucidated the structure of the "Yellow Enzyme" as an association of polypeptide and phosphorylated riboflavin. Three years later, O. Warburg discovered FAD as the cofactor of D-amino acid oxidase (2). He also observed that the coenzyme does not dissociate from the holoenzyme in the physiological range of pH and ionic strength but could be removed by treatment with acid. Many more FPs were described in the following years, but it was not until 1955 that indications of a flavin covalently bound to protein were obtained. Succinate dehydrogenase from beef heart mitochondria was the first enzyme in which covalently-bound flavin was unambiguously identified (3). The combined efforts of several laboratories, especially in San Francisco and Konstanz, led to the development of analytical and synthetic methods that were instrumental in elucidating the type of binding between the flavin and the polypeptide moieties; these achievements have been reviewed extensively (4–6). This report will deal mainly with the structural and cellular aspects of cFPs, the acceptor and donor molecules participating in their biosynthesis, and with mechanisms involved in these processes.

Structural and Cellular Aspects of Covalent Flavoproteins

When the ground was broken by the succinate dehydrogenase work, other FPs were reported to contain covalently-bound coenzyme. Within 10 years, their number rose to 18; 5 different types of flavin-amino acid linkage were identified (Figure 1).

Can we expect to find many more cFPs? The answer is most likely yes as far as microorganisms are concerned; their potential for oxidative degradation of organic compounds is almost without limit; several of the enzymes listed in Table 1 bear witness to this fact. In cells of higher organisms, however, the number of cFPs appears to be limited. Four cFPs were distinguished in rat liver mitochondria and associated with succinate dehydrogenase, monoamine oxidase, sarcosine dehydrogenase, and oxidase (21, 29, 30).

It is still difficult to rationalize the existence of cFPs in terms of structural or functional peculiarities. Do FPs catalyzing a given reaction have the same kind

[1]RF = riboflavin; FP = flavoprotein, cFP = flavoprotein with covalent flavin-protein linkage.

FAD 8α-N_1(His)

FAD 8α-N_3(His)

FAD 8α-S (Cys)

FAD 8α-O (Tyr)

FMN 6-S (Cys)

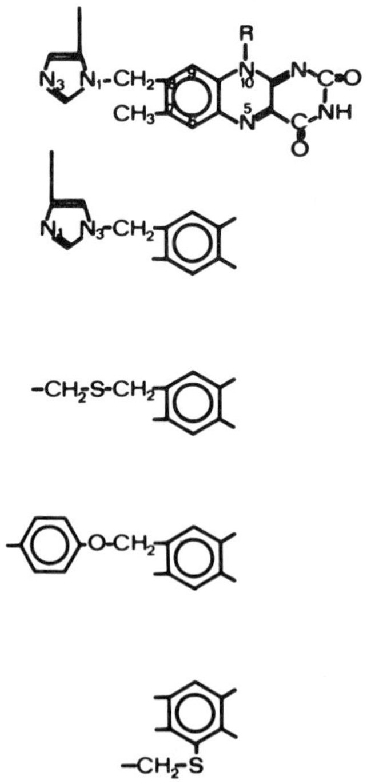

Figure 1. The flavin-amino acid bonds observed in flavoproteins.

of coenzyme attachment in different cells? So far, only succinate dehydrogenase, monoamine oxidase, and, perhaps, L-gulonolactone oxidase (31) indicate such connections. Riboflavin-8α attachment to His or Cys shifts the E'_o by some 20–30 mV (32); such an effect, however, is not specific for covalent binding, as a much wider range of biological redox potentials is found with noncovalent FPs.

The amino acid sequence adjacent to the binding residue has been determined for several cFPs; however, there are no particular features evident to suggest a preferential covalent flavin binding. The subunit structures of the known cFPs also allow no deductions about the necessity or feasibility for covalent coenzyme attachment. It is, nevertheless, remarkable that in one organism, *Arthrobacter oxidans*, D, L-nicotine induces a L-specific 6-hydroxynicotine oxidase (6-HLNO) composed of 2 identical subunits with 1 mole each of noncovalent FAD as well as a D-specific enzyme (6-HDNO) consisting of only 1 subunit (M_r 50,000) with one covalently bound FAD (33).

These oxidases belong to a group of enzymes specifically involved in the degradation of nicotine. Their structural genes have recently been associated with a large plasmid of about 160 kb by Dr. Brandsch in Freiburg, Through conjugation, it could be transferred into other strains of the genus *Arthrobacter* with a frequency of ca 10^{-7} conferring nicotine-dependent enzyme synthesis. The isolated plasmid DNA could also transform *E.coli* SF-8; this transformant produced the same 6-hydroxynicotine oxidase as *A.oxidans* (Figure 2).

Table 1. The Flavoproteins With Established Covalent FAD-Apoprotein Attachment.

FAD-8α-N1(His)	**Sources**	**Ref.**
Thiamine dehydrogenase (EC 1.1.3)	Soil bacterium (ATCC 25589)	7
Cyclopiazonate oxidocyclase (EC 1.3.99.9)	*Penicillium cyclopium*	7
L-Gulonolactone oxidase (EC 1.1.3.8)	Liver microsomes	8
L-Galactonolactone oxidase (EC 1.1.3)	Yeast	9
Cholesterol oxidase (EC 1.1.3.6)	Schizophyllum commune	10
FAD-8α-N3(His)		
Succinate dehydrogenase (EC 1.3.99.1)	Beef heart mitochondria	11
	Yeast	12
(Fumarate reductase)	*Vibrio succinogenes*	13
	E. coli (anaerobic)	14
6-Hydroxy-D-nicotine oxidase (EC 1.5.3.6)	*Arthrobacter oxidans*	15
D-Gluconolactone dehydrogenase (EC 1.1.1)	*Penicillium cyaneofulvum*	16
Choline oxidase (EC 1.1.3.17)	*Arthrobacter globiformis*	17
	Cylindrocarpon didymum M-1	18
	Alcaligenes sp.	19
Sarcosine dehydrogenase (EC 1.5.99.1)	*Pseudomonas* sp. WRF	20
	Liver mitochondria	21
Sarcosine oxidase (EC 1.5.3.1)	*Corynebacterium* sp. U-96	22
FAD-8α-S(Cys)		
Monoamine oxidase (EC 1.4.3.4)	Liver	23
Flavocytchrome C_{552}	Chromatium	24
Flavocytochrome C_{553}	*Chlorobium thiosulfatophilum*	25
FAD-8α-O(Tyr)		
p-Cresol methylhydroxylase (EC 1.17.)	*Pseudomonas putida*	26
FMN-6-S(Cys)		
Trimethylamine dehydrogenase (EC 1.5.99.7)	*Bacterium* sp. $W_3A_4$1	27
Dimethylamine dehydrogenase (EC 1.5.99)	Hyphomicrobium X	28

Peptides in Covalent Flavoprotein Synthesis

Flavinylation may be a posttranslational process using the apoprotein as an acceptor. Alternatively, a cotranslational process can be envisaged by analogy to the mechanism of glycoprotein synthesis (34). Posttranslational FAD attachment was deduced from *in vivo* measurements of the covalent Rf incorporation into proteins of the outer mitochondrial membrane of rat liver in presence and absence of cycloheximide (35).

The search for an apoenzyme or a flavin-free precursor of 6-HDNO in *A.oxidans* yielded negative results (36). In wild-type cells, addition of inhibitors of translation or transcription to nicotine-induced cells stopped the formation of the active enzyme immediately. As with most bacterial enzymes, 6-HDN synthesis in *A.oxidans* appears to be transcription-limited. The flavinylation step was made the rate-determining process of 6-HDNO synthesis using a riboflavin-requiring (rf^-) mutant of *A.oxidans*; even in these cells, no acceptor for posttranslational flavinylation could be detected.

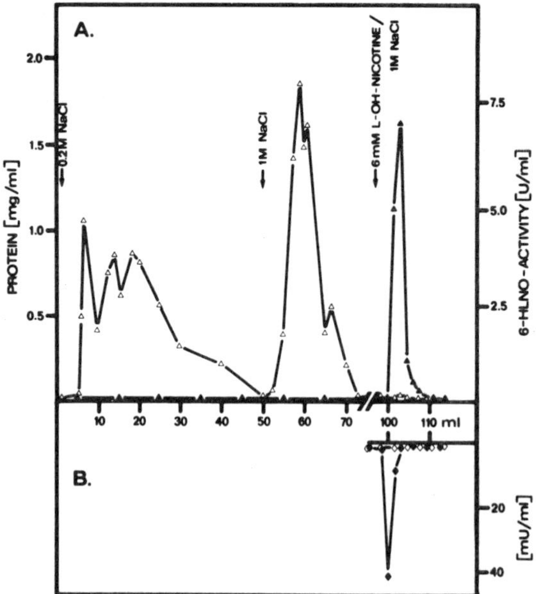

Figure 2. Identity of 6-HLNO synthesized in *A.oxidans* and in transformed *E.coli* SF-8. The enzymes were purified from sonicated cells by the same procedure. Extracts in 100 mM phosphate buffer, pH 7.0, were applied to Aminohexyl Sepharose 4 B and eluted with increasing NaCl concentrations. Addition of 6-hydroxy-L-nicotine specifically elutes 6-HLNO which is homogeneous in SDS PAGE (Hinkkanen, A., Lilius, E., Brandsch, R. and Decker, K., unpublished).

The possibility of cotranslational flavinylation was investigated by *in vitro* and *in vivo* techniques. A mixture of tRNA-bound peptides was isolated by the procedure of Kiely et al. (34) from a cell-free system synthesizing 6-HDNO and from nicotine-induced whole cells. After cleavage of the RNA-peptide bond, the liberated peptides produced a pattern in SDS PAGE (Figure 3) consistent with cotranslational flavinylation (37). These observations are taken to indicate that in 6-HDNO synthesis by *A.oxidans*, the acceptor for the flavin moiety is peptidyl-tRNA and thus flavinylation is a cotranslational event.

Flavins in Covalent Flavoprotein Synthesis

The question of the donor molecule in cFP synthesis is easier to answer but requires—due to the impermeability of cell membranes for most nucleotides—again a cell-free system for final proof. No indication was ever obtained in Rf autotrophs for the obligatory covalent attachment of a Rf precursor to the apoprotein: every cell so far tested readily incorporated extracellular Rf. Auxotrophic cells possess an efficient uptake system for the vitamin (38). A remarkable finding was reported by Grossman et al. (12): in yeast cells, some structural analogs of riboflavin, e.g., 7α- and 8α-methylriboflavin and 5-deazaflavin, can replace Rf to some extent in the covalent cofactor of succinate dehydrogenase and sustain the specific activity of that enzyme.

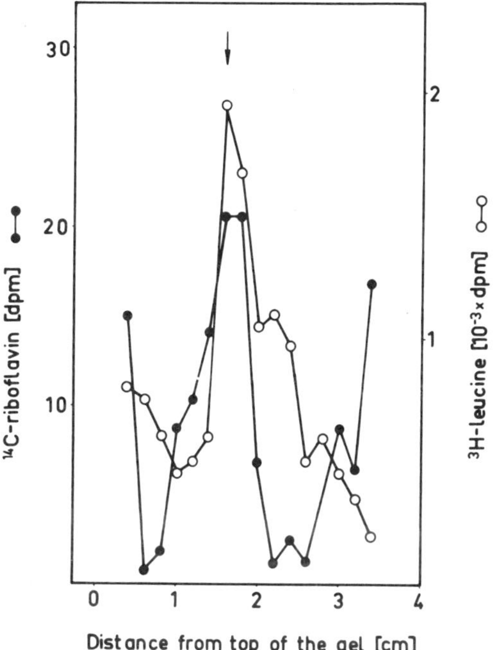

Figure 3. SDS PAGE of the peptides released from tRNA of intact cells. rf$^-$ Mutant cells of *A.oxidans* were grown in presence of (2-^{14}C) riboflavin. Rapidly prepared cell extracts (French press) were chromatographed on DEAE cellulose (34). The pooled fractions eluted with 0.25 M and 0.4 M NaCl were precipitated with 2 vol of 1% TCA in ethanol, treated with 3% SDS and analyzed in SDS PAGE as described previously (37).

Neither in these nor in other eucaryotic cells has it been possible so far to decide which form of Rf is actually incorporated into cFPs, e.g., Rf itself, FMN, or FAD, and whether the oxidized or reduced flavin participates in the flavinylation reaction. Phosphorylation and adenylylation of Rf, where investigated, do not seem to be limiting steps in cFP synthesis. In fact, no measurable amount of free intracellular Rf could be found in *A.oxidans* even in the presence of 15 μM extracellular Rf (39).

All the cFPs known contain FAD as coenzyme. The likelihood of the incorporation of this species was indicated in studies of cFP synthesis in rat liver and cerebrum (40). Direct evidence of FAD attachment to the acceptor polypeptide was obtained with the cell-free system of *A.oxidans* (36) and supported by *in vivo* studies using rf$^-$ mutant cells of this organism (39). FAD highly labelled in the adenine moiety (41) was used in cell-free translation of 6-HDNO. The results (Table 2) clearly indicate that FAD as such is the flavin donor of the covalent flavinylation step of 6-HDNO formation; it was also shown that *in vitro* only the 8α-His(N3) bond was synthesized (42). The covalent binding of FAD during 6-HDNO translation appears to require higher concentrations of the coenzyme than the noncovalent FAD attachment to 6-HLNO. The studies with intact rf$^-$ mutant cells of *A.oxidans* (39) showed that ca 5 μM intracellular FAD is necessary to allow half-maximal rate of

Table 2. Incorporation of (^{3}H-ado)FAD, (^{14}C)Histidine and (^{3}H)Leucine into 6-Hydroxy-D-Nicotine Oxidase During Biosynthesis *in Vitro* and *in Vivo*.

	FAD : Leu : His (mol : mol : mol)	**Ref.**
In vitro	1 : 34 : 13.7	42
In vivo	1 : 34 : 13	44

6-HDNO synthesis. The reconstitution of enzymatically active 6-HLNO from apoenzyme and FAD proceeded at an $S_{(0.5\ V)}$ of 80 nM (43).

Mechanisms of Covalent Flavin Attachment

Although structures of cFP have been known for over 10 years, very little has been learned about the chemical mechanisms of covalent binding. One still has to consider nonenzymatic as well as truly enzyme-catalyzed processes. The

Figure 4. Proposed mechanism of FAD-8α-His coupling.

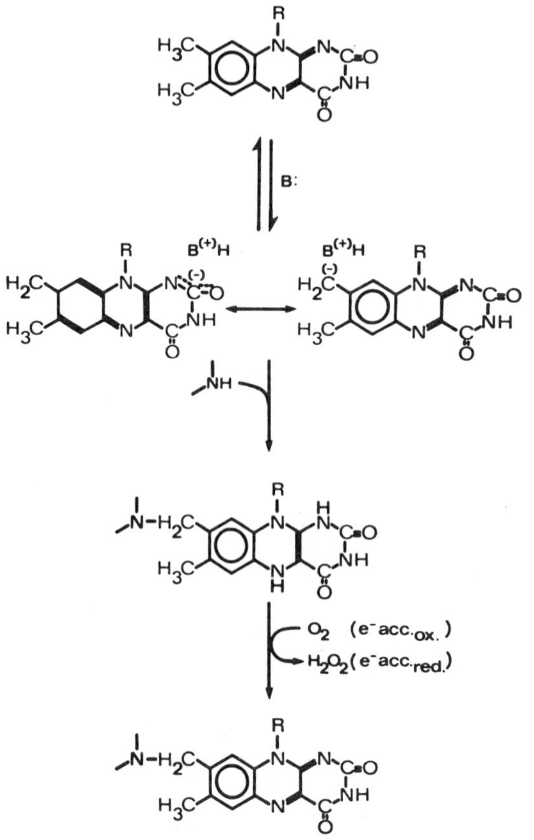

involvement of catalysts would be expected, if one of the participants, the flavin or the amino acid residue, has to be activated prior to coupling. The activation of the 8α-CH_3 group can be visualized through a monooxygenase-type hydroxylation followed immediately by (pyro)phosphorylation of the 8α-CH_2OH group. Pyrophosphate would be a good leaving group in this position facilitating the coupling with his-NH, cys-SH or tyr-OH; this process may or may not require a synthase.

Alternatively, activation of the amino acid residue can be considered: phosphorylated histidyl (45) and tyrosyl (46) residues of polypeptides are well documented. Thiophosphates, however, have never been observed in biological systems. From a chemical point of view, the activation of the amino acid residue appears to be less likely than that of the flavin moiety.

The prerequisite for a "nonenzymatic" coupling would be a high degree of conformational specificity of the binding region, i.e., the incoming flavin cofactor must be arranged and held for a finite time in a position which puts the 8α-C in exact and close proximity to the binding atom of the proper amino acid residue; furthermore, a base would have to be in a position to facilitate proton abstraction (Figure 4).

The driving force of the reaction could be the oxidation of the reduced histidyl-FAD. This mechanism would be equally feasible with histidine and cysteine and with the flavin positions 6 and 8α.

Taking this "nonenzymatic" coupling mechanism for granted, one could question any particular biological role of the covalent flavin-protein linkage. The formation of cFPs would occur spontaneously whenever the binding site for the flavin coenzyme happens to contain a proper nucleophilic amino acid residue in close proximity to the coupling site of the dimethyl-isoalloxazine ring and a base in a position to facilitate proton abstraction. Such a chance process might explain the hitherto futile attempts to assign a particular function to cFPs in cellular metabolism or evolution.

ACKNOWLEDGMENT

The work of the author was supported by grants from the Deutsche Forschungsgemeinschaft, Bonn-Bad Godesberg, through SFB 46.

References

1. Theorell, H. (1935) *Biochem Z* 275:37.
2. Warburg, O. and Christian, W. (1938) *Biochem Z* 298:150–168.
3. Singer, T.P., Kearney, D.B., and Massey, V. (1955) *Arch Biochem Biophys* 60:255–257.
4. Kenney, W.C. and Walker, W.H. (1972) *FEBS Lett* 20:297–301.
5. Singer, T.P. and Edmondson, D.E. (1974) *FEBS Lett* 42:1–14.
6. Edmondson, D.E. and Singer, T.P. (1976) *FEBS Lett* 64:255–265.
7. Edmondson, D.E. and Kenney, W.C. (1976) *Biochem Biophys Res Comm* 68:242–248.
8. Kenney, W.C., Edmondson, D.E., Singer, T.P., Nakagawa, H., Asano, A., and Sato, R. (1976) *Biochem Biophys Res Comm* 71:1194–1200.
9. Kenney, W.C., Edmondson, D.E., Singer, T.P., Nishikimi, M., Noguchi, E., and Yagi, K. (1979) *FEBS Lett* 97:40–42.
10. Kenney, W.C., Singer, T.P., Fukuyama, M., and Miyake, Y. (197() *J Biol Chem* 254:4689–4690.

11. Walker, W.H., Singer, T.P., Ghisla, S., and Hemmerich, P. (1972) *Eur J Biochem* 26:279–289.
12. Grossman, S., Goldenberg, J., Kearney, E.B., Oestreicher, G., and Singer, T.P. (1976) *Flavins and Flavoproteins*. Singer, T.P. (ed.) Amsterdam-Oxford-New York: Elsevier, pp. 318–322.
13. Kenney, W.C. and Kröger, A. (1977) *FEBS Lett* 73:239–243.
14. Weiner, J.H. and Dickie, P. (1979) *J Biol Chem* 254:8590–8593.
15. Möhler, H., Brühmüller, M., and Decker, K. (1972) *Eur J Biochem* 29:143–151.
16. Harada, Y., Shimizu, M., Murakawa, S., and Takahashi, T. (1979) *Agric. Biol Chem* 43:2635–2636.
17. Ohishi, N. and Yagi, K. (1979) *Biochem Biophys Res Comm* 86:1084–1088.
18. Mori, N., Tani, Y., Yamada, H., and Hayashi, R. (1981) *Agric Biol Chem* 45:539–540.
19. Ohta-Fukuyama, M., Miyake, Y., Emi, S., and Yamano, T. (1980) *J Biochem* 88:197–203.
20. Pinto, J.T. and Frisell, W.R. (1975) *Arch Biochem Biophys* 169:483–491.
21. Sato, M. et al., personal communication.
22. Hayashi, S., Nakamura, S., and Suzuki, M. (1980) *Biochem Biophys Res Comm* 96:924–930.
23. Walker, W.H., Kearney, E.B., Seng, R., and Singer, T.P. (1971) *Biochem Biophys Res Comm* 44:287–292.
24. Kenney, W.C. and Singer, T.P. (1977) *J Biochem Chem* 252:4767–4772.
25. Kenney, W.C., McIntire, W., and Yamanaka, T. (1977) *Biochim Biophys Acta* 483:467–474.
26. McIntire, W., Edmondson, D.E., and Singer, T.P. (1980) *J Biol Chem* 255:6553–6555.
27. Steenkamp, D.J., McIntire, W., and Kenney, W.C. (1978) *J Biol Chem* 253:2818–2824.
28. Steenkamp, D.J. (1979) *Biochem Biophys Res Comm* 88:244–250.
29. Sato, M., Ohishi, N., Nishikimi, M., and Yagi, K. (1977) *Biochem Biophys Res Comm* 78:868–873.
30. Addison, R. and McCormick, D.B. (1978) *Biochem Biophys Res Comm* 81:133–138.
31. Nishikimi, M., Yamauchi, N., Kiuchi, K., and Yagi, K. (1981) *Experientia* 37:479–480.
32. Edmondson, D.E. and Singer, T.P. (1973) *J Biol Chem* 248:8144–8149.
33. Decker, K. (1976) *Trends in Biochem Sciences* 1:184–185.
34. Kiely, M.L., McKnight, S., and Schimke, R.T. (1976) *J Biol Chem* 251:4590–4595.
35. Martinez, P. and McCauley, R. (1977) *Biochem Biophys Acta* 497:347–446.
36. Decker, K., and Hamm, H.-H. (1980) *Flavins and Flavoproteins*. Yagi, K., and Yamano, T. (eds.) Tokyo: Jap Sci Soc Press, pp. 251–263.
37. Hamm, H.-H. and Decker, K. (1978) *Eur J Biochem* 93:449–454.
38. Cecchini, G., Perl, M., Lipsick, J., Singer, T.P., and Kearney, E.B. 91979) *J Biol Chem* 254:7295–7301.
39. Hinkkanen, A. and Decker, K., this volume.
40. Pinto, J. and Rivlin, R.S. (1979) *Arch Biochem Biophys* 194:313–320.
41. Decker, K. and Hamm, H.-H. (1980) *Meth Enzymol* 66E;227–235.
42. Hamm, H.-H. and Decker, K. (1980) *Eur J Biochem* 104:391–395.
43. Decker, K. and Dai, V.D. (1967) *Eur J Biochem* 3:132–138.
44. Schimz, A. and Decker, K. (1978) *Arch Microbiol* 116:175–180.
45. Walinder, O. (1968) *J Biol Chem* 243:3947–3952.
46. Barbacid, M., Beemon, K., and Devare, S.G. (1980) *Proc Natl Acad Sci USA* 77:5158–5162.

Published 1982 by Elsevier North Holland, Inc.
Vincent Massey and Charles H. Williams, Editors
Flavins and Flavoproteins

CHAPTER 77

Unusual Properties of Flavin Prosthetic Group in Rat Liver Choline Dehydrogenase

Haruhito Tsuge, Yoshihiro Futamura, and Kazuji Ohasi

Department of Agricultural Chemistry, Gifu University, Kagamigahara, Gifu 504, Japan

Summary

The prosthetic group of rat liver mitochondrial choline dehydrogenase (EC 1.1.99.1) was studied. Choline dehydrogenase seemed to contain some type of modified FAD as the prosthetic group which gave rise to a visible absorption spectrum similar to that of the reduced flavin in the intact enzyme. Although a possibility of covalently-bound flavin easily cleaved by treatment with a reducing agent could not be ruled out completely, the flavin prosthetic group seemed to bind noncovalently to the enzyme on the molar basis of 1 : 1.

Introduction

Choline dehydrogenase (EC 1.1.99.1) has recently been purified from rat liver mitochondria to apparent homogeneity with a modification by thionitrobenzoate (1) and DEAE-Sepharose CL-6B column (2). This enzyme belongs strictly to the dehydrogenase type, due to a failure in direct electron transfer to molecular oxygen, and choline can be dehydrogenated to betaine via betainealdehyde in the presence of an appropriate electron acceptor. The molecular weight of the active form, where the enzyme is monomeric, is estimated to be ~75,000 (3).

Thus, the identification of the prosthetic group of the enzyme was an important task. We dealt with the properties of the prosthetic group, because it is questionable whether the coenzyme is a flavin. Only a few reports on this have been published in the past (4, 5).

Experiments

Choline dehydrogenase was purified as reported (1–3). The preparation was almost pure, based on a 7.5% polyacrylamide disc gel electrophoresis (PAGE) containing 0.1% Triton X-100 (TX-100) and 0.5% sodium dodecylsulfate (SDS). The preparation having a specific activity of more than 4.0 μmol O_2 consumed/min/mg protein was used. For the choline dehydrogenase assay, a solution containing 16.7 mM choline chloride, 1 mM phenazinemethosulfate, 1 mM KCN, and 17 mM KH_2PO_4-KOH buffer (pH 7.6) in a total volume of 3.0

ml was incubated at 30°C, and oxygen consumption was recorded oxygraphically (1–3). Protein was determined by the modified-Lowry method (6).

All absorption spectra were recorded with a Hitachi 200-10 double-beam spectrophotometer with a 1 cm cuvette. Fluorescence spectra were measured with a Hitachi 203 fluorescence spectrophotometer with a Xenon lamp as a light source.

Results

Visible absorption spectra of purified choline dehydrogenase (0.51 mg) were measured in the presence of 0.1% TX-100 and 0.1 M KCl at pH 8.0. As shown in Figure 1, a fresh preparation obtained through a DEAE-Sepharose CL-6B column gave a small peak around 410 nm. However, the absorption spectrum changed rather easily as indicated, whereas the activity to oxidize the substrate did not alter at all either before or after the spectral change. The time-dependent change of the absorption spectrum was remarkable: (a) The small peak at 410 nm disappeared with the hyperchromic changes, and (b) there were slight twin peaks centering at 370 and 450 nm that overlapped in a strong absorbance at a shorter wavelength. The addition of choline (60 mM) did not induce any detectable change, and an excess amount of dithionite caused the disappearance of the peak at 450 nm, suggesting that the preparation contained a flavin chromophore. However, it was uncertain that the chromophore functioned as a cofactor.

Because of our preconception that choline oxidase contains a type of covalently-bound flavin (7,8), as well as our findings that one fluorescence band corresponding to the protein-staining band under a UV lamp was

Figure 1. Visible absorption spectra of choline dehydrogenase. Each spectrum was measured using 0.51 mg protein (6.6 nmol). (A) Freshly prepared enzyme (1). After performic acid oxidation (2), the enzyme preparation was allowed to stand for 2 hr at 44°C as a solution state (3), to which an excess amount of dithionite was added (4). (B) The enzyme preparation (1) was treated with 0.1 N HCl (2), 0.8 M guanidine hydrochloride (3), and 0.1 N NaOH (4). These spectra were measured without centrifugation. The enzyme preparation (1) was heated at 100°C (5) and treated with 5% TCA (6), which were measured after centrifugation. Freshly prepared enzyme as control (7). (C) Time-dependent change of absorption spectrum of freshly prepared preparation. Time after 0 hr (1); 12 hr at 25°C (2); 4 hr at 30°C (3); 1 hr at 44°C (4) and 2 hr at 44°C (5).

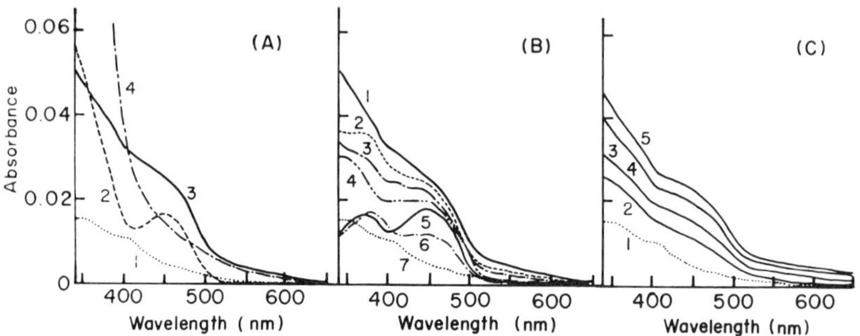

observed on a SDS PAGE oxidized by performic acid, the enzyme (9.36 mg) was precipitated by 5% trichloracetic acid (TCA) and then subjected to tryptic-chymotryptic digestion for 17 hr at 37°C. It was noteworthy, however, that the supernatant emitted strong fluorescence, while the digest was essentially nonfluorescent. It was also found that heating at 100°C or performic acid oxidation could release the chromophore from the choline dehydrogenase preparation.

Thus, the supernatant solution with 5% TCA was applied to a column of Florisil, from which three peaks with respect to the absorbance at 445 nm were obtained. These fractions, fluorescent under a UV lamp, were combined and lyophilized to dryness, and then applied to a column of DEAE-Sephadex A-25. Fluorescent compounds were also fractionated into three, where these peaks were designated as F-1, F-2, and F-3, in that order (see Figure 2). Figure 3 shows the absorption and fluorescence spectra of the lyophilized samples in 20 mM phosphate buffer (pH 7.0). The absorption spectrum of F-3 resembled that of FMN or FAD, while spectra of F-1 and F-2 were different in two points: (a) there was no peak at 370 nm, and (b) in spite of the relatively low intensity at 450 nm, a significant absorbance at a longer wavelength (>500 nm) was noted in comparison to that of authentic FAD. The fluorescence excitation and emission spectra of one of the three compounds agreed well with those of FAD as shown in the inset of Figure 3. However, it must be noted that the fluorescence efficiency of F-1 and F-2 seemed to be high.

The F-3 fraction was identified as FAD, judging from its ability to regain glucose oxidase activity (9), as well as the results obtained from thin layer or paper chromatography (10). The structural identification of F-1 and F-2 remains to be established, but it is noted that the Rf values of these two resembled that of FMN in all solvent systems used. These compounds would be derived from the intact chromophore present in the enzyme during the purification process.

Figure 2. Separation of three compounds by DEAE-Sephadex A-25 column. Pyridine-acetate (pH 5.5) was used as the elution buffer.

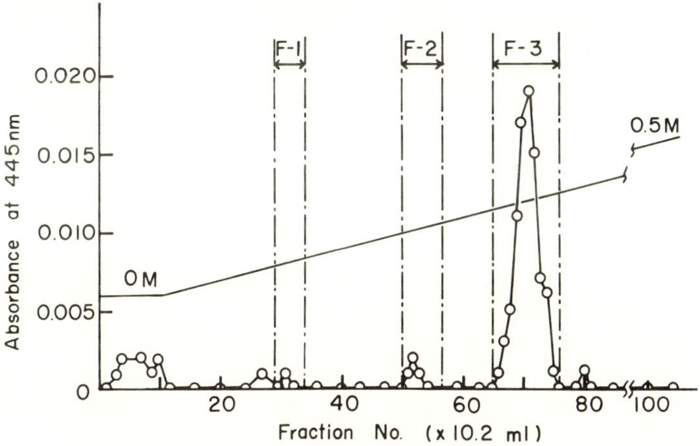

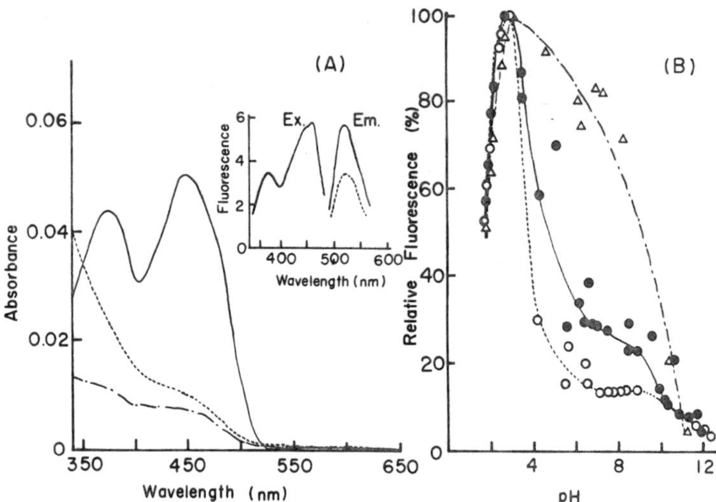

Figure 3. Absorption spectra and fluorescence properties of purified chromophores. (A) Visible absorption spectra of F-1 (---); F-2 (–·–); and F-3 (—) in 20 mM phosphate buffer (pH 7.0). (Inset) Fluorescence spectrum of F-3, emitted at 520 nm (—), and excited at 465 nm (—) or at 375 nm (---). (B) pH-Fluorescence curve of F-3, excited at 465 nm and emitted at 520 nm. (●) authentic FAD; (○) supernatant of heat treatment; and (△) F-3.

The recovery of flavin chromophore, as summarized in Table 1, was calculated in terms of absorbance at 450 nm using the known molar extinction coefficient (11). The chromophore in toto recovered from DEAE-Sephadex A-25 was about 90% and the major fraction (F-3) was 50% or more. Other methods enabling the release of the chromophore were adopted and overall recovery, when calculated as FAD, ranged from 50% to 70%.

Table 1. Recovery of Prosthetic Group and Rf Value on Thin Layer Chromatography.

Method	Flavin contents (nmol)	Recovery (%)		Rf value (I)[d]	(II)[e]
5% TCA sup[a]	61	51		—	—
DEAE-Sephadex[a]					
F-1	29	24[c]	} 91	0.52	0.12
F-2	14	12[c]		0.52	0.06
F-3	66	55		0.37	0.02
Performic acid oxidation[a]	72	60		—	—
Heat at 100°C[b]	4.8	73		0.31	0.00

[a]9.36 mg (S.A. 4.4 unit/mg; 120 nmol) was used.

[b]0.51 mg (S.A. 4.0 unit/mg; 6.6 nmol) was used.

[c]Determined as FMN.

[d]5% Na_2HPO_4 with cellulose plate.

[e]Butanol:acetic acid:water (4:1:5, V/V/V) with silica gel plate.

Discussion

The findings described above may imply that mammalian liver choline dehydrogenase contained a modified flavin skeleton as the prosthetic group, and presumably, the flavin group was attached to apoprotein noncovalently. However, we have to consider the fact that monoamine oxidase purified from beef liver (12) contains 8α-cysteinylflavin cleaved by a treatment with 100 mM dithiothreitol. We used a similar treatment (2) to release thionitrobenzoate. Thus, covalently-bound flavin might be released. It was certain, however, that choline dehydrogenase did not contain 8α-(n-3)-histidylflavin.

In conclusion, the enzyme contained 1 mol flavin per mol of enzyme, whose structure in the intact choline dehydrogenase probably was stabilized as an FAD-adduct or its reduced form (13). During the purification of the prosthetic group, some part decomposed to FMN-like substances and another part changed to FAD.

References

1. Tsuge, H., Onishi, H., Futamura, Y., Nakano, Y., and Ohashi, K. (1980) *J Jpn Soc Food Nutr* 33:143–150 (in Japanese).
2. Tsuge, H., Nakano, Y., Onishi, H., Futamura, Y., and Ohashi, K. (1980) *Biochim Biophys Acta* 614:274–284.
3. Tsuge, H., Futamura, Y., Nakano, Y., and Ohashi, K. (1980) *Biochem Int* 1:519–525.
4. Rothschild, H.A., Cori, O., and Barron, E.S.G. (1954) *J Biol Chem* 208:41–53.
5. Kimura, T. and Singer, T.P. (1962) *Methods Enzymol* 5:562–570.
6. ChandraRajan, J. and Klein, L. (1975) *Anal Biochem* 69:632–636.
7. Ohishi, N. and Yagi, K. (1979) *Biochem Biophys Res Commun* 86:1084–1088.
8. Ohta-Fukuyama, M., Miyake, Y., Emi, S., and Yamano, T. (1980) *J Biochem* (*Tokyo*) 88:197–203.
9. Tsuge, H. and Mitsuda, H. (19710 *J Vitaminol* (*Kyoto*) 17:24–31.
10. Fazekas, A.G. and Kokai, K. (1971) *Methods in Enzymol* 18(B):385–398.
11. Weimar, W.R. and Neims, A.H. (1975) In *Riboflavin*. Rivlin, R.S. (ed.), p. 1-47, New York: Plenum Press.
12. Watanabe, K., Minamiura, N., and Yasunobu, K.T. (1980) *Biochem Biophys Res Comm* 94:579–585.
13. Iwatsuki, N., Joe, C.O., and Werbin, H. (1980) *Biochemistry* 19:1172–1176.

Published 1982 by Elsevier North Holland, Inc.
Vincent Massey and Charles H. Williams, Editors
Flavins and Flavoproteins

CHAPTER 78

The Intracellular FAD Pools of *Arthrobacter oxidans* and Their Correlation to 6-Hydroxy-D-Nicotine Oxidase Synthesis

A. Hinkkanen and K. Decker

Biochemisches Institut, Albert-Ludwigs-Universität, Hermann-Herder-Str. 7, D-7800 Freiburg im Breisgau, Federal Republic of Germany

6-Hydroxy-D-nicotine oxidase (6-HDNO) is an inducible enzyme in *Arthrobacter oxidans*. It has been shown to consist of one polypeptide chain of M_r 50,000 to which FAD is is covalently linked through a 8α-His(N3) bond. Experiments *in vitro* and *in vivo* indicate that flavinylation is a cotranslational process in which FAD is the donor substrate (reviewed in Reference 1).

In a riboflavin-requiring (rf^-) mutant strain of *A.oxidans*, growth and 6-HDNO synthesis displayed different riboflavin (Rf) dependencies (Figure 1). With these cells it was possible to study 6-HDNO formation under conditions where flavinylation is the rate-limiting step.

Flavin coenzymes occur in at least 3 distinct pools: (a) as free FAD (FMN); (b) in noncovalent attachment to proteins but with restricted exchangeability; (c) covalently bound to a polypeptide chain. In this study, these three pool sizes were quantitatively measured, their levels correlated with the extracellular Rf supply and an approximate half-saturation concentration of FAD for 6-HDNO synthesis was derived.

The relationship of intracellular Rf, FMN, and FAD was established using [2-^{14}C] labelled Rf in the growth medium; chromatography of an acid cell extract (cellulose plates, 5% aqueous Na_2HPO_4 as solvent) showed undetectable Rf, traces of FMN and >95% of the radioactivity as FAD. FAD was, therefore, considered to be the only flavin coenzyme present in *A.oxidans*.

The FAD contents of the 3 pools (Table 1) were measured by two independent methods: (a) rf^- mutant cells were grown in the presence of labelled Rf of defined specific radioactivity. After separation of the individual coenzyme pools, the amount of FAD was calculated from the incorporated label; the radiochemical homogeneity of the samples was ascertained by chromatographic analysis. (b) The FAD content of the separated pools was also measured enzymatically in a catalytic assay (2); spectroscopically-defined FAD served to calibrate the assay and as internal standard.

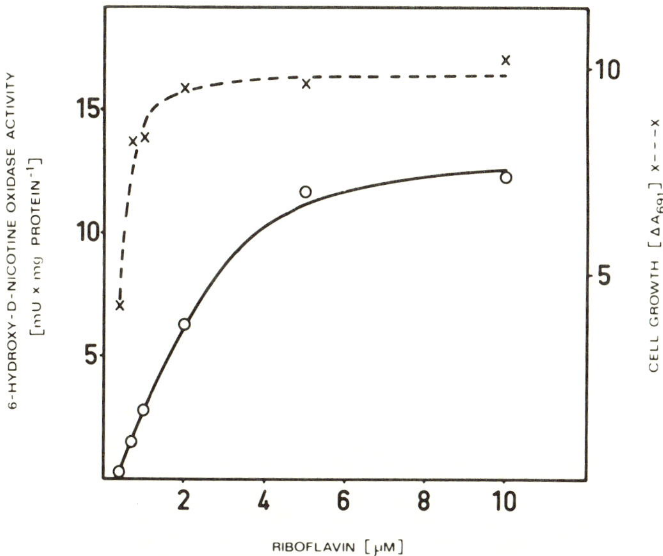

Figure 1. Dependence of growth and of 6-HDNO synthesis on the riboflavin supply of riboflavin-requiring mutant cells of *A. oxidans*.

The carefully washed cells were sonicated for 3.5 min (Branson sonifier, power stage 1) at 0°C and centrifuged. Experiments using [2-^{14}C]Rf ascertained that the pellet contained covalently-bound flavin only. From the supernatant a protein-rich and a protein-free fraction were separated by gel filtration (Figure 2). From the former, FAD was liberated by acid treatment; it represents the FAD which does not dissociate from its apoprotein at neutral pH and 0°C. The denatured protein was centrifuged and rehomogenized in 5% $HClO_4$ several times; the resulting pellet contained the cytosolic covalently-bound flavin.

A. oxidans contains, besides 6-HDNO, at least one more covalent flavoprotein which is independent of nicotine induction. The content of covalently-bound FAD was determined from the data of Figure 3.

Table 1. Content and Distribution of FAD Under Optimal Riboflavin Supply.

FAD		nmol/g dry wt.	% of total
Free		42	16
Noncovalently bound to proteins		188	71
Covalently bound			
6-HDNO	19		
Other	15		
Subtotal		34	13
Total		264	100

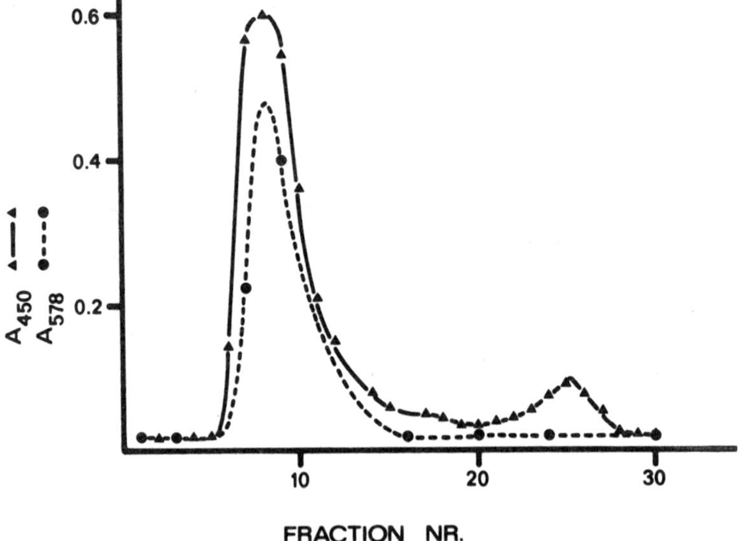

Figure 2. Separation of flavoproteins and free FAD by gel filtration. Sonicated extracts of 500 mg wet wt of rf^- cells (in 2 ml 0.1 M phosphate buffer, pH 7.0) grown in the presence of [2-^{14}C]Rf were passed through an Ultrogel AcA 44 column (25 ml). Fractions of 1 ml were collected. FAD content was followed spectrophotometrically at 450 nm, protein (after reaction of an aliquot with Coomassie blue) at 578 nm. All steps were carried out in the dark or under red light.

Figure 3. Cell growth, 6-HDNO synthesis and covalent incorporation of [2-^{14}C] riboflavin in D,L-nicotine-induced and in noninduced cells of *A.oxidans*. (△) enzyme activity (in induced cells only); (×) absorbancy at 691 nm (d=1 cm) (the same for induced and noninduced cells); protein-bound radioactivity in noninduced (○) and in induced (●) cells.

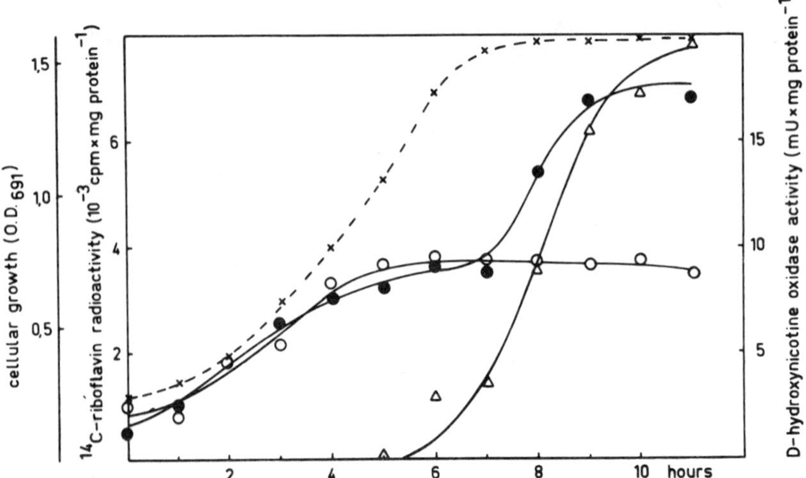

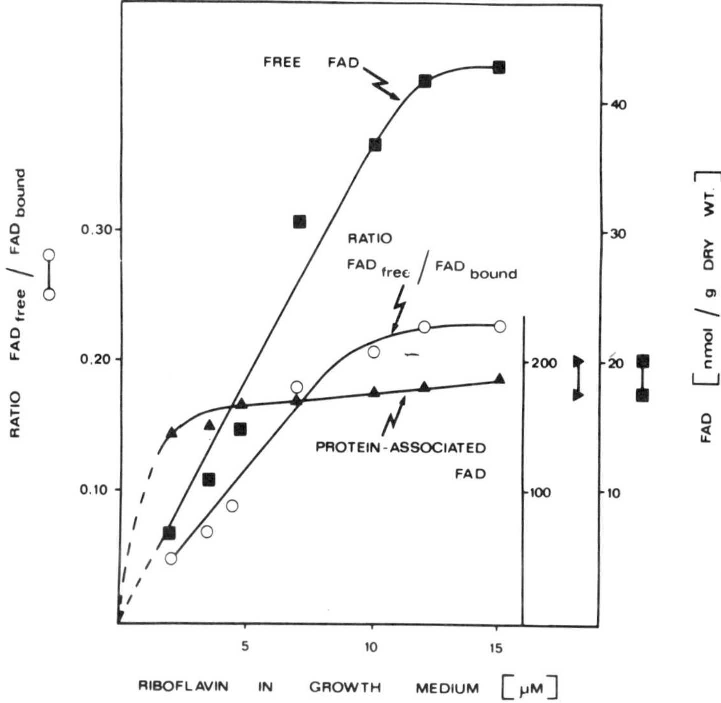

Figure 4. Free and protein-associated FAD of rf^- cells of *A. oxidans* as a function of the external riboflavin concentration.

Table 2. Growth, Content of Free FAD and Rate of 6-Hydroxy-D-Nicotine Oxidase Synthesis in the rf^- Mutant of *A. Oxidans*.

	Extracellular riboflavin (μM)		Ratio $\left(\frac{\text{at } 15\ \mu M}{\text{at } 2\ \mu M}\right)$
	2	15	
Increase of cell density (A_{691}/hr)[a]	0.82	0.83	1
Content of free FAD (nmol/g dry wt)	7	43	6.1
Increase of 6-HDNO activity (mU/mg protein·hr)[a]	1.1	5.8	5.3

[a]Measured during phase of quasi-linear increase.

Finally, the pool sizes of free and protein-associated FAD as a function of the extracellular Rf concentration were determined using the rf$^-$ mutant cells (Figure 4). Obviously, 2 μM Rf not only allowed optical cell growth but was also sufficient for the synthesis of the bulk of the flavoproteins (except 6-HDNO). The free FAD pool, however, reached its maximal level only at about 12 μM Rf. The rate of 6-HDNO synthesis did not follow the majority of the flavoproteins but paralleled the content of the free FAD (Table 2). Since 6-HDNO formation is flavin-limited up to about 10 μM external Rf, it can be deduced that the flavinylation step attains its half-maximal rate at an intracellular FAD concentration of ca 5 μM, assuming an aqueous phase of 2.5 ml per g cell dry weight (3).

ACKNOWLEDGMENT
This work was supported by a grant of the Deutsche Forschungsgemeinschaft, Bonn-Bad Godesberg, through SFB 46.

References

1. Decker, K. (1981) Biosynthesis of Covalent Flavoproteins, this volume.
2. Friedemann, H.C. (1974) In *Methoden der Enzymatischen Analyse*. Bergmeyer, H.U. (ed.) Weinheim: Verlag Chemie, pp. 2232–2235.
3. Riebeling, V., Thauer, R.K., and Jungermann, K. (1975) *Eur J Biochem* 55:445–453.

Published 1982 by Elsevier North Holland, Inc.
Vincent Massey and Charles H. Williams, Editors
Flavins and Flavoproteins

CHAPTER 79

Chemical and Biochemical Properties of p-Cresol Methylhydroxylase

W.S. McIntire, D.J. Hopper, D.E. Edmondson, and T.P. Singer

Molecular Biology Division, Veterans Administration Medical Center, San Francisco, California, Department of Biochemistry and Biophysics, University of California, San Francisco, California, and Department of Biochemistry, University College of Wales, Aberystwyth, Dyfed SY3 3DD, Wales, United Kingdom

The number of enzymes containing covalently-bound flavin has increased rapidly in the past few years. Since the identification of the mode of FAD linkage in monoamine oxidase and succinate dehydrogenase in the early 1970s, the flavin-amino acid bond for about 20 enzymes has been determined. All these flavin linkages fall into four categories; (1) 8α-S-cysteinyl-, (2) 8α-N(1)-histidyl-, (3) 8α-N(3)-histidyl-FAD, and (4) 6-S-cysteinyl-FMN (1). Recently we discovered a fifth type of flavin covalently bound to a protein.

The three p-cresol methylhydroxylases from *Ps. putida* strains 9866 and 9869 are all flavocytochromes of approximately equal molecular weight (2,3). Each is dimeric and contains 1 mol of covalent flavin and 1 mol of c type heme per mol of enzyme. All three enzymes catalyze a 2-electron oxidation of p-cresol, followed by addition of water, to form p-hydroxybenzylalcohol. Each of the enzymes will then oxidize the alcohol to p-hydroxybenzaldehyde.

Pure peptic flavin peptides were isolated from each of these enzymes and the amino acid analyses were identical (1 Asn, 1 Ser, 4 Gly, 1 Met, 1 Arg, 1 Tyr, 1 Trp). Cysteine was not present and amino acid analysis of the acid-hydrolyzed, performic acid-oxidized peptides did not yield cysteic acid. The absorption spectra did indicate 8α-substitution of the flavin (Figure 1), but the amino acid analysis and a pH fluorescence study ruled out both the 8α-S-cysteinyl and the 8α-histidyl derivatives (4,5).

The first indication as to the nature of the aminoacyl flavin came as a result of peptide sequencing work. An unusual N-dansyl-tyrosine was obtained as the N-terminal amino acid derivative (6). Normally, an N-terminal tyrosine would produce N,O-bisdansyl-tyrosine. The only explanation for this observation is that the phenolic oxygen of tyrosine is blocked, possibly the flavin. The resulting sequence for one of the peptides was NH_2-Tyr-Asn-Trp-Arg-Gly-Gly-Gly-(Gly, Ser, Met)-COOH.

The peptide was converted from the FAD to the riboflavin form and digested with amino peptidase M. The pure aminoacyl flavin was dansylated before and after acid hydrolysis. N-dansyl- and N,O-bisdansyl-tyrosine, respectively, were found. Acid hydrolysis and amino acid analysis indicated 1 mol of tyrosine per mol of flavin.

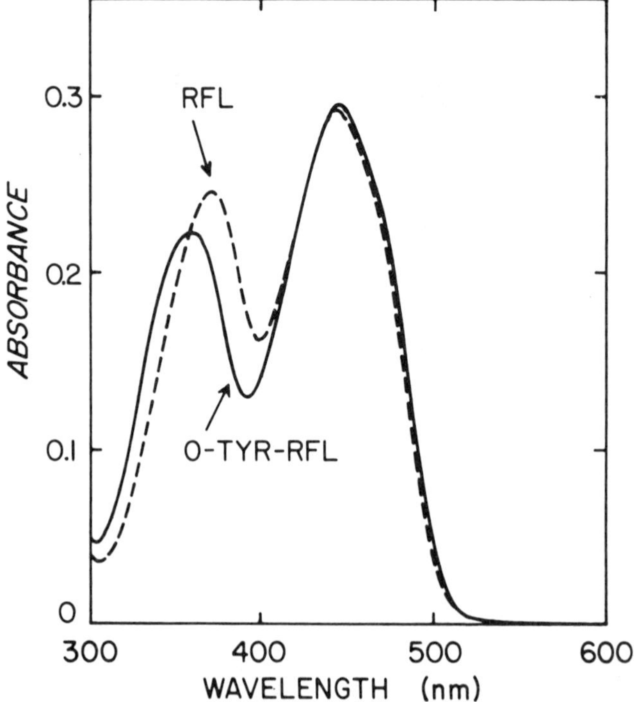

Figure 1. Spectra of riboflavin (RLF) and 8α-O-tyrosylriboflavin at neutral pH. A shift of the near UV band, relative to riboflavin, is indicative of 8α substitution.

Further proof of the structure was provided by synthesis of 8α-N-tyrosylriboflavin (Figure 2). The Cu^{2+} complex of tyrosine blocks both the α-amino- and carboxyl-groups of tyrosine, allowing only the phenolic oxygen to form the bond with 8α-Br-tetraacetylriboflavin (7). The purified deacetylated material was identical with the natural aminoacyl riboflavin, as judged by their spectra, dansylation experiments, amino acid analysis, and thin layer chromatography in 4 solvent systems. The NMR spectrum in 99.5% [U-^{2}H]-methanol also supports the 8α-O-tyrosyl-riboflavin structure (1).

The fluorescence of this new flavin is 1.5% that of riboflavin at pH 7.0 and 3.0. The quantum yield increases to 6.6% in methanol and to 12.3% in dimethylformamide. These fluorescence changes are similar to those found by Falk and McCormick for 8α-cysteinyl flavins (8). The intense quenching of these flavins (and of the unprotonated 8α-histidyl flavins) can be attributed to the interaction of the nonbonding electrons of the atom adjacent to the 8α-carbon, with the π system of the flavin. Organic solvents (or a proton, as in the case of the histidyl flavins) will interact with these electrons and thus cause a fluorescence increase (8).

Another important property of 8α-O-tyrosyl flavin is its oxidation-reduction chemistry. Sodium dithionite titration produced surprising results. Figure 3A shows that little change occurred at 446 nm during the uptake of the first 2 electrons by the near UV maximum shifted from 359 nm to 371 nm. Full

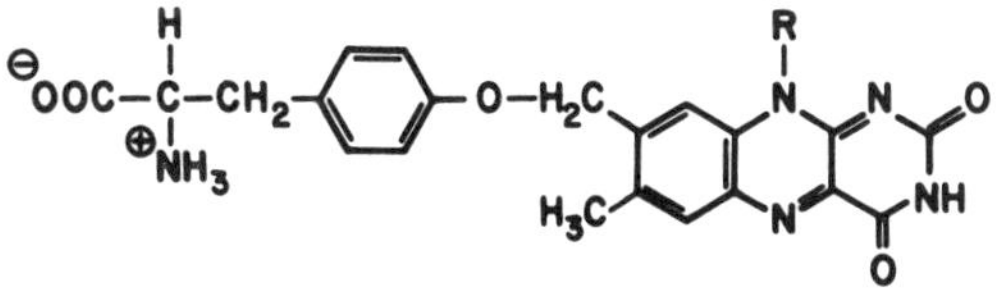

Figure 2. Structure of 8α-O-tyrosylriboflavin.

reduction of the flavin required a second mol of dithionite (2 more e^-). After reoxidation, the flavin was identified as riboflavin and amino acid analysis indicated that 1 mol of free tyrosine had been set free. This is the first example of an unmodified natural aminoacyl flavin undergoing reductive 8α elimination. Previous examples of this elimination are provided by 8α-S-cysteinylsulfone- and 8α-S-cysteinyl-8α-hydroxyriboflavin, where cysteinesulfinic acid and a hydroxyl group, respectively, are lost on reduction of the flavin (9, 10).

The reductive elimination precludes a direct determination of the midpoint potential of this flavin. From sulfite binding (11), we calculated an $E_{m,7}$ value of -169 mV ($K_D = 0.107$ M at 25°C) for 8α-O-tyrosylriboflavin.

These observations raised the interesting question whether the flavin undergoes the same reductive cleavage in the native enzyme and, if so, how this can be reconciled with its redox function. The absorption spectrum of one of these hydroxylases is shown in Figure 3B and the source of dithionite titration, monitored at selected wavelengths, in its inset. Absorbance changes at 418 nm represent reduction of the heme and require ~1 electron equivalent of dithionite. Changes at 450 nm reflect reduction of both the heme and the flavin: most of the bleaching is complete on addition of 3 e^- equivalents. Changes at 385 nm are more complex: the initial decrease is due to heme reduction, the increase during the uptake of the second e^- represents formation of the red flavin radical and the decrease during the third e^- uptake is due to further reduction of the flavin. The EPR spectrum of the 2 e^- reduced enzyme gave a peak-to-peak linewidth of 13 Gauss, in line with that given by the red (anionic) radicals of other 8α-substituted flavins but much less than the linewidth (19–23 Gauss) of blue (neutral) radicals (12). Precipitation with trichloroacetic acid in air at the end of the dithionite titration did not release any flavin, showing that no reductive cleavage had occurred. Thus, the tyrosyl flavin linkage is stabilized in the structure of the holoenzyme. Similarly, tryosylriboflavin bound to apoflavodoxin did not undergo cleavage on dithionite titration but, instead, large absorption maxima developed at 610 and 665 nm, indicating the formation of a blue (neutral) radical with this enzyme. Titration of the hydroxylase with p-cresol gave virtually the same results as dithionite. Since the substrate donates electrons in pairs but the enzyme accepts 3 e^-, the possibility of interenzymic electron transfer must be considered.

Steady-state kinetic assays using the enzyme from *Ps. putida* 9869, form A, and PMS-DCIP as oxidant gave the following values: K_M (p-cresol) $= 12.1 \pm 1.1$ μM; K_M (PMS) $= 1.58 \pm 0.16$ mM; $V_{max} = 24.7 \pm 0.9$ μmols/min/mg at 30°C. The parameters for p-hydroxybenzylalcohol were: K_M (substrate) $= 53.8 \pm 5.9$ μM; K_M (PMS) $= 0.56 \pm 0.09$ mM; and $V_{max} = 23.2 \pm 1.6$ μmol/min/mg. Pre-steady-state kinetics are under current study.

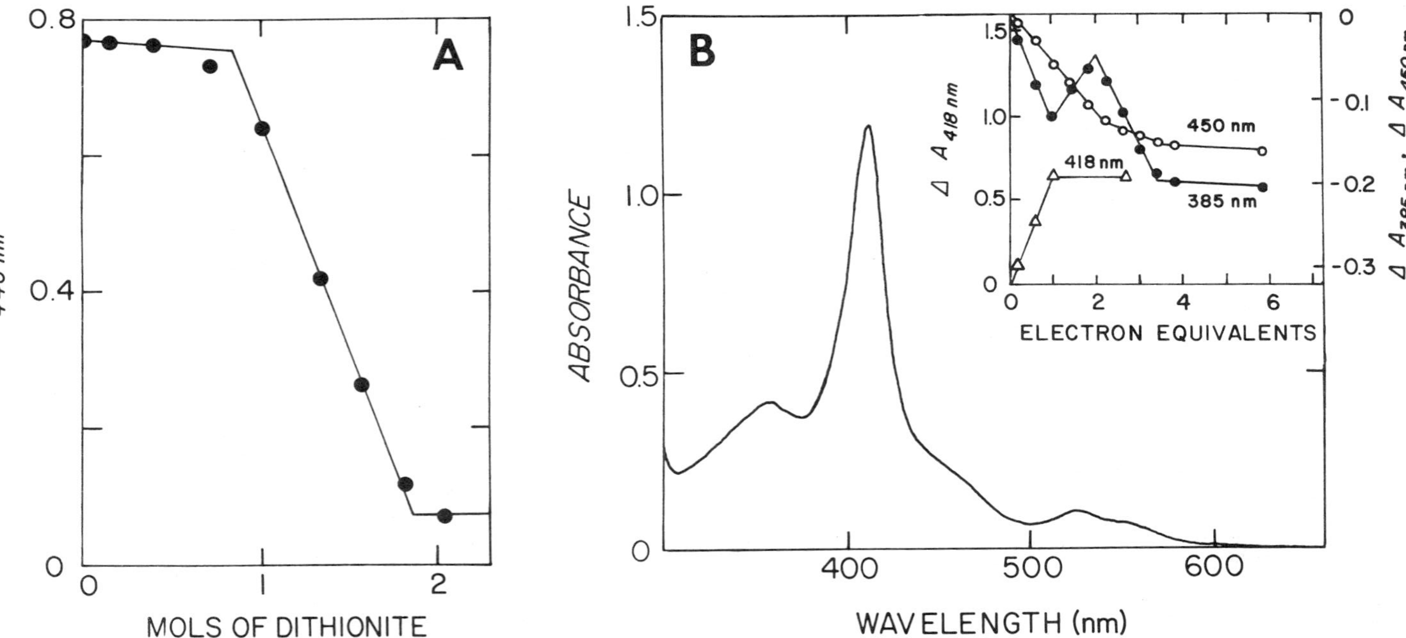

Figure 3. Anaerobic dithionite titrations. (A) Graph of A_{446nm} for the tyrosyl flavin versus dithionite added. 150 nmols of flavin in 2.17 mL of 50 mM phosphate pH 7.0. Points are corrected for volume change. Note points in first phase were recorded 15 minutes after the addition of dithionite. (B) Absorption spectrum of oxidized p-cresol methylhydroxylase from *Ps. putida* strain 9869 form A. Inset shows the changes that occur at 385, 418, and 450 nm on addition of dithionite. Enzyme: ~0.86 mg in 0.73 mL of 50 mM phosphate, pH 7.0. Points for the first and second electron equivalents were recorded immediately after addition of dithionite and for the third phase were 15 to 20 min after dithionite addition. Points are corrected for volume change and are plotted as Δ absorbance (measured absorbance minus absorbance of fully oxidized enzyme).

References

1. McIntire, W., Edmondson, D.E., Hopper, D.J., and Singer, T.P. (1981) *Biochemistry* 20:3068–3075.
2. Hopper, D.J. and Taylor, D.G. (1977) *Biochem J* 167:155–162.
3. Keats, M.J. and Hopper, D.J. (1978) *Biochem J* 175:649–658.
4. Kenney, W.C., McIntire, W., and Yamanaka, T. (1977) *Biochem Biophys Acta* 483:467–474.
5. Singer, T.P. and Edmondson, D.E. (1980) *Methods Enzymol* 66:253–264.
6. McIntire, W., Edmondson, D.E., Singer, T.P., and Hopper, D.J. (1980) *J Biol Chem* 255:6553–6555.
7. Yamashire, D. and Li, C.H. (1973) *J Am Chem Soc* 95:1310–1315.
8. Falk, M.C. and McCormick, D.B. (1976) *Biochemistry* 15:646–653.
9. Edmondson, D.E. and Singer, T.P. (1973) *J. Biol Chem* 248:8144–8149.
10. Kenney, W.C. and Singer, T.P. (1977) *J. Biol Chem* 252:4767–4772.
11. Müller, F. and Massey, W. (1969) *J Biol Chem* 244:4007–4016.
12. Edmondson, D.E., Ackrell, B.A.C., and Kearney, E.B. (1981) *Arch Biochem Biophys* 192:769–781.

Published 1982 by Elsevier North Holland, Inc.
Vincent Massey and Charles H. Williams, Editors
Flavins and Flavoproteins

CHAPTER 80

The Semiquinone Form of Succinate Dehydrogenase and Other 8α-Substituted Flavoenzymes

B.A.C. Ackrell, W. McIntire, D.E. Edmondson,[1] and E.B. Kearney

Veterans Administration Medical Center and Department of Biochemistry and Biophysics, University of California, San Francisco, California

Cardiac succinate dehydrogenase appears to be an exception to the generalization that class 4 dehydrogenases (1) form neutral semiquinones, since its flavin radical shows an ESR spectral linewidth characteristic of anionic semiquinones, with additional narrowing resulting from elimination of hyperfine coupling to the 8-CH_3 group by the 8α covalent linkage to the protein (2). Ohnishi and her colleagues have suggested, however, on the basis of pH dependence of one-electron redox potential monitored by ESR, that the flavin radical in the enzyme, which amounted to ~7% of the total flavin in their experiment, is the neutral form, ionizing to the anionic form with a pK_a of 8.0 (3). No change in linewidth of the ESR signal as a function of pH was evident. Confirmation of the form of the semiquinone from optical spectra is complicated by spectral contributions from the three nonheme iron-sulfur clusters of the enzyme. The low amount of radical observed is in contrast to the large signal obtained by us previously (~70% of the flavin) on reduction with succinate (4). It was suggested that this high proportion of radical was an artifact resulting from modifications to iron-sulfur center 3 of the enzyme (3).

In view of these uncertainties, we have studied further the nature of the semiquinone in succinate dehydrogenase, and compared its ESR spectral characteristics with those of other protein-bound 8α-substituted flavins. Succinate dehydrogenase with fully active center 3 has been used for these experiments (5).

It is evident from the examples cited in Table 1 that 8α-substituted flavoenzymes, whether the linkage be 8α-N(3)-histidyl, 8α-N(1)-histidyl, 8α-S-cysteinyl, or 8α-O-tyrosyl FAD, exhibit around neutral pH the anionic (red) form of the flavin semiquinone, and, where examined, an ESR signal linewidth of ~12G. In the case of succinate dehydrogenase (8α-N(3)-histidyl FAD), where the signal linewidth of 12 G is apparent, direct confirmation of the type of radical by absorption spectroscopy is complicated and difficult. Evidence for the anionic form was obtained, however, by comparison of the radical signals in

[1]Present address: Department of Biochemistry, Emory University, Atlanta, Georgia 30322.

Table 1. Flavin Semiquinone Forms of Flavoenzymes Containing 8α-Substituted Flavin.

Enzyme	Flavin	Signal linewidth	From absorption spectra	Reference
Thiamine dehydrogenase	8α-N(1)-histidyl FAD	12G	Anionic	6
β-Cyclopiazonate oxidocyclase	8α-N(1)-histidyl FAD	—	Anionic	7
D-6-Hydroxynicotine oxidase	8α-N(3)-histidyl FAD	—	Anionic	8
Monoamine oxidase	8α-S-cysteinyl FAD	12[a]	Anionic	7
Chromatium cytochrome c_{552}	8α-S-cysteinyl FAD	—	Anionic	9
Choline oxidase	8α-N(3)-histidyl FAD	11–12	Anionic	10
Cholesterol oxidase	8α-N(1)-histidyl FAD	11–12	Anionic	10
p-Cresol methylhydroxylase	8α-O-tyrosyl FAD	12	Anionic	This work
Succinate dehydrogenase	8α-N(3)-histidyl FAD	12	—	7

[a]Hemmerich, P., personal communication.

H_2O and D_2O. With both succinate dehydrogenase and thiamine dehydrogenase (which is known from optical data to exhibit the anionic form [6]), the signal was unaffected by D_2O (Table 2). Moreover, the linewidth of the succinate dehydrogenase signal was invariant in the pH range 6.1 to 9.1. These data are consistent, therefore, with the original conclusion of Palmer et al. (2), but in disagreement with the conclusion of Ohnishi and colleagues (3) that the enzyme exhibits the neutral (blue) form, with the anionic form being observed at higher pH ($pK_a \sim 8.0$).

Ohnishi and colleagues (3) have attributed the high levels of flavin radical previously observed by us (4) in modified preparations to loss of center 3 integrity, with the two electrons derived from succinate oxidation residing in

Table 2. The Effect of D_2O and pH on the ESR Spectral Linewidths of 8α-Substituted Flavin Radicals.

		Linewidth (G)	
	pH	H_2O	D_2O
Succinate dehydrogenase	6.1	12	—
	7.6	12	12
	9.1	12	—
Thiamine dehydrogenase	7.0	12	12
8α-N(3)-histidyl FMN peptide of succinate dehydrogenase + apoflavodoxin	7.0	19–20	14–15
8α-N(1)-histidyl FMN peptide of thiamine dehydrogenase + apoflavodoxin	7.0	23	15
8α-S-cysteinyl FMN peptide of monoamine oxidase + apoflavodoxin	7.0	23	14
Synthetic 8α-O-tyrosyl-riboflavin + apoflavodoxin[a]	No ESR data; gives the neutral (blue) radical by absorption spectroscopy		

[a]McIntire et al., this volume.

nonheme iron center 1 of the enzyme and the flavin semiquinone. Nonheme iron center 2 in this preparation is of low redox potential and not reducible by succinate. With center 3 active, the enzyme would have the capacity to accept four electrons with full reduction of the flavin. We have observed, however, relatively high levels of radical (Table 3) in enzyme preparations that are unmodified, as determined by the diagnostic "low K_m" ferricyanide reductase activity (11). This criterion of intactness is based on our previous observation that enzyme preparations exposed to air lose concomitantly their "low K_m" ferricyanide reductase activity, their reconstitutive capacity, and the ability to exhibit the signal of center 3, while retaining their phenazine methosulfate reductase activity (11). Such high levels of radical may thus reflect the uptake and redistribution of only two electrons by a significant proportion of molecules in intact preparations, since input of three electrons is not possible. It must be recalled, however, that in this type of intact preparation, center 2, whose presence has been verified by core extrusion techniques (13), does not exhibit an ESR signal. Hence, the possibility that succinate can reduce this center cannot be tested by EPR and thus cannot be eliminated at this time, although this would imply a higher midpoint potential than usually accorded center 2. If such were true, the enzyme could take up four electrons to be distributed between nonheme iron centers 1, 2, and 3 and the flavin semiquinone.

It is of interest that the aminoacyl FMN peptides of 8α-substituted flavoenzymes produce typical neutral radicals on binding to apoflavodoxin (Table 2). These data are in agreement with the findings of Palmer et al. (2) and their suggestion that the dominant anistropic coupling constants are at positions other than the 8 position.

Table 3. Semiquinone Formation in Substrate-Reduced Succinate Dehydrogenase.

	pH	Buffer	Reductase activities[a]		% radical
			PMS	$Fe(CN)_6^{3-}$	
Expt. I	7.0	PIPES	61	50	39
	7.5	HEPES	67	46	38
	9.3	BICINE	61	45	48
Expt. II	6.1	MES	—	—	32
	9.1	CHES	—	—	43

[a]Activities are V_{max} values with respect to acceptor at 38°, expressed as μmol min^{-1} mg^{-1}; PMS, phenazine-methosulfate-dichlorophenolindophenol reductase; $FE(CN)_6^{3-}$, "low K_m"-ferricyanide reductase activity. Enzyme samples, stored in liquid N_2 as ammonium sulfate pellets were redissolved anaerobically in argon-saturated, succinate-containing 50 mM buffers, as listed, and the ammonium sulfate was removed by passage through Sephadex G-25 by the Penefsky method (12) under anaerobic conditions. Samples were transferred under a stream of argon to ESR tubes for measurement of semiquinone concentrations and for assay of activity. The initial activities of a control sample of enzyme were 67 (PMS) and 52 ($Fe(CN)_6^{3-}$). The pH values given are those of the buffers at 22°; values in liquid N_2, estimated from the color of added indicators, were PIPES, 7.0, HEPES, ~8.0, and BICINE, 9.0. ESR spectra were measured at −46°C, microwave power, 0.2 mW, modulation amplitude, 3.2 Gauss, microwave frequency, 9.1 GHz, in a Varian E-4 spectrometer.

ACKNOWLEDGMENTS
This investigation was supported by National Institutes of Health Grant HL-16251, and the Veterans Administration.

References

1. Massey, V. and Hemmerich, P. (1980) *Biochem Soc Trans* 8:246–257.
2. Palmer, G., Müller, F., and Massey, V. (1971) in *Flavins and Flavoproteins*. Kamin, H. (ed.) Baltimore: University Park Press, pp. 123–140.
3. Ohnishi, T., King, T.E., Salerno, J.C., Blum, H., Bowyer, J.R., and Maida, T. (1981) *J Biol Chem* 256:5577–5582.
4. Beinert, H., Ackrell, B.A.C., Kearney, E.B., and Singer, T.P. (1975) *Eur J Biochem* 54:185–194.
5. Ackrell, B.A.C., Kearney, E.B., and Coles, C.J. (1977) *J Biol Chem* 252:6963–6965.
6. Gomez-Moreno, C., Choy, M., and Edmondson, D.E. (1979) *J Biol Chem* 254:7630–7635.
7. Edmondson, D.E., Ackrell, B.A.C., and Kearney, E.B. (1981) *Arch Biochem Biophys* 208:69–74.
8. Brühmüller, M., Möhler, H., and Decker, K. (1972) *Eur J Biochem* 29:143–151.
9. Cusanovich, M.A. and Tollin, G. (1980) *Biochemistry* 19:3343–3347.
10. Ohta-Fukuyama, M., Miyake, Y., Shiga, K., Nishina, Y., Watari, H., and Yamano, T. (1980) *J Biochem* (*Tokyo*) 88:205–209.
11. Beinert, H., Ackrell, B.A.C., Vinogradov, A.D., Kearney, E.B., and Singer, T.P. (1977) *Arch Biochem Biophys* 182:95–106.
12. Penefsky, H.S. (1977) *J Biol Chem* 252:2891–2899.
13. Coles, C.J., Holm, R.H., Kurtz, D.M., Sr., Orme-Johnson, W.H., Rawlings, J., Singer, T.P., and Wang, G.B. (1979) *Proc Natl Acad Sci USA* 76:3805–3808.

PART VII:

Flavin Biosynthesis and Flavin Binding

Published 1982 by Elsevier North Holland, Inc.
Vincent Massey and Charles H. Williams, Editors
Flavins and Flavoproteins

CHAPTER 81

Biosynthesis of Riboflavin—Preparation and Enzymatic Conversion of Phosphorylated Pyrimidines Intermediates

A. Bacher, P. Nielsen, P. Rauschenbach, and G. Klein

Lehrstuhl für Organische Chemie und Biochemie, Technische Universität München, Lichtenbergstrasse 4, 8046 Garching, German Federal Republic

The involvement of several pyrimidine type intermediates in the biosynthesis of riboflavin has been recognized for some time (Figure 1). Studies in yeast suggested the following sequence of events: (a) removal of carbon atom 8 from a guanine precursor (**I**); (b) reduction of the ribosyl side chain of the pyrimidine (**II**); (c) deamination of the pyrimidine (**III**) in position 2 (1). On the other hand, studies in *Escherichia coli* suggested a different pathway with formation of (**IV**) by deamination of (**II**) (2). So far, the suggested pyrimidine phosphate intermediates had not been isolated, and the arguments were based on circumstantial evidence.

Synthetic Procedures

We have prepared the pyrimidine phosphates (**III**) and (**V**) by the sequence of reactions shown in Figure 2. No attempts were made to isolate the highly unstable compounds in solid form. They were prepared as needed and used in enzyme studies or for the preparation of pteridines. Condensation of (**III**) and (**V**) with diacetyl yields the lumazine (**X**) and the pteridine (**XI**), respectively. By the use of appropriate dicarbonyl compounds, the preparation of a variety of lumazine and pteridine derivatives should be possible, which may deserve interest as analogs of FMN.

It was our aim to develop procedures for the synthesis of pure 5′-isomers. This aim has not been fully attained. Although all synthetic operations were performed at room temperatures, the synthetic intermediates (**VIII**) and (**IX**) were found to contain about 20% of non-5′-isomers. The problem of isomer formation seems to be quite common with phosphorylated ribityl derivatives and has been studied in considerable detail in the case of FMN (3). The isomers of the nitropyrimidines and of the pteridines can be separated by reverse phase HPLC (Figure 3). Pure isomers were obtained in mg quantities by HPLC.[1] For the routine enzyme test, the synthetic product containing about 20% isomers can be used with no disadvantage.

[1] We have found in preliminary studies that the isomers of FMN can also be separated at the mg level by reverse phase HPLC using an eluent of 100 mM ammonium formate buffer pH 3.7 and 17% methanol.

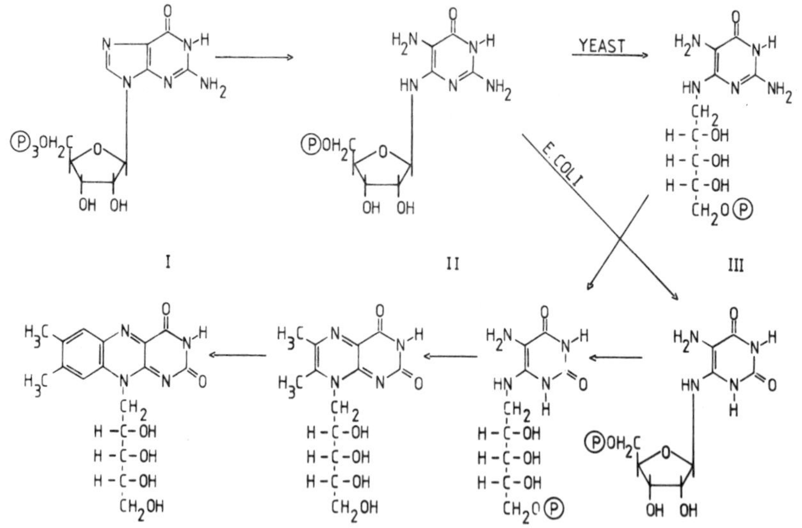
YEAST
E. COLI
I
II
III

Figure 1. Biosynthesis of riboflavin.

Figure 2. Synthesis of pyrimidine and pteridine phosphates. Experimental conditions: (a) dimethoxytrityl chloride, pyridine, dimethyl sulfoxide; (b) pyridine, acetic anhydride; (c) 80% acetic acid; (d) cyanoethyl phosphate, pyridine, dicyclohexyl carbodiimide; (e) 25% aqueous NH_3; (f) Pd/charcoal, water; (g) diacetyl, water. All operations were performed at room temperature.

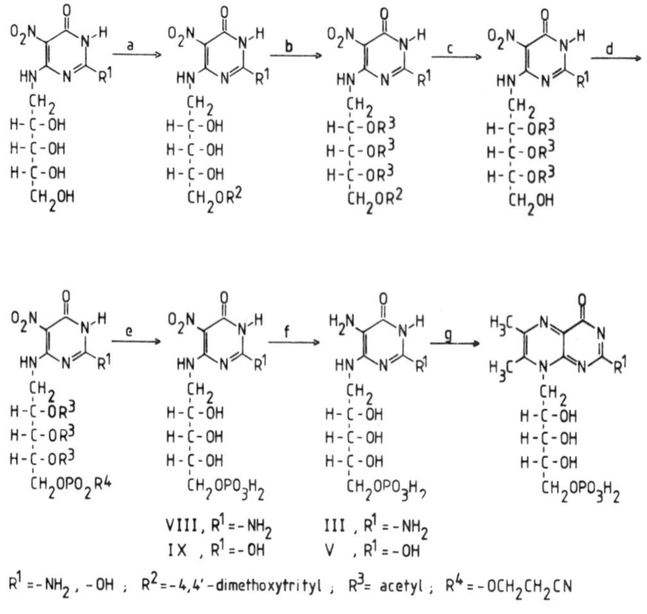
VIII, R1 = -NH2
IX, R1 = -OH
III, R1 = -NH2
V, R1 = -OH
R1 = -NH2, -OH ; R2 = -4,4'-dimethoxytrityl ; R3 = acetyl ; R4 = -OCH2CH2CN

4'-phosphate(?)
0,01 AU
335 nm
min 75
50
25
0

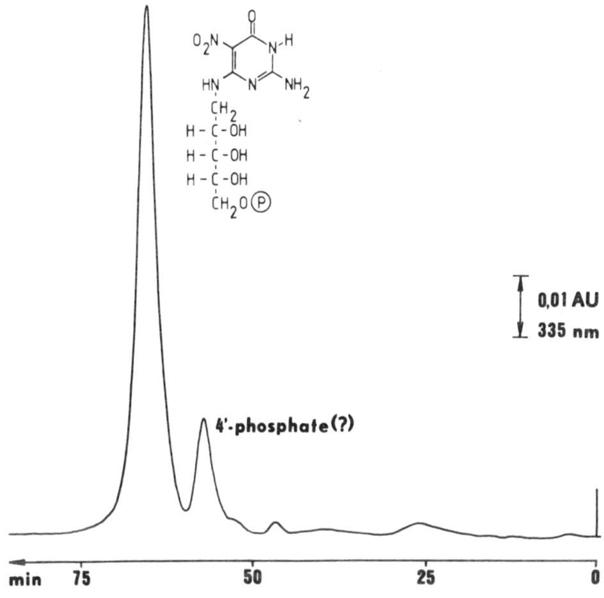

Figure 3. HPLC of (VIII); column: RP-18 Nucleosil; eluent: 10 mM ammonium formate buffer; pH 3.7.

Figure 4. Identification of the product of deaminase.

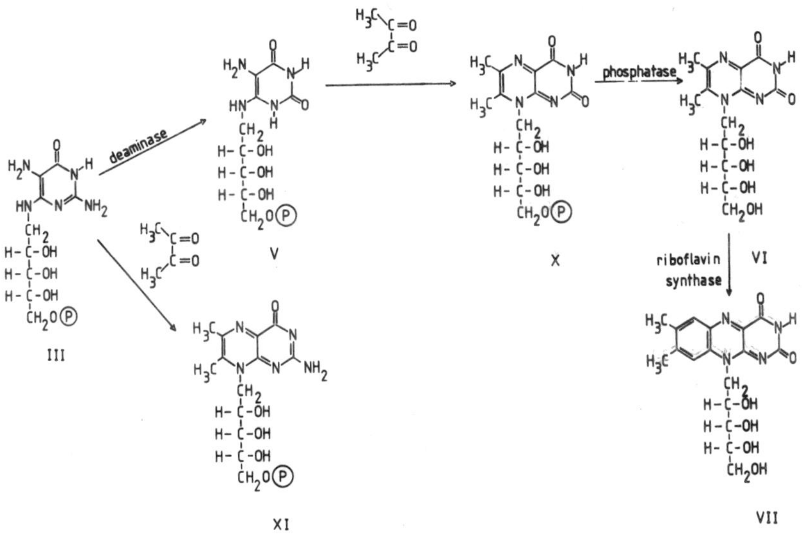

deaminase
phosphatase
riboflavin synthase
III
V
X
VI
XI
VII

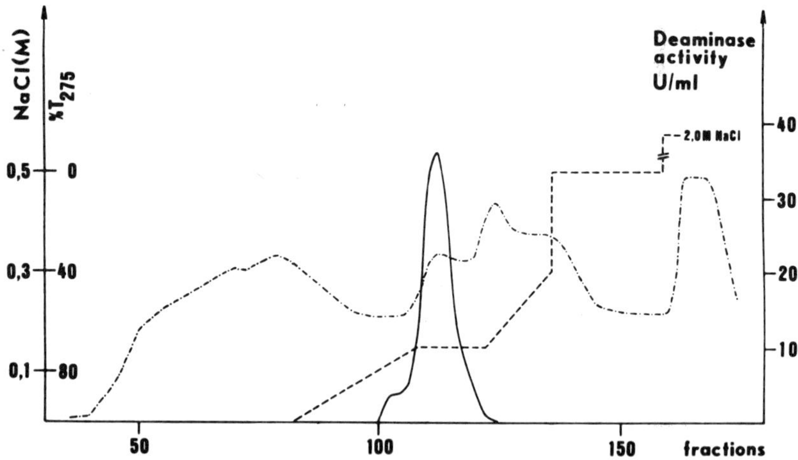

Figure 5. Chromatography of pyrimidine deaminase from *Candida guilliermondii* on DEAE Sephacel. Deaminase activity (—); protein concentration (–··–); NaCl gradient (---); 33 mM phosphate buffer pH 7.0, 5 mM $MgCl_2$.

Enzyme Studies

We have studied the enzymatic deamination of (**III**) by a deaminase from yeast. The expected product of this reaction, the 5-amino-pyrimidine phosphate (**V**), is not easily isolated from the reaction mixture. It was characterized as shown in Figure 4: The labile (**V**) was converted to (**X**) by diacetyl. (**X**) was isolated by chromatography and identified by comparison with a synthetic sample by HPLC, thin layer chromatography, electrophoresis, absorption spectra, and by fluorometric titration with apoflavodoxin (courtesy of S. Ghisla). For routine assay, (**X**) was converted to riboflavin (Figure 4) which was monitored by bioassay (4).

Rib 2 mutants of the yeast *Saccharomyces cerevisiae* were deficient in deaminase activity thus confirming the hypothesis that the gene rib 2 codes for deaminase (1).

Deaminase was partially purified from cell extracts of the flavinogenic yeast *Candida guilliermondii* by the procedures summarized in Table 1. The enzyme requires Mg^{++} ions and a sulfhydryl compound for activity. No other cofactors are required.

Table 1. Purification of Pyrimidine Deaminase from Cell Extracts of *C. guilliermondii*.

Procedure	Total protein (mg)	Activity (nmol/hr)	Specific activity (nmol/mg hr)	Purification
Cell extract	2 140	89 300	41.8	1.0
Ammonium sulfate	477	48 000	101	2.4
DEAE cellulose	54	31 206	578	13.8

ACKNOWLEDGMENTS
This work was supported by the Deutsche Forschungsgemeinschaft. We thank Margit Bunz and Fritz Wendling for excellent technical assistance.

References

1. Oltmanns, A. and Bacher, A. (1972) *J Bacteriol* 110:818–822.
2. Burrows, R.B. and Brown, G.M. (1979) *J Bacteriol* 136:657–667.
3. Scola-Nagelschneider, G. and Hemmerich, P. (1976) *Eur J Biochem* 66:567–577.
4. Klein, G. and Bacher, A. (1980) *Z Naturforsch* 35 b:482–484.

Published 1982 by Elsevier North Holland, Inc.
Vincent Massey and Charles H. Williams, Editors
Flavins and Flavoproteins

CHAPTER 82

Flavin Mononucleotide Binding by *Desulfovibrio gigas* Flavodoxin

J. Lawrence Fox[1] and Hermann Heumann

Max-Planck-Institute für Biochemie, D-8033 Martinsried bei München, West Germany

Introduction

Flavodoxins are a low molecular weight class of flavoenzymes which belong to the dehydrogenase classification (1,2) and have been isolated from a variety of microorganisms (3). It has been suggested that the flavodoxins may occur in bacteria that are among the oldest extant organisms (4–8).

The great similarities of the amino acid compositions (8), magnetic resonance properties (9), and Raman resonance properties (10) suggested that the flavodoxins from *Desulfovibrio gigas* and *vulgaris* were highly homologous in nearly all respects. The kinetics of FMN binding to the *D.vulgaris* apoenzyme indicated a single step process and possessed an anomalous enthalpy of binding, in comparison to other flavodoxins (11). Preliminary studies of the proteolytic peptide maps and subsequent sequence studies of the *D.gigas* flavodoxin (Brune and Fox, unpublished) indicated a very poor level of homology to the *D.vulgaris* protein. We therefore decided to examine the FMN binding to the *D.gigas* apoflavodoxin.

Methods and Materials

Desulfovibrio gigas flavodoxin was kindly supplied by Drs. LeGall and Bruschi of C.N.R.S., Marseille, France. The FMN was removed by treatment with 5% trichloroacetic acid or by titration to pH 5.0 with pH 1.9 saturated ammonium sulfate. The apoenzyme was dialyzed, the protein concentration was determined assuming $a_{m280} = 25{,}000$, diluted to 2.92×10^{-5} M and frozen in 250 microliter aliquots at $-20°C$. The FMN released into the soluble phase during the first precipitation was spun, neutralized with 1 M KH_2PO_4, and lyophilized in the dark. FMN resuspended in buffer was diluted to 4.52×10^{-8} M and 3.0 ml aliquots were frozen.

The fluorescence of the FMN was followed in a photon-counting spectrofluorimeter (Heumann and Zillig, unpublished) which possesses a rhodamine sulfate internal standard. For the titration experiments, an automated syringe and magnetic stirring bar driven by a controlled motor were used. A shutter was incorporated into the light path, so that the sample was only illuminated during measurement, thus reducing the problems of the photochemistry of this

[1]Permanent address: Department of Zoology, University of Texas at Austin, Austin, Texas 78712.

system. Kinetic measurements were made by manual addition of titrant. Sample to rhodamine sulfate standard ratios were recorded.

The fluorescence was corrected by subtracting the background noise. A similar approach was used for the titration data, but volume corrections were also made. The binding constants, K_a, were fit by computerized techniques to the equations:

$$\Delta F = \frac{k\left(B - \sqrt{B^2 - x}\right)}{2n}$$

where $x = 4\ nc_Fc_A$, $B = 1/K_a + c_F + nc_A$, k = a proportionality constant for bound flavin fluorescence, n = the proportion of active apoenzyme, c_A = the changing apoenzyme concentration, c_F = the constant flavin concentration, and ΔF = the fluorescence difference using a least-squares fit (12).

Results

Titration of the FMN by *D. gigas* apoflavodoxin yielded binding as typified by Figure 1. Association constants of $6.0 \times 10^8\ M^{-1}$ at 24.5°C and $4.8 \times 10^8\ M^{-1}$ at 37°C were determined. This yields a standard enthalpy of −3.3 kcal/mole, a standard free energy of −11.8 kcal/mole and a standard entropy of 48.0 e.u.

A typical set of kinetic analyses are shown in Figure 2. A pseudo-first-order on rate constant of 0.09 sec^{-1} was obtained in buffer at 24.5°C and 0.18 sec^{-1}

Figure 1. Titration of FMN by *Desulfovibrio gigas* apoflavodoxin. A 2.7 ml solution of 4.52×10^{-8} M FMN was titrated with 3 microliter aliquots of a 2.92×10^{-5} M solution of apoenzyme. Additions were allowed to stir for 10 sec and then sat for 40 min before measurement. Fluorescence was followed at 540 nm following 445 nm excitation.

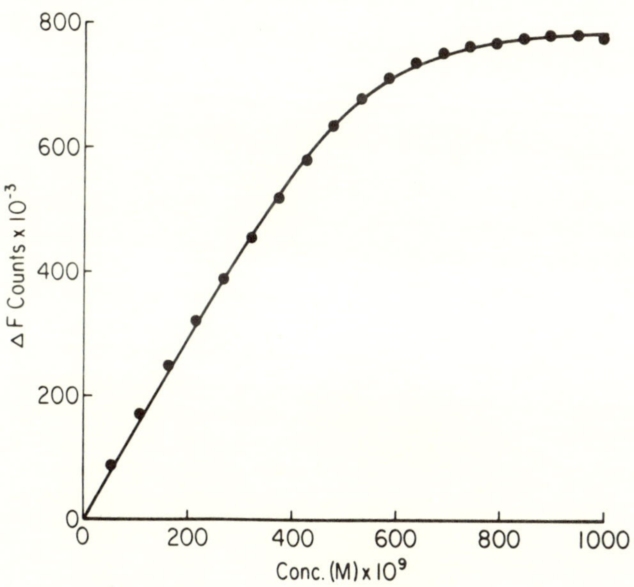

C.P.C. ABERYSTWYTH LLYFRGELL

at 37°C. An activation energy of 10 kcal/mole was calculated. The second-order rate constants calculated from these would be 2.0×10^{6} M^{-1} sec^{1} and 4.0×10^{6} M^{-1} sec^{-1}, respectively. In 0.2 M NaCl, the rate constants were 50% lower, but yielded the same activation energy.

Discussion

The results indicate that considerable differences between the *D.vulgaris* and *D.gigas* flavodoxins are found. The *D.gigas* association constant is an order of magnitude tighter than the value reported for the *D.vulgaris* flavodoxin (13), although the on kinetic rates are probably within the experimental error of comparing results from two different types of experiment. The *D.gigas* protein does not possess the anomalous enthalpy for binding reported for the *D.vulgaris* protein (11). The entropic term is high, but not as pronounced as for the *D.vulgaris* protein. How this is related to the structural features of the active site is not yet fully clear, but the *D.gigas* protein appears to have a lower aromatic environment in its active site (14, 15) according to our incomplete amino acid sequence studies (Brune and Fox, unpublished) which show very low levels of homology between the *D.vulgaris* and the *D.gigas* flavodoxin

Figure 2. Kinetics of FMN binding by *D.gigas* apoflavodoxin. Conditions were essentially as in Figure 1. 250 microliter aliquots of the apoenzyme were added by syringe to initiate the reaction and photon count ratios were recorded at 1-sec intervals. The insert shows a semilog plot of the kinetic data.

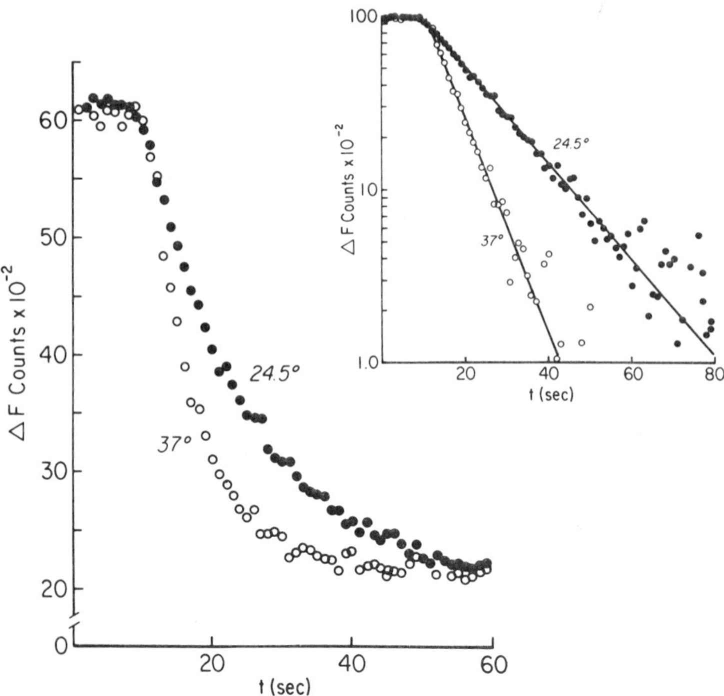

active sites. The lone tryptophan does not appear to be in the flavin binding site, but rather is placed near the C-terminus, about as far from the flavin binding site as it can get. This may be the most important difference between the *D. gigas* flavodoxin and the others known to date, as it appears to break the rule that at least one tryptophan is required for flavin binding to the apoenzyme.

ACKNOWLEDGMENTS
This research was supported by fellowships from the Alexander von Humboldt Stiftung and the University Research Foundation and Grant support from the National Institutes of Health (GM-21688) to J.L. Fox and the Max-Planck-Gesellschaft to H. Heumann.

References

1. Mayhew, S.G. and Ludwig, M.L., (1975) In *The Enzymes*, Vol. XII, 3rd Ed. P.D. Boyer, (Ed.) New York: Academic Press, pp. 57–118.
2. Edmondson, D.E. and Tollin, G. (1971) *Biochem* 10:113–124.
3. Fox, J.L., Smith, S.S., and Brown, J.R. (1972) *Z Naturforsch* 27b:1096–1100.
4. Fox, J.L. (1976) In *Flavins and Flavoproteins*. T.P. Singer, (Ed.) Amsterdam: Elsevier Scientific Publishing Company, pp. 439–446.
5. Fox, J.L. (1976) In *Protein Structure and Evolution*. J.L. Fox, Deyl, Z., and Blazej, A. (eds.) New York: Marcel Dekker, Inc., pp. 261–272.
6. Rossmann, M.G., Moras, D., and Olsen, K.W. (1974) *Nature* 250:194–199.
7. Dayhoff, M.O. and Schwartz, R.M. (1977) In *Evolution of Protein Molecules*. H. Matsubara and T. Yamanaka, (eds.) Tokyo: Japan Scientific Societies Press and Center for Academic Publications Japan, pp. 323–342.
8. Dubourdieu, M. and Le Gall, J. (1970) *Biochem Biophys Res Commun* 38:965–972.
9. Favaudon, V., Le Gall, J., and Lhoste, J.-M., (1976) In *Flavins and Flavoproteins*. T.P. Singer (ed.), Amsterdam: Elsevier Scientific Publishing Co., pp. 434–438.
10. Irwin, R.M., Visser, A.J.W.G., Lee, J., and Carreira, L.A. (1980) *Biochem* 19:4639–4646.
11. Dubourdieu, M., MacKnight, M.L., and Tollin, G. (1974) *Biochem Biophys Res Commun* 60:649–655.
12. Marquardt, D.W. (1963) *J Soc Indust Appl Math* 11:431–444.
13. Dubourdieu, M., Le Gall, J., and Favaudon, V. (1975) *Biochim Biophys Acta* 376:519–532.
14. Kobayashi, K. and Fox, J.L. (1977) In *Evolution of Protein Molecules*. H. Matsubara and T. Yamanaka (eds.), Tokyo: Japan Scientific Societies Press and Center for Academic Publications Japan, pp. 233–242.
15. Teitell, M.F. and Fox, J.L. (1980) *Int J Quantum Chem* 18:449–456.

Published 1982 by Elsevier North Holland, Inc.
Vincent Massey and Charles H. Williams, Editors
Flavins and Flavoproteins

CHAPTER 83

Thermodynamics of the Association of 8-Substituted Riboflavin with Egg White Riboflavin-Binding Protein

Kunio Matsui, Kyoko Sugimoto, and Sabu Kasai

Division of Biology, Research Institute for Atomic Energy, Osaka City University, Sumiyoshi-ku, Osaka 558, Japan

Riboflavin-binding protein (RBP) from hen egg white was investigated by many people to elucidate the mechanism of flavocoenzyme-apoprotein interaction. They reported hydrophobic and stacking interactions between flavin and RBP (1–5), and burying of 8-substituent in a "crevice" of the protein (2,4,5). We measured the association constant (K) of 14 species of riboflavin derivatives by using flavin and protein fluorescence. The derivatives were riboflavins in which 8-CH_3 was substituted by $(CH_3)_2N$- (MM), $CH_3 \cdot C_2H_5N$- (ME), $(C_2H_5)_2N$- (EE), H_2N- (HH), $CH_3 \cdot HN$- (HM), $C_2H_5 \cdot HN$- (HE), CH_3O- (MO), C_2H_5O- (EO) groups, and F- (F), Cl- (Ch), Br- (B), I- (I), H- (Hy) atoms. The thermodynamic parameters at 25°C were obtained by the ordinary method from K in a range of 10–42°C. The flavins, except riboflavin (Me) and Hy, were classified in 4 groups from $\Delta H^\circ - \Delta S_u$ relation, i.e., RR′N, HRN, RO, and halogen groups. In each group, a linear free energy relationship was found. The comparison of the ΔS_u with the bulkiness of the substituents showed a capacity of the binding site (8S-site), and suggested electric repulsion of halogen substituents from the site. The correlation between the constant term of the linear free energy relationship in each group and the wavelength of the visible absorption peak of the flavins suggested the electronic nature of the association.

Results and Discussion

ΔG° and ΔS° were corrected for cratic values, and the unitary values (ΔG_u and ΔS_u) were used. $-\Delta H^\circ$ of the flavins was plotted against $-\Delta S_u$ in Figure 1, which showed linear relations between 2 parameters in the 4 groups. From the linear relations, Equations (83.1) and (83.2) were obtained.

$$\Delta H^\circ = u \cdot \Delta S_u + v \quad (83.1)$$

$$-\Delta G_u = -(u - T) \cdot \Delta S_u - v \quad (83.2)$$

where u and v were peculiar constants for each group, and u is the so-called isokinetic temperature. The values of u were estimated as 425, 437, 426, and 321°K for RR′N, HRN, RO, and halogen group, respectively, and the values

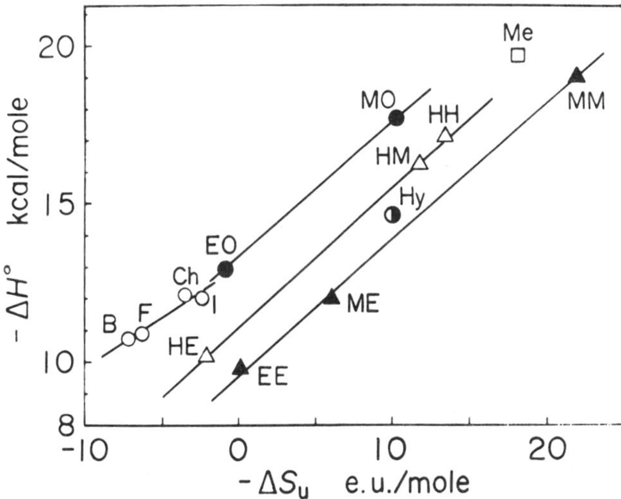

Figure 1. Correlation of $-\Delta H^\circ$ with $-\Delta S_u$ in the flavin-RBP association. (▲) Flavins of RR′N group; (△) of HRN; (●) of RO group; (○) of halogen group.

of $-v$ as 9.6, 11.1, 13.3, 13.0 kcal/mole, respectively. The similarity of u of 8-N, O groups suggested the similarity of their association mode, and the distinction of u of the halogen group suggested a different mode from those of the 8-N, O.

In Figure 2, ΔS_u of the flavins was plotted against the bulkiness of the substituents, which was the difference of the molecular volume of substituted benzene from that of benzene itself. In the 8-N, O groups, the more bulky the substituents were, the larger the ΔS_u of the flavins were. The broken line m was the regression line of MM, HM, and MO (m group), which have only methyl group as alkyl group, and the line e was that of EE, HE, and EO (e group), which have only ethyl group as alkyl. The slopes of the lines m and e were 0.62 and 0.04 e.u./(mole·ml), respectively. This suggested that the substituents of the m group were in a crevice of the 8S-site and the association was affected by the bulkiness of the substituents, and those of the e group were hardly captured in the crevice, and their association was hardly affected by the bulkiness. Therefore, the crevice has a capacity to accept O- or N-CH_3 group, but not a capacity to accept O- or N-C_2H_5 groups. These were more quantitative evidences of burying in the 8S-site of the 8-substituents in the association, than those of the previous reports (2, 4, 5).

The behavior of the flavins of the halogen group was different from that of the other 3 groups; e.g., they had low values of $-\Delta S_u$ in spite of the low bulkiness of the substituents. On the other hand, Hy, of the analogous structure, has a rather high value of $-\Delta S_u$ as expected from the low bulkiness. This was explained by an electric repulsion between the polarized negative charge in halogen atoms and an electric field in the 8S-site. An implication of an acidic group (pK_a 3.8, probably carboxyl group) in the association was reported (6, 7). So the negative group may be in or near the surface of the 8S-site and repel the halogen substituents. As the negative pole in the 8S-site,

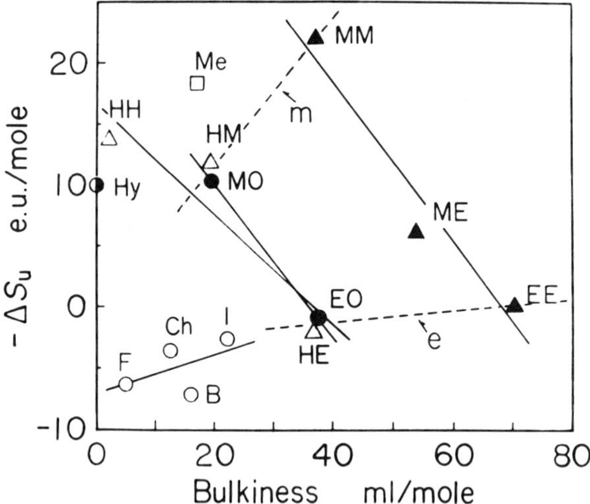

Figure 2. Correlation of $-\Delta S_u$ with bulkiness of the 8-substituents in the flavin-RBP association. (-▲-▲-) RR′N group; (-△-△-) HRN group; (-●-●-) RO group; (-○-○-) halogen group. Broken lines, regression lines of m and e groups.

tyrosyl residue is improbable, because, in the association with F, such a tyrosyl residue must be converted to tyrosyl-O(4α)-flavin (Kasai, S., Sugimoto, K., Miura, R., Yamano, T., and Matsui, K., unpublished results).

As the Equations (83.1) and (83.2) show, v is the constant term of ΔH^o and ΔG_u of the association, and is peculiar to each group. We wondered about a relation between v and the electronic state of α-atoms of the substituents. In

Figure 3. Correlation of the wavelength of the visible absorption peak of the flavins with $-$v. (▲) Flavins of RR′N group; (△) of HRN group; (●) of RO group; (○) of halogen group. pH 7.0.

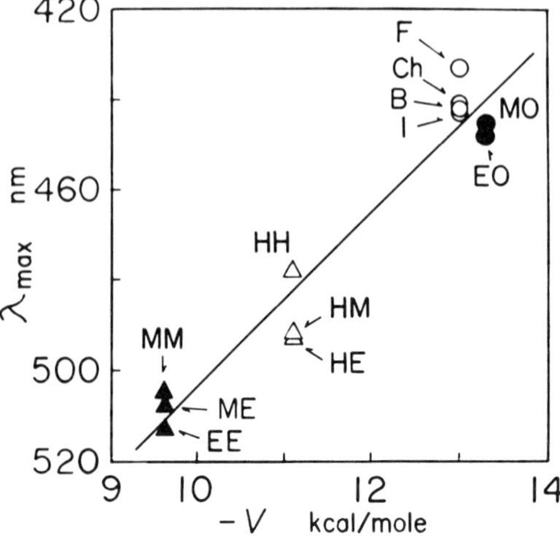

Figure 3, the wavelength of the visible absorption peaks of the flavins was plotted against v. A very good linear correlation was found. This showed clearly that v contained the energy of electronic or stacking interaction between the isoalloxazine ring and an aromatic amino acid residue of RBP.

ACKNOWLEDGMENTS
The authors thank Prof. Y. Yukawa of Osaka Women's University, Prof. K. Otozai of Osaka University, and Prof. K. Nishimoto of Osaka City University for their useful discussions and advice.

References

1. Nishikimi, M. and Kyogoku, Y. (1971) *J Biochem* 73:1233–1242.
2. Becvar, J.E. (1973) PhD Thesis, Univ. of Michigan, Ann Arbor, Michigan.
3. Blankenhorn, G. (1978) *Eur J Biochem* 82:155–160.
4. Walsh, C., Fisher, J., Spencer, R., Graham, D.W., Ashton, W.T., Brown, J.E., Brown, R.D., and Rogers, E.F. (1978) *Biochemistry* 17:1942–1951.
5. Choi, J.-D. and McCormick, D.B. (1980) *Arch Biochem Biophys* 204:41–51.
6. Farrell, H.M., Jr., Mallette, M.F., Buss, E.G., and Clagett, C.O. (1969) *Biochim Biophys Acta* 194:433–442.
7. Murthy, U.S., Podder, S.K., and Adiga, P.R. (1976) *Biochim Biophys Acta* 434:69–81.

Published 1982 by Elsevier North Holland, Inc.
Vincent Massey and Charles H. Williams, Editors
Flavins and Flavoproteins

CHAPTER 84

Recent Findings Concerning Mammalian Riboflavin-Binding Proteins

Alfred H. Merrill, Jr., Gary Shapira, and Donald B. McCormick

Department of Biochemistry, Emory University School of Medicine, Atlanta, Georgia

Micronutrients such as riboflavin frequently require binding (carrier) proteins for proper transport and utilization. In plasma, these proteins serve to sequester the nutrient from glomerular filtration, as well as to target its delivery to specific tissues. Cellular binding proteins play analogous roles, acting as traps and influencing intracellular distribution and metabolism. As yet, the macromolecules which might be involved in these processes for riboflavin have not been fully identified; at least in part, because they, like riboflavin, are present in amounts too small for easy detection.

Our approach has been to use flavin affinity chromatographic materials (1) to isolate proteins that bind riboflavin, as a prerequisite to elucidating the importance of protein binding to the utilization of this vitamin. Because flavoproteins exhibit differing flavin specificities, flavinyl agaroses with the 7,8-dimethylisoalloxazine ring immobilized via various positions (N^3, N^{10} and 8α) have been used, and have successfully yielded homogeneous riboflavin-binding proteins from laying-hen liver, blood, and eggs (2,3), bovine (4) and human (5) plasma, as well as flavokinase, which binds riboflavin as substrate, from rat liver (6). Some of these findings, plus others, are summarized below to present current knowledge concerning the mammalian riboflavin-binding proteins and their possible physiological functions.

Riboflavin-Binding Proteins in Plasma

Albumin

Purified human serum albumin (Cohn fraction V) binds riboflavin loosely (K_d-0.77 mM at 30° and 0.31 mM at 5°), but because this protein is present in high concentrations, this binding is sufficient to account for the fraction of riboflavin that is bound in whole plasma (42%) (7). Binding can also be demonstrated by polyacrylamide gel electrophoresis of [2-^{14}C]riboflavin plus purified albumin or whole plasma; however, the radiolabel is predominantly associated with the cathodal edge of the albumin band (8). To investigate whether this anomaly reflects binding by albumin or by a copurifying contaminant, commercially obtained albumin was further purified by preparative isoelectric focusing in granulated gel (Table 1). Albumin subspecies with various isoelectric pHs were found, as has been reported previously and

Table 1. Riboflavin-Binding by Human Serum Albumin.

Fraction number	pH of isoelectric focusing fraction	Protein (mg/fraction)	Riboflavin bound (μM)
Unfractionated human serum albumin	—	—	0.07
7	4.0	10.8	0.10
8	—	15.1	0.10
9	4.3	17.2	0.09
10	—	12.9	0.11
11	4.6	10.8	0.10
12	—	25.8	0.08
13	4.9	23.7	0.09
14	—	17.2	0.08
15	5.3	4.3	0.08

Human serum albumin (fraction V, Sigma Chemical Co., St. Louis, Missouri) was resolved into various subfractions by isoelectric focusing in granulated gel (LKB multiphor, LKB Instruments Inc., Rockville, Maryland) using Pharmalyte 3–10 ampholines (Pharmacia Fine Chem., Uppsala, Sweden). Riboflavin-binding was measured at 4°C by equilibrium dialysis using 1 mg/ml (15 μM) of each protein fraction (after removal of ampholines) and 1.2 μM[2-^{14}C]riboflavin (Amersham Corp., Arlington Heights, Illinois). Fractions not listed did not contain detectable levels of protein.

interpreted to reflect variously liganded forms (9). Each fraction was analyzed for riboflavin-binding by equilibrium dialysis and electrophoresis; both methods failed to detect preferential binding by any of the various forms. Hence, it appears more likely that riboflavin-binding alters the electrophoretic mobility of albumin, than that a contaminating protein or an albumin subspecies is responsible for the observed phenomena.

Photodegradation of [2-^{14}C]riboflavin by exposure of a neutral solution to direct sunlight yields products which are more tightly bound (i.e., 0.86 μM under the conditions described in Table 1) than the parent molecule, and which comigrate with albumin upon electrophoresis. Albumin, which binds hydrophobic molecules in general, would be expected to bind various photodegradation products of riboflavin, many of which arise from photolytic cleavage of the ribityl side chain (10). This tighter binding of these molecules should protect them from glomerular filtration and, thus, help explain why catabolites with altered side chains are usually not found in urine (11). Neither these nor radiochemical contaminants were wholly responsible for the binding previously attributed to riboflavin, as similar results are obtained when [2-^{14}C]riboflavin is freshly purified by paper chromatography and protected from light.

Riboflavin-Binding Immunoglobulins

A riboflavin-binding subfraction of human plasma proteins has been obtained using N^3-flavinyl agarose beads with elution of bound proteins by riboflavin (5). These proteins had electrophoretic mobilities on cellulose acetate and polyacrylamide gels corresponding to γ-globulins, and apparent molecular weights on sodium dodecylsulfate-polyacrylamide gels of 155,000 and (including β-mercaptoethanol) 55,000 and 25,000. They are identified as IgGs by quantitative binding to immobilized protein A from *Staphylococcus aureus* and reaction with antihuman IgG antisera. A maximum of 7 mg of these immunoglobulins were obtained from 100 ml of plasma; hence, they probably represent about 1% of the total IgGs. Heterogeneity in this IgG subfraction was observed as multiple bands on electrophoresis (including a tight doublet upon SDS polyacrylamide gels) and curved Scatchard plots. Approximately 20% of these IgGs bound riboflavin with $K_d \leq 4\ \mu M$.

The contribution of these proteins, relative to albumin, to the protein-binding of riboflavin in human plasma is small (5 to 6%) (5). This may not be the case, however, under all physiological conditions. In hypoanalbuminemia, for example, the concentrations of γ-globulins and other plasma proteins increase to compensate for the lower levels of albumin (12).

Other riboflavin-binding IgGs have been isolated from the serum of a patient with multiple myeloma (13). This IgG^{Gar} consists of two types of binding sites, one of which binds riboflavin essentially irreversibly while the other binds riboflavin tightly ($K_d = 0.6$ nM) but reversibly (14). Whether or not these IgGs are present among those isolated by affinity chromatography remains to be determined. A monoclonal IgA produced by mouse plasmacytoma MOPC-315 also binds riboflavin, but less tightly ($K_{apparent} = 36\ \mu M$) (15). For at least some of the riboflavin-binding immunoglobulins, these interactions probably reflect a hydrophobic site which coincidentally accommodates the flavin ring. Some IgGs, for example, can be isolated using phenyl agaroses (16).

Pregnancy-Specific Riboflavin-Binding Proteins

Chromatography of bovine plasma of flavinyl agarose beads yielded flavokinase, γ-globulins, and a lower molecular weight protein that bound riboflavin tightly ($K_d \leq 1\ \mu M$) (4). This latter protein was only found in plasma from pregnant cows, which indicates that it may have a role in the nutrition of the conceptus. Plasma carrier proteins are thought to be involved in the transport of some other vitamins (i.e., folate and vitamin B_{12}, inter alia) (17, 18) and are known to be crucial for the transport of riboflavin to the ovary for incorporation into egg yolk by laying hens (19). If they are also utilized during mammalian reproduction, their lack or malfunction would lead to the same results as inadequate dietary consumption of riboflavin, which are fetal death and congenital malformations (20).

Antibodies prepared against the avian riboflavin-binding protein have been reported to cross-react with a component of rat plasma which is induced during pregnancy, or by estrogens (21). Furthermore, injection of these antibodies into pregnant rats caused fetal death. This observation, plus the subsequent purification of a riboflavin-binding protein from pregnant rat serum (22) strengthens the likelihood that these proteins have a physiological function related to their riboflavin-binding capacities.

Analysis of small amounts of plasma from pregnant humans by flavin affinity chromatography yielded only the riboflavin-binding IgGs (5); although, the pregnancy-specific proteins might have been present at levels too low for detection. Identical treatment of fetal (cord) blood, however, yielded both the immunoglobulins and small amounts (2 $\mu g/ml$ of fetal blood) of additional riboflavin-binding proteins. Hence it appears that not only are these proteins found in humans, but they are more concentrated in fetal blood than maternal. This latter finding might be a clue to understanding why the concentration of riboflavin in fetal blood is 1.7- (23) to 5.0-fold (24) higher than that in maternal circulation.

To estimate the contribution to binding by fetal albumin, which may or may not be functionally similar to maternal albumin (25), [2-^{14}C]riboflavin-binding by fetal blood proteins was examined by electrophoresis. Unlike the previous findings with maternal plasma, little or no radiolabel was detected in the fetal albumin band. However, this finding was not confirmed upon purifying fetal albumin by chromatography on DEAE Affi-Gel Blue (Bio-Rad Laboratories, Richmond, California) and comparing [2-^{14}C]riboflavin-binding by this fraction with adult albumin by equilibrium dialysis. Fetal and adult albumins bound riboflavin equally well; the electrophoretic results may be artifacts caused by quenching.

Riboflavin-Binding Proteins in Kidney

Dog kidneys (60 g) were decapsulated and dissected to yield primarily the cortex. A 20% (w/v) homogenate was prepared by adding 50 mM potassium phosphate buffer (pH 7.2) containing 1 mM β-mercaptoethanol and blending at low speed for 3 min. After centrifugation at 24,000×g for 30 min, the supernatant was mixed with solid NaCl (to 0.5 M), 10% (w/v) Tween 20 (to 0.2%), and 2 ml of N^3- or N^{10}-flavinyl agarose beads. The mixture was stirred for 1 hr and poured into a small chromatographic column. The beads were washed with 50 mM potassium phosphate buffer (pH 7.2) containing 0.5 M NaCl until protein was not detected in the washes, then 1 mM riboflavin was included in the buffer. The initial yellow fractions were analyzed for protein and, after dialysis to remove unlabelled riboflavin, riboflavin-binding was measured by equilibrium dialysis using 2 μM [2-^{14}C]riboflavin. The protein contents of the eluates were found to parallel riboflavin-binding; hence, they were pooled for subsequent analyses.

Small amounts of protein (200 to 400 μg) were obtained using either N^3- or N^{10}-flavinyl agarose beads. Electrophoresis under nondissociating conditions revealed that both beads yielded the same two major bands, which corresponded to none of the plasma riboflavin-binding proteins observed previously. Upon concentration of storage for several days at 4°C, the proteins tended to precipitate from solution.

To test the specificity of riboflavin-binding, the competition of excess nonradiolabelled riboflavin (50 μM) or equivalent concentrations of FMN or FAD, versus [2-^{14}C]riboflavin (1.5 μM) was determined by equilibrium dialysis. Excess riboflavin decreased the amount of bound radiolabel by 81%, establishing that the apparent binding of riboflavin which is detected as bound radioactivity is not due solely to the binding of a radiochemical contaminant. Neither FMN nor FAD decreased the amount of bound [2-^{14}C]riboflavin; hence, these proteins are probably not apoflavoenzymes which normally bind a flavin coenzyme, but will also bind riboflavin less avidly. An estimate of the K_d for the radioflavin-protein complex was obtained by varying the concentration of [2-^{14}C]riboflavin and determining the cpm_{inside} versus $cpm_{outside}$ of the dialysis membrane. Assuming a 1:1 binding stoichiometry, these data were plotted as:

$$\frac{[Rf]}{[Rf \cdot P]} = [Rf]\left(\frac{1}{[P]_t}\right)\frac{K_d}{[P]_t}$$

where [Rf], [Rf·P], and $[P]_t$ represent the equilibrium concentrations of free riboflavin, bound riboflavin, and total riboflavin-binding protein, respectively, and can be approximated as $[Rf] = cpm_{outside}$ and $[Rf \cdot P] = cpm_{inside} - cpm_{outside}$. From the slope and intercept of the resulting line, a K_d of approximately 2 μM was calculated.

Examination of the absorbance spectrum of this protein fraction revealed major absorbance maxima at 270 and 406 nm, and much smaller maxima at approximately 534 and 570 nm, features which are typical of heme compounds or hemoproteins. Chromatography of this material on Sephadex G-100 yielded two major fractions, one eluting in the void volume and apparently a protein aggregate, and another eluting in the volume expected for a small protein ($\sim 10^4$ MW). This latter fraction both retained the absorbance spectrum of the starting sample and bound riboflavin.

The available evidence does not rule out the possibility that this riboflavin-binding protein is a familiar protein (such as a form of cytochrome c) that merely interacts with hydrophobic ligands like riboflavin. It is equally reasonable that it might function in the trapping of this vitamin during renal reabsorption. The finding of heme associated with this protein does not diminish this possibility because Z-protein (or fatty acid-binding protein), which is the cytosolic equivalent of albumin for the binding of numerous molecules and has a low molecular weight (12,000) (26), has been reported to contain heme.

Other plasma and cellular riboflavin-binding proteins await purification and characterization (27), including the membrane proteins responsible for transport (28). It is hoped that these, too, will be amenable to isolation by affinity chromatography so that riboflavin transport and metabolism can be understood in greater biochemical and physiological detail.

ACKNOWLEDGMENTS
This work was supported by NIH Grant AM-26746 (to D.B.M.) and funds from the Emory University Research Committee (to A.H.M.).

References

1. Merrill, A.H., Jr. and McCormick, D.B. (1980) In *Methods in Enzymology*. McCormick, D.B. and Wright, L.D. (eds) Vol. 66, New York: Academic Press, pp. 338–344.
2. Merrill, A.H., Jr. and McCormick, D.B. (1978) *Anal Biochem* 89:87–102.
3. Froehlich, J.A., Merrill, A.H., Jr., Clagett, C.O., and McCormick, D.B. (1980) *Comp Biochem Physiol* 66B:397–401.
4. Merrill, A.H., Jr., Froehlich, J.A., and McCormick, D.B. (1979) *J Biol Chem* 254:9362–9364.
5. Merrill, A.H., Jr., Froehlich, J.A., and McCormick, D.B. (1981) *Biochem Med* 25:198–206.
6. Merrill, A.H., Jr. and McCormick, D.B. (1980) *J Biol Chem* 255:1335–1338.

7. Jusko, W.J. and Levy, G. (1969) *J Pharm Sci* 58:58–62.
8. Cavalli-Sforza, L.L., Daiger, S.P., and Rummel, D.P. (1977) *Am J Hum Genet* 29:581–592.
9. Evenson, M.A. and Deutsch, H.F. (1978) *Clin Chim Acta* 89:341–354.
10. Cairns, W.L. and Metzler, D.E. (1971) *J Am Chem Soc* 93:2772–2777.
11. Yang, C. and McCormick, D.B. (1967) *J Nutr* 93:445–453.
12. Cormonde, E.J., Lyster, D.M., and Israels, S. (1975) *J Pediat* 86:862–867.
13. Farhangi, M. and Osserman, E.F. (1976) *N Engl J Med* 294:177–183.
14. Chang, M.Y., Friedman, F.K., and Beychok, S. (1981) *Biochemistry* 20:2922–2926.
15. Eisen, H.N., Michaelides, M.C., Underdown, B.J., Schulenburg, E.P., and Simms, E.S. (1970) *Fed Proc* 29:78–84.
16. Hofstee, B.H.J. (1979) *Biochem Biophys Res Comm* 91:312–318.
17. Fernandes-Costa, F. and Metz, J. (1979) *Brit J Haematol* 41:335–342.
18. Fernandes-Costa, F. and Metz, J. (1979) *Brit J Haematol* 43:625–630.
19. Farrell, H.M., Buss, E.G., and Clagett, C.O. (1970) Int J Biochem 1:168–172.
20. Warkany, J. (1975) In *Riboflavin*. Rivlin, R.S. (ed.) New York: Plenum Press, pp. 279–302.
21. Muniyappa, K. and Adiga, P.R. (1980) *FEBS Lett* 110:209–212.
22. Muniyappa, K. and Adiga, P.R. (1980) *Biochem J* 187:537–540.
23. Baker, H., Frank, O., Thomson, A.D., Langer, A., Munves, E.D., DeAngelis, B., and Kaminetsky, H.A. (1975) *Am J Clin Nutr* 28:59–65.
24. Rosso, P. and Cramoy, C. (1979) In *Human Nutrition: A Comprehensive Treatise*. Winick, M. (ed.) Vol. 1. New York: Plenum Press, pp. 133–228.
25. Gitzelmann-Cumarasamy, N., Gitzelman, R., Wilson, K.J., and Kuenzle, C.C. (1979) *Proc Natl Acad Sci USA* 76:2960–2963.
26. Billheimer, J.T. and Gaylor, J.L. (1980) *J Biol Chem* 255:8128–8135.
27. Frank, O., Luisada-Opper, A.V., Feingold, S., and Baker, H. (1970) *Nutr Rep Internat* 1:161–168.
28. Spector, R. and Boose, B. (1979) *J Biol Chem* 254:286–289.

PART VIII:

Physical Techniques in Flavin Research

Published 1982 by Elsevier North Holland, Inc.
Vincent Massey and Charles H. Williams, Editors
Flavins and Flavoproteins

CHAPTER 85

Physical Techniques in Flavin Research

Franz Müller and Chrit T.W. Moonen

Department of Biochemistry, Agricultural University, Wageningen, The Netherlands

Introduction

It is a very difficult task to describe and to discuss in a short review the potentials and limitations of different physical techniques useful in obtaining structural information on flavins and flavoproteins. Therefore, a few modern techniques are selected and the most recent data published presented. Moreover, the limitation of space does not allow us to cover the corresponding theories; a few references will ease finding the appropriate literature.

A few flavoproteins are presently studied by X-ray diffraction (see elsewhere in this volume). This technique provides the static structure of a biomolecule but it cannot reveal the dynamic structure. In catalysis, the dynamic structure is of fundamental importance and it may deviate considerably from the static structure. For this reason three-dimensional, structural data should always be complemented by dynamic, structural data. Only the combination of both approaches can provide us with the information needed to reveal the catalytic mechanism of a particular enzyme. In this light, a few techniques have been selected which are either very versatile or very specific in the content of information. For each technique, a few general remarks will be made as a help for the less physically oriented reader.

Techniques

Magnetic Resonance

Nuclear magnetic resonance (NMR) is the most versatile technique able to yield detailed information on the molecular and submolecular level of molecules. Modern instruments allow the investigation of a large number of different nuclei, including metal ions (1,2). NMR studies on biomolecules are, however, hampered by a few instrumental and nuclei specific factors, i.e., resolution and sensitivity. The sensitivity is determined by the gyromagnetic ratio (γ), the natural abundance of the nucleus in question and the magnetic field strength. For instance ^{13}C is about a factor of 6000 and ^{15}N about a factor of 250,000 less sensitive than ^{1}H. These shortcomings of the corresponding nuclei can, in part, be compensated by isotopic enrichments. The enrichment is in fact the only possible way to study protein-bound flavocoenzymes. The resolution of an NMR spectrum depends mainly on two factors: (a) the magnetic field strength and (b) the molecular weight of the protein studied. The superconducting high-field magnets have aided tremendously to improve the resolution. On the other hand, the limitation of the resolution of a biomolecule is not only determined by the wealth of a particular nucleus present, e.g., ^{1}H, but also by the molecular weight of the molecules studied, since the rotational correlation time influences the spin-spin relaxation time, which in turn determines the linewidth of the NMR signals. It should also be noted that NMR is an inherently insensitive technique due to the fact that the energetic levels of the two spin states are separated by only ~0.1 cal, in ESR this separation is ~10 cal which makes the latter technique more sensitive.

The *chemical shift* values of a particular nucleus give information on its surrounding or on its electronic state. With respect to the latter ^{13}C, ^{15}N, and ^{31}P are particularly valuable nuclei. In studying protein-bound flavin, these three atoms are the nuclei of choice. In Figure 1, the ^{13}C and ^{15}N chemical shifts of free and protein-bound flavins are presented in a correlation diagram. The ^{13}C chemical shifts of C(2), C(4), and C(10a) of oxidized FMN appear in aqueous solution at lower field than those of TARF in chloroform. The chemical shifts of these atoms, except that for C(4) in *A.vinelandii*, exhibit similar trends when FMN is bound to the apoproteins of *M.elsdenii* and *A.vinelandii* flavodoxin. These downfield shifts strongly suggest that C(2α) is hydrogen-bonded to the apoprotein. The same holds for C(4α) in *M.elsdenii* flavodoxin. In contrast, C(4α) in *A.vinelandii* flavodoxin seems not to be hydrogen-bonded and its environment is similar to that of TARF in chloroform. These results are supported by ^{15}N NMR data (3) (Franken, Rüterjans, and Müller, unpublished results). In the reduced state, comparing the molecules in polar and apolar solvents, the largest difference in chemical shifts is shown by C(4a) and C(10a). These shifts have to be interpreted in terms of the conformation of the molecule in the two solvents, i.e., in apolar solvents less planar than in polar solvents. In forming the ionized flavohydroquinone, the chemical shifts of C(2) and C(10a) are drastically influenced (downfield shift) due to the negative charge on N(1), which seems rather localized. These two chemical shifts can be used to determine the ionization state and ionization constant ($pK_a = 6.7$) of flavohydroquinone. Therefore, it is concluded from Figure 1 that the reduced flavin in *M.elsdenii* and *A.vinelandii* flavodoxin is ionized. Moreover, the chemical shifts of C(4a) and C(10a) are good indicators of the planarity of the molecule and the total negative charge on N(1). Thus the negative charge on N(1) in *M.elsdenii* flavodoxin is somewhat neutralized (by a positively charged group?) as compared to that in *A.vinelandii* flavodoxin, whereas both prosthetic groups possess a planar, or almost planar structure. These data are supported by ^{15}N studies on *M.elsdenii* flavodoxin (Franken, Rüterjans, and Müller, unpublished data).

In addition to information obtained from chemical shifts and coupling constants, relaxation studies can yield valuable information with respect to internuclei distances and dynamics in a protein. Two different relaxation processes can be distinguished, i.e., the *spin-lattice relaxation* (T_1) and the *spin-spin relaxation* (T_2). The spin-spin relaxation is directly manifested in the linewidth in a normal NMR spectrum. To obtain the T_1 relaxation time, a more complicated pulse sequence is needed than in measuring the chemical shifts. Two examples will be discussed shortly to demonstrate the information accessible by these two relaxation processes.

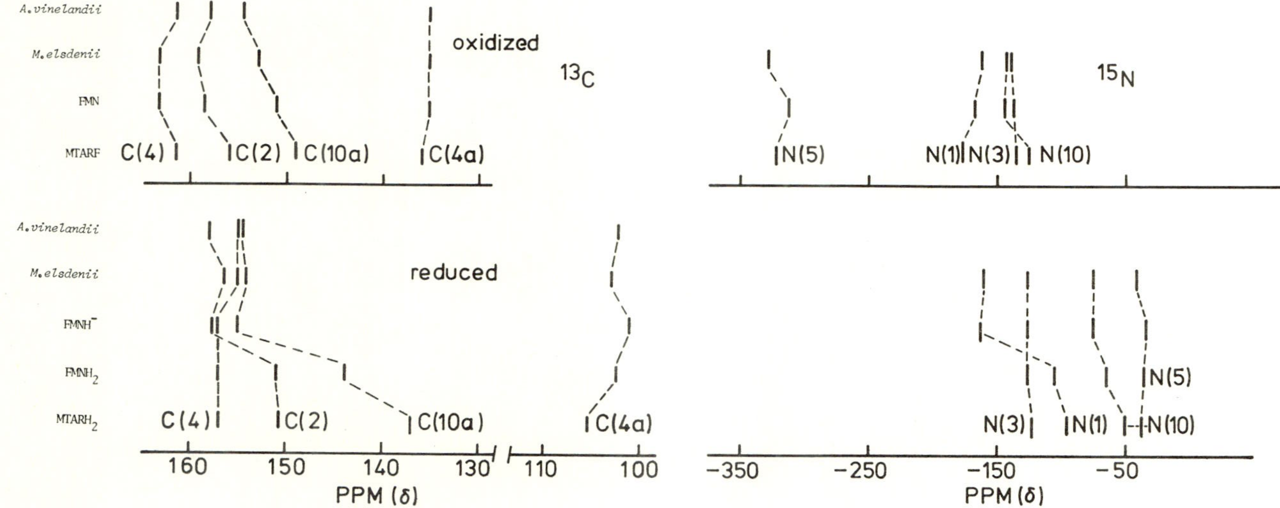

Figure 1. Correlation diagram of ^{13}C and ^{15}N chemical shifts of free and protein-bound flavins in the oxidized and 1,5-dihydro state.

Figure 2 demonstrates the analysis of the ^{31}P resonance of the phosphate group of FMN bound to *M.elsdenii* flavodoxin. The spectra represent different redox states of the flavodoxin, i.e., A, C, E are the oxidized, the semiquinone, and the hydroquinone state, respectively. The solution measured in Figure 2B consisted of about 50% oxidized and 50% semiquinone whereas in D about 98% hydroquinone and 2% semiquinone was present. The increased linewidth in spectrum C is a typical T_2 effect due to the semiquinone state. From this, direct information on the distance of the phosphorus atom from the paramagnetic center (i.e., ~8 Å) is obtained. T_2 analysis (Figure 2B and D) yields information on the rate of exchange of one electron between two molecules of flavodoxin in different redox states. From the spectra, one can calculate that this exchange is slow ($<2\ s^{-1}$) between the oxidized and the semiquinone state and fast ($>100\ s^{-1}$) between the semiquinone and the reduced state. These data can be correlated directly to the biological function of flavodoxins, i.e., shuttling between semiquinone and reduced state is kinetically more favorable than that between oxidized and semiquinone state.

An example of a T_1 experiment is shown in Figure 3. In this series of spectra, we observe the time needed to recover the equilibrium magnetization of the phosphorus atom in *M.elsdenii* flavodoxin after a 180° pulse. This return of the magnetization is described by an exponential function $\int \infty \exp(-t/T_1)$. The T_1 can be calculated by fitting the measured curve with an exponential function. The T_1 value tells us about the mobility of the phosphate group, on the one hand, and the number of hydrogen bonds formed by the phosphate, on the other hand. The latter information must be deduced from T_1 experiments using native and in deuterium oxide-reconstituted flavodoxin. In the case of *M.elsdenii* flavodoxin, about five hydrogen bonds are formed by the phosphate group which is in the dianionic state.

A phenomenon which is in fact related to the spin-lattice relaxation is the *Nuclear Overhauser Effect* (*NOE*). The NOE is the increase or decrease of the intensity of a particular resonance line irradiating a nucleus at another frequency. Here we will pay special attention to the possibility of obtaining structural and dynamic information using homonuclear (^1H-^1H) NOE measurements. In small molecules, ^1H-^1H NOE can give information on distances (1). However, ^1H spin diffusion occurs in protein at high magnetic fields (4) (except in the case of a high internal mobility). This spin diffusion makes the conventional ^1H-^1H NOE measurements useless. However, theory shows (16) that the initial build-up rates of NOEs are simply related to the inverse sixth power of the distance between the observed and the preirradiated proton. Thus if we irradiate a proton for only a short time, the NOE will build up before spin diffusion will become important. This is demonstrated in Figure 4 where the so-called Driven NOE technique was used. Irradiation of the high field resonance at −0.74 ppm (methyl group of Leu-62) gives (among others) a NOE at 6.32 ppm. This peak must belong to Trp-91 as revealed by photo-CIDNP NMR (5). This experiment shows that Leu-62, Trp-91 and the isoalloxazine are very close to each other in three-dimensional structure. This shows that NOE measurements are a very powerful tool in obtaining structural information. It should be noted that in the near future 2-dimensional NOE measurements will facilitate this kind of measurement.

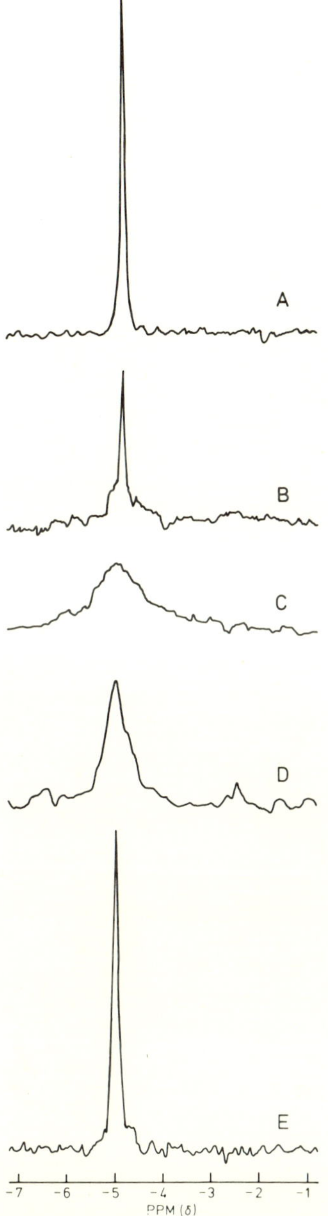

Figure 2. ^{31}P NMR spectra of *M.elsdenii* flavodoxin in the oxidized, the semiquinone and the hydroquinone state (see text).

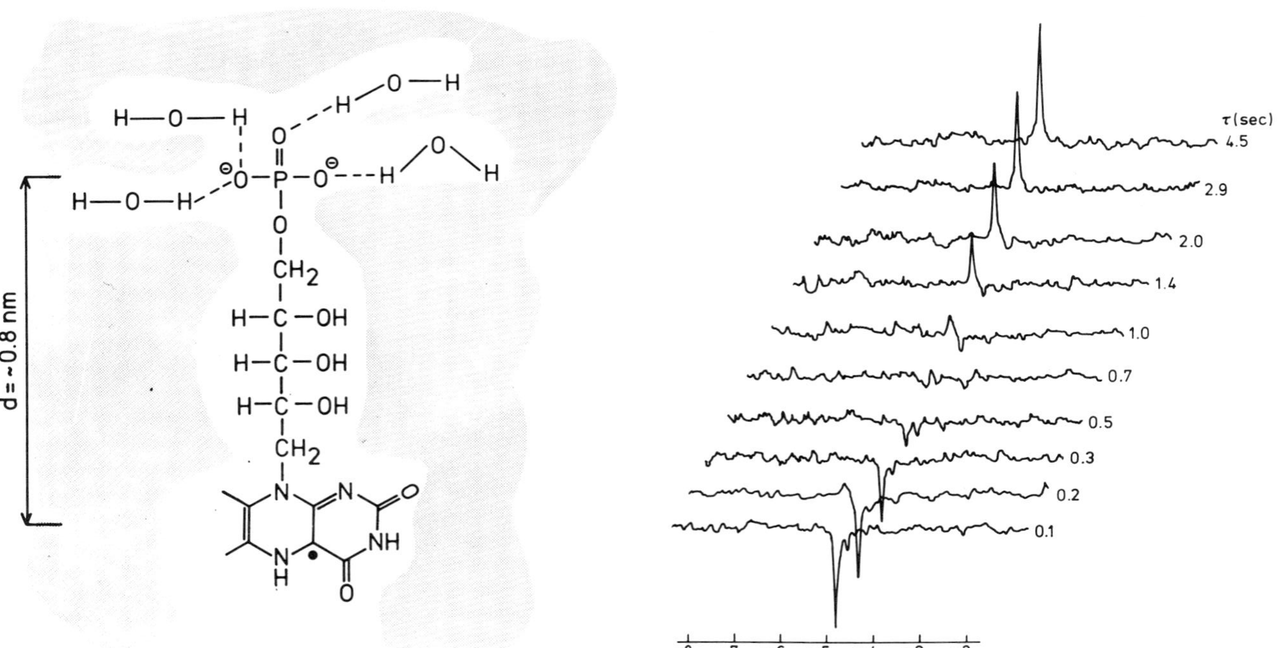

Figure 3. Partially relaxed ^{31}P NMR spectra of *M. elsdenii* flavodoxin (see text). The structure given is an illustration of the information obtained from Figures 2 and 3.

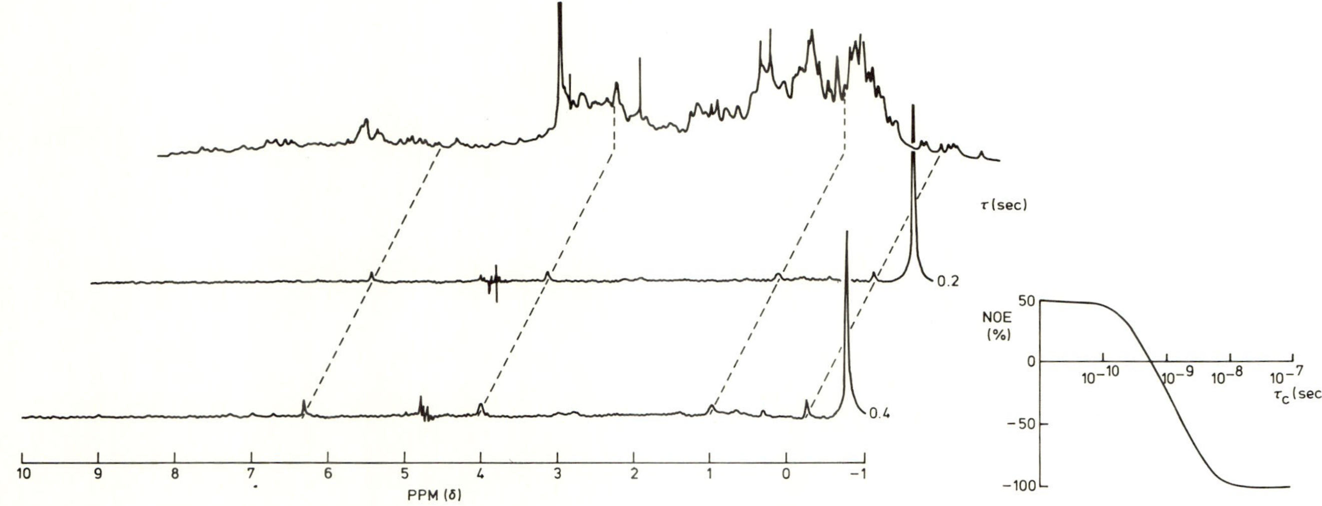

Figure 4. Double resonance homonuclear ^1H NMR experiments (NOE) on *M. elsdenii* flavodoxin (see text).

The *photo-chemically induced dynamic nuclear polarization* (CIDNP) technique has proven a powerful tool for easy assignment of proton resonances in complex protein NMR spectra. The experiment starts with photo excitation of a dye, thus populating its reactive triplet state. This triplet abstracts a hydrogen atom from an aromatic residue in a protein, forming a biradical, which upon recombination gives a different population of the spin states than under usual magnetization (6). It was found that protein-bound flavins do not yield a CIDNP signal because they cannot fulfil the requirement of fast dissocation of the biradical (molecular restrictions). The observation of a signal of a particular amino acid residue can depend on the structure of the dye used. In our cases, we used various lumiflavin (Lfl) derivatives as external dyes. For example, using Lfl-CH_2COO^- at a molar ratio lower than that of the protein, three lines could be observed at high field (Figure 5B, peaks 1). Increasing the concentration of the dye led to the appearance of a second set of resonance lines (Figure 5B). Both sets of lines originate from tryptophan residues, i.e., set 1 is assigned to Trp-100 and set 2 to Trp-91. Using a Lfl-derivative carrying a positively charged group on a side chain yielded a spectrum where an additional line could be observed (Figure 5C). This latter is assigned to Tyr-6.

Besides NMR, the *Raman spectroscopy* technique is of considerable interest for researchers in the field. This technique should specifically reveal the subtle interactions between the prosthetic group and the apoprotein of a particular flavoprotein. The information obtainable from Raman spectra is comparable to that from infrared spectra, i.e., vibrational modes are measured. The vibrational frequencies depend on intra- and intermolecular factors. Theoretically, if all vibrational frequencies of free flavin can be assigned, which seems to be a problem, then the specific interactions between the apoprotein and the flavocoenzyme could be extracted from the spectra. However it should be emphasized that the Raman spectrum is not a variety of the fluorescence spectrum. In Raman experiments, the molecules are never actually in an excited electronic state. The Raman effect originates from inelastic scattering of light by molecules. If the molecule is to show a vibrational Raman spectrum, the polarizability must change as the molecule vibrates. It is not known what transitions the Raman lines represent, apart from the qualitative remark that they represent modes involving a change in polarizability. The theoretical physical background of Raman scattering is rather complex. Therefore the reader interested in this technique is referred to the literature (7–8). Depending on the photophysical properties of a molecule, i.e., fluorescent or nonfluorescent, different techniques are used. If a molecule is fluorescent, only coherent anti-Stokes Raman spectroscopy (CARS) can be used (8, 9).

At this place, a combined discussion of NMR and Raman data would be appropriate. This is not yet possible because of the limitation of NMR data. In Table 1, a few recently published Raman frequencies for various flavoproteins are collected. The assignments given are still tentative, indicating the difficulties to assign the wealth of frequencies in a Raman spectrum. At this moment, one has to be satisfied with the crude assignment of a particular vibrational mode to one of the three rings of the isoalloxazine ring. Probably the least interesting frequencies for our purpose are those at $\sim$1630 and $\sim$1230 cm^{-1} belonging to a vibrational mode of the benzene subnucleus. As a matter of

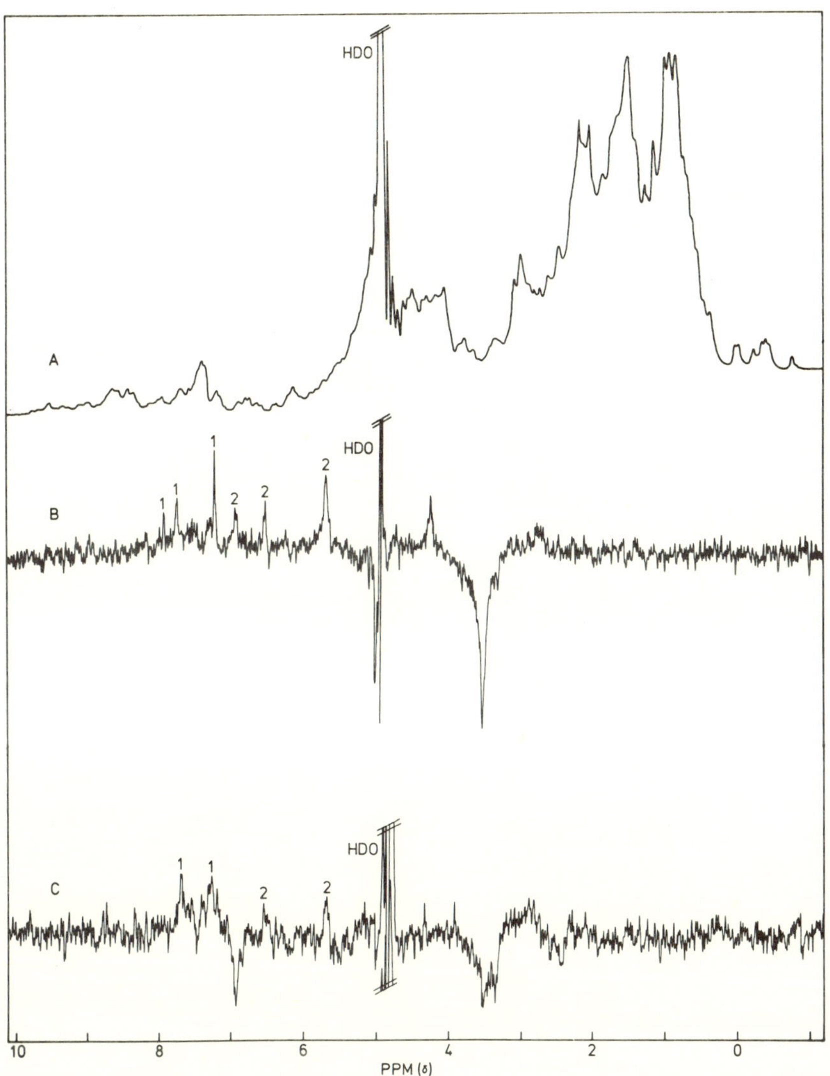

Figure 5. Photo-CIDNP spectra of *M. elsdenii* flavodoxin (see text).

fact, the frequency at ~1230 cm^{-1} was also identified from visible light absorption spectra corresponding to the resolved vibrational structure of the first light absorption band (16). The most interesting frequencies are those at ~1585, ~1355, and ~1260 cm^{-1}, if their assignment is correct. These frequencies should reveal whether or not hydrogen bonds to the apoprotein are formed. The above-mentioned ^{13}C chemical shifts of *M.elsdenii* flavodoxin indicate that hydrogen bonds are formed between C(2α) and C(4α) and the apoprotein. Although not identical (but probably similar) interactions could be expected for the other flavodoxins, a cautious comparison between the two sets

Table 1. Some Vibrational Frequencies (cm^{-1}) of Free and Protein-Bound Flavins.

		C.MP(12)				Ribofl.bind	Glucose	
FMN(10)	**FAD(11)**	**Ox.**	**Sq.**	**D.gigas(10)**	**D.vulgaris(10)**	**protein(13)**	**oxidase(14)**	**Assignment (11, 13, 15)**
1631	1635	1628	1611	1624	1626	1631	1626	Ring I
1585	1584	1577	—	1573	1571	1584	1578	Ring II (C(4a)-N(5))(C(10a)-N(1)
1551	1540	—	—	1545	—	1548	—	Ring II, III
1503	1507	1505	—	1501	1501	1503	1501	----
1413	1416	1409	1391	1407	1407	1407	1404	Ring III
1355	1359	1356	1378	1352	1354	1355	1364	(N(5))-C(10a)-N(1)-C(2)
1261	1260	1257	1269	1251	1254	1252	1260	Ring III C(2)-N(3)-C(4)
1233	1231	1231	1232	1227	1228	1229	1230	Ring I
1187	1185	1197	—	1173	1173	1179	1182	Ring III
1166	1164	1163	—	1163	1164	1161	1158	Ring III

of data is feasible. Thus the frequencies at ~1585 and ~1250 cm^{-1} in *D.gigas* and *D.vulgaris* flavodoxin are shifted to lower frequencies as compared to those in FMN, while that at ~1335 cm^{-1} remains almost unaffected. Relating these observations to the NMR results, it seems that the frequencies at ~1250 cm^{-1} (~1260 cm^{-1} in free FMN) indicate hydrogen bond formation between the apoprotein and C(2α) and C(4α) and/or N(3). It will be interesting to await the somewhat refined assignment by Bowman and Spiro (15) and the subsequent improvements, making Raman spectroscopy an interesting tool to study flavoapoprotein-prosthetic group interactions.

ACKNOWLEDGMENTS

We are grateful to Mrs. J.C. Toppenberg-Fang, Mr. M.M. Bouwmans, and Mr. B.J. Sachteleben for assistance in the preparation of this paper.

The unpublished work described in this paper was supported by the Netherlands Foundation for Chemical Research (SON) with financial aid from the Netherlands Organization for the Advancement of Pure Research (ZWO).

References

1. Wüthrich, K. (1976) NMR. In *Biological Research Peptides and Proteins*. Amsterdam: Elsevier/North Holland.
2. James, T.L. (1975) *Nuclear Magnetic Resonance in Biochemistry*. New York: Academic Press.
3. Kawano, K., Ohishi, N., Suzuki, A.T., Kyogoku, Y., and Yagi, K. (1978) *Biochemistry* 17:3854–3859.
4. Kalk, K. and Berendson, H.J.C. (1976) *J Magn Reson* 24:343–366.
5. Müller, F., Van Schagen, C.G., and Kaptein, R. (1980) In *Methods in Enzymology*, Vol. 66. New York: Academic Press, pp. 385–416.
6. Turro, N.J. (1978) *Modern Molecular Photochemistry*. Menlo Park: Benjamin/Cummings, pp. 275–288.
7. Tobin, M.C. (1972) In *Methods in Enzymology*, Vol. 26. New York: Academic Press, pp. 473–497.
8. Nestor, J.R. (1978) *J Raman Spectr* 7:90–95.
9. Begley, R.F., Harvey, A.B., and Byer, R.L. (1974) *Appl Phys Lett* 25:387–390.
10. Irwin, R.M., Visser, A.J.W.G., Lee, J., and Carreira, L.A. (1980) *Biochemistry* 19:4639–4646.
11. Dutta, P.K., Spencer, R., Walsh, C., and Spiro, T.G. (1980) *Biochim Biophys Acta* 623:77–83.
12. Dutta, P.K. and Spiro, T.G. (1980) *Biochemistry* 19:1590–1593.
13. Kitagawa, T., Nishina, J., Kyogoku, Y., Yamano, T., Ohishi, N., Takai-Suzuki, A., and Yagi, K. (1979) *Biochemistry* 18:1804–1808.
14. Dutta, P.K., Nestor, J.R., and Spiro, T.G. (1978) *Biochem Biophys Res Commun* 83:209–216.
15. Bowman, W.D. and Spiro, T.G. (1981) *Biochemistry* 20, in press (personal communication).
16. Dubs, A., Wagner, G., and Wüthrich, K. (1979) *Biochim Biophys Acta* 577:177–194.
17. Eweg, J.K., Müller, F., Visser, A.J.W.G., Veeger, C., Bebelaar, D., and Van Voorst, J.D.W. (1979) *Photochem Photobiol* 30:463–471.

Published 1982 by Elsevier North Holland, Inc.
Vincent Massey and Charles H. Williams, Editors
Flavins and Flavoproteins

CHAPTER 86

Application of an Improved Spectroelectrochemical Cell to the Study of the Redox Properties of Lumiflavin 3-Acetate, Tetraammine Ruthenium(II) 10-Methylisoalloxazine, and 5-Deazaflavin

David Condit and Marian Stankovich

Department of Chemistry, University of Massachusetts, Amherst, Massachusetts and Department of Chemistry, University of Minnesota, Minneapolis, Minnesota

The goal of our research is the study of the redox properties of selected flavoproteins using combined spectral and electrochemical techniques. In order to perform these studies, spectroelectrochemical cells suitably anaerobic for flavoprotein reduction were developed. Two versions of the spectroelectrochemical cell have been developed to date. The first successful version (1) was tested on glucose oxidase and flavodoxin. An improved version of this cell is described in this work (Figure 1). The modifications enable experiments to be performed more quickly and enable the reduction of compounds with more negative redox potential ($E^{\circ\prime}$) values. The two major improvements are: (a) the glass stirrer was replaced by a propeller-shaped polyethylene-encased magnetic stirrer which provided more efficient stirring; and (b) the negative potential limit of the gold wire working electrode was extended by coating it with mercury. This electrode enabled the use of mediators with potentials more negative than methyl viologen (MV) ($E^{\circ\prime} = -0.448$V vs SHE) (2) such as 4,4′-dimethyl-1,1′-trimethylene-2,2′-dipyridinium bromide ($E^{\circ\prime} = -0.548$V) (3).

This improved cell was tested by reductive titration of lumiflavin-3-acetate using both mediator dyes previously mentioned. Electrons were transferred with 100% current efficiency to lumiflavin-3-acetate by both mediator dyes. Figure 2 is a plot of absorbance vs time for the methyl viologen-mediated reduction of lumiflavin-3-acetate. As each successive charge increment (3 millicoulombs) was added, the decrease in absorbance of oxidized lumiflavin-3-acetate was rapid. This indicates rapid electron transfer from the reduced methyl viologen to the oxidized lumiflavin. Only after the endpoint of the titration does the spectrum of the reduced methyl viologen appear. The reduced viologen then undergoes dimerization (2,3).

An attempt was made to reduce 5-deazaflavin using an excess of the mediator dye 4,4′-dimethyl-1,1′-trimethylene-2,2′-dipyridinium bromide. Fig-

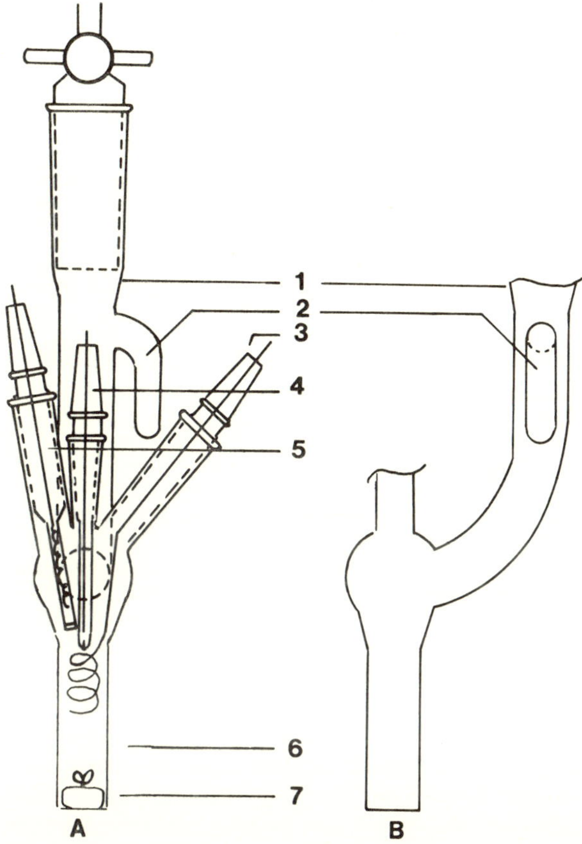

Figure 1. Spectroelectrochemical cell: main body is constructed of pyrex. (a) Front view, electrodes are in the plane of the paper; (1) main connection of nitrogen line; (2) sidearm; (3) mercury-coated gold coil working electrode; (4) silver-silver chloride reference electrode; (5) silver-silver chloride auxiliary electrode; (6) standard 3-ml pyrex cuvette; (7) modified stirrer. (B) Side view, cell is rotated 90° to give a clearer view of (1) and (2).

ure 3 shows the spectral progress of the reductions under 3 degrees of anaerobiosis. The spectra in Figure 3a indicate production of two electron reduced deazaflavin whereas spectra in Figure 3c indicate the dominant reduction product is deazaflavin dimer (4). In all experiments, the current efficiency is considerably less than 100%.

A possible explanation for the product mixtures of deazaflavin obtained under different anaerobic conditions is suggested by the following. The plots of absorbance at 398 nm versus time are shown in Figure 4. Both dFl_{ox} (oxidized deazaflavin) and $MV^{+\bullet}$ (one electron reduced viologen) absorb at this wavelength. In the most anaerobic case (Figure 4c), absorbance at 398 nm initially increases sharply as the system is reduced. Then the A_{398} falls off. This initial increase is due to production of $MV^{+\bullet}$ radical. The $MV^{+\bullet}$ does not react with

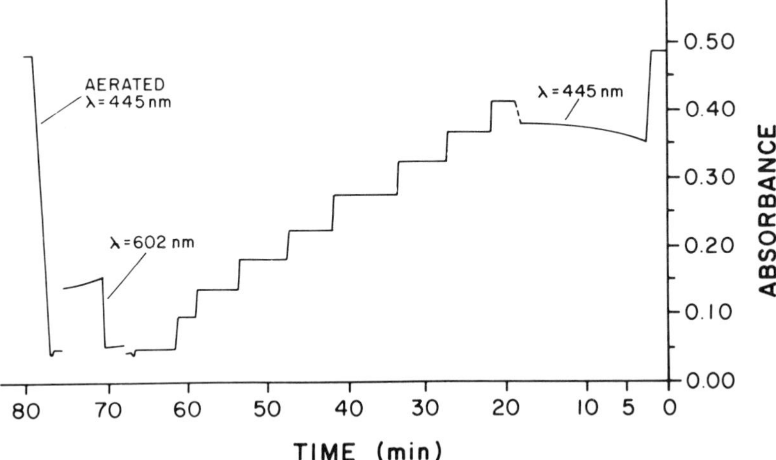

Figure 2. Reductive titration of 3.88×10^{-5} M lumiflavin-3-acetate mediated by 0.45 mM methyl viologen with 0.1 N KCl and 0.09 N phosphate buffer at pH 7.0. Initial charge increment is 9 millicoulombs. About 45 minutes was given (---) to let absorbance stabilize after first charge increment. Subsequent increments 3 millicoulombs. Decreasing absorbance at 602 nm after the endpoint is due to methyl viologen dimerization.

solution species rapidly. The radical can react in 3 ways: (a) form $(MV)_2$ dimer, (b) reduce dFl_{ox} to $dFl^{\bar{\bullet}}$ (one electron reduced deazaflavin) which can react further to dimerize to accept another electron, and (c) react further with $dFl^{\bar{\bullet}}$ to form fully reduced deazaflavin. The product mixtures suggest that dimerization of both deazaflavin and $MV^{+\bullet}$ are the most probable reactions under anaerobic conditions. In solutions which were not rigorously anaerobic (Figure 4a), there were no large initial increases in absorbance at 398 nm. The

Figure 3. Titration of 1.5×10^{-5} M 5-deazaflavin, mediated by 5×10^{-4} M, 4,4′-dimethyl-1,1′-trimethylene-2,2′-dipyridinium bromide. (a) Least anaerobic conditions; (c) most anaerobic conditions.

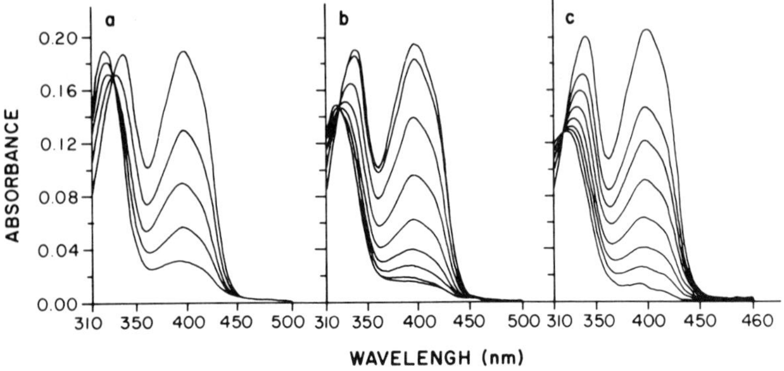

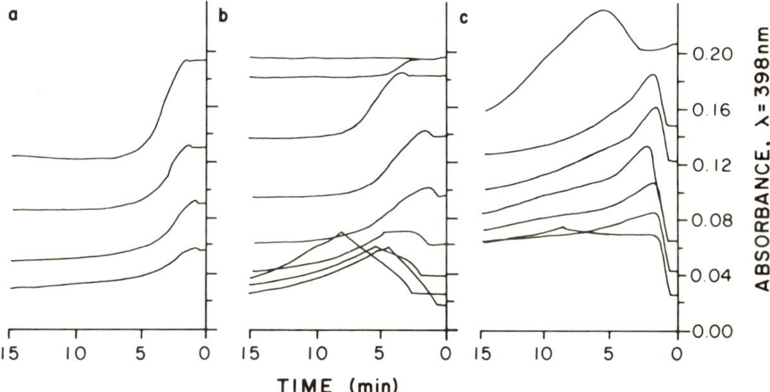

Figure 4. Absorbance vs time plots for the spectral titrations of 5-deazaflavin shown in Figure 3. Note the absorbance increase due to viologen radical in Figure c.

$MV^{+\bullet}$ which was generated was not stable in O_2 containing solutions. In this case, oxygen reacts with $MV^{+\bullet}$ as well as deazaflavin radical and dimer. Therefore, the only stable product generated is the 2 electron reduced deazaflavin which is oxygen stable (5). This is supported by the spectral evidence (Figure 3).

Direct (unmediated) reductive titrations of the ruthenium(II) 10-methylisoalloxazine complex were performed at pH 7.0 and 8.5. A single electron transfer to the complex (n_1) occurs at $E^{o\prime} = -350$ mv vs SHE (pH 8.5), producing a stable flavin radical. The second electron transfer (n_2) occurs at potentials 250 mv more negative. The titration yielded $n_1 = 1.09$ and $n_2 = 1.06$, confirming the predictions of the cyclic voltammetric experiments of Clark and Dowling (6). Thus current efficiency of 91 to 94% was obtained for this compound.

Conclusion

Anaerobic conditions are an obvious requirement for obtaining 100% current efficiency in any reductive titration. However, for titrations in which the viologen dyes are the electron transfer mediators, there is an additional requirement for 100% efficiency. The rate of electron transfer between $MV^{+\bullet}$ and oxidized flavin or flavoprotein must be faster than the rate of dimerization of $MV^{+\bullet}$. If this requirement is met, there is no $MV^{+\bullet}$ dimerization until after the endpoint of the reaction. Preliminary results for anaerobic deazaflavin indicate that both $MV^{+\bullet}$ and deazaflavin radical dimerize because electron transfer between the 2 species is extremely slow. Current efficiency was well below 100%. In contrast, the methyl viologen mediated reduction of lumiflavin-3-acetate to the two electron reduced form was rapid (Figure 2) and occurred with 100% current efficiency.

ACKNOWLEDGMENTS
We wish to thank M. Clarke for the gift of ruthenium complex; V. Massey and P. Hemmerich forthe 5-deazaflavin and lumiflavin 3-acetate; F. Hawkridge for the modified viologen. This work was supported by an NIH Biomedical Research Grant from the University of Massachusetts and by a Research Corporation starter grant.

References

1. Stankovich, M.T. (1980) *Analytical Biochemistry* 109:295.
2. Szentrimay, R., Yeh, P., and Kuwana, T. (1977) In *Electrochemical Studies of Biological Systems, ACS Symposium No. 38.* Sawyer, D.T. (ed.), Washington, D.C.: Amer. Chem. Soc., pp. 142–169.
3. Bowden, E.F. and Hawkridge, F.M. (1981) In *Extended Abstracts, Electrochem. Soc., 159th Meeting.* Princeton, New Jersey: Electrochem. Soc., pp. 1066–1067.
4. Massey, V. and Hemmerich, P. (1978) *Biochemistry* 17(1):9.
5. Duchstein, H., Fenner, H., Hemmerich, P., and Knappe, W. (1979) *Eur J Biochemistry* 98:167.
6. Clarke, M.T. and Dowling, M.G., Dept. of Chem., Boston College, submitted.

Published 1982 by Elsevier North Holland, Inc.
Vincent Massey and Charles H. Williams, Editors
Flavins and Flavoproteins

CHAPTER 87

Oxidation and Reduction of Flavin Compounds Studied by Radiation Chemical Techniques

A.J. Elliot, R. Ahmad, L. McIntosh, D. Burke, K.J. Stevenson, and D.A. Armstrong

Chemistry Department and Biochemistry Division, University of Calgary, Calgary, Alberta

Summary

One-electron (a) and overall two-electron (b) redox reactions of flavins, viz: (a) $F+H^+ +e \rightleftharpoons FH\cdot$ and (b) $F+2H^+ +2e \rightleftharpoons FH_2$, have been studied with $\cdot CO_2^-$ and cyclic disulphide anions as reducing agents, and thiyl radicals ($RS\cdot$) as oxidizing agents. The reductions and oxidations occurred with efficiency and were reversible. Transient $FH\cdot$ intermediates were observed by pulse radiolysis. The flavoprotein lipoamide dehydrogenase (E) was also reduced by $\cdot CO_2^-$. The first electron added is located primarily on the flavin. Continued treatment with $\cdot CO_2^-$ produced mainly EH_2 but also some EH_4.

Introduction

When suitable concentrations of solutes are present, the exposure of nitrous oxide saturated aqueous solutions to fast electrons (1 to 10 MeV) can be used to produce $\cdot OH$ radicals (1), viz:

$$H_2O \rightsquigarrow 2.7\ e^-(aq)+2.8\ \cdot OH+0.6\ H\cdot \qquad (1)^1$$

$$e^-(aq)+N_2O+H_2O \rightarrow N_2+OH^- + \cdot OH \qquad (2)$$

or other one-electron oxidants, such as thiyl radicals ($RS\cdot$):

$$\cdot OH+RSH \rightarrow RS\cdot +H_2O \qquad (3)$$

For studies of reduction 0.01 to 0.2 M formate ion is added to produce $\cdot CO_2^-$:

$$\cdot OH \text{ or } \cdot H+HCO_2^- \rightarrow H_2O \text{ or } H_2 + \cdot CO_2^- \qquad (4)$$

The fast electrons can be introduced in a short pulse ($\sim 1\ \mu s$) as a beam of particles from an accelerator or they can be set in motion in situ in the solution by exposing it to a continous beam of Co^{60} gamma rays (1). The former

[1]Coefficients are yields per 100 eV of fast electron energy expended. The small yield of $H\cdot$ is often neglected. $e^-(aq)$ is the solvated electron.

method, known as pulse radiolysis, has already been applied to the investigation of one-electron reductions (2) of flavins (F), viz:

$$\cdot CO_2^- + F(+H^+) \rightarrow CO_2 + FH\cdot \qquad (5)$$

and to the electron transfer reactions [4–7] of flavin (FH·) radicals. Here we report a study of reaction (**5**) for flavin adenine dinucleotide (RAD), lumichrome (LC), lumiflavin (LF), and pig heart lipoamide dehydrogenase (LPDH). Also we describe the use of the continous gamma radiolysis method to investigate two-electron reduction and oxidation:

$$F + 2H^+ + 2e \rightleftharpoons FH_2 \qquad (6)/(-6)$$

Special attention has been paid to the reactions:

$$FH_2 + RS\cdot \rightarrow FH\cdot + RSH \qquad (7)$$

$$FH\cdot + RS\cdot \rightarrow F + RSH \qquad (8)$$

$$F(+H^+) + DS_2^{\bar{\cdot}} \rightarrow FH\cdot + DS_2 \qquad (9)$$

The disulphide anions (1,8) of dithiothreitol (DS_2) and lipoamide (LS_2) were prepared by reaction (**10**) (1).

$$\cdot CO_2^- + DS_2 \text{ or } LS_2 \rightarrow CO_2 + DS_2^{\bar{\cdot}} \text{ or } LS_2^{\bar{\cdot}} \qquad (10)$$

Experimental

The flavins and LPDH were obtained from Sigma. Descriptions of the Co^{60} source and pulsed radiolysis equipment and of all procedures may be found in reference 9.

Results and Discussion

Flavins

The difference spectrum observed with FAD when the formation of FH· in reaction (**5**) was complete (and before significant decay had occurred, see solid line in Figure 1a), agreed with that previously reported by Anderson (5). For FAD, LC and LF k_5 was 1.2±0.2, 1.8±0.2, and 2.6±0.4 respectively in units of $10^9\ M^{-1}s^{-1}$.

In the case of FAD the FH· radicals were observed to decay by a second order reaction:

$$2\ FH\cdot \rightarrow F + FH_2 \qquad (11)$$

Also, in reaction (**9**), the initial absorbance due to $DS_2^{\bar{\cdot}}$ (Figure 1a) decayed in about 50 μs for our conditions, and at 100 μs the difference spectrum of F and FH· was the same as that obtained with $\cdot CO_2^-$. This shows that reaction (**9**), which was originally observed for $LS_2^{\bar{\cdot}}$ (8), is likely to be a general reaction of cyclic disulphides.

Figure 2a provides examples of the progress of reaction (**6**) followed by measuring the absorption at 360 and 450 nm and calculating F concentration with the known values of ϵ_{FH} and ϵ_F. The reduction yields were over 95% and

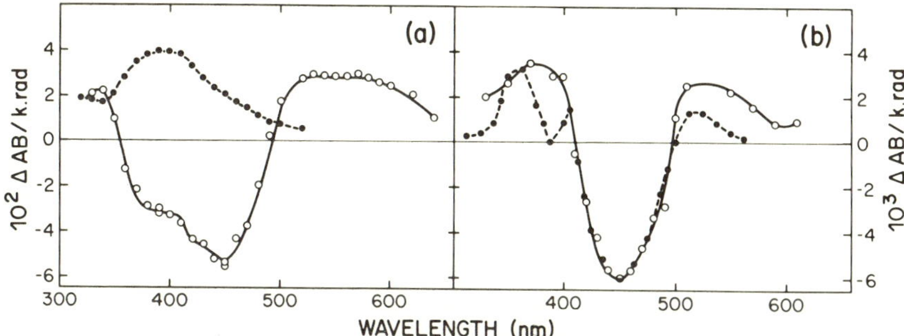

Figure 1. Change in absorbance (Δ AB) for one-electron reductions of FAD and LPDH in nitrous oxide saturated 0.005 M phosphate buffer containing 0.01 to 0.2 M sodium formate. Dose 0.4 k rad/pulse. (a) FAD 24 to 70 μM, at pH 6.8~100 μs after pulse: (—) $\cdot CO_2^-$; (O) $DS_2^{\dot{-}}$ made by adding 3 mM DS_2 to the solutions; (-●-●-) initial spectrum of $DS_2^{\dot{-}}$. (b) LPDH 38 μM at pH 7.0: (O) $\cdot CO_2^-$ 700 μs after pulse. This spectrum improved over that in ref. 9; (-●-●-) glucose oxidase semiquinoid anion (10) normalized to present spectrum at 450 nm.

the number of reducing equivalents used per mole of FAD was close to 2.0, indicating that reactions (**5**), (**9**) and (**11**) were occurring with very high efficiency. Results similar to those described above were obtained for FAD and F over the pH range 6 to 8. For LC, however, only one electron reduction was observed at pH 6 to 7.

Quantitative two-electron reoxidation of the reduced flavins was effected in every case by O_2, and for the dihydro form of FAD, also by thiyl radicals produced in reaction (**3**) from β-mercapto propionic acid at pH 7. The conclusion from this observation is that, whereas the natural tendency of *free* disulphide anions is to transfer electrons to FAD, (reaction **9**), for RS· radicals hydrogen atom or electron transfer from FADH· is the preferred direction (reaction **8**).

Lipoamide Dehydrogenase

The slope of the initial part of the titration plot for the enzyme (E) in Figure 2b shows that 8.6 CO_2^- are consumed per EH_2 produced. The reduction is therefore only about one-fourth as efficient as for the free flavins and we found evidence for some EH_4 production. The reduction was reversible and, at the doses used here, $\cdot CO_2^-$ did not inactivate the enzyme (9).

If the first electron added to E resided on the disulphide at the active site, there would be no loss of the flavin 450 nm absorbance, but an increased absorbance would be observed in the 350 to 450 nm range as exhibited by the $DS_2^{\dot{-}}$ spectrum in Figure 1a and the $\dot{S}S^-$ spectra of protein disulphides (1). In actual fact, the one-electron reduced form exhibited the decrease in absorbance from 410 to 500 nm which is characteristic of loss of oxidized flavin (Figure 1b). This in our view is unequivocal evidence that the flavin is involved in the one-electron reduction. However, the increased (rather than decreased) absorbance from 360 to 410 nm resembles an anionic flavin radical (see Figure

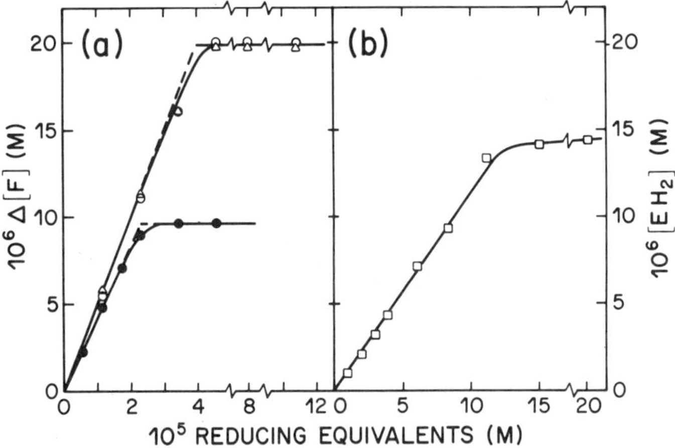

Figure 2. Redox titrations of FAD and LPDH in 0.005 M phosphate buffer at pH 7. (a) FAD→$FADH_2$ in 0.2 M $NaHCO_2$: (Δ,o)·CO_2^- +2O μM FAD followed at 370 and 450 nm respectively; (●) $LS_2^{\cdot-}$+9.5 μM FAD followed at 450 nm. (b) E→EH_2 in 0.02 M $NaHCO_2$: (□) E=(LPDH)+$CO_2^{\cdot-}$. [EH_2] calculated from absorbance at 550 nm (7), initial [E]=25 μM.

2b, glucose oxidase spectrum) as opposed to a neutral flavin radical. It is not clear why this should be so at pH 7 unless protonation, which was too fast to observe with free flavins at our buffer concentrations (3), is very slow for $E^{\cdot-}$. Another explanation is the involvement of both flavin and disulfide, viz: S^- S—FAD· . Investigation of these alternatives is continuing.

References

1. Adams, G.E. and Wardman, P. (1977) *Free Radicals in Biology*. Pryor, W.A. (ed.) Vol. 3. New York: Academic Press, pp. 53–59.
2. Land, E.J. and Swallow, A.J. (1969) *Biochemistry* 8:2117.
3. Meisel, D. and Neta, P. (1975) *J Phys Chem* 79:2459.
4. Faraggi, M., Hemmerich, P., and Pecht, I. (1975) *FEBS Lett* 51:47.
5. Anderson, R.F. (1976) *Ber Bunsen-Gesellschaft* 80:969.
6. Williams, C.H., Jr. (1976) *Enzymes, 3rd Ed.* 13:89.
7. Matthews, R.G. and Williams, C.H., Jr. (1976) *J Biol Chem* 251:3956.
8. Chan, S.W., Chan, P.C., and Bielski, B.H.J. (1974) *Biochim Biophys Acta* 338:213; Chan, P.C. and Bielski, B.H.J. (1973) *J Am Chem Soc* 95:5504.
9. Elliot, A.J., Munk, P.L., Stevenson, K.J., and Armstrong, D.A. (1980) *Biochem* 19:4945.
10. Massey, V. and Ghisla, S. (1974) *Ann NY Acad Sci* 277:446.

Published 1982 by Elsevier North Holland, Inc.
Vincent Massey and Charles H. Williams, Editors
Flavins and Flavoproteins

CHAPTER 88

Ab Initio MO Study of the Stacking Complexes, Flavin-Tyrosine, Flavin-Tryptophan, and Flavin-NADH

Yoshitaka Watanabe,* Kichisuke Nishimoto,* and Hiroshi Kashiwagi†

**Department of Chemistry, Faculty of Science, Osaka City University, Osaka; †Institute for Molecular Science, Okazaki, Japan*

Recently, the three-dimensional structure of several flavoenzymes containing FAD or FMN has been determined by X-ray diffraction analysis and the binding position of FAD or FMN in the active sites has been clarified.

In FMN-containing enzymes, the structures of two flavodoxins, *Clostridium* MP (1) and *Desulfovibrio vulgaris* (2), were analyzed and discussed with respect to the difference and the resemblance of the two enzymes. For an FAD-containing enzyme, glutathione reductase (EC 1.6.4.2) has been studied at 3.0 Å resolution (3). It was recently demonstrated that Resonance Raman spectroscopy could give data about flavin-ligand interaction (charge transfer interactions of phenols with the flavin mononucleotide of Old Yellow Enzyme (EC 9.6.99.1)) (4).

Figure 1. The geometries of the complexes of lumiflavin with the aromatic compounds: (I) FMN-tyrosine complex model in *Desulfovibrio vulgaris* flavodoxin (lumiflavin-p-cresol). (II) FMN-p-cresol complex model in Old Yellow Enzyme (lumiflavin-p-cresol). (III) FMN-tryptophan complex model in *Desulfovibrio vulgaris* flavodoxin (lumiflavin-skatole). (IV) FMN-tryptophan complex model in *Clostridium* MP flavodoxin (lumiflavin-skatole). (V) FAD-NADH complex model in glutathione reductase (lumiflavin-N-methyl-1,4-dihydronicotinamide).

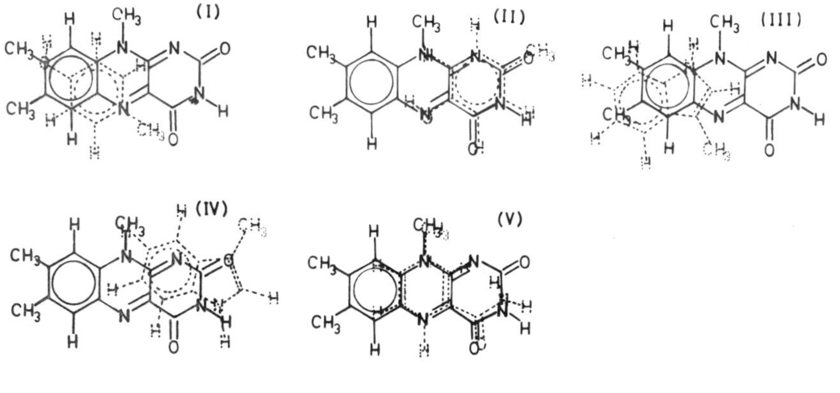

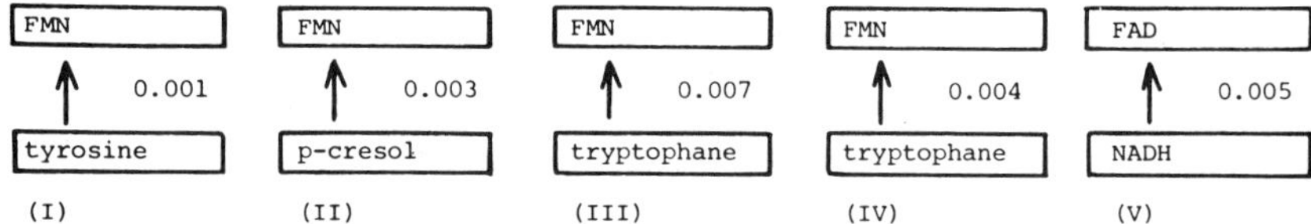

Figure 2. The charge flux diagram in flavoprotein.

Figure 3. The difference electron density of lumiflavin-p-cresol, lumiflavin-skatole, and lumiflavin-N-methyl-1,4-dihydronicotinamide complexes. The full lines indicate density increases and the dotted lines decreases. Values of these lines are ± 0.0004, ± 0.0008, ± 0.0016, and ± 0.0020 (e/Å^3), successively. Y=0.0: on the flavin plane. Y=0.5; 0.5 a.u. plane above flavin plane between the stacking complexes.

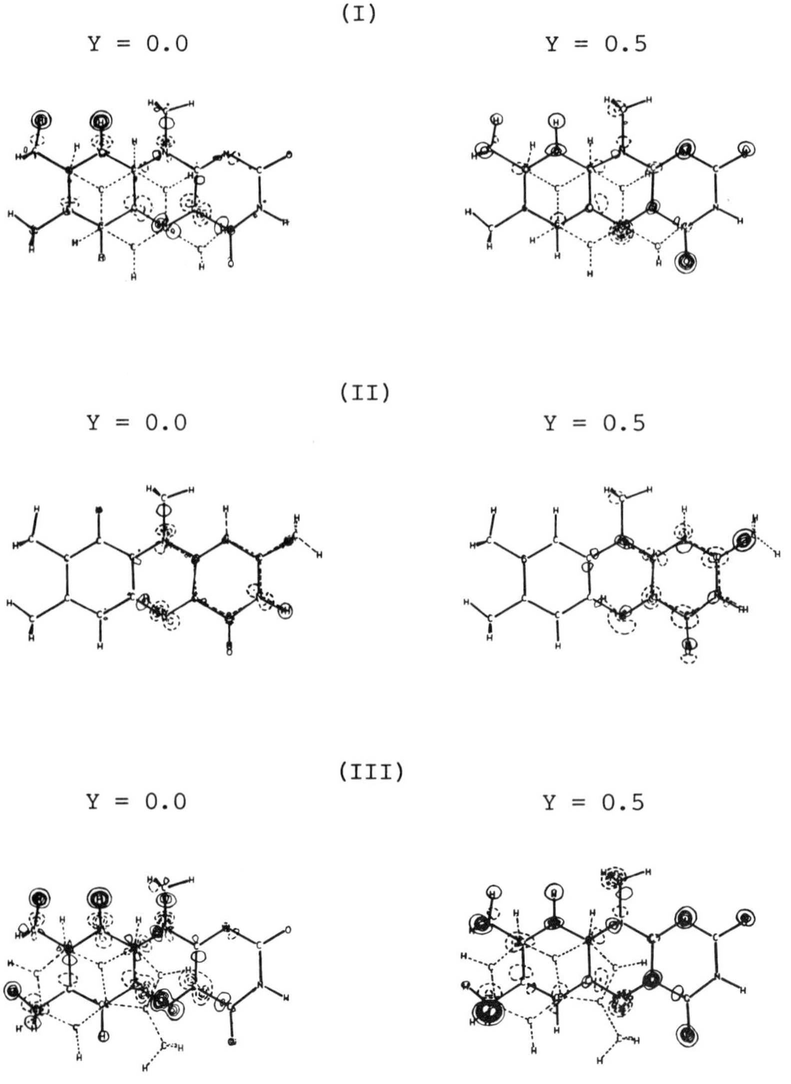

(I)

Y = 0.0

Y = 0.5

(II)

Y = 0.0

Y = 0.5

(III)

Y = 0.0

Y = 0.5

In order to elucidate the nature of stacking interactions in flavodoxins, we carried out ab initio MO calculations. We consider the stacking of the isoalloxazine nucleus with the aromatic groups in flavodoxins, with the inhibitor in Old Yellow Enzyme (p-cresol etc.), and with NADH in glutathione reductase.

It is pointed out that the ab initio SCF MO method (5–8) based on STO-3G basis set is not appropriate for the calculations of the charge transfer complexes. In order to investigate the charge transfer interaction, an extended basis set such as the 4-31G must be used.

The charge transfer interaction between aromatic systems mainly comes from π-electron interaction. Therefore, one approach is to use a split basis set for π-orbitals and STO-3G for σ-orbitals. We found that this approach works satisfactorily for benzene+TCNE (tetracyanoethylene) and naphthalene+TCNE complexes (6).

STO-3G π-split basis set might be recommended for the ab initio SCF MO calculation of very large stacking complexes. We calculated the planar stacking interaction with the maximum phase overlapping geometry of HOMO of electron donor, and LUMO of electron acceptor (9) (Figure 1). Since FMN, FAD, tryptophan, tyrosine, and NADH are molecules too large for the ab initio MO calculation, we used lumiflavin, skatole, p-cresol, and N-methyl-1,4-dihydronicotinamide instead of them. The distance between the flavin plane and the nearest atoms of the stacking molecule is put to be 3.5 Å.

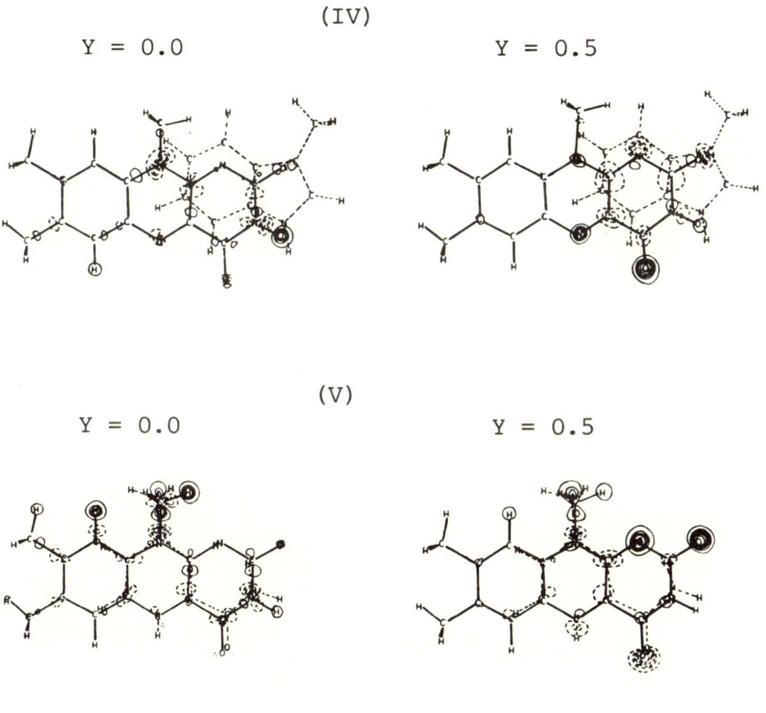

(IV)

Y = 0.0 Y = 0.5

(V)

Y = 0.0 Y = 0.5

Table 1. Lumiflavin-p-Cresol, Lumiflavin-Skatole, and Lumiflavin-N-Methyl-1, 4-Dihydronicotinamide Complexes (Intermolecular Distance R = 3.5 Å).

Basis set	Geometry	Binding energy (Kcal/mol)	Total differential dipole moment (Debye)
STO-3G	I	−2.489	0.393
π-split	II	−2.071	0.564
	III	−2.358	0.533
	IV	−1.850	0.805
	V	−0.415	0.610

A threshold for overlap integral $|S_{ij}|$; 1×10^{-3}.
A threshold for exchange integral $(ij|ij)^f$; 1×10^{-6}.

The calculated results are summarized as follows: The aromatic residue behaves as a weak electron donor and the flavin as a weak electron acceptor in the stacking interaction (Figure 2).

When the aromatic residue of a protein has a stacking interaction with the hydrophobic part of flavin in models (I) and (III), the π-electron population of the hydrophilic part in the flavin nucleus is varied (Figure 3. However, if the aromatic residue interacts with the hydrophilic part of the flavin in models (II) and (IV), the hydrophobic part is not influenced by this interaction.

The stacking interaction produces some amount of induced dipole moment in the flavin complex as shown in Table 1.

References

1. Burnett, R.M., Darling, G.D., Kendall, D.S., Lequesne, M.E., Mayhew, S.G., Smith, W.W., and Ludwig, M.L. (1974) *J Biol Chem* 249:4383.
2. Watenpagh, K.D., Sieker, L.C., and Jensen, L.H. (1973) *Proc Natl Acad Sci US* 70:3857.
3. Schulz, G.E., Schirmer, R.H., Sachsenheimer, W., and Pai, E.F. (1978) *Nature* 273:120.
4. Kitagawa, T., Nishina, Y., Shiga, K., Watari, H., Matsumura, Y., and Yamano, T. (1979) *J Am Chem Soc* 101:3376.
5. Osanai, Y. and Kashiwagi, H. (1980) *Int J Quantum Chem* 17:1031.
6. Watanabe, Y. and Kashiwagi, H., to be published.
7. We have used the Program JAMOL3 (Kashiwagi, H., Takada, T., Miyoshi, E., Obara, S., and Sasaki, F.).
8. Watanabe, Y., Nishimoto, K., and Kashiwagi, H. (1980) *IMS Ann Rev* 119.
9. Fukui, K. (1975) *Theory of Orientation and Stereoselection*. New York: Springer-Verlag.

Published 1982 by Elsevier North Holland, Inc.
Vincent Massey and Charles H. Williams, Editors
Flavins and Flavoproteins

CHAPTER 89

Ab Initio MO Study on Hydrogen Bonding in Lumiflavin-Water Complex

Yoshitaka Watanabe,* Kichisuke Nishimoto,* Hiroshi Kashiwagi,** and Kunio Yagi†

**Department of Chemistry, Faculty of Science, Osaka City University, Osaka; **Institute for Molecular Science, Okazaki; †Institute of Biochemistry, Faculty of Medicine, University of Nagoya, Nagoya, Japan*

In our previous work (1,2), the effect of hydrogen bonding on the transition energy and the oscillator strength of the isoalloxazine nucleus of flavins was studied by P-P-P (Pariser-Parr-Pople) method using model potential. The data were used to explain the experimental results (3).

In order to elucidate further the detailed effects of hydrogen bonding on the electronic structure of the isoalloxazine nucleus of flavins, we carried out ab initio MO calculations of lumiflavin-water complex, a model of hydrogen-bonded flavin in flavoprotein. Since the geometry of hydrogen-bonded lumiflavin has never been determined experimentally, the equilibrium structure was calculated by partial optimization. Hydrogen bonding can be described by a resonance hybrid of the following structure:

$$X-H\cdots Y \quad \textbf{(I)} \qquad (X-H)^{-}\cdots Y^{+} \quad \textbf{(II)} \qquad X^{-}\cdots(H-Y)^{+} \quad \textbf{(III)}$$

where structure **(II)** and **(III)** represent the charge transfer interaction in hydrogen bonding. Accordingly, hydrogen bonding formation brings about the transfer of charge from an electron donor to a flavin of oxidized form, and induced electronic polarization in the flavin nucleus.

The geometry of lumiflavin-water complexes is shown in Figure 1. Ab initio MO calculation was made on these complexes by using the program JAMOL 3 (H. Kashiwagi, T. Takada, E. Miyoshi, S. Obara, and F. Sasaki) and semiorthogonalized orbitals (4). The data are shown in Table 1. The charge transfer in hydrogen bonding at each site can be quantitatively represented by charge migration (ΔQ). The strongest charge transfer interaction is occurring at O(14). It is noted that the values ε(LUMO) $-$ ε(HOMO) obtained in the present study run parallel with the value obtained previously by P-P-P method. The order of the magnitude of the stabilization energy of hydrogen bonding was N(3)H > O(12) > N(5) > N(1) > O(14). The magnitude of the effect of the occurrence of hydrogen bonding on the electron acceptability of N(5) is in the order of N(5) > O(14) > N(1) > O(12) > N(3)H.

The calculated results were analyzed and difference charge density maps were delineated as shown in Figure 2. From these data, the following conclusions were obtained: (a) The polarization is delocalized over the region of the

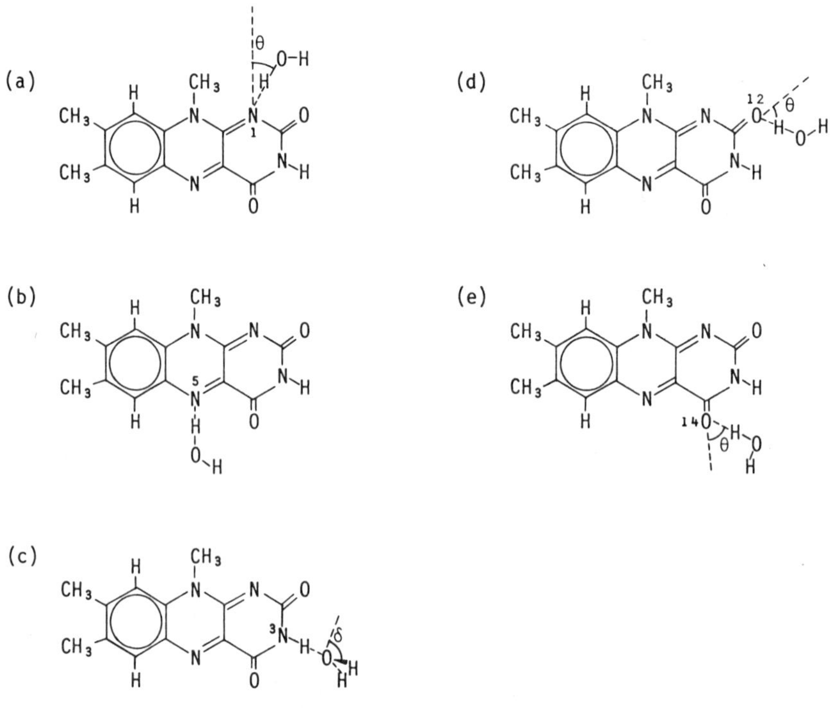

(a) (b) (c) (d) (e)

Figure 1. Possible structure of lumiflavin-water complexes. (a): N(1)-hydrogen bonding (R=3.1 Å, $\theta=30°$); (b): N(5)-hydrogen bonding (R=3.0 Å, $\theta=0°$); (c): N(3)H-hydrogen bonding (R=2.6 Å, $\delta=37.7°$); (d): O(12)-hydrogen bonding (R=2.77 Å, $\theta=60°$); (e): O(14)-hydrogen bonding (R=2.7 Å, $\theta=60°$).

Table 1. The Calculated Results on Lumiflavin-Water Complexes.

Hydrogen bonding	NO	N(1) (R = 3.1 Å) $\theta = 30°$	N(5) (R = 3.0 Å) $\theta = 0°$	N(3) (R = 2.6 Å) $\delta = 37.7°$	O(12) (R = 2.77 Å) $\theta = 60°$	O(14) (R = 2.7 Å) $\theta = 60°$
ΔQ	—	−0.021	−0.026	+0.074	−0.038	−0.046
ΔE	—	5.64	5.78	7.81	6.25	5.51
LUMO energy	0.1345	0.1271	0.1237	0.1478	0.1283	0.1258
HOMO energy	−0.2209	−0.2306	−0.2275	−0.2079	−0.2292	−0.2267
HOMO-LUMO energy	−0.3553	−0.3577	−0.3511	−0.3557	−0.3575	−0.3525
Coefficient of LUMO at						
N(5)	0.5426	0.5478	0.5524	0.5323	0.5452	0.5504
C(4a)	−0.4214	−0.4123	−0.4299	−0.4284	−0.4107	−0.4169

ΔQ: calculated charge migration, ΔQ = Q(lumiflavin, hydrogen bonding) − Q(lumiflavin, free).

ΔE: hydrogen bonding energy (kcal/mol).

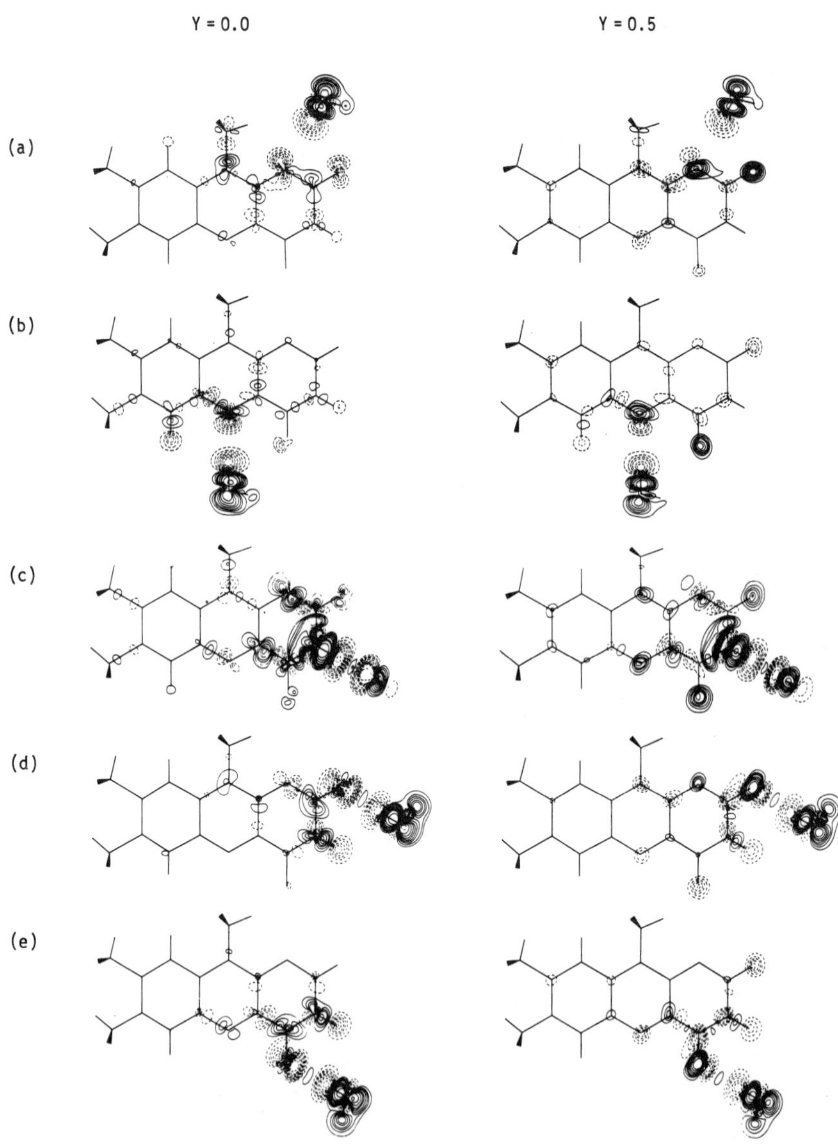

Figure 2. The difference electron density of lumiflavin-water complexes. (a): N(1)-hydrogen bonding; (b): N(5)-hydrogen bonding; (c): N(3)h-hydrogen bonding; (d): O(12)-hydrogen bonding; (e): O(14)-hydrogen bonding. Y=0.0: lumiflavin-water plane; Y=0.5: 0.5 a.u. plane above lumiflavin-water plane. The full lines indicate density increase and dotted lines decrease. Values of these lines are ±0.0005, ±0.0010, ±0.0015, ±0.0020, ±0.0030, ±0.0040, ±0.0060, ±0.0100, ±0.0200 (e/Å^3), successively.

polar part of flavin. An interesting fact is that C(8) is rather sensitive to delocalization through π-type polarization. (b) The hydrogen bonding at O(14) or N(5) promotes significantly the electron-acceptability of N(5). (c) The electron-acceptability of C(4a) increases upon formation of hydrogen bonding at N(3)H or N(5), but decreases upon formation of hydrogen bonding at N(1) or O(12).

References

1. Nishimoto, K., Watanabe, Y., and Yagi, K. (1978) *Biochim Biophys Acta* 526:34–41.
2. Nishimoto, K., Watanabe, Y., and Yagi, K. (1980) In *Flavins and Flavoproteins*. Yagi, K. and Yamano, T. (eds.) Tokyo: Japan Scientific Societies Press, pp. 493–500.
3. Yagi, K., Ohishi, N., Nishimoto, K., Choi, J.D., and Song, P.-S. (1980) *Biochemistry* 19:1553–1557.
4. Osanai, Y. and Kashiwagi, H. (1980) *Int J Quantum Chem* 17:1031–1037.

Published 1982 by Elsevier North Holland, Inc.
Vincent Massey and Charles H. Williams, Editors
Flavins and Flavoproteins

CHAPTER 90

Picosecond Fluorescence Lifetime of FAD in D-Amino Acid Oxidase-Benzoate Complex

Kunio Yagi,* Fumio Tanaka,** Nobuaki Nakashima,† and Keitaro Yoshihara†

*Institute of Biochemistry, Faculty of Medicine, University of Nagoya, Nagoya; **Mie Nursing College, Tsu; and †Institute for Molecular Science, Okazaki, Japan*

Evidence has been given that the monomer and dimer of D-amino acid oxidase are different in their various properties. The equilibrium between the monomer and the dimer of D-amino acid oxidase shifted toward the dimer upon complex formation with benzoate (1). In this work, we examined the dependence of the dissociation constant of FAD and the relative quantum yield of fluorescence on the concentration of the enzyme-benzoate complex. The fluorescence lifetime of FAD was measured by using a mode-locked Nd:YAG laser.

Dissociation Constant of FAD and Relative Quantum Yield

The apparent dissociation constant of FAD (K) and the relative quantum yield of FAD combined with the protein moiety to that of free FAD (r) were determined by measuring fluorescence intensity and polarization anisotropy under steady-state excitation, according to the following procedure. Relative intensity of the fluorescence of FAD (R_1) was measured at various concentrations of the complex. Relative intensity of the fluorescence of FAD in the complex to that of free FAD in the complex solution (R_2) was determined by measuring fluorescence polarization of the complex solution (2–4). K and r were evaluated from R_1 and R_2 as follows (2–4):

$$K = \frac{R_1^2 [P]_0}{(1 + R_2 - R_1)(1 + R_2)} \tag{90.1}$$

$$r = \frac{R_1 R_2}{1 + R_2 - R_1} \tag{90.2}$$

where $[P]_0$ in Equation (90.1) represents total concentration of the complex.

Observed values are shown in Figure 1. K value was relatively constant in the range of the concentration from 50 μM to 10 μM (0.14–0.15 nM).

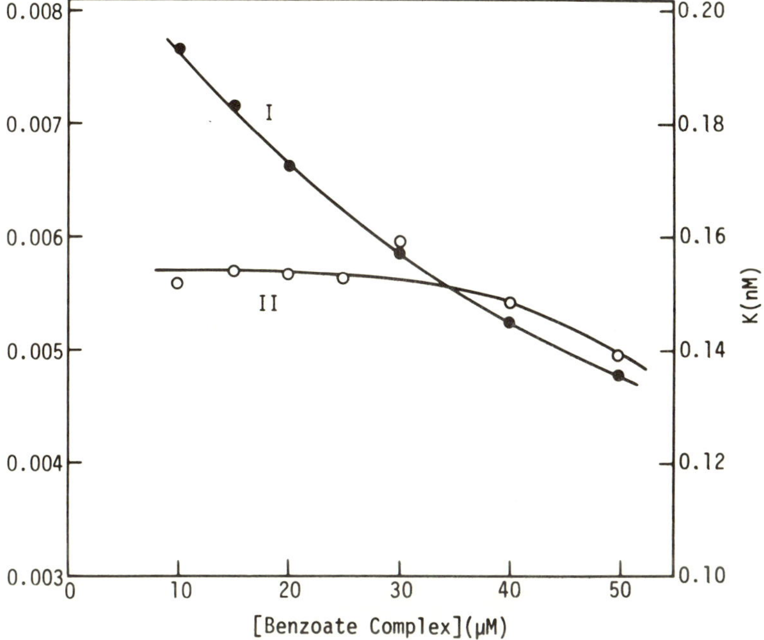

Figure 1. Dependence of r and K on the concentration of D-amino acid oxidase-benzoate complex. I: relative quantum yield, r; II: dissociation constant, K.

Picosecond Fluorescence Lifetime

The fluorescence decay was measured with a Hamamatsu streak camera (HTV 1000) by exciting with the 3rd harmonic (355 nm) of a passively mode-locked laser (1.06 μm) (4,5). The pulse width of the 3rd harmonic was measured to be 30 ps. Distortions of the camera and streak speed were fully corrected. Scattered light from the exciting pulse was completely rejected by yellow, sharp cutoff filters. After distortion of the observed decay curves and exciting pulses was corrected, the time scale was linearized at intervals of 5 ps, averaging over the signal points existing at time between $(5n-2.5)$ and $(5n+2.5)$ ps $(n=1,2,\cdots)$. Then, several decay data were accumulated (6–8 times) and used as an observed decay curve, $I_o(t)$. Decay parameters were determined by minimizing a root of mean square (RMS) between $I_o(t)$ and a calculated decay curve, $I_c(t)$. $I_c(t)$ was obtained by convoluting a decay function with two or three-exponential terms with exciting pulse.

The decay curves were obtained at 5–100 μM. In Figure 2, one of the decay curves is shown. Every RMS with three fluorescence species was always smaller than that with two fluorescence species. Decay parameters with the minimum values of RMS are listed in Table 1. Since the D-amino acid oxidase-benzoate complex exists mostly as the dimeric form (1), the fluorescent species with τ_1 (1.0 ± 0.5 ps) was assigned to the dimer. The species with the longest lifetime, τ_3, was assigned to FAD dissociated from the complex ($\tau_3=2300$ ps). Taking

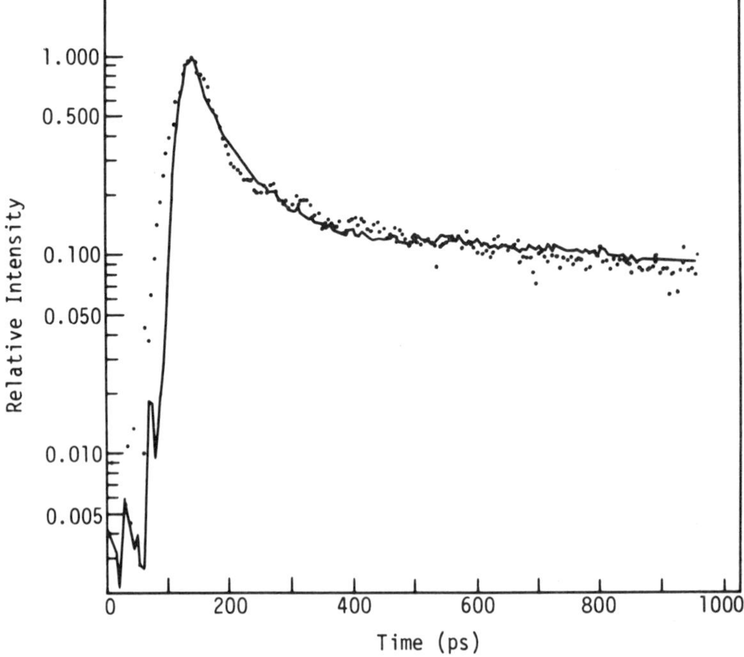

Figure 2. Fluorescence decay curve of FAD in D-amino acid oxidase-benzoate complex solution. Fluorescence decay curves were recorded on a streak camera. Half-width of excitation pulse was about 30 ps. Observed data are indicated by dots. Calculated decay curve is shown by a solid curve. Concentration of the complex was 50 μM in 0.017 M pyrophosphate buffer (pH 8.3) containing 0.1% benzoate. Decay parameters are listed in Table 1.

into account the dissociation constant of benzoate from the complex (3–12 μM), the species with τ_2 was assigned to the monomer. From the component fractions, α_1 and α_2, corresponding to τ_1 and τ_2, the dissociation constant of the dimer into the monomer was evaluated to be 0.4 ± 0.3 μM.

Results obtained by fluorescence lifetime measurements were coincident with those of steady-state excitation, since the monomer of the complex has a much longer lifetime than the dimer and therefore the relative quantum yield could increase as the concentration is lowered. The value of τ_1 may not be so accurate as 1.0 ± 0.5 ps, if we consider that the excitation laser pulse had ca 30 ps of half-width. However, we are confident that the lifetime of the dimer is less than 5 ps.

Upon complex formation with benzoate, the values of fluorescence lifetime of the dimer and of the monomer were reduced to ca 1/40 and 1/2. K value decreased to ca 1/1000 and the dissociation constant of the dimer into the monomer decreased to ca 1/10. The remarkable changes in these parameters suggest that the molecular interaction of FAD with electron donating amino acid residues was markedly enhanced by the complex formation, which contributes to the great stability of the complex.

Table 1. Fluorescence Decay Parameters of FAD in D-Amino Acid Oxidase-Benzoate Complex Solution.

	Concentration (μM)	τ_1 (ps)	α_1	τ_2 (ps)	α_2	τ_3 (ps)	α_3	RMS
I	5	13	0.86	530	0.14			0.0672
	10	15	0.90	620	0.10			0.0533
	50	14	0.90	460	0.10			0.0635
	100	12	0.90	350	0.10			0.0544
II	5	1.2	0.915	70	0.067	2300	0.018	0.0568
	10	0.4	0.910	50	0.076	2300	0.014	0.0442
	50	0.8	0.920	60	0.069	2300	0.011	0.0472
	100	0.6	0.925	60	0.066	2300	0.009	0.0414

Buffer system: 0.017 M pyrophosphate-HCl containing 0.1% benzoate (pH 8.3). I: 2-components analysis: II: 3-components analysis. Fluorescent species with lifetime τ_1, and τ_2, and τ_3 were assigned to the dimer, monomer, and FAD dissociated from the complex, respectively. α_1, α_2, and α_3 are component fractions corresponding to τ_1, τ_2, and τ_3, respectively. RMS: root of mean square between the observed and calculated fluorescence intensities.

ACKNOWLEDGMENT
The authors are indebted to the Computer Center, Institute for Molecular Science, Okazaki, National Research Institutes for the use of the HITAC M-200 H computer.

References

1. Yagi, K., Naoi, M., Harada, K., Okamura, K., Hidaka, H., Ozawa, T., and Kotaki, A. (1967) *J Biochem (Tokyo)* 61:580–597.
2. Tanaka, F. and Yagi, K. (1979) *Biochemistry* 18:1531–1536.
3. Tanaka, F. and Yagi, K. (1980) In *Flavins and Flavoproteins*. Yagi, K. and Yamano, T. (eds.) Tokyo: Japan Scientific Societies Press, pp. 387–394.
4. Nakashima, N., Yoshihara, K., Tanaka, F., and Yagi, K. (1980) *J Biol Chem* 255:5261–5263.
5. Sumitani, M., Nakashima, N., Yoshihara, K., and Nagakura, S. (1977) *Chem Phys Lett* 51:183–185.

Published 1982 by Elsevier North Holland, Inc.
Vincent Massey and Charles H. Williams, Editors
Flavins and Flavoproteins

CHAPTER 91

A Study of D-Amino Acid Oxidase-Inhibitor Complexes by Resonance Raman Spectroscopy

Yasuzo Nishina,* Kiyoshi Shiga,* Hiromasa Tojo,** Retsu Miura,** Toshio Yamano,** and Hiroshi Watari*

**National Institute for Physiological Sciences, Okazaki, Aichi; **Department of Biochemistry, Osaka University Medical School, Osaka, Japan*

Hog kidney D-amino acid oxidase (DAO) is one of the most extensively investigated flavoenzymes. Many benzoate derivatives are known to be competitive inhibitors for the enzyme. The complexes of the enzyme with aminobenzoates or hydroxybenzoates have charge-transfer bands, and the investigation of these complexes has been carried out systematically measuring visible absorption (1) and circular dichroism spectra (2, 3).

Recently, many investigations using resonance Raman (RR) technique in flavins and flavoproteins were carried out, which revealed that RR spectroscopy is useful for studying the structure of flavin, and flavin-protein and flavin-ligand interactions (4–11).

In the present paper, we report the RR spectra of complexes with DAO and aminobenzoates or hydroxybenzoates. For comparison with DAO-*o*-OH-benzoate complex, the RR spectra of quasi D-amino acid oxidase (Q-DAO)-*o*-OH-benzoate and Old Yellow Enzyme (OYE)-*o*-OH-benzoate complexes were also observed.

Materials and Methods

The purification procedures for DAO, Q-DAO, and OYE are described elsewhere (12–14). Each enzyme concentration was determined spectrophotometrically for FAD or FMN bound to the enzymes. Experiments were carried out in 0.1 M sodium pyrophosphate buffer, pH 8.3.

RR spectra were obtained by a JASCO R-800 spectrometer (Japan Spectroscopic Co.) using an argon ion laser (Spectra Physics model 164) or a He-Ne laser (Kinmon Electrics, model KLG-103). Each sample contained ca 2%(w/w) of $(NH_4)_2SO_4$ as an internal standard, which gave the Raman line at 981 cm^{-1} (marked by SO_4^{2-}). The frequency calibration of Raman spectrometer was performed with indene (15). Visible and ultraviolet absorption spectra were measured with a Hitachi 340 spectrophotometer.

Results and Discussion

Resonance Raman Spectra of D-Amino-Acid Oxidase-Aminobenzoate Complexes

The RR spectra of the complexes of DAO with aminobenzoates (*o*-, *m*-, and *p*-) excited at 514.5 nm, were similar to one another and also to that of oxidized flavin. In the case of DAO-*o*-NH_2-benzoate complexes, however, the line at 568 cm^{-1} derived from *o*-NH_2-benzoate was intensified (Figure 1). The RR spectrum of the complex was observed with excitation at 632.8 nm within the charge-transfer band (Figure 1). The Raman lines at 1583 and 569 cm^{-1} which are derived from flavin and *o*-NH_2-benzoate, respectively, are intensified. This shows that the charge-transfer band is induced by the direct interaction between flavin and *o*-NH_2-benzoate. The observed 1583 cm^{-1} line is known to involve the vibrational displacements of N(5) and C(4a) atoms of isoalloxazine (8). This suggests that charge-transfer interaction occurs in the C(4a)-N(5) region. As the line observed at 569 cm^{-1} derived from the ring deformation mode of *o*-NH_2-benzoate is specifically intensified, an interaction in the parallel arrangement of flavin and ligand may be suggested in a similar manner as OYE-phenol complexes (10, 11).

Figure 1. Resonance Raman spectra excited at 632.8 nm or 514.5 nm of a complex of DAO with *o*-NH_2-benzoate. DAO (6.5×10^{-4} M) + *o*-NH_2-benzoate (2.2 mM), 2.2%(w/w) $(NH_4)_2SO_4$. The RR spectra of the complex were observed in 0.1 M sodium pyrophosphate buffer, pH 8.3

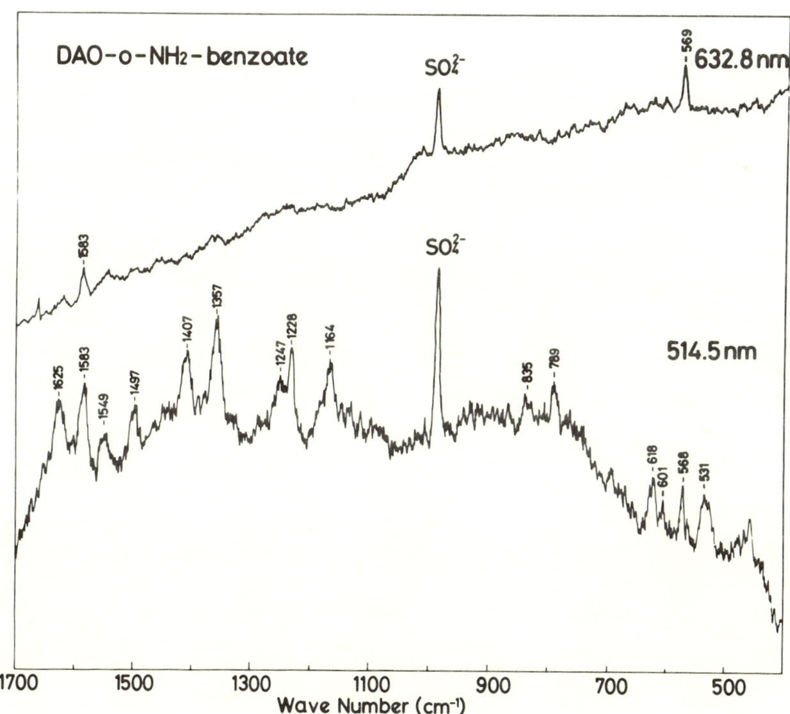

The Raman line derived from aminobenzoate was not observed in the RR spectra of complexes of DAO with *m*- or *p*-NH_2-benzoate (results not shown). This suggests that these interactions of flavin-aminobenzoate are very different from those in the DAO-*o*-NH_2-benzoate complex. In the *meta*- and *para*-substituted benzoates, the amino group in the compound may either be the electron donating site or, at least, very important for the charge-transfer interaction (2).

Resonance Raman Spectra of D-Amino Acid Oxidase-Hydroxybenzoate Complexes

The RR spectra of complexes of DAO with hydroxybenzoates (*o*- and *m*-) excited at 514.5 nm were similar to those of DAO-NH_2-benzoate complexes (Figure 2). However, it is noteworthy that the Raman line at 565 cm^{-1} is observed only in the RR spectrum of DAO-*o*-OH-benzoate, but not in the DAO-*m*-OH-benzoate complex. This line is not derived from the oxidized flavin, but from *o*-OH-benzoate, whose Raman spectrum has the line at 570 cm^{-1} in the same buffer solution. This fact strongly suggests that *o*-OH-benzoate makes a great contribution at 514.5 nm, namely it joins the charge-

Figure 2. Resonance Raman spectra excited at 514.5 nm of complexes of DAO with hydroxybenzoates. DAO (3.5×10^{-4} M) + *o*-OH-benzoate (1.5 mM), 2.2%(w/w) $(NH_4)_2SO_4$; DAO (3.1×10^{-4} M) + *m*-OH-benzoate (3.6 mM), 2.1%(w/w) $(NH_4)_2SO_4$. The RR spectra of the complexes were observed in 0.1 M sodium pyrophosphate buffer, pH 8.3.

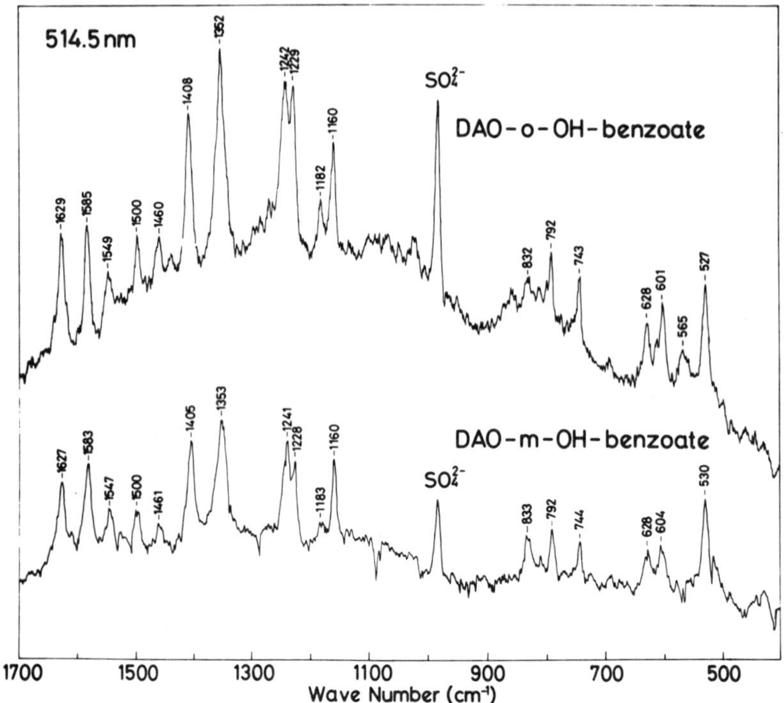

transfer interaction and that the absorption band around 514.5 nm has the character of a charge-transfer band. Therefore, this supports the indication by Massey and Ganther (1), that *o*-OH-benzoate forms a charge-transfer complex with the enzyme.

In the RR spectrum of DAO-*o*-OH benzoate complex in the buffer solution containing 50% D_2O (results not shown), the line at 1242 cm^{-1} shifted to 1284 cm^{-1}. The line is known to involve large vibrational displacements of C(2) and N(3) atoms and to be strongly coupled with the N(3)-H bending mode (10, 11). The shift mentioned above is very different from that in FAD (1261–1300 cm^{-1}) (4, 6, 7, 9), riboflavin-binding protein (1250–1295 cm^{-1}) (4, 8), or *D. gigas* flavodoxin (1251 – 1280 cm^{-1}) (9). Therefore, the environment around N(3)-H and C(2)=0 of FAD was different between DAO and others.

It is very interesting to compare the RR spectra of DAO, OYE, and Q-DAO binding with the same ligand, *o*-OH-benzoate. The RR spectrum of the OYE-*o*-OH benzoate complex excited at 514.5 nm had characteristics of oxidized flavin (results not shown). Noteworthy is the fact that the line at 545 cm^{-1} derived from *o*-OH-benzoate was observed. The result suggests that the complex has an absorption band around 514.5 nm due to a charge-transfer interaction similar to that of DAO, although it does not have an absorption band in the long wave-length region.

The Raman line derived from *o*-OH-benzoate in the complexes of DAO and OYE is 565 and 545 cm^{-1}, respectively. Therefore, the mode of the interaction between *o*-OH-benzoate and these flavoenzymes is different.

The RR spectrum of Q-DAO-*o*-OH-benzoate complex excited at 514.5 nm is identical with that of DAO-*o*-OH-benzoate complex. The result supports that the active sites of these two enzymes have similar structures (13).

References

1. Massey, V. and Ganther, H. (1965) *Biochemistry* 4:1161.
2. Shiga, K., Horiike, K., Isomoto, A., and Yamano, T. (1976) *J Biochem* 80:1101.
3. Shiga, K., Horiike, K., Nishina, Y., Isomoto, A., and Yamano, T. (1977) *J Biochem* 81:1465.
4. Dutta, P.K., Nestor, J.R., and Spiro, T.G. (1977) *Proc Natl Acad Sci US* 74:4146.
5. Nishina, Y., Kitagawa, T., Shiga, K., Horiike, K., Matsumura, Y., Watari, H., and Yamano, T. (1978) *J Biochem* 84:925.
6. Nishimura, Y. and Tsuboi, M. (1978) *Chem Phys Lett* 59:210.
7. Benecky, M., Li, T.Y., Schmidt, J., Frerman, F., Watters, K.L., and McFarland, J. (1979) *Biochemistry* 18:3471.
8. Kitagawa, T., Nishina, Y., Kyogoku, Y., Yamano, T., Ohishi, N., Takai-Suzuki, A., and Yagi, K. (1979) *Biochemistry* 18:1804.
9. Irwin, R.M., Visser, A.J.W.G., Lee, J., and Carreira, L.A. (1980) *Biochemistry* 19:4639.
10. Kitagawa, T., Nishina, Y., Shiga, K., Watari, H., Matsumura, Y., and Yamano, T. (1979) *J Am Chem Soc* 101:3376.
11. Nishina, Y., Kitagawa, T., Shiga, K., Watari, H., and Yamano, T. (1980) *J Biochem* 87:831.
12. Curti, B., Ronchi, S., Branzoli, U., Ferri, G., and Williams, C.H., Jr. (1973) *Biochim Biophys Acta* 327:266.
13. Shiga, K., Nishina, Y., Horiike, K., Tojo, H., Watari, H., and Yamano, T. (1980) *Medical J Osaka Univ* 30:71.
14. Abramovitz, A.S. and Massey, V. (1976) *J Biol Chem* 251:5321.
15. Hendra, P.J. and Loader, E.J. (1968) *Chem Ind* 718.

Published 1982 by Elsevier North Holland, Inc.
Vincent Massey and Charles H. Williams, Editors
Flavins and Flavoproteins

CHAPTER 92

Normal Mode Analysis of Lumiflavin and the Interpretation of Flavoprotein Resonance Raman Spectra

Thomas G. Spiro, W. David Bowman,[1]
Prabir K. Dutta,[2] and Michael J. Benecky

Department of Chemistry, Princeton University Princeton, New Jersey

Introduction

Resonance Raman (RR) spectroscopy has great potential application in the study of flavoprotein structure and dynamics. This technique allows in situ monitoring of flavin vibrational modes without interference from the protein backbone or solvent via the enhancement observed when the laser excitation source is tuned to flavin's π-π^* electronic transitions. The flavin RR spectrum may be expected to be sensitive to interactions between the isoalloxazine ring and the protein matrix which determine flavoenzyme reactivity.

The intense fluorescence exhibited by flavins has been a major obstacle to RR studies. The first flavin RR spectra were obtained by the use of coherent anti-Stokes Raman scattering (CARS) (1), a nonlinear optical technique which generates the Raman signal as a coherent beam of light which can be spatially filtered from the isotropic flavin fluorescence. Nishina et al. (2) subsequently showed that the apoprotein of riboflavin binding protein (RBP) quenches the fluorescence of bound riboflavin sufficiently to use spontaneous Raman techniques to obtain good quality RR spectra. Most of the available flavin RR data in the literature are on oxidized flavin (1–7) but spectra of semiquinone forms (8) and charge-transfer complexes (9) have been obtained.

Attention has focused initially on the assignment of the observed flavin RR bands to specific structural elements of the isoalloxazine ring. A variety of chemically and isotopically substituted flavins has been examined (3–7), and qualitative interpretations of the resulting vibrational frequency shifts have been offered. These suggestions need to be extended, however, and to this end we have carried out a normal mode analysis of the in-plane vibrations of lumiflavin, in which the ribose substituent at N_{10} is replaced by a methyl group.

Results

Figure 1 shows the numbering system used in the normal mode analysis. Bond length-stretching force constant correlations were used, and bending and

[1]Present address: Procter and Gamble, Cincinnati, Ohio.
[2]Present address: Exxon Fuel and Refining Corporation, Linden, New Jersey.

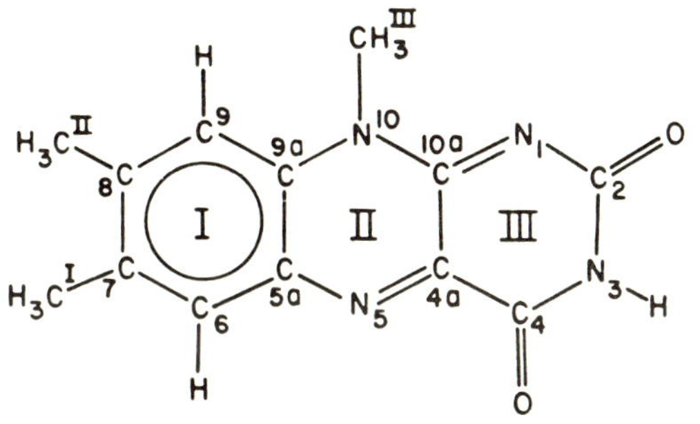

Figure 1. Structure and numbering for lumiflavin.

interaction force constants were transferred from smaller conjugated molecules to construct a reasonable initial force field. Details of the computation are given elsewhere (10). The calculated frequencies and isotope shifts lead to satisfactory assignments for the flavin RR bands and provide insights into the experimentally observed chemical effects.

Table 1 compares calculated and observed (for RBP) frequencies for the 13 RR bands, labelled I-XIII detected above 1100 cm^{-1}. Figure 2 shows a representative RR spectrum, obtained by excitation in the long wavelength absorption band of RBP. Potential energy distributions (P.E.D.) for the calculated bands are also given in Table 1. An underlined P.E.D. entry means that the internal coordinate displacement is opposite in phase to those which are not underlined.

Table 1. Results of the Normal Mode Analysis of Flavin.

	Frequency (cm^{-1})		
Mode	Observed	Calculated	Potential energy distribution (%)
I	1631	1731	νC_{5a}-$C_6(10)$, νC_7-$C_8(18)$, νC_8-$C_9(13)$, νC_{5a}-$C_{9a}(17)$
II	1584	1510	νC_{4a}-$N_5(15)$, νN_{10}-$C_{10a}(18)$, $\underline{\nu C_{10a}\text{-}N_1(12)}$, $\underline{\nu C_{4a}\text{-}C_{10a}(13)}$
III	1548	1451	$\underline{\nu C_{4a}\text{-}N_5(28)}$, νC_{10a}-$N_1(14)$, $\delta^{as}CH_3^{III}(72)$, $\delta^{s}CH_3^{III}(10)$
IV	1503	1424	$\delta^{as}CH_3^{III}(72)$, $\delta^{s}CH_3^{III}(10)$
V	1465	1407	$\underline{\nu C_6\text{-}C_7(14)}$, $\underline{\nu C_8\text{-}C_9(13)}$, νC_9-$C_{9a}(10)$, $\delta^{s}CH_3^{III}(10)$
VI	1407	1394	νN_1-$C_2(12)$, $\underline{\nu C_{5a}\text{-}C_6(17)}$, $\nu C_8 0 C_9(23)$, νC_{5a}-$C_{9a}(19)$
VII	1355	1379	νN_{10}-$C_{10a}(22)$, νC_{5a}-$C_{9a}(12)$, $\underline{\delta^{s}CH_3^{III}(19)}$
VIII	1302	1344	νN_5-$C_{5a}(14)$, $\underline{\delta^{s}CH_3^{II}(73)}$
IX	1282	1291	$\underline{\delta^{s}CH_3^{I}(59)}$
X	1252	—	Analogous to 1236 cm^{-1} uracil mode (see text)
XI	1229	1204	$\underline{\nu N_3\text{-}C_4(20)}$, νC_8-$CH_3(19)$, $\underline{\delta^{rock}CH_3^{III}(14)}$
XII	1179	1147	$\underline{\nu N_1\text{-}C_2(12)}$, νC_2-$N_3(16)$, $\underline{\nu N_3\text{-}C_4(17)}$, νC_4-$C_{4a}(14)$, $\underline{\nu C_{4a}\text{-}C_{10a}(12)}$, $\underline{\nu C_7\text{-}CH_3(16)}$, $\delta^{rock}CH_3^{III}(12)$
XIII	1161	1124	νC_2-$N_3(11)$, νC_{4a}-$C_4(20)$, $\underline{\nu C_{4a}\text{-}C_{10a}(11)}$, νC_7-$CH_3(14)$, $\delta C_6H(20)$.

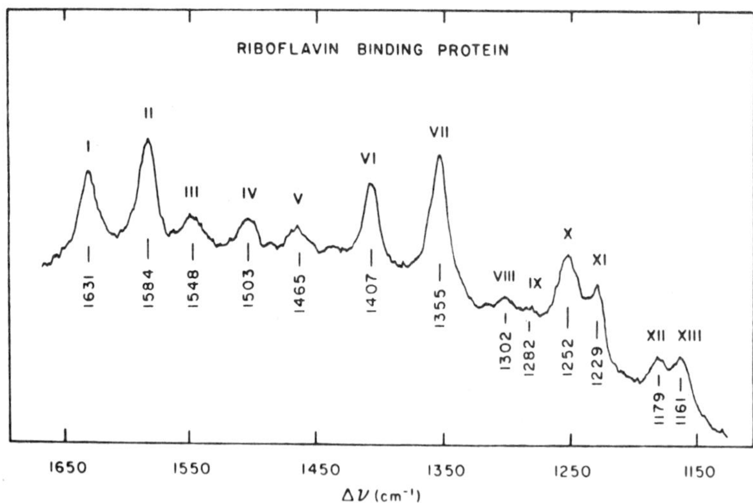

Figure 2. Resonance Raman spectrum of riboflavin (0.5 mM) bound to egg white riboflavin binding protein (~1 mM). Excitation wavelength 488 nm, 100 mW. Spectral slit width=10 cm^{-1}. Scan rate=.5 cm^{-1}/sec. Time constant=5 sec. Sensitivity=10^4 counts.

Twenty-five frequencies are calculated to fall between 1100 and 1800 cm^{-1}. The assignment of 13 of these to the observed RR bands is based on the general frequency order, and a detailed match of observed isotope shifts and chemical considerations is discussed elsewhere (10). The disparity between the observed and calculated frequencies is as high as 100 cm^{-1}, and some of the calculated isotope shifts are significantly in error. Many of the discrepancies could have been removed by judicious adjustment of interaction force constants. However, refinement of the force field would be premature because of the incompleteness of the vibrational data.

Discussion

The present calculation must be considered only a starting approximation to the analysis of the flavin normal modes. It nevertheless does provide a framework for understanding the selective effects observed for chemical substitutions. Although the internal coordinates mix extensively, as expected, they segregate themselves into identifiable regions of the molecule in certain modes. Attention naturally focuses on the most prominent RR bands, I, II, and X which are readily identified and monitored.

Band I is a mode associated with the xylene portion, ring I, of the isoalloxazine skeleton involving in-phase stretching of bonds para to one another It has been shown to shift down upon 7,8-dichloro substitution by Nishina et al. (2) who associated it with the 1607 cm^{-1} o-xylene mode. This band shifts up when position 8 of the isoalloxazine ring is substituted with an electron donating group which shifts the flavin resonance structure towards the quinoid form shown in Figure 3 [e.g., band I occurs at 1641 and 1638 cm^{-1} for the 8-CH_3NH and 8-O^- flavin derivatives respectively (6)]. Schopfer and Morris (7) report that the band I frequency shows an inverse linear correlation with Hammet sigma values for 7,8 substituents. The slight upshift in band I for FAD in water (1635 cm^{-1}) has been suggested (8) to result from the stabilization of the 8 H^+CH_2 =resonance form responsible for the appreciable 8-CH_3 H-D exchange rate (11).

Kitagawa et al. (4) have shown that band II shows sizable isotope shifts on $^{15}N_5$ and $^{13}C_{4a}$ substitution. Our calculation shows that this mode has a major component of $C_{4a}=N_5$ stretching, and also of C_{4a}-C_{10a} and $C_{10a}=N_{10}$ stretching in alternating phase. In this, it resembles a pyrazine mode observed at 1584 cm^{-1} (12). This correspondence has been noted by Dutta and Spiro (8) along with the remarkable parallel in the upshift on pyrazine protonation (1612 cm^{-1}) and protonated semiquinone formation (1611 cm^{-1}). Moreover, the 1584 cm^{-1} and 1611 cm^{-1} modes of oxidized flavin and semiquinone are the only strongly enhanced bands in the second π-π^* transitions of these two species (8). It appears that the molecular distortion in the second π-π^* state (13) involves a large displacement along this normal mode.

This mode also involves a significant contribution of $C_{10a}=N_1$ stretching, out of phase with $C_{4a}=N_5$. Thus it strongly involves the $N_5=C_{4a}$-$C_{10a}=N_1$ conjugation pathway. This accounts for the large downshift of band II when the conjugation pathway is altered by substitution at position 8 with electron donor groups which tend to stabilize the quinoid resonance structure shown in Figure 3 [e.g., 8-CH_3NH (1567 cm^{-1}), 8-O^- (1555 cm^{-1}) (6)]. Band II is observed at a lower frequency (1577 cm^{-1}) in flavodoxin and glucose oxidase compared to 1584 cm^{-1} in FAD and RBP. H-bonding to either N_5 or N_1 might be expected to lower the band II frequency, since these heteroatoms are at either end of the conjugated double bond system. Flavodoxin has a N_5 H-bond in the semiquinone but not in the oxidized form (14). However, there is a potential H-bond donor (Gly-89 NH) in the vicinity of N_1 and $C_2=O$ (14). H-bonding to this region might be responsible for the lowered band II frequency in flavodoxin and glucose oxidase.

Band X (1252 cm^{-1} in RBP) has been the focus of much attention because of its large upshift upon N_3 deuteration to $\sim$1295 cm^{-1} (2,3,5). A similar mode is observed in uracil at 1236 cm^{-1}, shifting to 1258 cm^{-1} upon N_1-N_3 dideuteration (15). This effect has been explained (2) as resulting from an interaction of a C-N_3 stretching mode with an unobserved N_3-H bending mode, whose frequency is expected to be somewhat higher; deuteration would

Figure 3. Quinoid resonance structure for 8-substituted isoalloxazines.

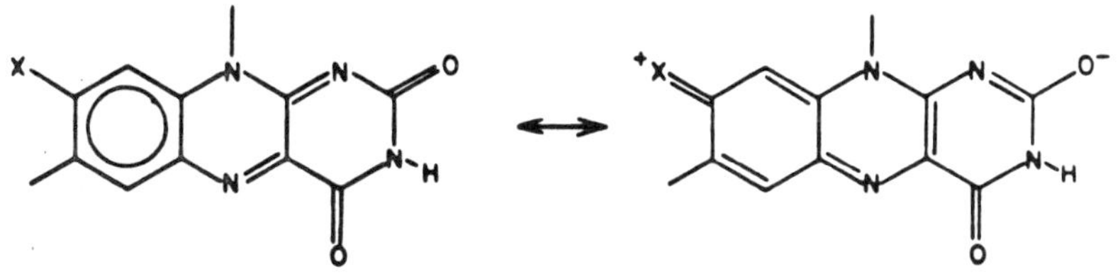

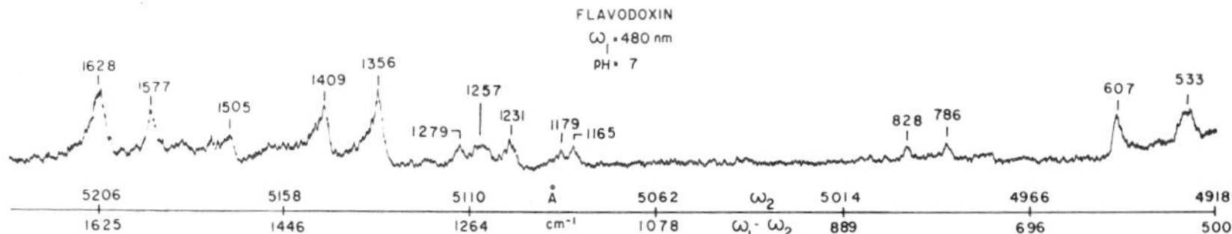

Figure 4. Resonance CARS spectrum of Flavodoxin from *Clostridium* MP. ~1 mM Flavodoxin in 5 mM. Tris buffer pH 7.3 in a capillary at 10°C. Instrumental conditions $\omega_1 = 480$ nm; ω_2 scan speed = .6 nm/min; 30 pulses averaged; repetition rate 10 pulses/sec.

greatly lower this frequency and relieve the interaction allowing the 1252 cm^{-1} to rise. Our calculation produced two modes at 1269 and 1258 cm^{-1} with substantial N_3-H bending contributions. But neither one is calculated to shift up on deuteration or give a large $^{13}C_2$ isotope shift which is experimentally observed (4) for this mode. Likewise standard valence force field calculations on uracil fail to show a deuteration upshift (15). A molecular orbital-constrained force field did give an upshift of the uracil 1236 cm^{-1} mode (16). It was then found to have only a small contribution from NH bending, but a large contribution from C=O bending, which decreased on deuteration, leaving a relatively pure C-N stretching mode at a higher frequency. A similar change in normal mode composition is likely to account for the deuteration upshift in the flavin 1252 cm^{-1} mode.

Band X is believed to be sensitive to H bonding from N_3-H and also possibly $C_2=O$ and $C_4=O$. This band shows the greatest variability among the proteins studied to date. Figure 4 shows the resonance CARS spectrum of flavodoxin from *Clostridium* MP. Flavodoxin shows essentially the same band X frequency (1257 cm^{-1}) as FAD in H_2O (1260 cm^{-1}), while in RBP it is 8 cm^{-1} lower (1252 cm^{-1}). Figure 5 shows the resonance CARS spectrum of the benzoate complex of D-amino acid oxidase which has the resolved 450 nm absorption band characteristic of flavin in a hydrophobic

Figure 5. Resonance CARS spectrum of the Benzoate complex of D-amino acid oxidase ~.1 mM D-amino acid oxidase in a 5 mm pathlength cell in pyrophosphate buffer at pH 8.3. ω_1 =495 nm. All other conditions same as Figure 4.

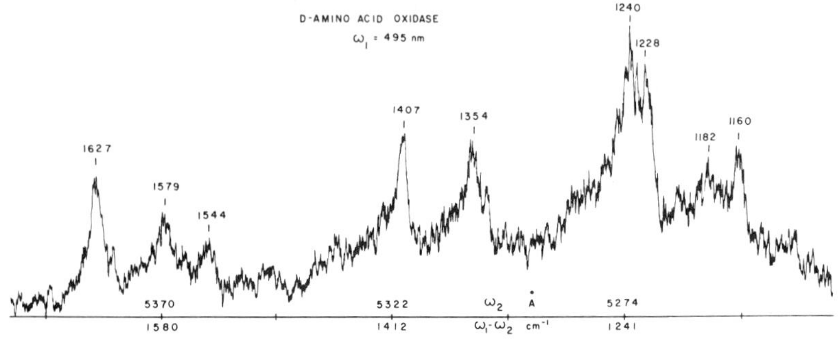

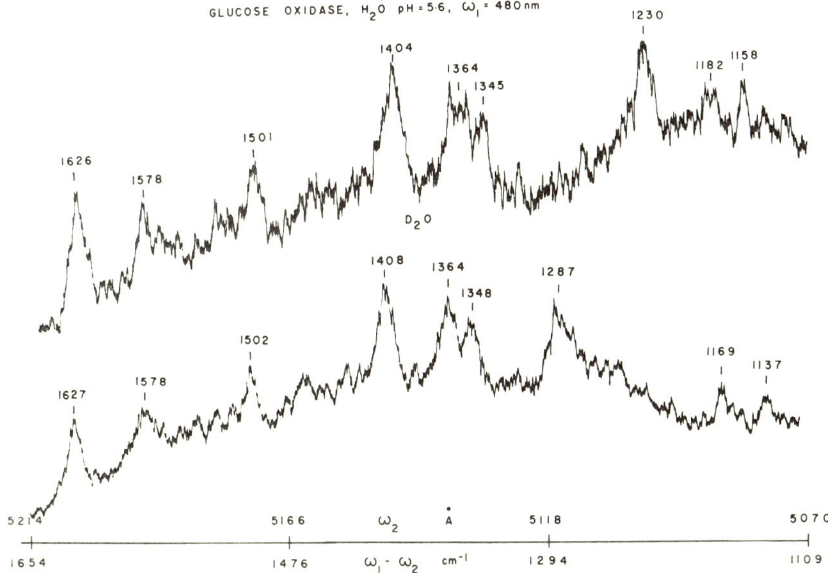

Figure 6. Resonance CARS spectra of glucose oxidase in acetate buffer at pH 5.6 in H_2O and D_2O. Spectra recorded in a 5 mm cell. Conditions identical to those in Figure 4.

environment. In this protein, band X occurs at a very low frequency 1240 cm^{-1}. Figure 6 shows that in glucose oxidase this band is missing, but its upshifted counterpart in D_2O is observed at 1287 cm^{-1}. Thus, the H-bond pattern appears to be appreciably different for these proteins as monitored by RR spectroscopy.

References

1. Dutta, P.K., Nestor, J.R., and Spiro, T.G. (1977) *Proc Natl Acad Sci* 76:4146.
2. Nishina, Y., Kitagawa, T., Shiga, K., Horiike, K., Matsumura, Y., Watari, H., and Yamano, T. (1978) *J Biochem* (*Tokyo*) 84:925.
3. Dutta, P.K., Nestor, J.R., and Spiro, T.G. (1978) *Biochim Biophys Res Comm* 83:209.
4. Kitagawa, T., Nishina, Y., Yoshimasa, K., Yamano, T., Ohishi, N., Takai-Suzuki, A., and Yagi, K. (1979) *Biochemistry* 18:1804.
5. Benecky, M., Li, T.Y., Schmidt, J., Frerman, F., Watters, K.L., and McFarland, J.T. (1979) *Biochemistry* 18:3471.
6. Dutta, P.K., Spencer, R., Walsh, C., and Spiro, T.G. (1980) *Biochim Biophys Acta* 623:77.
7. Schopfer, L.M. and Morris, M.D., (1980) *Biochemistry* 19:4932.
8. Dutta, P.K. and Spiro, T.G. (1980) *Biochemistry* 19:1590.
9. Kitagawa, T., Nishina, Y., Shiga, K., Watari, H., Matsumura, Y., and Yamano, T. (1979) *J Am Chem Soc* 101:3376.
10. Bowman, W.D. and Spiro, T.G. (1981) *Biochemistry*, in press.
11. Bullock, F.J. and Jardetzky, O. (1965) *J Org Chem* 30:2056.

12. Lord, R.C., Marson, A.J., and Miller, F.A. (1957) *Spectrochim Acta* 9:113.
13. Eaton, W.A., Ofrichter, J., Makinen, M.W., Andersen, R.W., and Ludwig, M.L. (1975) *Biochemistry* 14:2146.
14. Smith, W.W., Burnett, R.M., Darling, G.D., and Ludwig, M.L. (1977) *J Mol Biol* 117:195.
15. Susi, H. and Ard, J.S. (1971) *Spectrochim Acta* 27A:1549.
16. Bowman, W.D. and Spiro, T.G. (1980) *J Chem Phys* 73:5482.

Published 1982 by Elsevier North Holland, Inc.
Vincent Massey and Charles H. Williams, Editors
Flavins and Flavoproteins

CHAPTER 93

High Resolution Optical Spectroscopy of Lumiflavin and Lumichrome Derivatives in a *N*-Decane Matrix at 4.2 K

R.J. Platenkamp,* A.J.W.G. Visser,** and J. Koziol†

**Center for the Study of Excited States of Molecules, Huygens Laboratory, Leiden, The Netherlands; **Biochemistry Department, Agricultural University, Wageningen, The Netherlands; †Department of Instrumental Analysis, School of Economics, Poznan, Poland*

Summary

Lumiflavin and lumichrome derivatives equipped with an undecyl "tail" on nitrogen 3 yield highly resolved emission and excitation spectra when incorporated in single crystals of *n*-decane at 4.2 K. The best spectra are obtained using laser site-selection. Vibrational frequencies are similar in ground and excited states, the O-O transitions being the most intense. We conclude that the geometry of the first excited singlet state of these quasi-isolated molecules differs only slightly from that of the ground state. Characteristic photochemistry occurs in both isoalloxazine and alloxazine derivatives even at 4.2 K. High resolution optical spectroscopy applied to flavins enables a number of exciting possibilities which will be briefly outlined.

Introduction

Flavins and flavoproteins are generally characterized by broad absorption and emission spectra. Lowering of the temperature often results in sharpening of the band or the development of some structure. A vibrational analysis, however, is out of the question.

We have recently shown that spectra of very good quality could be obtained from flavin derivatives equipped with a long aliphatic chain of about the same length as the parent molecule (1). The aliphatic chain enhances the solubility of the flavin in the apolar solvent and functions as a kind of anchor to lock the flavin within the matrix. Having obtained these sharplined Shpolskii spectra (2), its quality (i.e., reduction of the background and spectral resolution) can still be improved by varying conditions such as choice of solvent and location of attachment of the aliphatic chain in the isoalloxazine ring. Nitrogen 3 (and not nitrogen 10) proves to be the best position of attachment. We could rationalize this observation by the much nicer incorporation of N_3-undecyllumiflavin in the host solvent (*n*-decane) (1). Even in the latter case, the molecules were not uniquely oriented, as they occupy quite a number of sites. Therefore site-selection experiments were performed both in excitation and in emission. The high resolution spectra obtained as described in reference 1 enabled us to obtain the vibrational frequencies in ground and excited states of the lumiflavin.

After establishment of the appropriate experiment conditions, one can obviously investigate a wide range of flavin derivatives. We have chosen N_3-undecyllumichrome, since this alloxazine is isomeric with isoalloxazine and has completely different electronic and optical properties. N_3-undecyllumichrome was synthesized from N_3-undecyllumiflavin according to published procedures (3).

The main features of these investigations will be summarized and other applications of high resolution spectroscopy in flavin research are discussed.

Results and Discussion

Site-Selection Experiments on N_3-Undecyllumiflavin (N_3-ULF)

The broad band excited fluorescence spectrum of N_3-ULF in a single crystal of *n*-decane is shown in Figure 1 (middle trace). The upper trace shows the room temperature fluorescence spectrum for comparison. A number of origins, denoted by the capitals A, B, etc., are present in the single crystal spectrum. Exciting a single vibronic line of site A leads to the site-selected fluorescence spectrum shown in the lower trace (excitation at 21140 cm^{-1}, bandwidth 0.5 cm^{-1}). The spectral background is completely reduced and a rich vibronic spectrum has appeared from which vibrational frequencies in the ground state can be determined (see Table 1). Note that the O-O transition at 20956 cm^{-1} is very intense.

Applying the same protocol, broad-band detection followed by detection at a single vibronic transition, single-site absorption or excitation spectra can be obtained. This is illustrated in Figure 2. The upper trace is a broad-band detected excitation spectrum, in which a number of sharp lines are present, superimposed on a broad background. Several lines coincide with lines in the emission spectrum of Figure 1 (lower trace). These lines should be origins and are labelled by the same symbol. The lower trace of Figure 2 is a site-selected excitation spectrum (site A) with the detection monochromator set on the vibronic line at 20360 cm^{-1}. The vibrational frequencies in the first excited singlet are labelled on top of the peaks. The strongest line is the O-O transition at 20956 cm^{-1} as in the fluorescence spectrum. The ground and excited state vibrational frequencies are similar and this lends credence to the idea that practically no geometry alternations are involved upon excitation of the quasi-isolated flavins.

In Table 1, several resonance Raman frequencies (4,5) are incorporated and it is gratifying to see that many frequencies correspond very well to the ones found here. In the high resolution spectra, the most intense bands occur in the low frequency region, especially the mode at 156 cm^{-1} is very strong. The modes at lower frequency are probably related to movements of the lumiflavin moiety in its cage, whereas the relatively strong 156 cm^{-1} vibration might be a butterfly motion of the molecule ($N_{10}-N_3$-axis).

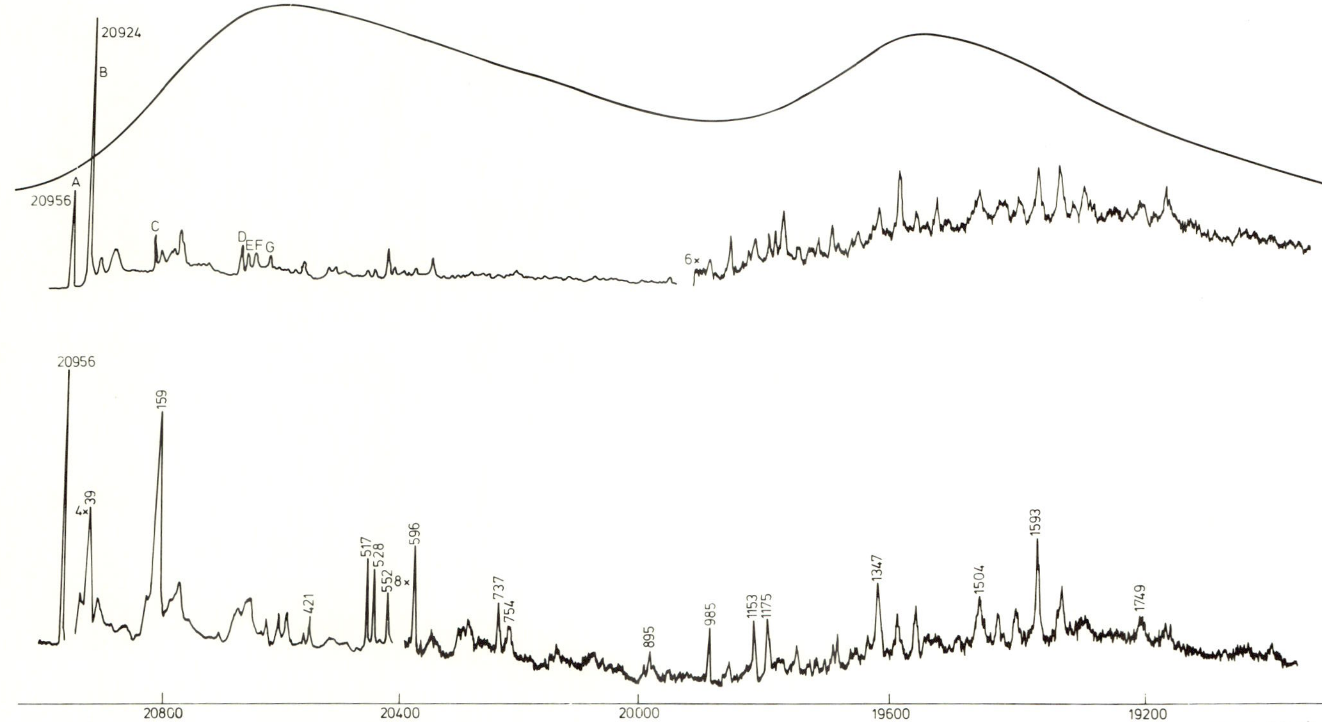

Figure 1. Fluorescence spectra of N_3-undecyllumiflavin in n-C_{10}. Upper trace: solution at room temperature. Middle trace: single crystal with broad-band excitation (4.2 K). Lower trace: site-selected fluorescence of site A (4.2 K) with laser excitation at 21110 cm^{-1}.

Table 1. Emission Spectra of N_3-Undecyllumiflavin and N_3-Undecyllumichrome at 4.2 K. Peak Positions in cm^{-1}. Raman Frequencies for Flavin are Given for Comparison.[a]

N_3-undecyllumichrome (19305 = 0–0)	N_3-undecyllumiflavin (20956 = 0–0)	Raman
	39	
	58	
161	159	
276		
306		300
	323	
	353	
381	381	
	421	429
436		
454		469
496		
	517	521
	528	
	552	
600	596	605
		633
687	686	
	737	740
	754	
	834	834
872		
	895	
988	985	994
1095	1081	
	1153	
	1175	1170
1200		1192
		1233
	1284	1261
	1302	
	1347	1355
1370		
	1406	1412
1465		
	1504	1503
	1533	
	1593	1582
		1629

[a]Raman frequencies <1000 cm^{-1} from reference 4; >1000 cm^{-1} from reference 5.

High Resolution Emission Spectra of N_3-Undecyllumichrome (N_3-ULC)

Next to fluorescence spectra, we have obtained high resolution phosphorescence spectra of N_3-ULC in a single crystal of *n*-decane. N_3-ULC was excited in the O-O transition of the first excited singlet (24761.7 cm^{-1}). The single-site phosphorescence spectrum was acquired by gating the boxcar after the fluorescence (Figure 3). As with N_3-ULF, the O-O transition is the most prominent

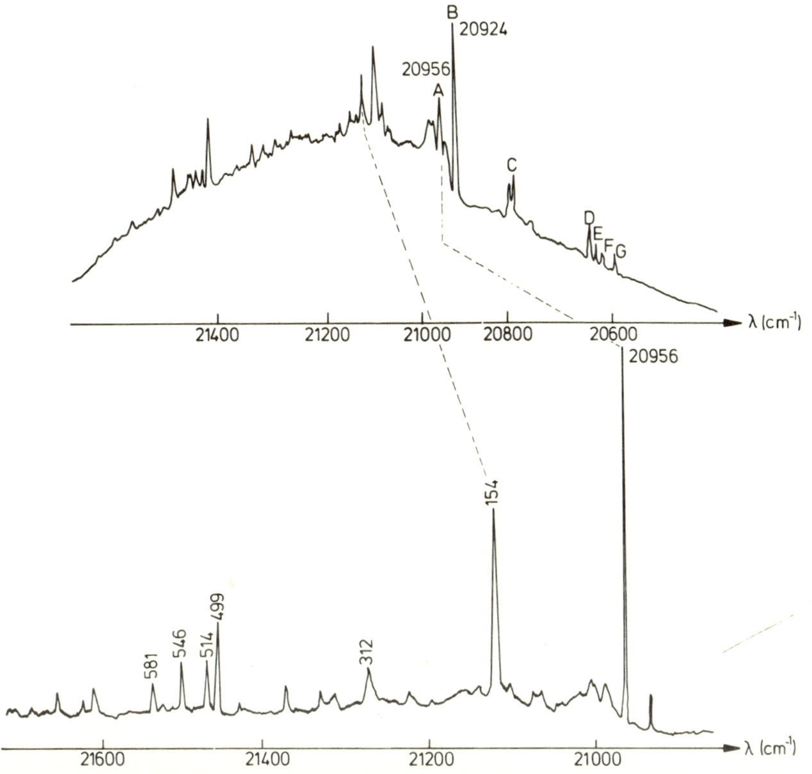

Figure 2. Absorption spectra of N_3-undecyllumiflavin in n-C_{10} at 4.2 K. Upper trace: broadband detection. Lower trace: site-selected absorption spectrum of site A, detection at 20360 cm^{-1}.

one. With the transition at 19385 cm^{-1} as origin, the complete vibronic structure can be obtained as shown in Figure 3. The frequencies are listed in Table 1. The same frequencies are obtained from the fluorescence spectrum (not shown), since both fluorescence and phosphorescence spectra provide information on ground state vibrations. The fluorescence spectrum, however, exhibits additional photochemical aspects to be discussed below. The frequencies of N_3-ULF and N_3-ULC are evidently different (see Table 1). This is due to the difference in electronic structure of both compounds. These experiments show that one must be cautious to extrapolate the vibrational properties of part of the flavin (e.g., the uracil part) to the entire molecule (6).

Photochemical Aspects

In N_3-ULF, we were able to detect photochemical site conversion between flavins occupying different sites (A and B) in the host crystal (1). We found that after irradiation into the O-O transition of site A, part of its population was transferred to site B. This process proved to be reversible since irradiation of site B into a vibronic band resulted in transfer of population from site B to

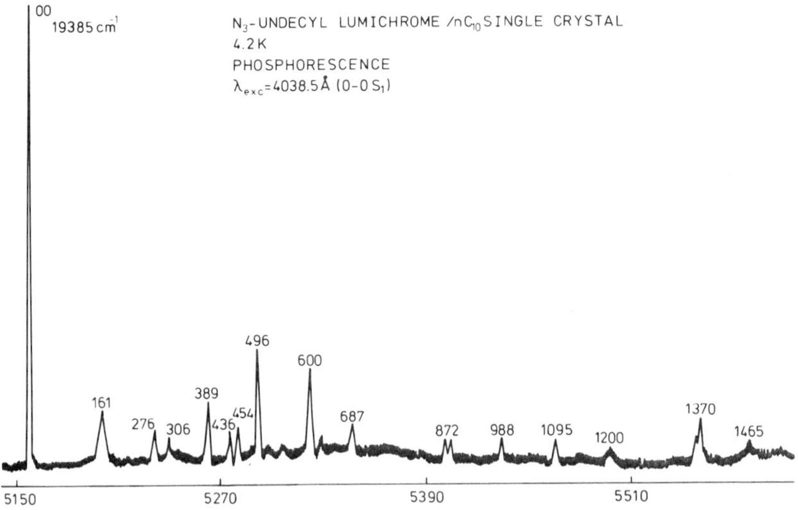

Figure 3. Site-selected phosphorescence spectrum of N_3-undecyllumichrome in n-C_{10} single crystal with laser excitation at 24762 cm^{-1} (O-O S_1).

site A. These phenomena probably reflect a reorientation of the guest in the host. We hoped for the observation of photochemical tautomerism in the N_3-ULC experiments. However, no production of N_3-ULF was found. On the other hand, the O-O fluorescence transitions appeared as intense sharp peaks ($\Delta\upsilon \approx 1$ cm^{-1}) with a large number of satellites spaced a wavenumber apart up to 50 cm^{-1} to high energy. Although not yet confirmed in a deuterium experiment, we believe to see the emission from lumichrome molecules that have lost a proton at N_1. This proton hides in various ways in the matrix, causing the satellite structure, and recombines with the parent molecule in the ground state.

Perspectives

The results shown are innovative in flavin research, as they contain the first high resolution optical spectra of this asymmetric molecule. The crucial point is the presence of the long aliphatic chain, which locks the flavin in the matrix in a well-defined manner. A number of potential applications using oriented flavin molecules can be considered:

1. Substitution of isotopes in the isoalloxazine ring (^{13}C, ^{15}N, ^{2}H) may establish the assignment of the vibrational modes.
2. Application of a uniform electric field may affect the positions of the O-O transitions (Stark Effect), leading to a determination of dipole moments (direction and magnitude) in ground and excited states.
3. Oriented lumiflavin radicals or photoexcited triplets can be used in EPR-spectroscopy to determine the distribution of spin densities and spin axes in

the flavin nucleus. The results can be compared with the calculated spin densities (7).

4. Flavoprotein crystals with the prosthetic group in a fixed orientation are appropriate subjects for high resolution optical spectroscopy. Interactions of the flavin with the protein matrix might be traced down.

ACKNOWLEDGMENTS

The investigations were supported by the Netherlands Foundation for Chemical Research (S.O.N.) with financial aid from the Netherlands Organization for the Advancement of Pure Research (Z.W.O.).

References

1. Platenkamp, R.J., van Osnabrugge, H.D., and Visser, A.J.W.G. (1980) *Chem Phys Lett* 72:104.
2. Shpolskii, E.V., Il'ina, A.A., and Klimova, L.A. (1952) *Dokl Adad Nauk SSR* 17:235; Shpolskii, E.V. (1963) *Usp Fiz Nauk* 80:225.
3. Goldner, H., Dietz, G., and Carstens, E. (1966) *Liebigs Ann Chem* 694:142; Müller, F. and Dudley, K.H. (1971) *Helv Chim Acta* 54:1487.
4. Kitagawa, T., Nishina, J., Yamano, T., Ohishi, N., Takai-Suzuki, A., and Yagi, K. (1979) *Biochemistry* 18:1804.
5. Irwin, R.M., Visser, A.J.W.G., Lee, J., and Carreira, L.A. (1980) *Biochemistry* 19:4639.
6. Bowman, W.D. and Spiro, T.G. (1980) *J Chem Phys* 73:5482.
7. Platenkamp, R.J., Palmer, M.H., and Visser, A.J.W.G. (1980) *J Mol Struc* 67:45.

Published 1982 by Elsevier North Holland, Inc.
Vincent Massey and Charles H. Williams, Editors
Flavins and Flavoproteins

CHAPTER 94

Resolution of Flavin EPR Spectra Using Fourier Transforms

W.R. Dunham,* L.J. Harding,** H.J. Grande,† and F. Müller†

**Biophysics Research Division, Institute of Science and Technology; **Computing Center, The University of Michigan, Ann Arbor; and †Department of Biochemistry, Agricultural University, Wageningen, The Netherlands*

Introduction

Flavin-radicals play an important role in nature. When bound to proteins, they function in many redox reactions and electron-transfer systems, but little is known about the electron distribution determined by the protein. Electron spin resonance spectra of isoalloxazine radicals substituted at different positions in the molecule can provide valuable information on the potential interaction positions with the protein. The complexity of the molecules, which lack symmetry and possess relatively large linewidths, have made the analysis (1–3) of EPR spectra of isoalloxazine radicals difficult. However, improved methods of spectral simulation have provided some definitive results. Westerling et al. (4) were able to determine the larger hyperfine coupling constants of a series of isoalloxazine cation radicals but were not able to obtain accurate values for the smaller (<2 G) coupling constants or to identify unambiguously their positions. Using the Fourier transform method delineated in Dunham et al. (5), Müller et al. (6) were able to determine the smaller coupling constants unambiguously. From this study, it is apparent that a good fit can be obtained only if the smaller (<2 G) coupling constants are included and if the fit is based on precise digitization of the spectra. For this study, the spectra were obtained by direct computer digitization to 4096 points; presumably the digitization used by Westerling et al. (4) was too coarse to show the discrepancies between the experimental and theoretical spectra that are engendered by the smaller coupling constants.

The additional results delineated here illustrate that the Fourier transform method can be used to detect relatively small differences between isoalloxazine cation radicals. Specifically, we show the influence of the solvent on these radicals and the effects of various substitutions at the N(10) position to provide a connection with our earlier study of flavin radicals.

Results and Discussion

Because the cations are stable only in acids, it is not possible to use a wide variety of solvents. In the solvents used, the polarization and the acidity of the anion can influence the interaction with the isoalloxazine radical. The results presented in Table 1, which is arranged so that the coupling constant (a-value) of the proton at position 5 is increasing, show no straightforward correlation between the solvent and the a-values. The proton at position 5 is affected most; the a-value of the N(5) shows the same tendency but to a smaller degree. The a-values of the N(10) and CH3(10) vary only slightly. The a-value of the H(8) correlates with the presence of trifluoracetic acid, while the H(6) is effectively constant. The variations in the H(7) and H(9) are not regarded as significant because they are within the accuracy that we have generally found appropriate for these small (<.5 G) values.

Based on the the width of the spectra and the a-values at N(5), there is a change in the polarization of the NH-bond that is caused by the solvent and that results in both a-values at the N(5) position increasing together. Apparently, the trifluoroacetic acid ion is the most polarizable anion, while the chloride anion is the least polarizable. Adding toluene to trifluoroacetic acid decreases the polarity of the solvent.

The aromatic ring protons, however, behave as if only the acidity is important. Our NMR article (7) showed that a change of the 2 carbonyl into an iminol function shows up at positions 8 and 9 (compare compounds XXII and XXV of Table 5 in reference 7). The strongest acid, trifluoroacetic, shows this influence of protonation also. Since the N(10) position varies only slightly, there is an apparent lack of polarizability at this position.

Replacement of the CH3 group at N(10) by bulkier groups supports the suggestion of Ehrenberg (8) for FMN that the protons of the N(10)-CH2 group are not magnetically equivalent. Although the spectra of the N(10)-substituted ethyl and acetyl-butyl radicals can be fitted with equivalent N(10) protons, the a-value obtained for the H(8) proton is inconsistent with all previously analyzed isoalloxazine radicals. More specifically, the H(8) proton is not in its normal range (2.5 to 2.9 G), while the N(10) protons are both within this normal range. It appears, therefore, that the equivalent proton a-values actually correspond to the H(8) proton and one of the N(10) protons, i.e., the N(10) protons are not equivalent. This conclusion is further supported by the results for the N(10)-substituents isopropyl, phenyl, and the ethyl bridge; these compounds possess an a-value in the normal range of the H(8) proton that cannot be confused with the a-value for any other position. Thus, we may conclude that one of the pseudo-doublets belongs to the C(8) and that the protons at the N(10) position are slightly inequivalent.

Additional spectra of these same N(10)-substituted compounds, which are in formic acid rather than HC1/acetonitrile (this solvent change was made to improve the resolution by decreasing the linewidth) have been examined. Our preliminary analysis indicates that they are consistent with the results in Table 2 with respect to the inequivalence of the N(10) protons but further analysis is necessary to delineate the solvent effects.

The ribityl-substituted cation in HC1/acetonitrile provides further confirmation of the proton inequivalence. Two fits to this compound are given in Table 2: the first with the protons forced to be equivalent and the second with the protons allowed to be inequivalent. Quantitatively, the relative squared error is larger than normally accepted in both cases; however, qualitatively, the

Table 1. Coupling Constants (in Gauss) of 10-Methyl Isoalloxazine Cation Radical as Dependent on Solvents.

Solvent	N5	H5	N10	10CH3	H8	H6	H9	H7	W[a]	LW[b]	RSE[c]
HCl/CH3CN[d]	7.39	7.80	4.87	5.07	2.70	1.77	0.41	0.39	53.32	0.47	0.90
	7.36	7.74	4.84	5.04	2.70	1.76	0.50	0.42	53.28	0.55	0.96
TOL/TFA[e]	7.41	7.95	4.87	5.10	2.90	1.70	0.55	0.45	54.51	0.95	5.71
HCOOH	7.47	8.16	4.92	5.11	2.68	1.74	0.41	0.41	53.86	0.33	1.15
	7.47	8.17	4.95	5.11	2.69	1.74	0.46	0.37	54.05	0.42	2.19
TFA[f]	7.58	8.33	4.94	5.07	2.87	1.72	0.52	0.45	54.95	0.71	2.86

[a]The total width of the spectrum in Gauss.

[b]Linewidth in Gauss.

[c]RSE is the relative square error, which is 1000 times the sum of the squares of the error divided by the sum of the squares of the data. For two solvents, the multiple solutions correspond to samples prepared at at different times and show the reproducibility of our method.

[d]50% 4N HCl/50% acetonitrile.

[e]80% toluene/20% trifluoroacetic acid.

[f]Trifluoroacetic acid.

Table 2. Coupling Constants (in Gauss) of Isoalloxazine Cation Radicals with Different Substitutents in 50% 4N HCl/50% Acetonitrile.

N(10)-R	N5	H5	N10	H10′	H10″	H8	H6	H9	H7	LW[a]	RSE[b]
Ethyl	7.41	7.66	4.81	2.47	2.34	2.65	1.76	0.52	0.43	0.87	2.51
Acetyl-butyl	7.41	7.70	4.80	2.76	2.14	2.76	1.75	0.52	0.53	0.73	2.73
Isopropyl	7.34	7.80	4.92	0.91	—	2.58	1.74	0.47	0.43	0.66	2.96
Phenyl	7.51	8.08	4.48	2 × 0.42[c]		2.63	1.65	0.46	0.44	0.82	6.19
Ethyl bridge[d]	7.50	7.93	4.60	7.06	7.06	2.63	1.67	0.68	0.45	0.99	3.61
Ribityl Eq[e]	7.46	7.36	5.23	2.74	1.97	2.74	1.63	0.40	0.00	0.75	14.76
Ineq[f]	7.48	7.78	4.75	3.03	2.44	2.63	1.71	0.00	0.00	0.89	7.12

[a]Linewidth in Gauss.

[b]RSE is the relative squared error, see Table 1.

[c]The orthoprotons on the phenyl residue.

[d]Ethyl bridge between N(10) and N(1).

[e]Fit with protons at N(10) forced to be equivalent.

[f]Fit with protons at N(10) allowed to be inequivalent.

fit with inequivalent protons is clearly superior. Accepting the inequivalence, it can be concluded from Table 2 that the inequivalence is larger, the larger the N(10) side chain and that the electron distribution at the other positions is hardly affected by the substitutions. This inequivalence at N(10) implies restricted rotation about the N(10)-R bond for the ethyl, acetyl-butyl, and ribityl substituents.

The higher a-values of the aromatic ring [H(9), the N(5), and the H(10) itself] indicate that the ethyl bridge is flatter than the other isoalloxazines (cf ref. 6). Comparison of the isoalloxazines in Table 1 and the ethyl and isopropyl substituents in Table 2 indicate that methylation at C(10′) strongly influences the a-values of the protons at this position. These protons are significantly different in the ethyl bridge substituent also. This implies that the bonding at the C(10′) position strongly affects the spin density at this position. In contrast, the N(10) is not readily polarizable as shown by the solvent study. This apparent contradiction would be an interesting subject for theoretical examination.

In the near future, we plan experiments with 13C-enriched isoalloxazine to study the complete spin density distribution in the neutral and anionic radicals.

ACKNOWLEDGMENTS

This investigation was supported by a NATO Grant (1854), by the National Institute of Health under GM12176, and by The University of Michigan Computing Center.

References

1. Guzzo, A.V. and Tollin, G. (1963) *Arch Biochem Biophys* 103:231–243 and 244–148.
2. Müller, F., Hemmerich, P., Ehrenberg, A., Palmer, G., and Massey, V. (1970) *Eur J Biochem* 14:185–196.
3. Müller, F., Erikssòn, L.E.G., and Ehrenberg, A. (1970) *Eur J Biochem* 12:93–103.
4. Westerling, J., Mager, H.I.X., and Berends, W. (1975) *Tetrahedron* 31:437–440.
5. Dunham, W.R., Fee, J.A., Harding, L.J., and Grande, H.J. (1980) *J Mag Res* 40:351–359.
6. Müller, F., Grande, H.J., Harding, L.J., Dunham, W.R., Visser, A.J.W.G., Reinders, J.H., Hemmerich, P., and Ehrenberg, A. (1981) *Eur J Biochem* 116:17–25.
7. Grande, H.J., Van Schagen, C.G., Jarbandhan, T., and Müller, F. (1977) *Helv Chim Acta* 60/2:348–366.
8. Ehrenberg, A. (1962) *Arkiv Kemi* 19:97–117.

Published 1982 by Elsevier North Holland, Inc.
Vincent Massey and Charles H. Williams, Editors
Flavins and Flavoproteins

CHAPTER 95

Separation and Characterization of Flavin Derivatives by Means of Gel Chromatography and High Performance Thin-Layer Chromatography

Janos Nagy,* Joseph Knoll,* William C. Kenney,** and Thomas P. Singer**

**Department of Pharmacology, Semmelweis University of Medicine, Budapest, Hungary; **Liver Studies Unit and Molecular Biology Division, Veterans Administration Medical Center, San Francisco, California*

Introduction

Gel chromatography is probably one of the most popular methods in biochemistry. In classical terms, it is a separation based on differences in molecular dimensions. Besides the molecular sieving action of the gel, however, some other side effects may occur, especially, if tightly cross-linked gels are used, e.g., Sephadex G-10, G-15, G-25, Bio-Gel, P-2, P-4, and when no correlation can be found between elution behavior and molecular weights. Small molecules can penetrate into the gel phase and approach the cross-links more closely, so they are consequently distributed between the swollen gel and mobile phase. This fact ensures the separation of small, structurally very similar compounds. These observations have first been reported by Gelotte (1) and Marsden (2), but systematic studies have been done in our laboratory in the last decade, focusing on the possibility of separation of low molecular weight substances (3).

In our previous work, we studied the gel chromatographic behavior of biologically active amines (4), amino acids (5), and some investigations were also extended to the purification of covalently-bound flavin adducts (6,7). Therefore, in this paper, we briefly review the results obtained on separation of flavin derivatives by these techniques improving various buffer systems at different pH. The rapid characterization and detection of the separated substances were carried out by high performance thin-layer chromatography (HPTLC).

Materials and Methods

Gel chromatographic model experiments were performed on Pharmacia glass columns at room temperature. The separations were monitored by means of

LKB Uvicord II system at 254 nm. In some cases, the absorbance values were measured on a Cary 219 spectrophotometer. High performance thin-layer plates (HPTLC-silica gel 60) were purchased from E. Merck, Darmstadt. Thin-layer chromatography was carried out with the following solvents: isopropanol : water (9 : 3, v/v); n-butanol : acetic acid : water (4 : 2 : 2, v/v); chloroform : methanol : concentrated ammonia (2 : 1 : 1, v/v/v); FMN was from Sigma Chemical Co., and Riboflavin from Bio-Rad Laboratories. Other chemicals and experimental procedures were as previously reported (6, 7).

Results and Discussion

Separation of flavin derivatives is also based on the gel-solute interaction where K_d values of substances to be separated are definitely higher than 1.00. Specific interactions may occur because of the aromatic character of the isoalloxazine ring, and the polyhydroxylic side chain at N(10) position, and they have been found to be retarded on gel more than would be expected from their molecular size, i.e., their K_d values are increased. What factors do play important roles in the separation? Obviously, the characteristic feature of the molecules themselves, because of the correlation between the number of aromatic rings, hydroxyl and amino groups, and the increased retardation. Two other effects can be considered: differences in charge or in phosphorylated level, and the type of eluent. In addition, the selection of optimal gel type can also be advantageous.

The gel chromatographic behavior of FMN and riboflavin was basically studied on Sephadex G-15 and Bio-Gel P-2 columns using different kinds of eluents. The distribution coefficient (K_d) and resolution (R_s) values are listed in Table 1 and K_d was calculated according to the next formula:

$$K_d = \frac{V_e - V_{albumine}}{V_{acetone} - V_{albumine}}$$

where V_e is the elution volume of the substance investigated, albumine and acetone as internal standards determine the void volume (V_o) and the total permeation volume (V_∞). Resolution was calculated as usual.

The separation of FMN and riboflavin is excellent in most cases, although resolution values are generally better on Bio-Gel P-2 (great peak distances were obtained) than on Sephadex G-15 gels. Further advantage of the former gel type is that the increased retardation of flavins ($K_d > 1.00$) assures the removal of salts from the low molecular weight flavin mixtures, i.e., a complete desalting procedure can be reached. Basically, the data shown in Table 1 demonstrate how the flavins can be moved by applying different kinds of buffers (at various pH range) and selecting the optimal gel type.

All the advantages of these types of gels can be utilized if a two-column system is constructed (Figure 1). This set-up consists of Column I and Column II packed with the same or different types of gels, plus accessories. Figure 2 shows a model separation of FMN and riboflavin performed on coupled columns. The increased K_d values and great peak distance create the possibility of increasing the number of separable components. One other instance is when Column I is packed with Bio-Gel P-2 and Column II with Sephadex

Table 1. K_d and R_s Values of FMN and Riboflavin on Sephadex G-15 and Bio-Gel P-2 Columns.

	Sephadex G-15			Bio-Gel P-2		
	K_d			K_d		
	FMN	Riboflavin	R_s	FMN	Riboflavin	R_s
0.5 M HCOOH, pH 2.2	1.70	2.87	1.38	6.00	6.00	0.00
0.1 M NaH_2PO_4, pH 5.0	1.88	3.28	1.24	4.02	7.30	1.37
0.5 M Pyr-Ac., pH 5.0	1.00	1.87	1.58	2.06	3.00	1.15
0.1 M CH_3COONH_4, pH 6.6	1.43	3.06	1.82	1.74	5.65	3.22
0.15 M NaCl, pH 6.6	1.16	3.30	2.11	2.36	5.13	2.15
0.1 M Na-phosphate, pH 7.5	1.45	2.78	1.23	1.48	5.91	2.18
0.1 M NH_4HCO_3, pH 8.4	1.00	3.02	2.33	1.00	5.65	3.39
0.1 M Na_2HPO_4, pH 9.1	1.58	3.12	1.50	1.60	5.64	2.37

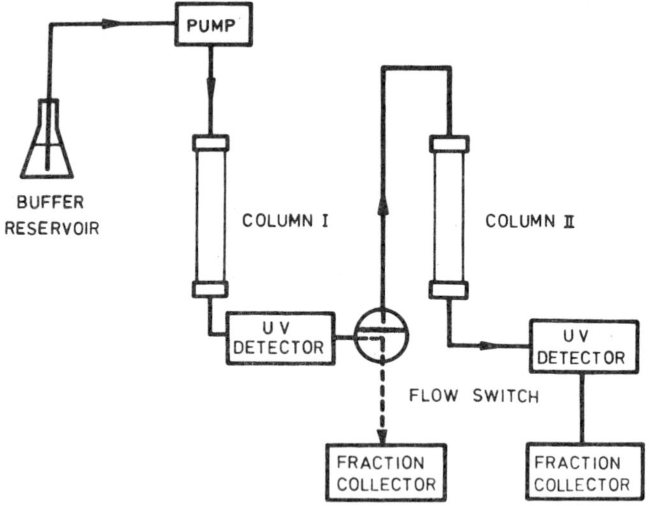

Figure 1. Schematic representation of two column gel chromatographic system. Column sizes: 2.5×90 cm. Buffer is delivered by peristaltic pump, separation is monitored by LKB Uvicord II instrument. Flow switch can be manually or automatically operated according to whether one or both columns are used at the same time.

G-15 gels and 0.5 M formic acid is used as eluent. No separation occurs on the primary column, FMN and riboflavin elute together at $K_d = 6.00$ (Table 1). Nevertheless, the second chromatographic step gives fine resolution of the strongly retarded flavins on Column II.

In particular, this system is suitable for separation of very complex mixtures of flavins or flavin peptides, where besides the group separation—removal of the high molecular weight peptide fragments and salts—a fine differentiation of flavin derivatives is obtained.

The most important gel chromatographic applications are the purification processes of flavin coenzymes and flavin peptide adducts. Flavin peptides obtained by proteolytic digestion of bovine "B" form and human "A" form

Figure 2. Separation of bovine serum albumine, acetone, FMN, and riboflavin on coupled columns of Sephadex G-15 and Bio-Gel P-2. The UV extinction was recorded on an Uvicord II instrument. Columns: 2.5×90 cm, packed with Sephadex G-15 (Column I) and Bio-Gel P-2 (Column II). Eluent: 0.9% sodium chloride solution.

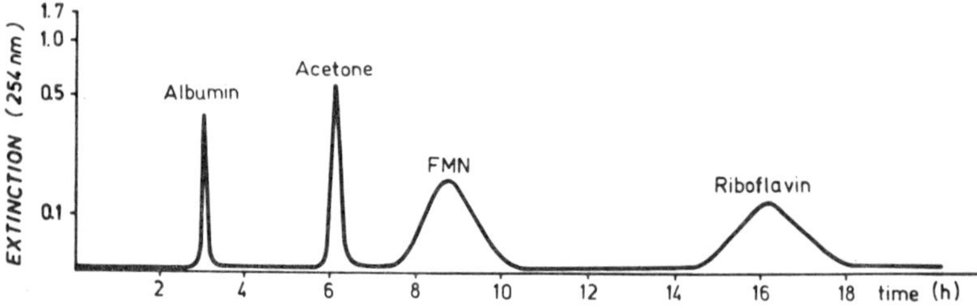

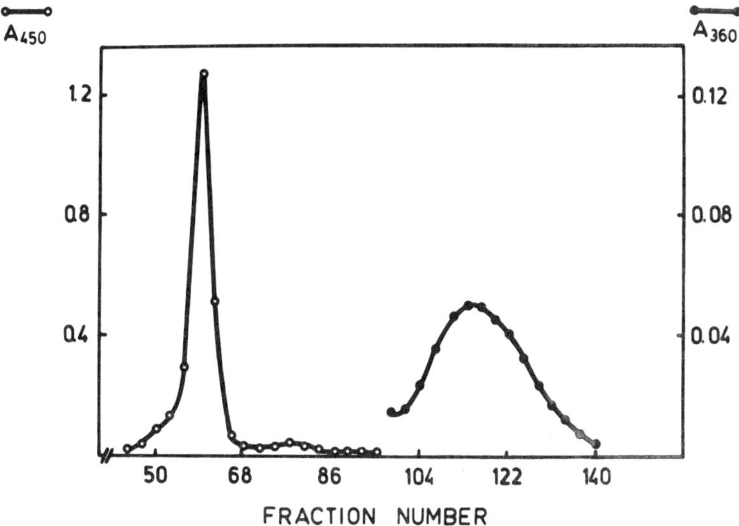

Figure 3. Isolation of 8α-S-(N-acetylcysteinyl)-4a-phenyldihydrotetraacetyl-riboflavin (2nd peak) from a reaction mixture of 8α-S-(N-acetylcysteinyl)-tetraacetylriboflavin (1st peak) and "in situ" generated phenyldiazene. Column: 2.5×84 cm, packed with Sephadex G-15. Eluent: 0.5 M pyridinium acetate, pH 5.2. Fractionation was followed by measuring the absorbances at 360 and 450 nm.

monoamine oxidase were also purified by this method (8). Moreover, after suicide inactivation of flavoenzymes by arylhydrazines, the cofactor flavin peptide adducts and parent peptides could be successfully separated on gel columns (7,9). An interesting aspect of this method is that the main interaction between FMN and riboflavin and the gel matrix can also be employed in the separation of flavocyanine products, e.g., in separation of 4a-phenyldihydro-FMN (K_d =2.65) and 4a-phenyldihydroriboflavin (K_d =5.41). Naturally, the K_d values of these adducts are higher than those of their parent compounds. Another closely related application is the isolation of phenyl adduct formed on incubation of 8α-S-(N-acetylcysteinyl)-tetraacetylriboflavin with phenyldiazene (9). It can be seen in Figure 3, that the "phenyl adduct" is eluted as the 2nd peak, because of the difference in one benzene ring, and serves as model compound in the suicide inactivation studies of monoamine oxidase.

Characterization of flavin derivatives separated on preparative gel columns was performed by high performance thin-layer chromatography with high sensitivity. The most convenient way of doing this is to apply tiny aliquots from the fractions and develop in solvent systems as given in the Methods section. Short running time with high resolution was achieved.

References

1. Gelotte, B. (1960) *J Chromatogr* 3:330.
2. Marsden, N.V.B. (1965) *Ann N Y Acad Sci* 125:428.

3. Nagy, J. (1974) PhD Thesis, Semmelweis Univ. of Medicine.

4. Kalász, H., Nagy, J., Magyar, K., and Knoll, J. (1976) In *Symposium on Pharmacology of Catecholaminergic and Serotoninergic Mechanisms*. J. Knoll and K. Magyar (eds.), Budapest: Akademiai Kiado, pp. 91–94.

5. Miklos, P., Nagy, J., Kalász, H., and Knoll, J. (1975) *Proc. 15th Hung. Ann. Meet. Biochem.*, pp. 77–78.

6. Nagy, J., Kenney, W.C., and Singer, T.P. (1979) *J Biol Chem* 254:2684.

7. Nagy, J., Kenney, W.C., Salach, J.I., and Singer, T.P. (1980) In *Monoamine Oxidases and their Selective Inhibition*. J. Knoll and K. Magyar (eds.) Budapest: Akademiai Kiado, pp. 85–93.

8. Nagy, J., and Salach, J.I. (1981) *Arch Biochem Biophys*, in press.

9. Kenney, W.C., Nagy, J., Salach, J.I., and Singer, T.P. (1979) In *Monoamine Oxidases: Structure, Function, and Altered Functions*. T.P. Singer, R.W. Von Korff, and D.L. Murphy (eds.) New York: Academic Press, pp. 25–37.

Published 1982 by Elsevier North Holland, Inc.
Vincent Massey and Charles H. Williams, Editors
Flavins and Flavoproteins

CHAPTER 96

Electrochemistry of N^5-Metallated Flavins in Aqueous and Nonaqueous Media

M.J. Clarke and M.G. Dowling

Department of Chemistry, Boston College, Chestnut Hill, Massachusetts

Several lines of research have recently shown that flavins coordinated with a Lewis acid at the N^5 position tend to undergo electron transfer in discreet 1e-steps rather than through the single 2e-transfer exhibited by free flavins in aqueous media (1–3). Lewis acids capable of inducing this transformation include: H^+, alkyl groups (R^+), and metal cations. Moreover, 1e-transfer flavodoxins relegate the flavin to cycling between $Fl^{\cdot}$ and Fl_{red} through protein controlled protonation at the N^5 site. Unfortunately, attempts to investigate the electrochemistry of N^5 coordinated flavin systems are complicated by dimerization (1) of $RFl^{\cdot}$, protein interference with electron transfer from the electrode surface (2), or the lability of most metalloflavin complexes (3).

Using a metal cation to coordinate the N^5 position can prevent flavin dimerization on both steric and electrostatic grounds, providing that the metalloflavin complex remains stable in all flavin oxidation states. Such coordination can also overcome the solubility and adsorption problems encountered in studying many flavin derivatives in both aqueous and nonaqueous media (3). Hemmerich had pointed the way in the synthesis of such complexes by suggesting that stable metalloflavins result only when extensive metal to ligand backdonation of electron density occurs (5). The $(NH_3)_5Ru(II)$ ion is well known for being able to enter into this type of strong $d\pi - \pi^*$ retrodative bonding (6). A series of stable metalloflavin complexes incorporating this ion have been prepared and thoroughly characterized by X-ray diffraction, electronic spectroscopy, and electrochemical methods (7). The structure of these complexes reveals chelation via the N^5 and O^4 sites and significant bending of the flavin ring, which is at least partially due to a substantial increase in the electron density on the ring resulting from metal backbonding (8). This π-interaction yields complexes whose spectra are reminiscent of neutral flavin radicals. Several lines of evidence indicate that the complexes persist in solution (in appropriate pH ranges) without dimerization and are related by simple acid-base and redox equilibria (7).

Experimental

Details of experimental methods in aqueous solution have been reported elsewhere (8). Potentiometric (pH*) measurements in DMF were made with a

Tacussel DMF-B10 glass electrode against a DMF-C10 calomel reference electrode (10, 11). This electrode system was calibrated between pH* 0 and 3 with freshly prepared trifluoromethylsulfonic acid (HTFMS) solutions in DMF, in which the HTFMS is fully ionized (12). At higher pH*, 2,4-dinitrophenol and p-nitrophenol were used as calibrants (14).

Cyclic voltammetry (CV) measurements were performed using an Ag, AgCl reference electrode filled with 0.4 M $(Et_4N)Cl$ in AgCl saturated water (15). Working electrodes were either a Metrohm HMDE or a Beckman Pt disk with a Pt wire auxiliary electrode. Sample solutions were purged with Ar or N_2 and blanketed with the inert gas. Ionic strengths were adjusted to 0.1 with TEAP. Scan rates were 125 mV/sec. Voltage calibration was obtained using the ferrocene-ferrocinium couple (0.400 V) (16); however, E_{cv} values are reported relative to SHE.

Results

A thorough consideration of the electrochemistry of the various Ru-Fl complexes appears elsewhere (8); however, for comparison purposes the E_{cv} vs pH diagram for the 10-methylisoalloxazine (10-MeIAlo) complex is presented in Figure 1. The diagram for (3,10-Me_2IAlo) $(NH_3)_4$Ru(II) is similar but does not exhibit N^3-protonation equilibria, so that the Ru-Fl_{ox} → Ru-$Fl^{\cdot}$ couple remains constant at high pH.

Figure 2 depicts a typical CV scan for the Ru-Fl complexes in neutral DMF. For the 3,10-Me_2IAlo complex all three couples are essentially reversible, while for the 10-MeIAlo complex only the two more cathodic couples are reversible. By analogy to the results in aqueous solution, the couple occurring

Figure 1. E_{cv} vs pH plot for (10-MeIAlo) $(NH_3)_4$Ru(II) in aqueous media.

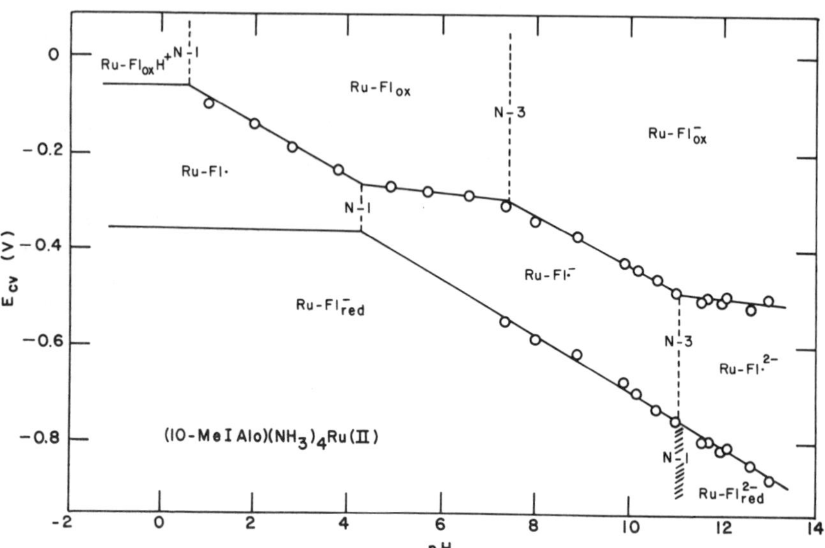

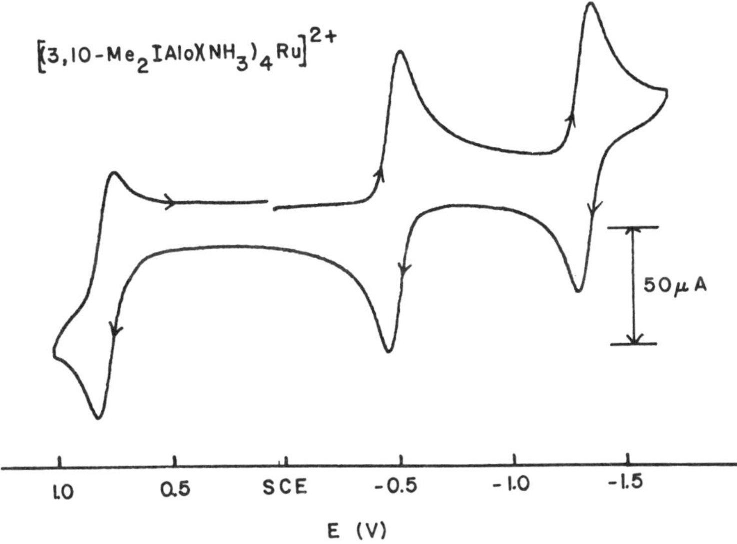

Figure 2. Cyclic voltammetry scan of (3,10-Me_2IAlo) $(NH_3)_4$Ru(II) in neutral DMF, 0.1 M TEAP on Pt disk electrode.

at highest potential is assigned to Ru(II-III), the middle couple corresponds to adding 1e to the flavin ligand and the couple at lowest potential is attributed to forming the fully-reduced flavin species. In contrast to the results in aqueous media, there appears to be a complete absence of electrode adsorption phenomena in DMF. The Ru(II-III) couple is 0.67 V between pH* 1 and 8.

Figure 3 summarizes the reversible coordinated flavin E_{cv} potentials as a function of pH* for the 10-MeIAlo complex. The results for the 3,10-MeIAlo species are essentially identical. The pK_a values estimated from the discontinuities in these plots correlate well with those independently determined by spectrophotometric titration. Similar spectra are obtained for the corresponding protonated and deprotonated Ru-Fl_{ox} forms of the 10-MeIAlo and 3,10-MeIAlo complexes so that they share a common protonation site, which is assumed to be N^1.

Discussion

Relative to the analogous results in aqueous media the 1e-reduction potentials of Ru-FlH_{ox}^+ and Ru-Fl_{ox} are shifted negatively in DMF by 0.2 and 0.3 V, respectively. The E_{cv} of Ru-$Fl^{\cdot}$ is estimated to be approximately 0.6 V more negative in DMF. This relative destabilization of the reduced form can be attributed to the better electron-pair donor characteristics of DMF relative to H_2O. (Their respective Gutman donor numbers (13) are 26.6 and 18.) As a result, the enthalpies of cation solvation are generally more negative in DMF than in H_2O (13). Conversely, DMF is less good as a H-bonding solvent, so that the negative charge residing on the ligand as a result of reduction or

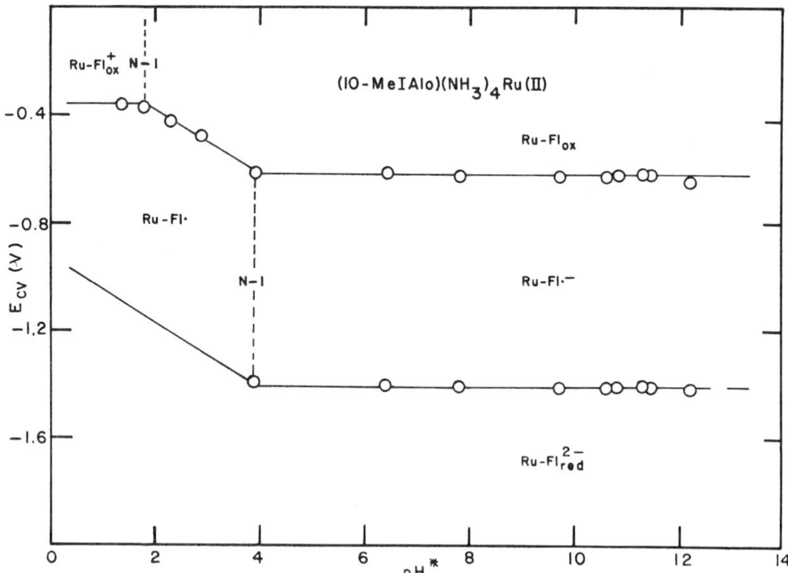

Figure 3. E_{cv} vs pH* plot for (10-MeIAlo) $(NH_3)_4Ru(II)$ in DMF.

deprotonation is less well stabilized by proton interactions. Thus, the more cationic forms of the complex ion are stabilized and the anionic ligand forms destabilized in going from water to DMF, so that the complexes become more difficult to reduce.

Comparison with the reduction potentials for free flavins in DMF reveals that the metal ion exerts a substantial influence on the flavin redox chemistry. While the addition of the first electron to Ru-Fl_{ox} occurs at a potential similar to that for tetraacetylriboflavin (Ac_4Rib_{ox}) in DMF (10), the E^0 value for the addition of the second electron is approximately 1.1 V more negative. An analogous situation is seen for Et-Fl_{ox} (5-ethyl-3-methylflavinium) for which the first electron adds at a calculated potential of 0.42 V and the second at approximately -0.9 V (1). These results clearly indicate that it is the destabilization of the lowest flavin oxidation state following N^5-coordination that causes the two electrons to be passed in sequence rather than simultaneously.

The first-electron reduction potential for Ru-Fl_{ox} is 0.78 V more negative than for Et-Fl^+ in DMF (1). This shift is in the opposite direction from what might be expected on changing from Et^+ to a dipositive metal ion. The difference can be attributed to the fact that Ru(II) is also a good Lewis π-base and donates considerable electron density into the same orbital which accepts the additional electron in the redox process. On the other hand, coordination of this dipositive ion does make it more difficult to add a proton at a site on the ligand which receives little of the added electron density, especially in a relatively poor proton donor solvent. For these reasons, the fully-reduced flavin is not stabilized by proton addition as occurs with the free ligand. This results in sequential electron transfer with no pH dependence over a broad range and the occurrence of Ru-Fl_{red} only at quite negative potentials.

While the Ru-Fl system is clearly nonbiological, it does represent a stable, tractable system for probing unusual electronic effects in flavin coenzymes. Moreover, the ^{3}-donor and proton-controlled effects illustrated here are among those that can be easily generated by proteins, albeit in different fashions.

ACKNOWLEDGMENT
This work was supported by PHS grant GM-26390.

References

1. Nanni, E.J., Sawyer, D.T., Ball, S.S., and Bruice, T.C. (1981) *J Am Chem Soc* 103:2797–2802.
2. Stankovich, M.T. (1980) *Analytical Biochem* 109:295–308.
3. Hemmerich, P. (1977) In *Bioinorganic Chemistry-II*. Raymond, K.N. (ed.) Washington, D.C.: American Chemical Society, pp. 312–329.
4. Dryhurst, G. (1977) *Electrochemistry of Biological Molecules*. New York: Academic Press, pp. 365–391.
5. Hemmerich, P. and Lauterwein, J. (1973) In *Inorganic Biochemistry*. Eichhorn, G.L. (ed.), Vol. 2, New York: Elsevier, pp. 1168–1190.
6. Taube, H. (1973) *Survey of Progress in Chemistry* 6:1.
7. Clarke, M.J., Dowling, M.G., Garafalo, A.R., and Brennan, T.F. (1980) *J Biol Chem* 225: 3472–3481.
8. Clarke, M.J., Dowling, M.G., Garafalo, A.R., and Brennan, T.F. (1979) *J Am Chem Soc* 101:223–225.
9. Clarke, M.J. and Dowling, M.G. (1981) *Inorg Chem* 20, in press.
10. Favaudon, V. and Lhoste, J.M. (1975) *Biochemistry* 14:4731–4738.
11. Favaudon, V. (1977) *Biochemistry* 16:293–307.
12. "Fluorad FC-24", Technical Bulletin, 3M Co., St. Paul, Minnesota, and references therein.
13. Burgess, J. (1978) *Metal Ions in Solution*. New York: Wiley, pp. 32, 202, 206.
14. Breant, M. and Demange-Guerin, G. (1969) *Bull Chim Soc Fr* 8:2935–2945.
15. Sawyer, D.T. and Roberts, J.L. (1974) *Experimental Electrochemistry for Chemists*. New York: Wiley.
16. Gagne, R.R., Koval, C.A., and Lisensky, G.C. (1980) *Inorg Chem* 19:2854–2855.

Published 1982 by Elsevier North Holland, Inc.
Vincent Massey and Charles H. Williams, Editors
Flavins and Flavoproteins

CHAPTER 97

The Electronic Structure of Flavins by Ab Initio Molecular Orbital Calculations and X-PES

M.H. Palmer,* J.R. Wheeler,* R.J. Platenkamp,** and A.J.W.G. Visser†

**University of Edinburgh, Scotland; **Leiden University, The Netherlands; †Agricultural University, Wageningen, The Netherlands*

Introduction

Recent results of ab initio molecular orbital calculations on isoalloxazines and related compounds have put the quantum mechanical aspect of flavin reactivity on a more rigorous footing (1,2). The most pressing experimental need is to obtain reliable experimental information concerning the total electron density for the flavin ring nuclei.

The best estimates that can be obtained are from X-ray photoelectron spectra (X-PES or ESCA). For any nucleus with localized core electrons (e.g., 1s of C, N, O or 2s, 2p of P, S) the binding energy of these electrons (E_i) can be found by X-PES (3). For organic molecules, the absence of core electrons for H means that not all the electron densities can be observed; furthermore, the line width of the X-ray (0.4~0.8 eV depending on monochromation or not) means that not all the separate binding energies will be observed. Experimentally the relationship between the electron binding energy ($E_i = -IP_i$), the irradiation energy ($h\nu$), kinetic energy (K.E.), electron densities (q_i, q_j, etc.) and internuclear distances r_{ij} are given in Equations (97.1) and (97.2).

$$K.E_i = h\nu - E_i \tag{97.1}$$

$$E_i - E_o = kq_i + \sum_j q_j/r_{ij} \tag{97.2}$$

In Equation (97.2) E_o represents some reference level, e.g., C_{1s} in methane or other alkanes (290.7 eV), q_i is the appropriate nuclear charge on center i, with other charges q_j at distances r_{ij}. The proportionality constant k has different values for different nuclei C, N, O, etc. In practice, owing to the inability to observe sufficient E_i, the experiment is normally interpreted by means of recourse to electronic structure calculation, and seeing whether the latter will reproduce the experimental envelope to the X-PES. Unfortunately this leads to differing values for k for different computational methods.

Experimental Methods and Deconvolution

X-PES of the flavins **1**a, b, c and of the lumazine derivatives **2**, **3**a, **3**b were obtained using either an AEI ES200B (Kratos) or an ESCA-3 (Vacuum Generators), using unmonochromated AlK_α radiation (1486.6 eV) or MgK_α (1253.6 eV). It was not possible to cast conducting films of the samples from solution and hence minimize sample charging, of critical importance to monochromatic usage. Repetitive scanning and averaging was done. The effective X-ray line width was about 0.8 eV (FWHH). The spectra for O, N, and C 1s levels are shown in Figures 1–3. Data reduction before curve fitting was carried out in two stages. A weighted average procedure was used to smooth the data (4); base-line correction to remove the absorption edge phenomenon was carried by subtraction of the proportion of area from the low binding side to total area (4). Deconvolution of the envelopes was then carried out using a set of Gaussian lines. Each core level IP was assigned the same peak shape and area, so that the N_{1s} region was thus a 1 : 1 : 1 : 1 quartet with variable spacing in the general case of the present systems. The principal results are shown in Table 1.

Discussion

All of the compounds showed broad N_{1s} peaks with FWHH in the range 2.0 to 3.5 eV; no completely resolved subsidiary maxima were observed for any of the 4 N-atoms present. The sharpest spectrum was produced by 1H-lumazine (**2**), which is known to be intermolecularly H-bonded across the 3-NH and 2-CO

Figure 1. Nitrogen is core level (X-PES spectra).

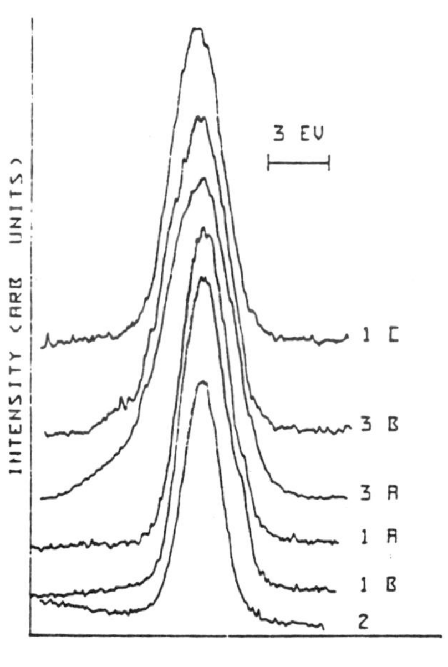

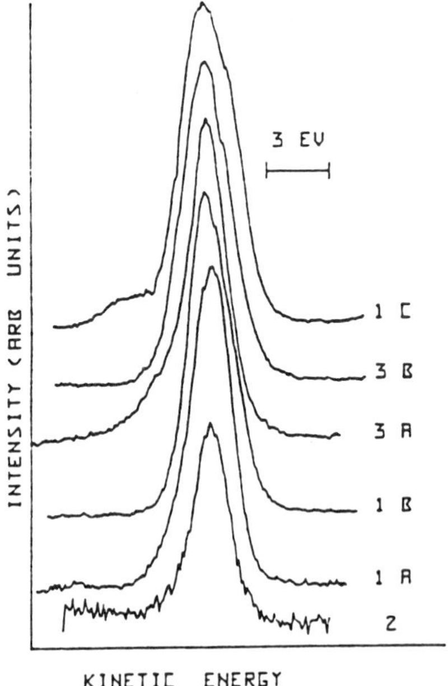

Figure 2. Oxygen is core level (X-PES spectra).

groups in the solid state (5). Thus the N-atoms fall into two classes: pyrrolic and pyridine-like. Previous workers have attributed an increased binding energy to -NH- relative to -N=groups (6). The deconvolution suggests a 1:2:1 arrangement with separations of about 0.5 and 1.0 eV. The Madelung potential calculations (Table 2), for an isolated molecule, indicate two pairs separated by about 3 eV, and corresponding to pyridine-like (higher binding) and pyrrolic (lower binding). The H-bonding on N-3 will convert the center to a more pyridine-like one, and a shift to lower binding energy. This suggests the assignment (high to low binding) N_5, N_8+N_3, N_1.

In the 8-substituted compounds (3a, b) the N_{1s} peak is broadened to 3.6 ± 0.2 eV, and markedly asymmetric with a clear shoulder on the low KE (high binding) side. Deconvolution of both these spectra suggests that two levels are nearly degenerate ($\Delta E\sim0.4$ eV) with neighbors about 1.5 eV and 1.0 eV to low and high KE respectively. The calculations of 8-methyllumazine (3a) yield the Madelung potentials (Table 2) from the Mulliken populations. The low KE band is attributed to N-5 on this basis, with N-1/N-8 nearly degenerate.

Table 1. Deconvoluted Kinetic Energy Maxima From Selected Spectra.

	Nitrogen (eV)				Oxygen (eV)	
1-H lumazine	1084.90	1085.40	1085.40	1086.30	953.84	954.60
8-Methyl lumazine	1083.33	1084.93	1085.51	1086.66	952.96	953.39
8-Ribityl lumazine	1083.36	1084.72	1085.36	1086.38	—	—
Riboflavin	1085.24	1085.47	1086.68	1086.99	—	—
Lumiflavin	1085.26	1085.60	1086.28	1087.44	952.54	953.55

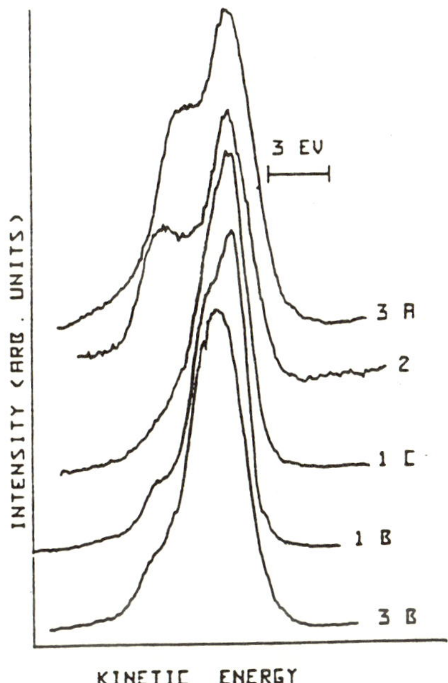

Figure 3. Carbon is core level (X-PES spectra).

The close similarity between the N_{1s} spectra for 8-methyl- and 8-ribityl-lumazines is an indication that intra- and intermolecular H-bonding between the carbohydrate group and the rings is unimportant. This conclusion is less certain for the corresponding flavins (1a, b). In flavins with a 3-NH and 2-CO, intermolecular H-bonding occurs with the corresponding units of adjacent molecules (7) as in lumazine (5); 3-NMe compounds do not H-bond (8). We thus anticipate that both riboflavin and lumiflavin will be H-bonded across the 3-NH/2-CO groups, and that a high binding energy shift of N-3 will occur relative to the free molecule. For the latter, 10H- and 10Me-isoalloxazines (1a) are models, and there the Madelung potentials suggest the binding energy order $N_5 > N_1 > N_{10} > N_3$ in the NH compound, and $N_5 > N_{10} > N_1 > N_3$ in the NMe compound. Thus replacement of NH by NMe leads to an increase in binding energy of the latter. This arises from the effect of Me being a weaker *donor* than H, not to an overall Me electron withdrawing effect (cf 9). Both flavins (1a, b) show a shoulder on the low binding side of the N_{1s} peak, suggesting a 2 : 1 : 1 grouping. In the light of the previous discussion, it seems likely that N_1 and N_3 are reversed in position relative to the Madelung potential values shown in Table 2.

The Madelung potentials for 1H-alloxazine and reduced alloxazine (Table 2) show the same effects as previously, namely pyridine-like N to higher binding energy, and show all 4 N-atoms of the latter molecule nearly degenerate. However, in the crystalline state, 5-H/4-CO H-bonding is known to occur in reduced alloxazines (10) and will compete with 3-H/4-CO H-bonding. Thus we could anticipate a splitting of the N_{1s} levels with N_5 and N_3 to higher

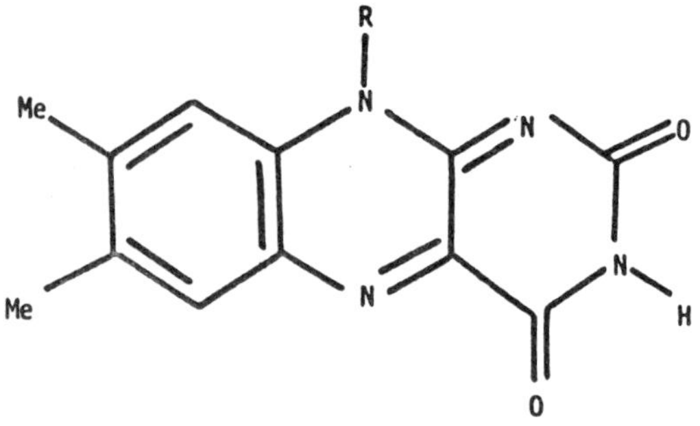

1 — 1, R
(a) $-CH_3$, (LUMIFLAVIN)
(b) $-CH_2(CHOH)_3CH_2OH$ (RIBOFLAVIN)
(c) $-CH_2(CHOH)_3CH_2OPO_3H$ Flavin (FMN) mono-nucleotide

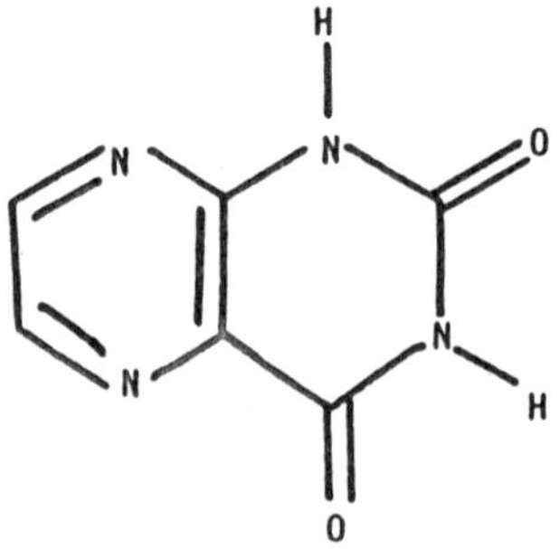

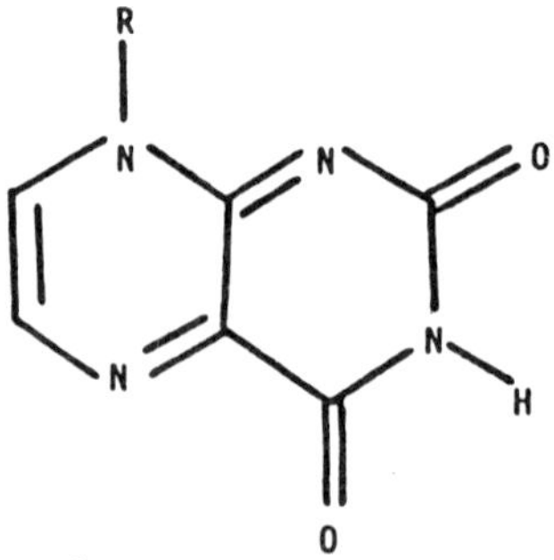

2 — 1H-LUMAZINE

3
(a) R=Me
(b) $R=-CH_2(CHOH)_3CH_2OH$

Structure 1.

binding energy. The O_{1s} peaks clearly differentiate between the carbonyl carbohydrate and phosphate oxygen; comparison of lumiflavin and riboflavin, with the high KE shoulder on the latter, suggests that the CO groups are less strongly bound than the carbohydrate hydroxylic portion.

The C_{1s} spectra of the flavins clearly show the CO groups to high binding energy, as predicted by the Madelung potentials. In the case of 1H- (2) and 8Me-lumazines (3a) the improved proportions of CO to other C atoms lead to nearly separated IPs, and in the former case a ratio 2 : 1 : 3 (high to low binding) seems possible.

In all of these calculated potentials (Table 2), the spread of values is too large, and only the relative order is important. It is also important to note that

Table 2. Relative Binding Energy from Ab Initio Calculations.

	N_1	N_3	N_5	$N_8/_{10}$	O_{12}	O_{14}
Uracil	−6.69	−6.61	—	—	−8.42	−8.05
1-H lumazine	−6.43	−6.36	−2.89	−3.82	−8.41	−7.57
1-H alloxazine	−6.47	−6.39	−2.87	−3.99	−8.49	−7.63
8-Methyl lumazine	−5.33	−6.55	−2.76	−4.04	−8.27	−7.93
10-H isoalloxazine	−5.05	−6.53	−2.53	−5.82	−8.05	−7.80
10-Me isoalloxazine	−5.21	−6.53	−2.57	−4.11	−8.14	−7.83
Reduced alloxazine	−6.29	−6.36	−6.51	−6.39	−8.76	−8.41

Nitrogen data from Tables 1 and 2 are brought to same reference level by $E_0 = 1083.2$ eV.

while Equation (97.2) allows for the variation in 1-center charges with method of calculation through k (see [11] for the present values), this is not the case with the 2-center portion, whose value is fixed irrespective of the method of calculation. In more accurate work, a second constant for this term may be necessary.

Conclusions

A great deal of information remains to be extracted from the core level spectra of flavins. The accuracy of the fitting procedure is greatly diminished by the low resolution (high linewidth) of the X-ray source; the use of a narrow source, e.g., a synchrotron, would allow more complex and interesting molecules to be studied, e.g., FMN.

ACKNOWLEDGMENTS

Thanks are due to Drs. D. Shuttleworth and P. Swift (Shell Research, Thornton) and Dr. H. Bishop (A.E.R.E., Harwell) for running the spectra, and to the Agricultural University and NATO for research grants.

References

1. Palmer, M.H., Simpson, I., and Platenkamp, R. J. (1980) *J Mol Struct* 66:243.
2. Platenkamp, R.J., Palmer, M.H., and Visser, A.J.W.G. (1980) *ibid.* 67:45.
3. Siegbahn, K. et al. (1971) *ESCA Applied to Free Molecules*. Amsterdam: Elsevier North-Holland.
4. Shirley, D.A. (1972) *Phys Rev B* 5:4709.
5. Norrestam, R., Stensland, B., and Söderberg, E. (1972) *Acta Cryst* B28:659.
6. (a) Clark, D.T., Peeling, J., and Colling, L. (1976) *Biochem Biophys Acta* 453:533; (b) Peeling, J., Hruska, F.E., and McIntyre, N.S. (1978) *Can J Chem* 56:1555.
7. Wang, M. and Fritchie, C.J. (1973) *Acta Cryst* B29:2040.
8. von Glehn, M. and Norrestam, R. (1972) *Acta Chem Scand*, 26:1490; Norrestam, R. and Stensland, B. (1972) *Acta Cryst* B28:440.
9. Peeling, J., Hruska, F.E., McKinnon, D.M., Chauhan, M.S., and McIntyre, N.S. (1978) *Can J Chem* 56:2405.
10. Kierkegaard, P., Norrestam, R., Werner, P-E., Csöregh, I., von Glehn, M., Karlsson, R., Leijonmark, M., Rönnquist, O., Stensland, B., Tillberg, O., and Torbjörnsson, L. (1971) *Flavins and Flavoproteins*. London: Butterworths, p. 1.
11. Snyder, L.C. (1971) *J Chem Phys* 55:95.

PART IX:

Flavoproteins and Redox Systems

PART IX A:

Acyl Coenzyme Dehydrogenases

Published 1982 by Elsevier North Holland, Inc.
Vincent Massey and Charles H. Williams, Editors
Flavins and Flavoproteins

CHAPTER 98

Coenzyme A Persulphide, the Tightly-Bound Ligand in the Green Form of Butyryl CoA Dehydrogenase

Gary Williamson,* Paul C. Engel,* John P. Mizzer,** Colin Thorpe,** and Vincent Massey†

**Department of Biochemistry, University of Sheffield, England; **Department of Chemistry, University of Delaware, Maryland; †Department of Biological Chemistry, University of Michigan, U.S.A.*

Native butyryl CoA dehydrogenase (BCD) from various sources is green owing to a broad absorption band centered at 710 nm (1–3). After copper involvement (1) was ruled out (2), an interaction between the FAD and a cysteine residue was proposed (4). The greenness of bacterial BCD (*M. elsdenii*) was variable, however. Addition of various acyl CoA compounds to yellow BCD [formed by reoxidizing dithionite-reduced green BCD (2,3)] produced different long-wavelength bands (5,6). Amino acid analysis showed that green BCD contained CoA in some form, and an unstable CoA-containing "re-greening factor" was released by trichloroacetic acid (5,6). Engel and Massey (5) concluded that an unknown acyl CoA species gave rise to the 710 nm band by charge-transfer interaction with FAD. It remained difficult, however, to explain irreversible de-greening by dithionite (2,3), the effects of some sulphydryl reagents (3,7) and the instability of the released "re-greening factor" (6). Attempts to identify the putative acyl group by GLC/MS were unsuccessful. Recent work has implicated S^{2-} in the "re-greening" of yellow BCD. The experiments described here suggest that the true "re-greening factor" is a persulphide derivative of Coenzyme A.

Pig Kidney GAD: Effect of S^{2-} and CoASH

Pig kidney general acyl CoA dehydrogenase (PKGAD) is yellow when isolated (8), but, like bacterial BCD (5), forms a complex with acetoacetyl CoA with a broad absorption band at 580 nm (C. Thorpe, unpublished results). Incubation of PKGAD with Na_2S and CoASH slowly gave rise to a 710 nm band. Preincubation of the CoASH with Na_2S led to much faster greening. Reaction between these two species could conceivably involve oxidation of S^{2-} to an S_0 species and subsequent production of a CoA persulphide. Significantly, elemental sulphur, S_0, dissolved in Na_2S gave rapid greening when added to PKGAD and CoASH. With PKGAD, the 710 nm band declines with time, and the maximum $A_{710}:A_{430}$ ratio was fairly low (0.22). Further experiments were therefore performed on bacterial BCD.

Bacterial BCD: Full Greening by CoASH with S_0

The yellow form of BCD (Figure 1) is also "greened" by CoASH and Na_2S, in this case to a maximum $A_{710}:A_{430}$ ratio of 0.53, very close to the value of 0.54 predicted from anaerobic titrations of partially green BCD (3). The green complex was stable, and neither dialysis nor storage in 75% $(NH_4)_2SO_4$ significantly decreased $A_{710}:A_{430}$. Addition of Na_2S without CoASH to yellow BCD also resulted in some greening but gave low $A_{710}:A_{430}$ values (≤ 0.3). Amino acid analysis showed this was due to residual CoA in the yellow BCD. Addition of exogenous CoASH produced fuller greening.

As with PKGAD, greening of bacterial BCD by CoASH and Na_2S is slow (30–60 min). Rapid greening (within 5 min) has been observed in three cases (Table 1): (a) addition of a pre-incubated mixture of CoASH and Na_2S; (b) addition of a solution of S_0 in Na_2S to BCD solution already containing added CoASH; (c) addition of colloidal sulphur to BCD solution already containing added CoASH. In the absence of enzyme, incubation of CoASH with Na_2S, in either phosphate, pH 7 or 0.2 N NaOH, gradually produced an absorption band at 335–340 nm. This was destroyed by acid, and is consistent with formation of a persulphide (10).

Incorporation of ^{35}S and ^{14}C-CoA

$^{35}S^{2-}$ and ^{14}C-labelled CoA were incorporated into yellow BCD in order to determine the stoichiometry of greening. Parallel incubations contained yellow BCD with either unlabelled CoASH and $^{35}S^{2-}$ or [adenine-8-^{14}C]-CoASH and unlabelled S^{2-}. Unbound material was removed by gel filtration followed by dialysis. Results showed the following: (a) close correlation between increasing

Figure 1. Green and yellow forms of BCD.

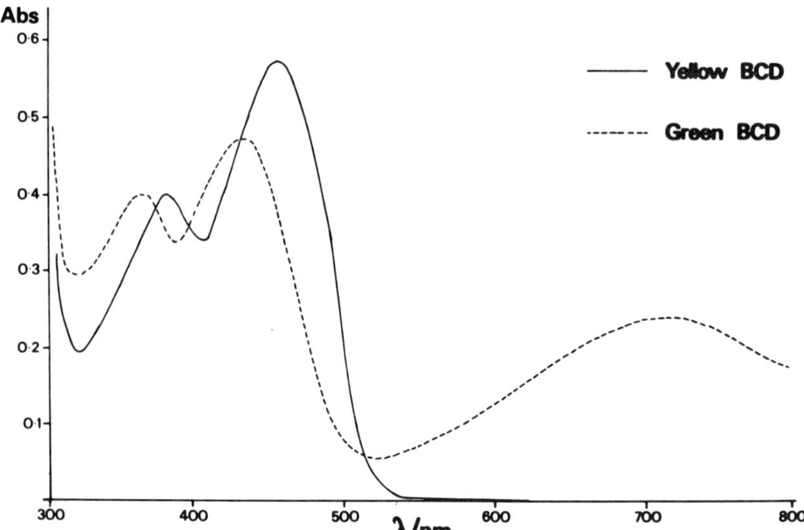

Table 1. Greening of BCD by Various Sulphur Species.

Molar ratio of CoASH to BCD flavin	Sulphur species	A_{710}:A_{430} after time shown	Time (min)
5	S^{2-}	0.43[a]	30
5	S^{2-} after 15 min preincubation with CoASH	0.43[a]	2
1	S^{2-}/S_0 mixture	0.48	2
3	S_0 (colloidal)	0.39	5

[a]In the first two incubations shown, with a 40-fold molar excess of S^{2-} over BCD flavin in both cases, further greening occurred after more prolonged incubation.

greenness and incorporation of ^{14}C-CoASH (Figure 2), although the stoichiometry was complicated both by the presence of residual CoASH in the yellow BCD and by some nonspecific binding, (b) poor correlation between *initial* ^{35}S incorporation and greening (Figure 3). In all cases, 1–2 mol ^{35}S per mol bound FAD was initially incorporated, for a range of A_{710}:A_{430} of 0.23–0.50. Samples were then either kept at 4°C for 10 days in 0.1 M phosphate, pH 7, or treated with 5 mM KCN at 4°C for 48 hours. Both treatments decreased the bound ^{35}S *without any significant de-greening*, and resulted in samples containing 1.12–1.26 mol residual ^{35}S per mol fully green enzyme (Table 2). Gel filtration of the KCN-treated sample on a calibrated Sephadex G-25 column indicated release of both $^{35}S^{2-}$ and $CN^{35}S^{-}$. These results suggest that 1 mol sulphur is required to "green" 1 mol enzyme-bound FAD, and also show substantial nonspecific labelling, some of this probably as cysteine persulphide. Nonspecific incorporation of $^{35}S^{2-}$ was also found with lactate dehydrogenase and bovine serum albumin run as controls.

Persulphide Analysis

Total persulphide in green BCD was analyzed by alkaline denaturation in the presence of 5 mM KCN, followed by assay of thiocyanate (11). Sample 3 in Table 2 showed an initial content of 1.35 mol persulphide/mol FAD, i.e., 72% of total bound ^{35}S, but, after two treatments with KCN (see Table 2), contained only 1.2 mol persulphide/mol FAD, accounting for all remaining ^{35}S.

Displacement of Label

Addition of a 7-fold molar excess of acetoacetyl CoA to ^{35}S-labelled green BCD (A_{710}:A_{430} =0.40, 0.98 mol ^{35}S per mol FAD) produced the expected increase in A_{580} and decrease in A_{710}, but displaced only 35% of the ^{35}S. A parallel experiment with ^{14}C-CoA-labelled enzyme resulted in displacement of 81% of ^{14}C. On the other hand, treatment with dithionite removed 85–90% of ^{35}S label but only 50% of ^{14}C-CoA.

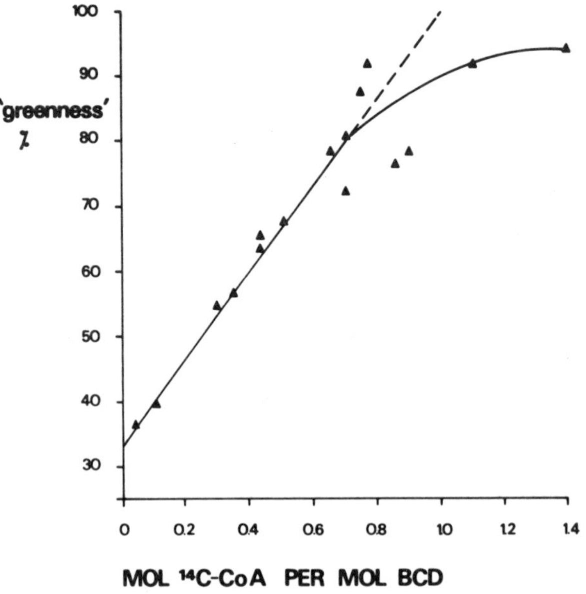

Figure 2. Correlation of "greenness" with bound ^{14}C-CoA.

Denaturation

These results suggest that the incorporated sulphur is not inescapably associated with the CoA. This was followed up by denaturation experiments: (a) with 5M guanidinium chloride at pH 7, (b) with NaOH to give pH 9.5, and (c) with 3% trichloracetic acid. Iodoacetic acid was included to minimize transfer of S between -SH groups. In all cases, elution from a calibrated Sephadex G-25

Figure 3. Correlation of "greenness" and quantity of ^{35}S incorporated.

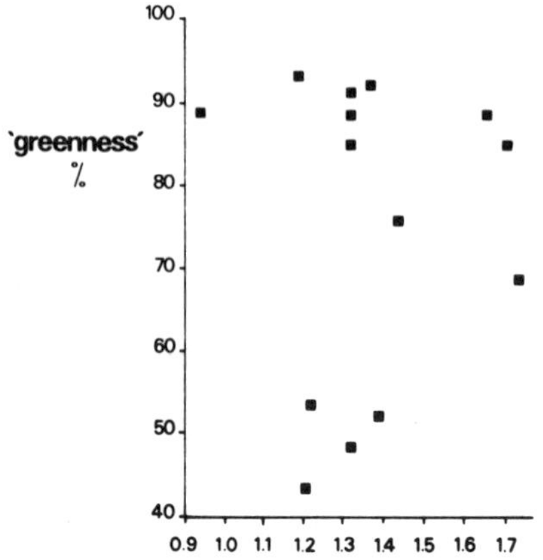

Table 2. Decline in ^{35}S Content of Greened BCD Upon Storage or 5 mM KCN Treatment.

	A_{710}:A_{430}	% Greenness	Mol ^{35}S per mol BCD	Mol ^{35}S per mol BCD extrapolated to A_{710}:A_{430} = 0.54	
Sample No. 1	0.49	91	1.32	1.45	Initial incorporation
	0.47	87	0.97	1.12	After storage
Sample No. 2	0.47	87	1.85	2.13	Initial incorporation
	0.47	87	1.02	1.17	After storage
Sample No. 3	0.53	98	1.87	1.91	Initial incorporation
	0.53	98	1.40	1.43	After first KCN treatment
	0.52	96	1.21	1.26	After second KCN treatment

column gave a protein peak at 14 ml and an FAD peak at 30–35 ml (see Figure 4). Free CoASH and $^{35}S^{2-}$, run separately, peaked at 27 ml. All three methods of denaturation gave 74–78% of loaded ^{14}C-CoA under the CoASH peak and 22–26% under the protein peak (100% recovery). The distribution of ^{35}S varied with conditions of denaturation (Table 3), but two important points are the loss of 50% of the label under acid conditions, and the association of 42% of the counts with the protein peak in guanidinium chloride at pH 7.

Summary

It is probable that the charge-transfer donor in green BCD is CoA-S-S^-. This novel species would be unstable when released under acid conditions, and its properties appear to explain the previously puzzling effects of dithionite, -SH reagents, etc. (2,3,6,7).

Figure 4. Elution profile from Sephadex G-25 of denatured green BCD.

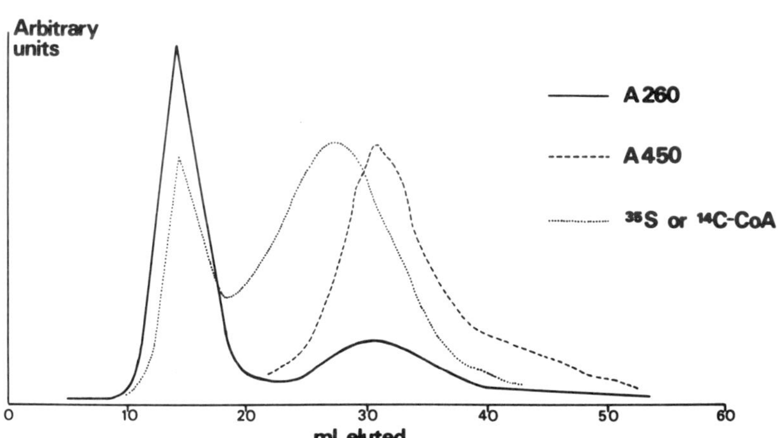

Table 3. Recovery and Distribution of ^{35}S.

	Method of denaturation		
	TCA*	**Guanidinium chloride pH 7**	**NaOH pH 9.5**
Protein peak or* pellet	19%	42%	34%
CoA/S^{2-} peak	81%	58%	66%
Recovery of label	50%	106%	105%

References

1. Mahler, H.R. (1954) *J Biol Chem* 206:13.
2. Steyn-Parvé, E.P. and Beinert, H. (1958) *J Biol Chem* 223:853.
3. Engel, P.C. and Massey, V. (1971) *Biochem J* 125:879.
4. Palmer, G. and Massey, V. (1968) In *Biological Oxidations*. T. P. Singer (ed.) Interscience.
5. Engel, P.C. and Massey, V. (1971) *Biochem J* 125:889.
6. Engel, P.C. (1972) *Z f Naturforsch* 27b:1080.
7. Engel, P.C. and Jones, J.B. (1978) *Biochem J* 171:51.
8. Thorpe, C., Matthews, R.G., and Williams, C.H., Jr. (1979) *Biochem* 18:331.
9. McKean, M.C., Frerman, F.E., and Mielke, D.M. (1979) *J Biol Chem* 254:2730–2735.
10. Rao, G.S. and Gorin G., (1959) *J Org Chem* 24:749.
11. Sörbo, B. (1957) *Biochim Biophys Acta* 23:412.

Published 1982 by Elsevier North Holland, Inc.
Vincent Massey and Charles H. Williams, Editors
Flavins and Flavoproteins

CHAPTER 99

On the Dehydrogenation of Octanoyl CoA by General Acyl CoA Dehydrogenase and Electron Transfer Flavoprotein

Carole L. Hall

School of Chemistry, Georgia Tech, Atlanta, Georgia

The first step of β-oxidation in mammalian mitochondria requires the presence of two different flavoproteins, a substrate-specific dehydrogenase and electron transfer flavoprotein (ETF) (1,2). Recent work has shown that acyl CoA substrates bind rapidly and tightly to general acyl CoA dehydrogenase (G-AD) from pig liver, and rapidly (100 sec^{-1}) reduce the flavin (3). G-AD-octanoyl CoA (C_8CoA) complex rapidly reduces ETF in two one-electron steps, via anionic semiquinone of ETF. The transfer of the first electron occurs with a bimolecular rate constant of $\sim 6 \times 10^6$ M^{-1} sec^{-1}. Transfer of the second electron is an order of magnitude slower and may or may not be concentration dependent (4). Thus there is evidence for a kinetic ternary complex, but the point at which the enoyl CoA dissociates is unclear.

McKean et al. (5) have reported that when the oxidation of butyryl CoA by G-AD and ETF is treated as a bisubstrate reaction, the results are consistent with an ordered sequential mechanism in which reduced ETF dissociates from the ternary complex before (oxidized) enoyl CoA dissociates from the dehydrogenase. However, butyryl CoA is an atypical substrate for this dehydrogenase in that it is the only even-numbered straight chain acyl CoA which does not bind tightly with a 1 : 1 stoichiometry with dehydrogenase flavin, and the K_m is an order of magnitude greater than the others (3,6,7). Earlier attempts to treat the G-AD--C_8CoA-ETF reaction as a bisubstrate reaction led to converging lines in all combinations of fixed and variable components (C.L.H. unpublished, see ref. 2).

ETF has always been thought to function solely in transferring electrons from the dehydrogenases to the acceptor in the electron transport chain or to dyes. However, spectral evidence has been presented that reduced ETF can bind enoyl CoA derivatives accompanied by apparent reoxidation of the flavin, possibly suggesting reduction of the enoyl CoA (3). Thus the role of ETF might be broader and include some interaction with acyl CoA .

ETF which has been reduced by C_8CoA via catalytic amounts of G-AD showed changes in the UV and near UV which could be correlated with changes in the visible absorbance previously (3) shown to reflect oxidation-reduction states of ETF flavin (Figure 1). All of the ETF was converted to the anionic semiquinone (see ref. 3) and fully reduced forms with rate constants

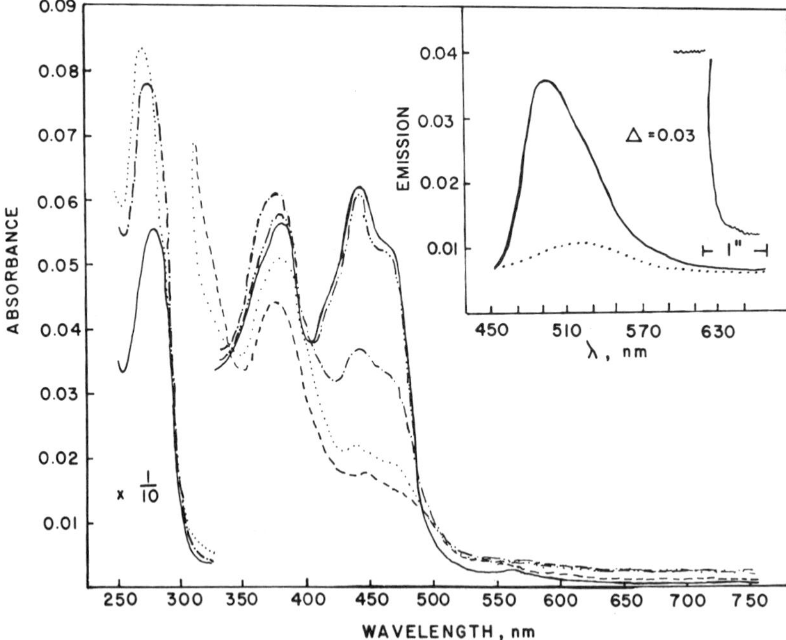

Figure 1. Absorbance and fluorescence changes in ETF during catalysis. ETF (10.5 nmoles flavin) and G-AD (0.12 nmoles flavin) were mixed in 2 ml KPi buffer (20 mM, pH 7.6) and absorbance and fluorescence (inset) spectra were recorded (—). The trace to the right of the emission spectrum (inset) shows the decrease in fluorescence caused by addition of 36 nmoles C_8CoA. (The excitation λ was 380 nm. Slits were 10 nm each for excitation and emission. Emission is in arbitrary units). The emission spectrum was then immediately recorded (· · ·) followed by the absorption spectrum (. · · ·), starting about 4 min after addition of C_8CoA. Spectra started 8 min (– –), 12 min (–·–·), and 16 min (–··–··) after addition of C_8CoA are shown. No further changes occurred up to 12 hr later. The spectrum in the UV shown by (· · ·) represents both the 4 and 8 min traces, while that shown by (–·–·) represents both the 12 and 16 min traces. The increased (and blue-shifted) absorbance in the UV seen at 4 and 8 min is probably due to the reduced ETF since similar blue-shifted hyperchromicity is seen in ETF photoreduced anaerobically via EDTA with or without deazariboflavin (not shown). Virtually identical visible absorbance changes were seen when all components were 10× more concentrated (not shown).

closely approximating those observed previously when both enzymes were present in "equal" amounts. (Rate constants were estimated by assuming the end points of semiquinone formation and t_0 for full reduction as in ref. 4). Absorbance increase at 280 nm, suggestive of appearance of enoyl CoA, appeared slowly, mostly appearing only as 440 nm absorbance (reoxidation) increased. Changes at 280 nm suggested that 75–90% of the substrate was dehydrogenated (Figure 2).

To try to determine if changes at 280 nm, often used for assay of hydratase (8), could be correlated with production and release of octenoyl CoA ($C_8^=$ CoA), experiments were designed for GLC analysis of the acyl CoAs present during the time course of the reoxidation of ETF after reduction by C_8CoA via

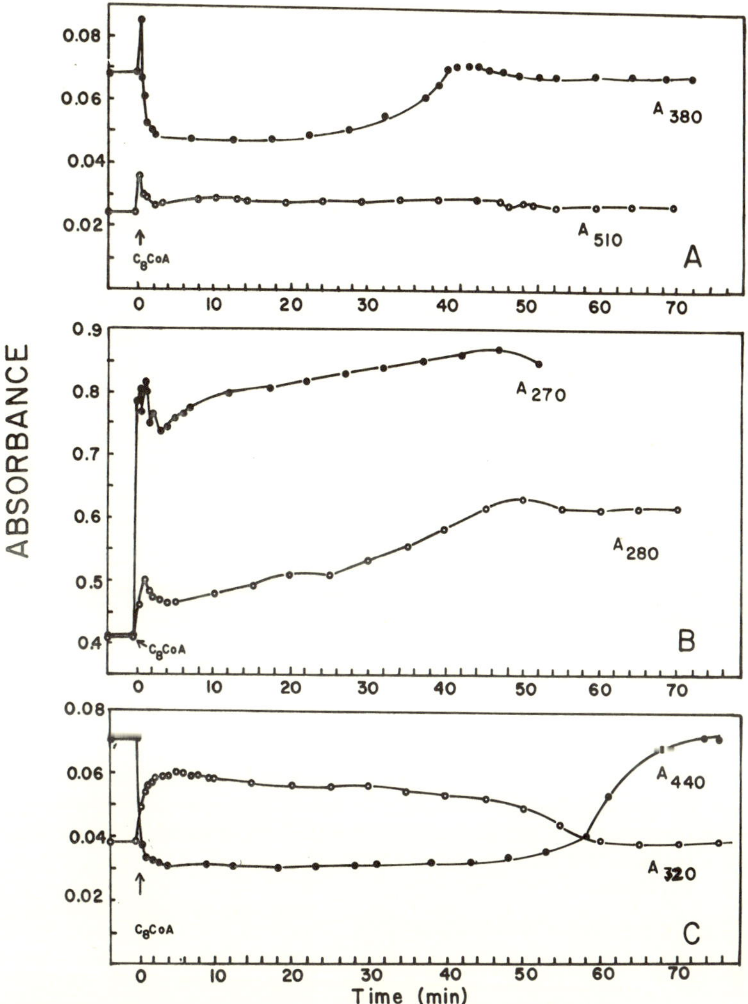

Figure 2. Absorbance changes in ETF during reduction by C_8CoA via G-AD and subsequent reoxidation by air. ETF (5.4 nmoles flavin) and G-AD (0.12 nmoles flavin) were mixed in 1 ml KPi buffer. The absorption at each wavelength prior to addition of C_8CoA is shown to the left of the plots. C_8CoA (36 nmoles) was added at t=0. The absorbance changes were recorded continuously at each wavelength and representative points from each trace were plotted. Lines were drawn to fit the points. (Different chart speeds were used at different times in different experiments. Data at each wavelength were obtained on a fresh preparation).

catalytic G-AD as in Figures 1 and 2. Conditions were similar to those of Figures 1 and 2, but scaled up proportionately several-fold. Reoxidation times were similar, i.e., 40–60 min for full reoxidation when C_8CoA was in 5–10-fold molar excess over ETF flavin. Other experiments have shown that reoxidation time is dependent on concentration of substrate and ETF concentration (not shown).

U.C.W. ABERYSTWYTH LLYFRGELL

Samples were ultrafiltered at different times after addition of C_8CoA to separate "free" from "bound" acyl CoA. The retained protein was resuspended in an equal volume of buffer, acidified and boiled to denature the protein, and centrifuged. The ultrafiltration effluents, denatured protein supernatants, and resuspended denatured protein were treated with 10 N NaOH to cleave the thiolesters, acidified, extracted into n-hexane, and methylated with 10% BF_3 in methanol. The methyl esters were extracted into 1 ml n-hexane for GLC analysis. In some cases, the samples were extracted with hexane prior to thiolester cleavage, analyzed for volatile components, and methylated to analyze for free fatty acids.

Table 1 shows estimates (based on peak areas of esters from known amounts of acyl CoAs) of methyl octanoate (C_8OCH_3) and methyl octenoate ($C_8^= OCH_3$) present in the reaction mixtures at different times after start of the reaction. ETF was mostly reduced (judged by eye) at t=2.6 and t=11 min and remained so through the subsequent work-up, only becoming noticeably yellow when boiled. The ETF appeared partly reoxidized at 31 min (data not shown) and fully reoxidized at 51 min. Other experiments which were followed spectrophotometrically showed similar time courses for reoxidation for the relative concentrations of components used. The table shows that within 1–20 min after addition of C_8CoA perhaps an equimolar amount (to ETF flavin) of $C_8^=$ CoA was associated with the protein but not in the ultrafiltration effluent. The amounts of methyl octanoate recovered from the ultrafiltration effluent decreased with time in the early time points, but never could an amount of methyl octenoate much greater than 1–3 x the ETF flavin be found, even when the amount of C_8CoA was 10 times in excess of ETF flavin (data not shown) and this was not present in the ultrafiltration effluent until the ETF was reoxidized. When C_8CoA was present in only 2-fold excess to ETF flavin, only one-half equivalent of $C_8^= OCH_3$ could be found, but otherwise results were very similar to those reported in the table.

These results suggest that octenoyl CoA and also octanoyl CoA are bound either specifically or nonspecifically to ETF and appear at times soon after addition of C_8CoA in the presence of catalytic G-AD in amounts which seem stoichiometric to the flavin. The results are consistent with the conclusions drawn from the spectral studies in that enoyl CoA free in solution does not seem to appear until full reoxidation has occurred. The results also show that stoichiometric conversion of C_8CoA to $C_8^=$ CoA does not occur under these conditions. In some experiments, where the reoxidized mixture was allowed to stand longer, no $C_8^= OCH_3$ could be found. When authentic $C_8^=$ CoA was mixed with ETF at close to 1 : 1 or 1 : 2 stoichiometries and incubated for 40–60 min, no $C_8^= OCH_3$ could be found, and at higher ratios only about 75% was found. These results suggest there is some destruction of either the saturated or unsaturated acyl CoAs, either concomitant with denaturation or independent of it. Finally, the results suggest that ETF participates in the dehydrogenation of acyl CoA in a way which involves the acyl CoA as well as the electrons. Recent studies with tritium-labelled butyryl and iso-valeryl CoA have shown that 30–60% of the tritium is bound to ETF when only purified enzymes are used, in contrast to experiments using only sonicated mitochondria, where all the tritium was releasable with acid (9).

Table 1. Recovery of Methyl Octanoate and Methyl Octenoate from ETF Reduced by C_8CoA and G-AD.

	t = 2.6 min		t = 11 min		t = 20 min		t = 51 min	
Sample	C_8OCH_3	$C_8^{=}OCH_3$	C_8OCH_3	$C_8^{=}OCH_3$	C_8OCH_3	$C_8^{=}OCH_3$	C_8OCH_3	$C_8^{=}OCH_3$ [a]
PM 10[b]	150	0	14	0	34	0	174	35
DPSH[c]	24	0	0	0	0	0	8	0
DPS[d]	0	0	30	0	14	6	0	0
DP[e]	0	16	0	0	0	0	0	0

[a] nmoles/ml (1 ml original volume, worked up to 1 ml n-hexane).

[b] Ultrafiltration effluent, Amicon PM 10 membrane, 22 mm. Most of the sample passed through the membrane in 1–2 min.

[c] Denatured protein supernatant, extracted prior to thiolester cleavage, methylated.

[d] Denatured protein supernatant, cleaved, extracted, and methylated.

[e] Denatured protein, resuspended in 1 ml buffer, cleaved, extracted, and methylated.

Each experimental tube contained 22 nmoles ETF flavin, 0.1 nmoles G-AD flavin, and 250 nmoles C_8CoA in one ml 20 mM KPi, pH 7.6 (See text).

If enoyl CoA dissociated from the ternary complex concomitant with transfer of the second electron to ETF (to form fully reduced ETF), one equivalent of enoyl CoA should have been found in the ultrafiltration effluent within 1–2 min after addition of substrate in these experiments. This has not been found in several experiments utilizing different amounts and ratios of C_8CoA to ETF flavin, but one-half to 3 equivalents can be found in solution after full reoxidation of ETF.

ACKNOWLEDGMENTS
The author wishes to thank Teresa Detar and Eva Topfl for technical assistance The octenoyl CoA was synthesized by Dr. Robert Conway and was the kind gift of Dr. Robert Waterson. This work was supported by USPHS Grant #Gm25494.

References

1 Beinert, H. (1963) In *The Enzymes* 7. Boyer, P., Lardy, H., and Myrback, K., (eds). N.Y.: Academic Press, pp. 447–476.
2 Hall, C.L. and Kamin, H. (1975) *J Biol Chem* 250:3476–3486.
3. Hall, C.L., Lambeth, J.D., and Kamin, H. (1979) *J Biol Chem* 254:2023–2031.
4. Hall, C.L. and Lambeth, J.D. (1980) *J Biol Chem* 255:3591–3595.
5. MeKean, M.C., Frerman, F.E., and Mielke, D.M. (1979) *J Biol Chem* 254:2730–2735.
6. Hauge, J.G. (1956) *J Am Chem Soc* 78:5266–5272.
7. Thorpe, C., Matthews, R.C., and Williams, C.H., Jr. (1979) *Biochem* 18:331–337.
8. Steinman, H.M. and Hill, R.L. (1977) *Methods in Enzymology* 35:136–151.
9. Rhead, W.J., Hall, C.L., and Tanaka, K. (1981) *J Biol Chem* 256:1616–1624.

Published 1982 by Elsevier North Holland, Inc.
Vincent Massey and Charles H. Williams, Editors
Flavins and Flavoproteins

CHAPTER 100

Reaction of General Acyl-CoA Dehydrogenase with 3,4-Pentadienoyl-CoA

Alexandra Wenz,[†] Sandro Ghisla,[†] and Colin Thorpe*

[†]*Facultät für Biologie der Universität Konstanz, Postfach 5560 D-7750 Konstanz, West Germany;* **Department of Chemistry, University of Delaware, Newark, Delaware*

Recently several thioester inhibitors of acyl-CoA dehydrogenases have been described whose mode of action probably involves an initial removal of a C-2 proton followed by isomerization of the thioester to the active species. Thus general acyl-CoA dehydrogenase is inhibited via irreversible covalent modification of the protein using 3-alkynoyl-CoA derivatives (1), and a similar inhibition has been reported for butyryl-CoA dehydrogenase from *Megasphaera elsdenii* and glutaryl-CoA dehydrogenase from *Pseudomonas fluorescens* using 3-alkynoyl-pantetheine thioesters (2). Methylenecyclopropylacetyl-CoA, a metabolite of hypoglycin, effects irreversible flavin modification on incubation with *M. elsdenii* butyryl-CoA dehydrogenase and pig kidney general acyl-CoA dehydrogenase (3). This paper deals with 3,4-pentadienoyl-CoA, a novel inhibitor of general acyl-CoA dehydrogenase. The mode of action of this allenic thioester is also consistent with an initial proton abstraction.

A titration of pig kidney general acyl-CoA dehydrogenase with 3,4-pentadienoyl-CoA is shown in Figure 1. The reduced spectrum obtained at one equivalent of thioester is strikingly different from that obtained on reduction with octanoyl-CoA (4); it exhibits neither a 350 nm transition nor detectable long wavelength absorbance, but shows a very high extinction at 316 nm (Figure 1). The spectrum resembles N-(5)-acetyl-1,5-dihydroflavins (5,6). Reduction with equimolar inhibitor is rapid and biphasic, with a rate constant of $\geqslant 200\ sec^{-1}$ at 25°C for the fast phase, accounting for >50% of the bleaching at 450 nm and a 4-fold slower second phase.

A slow regeneration of up to ~80% oxidized enzyme occurs on incubation of the reduced enzyme species. This process is independent of O_2 or the effective mediator phenazine methosulfate ($t_{1/2}$ 30 min, 25°; activation energy 19 kcal/mole). Increased concentrations of 3,4-pentadienoyl-CoA decrease the extent of A_{450} nm recovery and lengthen the lag phase preceding this apparent reoxidation, Figure 2. 3,4-pentadienoyl-CoA is not a substrate for general acyl-CoA dehydrogenase in the standard assay system (4), but is a potent inhibitor of the enzyme. Assays performed in the presence of 6 μM compound show progressive inhibition (exhibiting 60% and 11 % of control rates after 15 and 120 sec respectively). In contrast, preincubation of enzyme with inhibitor yields an initially inactive reduced species (see Figure 1), which regains up to 75% activity after 2 minutes exposure to octanoyl-CoA in the assay mixture.

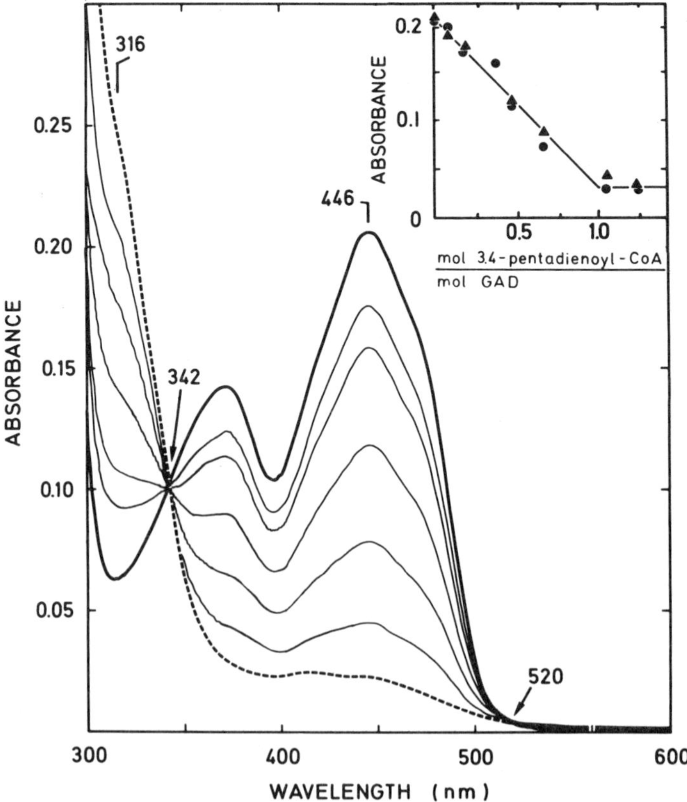

Figure 1. Spectral course of the reaction of general acyl-CoA dehydrogenase with 3,4-pentadienoyl-CoA. The oxidized enzyme, 1.3×10^{-5} M in 0.1 M phosphate buffer, pH 7.6 (curve:—), was titrated at 25° with the inhibitor concentrations shown in the inset, and the intermediate spectra were recorded within 5 min. The arrows denote the isobestic points of the conversion. Curve (---) represents the spectrum of the inactive enzyme after addition of 1.5 equivalents of inhibitor and before noticeable oxidation had occurred. The inset shows the absorbance changes at 450 nm measured under aerobic (●–●), and anaerobic (▲–▲–) conditions.

This work demonstrates that 3,4-pentadienoyl-CoA is a potent, substantially reversible inhibitor of general acyl-CoA dehydrogenase. The spectrum of the reduced enzyme and its resistance to reoxidation suggest that it may represent a covalent adduct. A complex of reduced enzyme and the corresponding cumulene seems much less likely, and would be expected to react with phenazine methosulfate and exhibit a long wavelength band. Adduct formation could reasonably involve attack of a C(2)- or C(4)- carbanion on oxidized flavin (Scheme 1).

Scheme 1 provides for the ready reversal of adduct formation (lower left) by octanoyl-CoA, since this substrate binds tightly to the oxidized enzyme. The apparent reoxidation of the reduced species in Figure 1 and the lag phase in Figure 2 suggest that GAD catalyzes turnover of the inhibitor. Indeed prolonged preincubation of 3,4-pentadienoyl-CoA with GAD yields 2,4-

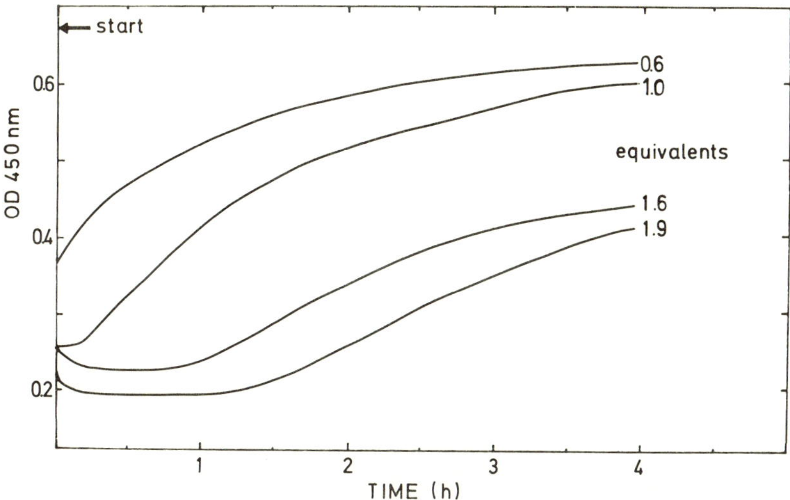

Figure 2. Time dependence of the oxidation state of general acyl-CoA dehydrogenase upon incubation with varying concentrations of 3,4-pentadienoyl-CoA. The enzyme, 4.4×10^{-6} M (0.8 ml) was incubated at 25° with the equivalents of inhibitor indicated. After a very rapid decrease of 450 nm absorption, a slow reoxidation occurs, which leads to a maximal 80% reappearance of the original 450 nm absorbance. This process is unaffected by the presence of oxygen or PMS. The recovery of catalytic activity parallels the spectral changes.

Scheme 1. Proposed mechanism for the reaction between oxidized general acyl-CoA dehydrogenase and 3,4-pentadienoyl-CoA. (B=enzyme active site base, $E{\sim}Fl_{ox}$= oxidized GAD). In the lower left hand structure, the covalent bond could alternatively be to the α-position of the inhibitor.

[~ Fl_{ox}

H_2C=CH-CH=CH-COS-

H_2C=C=CH-CH_2-COS-

~BI

H_2C=CH-CH=CH-COS-

?

~BI

H_2C=C=C-C-COS-

H H

~B-H

H_2C=C-C=C-COS-

H H

~B-H

H_2C=C-C-C-C-S-

H H

pentadienoyl-CoA as evident by the immediate formation of the characteristic spectrum of the reduced enzyme 2,4-pentadienoyl-CoA complex (7) on the addition of dithionite.

ACKNOWLEDGMENTS
This work was supported in part by Grants from D.F.G. (to S.G.) and NIH (GM 26643 to C.T.)

References

1. Frerman, F.E., Miziorko, H.M., and Beckman, J.D. (1980) *J Biol Chem* 255:11192–11198.
2. Gomes, B., Fendrich, G., and Abeles, R.H. (1981) *Biochemistry* 20:1481–1490.
3. Ghisla, S., Wenz, A., and Thorpe, C. (1980) *Enzyme Inhibitors*. Brodbeck, U. (ed.), pp. 43–60.
4. Thorpe, C., Matthews, R.G., and Williams, C.H., Jr. (1979) *Biochemistry* 18:331–337.
5. Ghisla, S. and Massey, V. (1975) *J Biol Chem* 250:577–584.
6. Brüstlein, M. and Hemmerich, P. (1968) *Fed Eur Biochem Soc Lett* 1:335–338.
7. Engel, P.C. and Massey, V. (1971) *Biochem J* 125:889–902.

Published 1982 by Elsevier North Holland, Inc.
Vincent Massey and Charles H. Williams, Editors
Flavins and Flavoproteins

CHAPTER 101

Physical and Chemical Probes of General Acyl CoA Dehydrogenase Structure

Frank Frerman and Vladimir Kushnaryov

The Medical College of Wisconsin, Milwaukee, Wisconsin

As an approach to understanding the reaction between the substrate reduced general acyl CoA dehydrogenase (GAD) and electron transfer flavoprotein (ETF) and the electron transfer reaction, the structure of the GAD was investigated in negatively stained preparations with the electron microscope. Also, the exposure of the flavin in the GAD was estimated by solvent perturbation difference spectroscopy with solvents of different molecular diameters.

Electron Microscopic Studies

The GAD (0.2 mg/ml) in 0.05 M NH_4^+ acetate, pH 7.4, was stained with 1% NH_4^+ phosphomolybdate or phosphotungstate, pH 7.0, and examined in the electron microscope. Both the oxidized and the stable substrate-reduced binary complex were studied. Neither negative stain inhibited the catalytic activity of the dehydrogenase. Examination of both states of the enzyme showed uniform fields of triangular particles. Slightly more detail is observed in magnified images of the substrate reduced enzyme (Figure 1). Table 1 summarizes the results of measurements made on 200 particles of each enzyme form and these show no significant differences in the sizes of the two forms. The mean height: base ratio of the particles, 1.02, indicates that the faces of the particles are isosceles triangles and suggests that the tetrameric enzyme is an irregular tetrahedron. The calculated volume for a protein with a molecular weight of 1.8×10^5 (1) agrees with the volume of the particles calculated from the dimensions given in Table 1. The lack of significant difference between the dimensions of the oxidized and octanoyl CoA reduced binary complex agrees with the results of immunochemical experiments which showed that binding of an acyl CoA ligand results in a limited conformational change (1).

Solvent Perturbation Spectroscopy

The environment of the flavin in the GAD and its complexes with acyl CoAs has been investigated by visible absorption, circular dichroism, resonance Raman, and electron spin resonance spectroscopy (2–4). However, nothing is known about the exposure of the flavin in different states of the enzyme. Such information will be important in understanding the flavin-flavin electron

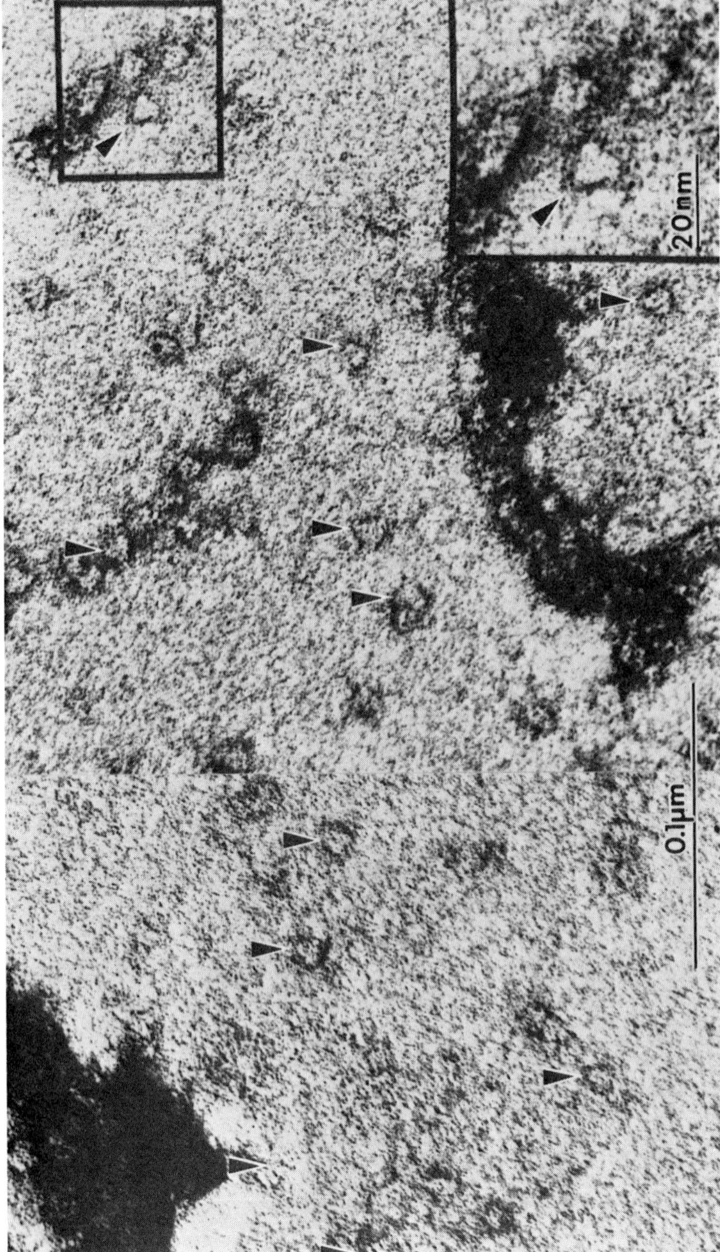

Figure 1. Electron micrographs of the GAD negatively stained with 0.1% sodium phosphotungstate, pH 7.1.

Table 1. Analyses of Negatively Stained Oxidized and Substrate-Reduced Forms of the General Acyl CoA Dehydrogenase.

Parameter	Oxidized	Substrate reduced complex
Base	116.2 ± 0.7 A	107.5 ± 1.4 A
Height	117.8 ± 0.7 A	116.2 ± 1.5 A
Maximum side	128.2 ± 0.6A	128.1 ± 4.1 A
Volume, from micrographs	$2.18 \pm 0.03 \times 10^5$ A^3	—
Volume, calculated	2.19×10^5 A^3	—

Measurements were made on 200 oxidized molecules and 200 reduced molecules.

transfer reaction with ETF. The exposure of the flavins in the free GAD and the GAD:S-octyl CoA complex was determined by solvent perturbation difference spectroscopy (5). FMN was used to determine 100% exposure of the isoalloxazine chromophore to the solvents. FMN was used to avoid possible complications due to stacking of the adenine and isoalloxazine rings in FAD, the prosthetic group of GAD. FAD is probably in the extended conformation in the GAD (2). Egg white riboflavin binding protein (RBP) was used as a flavoprotein model system since much is known about the physical and chemical properties of the flavin site. The data were in qualitative agreement with models of the RBP flavin site (6–7) in which the dimethylbenzene ring of the flavin is buried in the protein and the pyrimidinoid ring is relatively exposed. Figure 2 shows the solvent perturbation spectra of the free GAD using 20% ethylene glycol and 45% ^2H_2O as the perturbants. The exposure of the GAD flavin to these perturbants with molecular diameters of 2.0 A and 4.5 A is about 90%. Other commonly used, "mild" perturbants significantly alter dehydrogenase and ETF structures as determined from far ultraviolet circular dichroism spectra of the proteins in those solvents. The solvent perturbation difference spectra of the GAD and RBP flavins are qualitatively very similar to those of the FMN model. Perturbation of the dehydrogenase flavin in the enzyme saturated with the dead-end inhibitor, S-octyl CoA, with 20% ethylene glycol shows that the exposure of the flavin to ethylene glycol falls to about 40% (Table 2).

This decrease is likely due to restriction of approach to the flavin in the region of N5−C4a (2,3). That S-octyl CoA blocks access to the flavin, and very likely to this region of the flavin, is supported by studies on the chemical reactivity of the GAD flavin. The first-order rate constants for photochemical reduction (8) of the free and S-octyl CoA complexed enzymes to the neutral semiquinones were 0.17 min^{-1} and 0.30 hr^{-1}, respectively, at 4°C. Reoxidation of the semiquinone form of the free enzyme by O_2 was too fast to measure accurately; the first-order rate constant for reoxidation of the semiquinone form of the complex was 0.67 min^{-1} at 25°C. Since reoxidation probably occurs through the C4a position (9), the chemical reactivity data support the hypothesis that the acyl CoA analog blocks access to the flavin by binding at or near the C4a region of the GAD flavin.

In contrast to the relatively high exposure of the flavin in the uncomplexed dehydrogenase, the ETF flavin is only 56% accessible to 45% ^2H_2O.

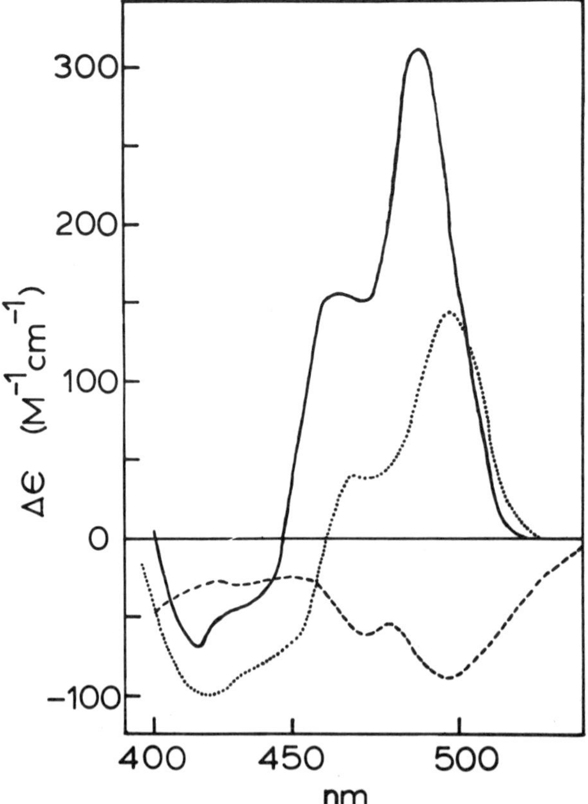

Figure 2. Solvent perturbation difference spectra of GAD. The spectra were determined in 0.1 M potassium phosphate, pH 7.1, at 20°C using tandem cuvettes. The difference spectra are those of free GAD in 20% ethylene glycol (—) or 45% ^2H_2O (---) and the GAD-octyl CoA complex in 20% ethylene glycol (· · ·). The concentrations of GAD flavin ranged from 9×10^{-5} M to 1.5×10^{-4} M in these experiments.

Table 2. Summary of Solvent Perturbation Difference Spectra.

Protein	Perturbant (diameter, A°)	$\Delta \lambda_{max}$ (nm)	$\Delta \varepsilon_{max}$ (M^{-1} cm^{-1})	Exposure relative to FMN (%)
RBP	Ethylene glycol (4.4)	486	164.0	51
	Glycerol (5.4)	488	153.0	51
	Sucrose (9.4)	485	92.1	32
GAD	Ethylene glycol	486	312.0	92
	^2H_2O (2.0)	485	−86.0	87
+octyl CoA	Ethylene glycol	496	143.7	44
ETF	^2H_2O	475	65.7	56

All measurements were made in 0.1 potassium phosphate, pH 7.1.

References

1. Frerman, F.E., Kim, J.J.P., Huhta, K. and McKean, M.C. (1980) *J Biol Chem* 255:2195–2198.
2. Auer, H.E. and Frerman, F.E. (1980) *J Biol Chem* 255:8157–8163.
3. Benecky, M., Li, T.Y., Schmidt, J., Frerman, F.E., Waters, K.L., and McFarland, J.T. (1979) *Biochemistry* 18:3471–3476.
4. McKean, M.C., Sealy, R.C., and Frerman, F.E., this volume.
5. Herskovitz, T.T. and Laskowski, J., Jr. (1962) *J Biol Chem* 237:2481–2492.
6. Choi, J.D. and McCormick, D.B. (1980) *Arch Biochem Biophys* 204:41–51.
7. Blankenhorn, G. (1978) *Eur J Biochem* 82:155–160.
8. Massey, V. and Hemmerich, P. (1977) *J Biol Chem* 252:5612–5614.
9. Bruice, T.C. (1976) In *Progress in Bioorganic Chemistry*. New York: John Wiley and Sons, pp. 1–87.

Published 1982 by Elsevier North Holland, Inc.
Vincent Massey and Charles H. Williams, Editors
Flavins and Flavoproteins

CHAPTER 102

ESR and ENDOR Studies of Sarcosine Dehydrogenase, Acyl CoA Dehydrogenase, and Electron Transfer Flavoprotein

Martha C. McKean, Roger C. Sealy, and Frank E. Frerman

The Medical College of Wisconsin, Milwaukee, Wisconsin

The flavin radicals of sarcosine dehydrogenase, acyl CoA dehydrogenase (GAD), and electron transfer flavoprotein (ETF) were studied by direct ESR techniques using both photoreduction and enzymatic reactions to generate semiquinones. ETF is an acceptor of reducing equivalents from both dehydrogenases and has been implicated as a one-electron acceptor by observations of optical spectra (1,2). This suggests that there may be other radical species generated in the dehydrogenation and electron transfer catalyses of these enzymes. Such a species has been reported for GAD but until now has remained unconfirmed (2,3).

Photochemically Generated Semiquinones

Anaerobic photoreduction of GAD produced an ESR signal of linewidth = 20.5 ± 0.1 G, g = 2.003 (pH 7.6), consistent with the earlier assignments (based on optical spectra) of a neutral semiquinone (Figure 1A) (4,5). The ENDOR spectrum exhibited the 8-methyl resonance at 18 MHz, as expected for a neutral, blue semiquinone (6). The ESR spectrum of photoreduced ETF had a narrower linewidth of 14.5 ± 0.1 G, g = 2.004 (pH 8.1), in keeping with its red anionic semiquinone optical spectrum (Figure 1B). The ENDOR spectra of ETF were typical of anionic flavin semiquinones with the matrix signal at 13.6 MHz and absorption at 19 MHz due to the 8-methyl group of the flavin. When this experiment was carried out in ^2H_2O, there was little decrease in the matrix ENDOR signal relative to the 19 MHz signal, suggesting that the immediate solvent environment, i.e., the matrix, of the radical does not exchange freely with bulk solvent. Lowering the pH to 7.3 did not affect the position of the 8-methyl resonance. If the pK of the anion semiquinone is 8.4, as with free flavin, and/or if there were free access of bulk solvent to the flavin in the holoenzyme, one would expect a change in position of the methyl proton resonance due to protonation of the flavin radical at N-5 to yield the neutral species. The photochemically generated semiquinone of sarcosine dehydrogenase had a linewidth of only 11 G and its ENDOR spectrum had very little

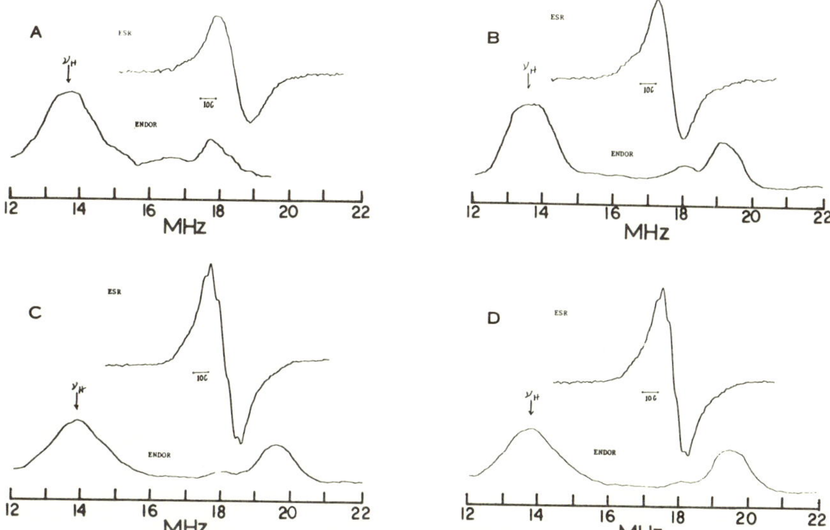

Figure 1. ESR and ENDOR spectra of GAD and ETF. Temperature and microwave power conditions for spectra are given. (A) Photoreduction of GAD: −150°C, 0.06 mW (ESR); −160°C, 0.3 mW (ENDOR). (B) Photoreduction of ETF: −160°C, 0.02 mW (ESR), 0.3 mW (ENDOR). (C) Octanoyl CoA-GAD: −160°C, 0.02 mW (ESR), 0.2 mW (ENDOR). (D) GAD, ETF, octanoyl CoA reaction: −160°C, 0.02 mW (ESR), 1 mW (ENDOR).

absorbance in the 8-methyl region (Figure 4). These findings confirm the 8α-covalent linkage of the flavin (7).

Enzymatically Generated Semiquinones

In marked contrast to the radical produced by photochemical reduction, the ESR spectrum of GAD reduced with octanoyl CoA under anaerobic conditions (pH 7.6), had a linewidth of 14.8±0.7 G, g=2.004 and was similar in linewidth and lineshape to the radical produced from ETF by photochemical reduction, with the exception of a greater degree of resolution of hyperfine structure (Figure 1C). The radical yield of this reaction was found to be 0.8 radicals per dehydrogenase tetramer. The ENDOR spectrum of this semiquinone had the 8-methyl resonance positioned at 19 MHz, characteristic of a red anion semiquinone. This result is an exception to the usual correlation between optical and ESR spectra of neutral and anionic semiquinones of flavoproteins. The optical spectrum of octanoyl CoA-reduced dehydrogenase (anaerobic) exhibits the usual "charge transfer" absorbance band at 565 nm which is also found in blue semiquinones and has no increased absorbance at 370 nm, where anion semiquinones show spectral changes. A parallel circular dichroism (CD) experiment was performed and the resulting spectrum (Figure 2A) differed from CD spectra of both GAD neutral semiquinone and aerobic octanoyl CoA-reduced GAD (Figure 2B) (8). The CD spectrum of the anaerobic

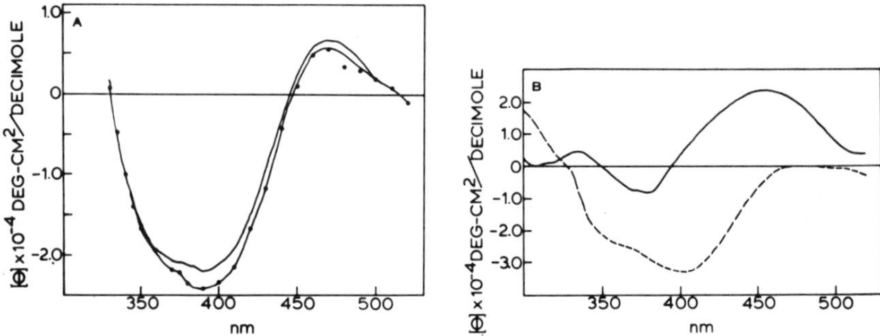

Figure 2. CD spectra of GAD. (A) Octanoyl CoA-reduced GAD, anaerobic (—). Simulated spectrum (-●-●-). (B) Neutral Semiquinone (—). Octanoyl CoA-reduced GAD, aerobic (---).

octanoyl CoA-GAD complex was well-simulated (Figure 2A) by summing spectra of Figure 2B assuming 1 part GAD semiquinone and 3 parts substrate reduced-GAD.

The ESR spectrum of GAD and ETF in the presence of octanoyl CoA was identical to that of the octanoyl CoA-reduced dehydrogenase and experimental conditions were such that ETF should have been reduced to hydroquinone (Figure 1D) (2). The radical yield for this reaction was 1.8 radicals per dehydrogenase holoenzyme. Interaction with ETF apparently increases the capacity of dehydrogenase to form free radicals. The ENDOR spectrum of the two enzyme plus substrate reaction again showed the 8-methyl resonance at 19 MHz, typical of an anion semiquinone. To assess the dependence of radical yield on substrate in this reaction, the octanoyl CoA concentration was varied while GAD and ETF were held constant at 37.5 μM and 84.3 μM respectively (Figures 3 and 4). Radical yield reached a maximum when substrate was equimolar with ETF flavin and did not diminish at substrate concentrations as

Figure 3. Dependence of free radical yield on substrate concentration. Enzyme concentrations were constant. Radical concentrations measured by ESR.

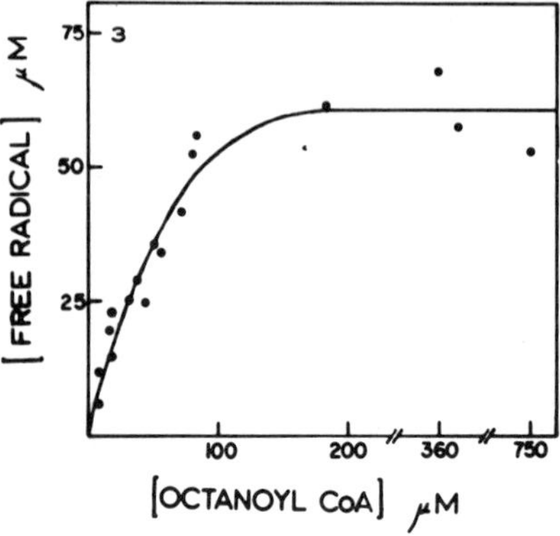

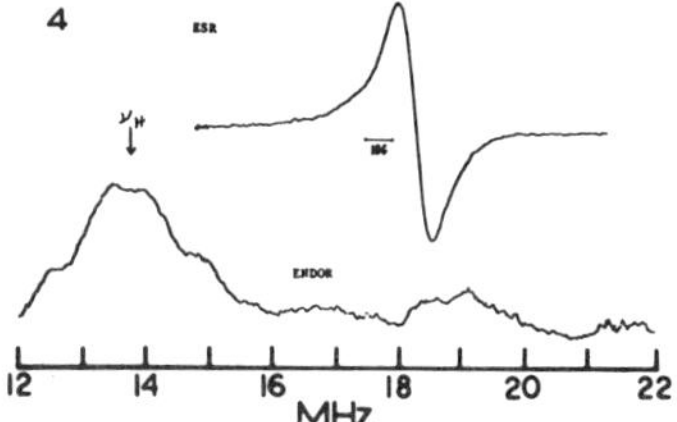

Figure 4. ESR and ENDOR spectra of sarcosine dehydrogenase: −160°C, 0.2 mW.

high as 750 μM. The maximum radical yield was also coincident with a substrate/GAD ratio of 2. The radical yield/GAD of this reaction did not increase when ETF was at 7-fold molar excess over GAD and substrate was saturating. Sarcosine dehydrogenase and sarcosine did not produce an ESR signal. The optical spectrum of this complex is similar to that of fully reduced flavin with no long wavelength absorbance.

Substrate reduced GAD has been shown to form a free radical in significant yield. Furthermore, when ETF is present as an electron acceptor and with excess substrate, the yield of this radical is increased 2-fold. Yield of flavin semiquinone is dependent on substrate concentration in the reaction, octanoyl CoA+GAD+ETF, where ETF concentration is limiting as a terminal electron acceptor. Radical yield is proportional to GAD concentration when ETF is in excess of substrate reducing equivalents.

References

1. Beinert, H. and Frisell, W.R. (1962) *J Biol Chem* 237:2988.
2. Hall, C.L. and Lambeth, J.D. (1980) *J Biol Chem* 255:3591.
3. Beinert, H. and Sands, R.H. (1961) In *Free Radicals in Biological Systems*. Blois, M.S., Brown, H.W., Lemmon, R.M., Lindbloom, R.Q., and Weissbluth, M. (eds.) New York: Academic Press, pp. 17–52.
4. Massey, V. and Palmer, G. (1966) *Biochemistry* 5:3181.
5. Palmer, G., Müller, F., and Massey, V. (1971) In *Flavins and Flavoproteins* (Third International Symposium). Kamin, H. (ed.) Baltimore, Maryland: University Park Press, pp. 123–140.
6. Eriksson, L.E.G., Hyde, J.S., and Ehrenberg, A. (1969) *Biochem Biophys Acta* 192:211.
7. Walker, W.H., Salach, J., Gutman, M., Singer, T.P., Hyde, J.S., and Ehrenberg, A. (1969) *FEBS Letters* 3:237.
8. Auer, H.E. and Frerman, F.E. (1980) *J Biol Chem* 255:8157.

Published 1982 by Elsevier North Holland, Inc.
Vincent Massey and Charles H. Williams, Editors
Flavins and Flavoproteins

CHAPTER 103

Electrostatic Interaction Between General Acyl CoA Dehydrogenase and Electron Transfer Flavoprotein

Joe D. Beckman and Frank E. Frerman

The Medical College of Wisconsin, Milwaukee, Wisconsin

Chemical modification of six carboxyl groups per subunit in the general acyl CoA dehydrogenase (GAD) results in a 10-fold increase in the K_m for the electron transfer flavoprotein (ETF) with little effect on the turnover number of the GAD (1). This suggested that the two proteins may react by complementary charge interaction in an analogous fashion to cytochrome c and cytochrome b_5, for example (2). To support this hypothesis, the effects of ionic strength, pH, and chemical modifications of ETF upon the kinetic constants for the reaction of the GAD with the ETF have been investigated.

Kinetic-Ionic Strength Studies

Figure 1 shows the effect of ionic strength on the K_m of the GAD for ETF and the turnover number. While k_{cat} is relatively unaffected, the K_m^{ETF} increases 120-fold over the range of ionic strength examined. A replot of the data according to the Bronsted equation ($\ln k_{cat}/K_m = \ln k_o + 2.34 \cdot Z_1 Z_2 I^{1/2}$) indicates $-Z_1Z_2 = 9.3$ and $k_o = 2.5 \times 10^9 M^{-1} sec^{-1}$. A second-order rate constant of $8.2 \times 10^9 M^{-1} sec^{-1}$ may be calculated using the approximate molecular radii of 57Å and 28Å for GAD and ETF, respectively. The data are consistent with 3–9 complementary charge interactions between the two proteins which must approach each other in a specific orientation.

Kinetic-pH Studies

Figure 2 shows the effect of pH on K_m and k_{cat} of the ETF reduction. The error in k_{cat} is a consequence of the method of analysis of the decreasing ETF fluorescence single progress curves (3). No significant change in k_{cat} is observed; K_m increases 4-fold as the pK of ε-amino groups is approached. Both GAD and ETF are stable in the pH range investigated.

ETF Chemical Modifications

Table 1 summarizes the results of chemical modification experiments which indicate that obliteration of ETF positively-charged groups dictates loss of

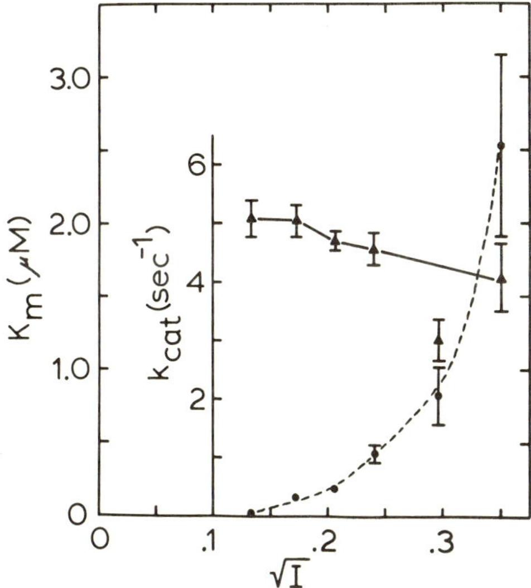

Figure 1. Effect of ionic strength on the K_m of general acyl CoA dehydrogenase for ETF and turnover number of the reaction. Enzyme assays were conducted at 25°C by monitoring the initial rates of ETF flavin fluorescence decrease which result from reduction by dehydrogenase. The Aminco-Bowman ratio spectrophotofluorimeter excitation and emission wavelengths were 340 nm and 495 nm, respectively. Excitation was into an isobestic point for ETF absorbance when reduced in order to avoid an inner filter effect. All assays were in triplicate and contained 225 µM butyryl CoA, 1.11 nM dehydrogenase flavin, 5 mM Tris·HCl pH 8.0, and KCl to yield the desired ionic strength in a total volume of 1.5 ml. ETF flavin was varied from 0.11–1.38 µM, and the reactions were started by addition of dehydrogenase.

enzyme activity by increasing the K_m of dehydrogenase for the ETF. Trinitrophenylation with TNBS and acetylation generate ETF charge isomers of increasing mobility on polyacrylamide gels. Acetylation indicates no change in the ETF flavin absorbance and that 14 amino groups are available to modification. Both ETF subunits are modified by TNBS to a similar extent, thereby suggesting that both subunits may interact with GAD.

The data indicate that specific complementary charge interactions are involved in the reaction of the GAD with ETF. However, no long-lived complex is formed even anaerobically where ETF cannot turn over (4). Previous

---

Figure 2. Effect of pH on the K_m of dehydrogenase for ETF and turnover number of the reaction. k_{cat} is in (sec^{-1}). Assays were conducted at 27°C on an Aminco-Bowman fluorimeter. Each sample contained 20 mM Tris·HCl of the appropriate pH made to 50 mM ionic strength with KCl, 190 µM butyryl CoA, 0.86 nM GAD flavin, 15 units of glucose oxidase and 0.75–3 µM ETF flavin. Assays were made anaerobic by repeated evacuation and purging with scrubbed nitrogen followed by addition of 20 µl of anaerobic 10% glucose. The reactions were started by the addition of anaerobic GAD. The decreasing ETF fluorescence single progress curves were analyzed according to ref. 3.

Table 1. Kinetic and Electrophoretic Properties of the Native ETF, Trinitrophenylated ETF, and Amidinated ETF Derivatives.

ETF derivative	Lysines modified per 61,000 g[a]	ETF K_m[b] ($M \times 10^7$)	Turnover no.[b] (min^{-1})
Native	0	1.13 ± 0.32	109 ± 5
TNP-ETF-1[c]	—	5.60 ± 0.91	142 ± 10
TNP-ETF-2[c]	—	9.84 ± 0.83	143 ± 5
TNP-ETF-3[c]	—	27.3 ± 3.7	197 ± 18
TNP-ETF-4[c]	—	62.8 ± 15.03	182 ± 36
Native	0	1.33 ± 0.25	168 ± 13
Amidinated (40 mM)[d]	11.1	1.86 − 0.39	207 ± 19
Amidinated (140 mM)[d]	13.4	1.83 ± 0.12	202 ± 7
Amidinated (230 mM)[d]	13.9	1.56 ± 0.15	181 ± 9

[a]The number of lysine residues modified was determined by reaction of the native and amidinated forms of ETF with TNBS. The number of lysine residues determined in the native protein was 45.9 ± 4.0 based on the flavin content of the protein.

[b]Kinetic constants were determined at ambient temperature with the DCPIP coupled assay. All points were in triplicate and the data treated with a TI-59 calculator program which provides the weighted constants and their standard errors.

[c]These charge isomers were electrophoretically purified to approximately 80% homogeneity and were contaminated with the next two higher mobility isomers. The major charge isomer electrophoretic mobilities in pH 9.5 nondenaturing polyacrylamide gels were: Native, 0.065; TNP-ETF-1, 0.083; TNP-ETF-2, 0.103; TNP-ETF-3, 0.122; TNP-ETF-4, 0.141.

[d]The numbers in parenthesis refer to the initial concentration of methyl acetimidate in the reaction mixture.

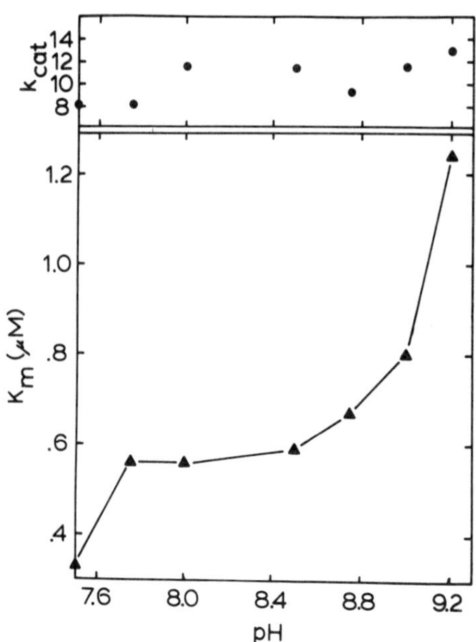

experiments have shown that electron flow may be reversed in the butyryl CoA dehydrogenase-ETF couple (5). This result suggests that reduced ETF could function as product inhibitor of the reoxidation of the dehydrogenase. Indeed, it was found in the kinetic-pH analyses (see legend to Figure 2) that ETF semiquinone acts as a product inhibitor with $K_i \leqslant K_m$. Therefore, the steady-state level of ETF semiquinone can limit flux through the GAD-catalyzed reaction. Since there are at least seven ETF-linked dehydrogenases in liver, the level of ETF semiquinone, as well as the K_m for ETF of a given dehydrogenase, can be controlling factors in the oxidations of fatty acyl CoAs, acyl CoAs derived from amino acid catabolism, sarcosine, and dimethyl glycine.

References

1. Frerman, F.E., Mielke, D., and Huhta, K. (1980) *J Biol Chem* 255:2199–2202.
2. Stonehuerner, J., Williams, J.B., and Millett, F. (1979) *Biochem* 18:5422–5427.
3. Foster, F.J. and Niemann, C. (1953) Proc Natl Acad Sci USA 39:999.
4. Auer, H.E. and Frerman, F.E., unpublished studies.
5. Crane, F.L. and Beinert, H. (1956) *J Biol Chem* 218:717–731.

Published 1982 by Elsevier North Holland, Inc.
Vincent Massey and Charles H. Williams, Editors
Flavins and Flavoproteins

CHAPTER 104

The Reaction of a New Pseudo Substrate, β-2-Furylpropionyl CoA, with "General" Fatty Acyl CoA Dehydrogenase

James T. McFarland, Mei-Young Lee, James Reinsch, and Wanda Raven

Department of Chemistry, University of Wisconsin - Milwaukee, Milwaukee, Wisconsin

We have developed a new reagent for the study of the "general" fatty acyl CoA dehydrogenase; the reagent, furylpropionyl CoA (FPCoA), yields a chromophoric product, Equation (104.1a), furylacryloyl CoA (FACoA), which is identical to the product of the reaction of β-2-furylacryloylchloride and CoASH.

$$\text{Fur-}CH_2\text{-}CH_2\text{-}\overset{O}{\overset{\|}{C}}\text{-SCoA} + \text{E-FAD} \rightleftharpoons \text{Fur-}CH\text{=}CH\text{-}\overset{O}{\overset{\|}{C}}\text{-SCoA} + \text{E-FADH}_2 \qquad (104.1a)$$

$$\text{Fur-}CH_2\text{-}CH_2\text{-}\overset{O}{\overset{\|}{C}}\text{-SCoA} + 2\text{ETF-FAD} \xrightleftharpoons{\text{E-FAD}} 2\text{ETF-FAD}\cdot + \text{Fur-}CH\text{=}CH\text{-}\overset{O}{\overset{\|}{C}}\text{-SCoA} \qquad (104.1b)$$

Steady state kinetic investigation (with the dye acceptor dichloroindophenol) of Equation (104.1b) indicates that the new substrate binds tightly, $K_m = 9.5$ μM, and reacts at a rate $V_{max}/E_o = 13.6\ sec^{-1}$ which is intermediate between that of butyryl CoA and octanoyl CoA. Furthermore, steady state kinetic investigation, observing double bond formation, yields $V/E = 14.7\ sec^{-1}$ while the same kinetic investigation, observing ETF reduction, yields $V/E = 27.7$ sec^{-1} and $K_m = 9.1\ \mu M$. These rates are consistent with the 2:1 stoichiometry shown in Equation (104.1b).

Investigation of the transient kinetic reaction of Equation (104.1a) gives the results shown in Figure 1. Two kinetic processes are observed at 450 nm where flavin reduction in the general fatty acyl CoA dehydrogenase can be observed. Similarly, the formation of FACoA proceeds in two kinetic steps at short time intervals with the addition of a third reaction at longer time intervals. Formation of the charge transfer product complex occurs in two rapid kinetic steps while dissociation of this complex occurs on the time scale of the third kinetic step for FACoA formation. The dissociation of the product charge transfer complex of this new pseudo substrate is complete, unlike that for other substrates where only partial dissociation occurs. Table 1 shows data demonstrating that the third kinetic step in double bond formation is a steady state process linearly increasing with enzyme concentration whereas product charge transfer complex dissociation is first-order, exhibiting approximately the same

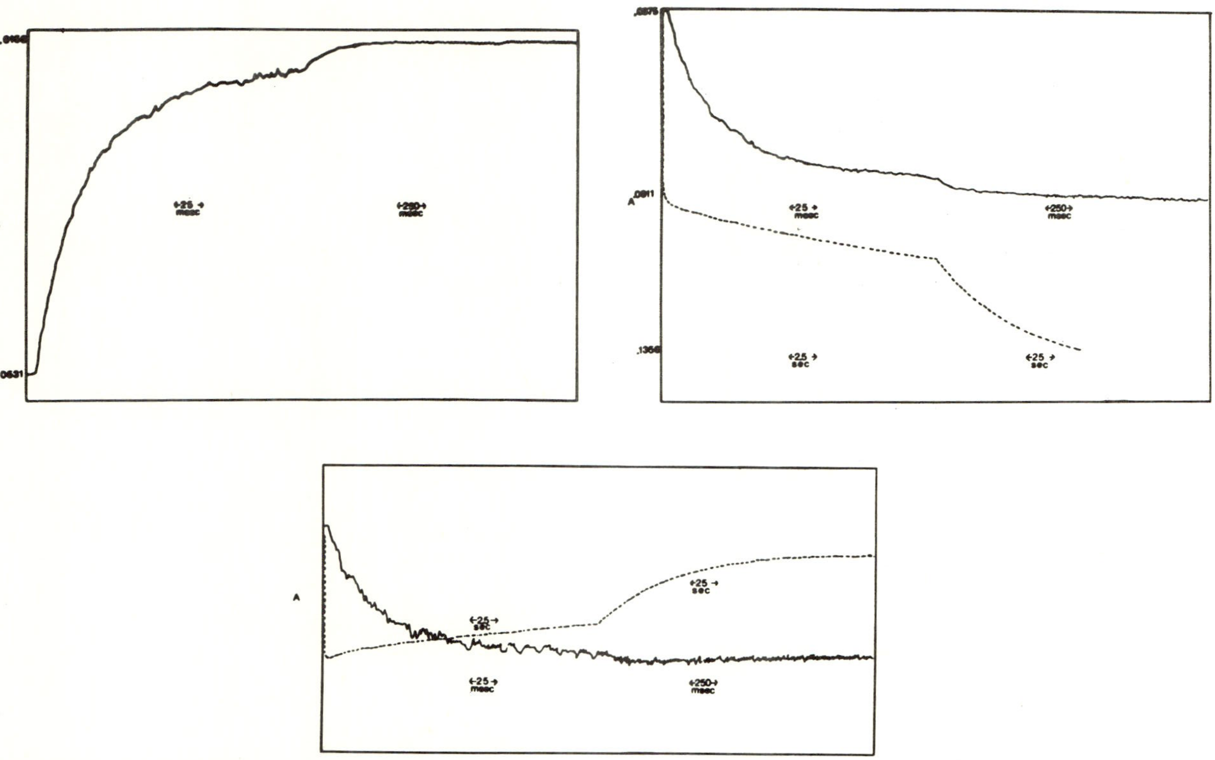

Figure 1. Reaction of FPCoA with fatty acyl CoA dehydrogenase (pH 8.5 20 mM Tris)—upper left: 450 nm; upper right: 340 nm; lower: 640 nm.

Table 1. Kinetic Parameters for FPCoA Reactions.

[S] μM	[E] μM	k_1 650 nm sec^{-1}	V340 nm M/ℓsec	V/E_0 340 nm sec^{-1}
30	7.0	1.3×10^{-2}	8.4×10^{-8}	1.2×10^{-2}
30	3.4	1.1×10^{-2}	4.4×10^{-8}	1.3×10^{-2}
30	1.7	1.0×10^{-2}	2.2×10^{-8}	1.3×10^{-2}
30	.85	—	0.9×10^{-8}	1.1×10^{-2}
5.3	2.8	1.4×10^{-2}	3.3×10^{-8}	1.2×10^{-2}
49.3	2.8	1.2×10^{-2}	3.5×10^{-8}	1.3×10^{-2}

rate constant as the zero-order formation of double bonds. Double bond formation in the slow, third step, is dependent upon the oxygen concentration in solution, decreasing to zero under anaerobic conditions. Figure 2 shows the spectrum of the product of this "oxidase" reaction. The spectrum is identical to that of our synthetic trans-FACoA and the extinction coefficient, $\varepsilon = 22{,}964$ ($\lambda_{max} = 340$ nm), was used to calculate the data shown in Table 1.

Table 2 shows that H_2O_2 is formed during the oxidase reaction; however, we account for only $\simeq 50\%$ of the H_2O_2 which should be formed in the reaction. Addition of superoxide dismutase during reaction raises the amount of per-

Figure 2. The product of the "oxidase" reaction of FPCoA; (a) and (b) are intermediate reaction times.

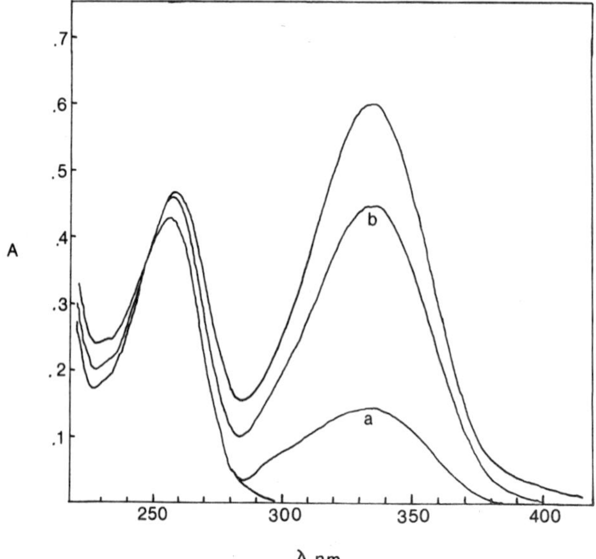

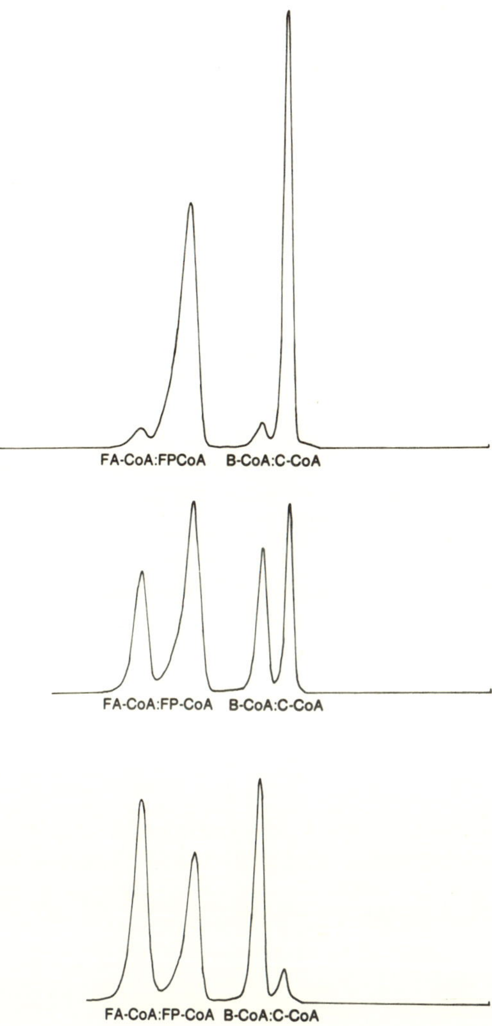

Figure 3. High performance liquid chromatograms during the reaction of FPCoA and crotonyl CoA catalyzed by acyl CoA dehydrogenase. Detection of constituents employs 260 nm absorbance of CoA. (FPCoA) = 116 μM; (crotonyl CoA) = 73 μM; (Enzyme) = 1.0 μM.

oxide formed only slightly, eliminating the possibility that the product of the reaction is primarily superoxide.

At this point, we were interested in determining if another electron acceptor would react with the reduced flavin generated by dissociation of the product C.T. complex. Figure 3 shows the high performance liquid chromatograms at various times during a "transhydrogenation reaction" between FPCoA and crotonyl CoA catalyzed by acyl CoA dehydrogenase. Notice that in the first frame we start out with crotonyl CoA and FPCoA with very little product

Table 2. Peroxide Assay for "Oxidase" Reaction.

[FPCoA] = 29 μM; [Enzyme] = 1.0 μM; pH 8.0; 20 mM Tris 25°C

C = C (total reaction) μM	**C = C (oxidase step) μM**	**H_2O_2 μM**	**% recovery of H_2O_2**
1.00	0.00	<0.26	
2.17	1.17	0.5	43
3.14	2.14	1.0	47
4.04	3.04	1.2	40
5.16	4.16	2.2	53
		2.5^a	60
5.71	4.71	2.0	42

[a] Superoxide dismutase added (SOD) = 5×10^{-2} mg/ml.

formed. At longer times, butyryl CoA and FACoA are formed until at equilibrium there is very little crotonyl CoA; ($\Delta G_0{}' = -1.41$ kcal/mole). These experiments were performed under conditions of excess FPCoA so that the fraction of FPCoA converted to FACoA is smaller than that for crotonyl CoA converted to butyryl CoA. The data suggest that the following mechanism is operating during the reactions of FPCoA:

$$E\begin{matrix}FAD\\FPCoA\end{matrix} \underset{\text{fast}}{\rightleftharpoons} E\begin{matrix}FADH_2\\FACoA\end{matrix} \xrightarrow{\text{slow}} FACoA + E{-}FADH_2$$

$$E{-}FADH_2 \xrightarrow{O_2} H_2O_2 + E{-}FAD$$

$$E{-}FADH_2 \xrightarrow{\text{crotonyl CoA}} CH_3{-}CH_2{-}CH_2{-}\overset{\displaystyle O}{\overset{\|}{C}}{-}SCoA + E{-}FAD$$

In conclusion, we have found that fatty acyl CoA dehydrogenase catalyzes two unusual reactions at the rate at which the charge transfer complex of reduced enzyme and FACoA dissociates. The first of these is an "oxidase" reaction producing H_2O_2 while the second is a transhydrogenase reaction in which crotonyl CoA is reduced to butyryl CoA. One point that these data make in a particularly graphic fashion is the importance of the charge transfer product complex in controlling the β-oxidation pathway. O_2 and crotonyl CoA cannot react with reduced dehydrogenase in the charge transfer complex; indeed, the complex must dissociate to allow this reaction to proceed. On the other hand, the rate of reaction of ETF with FPCoA is 30 sec^{-1}, indicating that ETF reacts directly through the charge transfer product complex to pick up a single electron to form semiquinone. Thus, the charge transfer product complex is primarily responsible for storing electrons for transfer to the electron transfer chain and for prohibiting the electrons from being expended upon O_2 reduction.

Published 1982 by Elsevier North Holland, Inc.
Vincent Massey and Charles H. Williams, Editors
Flavins and Flavoproteins

CHAPTER 105

Resonance Raman Studies on Enzymes Involved in Fatty Acid Oxidation

Jack Schmidt and James T. McFarland

Department of Chemistry, University of Wisconsin-Milwaukee, Milwaukee, Wisconsin

Our current studies continue to focus on the small shifts in the RR spectrum of FAD upon binding to proteins involved in the oxidation of fatty acyl CoA esters. Figures 1 and 2 compare the spectra of FAD in aqueous solution and FAD bound to the active site of the general fatty acyl CoA dehydrogenase and a yeast fatty acyl CoA oxidase.

There are four regions of the spectra which exhibit changes upon binding of FAD to protein. Two bands located at 1250 cm^{-1} and 1280 cm^{-1} in aqueous FAD, shift upon binding to the two proteins. The dehydrogenase shows a single band at 1253 cm^{-1} and the oxidase at 1251 cm^{-1} while FAD shows two bands (a major band at 1256 cm^{-1} and a shoulder at 1280 cm^{-1}). We have been successful in modeling this behavior, in part, using a FMN solution of 2/3 DMSO, 1/3 H_2O, 4 M KI. The FAD RR spectrum in this mixed solvent system shows a greatly reduced intensity in the shoulder at 1280 cm^{-1}, although the main band does not show the frequency shift which occurs on binding FAD to enzyme. The visible spectrum of the mixed solvent flavin changes in a manner consistent with reduced hydrogen bonding between flavin and water (1). These changes consist of a red shift in the 370 nm electronic transition and a loss of vibronic resolution in the 450 nm band upon going from DMSO-water to aqueous solution. Furthermore, the 1250 cm^{-1} band and the 1280 cm^{-1} band show the only large spectral shifts in D_2O indicating that these bands have a δN-H component at the N-3 (D_2O exchangeable) position of the flavin ring. Reduced hydrogen bonding would a priori be expected to shift δN-H bands to lower frequency. On the basis of these arguments, we have come to the conclusion that there is less hydrogen bonding at N-3 between FAD and the two enzymes' active sites than between FAD and water in aqueous solution.

The second spectral region which changes upon complexation of FAD at fatty acyl CoA oxidase is the 1407 cm^{-1} region. Although this band remains unchanged upon complexation to dehydrogenase, it is split into two bands at 1403 cm^{-1} and 1410 cm^{-1} in the RR spectra of the oxidase. The 1407 cm^{-1} band of riboflavin has been assigned to a ring vibration in the uracil ring (III) of the flavin molecule (2, 3). Figure 1 shows the comparison of the RR spectra of oxidase at pH 6.7 and pH 8.0. Notice that it appears as if the band at low frequency increases at pH 8.0 while the higher frequency band decreases. Furthermore, the yeast fatty acyl CoA oxidase stabilizes a red anionic semi-

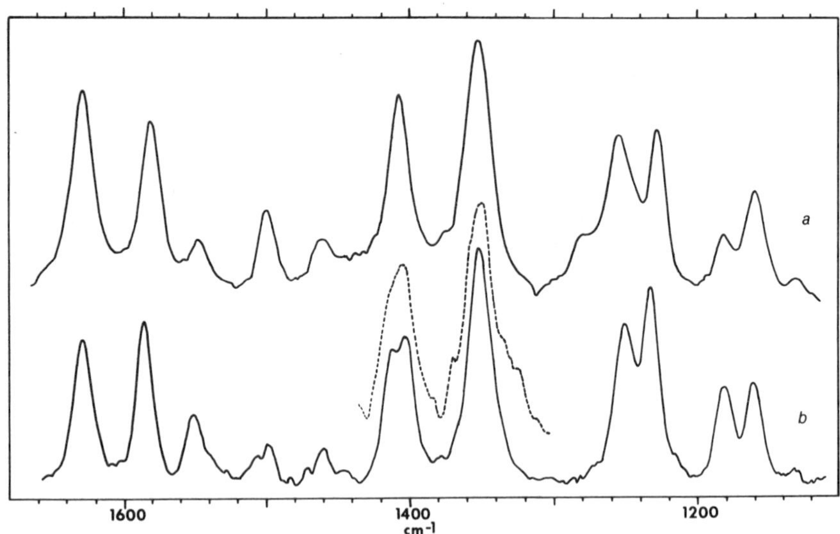

Figure 1. RR spectrum of: (a) FAD in 4 M KI. (b) Fatty acyl CoA oxidase pH 6.7 and pH 8.0 (dashed insert).

quinone at pH 8.0 but not at pH 6.7 (4). It is appealing to suggest that there are two forms of enzyme bound flavin in equilibrium with each other and that lowering the pH perturbs this equilibrium in favor of the form which stabilizes the anionic semiquinone. Accordingly, Massey and coworkers (5) have suggested that there is a specific hydrogen bond formed between the N-1 position

Figure 2. RR spectrum of (a) Fatty acyl CoA dehydrogenase pH 8.5. (b) Fatty acyl CoA dehydrogenase-acetoacetyl CoA complex.

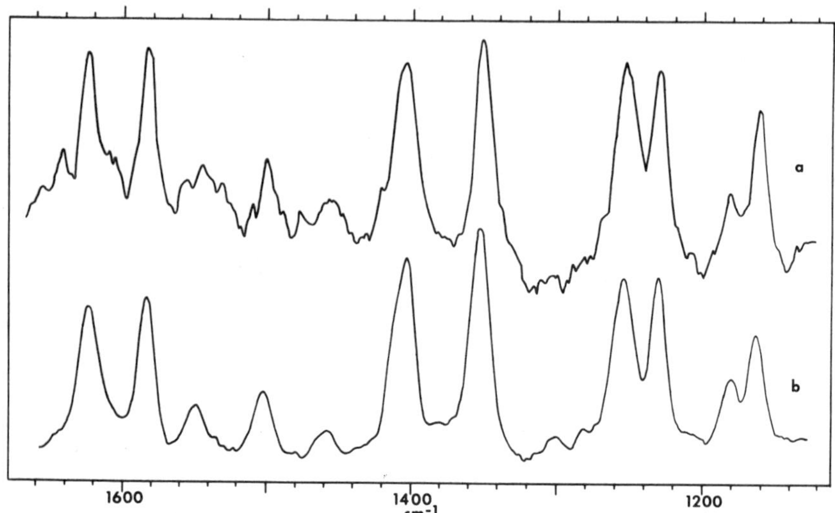

Scheme 1.

of the isoalloxazine ring and a protein residue, and it is this hydrogen bond which stabilizes red anionic semiquinone formation. Our data may reflect the following equilibrium partitioning of hydrogen bonded forms in FAD bound to the oxidase. It should be pointed out that this explanation is a preliminary hypothesis and that we are currently actively involved in testing this hypothesis (Scheme 1).

There are two other spectral regions which show significant shifts. Table 1 shows that FAD bound to dehydrogenase, but not oxidase, undergoes a shift in the 1629 cm^{-1} band to lower frequency in the enzyme complex. This vibration is a ring vibration and is assigned to Ring I of the flavin molecule (2,3). It can also be seen that the 1582 cm^{-1} band of FAD shifts to 1586 cm^{-1} upon binding to oxidase. This band is assigned to $\nu C{=}N$ and shows the largest isotope shift upon substitution of N-14 and C-13 at N-5 and C-4a (2). At

Table 1.

FAD		FATTY ACYL OXIDASE pH 6.7		GENERAL DEHYDROGENASE pH 8.5		GENERAL DEHYDROGENASE: AcAcCoA pH 8.5		PRELIMINARY ASSIGNMENTS
$\Delta\nu$ CM^{-1}	I	$\Delta\nu$ CM^{-1}	I	$\Delta\nu$ CM^{-1}	I	$\Delta\nu$ CM^{-1}	I	
1160	.44	1161	.43	1162	.66	1164	.47	RING III (A)
1182	.24	1182	.40	1182	.32	1181	.27	
1228	.68	1233	.82	1231	.87	1231	.74	RING I (A)
1255	.66	1251	.67	1253	.91	1255	.75	RING III δN-H AT N3 (A,B)
1280±2	.21	--		--		--		RING III δN-H AT N3 (C)
1353	1.0	1351	1.0	1352	1.0	1353	1.0	N5-C4a-C10a-N1-C1 (A,B)
1408	.68	1403 1412	.61 .57	1407	.75	1405±2	.88	RING III (A)
1461±2	.13	1460±2	.14	1454±2	.18	1459±2	.13	
1500	.30	1498±2	.15	1496	.30	1501±2	.25	
1548	.20	1551	.26	1544±4	.16	1550±2	.22	
1582	.74	1586	.67	1583	.49	1585	.70	νC=N RING II (A,B)
1629	.88	1629	.59	1624	.39	1625	.63	RING I (A,B)

[A] DUTTA, P.K., SPENCER, W.C., AND SPIRO, T.G. (1980) BIOCHEM. BIOPHYS. ACTA 623 77-83. [B] KITAGAWA, T., NISHINA, Y., KYOGOKA, Y., YAMANO, T., OHISHI, N., TAKAI-SUZUKI, A., AND YAGI, K. (1979) BIOCHEM. 18, 1804-1808. [C] SCHMIDT, J., THOMPSON, A. W., YU, T.-S, WATTERS, K. L., AND McFARLAND, J. T. (1981) SUBMITTED FOR PUBLICATION.

present, we do not understand the basis for these small shifts in the RR spectrum, but we are continuing to study chemical models in order to interpret these changes.

Previous work in our laboratory has indicated that reaction of butyryl CoA with the general fatty acyl CoA dehydrogenase proceeds through an obligatory intermediate between the C-2 anion of substrate and the flavoprotein (6). Figure 2 shows the RR spectrum of the acetoacyl CoA (AcAcCoA) complex of dehydrogenase; visible spectroscopy of the complex indicates that it too involves the C-2 anion of inhibitor (7). We have employed this complex as a nonreactive analogue of the substrate anion C.T. complex. Figure 3 shows the excitation profile of several of the strongest RR bands shown in Figure 2, the RR spectrum of the AcAcCoA complex. Notice that most of the bands are coupled only to the 450 nm $\pi \rightarrow \pi^*$ transition; however, the 1585 cm^{-1} band (νC=N involving C-4a and N-5) and the unassigned 1540 cm^{-1} band are markedly enhanced at the wavelength of the C.T. electronic transition. Since our kinetic data indicate that electron transfer takes place through an anionic C.T. complex and since both RR excitation profile and charge transfer are largely first electronic excited state phenomena, it follows that the RR excitation profile should yield information pertaining to those parts of the molecule involved in electron transfer. Figure 3 would seem to indicate that C.T.

Figure 3. Excitation profile of fatty acyl CoA dehydrogenase-acetoacyl CoA complex.

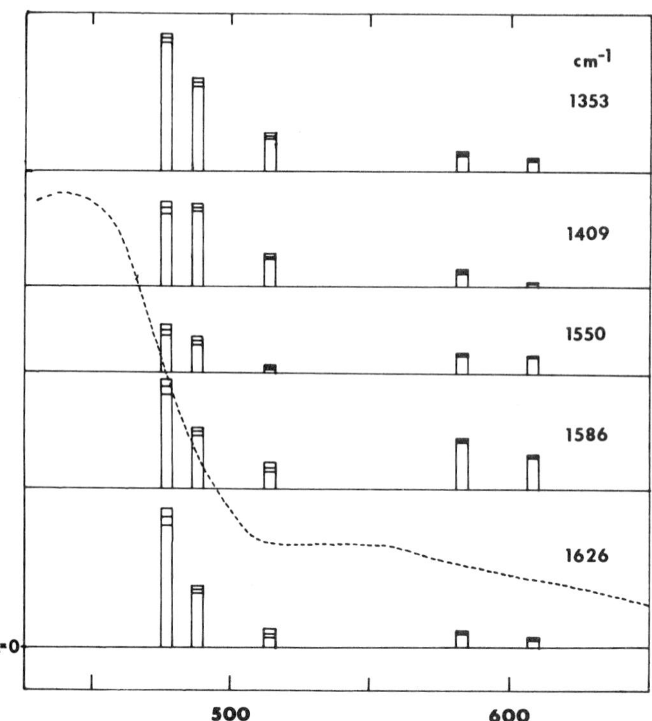

complexation with AcAcCoA involves N-5 and C-4a and on this basis we would suggest that this is a likely site for electron transfer. Molecular orbital calculations indicate that the LUMOs are also located at N-5 and C-4a of oxidized flavin (8) in accord with our RR interpretation.

References

1. Yagi, K., Ohishi, N., Nishimoto, K., Choi, T., and Song, P.-S. (1980) *Biochem* 19:1553–1557.
2. Kitagawa, T., Nishina, Y., Kyogoku, Y., Yamano, T., Ohishi, N., Takai-Suzuki, A., and Yagi, K. (1979) *Biochem* 18:1804–1808.
3. Dutta, P.K., Spencer, R., Walsh, C., and Spiro, T.G. (1980) *Biochem Biophys Acta* 623:77–83.
4. Frerman, F.E., personal communication.
5. Massey, V., Ghisla, S., and Moore, E.G. (1979) *J Biol Chem* 254:9640–9650.
6. Reinsch, J., Katz, A., Aprahamian, G., and McFarland, J.T. (1980) *J Biol Chem* 255:9093–9097.
7. Auer, N. and Frerman, F. (1980) *J Biol Chem* 255:8157–8163.
8. Dixon, D.A., Lindnor, D.L., Branchand, B., and Lipscomb, W.N. (1979) *Biochemistry*, pp. 5770–5775.

Published 1982 by Elsevier North Holland, Inc.
Vincent Massey and Charles H. Williams, Editors
Flavins and Flavoproteins

CHAPTER 106

Spectroelectrochemical Redox Titrations of the General Fatty Acyl-CoA Dehydrogenase Electron Transport Chain Components

Benjamin A. Feinberg and William G. Gustafson

Department of Chemistry, The University of Wisconsin-Milwaukee, Milwaukee, Wisconsin

The mammalian fatty acyl-CoA dehydrogenase electron transport chain consists of the fatty acyl-CoA dehydrogenase, ETF (electron transfer flavoprotein) and ETF dehydrogenase, a flavin-iron-sulfur redox protein. In the initial reactions fatty acyl-CoA binds to the dehydrogenase and then undergoes β-oxidation to form the trans-2,3-enoyl-CoA thioester. Although there has been considerable work done on these components, one aspect of mechanism that is unexplored is the determination of the midpoint reduction potentials of the flavins in each of the components. In considering the general fatty acyl-CoA dehydrogenase, there are four flavins (one/subunit of the tetramer) thus providing eight electron equivalents and eight possible E°s. Each flavin has of course two possible redox potentials: one for the Fl/FlH couple where Fl is the fully oxidized flavin and FlH is the one-electron reduced semiquinone, and the other for the FlH/FlH_2 redox couple, where FlH_2 is the two-electron fully reduced flavin. The ability to distinguish between the various E°s depends upon how large the potential difference is between them and also upon the observable spectroscopy of the flavins and their intermediates.

In this work, we have applied and further developed the methods of spectroelectrochemistry for the study of the redox behavior of the porcine liver general fatty acyl-CoA dehydrogenase electron transport chain components. The spectroelectrochemical cell used in this work is similar to that described and developed by Stankovich (1), but with modifications including a double gold electrode to provide greater surface area and stirring efficiency. In this method, the dye redox mediator, which necessarily has to have its redox potential in the region of the potential of the redox center being investigated ($\pm$100 mV), responds in a Nerstian fashion to the potential applied at the electrode with a potentiostat. The mediator also comes into equilibrium with the redox enzyme, and thus the redox enzyme is brought to the same potential as the applied potential. At the same time the visible absorption spectroscopy is measured for both the mediator and redox enzyme at different applied potentials and/or after different numbers of electrons have been added. Two important advantages of this method are: (1) no chemical titrants are added which might react with the flavins and (2) the high degree of anaerobicity that can be maintained throughout the titration.

The general porcine liver fatty acyl-CoA dehydrogenase was spectroelectrochemically titrated with the mediators indigo disulfonate ($E^{o\prime}_{7.0} = -134$ mV vs S.H.E.) and methylviologen ($E^{o\prime}_{7.0} = -440$ mV). Methylviologen was used to facilitate equilibrium and is spectroscopically undetectable until all other species are reduced. In Figure 1 are shown the spectra of the fatty acyl-CoA dehydrogenase with and without mediators, but all in 15% glycerol. The glycerol was added since it appears to stabilize the enzyme under the experi-

Figure 1. Spectrum of general fatty acyl-CoA dehydrogenase (.022 mM) in 100 mM phosphate, pH 7, 15% glycerol. (2) Same as in (1) but with indigo disulfonate and methylviologen added. (3) After spectroelectrochemical reduction and oxidation.

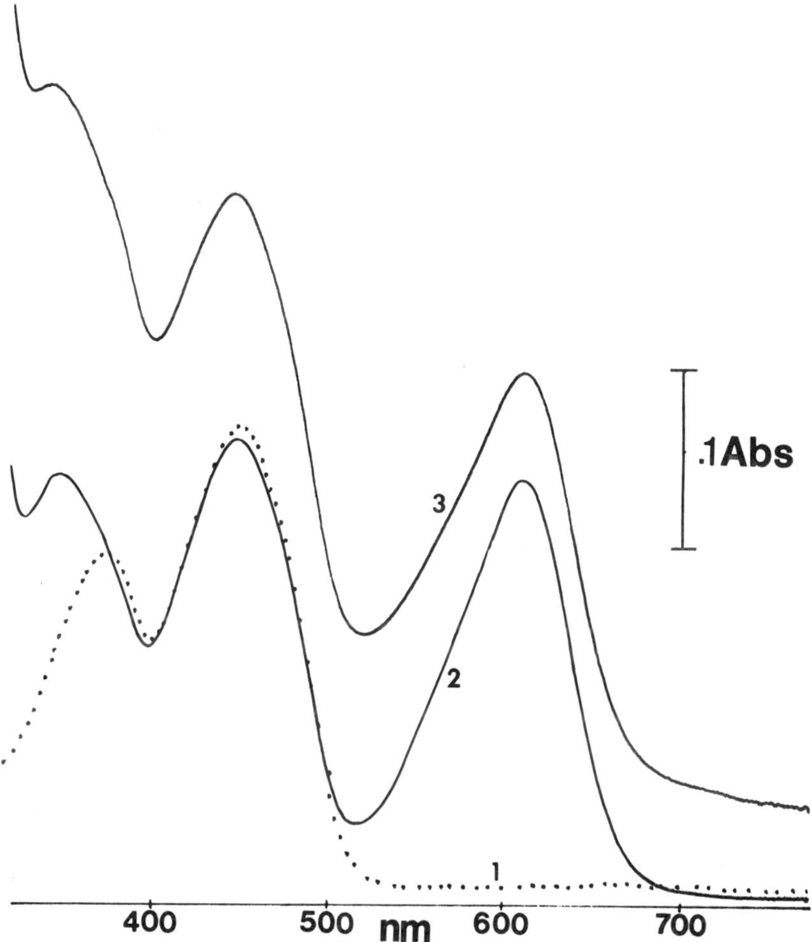

mental conditions used, which include stirring throughout the course of the titration. In Figures 2 and 3 are shown the spectroelectrochemical reduction and oxidation titrations respectively of the fatty acyl-CoA dehydrogenase. When the applied potential (determined either potentiometrically or, as here, by the ratio of IDS_{ox}/IDS_{red}) is plotted versus the spectrophotometrically determined log Fl/FlH_2, the titration curve revealed an $E^{o\prime} = -128$ mV vs S.H.E. The number of electron equivalents for the redox process was n=2 electrons per flavin (with all four flavins being apparently identical). This is a "population" $E^{o\prime}$ and at this time the four flavins can not be differentiated from each other. Although the blue, neutral flavin semiquinone is observed with dithionite titrations, the spectrum of IDS prevents us from clearly seeing this; however, an examination of the 557 nm (semiquinone)/610 nm (IDS) ratio did not indicate its formation. Coulometric titrations of the dehydrogenase with methylviologen are in progress to clarify this issue (Figure 4).

Preliminary results for the titration of ETF with the mediator pyocyanine ($E^{o\prime} = -34$ mV) give an $E^{o\prime}$ in the region of 0 to -25 mV.

Figure 2. Spectroelectrochemical reduction titration of fatty acyl-CoA dehydrogenase under conditions described in Figure 1.

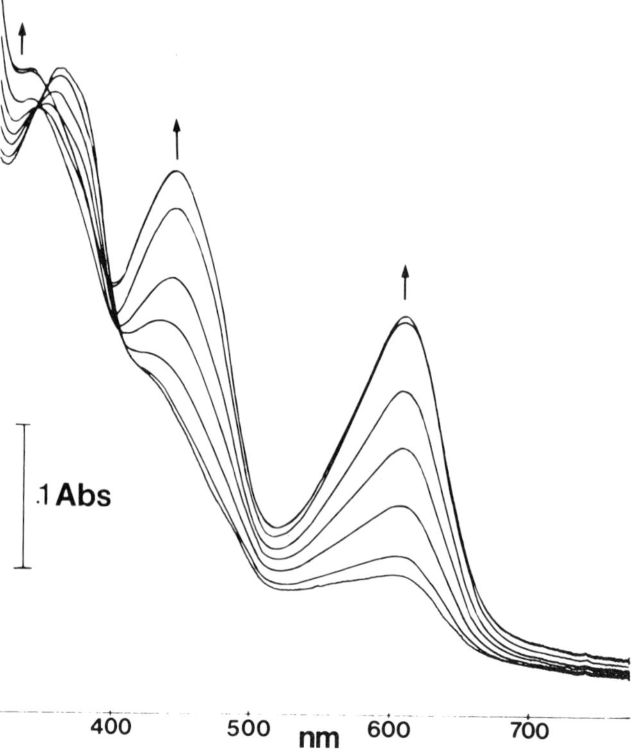

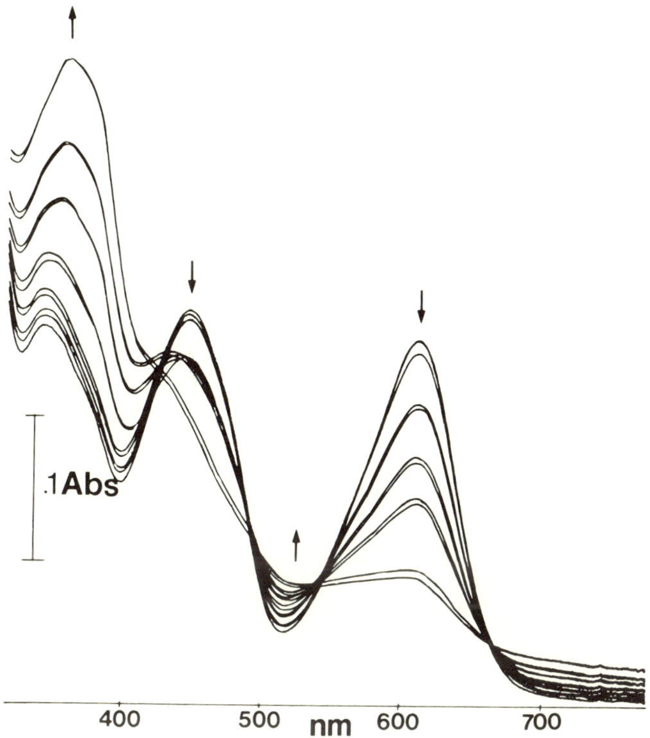

Figure 3. Spectroelectrochemical oxidation titration of fatty acyl-CoA dehydrogenase under conditions described in Figure 1.

Figure 4. Redox titration curve ($-E_{applied}$ vs log Fl/FlH_2) for general fatty acyl-CoA dehydrogenase. Conditions as in Figure 1.

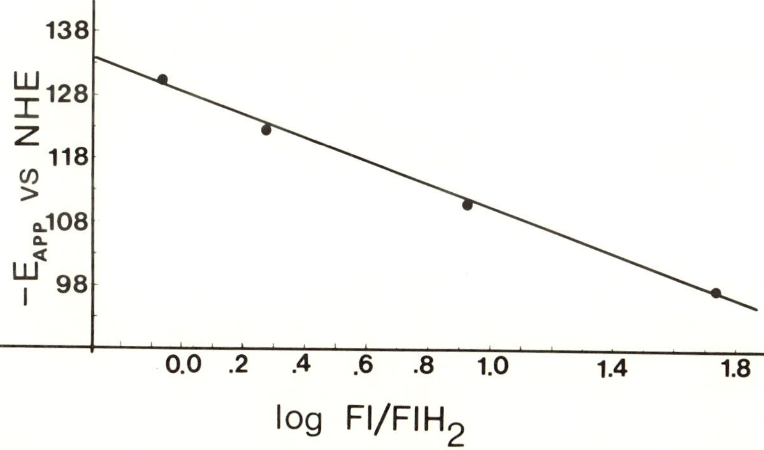

ACKNOWLEDGMENTS
We gratefully acknowledge the National Institutes of Health (GM 25486) and the Competitive Grants Office of the U.S. Department of Agriculture (No. 78-5901-0410-0156-1). We also thank Wanda Raven, George Aprahamian, and Dr. Jing Fong Wei for their excellent technical assistance and Dr. James McFarland for many stimulating discussion of this work.

Reference

1. Stankovich, M.T. (1980) *Anal Biochem* 109:295–308.

Published 1982 by Elsevier North Holland, Inc.
Vincent Massey and Charles H. Williams, Editors
Flavins and Flavoproteins

CHAPTER 107

Stabilization of the Red Semiquinone Form of Pig Kidney General Acyl-CoA Dehydrogenase by Acyl-CoA Derivatives

John P. Mizzer and Colin Thorpe

Chemistry Department, University of Delaware, Newark, Delaware

One aspect of the acyl-CoA dehydrogenases which deserves further study is their interaction with acyl-CoA derivatives. The work reported here shows that a variety of saturated and Δ^2-*trans*-enoyl-CoA derivatives profoundly stabilize the red semiquinone form of general acyl-CoA dehydrogenase (1), although dithionite or photochemical reduction yields the blue radical (1–4) as would be expected for a flavoprotein dehydrogenase (5). This finding may be relevant to the interflavin electron transfer between the dehydrogenase and the physiological oxidant, electron transferring flavoprotein (3), and possibly to the mechanism of substrate dehydrogenation.

Dithionite titration of pig kidney general acyl-CoA dehydrogenase at pH 6.7 yields very similar levels of blue semiquinone to that obtained at pH 7.6 (2). However, reduction in the presence of 190 μM crotonyl-CoA generates a much higher yield of blue radical (Figure 1; $\varepsilon_{560} = 4.9$ vs 2.2 mM^{-1} cm^{-1}). The sizable level of 2e-reduced enzyme generated in the absence of crotonyl-CoA is recycled to oxidized enzyme in its presence, allowing another round of reduction. Thus the stoichiometry of 2.6 electron/FAD for maximal radical formation is significantly greater than unity. Eventually, a high proportion of the enzyme is sequestered in the semiquinone form, no longer capable of reoxidation by product. The stability of the radical spectra shown in Figure 1 underscores the extreme slowness of semiquinone disproportionation in this enzyme (2) and suggests that there is not effective intraflavin redox communication within the tetramer. In marked contrast to the above, titration at pH 8.6 in the presence of crotonyl-CoA reveals the red anion radical as the predominant species, although this form is not evident during reduction of the free enzyme at pH 8.6 or 9.6. The dashed spectrum (Figure 1) exhibits an apparent $\varepsilon = 16.0$ mM^{-1} cm^{-1} at 385 nm, and despite a clear resemblance to other red radical spectra (4), displays an additional long wavelength tail extending beyond 700 nm.

Figure 2 clearly shows that the conversion of the preformed neutral semiquinone crotonyl-CoA complex to the corresponding red anion complex exhibits an apparent pK of 7.3. Crotonyl-CoA binding thus lowers the pK of the blue semiquinone by at least 2.5 pH units implying a strong preferential binding to the red anion form. Indeed crotonyl-CoA binds tightly to the

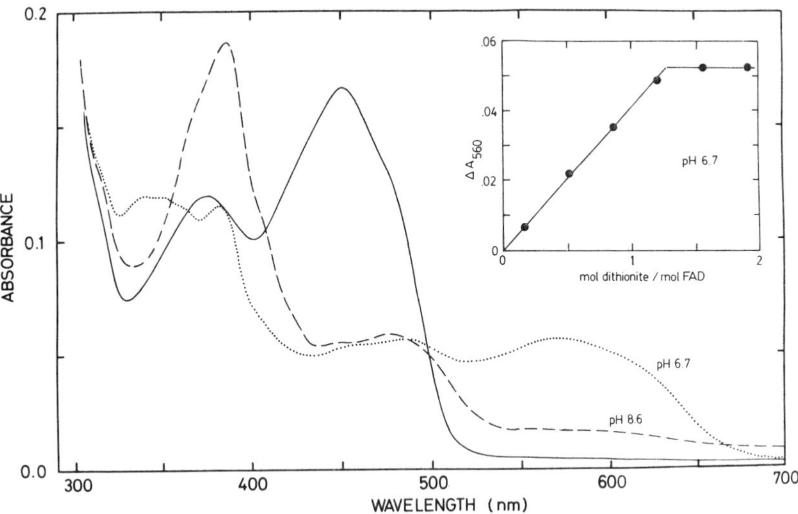

Figure 1. Dithionite reduction of pig kidney general acyl-CoA dehydrogemase at pH 6.7 and 8.6 in the presence of crotonyl-CoA. Spectra: (···), maximal semiquinone formed during dithionite reduction of enzyme in phosphate buffer at pH 6.7 containing 190 μM crotonyl-CoA (also see inset); (---), maximal semiquinone formed in tris buffer at pH 8.6 in the presence of 177 μM crotonyl-CoA; (—), corresponding oxidized enzyme in the presence of 177 μM crotonyl-CoA at pH 8.6.

Figure 2. pH titration of the flavosemiquinone crotonyl-CoA complex. An anaerobic solution of enzyme (11.8 μM in 10 mM phosphate buffer, pH 6.7, containing 190 μM crotonyl-CoA) was titrated to maximal blue semiquinone formation as in Figure 1. The spectrum was recorded (curve 1) and then the pH of the solution was raised by the addition of deoxygenated tris buffer (2 M, pH 9.1). Curves 1–5 correspond to pH values of 6.8, 7.10, 7.34, 7.80, and 8.44 respectively.

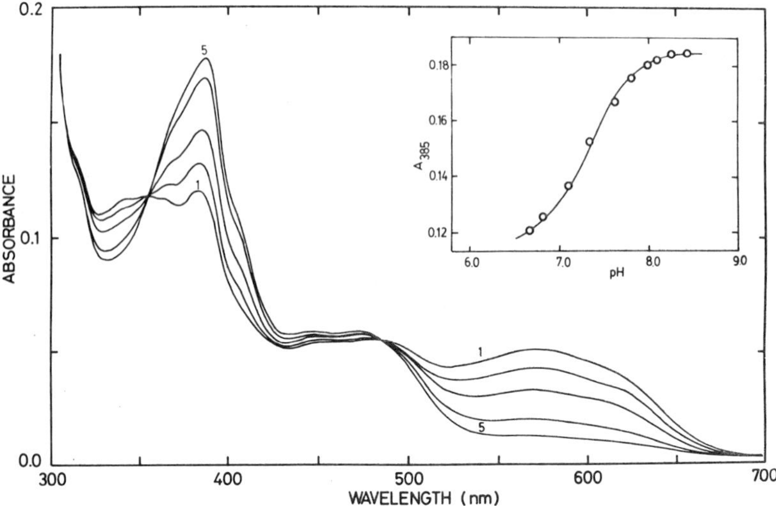

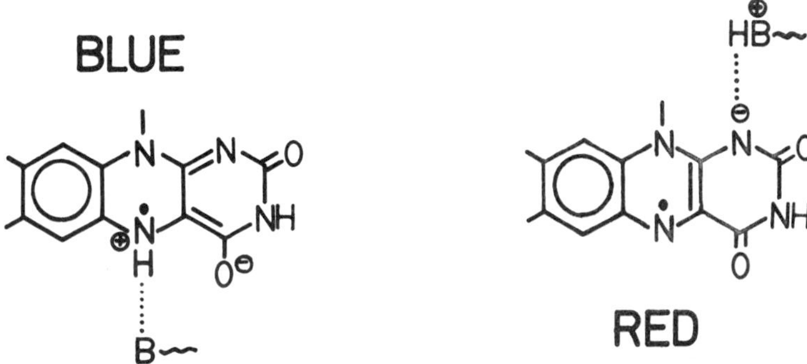

Figure 3. Possible stabilization of blue and red radical species by protein interactions (5).

preformed dehydrogenase semiquinone inducing anion radical formation (1). Octenoyl-CoA similarly stabilizes the anion radical.

The effects described above are not limited to enoyl-CoA derivatives, since butyryl-, octanoyl-, and palmitoyl-CoA, when added to the preformed neutral semiquinone, induce deprotonation. In contrast, octyl- (6), heptadecyl- (7), acetyl-, and acetoacetyl-CoA all fail to stabilize the anion radical. Massey and Hemmerich have suggested that flavin-protein interactions of the type shown in Figure 3 may be responsible for the stabilization of red and blue radical forms (5). However, binding CoA-derivatives per se is not a sufficient condition for red radical stabilization, since thioether-SCoA analogs, acetyl- and acetoacetyl-CoA are exceptions. While it is not yet clear what aspects of the ligand-protein interaction are important in lowering the pK of the blue radical, the ineffectiveness of compounds which are neither substrates nor products suggests that the stabilization of the red radical may be intimately involved in catalysis in the acyl-CoA dehydrogenases. Possibly the strong preferential binding of substrates or products to the anion radical reflects a contribution of this form to the transition state for oxidation of the substrate carbanion. Whether the present example and related observations involving radical anion stabilization in, e.g., glucose oxidase (8), cytochrome b_5 reductase (9), and glutathione reductase (10) have general significance for flavoprotein catalysis remains to be seen. A second speculation is that deprotonation of the acyl-CoA dehydrogenase semiquinone might ensure the efficient delivery of both dehydrogenase electrons to ETF since the anion radical would be expected to be a more powerful reductant than the corresponding neutral species (11).

ACKNOWLEDGMENT
This work was supported in part by NIH GM 26643.

References

1. Mizzer, J.P. and Thorpe, C. (1981) *Biochemistry*.
2. Thorpe, C., Matthews, R.G., and Williams, C.H. Jr. (1979) *Biochemistry* 18:331–337.

3. Beinert, H. (1963) *The Enzymes* 7:447–466.
4. Massey, V. and Palmer, G. (1966) *Biochemistry* 5:3181–3189.
5. Massey, V. and Hemmerich, P. (1980) *Biochem Soc Trans* 8:246–257.
6. Frerman, F.E., Miziorko, H.M., and Beckmann, J.D. (1980) *J Biol Chem* 255:11192–11198.
7. Thorpe, C., Ciardelli, T.L., Stewart, C.J., and Wieland, T. (1981) *EJB*, in press.
8. Massey, V., Palmer, G., Williams, C.H. Jr., Swoboda, B.E.P., and Sands, R.H. (1966) *Flavins and Flavoproteins* 8:133–158.
9. Iyanagi, T. (1977) *Biochemistry* 16:2725–2730.
10. Williams, C.H. Jr. (1976) *The Enzymes* 13:89–173.
11. Stankovich, M.T., Schopfer, L.M., and Massey, V. (1978) *J Biol Chem* 253:4971–4979.

Published 1982 by Elsevier North Holland, Inc.
Vincent Massey and Charles H. Williams, Editors
Flavins and Flavoproteins

CHAPTER 108

Pig Kidney General Acyl-CoA Dehydrogenase: Flavin Analog Studies

Colin Thorpe and Vincent Massey

Chemistry Department, University of Delaware, Newark, Delaware and Department of Biological Chemistry, University of Michigan, Ann Arbor, Michigan

Mammalian acyl-CoA dehydrogenases represent an important class of flavoproteins which catalyze the first step of β-oxidation with the introduction of *trans*-double bond into their thioester substrates (1). The general acyl-CoA dehydrogenase exhibits a broad specificity, and has received the most attention starting with the pioneering work of Beinert and coworkers (1). Despite a resurgence of interest in these enzymes (for example, see other papers in this volume) many aspects of their mechanism require clarification. We have recently developed a satisfactory method for the preparation of pig kidney general acyl-CoA dehydrogenase apoprotein (2) and have reconstituted the enzyme with a variety of flavin analogs including 8-Cl, 5-deaza, and 6-OH-FAD. This report summarizes some of our findings.

Reconstitutions were performed by incubating holoenzyme and the appropriate analog and gel filtration to remove excess flavin. Figure 1 shows the visible spectrum of the 8-Cl FAD enzyme at pH 7.6, 4° C. It is very similar to the spectrum of the native enzyme with little resolution of the main absorbance peaks (3). Octanoyl-CoA effects rapid and extensive bleaching at 442 nm with generation of a long wavelength band at 550 nm. In the native enzyme, this band has been ascribed to a charge transfer complex between reduced flavin as the donor and the bound enoyl-CoA as the acceptor (4,5). The 20 nm blue shift incurred on replacement of the natural flavin by the 8-Cl derivative is consistent with this ascription, since the latter has a redox potential 60 mV higher than that of FAD (6). The 8-Cl enzyme is appreciably active (~8% that of native enzyme) in the standard assay system using phenazine methosulfate to replace the physiological mediator electron transferring flavoprotein (1). Very similar results to those shown in Figure 1 are obtained with butyryl- and palmitoyl-CoA substrates. This behavior contrasts sharply with the native enzyme where the internal equilibrium:

$$EFl_{ox} \cdot SH_2 \overset{K_2}{\rightleftharpoons} EFl_{red} \cdot S$$

lies to the right with octanoyl-CoA, but to the left with palmitoyl-CoA (1,3). Obviously this discrimination is suppressed using the higher redox potential analog and in each case the reduced flavin species is favored.

The stability of the 8-Cl enzyme, even after prolonged storage, indicates that none of the 6 cysteine residues/subunit are capable of nucleophilic displace-

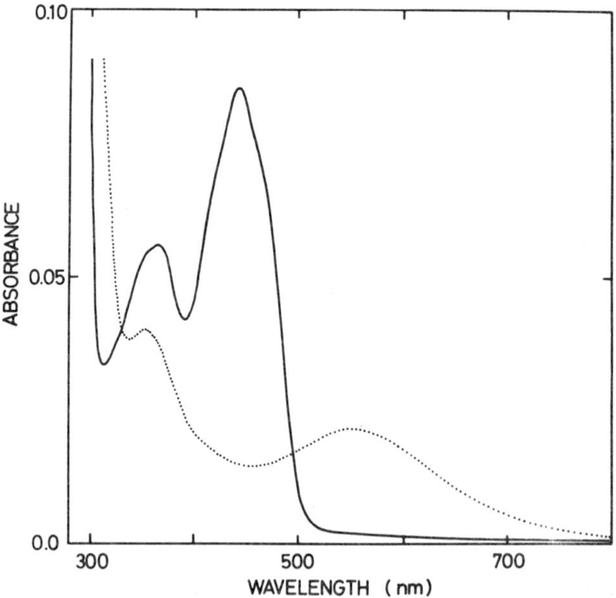

Figure 1. The effect of octanoyl-CoA on the absorption spectrum of 8-Cl FAD general acyl-CoA dehydrogenase. The spectrum of the enzyme (0.9 ml in 50 mM potassium phosphate buffer, pH 7.6, 4°) was recorded (—) and 8.7 μM octanoyl-CoA was added (· · ·). The spectral changes were completed very rapidly.

ment at the 8-position of the isoalloxazine ring as was observed with lipoamide dehydrogenases (6). Similarly, neither sodium sulfide nor thiophenol effect the characteristic changes expected for 8-S substitution (7–8). The 8-position of the flavin nucleus is thus either shielded from solvent or perhaps reaction of sulfur nucleophiles is electrostatically unfavorable (see later).

In contrast to the above, reconstitution with 5-deaza-FAD yields an enzymatically inactive derivative. Octanoyl-CoA merely perturbs the chromophore without significant reduction of bound deaza-FAD (Figure 2). The equilibrium K_2 (see above) lies far to the left with this strongly reducing analog. Accordingly, apoenzyme was reconstituted with 5-deaza-$FADH_2$ and a rapid reoxidation of this reduced enzyme was observed on the addition of crotonyl-CoA together with a disappearance of the strong fluorescence due to enzyme-bound deaza-$FADH_2$. The extreme thermodynamic destabilization of the 5-deazaflavin radical in comparison to normal flavin (9) suggests that reoxidation of the reduced enzyme derivative is unlikely to proceed via obligatory 1-electron steps.

The profound spectral changes accompanying the ionization of 6-OH flavins have been used as probes for the active site of several enzymes (8). Figure 3 shows that the apoprotein exhibits an approximately 40-fold preferential binding of the yellow neutral form, elevating the pK of the bound flavin from

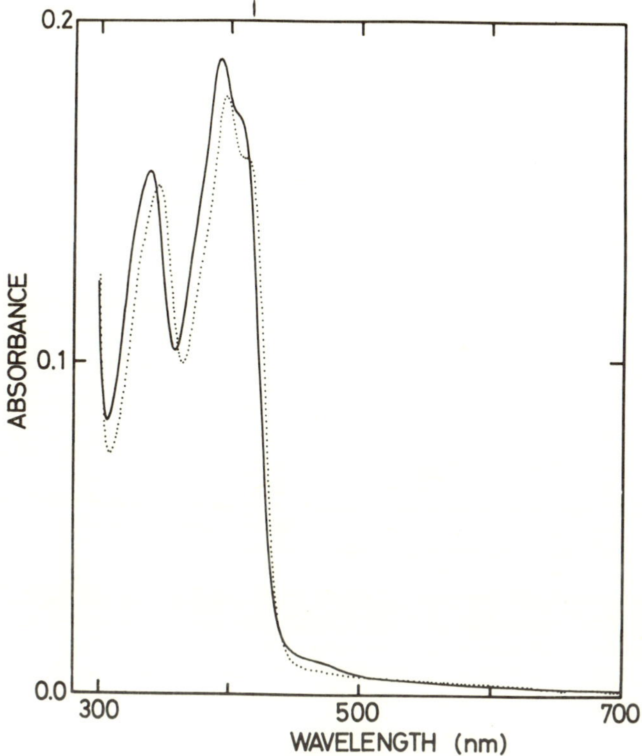

Figure 2. The effect of octanoyl-CoA on the absorption spectrum of 5-deaza-FAD enzyme. A solution of enzyme in 0.8 ml 50 mM phosphate buffer, pH 7.6, 4° (—) was treated with 39 μM octanoyl-CoA (· · ·).

7.1 (10) to 8.7 (see inset). Octanoyl-CoA does not reduce the enzyme, but promotes ionization of the chromophore to yield a spectrum similar to that found at high pH (Figure 3). Crotonyl-CoA induces very similar spectral changes. These two thioesters thus stabilize the green benzoquinoid species which carries a negative charge at the N-1-C-2=0 locus (8). These results are of great interest since crotonyl- and octanoyl-CoA also promote the ionization of the blue neutral semiquinone of native general acyl-CoA dehydrogenase yielding red anion-ligand complexes (11, 12). Thus ionizations of the oxidized 6-OH analog mirror those of the normal flavin semiquinone. The molecular basis for the stabilization of a negative charge localized in the N-1-C-2=0 position is not clear, but it might reflect a favorable electrostatic interaction with an adjacent amino acid side chain (8). This interaction is clearly not present in the absence of substrate, since the pK of 6-OH FAD is raised on binding to apoprotein, rather than being lowered as would be expected if the benzoquinoid form were stabilized.

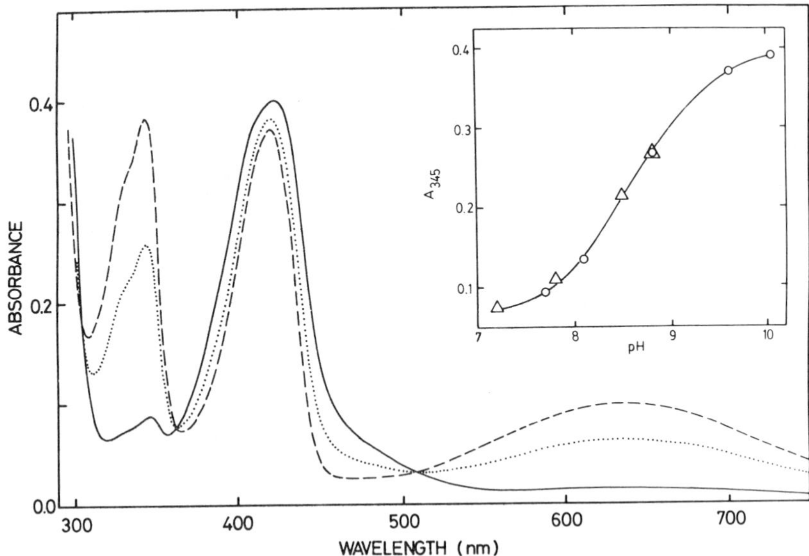

Figure 3. pH dependence of the spectrum of 6-OH FAD enzyme. The pH of the derivative (0.9 ml in 50 mM phosphate, 4°) was carefully adjusted by the addition of increments of solid sodium carbonate to an upper limit of 10.05 (○, see inset) and then lowered to 7.2 using solid KH_2PO_4(Δ). The spectra shown were recorded at pH 7.7 (—), pH 8.8 (· · ·), and pH 10.05 (---) respectively.

ACKNOWLEDGMENTS
This work was supported in part by NIH GM26643 (C.T.) and GM11106 (V.M.).

References

1. Beinert, H. (1963) *The Enzymes* 7:447–466.
2. Mayer, E.J. and Thorpe, C. *Anal Biochem*, in press.
3. Thorpe, C., Matthews, R.G., and Williams, C.H., Jr. (1979) *Biochemistry* 18:331–337.
4. Engel, P.C. and Massey, V. (1971) 125:879–887.
5. Massey, V. and Ghisla, S. (1974) *Ann NY Acad Sci* 227:446–465.
6. Moore, E.G., Cardemil, E., and Massey, V. (1978) *J Biol Chem* 253:6413–6422.
7. Moore, E.G., Ghisla, S., and Massey, V. (1979) *J Biol Chem* 254:8173–8178.
8. Massey, V. and Hemmerich, P. (1980) *Biochem Soc Trans*, 246–257.
9. Blankenhorn, G. (1976) *Eur J Biochem* 67:67–80.
10. Mayhew, S.G., Whitfield, C.D., Ghisla, S., and Schuman-Jorns, M. (1974) *Eur J Biochem* 44:570–591.
11. Mizzer, J.P. and Thorpe, C., this volume.
12. Mizzer, J.P. and Thorpe, C. (1981) *Biochemistry*.

PART IX B:

Interactions of Flavoproteins and Iron Sulfur Proteins

Published 1982 by Elsevier North Holland, Inc.
Vincent Massey and Charles H. Williams, Editors
Flavins and Flavoproteins

CHAPTER 109

Interaction of Adrenodoxin Reductase with the Electron Donor and Acceptor

Toshio Yamano,* Toshihiro Sugiyama,*
Yasuki Nonaka,* Yoshiyuki Ichikawa,*[1]
Teizo Kitagawa,** Hiroshi Sakomoto,† and
Yoshihiro Miyake†

**Departments of Biochemistry and **Molecular Physiological Chemistry, Osaka University Medical School, Kitaku, Osaka 530, Japan; †Department of Biochemistry, National Cardiovascular Center Research Institute, Osaka 565, Japan*

Summary

A short review is introduced on the interaction between NADPH-adrenodoxin reductase and adrenodoxin and their functions in electron transfer. From a mechanistic point of view, the complex formation of the reductase with adrenodoxin was regarded as a key step and the properties of the complex were extensively studied. Our recent results with circular dichroic study gave a dissociation constant much larger than the one regarded as almost established. The interaction of the reductase with $NADP^+$ was found to be physiologically important, since $NADP^+$ was found to prevent electron-leakage from the electron transfer pathway by stabilizing flavin semiquinone of the reductase. Chemical modification experiments on the reductase suggest arginine residue(s) as a candidate for the essential binding site of the reductase for $NADP^+$ nucleotides. Some properties of the reductase were studied in the electron transfer to free flavins added for the estimation of the redox potential. Identification of the flavin semiquinone of the reductase was demonstrated by resonance Raman spectroscopy.

NADPH-adrenodoxin reductase (AdR) is a component in the cytochrome P-450-linked steroid hydroxylase system in adrenocortical mitochondria. Its function is well-known to transfer electrons from NADPH to adrenodoxin and then to cytochrome P-450_{scc} or P-$450_{11\beta}$. We have succeeded in the crystallization of the reductase by using adrenodoxin-affinity chromatography (1) or conventional chromatography (2). Our main theme here on AdR is concerned with its interaction with adrenodoxin, the electron acceptor, and with pyridine nucleotides, the electron donor.

Since the pioneering work of Chu and Kimura (3,4) was published on the preparation of the reductase and the 1 : 1 complex formation between the reductase and adrenodoxin with a low dissociation constant of the order of 10^{-9} M, several papers have reported development of research on the nature

[1]Present address: Department of Biochemistry, Kagawa Medical University.

and physicochemical properties of the complex. Complex formation between these two proteins was observed spectroscopically as the enhancement of the absorbance at 450 nm over the sum of the two components. This fact had been recognized by Shin and San Pietro in the interaction between ferredoxin and ferredoxin reductase (5).

Lambeth et al. also reported a low value of dissociation constant of 10^{-8} M (8). They first speculated that the electron transfer from the flavoprotein to cytochrome P-450 or to cytochrome c may proceed without dissociation of the complex (7). In the same paper and also later (8), they reported that the redox potential of adrenodoxin is lowered when it is complexed with the reductase. Although the dissociation constant they obtained between oxidized AdR and adrenodoxin was quite small, they admitted that the complex formation did not suffice for cytochrome $P\text{-}450_{11\beta}$-mediated hydroxylations and that adrenodoxin in excess over the reductase was required. They proposed a shuttle mechanism (8), in which the oxidation-reduction potential of adrenodoxin was shifted by about -100 mV when bound to the reductase and the reduction of adrenodoxin promoted dissociation of the complex. For proof of this proposal, they carried out experiments using acetylated cytochrome c as a probe for electron transfer (9).

We initially prepared the reductase from bovine adrenocortical mitochondria by the procedure of Suhara et al. (6). We later found that the reductase is easily crystallized by using the adrenodoxin-Sepharose affinity column (1). For exploration of the microenvironment surrounding the chromophores of the interacting two proteins, we carried out fluorescence measurements on the reductase complexed with adrenodoxin (2). The fluorescence of bound FAD in AdR from bovine mitochondria is 12% that of free FAD at 520 nm on excitation at 450 nm. Figure 1 shows that further quenching of the fluorescence was brought about by the presence of bovine adrenodoxin, apoadrenodoxin, or other nonheme iron proteins, which is designated by Q on the abscissa. The ordinate represents the ratio of fluorescence intensity without Q to that with Q. Similarly, when the intensity of the tryptophan fluorescence of AdR was plotted against the concentration of the quencher, quenching of the tryptophan fluorescence at 337 nm with excitation at 285 nm was observed. In both cases, the ratio of the quencher to the reductase at the equivalence point was 1 : 1 (2).

Another approach to the analysis of interaction between the two proteins was to use spin-labelled adrenodoxin (10). Adrenodoxin was spin-labelled with N-(2, 2, 5, 5-tetramethyl-3-carbonylpyrroline-1-oxyl)imidazole (I) which is known to be specific for a tyrosine residue, or with N-(1-oxyl-2,2,6,6-tetramethyl-4-piperidinyl)maleimide(II), a cysteine specific reagent. The electron paramagnetic resonance (EPR) signal of the labelled adrenodoxin showed a pattern typical of a moderately immobilized spin-label. Addition of the spin-labelled adrenodoxin to AdR resulted in the appearance of a more immobilized component in the EPR spectrum. Figure 2 shows EPR spectra of adrenodoxin spin-labelled with (I) at 25°C in the presence of AdR at low (a) and high (b) ionic strengths and in the presence of bovine serum albumin (c).

Using this technique, the interaction of spin-labelled adrenodoxin and cytochrome $P\text{-}450_{scc}$ was investigated. The result was similar to that represented in Figure 2. However, addition of both the reductase and cytochrome

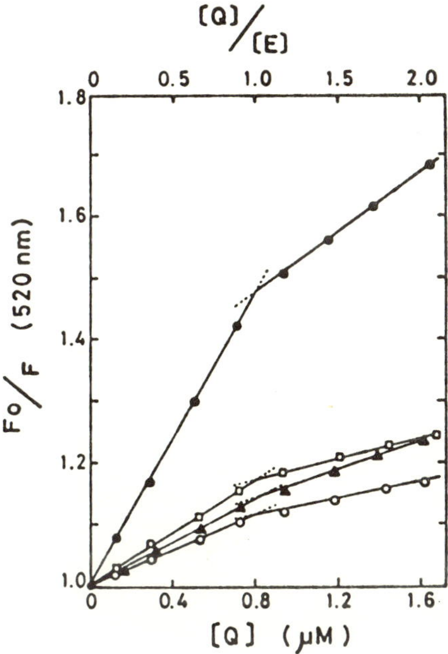

Figure 1. Stern-Volmer plots of fluorescence of flavin of AdR in the complex formation. (●) bovine adrenodoxin; (○) bovine apoadrenodoxin; (▲) spinach ferredoxin; (□) pig adrenodoxin.

$P\text{-}450_{scc}$ at the same time to spin-labelled adrenodoxin did not produce superposition of the more immobilized components. Kido and Kimura (11) demonstrated the occurrence of ternary complex formation in the presence of the three proteins in a system containing cholesterol and Emulgen. We have failed to obtain clear evidence for this ternary complex formation in our simple system.

Significant difference circular dichroism (CD) was observed around the region of the absorption of flavin and Fe-S core chromophores between the CD of the mixture and that of the algebraic sum of the two proteins when the reductase was complexed with adrenodoxin (12). The difference CD spectrum showed a broad trough between 500 and 600 nm and CD extrema at 452, 383, and 350 nm. This spectrum may reflect a change in the environment of one or both of the protein chromophore(s). By applying a curve-fitting procedure to a titration profile, where the difference CD change at 452 nm was plotted against the concentration of adrenodoxin while the concentration of the reductase was held constant, we directly obtained 4×10^{-7} M as the dissociation constant for the complex between the proteins in their oxidized states (Figure 3). The theoretical curves were obtained by assuming several values of K_d, which is a function of total adrenodoxin added or the ratio of total adrenodoxin to total AdR used. We also studied the salt-induced change of the difference ellipticity to obtain the dissociation constants for the complex at various concentrations of NaCl.

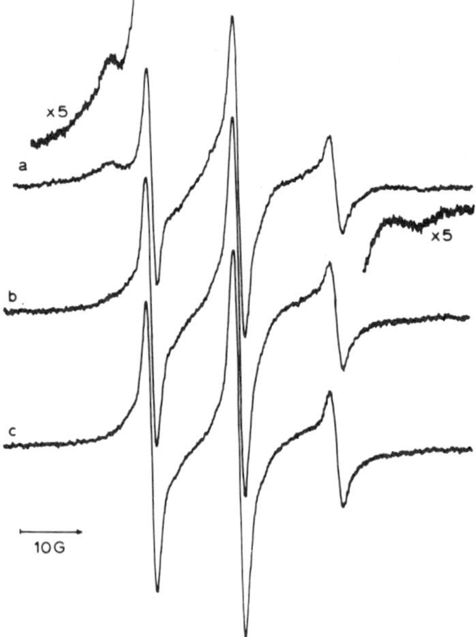

Figure 2. EPR spectra of spin-labelled adrenodoxin in the presence of the reductase at low (a) and high (b) ionic strengths and of serum albumin (c).

The dissociation constants mentioned above were larger than those obtained previously by kinetic methods, measurements of NADPH-cytochrome c reductase or NADPH-DCPIP reductase activity. To analyze this discrepancy, we examined the relationship between salt-induced changes of the NADPH-

Figure 3. Curve fitting to obtain a reasonable K_d, dissociation constant. On the ordinate is the fractional saturation, α, of AdR complexed with adrenodoxin calculated from the difference ellipticity. (A) is an expansion of (B). K_d assumed were 1×10^{-7}, 4.56×10^{-7}, 1×10^{-6} M from top to bottom.

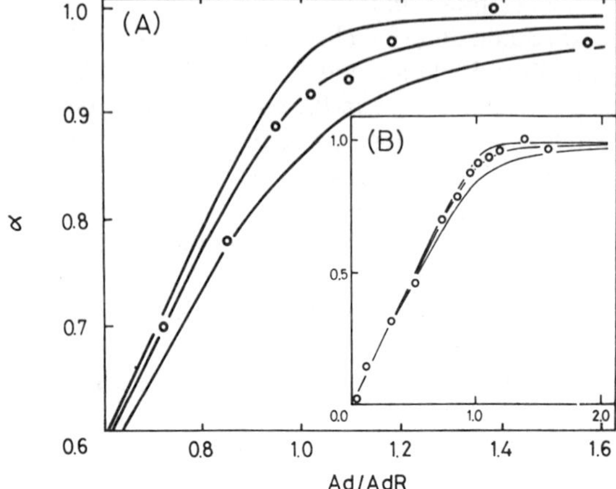

cytochrome c reductase activity or the NADPH-DCPIP reductase activity in the presence of adrenodoxin and the complex concentration calculated using the dissociation constant found from the difference CD study. There seems to be no correspondence between the activities and the calculated complex concentrations.

We noticed not only the difference in the values of K_m of the reductase for the pyridine nucleotides, but also, more importantly, the effect of $NADP^+$ to prevent the reduced reductase from being autooxidized (13). Lambeth and Kamin reported earlier that characteristic absorption spectra were obtained when the reduced AdR was titrated with $NADP^+$. They attributed the spectra to a charge-transfer complex formed between reduced AdR and $NADP^+$ (14). We have demonstrated that flavin semiquinone in much greater amount was observed in the reoxidation process of the reductase reduced with NADPH than that when NADH was used to reduce the reductase. These facts suggest that NADPH or $NADP^+$ induces some additional influence on the reductase when it binds to flavoprotein and that NADH induces no more than simple binding.

To elucidate the function of NADP nucleotide, we have attempted, first of all, chemical modification of the reductase to search for the essential amino acid as the binding site. p-Hydroxyphenylglyoxal was found to be a useful reagent for arginine residues because the reaction takes place under mild conditions and the extent of modification could be determined by the absorbance change at 340 nm as reported by Yamasaki et al. (15).

Pseudo-first-order reaction rate constants were obtained for the inactivation of the AdR at various concentrations of the reagent. Figure 4 shows the time course of the inactivation. When $NADP^+$ was added to the reaction mixture at the concentration of 50 μM, the k_{obs} was 70%. On the contrary, a thousand times higher concentration of NAD^+ was required to effect the same extent of prevention (Figure 5). The concentration of $NADP^+$ to prevent the inactiva-

Figure 4. Chemical modification of the reductase with p-hydroxyphenylglyoxal (PHPG). The concentrations of PHPG were (□) 0 mM; (△) 25.3 mM; (○) 42.1 mM in 50 mM phosphate buffer, pH 7.4, at 25°C.

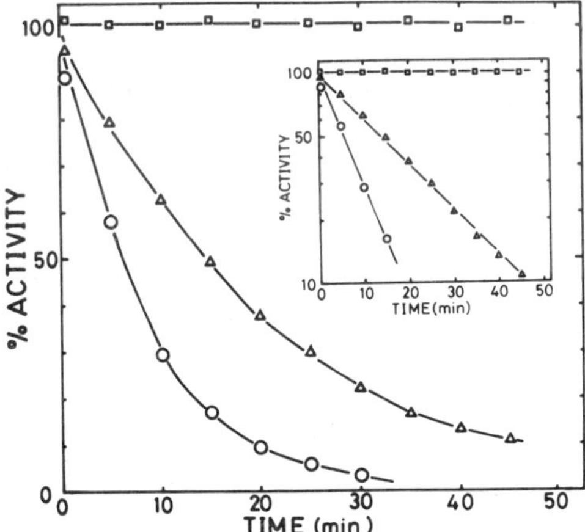

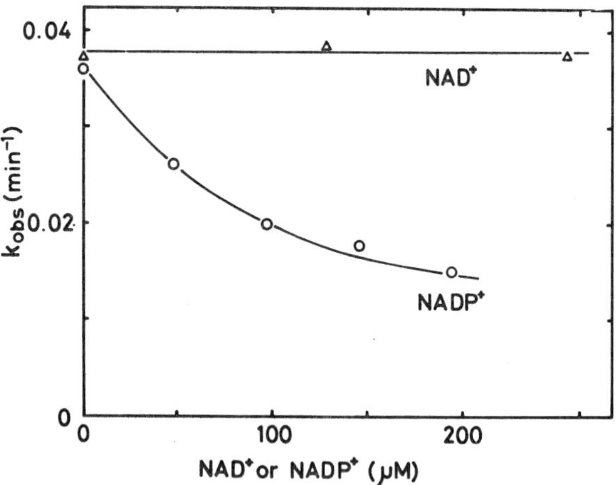

Figure 5. Effect of $NADP^+$ or NAD^+ on the k_{obs} (inactivation rate constant) at 25.3 mM PHPG.

tion compared well with the dissociation constant of $NADP^+$ for the reductase (14). The experiments with 2,3-butanedione also showed that inactivation of AdR took place and that $NADP^+$ prevented the inactivation while NAD^+ did not. 1,2-Cyclohexanedione, another arginine modifying reagent, showed similar inactivation of the reductase which was prevented by $NADP^+$, although the latter two reagents reacted with the enzyme more slowly than p-hydroxyphenylglyoxal.

Another remarkable difference between the electron donating nucleotides was observed in the behavior of the reduction rate of free flavins added as acceptors in the solution. Figure 6 shows that in the presence of NADPH, free FMN, FAD, or riboflavin accepts electrons from the reductase which in turn are transferred to oxygen, causing enhanced oxygen consumption, whereas no enhancement of oxygen consumption was observed in the presence of NADH. When quinones were used as electron acceptors in place of free flavins, there was no difference between NADPH and NADH. This may be explained in terms of the redox potentials of these free compounds and of the flavin semiquinone-$NADP^+$/fully reduced flavin of the reductase.

A quantitative study on the flavin semiquinone concentration during the reoxidation of the reductase reduced with NADPH was performed by EPR spectrometry. The flavin semiquinone was highly stabilized in the presence of $NADP^+$. This does not apply in the course of reoxidation of the reductase reduced with NADH. As has already been published by Sugiyama et al. (13), the autooxidation rate of the reduced pyridine nucleotides by the reductase is much higher in the case of NADH than NADPH. The observation is attributable to the stabilized complex between the flavin semiquinone of the reductase and $NADP^+$, thereby suppressing electron leak from the flavin moiety to oxygen. In addition to the complexes known hitherto, we emphasize the presence of the complex of the flavin semiquinone and $NADP^+$.

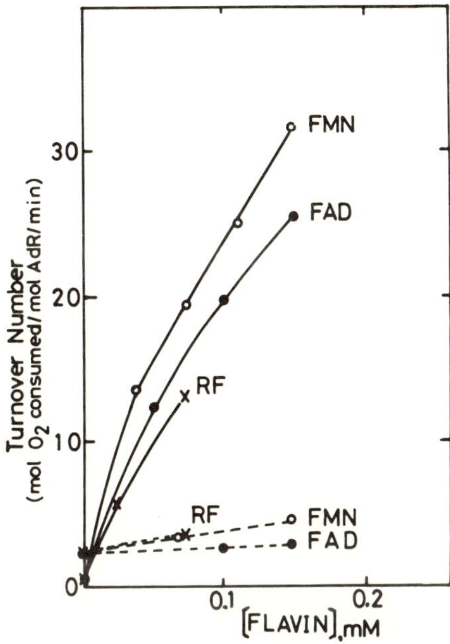

Figure 6. Electron flow from the reductase to free flavins added in the presence of pyridine nucleotides. Solid lines, NADPH; broken lines, NADH. The reductase, 3.3 μM; potassium phosphate, 50 mM, pH 7.5, at 25°C.

Laser resonance Raman study was performed with the reductase. As stated elsewhere in this volume, a Raman line of 1611 cm^{-1} was characterized as derived from the flavin semiquinone of the reductase. This Raman line is observable only when NADPH was added to the reductase solution, but not when NADH was added. Its appearance was always paralleled with the distinct EPR signal of g-value around 2.00 and the absorption spectrum of blue semiquinone type.

Figure 7. Proposed mechanism for electron transport in adrenodoxin reductase system.

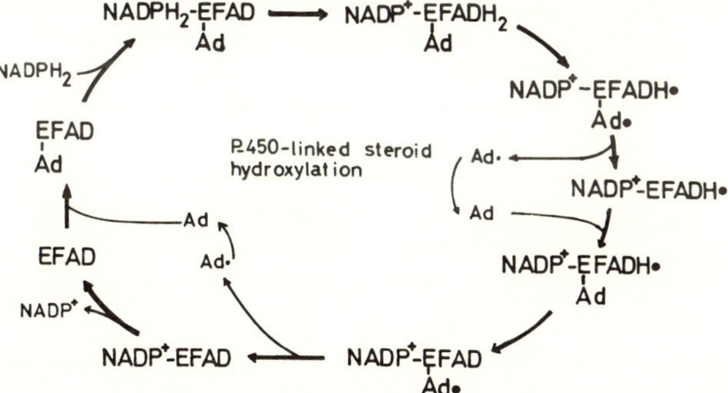

Lambeth et al. demonstrated the occurrence of E-$FADH_2$-$NADP^+$ from their spectroscopic data (14). We encountered a species with a similar absorption spectrum which exhibited diminished or no EPR signal.

Taking these results as well as specificity of $NADP^+$ toward the reductase into account, we propose the following the scheme for electron transport from NADPH to cytochrome P-450(s) (Figure 7). One electron is transferred from reduced AdR to adrenodoxin, partitioning the two electrons in reduced AdR into one each for AdR and adrenodoxin. Reduced adrenodoxin may be detached from the reductase by the shuttle mechanism of Lambeth and Kamin (8). After the electron on the flavin semiquinone is transferred to another molecule of adrenodoxin, $NADP^+$ would be replaced by another NADPH for the subsequent cycle of electron transport to proceed.

References

1. Sugiyama, T. and Yamano, T. (1975) *FEBS Letters* 52:145–148.
2. Hiwatashi, A., Ichikawa, Y., Maruya, N., Yamano, T., and Aki, K. (1976) *Biochemistry* 15:3082–3090.
3. Chu, J.-W. and Kimura, T. (1973) *J Biol Chem* 248:2089–2094.
4. Chu, J.-W. and Kimura, T. (1973) *J Biol Chem* 248:5183–5187.
5. Shin, M. and San Pietro, A. (1968) *Biochem Biophys Res Commun* 33:38–42.
6. Suhara, K., Ikeda, Y., Takemori, S., and Katagiri, M. (1972) *FEBS Letters* 28:45–47.
7. Lambeth, J.D., McCaslin, D.R., and Kamin, H. (1976) *J Biol Chem* 251:7545–7550.
8. Lambeth, J.D., Seybert, D.W., and Kamin, H. (1979) *J Biol Chem* 254:7255–7264.
9. Lambeth, J.D., Lancaster, J.R., and Kamin, H. (1981) *J Biol Chem* 256:3674–3678.
10. Miura, R., Sugiyama, T. and Yamano, T. (1979) J. Biochem. Tokyo 85, 1107–1113.
11. Kido, T. and Kimura, T. (1979) *J Biol Chem* 254:11809–11815.
12. Sakamoto, H., Ichikawa, Y., Yamano, T., and Takagi, T. (1981) *J Biochem Tokyo*, in press.
13. Sugiyama, T., Miura, R., and Yamano, T. (1979) *J Biochem Tokyo* 86:213–223.
14. Lambeth, J.D. and Kamin, H. (1976) *J Biol Chem* 251:4299–4306.
15. Yamasaki, R.B., Vega, A., and Feeney, R.E. (1980) *Anal Biochem* 109:32–40.

Published 1982 by Elsevier North Holland, Inc.
Vincent Massey and Charles H. Williams, Editors
Flavins and Flavoproteins

CHAPTER 110

Role of Flavins and Iron-Sulfur Proteins in the Reduction of Cytochrome P-450

Henry Kamin and J. David Lambeth

Department of Biochemistry, Duke University Medical Center, Durham, North Carolina and the Department of Biochemistry, Emory University School of Medicine, Atlanta, Georgia

Cytochrome P-450, operating as an oxgenase, takes electrons derived from metabolism and transmitted via an electron-transport chain, and utilizes them to insert molecular oxygen into substrate. This process is one example of a general oxidation-reduction sequence wherein substrates, which give up electrons two at a time, transmit them to final acceptors via enzymatic machinery which, at one point or other, deals with these electrons one at a time. Before proceeding with a description of how flavins and iron-sulphur proteins operate in cytochrome P-450 reduction, it is useful to deal first with the general problem of 2-electron→1-electron "step-down" processes. Some aspects of this subject are dealt with in a previous publication (1).

It has long been known that donor substrates are oxidized by the removal of 2-electron (or hydrogen) equivalents; since hemoproteins operate by 1-electron redox steps, the requirement for a "step-down" process became apparent when Keilin (2) demonstrated the substrate→hemoprotein→oxygen nature of the respiratory chain. How did this process occur? Michaelis and his coworkers, in the course of studies of semiquinones as intermediate oxidation forms in organic-chemical redox reactions, suggested that flavin semiquinones could be intermediates in redox reactions catalyzed by flavoenzymes (3,4) and Haas (5) observed a red form, presumably semiquinone, in "Old Yellow Enzyme" of yeast. Haas, Horecker, and Hogness in 1940 specifically suggested a flavin semiquinone as an obligatory intermediate during catalysis by yeast NADPH-cytochrome reductase, recognizing that the cytochrome must undergo a valence change of 1 while the reductase must accept two electrons from the NADPH (6). The first semiquinone actually observed in the course of a flavin-catalyzed reaction was by Beinert [see spectrum (7)] in the fatty acyl CoA dehydrogenase system. But semiquinones were considered unstable and transient redox forms until Masters, Kamin, and their colleagues (8,9) isolated and worked with the air-stable semiquinone form of microsomal NADPH-cytochrome P-450 (then "cytochrome c") reductase.

Electron-Pair Splitting

A "simple" step-down mechanism was first described by Strittmatter (10), who showed that the electron donor for cytochrome b_5 reduction could be either the flavin hydroquinone or semiquinone form of the cytochrome b_5 reductase. Although occurring at different potentials, each reductive step was sufficiently rapid to sustain catalysis. An analogous reaction, as studied in our laboratory, is the reduction of ferricyanide catalyzed by NADPH-adrenodoxin reductase (11); here again, both the fully reduced and semiquinone forms of the flavin are competent to reduce ferricyanide at rates commensurate with overall catalytic activity.

However, these apparently simple $2 \rightarrow 1$ electron transformations may be the exception rather than the rule; in a number of flavin enzymes a second redox group, which acts as a 1-electron carrier, serves as a transmitting group to provide single, constant-potential electrons to the acceptor. Some systems which serve as examples, and which have been studied in Durham, are shown in Figure 1; a more complete tabulation, including variants on this pattern, has been presented previously (1). The examples selected in Figure 1 consist of two types of P-450 reductases, (which utilize either a general or specific cytochrome P-450), and two 6-electron reductases: sulfite reductase of enteric bacteria and plants, and nitrite reductase of plants. These reductases utilize a protein subunit containing siroheme and a tetranuclear iron-sulfur center.[1]

This figure suggests a striking analogy: FMN and Fe_2S_2 seem to play similar roles. They both undergo one-electron transformations, receiving electrons from FAD (which serves as a two-electron acceptor from NADPH), and donate a single electron to the terminal one-electron acceptor. In the case of the microsomal reductase and sulfite reductase, the FAD and the FMN are on the same subunit;[2] in the case of nitrite reductase the ferredoxin donor is a separate molecule; one might wonder whether flavodoxin might serve as a FMN analogue to the ferredoxin.[3] Although the Fe_2S_2 of adrenodoxin is written in Figure 1 as a separate protein, it actually functions by shuttling (13) between complexes with the flavoprotein (14, 15) and with the reaction-specific cytochrome P-450 (16, 17). Ferredoxin certainly forms complexes with FNR (18, 19), but we do not yet know whether it forms complexes with nitrite reductase.

Figure 2A describes the electron-pair splitting in the six-electron reductases, and Figure 2B describes this process in the P-450 reductases. Studies of Siegel et al. (20) have shown that the flavin moiety contains three electrons in its "most oxidized" state and one electron in its "most reduced" state. We presume, but have not rigorously demonstrated, that the same holds for the flavin-Fe_2S_2 system which donates electrons to nitrite reductase. The same number of electrons reside in the "most oxidized" and "most reduced" forms of the P-450 reductases (Figure 2B); Lambeth and Kamin (1) have described

[1]Joint studies from Siegel and Münck's laboratories (12) have shown that the siroheme and the iron-sulfur center interact closely and are probably in close spatial proximity; addition or subtraction of a single electron affects the magnetic properties of all five iron atoms.

[2]The structure of the flavoprotein moiety of sulfite reductase remains a puzzle. It contains 8 protein subunits, four FAD, and four FMN. We have not yet succeeded in preparing a functional monomer.

[3]At the 1978 Flavin Symposium, several of us discussed the possible analogies between FMN and iron-sulphur groups, and wondered whether there might be amino acid homologies. At the present symposium, Veeger specifically asked the question whether there might be homologies between flavodoxin and the FMN domains of the FAD-FMN enzymes.

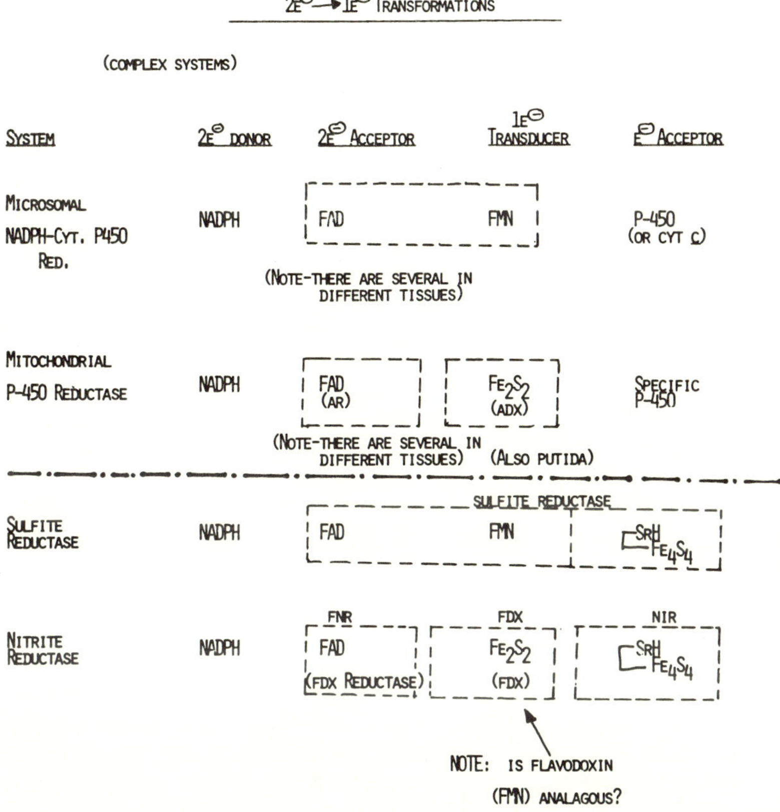

Figure 1. Flavoproteins in selected 2 electron-to-1 electron transformations.

this catalytic cycle in the mitochondrial adrenal steroid hydroxylating system, and Vermilion et al. (21) have shown that the microsomal FAD-FMN enzyme follows a reaction cycle strikingly similar to that of the flavoprotein moiety of sulfite reductase. The solution of the microsomal flavoprotein mechanism was made possible by the observation of Iyanagi and Mason (22) [see also the footnote, page 346 of (1)], that microsomal NADPH-cytochrome reductase contained equimolar quantities of FAD and FMN rather than the two moles of FAD proposed by previous workers.

Steroidogenic Electron Transport in Adrenal Mitochondria

The early studies of Estabrook, Omura, and their colleagues (23) provide the basic information which describes the adrenal system. The system consists of NADPH adrenodoxin reductase (a non-FAD enzyme, designated AR), adrenodoxin, (an Fe_2S_2 protein, designated ADX), and cytochromes P-450 which were later shown to be reaction-specific. All components are needed for

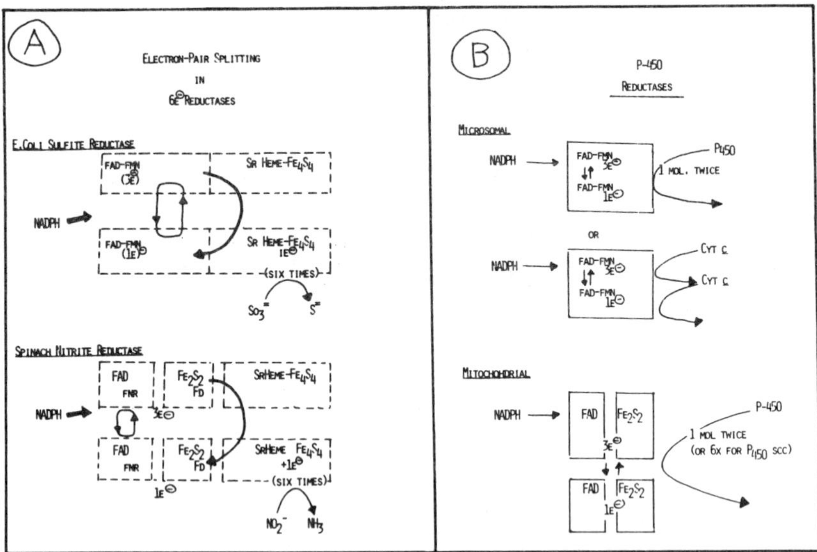

Figure 2. Electron-pair splitting in 6-electron reductases (A), and in P-450 reductases (B).

steroid hydroxylation; ferricyanide reduction requires only the flavoprotein, but cytochrome c reduction requires both AR and ADX. DCIP reduction does not require ADX but our laboratory (24) and Kimura's (25) have shown stimulation of DCIP reduction by the iron-sulphur protein.

Our laboratory has used the general strategy of studying the reactions "from left to right," studying first the interaction of pyridine nucleotide with flavoprotein, and then proceeding down the line toward cytochrome P-450 and hydroxylation. Our initial studies utilized the highly purified, soluble system; more recent studies have examined the system reconstituted in phospholipid vesicles. Our first efforts (24) showed the formation of a 1 : 1 complex between reduced flavoprotein and $NADP^+$. This species, formed rapidly upon anaerobic reaction of NADPH with AR, was a complex with absorbance at higher wavelengths. Spectrophotometric studies plus measurement of redox potentials and equilibrium titration showed that complex formation facilitates the reduction of the flavoprotein ($E_m ARH_2/AR = .29V$; $E_m ARH_2 \cdot NADP^+/AR = -.20$ V), and that dissociation of $NADP^+$ is facilitated by oxidation of the reduced flavoprotein. It was clear that the K_ds of pyridine nucleotide-flavoprotein complexes depended upon the redox state of the flavoprotein. This study, which described a relationship between complex formation and redox potential, established the general pattern for our subsequent studies, which described the various redox and ligated states of the components, and examined the relationship between complex formation and oxidoreduction reactions under a variety of conditions. When combined with appropriate studies of reaction rates, the observed relationships served to define the reaction mechanism. In brief, we describe a "shuttle" mechanism, where the ADX molecule shuttles electrons from a bimolecular complex with AR to a bimolecular complex with cytochrome P-450. At the cytochrome end of the reaction chain, formation of the ADX-P-450 complex facilitates the binding of cholesterol and vice versa, thus providing a control device at the hydroxylation step (26).

Catalytic studies varying the ratio of ADX to AR showed that a 1 : 1 complex was the effective species for cytochrome c reduction; the reaction rate of cytochrome c with flavoprotein alone was essentially zero.[4] Our studies of the stoichiometry of cytochrome c reduction under a variety of conditions have indicated that this rate is a valid measure of the concentration of the AR : ADX complex.

Our next task was to establish a system for the spectrophotometric assessment of the redox states of flavoprotein, iron-sulfur protein, and mixtures of both during titration with reduced pyridine nucleotide. Our laboratory's earlier studies on the microsomal flavoprotein (8, 9) had utilized three wavelengths to establish and quantify the three redox states of the flavoprotein; this involved relatively simple arithmetic. However, the addition of a second redox protein, ADX, to flavoprotein increases the complexities enormously and analysis requires computerized solutions. Figure 3 presents the spectrophotometric data obtained on titrating a 1 : 1 AR-ADX complex with NADH; NADH was used in this case rather than NADPH to remove the additional complexities of the $NADP^+$-ARH_2 complex. Lambeth et al. (15), using the wavelengths designated by the closed circles on the abscissa of Figure 3, designed a computer program which provided a good inventory of the concentrations of the redox states of the enzymatic components in mixture.

When a 1 : 1 AR-ADX mixture is titrated, the first electrons go into the flavoprotein; the ADX becomes reduced only after most of the flavoprotein has reacted. These types of data provide potentials for the components. We have found that the E_m of the flavoprotein is affected but little by complexation to adrenodoxin, but the E_m of adrenodoxin is profoundly affected by complexation. ADX has an E_m of -0.291 V, whereas, in the presence of the reductase, its potential shifts to -0.360 V. Thus, *complexation makes bound ADX a stronger reducing agent; reduction of ADX facilitates dissociation of the complex.* The facilitation of dissociation upon reduction of ADX is central to the shuttle mechanism we propose.

What are the redox and complexation states of the enzymes during catalysis? We first addressed this question to catalysis of ferricyanide and cytochrome c reduction (11). The mechanism for ferricyanide reduction seems relatively straightforward: ferricyanide reduction is extremely rapid with both the FH_2 and the $FH\cdot$ forms of the flavoprotein. All of the reducing power of the enzyme is transferable to the electron acceptor. Thus, the reaction follows the type of mechanism proposed by Strittmatter (10) for cytochrome b_5 reduction. For cytochrome c, the mechanism is more complex. When oxidized AR-ADX complex is reacted in the stopped-flow with different quantities of a 1 NADPH:2 cytochrome c mixture, the reaction starts in a rapid, first-order manner, but makes a sharp transition to a much slower rate constant at the precise point where one electron is left in the AR-ADX complex. The latter

[4]A 1 : 1 complex is also required for rapid DCIP reduction; however the rate constant for DCIP reduction by flavoprotein alone is not zero, but is about 60% of the rate catalyzed by the AR : ADX complex.

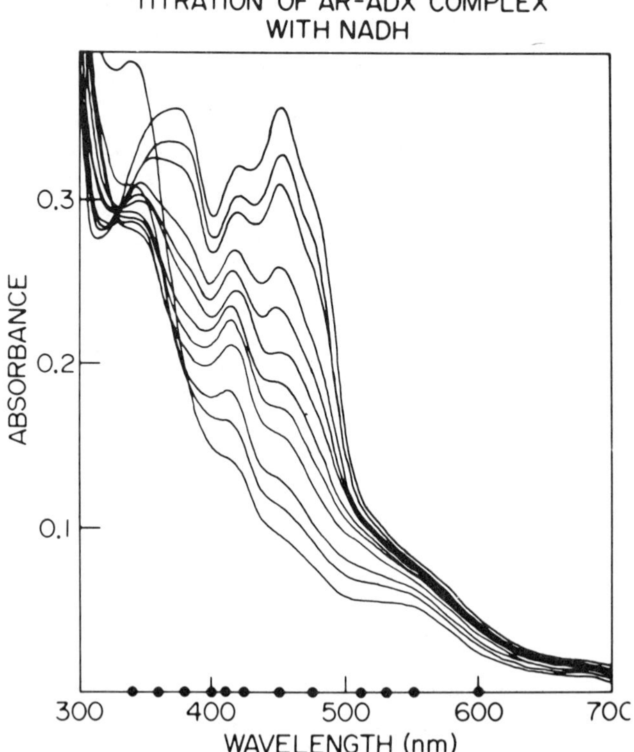

Figure 3. Titration of AR-ADX complex with NADH (15).

rate is too slow for catalysis, and it can therefore be concluded that the "most oxidized" state of AR-ADX complex contains one electron. This is the same number of electron equivalents as in the "most oxidized" states of the FAD-FMN flavoproteins, microsomal NADPH-cytochrome P-450 reductase (21) and NADPH-sulfite reductase (20). Figure 4 shows the catalytic cycle for cytochrome c reduction based on the above and other lines of evidence. The equilibrium of step c is unfavorable (about 1:60) for producing the catalytically-effective $\dot{A}\dot{R}$-A$\dot{D}$X form, but calculations show that even this small quantity is sufficient for catalysis. In the one-electron form, $\overline{\text{AR-ADX}}$, the electron appears to reside in the ADX about 3 times as much as in the AR.

Thus for cytochrome c reduction, the AR-ADX complex cycles between a 3-electron "most reduced" form and a one-electron "most oxidized" form. The immediate electron donor of cytochrome c is reduced ADX. The analogy between FAD-FMN enzymes and FAD-Fe_2S_2 enzymes is shown in Figure 5 (1). In both systems, oxidation and reduction are shown in the vertical direction, from top to bottom. Internal rearrangement of electrons ("electron shuffle") is shown in the horizontal direction. In all cases the acceptor flavin, FAD, takes two electrons from NADPH while in its fully oxidized form; it transfers one electron into a "vacancy" created by having the FMN in a

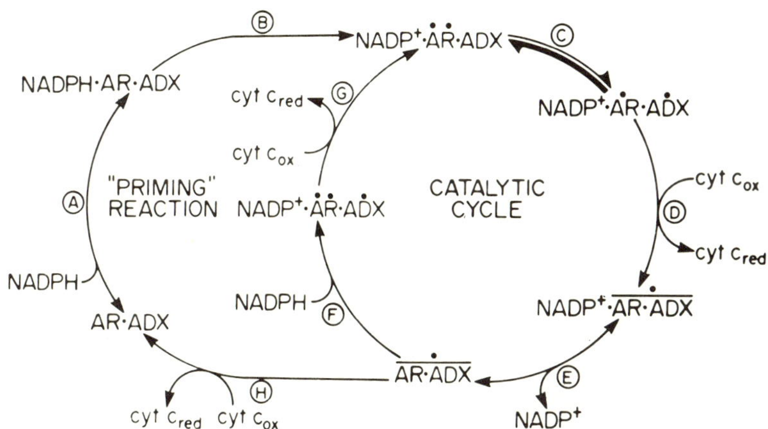

Figure 4. Mechanism of cytochrome c reduction catalyzed by the AR·ADX complex (11).

semiquinone form, or the iron-sulfur protein in its oxidized form. The latter groupings act in each case via a 1-electron transition, FMN cycling between the $2e^-$ and $1e^-$ forms, and the iron-sulfur grouping between one-e^- and zero-e^- forms. This provides for electron transfer to the final acceptor at a uniform potential.

Electron Transfer to Cytochrome P-450

We then turned our attention to the mechanism of reduction of the natural substrates of the AR-ADX system, the cytochromes P-450. The literature hinted that cytochrome P-450 reduction may involve some substantial differences from that of cytochrome c. Adrenal mitochondria contain a large excess of ADX and P-450 over AR (27) and Katagiri (17) showed the presence of an ADX-P-450 complex.

Even our first experiments revealed important differences in the behavior of cytochrome P-450 to that of cytochrome c. The most striking observation was that a 1:1 mixture of AR and ADX did not sustain hydroxylation; extra ADX was needed (16). When ADX was titrated into an AR-P-450 mixture, little if any hydroxylation occurred until the ADX:AR ratio *exceeded* 1:1. Figure 6 shows the spectrophotometric changes observed in such a titration. The wavelengths of the curve with solid circles represent the AR-ADX complex; the curve with open circles represents the ADX-P-450 complex. It is clear that the AR "sops up" the ADX first, as can be expected from the dissociation constants of the AR-ADX complex ($K_d = 5 \times 10^{-9}$; ADX-P-450 $K_d = 1 \times 10^{-7}$) (17). Since the 1:1:1 ratio of proteins failed to catalyze hydroxylation, it was clear that either there had to be an interaction of AR-ADX with yet another molecule of adrenodoxin, or that the ADX required prior release from its AR

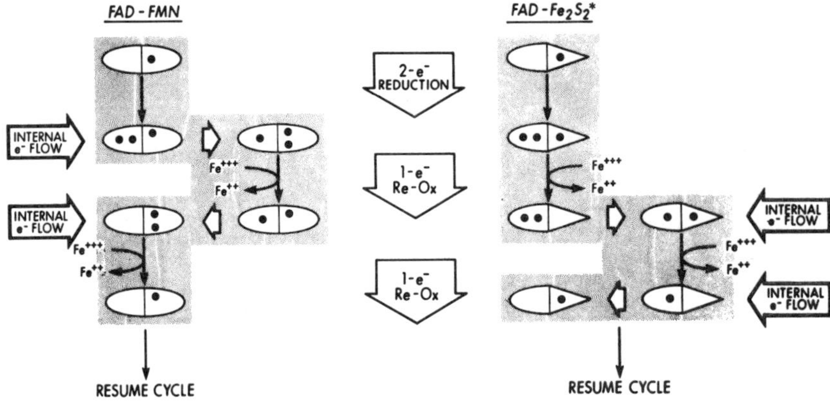

Figure 5. Topology of electron transferase cycles (1).

complex to provide an ADX-P-450 complex whose electrons could be utilized in hydroxylation. We were never able to find any evidence for a ternary AR-ADX-P-450 complex.

Since our previous data showed that the K_d of the AR-ADX complex was much higher upon reduction of ADX (15), it seemed plausible that ADX might be serving as a "shuttle." We tested this possibility both by utilizing ADX as an electron acceptor for the AR-ADX system and by examining the behavior of the AR-ADX complex at various salt concentrations (13). We had observed that increasing salt concentration, in a cation-specific manner, caused the dissociation of the AR-ADX complex, as measured by cytochrome c reduction. At low salt concentrations, cytochrome c reduction was maximal; but decreased to zero at high salt concentrations. The 50% inhibition point was different for each of a number of cations tested; divalent ions were effective. This differential salt-dependent dissociation permitted us to manipulate the dissociation constant of the AR-ADX complex. Testing the effect of different salts upon the 11-β cytochrome P-450-dependent hydroxylations, we found that for each salt, there was a bell-shaped curve, variably positioned; no activity was observed at either low or high concentrations. In each case, the shape of the curve could be accounted for by the supposition that, at low concentration, failure to dissociate reduced ADX prevented hydoxylation; at high salt concentrations, reaction was inhibited because dissociation of the AR-ADX complex prevented electron transfer between flavoprotein and iron-sulfur protein, a step which we had shown was rate-limiting for reduction of cytochrome c (28). Quantitation of these data and matching against prediction was excellent; the actual maximum did indeed represent, for each salt tested, the largest molar differences between the associated-oxidized state and the dissociated reduced state.

The "shuttle" mechanism can be envisioned as follows: a reduced flavin transfers an electron to the oxidized ADX which it holds in complex; the reduced ADX dissociates (finding its way to a cytochrome P-450), and is

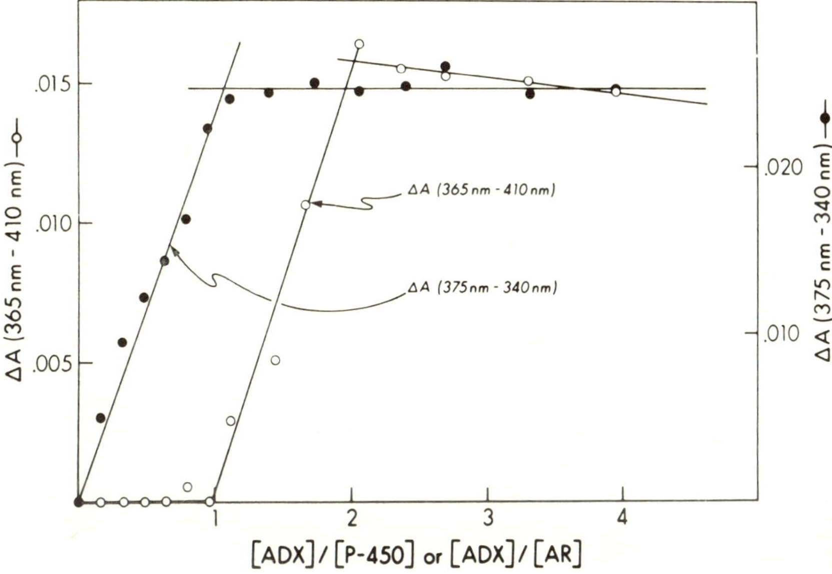

Figure 6. Difference spectra of titrations for interaction of ADX with AR and P-450 (16).

quickly replaced by another ADX (oxidized) to form a 1-electron-containing AR-ADX complex. The new complex accepts an electron pair from NADPH to form a 3-e^- containing complex. Again, the reduced ADX dissociates and is replaced by an oxidized ADX from the pool of excess mitochondrial ADX, and the cycle continues. Additional studies from our laboratory, including studies of electron transfer to acetyl cytochrome c, have given additional support to this proposed "shuttle" mechanism (29).

Studies in Liposomes

The cytochromes P-450 of adrenal mitochondria are intrinsic membrane proteins, and a proper understanding of their mechanism requires recognition of their normal milieu. Drs. Jack Lancaster and David Seybert brought to our laboratory both technical and intellectual expertise in this area, and we found that cytochrome $P\text{-}450_{scc}$ could be readily incorporated into phospholipid vesicles (30). The preparations were highly active and retained the expected mechanistic features of requiring both AR and ADX, with an excess of ADX over that of the flavoprotein, indicating that the "shuttle" mechanism operated in liposomes as well as in solution. Cytochrome P-450 was in the lipid phase of the membrane and could not be removed by salt treatment, but was reducible by soluble ADX. The substrate, cholesterol, was dissolved in the lipid phase and interacted with lipid-phase P-450. This was shown in two ways: (1) the cholesterol concentration-dependence of hydroxylation followed that of cholesterol to phospholipid ratio in the vesicle rather than the ratio of cholesterol to the total reaction volume. (2) Addition of enzyme to cholesterol-free vesicles caused a partial conversion of the high spin (cholesterol complex) form of P-450 to the low-spin (dissociated) form; this could be reversed by the addition of larger amounts of cholesterol. The spectrophotometrically-determined high spin → low spin conversion could serve to obtain a dissociation constant between substrate and P-450 (26).

The catalysis of hydroxylation was strikingly sensitive to the nature of the fatty acid side chain in the phospholipid vesicle. Figure 7 (31) shows that there is excellent correlation, for each of several phospholipids, between the spectrophotometrically-established dissociation constant for the cholesterol-P-450 complex and the observed rate of catalysis. Dr. John Salerno, then at our laboratory, suggested experiments utilizing the paramagnetic ion dysprosium (which causes, upon interaction, broadening or desaturation of EPR signals of other paramagnetic centers) to ascertain the nature of the juxtaposition of the prosthetic groups of the components (32). The ferric heme signal of cytochrome P-450 in the vesicle is affected by aqueous Dy-EDTA, showing that the heme group is accessible to the aqueous phase as well as to the lipid phase. Dy effects were abolished by ADX, showing that this aqueous phase material can interact with a heme accessible to both lipid-soluble and aqueous ligands. Similar results were seen for Dy effects on the reduced adrenodoxin signal. Complex formation with cytochrome P-450 abolished the Dy-Iron-sulfur interactions.

Figure 7. Correlation between cholesterol binding to P-450_{scc} and catalytic activity in the presence of different phospholipids (31).

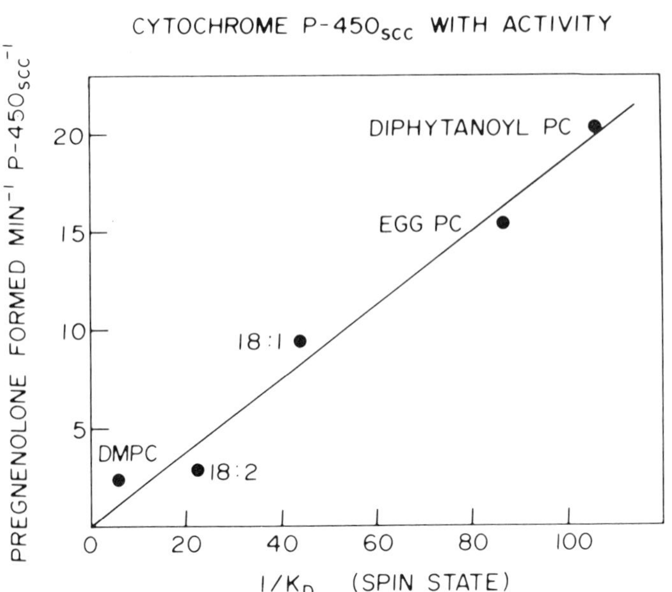

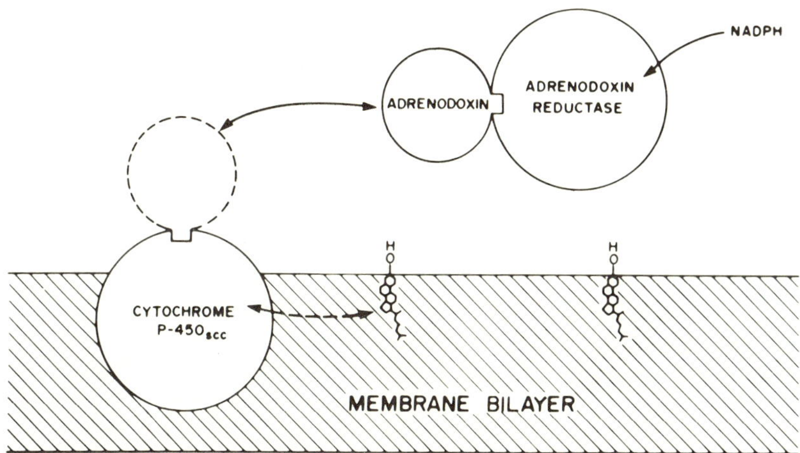

Figure 8. Interaction of components of P-450_{scc} system in phospholipid vesicles.

A model for these interactions is presented in Figure 8, which shows cytochrome P-450 in a membrane interacting with membrane cholesterol and with a reduced adrenodoxin dissociated from an aqueous-phase AR:ADX complex. Studies by Salerno et al. (32), using oriented multilayers of vesicles and centrifuged submitochondrial particles, showed that the heme plane is largely oriented parallel to the plane of the vesicle or membrane.

Studies of interactions between vesicle cytochrome P-450 and its substrates within the bilayer have shown an important control process; measuring reaction rates (as well as the extent of the low spin-high spin transformation upon addition of cholesterol) as a function of added adrenodoxin, has shown a mutually-facilitating effect: the addition of ADX markedly enhances the binding of cholesterol, and the binding of cholesterol facilitates the formation of ADX-P-450 complex (20). Thus, this hydroxylase, like the flavoprotein hydroxylases operating on organic phenols (33, 34), is substrate controlled, and futile oxygen consumption does not occur: both hydroxylatable substrate and reducing power must be simultaneously present before reaction with oxygen can occur. It remains to be seen whether the extraordinary sensitivity of this system to the fatty acid composition of the membrane could serve as one of the devices which can translate the stimuli of ACTH secretion into the enzymatic rates of steroid biosynthesis.

ACKNOWLEDGMENTS
Supported by NIH Grants GM21226 (HK) and AM27373 (JDL) and NSF Grant PCM 7924877 (H.K. and Lewis M. Siegel).

References

1. Kamin, H., Lambeth, J.D., and Siegel, L.M. (1980) In *Flavins and Flavoproteins*. Yagi, K. and Yamano, T. (eds.) vol. 6. Baltimore: University Park Press, pp. 341–348.
2. Keilin, D. (1925) *Proc Roy Soc B* 98:312–339.

3. Michaelis, L., Schubert, M.P., and Smythe, C.V. (1936) *Science* 84:138–139.
4. Michaelis, L., Schubert, M.P., and Smythe, C.V. (1936) *J Biol Chem* 116:587–607.
5. Haas, E. (1937) *Biochem Z* 290:291–292.
6. Haas, E., Horecker, B.L., and Hogness, T.R. (1940) *J Biol Chem* 136:747–774.
7. Hauge, J.G., Crane, F.L., and Beinert, H. (1956) *J Biol Chem* 219:727–733.
8. Masters, B.S.S., Kamin, H., Gibson, Q.H., and Williams, C.H., Jr. (1965) *J Biol Chem* 240:921–931.
9. Masters, B.S.S., Bilimoria, M.H., Kamin, H., and Gibson, Q.H. (1965) *J Biol Chem* 240:4081–4088.
10. Strittmatter, P. (1965) *J Biol Chem* 240:4481–4487.
11. Lambeth, J.D. and Kamin, H. (1977) *J Biol Chem* 252:2908–2917.
12. Christner, J.A., Münck, E., Janick, P.A., and Siegel, L.M. (1981) *J Biol Chem* 256:2098–2101.
13. Lambeth, J.D., Seybert, D.W., and Kamin, H. (1979) *J Biol Chem* 254:7255–7266.
14. Chu, J.-W., and Kimura, T. (1973) *J Biol Chem* 248:5183–5187.
15. Lambeth, J.D., McCaslin, D.R., and Kamin, H. (1976) *J Biol Chem* 251:7545–7550.
16. Seybert, D.W., Lambeth, J.D., and Kamin, H. (1978) *J Biol Chem* 253:8355–8358.
17. Katagiri, J., Takikawa, O., Sato, H., and Suhara, K. (1979) *Biochem Biophys Res Comm* 77:804–809.
18. Batie, C.J. and Kamin, H. (1981) *J Biol Chem* 256:7756–7763 (See also this volume).
19. Foust, G.P., Mayhew, S.G., and Massey, V. (1969) *J Biol Chem* 244:964–970.
20. Siegel, L.M., Faeder, E.J., and Kamin, H. (1972) *Z Naturforsch* 276:1087–1089.
21. Vermilion, J.L., Ballou, D.P., Massey, V., and Coon, M.J. (1981) *J Biol Chem* 256:266–277.
22. Iyanagi, T. and Mason, H.S. (1973) *Biochemistry* 12:2297–2308.
23. Omura, T., Sanders, E., Estabrook, R.W., Cooper, D.Y., and Rosenthal, O. (1966) *Arch Biochem Biophys* 177:660–673.
24. Lambeth, J.D. and Kamin, H. (1976) *J Biol Chem* 241:4299–4306.
25. Chu, J.-W. and Kimura, T. (1973) *J Biol Chem* 248:2089–2094.
26. Lambeth, J.D., Seybert, D.W., and Kamin, H. (1980) *J Biol Chem* 255:138–143.
27. Ohashi, J. and Omura, T. (1978) *J Biochem* (*Tokyo*) 83:249–260.
28. Lambeth, J.D. and Kamin, H. (1979) *J Biol Chem* 254:2766–2774.
29. Lambeth, J.D., Lancaster, J.R., and Kamin, H. (1981) *J Biol Chem* 256:3674–3678.
30. Seybert, D.W., Lancaster, J.R., Lambeth, J.D., and Kamin, H. (1979) *J Biol Chem* 254:12088–12098.
31. Lambeth, J.D., Kamin, H., and Seybert, D.W. (1980) *J Biol Chem* 255:8282–8288.
32. Salerno, J.D., Lancaster, J.R., Lambeth, J.D., and Kamin, H. (1980) *Fed Proc* 39:1824.
33. Kamin, H., White-Stevens, R.H., and Presswood, R.P. (1978) In *Methods in Enzymology*, 53. S. Fleischer and L. Packer (eds.) New York: Academic Press, pp. 527–543.
34. Husain, J., Schopfer, L.M., and Massey, V. (1978) In *Methods in Enzymology*, 53. S. Fleischer and L. Packer (eds.) New York: Academic Press, pp. 543–558.

Published 1982 by Elsevier North Holland, Inc.
Vincent Massey and Charles H. Williams, Editors
Flavins and Flavoproteins

CHAPTER 111

Interactions of Ferredoxin-NADP$^+$ Reductase With One- and Two-Electron Carriers

G. Zanetti and B. Curti

Institute of General Physiology and Biochemistry, Faculty of Science, University of Milano, Milano, Italy

Ferredoxin-NADP$^+$ reductase (FNR) functions in photosynthesis, as a part of the thylakoid membrane redox chain, transferring electrons from one-electron donors (i.e., ferredoxin, Fd) to two electron acceptors (i.e., NADP$^+$). The 1 to 2 e^- transformation is unique to such flavoproteins as FNR. In order to gain a better insight into this process, we have investigated the complex formation between FNR and one- or two-electron carriers at different stages of reduction. The existence of a complex between oxidized FNR and oxidized Fd has long been recognized (1). The same complex has been shown to be involved in the photoreduction of NADP$^+$ (2). The study of the complex between the two proteins at the reduced stage has been hampered by the difficulty of obtaining complete reduction of FNR and Fd by the usual method of EDTA-light irradiation (3), even after addition of catalytic amounts of free flavin (see Figure 1a for FNR reduction). In contrast, light irradiation in the presence of deazaflavin (4) brings about complete reduction of the enzyme in a few minutes (Figure 1b). The characteristic absorption spectrum of the neutral blue semiquinone can be measured transiently during the photoreduction. The reduction of Fd can be obtained almost as rapidly under the same conditions. The use of deazaflavin then made it possible to reduce either or both the proteins in separate compartments of an anaerobic cell; then we mixed the two solutions in order to have different numbers of electrons in the two-protein system. The spectrum was recorded. NaCl crystals were then added anaerobically to give 0.5 M salt to dissociate the complex and the spectrum was recorded again. Difference spectra were obtained either by algebraic subtraction (Figure 2a) or through the use of a computer (Figure 2b). Figure 2a shows the difference spectra obtained by mixing one mol of Fd_{ox} with one mol of completely reduced FNR (lower spectrum) or by mixing two moles of Fd_{red} with one mol of FNR_{ox} (upper spectrum). The resulting difference spectrum should be characteristic of the interactions between FNR_{red} and Fd_{ox}, considering that both electrons are donated by Fd_{red} to FNR. The difference spectrum shown in Figure 2b was obtained by mixing one mol of Fd_{red} with one mol of FNR_{ox}. This type of spectrum, which is obtained with 0.5–1 e^- in the system, seems to sum up the features of both the oxidized complex and the

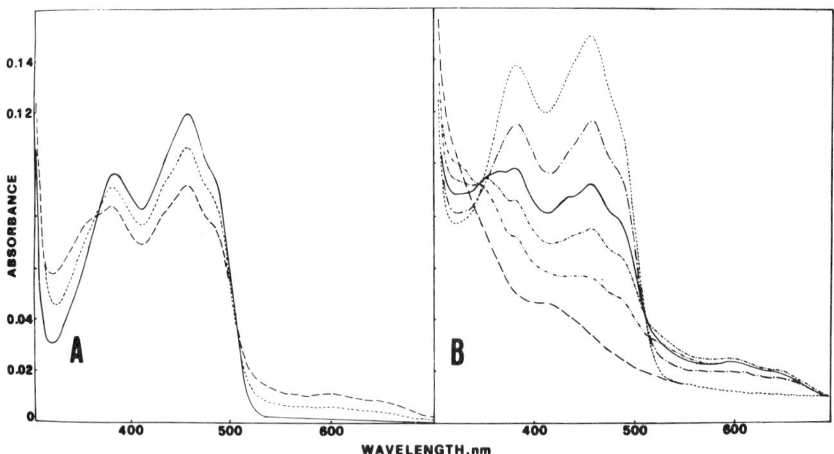

Figure 1. Photochemical reduction of FNR in the presence of EDTA. Conditions: 10 mM Hepes, pH 7, 10°C. (A) 11.2 μM FNR, 40 mM EDTA, and 1 μM lumiflavin-3-acetate. (Curves 1–3) oxidized and after 22 or 67 min of irradiation, respectively. (B) 12.2 μM FNR, 15 mM EDTA, and 1.2 μM 5-deazaflavin. (Curves 1–6) after zero, 25, 50, 75, 105, and 250 s of illumination.

two-electron reduced species. By mixing the two completely reduced proteins (a total of 3 e^-) again, a difference spectrum is obtained (5). These results indicate that a complex between Fd and the reductase which involves a perturbation of the protein chromophores is formed regardless of the oxidation-reduction states of the two components. Thus, it is tempting to postulate that ferredoxin could also form a stable complex with the flavoprotein *in vivo*.

The involvement of the same complex between FNR and Fd in the reduction of cytochrome c has been challenged by Davis and San Pietro (2) who postulated a different kind of interaction between Fd and FNR in the cyt c reaction. According to their hypothesis, a (Fd-cyt c) complex is the true substrate for the flavoprotein. We have indeed verified by difference spectroscopy the formation of a complex between Fd_{ox} and oxidized cyt c in the absence of FNR: the results are partially obscured by a concomitant cyt c reduction which seems to take place following the oxidation of the free sulfhydryl group of Fd. An interaction between these two proteins is also suggested by isoelectrofocusing experiments, where the appearance of a new band at short running times is observed when a mixture of Fd and cyt c is loaded on the gel. Nevertheless, in spectral experiments, the addition of cyt c to the complex between Fd_{ox} and FNR_{ox}, as shown in Figure 3, does not promote the dissociation of the complex, whereas an additional band is observed in the 400 nm region. These results suggest that a ternary complex of the three proteins is formed, and since no spectral perturbations were detected by mixing FNR and cyt c (1), complex formation must therefore be mediated by Fd. Thus, it seems that no mutual exclusion exists between cyt c and the reductase. The NADPH-cyt c reductase reaction should be considered a valid model for the *in vivo* reduction of $NADP^+$, in comparison to other diaphorase assays.

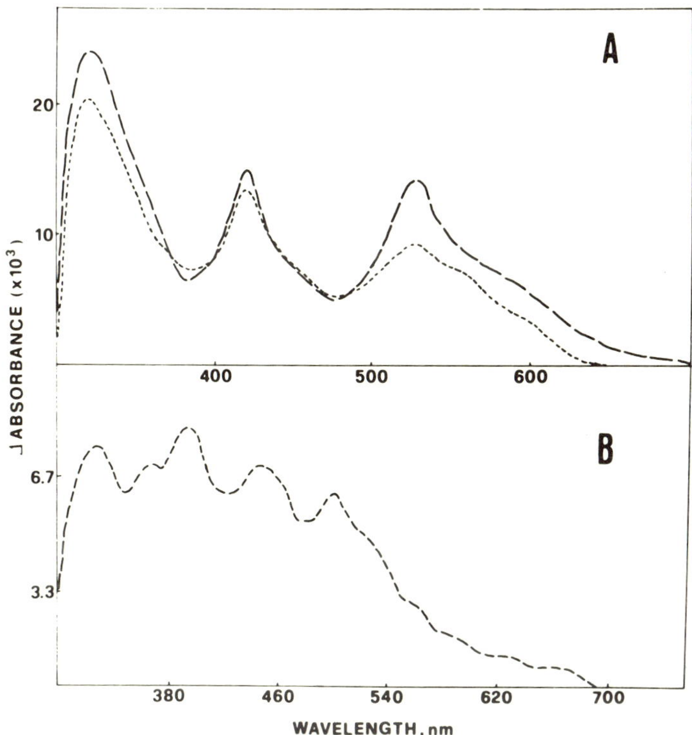

Figure 2. Difference spectra of FNR-Fd complexes obtained at different reduction levels of the two proteins. (A) Two electrons in the protein system: 50 mM Hepes, pH 7, 10°C. (Lower curve) FNR and Fd were 10 μM. (Upper curve) FNR was 10 μM and Fd 20 μM. (B) One electron in the protein system: 10 mM Hepes, pH 7, 10°C. FNR was 7.2 μM and Fd 8 μM. This difference spectrum was recorded by an Apple II plus computer interfaced to a Cary 219.

Finally, the interaction of FNR with the two-electron donor NADPH has been re-examined, in an attempt to resolve the discrepancies existing in the literature about semiquinone formation and its extinction coefficient values (6). We used a stoichiometric amount of NADPH in the presence of a regenerating system (glucose-6-phosphate and glucose-6-phosphate dehydrogenase). Immediate appearance of semiquinone was observed followed by a slow conversion to the fully reduced form. In the light of the hydride transfer mechanism of NADPH and considering the low levels of semiquinone produced by photoreduction with either lumiflavin or deazaflavin (Figure 1), we thought that the amount of semiquinone found could be due to traces of oxygen present in the system. Indeed, as shown in Figure 4b, admission of a small amount of air to the anaerobic cell considerably increased the formation of semiquinone. The semiquinone was then very slowly reduced, at a rate which was possibly limited by dismutation. These results confirm earlier findings of a rapid reoxidation to the semiquinone level of reduced FNR by traces of O_2 (7); however, in experiments similar to that of Figure 4a, the presence of an O_2 trap

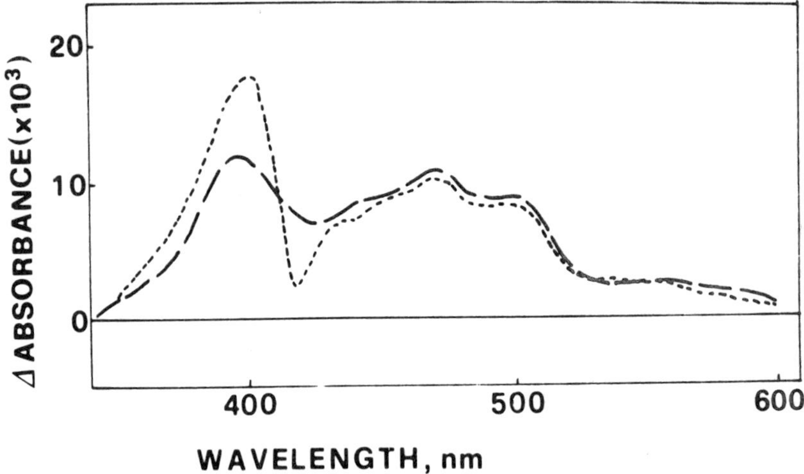

Figure 3. Effect of cytochrome c addition to the $(FNR\text{-}Fd)_{ox}$ complex. Conditions: 10 mM Hepes, pH 7, 20°C. The three proteins were each 4.5 μM. The difference spectra were recorded by using matched split-compartment cells with a mid-partition and an optical path length of 0.9 cm. (– –) $(FNR\text{-}Fd)_{ox}$ complex; (---) $(FNR\text{-}Fd)_{ox}$ complex plus cyt c.

Figure 4. Reduction of FNR by a NADPH regenerating system. (A) Conditions: 50 mM Hepes, pH 7, 8 μM FNR, 10 μM NADPH, 2 mM glucose-6-phosphate, and 10 μg glucose-6-phosphate dehydrogenase. (Curves 1–6) oxidized and 6, 13, 41, 70, 100 min after NADPH addition. (B) Same conditions as in (A) except that an O_2 trap was added to the medium: 0.12 M glucose and after anaerobiosis was established, 50 μg glucose oxidase. (Curves 1–5) oxidized, 75 min after NADPH addition and 10, 20, 50 min after a limited amount of air was admitted.

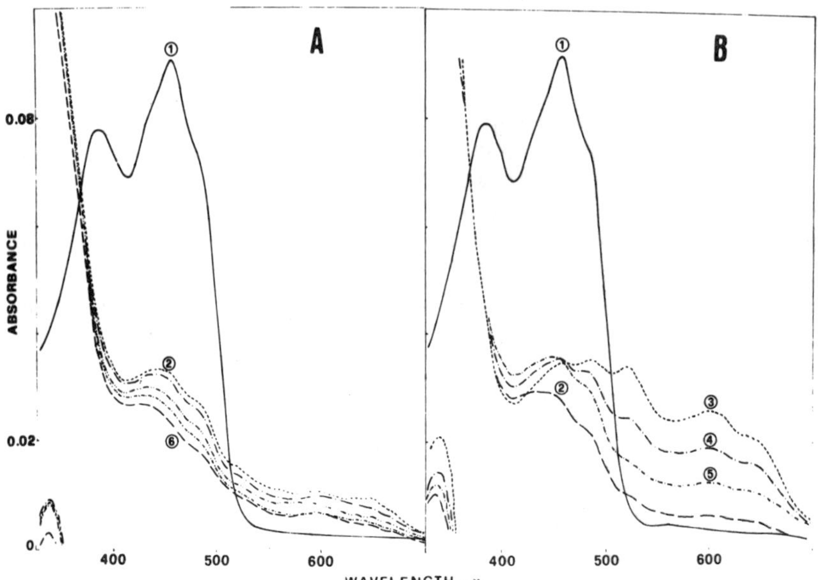

(glucose and glucose oxidase) did not diminish the extent of semiquinone formation (Figure 4b). From the maximal absorbance values obtained in several experiments, an $\varepsilon_{600\ nm}$ of 2,750 $cm^{-1}\ M^{-1}$ was calculated. Conversely, extrapolation of the data obtained from experiments similar to that of Figure 1b yields a much lower value ($\varepsilon_{600\ nm} = 1,600\ cm^{-1}\ M^{-1}$). In conclusion, although thermodynamic semiquinone stabilization (8,9) has been proposed for this type of flavoprotein, the radical we observe is due mainly to kinetic stabilization.

ACKNOWLEDGMENTS
This work has been supported by Grants from the Consiglio Nazionale delle Richerche of Italy. We thank Dr. Massey for supplying us with a sample of deazaflavin.

References

1. Foust, G.P., Mayhew, S.G., and Massey, V. (1969) *J Biol Chem* 244:964–970.
2. Davis, D.J. and San Pietro, A. (1977) *Arch Biochem Biophys* 182:266–272.
3. Massey, V. and Palmer, G. (1966) *Biochemistry* 5:3181–3189.
4. Massey, V. and Hemmerich, P. (1977) *J Biol Chem* 252:5612–5614.
5. Zanetti, G. and Curti, B. (1981) *FEBS Lett*, in press.
6. Zanetti, G. and Curti, B. (1980) *Methods Enzymol* 69:250–255.
7. Massey, V., Matthews, R.G., Foust, G.P. Howell, L.G., Williams, C.H., Jr., Zanetti, G., and Ronchi, S. (1970) In *Pyridine-Nucleotide-dependent Dehydrogenases*. H. Sund (ed.) New York: Springer-Verlag, pp. 393–409.
8. Keirns, J.J. and Wang, J.H. (1972) *J Biol Chem* 247–7374–7382.
9. Massey, V. and Hemmerich, P. (1980) *Biochem Soc Trans* 8:246–257.

Published 1982 by Elsevier North Holland, Inc.
Vincent Massey and Charles H. Williams, Editors
Flavins and Flavoproteins

CHAPTER 112

A Quantitative Analysis for Complex Formation Between Ferredoxin and Ferredoxin-NADP$^+$ Reductase by Using the Immobilized Proteins

Masateru Shin,* Naoko Sakihama,* Kazunori Niimi,*[1] and Reiko Oshino**

**Department of Biology, Faculty of Science, Kobe University, Nada-ku, Kobe 657 and **Laboratory of Biochemistry, Kobe Yamate Women's College, Chuoh-ku, Kobe 657, Japan*

Introduction

Ferredoxin-NADP$^+$ reductase (FNR, EC 1.18.1.2, NADPH: ferredoxin oxidoreductase) forms a tightly bound complex with ferredoxin at low ionic strengths in the molar ratio of one to one. The phenomenon was first observed by measuring changes in absorption spectra when these two proteins were mixed (1–3). Measurements of circular dichroism (4,5) or fluorescence intensity (5) were also useful for detecting the complex formation. However, these spectrophotometric measurements did not always prove that the two proteins actually conjugated, so that ultracentrifugal (3) or gel filtration analyses (5) had to be carried out. We here describe a new method for analyzing the complex formation by using immobilized ferredoxin or FNR on Sepharose 4B. The new method is not only as sensitive and convenient as spectrophotometric analyses, but also gives direct evidence that the two proteins actually conjugate.

Methods, Results, and Discussion

Highly purified ferredoxin or FNR was immobilized by a coupling reaction with CNBr-Sepharose 4B by the methods described previously (6). The complex formation of ferredoxin with immobilized FNR was measured as follows: The FNR-Sepharose 4B, 0.2 ml, was suspended with 20 mM Tris-HCl buffer, pH 7.5, in 1.5 ml total volume. Concentrated ferredoxin was added with mixing to the suspension in a spectrophotometric cell. On each addition the cell was directly centrifuged in a Sakuma centrifuge (Microlabofuge, Model M-15) equipped with a cell holder at 2,000 rpm for 5 min. The absorbance of the supernatant was measured and then plotted against concentrations of

[1]Present address: Otsuka Pharmaceutical Co., Ltd., 2-9 Kanda, Tsukasa-cho, Chiyoda-ku, Tokyo, Japan.

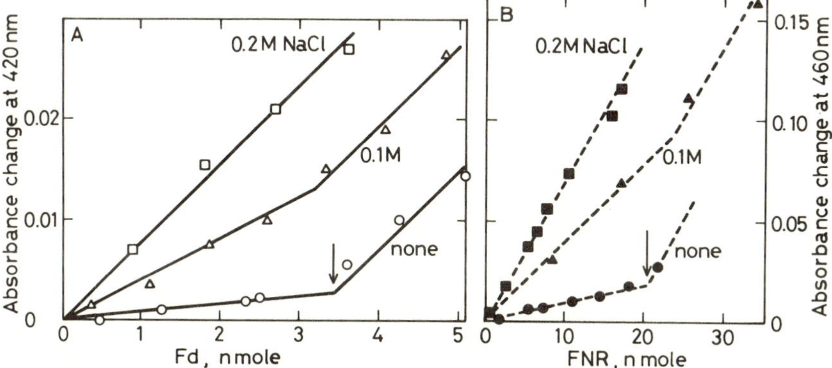

Figure 1. Titration curve of immobilized FNR or ferredoxin with the partner protein in various concentrations of NaCl. (A) Titration of FNR-Sepharose 4B with ferredoxin. (B) Titration of ferredoxin-Sepharose 4B with FNR. For experimental conditions, see text. Arrows indicate saturation point.

ferredoxin added (Figure 1A). The complex formation of FNR with immobilized ferredoxin was also achieved in the same manner except that concentrated FNR was added to ferredoxin-Sepharose 4B (Figure 1B).

The experimental data obtained were treated quantitatively as follows: The slopes before and after the saturation point were defined as initial slope, G_i, and saturation slope, G_{sat}, respectively. The initial slope varied depending on ionic strength, whereas the saturation slope was constant regardless of ionic strength. Therefore, the percentage of ferredoxin (or FNR) bound in various conditions was estimated from the slopes by using Equation (112.1).

$$\text{Ferredoxin (or NNR) bound, } \% = \left(1 - \frac{G_i}{G_{sat}}\right) \times 100 \qquad (112.1)$$

Table 1 summarizes the fate of ferredoxin or FNR bound on the immobilized partner protein at various concentrations of NaCl. The dissociation constant of the complex, K, is expressed by Equation (112.2) in the titration experiment,

$$\frac{[\mathrm{Fd}]_{free}}{[\mathrm{Fd}]_{total} - [\mathrm{Fd}]_{free}} = \frac{K + [\mathrm{Fd}]_{free}}{[\mathrm{FNR}]_{gel\text{-}bound}} \qquad (112.2)$$

where $[\mathrm{Fd}]_{free}$ is the concentration of ferredoxin present in the medium; $[\mathrm{Fd}]_{total}$, that of ferredoxin added; and $[\mathrm{FNR}]_{gel\text{-}bound}$, that of FNR immobi-

Table 1. Analyses for Complex Formation of Ferredoxin or FNR With the Immobilized Partner Protein at Various Concentrations of NaCl.

NaCl concentrations	(M)	0	0.1	0.2 or more
Ferredoxin bound on FNR-Sepharose 4B	(%)	89.6	48.4	0
FNR bound on ferredoxin-Sepharose 4B	(%)	87.5	43.1	0

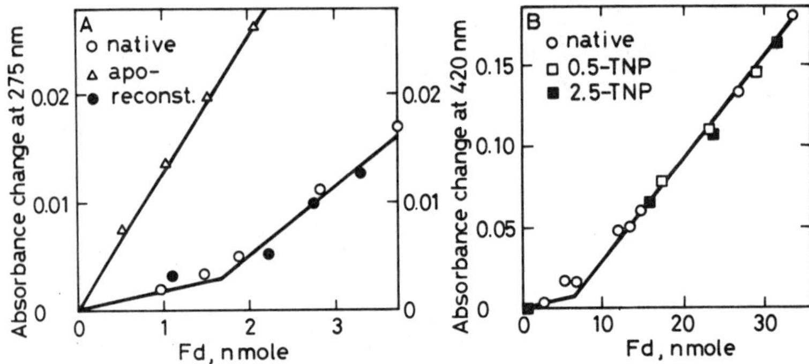

Figure 2. Analyses of complex formation of chemically modified ferredoxins. (A) native-, apo-, and reconstituted ferredoxins. (B) trinitrophenylated ferredoxin. For experimental conditions, see text.

lized on the Sepharose. Based on Equation (112.2), K was evaluated graphically at 5×10^{-7} M by plotting $[Fd]_{free}/[Fd]_{total} - [Fd]_{free}$ against $[Fd]_{free}$. This value was identical with that observed in the free system (3).

The new method proved useful in analyzing the complex formation of chemically modified ferredoxins as well. Apoferredoxin, prepared according to Hong and Rabinowitz (7), no longer had this activity, but it was recovered by reconstituting the iron-sulfur center (Figure 2A). Adrenodoxin, kindly supplied by Prof. Dr. T. Yamano, Osaka University, was confirmed to form the complex with spinach FNR, but apoadrenodoxin did not. Trinitrophenylated ferredoxin, prepared according to Davis and San Pietro (8), formed the complex just as the native one did (Figure 2B).

References

1. Nelson, N. and Neumann, J. (1968) *Biochem Biophys Res Commun* 30:142–147.
2. Shin, M. and San Pietro, A. (1968) *Biochem Biophys Res Commun* 33:38–42.
3. Foust, G.P., Mayhew, S.G., and Massey, V. (1969) *J Biol Chem* 244:964–970.
4. Cammack, R., Neumann, J., Nelson, N., and Hall, D.O. (1971) *Biochem Biophys Res Commun* 42:292–297.
5. Shin, M. (1973) *Biochim Biophys Acta* 292:13–19.
6. Shin, M. and Oshino, R. (1978) *J Biochem* 83:357–361.
7. Hong, J.-S. and Rabinowitz, J.C. (1967) *Biochem Biophys Res Commun* 29:246–249.
8. Davis, D.J. and San Pietro, A. (1977) *Biochem Biophys Res Commun* 74:33–40.

Published 1982 by Elsevier North Holland, Inc.
Vincent Massey and Charles H. Williams, Editors
Flavins and Flavoproteins

CHAPTER 113

Steady State and Transient Kinetics of Oxidation of Reduced Ferredoxin With Ferredoxin-NADP$^+$ Reductase

Ryuichi Masaki,* Midori Matsumoto,**
Shinya Yoshikawa,** and Hiroshi Matsubara*

**Department of Biology, Faculty of Science, Osaka University, Toyonaka, Osaka and*
***Department of Biology, Konan University, Higashinada, Kobe, Japan*

Introduction

Ferredoxin-NADP$^+$ reductase (FNR, EC1.18.1.2) functions as the terminal enzyme in the photosynthetic NADP$^+$ reduction. It transfers electrons from reduced ferredoxin (Fd) to NADP$^+$. Due to the strong autooxidizability of reduced Fd, the kinetic studies have mainly been carried out on the electron transfer reaction from NADPH to nonphysiological electron acceptors such as DCIP, NAD$^+$, and cytochrome c (1). This system will be referred to as the *artificial reaction* in this paper. We studied the kinetics of the *physiological reaction*, namely, the electron transfer from reduced Fd to NADP$^+$ mediated by FNR. Our results showed that the reaction was one of a sequential mechanism and the active intermediate of FNR was the fully reduced FNR. These results are in conflict with those obtained from the *artificial reaction* (1).

Materials and Methods

Ferredoxin and FNR were prepared from *Spirulina platensis*, a blue-green alga (2,3). The reduced Fd was prepared by adding sodium dithionite solution stoichiometrically under an atmosphere of argon.

Results and Discussion

Anaerobic oxidation of reduced Fd with a catalytic amount of FNR was followed by an absorbance increase at 420 nm. The reaction was of apparent first-order with respect to reduced Fd concentration. The first-order rate constant was dependent on the total concentration of Fd. Therefore, the initial steady state rate was obtained by multiplying the total concentration of Fd by the first-order rate constant. Figure 1A shows the Lineweaver-Burk plots at various concentrations of NADP$^+$. Remarkable deviations from the straight line at low NADP$^+$ concentrations suggest substrate inhibition. In the Fd concentration range in which the apparent substrate inhibition was not promi-

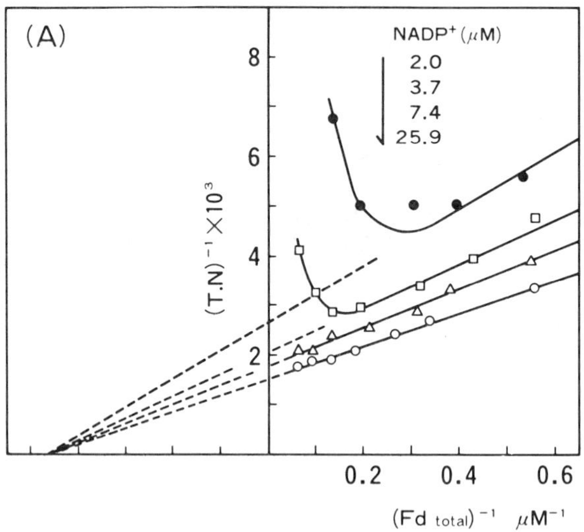

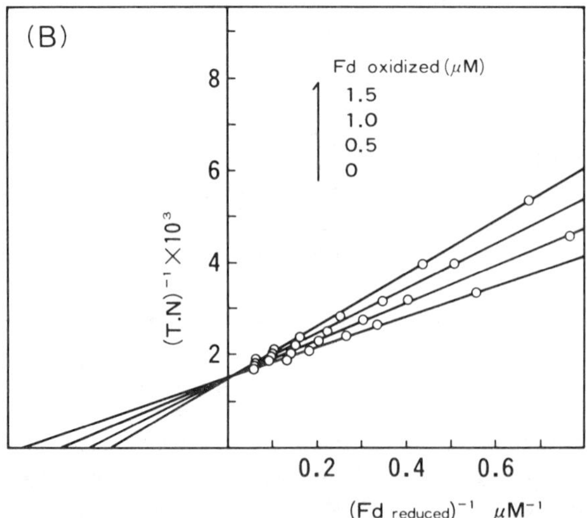

Figure 1. Lineweaver-Burk plots of FNR activity. The steady state velocities in 20 mM Tris-HCl buffer, pH 8.0, containing 75 mM NaCl are expressed in terms of turnover number (T.N.) by dividing the initial velocities (μM/sec) by the enzyme concentration. (A) The initial velocity patterns in the presence of various concentrations of $NADP^+$. (B) The effect of oxidized Fd on the enzyme activity in the presence of 25.9 μM $NADP^+$.

nent, the relation between the activity and substrate concentration showed a set of straight Lineweaver-Burk plots with an intersecting point on the abscissa (Figure 1A). This relationship is expressed as $V_{max}/v_0 = [1 + K_m{}^F/(Fd)_t][1 + K_m{}^D/(NADP^+)]$ where V_{max}, $K_m{}^F$, and $K_m{}^D$ denote maximum velocity and Michaelis constants for Fd and $NADP^+$, respectively. The experimental values were $V_{max} = 700\ s^{-1}$, $K_m{}^F = 2.2\ \mu M$, and $K_m{}^D = 1.7\ \mu M$. This result suggests a sequential mechanism including a ternary complex of the two substrates and FNR, and excludes a ping-pong mechanism (4) which would give a set of parallel double reciprocal plots obtained in the *artificial reaction* (3).

The effect of oxidized Fd on the enzyme activity was examined by measuring the rate at an appropriate time during the enzymic oxidation of reduced Fd. The effect of oxidized Fd on the Lineweaver-Burk plots (Figure 1B) indicates that the oxidized Fd acts as a competitive inhibitor to the reduced Fd. In a sequential mechanism including two substrates (A, B) and the products (P, Q), one of the products (Q) which is to dissociate from the enzyme at the last step in the reaction inhibits competitively the association of the first substrate to react with the free enzyme (4). In the present system, the reduced and oxidized Fd act as A and Q, respectively. The above results and consideration give a possible reaction scheme of FNR with the substrate as shown in Figure 2A.

When FNR (4 μM) was reacted anaerobically with dithionite-reduced Fd (8 μM) in the presence of $NADP^+$ (80 μM) in a stopped flow apparatus, a rapid increase in absorbance ($t_{1/2}$=6 msec at 20°C) was observed. The absorbance difference at the end of the reaction and that at 4 msec after mixing was determined at different wavelengths, as given in Figure 3 (–●–). This spectrum (two maxima at 470 and 390 nm) resembles the difference spectrum of oxidized versus reduced FNR. Therefore, the electron transfer from reduced Fd to FNR seems to be completed within the dead time of the apparatus (4 msec) and the observed spectral change is due to the oxidation of the fully reduced FNR with $NADP^+$. This result is also in conflict with those of the

Figure 2. (A) A reaction scheme for the *physiological reaction*. (B) A ping-pong mechanism proposed for the *artificial reaction*. D denotes the artificial electron acceptor.

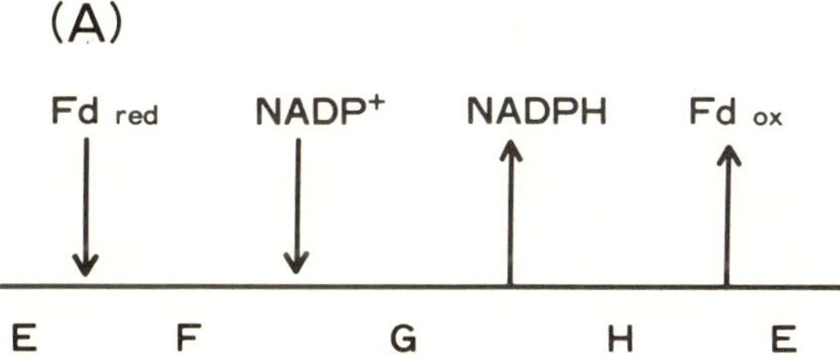

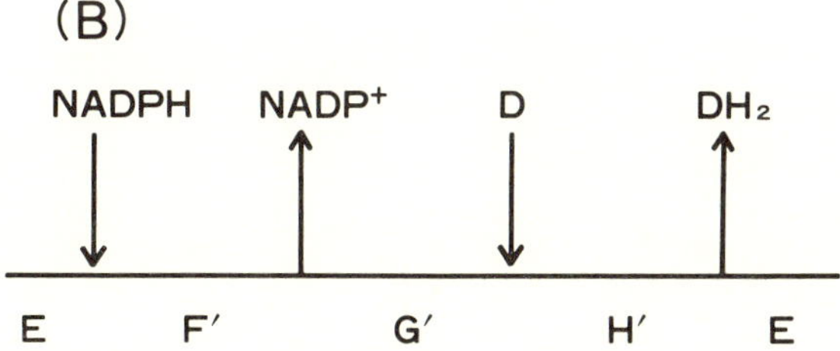

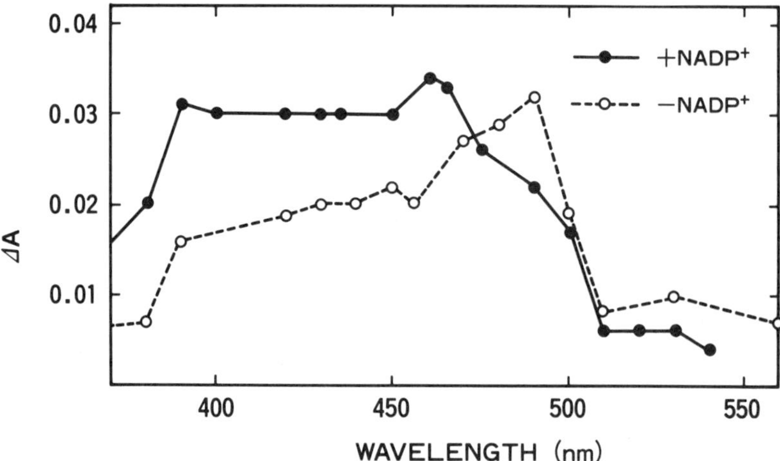

Figure 3. The difference spectra at the times of infinite and 4 msec after mixing the dithionite reduced Fd with FNR in a stopped flow apparatus in the presence (-●-) and absence (--○--) of $NADP^+$ in 25 mM Tris-HCl buffer, pH 8.0. Final concentrations of Fd, FNR, and $NADP^+$ are 8, 4, and 80 μM, respectively.

artificial reaction system ($NADP^+$ diaphorase reaction with ferricyanide), in which a semiquinone form was detected as an active intermediate (5).

Both the steady state and transient kinetic studies suggest that the *artificial reaction* is not merely the reverse reaction of the *physiological reaction*; in other words, it is unlikely that the *artificial reaction* proceeds via the same reaction path as that of the *physiological reaction* in the opposite direction.

When reduced Fd was mixed with oxidized FNR in the absence of $NADP^+$, no significant absorbance change was observed, except for a slow absorbance increase due to the formation of fully oxidized FNR as a result of oxygen leakage into the stopped flow apparatus. The slow spectral change characterized by a maximum at 490 nm (--○--) in Figure 3 is clearly different from the rapid spectral change in the presence of $NADP^+$, suggesting that the intermediate species produced within the first 4 msec in the absence of $NADP^+$ is different from that in the presence of $NADP^+$. Namely, $NADP^+$ seems to affect the reaction of oxidized FNR with reduced Fd.

References

1. Shin, M. (1968) In *Flavins and Flavoproteins*. Yagi, K. (ed.) Tokyo: University of Tokyo Press and Baltimore: University Park Press, pp. 1–12.
2. Wada, K., Hase, T., Tokunaga, H., and Matsubara, H. (1975) *FEBS Lett* 55:102–104.
3. Masaki, R., Wada, K., and Matsubara, H. (1979) *J Biochem* 86:951–962.
4. Cleland, W.W. (1970) In *The Enzymes*. Boyer, P.D. (ed.) Vol. 12. New York and London: Acad. Press, pp. 1–56.
5. Massey, V., Matthews, R.G., Foust, G.P., Howell, L.G., Williams, C.H., Jr., Zanetti, G., and Ronchi, S. (1970) In *Flavins and Flavoproteins*. Amsterdam: Elsevier Sci. Publ. Co., pp. 393–411.

Published 1982 by Elsevier North Holland, Inc.
Vincent Massey and Charles H. Williams, Editors
Flavins and Flavoproteins

CHAPTER 114

Equilibria of the Complexes of Ferredoxin : NADP$^+$ Reductase With Ferredoxin and NADP(H)

Christopher J. Batie and Henry Kamin

Duke University Medical Center, Durham, North Carolina

Ferredoxin: NADP$^+$ reductase (FNR) catalyzes photosynthetic NADP$^+$ reduction in green plants; electrons flow from ferredoxin (Fd) to FNR, an FAD protein, and thence to NADP$^+$. This system of electron transport is chemically analogous to the adrenodoxin-adrenodoxin reductase system which has been studied in our laboratory (1). In the spinach system, as in the adrenal, the flavoprotein forms complexes with each of its substrates; these complexes have characteristic difference mixing spectra (2). Our studies deal with the relation of these complexes to each other, and with the role that complex formation plays in electron transfer. Our goals are (1) to determine whether the reduction state of the components affects the association of the Fd: FNR and NADP(H): FNR complexes and, conversely, whether complex formation affects the redox behavior of any component, and (2) to determine whether NADP$^+$ reduction proceeds via only bimolecular complexes or via a ternary complex of all three components.

Effects of pH and Redox State on the Fd : FNR Complex

Initial studies of the relation of pH to the association of the Fd : FNR complex revealed that from pH 6 to pH 8 the K_d of the complex increased with pH; K_d was constant above pH 8.5 (3). We explained this pH-dependence as the consequence of binding one proton upon complex formation to a group with a pK_a of 8.2 in the complex, and with a pK_a of 6.0 on the free protein, which could be either Fd or FNR.

We have previously used Dy^{3+} to probe paramagnetic centers for changes in probe accessibility accompanying complex formation (4). The Dy^{3+} ion has an effective spin of 15/2 and will strongly affect nearby paramagnetic centers (reduced Fd and FNR semiquinone here) by broadening their spectra through dipolar interactions. We used Dy^{3+} chelated by EDTA, a negatively charged complex, or Dy^{3+} in the positively charged o-phenanthroline complex. Reduced Fd interacted strongly with Dy^{3+} (Figure 1A) as shown by the broadening of the Fd epr spectrum when Dy^{3+} was included in the dithionite-reduced sample (--- Fd; ·–·–·– · Fd+o-phenanthroline-Dy). The o-phenanthroline chelate broadened the Fd signal more than did the EDTA chelate, a result compatible with expected Coulomb interactions with negatively charged Fd.

Inclusion of FNR in the sample substantially shielded the Fd from Dy induced line broadening (— Fd : FNR + o-phenanthroline-Dy). The epr spectrum of the FNR semiquinone produced by aerobic addition of excess NADPH was studied by similar means (Figure 1B). FNR semiquinone has a 19 gauss line (—) which was only slightly broadened by o-phenanthroline-Dy (---), but was broadened to a 39 gauss line by EDTA-Dy (·—·—); addition of Fd to this last sample (···) narrowed the line to 32 gauss. These results show that Fd and FNR form reduced complexes (the one- and three-electron reduced complexes in this case) in which the accessibility of both the iron-sulfur center and the flavin to the dysprosium probes is substantially reduced.

If reduction of one component of a complex affects the extent of association of the complex, then complex formation will alter the redox behavior of that component. Potentiometric titrations of the Fd : FNR complex were used to test this possibility. When titrated separately, measuring spectra and E_h simultaneously in a Dutton cell, at pH 8.4, Fd and FNR had similar E_ms, −420 mV and −442 mV respectively. Figure 2 displays data from oxidative and reductive titrations of the Fd : FNR complex. The E_m of FNR in the complex was 17 mV more positive than when measured separately, −425 mV. Though the data for Fd reduction deviate substantially from the Nernstian n = 1 curve drawn through the points, the best fit was for an E_m of −510 mV, 90 mV more negative than uncomplexed Fd. The large change in the Fd potential was also observed when neutral red, methyl viologen, or 10-fold excess FNR was used to indicate E_h; thus formation of the Fd : FNR complex makes electron transfer from Fd to FNR highly favored. The change in the potential of the Fd couple on binding to FNR can be interpreted as a 30-fold increase in K_d on Fd reduction. Quantitative differences between these data and those recently reported by Smith et al. (5) may arise from differences of pH and of the techniques to determine the redox state of Fd and FNR.

Figure 1. Dy^{3+} effects on the epr signals of FNR_{sq} and Fd_{red}, separately and in complex. (A) Samples were prepared with 100 μM Fd, excess dithionite, and as indicated 100 M FNR, 1 mM o-phenanthroline-Dy, or 1 mM EDTA-Dy. epr spectra were recorded at 10 K, and 20 mW. (B) Samples prepared by aerobic addition of 500 μM NADPH to 100 μM FNR, and as indicated 100 μM Fd, 1 mM o-phenanthroline-Dy, or 2 mM EDTA-Dy. Spectra recorded at 20 K, and 1 mW.

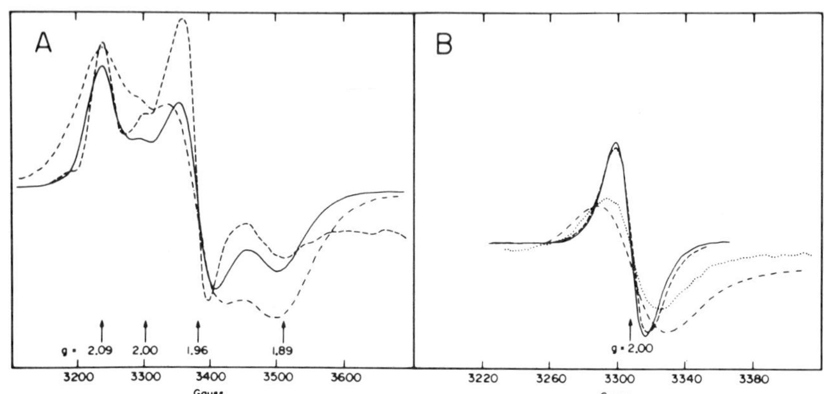

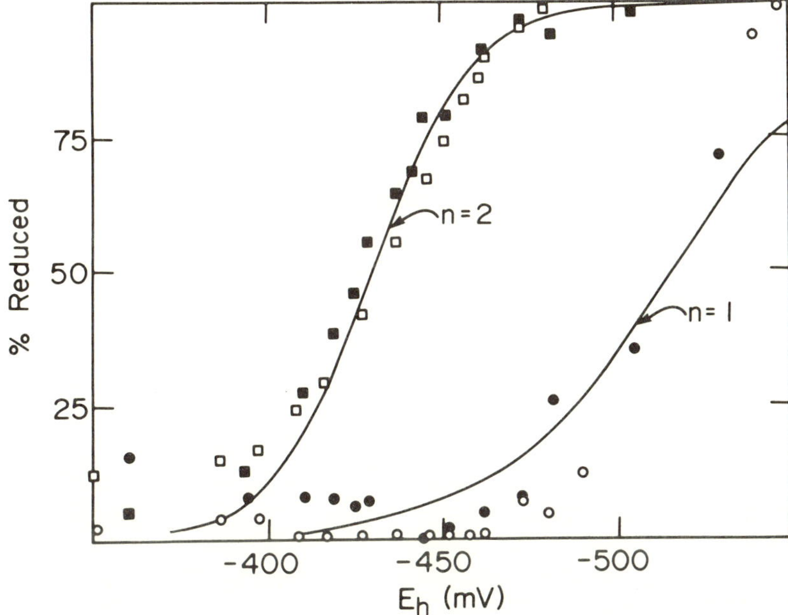

Figure 2. Potentiometric titration of Fd:FNR complex. Titrations conducted in a Dutton cell using dithionite as reductant (closed symbols), and ferricyanide as an oxidant (open symbols). E_h measured using platinum and calomel electrodes with benzyl and methyl viologen as mediators. Reduction of each species calculated from spectra. (□ ■) FNR; (○ ●) Fd.

Complexes of NADP(H) with FNR

We also investigated complexes of $NADP^+$ and NADPH with oxidized FNR. Mixing $NADP^+$ with FNR produced a mixing spectrum (2), which was stable in our hands. From pH 6.5 to pH 9 the mixing spectrum remained unchanged, as did the K_d of 60 μM when measured in 50 mM buffer. When we titrated NADPH into FNR anaerobically, we produced a green species with a peak at 460 nm and a broad long-wavelength band (Figure 3A, line b) similar to that previously reported (8). Titration of 150 μM FNR indicated a 1:1 NADPH:FNR complex, epr silent, and stable in the absence of O_2. During titration of 25 μM FNR the A_{600} changed as shown in Figure 3B. The line is the best least-squares fit to the date, $K_d = 6.08$ μM, and $\Delta\varepsilon_{600} = 1500$ M^{-1} cm^{-1}; NADPH seems to bind to FNR with an apparent K_d of about one tenth that of the $NADP^+$:FNR complex. The two-electron reduced species can also be produced by mixing photoreduced FNR with $NADP^+$, and thus may be an intermediate in the *in vivo* reaction. NADH or an NADPH-generating system did not produce the species.

The Relation of the Fd:FNR and $NADP^+$:FNR Complexes

To investigate the possibility that FNR forms a ternary Fd:FNR:$NADP^+$ complex, we used both FNR binding to immobilized ligands, Fd-Sepharose

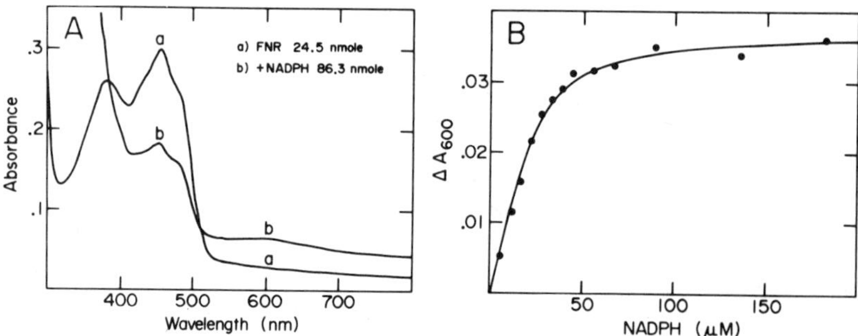

Figure 3. Anaerobic titration of FNR with NADPH. (A) Spectra of: (a) initial FNR solution, (b) after addition of 3.5 fold molar excess NADPH. (B) The plot of ΔA_{600} vs NADPH added.

(6), and 2′,5′ADP-Sepharose, and difference mixing spectra. We found that FNR bound to an Fd-Sepharose column was eluted by 3 mM $NADP^+$ (Figure 4B), but not by a buffer of equal ionic strength. This suggests that $NADP^+$ and Fd compete for complexation with flavoprotein. A 0.1 mM pulse of $NADP^+$ eluted from the Fd-Sepharose column in the same position whether or not FNR was bound to the column, and thus was not retarded by formation of a ternary complex Conversely, FNR bound to the $NADP^+$ analogue, 2′,5′ADP-Sepharose was eluted by 5 μM Fd in a sharp peak (Figure 4A). When we measured formation of Fd:FNR and $NADP^+$:FNR complexes by difference mixing spectra, we found that the spectrum of the Fd:FNR complex could be changed to that of the $NADP^+$ complex by addition of

Figure 4. Elution of FNR from affinity columns. (A) FNR was applied to a 2′,5′ADP-Sepharose column washed with 10 mM MOPS buffer, pH 7. After washing, 5 μM Fd was applied to elute the FNR. FNR assayed by ferricyanide reductase activity, Fd by A_{420}. (B) FNR applied to an Fd-Sepharose column and washed with 10 mM Tris buffer, pH 8.0. 3 mg/ml $NADP^+$ was applied; after elution of 20 ml, 0.5 M NaCl was applied to elute any remaining FNR. FNR assayed as before, $NADP^+$ as A_{260}.

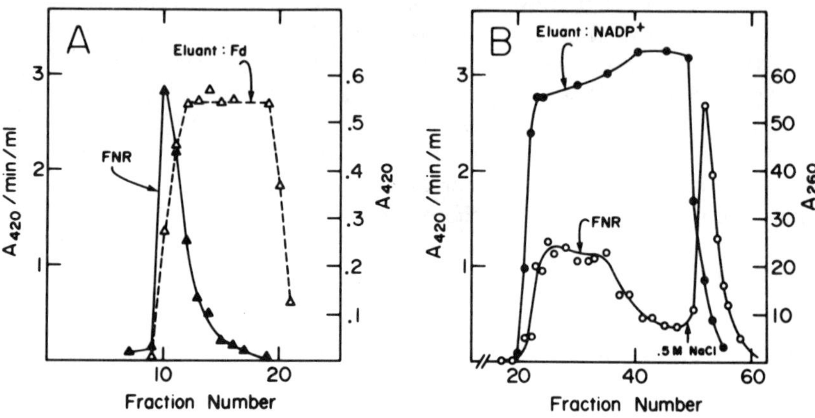

sufficient $NADP^+$ and vice versa. Intermediate difference spectra could be accounted for as the sum of Fd : FNR and $NADP^+$: FNR complexes. Thus, it appears that FNR will only form bimolecular complexes with its substrates under equilibrium conditions, and the Fd and $NADP^+$ compete for binding to the $NADP^+$ reductase. These results appear to be in conflict with the conclusions of Ricard et al., (7), who suggest a ternary complex.

ACKNOWLEDGMENTS
Supported by NIH Grant GM 21226 and NSF Grant PCM 7924877.

References

1. Lambeth, J.D. and Kamin, H. (1976) *J Biol Chem* 252:2908–2917.
2. Foust, G.P., Mayhew, S.G., Massey, V. (1969) *J Biol Chem* 244:964–970.
3. Batie, C.J. and Kamin, H. (1981) *J Biol Chem*, in press.
4. Salerno, J.C., Lancaster, J.R., Lambeth. J.D., and Kamin, H. (1980) *Fed Proc* 39:1824.
5. Smith, J.M., Smith, W.A., and Knaff, D.B. (1981) *Biochem Biophys Acta* 635:405–411.
6. Shin, M. and Oshino, R. (1978) *J Biochem* (*Tokyo*) 83:357–361.
7. Ricard, J., Nari, J., and Diamantidis, G. (1980) *Eur J Biochem* 108:55–66.
8. Massey, V., Matthews, R.G., Foust, G.P., Howell, L.G., Williams Jr., C.H., Zannetti, G., and Ronchi, S. (1969) In *Pyridine Nucleotide-Dependent Dehydrogenases*. (ed. Sund, H.) Berlin: Springer-Verlag, pp. 393–411.

Published 1982 by Elsevier North Holland, Inc.
Vincent Massey and Charles H. Williams, Editors
Flavins and Flavoproteins

CHAPTER 115

Resonance Raman Study on the Reaction Intermediates of Adrenodoxin Reductase and NADPH

H. Sakamoto,* T. Kitagawa,** T. Sugiyama,***
Y. Miyake,* and T. Yamano***

**Department of Biochemistry, National Cardiovascular Center Research Institute, Suita, Osaka, Japan; **Department of Molecular Physiological Chemistry, Osaka University Medical School, Nakanoshima, Kitaku, Osaka, Japan; and ***Department of Biochemistry, Osaka University Medical School, Nakanoshima, Kitaku, Osaka, Japan*

Summary

The stable complexes formed in anaerobic reduction of adrenodoxin reductase (AdR) by NADPH were investigated with resonance Raman, EPR, and absorption spectroscopies. There existed at least two molecular species assignable to a blue semiquinone and the $AdRH_2$-$NADP^+$ CT complex. Their relative populations depended upon pH, temperature, and the concentration of NADPH. The identical complexes were also obtained from photoreduction of AdR and reduction of AdR by NADH, both in the presence of $NADP^+$. We note that the isoalloxazine ring of the $AdRH_2$-$NADP^+$ CT complex adopts practically the structure of the oxidized form.

Adrenodoxin reductase (NADPH: adrenal ferredoxin oxidoreductase, EC 1.6.7.1) contains one FAD coenzyme per molecule ($M_r = 54{,}000$) and functions as an electron transferase in the adrenal mitochondrial steroid hydroxylase system (1). Lambeth and Kamin demonstrated that adrenodoxin reductase (AdR) and NADPH in anaerobic conditions form a tightly bound one-to-one complex having no EPR signal but distinctive long wavelength absorption, and assigned it to a charge transfer (CT) complex of $AdRH_2$ and $NADP^+$. Sugiyama et al. (3), on the other hand, found an enzyme-substrate complex (AdR-NADPH) in their rapid-scan stopped-flow spectrophotometry during reduction of AdR by NADPH. The intermediate displayed a broad absorption around 520 nm.

Resonance Raman scattering from flavoproteins reveals the vibrational spectrum of a flavin coenzyme interacting with its immediate environment (4, 5). The spectra of the isotope-labelled riboflavin allowed the assignments of representative Raman lines (6), and the characteristic Raman spectra of the Old Yellow Enzyme-phenol CT complexes enabled us to deduce the arrangement of FMN and phenol (7, 8). Accordingly, the technique may provide some structural information on the intermediates present in the anaerobic reduction of AdR by NADPH. In the present study, we adopted resonance Raman as

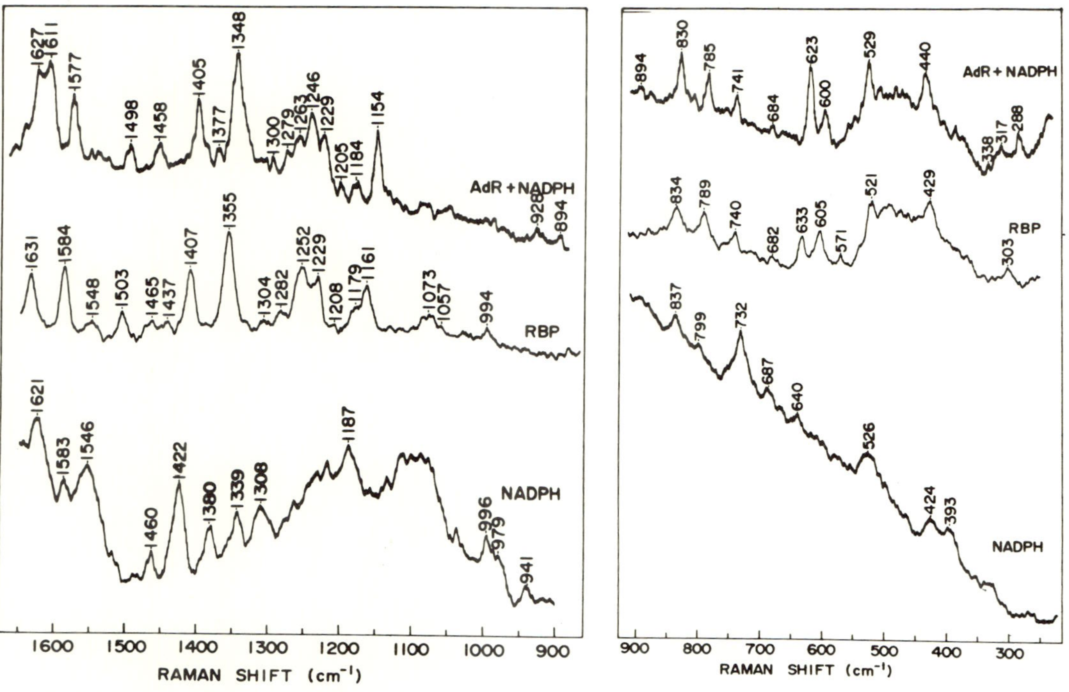

Figure 1. The resonance Raman spectra of the AdR-NADPH mixture (top), oxidized riboflavin bound to egg riboflavin binding protein (middle), and free NADPH (bottom) excited at 488.0 nm.

well as EPR and absorption spectroscopies to study the flavin structure of the redox intermediates.

Materials and Methods

AdR was isolated from bovine adrenocortical mitochondria and purified with immobilized adrenodoxin-Sepharose affinity chromatography as reported elsewhere (9). The concentration of AdR was determined spectrophotometrically with $\varepsilon_{mM} = 11.3$ at 450 nm, pH 7.4. For the measurement of Raman spectra, 50 μL of 0.35 mM AdR solution in 50 mM phosphate buffer, pH 7.4, was put into a cylindrical cell with a stopcock and was degassed by repeated evacuation and flushing with Ar gas. After addition of NADPH, the cell was evacuated to 0.01 Torr, the stopcock closed, and the cell placed in a temperature-controlled cell holder. For the measurement of EPR, the AdR solution in the EPR tube with an air-tight septum cap was degassed by evacuation to 0.03 Torr first and then NADPH was added into the tube without incorporation of air.

Results and Discussion

A typical resonance Raman spectrum of the AdR-NADPH reaction mixture excited at 488.0 nm is compared with the spectra of oxidized riboflavin bound to riboflavin binding protein and of free NADPH in Figure 1. The oxidized AdR alone showed very strong fluorescence and reduced AdR alone showed no Raman line under the present conditions. The Raman spectrum was never observed for the reduction of AdR by NADH. The whole spectrum of the

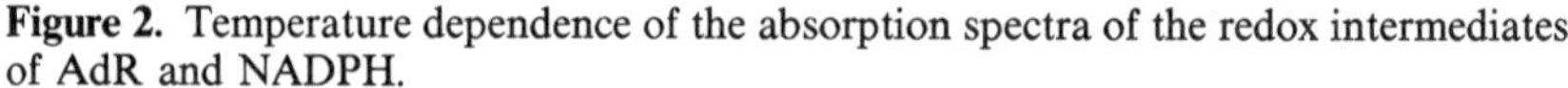
Figure 2. Temperature dependence of the absorption spectra of the redox intermediates of AdR and NADPH.

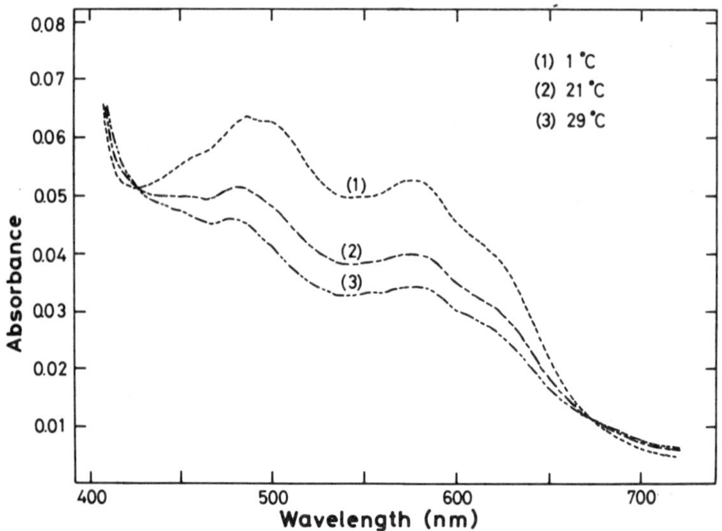

AdR-NADPH reaction mixture closely resembled the spectrum of the oxidized riboflavin except for the Raman line at 1611-cm^{-1}. The behavior of the 1611-cm^{-1} line was distinct from that of the 1627-cm^{-1} line and the feature was most apparent in the relative intensity of two lines at 1611- and 1627-cm^{-1}. The 1611-cm^{-1} line was remarkably intensified upon excitation at 514.5 nm while the 1627-cm^{-1} line was intensified on excitation at 460–480 nm. The 1611-cm^{-1} line was more intense at acid pH and the 1627-cm^{-1} line more at alkaline pH. Their relative intensity changed reversibly with temperature, the 1627-cm^{-1} line being more intense at higher temperature. The 1611-cm^{-1} line was intense when higher concentrations of NADPH were used.

When NADPH was added to the AdR solution, the sample became colorless first and violet second, and then remained unaltered for a few days. Figure 2

Figure 3. The EPR and the resonance Raman spectra of the AdR-NADPH mixture with the higher concentration of NADPH.

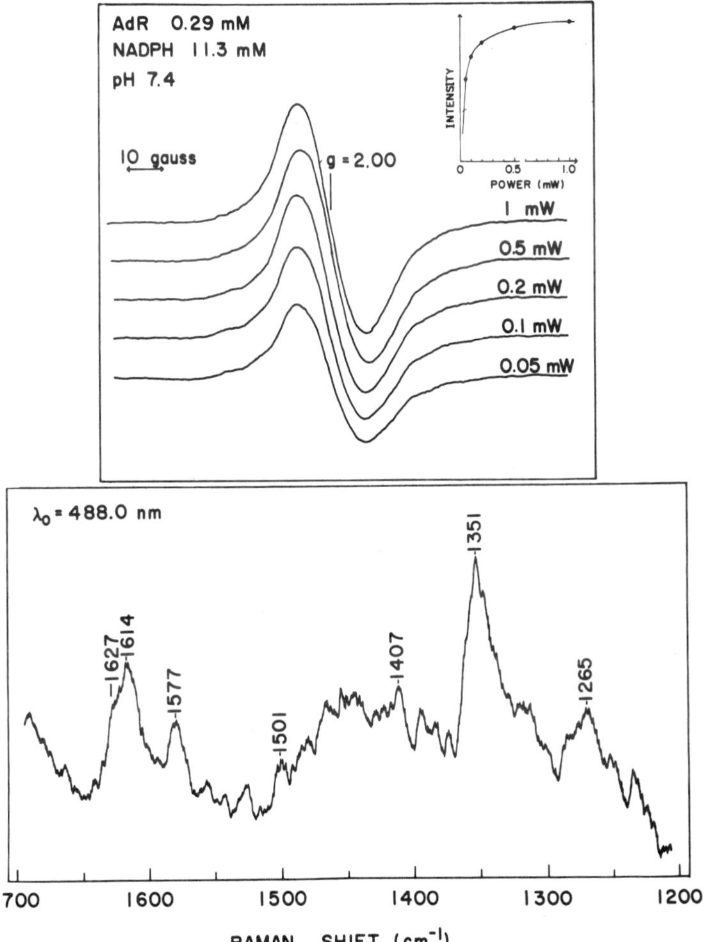

shows the temperature dependence of the absorption spectra after becoming stationary. The spectrum changed reversibly with isobestic points at 655 and 427 nm. Therefore it became evident that two molecular species are in equilibrium in which one has absorption maxima around 510 and 580 nm like a blue semiquinone and the other around 450, 480, and 700 nm like the $AdRH_2$-$NADP^+$ CT complex.

The AdR and $NADP^+$ mixture showed intense fluorescence and no Raman line was observed. However, when NADH was added to the solution, or when EDTA was added and irradiated by the 457.9 nm line in the presence of $NADP^+$, the same spectrum as shown in Figure 1 was observed. On the contrary, in the absence of $NADP^+$, the spectrum showed the reduced form. Therefore, neither 1611- nor 1627-cm^{-1} species can be ascribed to the AdR-$NADP^+$ CT complex.

Figure 3 shows the EPR spectrum of the AdR and NADPH mixture at 77 K and the Raman spectrum of the identical sample at room temperature. The $g=2.00$ signal was clearly observed and this signal was very stable and could be observed even after 8 days. This sample exhibited an intense 1611-cm^{-1} line but a very weak 1627-cm^{-1} line. The integrated intensity was larger at pH 6.6 than at pH 7.7. Thus, the 1611-cm^{-1} line was more intense when the $g=2.00$ signal was more intense. Consequently, we conclude that the 1611-cm^{-1} species corresponds to a blue semiquinone with absorption maxima around 520 and 580 nm. The 1627-cm^{-1} species may be assigned to the $AdRH_2$-$NADP^+$ CT complex (2) judging from the similarity in the absorption spectra of the two complexes. Nevertheless, we emphasize the resemblance of its Raman spectrum to those of oxidized flavoproteins. Presumably a charge transfer actually took place in the ground state of the complex and hence the isoalloxazine ring adopted practically the structure of the oxidized form. The specific binding of $NADP^+$ to half reduced flavin may cause the stability of the semiquinone form.

References

1. Chu, J.-W. and Kimura, T. (1973) *J Biol Chem* 248:2089–2094.
2. Lambeth, J.D. and Kamin, H. (1976) *J Biol Chem* 251:4299–4306.
3. Sugiyama, T., Miura, R., and Yamano, T. (1979) *J Biochem (Tokyo)* 86:213–223.
4. Dutta, P.K., Nestor, J.R., and Spiro, T.G. (1977) *Proc Natl Acad Sci USA* 74:4146–4149.
5. Nishina, Y., Kitagawa, T., Shiga, K., Horiike, K., Matsumura, Y., Watari, H., and Yamano, T. (1978) *J Biochem (Tokyo)* 84:925–932.
6. Kitagawa, T., Nishina, Y., Kyogoku, Y., Yamano, T., Ohishi, N., Takai-Suzuki, A., and Yagi, K. (1979) *Biochemistry* 18:1804–1808.
7. Kitagawa, T., Nishina, Y., Shiga, K., Watari, H., Matsumura, Y., and Yamano, T. (1979) *J Am Chem Soc* 101:3376–3378.
8. Nishina, Y., Kitagawa, T., Shiga, K., Watari, H., and Yamano, T. (1980) *J Biochem (Tokyo)* 87:831–839.
9. Sugiyama, T. and Yamano, T. (1975) *FEBS Lett* 52:145–148.

Published 1982 by Elsevier North Holland, Inc.
Vincent Massey and Charles H. Williams, Editors
Flavins and Flavoproteins

CHAPTER 116

Flavin to Iron-Sulfur Electron Transfer: Stopped Flow Studies of D_2O Effects on NADPH-Cytochrome c Reduction by the Adrenodoxin Reductase-Adrenodoxin Complex

J. David Lambeth

Emory University School of Medicine, Atlanta, Georgia

Summary

Stopped flow studies at two pH values demonstrate that the rate of flavin to iron-sulfur one-electron transfer from the fully-reduced flavin is not affected by D_2O, while the rate of transfer from the neutral flavin semiquinone occurs with a minimum isotope rate effect of 1.5 to 1.8: semiquinone to iron-sulfur electron transfer is rapid only in the presence of excess NADPH and the isotope effect is expressed only under these conditions. Data suggest the possibility of different one-electron transfer mechanisms for reduction of the iron-sulfur center by fully-reduced versus semiquinone flavin, and binding of NADPH appears to facilitate the latter electron transfer.

Introduction

Cytochrome c reduction has been used previously to study the electron transferring properties of the FAD-protein NADPH-adrenodoxin reductase (AR) and the Fe_2S_2 protein adrenodoxin (ADX) (1–4). During catalysis of cytochrome c reduction, the 1-electron containing 1 : 1 flavoprotein : iron-sulfur protein complex (1) reduces cytochrome c too slowly to be involved in catalysis, and the complex cycles between its 1- and 3-electron reduced[1] states (via a 2-electron-reduced intermediate during the reoxidative half reduction) (2). NADPH-reduction of flavin is rapid (20–30 s^{-1}) compared with turnover (~2 s^{-1}), and reduced ADX reduces cytochrome c very rapidly (>300 s^{-1}) both in the presence and absence of flavoprotein. While electron transfer to the native acceptor, cytochrome P-450, appears to require dissociation of the AR-ADX complex, transfer to cytochrome c probably does not, and in any case, this process is not rate-limiting (4–6). These and other data (2,4) indicate

[1] The distribution of electrons among redox centers includes the possibility of electrons residing on bound pyridine nucleotide (see below).

that flavoprotein to iron-sulfur electron transfer is rate-limiting in cytochrome c reduction, and that the rate of hemoprotein reduction is thus a measure of flavin to iron-sulfur electron transfer. We have previously reported that turnover in NADPH-cytochrome c reduction occurs at half the normal rate when the reaction is carried out in D_2O (2). In the present studies, stopped flow spectrophotometry has been used to investigate the specific step in the catalytic cycle at which D_2O exerts its effect.

Results and Discussion

Reduction of cytochrome c by the adrenodoxin reductase-adrenodoxin complex was measured in the stopped flow spectrophotometer using H_2O versus D_2O as solvent. Figure 1 shows sample kinetic records, each the average of at least four separate reactions. The upper record was obtained by reacting

Figure 1. Reaction of AR-ADX complex with NADPH plus cytochrome c. Adrenodoxin reductase-adrenodoxin complex (5 μM) was reacted with NADPH and cytochrome c (5 μM and 10 μM respectively for the upper record, and 10 μM and 20 μM respectively for the lower) using an Update Instruments stopped flow spectrophotometer. The absorbance increase at 550 nm was recorded using an interfaced Northstar computer. Reaction was carried out in 10 mM K Pi pH 7.0; each datum number represents 10 msec.

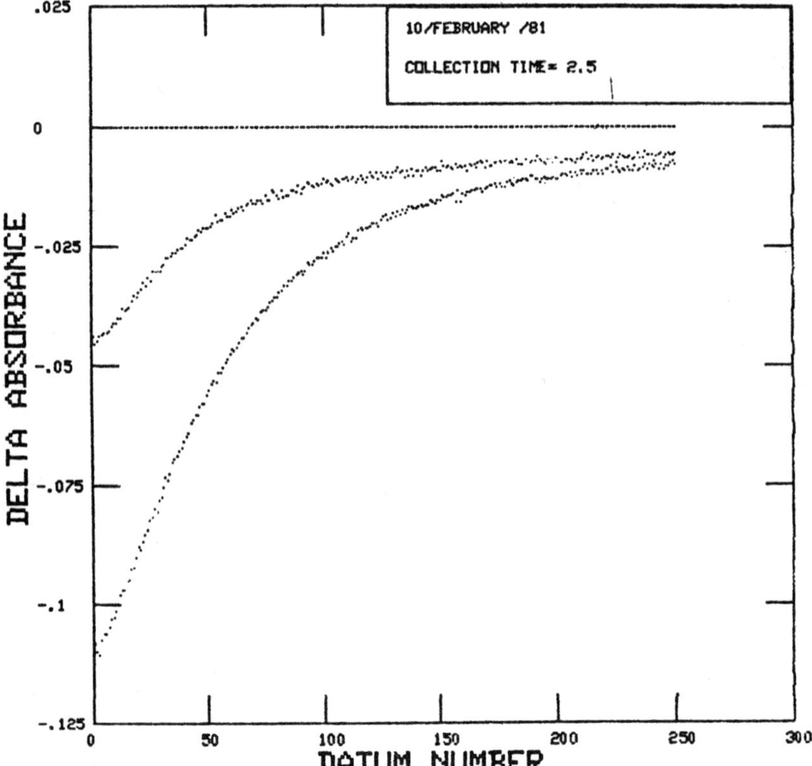

complex with one equivalent of NADPH plus two of cytochrome c. Thus, the 1-electron reduced complex formed initially, and can be reoxidized sequentially by two molecules of cytochrome c. As reported previously (2), one equivalent of cytochrome is reduced at a rate consistent with the slowest step in turnover, while the second is reduced much too slowly to be involved catalytically (i.e., the 1-electron reduced complex reduces cytochrome c slowly). The lower kinetic record was obtained using two and four equivalents respectively of NADPH and cytochrome per equivalent of enzyme complex. Following the first electron transfer to cytochrome c, the complex reacts with the second NADPH to form the 3-electron reduced complex. This species probably initially represents a complex in which one electron resides on flavin and an electron pair resides on pyridine nucleotide, since if an electron was present on adrenodoxin, cytochrome c should be reduced instantaneously (3). The second and third cytochrome c reductions then occur relatively rapidly from the 3- and 2-electron reduced forms of the complex, followed finally by the slow reduction of the fourth cytochrome c by the one electron-containing complex. For the first set of conditions (1 NADPH and 2 cytochromes per complex), D_2O had no effect on the rate of either phase of cytochrome reduction (data not shown). However, when the complex was reacted with 2 NADPH and 4 cytochromes per complex, a small but reproducible decrease in initial rate was seen in D_2O (see Figure 2). Apparent first-order rate constants for each phase obtained from first-order plots of the data are summarized in Table 1. At both pH 7.0 and 7.6, D_2O rate effect was seen only when the 3-electron reduced complex participated in cytochrome reduction.

In order to estimate the magnitude of the isotope effect, the reaction records at 1 NADPH per complex (e.g., Figure 1, upper record) were subtracted from those at 2 NADPH per complex (e.g., Figure 1, lower record). The resulting rate (see Figure 3) should, to a first approximation, eliminate the contribution due to the first and last cytochrome c reductions in the 4 cytochrome reduction experiment. The remaining initial rate should thus reflect the reduction of cytochrome c by the 3-electron-reduced complex. This rate demonstrates an isotope effect of 1.8 at pH 7 and 1.6 at pH 7.6 (see Table 2), consistent with the

Table 1. Effect of D_2O on the Rate of Reduction of Cytochrome c.

Experiment		Solvent	Phase 1 s^{-1}	Phase 2 s^{-1}
pH 7.06				
1. AR-ADX (5 μM)	550 nm	H_2O	1.73	.33
NADPH (5 μM) + CYTC (10 μM)		D_2O	1.68	.39
2. AR-ADX (5 μM)	550 nm	H_2O	2.00	.45
NADPH (10 μM) + CYTC (20 μM)		D_2O	1.41	.55
pH 7.6				
3. AR-ADX (5 μM)	550 nm	H_2O	1.33	.31
NADPH (5 μM) + CYTC (10 μM)		D_2O	1.25	.33
4. AR-ADX (5 μM)	550 nm	H_2O	1.42	.42
NADPH (10 μM) + CYTC (20 μM)		D_2O	1.15	.53

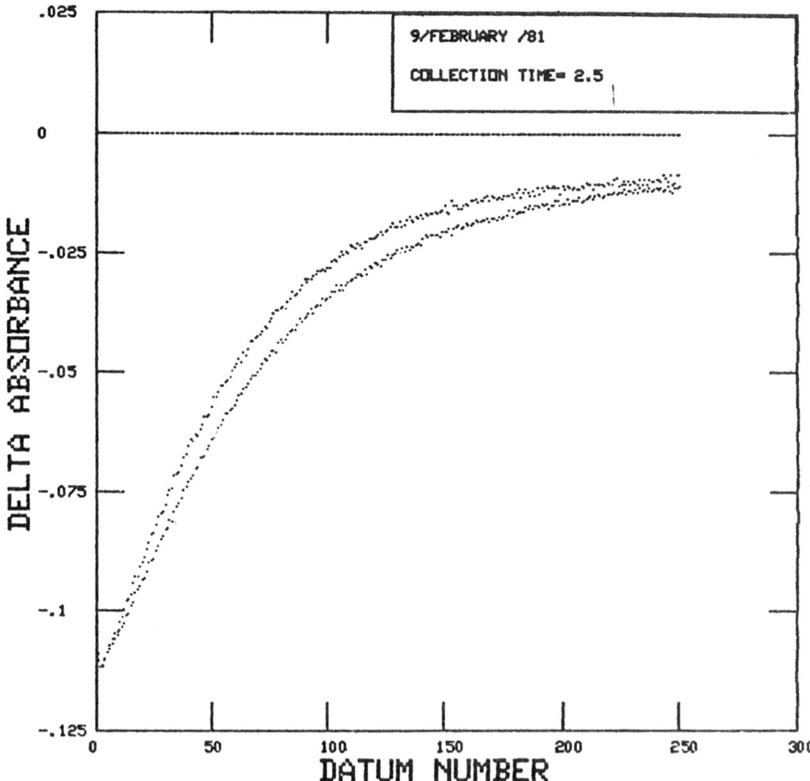

Figure 2. Effect on D_2O on the rate of reduction of cytochrome c by the adrenodoxin reductase-adrenodoxin complex. The kinetic records (pH 7.6) for reaction of the AR-ADX complex with two equivalents of NADPH and four equivalents of cytochrome c are shown for reaction in H_2O (upper) versus D_2O (lower).

isotope effect of 2 seen during turnover at pH 7 (3). This effect is considered to be a minimum, since there is also a rate contribution due to the third cytochrome c reduction (i.e., from the 2-electron reduced complex).

Since cytochrome c is reduced very rapidly once adrenodoxin becomes reduced (3), one can equate rates of cytochrome c reduction with rates of iron-sulfur center reduction. Thus, the simplest explanation for the above experiments is as follows. Once formed, the 1-electron-containing complex (with the electron residing in a flavin semiquinone) has two possible fates: First, in the absence of a second NADPH, the flavin to iron-sulfur electron transfer occurs very slowly and is D_2O-independent. However, in the presence of a second NADPH, the bound pyridine nucleotide functions as an effector to allow rapid flavin semiquinone to iron-sulfur electron transfer. It is this latter electron transfer which appears to be sensitive to D_2O. Thus, the semiquinone to iron-sulfur electron transfer occurs rapidly only in the presence of bound pyridine nucleotide, and the D_2O effect on this step is expressed only under

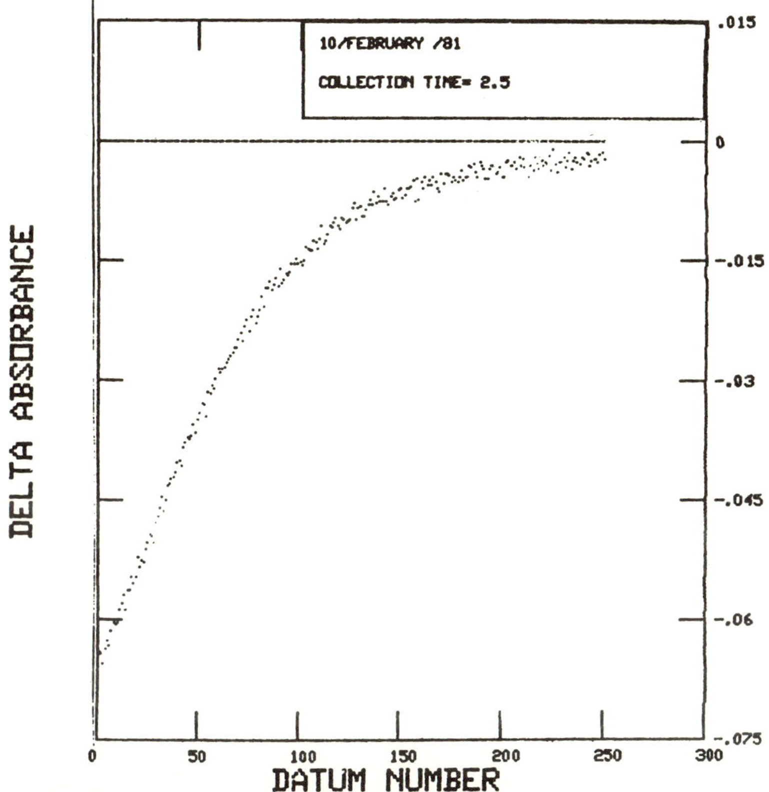

Figure 3. Subtracted absorbance changes. Kinetic records using 5 μM NADPH and 10 μM cytochrome c were subtracted from those using 10 μM NADPH and 20 μM cytochrome c in order to minimize the contributions of the first and last electron transfer to the latter reaction. Sample data in D_2O are shown.

these conditions. The finding of a D_2O effect on flavin semiquinone to iron-sulfur electron transfer but no effect on fully reduced flavin to iron-sulfur electron transfer suggests the possibility of different electron transfer mechanisms for reduction of the iron-sulfur protein by the fully-reduced versus the semiquinone flavin in adrenodoxin reductase.

Table 2. Effect of D_2O on Cytochrome c Reduction: Estimation of Isotope Rate Effect.

Equivalent pH	Subtracted rates H_2O s^{-1}	Subtracted rates D_2O s^{-1}	Minimum isotope rate effect
pH 7.0	1.94	1.07	1.81
pH 7.6	1.36	0.88	1.55

ACKNOWLEDGMENT
Supported by NIH Grant AM 27373.

References

1. Lambeth, J.D., McCaslin, D.R., and Kamin, H. (1976) *J Biol Chem* 251:7545–7550.
2. Lambeth, J.D. and Kamin, H. (1977) *J Biol Chem* 252:2908–2917.
3. Lambeth, J.D. and Kamin, H. (1979) *J Biol Chem* 254:2766–2774.
4. Lambeth, J.D., Lancaster, J.R., Jr. and Kamin, H. (1981) *J Biol Chem* 256:3674–3678.
5. Lambeth, J.D., Seybert, D.W., and Kamin, H. (1979) *J Biol Chem* 254:7255–7264.
6. Lambeth, J.D., Seybert, D.W., and Kamin, H. (1980) *J Biol Chem* 255:4667–4672.

Published 1982 by Elsevier North Holland, Inc.
Vincent Massey and Charles H. Williams, Editors
Flavins and Flavoproteins

CHAPTER 117

Further Studies of Interactions Between Adrenodoxin Reductase and Adrenodoxin: Circular Dichroism, Laser Raman Spectroscopy and the Prediction of the Binding Structure

Tokuji Kimura, Ellen Bicknell-Brown, Bee T. Lim, Satoshi Nakamura,* Hideyo Hasumi,* Kunimasa Koga,** and Hajime Yoshizumi**

*Department of Chemistry, Wayne State University, Detroit, Michigan; *Department of Biophysical Chemistry, Kitasato University, Sagamihara, Kanagawa, Japan: and **Suntory Central Research Institute, Shimamoto, Osaka, Japan*

Introduction

Adrenodoxin reductase (an FAD-containing dehydrogenase) interacts with adrenodoxin (an Fe_2S_2 protein) and the complex is catalytically important for the electron transfer system of the steroid hydroxylation reactions in adrenocortical mitochondria. Chu and Kimura (2) showed that the interacting forces between the flavoprotein and the iron-sulfur protein are mainly ionic with a dissociation constant of 10^{-9} M in low ionic strength medium. Although there are many works on this complex (4,6), the structure of the binding site is unclear because crystallographic information is lacking. In this study, we have attempted to gain a better understanding of the complex by the use of a computer-assisted circular dichroism (CD) spectrometer, differential scanning calorimeter, and laser Raman spectrometer. We have predicted the secondary structure of the binding domain based on the method of Chou and Fasman (1).

Results and Discussion

As shown in Figure 1, adrenodoxin (curve a), adrenodoxin reductase (curve b), and their complex (curve c) display the characteristic CD spectra in the visible region. The difference spectrum between the complex and the sum of its components showed positive peaks at 450 and 350 nm, and a negative peak at 375 nm (curve d). From these results, it can be concluded that both flavin and iron-sulfur centers are perturbed upon the complex formation. In addition to the previous implication that the flavin site is perturbed by complexing, we emphasize here the changes in the iron-sulfur dichroism as determined by the minimum at 375 nm.

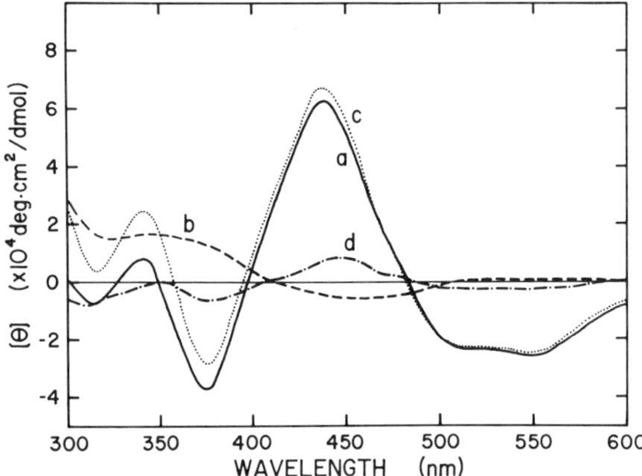

Figure 1. CD spectra of adrenodoxin (a), adrenodoxin reductase (b), their complex (c), and the difference CD spectrum (d). The difference spectrum was obtained by subtracting the additive spectra of the component proteins from the spectrum of the complex. The average spectrum was obtained after 2 runs.

Figure 2 represents the CD spectra in the ultraviolet region. The spectra of adrenodoxin, adrenodoxin reductase, the complex, and the difference are shown as curves a, b, c, and d, respectively. From these results, we conclude that upon the complex formation small changes in the protein structures occur.

Perhaps the content of the α-helical region increases upon complexing. No large structural changes were ever observed, indicating that the gross structures of both proteins in their free forms were retained in the complex. We have carried out similar CD experiments with spinach ferredoxin and its reductase. The results from both visible and UV regions were found to be largely similar to the adrenal system. Thus, the binding mode of the spinach system resembles that of the adrenal system.

Figure 3 indicates the differential scanning calorimetric thermograms of the spinach system. The apparent denaturation temperatures of ferredoxin, ferredoxin reductase, and their complex were found to be 53.3°, 66.3°, and 71.1°C, respectively. These results confirm the tight binding found between iron-sulfur protein and flavoprotein. The thermogram of the adrenal system is predicted to be closely similar to that of the spinach system as judged from their CD properties of protein-protein interactions.

The Raman spectrum of apoadrenodoxin was measured with particular emphasis on the Raman doublet due to ^{82}Tyr, which is the sole tyrosine residue in the polypeptide and located on the binding site of adrenodoxin to its reductase (Taniguchi and Kimura, 1976). The tyrosine doublet due to ^{82}Tyr was observed at 860 and 832 cm^{-1} and the relative intensity ratio was approximately 2:10. This unusual ratio suggests that the tyrosine residue is strongly hydrogen-bonded to a neighboring carboxylic acid residue (Table 1).

The secondary structure of the adrenodoxin polypeptide chain was predicted empirically using the method of Chou and Fasman (1). From these predictions, we concluded that the carboxylic acid-rich domain from residue 65 through 84 consists of a loop: two α-helical coils are connected by a short β-turn

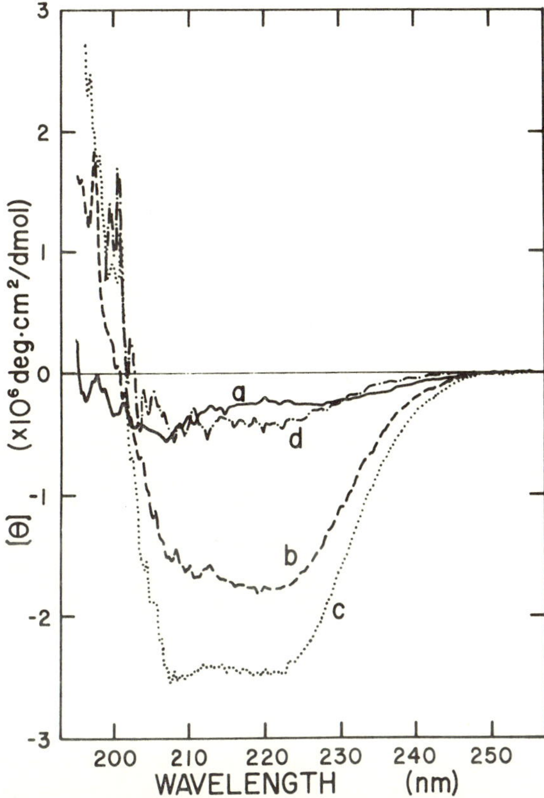

Figure 2. Ultraviolet CD spectra of adrenodoxin (a), adrenodoxin reductase (b), their complex (c), and the difference spectrum (d). The average spectrum was obtained after 16 runs.

sequence. In this conformation, the tyrosine residue at position 82 is proximal to glutamic residues at positions 65 and 68. ^{73}Glu, ^{74}Glu, and ^{79}Asp are also close to ^{82}Tyr. However, the steric arrangement for the interaction is less favorable. In this context, we propose that the binding site of adrenodoxin to its reductase is located within the peninsular domain from the residue No. 64

Table 1. Comparison of Frequencies and Intensity Ratios of the Tyrosine Raman Doublet in Apoadrenodoxin With Other Proteins.

	Doublet Frequencies (cm^{-1})	I_{850}	I_{830}	H-bonding
Apoadrenodoxin	860, 832	2	10	Strong donor
Anhydrous tyrosinate	856, 828	9	10	Ionized
Erabutoxin a	850, 830	sh	10	Donor
Erabutoxin b	850, 830	sh	10	Donor
Toxin B	850, 827	10	8	Weak
Myotoxin a	858, 826			Weak
Fd phage	850, 830	10	2.7	"Excess" H accepting

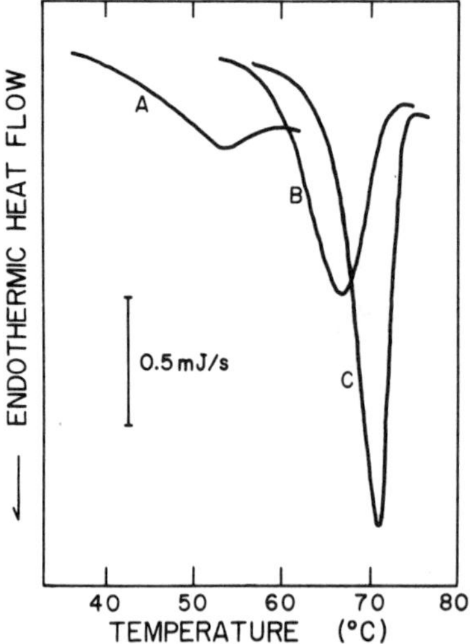

Figure 3. DSC thermograms of spinach ferredoxin (a), ferredoxin-NADP reductase (b), and their complex (c) in 10 mM Tris-HCl buffer, pH 7.4. The heating rate was 5°C/min.

to No. 84. A similar arrangement was seen in pituidaredoxin and spinach ferredoxin. It was reported by Hasumi et al. that in spinach ferredoxin the tyrosine residue of the polypeptide chain participates in the binding to the reductase. In spite of these similarities between adrenodoxin and spinach ferredoxin, adrenodoxin reductase does not give electrons to spinach ferredoxin effectively. This can be accounted for by considering the size of the binding domains: the spinach domain appears to be much larger than the adrenal one. Additionally, *E. coli* thioredoxin is known to have a peninsular domain (a male protein) for its redox active site, as seen from the X-ray crystallographic analysis (5). It is of interest that these tight interactions between these proteins occur in a key and lock interaction rather than a contact surface one.

ACKNOWLEDGMENTS

We are pleased to acknowledge Mr. Jeffrey H. Parcells for his preparation of adrenodoxin reductase from bovine adrenals. This study was partially supported by NIH Grants (GM-27243, GM-27744, and AM-12713).

References

1. Chou, P.Y. and Fasman, Ct.D (1978) *Ann Rev Biochem* 47:251–276.
2. Chu, J.W. and Kimura, T. (1973) *J Biol Chem* 248:5183–5187.
3. Hasumi, H. and Nakamua, S. (1978) *J Biochem* 84:707–717.
4. Hiwatashi, A., Ichikawa, Y., Maruya, V., Yamano, T., and Aki, K. (1976) *Biochemistry* 15:3082–3090.
5. Holmgren, A. Söderberg, B-O., Eklund, H., and Branden, C-I (1975) *Proc Natl Acad Sci US* 72:2305–2309.
6. Lambeth, J.D., McCaslin, D.R., and Kamin, H. (1976) *J Biol Chem* 251:7545–7550.

PART IX C:

Interactions of Flavoproteins and Cytochromes

Published 1982 by Elsevier North Holland, Inc.
Vincent Massey and Charles H. Williams, Editors
Flavins and Flavoproteins

CHAPTER 118

Circular Dichroism Studies on NADH-Cytochrome b_5 Reductases Purified from Human and Rabbit Erythrocytes

Toshitsugu Yubisui, Masazumi Takeshita, and Yoshimasa Yoneyama*

*Department of Biochemistry, Medical College of Oita, Hazama-cho, Oita, and *Department of Biochemistry, Kanazawa University School of Medicine, Kanazawa, Japan*

NADH-cytochrome b_5 reductase in human erythrocytes (1,2) is known to function as a methemoglobin reducing enzyme, unlike the function of liver microsomal enzyme (3,4). Although the function of the enzyme in the erythrocytes is different from that of the microsomal enzyme, the purified enzyme from human erythrocytes has properties (5–7) very similar to those of the liver microsomal enzyme (8). As the enzyme in the erythrocytes can be obtained as a soluble form from hemolysate without any treatment such as proteolytic digestion or extraction with detergent, the soluble enzyme is suitable for investigating the protein structure or conformation of the enzyme in relating to the flavin-binding. In this paper, we describe CD spectra of the purified cytochrome b_5 reductases of human and rabbit erythrocytes which were measured under various conditions.

NADH-cytochrome b_5 reductase of human erythrocytes was purified by procedures described previously (7). The enzyme of rabbit erythrocytes was purified by simple procedures developed in our laboratory (9); fractionation with ammonium sulfate, gel filtration on Sephadex G-75, and an affinity chromatography on 5′-AMP-Sepharose 4B. The enzyme of rabbit erythrocytes was purified about 12,000-fold from hemolysate in respect to the NADH-cytochrome b_5 reductase activity (9). By the affinity chromatography, the enzyme was highly purified and the enzyme was obtained as a homogeneous protein judging from SDS-polyacrylamide gel electrophoresis.

The purified cytochrome b_5 reductases of human and rabbit erythrocytes showed very similar CD spectra as well as absorption spectra. In the region from 250 to 600 nm, both the enzymes (oxidized form) showed negative CD at 285, 460, and 490 nm, and positive CD at 310, 370, and 390 nm, respectively.

As the flavin-binding in the enzyme of human erythrocytes is weak, the flavin content in the purified enzyme varied from 0.4 to 1.0 mol FAD/mol of enzyme in different batches of preparations. However, we found a good correlation between the flavin content, intensity of CD at 460 nm (or at 285 nm), and enzyme activity as shown in Table 1. The intensity of CD at 460 nm

U.C.W. ABERYSTWYTH LIBRARY

Table 1. Flavin Content, Molar Ellipticity of CD Spectra at 460 nm, and Enzyme Activity of the Purified NADH-Cytochrome b_5 Reductase of Human Erythrocytes.

Enzymes	Flavin content mol/mol enzyme	CD ($[\theta] \times 10^{-4}$) $deg \cdot cm^2/dmol$ FAD	Enzyme activity mol/min/mol FAD
1[a]	1.02	−3.43	2,935
2[a]	0.63	−3.33	2,436
3[a]	0.48	−3.23	2,636

[a] Different batches of enzymes purified by our method (7).

and enzyme activity based on the flavin content was constant even though the flavin content varied in different batches of preparations. These results apparently indicate that the enzyme of human erythrocytes contains FAD as a prosthetic group, and also that the CD can be used as a measure of the active enzyme species, or the content of enzyme-bound flavin.

To investigate further the flavin-binding in the soluble cytochrome b_5 reductases in erythrocytes, CD spectra of the enzyme purified from rabbit erythrocytes were measured under various conditions. Figure 1 shows the CD spectra of oxidized and NADH-reduced forms of the enzyme. Upon reduction the enzyme with a slight excess of NADH under anaerobic conditions, the sign of the CD spectra of the oxidized form reversed in the region from 300 to 500

Figure 1. CD spectra of the purified NADH-cytochrome b_5 reductase of rabbit erythrocytes. The CD spectra were measured in 0.05 M phosphate buffer (pH 7.6) containing 1 mM EDTA and 0.1 mM DTT at 25°C with a Union Dichrograph, model Mark III-J fitted with a computer system 77 (Union Giken, Osaka, Japan). Each spectrum was recorded as an average trace from 3–5 times accumulated scannings. Solid line, oxidized; broken line, NADH-reduced; dotted line, air-reoxidized from the NADH-reduced form of the enzyme.

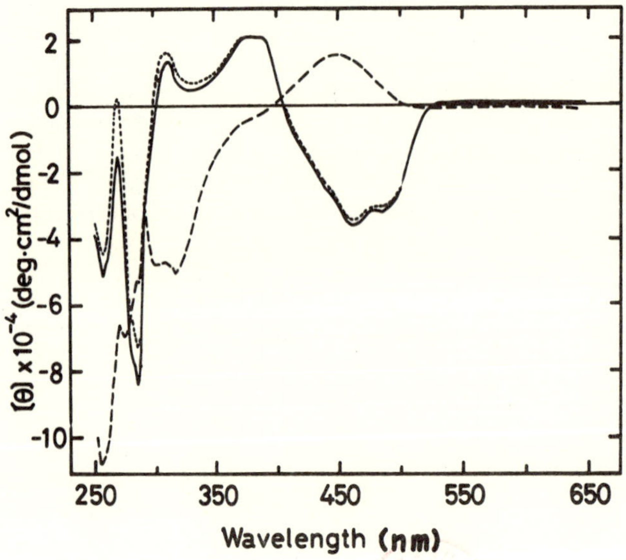

nm. This enzyme species is considered to be a reduced enzyme (flavin)-NAD^+ (charge transfer) complex as confirmed from the absorption spectra which showed a low and broad absorption band extending from 500 to 650 nm. The NADH-reduced enzyme showed positive CD at 450 nm, and negative CD at around 320 nm. A similar reduced enzyme-NAD^+ complex of the liver microsomal cytochrome b_5 reductase was reported by Strittmatter (8), and by Iyanagi and Anan (10). Introducing air to the NADH-reduced enzyme, the CD spectrum of the NADH-reduced form returned almost completely to that of the oxidized form. In the UV region from 250 to 300 nm, however, a slight difference was observed, and this may reflect a weak binding of NAD^+ to the oxidized form of the enzyme (10). When the reactive cysteines were titrated with PCMB, a significant decrease of CD at 370–390 nm was observed, and the extent of the decrease reached near maximum with the binding of 1 mole PCMB to the first reactive cysteine accompanied with almost complete loss of enzyme activity (Figure 2). By the addition of excess dithiothreitol (3 mM) to the PCMB-bound enzyme, a partial recovery (about 40%) of enzyme activity was obtained, while no significant recovery of the decreased CD at 390 nm was observed. The PCMB titration did not cause any change of the negative CD at 460–490 nm, but when the enzyme was treated with tetranitromethane to modify tyrosine, the CD in the UV region as well as the visible region decreased significantly in its intensity and the sign of CD at 260 nm was reversed, depending on the concentration of the reagent (Figure 3). These results suggest that the environment around the flavin-binding site in the

Figure 2. Effect on PCMB on the CD spectra of the purified NADH-cytochrome b_5 reductase of rabbit erythrocytes. Conditions are the same as used in Figure 1, except that the enzyme is in the buffer without DTT. Solid line, oxidized; broken line, oxidized+3 mol PCMB/mol of enzyme; dotted line, oxidized+3 mol PCMB/mol of enzyme+DTT (3 mM).

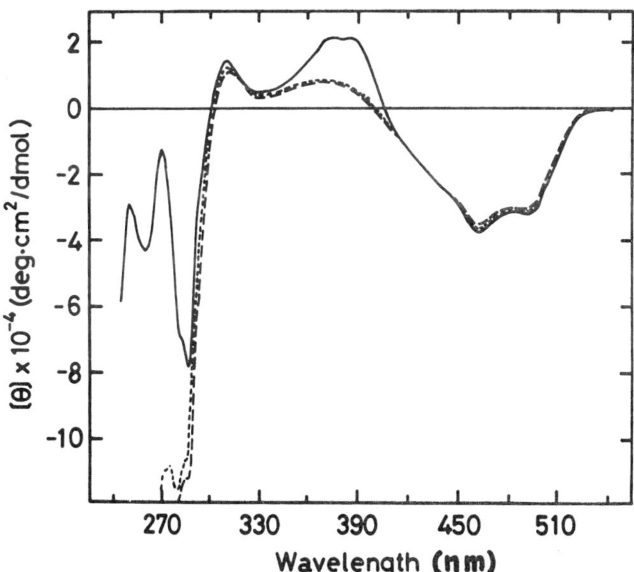

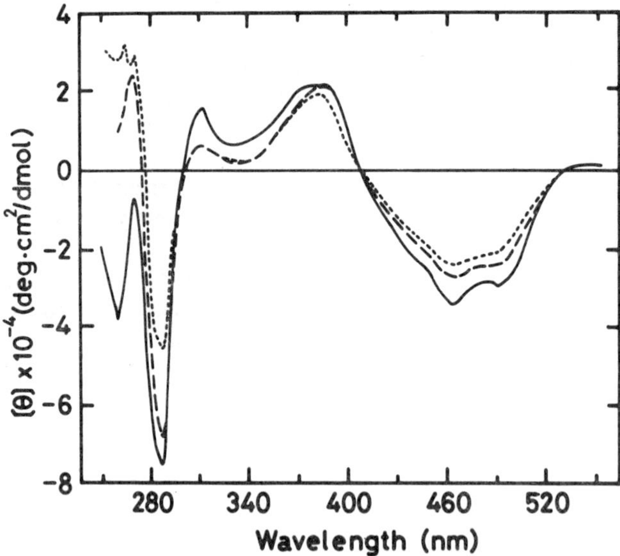

Figure 3. Effect of tetranitromethane on the CD spectra of the purified NADH-cytochrome b_5 reductase. The spectra were measured in 0.1 M Tris-HCl buffer (pH 8.0), and other conditions were the same as used in Figure 1. Solid line, oxidized; broken line, oxidized + 6 mol excess tetranitromethane/mol of tyrosine in the enzyme, 1 hr; dotted line, same as the broken line, 2 hr.

erythrocyte enzyme essentially coincides with that proposed by Strittmatter (8) where the first reactive cysteine in the enzyme is related to the NADH-binding, and tyrosine is related to the flavin-binding.

References

1. Hultquist, D.E. and Passon, P.G. (1971) *Nature (New Biol)* 229:252–254.
2. Sugita, Y., Nomura, S., and Yoneyama, Y. (1972) *J Biol Chem* 246:6072–6078.
3. Oshino, N., Imai, Y., and Sato, R. (1971) *J Biochem (Tokyo)* 69:155–167.
4. Enoch, H.G., Catala, A., and Strittmatter, P. (1976) *J Biol Chem* 251:5059–5103.
5. Passon, P.G. and Hultquist, D.E. (1972) *Biochim Biophys Acta* 275:62–73.
6. Kuma, F. and Inomata, H. (1972) *J Biol Chem* 247:556–560.
7. Yubisui, T. and Takeshita, M. (1980) *J Biol Chem* 255:2454–2456.
8. Strittmatter, P. (1966) In *Flavins and Flavoproteins*. Slater, E.C. (ed.) Amsterdam: Elsevier Publishing Co., pp. 325–339.
9. Yubisui, T. and Takeshita, M. (1981) Submitted for publication.
10. Iyanagi, T. and Anan, K. (1979) In *Flavins and Flavoproteins*. Yagi, K. and Yamano, T. (eds.) Tokyo: Japan Scientific Societies Press and Baltimore: University Park Press, pp. 725–734.

Published 1982 by Elsevier North Holland, Inc.
Vincent Massey and Charles H. Williams, Editors
Flavins and Flavoproteins

CHAPTER 119

NADPH-Cytochrome P-450 Reductase: Redox States of FMN and FAD During Electron Transfer From NADPH

Daniel D. Oprian and Minor J. Coon

Department of Biological Chemistry, The University of Michigan, Ann Arbor, Michigan

We have developed a model for the interaction of NADPH-cytochrome P-450 reductase with NADPH that extends the postulate of rapid interflavin electron transfer originally proposed by Masters and coworkers (1). The fundamental concept is that, given the rapid exchange of reducing equivalents between flavin moieties at any given time during the reaction of NADPH with the reductase, the distribution of electrons among the various oxidation states of each flavin is precisely determined by the reduction potentials for the individual flavin half-reactions. Quantitation of this model was approached using the mathematical formulation developed by Olson et al. (2) for the interaction of the multiple electron-accepting centers of xanthine oxidase. On the basis of a multiwavelength analysis of the reaction of NADPH with the reductase under anaerobic conditions by stopped flow spectrophotometry, the model was found to predict successfully the spectral course of each phase of the entire reaction.

The reaction of the reductase, purified from detergent-solubilized rabbit liver microsomes (3), with NADPH was found to exhibit triphasic kinetics. The first phase, $k_1 = 70\ sec^{-1}$, was characterized by an absorbance increase at 585 nm and decrease at 502 nm. The second phase ($k_2 = 7\ sec^{-1}$), for which the absorbance at 585 nm was found to decrease, was observed only in the presence of an excess of NADPH. Irrespective of the NADPH concentration employed, a very slow third phase was observed which, in the presence of a 10-fold excess of reduced pyridine nucleotide, displayed an absorbance increase at 585 nm with a first-order rate constant of $0.05\ sec^{-1}$. Spectra for each phase were constructed from a multiwavelength analysis as is shown in Figure 1A and B. That no species spectrally distinct from oxidized enzyme was formed in the dead time of the stopped flow spectrophotometer is demonstrated by the results shown in Figure 1A (solid triangles).

The observation of three phases for the reaction may be explained as follows. The first two phases arise from the sequential oxidation of two equivalents of reduced pyridine nucleotide. Since NADPH is an obligatory 2-electron donor, the equilibrium established at the end of the second phase is metastable, and thermodynamic relaxation gives rise to the very slow third phase. As was first suggested by Masters and coworkers (1), we have interpreted the absorbance changes observed for the first phase of the reaction as

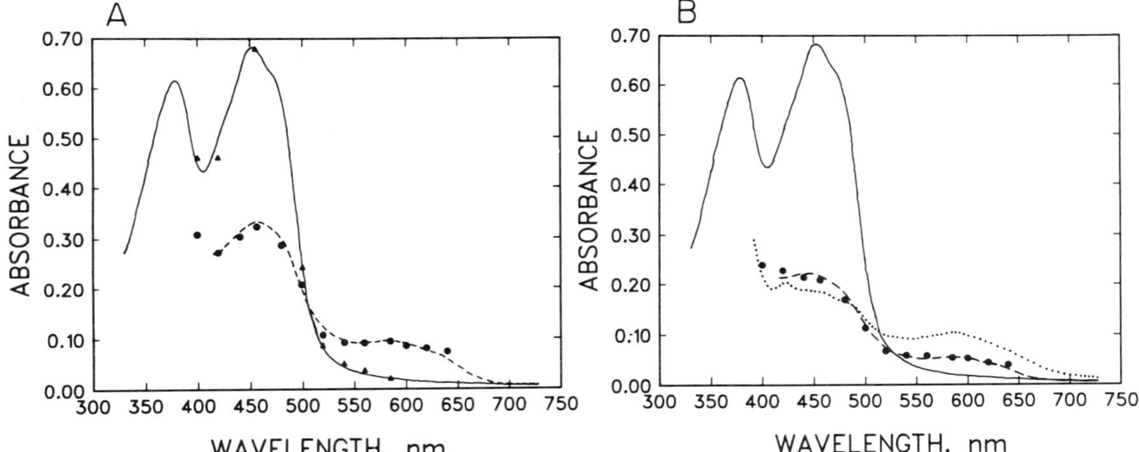

Figure 1. Absorption spectra obtained after first, second, and third phases of reduction of the reductase by NADPH. (A) an anaerobic solution of the reductase (32.2 μM) was rapidly mixed with an anaerobic solution of NADPH (332 μM), and the resulting absorbance change was monitored as a function of time at each of thirteen wavelengths. The data were analyzed as first-order triphasic reactions, and the absorbance change calculated for each phase was referenced to the final spectrum at equilibrium to obtain the absolute absorbance. (Solid line) oxidized reductase; (solid circles) absorbance at the end of the first phase (about 50 msec after mixing); (dashed line) spectrum calculated assuming a mixture of 70% ($FMNH_2$,FAD) and 30% (FMNH·,FADH·); (solid triangles) calculated absorbance at time t=0. (B) conditions and analysis of the data were as in Part A. (Solid line) oxidized reductase; (solid circles) absorbance at the end of the second phase (about 1 sec after mixing); (dashed line) spectrum calculated for a mixture of 50% ($FMNH_2$,$FADH_2$), 35% ($FMNH_2$,FAD), and 15% (FMNH·,FADH·); (dotted line) spectrum recorded at the end of the third and final phase of the reaction.

arising from the simultaneous formation of flavin semiquinone and fully reduced flavin. Accordingly, the transfer of reducing equivalents between the two flavin moieties is postulated to be much more rapid than that between reduced pyridine nucleotide and flavin. Reduction of the oxidized protein by a single equivalent of NADPH, therefore, will generate an equilibrium mixture of the three 2-electron-reduced forms of the enzyme: ($FMNH_2$,FAD), (FMNH·,FADH·), and (FMN,$FADH_2$). Reaction of the 2-electron-reduced flavoprotein with a second equivalent of NADPH will result in a mixture of the three 2-electron-reduced forms, fully reduced enzyme, and oxidized and reduced pyridine nucleotide. All nine possible oxidation states of the reductase will be present after thermodynamic relaxation in the third phase.

Simulation of the absorption spectrum resulting from each phase of the reaction required that we first accumulate reference spectra for each of the nine oxidation states of the flavoprotein from which composite spectra could be constructed, and then determine according to the above scheme the quantity of each species present at any point in the reaction. If we assume that the presence of FAD does not perturb the spectral properties of the FMN moiety, the spectra of enzyme-bound FMN, FMNH·, and $FMNH_2$ may be computed from the difference between spectra for the appropriate forms of the native and FMN-free reductase (4,5). With component spectra for bound FMN, FMNH·, $FMNH_2$, FAD, FADH·, and $FADH_2$ the absorption spectrum for each of the nine possible oxidation states of the reductase may be obtained by linear combination.

The quantitative expression of the reaction scheme described above was arrived at in the following manner. A set of four equilibria are defined,

$$\text{FADH}\cdot + \text{FMN} \overset{K_1}{\rightleftharpoons} \text{FAD} + \text{FMNH}\cdot$$

$$\text{FADH}\cdot + \text{FMNH}\cdot \overset{K_2}{\rightleftharpoons} \text{FAD} + \text{FMNH}_2$$

$$\text{FADH}\cdot + \text{FADH}\cdot \overset{K_3}{\rightleftharpoons} \text{FAD} + \text{FADH}_2$$

$$2\text{FADH}\cdot + \text{NADP}^+ \overset{K_4}{\rightleftharpoons} 2\text{FAD} + \text{NADPH} + \text{H}^+$$

the equilibrium constants of which may be computed from the reduction potentials determined by Iyanagi et al. (6). The reduction potentials were determined, however, at pH 7.0, whereas the present study was done at pH 7.4. The equilibrium constants corresponding to the new pH value may be calculated if we assume that each of the flavin half-reactions involves a single proton. The values arrived at for the equilibrium constants are: $K_1 = 1097$, $K_2 = 2.18$, $K_3 = 0.054$, and $K_4 = 0.247$. Significantly, only the value of K_4 differs from that obtained at pH 7.0. As detailed by Olson et al. (2), these four equilibrium constants may be used to determine the percentage of each species present at any point in the reaction. For example, consider the nine possible oxidation states of the reductase. The probability of formation for an intermediate state, i, is: $P_{(i)} = (\exp(-\Delta G_i/RT))/Z$, where ΔG_i is the free energy of formation for the ith state, and Z, the partition function, is the sum of all states containing the same number of reducing equivalents as the ith intermediate, $Z=\Sigma_i \exp(-\Delta G_i/RT)$. As $\Delta G_i = -RT\ln(K_i)$, $P_{(i)} = K_i/Z$; for those states containing more than one reducing equivalent, K_i is the product of all the individual constants involved in the formation of the state. As an example, the fractional amounts of ($FMNH_2$, FAD), (FMNH·, FADH·), and (FMN, $FADH_2$) for the 2-electron-reduced enzyme are given by K_1K_2/Z, K_1/Z, and K_3/Z, respectively, where Z is equal to $K_1K_2+K_1+K_3$. This is precisely the situation prevailing after reaction of the reductase with a single equivalent of NADPH. Since the midpoint potential for $NADP^+$/NADPH is over 100 mV more negative than that of the first two states, the equilibrium between oxidized and reduced pyridine nucleotide need not be considered. According to this treatment, (FMN, $FADH_2$) is calculated to be present in negligible quantities, whereas ($FMNH_2$, FAD) and (FMNH·, FADH·) account for 68.5 and 31.5% of the enzyme population, respectively. The spectrum represented by the *dashed line* in Figure 1A was generated from the reference spectra described above for an equilibrium mixture consisting of 30% di-semiquinone (FMNH·, FADH·) and 70% ($FMNH_2$, FAD).

In the presence of excess NADPH, the second phase may be represented by the following equilibria:

$$(\mathrm{FMNH\cdot,FADH\cdot}) \overset{K_2}{\rightleftharpoons} (\mathrm{FMNH_2,FAD})$$

$$(\mathrm{FMNH_2,FAD}) + \mathrm{NADPH} + \mathrm{H^+} \overset{K_e}{\rightleftharpoons} (\mathrm{FMNH_2,FADH_2}) + \mathrm{NADP^+}$$

where $K_e = K_3/K_4$. The percentage of each species present is calculated to be 36.5% ($FMNH_2$, FAD), 16.8% (FMNH·, FADH·), and 46.7% ($FMNH_2$, $FADH_2$). Of the 10 equivalents of NADPH present before mixing of pyridine nucleotide and reductase, 8.53 remain after the second phase. The spectral simulation representing this mixture is shown in Figure 1B.

The third phase of the reaction results in thermodynamic equilibration of the system and gives rise to the spectrum in Figure 1B. This spectrum could be simulated from the following distribution:

FMN	0	FAD	0.141
FMNH·	0.087	FADH·	0.681
$FMNH_2$	0.913	$FADH_2$	0.178
$NADP^+$	1.5	NADPH	8.5

where total flavin is equal to 2 and pyridine nucleotide is present at a concentration 5 times that of total flavin. Although the equilibrium position is unchanged, as evidenced by the constant $NADP^+$/NADPH ratio, the amount of $FADH_2$ decreases from 50% of the total FAD at the end of the second phase to 18% at the end of the third phase.

ACKNOWLEDGMENTS

Dr. Janice L. Vermilion deserves special thanks for invaluable discussions and suggestions. This work was supported by NSF Grant PCM8109394.

References

1. Masters, B.S.S., Kamin, H., Gibson, Q.H., and Williams, C.H. Jr. (1965) *J Biol Chem* 240:921–931.
2. Olson, J.S., Ballou, D.P., Palmer, G., and Massey, V. (1974) *J Biol Chem* 249:4362–4382.
3. French, J.S. and Coon, M.J. (1979) *Arch Biochem Biophys* 195:565–577.
4. Vermilion, J.L. and Coon, M.J. (1978) *J Biol Chem* 253:8812–8819.
5. Vermilion, J.L., Ballou, D.P., Massey, V., and Coon, M.J. (1980) *J Biol Chem* 256:266–277.
6. Iyanagi, T., Makino, N., and Mason, H.S. (1974) *Biochemistry* 13:1701–1710.

Published 1982 by Elsevier North Holland, Inc.
Vincent Massey and Charles H. Williams, Editors
Flavins and Flavoproteins

CHAPTER 120

Hydrophobic and Hydrophilic Domains of NADPH-Cytochrome P-450 Reductase

S.D. Black and M.J. Coon

Department of Biological Chemistry, The University of Michigan, Ann Arbor, Michigan

NADPH-cytochrome P-450 reductase, a unique component of the hepatic microsomal P-450-containing monooxygenase system (1), is known to have a two-domain structure (2, 3). The domains of the native reductase are physically separable by the action of various hydrolytic enzymes (4, 5), which produce a large soluble peptide (MW $\approx$71,000) and a small hydrophobic fragment (MW $\approx$6,000) (2). The large hydrophilic peptide exists as a monomer in aqueous solution, retains both flavins, and is functional in the reduction of cytochrome c. However, it can no longer effect the reduction of cytochrome P-450 (6). The small fragment is highly aggregated in aqueous solution (2) and has also been shown to be hydrophobic by its amino acid composition (2, 3). Addition of this peptide to a reconstituted hydroxylation system causes a marked inhibition of NADPH oxidation, providing further evidence for its role as an important moiety in the reductase (2). Our studies have been aimed at the further characterization of the hydrophobic and hydrophilic domains of the rabbit liver reductase (7).

The hydrophilic polypeptide has been obtained by treatment of the native enzyme with either trypsin or bromelain, as well as through adventitious cleavage by an endogenous, presumably lysosomal, endopeptidase during purification of the detergent-solubilized form. Each form of the reductase has been purified to apparent homogeneity, as can be seen in Figure 1. The tryptic and "adventitious" reductases exhibit the same electrophoretic mobilities while the bromelain-solubilized enzyme is approximately 1,000 smaller than these in molecular weight. Although the detergent-solubilized reductase has been shown to be blocked to various methods of amino-terminal analysis, each hydrophilic peptide proved to have a free N-terminus; the tryptic and "adventitious" forms have isoleucine, while the bromelain form has an N-terminal serine. This evidence and previous C-terminal work (2) allow an unambiguous assignment of the hydrophilic domain as the C-terminal peptide.

Tryptic digests of the detergent-solubilized reductase have been examined with various chromatographic techniques. A time-course of digestion was studied as shown in Figure 2 with reversed-phase high performance liquid chromatography. Peptide components are not eluted until the end of the gradient, at 50% n-propanol. Peaks a and b appear to be well correlated with the bands seen in the electrophoretic experiment (inset), thus leading to their identification as the tryptic and native enzymes, respectively. This elution

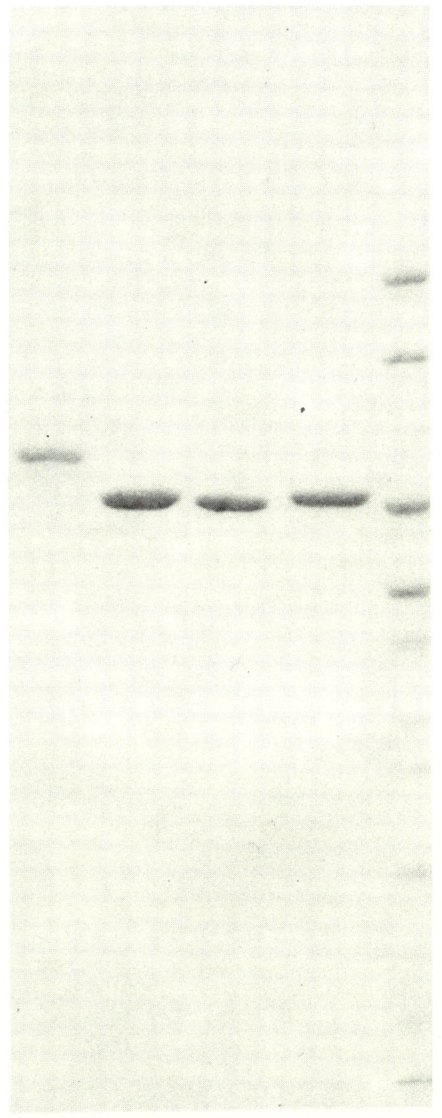

Figure 1. Electrophoretic behavior of purified detergent-solubilized reductase and various large proteolytic fragments. The proteins (0.5 μg samples) were submitted to SDS-polyacrylamide gel electrophoresis with migration from top to bottom, using a 7.5% separating gel. (a) native reductase; (b) trypsin-solubilized reductase; (c) bromelain-solubilized reductase; (d) "adventitious" reductase; (e) molecular weight standards: β-galactosidase, phosphorylase a, bovine serum albumin, catalase, glutamate dehydrogenase, ovalbumin, aldolase, and chymotrypsinogen.

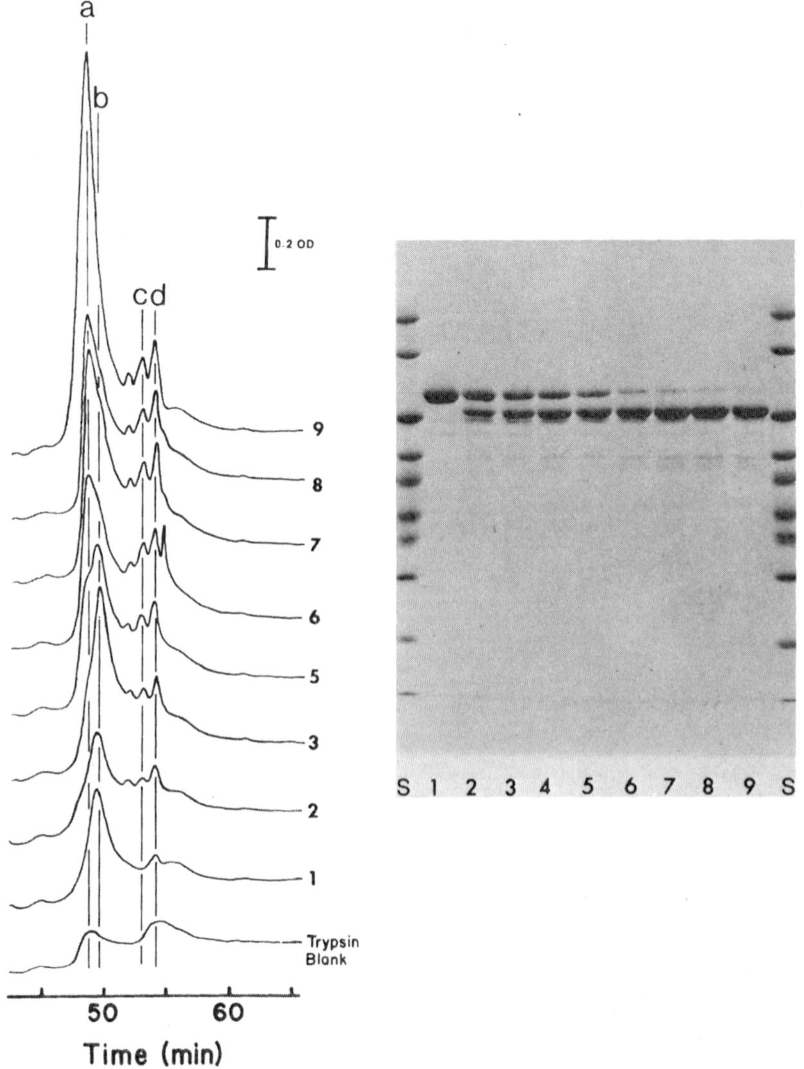

Figure 2. Time-course of tryptic digestion of native reductase examined by high performance liquid chromatography and SDS-polyacrylamide gel electrophoresis. Samples at various times of digestion (0°C; 1:50 trypsin:reductase, w/w) were acidified with phosphoric acid to pH 2.2, and aliquots containing the equivalent of 3.1 nmol of starting material injected onto the HPLC. The gradient from 1%-phosphoric acid, pH 2.2, to a 1:1 mixture of 1%-phosphoric acid, pH 2.2, and n-propanol (v/v) was developed at a rate of 1% n-propanol per minute, with a total flow rate of 0.5 ml per min. The separation was achieved using an Altex Ultrasphere ODS 5 μ reversed phase column at ambient temperature. Samples 1 through 9 correspond to the following times: 0 min, 30 min, 1hr, 1.5 hr, 2 hr, 4 hr, 6 hr, 8 hr, and 11 hr. In the electrophoretic experiment (see inset), each of the lanes 1 through 9 contained 1.0 μg total protein. Standards (labeled S) and conditions were as described in Figure 1.

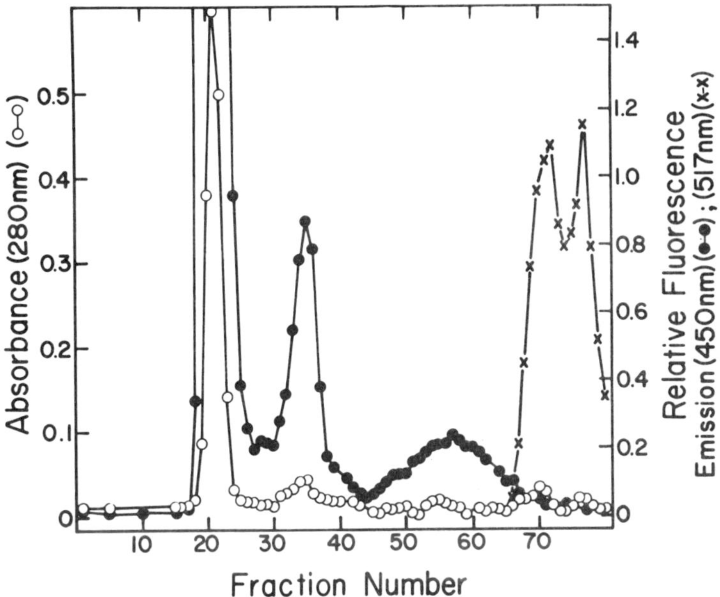

Figure 3. Chromatographic behavior of a tryptic digest of native reductase of Sephadex LH-60. Digestion (0°C; 1:50 trypsin:reductase, w/w) was allowed to proceed for 11 hr. The sample (52 nmol) was then applied to the column (1.75×100 cm) at 0.8 cm/hr and eluted with a solvent containing 50% formic acid: 50% ethanol. The eluate was monitored by the absorbance at 280 nm, by fluorescence emission at 517 nm (excitation, 360 nm), and by submitting aliquots of each fraction to alkaline hydrolysis followed by derivatization with o-phthaldialdehyde and determination of the resultant fluorescence (excitation, 340 nm; emission, 450 nm).

order is also in agreement with that predicted by the hydrophobicities of these peptides as determined by amino acid analysis (2). The late eluting and most hydrophobic peptides, c and d, appear to have resulted from the hydrophobic domain. Peak d is the primary product; peak c appears later in the course of the digest and is likely to have resulted from a further cleavage of d.

Tryptic digests have also been examined using calibrated gel filtration in nonaqueous solution. The Sephadex LH-60 profile detailed in Figure 3 shows the components of an 11-hr digest which were eluted in a solvent containing 50% formic acid (88% reagent): 50% ethanol (95% azeotrope) (v/v). This modification of the solvent system described by others (8) has proven especially useful in maintaining the solubility of highly hydrophobic peptides. The hydrophobic peptide is eluted at a molecular weight of 4,800 (fraction 35), with a yield of 77%; this represents a considerable improvement with respect to previous methods. N-terminal analysis of this peptide shows that it is blocked; the amino acid composition agrees well with that determined previously (2). These data provide strong evidence that the hydrophobic domain represents the actual amino-terminal peptide of the native reductase.

ACKNOWLEDGMENTS
This work was supported by NIH Grant AM-10339. S.D.B. was the recipient of a Merck Predoctoral Fellowship.

References

1. Lu, A.Y.H. and Coon, M.J. (1968) *J Biol Chem* 243:1331–1332.
2. Black, S.D., French, J.S., Williams, C.H., Jr., and Coon, M.J. (1979) *Biochem Biophys Res Comm* 91:1528–1535.
3. Gum, J.R. and Strobel, H.W. (1979) *J Biol Chem* 254:4177–4185.
4. Iyanagi, T. and Mason, H.S. (1973) *Biochemistry* 12:2297–2308.
5. Welton, A.F., Pederson, T.C., Buege, J.A., and Aust, S.D. (1973) *Biochem Biophys Res Comm* 54:161–167.
6. Lu, A.Y.H., Junk, K.W., and Coon, M.J. (1969) *J Biol Chem* 244:3714–3721.
7. French, J.S. and Coon, M.J. (1979) *Arch Biochem Biophys* 195:565–577.
8. Gerber, G.E., Anderegg, R.J., Herlihy, W.C., Gray, C.P., Biemann, K., and Khorana, H.G. (1979) *Proc Nat Acad Sci USA* 76:227–231.

Published 1982 by Elsevier North Holland, Inc.
Vincent Massey and Charles H. Williams, Editors
Flavins and Flavoproteins

CHAPTER 121

Studies on NADPH-Cytochrome P-450 Reductase From Bovine Adrenocortical Microsomes

Shiro Kominami, Tadashi Ogishima, and
Shigeki Takemori

Department of Environmental Sciences, Faculty of Integrated Arts and Sciences, Hiroshima University, Hiroshima 730, Japan

Introduction

Two types of the steroidogenic electron transfer system are present in adrenal cortex. One of these is located in the mitochondria and functions in cholesterol side chain cleavage and in steroid 11β or 18 hydroxylation. The other is present in the endoplasmic reticulum and catalyzes steroid 17α or C-21 hydroxylation. Immunochemical studies of Masters et al. (1) showed that the microsomal electron transfer system was composed of the two components: NADPH-cytochrome P-450 reductase and cytochrome P-450. We have recently succeeded in purifying these components from bovine adrenal microsomes and reconstituting the steroid C-21 hydroxylation system with the two purified components (2,3). The present paper describes the purification and steady state kinetics of NADPH-cytochrome P-450 reductase from bovine adrenal cortex microsomes.

Results

Purification of NADPH-Cytochrome P-450 Reductase

The bovine adrenal microsomes (1.5 g) were suspended (to 10 mg/ml) in 50 mM potassium phosphate buffer, pH 7.2, containing 0.1 mM EDTA, 0.1 mM dithiothreitol, 20% (v/v) glycerol, 1 μM FAD, 1 μM FMN and 1.5% (v/v) Emulgen 913. After stirring for 1 hr, the suspension was centrifuged at 105,000$\times$g for 1 hr. The supernatant was applied to a DEAE-Sephacel column (2.6$\times$12.5 cm) previously equilibrated with the basal buffer which was composed of 50 mM potassium phosphate buffer, pH 7.2, 0.1 mM EDTA, 0.1 mM dithiothreitol, 1 μM FAD, 1 μM FMN, 20% glycerol and 0.1% Emulgen 913. After washing the column with 80 ml basal buffer containing 50 mM KCl, the reductase was eluted with 600 ml basal buffer in which the KCl concentration was increased from 50 mM to 400 mM in a linear gradient. The fractions with the reductase were detected routinely by the ability to catalyze cytochrome c reduction. The pooled fractions were dialyzed overnight against 3 liters basal buffer and were concentrated to about 1/5 the original volume by ultrafiltration on an Amicon CF-25 membrane. The concentrated solution was passed through a Matrix gel red A column (1.2×4.5 cm) equilibrated with the basal buffer and was loaded on a 2′, 5′-ADP-Sepharose column (1.2×4.5 cm) previously equilibrated with the basal buffer. After washing the column with 70 ml basal buffer, the reductase was eluted with 50 mM potassium phosphate buffer, pH 7.2, containing 0.7 mM $NADP^+$, 0.05% Emulgen 913, 20% glycerol, 0.1 mM dithiothreitol, and 0.1 mM EDTA. The reductase fractions were pooled and concentrated on the Amicon membrane. In order to remove $NADP^+$ from the reductase, the concentrated reductase on the membrane was further washed more than ten times with 50 mM potassium phosphate buffer, pH 7.2, containing 20% glycerol, 0.1 mM dithiothreitol, and 0.1 mM EDTA. The specific activity of the final preparation was 77 μmol cytochrome c reduced min^{-1} mg^{-1} protein with an overall yield of 30%. The overall purification from the solubilized microsomes was 800-fold.

Properties of NADPH-Cytochrome P-450 Reductase

The purified reductase was electrophoretically homogeneous and its molecular weight was estimated to be 78,000±4,000. The purified reductase showed a visible absorption spectrum which had absorption maxima at 380 and 455 nm, and contained 0.8±0.1 mol of FAD and 0.7±0.1 mol of FMN per 78,000 g of protein by fluorimetric analysis (4). The aerobic titration of the reductase with NADPH produced a spectral species with absorption around 600 nm which could be attributed to the air-stable semiquinone radical (5). The ESR spectrum was obtained at 77° K for this species which was similar to that reported for the hepatic microsomal reductase (5).

Steady State Kinetics of Reduction of Cytochromes c and P-450

The activity for electron transfer from NADPH to cytochrome c was dependent on the pH and the ionic strength of the assay system. The maximum reactivity was observed in 0.3 M potassium phosphate buffer, pH 7.7. No detectable difference between anaerobic and aerobic systems was observed for the reduction rate of cytochrome c. Figure 1(a) shows double-reciprocal plots of the rate of cytochrome c reduction vs the concentration of NADPH at various concentrations of oxidized cytochrome c. It is quite obvious that all the lines are parallel. Figure 1(b) shows the double-reciprocal plots in the presence of various concentrations of $NADP^+$. The slope of the lines increased with increasing concentration of $NADP^+$ and all the lines crossed at a certain point. In order to obtain the effect of the concentration of oxidized cytochrome c against the inhibitory action of $NADP^+$, the activity was measured at various concentrations of oxidized cytochrome c taking the ratio of the concentration of $NADP^+$ to that of oxidized cytochrome c to be constant. Parallel lines were obtained as shown in Fig. 1(c).

The electron transfer reaction from NADPH to cytochrome P-450 (P-450_{C21}), specific for steroid C-21 hydroxylation, was analyzed by measuring the formation of a ferrous carbonyl P-450. The rate of P-450_{C21} reduction was markedly stimulated by increasing the concentration of a nonionic detergent, Emulgen 913, and reached a plateau at a concentration over 0.5%. The rate did not show so much dependence on the pH and the ionic strength. The rate was 30

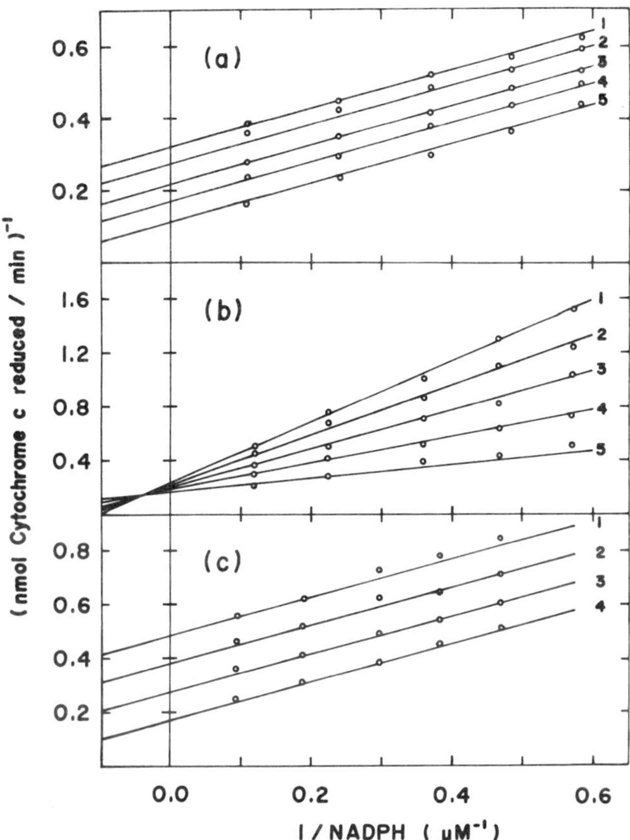

Figure 1. Double-reciprocal plots of the initial velocities of cytochrome c reduction vs NADPH concentration. (a) Concentrations of oxidized cytochrome c from lines 1 to 5 were 4.50, 5.57, 7.41, 11.41, and 23.9 μM, respectively. (b) Concentrations of $NADP^+$ from lines 1 to 5 were 59.11, 43.40, 28.87, 15.09, and 0.0 μM, respectively. The concentration of oxidized cytochrome c was fixed at 12.08 μM. (c) The ratio of the concentration of $NADP^+$ to that of oxidized cytochrome c was fixed at 0.54. Concentrations of $NADP^+$ from lines 1 to 4 were 1.52, 2.06, 3.11, and 6.14 μM, respectively. Concentrations of oxidized cytochrome c from lines 1 to 4 were 2.81, 3.81, 5.76, and 11.38 μM, respectively.

times faster in the presence of the steroid substrate than in its absence. The kinetic studies were performed in 50 mM potassium phosphate buffer, pH 7.2, containing 0.5% Emulgen 913, 20% glycerol, and 30 μM 17α-hydroxyprogesterone. The rate of P-450_{C21} reduction showed a similar dependence on the concentration of NADPH and $NADP^+$ to Figure 1.

Discussion

The following scheme which is usually referred to as Hexa-Uni-Ping-Pong mechanism (6, 7) can be deduced for the reduction of cytochromes c and P-450: Here, E, A, P, B, and Q represent the reductase, NADPH, $NADP^+$, the

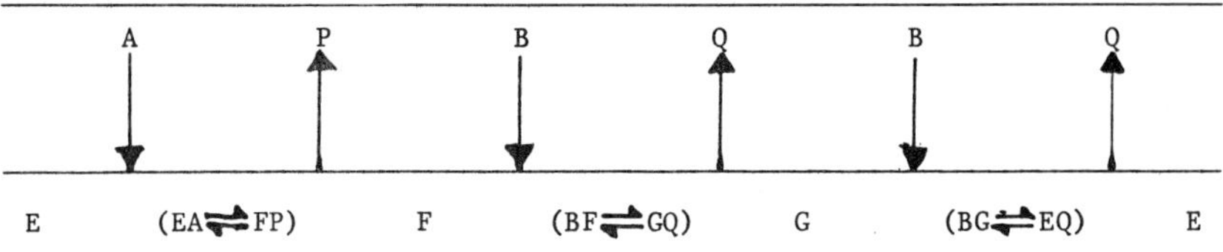

Scheme 1.

ferric cytochrome, and its ferrous form, respectively. F and G indicate some reduced intermediates of the reductase. The dependence of the activity on each component has been shown theoretically by Cleland (6) and Segel (7). In the presence of $NADP^+$ but not the reduced cytochrome, the rate equation becomes as follows:

$$1/v = ((K_{m,A}/V_{max}) + C \times K_{d,A} \times ([P]/[B]))/[A] + (K_{m,B}/V_{max})/[B] + C \times ([P]/[B]) + 1/V_{max} \qquad (121.1)$$

where $K_{m,A}$ and $K_{m,B}$ are Michaelis constants for NADPH and the ferric cytochrome, respectively. $K_{d,A}$ represents the dissociation constant for NADPH and C is some constant. The lines in Figure 1(a), (b), and (c) are the best fit lines for the above equation. In the case of cytochrome c reduction, the values for $K_{m,NADPH}$, $K_{m,cytochrome\ c}$ and $K_{d,NADPH}$ were calculated to be 8.0 μM, 17.0 μM, and 22.3 μM, respectively. The maximum molecular activity was estimated to be 6,000 mol cytochrome c reduced/min/mol of the reductase. In the case of P-450_{C21} reduction, the values for $K_{m,NADPH}$ and $K_{m,P\text{-}450}$ were calculated to be 0.37 μM and 2.4 μM, respectively. The maximum molecular activity was estimated to be 140 mol P-450_{C21} reduced/min/mol of the reductase.

It is concluded that the ternary complex, NADPH : reductase : cytochrome, is not involved in the electron transfer reaction from NADPH to cytochrome c or to P-450_{C21}.

References

1. Masters, B.S.S., Baron, J., Taylor, W.E., Isaacson, E.L., and LoSpalluto, J. (1971) *J Biol Chem* 246:4143–4150.
2. Kominami, S., Mori, S., and Takemori, S. (1978) *FEBS Lett* 89:215–218.
3. Kominami, S., Ochi, H., Kobayashi, Y., and Takemori, S. (1980) *J Biol Chem* 255:3386–3394.
4. Faeder, E.J. and Siegel, L.M. (1973) *Anal Biochem* 53:332–336.
5. Iyanagi, T. and Mason, H.S. (1973) *Biochemistry* 12:2297–2308.
6. Cleland, W.W. (1963) *Biochim Biophys Acta* 67:104–137.
7. Segel, I.H. (1975) In *Enzyme Kinetics*. New York: John Wiley & Sons, pp. 727–736.

Published 1982 by Elsevier North Holland, Inc.
Vincent Massey and Charles H. Williams, Editors
Flavins and Flavoproteins

CHAPTER 122

Transient Kinetics of Electron Transfer Reactions of Flavodoxin: Ionic Strength Dependence of Semiquinone Oxidation by Cytochrome c, Ferricyanide, and Ferric EDTA

Royce P. Simondsen and Gordon Tollin

Departments of Chemistry and Biochemistry, University of Arizona, Tucson, Arizona

The examination of the ionic strength dependence of the reaction between flavodoxin semiquinone and cytochrome c offers an opportunity to investigate the importance of electrostatic interactions between two proteins of known structure (1,2). Flavodoxin has a region of uncompensated negative charge in the vicinity of the FMN prosthetic group. It is not unreasonable that this could participate in electrostatic interactions which facilitate the attainment of the proper orientation for the transfer of electrons between prosthetic groups within a bimolecular reaction complex. As controls, we have also investigated two negatively-charged nonprotein reactants, ferricyanide and ferric EDTA.

C. pasteurianum flavodoxin (Fd) was isolated and purified according to published procedures (3). Horse heart cytochrome c (type VI) (cyt. c) was purchased from Sigma Chemical Co. and used without further purification.

The experiments were carried out in 0.005 M potassium phosphate buffer, pH=7.2. Ionic strength was varied by addition of appropriate amounts of NaCl. EDTA and deazariboflavin were used to generate Fd semiquinone. Stopped-flow measurements were conducted at a wavelength of 580 nm.

The calculation of the Marcus and Debye-Hückel parameters used an iterative program that minimized the least-square errors for the observed second-order rate constants.

Computer graphics analysis was done in collaboration with Drs. P.C. Weber and F.R. Salemme. Coordinates for *C. MP* Fd semiquinone (1) and tuna cyt. c (2) were obtained from the Brookhaven Protein Data Bank (4). Images of each protein were simultaneously displayed and could be independently translated and rotated in the MMSX computer graphics system (5). Molecules, initially oriented with their prosthetic groups toward each other, were moved in order to maximize the number of intermolecular charge pairs. In some cases, side chains were animated (6) in order to bring atoms of opposite charge within hydrogen-bonding distance (2.7 to 3.0 Å).

At low ionic strength, k_{obs} for the reduction of oxidized cyt. c by Fd semiquinone saturates at high cyt. concentration. Such behavior is evidence for complex formation between cyt. and Fd ($Ka \sim 6 \times 10^5\ M^{-1}$) and is not seen at higher ionic strengths or for the ferricyanide or ferric EDTA reactions (these latter reactions are monophasic at all ionic strengths).

The calculated second-order rate constants have been plotted vs $\sqrt{I}$ for the Fd-cyt. c reaction (Figure 1). As can be seen, the rate constant increases with decreasing ionic strength, consistent with a reaction occurring between two oppositely-charged molecules. This is in agreement with other work on cyt. c (7) which demonstrates a positively-charged reaction site. The lines are calculated from the Marcus theory equation derived by Wherland and Gray (8).

The reaction of Fd semiquinone with ferricyanide is slower than for cyt. c. This is due, in part, to the fact that both of the reactants are negatively charged and thus exert a repulsive force on each other. This is demonstrated in Figure 1 where the second-order rate constants vs $\sqrt{I}$ are plotted. The increase of the rate constant with increasing ionic strength is expected for a reaction occurring between two reactants of like charge. Similar results are obtained for ferric EDTA (Figure 1).

Table 1 lists charge productions and k_0 values for all three reactions calculated from both Marcus theory and the extended Debye-Hückel equation. Both sets of calculations yield comparable values for these parameters. For cyt. c the values calculated are unreasonable when full radii are used, i.e., the charge product is greater than the total of the charges on both of the reactants and k_0 is much greater than the diffusion-controlled limit. By decreasing the radii, the calculation yields more reasonable values. Thus, when the radii of the reactants are assumed to be 3-4 Å, a charge product of -15 to -30 is predicted and a k_0 value which is close to the diffusion limit ($\sim 4 \times 10^{10} M^{-1} s^{-1}$ calculated from the Debye-Smoluchowski equation) is obtained. For ferricyanide, the charge product indicates an interaction between like charges with a value of between 3 and 4 for the charge on flavodoxin. A similar result is obtained for the ferric EDTA reaction.

Figure 2 shows an hypothetical electron transfer complex between cyt. c and Fd. Four intermolecular charge pairs are formed between lysine side chains of cyt. c and aspartate and glutamate side chains of Fd. These are Lys 13-Glu 120, Lys 25-Glu 65, Lys 27-Glu 62, and Lys 79-Asp 58. Formation of these salt-links orients each molecule with the exposed edge of their respective prosthetic groups toward the intermolecular interface, and also results in a complementary fit of their respective molecular surfaces. In addition, the ferriheme introduces another positive charge and flavodoxin has a negative charge at Glu 63 which lies within the contact area. Thus, we would predict a charge product of -25, which is within the range of the experimental results.

In the electron transfer complex, the porphyrin ring of cyt. c and the flavin ring in Fd are nearly coparallel. The angle between the planes is 24°. The solvent-exposed portions of each prosthetic group are nearly within van der Waals contact of each other. The closest distance between atoms of each prosthetic group is 3.5 Å between the exposed vinyl of the porphyrin and C-8 CH_3 of the flavin macrocycle. The calculated area which encloses both the prosthetic groups and the charge-paired groups is 186 Å^2.

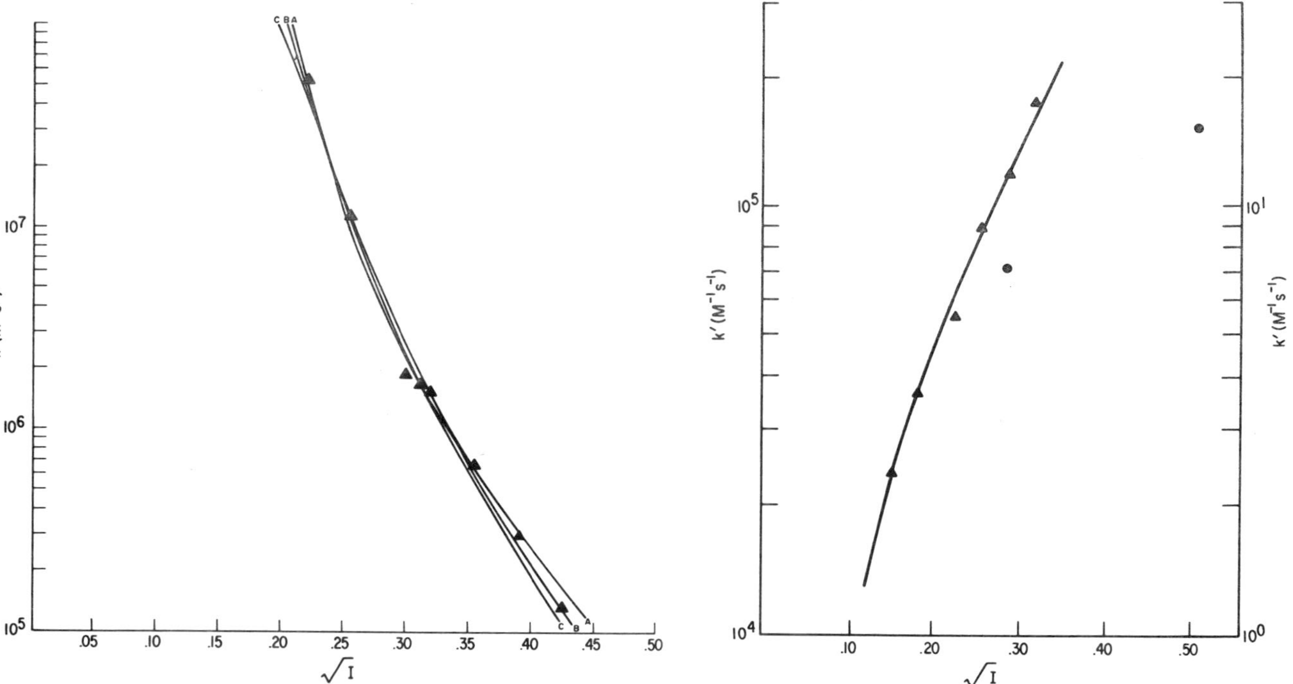

Figure 1. Ionic strength dependence of second-order rate constant for oxidation of flavodoxin semiquinone by cytochrome c. Solid lines represent theoretical curves obtained from Marcus theory for three different radii. (Left) (A) flavodoxin: 17.5 Å; cytochrome: 16.6 Å; (B) flavodoxin: 8 Å; cytochrome: 8 Å; (C) flavodoxin: 4 Å; cytochrome: 4 Å. Ionic strength dependence of second-order rate constant for oxidation of flavodoxin semiquinone. (Right) (▲) ferricyanide; (●) ferric EDTA. Solid line is theoretical curve obtained from Marcus theory with following radii: ferricyanide: 4.5 Å; flavodoxin: 4 Å.

Table 1. Theoretical Calculations of Ionic Strength Effects.

Oxidant	R_{Fld} (Å)	R_O (Å)	Marcus equation Z_1Z_2	Marcus equation k_0 ($M^{-1}s^{-1}$)	Marcus equation k_i ($M^{-1}s^{-1}$)	Debye-Hückel equation Z_1Z_2	Debye-Hückel equation k_0 ($M^{-1}s^{-1}$)	Debye-Hückel equation Z_2
Cytochrome c	17.5	16.6	−237.0	2.80×10^{27}	5.98×10^{4}	−234.5	1.72×10^{28}	7.5
	8.0	8.0	−62.2	7.07×10^{15}	1.40×10^{3}	−50.6	8.82×10^{14}	5.0
	4.0	4.0	−29.4	4.19×10^{12}	4.46	−24.9	8.36×10^{11}	5.0
	3.5	3.5	−24.1	5.94×10^{11}	3.48	−22.8	4.24×10^{11}	5.0
	3.0	3.0	−19.5	1.09×10^{11}	2.69	−20.3	1.59×10^{11}	5.0
$Fe(CN)_6^{3-}$	17.5	4.5	25.7	100	6.43×10^{5}	38.6	19.8	−3.0
	8.0	4.5	13.3	676	2.05×10^{6}	14.9	620	−3.0
	4.0	4.5	9.7	1380	7.42×10^{6}	8.5	1880	−3.0
	3.0	3.0	8.0	1980	4.31×10^{7}	7.7	2200	−3.0
$Fe(EDTA)^{1-}$	17.5	4.0	15.9	9.12×10^{-2}	23.3	14.6	.21	−1.0
	8.0	4.0	5.5	.89	27.4	5.5	.98	−1.0
	4.0	4.0	3.6	1.42	41.2	2.9	1.88	−1.0
	3.0	3.0	2.8	1.83	59.4	2.4	2.17	−1.0

Total net charge estimated from sequence data: Cytochrome c 7.5; flavodoxin −25.0.

R_{Fld} = radius of flavodoxin; R_O = radius of oxidant; k_0 = rate constant at zero ionic strength; k_i = rate constant at infinite ionic strength; Z_1Z_2 = calculated charge product; Z_2 = selected charge for oxidant.

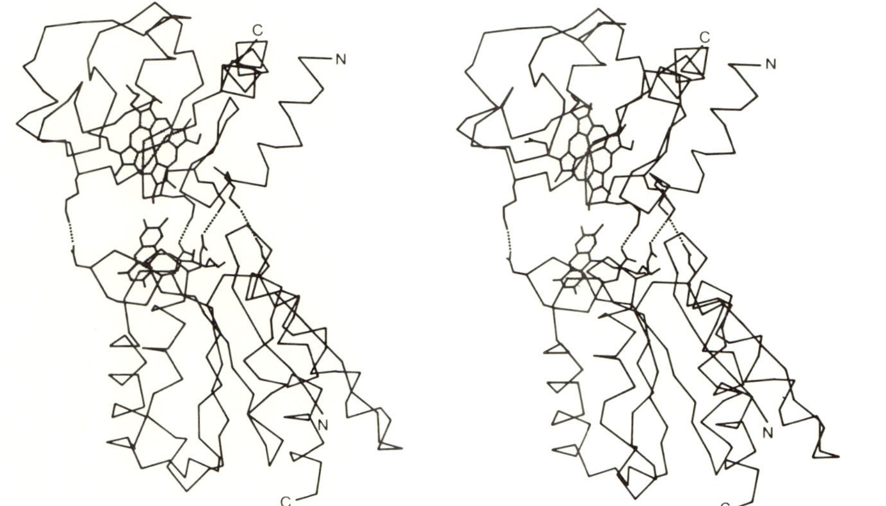

Figure 2. Stereoscopic views of an hypothetical electron transfer complex between cyt. c (upper molecule) and Fd. Only side chain residues involved in intermolecular charge pairs (dashed lines) are shown. Side chains which form covalent interactions with the heme group (shaded) of cyt. c, and a tryptophan residue (trp 90) which packs against the front side of the flavin (shaded) are omitted for clarity.

In their derivation of ionic strength effects, Wherland and Gray (8) used a Coulomb potential between uniformly charged spheres at a distance given by the sum of their radii. Given this and the equation which defines the potential between two parallel plates, it is possible to relate the contact surface area (assuming planarity) between Fd and cyt. c, and the distance between surfaces, which are calculated from the model structure, to the sum of the spherical radii given by the Marcus equation. This yields a value of 9.9 Å for the sum which is in reasonable agreement with the ionic strength data. Another simpler method of testing the model is to calculate the radius of a sphere which has the same surface area as that obtained from the model. This yields a value of 3.9 Å for the radius (of each molecule), which is again in good agreement with the data. We conclude from these calculations that the structural model provides a satisfactory means of accounting for the electrostatics of the Fd-cyt. c interaction.

A computer graphics study of the Fd-ferricyanide interaction indicates that approximately three negatively charged groups on Fd can interact with the ferricyanide within an electron transfer complex. This is also in good agreement with the theoretical calculations of the ionic strength effects (Table 1).

In conclusion, this investigation indicates that electrostatic interactions are probably important in orienting Fd for its role as an electron-transferring protein. The theoretical and computer analyses suggest that charges which are localized in the vicinity of the prosthetic groups, rather than the overall charge on the protein, are important in this regard. A more complete account of this research will be published elsewhere.

ACKNOWLEDGMENT

Work supported in part by a Research Grant from the National Institutes of Health (AM15057).

References

1. Smith, W.W. et al. (1977) *J Mol Biol* 117:195–225.
2. Swanson, R. et al. (1977) *J Biol Chem* 252:759–775.
3. Mayhew, S.G. (1971) *Biochim Biophys Acta* 235:289–302.
4. Bernstein, F.C. et al. (1977) *J Mol Biol* 112:535–542.
5. Barry, C.D. et al. (1974) *Fed Proc* 33:2368–2372.
6. Lederer, F. et al. (1981) *J Mol Biol* 148, in press.
7. Cusanovich, M.A. (1978) In *Bioorganic Chemistry Vol. IV Electron Transfer and Energy Conversion*; *Cofactors*; *Probes*. van Tamelen, E.E. (ed) New York: Academic Press, pp. 117–145.
8. Wherland, S. and Gray, H.B. (1976) *Proc Natl Acad Sci USA* 73:2950–2954.

PART X:

Complex Flavoproteins

PART X A:

Iron-Sulfur Flavoproteins

Published 1982 by Elsevier North Holland, Inc.
Vincent Massey and Charles H. Williams, Editors
Flavins and Flavoproteins

CHAPTER 123

Dehydrogenases of Methylotrophic Bacteria as Possible Models for Electron Transfer in Iron-Sulfur Flavoproteins

Helmut Beinert,* Robert W. Shaw,*
Daniel J. Steenkamp,† Thomas P. Singer,†
Randy Stevenson,¶ W. Richard Dunham,¶ and
Richard H. Sands¶

Institute for Enzyme Research, University of Wisconsin, Madison, †Department of Biochemistry and Biophysics, University of California and Molecular Biology Division, Veterans Administration Hospital, San Francisco, and ¶Biophysics Research Division, Institute of Science and Technology, University of Michigan, Ann Arbor Michigan

Trimethylamine (TMA) dehydrogenase from the methylotrophic bacterium W3A1 is an iron sulfur-flavoprotein (Fe-S-Fp) which catalyzes the conversion of TMA to dimethylamine (DMA) and formaldehyde with artificial electron (e^-) acceptors or with an ETF-like flavoprotein from W3A1 (1–5). Direct experimental evidence for the mode of e^- transfer in Fe-S-Fps where flavin is the e^- acceptor on the substrate side (as, e.g., in the dehydrogenases of the mitochondrial respiratory chain) has been scarce. While enhanced e^- spin relaxation of the semiquinones of these enzymes (6,7) and observation of half-field signals (8) suggested interaction with a neighboring Fe-S cluster, kinetic resolution of the reaction sequence was never achieved. Bacterial TMA dehydrogenase, which has a single flavin and [4Fe-4S] cluster per molecule and an overall turnover of 0.1 to 1 per sec, offers advantages for collecting information on this point.

Reduction of Trimethylamine Dehydrogenase and Kinetic Properties

When TMA dehydrogenase is reduced by dithionite, 3 e^- are taken up per molecule of enzyme to produce $FMNH^\cdot$ and $[4Fe\text{-}4S]^+$. On addition of TMA, only 2 e^- are taken up; but neither a significant $FMNH^\cdot$ nor $[4Fe\text{-}4S]^+$ signal is observed. Instead a set of signals characteristic of a triplet state appears. The optical absorption spectrum has features of $FMNH^\cdot$ spectra. Full development of this characteristic spectrum requires more than one mole of TMA per mole of dehydrogenase, suggesting the presence of an allosteric site (3). This is shown by the effect of the inhibitor $N(CH_3)_4Cl$ on the pattern of reduction of TMA dehydrogenase by dithionite. In the presence of this inhibitor, only 2 e^- enter the enzyme and the triplet state is formed as it is with TMA.

Kinetic analysis of the reaction of TMA dehydrogenase of W3A1 with TMA (2–4), carried out by the stopped flow and rapid freeze-quench techniques, both with optical monitoring in the liquid and frozen states and EPR monitoring in the frozen state, showed that 2 e^- reduction of FMN is a rapid process ($t_{1/2}$ <2 msec at 18°). No significant EPR signal of $FMNH^{\cdot}$ is detected at any time. Nevertheless, the optical spectra as well as appearance of the triplet EPR signal following the initial fast bleaching process clearly indicate that $FMNH^{\cdot}$ must be formed. The e^- transfer indicated by these spectral changes, namely from $FMNH^{\cdot}$ to the Fe-S cluster, is one to two orders of magnitude slower than initial formation of $FMNH_2$. However, the rates measured on liquid and freeze-quenched samples showed unsatisfactory agreement. Among the techniques used, only EPR can unambiguously demonstrate that e^- transfer has indeed taken place and what species are involved. The slow rates observed by EPR, therefore, raised the question as to whether the spin-coupled state could be an intermediate in the overall reaction.

Since the aforementioned work, three developments made a renewed study seem promising: similar TMA and DMA dehydrogenases of another organism (Hyphomicrobium X) (9) have been purified which, by comparison, furnished clues to the behavior of the W3A1 enzyme; considerable progress has been made in the understanding of the observed triplet EPR signals (see below); and the possibility of freezing artifacts with the pyrophosphate buffer previously used was recognized (cf 10) and minimized by substituting barbital. New kinetic data are shown in Figures 1 and 2; in Figure 1, a comparison of the rates of reduction of the three enzymes by their substrates as measured at 365 nm and in Figure 2 a comparison of the EPR (points) with the optical data (lines). While the progress curve for DMA dehydrogenase corresponds to a single first-order reaction, those of the TMA dehydrogenases are biphasic. The slow phase is due to reaction at an allosteric site of additional TMA with the reduced enzymes (3). Figure 3 shows EPR spectra obtained with the three enzymes at 320 msec reaction time. It is obvious from Figure 3 that the splitting of the triplet signal at g~2 and the half-field signal at g=4 are largest for W3A1 and smallest for DMA dehydrogenase. The agreement of kinetic optical and EPR data is, however, best for DMA dehydrogenase and poorest (not shown) for TMA dehydrogenase of W3A1. It appears that the allosteric effect manifest in the slow reaction phase of the TMA dehydrogenases changes the orientation of $FMNH^{\cdot}$ with respect to $[4Fe\text{-}4S]^+$ (see below) thus introducing changes in the triplet EPR signal which are not reflected in the light absorption measurements. The agreement of optical and EPR measurements is best when there are no changes in the splitting during the reaction. With each of the enzymes, however, the formation of the spin-coupled form in the rapid phase is well within the limits of the overall reaction rate thus allowing the conclusion that this form is an intermediate in the dehydrogenation reaction.

Can we consider the mode of reaction of these dehydrogenases with their substrates as a model for Fe-S-Fps in general? Is the reaction sequence the same and is it only the relative rates of the component reactions that are sufficiently different so that the $FMNH^{\cdot}-[4Fe\text{-}4S]^+$ intermediate never accumulates and that we cannot discern the individual steps as we can with these bacterial dehydrogenases? We have no definitive answers to these questions.

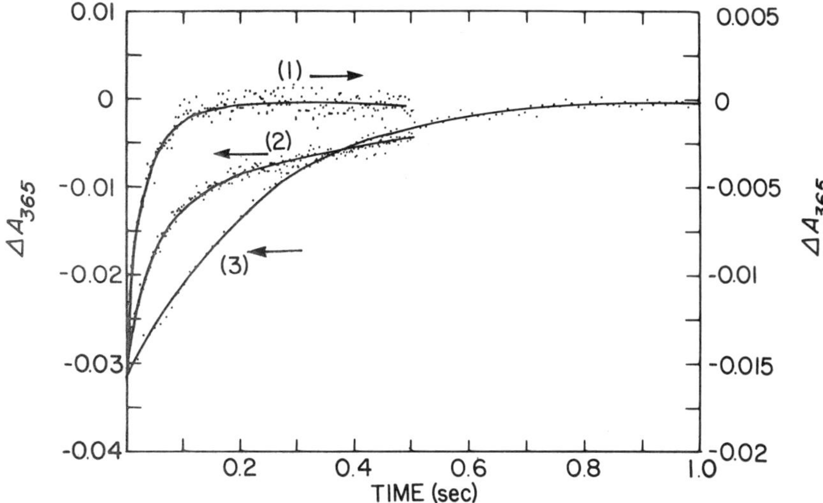

Figure 1. Stopped flow spectrophotometric determination of the rate of appearance of absorption at 365 nm following mixing to give final concentrations of (1) 7.76 μM TMA dehydrogenase from bacterium W3A1 with 100 μM TMA; (2) 6 μM TMA dehydrogenase from Hyphomicrobium X with 100 μM TMA; and (3) 7 μM DMA dehydrogenase with 100 μM DMA in 40 mM potassium barbital (pH 7.7) at 17°. Arrows indicate the appropriate ordinate.

Figure 2. Time course of the reaction of TMA and DMA dehydrogenases of Hyphomicrobium X with their respective substrates (0.1 mM) at 17° (points) compared to the reaction measured optically at 365 nm (lines) from Figure 1. The enzymes and substrates were mixed and rapidly freeze-quenched. The enzyme concentrations after mixing were: 102 μM for TMA dehydrogenase (TMAD) (▲) and 65 μM for DMA dehydrogenase (DMAD) (●). The ordinate shows maximal development of A_{365} nm in 1 sec adapted from Figure 1, the abscissa time of reaction in sec. Progress of the reaction was determined by EPR from integration of the signals at g=4. These data were normalized to the optical data by superimposing the points at 0.32 sec for TMAD and 0.4 sec for DMAD. Note that two of the observed EPR signals are shown in Figure 3. The conditions of recording are given there.

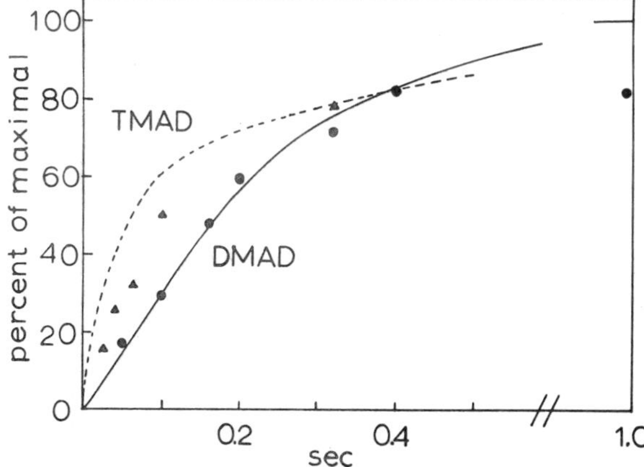

The TMA and DMA dehydrogenases may be simple models—and intuitively satisfying models—that nature has provided for e^- transfer in Fe-S-Fps.

Interpretation of the EPR Signals Observed With TMA and DMA Dehydrogenases

From the composition of these enzymes, one would expect to observe no EPR signal in the oxidized state, a signal typical of $[4Fe\text{-}4S]^+$ in the fully reduced state (g-values between 2.1 and 1.8), and a signal for FMNH$^{\cdot}$ (g=2.00) at intermediate states of oxidation. This behavior is observed on titration with dithionite. The additional pattern of sharp and broad lines in the g~2 region and the signal at g=4 which arise on addition of TMA (DMA) to the enzymes must, however, also be explained in terms of the $[4Fe\text{-}4S]^+$ cluster and FMNH$^{\cdot}$. The fact that on addition of substrate the expected signals typical for these components are not observed to any significant extent suggests that they interact with each other so as to produce the unusual features of the EPR spectra. Variable frequency EPR on TMA and DMA dehydrogenases shows that the characteristic broad outer splitting at g~2 seen on addition of substrate does, to a first approximation, not increase with frequency. This

Figure 3. Comparison of EPR signals of TMA and DMA dehydrogenases of W3A1 and Hyphomicrobium X as indicated in the figure when these enzymes were exposed to their respective substrates (0.1 mM) for 0.32 sec at 17°. The EPR spectra were recorded at 9.227 GHz, 0.8 mT modulation amplitude and 12 K and with 9 mWatt power at low-field (left) and 2.7 mWatt at high-field (right). Up to 9 scans were averaged by a Varian Data System. Note that the amplitudes of the signals at high-field would be 2.74-fold higher if microwave power and gain had been identical for low- and high-field. Note also that the amplification for the signals of W3A1 was 3.2 times that used for the signals of the Hyphomicrobium X enzymes. The samples were the same as those used for the experiments of Figure 2.

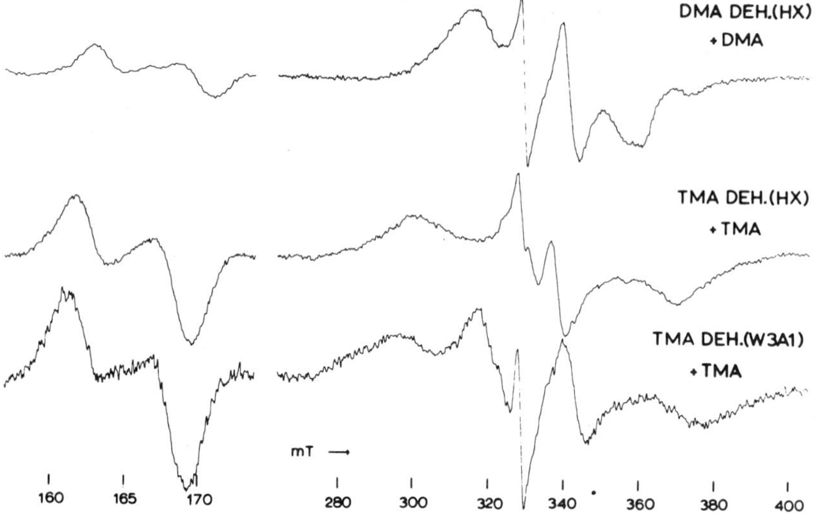

behavior is typical of spin-spin interaction and is familiar from "Hyperfine" interactions, viz., electron-nuclear spin-spin interactions. The signal at $g=4$, however, is only observable for dipolar or pseudodipolar electron-electron spin interactions. It arises from a transition between the upper and lower energy levels of the three levels of a triplet state.

A triplet state would arise between the two paramagnets because of an isotropic (orientation independent) exchange interaction which may be thought of as a correction to the electrostatic interaction in the system, due to the Pauli Exclusion Principle. The large splittings observed in the $g\sim 2$ region of the EPR spectra of the TMA and DMA dehydrogenases could originate from some combination of two interactions: magnetic dipole-dipole interaction or from anisotropic exchange, a third-order correction to the isotropic exchange which is due to spin-orbit coupling. Both interactions, dipolar and anisotropic exchange, can be parameterized by the zero field splitting (ZFS) tensor.

For TMA dehydrogenase of W3A1, e.g., we find a separation of the broad lines in the $g\sim 2$ region of 120 mT at X-band. We call this separation 2D. Assuming that the interaction giving rise to this separation is pure dipole-dipole, we estimate from Equation (123.1) (see Appendix) that the dipoles would have to be separated by ~ 3 Å. Such a center to center distance for a [4Fe-4S] cluster and $FMNH^{\cdot}$ seems unrealistic, even if the plane of the flavin were perpendicular to the vector connecting the two centers. Could we then explain the features of the spectra just in terms of anisotropic exchange interaction? Equation (123.2) provides an estimate for the size of this interaction in terms of the anisotropy of g-factors, i.e., quantities measurable from our spectra. From Equation (123.2) we estimate J, the strength of the exchange interaction, as $J=120\ cm^{-1}$ if we take the center of the two broad lines as $g=1.96$. Such a value is large but not unreasonable. However, a linear combination of the dipole-dipole and pseudodipolar interactions would seem to us the most likely interpretation.

The triplet Hamiltonian with all degrees of freedom is given by Equation (123.3). The ZFS Hamiltonian is usually given by Equation (123.4), but since we cannot a priori assume that the g-tensor and the ZFS tensor are parallel we must let the interacting structures rotate with respect to each other in all three dimensions. In attempts to simulate the observed spectra, we have to consider for each orientation three resonances: a pair at $g\sim 2$ split about an effective g-value and a single line at $g=4$ (the half-field signal). For the simulation of powder spectra, one must superimpose spectra from all orientations. The resulting spectrum at $g\sim 2$ displays three sets of two turning points each; these pairs are split by 2D, D+3E, and D−3E, respectively. The position in the field of the $g=4$ resonance is mainly determined by g anisotropy with second-order shifts downfield with increasing ZFS. Since the values of the ZFS tensor are not known, the g factors are calculated from spectra obtained at two frequencies (9 and 15 GHz), which allows us to eliminate the contributions from the ZFS parameters. The values of the ZFS parameters D and E are extracted from the spectra in the $g\sim 2$ region. In our case, the ZFS anisotropy is dominant over the anisotropy of the g tensor so that with large ZFS the g-value anisotropy is essentially masked. Note that the broad outer resonances in the $g\sim 2$ region in Figures 3 and 4 are composed of two separate but

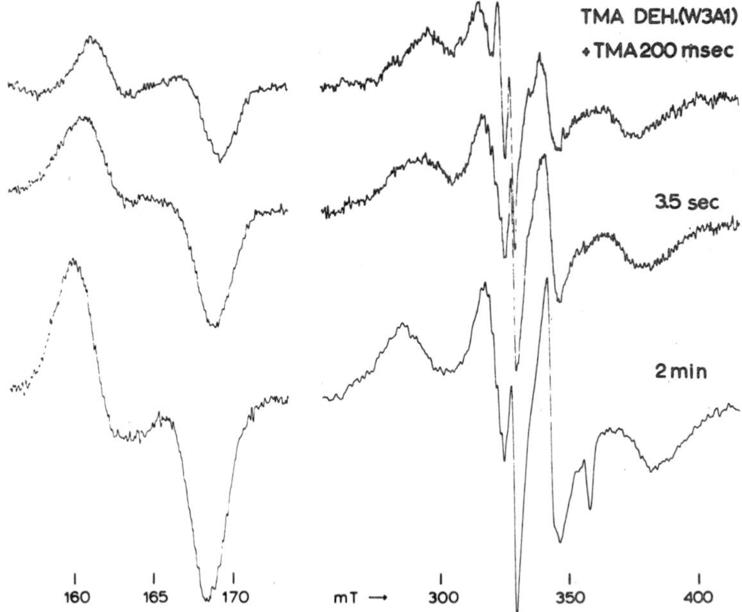

Figure 4. Comparison of EPR signals of TMA dehydrogenase of W3A1 when exposed to TMA for various lengths of time at 17°. (Top and center) with 0.1 mM TMA for 0.2 sec and 3.5 sec, respectively; (bottom) with 2 mM TMA for 2 min. The EPR spectra were recorded as for Figure 3. Note that the amplitudes of the signals at high-field would be 1.5-fold higher if microwave power and gain had been identical for low- and high-field.

adjacent resonances each. These correspond to two pairs of lines split by 2D and D+3E, respectively. Thus we arrive at the three principle g-values and the ZFS parameters D and E.

So far we have drawn only on the field positions of the lines, not their intensities. The shape of the resonance at $g=4$, however, can provide useful information since it is usually quite sensitive to variations in the angles of rotation (see above). The $g\sim2$ region generally shows the opposite behavior. The procedure for fitting these spectra then is to move back and forth between the $g\sim4$ and $g\sim2$ regions in refining parameters. The $g=4$ region is more sensitive to changes in the principal g values and the angles of rotation than the $g\sim2$ region, whereas the latter is more useful for refining D and E.

Although the broad outer resonances are obviously made up by two separate pairs of resonances, they can only be generated in the shape as observed by invoking a distribution in the splittings. We assume that for each powder orientation the splitting, or its equivalent, viz., the projection of the ZFS tensor along the direction of the magnetic field, has a Gaussian distribution with a spread proportional to that projection.

Figure 5 shows simulation of the EPR spectrum of DMA dehydrogenase plus DMA with the effect of the assumed distribution (note that minor contributions from $FMNH^{\cdot}$ and a Fe-S signal in the original spectrum are not

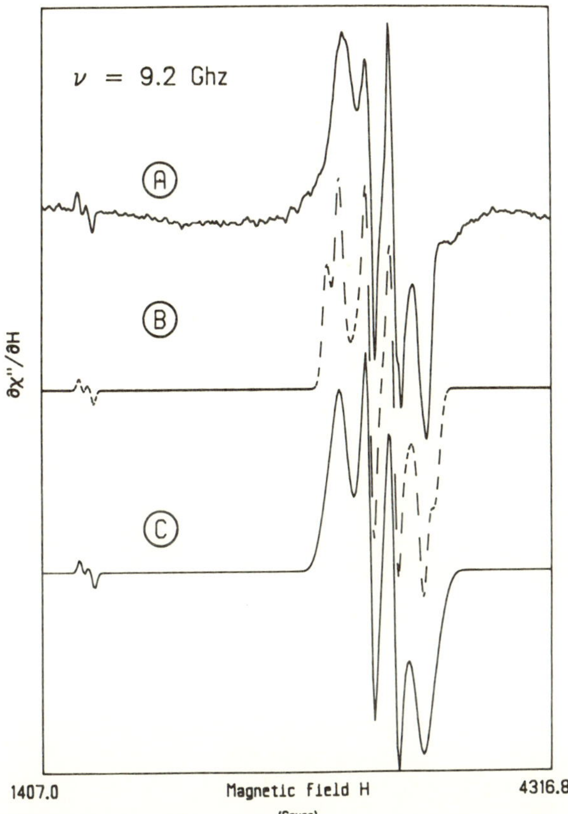

Figure 5. Computer simulation of the EPR spectrum of DMA dehydrogenase as recorded after reaction with 2 mM DMA for 2 min. (Top) spectrum as recorded; (center) simulation without assuming ZFS distribution; (bottom) simulation assuming ZFS distribution. The parameters used in the simulation were $g_x = 1.921$, $g_y = 1.958$, $g_z = 2.015$; rotation angles $\alpha = 0°$, $\beta = 45°$, $\gamma = 60°$ and ZFS parameters D = 287 gauss, E = 54 gauss.

incorporated into the simulation). It is obvious from these spectra and a comparison to the spectra of TMA dehydrogenase plus TMA (Figure 6) that the features and also the relative intensities at g~2 and g=4 can be reproduced very well by our simulations. Obviously, for DMA dehydrogenase D is much smaller than for TMA dehydrogenase and, consequently, also the spread of the splitting. Furthermore, as shown in Figure 5, as TMA dehydrogenase reacts with TMA, the separation of the broad outer lines increases as the reaction proceeds, i.e., D increases and so does the spread of the splitting. Since we know (3) that TMA dehydrogenase is subject to allosteric effects exerted by substrate or analogs, we interpret this observation to indicate that as TMA is bound at the allosteric site, changes take place in the active center leading to an increase in the parameter D.

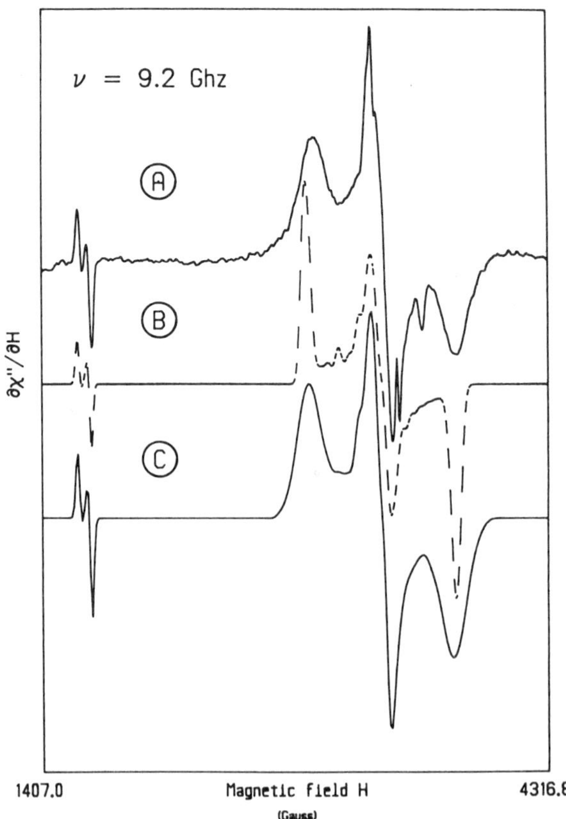

Figure 6. Computer simulation of the EPR spectrum of TMA dehydrogenase from Hyphomicrobium X as recorded after reaction with 2 mM DMA for 2 min. Curves as in Figure 5. The parameters used in the simulation were $g_x = 1.927$, $g_y = 1.960$, $g_z = 2.015$; rotation angles $\alpha = 0°$, $\beta = 45°$, $\gamma = 60°$ and ZFS parameters D = 428 gauss, E = 127 gauss.

While we thus have arrived at a plausible physical model for the type of interactions prevailing between the redox-active centers of these enzymes, can we draw any conclusions that are of interest from the enzymologist's point of view? Unfortunately, in the presence of exchange interaction, a simple treatment is not feasible. J, the exchange parameter, would have to be independently determined by, e.g., magnetic susceptibility measurements, to confirm our conclusions concerning its magnitude. The separation of the interacting centers in terms of atoms intervening (super exchange) could then be estimated. However, from the information at hand, namely that we are dealing with exchange interaction, it is obvious that the separation of the *centers* (FMNH$^\cdot$ and $[4Fe\text{-}4S]^+$) must be small, presumably not exceeding 10 Å, (i.e., the edge-to-edge distance would be less than 5 Å).

Appendix

Equation (123.1). $D \sim 3/4\, g_e^2 \beta^2 1/r^3$ — g_e is the electron g value and β is the Bohr magneton of the electron.

Equation (123.2). $J \sim \left(\frac{g}{\Delta g}\right)^2 D$ — $\Delta g = g - g_e$; g is a typical g value associated with the 2D splitting.

Equation (123.3). $\mathcal{H} = \beta \vec{S} \cdot \overleftrightarrow{G} \cdot \vec{D} + \vec{S} \cdot \overleftrightarrow{D} \cdot \vec{S}$ — where $S = S_1 + S_2 = 1/2 + 1/2 = 1 \cdot \vec{S} \cdot \overleftrightarrow{D} \cdot \vec{S}$ is the symmetric zerofield pseudodipole and dipole interaction.

Equation (123.4). $ZFS = \begin{pmatrix} D/3 & O & E \\ O & -2D/3 & O \\ E & O & D/3 \end{pmatrix}$ — the unrotated zero field Hamiltonian.

References

1. Steenkamp, D.J. and Singer, T.P. (1976) *Biochem Biophys Res Commun* 71:1289–1295.
2. Steenkamp, D.J., Singer, T.P., and Beinert, H. (1978) *Biochem J* 169:361–369.
3. Steenkamp, D.J., Beinert, H., McIntire, W.C., and Singer, T.P. (1978) In *Mechanisms of Oxidizing Enzymes*. Singer, T.P. and Ondarza, R.N. (eds.) Vol. 1. New York: Elsevier/North-Holland Inc., New York: pp. 127–141.
4. Singer, T.P., Steenkamp. D.J., Kenney, W.C., and Beinert, H. (1979) In *Flavins and Flavoproteins*. Yagi, K. and Yamano, T. (eds.) Tokyo: Japan Scientific Societies Press, and Baltimore: University Park Press, pp. 277–287.
5. Steenkamp, D.J. and Gallup, M. (1978) *J Biol Chem* 253:4086–4089.
6. Beinert, H. and Hemmerich, P. (1965) *Biochem Biophys Res Commun* 18:212–220.
7. Ohnishi, T., King, T.E., Salerno, J.C., Blum, H., Bowyer, J.R., and Maida, T. (1981) *J Biol Chem*, in press.
8. Salerno, J.C., Ohnishi, T., Lim, J., Widger, W.R., and King. T.E. (1977) *Biochem Biophys Res Commun* 75:618–624.
9. Steenkamp, D.J. (1979) *Biochem Biophys Res Commun* 88:244–250.
10. Williams-Smith, D.L., Bray, R.C., Barber, M.J., Tsopanakis, A.D., and Vincent, S.P. (1977) *Biochem J* 167:593–600.

Published 1982 by Elsevier North Holland, Inc.
Vincent Massey and Charles H. Williams, Editors
Flavins and Flavoproteins

CHAPTER 124

The Role of Flavodoxins in the Reduction of Nitrogenous Compounds and in the Production of Hydrogen

C. Veeger

Department of Biochemistry, Agricultural University, De Dreijen 11, 6703 BC Wageningen, The Netherlands

Flavodoxin and Nitrogen Fixation

The only known electron carriers operating in the transfer of reducing equivalents to nitrogenase are ferredoxin and flavodoxin. In anaerobes like *C. pasteurianum* the sequence

$$\text{pyruvate} \rightarrow \genfrac{}{}{0pt}{}{\text{ferredoxin}}{\text{flavodoxin}} \rightarrow \text{nitrogenase} \rightarrow N_2$$

is able to generate reducing equivalents of sufficiently low redox potential to reduce nitrogenase (1,2). In aerobes however the situation is different. Beneman et al. (3) postulated a sequence

$$\text{NADPH} \rightarrow \genfrac{}{}{0pt}{}{\text{ferredoxin}}{\text{flavodoxin}} \rightarrow \text{nitrogenase} \rightarrow N_2.$$

It is difficult to visualize how *A.vinelandii* flavodoxin, of which the hydroquinone-semiquinone couple E_m is -495 mV, can be reduced, since Haaker et al. (4) showed that within the cell the ratio $NAD(P)H/NAD(P)^+$ was around 2–3 : 1. Furthermore, they showed that nitrogen fixation comes to full inhibition by using uncouplers of oxidative phosphorylation or even by depolarizing the membrane (5,6).

Table 1 gives the activity of nitrogenase from *A.vinelandii* and from *R.leguminosarum* bacteroids. In agreement with Yates (7), but in contrast with Beneman et al. (8), flavodoxin hydroquinone serves as a 1-electron donor. *A.vinelandii* flavodoxin hydroquinone is a more efficient electron donor than reduced ferredoxin I; its K_m is around 5 μM. In fact, at infinite carrier concentration the activity with flavodoxin hydroquinone is around 1.5 times the activity with $Na_2S_2O_4$. The flavodoxins of *Megasphaera elsdenii* and *Desulfovibrio vulgaris* have a maximum velocity of about 30% and 15% respectively. The hydroquinone of *A. vinelandii* protein is extremely reactive with O_2 and plays in our opinion an essential role in protecting nitrogenase against damage by O_2 (9).

Scherings et al. (9) showed that the nitrogenase reaction comes to a stop when the redox potential of the donor exceeds -450 mV. However, under the conditions of these experiments, the flavodoxin hydroquinone concentration

Table 1. Comparison of the Efficiency of Several Electron Donors for the Nitrogenase from *A. vinelandii* and *R. leguminosarum* Bacteroids.

		A. vinelandii		*R. leguminosarum*
		Chloroplasts/2, 6 dichlorophenol indophenol/ascorbate	Photoreductant Deazaflavin/tricine	Deazaflavin/tricine
A.v.	flavodoxin	0	80	75
M.e	flavodoxin	7	16	
D.v	flavodoxin	N.d.	15	
A.v+M.e	flavodoxins	16	111	
A.v	ferredoxin I	33	13	58
A.v	FeS prot II	11	6	
A.v	ferredoxin +			
A.v	flavodoxin	59	89	127
None		0	3	

Data are expressed as percentage rate with dithionite as electron donor. The donors (10 μM) are photoreduced by two different methods. A.v: *A. vinelandii*; M.e: *M. elsdenii*; D.v: *D. vulgaris*; N.d.: not determined.

was below its K_m-value (Figure 1). In fact, recent work shows that the switch-off potential of the nitrogenase reaction under carefully controlled conditions of pH and ionic strength with different electron donors is around −400 mV, about 50 mV higher than the value reported earlier (Figure 2).

Figure 1. Activity-redox potential relation of flavodoxin hydroquinone oxidations (cf. ref. 9). (A) without treatment; (B) treatment of nitrogenase: (O—O, □—□) no treatment; (Δ—Δ, ×—×) prereduced by photoreduction.

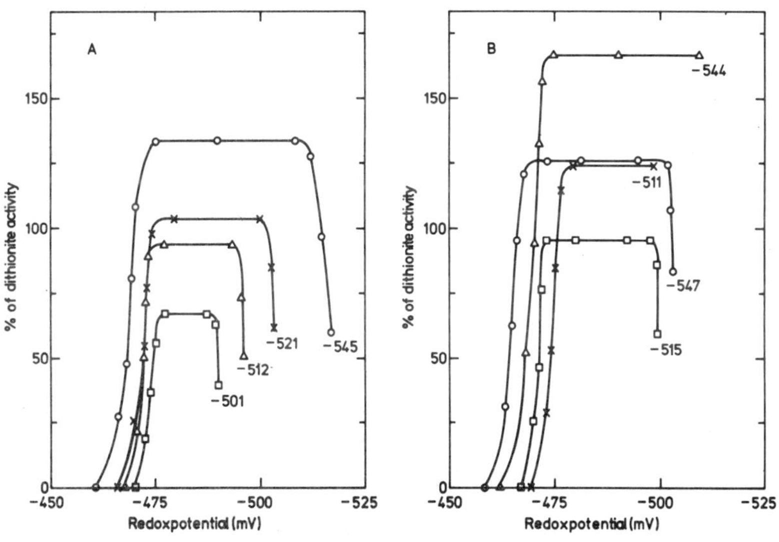

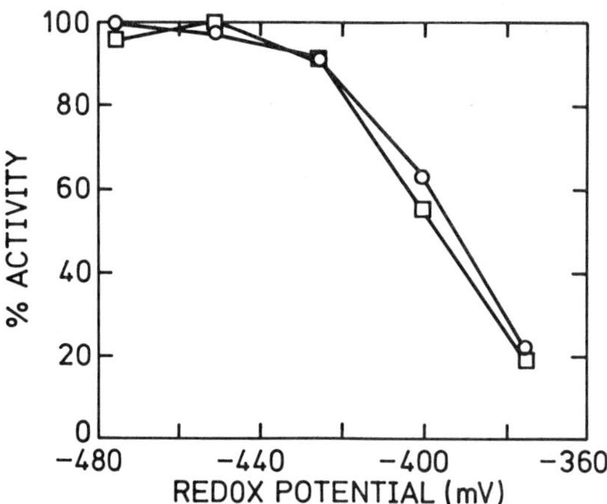

Figure 2. Activity-redox potential relation of nitrogenase. The redox potential of the donor was fixed by the method of ref. 14.

However, the lag in the activity, when the enzyme is not reduced prior to the assay, is always observed.

An enzyme which reduces *A.vinelandii* flavodoxin initially to the hydroquinone in the presence of NAD(P)H, was purified 200 times (cf. ref. 10). The specific activity of the pure FAD-containing protein is not very high, and varies between 6 and 30 nmoles·min^{-1}·mg^{-1}. In addition, this reductase reacts with the flavodoxins from *M.elsdenii* > *D.vulgaris* > *A.vinelandii* under anaerobic conditions, an order which corresponds with the redox potentials of the semiquinone-hydroquinone couple.

Although catalyzing initially the formation of flavodoxin hydroquinone, the soluble enzyme does not meet the requirements needed for electron donation to nitrogenase. The tight coupling of the nitrogenase activity with the membrane potential necessitates coupling of flavodoxin hydroquinone formation to reversed electron flow in the membrane. Reducing equivalents, generated via oxidation of NADH in the respiratory chain, reduce the flavodoxin semiquinone to its hydroquinone. It is possible that membrane-bound flavodoxin reductase catalyses this reaction, while in addition this membrane process makes use of the pH-dependence of the flavodoxin hydroquinone/semiquinone couple below the pK of protonation of the hydroquinone (10, 11) (Figure 3).

Flavodoxin and Nitrate Reductase

Bothe (12) recently succeeded in solubilizing nitrate reductase present in *A. vinelandii* membranes. Table 2 shows that flavodoxin hydroquinone (1–4 μM) is a much better "natural" electron donor, compared with higher concentration (25–100 μM) reduced ferredoxin. The artificial electron donors methyl viologen and benzyl viologen are more efficient.

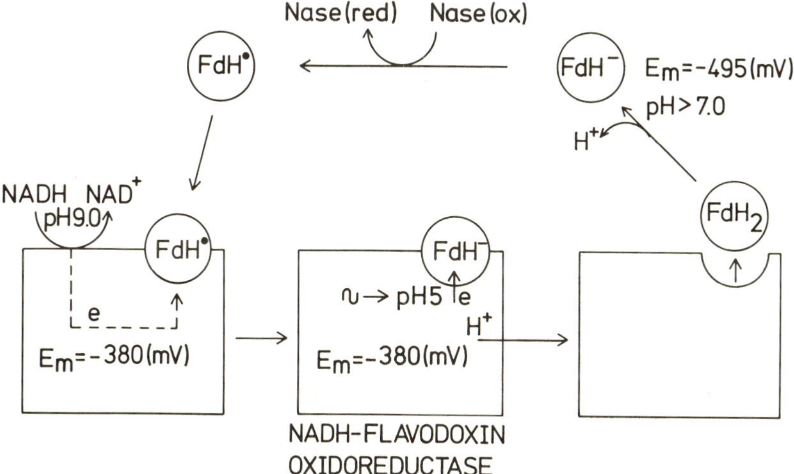

Figure 3. Proposed scheme for electron transport from NADH to N_2-ase in *A.vinelandii* (cf. ref. 10).

Effect of pH and Redox Potential on Hydrogen Production Activity of Hydrogenase

As has been shown by Mayhew et al. (13), at low pH *M.elsdenii* flavodoxin hydroquinone becomes protonated according to

$$FlH_2^- + H^+ \rightleftharpoons FlH_3$$

with a value for the ionization constant at 25°C of pK'=5.8. We determined the effect of temperature on the midpoint potential and ionization constant of flavodoxin hydroquinone and found (14) that the change in midpoint potential per degree Kelvin of the protonated species is greater than that of the unprotonated species. Above 15°C $\Delta E_B \cdot K^{-1}$ is -0.6 mV and $\Delta pK' \cdot K^{-1}$ is -9.7×10^{-3}. From these data, it can be calculated that at 30°C the value for $E_B = -378$ mV, with an ionization constant pK'=5.75. It was necessary to know these values, since our hydrogen production activities were determined at 30°C.

The effect of the pH and redox potential on the hydrogen production activity with either methyl viologen semiquinone (0.3 mM) or flavodoxin hydroquinone (50 μM) as donor is shown in Figure 4A and B respectively. From both figures, it appears that, first, the activity increases at decreasing pH; second, at a given pH the activity is independent of the redox potential up to a certain value, after which the activity declines; third, at decreasing pH values, this redox potential-independent part of the hydrogen production activity shifts to more positive potentials; fourth, the redox potential at which hydrogenase does not produce hydrogen any more increases with decreasing pH; fifth, in the range of redox potentials where the hydrogen production activity is constant, at less negative redox potentials, the concentration of the reduced electron carrier is constant while the concentration of the oxidized

Table 2. Nitrate Reduction by the Solubilized Nitrate Reductase (cf ref. 12).

Electron donor system	nmol NO_3^- reduced/hr × mg protein
1. $Na_2S_2O_4$	0
2. $Na_2S_2O_4$ + BV	3480
3. $Na_2S_2O_4$ + MV	2800
4. $Na_2S_2O_4$ + ferredoxin spinach	0
5. Spinach chloroplasts	2
6. Spinach chloroplasts + ferredoxin spinach	85
7. Spinach chloroplasts + ferredoxin *Azotobacter*	114
8. Spinach chloroplasts + flavodoxin *Azotobacter*	63
9. Deazaflavin EDTA	0
10. Deazaflavin EDTA + ferredoxin *Azotobacter*	51
11. Deazaflavin EDTA + flavodoxin *Azotobacter*	1390
12. NADH + BV or MV	0
13. NADPH	51
14. NADPH + MV	165
15. NADPH + BV	369

Solubilized nitrate reductase was obtained by treating the large particle fraction with Triton-X100 and DE-cellulose. Protein concentration of nitrate reductase was 0.3–0.5 mg in the assays.

form of the carrier increases. This latter observation suggests that, firstly, the oxidized forms of the electron carrier used do not inhibit the hydrogenase and secondly, that flavodoxin semiquinone cannot serve as electron donor for hydrogenase to produce hydrogen. Further, it is noticeable (14), that at a given pH the redox potential at which half of the maximal activity at that pH is observed is at the potential of the hydrogen electrode at that pH. (Bi)sulphate affects the hydrogen production activity with methyl viologen semiquinone and flavodoxin hydroquinone. However, since this effect is rather small and because at pH values below pH 7 the (bi)sulphite concentration did not exceed 100 mM, the effect of (bi) sulphite on the patterns observed may be assumed to be negligible.

The pH-dependence of the redox potentials at which hydrogen is no longer produced, the end potentials, was determined independently with both electron donors. The end potentials become more positive at decreasing pH. The slope of Δ(end potential)/ΔpH is -61 mV and thus parallels the hydrogen electrode; the end potentials are, on average, 58 mV more positive than the hydrogen electrode at the corresponding pH. Further, it is noticeable that the redox potentials at which the hydrogen production activity ceases to be redox potential-independent are, at the pH values tested, about 45 mV more negative than the hydrogen electrode at the corresponding pH values.

If it is assumed that the midpoint potentials of the ferredoxin-type clusters within the enzyme, like nearly all other electron-transferring ferredoxins, are pH-independent and transfer one electron per cluster at an average midpoint potential of -400 mV, it can be calculated that at -350 mV only 13% of the clusters are in the reduced form; nevertheless, at pH 5.25 at this potential nearly maximal activity is observed. Furthermore, at pH 8 and -430 mV, 76% of the clusters are in the reduced form, but no activity is observed (Figure 4A).

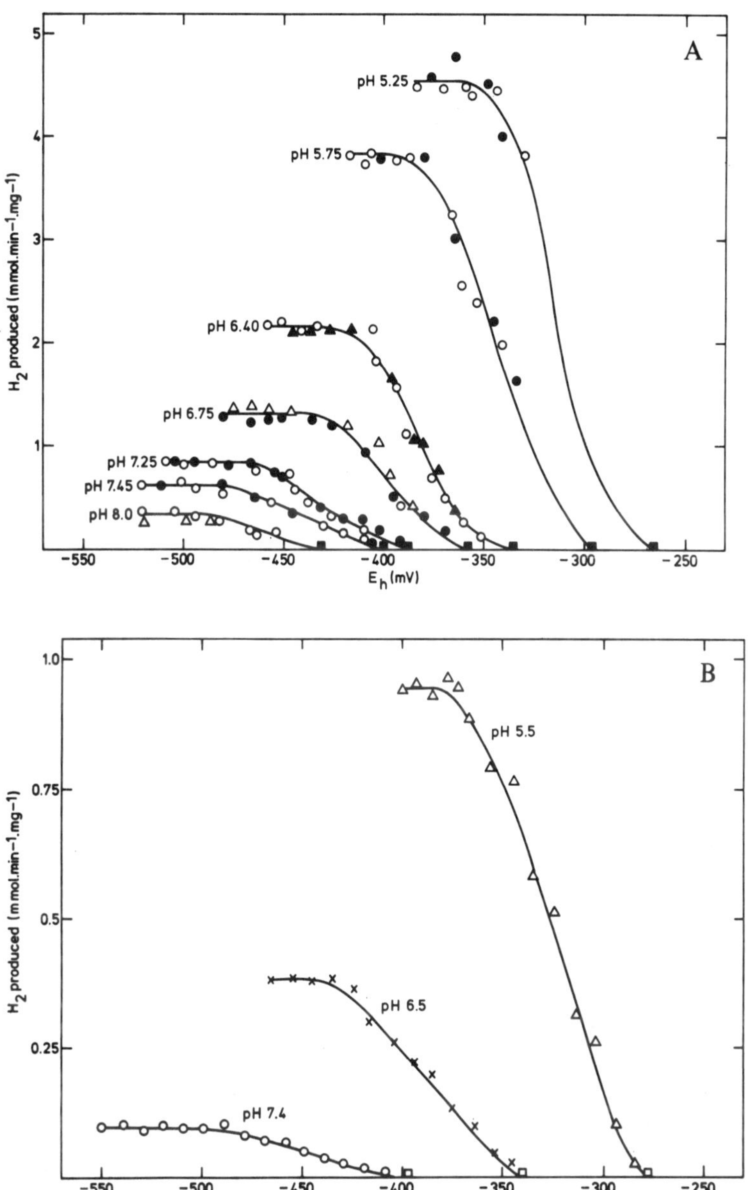

Figure 4. Effects of pH and redox potential on the hydrogen production activity with methyl viologen [A] and flavodoxin hydroquinone [B], (cf. ref. 14).

This suggests that the midpoint potential(s) of a(the) ferredoxin-type cluster(s) is(are) pH-dependent.

If, at a given pH, the activities are considered to express the degree of reduction of the enzyme, with the redox potential-independent part of the activity representing full (100%) reduction and the end potentials representing complete oxidation (0% reduction), an n=2-type of redox titration curve is obtained (Figure 5). These n=2-type redox titration curves are observed at the pH range tested and for both electron donors and are thus a property of the enzyme.

The mechanistic model of Figure 6, explains the n=2-type redox behavior together with the observed change of the apparent midpoint potential of −60 mV/pH. It is supposed that the two ferredoxin-type clusters are each transferring one electron. The kinetic model proposes that both ferredoxin-type clusters need to be reduced in two one-electron reduction steps after which two electrons are transferred to the catalytic center.

FlH_3 is a better substrate than FlH_2^-, though the $[S]_{0.5}$ values for both substrates are almost the same (Figure 7). The reason for this might be that FlH_3 carries, besides an electron, also a proton.

If it is assumed that the kinetics of the hydrogenase with methyl viologen semiquinone are unaltered at low pH values, a turnover number of $0.8-1.7\times 10^5$ mol H_2 produced$\cdot s^{-1}\cdot$(mol hydrogenase)$^{-1}$ can be calculated. The fact that the extrapolated turnover number with flavodoxin hydroquinone of $0.4-1.6\times 10^4$ mol H_2 produced$\cdot s^{-1}\cdot$(mol hydrogenase)$^{-1}$, is an order of magnitude lower indicates the existence of a flavodoxin hydroquinone hydrogenase enzyme-substrate complex.

Figure 5. Redox potential-dependent hydrogen production as related to redox titrations (cf. ref. 14).

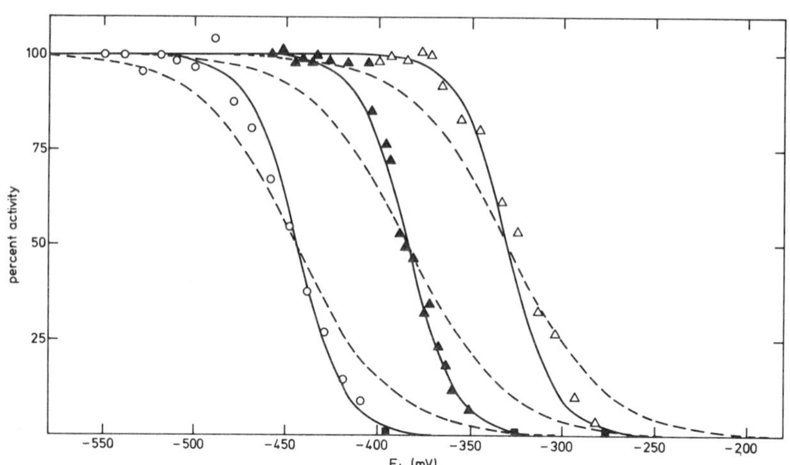

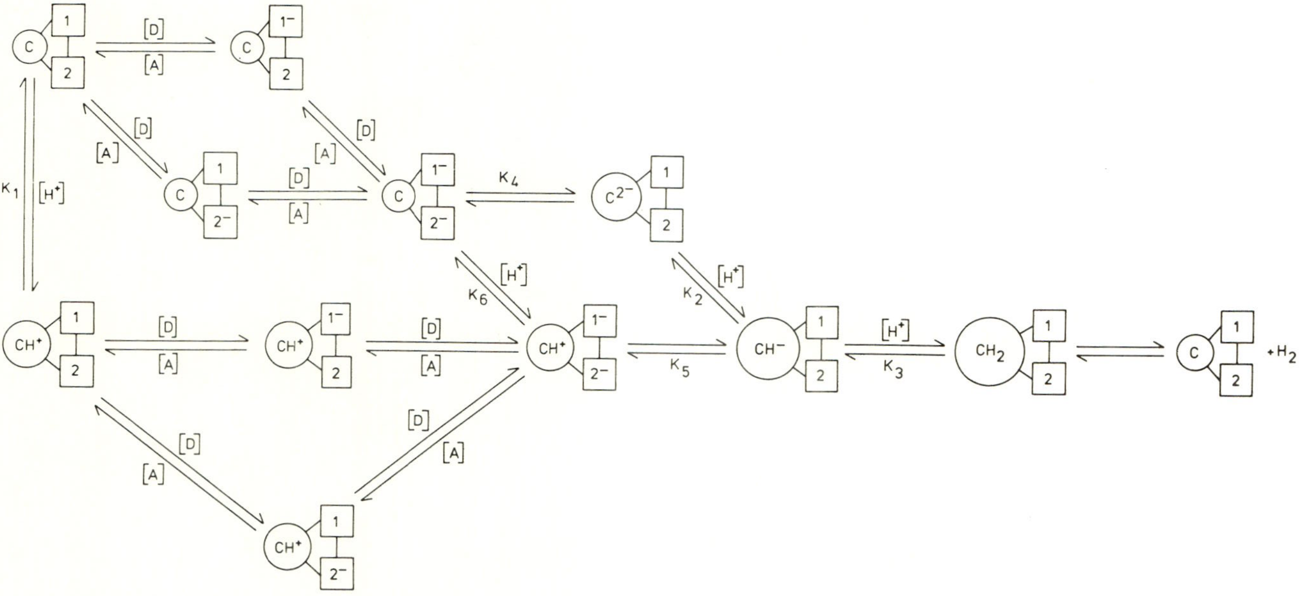

Figure 6. Model for *M. elsdenii* activity (cf. ref. 14).

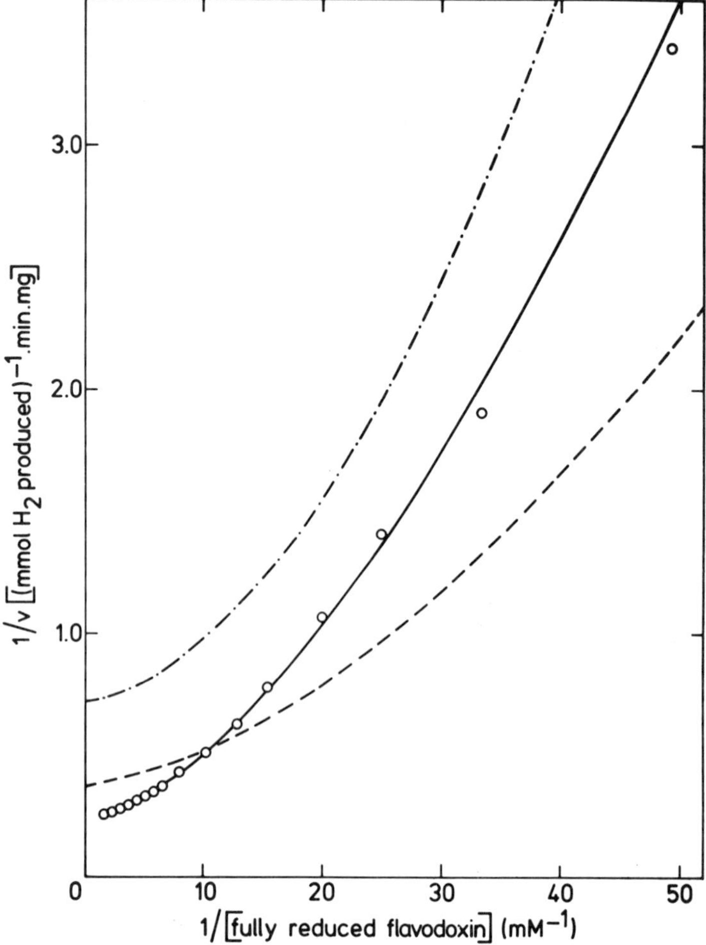

Figure 7. Efficiency of H_2-production by FlH_2^- (–·–·) and FlH_3 (---), compared to observed activity at pH 5.5 (cf. ref. 14).

ACKNOWLEDGMENTS

I express my sincere thanks to Drs. C. van Dijk, H.J. Grande, H Haaker, and C. Laane for their contributions to the data and in the ideas presented here. I express my thanks to Prof. H. Bothe for the permission to cite some of his unpublished data. This work was supported by the Netherlands Foundation for Chemical Research (S.O.N.) with financial aid from the Netherlands Organization for the Advancement of Pure Research.

References

1. Mortenson, L.E. (1964) *Proc Natl Acad Sci USA* 52:272–279.
2. Knight, E. and Hardy, R.W.F. (1966) *J Biol Chem* 241:2752–2756.
3. Beneman, J.R., Yoch, D.C., Valentine, R.C., and Arnon, D.I. (1971) *Biochim Biophys Acta* 226:205–212.

4. Haaker, H., de Kok, A., and Veeger, C. (1974) *Biochim Biophys Acta* 357: 344–357.
5. Laane, C., Krone, W., Konings, W.N., Haaker, H., and Veeger, C. (1979) *FEBS Lett* 103:53–57.
6. Laane, C., Krone, W., Konings, W.N., Haaker, H., and Veeger, C. (1980) *Eur Biochem* 103:39–46.
7. Yates, M.G. (1972) *FEBS Lett* 27:63–67.
8. Beneman, J.R., Yoch, D.C., Valentine, R.C., and Arnon, D.I. (1969) *Proc Natl Acad Sci USA* 64:1079–1086.
9. Scherings, G., Haaker, H., and Veeger, C. (1977) *Eur J Biochem* 77:621–630.
10. Veeger, C., Laane, C., Scherings, G., Matz, L., Haaker, H., and Van Zeeland-Wolbers, L. (1980) *Nitrogen Fixation*. W.E. Newton and W.H. Orme-Johnson (eds) Baltimore: University Park Press, Volume I, pp. 111–137.
11. Barman, B.C. and Tollin, G. (1972) *Biochemistry* 11:4755–4759.
12. Bothe, H. and Häger, K.P., submitted for publication.
13. Mayhew, S.G., Foust, G.P., and Massey, V. (1969) *J Biol Chem* 144: 803–810.
14. Van Dijk, C. and Veeger, C. (1981) *Eur J Biochem* 114:209–219.

Published 1982 by Elsevier North Holland, Inc.
Vincent Massey and Charles H. Williams, Editors
Flavins and Flavoproteins

CHAPTER 125

Resolution and Partial Characterization of Electron Transport Components of *Thermus thermophilus*: NADH → Flavoprotein → Menaquinone → Rieske Fe/S Protein

M.G. Choc, R.M. Lorence, T. Yoshida, W.R. Dunham, and J.A. Fee

Biophysics Research Division, The University of Michigan, Ann Arbor, Michigan

We have noted significant spectral similarities between the membrane associated electron transport components of the extremely thermophilic bacterium, *Thermus thermophilus*, and those of mitochondria (1). This observation was extended by the purification of a copper-containing cytochrome oxidase from this organism which possessed spectral and magnetic properties virtually identical to those of purified mitochondrial cytochrome oxidase, although the subunit composition was different (2). We report here a preliminary characterization of a flavin-containing system having menaquinone reductase activity which effectively transfers electrons to an Fe/S protein having properties in common with the Rieske protein of mitochondria (3–5).

Optical spectra of the samples used in this study are shown in Figure 1. As judged from SDS gel electrophoresis of the proteins (Figure 2), the flavoprotein sample contains major polypeptides of 20 and 35 kdalton while the Fe/S protein sample contains primarily one polypeptide of 25 kdalton. The menaquinone sample behaved as a single substance during thin layer chromatography (cf Table 1).

Extraction of *Thermus* membranes with a nonionic detergent followed by chromatography on DE-52 yielded a brown colored fraction containing flavin, Fe/S, and b and c cytochromes. Addition of NADH to this material resulted in the appearance of an EPR signal identical to that reported earlier for the Rieske protein with g values: 2.02, 1.90, 1.79. Further purification separates the flavin-containing moiety from the Fe/S protein. The flavoprotein can be reduced (Figure 1) by NADH but no EPR signals are elicited. In contrast, the partially purified Fe/S protein can no longer be reduced by added NADH either in the absence or presence of the separated flavoprotein. However, dithionite elicits the signal shown in Figure 3 which is identical to that observed in the whole membranes of *Thermus* and in the above described extract.

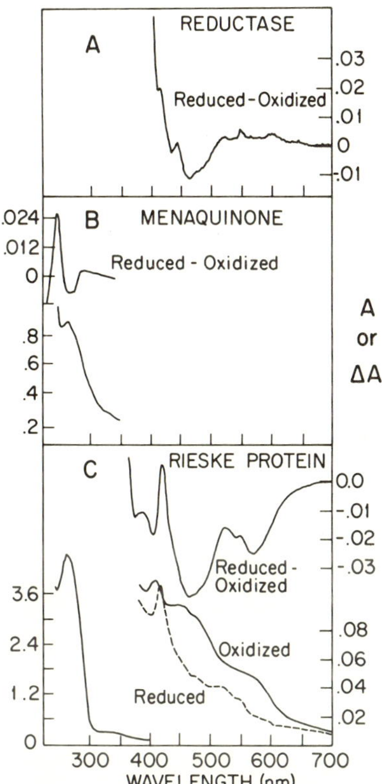

Figure 1. Optical spectra of the redox components used in this study: (A) Difference spectrum (reduced-oxidized) of NADH: Menaquinone reductase. The sample was reduced under anaerobic conditions with 0.8 mM NADH. (B) Menaquinone purified from plasma membranes of *Thermus*. Top: Reduced-oxidized. The sample was reduced with excess N_aBH_4. Bottom: Oxidized. (C) Top: Reduced-oxidized. The protein was reduced with dithionite under anaerobic conditions. Bottom: Oxidized (solid line) and reduced (dashed line).

The reductase activity, as monitored by NADH absorption at 340 nm, was quite specific for menadione and the extracted menaquinone as shown in Table 1, and activity toward the former can be completely inhibited by rotenone.

The results of our attempts to reconstitute the electron transfer segment present in the original extract are shown in Figure 4. In the presence of catalytic amounts of the flavoprotein and menaquinone, electrons are transferred from NADH to the Fe/S protein, while ubiquinone (Q-6), which must be present at very low levels in the membranes of *Thermus*, was a very poor mediator of electron transfer to the Fe/S protein. These results suggest a minimal scheme of NADH oxidation by the respiratory system of *Thermus*:

$$\text{NADH} \rightarrow \text{Flavoprotein} \rightarrow \text{Menaquinone} \rightarrow \text{Rieske protein.}$$

The use of these purified components, along with the pure cytochrome c_1aa_3 complex (2) should allow us to reconstruct a functional respiratory system which will transfer electrons from NADH to oxygen.

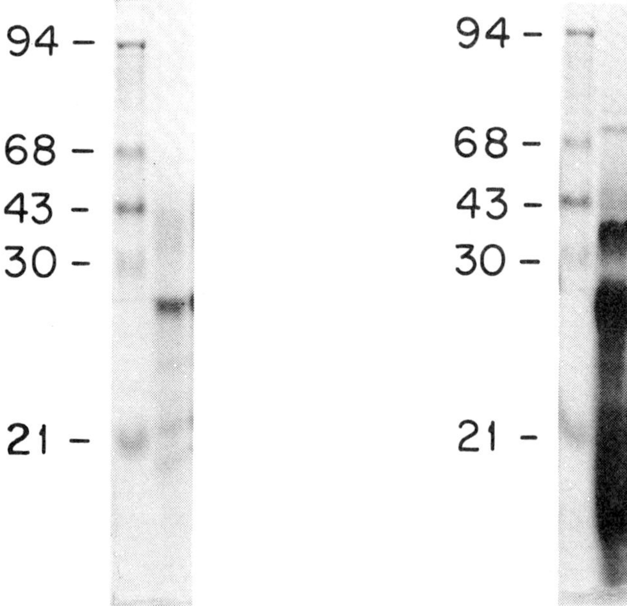

Figure 2. SDS gel electrophoresis of the partially purified fractions of *Thermus* Rieske protein (left) and NADH: Menaquinone reductase. The samples were denatured by heating at 70°C for 12 hr in 4% SDS and 4% 2-mercaptoethanol, and the gels were 7% acrylamide (bis 1:15) plus 8 M in urea. The molecular weight standards were as indicated.

Table 1. NADH: Acceptor Reductase Properties of Partially Purified Flavoprotein from Plasma Membranes of *Thermus thermophilus*.

Electron acceptor	Activity[a]
Ubiquinone-6[b]	1.4
DCPIP[c]	6.2
Menaquinone[b,d]	18.
Menadione[b]	26.
Menadione + Rotenone(40 μM)[b,e]	0.0

[a]The assay was carried out anaerobically or in the presence of 1 mM KCN using 0.9% deoxycholate, 30 mM Tris.HCl pH 7.9, 0.155 mM NADH, 20 μM electron acceptor, and 0.18 μM flavin. The approximate concentration of flavin was estimated from the difference spectrum and assuming $\Delta\varepsilon_{450} = 10\ \text{mM}^{-1}\ \text{cm}^{-1}$. Activities are expressed as millimoles NADH oxidized per minute per μmole of flavin.

[b]$\Delta\varepsilon_{340} = 6.8\ \text{mM}^{-1}\ \text{cm}^{-1}$ for NADH oxidation (6).

[c]$\Delta\varepsilon_{600} = 20.6\ \text{mM}^{-1}\ \text{cm}^{-1}$ for DCPIP reduction (7).

[d]The menaquinone was extracted from plasma membranes with a mixture of petroleum ether and methanol (8) and purified by silica gel chromatography (9). Spectroscopic measurements of the extracts suggested menaquinone to be the dominant if not only quinone present. The concentration of menaquinone was determined using $\varepsilon_{264} = 11.7\ \text{mM}^{-1}\ \text{cm}^{-1}$ for the oxidized material in cyclohexane (10).

[e]The flavoprotein was incubated with the rotenone 20 min prior to assay.

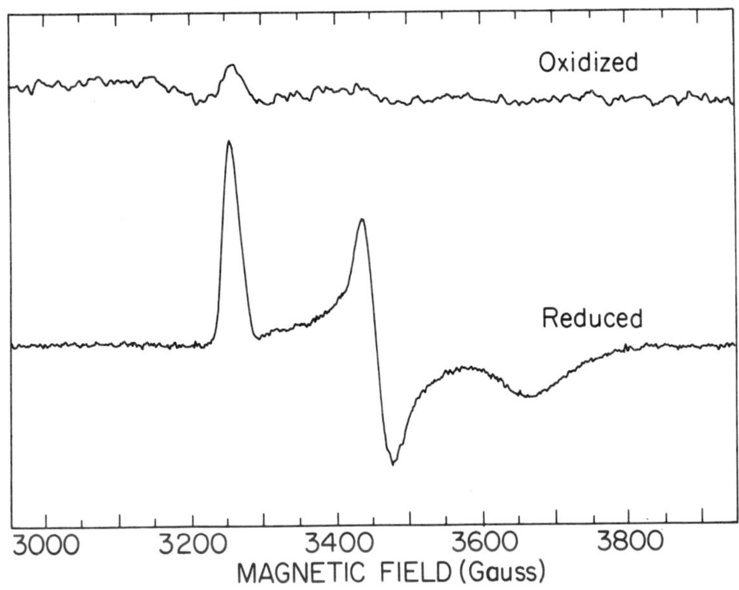

Figure 3. EPR spectra of the oxidized and reduced forms of *Thermus* Rieske protein. The conditions of recording were: 9.203 GHz, 22°K, 10 Gauss field modulation at 100 kHz, 250 Gauss/min field sweep at 0.25 s time constant. (Oxidized) relative gain =2.5 at 2 mW microwave power; (Reduced) relative gain = 1 at 2 mW microwave power. The sample was reduced with a small excess of dithionite.

Figure 4. Reduction of the Rieske protein from *Thermus* by NADH as measured by the decrease in absorbance at 460 nm. The reaction was carried out anaerobically, 1 mM KCN, 0.85% deoxycholate, 30 mM Tris. HCl, pH 7.8, in a final volume of 1 mL. The concentration of Rieske protein was 21 μM estimated using $\epsilon_{460} = 6.7\ mM^{-1}\ cm^{-1}$. The approximate final concentrations (cf Table 1) of other components were as follows: Reductase, 0.18 μM; NADH, 1 mM; Ubiquinone-6, 34 μM; and menaquinone, 60 μM.

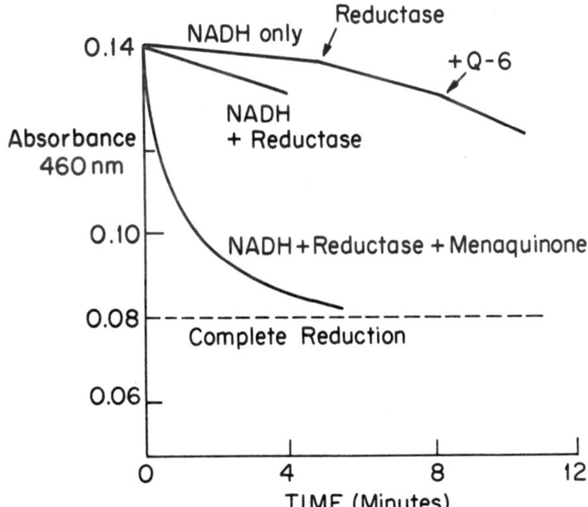

ACKNOWLEDGMENTS
We thank K.L. Findling for expert technical assistance. This work was supported by U.S.P.H.S. Grant GM12176.

References

1. Fee, J.A., Findling, K.L., Lees, A.C., and Yoshida, T. (1978) In *Frontiers in Biological Energetics*. Dutton, P.L., Leigh, J.S., and Scarpa, A. (eds) New York: Academic Press, Vol. I, pp. 118–126.
2. Fee, J.A., Choc, M.G., Findling, K.L., Lorence, R., and Yoshida, T. (1980) *Proc Natl Acad Sci* 77:147–151.
3. Rieske, J.S. (1976) *Biochem Biophys Acta* 456:195–247.
4. Green, D.E., Wharton, D.C., Tzagoloff, A., Rieske, J.S., and Brierley, G.P. (1965) In *Oxidases and Related Redox Systems*. King, T.E., Mason, H.S., and Morrison, M. (eds). New York: John Wiley and Sons, Vol. II, pp. 1032–1076.
5. Trumpower, B.L. and Edwards, C.A. (1979) *J Biol Chem* 254:8697–8706.
6. Schatz, G. and Racker, E. (1966) *J Biol Chem* 241:1429–1438.
7. King, T.E. and Howard, R.L. (1967) In *Methods in Enzymology*. Estabrook, R.W. and Pullman, M.E. (eds.) New York: Academic Press, Vol. X, pp. 275–294.
8. Kroger, A. and Klingenberg, M. (1966) *Biochem Z* 344:317–336.
9. Lester, R. and Crane, F.L. (1959) *Biochem Biophys Acta* 32:492–499.
10. Bishop, D.H.L., Pandya, K.P., and King, H.K. (1962) *Biochem J* 83:606.

Published 1982 by Elsevier North Holland, Inc.
Vincent Massey and Charles H. Williams, Editors
Flavins and Flavoproteins

CHAPTER 126

Regulation of Succinate Dehydrogenase Activity by Monovalent Inorganic Anions: Kinetic and Molecular Studies

Franco Bonomi, Silvia Pagani, and Paolo Cerletti

Department of General Biochemistry, University of Milan, Via G. Celoria, 2, I-20133 Milano, Italy

A number of literature reports analyze the role of monovalent inorganic anions as activators of succinate dehydrogenase (SDH). Their effect on the activated beef heart enzyme has however not been considered (1–3).

We studied the interaction of bromide with soluble, activated, substrate-reduced SDH. The enzyme at the gel eluate purification step (4) was precipitated with ammonium sulphate, redissolved, and desalted in 5 mM succinate-50 mM Tris acetate, pH 7.5. Small aliquots (4.12 nmoles protein-bound flavin mg^{-1}) were stored in liquid nitrogen and thawed just before each experiment.

Addition of bromide to the assay medium for SDH considerably lowers the activity of the enzyme. The inhibition is competitive at succinate concentrations in the range of the dissociation constant of the substrate from the catalytic site [0.11 mM(5)], and gradually becomes uncompetitive as the succinate concentration increases (Figure 1). This indicates that two different mechanisms are involved. Indeed the number of anions required for the inhibition, as measured from the slopes in the Hill plot of Figure 2, goes from two to one as the succinate concentration increases (inset A). Inhibition by a single Br^- is promoted by the binding of succinate to a site having a dissociation constant for succinate of 1.52 mM (inset B).

Inhibition by a single Br^- is pH-dependent and the pH for 50% inhibition increases with the Br^- concentration (Figure 3). This indicates that protonation of a group on the protein, with an apparent pK_a of 8.8, favors binding of the inhibitory anion according to the equation:

$$\text{succ}\,E_{\text{active}} + H^+ \rightleftharpoons \text{succ}\,E^+_{\text{active}} \underset{-Br^-}{\overset{+Br^-}{\rightleftharpoons}} \text{succ}\,EH^+Br^-_{\text{inhibited}} \qquad (1)$$

At low succinate concentration the inhibition is competitive and requires two Br^-. They probably interact with the same charged groups involved in the binding of succinate, replacing the two carboxylate anions. Dissociation of succinate leads to oxidation of the enzyme, as appears from the absorbance spectra of Figure 4. The spectra of the enzyme oxidized by Br^-, fumarate, or oxaloacetate show different features at wavelengths different from those typi-

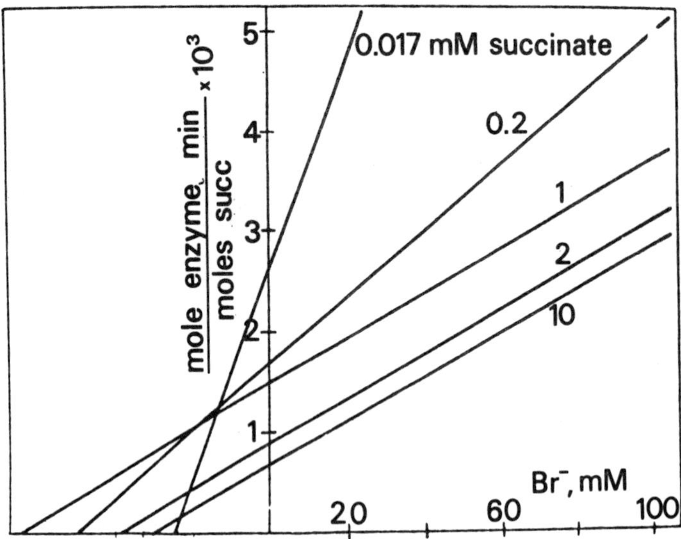

Figure 1. Dixon plot for the inhibition of SDH by Br^-. Enzyme activity was assayed at 13°C in a medium containing the indicated concentration of effectors.

cal of oxidized flavin. When Br^- acts as oxidant, flavin oxidation is accompanied by disappearance of the semiquinone signal, as confirmed by EPR spectra in Figure 5. The spectroscopic properties of the enzyme treated with bromide reproduce those observed when any kind of high affinity effector (succinate, malonate, oxaloacetate, etc.) is removed from the enzyme in phosphate buffer (6,7). The spectral changes induced by Br^- are instantaneously reversed by succinate even at 10°C; moreover once the Br^--treated enzyme is assayed at 15°C in a Br^--free medium, it immediately displays full activity. These observations indicate that the enzyme modified by Br^- is still in the activated form. Only association-dissociation equilibria of ligands appear therefore to be involved in the action of Br^- on the activated enzyme, different from the activation-deactivation process where high activation energies and slow reaction rates are implicated, suggesting conformational modifications (1–3, 8).

Titrations of flavin oxidation by increasing Br^-, as shown in the inset of Figure 4, in the presence of different succinate concentrations yield the results analyzed in Figure 6. They quantify the effect of succinate on the affinity of Br^- for the enzyme and allow to draw for the interaction with two Br^- with concomitant oxidation the following stoichiometry, which involves competitive inhibition:

$$\mathrm{E\,succ}^{\text{reduced}}_{\text{active}} + 2\mathrm{Br}^- \rightleftharpoons \mathrm{E(Br^-)}_2{}^{\text{oxidized}}_{\text{inhibited}} + \mathrm{succ} \qquad (2)$$

Reactions **1** and **2** describe the interaction of Br^- with SDH. They differ not only in the number of anions involved, but in the site of their action. In reaction **2**, two Br^- substitute the succinate bound at the catalytic site,

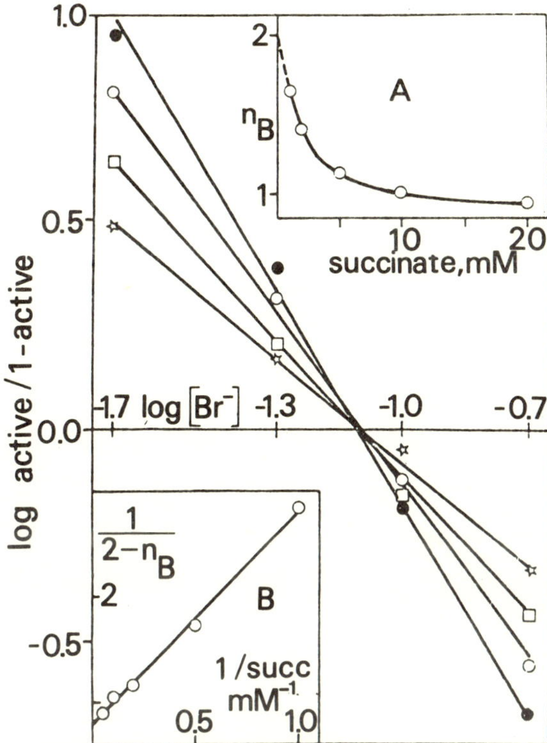

Figure 2. Active fraction of SDH as a function of the Br^- and succinate concentrations. Enzyme activity was tested in the presence of 1 (dots), 2 (circles), 5 (squares), or 20 (stars) mM succinate and the indicated concentrations of Br^-. For each succinate concentration, the activity of the uninhibited enzyme was measured in the absence of Br^-. Inset A: replot of the slopes of the lines in Figure 2 versus the succinate concentration. In inset B: the ratio between the enzyme inhibited by one and that inhibited by two bromides, given n_s and k_d^{succ}.

therefore the enzyme becomes oxidized and competitive inhibition results. Reaction **1** takes place only when the succinate concentration is high enough to saturate a site with $K_d^{succ} = 1.52$ mM and the Br^- present is not sufficient to compete for the catalytic site. The inhibition is uncompetitive and silent with respect to the oxidation state of SDH. Summarizing the prevailing type of interaction depends on the bromide to succinate ratio: at low ratios **1** is the favored scheme, whereas when the ratio increases reaction **2** (which depends on the square of Br^- concentration) will take place. At intermediate ratios, the two forms of inhibition coexist, and a nonlinear relationship between inhibition and oxidation results: inhibition is manifest long before oxidation begins.

Since respiring submitochondrial particles accumulate anions (9), and anions of physiological interest behave quite similarly to Br^- in other interactions with SDH, e.g., the activation process (2), this novel type of SDH inhibition may have some physiological interest.

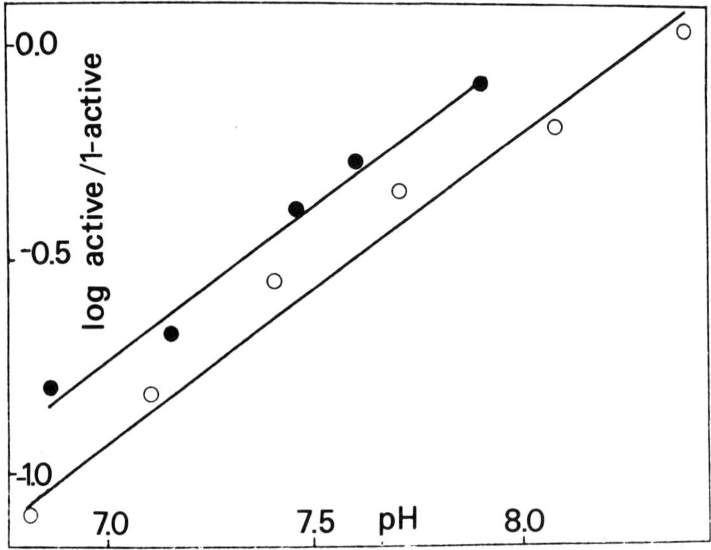

Figure 3. pH dependence of the inhibition of SDH by 100 (dots) or 200 (circles) mM KBr in the presence of 5 mM succinate. For each pH value, the activity of the uninhibited enzyme was measured in the absence of KBr.

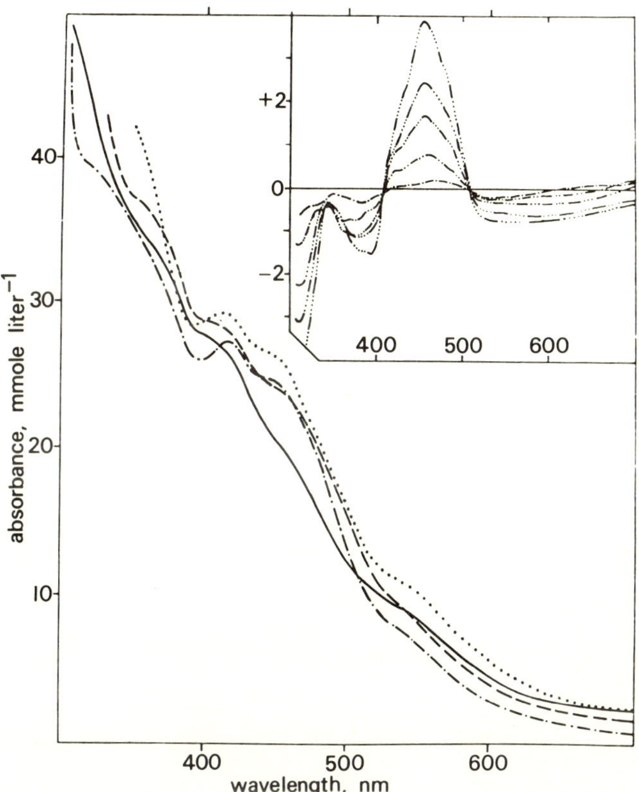

Figure 4. Spectral effects of KBr. Absorbance spectra of SDH in 5 mM succinate were recorded as such (full line) or in the presence of 0.8 M KBr (dashes and dots), 0.15 M fumarate (dashes), 50 mM fumarate and 2 mM oxaloacetate (dots). Inset: difference spectra. Aliquots of 5 M KBr were added to the sample cell. Succinate was 20 mM, KBr was 22, 174, 336, 487, and 758 mM from lowest to highest tracing at 450 nm.

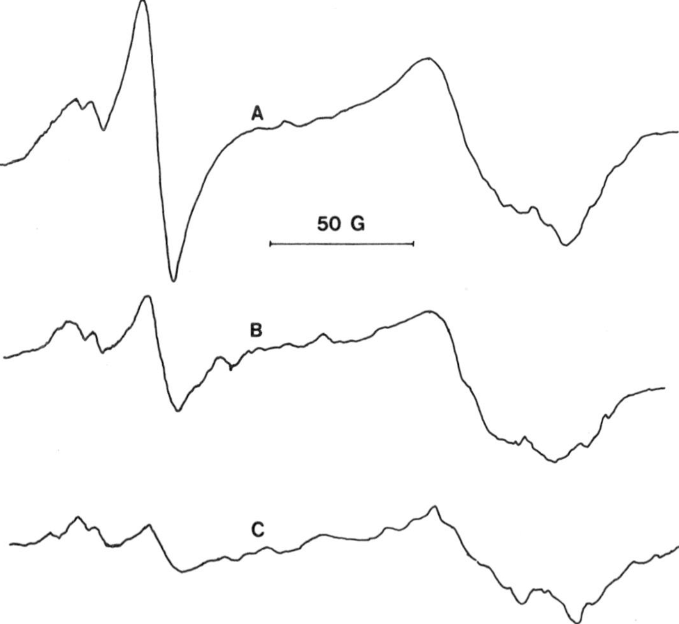

Figure 5. EPR spectra of SDH in 20 mM succinate were recorded at liquid N_2 temperature as such (A) or in the presence of 166 (B) or 666 (C) mM KBr. Enzyme concentration was the same throughout.

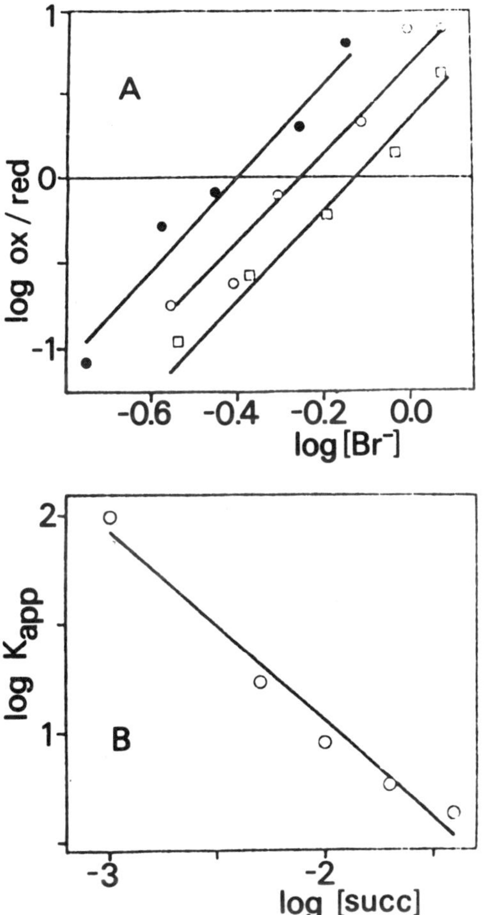

Figure 6. Oxidation state of flavin as a function of KBr and succinate concentration. SDH was titrated with KBr as in the inset of Figure 4 in the presence of 5 (dots), 10 (circles) or 40 (squares) mM succinate. When KBr addition did not further increase A_{455}, the flavin was considered fully oxidized. Dilution was accounted for. The reciprocal of the square of the KBr concentration giving 50% oxidation was taken as the K_{app} for oxidation. Its dependence on the succinate concentration is shown in B.

References

1. Kearney, E.B., Ackrell, B.A.C., Mayr, M., and Singer, T.P. (1974) *J Biol Chem* 249:2016–2020.
2. Ackrell, B.A.C., Kearney, E.B., and Mayr, M. (1974) *J Biol Chem* 249:2021–2027.
3. Gutman, M. (1976) *Biochemistry* 15:1342–1348.
4. Cerletti, P., Zanetti, G., Testolin, G., Rossi, C., Rossi, F., and Osenga, G. (1971) In *Flavins and Flavoproteins*. Kamin, H. (ed.) Baltimore: University Park Press, pp. 629–647.
5. Zeylemaker, W.P., Klaasse, A.D.M., and Slater, E.C. (1969) *Biochim Biophys Acta* 191:229–238.
6. Bonomi, F., Pagani, S., and Cerletti, P. (1977) In *Flavins and Flavoproteins, Physicochemical Properties and Functions*. W. Ostrowski (ed.) Warsaw: Polish Scientific Publ., pp. 151–164.
7. Bonomi, F., Pagani, S., and Cerletti, P. (1979) *Proceedings XII FEBS Meeting, Dresden, vol. 53.* Processing and turnover of proteins and organelles in the cell. S. Rapoport and T. Schewe (eds.) New York: Pergamon Press.
8. Gutman, M., Bonomi, F., Pagani, S., Cerletti, P., and Kroneck, P. (1980) *Biochim Biophys Acta* 591:400–408.
9. Liberman, E.A., Topaly, V.P., Tsofina, L.M., Jasaitis, A.A., and Skulachev, V.P. (1969) *Nature* 222:1076.

Published 1982 by Elsevier North Holland, Inc.
Vincent Massey and Charles H. Williams, Editors
Flavins and Flavoproteins

CHAPTER 127

A Possible Solution to Some Unexplained Observations on NADH Dehydrogenase From Beef-Heart Mitochondria

Simon P.J. Albracht

Laboratory of Biochemistry, B.C.P. Jansen Institute, University of Amsterdam, Plantage Muidergracht 12, 1018 TV Amsterdam, The Netherlands

Introduction

NADH:Q oxidoreductase [EC 1.6.99.3] is the least well-understood enzyme complex of the mitochondrial respiratory chain. The molecular weight of the purified enzyme, Complex I, is about 700,000 per FMN molecule (1), but the hydrodynamic properties indicate a dimeric structure (2). There are 4 different Fe-S clusters supposed to function in this complex (3–6), all of which are completely reduced by NADH within 5 ms (3,4). The stoichiometry of the clusters in Complex I, 0.5 : 1 : 0.5–1 : 1 for the clusters 1–4, respectively (5), is not understood. The ratio of the clusters 1 and 2 in submitochondrial particles (SMP) also has been found to equal 0.50 ± 0.05 ($n = 10$) (7).

The enzyme complex also reacts with NADPH (8). The rotenone-sensitive NADPH oxidation by SMP is not related to the trypsine-sensitive NADPH-NAD^+ transhydrogenase activity (9, 10) and occurs optimally at pH 6.0 (8). The clusters 1 and 3 are only partly reduced by NADPH (11), whereas the other clusters are fully reduced. According to the groups of Hatefi and Ragan (11, 12) this partial reduction is the result of the balance between the rate of hydrogenation of NADPH and the rate of autooxidation of the complex. However, since NADPH oxidation by SMP at pH 6.0 is rotenone sensitive (8,9), the rate of reduction of the complex must be at least 50 times greater than the autooxidation rate. This implies that part of the clusters 1 and 3 cannot be involved in NADPH oxidation. Freeze-quench experiments were carried out further to examine this problem.

Materials and Methods

SMP were prepared from beef-heart mitochondria by sonication and differential centrifugation and suspended in 0.25 M sucrose, 100 mM potassium phosphate buffer (pH 6.2), 30 μM rotenone and 1% (v/v) ethanol (Medium A) to a protein concentration of about 60 mg/ml. This suspension was mixed at room temperature (20–22°C) with the substrates NADH or NADPH (5 mM) dissolved in Medium A. Where indicated, NAD^+ or $NADP^+$ (2 mM) were added to the substrate solutions. EPR spectroscopy was performed as in (13). The relative amplitudes of the various Fe-S signals were obtained as follows: (A) for cluster 1 from the g=1.94 line at 45 K and 10 mW microwave power; (B) for cluster 2 from the g=1.92 line at 14 K and 2 mW; (C) for cluster 3 from the g=1.88 line at 9 K and 20 mW. The g=2.10 line showed the same behavior; (D) for cluster 4 from the 1.86 line at 9 K and 20 mW. Kinetic freeze-quench experiments were carried out using a mixing chamber and quenching bath as reported in (14). The quenching time (5 ms) is included in the times indicated in the figures. Syringes were driven by a hydraulic ram system (15) developing a force of 3000 N.

Results

The pattern of optical and EPR properties of Complex I reduced by NADPH at pH 6.5, 7.5, and 8.0 in air at room temperature was found to be essentially the same as reported by the group of Hatefi (11). Similar observations were made with SMP in the presence of rotenone. Especially the incomplete reduction of clusters 1 and 3 was noteworthy. The results of freeze-quench experiments with NADPH in the presence of rotenone at pH 6.2 are shown in Figure 1 (open symbols). Reaction of cluster 2 is biphasic. The rapid phase, representing half the signal, is completed within 10 ms. Reduction of clusters 1 and 3 is slow and, as expected, incomplete after 100 ms. The reduction of cluster 4 is not distinguishable from first-order kinetics with a half time of 10 ms.

Figure 1. Reduction by NADPH of the Fe-S clusters of NADH:Q oxidoreductase in submitochondrial particles in the presence of rotenone at pH 6.2. Left panel: cluster 1 (circles) and cluster 2 (triangles). Right panel: cluster 3 (circles) and cluster 4 (triangles). Open symbols and dashed lines: NADPH; closed symbols and solid lines: NADPH plus NAD^+. Signal amplitudes are relative to those obtained with NADH. The latter were independent of time between 5 and 100 ms.

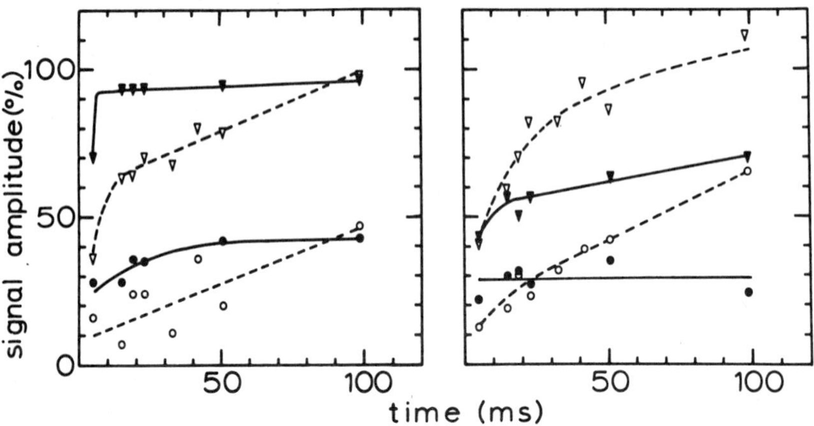

It is not likely that the rapid reduction of part of clusters 2 and 4 is due to contamination by NADH in the NADPH solution, since the initial extent of reduction of cluster 2 was found to be lower at higher pH. However, it could be due to contaminating NAD^+, from which NADH could be formed by transhydrogenase activity. Addition of NAD^+ brings about a marked change in the reduction pattern of the Fe-S clusters (Figure 1, closed symbols). Reduction of cluster 2 is now complete within 10 ms; one-third of clusters 1 and 3 is reduced also within 10 ms and only the extent of reduction of cluster 3 increases somewhat in the following 40 ms. Reduction of cluster 4 proceeds biphasically, half of its signal appearing in 10 ms.

Reduction of all the clusters was completed within the quenching time (5 ms) when NADH was the substrate. The presence of NADPH or $NADP^+$ did not affect this pattern. NAD^+ alone induced no signals and gave the same result as mixing with buffer.

Discussion

Hatefi et al. (9) reported NADPH oxidation activities up to 250 nmol/min/mg of SMP at 30°C and pH 6.0. Slightly lower values were obtained in this study. On the basis of a cytochrome c_1 content of 0.5 nmol/mg protein (4) and the stoichiometries of the prosthetic groups in the respiratory chain (16), it can be calculated that the turnover number can be 3600 min^{-1} per cluster 1 at 30°C. A cluster, involved in NADPH oxidation, should thus be reduced in about 17 ms. Corrected for the temperature used here, an upper limit of 50 ms is allowed. This eliminates clusters 1 and 3, but also half of clusters 2 and 4 (Figure 1).

When NAD^+ is added, the transhydrogenase activity will produce NADH at a rate of about 50 $\mu M/s$ and the redox potential of the resultant $NAD^+/NADH$ couple will first rapidly decrease and then slowly come to an equilibrium value. As seen in Figure 1, clusters 1 and 3 then show a rapid initial phase of reduction, followed by a stabilization. At 100 ms, cluster 1 is reduced by 43%. Assuming equilibrium between clusters 1 and 3 and using the redox potentials given in (17), cluster 3 should be 28% reduced, which is in agreement with what was found experimentally. However, cluster 4 is 70% reduced, instead of 16% calculated from the redox potentials. Also its initial reduction is not influenced by NAD^+ whereas the second phase in inhibited. It is concluded that half of cluster 4 is rapidly reduced by NADPH, whereas the other half is in equilibrium with clusters 1 and 3.

The following work hypothesis for the composition of NADPH:Q oxidoreductase is proposed (Scheme 1). The complex consists of two different molecules, A and B, each consisting of a flavoprotein moiety, reacting with substrate, and a Fe-S protein moeity, that takes care of energy generation and the reaction with Q-10. The scheme explains the kinetic experiments and the stoichiometry of the prosthetic groups, although the status of cluster 3 needs further study. Since no intense radical signals were observed in this study, reducing equivalents are considered to be transferred virtually pairwise to and

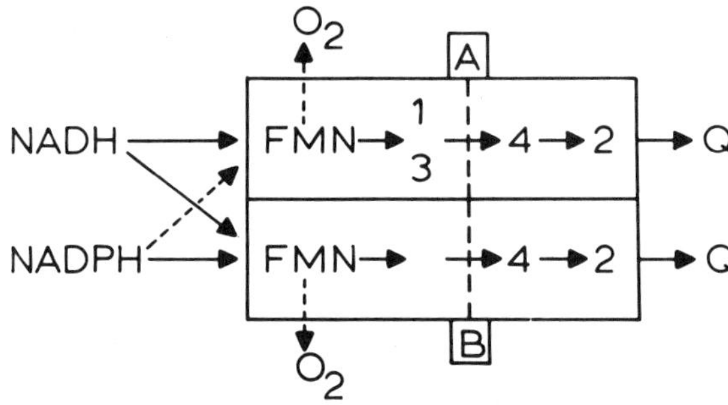

Scheme 1. The solid arrows indicate fast reactions, the dashed arrows stand for slow reactions.

from the flavin, requiring a minimum of two accepting Fe-S clusters. It is not excluded that the flavoprotein moiety in molecule B might contain Fe-S clusters.

ACKNOWLEDGMENT

The author thanks Prof. E.C. Slater for his stimulating criticism on the manuscript.

References

1. Hatefi, Y., Haavik, A.G., and Griffiths, D.E. (1962) *J Biol Chem* 237:1676–1680.
2. Dooijewaard, G., De Bruin, G.J.M., Van Dijk, P.J., and Slater, E.C. (1978) *Biochim Biophys Acta* 501:458–469.
3. Orme-Johnson, N.R., Hansen, R.E., and Beinert, H. (1974) *J Biol Chem* 249:1922–1927.
4. Orme-Johnson, N.R., Hansen, R.E., and Beinert, H. (1974) *J Biol Chem* 249:1928–1939.
5. Albracht, S.P.J., Dooijewaard, G., Leeuwerik, F.J., and Van Swol, B. (1977) *Biochim Biophys Acta* 459:300–317.
6. Ohnishi, T. (1979) In *Membrane Proteins in Energy Transduction*. Capaldi, R.A. (ed.) New York: Marcel Dekker, Inc., pp. 1–87.
7. Albracht, S.P.J., Leeuwerik, F.J., and Van Swol, B. (1979) *FEBS Lett* 104:197–200.
8. Hatefi, Y. and Hanstein, W.G. (1973) *Biochemistry* 12:3515–3522.
9. Hatefi, Y., Djavadi-Ohaniance, L., and Galante, Y.M. (1975) In *Electron Transfer Chains and Oxidative Phosphorylation*. Quagliariello, E., Papa, S., Palmieri, F., Slater, E.C., and Siliprandi, N. (eds.) Amsterdam: North-Holland Publishing Company, pp. 257–263.
10. Djavadi-Ohaniance, L. and Hatefi, Y. (1975) *J Biol Chem* 250:9397–9403.
11. Hatefi, Y. and Bearden, A.J. (1976) *Biochem Biophys Res Commun* 69:1032–1038.
12. Ragan, C.I. (1976) *Biochem J* 158:149–151.
13. Albracht, S.P.J. (1980) *Biochim Biophys Acta* 612:11–28.
14. Ballou, D.P. and Palmer, G. (1974) *Anal Chem* 46:1248–1253.
15. Bray, R.C. (1964) In *Rapid Mixing and Sampling Techniques in Biochemistry*. Chance, B., Eisenhardt, R.H., Gibson, Q.H., and Lonberg-Holm, K.K. (eds.) New York: Academic Press, pp. 195–203.
16. Albracht, S.P.J., Van Verseveld, H.W., Hagen, W.R., and Kalkman, M.L. (1980) *Biochim Biophys Acta* 593:173–186.
17. Ingledew, W.J. and Ohnishi, T. (1980) *Biochem J* 186:111–117.

Published 1982 by Elsevier North Holland, Inc.
Vincent Massey and Charles H. Williams, Editors
Flavins and Flavoproteins

CHAPTER 128

Methane Monooxygenase: An Iron-Sulfur Flavoprotein Complex

Howard Dalton and John Colby

Department of Biological Sciences, University of Warwick, Coventry, England

Methane monooxygenase (MMO) is the enzyme responsible for the initial hydroxylation of methane to methanol by bacteria which can grow on methane as their sole carbon and energy source. The enzyme complex from the organism *Methylococcus capsulatus* (Bath) incorporates one atom of dioxygen into the substrate in the presence of reduced pyridine nucleotide:

$$CH_4 + O_2 + NAD(P)H + H^+ \rightarrow CH_3OH + H_2O + NAD(P)^+$$

The soluble enzyme complex has been resolved into three components by DEAE-cellulose chromatography (1).

Proteins A and B have been purified to better than 90% homogeneity and protein C has been judged pure by gel electrophoresis.

Protein A has a molecular weight of about 200,000, and contains 2 g atoms of nonheme iron per mole of protein. Although determination of acid labile sulfide by colorimetric techniques has proved negative, core extrusion studies have indicated the presence of a 2Fe-2S* center within this protein.

Protein B is a colorless protein of mol wt about 15,000 and has to be purified in the presence of phenyl methyl sulfonyl fluoride. Protein C (mol wt 45,000) contains 2 g atoms of nonheme iron, 2 mol of acid labile sulfide and 1 mol FAD per mol of protein. Because of its inherent instability, purification of C is done in the presence of sodium thioglycollate.

All three proteins are necessary to catalyze the reductive oxygenation of the hydrocarbon substrate (i.e., methane to methanol, ethene to epoxyethane, cyclohexane to cyclohexanol or benzene to phenol). Neither proteins A nor B have any catalytic activity alone or in combination, but protein C will catalyze the $NAD(P)H_2$ – driven reduction of $K_3Fe(CN)_6$, dichlorophenol indophenol, cytochrome c and O_2.

The flavin prosthetic group from protein C could be released either by boiling or by treatment with hexamethyl phosphoramide at pH 8.5 (Figure 1). Further treatment of the protein with p-methoxybenzene thiol under anaerobic conditions (Figure 2) gives a difference spectrum characteristic of a 2Fe-2S* core (Figure 3) and was similar to that observed with other 2Fe-2S* proteins (3). Core extrusion studies at different protein concentrations gave a linear relationship between protein concentration and A_{510} (2). Using an E_{mM} value of 11.2 for the extruded core, protein C contained 0.96 (2Fe-2S*) centers per mole.

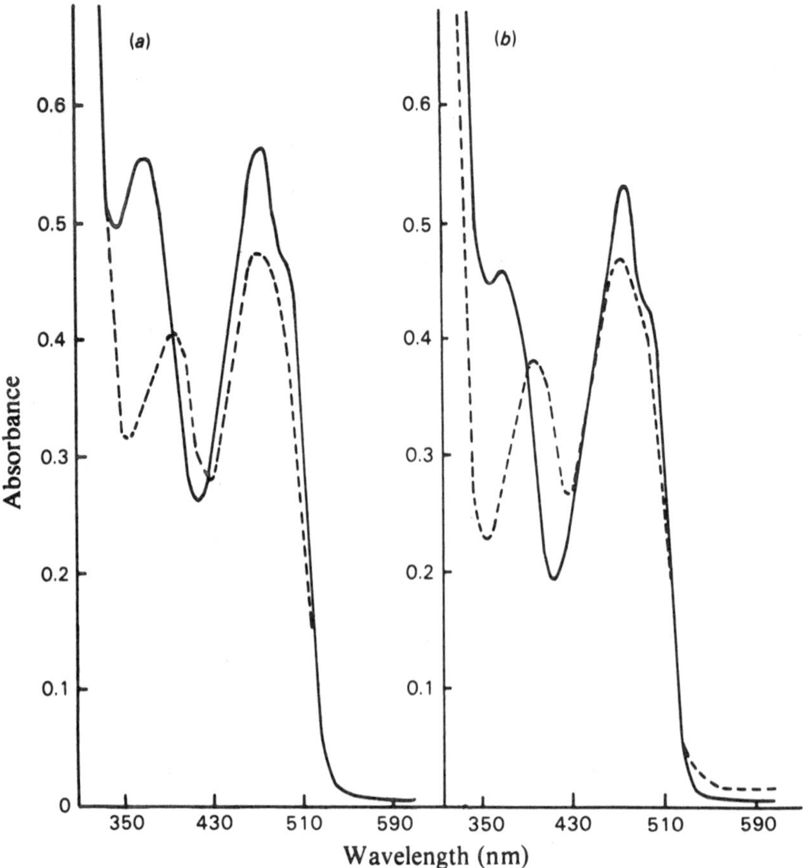

Figure 1. Effect of solvent on absorption spectra of FAD (a) and protein C (b) both at 0.043mM. (---) spectrum in 50 mM phosphate buffer pH 7+1% SDS (—) spectrum in hexamethyl phosphoramide in tris HCl buffer pH 8.5.

Reduction of protein C under anaerobic conditions with $NADH_2$, sodium dithionite or light/EDTA produced a new spectral species in the 570–630 nm region which was probably due to the flavin semiquinone (Figure 3). The neutral semiquinone could also be detected as a free radical signal at g=2.002 by epr spectroscopy when the protein was partially reduced with $NADH_2$ (Figure 4). Reoxidation of this protein was also effected by the addition, under anaerobic conditions, of protein A, suggesting that electrons could be transferred from protein C to protein A.

Since the hydrocarbon substrate appears to bind to protein A under anaerobic conditions (Dalton 1980), a possible scheme for electron transfer with the MMO complex can be envisaged as shown in Figure 5.

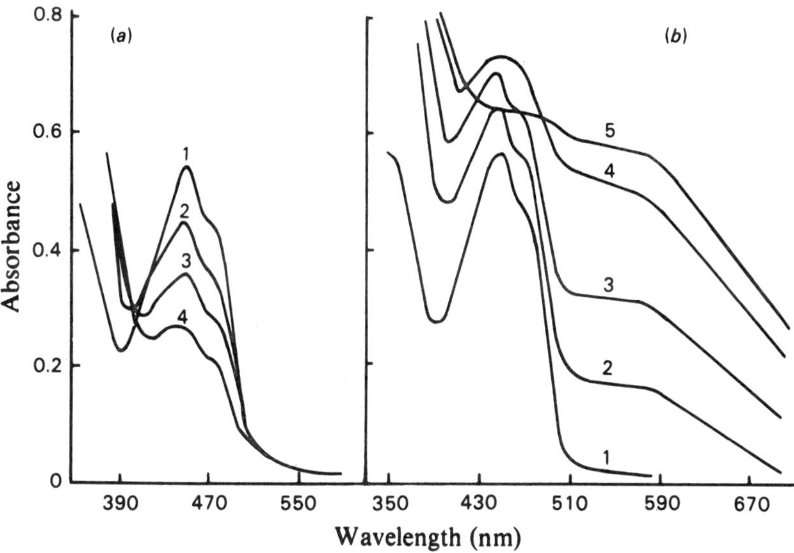

Figure 2. Absorption spectra of FAD (*a*) and protein C (*b*) in hexamethyl phosphoramide after treatment with p-methoxybenzene thiol. Spectra were recorded at various times after addition of thiol. (*a*): 1, before thiol; 2, 30 min.; 3, 60 min; and 4, 120 min. after adding thiol. (*b*): 1, before thiol; 2, 15 min.; 3, 30 min.; 4, 60 min; and 5, 120 min. after adding thiol.

Figure 3. Absorption spectrum of extruded Fe-S* core. This was obtained by subtracting 4a from 5b in Figure 2.

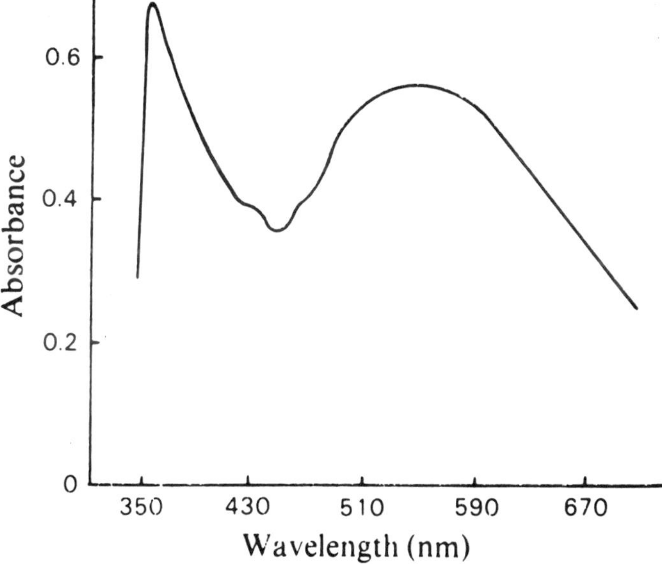

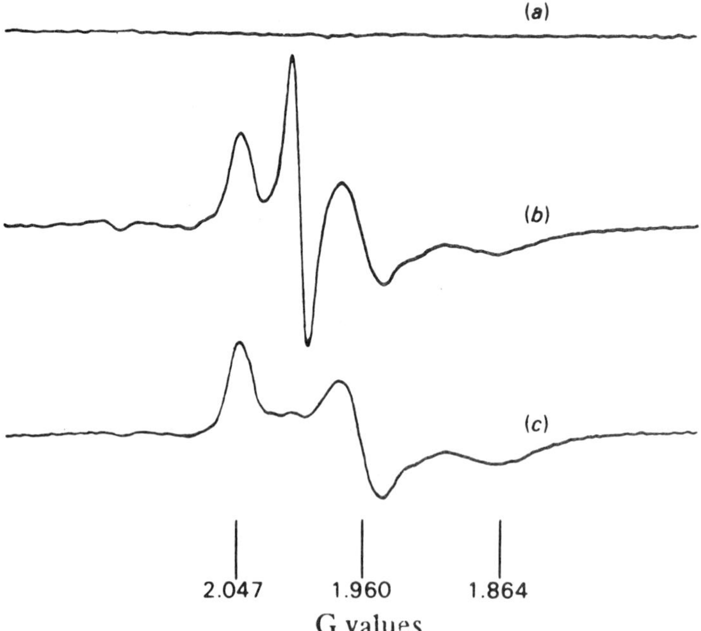

Figure 4. epr spectra of protein C (0.07/μmole) in 50 mM tris HCl pH 7.0. Spectrum (*a*) anaerobic untreated protein. (*b*) after incubation for 1 minute with 5 μmole NADH under anaerobic conditions. (*c*) 1 minute after addition of sodium dithionite to (*b*). Spectra were recorded at 19°K at 9.232 GH_z and 10 mW for (*a*) and 1 mW for (*b*) and (*c*).

Figure 5. Possible mechanism of electron transfer within the MMO complex of *M.capsulatus* (Bath). A_{ox}/red oxidized or reduced protein A; $C_{ox}/_{red}$ oxidized or reduced protein C.

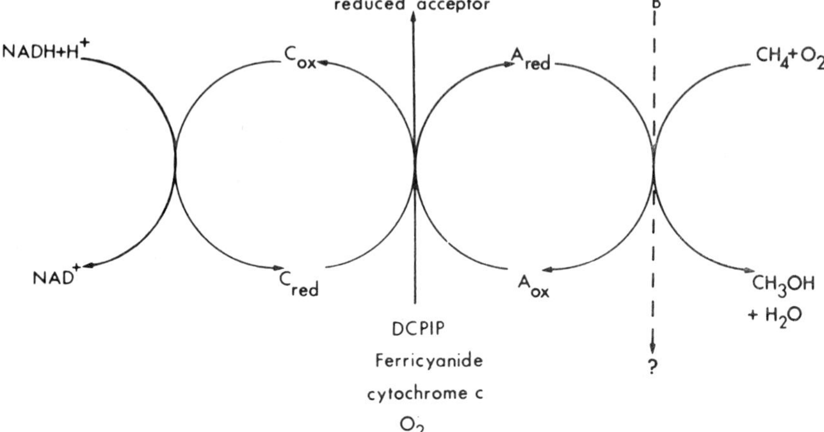

ACKNOWLEDGMENTS
Howard Dalton wishes to thank the Science Research Council for financial support and the Royal Society and the organizing committee for enabling this work to be presented at the Symposium.

References

1. Colby, J. and Dalton, H. (1978) *Biochem J* 174:461–468.
2. Colby, J. and Dalton, H. (1979) *Biochem J* 177:903–908.
3. Hill, C.G., Steenkamp, D.J., Holm, R.H., and Singer, T.P. (1977) *Proc Nat Acad Sci USA* 74:547–551.
4. Dalton, H. (1980) *Adv Appl Microbiol* 26:71–87.

Published 1982 by Elsevier North Holland, Inc.
Vincent Massey and Charles H. Williams, Editors
Flavins and Flavoproteins

CHAPTER 129

The Immediate Fate of Formaldehyde in Oxidative N-Demethylations by Dehydrogenases From Pig Liver Mitochondria and Methylotrophic Bacteria

Daniel J. Steenkamp

Molecular Biology Division, Veterans Administration Medical Center, San Francisco, California and Department of Biochemistry and Biophysics, University of California, San Francisco, California

Although oxidative N-demethylations are of widespread occurrence in biochemistry, the immediate fate of the methyl group which is oxidized to the oxidation level of formaldehyde is uncertain. The rapid incorporation of radioactive carbon from the methyl group of sarcosine, but not from free formaldehyde, into serine was reported by MacKenzie and coworkers (1). It was postulated that "active" rather than free formaldehyde is formed as a product of the oxidative demethylation of sarcosine. The recent identification of dimethylglycine (2, 3) and sarcosine dehydrogenase as the folate binding proteins of rat liver mitochondria suggested that these enzymes use tetrahydrofolate as a coenzyme in the direct synthesis of 5, 10-methylenetetrahydrofolate (MTHF) from sarcosine or dimethylglycine. However, since formaldehyde condenses nonenzymatically with tetrahydrofolate (THF) to form MTHF (4), positive identification of MTHF as the immediate reaction product in the oxidative demethylation of dimethylglycine was not possible (3).

In certain methylotrophic bacteria, formaldehyde produced by oxidative demethylation is, as in mitochondria, used for the synthesis of serine by the enzyme serine hydroxymethyl transferase (5). It is generally assumed that the MTHF required for the latter reaction is formed nonenzymatically by condensation of formaldehyde and THF (5). The data of Kallen and Jencks (6) suggest, however, that this condensation is too slow to be physiologically significant at the low concentrations of both formaldehyde and THF which may be expected to exist *in vivo* (7), a fact noted earlier by Osborn (8).

In view of these considerations, it was important to determine whether MTHF is indeed a direct product of the oxidative demethylation of dimethylglycine and sarcosine, as proposed by Wittwer and Wagner (2), and, if so, to establish if this is a hitherto unrecognized aspect of oxidative demethylations in general.

The Demethylation of Sarcosine and Dimethylglycine

The formation of MTHF rather than free formaldehyde can be established using the NADP-linked enzyme methylenetetrahydrofolate dehydrogenase, since the nonenzymatic condensation of formaldehyde with THF is severely rate-limiting in the conversion of these reactants to methenyl-THF, as shown in Figure 1. The formation of $NADPH_2$, due to the combined activities of MTHF dehydrogenase and methenyltetrahydrofolate cyclohydrolase is much slower when the reaction is initiated by the addition of formaldehyde to an assay mixture containing THF than when preformed MTHF is used as the substrate. To avoid complications which may arise from the presence of electron acceptors, it was decided to determine whether MTHF is released as a product upon reduction of sarcosine or dimethylglycine dehydrogenase with their substrates in the absence of electron acceptors.

For this purpose, sarcosine and dimethylglycine dehydrogenase were isolated from pig liver mitochondria by a modification of the procedure of Wittwer and Wagner (4). Both enzymes were homogeneous as judged by SDS polyacrylamide gel electrophoresis. The results of an analysis of the products formed upon reduction of these enzymes with their substrates are shown in Figure 2.

Figure 1. Rate of $NADPH_2$ formation by MTHF-dehydrogenase in the presence of MTHF compared to THF plus formaldehyde. Reaction mixtures comprised 50 μmols potassium barbital buffer, pH 7.5, 2 μmols NADP, 1 μmol cysteine, and 0.04 units MTHF dehydrogenase in a volume of 1.04 ml at 30°C. Reactions were started under anaerobic conditions by the addition of 6.6, 13.2, and 19.8, nmols d,1 (L)-MTHF to obtain curves (2), (3), and (4), respectively. Curve (1) was obtained upon the addition of 80 nmols formaldehyde to an assay mixture which contained, in addition, 92.5 nmols, d,1 (L)-THF. The traces represent absorbance changes at 340 nm.

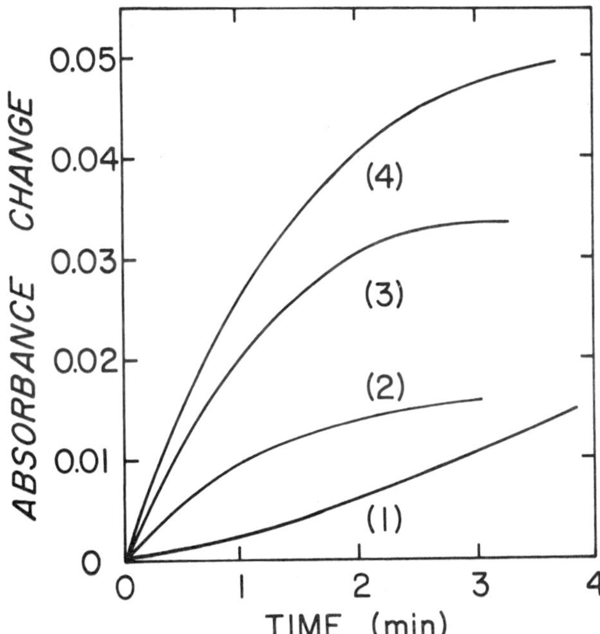

The release of formaldehyde could be detected by a NAD-linked formaldehyde dehydrogenase. It is evident that either of these enzymes release formaldehyde only in the absence of THF, and that MTHF is formed when THF is present. The yield of either formaldehyde or MTHF was quantitated by logarithmic extrapolation of the change in absorbance at 340 nm, due to the formation of the reduced pyridine nucleotides, to the time of addition of the respective substrates, and probably overestimated the actual yields. A difference molar extinction coefficient of 7.1 mM^{-1} cm^{-1} for the NADP-linked conversion of

Figure 2. Enzymatic analysis of the products formed in the reduction of sarcosine or dimethylglycine dehydrogenase. Reaction mixtures containing the enzymes were made anaerobic by the inclusion of 1 unit glucose oxidase and 2.3 μg catalase. After replacing the air in a 1 ml cuvette fitted with a silicone rubber stopper with Ar, 4 μmols glucose was added. Assay mixtures contained, in addition, (a) 10.15 nmols sarcosine dehydrogenase, 0.4 μmols NAD, 80 μg formaldehyde dehydrogenase, and 36 μmols potassium barbital, pH 7.7 in 0.955 ml; (b) as (a) but with 0.125 μmols d,l (L)-THF also present in a final volume of 1.0 ml; (c) 11.26 nmol sarcosine dehydrogenase, 2 μmols NADP, 0.04 units MTHF dehydrogenase, 1 μmol cysteine, 0.125 μmols d,l (L)-THF, and 36 μmols potassium barbital, pH 7.7 in 1.0 ml; (d) 6.8 nmols dimethylglycine dehydrogenase, 60 μg formaldehyde dehydrogenase, 0.4 μmols NAD and 30 μmols potassium barbital in 0.8 ml; (e) as (d) but with 0.1 μmol d,l (L)-THF present in a final volume of 0.84 ml; (f) as (c) but with 8.2 nmol dimethylglycine dehydrogenase instead of sarcosine dehydrogenase. Other additions to the mixtures are as shown. Spectral changes were recorded at 340 nm and the offset on the instrument was used to reset the baseline following the addition of d,l (L)-MTHF in (c) and (f) and prior to the addition of formaldehyde in (a) and (d). The experiment was conducted at room temperature.

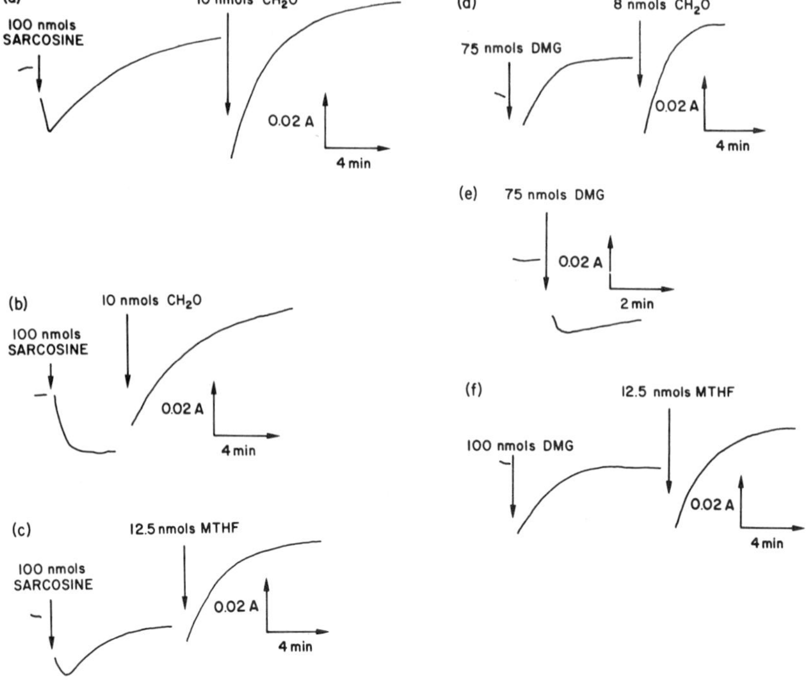

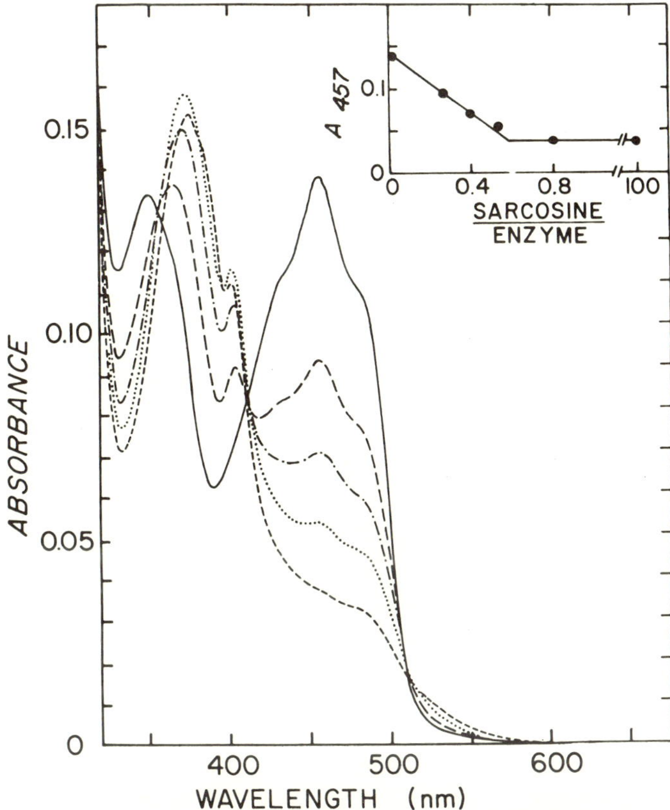

Figure 3. Anaerobic titration of sarcosine dehydrogenase with sarcosine. Anaerobicity was achieved as described in the legend of Figure 2. The buffer was 40 mM potassium barbital, pH 7.7. Additions of sarcosine were as shown in the inset.

MTHF to 10-formyl-THF was assumed. Sarcosine and dimethylglycine dehydrogenase released 0.66 and 0.95 moles formaldehyde and 0.33 and 0.64 moles MTHF respectively, per mole of the enzyme. The relatively low yield of MTHF in the reduction of sarcosine dehydrogenase by sarcosine appears to be consistent with the formation of a high proportion of the anionic semiquinone upon reduction, which required only 0.59 moles sarcosine (Figure 3).

The Dehydrogenases of Methylotrophic Bacteria

For the purpose of this investigation, the trimethylamine dehydrogenase of bacterium W_3A_1, an organism in which formaldehyde is assimilated into the ribulose-5-phosphate pathway by the enzyme hexulose phosphate synthetase, and the trimethylamine and dimethylamine dehydrogenases of *Hyphomicrobium X*, which utilize the serine pathway of formaldehyde assimilation, were chosen. Upon reduction with their substrates, these enzymes were found to release formaldehyde stoichiometric with the enzyme, both in the presence and absence of tetrahydrofolate. The yield of formaldehyde was dramatically decreased in the presence of thiols, such as cysteine or thioglycollic acid, probably due to the formation of cyclic adducts, such as the thiazolidine of cysteine, which are no longer substrates for formaldehyde dehydrogenase. In the presence of thiols such as β-mercaptoethanol or glutathione, the rate of $NADH_2$ formation due to the formaldehyde dehydrogenase reaction changed to that characteristic of the hydroxymethylated thiol where the reaction rate depends on dissociation of the adduct. By contrast, the inclusion of 1 mM glutathione in a reaction mixture caused only a slight deviation in the kinetics of $NADH_2$ formation observed upon the addition of formaldehyde. In view of the high levels of glutathione in living cells (9), it is likely that the oxidative demethylations of trimethylamine or dimethylamine in methylotrophic bacteria give rise to S-hydroxymethylglutathione rather than free formaldehyde.

ACKNOWLEDGMENT
This investigation was supported by program Project Grant HL-16251 from the National Institutes of Health.

References

1. MacKenzie, C.G. and Frisell, W.K. (1958) *J Biol Chem* 232:417–427.
2. Wittwer, A. and Wagner, C. (1981) *J Biol Chem* 256:4102–4108.
3. Wittwer, A. and Wagner, C. (1981) *J Biol Chem* 256:4109–4115.
4. Kisliuk, K.L. (1957) *J Biol Chem* 227:805–814.
5. Colby, J., Dalton, H., and Whittenbury, R. (1979) *Ann Rev Microbiol* 33:481–517.
6. Kallen, K.G. and Jencks, W.P. (1966) *J Biol Chem* 241:5851–5863.
7. Jaenicke, L. (1971) *Methods Enzymol* 18:605–614.
8. Osborn, M.J., Vercamer, E.N., Talbert, P.T., and Huennekens, F.M. (1957) *J Am Chem Soc* 79:6565–6566.
9. Meister, A. (1975). In *Metabolism of Sulfur Compounds*. D.M. Greenberg, (ed.) New York: Academic Press, pp. 101–185.

PART X B:

Iron-Sulfur Molybdoflavoproteins

Published 1982 by Elsevier North Holland, Inc.
Vincent Massey and Charles H. Williams, Editors
Flavins and Flavoproteins

CHAPTER 130

The Flavin and the Other Catalytic and Redox Centers of Xanthine Oxidase and Related Enzymes

R.C. Bray

School of Molecular Sciences, University of Sussex, Brighton, England

Introduction: Composition and Structure of the Enzymes

This review is concerned with the complex flavoproteins, xanthine oxidase, xanthine dehydrogenase, and aldehyde oxidase. These enzymes have been reviewed extensively by the present author (1); this review should be consulted where no reference is given here. Xanthine oxidase from cow's milk has been studied in more detail than have other members of the series and, unless specifically indicated to the contrary, all statements are to be taken as referring to this enzyme. All the enzymes contain molybdenum and iron-sulphur clusters in addition to FAD.

The enzymes have M_r about 300,000 but in most preparations, particularly of the milk enzyme, the molecule has been degraded by proteolysis. The native enzymes have two subunits only, each having M_r about 150,000 (2–5). Though incomplete incorporation of molybdenum can occur (6,7; see below), the normal composition per subunit is 1 Mo, 1 FAD, 4 Fe, and 4 acid-labile sulphur.

Figure 1a summarizes such structural information as is available concerning the enzyme half-molecule. The four redox-active centers are not in immediate contact with one another. Molybdenum and FAD are assumed to be near the surface of the molecule (see below) but the locations of the iron-sulphur centers, Fe-S 1 and Fe-S 2, are less certain. Treatment of chicken liver xanthine dehydrogenase with subtilisin at pH 10 has been reported (3) to split the molecule in a number of ways including the one indicated, to yield a molybdenum- and iron-sulphur-containing fragment assumed to be flavin-free, of about one-third the size of the original half-molecule. The only firm distance information available is the distance, (a) (Figure 1a), between molybdenum and Fe-S 1. From magnetic coupling between these centers, it was concluded (8,9) that (a) must be 8–14 Å. The rest of the molecule is drawn roughly to scale with this distance (though the shape of the molecule and the relative locations of the four centers are represented in a quite arbitrary manner). Since intramolecular electron transfer among the centers occurs rapidly (see below), they are likely to be within, say, 20 Å of one another. Thus, the centers must be confined within a relatively small part of the molecule, as is indicated in Figure 1a.

Mechanism of Action: Miscellaneous Constituents

Figure 1b shows in greater detail a schematic representation of the functional part of the molecule, and indicates the sites at which substrate molecules react. We note that overall kinetics are better fitted by a modified (10), than by a classical, ping-pong mechanism. Before considering either the molybdenum, or the flavin (together with the group or groups indicated by X (Figure 1b) which modulate its properties), we have to consider other constituents of the enzymes whose roles have not as yet been fully defined.

Figure 1. Schematic representation of the xanthine oxidase or dehydrogenase half-molecule. In (A), approximate relative sizes are indicated, though the shape of the molecule and the positions of the different centers are arbitrary. It is known, however, that molybdenum and flavin are on the surface, while the distance, a, has been estimated (8,9). In (B), reactions with substrates and intramolecular electron transfer reactions are indicated.

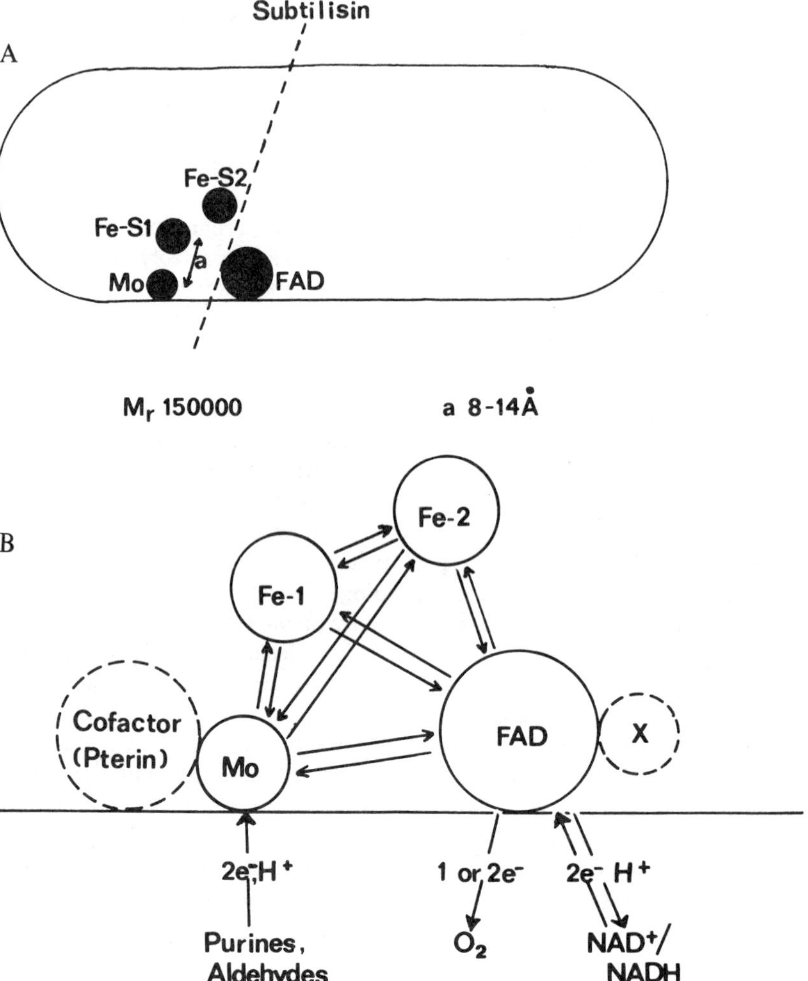

The iron-sulphur centers will be mentioned only briefly. The original characterization as two [2Fe-2S] clusters depended on EPR (11). This has now been confirmed (12) by iron-sulphur core extrusion with ^{19}F NMR characterization. In showing (11) two g-values above the free electron value, reduced Fe-S 2 is unusual among [2Fe-2S] clusters (13), though this is not necessarily incompatible with existing theory (14; J.F. Gibson, personal communication).

The intriguing possibility that selenium may be present in certain bacterial xanthine dehydrogenases has recently been raised (15). Further work is awaited with interest.

Finally we turn to the "molybdenum cofactor." Genetic evidence for a common cofactor associated with incorporation of molybdenum into molybdenum-containing enzymes is of long standing; this and later work has been reviewed (16). A more recent development is the proposal (17) that the cofactor is a reduced pterin derivative substituted in the 6-position. Further work is awaited eagerly. In the meantime, evidence (17) pertaining to the presence of this new constituent in milk xanthine oxidase may be summarized. Molybdenum-containing enzymes were found on denaturation with guanidinium chloride to yield a product, converted on oxidative degradation to pterin-6-carboxylic acid. This was obtained in unspecified yield and was identified by fluorescence and thin layer chromatography. No direct evidence that molybdenum in the enzyme is bound to the pteridine was presented, nor are pteridines particularly good metal chelators (18). Additional studies on the cofactor by Russian workers (19) are also of interest.

Figure 2 shows work, perhaps of relevance to the cofactor, carried out some years ago in the author's laboratory, and may serve as a reminder that xanthine oxidase, as isolated by conventional procedures from many milk samples, is deficient in molybdenum (6). De-molybdo enzyme may be removed from such samples by selective denaturation with sodium salicylate (6). The conditions used (temperature, buffer, and enzyme concentration) are critical. Figure 2 shows that with 0.2 M salicylate at 37°C, and 90–100-hr reaction time, the de-molybdo form is about 85% destroyed, in comparison with an activity loss of only about 50%. These conditions have been adopted for routine preparation of the enzyme in the author's laboratory. That the de-molybdo enzyme is more readily denatured than is the functional form presumably indicates that molybdenum and the cofactor serve, in addition to more specific roles, to stabilize the molecule.

The Flavin Site

Special interest in relation to the flavin site in these enzymes is twofold. First, some are oxidases utilizing oxygen effectively as reducing substrate but scarcely reacting with NAD^+, while others are dehydrogenases with this acceptor specificity pattern reversed. Additionally, interconversions between dehydrogenase and oxidase forms are in some cases possible (see 1 for review of earlier work). Despite study of these interconversion processes (2,3,20–22), the precise chemical natures of the group or groups (denoted by X in Figure 1b) which determine the properties of the flavin in the different enzyme forms remains obscure. However, thiol group modification and proteolysis clearly have a role in interconversions.

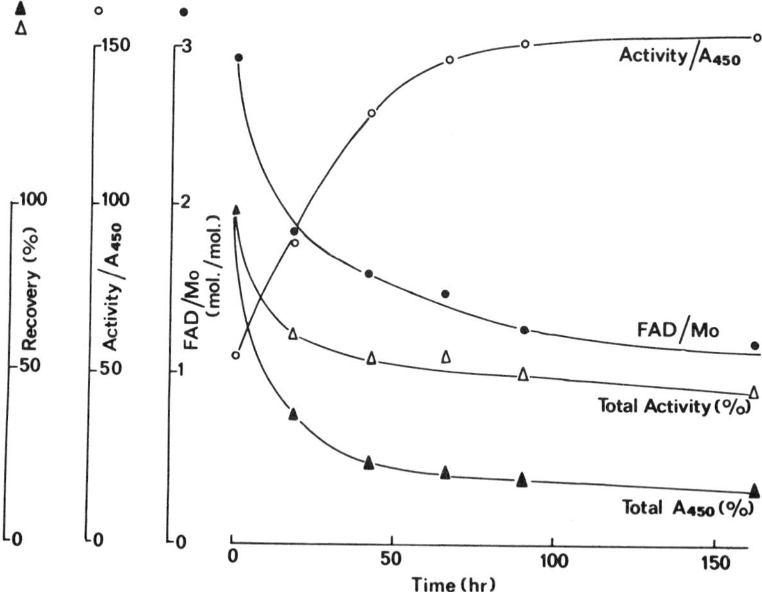

Figure 2. Selective denaturation of de-molybdo milk xanthine oxidase by sodium salicylate. The procedure was based on that of Hart et al. (6). The enzyme sample (having an unusually high FAD/Mo ratio) was incubated at 37° with 0.2 M sodium salicylate in 0.1 M sodium pyrophosphate, pH 7.0. Samples were withdrawn at the times indicated and analyzed after removal of precipitated material by centrifugation and of low molecular weight material by gel filtration. Selective destruction of the de-molybdo enzyme is indicated by FAD/Mo tending towards 1.0 and by Activity/A_{450} increasing. [Loss of activity was unusually high in this experiment, probably because of the low concentration of active enzyme (unpublished data of R.T. Pawlik and R.C. Bray).]

A second unusual (23) feature of the enzymes, which relates to the flavin, is that under some conditions oxygen is reduced to hydrogen peroxide, while under others is reduced to superoxide (24). Quantitative aspects have recently been reinvestigated (25) and will be mentioned below.

Differing reactivity of the flavin in the different enzymes is associated with, and no doubt in part dependent on, variations in its redox potentials (Figure 3). Thus, the FAD/FADH$^{\cdot}$ potentials are about -350 to -360 for both the milk oxidase and the turkey liver dehydrogenase, but the FADH$^{\cdot}$/$FADH_2$ potential is -236 mV in the milk enzyme whereas in the turkey enzyme it has the low value of -366 mV (26,27). Potentials in the chicken liver dehydrogenase (28) are comparable to those in the turkey enzyme. Though titration data are not available, oxidase-dehydrogenase interconversions appear to be accompanied by comparable changes of potential (2,29). No doubt (26) the prevalence of FADH$^{\cdot}$ over $FADH_2$, in dehydrogenases reduced by substrates, explains their low reactivity to oxygen. For the milk enzyme, $FADH_2$ in the reduced enzyme has clearly been shown to react much more rapidly with oxygen than does FADH$^{\cdot}$ (25).

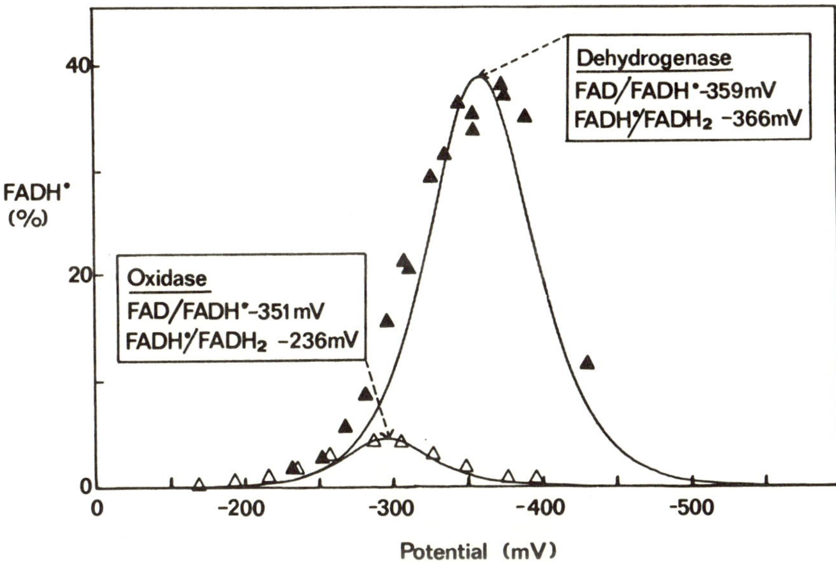

Figure 3. Comparison of the conversion of FAD into FADH during potentiometric titration of milk xanthine oxidase and turkey liver xanthine dehydrogenase. The titrations were carried out in the presence of mediator dyes in pyrophosphase buffer pH 8.2. Samples, adjusted to the appropriate potentials, were frozen for EPR measurement. Solid lines through the experimental points were calculated from the values of the potentials indicated [replotted from (26)].

We further note that at normal pH values the flavin semiquinone in both oxidases and dehydrogenases is of the "blue" type [EPR linewidth 1.9 mT (30)], while at high pH values the flavin semiquinone changes to the "red" type [linewidth 1.35 mT; pK 8.6 (31)]. Possibly connected with this, rates of reoxidation by oxygen of reduced milk xanthine oxidase molecules increase with increasing pH (25). In view of these findings, it might be of interest to test whether minor oxidase activity exhibited by the avian dehydrogenases becomes more significant if the pH value is changed.

In contrast to oxidase activity, NAD^+-reducing activity of the dehydrogenases may depend on the presence of a binding site for the coenzyme near to the flavin, this site being absent in the oxidases. Observations of NAD^+ difference spectra (22) in the dehydrogenases support this interpretation, as does the finding (21) that NADH inhibits oxidase activity in some forms of the enzymes only.

In relation to differences in flavin environment, it is interesting that there are slight absorption differences in the 300–500 nm region between the oxidases and the dehydrogenases [Figure 4a: (28); see also e.g. (6) and (32)]. Though Fe-S and possibly Mo provide features in this region, it is clear from the flavin difference spectra [Figure 4b, obtained by subtracting spectra of the deflavo enzymes (cf 33) from those of the native forms], that the environment of the flavin differs. Perhaps a more hydrophobic environment in the oxidase is

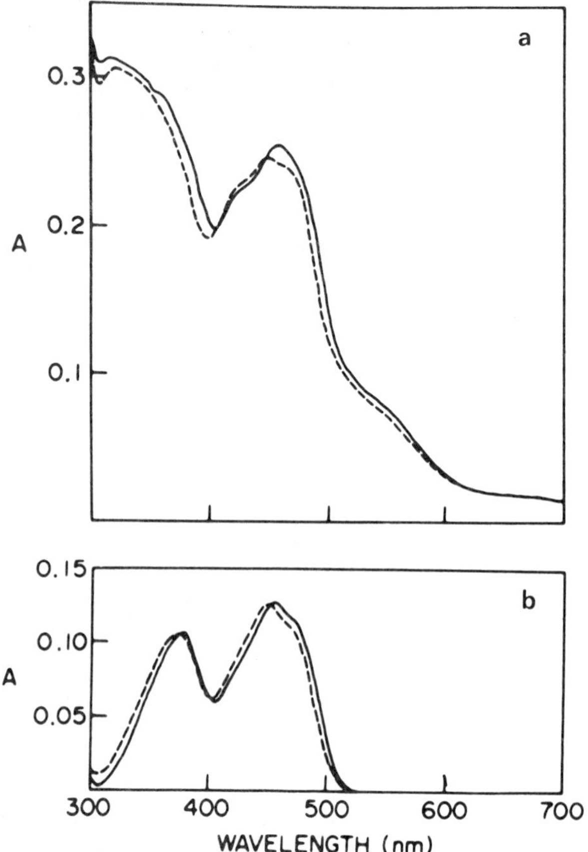

Figure 4. Comparison of the spectra of milk xanthine oxidase and chicken liver xanthine dehydrogenase. The solid line corresponds to the dehydrogenase and the dashed line to the oxidase. (a) Shows the enzyme as prepared and (b) shows flavin difference spectra [reproduced from (28)].

indicated (cf 34). Possibly the claim (20) that there is no change in the spectrum during the oxidase-dehydrogenase interconversions should be reinvestigated. However, it is quite possible that the hypothetical group, X (Figure 1b), endowing the flavin with dehydrogenase characteristics, is of a different nature in xanthine dehydrogenases from different sources.

A final point concerns inhibition of aldehyde oxidase by menadione (35). This phenomenon seems entirely unexplained though preliminary experiments from the author's laboratory (R.C. Bray, G.N. George, L. Norlander, and A. Zimmerman, unpublished work) may implicate the flavin. Thus, this enzyme may exhibit still further an as yet unknown difference in its flavin site.

Clearly much work remains to be done on the chemistry of the flavin site of the enzymes.

Intramolecular Electron Transfer to and From the Flavin Site

In reduction and reoxidation of xanthine oxidase by substrates, intramolecular electron transfer among the four redox-active constituents does not appear to be rate-limiting (36). Thus, we assume that all the electron transfer reactions represented by the arrows within the enzyme in Figure 1b are possible and occur rapidly [$k>100\ sec^{-1}$ (37)]. A feature of xanthine oxidase is that the redox potentials of all four constituents are independent of one another, that is reduction of one constituent does not affect the redox potential of another. The enzyme half-molecule can take up a maximum of six electrons on full reduction. In partly reduced forms, the distribution of electrons is governed by, and can be predicted from, the redox potentials of the different sites. Thus, it has been possible to predict the forms of stoichiometric reductive titrations of all the constituents of the enzyme, when using either 1-electron (36) or 2-electron (26) reducing agents.

It is important to note that 2-electron reduced xanthine oxidase is chemically (and spectroscopically) a distinct species from, for example, the 4-electron reduced molecule. In fact, it seems (25,36,37) that it is only the 1-electron and the 2-electron reduced forms whose flavin reacts with oxygen to give superoxide, in contrast to the more extensively reduced forms of the enzyme, which give hydrogen peroxide only.

Recently (38), FAD has been removed from xanthine oxidase and the enzyme reconstituted with FAD analogues having higher or lower redox potentials. It was found (38) that reductive titration curves could again be predicted from a knowledge of the potentials, with potentials for the different enzyme-bound flavin derivatives bearing a roughly constant relation to those of the free flavins, and with no changes in the potentials of the other constituents. Further, in stopped-flow experiments on the time-course of reduction of the modified enzyme by xanthine, there was evidence (38) for very rapid redistribution of electrons from molybdenum to the site of highest redox potential, whether this was flavin or iron-sulphur. These new results thus further support earlier conclusions (36,37) that intramolecular electron transfer within the enzyme is rapid.

The Molybdenum Site

The author's main interest in recent years has concerned the structure and functioning of the molybdenum site. Since this is scarcely relevant to a volume devoted to flavins, the work will be surveyed briefly, concentrating on developments since a review (39) was written.

Attention has been focused on the natures and structures of several different molybdenum(V) EPR signal-giving species which are obtainable from most of the enzymes, but particularly from milk xanthine oxidase. Each of the signals has been given a name (Table 1). The natures of the species are in principle deducible by essentially standard biochemical procedures and these have led to the conclusions given. Much progress towards elucidation of the structures has recently been made by EPR investigations involving substitutions with differ-

Table 1. Superhyperfine Coupling to Mo(V) in Reduced Xanthine Oxidase.

Enzyme form	Signal name	Nature	Couplings ^{1}H	^{13}C	^{17}O	$^{33}S^{a}$
Active	Very Rapid	Transient covalent intermediate	None	1 We[b]	1 St[c]	1 St
	Rapid type 1	Simple reduced forms	1 St[d] 1 We	—	1 St	1 We
	Rapid type 2	Simple reduced forms	2 St	—	2 St[e]	1 We
	Inhibited	Stable inhibitory complex (aldehydes)	1 We[f,g] or None	1 St[h]	2 We[f]	1 We
Desulpho	Slow	Simple reduced form	1 St	—	2 St[e]	None
	Desulpho Inhibited	Stable complex (glycol)	None	—	2 We[f]	None

[a] CN^--labile S atom.
[b] From C-8 carbon of xanthine.
[c] Atom used to follow ^{17}O exchange in oxidized enzyme ($k \sim 10^{-3} sec^{-1}$) or with substrate ($k \gg 10^{-1} sec^{-1}$).
[d] From C-8 proton of xanthine, but exchanges with solvent ($k = 27 sec^{-1}$).
[e] Might be 1 St, 1 We.
[f] Not exchangeable with solvent.
[g] From formaldehyde.
[h] From C = 0 of aldehyde.

The nature of each signal-giving species is indicated, together with the number of nuclei of each type which are coupled; We = weak coupling; St = strong coupling. References: ^{1}H, ^{13}C (39); ^{17}O (42, 43); ^{33}S (40, 41).

ent stable isotopes. These extensive studies, still incomplete at the time of writing, are illustrated in Figure 5 and summarized in Table 1. Figure 5 shows effects of substitution with ^{33}S (40, 41), or with ^{17}O (42), on one of the signals, Very Rapid. Current hypotheses as to the structures of the various species (depending in large measure on very recent data from oxygen-17 substitution work, to be published elsewhere (43)) are summarized in Figure 6. All these structures are to a greater or lesser extent tentative but studies in hand should serve further to confirm or refine them.

The proposed mechanism (39, 41–43) of substrate hydroxylation, based on these data, involves its coordination to molybdenum and the transfer from it, of a proton to an accepting group, and of two electrons to Mo(VI). Concerted with this, the substrate carbonium ion is stabilized by reaction with a nucleophile. The nucleophile is identified as a terminal oxygen ligand of molybdenum and the proton acceptor as either another terminal oxygen, or perhaps more likely, as a terminal sulphur ligand. Product release takes place by hydroxyl ion attack, and cleavage of the molybdenum oxygen bond.

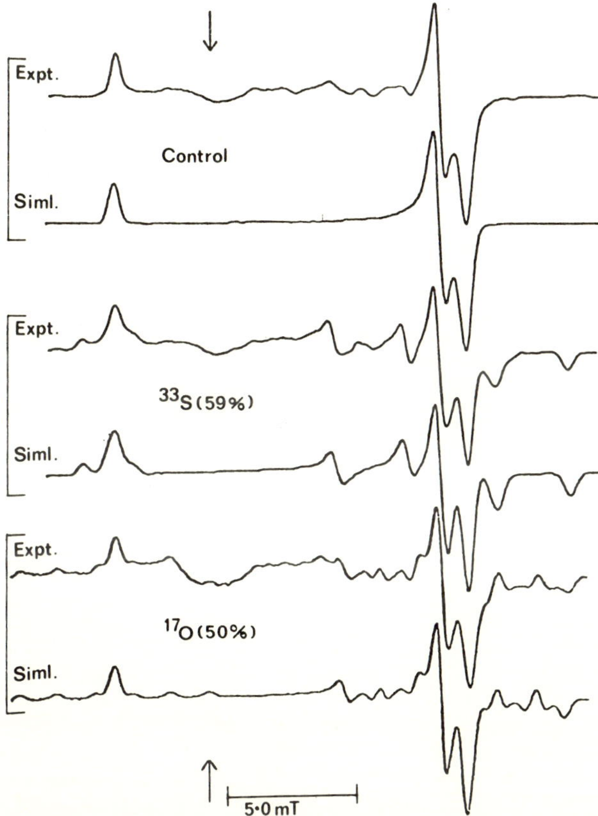

Figure 5. Hyperfine coupling of ^{33}S and of ^{17}O to Mo(V) in the Very Rapid epr signal from xanthine oxidase. Xanthine was used to generate the signal; experimental and computer-simulated spectra are shown. At the top, a control with unenriched enzyme is shown; in the center, ^{33}S was present in the CN^--labile site of the enzyme; and at the bottom, the water was enriched with ^{17}O [data are from (41) and (42)]. For both isotopes, coupling is strong, but whereas for ^{17}O it is isotropic, for ^{33}S it is highly anisotropic with noncoincident axes of g and A.

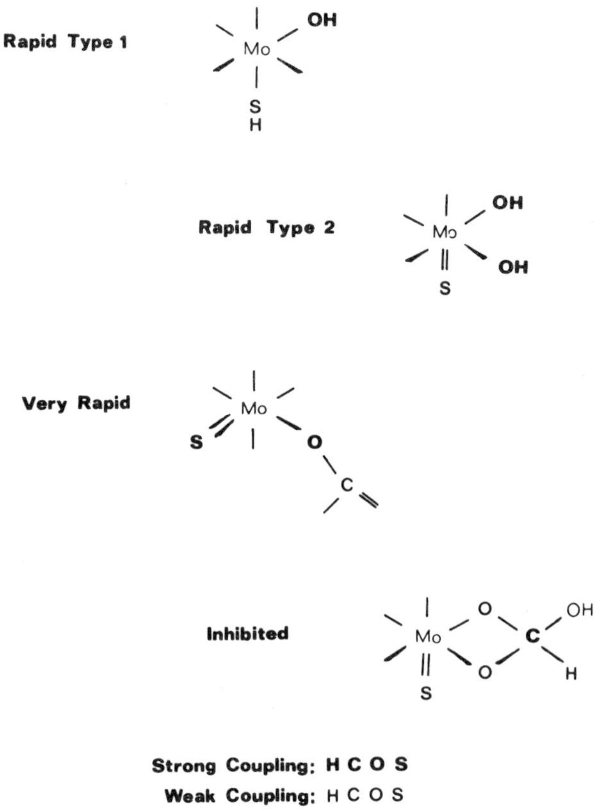

Rapid Type 1

OH
Mo
S
H

Rapid Type 2

OH
Mo
OH
S

Very Rapid

Mo
S
O
C

Inhibited

O
OH
Mo
C
O
H
S

Strong Coupling: H C O S

Weak Coupling: H C O S

Figure 6. Possible structures for the various Mo(V) EPR signal-giving species from milk xanthine oxidase. Structures are based largely on data in Table 1, generally with the convention that strong coupling (*heavy type*) denotes an equatorial ligand and weak coupling (*normal type*) an axial one. Coupling of ^{13}C to Mo(V) in Inhibited, is assumed analagous to transanular coupling of ^{31}P to the metal in dithiophosphinate complexes (44). (The $-OH$ group in Inhibited is not coupled.) Note that in Rapid Type 1, the strongly coupled H might be on the S and the weakly coupled one on O, rather than as illustrated.

ACKNOWLEDGMENTS

The author thanks his colleagues Dr. S. Gutteridge, Dr. J.P.G. Malthouse, and Mr. G.N. George for helpful discussions and Dr. V. Massey and Dr. G. Palmer for supplying manuscripts prior to publication.

References

1. Bray, R.C. (1975) In *The Enzymes*, Vol. XII, 3rd Ed. Boyer, P.D. (ed). New York: Academic Press, pp. 299–419.
2. Waud, W.R. and Rajagopalan, K.V. (1976) *Arch Biochem Biophys* 172:365–379.
3. Coughlan, M.P., Betcher-Lange, S.L., and Rajagopalan, K.V. (1979) *J Biol Chem* 254:10694–10699.
4. Mangino, M.E. and Brunner, J.R. (1977) *J Dairy Sci* 60:841–850.

5. Lyon, E.S. and Garrett, R.H. (1978) *J Biol Chem* 253:2604–2624.
6. Hart, L.I., McGartoll, M.A., Chapman, H.R., and Bray, R.C. (1970) *Biochem J* 116:851–864.
7. Johnson, J.L., Waud, W.R., Cohen, H.J., and Rajagopalan, K.V. (1974) *J Biol Chem* 249:5056–5061.
8. Lowe, D.J. and Bray, R.C. (1978) *Biochem J* 169:471–479.
9. Coffman, R.E. and Buettner, G.R. (1979) *J. Phys Chem* 83:2392–2400.
10. Coughlan, M.P. and Rajagopalan, K.V. (1980) *Eur J Biochem* 105(1):81–84.
11. Lowe, D.J., Lynden-Bell, R. and Bray, R.C. (1972) *Biochem J* 130:239–249.
12. Wong, G.B., Kurtz, D.M., Jr., Holm, R.H., Mortenson, L.E., and Upchurch, R.G. (1979) *J Am Chem Soc* 101:11, 3078–3090.
13. Sands, R.H. and Dunham, W.R. (1975) *Quart Rev Biophys* 7:443–504.
14. Gibson, J.F., Hall. D.O., Thornley, J.H.M., and Whatley, F.R. (1966) *Proc Nat Acad Sci USA* 56:987–990.
15. Wagner, R. and Andreesen, J.R. (1979) *Arch Microbiol* 121:255–260.
16. Johnson, J.L. (1980) In *Molybdenum and Molybdenum-Containing Enzymes*. Coughlan, M.P. (ed.) Oxford: Pergamon Press, pp. 347–383.
17. Johnson, J.L., Hainline, B.E., and Rajagopalan, K.V. (1980) *J Biol Chem* 255:1783–1786.
18. Sillén, L.G. and Martell, A.E. (1964) Stability Constants of Metal-Ion Complexes. London: Chemical Society, p. 722.
19. Alikulov, Z.A., L'Vov, N.P., Burikhanov, S.S., and Kretovich, V.L. (1980) *Iz Akad Nauk SSSR Ser Biol* 5:712–718.
20. Waud, W.R. and Rajagopalan, K.V. (1976) *Arch Biochem Biophys* 172:354–364.
21. Kaminski, Z.W. and Jezewska, M.M. (1979) *Biochem J* 181:177–182.
22. Nakamura, M., Kurebayashi, H., and Yamazaki, I. (1978) *J Biochem* 83:9–17.
23. Nakamura, S. and Yamazaki, I. (1969) *Biochim Biophys Acta* 189:29–37.
24. Fridovich, I. (1970) *J Biol Chem* 245:4053–4057.
25. Porras, A.G., Olson, J.S., and Palmer, G. (1981) *J Biol Chem* 256:9096–9103.
26. Barber, M.J., Bray, R.C., Cammack, R., and Coughlan, M.P. (1977) *Biochem J* 163:279–289.
27. Cammack, R., Barber, M.J., and Bray, R.C. (1976) *Biochem J* 157:469–478.
28. Barber, M.J., Coughlan, M.P., Kanda, M., and Rajagopalan, K.V. (1980) *Arch Biochem Biophys* 201:468–475.
29. Bray, R.C., Barber, M.J., Coughlan, M.P., Dalton, H., Fielden, E. M., and Lowe, D.J. (1976) In *Flavins and Flavoproteins*. Singer, T.P. (ed.) Amsterdam: Elsevier, pp. 553–565.
30. Barber, M.J., Bray, R.C., Lowe, D.J., and Coughlan, M.P. (1976) *Biochem J* 153:297–307.
31. Barber, M.J. and Siegel, L.M. (1980) *Fed Proc* 39:1677.
32. Cleere, W.F. and Coughlan, M.P. (1975) *Comp Biochem Physiol* 50B:311–322.
33. Komai, H., Massey, V., and Palmer, G. (1969) *J Biol Chem* 244:1692–1700.
34. Ohama, H., Sugiura, N., Tanaka, F., and Yagi, K. (1977) *Biochemistry* 16:126–131.
35. Rajagopalan, K.V., Fridovich, I., and Handler, P. (1962) *J Biol Chem* 237:922–928.
36. Olson, J.S., Ballou, D.P., Palmer, G., and Massey, V. (1974) *J Biol Chem* 249:4363–4382.
37. Olson, J.S., Ballou, D.P., Palmer, G., and Massey, V. (1974) *J Biol Chem* 249:4350–4362.
38. Hille, R., Fee, J.A., and Massey, V. (1981) *J Biol Chem* 256:8933–8940.
39. Bray, R.C. (1980) In *Advances in Enzymology and Related Areas of Molecular Biology*, Vol. 51. Meister, A. (ed.) New York: John Wiley & Sons, pp. 107–165.
40. Malthouse, J.P.G. and Bray, R.C. (1980) *Biochem J* 191:265–267.
41. Malthouse, J.P.G., George, G.N., Lowe, D.J., and Bray, R.C. (1981) *Biochem J* 199:629–637.
42. Gutteridge, S. and Bray, R.C. (1980) *Biochem J* 189:615–623.
43. Gutteridge, S. and Bray, R.C., unpublished work.
44. Stiefel, E.I., Newton, W.E., and Pariyadath, N. (1977) In *Chemistry and Uses of Molybdenum*. Mitchell, P.C.H. and Seaman, A. (eds.) London: Climax Molybdenum Company, pp. 265–270.

Published 1982 by Elsevier North Holland, Inc.
Vincent Massey and Charles H. Williams, Editors
Flavins and Flavoproteins

CHAPTER 131

The Oxidative Half-Reaction of Xanthine Oxidase

Russ Hille and Vincent Massey

Department of Biological Chemistry, University of Michigan, Ann Arbor, Michigan

The oxidation of fully reduced xanthine oxidase (i.e., enzyme containing six reducing equivalents) by O_2 produces both H_2O_2 and $O_2^{\overline{\cdot}}$ (1–3). The process is necessarily a sequential one involving the reaction of successive O_2 molecules with enzyme species at intermediate levels of reduction. It has been inferred from a variety of data that $O_2^{\overline{\cdot}}$ is produced only late in the oxidation sequence (4), but direct evidence has been lacking. Furthermore, the actual amounts of $O_2^{\overline{\cdot}}$ reported in rapid quench experiments by monitoring its EPR signal are twenty-fold lower than expected on the basis of the model proposed (4). We have reinvestigated the production of $O_2^{\overline{\cdot}}$ in the oxidation of xanthine oxidase by making use of the well-established ability of $O_2^{\overline{\cdot}}$ to reduce cytochrome c (1). We find that $O_2^{\overline{\cdot}}$ is in fact produced only in the last two steps of enzyme oxidation with an overall stoichiometry of two $O_2^{\overline{\cdot}}$ formed per enzyme molecule oxidized.

Xanthine oxidase was isolated by the method of Massey et al. (5), and was typically found to have AFR values between 140 and 160 (i.e., approximately 70% functional active sites). Reduced enzyme for kinetic experiments was prepared by carefully titrating the anaerobic enzyme with sodium dithionite in a tonometer equipped with a sidearm cuvette for spectrophotometric observation. Kinetic experiments were performed with a stopped-flow spectrophotometer designed and built by Dr. D.P. Ballou of this department and interfaced with a Data General Nova 2 minicomputer. Computer simulations of the kinetic data were performed with the same minicomputer using a fourth-order Runge-Kutta simulation package (6). All experiments were performed in 0.1 M pyrophosphate pH 8.5 at 25°C.

The reaction of fully reduced xanthine oxidase with O_2 as observed at 450 nm is shown in Figure 1. It can be seen that the reaction is markedly biphasic and that the fast phase of the reaction is nonlinear in a semilogarithmic plot. When 50 μM cytochrome c is included in the reaction, the time courses shown in Figure 2A are observed at 550 nm, depending on whether superoxide dismutase is included as well. The difference between the two traces reflects the superoxide-dependent reduction of cytochrome c. (As it turns out, no superoxide-independent reduction of cytochrome c was observed. In the presence of dismutase, the absorbance change at 550 nm is due entirely to the oxidation of xanthine oxidase.) A semilogarithmic plot of the difference between the two traces is shown in Figure 2B. Three phases can be distinguished: a pronounced

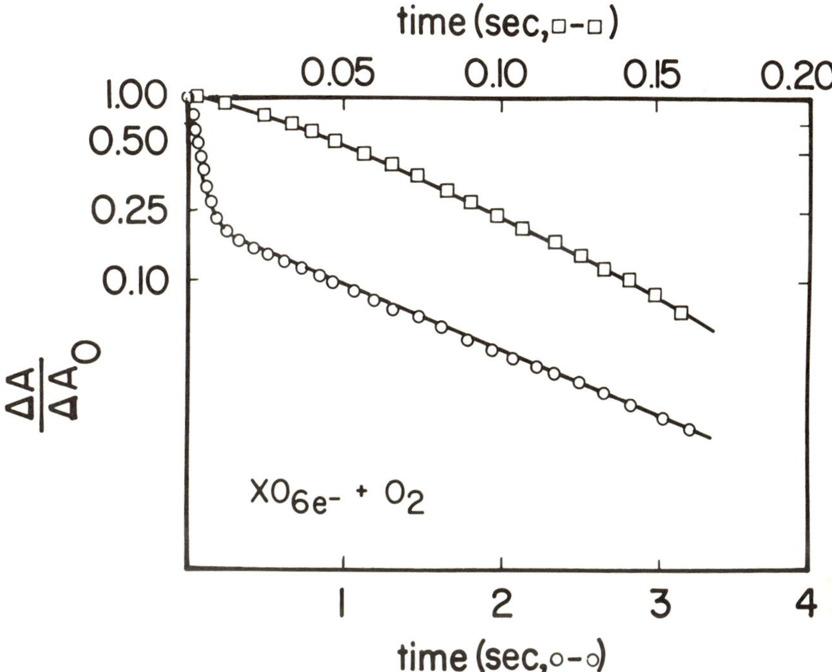

Figure 1. The reaction of fully reduced xanthine oxidase with 200 μM O_2 (before mixing). Circles: the absorbance change as observed at 450 nm. Squares: the fast phase of the reaction on an expanded time scale. The fast phase is nonlinear for reasons discussed in the text.

lag of approximately 150 ms followed by a burst phase and finally a slower phase with a rate identical to that for the slow phase of the reaction shown in Figure 1 (i.e., 0.9 s^{-1}). The stopped-flow apparatus with which these experiments were performed had the capability of scanning wavelength automatically, and from spectra recorded at the end of the two reactions shown in Figure 2A the amount of cytochrome c reduced by $O_2^{\cdot-}$ could be quantitated. It was found from several experiments that the stoichiometry of cytochrome c reduced to enzyme oxidized was 2.1±0.1. Because there was essentially no cytochrome c reduction during the fast phase of enzyme oxidation, a scheme very similar to that of Olson et al. (4) can be proposed:

$$XO_{6e}\text{-}\rightarrow XO_{4e}\text{-}\rightarrow XO_{2e}\text{-}\rightarrow XO_{1e}\text{-}\rightarrow XO_{ox}$$

where the subscripts refer to the number of reducing equivalents present in each intermediate. H_2O_2 is produced (rapidly) in each of the first two steps and $O_2^{\cdot-}$ in each of the last two. Extinctions for each of the intermediates can be calculated (7), and after including explicitly the second-order rate for the reduction of cytochrome c by $O_2^{\cdot-}$ ($k_2 = 1.6 \times 10^5$ $M^{-1}s^{-1}$, ref. 3) this scheme can be used to simulate the data. The lines in Figures 1 and 2B represent such simulations and it can be seen that the fits to the data are excellent. In these simulations, a single intrinsic rate constant for each of the first three steps was

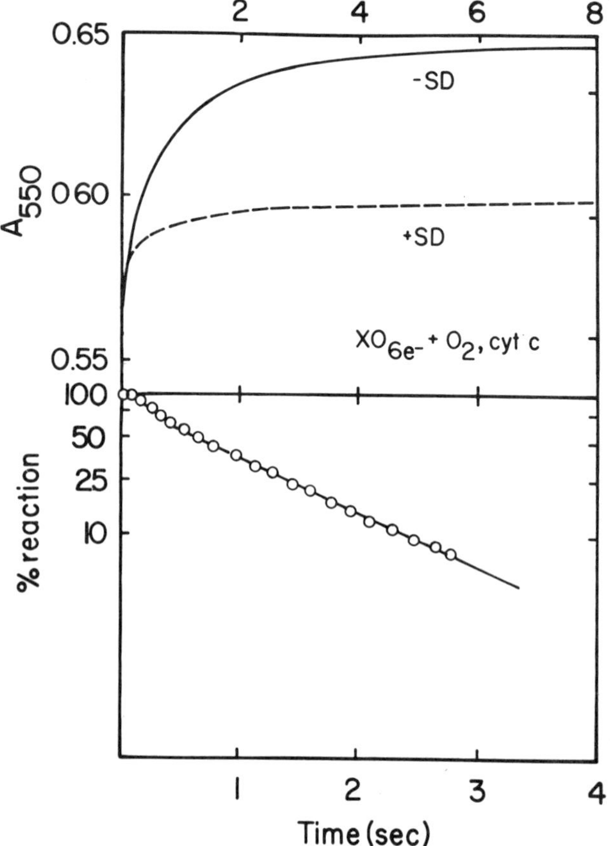

Figure 2. The reaction of fully reduced xanthine oxidase with 200 μM O_2 (before mixing) in the presence of 50 μM cytochrome c (before mixing). (A) solid and dashed lines reflect the absorbance changes observed at 550 nm in the absence and presence, respectively, of 5 μg/ml superoxide dismutase (before mixing). (B) a semilogarithmic plot of the *difference* between the two traces in panel A. The limiting rate of the reaction is identical to that for enzyme oxidation in the absence of cytochrome c (Figure 1).

used which was multiplied by the fraction of the flavin site present as $FADH_2$ in the appropriate intermediate to obtain the effective rate constant for each step. The rate constants used were 35, 33.5, 20, and 0.9 s^{-1}, respectively. It was found that the intrinsic rate required to fit adequately the fast phase of the reaction (35 s^{-1}) was nearly twice the rate for the fast phase determined graphically (17.5 s^{-1}). Both this and the apparently accelerating behavior of the fast phase (see Figure 1, squares) can be attributed to the sequential nature of the reaction (4).

To verify this scheme, experiments were carried out with XO_{2e^-} and the XO_{6e^-} alloxanthine complex, an analog (as regards oxidation) for XO_{4e^-}. XO_{2e^-}, generated by tipping a substoichiometric amount of xanthine into anaerobic, oxidized enzyme, is oxidized biphasically (Figure 3A), and the fast phase of the reaction exhibits no lag, as expected from the proposed scheme. The relative absorbance changes in each phase are those expected from calculations of extinctions for XO_{2e^-} and XO_{1e^-} (7). In the cytochrome c experiment, $O_2^{\cdot-}$ is produced in two phases of equal absorbance change strictly analogous to the two phases observed in the oxidation of XO_{2e^-} in the absence of cytochrome c (Figure 3B). No lag is observed and the stoichiometry is two $O_2^{\cdot-}$ per XO_{2e^-} oxidized, providing strong evidence in favor of the proposed scheme. The oxidation of XO_{6e^-} alloxanthine is found to be biphasic as well (Figure 4A), with a fast phase more bowed in a semilogarithmic plot than that for XO_{2e^-} but less so than that for XO_{6e^-}. $O_2^{\cdot-}$ production also exhibits behavior intermediate between that of XO_{2e^-} and XO_{6e^-}. The solid lines in Figures 3 and 4 represent simulations using the same rate constants as in Figures 1 and 2, only modifying the scheme as is appropriate for the level of

Figure 3. The reaction of two electron-reduced xanthine oxidase with 200 μM O_2 (before mixing). Top (A) circles: the absorbance change as observed at 450 nm. Squares: the fast phase of the reaction shown on an expanded time scale. Bottom (B) semilogarithmic plot of the differential absorbance change at 550 nm with cytochrome c in the absence and presence of superoxide dismutase (as in Figure 2).

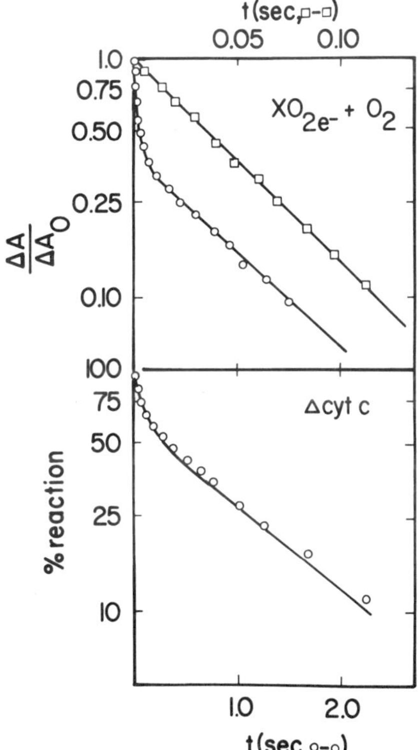

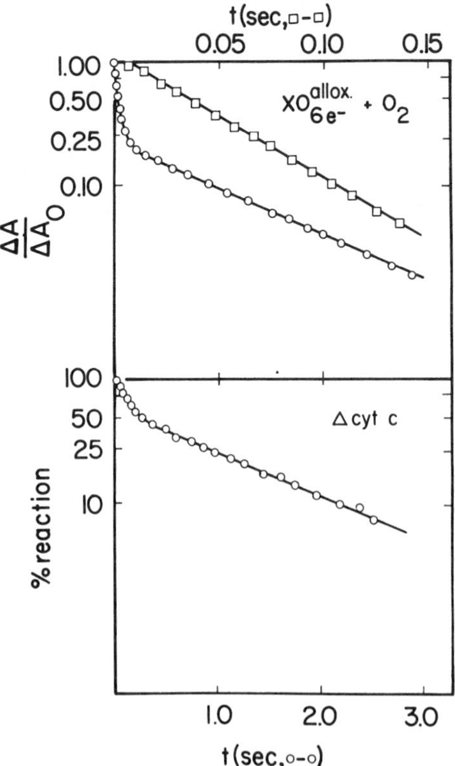

Figure 4. The reaction of fully reduced enzyme·alloxanthine complex with 200 μM O_2 (before mixing). Top (A) circles: the absorbance change as observed at 450 nm. Squares: the fast phase of the reaction shown on an expanded time scale. Bottom (B) semilogarithmic plot of the differential absorbance change at 550 nm with cytochrome c in the absence and presence of superoxide dismutase (as in Figure 2).

enzyme reduction at the outset of the experiment. Again, it can be seen that the fits to the data are excellent, providing conclusive evidence for the validity of the proposed scheme.

ACKNOWLEDGMENTS

Financial support from the National Institutes of Health (GM11106), the AMAX Foundation, and the Michigan Society of Fellows is gratefully acknowledged.

References

1. McCord, J.M. and Fridovich, I. (1968) *J Biol Chem* 243:5753–5760.
2. Knowles, P.F., Gibson, J.F., Dick, F.M., and Bray, R.C. (1969) *Biochem J* 111:53–58.
3. Ballou, D., Palmer, G., and Massey, V. (1969) *Biochem Biophys Res Commun* 36:898–904.
4. Olson, J.S., Ballou, D.P., Palmer, G., and Massey, V. (1974) *J Biol Chem* 249:4350–4362.

5. Massey, V., Brumby, P.E., Komai, H., and Palmer, G. (1969) *J Biol Chem* 244:6049–6055.

6. LaFara, R.L. (1973) *Computer Methods for Science and Engineering.* Rochelle, N.J.: Haydon Book Company, Inc., pp. 254–256.

7. Olson, J.S., Ballou, D.P., Palmer, G. and Massey, V. (1974) *J Biol Chem* 249:4363–4382.

Published 1982 by Elsevier North Holland, Inc.
Vincent Massey and Charles H. Williams, Editors
Flavins and Flavoproteins

CHAPTER 132

On the Nature of the Cyanolysable Sulfur in Xanthine Oxidase

Takeshi Nishino, Tomoko Nishino, and Keizo Tshushima

Department of Biochemistry, Yokohama City University School of Medicine, Yokohama, Japan

The molybdenum-containing hydroxylases, xanthine oxidase (1), xanthine dehydrogenase (2,3), and aldehyde oxidase (4) can be converted into an inactive form known as desulfo-enzyme on treatment with cyanide. The desulfo-enzyme is also known to be present more or less in these purified enzyme preparations as a degradation product during purification or storage (5). But in chicken liver, some amount of the desulfo-enzyme seems to be present in situ. The purified enzyme from the liver of chickens fed a high-protein diet had a higher specific activity (6). As shown in Figure 1, the activity-flavin ratio of xanthine dehydrogenase in the supernatant solution of liver homogenate seemed to be changed during the dietary adaptation, while molybdenum-flavin ratio was almost constant (data not shown). The value of activity-flavin ratio of the enzyme obtained on the 3rd day of adaptation to a high-protein diet was consistent with that of the enzyme purified from the livers of chickens adapted to a high-protein diet (6). However, the physiological significance of the presence of desulfo form in the liver is not clear.

Despite a great interest in the mechanistic role in enzyme action and physiological role of this labile sulfur in question, the nature of this sulfur is still hypothetical (1,8,9).

From experiments with affinity chromatography using folate, a competitive inhibitor of xanthine oxidase, as a ligand, it was known that hypoxanthine and folate could bind to the desulfo-enzyme, suggesting that the sulfur atom may not be involved in the formation of the enzyme-substrate complex. A milk xanthine oxidase preparation having the value of AFR; 130 and PFR; 11 was applied on the affinity column. Almost all the material having absorbance at 450 nm, was adsorbed on this column. After washing the impurity with the buffer, the enzyme was eluted with buffer containing hypoxanthine as shown in Figure 2. $AFR^{25°C}$ and PFR values of the eluted enzyme were 126 and 5.2 respectively, suggesting that the purified enzyme is almost pure but the content of desulfo-form is not changed by this affinity chromatography. As it has been known that only the active enzyme in a reduced form can bind pyrazolo[3,4-d]pyrimidine derivatives (10,11), we adopted allopurinol or oxipurinol as a counter reagent of this affinity column to elute the active enzyme exclusively. After treatment with oxipurinol of the enzyme that was eluted from the first affinity column and was reduced with xanthine, the

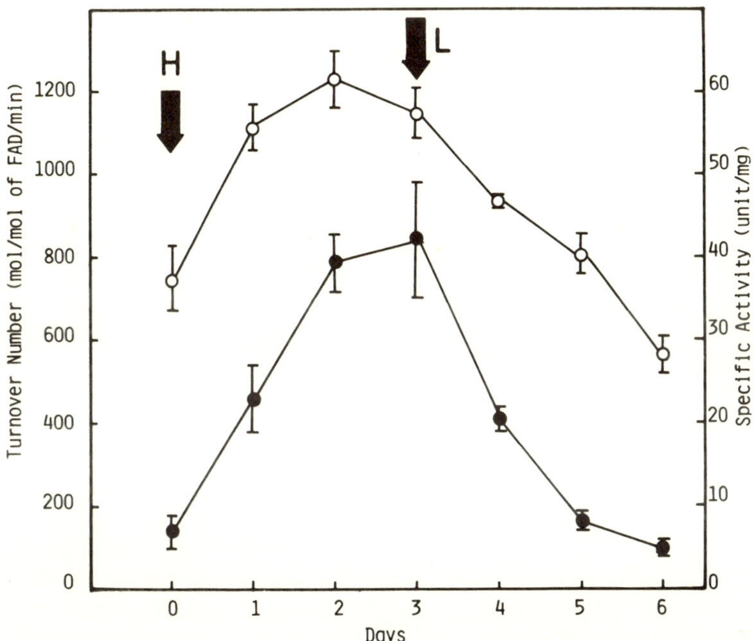

Figure 1. Changes in the activity-flavin ratio and the specific activity in the chicken liver extract during dietary adaptation. The diet was changed from a 20% to a 75% casein diet on the day indicated by arrow H. The casein content in the diet was changed from a 75% to a 5% on the day indicated by arrow L. The activity of xanthine dehydrogenase in the supernatant solution from liver extract was determined. An excess of the specific rabbit antiserum which was prepared against chicken liver xanthine dehydrogenase was added to the liver supernatant solution, and the flavin content in the immunoprecipitate was determined by the method of Yagi (7). Each point represents the mean ± SD of 5 observations.

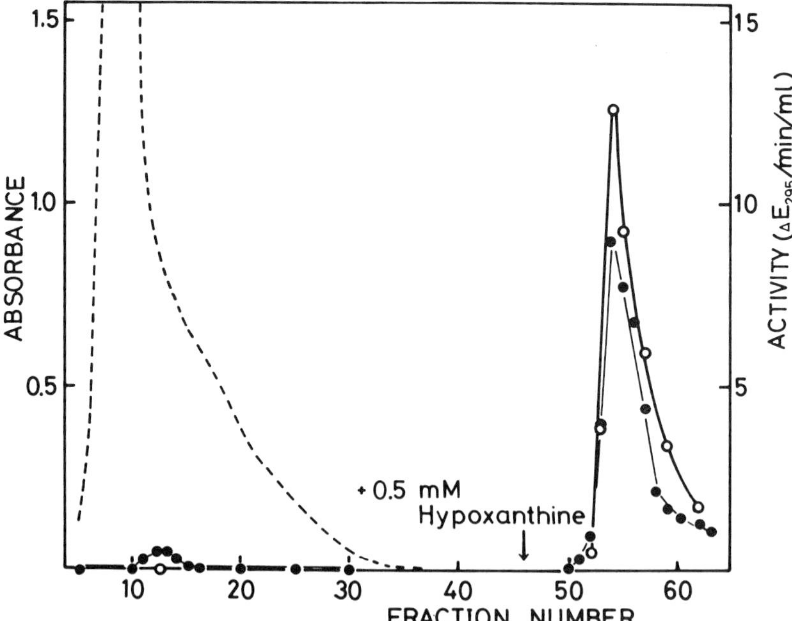

Figure 2. Purification of milk xanthine oxidase by affinity chromatography on Sepharose 4B/folate. After washing the impurities with the buffer, the enzyme was eluted with the buffer mixture containing 0.5 mM hypoxanthine. (---) absorbance at 280 nm; (●--●) absorbance at 450 nm; and (○--○) enzyme activity.

enzyme was passed through G-25 in order to remove excess oxipurinol and xanthine. Then the oxipurinol-inactivated enzyme was applied again on the same affinity column. As shown in Figure 3, only the sulfo-enzyme which bound oxipurinol passed through the column; on the other hand the desulfo-enzyme remained on the column and was eluted also with the buffer containing hypoxanthine. Highly active enzyme ($AFR^{25°C}$;205–190) was obtained after reactivation of the oxipurinol-bound enzyme by treatment with ferricyanide (11).

Effective reactivation of desulfo-enzyme by the system consisting of rhodanese, ^{35}S-thiosulfate and dithiothreitol was accompanied by reincorporation of more than one sulfur atom into apoenzyme. This reactivation might presumably be caused by the incorporation into apoenzyme of ^{35}S-sulfide produced in the system as above (12). DTT could be replaced by cysteine, GSH, and dihydrolipoate, but with less activity. After treatment of ^{35}S-containing enzyme with cyanide in the presence of allopurinol, which might release only ^{35}S other than the molybdenum-linked one, no trace amount of ^{35}S-cysteine could be detected by amino acid analysis in the acid hydrolysate of the enzyme.

The results presented here support the proposition that the sulfur is a terminal sulfur ligand of the molybdenum atom (Mo=S) (9,13).

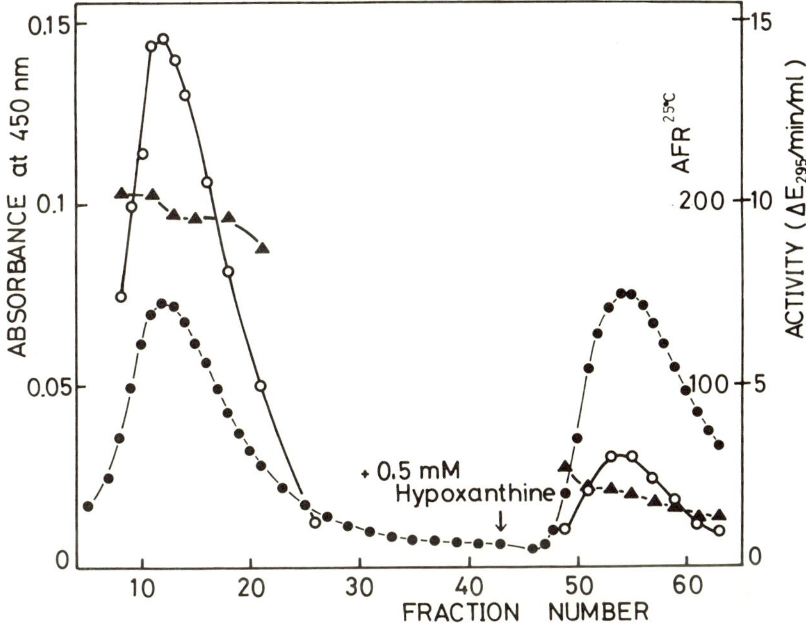

Figure 3. Resolution of active and inactive xanthine oxidase by the second affinity chromatography of Sepharose 4B/folate. The oxipurinol-treated enzyme was rechromatographed on the same affinity column as described in the text. The enzyme activity (○--○) was measured after treatment with ferricyanide. (●--●) absorbance at 450 nm and (▲--▲) AFR value at 25°C.

ACKNOWLEDGMENTS
This research was supported in part by a Grant-in-Aid for Scientific Research from the Japanese Ministry of Education, Science, and Culture (No. 17709 to T.N. and No. 458081 to K.T.)

References

1. Massey, V. and Edmondson, D. (1970) *J Biol Chem* 245:6595–6598.
2. Cleere, W.F. and Coughlan, M.P. (1974) *Biochem J* 143:331–340.
3. Nishino, T., Itoh, R., and Tshushima, K. (1975) *Biochim Biophys Acta* 403:17–22.
4. Branzoli, V. and Massey, V. (1974) *J Biol Chem* 249:4346–4349.
5. Bray, R.C. (1975) *The Enzymes* 3rd Ed. 12:299–419.
6. Nishino, T. (1974) *Biochim Biophys Acta* 341:93–98.
7. Yagi, K. (1971) *Methods in Enzymology* Vol. 18 part B:290–296.
8. Coughlan, M.P. (1977) *FEBS Letters* 81:1–6.
9. Gutteridge, S., Tanner, S.J., and Bray, R.C. (1978) *Biochem J* 175:887–897.
10. Massey, V., Komai, H., Palmer, G., and Elion, G.B. (1970) *J Biol Chem* 245:2837–2844.
11. Edmondson, D., Massey, V., Palmer, G., Beacham, L.M., III, and Elion, G.B. (1972) *J Biol Chem* 247:1597–1604.
12. Sörbo, B. (1975) *Metabolic Pathways* 3rd Ed. 7:433–456.
13. Malthouse, J.P.G. and Bray, R.C. (1980) *Biochem J* 191:265–267.

Published 1982 by Elsevier North Holland, Inc.
Vincent Massey and Charles H. Williams, Editors
Flavins and Flavoproteins

CHAPTER 133

Proton and Electron Affinities and Magnetic Interactions Associated With the Molybdenum, Flavin, and Iron-Sulfur Centers of Milk Xanthine Oxidase

Michael J. Barber and Lewis M. Siegel

Department of Biochemistry, Duke University School of Medicine, and the Veterans Administration Hospital, Durham, North Carolina

Bovine milk xanthine oxidase (XO) is a dimer of identical subunits, each of which contains 1 FAD, $2Fe_2S_2$ centers (termed Fe/S I and II), and a Mo center with a novel pterin cofactor (1,2). Stiefel (3) proposed that Mo functions as a simultaneous H^+ and e^- acceptor in XO, facilitating exchange of $-$OH for substrate-derived -H. Gutteridge et al. (4) demonstrated H-transfer from 1-methyl-xanthine to Mo^V, thus supporting Stiefel's hypothesis. Olson et al. (5,6) proposed that xanthine donates e^- to the Mo and O_2 accepts e^- from reduced FAD, with e^- transfer between these groups being so rapid that the e^- distribution among the centers during catalysis should be governed solely by the relative reduction potentials of the various centers. We report here thermodynamic data which may aid in the testing of these hypotheses concerning XO function, as well as data which may shed light on possible pathways of e^- transfer within XO.

Variation of XO Reduction Potentials with pH

Potentiometric titrations of XO were performed in the pH range 5.7–10.9. Enzyme and mediator dyes were poised at desired potentials, samples removed anaerobically and frozen in liquid N_2 for EPR analysis of the paramagnetic species Mo^V, FAD(H)$\cdot$, Fe/S I_{red}, and Fe/S II_{red}. E_m values were derived from these data (7). Figure 1 shows the variation of these observed E_m with pH. Figure 1A and B indicate the behavior of two species of Mo^V EPR signal (1), "Rapid" arising from the functional Mo center of native XO, and "Slow" arising from the nonfunctional Mo center of "desulfo" XO (in which a terminal S ligated to the functional Mo center has been replaced by an oxygen (8)). It is apparent that H^+ addition accompanies e^- addition to each of the prosthetic groups of XO. The complex curves indicate that there are H^+ ionizations associated with oxidized and/or reduced states of each center. These curves were fit to the data using the reactions and thermodynamic parameters shown in Figure 2.

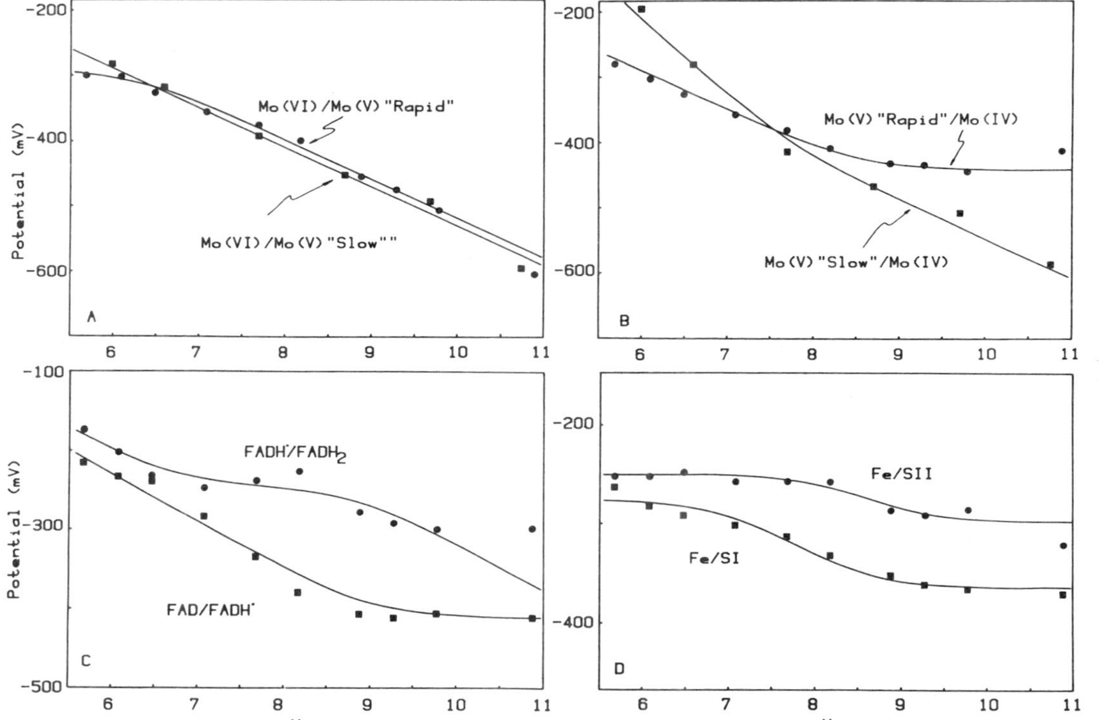

Figure 1. Variation of the potentials for XO prosthetic groups with pH. Buffers used (all at 50 mM and containing 1 mM EDTA): pH 5.7–6.1, Mes; 6.5–7.1, Pipes; 7.7 and 8.7, Bicine; 8.2, Tris-HCl; 8.9–9.8, Ches, 10.7 and 10.9, Caps. Theoretical curves follow equations based on the reactions and parameters of Figure 2. (A) E_m for Mo^{vi}/Mo^{v} "Rapid" in native XO (●) and Mo^{vi}/Mo^{v} "Slow" in desulfo XO (■). (B) E_m for Mo^{v} "Rapid"/Mo^{iv} in native XO (●) and Mo^{v} "Slow"/Mo^{iv} in desulfo XO (■). (C) E_m for FAD/FAD(H)$^{\cdot}$ (■) and FAD(H)$^{\cdot}$/FAD$(H)_2$(●). The terms FADH and $FADH_2$ used in the figure represent relative oxidation states and are not meant to indicate protonation states. (D) E_m for Fe/S I (■) and Fe/S II (●).

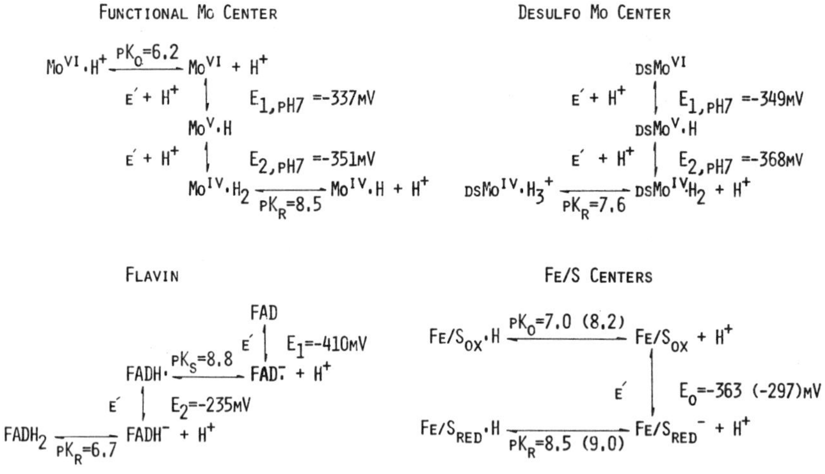

FUNCTIONAL Mo CENTER

$Mo^{VI}\cdot H^+ \xrightarrow{pK_O=6.2} Mo^{VI} + H^+$

$E' + H^+ \downarrow \quad E_{1,pH7} = -337mV$

$Mo^{V}\cdot H$

$E' + H^+ \downarrow \quad E_{2,pH7} = -351mV$

$Mo^{IV}\cdot H_2 \xrightarrow[pK_R=8.5]{} Mo^{IV}\cdot H + H^+$

DESULFO Mo CENTER

$DSMo^{VI}$

$E' + H^+ \downarrow \quad E_{1,pH7} = -349mV$

$DSMo^{V}\cdot H$

$E' + H^+ \downarrow \quad E_{2,pH7} = -368mV$

$DSMo^{IV}\cdot H_3^+ \xrightarrow[pK_R=7.6]{} DSMo^{IV}\cdot H_2 + H^+$

FLAVIN

FAD

$E' \downarrow \quad E_1 = -410mV$

$FADH\cdot \xrightarrow{pK_S=8.8} FAD^{\cdot-} + H^+$

$E' \downarrow \quad E_2 = -235mV$

$FADH_2 \xrightarrow[pK_R=6.7]{} FADH^- + H^+$

FE/S CENTERS

$Fe/S_{OX}\cdot H \xrightarrow{pK_O=7.0\ (8.2)} Fe/S_{OX} + H^+$

$E' \downarrow \quad E_0 = -363\ (-297)mV$

$Fe/S_{RED}\cdot H \xrightarrow[pK_R=8.5\ (9.0)]{} Fe/S_{RED}^- + H^+$

Figure 2. Thermodynamic parameters for reduction and protonation of XO centers. Parameters for Fe/S II are given in parentheses next to those for Fe/S I.

A number of conclusions can be derived from these data:

1. The potentials of the native and desulfo Mo are nearly identical when one considers reductions involving comparable changes in protonation state, i.e., substitution of terminal O for S on the Mo does not change its e^- affinity.
2. The two types of Mo center have markedly differing affinities for H^+, particularly in the Mo^{iv} state. Since the Mo^{v} state exists as $Mo^{v}\cdot H$ in both enzyme forms between pH 5.7 and 10.9, it is the differential protonation of the Mo^{iv} state which causes the observed Mo^{v}/Mo^{iv} reduction potentials to differ so markedly in native and desulfo-enzymes at acid and alkaline pH values.
3. The FAD/FAD(H)$\cdot$ and FAD(H)$\cdot$/FAD(H)$_2$ potentials are minimally separated when the hydroquinone is in its protonated form, governed by pK=6.7. Under such conditions, approximately 20% of the enzyme flavin can be converted to semiquinone. There is a pK = 8.8 for conversion of neutral to anionic semiquinone; this pK was confirmed by the observation that the linewidth of the FAD radical EPR signal decreased from 19 to 14 Gauss as the pH was raised from 5.7 to 10.9.
4. The relative order of e^- addition to the XO prosthetic groups should vary with pH due to the different pH-dependence of the FAD and Fe/S centers. At pH 6, the order should be FAD, FeS/II, Fe/S I; at pH 10, the order should be Fe/S II, FAD, Fe/S I. This variation could be used to advantage in experimental tests of the Olson et al. (5,6) model of e^- transfer within XO.

Coulometric Determination of XO Reduction Potentials at 25°

If XO is equilibrated with a given solution potential at 25°, is it valid to determine the degree of reduction (and hence the midpoint potentials) of XO centers (by EPR) in frozen samples? Palmer and Olson (6) have pointed out that the entropies of reduction of the various XO centers need not be identical ($dE_m/dT = \Delta S/nF$) and that redistribution of e^- within an XO molecule might well continue at temperatures below the freezing point. Williams-Smith et al. (9) showed that solution pH changes on freezing in several buffers and one might anticipate e^- redistribution when potentials are pH-dependent. In the zwitterionic buffers used in this work, however, pH shifts on freezing are minimal (9).

It is therefore important to determine potentials for the XO centers in experiments conducted entirely at one temperature. To achieve this, we performed coulometric titrations (10) of native XO and two of its modified forms in collaboration with J.T. Spence of Utah State University. Defined amounts of XO were added to a fixed volume of solution containing an excess of mediator dyes, the solution being maintained at a fixed potential with a potentiostat, and the number of e^- taken up measured. Figure 3 shows results of a titration at pH 7.7. Curve A shows that native XO (Mo centers 85% functional) required 5.7 e^- for full reduction. Curve B shows that 4 e^- were required for reduction of desulfo XO (<0.5% functional Mo). Curve C shows that 2 e^- were required for reduction of desulfo XO with its FAD rendered nonreducible by alkylation (11). Clearly, desulfo Mo is not reduced in these experiments during the time required for e^- uptake by functional Mo or the other centers. Thus, the 2 Fe/S centers (the only groups capable of electron uptake in sample C) take up 2 e^-, the FAD (reducible only in samples A and B) takes up 2, and the functional Mo (reducible only in sample A) takes up 2. The set of E_m which achieved the best simultaneous fit to all three curves were (values obtained by EPR analysis of frozen samples in parentheses): Fe/S centers, −230 and −320 (II: −255, I: −310) mV; $FAD/FADH_2$, −280 (−283) mV; Mo^{vi}/Mo^{v}, −375 (−373) mV; Mo^{v}/Mo^{iv}, −405 (−377) mV. At pH 8.9, the following E_m were obtained by coulometry (EPR): Fe/S centers, −255 and −355 (II: −285, I: −350) mV; $FAD/FADH_2$, −330 (−340) mV; Mo^{vi}/Mo^{v}, −450 (−453) mV; Mo^{v}/Mo^{iv}, −450 (−427) mV. Thus there is good agreement between potentials obtained by the two methods for Mo^{vi}/Mo^{v}, $FAD/FADH_2$, and one of the Fe/S center potentials, which corresponds to that of Fe/S I in the EPR analysis. The Mo^{v}/Mo^{iv} and Fe/S $II_{ox/red}$ potentials are shifted by about ±30 mV in frozen samples. Thus it seems probable that there are significant differences in the entropies for reduction of the different carriers in XO. Comparison of the coulometric and EPR-derived potentials at the two pH values studied shows, however, that the variation of each potential with pH over this range is identical within experimental error by the two methods. Thus, the relative protonation behavior of the oxidized and reduced XO centers determined by the EPR analysis method is probably valid. However, it is probably not valid to combine results obtained by EPR analysis of some centers at cryogenic temperatures with optical measurements of other centers at much higher temperatures to determine e^- distributions within the enzyme during titrations and during turnover (e.g., rapid freezing vs stopped flow). The Olson et al. (5,6) model of e^- transfer within XO, which is based, in part, on experimental data obtained in such a fashion, may therefore be in need of some reexamination.

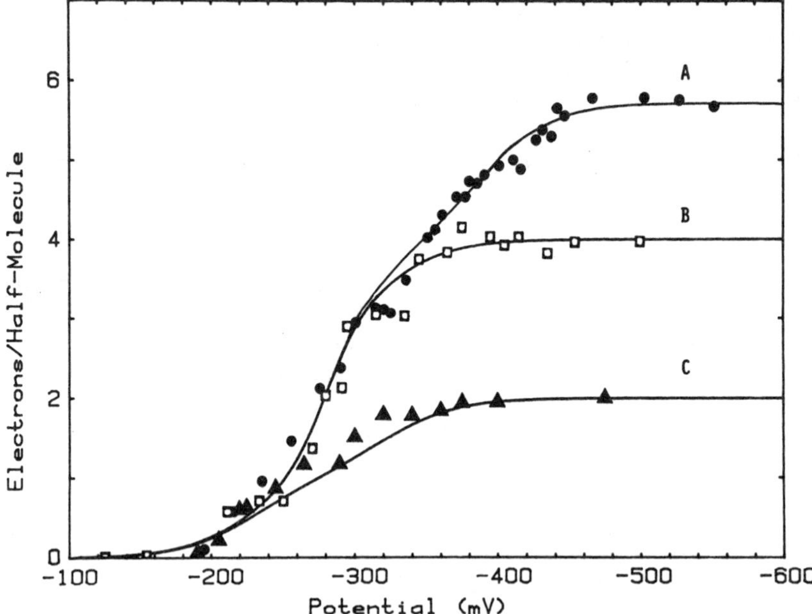

Figure 3. Coulometric titrations of native and modified form of XO in 50 mM Bicine-1 mM EDTA, pH 7.7. Each point represents the electron uptake following addition of the following amounts of XO subunits to the titration vessel: Curve A: native XO, 2.72 nmol; Curve B: desulfo XO, 2.08 nmol; Curve C: desulfo XO with alkylated FAD, 1.10 nmol. E_m values used for curve fitting are in the text.

Magnetic Interactions Between Centers in XO

We now turn to a series of experiments in which the variation of the reduction behavior of the various XO centers with pH is used in purely technical fashion. One can choose experimental conditions (by poising enzyme at appropriate potentials and pH values) so that specific pairs of centers are rendered paramagnetic. If such centers are physically close to one another they can exhibit spin-spin coupling which may be manifested in EPR analysis by: (a) splitting of the spectrum of one species by the presence of the other, and/or (b) an increase in the relaxation rate of the normally more slowly relaxing species in the presence of the more rapidly relaxing species. The latter effect is manifested as an increase in the microwave power saturation parameter $P_{1/2}$ defined by the semi-empirical equation of Beinert and Orme-Johnson (12):

$$S = k\sqrt{P}/\left[1 + P/P_{1/2}\right]^{0.5b} \tag{133.1}$$

where S is the signal amplitude, P is the incident microwave power, and b is a parameter which can vary from 1.0 for inhomogeneously-broadened to 4.0 for homogeneously-broadened lines. In plots of log $(S/\sqrt{P})$ vs log P, the curves tend to a slope of 0.5b under conditions of saturation.

Lowe and Bray (13) have shown the presence of mutually-induced splittings in the EPR spectra of Mo^{V} and Fe/S I_{red}, a result which was analyzed by Coffman and Buettner (14) to indicate a Mo--Fe/S I distance of 8–14 A. We have performed similar studies and confirmed the results of Lowe and Bray using the stable Mo^{V} species in the desulfo glycol inhibited form of XO. We also observed a 100-fold increase in the $P_{1/2}$ of the Mo^{V} EPR signal at 103K when Fe/S I was reduced vs oxidized. We could detect no effect of Fe/S II reduction on either splitting of the Mo^{V} spectrum at 30K or on the $P_{1/2}$ of the Mo^{V} signal at 103K. Since reduced Fe/S II is a more rapidly relaxing species than Fe/S I, we can consider absence of Mo^{V}-Fe/S II_{red} magnetic interaction to be significant. We therefore conclude that these centers are relatively far apart in XO.

By poising native XO at a series of potentials at pH 6.1, we were able to prepare samples containing both amounts of $FADH^{\cdot}$ adequate for microwave power saturation studies and with differing proportions of reduced Fe/S I and Fe/S II. Figure 4 shows the saturation behavior of the $FADH^{\cdot}$ EPR signal in such a series of XO samples. Curve A was obtained at a potential of −142 mV; this sample contained no reduced Fe/S I and no more than 5% reduced Fe/S II. The saturation behavior of this sample was fit by Equation (133.1) with $b=1.0$ and $P_{1/2}=0.25$ mW. Curve G was obtained at a potential of −327 mV; this sample contained 98% of Fe/S II and 85% of Fe/S I in the reduced form. The saturation behavior of this sample was fitted with $b=1.0$ and $P_{1/2}=18$ mW. Curves B–F, obtained at intermediate potentials, showed $FADH^{\cdot}$ saturating as a mixture of two components, one with $P_{1/2}=0.25$ mW, and the other ("relaxed") with $P_{1/2}=18$ mW. Figure 5A shows the fraction of $FADH^{\cdot}$ exhibiting the relaxed behavior of each of the potentials measured between −142 and −269 mV, together with the fractions of Fe/S I and Fe/S II found to be in the reduced state at each potential. It can be seen that, at most potentials, more of the $FADH^{\cdot}$ is relaxed than can be accounted for by the presence of either paramagnetic Fe/S center. However, if one calculates the fraction of XO subunits in which either (or both) Fe/S center is reduced at a given potential, Figure 5A shows a good correlation between this parameter and the amount of relaxed $FADH^{\cdot}$. In a second type of experiment, XO solutions at a number of pH values were poised at potentials which would lead to maximum formation of $FADH^{\cdot}$, and the power saturation properties of these samples examined. Again, all curves could be fit by a mixture of two components with $P_{1/2}$ of either 0.25 or 18 mW. Figure 5B shows that the amount of relaxed $FADH^{\cdot}$ at any of the pH values correlated well with the proportion of subunits in which either (or both) Fe/S center was reduced. These results indicate that $FADH^{\cdot}$ interacts magnetically with both reduced Fe/S I and Fe/S II. Analysis of the relaxation data (performed in collaboration with J.C. Salerno), together with information on relaxation properties of the Fe/S I centers themselves obtained from line-broadening studies as a function of temperature, has yielded approximate distances between FAD and either Fe/S center of 12–20 A. No magnetic interaction between $FADH^{\cdot}$ and Mo^{V} (studied in desulfo glycol inhibited XO) could be detected. No splittings of $FADH^{\cdot}$ spectra were seen in any of the experiments of this work.

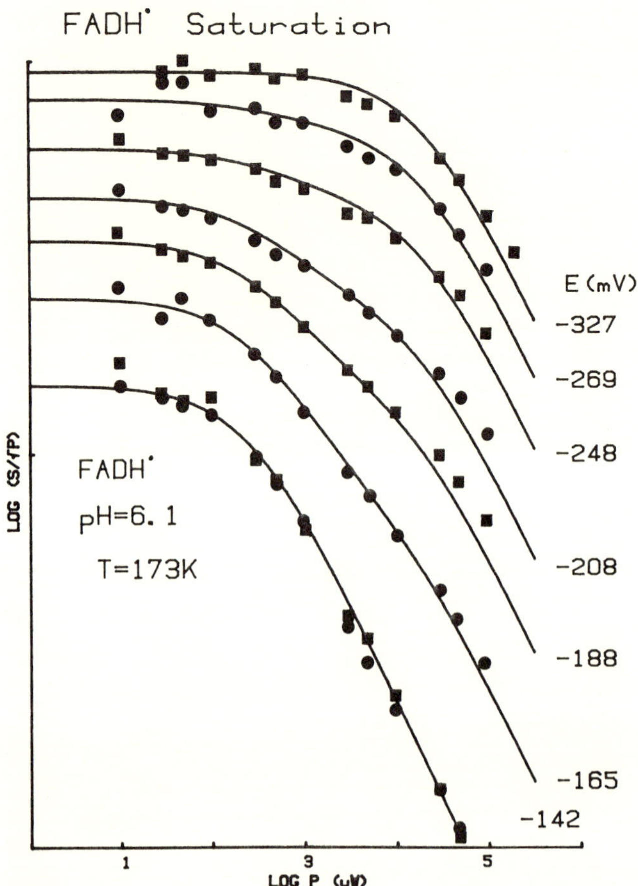

Figure 4. Microwave power saturation of the FADH˙ EPR signal in XO poised at different potentials, in 50 mM Mes−1 mM EDTA, pH 6.1. EPR spectra taken at 173K. Points were fit using 1 or 2 relaxing components, with $P_{1/2}$ = 0.25 or 18 mW.

Finally, studies were conducted on the relaxation behavior of Fe/S I_{red} in the presence and absence of Fe/S II_{red}. XO was prepared with 5% of the Fe/S I and 8% of the Fe/S II in the reduced state by poising the enzyme at −170 mV at pH 5.7, a pH at which the two Fe/S E_m are nearly identical. The saturation behavior of the sample at 20K was fit by Equation (133.1) with b = 2.0 and $P_{1/2}$ = 20 mW. A second XO sample was poised at −370 mV at pH 5.7; both Fe/S centers were over 95% reduced. Saturation data for this sample yielded b = 2.0 and $P_{1/2}$ = 50 mW. Thus there is significant relaxation of Fe/S I when Fe/S II is paramagnetic. No splitting of the EPR spectrum of either Fe/S center was detected in any of the experiments of this work. Although angular factors come into particular importance when analyzing couplings of the relatively anisotropic Fe/S centers, it is possible to estimate

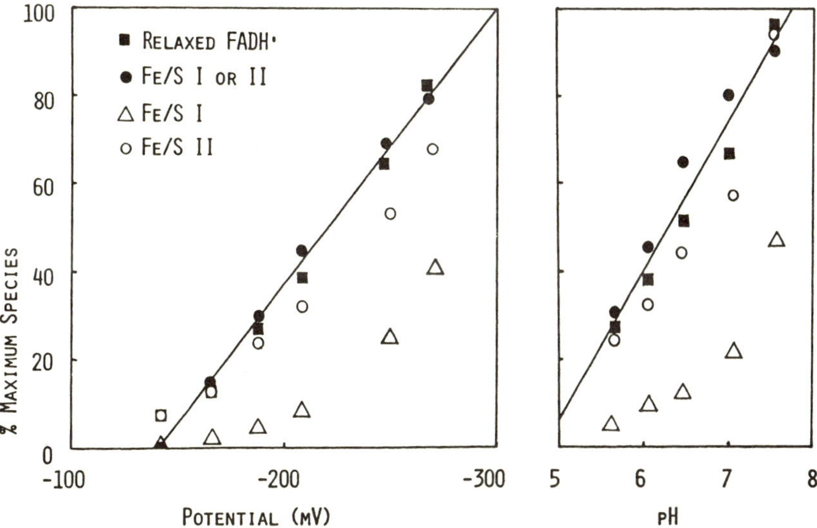

Figure 5. Correlation of fraction of FADH˙ in the "relaxed" state with the fraction of Fe/S centers reduced. (■) Fraction of FADH˙ EPR signal saturating with $P_{1/2} = 18$ mW. ((△) Fraction Fe/S I reduced (F_I). (○) Fraction Fe/S II reduced (F_{II}). (●) Fraction of XO subunits which contain either reduced Fe/S I, reduced Fe/S II, or have both centers reduced ($F_I + F_{II} - F_I F_{II}$).

the distance between these centers to be approximately 11–18 A. The distance estimates obtained from these studies suggest a spatial arrangement of prosthetic groups in XO subunits like that shown in Figure 6. The results indicate that the Mo, Fe/S I, and Fe/S II centers should be placed in a roughly linear array, while the FAD is placed in a position to interact with the Fe/S centers but not with the Mo center.

Figure 6. Distances between XO centers from magnetic interaction results.

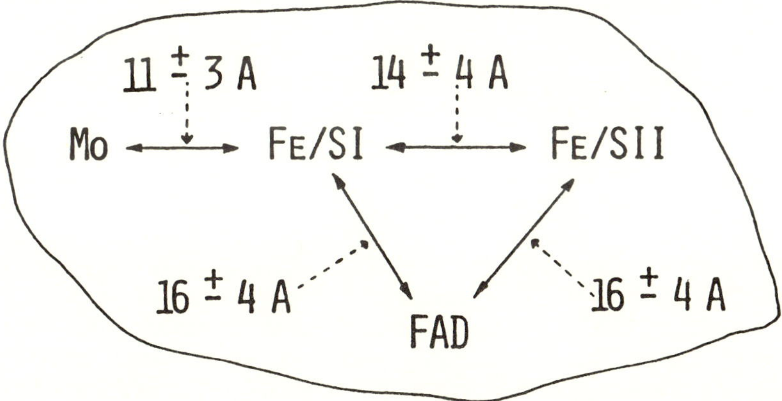

ACKNOWLEDGMENTS
This work was supported by Grant AM-13460 from the National Institutes of Health and Project Grant 7875-01 from the Veterans Administration.

References

1. Bray, R.C. (1980) *Adv Enzymol* 51:107–165.
2. Johnson, J.L., Hainline, B.E., and Rajagopalan, K.V. (1980) *J Biol Chem* 255:1783–1786.
3. Stiefel, E.I. (1973) *Proc Nat Acad Sci USA* 70:988–992.
4. Gutteridge, S., Tanner, S.J., and Bray, R.C. (1978) *Biochem J* 175:869–886.
5. Olson, J.S., Ballou, D.P., Palmer, G., and Massey, V. (1974) *J Biol Chem* 249:4350–4362.
6. Palmer, G. and Olson, J.S. (1980) In *Molybdenum and Molybdenum-Containing Enzymes*. M.P. Coughlan (ed.) Oxford: Pergamon Press, pp. 189–220.
7. Barber, M.J., and Salerno, J.C. (1980) In *Molybdenum and Molybdenum-Containing Enzymes*. M.P. Coughlan (ed.) Oxford: Pergamon Press, pp. 543–568.
8. Bordas, J., Bray, R.C., Garner, C.D., Gutteridge, S., and Hasnain, S.S. (1980) *Biochem J* 191:499–508.
9. Williams-Smith, D.L., Bray, R.C., Barber, M.J., Tsopanakis, A.S., and Vincent, S.P. (1977) *Biochem J* 167:593–600.
10. Watt, G.D. (1979) *Anal Biochem* 99:399–407.
11. McGartoll, M.A., Pick, F.N., Swann, J.C., and Bray, R.C. (1970) *Biochem Biophys Acta* 212:523–526.
12. Beinert, H. and Orme-Johnson, W.H. (1967) In *Magnetic Resonance in Biological Systems*. A. Ehrenberg, B.G. Malmstrom, and T. Vangaard (eds.) Oxford: Pergamon Press, pp. 221–247.
13. Lowe, D.J. and Bray, R.C. (1978) *Biochem J* 169:471–479.
14. Coffman, R.E. and Buettner, G.R. (1979) *J Chem Phys* 83:2392–2400.

Published 1982 by Elsevier North Holland, Inc.
Vincent Massey and Charles H. Williams, Editors
Flavins and Flavoproteins

CHAPTER 134

Rabbit Liver Aldehyde Oxidase: Reduction Potentials and Spectroscopic Properties

Michael J. Barber, Michael P. Coughlan,
K.V. Rajagopalan, and Lewis M. Siegel

Department of Biochemistry, Duke University School of Medicine and Veterans Administration Hospital, Durham, North Carolina and Department of Biochemistry, University College, Galway, Ireland

Although milk xanthine oxidase (XO) and avian liver xanthine dehydrogenase (XDH) have been extensively studied (1), rabbit liver aldehyde oxidase (AO), with a similar molecular structure (M=300,000; Mo, FAD, Fe-S^{2-} in the ratio 1 : 1 : 4) but differing specificity, has been largely neglected since the early work of Rajagopalan et al. (2). To facilitate a comprehensive comparison of these three Mo hydroxylases, we have examined a number of spectroscopic and potentiometric properties of AO. All studies were performed in 50 mM KP_i, 0.1 mM EDTA, pH 7.8.

Iron-Sulfur Centers

EPR spectra of fully reduced AO at 20 K show two distinct types of Fe/S. One (Figure 1A) is of near axial symmetry with $g_{av} = 1.96$, relatively easily saturated with power at 20 K ($P_{1/2} = 17$ mW), and is observable at 80 K. These properties are nearly identical to those of Fe/S I in XO and XDH (4). The other (Figure 1B) is rhombic with $g_{av} = 2.01$, resistant to saturation at 20 K ($P_{1/2} = 300$ mW), and is only observable at T<40 K. These properties are nearly identical to those of Fe/S II in XO and XDH (3). No magnetic interaction between the Fe/S centers was detected in AO, unlike the results found with XO (4).

Potentiometric titration of these centers (Figure 2) yielded $E_m = -207$ mV for Fe/S I and −310 mV for Fe/S II. The AO E_m for Fe/S II is similar to that in XO (−303 mV) and XDH (−275 mV) in P_i or PP_i buffers at pH 8 (5,6). The E_m for Fe/S I is much more positive in AO than in XO (−343 mV) or XDH (−280 mV).

Changes in optical spectra of AO seen during the course of reductive titrations could be fit rather well (at 450 and 550 nm) by extinction coefficients determined by Olson et al. (7) for the Fe/S and FAD centers of XO, except that the coefficients assigned to Fe/S I and Fe/S II, respectively, in XO, had to be assigned to Fe/S II and Fe/S I, respectively in AO. This result suggests a marked change in structure and/or environment of one or both Fe/S centers in AO as compared to XO.

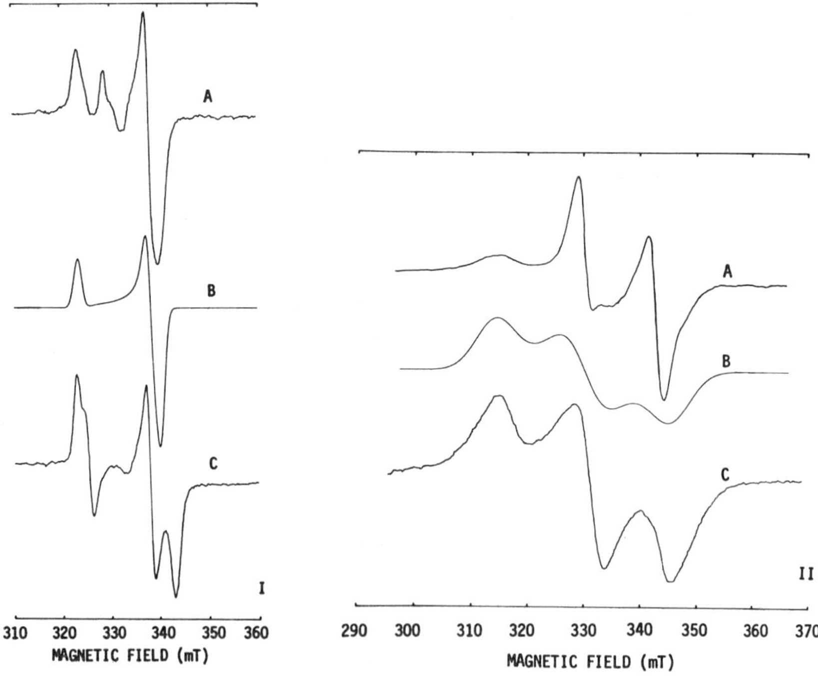

Figure 1. EPR spectra of reduced Fe/S centers in AO and XO. (A) Spectra of Fe/S I. A: AO poised at −500 mV in 50 mM KP_i, pH 7.8; B: computer simulation of spectrum A, using g=2.018, 1.930, 1.918.; C; native XO poised at −500 mV in 50 mM Bicine, pH 7.7. Spectra were recorded at 20 K and 1 mW power. (B) Spectra of Fe/S II. Same conditions as in A, except that the computer simulation used g=2.106, 2.003, 1.915, and spectra were recorded at 20 K and 400 mW power.

FAD

EPR signals at g=2.0036 with the 1.9 mT linewidth of neutral flavin radicals (8) were observed on reduction of AO with substrates, $S_2O_4^{2-}$, and during potentiometric titrations. In the titration of Figure 2, a maximum of 17% of the FAD was converted to FADH˙. E_m values were: FAD/FADH˙, −258 mV; FADH˙/$FADH_2$, −212 mV. The first E_m is 100 mV more positive than that for either XO or XDH at similar pH, while the second E_m is 160 mV more positive than that for XDH, but similar to that for XO. FAD is thus more easily reduced in AO than in XO or XDH.

Although we have not been able as yet to prepare AO samples with FADH˙ in the absence of reduced Fe/S I, we have obtained some evidence for magnetic interaction between FAD and one or both Fe/S centers in AO, in that $P_{1/2}$ for saturation of the FADH˙ EPR signal at 173 K increased in the following manner with the percent Fe/S I and Fe/S II reduced, respectively (in samples poised at various potentials): 0.6 mW—48%, 0%; 0.8 mW—72%, 0%; 1.0 mW—86%, 20%; 1.7 mW—100%, 52%. These relaxation effects are much weaker in AO than in XO, however.

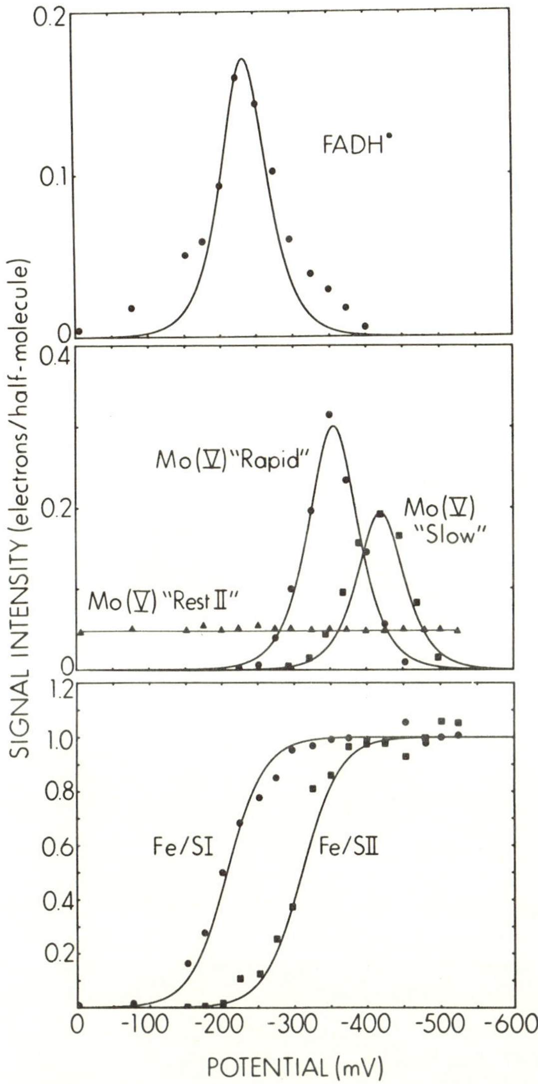

Figure 2. Behavior of FAD, Mo, and Fe/S during potentiometric titration of AO in 50 mM KP_i, pH 7.8. Concentrations of paramagnetic species were measured by EPR analysis of samples removed from the titration vessel and quickly frozen in liquid N_2 (6). Theoretical Nernst curves were derived using E_m values given in the text.

Molybdenum

Figure 3 shows a number of Mo^V EPR signals which have been observed after various treatments of AO; these signals are generally similar to those observed in XO after analogous treatment. The "resting II" (3) signal (Figure 3A) is seen in untreated AO; in the preparations used, it corresponded to 5% of the

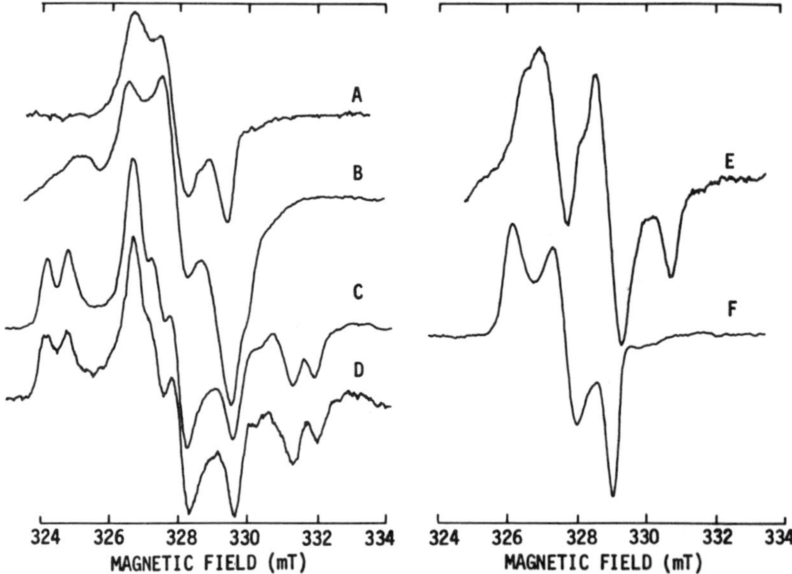

Figure 3. Mo^{V} EPR spectra from functional and nonfunctional forms of AO. (A) "Resting II" signal from oxidized native AO (g = 1.979, 1.972, 1.963). (B) Signal from native AO reduced anaerobically with 10 mM N-methyl nicotinamide for 10 ms and then incubated at −50° for 10 min. (C) AO reacted with 0.15 M formaldehyde for 20 min (g = 1.994, 1.979, 1.952; A = 0.60, 0.60, 0.62). (D) AO reacted with 2 mM benzaldehyde for 15 min in the presence of 0.75 M methanol (same g and A values as in C). (E) Cyanide-treated AO reduced with dithionite (20 mol/mol enzyme) (g = 1.971, 1.967, 1.955; A = 1.55, 1.43, 1.48). (F) Same as E, but reacted for 72 hr in the presence of 50% (v/v) ethylene glycol (g = 1.980, 1.971, 1.963).

enzyme Mo. No ^{1}H hyperfine splitting could be detected. A nearly identical signal of much greater intensity could be generated by treating AO with ethylene glycol (Figure 3F). Treatment of AO with methanol or formaldehyde results in production of nearly identical "inhibited" Mo^{V} signals (Figure 3C, D) with hyperfine splitting apparently due to a single ^{1}H. Reduction of CN^{-}-treated AO resulted in formation of a species (Figure 3E) similar to the "Slow" Mo^{V} of desulfo XO and XDH (3).

Only 30% of the AO flavin was associated with catalytically functional Mo centers in the enzyme preparation used. Given the presence of a constant background of "resting II" Mo^{V}, this has made observation of pure EPR spectra associated with the functional Mo^{V} (termed "Rapid" [3]) difficult. The spectrum shown in Figure 3B was produced by reaction of AO with N-methyl nicotinamide for 20 ms using the rapid freeze technique, followed by incubation of the frozen sample at −50° for 10 min, during which time the intensity of the Mo^{V} "Rapid" signal became maximal. (Further incubation resulted only in more "Slow" signal.)

Figure 2 shows the behavior of the Mo^{V} "Rapid" and "Slow" species on potentiometric titration of native and cyanide-treated AO. E_m for Mo^{VI}/Mo^{V}

and Mo^{v}/Mo^{iv}, respectively, were: −359 and −351 mV (native), and −439 and −401 mV (desulfo). These values are near to those reported for XO and XDH under similar conditions (5,6). Magnetic interaction between Mo^{v} ("resting II") and reduced Fe/S I was readily observed, both by splitting of the Mo^{v} spectrum at 30 K and a 100-fold increase in the $P_{1/2}$ for its saturation at 20 K when Fe/S I was reduced but not oxidized. Thus, as in XO and XDH (4,9), Mo and Fe/S I are close together in AO.

We can conclude that the Mo centers of AO, XO, and XDH are remarkably similar. Magnetic interactions involving Mo–Fe/S I and FAD–Fe/S are always present. However, significant differences exist in properties of the individual Fe/S centers and the magnetic interactions between them in AO and XO/XDH. The FAD e^- affinity varies considerably in the three enzymes. These differences may be of use in examining models of e^- transfer (8) within the Mo hydroxylases.

ACKNOWLEDGMENTS
This work was supported by Grants AM-13460 and GM-00091 from the National Institutes of Health and Project Grant 7875-01 from the Veterans Administration.

References

1. Coughlan, M.P. (1980) In *Molybdenum and Molybdenum-Containing Enzymes.* M.P. Coughlan (ed.) Oxford: Pergamon Press, pp. 221–239.
2. Rajagopalan, K.V., Handler, P., Palmer, G., and Beinert, H. (1968) *J Biol Chem* 243:3784–3796, 3797–3806.
3. Bray, R.C. (1975) *The Enzymes, 3d Ed.* 12:299–419.
4. Barber, M.J., Salerno, J.C., and Siegel, L.M. (1981) *Fed Proc* 40:1665.
5. Cammack, R., Barber, M.J., and Bray, R.C. (1976) *Biochem J* 157:469–478.
6. Barber, M.J., Coughlan, M.P., Kanda, M., and Rajagopalan, K.V. (1980) *Arch Biochem Biophys* 201:468–475.
7. Olson, J.S., Ballou, D.P., Palmer, G., and Massey, V. (1974) *J Biol Chem* 249:4350–4362.
8. Palmer, G., Muller, F., and Massey, V. (1971) In *Flavins and Flavoproteins.* H. Kamin (ed.) Baltimore: University Park Press, pp. 123–137.
9. Lowe, D.J., and Bray, R.C. (1978) *Biochem J* 169:471–479.

Published 1982 by Elsevier North Holland, Inc.
Vincent Massey and Charles H. Williams, Editors
Flavins and Flavoproteins

CHAPTER 135

Room Temperature Potentiometric Measurements on Xanthine Oxidase

Arturo G. Porras and Graham Palmer

Department of Biochemistry, Rice University, Houston, Texas

Introduction

Frequently, the characterization of the potentiometric behavior of electron transfer centers in flavoproteins and other redox enzymes utilizes low-temperature electron paramagnetic resonance (EPR) as the means of visualizing the oxidation state of the participating species.

Unfortunately this technique has some inherent limitations which can result in quantitations which are misleading. In particular, the requirement that the EPR measurement be made at temperatures that are very much below that at which the sample is prepared, can affect the experimental results in two ways. First, it is well-known that there are changes in pH which can occur upon cooling the sample and, in some buffer systems, these changes in pH can be substantial (1). The desirability of obtaining data in several different buffers is emphasized by this consideration. Second, we have recently stressed the fundamental point that reduction potentials are inherently temperature-dependent (2). Thus, changes in temperature are expected to lead to changes in midpoint potential, and hence the observed oxidation-reduction level of a redox center measured at low-temperature by EPR is not necessarily that specified by the system potential which was established at room temperature. This complication is particularly worrisome in those enzymes which contain multiple redox centers, for in this case, freezing of the sample will not necessarily lock-in the redox state, but intramolecular electron equilibration may occur all the way down to very low temperatures.

We are attempting to establish the extent to which these effects are present in xanthine oxidase. The available data in the potentiometric behavior of the FlH$^{\cdot}$, Mo(V), and Fe/S centers in this enzyme were obtained between 25–120°K (3), with the iron-sulfur centers being measured at the colder temperatures and the flavin and molybdenum characterized at the higher temperature where saturation of the EPR signals is less evident. Thus, in these measurements not only are the species evaluated at temperatures far away from ambient, but the different species are characterized at temperatures unrelated to each other.

The obvious solution to this dilemma is to prepare and characterize the enzyme samples at room temperature, but the implementation of this solution is complicated by the limitations of the spectroscopic techniques available. Thus, while both flavin and the iron-sulfur centers contribute substantially to the absorption spectrum, there is no simple objective way to establish the contribution of the three chromophores to the absorbance changes obtained during oxido-reduction, and we have found it desirable to use a combination of circular dichroism spectroscopy and room temperature EPR to obtain the relevant data.

Room Temperature CD Potentiometric Measurements

It has been known for many years that the visible circular dichroism of xanthine oxidase is dominated by the iron-sulfur chromophores (4); indeed there is no evidence that any of the other species are significant contributors to the optical activity (5). The CD spectrum resembles markedly that of common two-iron iron-sulfur proteins such as spinach ferredoxin, and the CD intensity [$\Delta\varepsilon$ (432 nm)=48] is about twice that observed with the plant protein, implying that both iron-sulfur centers in xanthine oxidase make comparable contributions to the observed CD spectrum.

The implementation of a potentiometric-CD assembly is straightforward and only requires the marriage of a standard potentiometric vessel (e.g., Metrohm EA875 vessel with EA874 cap) with a spectrophotometer cell. The potentials are measured with a Metrohm EA 234 combination platinum electrode with Ag/AgCl reference system; this electrode is equipped with a ground joint for use with the titration ensemble. The reaction mixture comprises the enzyme plus the usual complement of mediator dyes necessary for promoting electromotive equilibration between the enzyme and the electrode. These dyes are colored and thus compete for the circularly polarized light. However, they show insignificant optical activity. Thus, the dyes make no contribution to the measured CD spectra, but do degrade the quality of the data by lowering the signal-to-noise ratio. Fortunately, the CD of the enzyme is quite intense and the increase in the noise level is not serious.

Figure 1 displays the CD spectrum of xanthine oxidase as the potential is lowered from 150 millivolts (fully oxidized) to −400 millivolts (fully reduced). These changes in the CD spectra occur instantaneously, are stable with time, and are reversible.

A plot of the dependence of the CD intensity versus potential is shown in Figure 2 together with a theoretical line for a system comprising two $n=1$ redox centers. In fitting the data, four disposable parameters were assumed, the redox potential and spectral weight of each of the two chromophores. The fit shown in Figure 2 was obtained assuming equal spectral weights and using −200 and −300 mV respectively for the two potentials. Clearly the data are explained well by this scheme, in particular markedly different spectral weights are not allowed by the data. Delta-delta plots of the changes observed at other wavelengths are quite linear and thus the analysis (Figure 2) is independent of the wavelength selected. For comparison, the original (3) and very recent (6) data obtained with xanthine oxidase (3) by EPR are collected in Table 1. It is clear that there is a significant difference between our room temperature CD data and the earlier low-temperature EPR results. Thus, the original data (3) produced potentials of −343 and −303 mV, for Fe/S I and Fe/S II respectively, while our present measurements led to values of −200 and −300

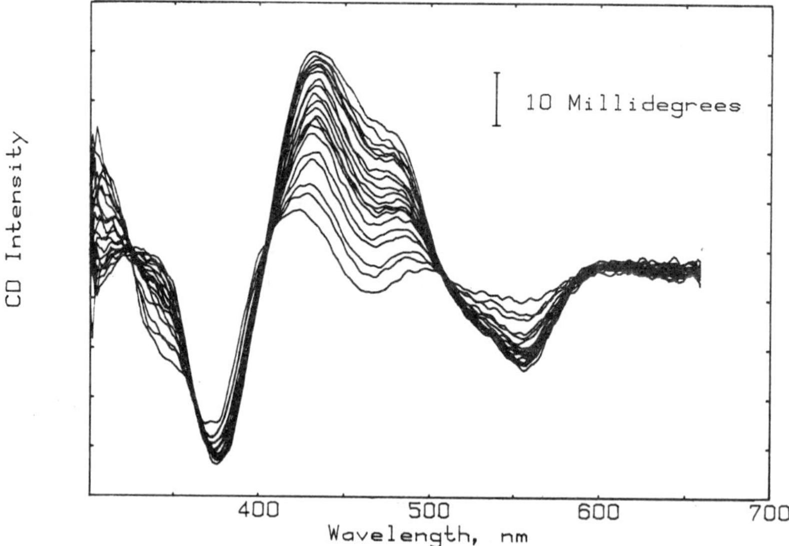

Figure 1. Circular dichroism spectral changes associated with reduction of xanthine oxidase during potentiometric titrations. The spectra shown in this figure were obtained using 21 μM enzyme in 0.1 M MOPS, pH 7.2. The fully oxidized spectrum shown was taken at a potential reading of +150 millivolts. The rest of the data range from −65 to −436 mV at intervals of approximately 20 mV.

mV, although the assignment of each of these values to a specific iron-sulfur center is not possible from our data. It is, however, tantalizing that the two potentials we deduce agree closely in value with those reported at this meeting by Barber et al. (6), who find values of −207 and −310 mV for Fe/S I and Fe/S II, respectively. In the absence of other information, we are thus inclined to correlate our two values with those of Barber et al. (6), as is detailed in Table 1.

The CD properties of the two centers are not dramatically different. This is confirmed by a comparison of the CD changes occuring during the first one-third of the titration with those occurring during the last third. These difference-CD changes are quite similar, although careful comparison establishes that they are not identical; this similarity is quite surprising in view of the marked differences in EPR (4) and visible spectra (7) reported for the two iron-sulfur centers (Figure 3).

It must be noted that potentiometric data of this kind can also be explained by a more complicated scheme in which Fe/S I and Fe/S II have comparable potentials and there is an allosteric linkage between them so that reduction of Fe/S I lowers the potential of Fe/S II by 100 MV, and vice versa. With this alternative, both species contribute to the spectral changes at all points along the titration curve, although the species with the more positive potential would dominate the early phase, etc. We have no means of excluding this alternative explanation with our results, although the possibility of this behavior does seem to be eliminated by the EPR studies (3,6).

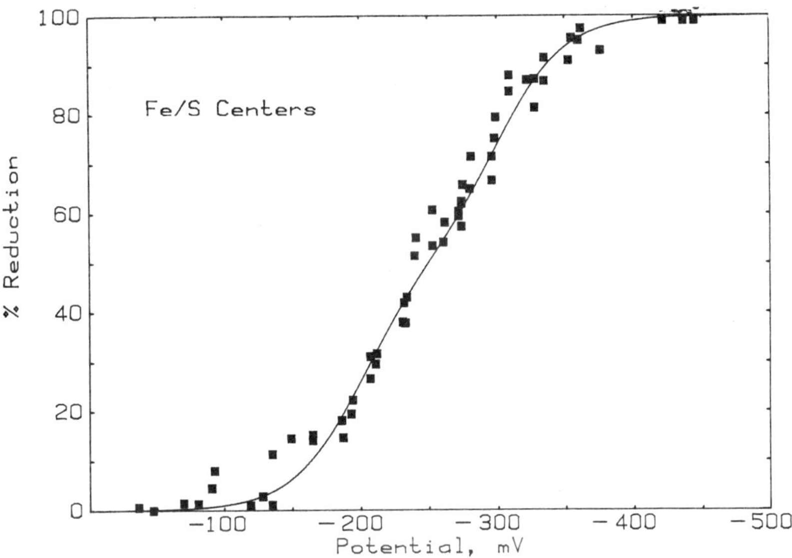

Figure 2. Potentiometric titration of the iron-sulfur centers of xanthine oxidase. The symbols represent the percent change in the intensity of the circular dichroism at 460 nm observed at each potential value with respect to fully oxidized enzyme. The data were taken in 0.1 M MOPS, pH 7.2 during both oxidative and reductive titrations. Since the CD spectrum of xanthine oxidase arises purely from the iron-sulfur centers, the redox potentials of the centers can be obtained in this way. The line represents the change expected for two 1-electron accepting centers with identical spectral characteristics and respective midpoint potentials of −200 and −300 mV. By comparison, the corresponding potentials obtained at pH 8.3 are −220 and −320 mV.

Room Temperature EPR Studies

Because both flavin radical and Mo(V) have relatively long electron spin relaxation times, the acquisition of room temperature EPR data is straightforward. However, sensitivity is maximized through the use of the so-called "flatcell" as opposed to narrow-bore, capillary sample tubes. Transfer of material in and out of the flat cell is tricky because of the potential for trapping pockets of gas in the flat zone of this cell. To achieve the reliable transfer of sample from the potentiometric vessel into the flat cell and still maintain acceptable anaerobiosis, we have employed flexible, narrow-bore (0.1 mm) glass tubing normally used in glc systems. This tubing connects the base of the Metrohm potentiometric vessel with the lower stem of the flat cell which is located in the Varian E-231 multi-purpose rectangular cavity. Liquid can be transferred between the potentiometric vessel and the flat cell by the application of either a gentle vacuum or a small N_2 pressure to the upper stem of the flat cell.

In these experiments, there are two complications: First, the mediator dyes exhibit large free radical signals when partially reduced. Fortunately, the contribution of these radicals is still quite small in the range of potentials where the flavin is titrating and this small contribution can be eliminated by

Table 1. Comparison of Xanthine Oxidase Potentials Obtained by Room Temperature and Low Temperature EPR. [All values are in millivolts (vs SHE)].

	Mo(VI)/Mo(V) (Fast) (Slow)	Mo(V)/Mo(IV) (Fast) (Slow)	FAD/FADH$^{\cdot}$	FADH$^{\cdot}$/FADH$_2$	Fe/S I	Fe/S II
Room temperature[a] pH 7.2, MOPS	−300 −344	−300 −306	−242	−227	−300	−200
Low temperature[b] pH 7.1, pipes	−355 —	−356 —	−280	−244	−300	−255
Low temperature[b] pH 8.2, tris	−397 —	−405 —	−378	−223	−330	−255
Low temperature[c] pH 8.3, pyrophosphate	−355 −440	−355 −380	−351	−236	−343	−303

[a]This work.

[b]Barber and Siegel, this volume.

[c]Reference 3.

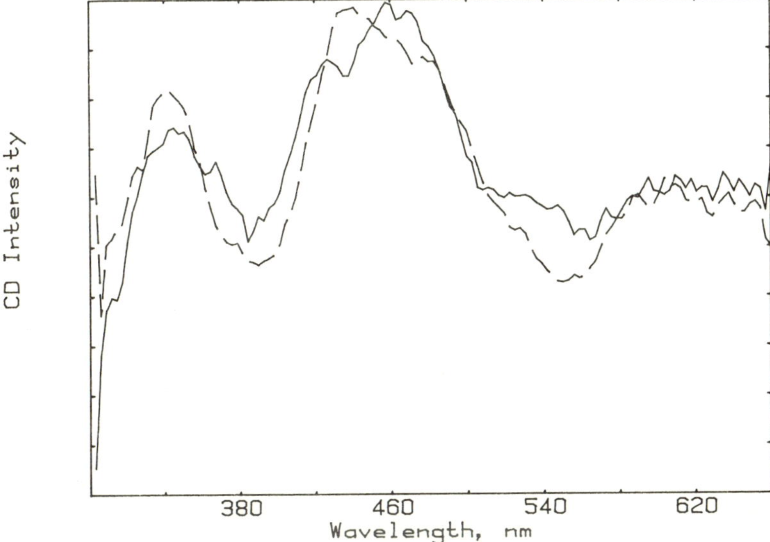

Figure 3. Comparison of the circular dichroism spectral changes associated with reduction of the iron-sulfur centers in xanthine oxidase. The spectra shown here were obtained by subtracting the spectrum obtained at 31.3% reduction from that of fully oxidized enzyme (solid line) and by subtracting that of fully reduced enzyme from the spectrum obtained at 64.4% reduction (broken line).

computer, taking advantage of the substantial difference in linewidth between the flavin and dye radical signals (Figure 4). (The dye signals have an apparent linewidth of 5 Gauss, the modulation amplitude employed in recording the data.) Second, the molybdenum region has contributions from both the "slow" and the "fast" species. These can be partially resolved but because the spectral resolution is somewhat poorer at room temperature the separation (Figure 5) is not as clean as that obtainable from spectra obtained at liquid nitrogen temperatures.

The dependence of the intensity of flavin radical on system potential is shown in Figure 6. The most striking feature of these data is the relatively large yield of radical, about 0.35 moles of $FlH^{\cdot}$, at this pH. However, the yield of radicals is strongly pH dependent, and at pH 8.3 we find about 0.13 moles of radicals. The data can be fit in the usual way assuming two consecutive one-electron processes

$$A(0) \overset{E_1}{\rightleftharpoons} A(1) \overset{E_2}{\rightleftharpoons} A(2)$$

obtaining values of -242 and -227 mV for E_1 and E_2 at pH 7.2 and -300 and -240 mV for E_1 and E_2 respectively at pH 8.3. The latter values are to be compared with the EPR data which yielded values of -351 and -263 mV for E_1 and E_2 at pH 8.3. It would appear that the flavin couple is not as sensitive to temperature as are the iron-sulfur centers and that the effect of temperature

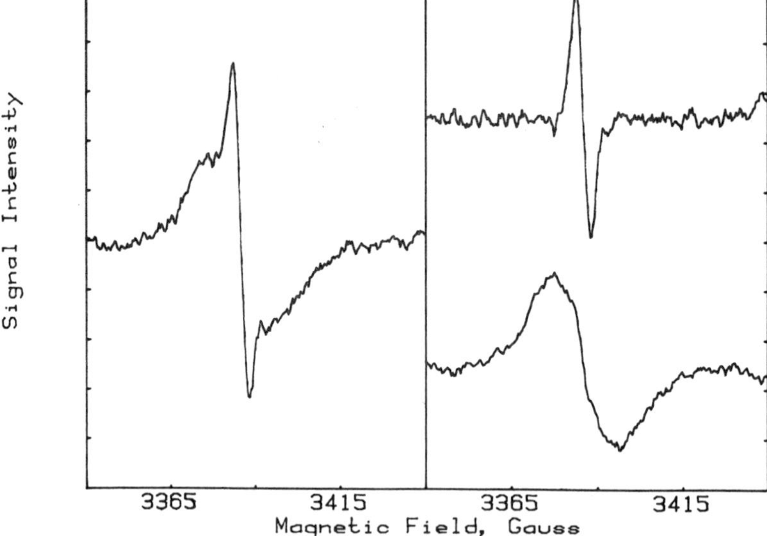

Figure 4. Deconvolution of radical EPR signals obtained during potentiometric titrations of xanthine oxidase. The measured spectrum is shown on the left, and was obtained under the following conditions: 100 μM enzyme in 0.1 M Bicine, pH 8.3; field modulation, 5 gauss; microwave frequency, 9.506 GHz; microwave power, 5 milliwatts; temperature, 23°C; applied potential −252 mV. The spectra on the right corresponded to the contributions of an unidentified radical in the mediator dye mixture (top) and the flavin semiquinone observed in the xanthine oxidase sample (bottom).

is asserted primarily through an increase in the affinity of the flavin for the first electron added, the affinity for the second electron being essentially unmodified.

The data on the two molybdenum species are shown in Figures 7 and 8. About 35% of the "rapid" molybdenum is maximally converted to Mo(V) and fitting the data to Scheme 1 yields the values $E_1 = E_2 = -300$ mV; the identity of the E_1 and E_2 follows automatically from the yield of the signal and the value from the location of the signal. These data are not in obvious contradiction to the EPR data obtained at pH 8.2 which also show $E_1 = E_2$, although the absolute values are lower by about 50 mV (Table 1). Our preliminary data at pH 8.3 confirm that the molybdenum potentials do decrease with increasing pH but we do not yet know the magnitude of this decrease.

Substantial amounts of slow molybdenum are also observed at room temperature, the maximum yield being 0.2 mole/mole total inactive Mo. Fitting the data to Scheme 1 suggests that $E_1 = -344$ mV and $E_2 = -306$ mV. These values are both qualitatively and quantitatively different from the EPR data which required E_2 to be less than E_1 (Table 1), and $1/2\ (E_1 + E_2)$ decreasing by about −140 mV under the conditions of the low-temperature EPR measurements on this enzyme.

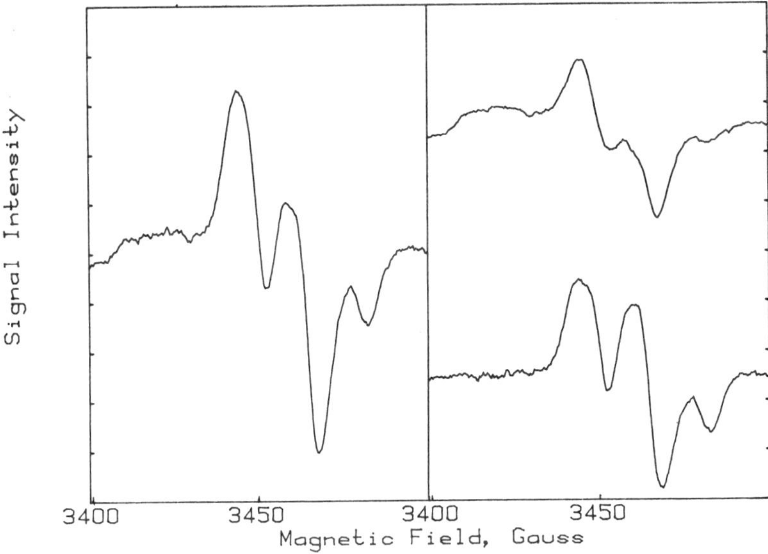

Figure 5. Deconvolution of rapid and slow molybdenum signals during potentiometric titrations of xanthine oxidase. The measured spectrum is shown on the left and was obtained under the conditions given in Figure 4, at −376 mV. This spectrum was deconvoluted into the contributing rapid (top right) and slow (bottom right) molybdenum signals by subtraction of the appropriate amount of pure "slow" signal.

Figure 6. EPR potentiometric titration of the FAD in xanthine oxidase. The symbols represent the integrated intensity of the flavin semiquinone radical signal obtained as described in the text. The intensity is presented here as (%) mole $FADH^{\cdot}$/mole total flavin. The data were obtained in 0.1 M MOPS, pH 7.2 during both oxidative and reductive titrations. The line represents the intensity expected for a 2-electron accepting system which is reduced in 1-electron steps, with midpoint potentials of −242 mV for the FAD/$FADH^{\cdot}$ couple and −227 mV for the $FADH^{\cdot}/FADH_2$ couple. The corresponding potentials obtained at pH 8.3 are −300 and −240 mV respectively.

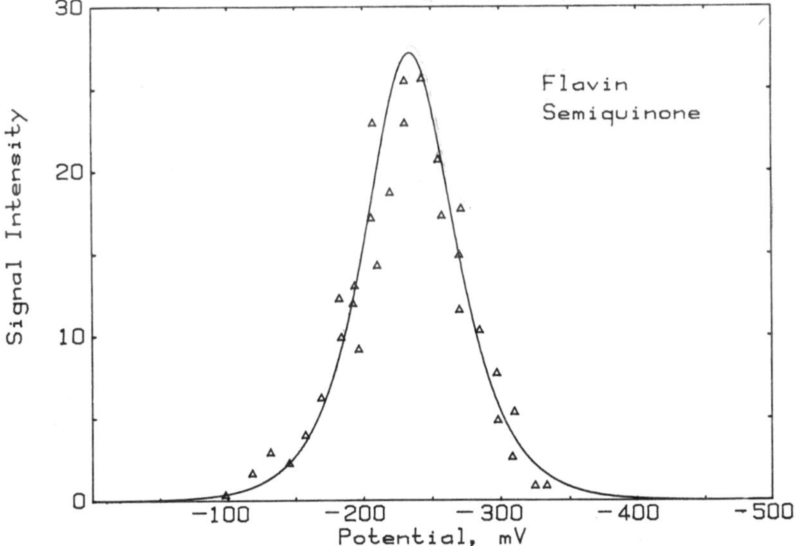

Conclusions

The preliminary results we have so far obtained indicate that small differences are to be found between electron affinities determined by measurements made at room temperature and at 20–100 K. Such differences imply that the temperature dependency of redox potentials is a significant phenomenon. We find deviations of 0 to about −60 mV between our data and those recorded by low temperature EPR at the same pH (6). For a one-electron process these variations translate into a minimum entropy change of about 0 to −8 eu for electron addition (equivalent to about 0 to −1 mV for every 3 degrees). This range of values should be compared to the value of −27 eu obtained from calorimetric measurements of cytochrome c (8). Clearly the values found are to be considered reasonable.

In assigning the potentials of the Fe/S centers, we are forced to use an arbitrary assignment of a center to each of the high and low potential phases (Figure 2). We elected to assign Fe/S I as the low potential center to maintain consistency with the EPR data of Michael Barber and Lewis Siegel in their contribution to this symposium. Interestingly, this center exhibits a potential which is independent of temperature. We find Fe/II to be the most positive redox center in the enzyme. This conclusion appears to be valid at all pH values between 6.2 and 9.2 and contradicts the picture we had drawn earlier (7) for the distribution of potentials among the four redox centers. It appears that the enzyme is not perfectly tuned to maximize the degree of reduction of the

Figure 7. EPR potentiometric titration of the rapid molybdenum in xanthine oxidase. The symbols represent the integrated intensity of the rapid molybdenum signal obtained as described in Figure 4. The intensity is presented as (%) mole Mo(V)/mole active molybdenum. The data were obtained in 0.1 M MOPS, pH 7.2 during both oxidative and reductive titrations. The line represents the intensity expected for a 2-electron accepting system which is reduced in 1-electron steps, with midpoint potentials of −30 mV for both the Mo(VI)/Mo(V) and the Mo(V)/Mo(IV) couples.

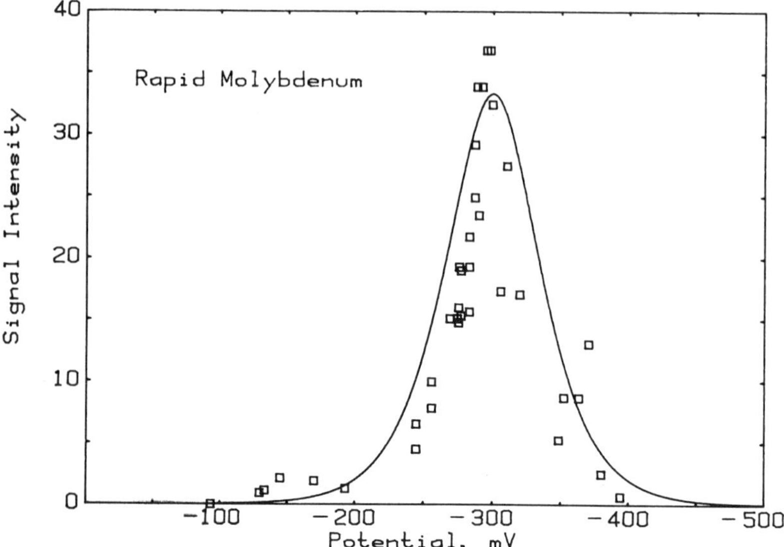

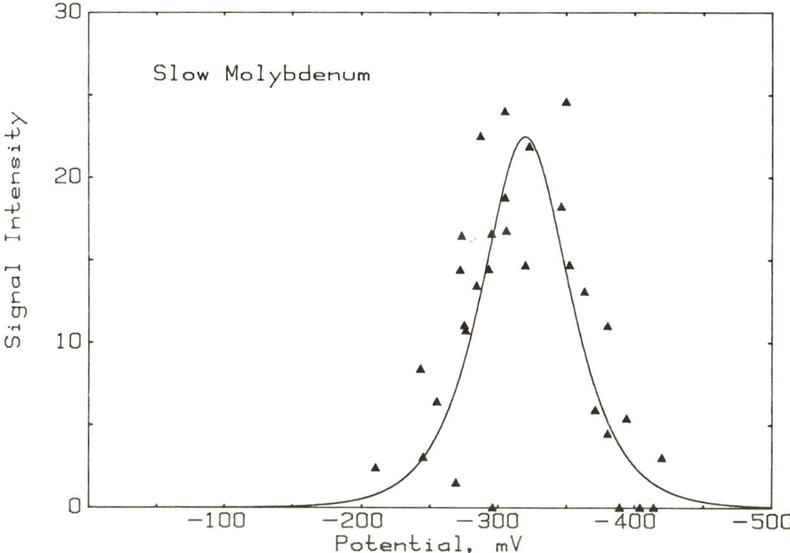

Figure 8. EPR potentiometric titration of the slow molybdenum in xanthine oxidase. The symbols represent the integrated intensity of the slow molybdenum signal obtained as described in Figure 4. The intensity is presented as (%) mole Mo(V)/mole inactive molybdenum. The data were obtained in 0.1 M MOPS, pH 7.2 during both oxidative and reductive titrations. The line represents the intensity expected for a 2-electron accepting system which is reduced in 1-electron steps, with midpoint potentials of −344 mV for the Mo(VI)/Mo(V) couple and −306 mV for the Mo(V)/Mo(IV) couple.

flavin nor the degree of oxidation of the molybdenum. Rather, it appears that each of these two centers is associated (thermodynamically) with an individual iron-sulfur center.

Substantial differences were also found for the potentials of "rapid" molybdenum; the flavin appears to occupy an intermediate position in its sensitivity to temperature.

The principal impact of our conclusions will be felt when attempting to provide a detailed picture of the kinetic behavior of this enzyme. We have previously described in detail (7) the important interplay of kinetic and thermodynamic factors which control the time dependence of the redox changes in the molybdenum, flavin, and iron-sulfur centers. Clearly, the accurate reproduction of such changes requires access to the electron affinities of these components under conditions pertinent to the kinetic data. We deem it rather unlikely that the description of the kinetic behavior of the enzyme provided previously will be qualitatively affected by revision in the potentials such as those we now report. However, it is to be expected that the quantitative aspects of the analysis will undergo some modification, hopefully in the direction of improving the agreement between theory and fact.

Finally, we note that the phenomenon of temperature-dependent potentials has an important impact on the interpretation of kinetic data obtained by the rapid-freezing technique. It is clear that for those enzymes which possess multiple redox centers, temperature-dependent intramolecular electron re-arrangement is a real problem and one that must be addressed in the evaluation of such data.

ACKNOWLEDGMENTS
We are most appreciative of the courtesy of Dr. Lewis Siegel and Dr. Michael Barber, allowing us to include some of the data from their contribution in this article. This work was supported by the NIH (GM21337) and the Welch Foundation (C636).

References

1. Williams-Smith, D.L., Bray, R.C., Barber, M.J., Tsopnakis, A.D., and Vincent, S.P. (1977) *Biochem J* 167:593–600.
2. Palmer, G. and Olson, J.S. (1980) In *Molybdenum and Molybdenum Containing Enzymes*. M. Coughlan (ed.) New York: Pergamon Press, pp. 187–220.
3. Cammack, R., Barber, M.M., and Bray, R.C. (1976) *Biochem J* 157:469–478.
4. Palmer, G. and Massey, V. (1969) *J Biol Chem* 244:2614–2620.
5. Komai, H., Massey, V., and Palmer, G. (1969) *J Biol Chem* 244:1692–1700.
6. Barber, M.J. and Siegel, L.M. (1981), this volume.
7. Olson, J.S., Ballou, D.P., Palmer, G., and Massey, V. (1974) *J Biol Chem* 249:4363–4382.
8. Watt, G.D. and Sturtevant, J.M.C. (1969) *Biochemistry* 8:4567–4571.

PART X C:

Flavocytochromes

Published 1982 by Elsevier North Holland, Inc.
Vincent Massey and Charles H. Williams, Editors
Flavins and Flavoproteins

CHAPTER 136

Interactions and Electron Transfer Between the Flavodehydrogenase and the Cytochrome b_2 Components of Yeast Flavocytochrome b_2

Françoise Labeyrie

Centre de Génétique Moléculaire du CNRS - 91190 Gif - sur - Yvette, France

Introduction

Its apparently simple complexity was the particularity of the flavocytochrome b_2 molecule that gave investigators the confidence to hope to attain a precise analysis and understanding of the relationships, at structural and functional levels, between the four elements: (a) electron donor, the dehydrogenable substrate, (b) the one-electron acceptor, cytochrome c, and (c), (d) the enzyme prosthetic groups, heme-bound as cytochrome b_2 and flavin mononucleotide in its proteic site.

The purpose of this review article is to examine the present state of knowledge. Many studies have been carried out since 1955, when excellent, pure preparations of enzyme, well-characterized as an L-lactate cytochrome c reductase, with a flavocytochrome b_2 structure, became available (1). The details of the evolution of ideas and the respective contributions of different laboratories during the period preceding detailed structural analysis and rapid kinetic studies will not be given. Nevertheless, homage must be paid to the considerable pioneer work of Appleby and Morton and coworkers and to important contributions by Boeri, Nygaard, Mahler, Ogura and colleagues, and Baudras in France. Contributions of these laboratories are reviewed in the following articles (2–6).

In 1965, however, the model proposed for the structure of the molecule was still incorrect and hypotheses on the mechanism of electron transfers were not supported by any reliable data. The progress made since then is due mainly to the development of new methods for characterization of polypeptide chains and for rapid kinetic studies.

At present, it is possible to summarize in two sentences the main knowledge of flavocytochrome b_2 structure and functional features: (a) *Flavocytochrome b_2 structure*: A block consisting of a tetrameric flavodehydrogenase domain carrying four cytochrome b_2 globules, capable of binding strongly four cytochrome c units. (b) *Flavocytochrome b_2 function*: A "transformase" 2e/1e in which electrons are able to tunnel from flavin to cyt b_2 and from cyt b_2 to cyt c.

The Two-Domain Structure

In order to distinguish flavocytochrome b_2 prepared from the yeasts *Saccharomyces cerevisiae* and *Hansenula anomala*, the symbols S and H are generally used.

Both enzymes consist of a tetramer assembly of four identical chains of 60,000 daltons (7, 8). The polypeptide chains of both have a triglobular folding [Gervais and coworkers (9–12)]. Globule 1, at the N terminus, is the "cytochrome b_2" unit (MW 12,000), a building block originating from the cytochrome b_5 family (13–16). Globules 2 and 3, together, form (9, 12) the flavodehydrogenase entity. Both are required for the formation of the flavin site (17, 18) the stability of which is also controlled by the overall tetrameric structure (personal observation).

In spite of all these similarities, the two enzymes present many differences due probably to many point substitutions of residues along the chain. Some of these differences have been characterized in the middle part of the chain (19). As far as their behavior is concerned, the differences effect (a) the stability of their quaternary structure at low ionic strength (20, 21, 7); (b) the precise pattern of peptide bonds, hypersensitive to various proteases, in the native chains, in the zones of junction between globular folding units (10); (c) the length and stability of proteolytic fragments that can be obtained under controlled proteolytic action (11, 12, 19) (a consequence of (b); (d) their molecular activities (1200 s^{-1} for H, 500 s^{-1} for S) (8, 22), partly controlled by the rate of the lactate-to-flavin electron transfer, as will be seen below. It should be underlined that, because of a proteolytic cleavage during the preparation, the enzyme preparation obtained by the Appleby-Morton procedure, consists of a nicked-form, active flavocytochrome b_2 known as S-cleaved (8). A similar cleaved enzyme can be obtained by trypsin or clostripain action from the H-enzyme (19).

Electron Transfers Within the Flavodehydrogenase Moiety

The lactate to flavin two-electron transfer, the first step (after lactate binding) in the overall catalytic reaction as well as in the partial reaction of enzyme reduction, is also the slowest of the steps involved in the turnover (23–25). Its rate constant varies markedly for the different forms of the enzyme that have been studied. The most rapid corresponds to the intact H-flavocytochrome b_2 (380 s^{-1} at 5°C) (26) being respectively 130 (27) and 40 s^{-1} for the intact and cleaved (Appleby-Morton) S-enzyme (28). Analysis of the isotopic effect, when the α-CH hydrogen of L-lactate is replaced by deuterium shows that this slowest step does not limit the rate of turnover but is one of the steps controlling the rate (25–27). This is true essentially for the more active forms of the enzyme ("intact"-S and H, "cleaved"-S). For much less active forms in which the flavin site has been markedly affected after cleavage and peptide deletion in the interglobule 2–3 zone, we might suspect that step 1 becomes entirely limiting at substrate and acceptor saturation.

One important question is the following: Does the absence of heme or of the cyt b_2 entity modify the functional parameters of the lactate binding and lactate to flavin transfer? As far as heme depletion is concerned, a clear answer has been afforded by Iwatsubo et al. (29): Values of $k_{red,max}$ and K_m lactate do not depend on the presence of heme. Current studies should afford a definite answer to the role of cyt b_2 core depletion (19).

Electron Transfers from the Flavodehydrogenase (Heme-Free Derivatives) to External Acceptors

Another important problem concerns the reactivity of flavin with monoelectronic external acceptors in derivatives deprived of the heme prosthetic groups or of the cyt b_2 globule.

In the case of ferricyanide. It has been shown with a heme-free derivative obtained from S-cleaved flavocyt b_2 that the fully reduced form is much less reactive than flavocytochrome b_2 (2 orders of magnitude in terms of bimolecular rate constants: $10^5 M^{-1}s^{-1}$ instead of $10^7 M^{-1}s^{-1}$), and that this hydroquinone form is at least 20 times less reactive than the corresponding semiquinone (29) (no saturation behavior has ever been detected in any case, the highest pseudo-first-order rate constant measured being 200 s^{-1}.

In the case of cytochrome c. Flavodehydrogenase, heme-free derivatives, of flavocytochrome b_2 have no lactate-cytochrome c reductase activity under conditions where ferricyanide reductase activities are high (30). Experiments were carried out by Gervais et al. to see whether or not an isolated flavodehydrogenase entity at high concentration (6 μM) was able to achieve cyt c reduction (30 μM) by lactate: No activity has been found (19). A parallel experiment in the case of cytochrome b_2 core, the isolated cytochrome globule, yielded an identical result (19); again no reduction took place.

These results, compared with those obtained for the reaction of the flavodehydrogenase covalently linked to its acceptor cyt b_2, within the flavocytochrome b_2 assembly, will be examined and interpreted in the discussion-conclusion section.

Intramolecular, Flavodehydrogenase Cytochrome b_2, One-Electron Transfer Within Flavocytochrome b_2

Information on the parameters of electron exchange between flavocytochrome b_2 prosthetic groups when reduction by lactate or reoxidation by external acceptors are studied, can be obtained only with great difficulty (31). The difficulties arise from two kinds of problems, those concerning the detection of species F_{SQ} and F_{ox}, and those resulting from the slowness of the first step, i.e., dehydrogenation of bound lactate, relative to certain electron transfers taking place afterwards. There are now only two studies being carried out by our research group (Blandin-Capeillère and Iwatsubo) in collaboration with R.C. Bray at Brighton (28,31,26), in which all species are measured, both during the time course of reduction by lactate and once the equilibrium is reached (titration data). The data concerning redox equilibrium, summarized in reference 26, show that the redox potentials of the different 1e-systems are close to each other, and that the semiquinone has a significant stability ($K = (F_{SQ})^2/(F_{ox})(F_{red}) \sim 2$).

Other studies where, apart from the heme, only the amount of F_{ox} was estimated have been carried out in several laboratories and concern the baker's yeast flavocytochrome b_2 in its "cleaved" (32, 33) and "intact" forms (27).

Aspect of the Reaction of Enzyme Reduction by L-Lactate

In all studies carried out on the S-cleaved at saturating substrate concentrations, the most striking observations were:

1. A synchrony of H_{red} and $(1\text{-}F_{ox})$ time courses that persists during the whole reduction (6e accepted per couple of protomers from L-lactate molecules). No lag or burst has ever been detected under such conditions at the level of absorption signals.
2. Biphasicity of H_{red} and $(1\text{-}F_{ox})$ time courses (85% of rapid phase, 15% of a phase slower by a factor of ~ 10) when working with saturating L-lactate concentrations. This biphasicity was recognized by Suzuki and Ogura (33) and fully analyzed by Blandin-Capeillère and coworkers (28, 26) who showed that neither of these phases were controlled by intermolecular electron exchange and that phase I corresponds to the entry of 2e/protomer while phase 2 corresponds to that of the third electron.
3. The time course of F_{SQ}, when determined, exhibits a rapid initial increase in phase I which however was at a level lower than H_{red} reaching a plateau near 50% in phase II.

Reaction Schemes

Initial Steps of Reduction Also Involved in the Turnover

The several reaction schemes that have been proposed (Morton and Sturtevant (23); Suzuki and Ogura (33); Blandin-Capeillère, Iwatsubo, Bray and Labeyrie (26); Pompon, Iwatsubo, and Lederer (27) in order to interpret the reduction kinetics of flavocyt b_2 by lactate, proposed the same order of succession for the initial steps. After reversible lactate binding, the following steps take place successively: *Step 1*: lactate to flavin 2e-transfer yielding F_{red}. *Step 2*: 1e-transfer from F_{red} to H_{ox} yielding F_{SQ}, the flavin semiquinone, and H_{red}.

In the two first studies, step 2 was considered to be an irreversible process. With such a hypothesis, Morton and Sturtevant found it impossible to explain their experimental results qualitatively. Because of the small differences observed in our laboratory between redox potentials of the heme and flavin 1e-systems, it became evident that all permitted intramolecular transfers have to be considered as reversible processes, with a precisely defined k_+/k_- ratio equal to the equilibrium constant. This reversibility was introduced in step 2 by Blandin-Capeillère et al. (26), and confirmed as a necessary assumption (27). There is now a general consensus concerning the nature of the role of these two basic initial steps. They explain satisfactorily the biphasic behavior (the final level in phase I being essentially controlled by the k_{+2}/k_{-2} ratio) and the presence of F_{SQ} in early reaction times. Step 2 therefore is a fast flavin-to-heme 1e-transfer taking place within an "active site" assembling these two related groups. This kind of transfer will be termed here "intra-site" transfer in contrast to other kinds of transfers, involving two different "active sites," called "inter-site" transfers. This terminology is preferred to "intra-protomer" and "inter-protomer" since it is not known whether or not an individual active site is formed by a single protomer or by a junction between two protomers (Figure 1).

In the presence of external acceptors, it seems that the reduction does not go further. Electrons are delivered by cyt b_2 and possibly F_{SQ}.

Final Steps Involved in Total Reduction and Not in the Turnover

In the absence of external acceptors, total reduction requires the introduction of a new couple of electrons for each pair of active sites. For this, regeneration of F_{ox} is needed. This regeneration must require an intersite 1e-transfer: An exchange from one F_{SQ} to another F_{SQ}, that is, a dismutation yielding F_{ox} plus F_{red}, was first envisaged (28), and the possibility of "inter-site" F_{SQ} to H_{ox} or H_{ox} to H_{red} (the latter allowing a new intrasite F_{SQ} to H_{ox}) was later proposed as an alternative (27).

In the first linear scheme proposed by Blandin-Capeillère et al. (designated scheme BC), *step 3*, consisting of a reversible disproportionation was followed by *steps 4*, *5*, *6* involving first rebinding of lactate and a conformational change as limiting step (or vice versa), followed by the final 2e-transfer to the regenerated F_{ox}.

A "square" three-redox-level scheme (referred to here as scheme P) was then proposed and simulated by Pompon (34). It involved all kinds of possible redox equilibria for each value of the number of electrons accepted per protomer (1,2) and irreversible processes (as previously) for each entry of a couple of electrons. Scheme P improved on scheme BC by the introduction of transient forms with one bound lactate or one reduced flavin per pair of active sites. Such transients were not considered for the sake of simplification in

Figure 1. Principal steps of electron exchange within flavocytochrome b_2. For details of reaction schemes BC and P (see text), original publications (34,35) should be consulted.

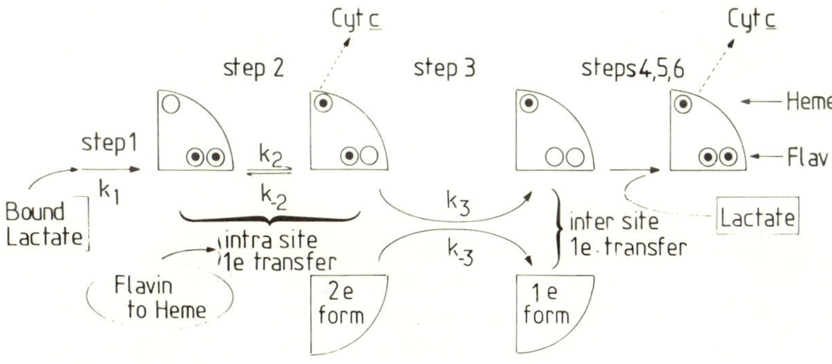

Schematic presentation of reactions involved in flavo Cyt. b_2

simulations of the linear scheme BC. In fact, as underlined in (34) such a simplification supposes a cooperativity of two associated active sites for their reduction. The major interest of Pompon's study was to propose three variants of his scheme differing by the mode of inter-site transfers. The first one assumed a dismutation as in scheme BC, the second a transfer from F_{red} to H_{ox} of another active site, and the third a transfer from H_{red} to H_{ox} of another active site, allowing a new rapid intra-site F_{red} to H_{ox} exchange.

Both schemes, BC and P (with its variants) were the object of thorough simulation studies with different objectives. The former (35) was aimed to achieve a satisfactory fit of all species (F_{SQ} and F_{red} together with F_{ox} and H_{ox}) at equilibrium and at maximal rates of electron entry. With F_{SQ}, great difficulties were encountered. Introduction of steps 4-5-6 with the rate constants that will be given in the next section solved the problem of fit. The scope of the challenger (34) was to provide a satisfactory fit of F_{ox} and H_{red} and of the $H_{red}/(1\text{-}F_{ox})$ ratio in the range of very low (down to 0.5 s^{-1}) rates of electron entry. This was possible with the three variants of "inter-site" transfers and the same set of k_{+3}/k_{-3} values (50 times lower) for all kinds of "inter-site" transfers. However, there were two constraints for a satisfactory fitting: First, an anticooperative factor of 0.5 had to be introduced for the entry of the second electron couple on each pair of active sites, in the case of inter-site transfers of the F to H or H to H kinds. Second, an anticooperative factor of similar intensity ($\times$0.3) also had to be introduced for the entry of the final electron couple, one per two active sites. Although such factors suppose the intervention of a conformation change, this process is no longer considered as an individual limiting step.

Rate Constants of "Intra-Site" Flavin to Cyt b_2 Electron Tunneling

With S-cleaved-flavocyt b_2 at 24°, at substrate saturation, k_{1max} is 120 s^{-1}. k_{+2} and k_{-2}, for step 2 can only be given minimal values, their ratio being precisely fixed; they were proposed to be 600/120 s^{-1}. They could not be evaluated more accurately because of the synchrony of F_{ox} loss and H_{red} appearance due to the slowness of step 1.

With H-intact-flavocyt b_2 at 5°, k_{1max} is much higher, 380 s^{-1} so that k_{+2}/k_{-2} can be precisely determined to be 350/120 s^{-1}. Only these values give satisfactory fit of simulated time courses presenting a detectable lag and different apparent rate constants for species H_{red} and F_{ox}.

With S-cleaved enzyme at 5° with suboptimal to low concentrations of L-lactate and L-deutero lactate (k_1 varying from 30 s^{-1} to 0.5 s^{-1}), values of 500/120 s^{-1}, taken as similar to those previously proposed (26), were tested in simulations of scheme P and proved to be valid (34).

Rate Constants of "Inter-Site" Transfers

There is at present no consensus of opinion concerning the values of the rate constants involved in such "inter-site" transfers and in the following steps allowing full reduction. Together the processes involved in the terminal reduction phase are slow ($k_{app,max}$ ~7 s^{-1} at 24°C and ~4 s^{-1} at 5°C) and appear to decrease in rate when lactate concentrations are decreased, remaining roughly 10–20 times lower than k_{+1} (27). Blandin-Capeillère (35), in order to fit the semiquinone time course, proposed best estimates of 80 and 120 s^{-1} for

k_{+3} and k_{-3} at 24° for the "inter-site" disproportionation reaction and $k_5 = 7$ s^{-1} for the limiting irreversible step (34). For any kind of "inter-site" exchanges, Pompon proposes and finds satisfactory fits of his scheme with values of 8 and 12 s^{-1} at 4° (same ratio as BC values but 3 times slower if identical temperatures were compared).

Thus *no conclusion has been reached* concerning the kind of "inter-site" transfers actually involved in this flavocytochrome b_2 system. From a structural point of view, it seems highly probable that only certain groups in the tetrameric assembly are close enough to each other to achieve efficient electron tunneling transfers.

Comparison of Relative Proportions of Redox Species at Equilibrium and in the Course of the Reaction

One important problem for the understanding of electron exchange within active sites and inter-active sites was to know whether or not: (a) Prosthetic groups reach their thermodynamic redox equilibrium during the course of the reaction and also (b) all possible kinds of exchange that might be reached, in exchanges involving bimolecular processes at the level of whole flavocyt b_2 molecules, are all permitted within a molecule by means of intra-site or inter-site exchanges.

The first attempts to compare redox states of heme with those of flavin was achieved by our group in 1968 (36) and detailed in further studies (26). It was shown that heme and flavin were not at equilibrium. Apparent equilibrium constants $\left(K_{app} = \frac{H_r}{H_{ox}} \times \frac{(1-F_{ox})}{F_{ox}}\right)$ calculated from the data of Pompon et al. at 5°C on S_x-cleaved flavocyt b_2 and compared to the true equilibrium value K_e (at 20°C) show how K_{app} values become closer to K_e when k_{+1} decreases to 0.5 s^{-1} (Figure 2A). The curve indicates that complete equilibration of redox species in this case results from processes that are so slow that they could be bimolecular. In the case of the H-intact enzyme (data in ref. 26) as shown in Figure 2B, the situation is strikingly different. True equilibrium between F_{SQ}/F_{red} and heme needs 100 msec for $\tau = 35$ msec and 14 msec for $\tau = 2$ msec. Apparent equilibrium between $F_{ox}/(1\text{-}F_{ox})$ and heme is reached within 20 msec for $\tau = 2$ msec and 10 msec for $\tau = 35$ msec. Equilibrium between all prosthetic groups seems to be reached in a period markedly shorter for the H-enzyme than for its S-cleaved or -intact homologue.

Conclusion

This article defines the present state of knowledge concerning the 1e-transfer taking place between the flavodehydrogenase and cyt b_2 entities associated together within flavocyt b_2 individual molecules. Two kinds of reversible transfers have been clearly distinguished: One takes place between flavin and heme associated in one "active center," exchanging electrons with "on" and "off" life times which have been precisely determined for the H-enzyme flavocyt b_2 ($\tau \sim 2$ msec). There remain unresolved problems, in particular, whether or not there is a difference in the rate constant when hydroquinone flavin or semiquinone flavin exchanges 1e with its heme acceptor. The other kind of transfer certainly takes place between two different active centers, in a

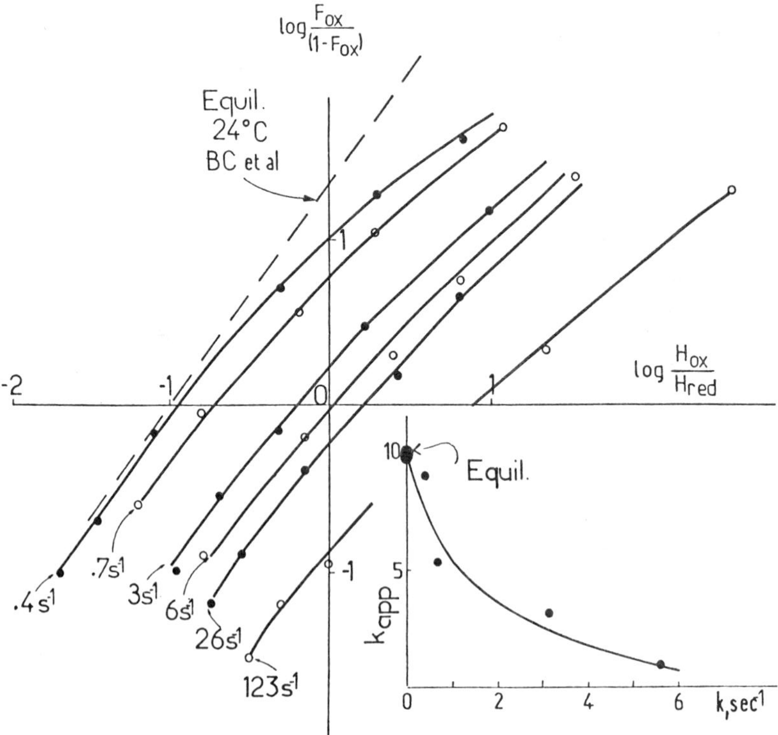

Figure 2A. Comparison of redox levels of heme and flavin groups during time courses (at various k_1) and at equilibrium. For S-flavocyt b_2 (cleaved and intact form). (Solid lines) Calculation from Pompon et al.'s kinetic data at 5°C (27). k_1 values are given for each curve. They are controlled by variation of substrate concentration by use of D or H isotopes at the α-CH of lactate, by use of cleaved or intact forms of the enzyme. Note that the expression $F_{ox}/(1\text{-}F_{ox})$ has no exact thermodynamic meaning since F_{SQ} is present in a significant amount. (Dotted lines) Equilibrium data, calculated for the same flavin species from data at 20°C on the S-cleaved enzyme (36). (Insert) $K_{app} = H_{red}/H_{ox}$ for half-flavin reduction, calculated from x_0 values of each curve corresponding to one k_1 value.

manner that cannot yet be assigned as either flavin-to-flavin, flavin-to-heme or even heme-to-heme, and with rate constants that can still be considered as speculative. This second kind of transfer appears to occur much more rapidly with the H-enzyme than with the S-enzyme since in the former it allows a rapid equilibrium of all redox species. It is not involved in the turnover.

The significance of a ~1 msec life-time for the e-transfer transition within an active site needs comment. Which is the limiting step in such a transfer? Is it the electron tunneling itself between the two partners retained together for a time long enough in a stable position, or is the electron tunneling rate constant multiplied by a probability factor depending on the time of residence of the two partners in a favorable position? Indeed the apparent absence of protein-protein binding interactions between the flavodehydrogenase and the cyt b_2

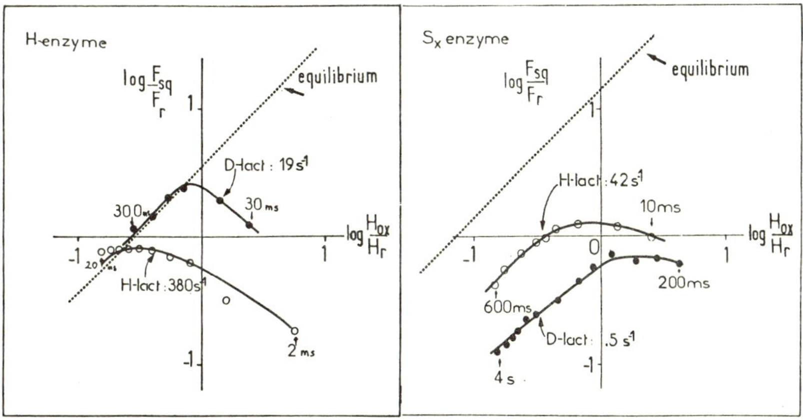

Figure 2B. For S-cleaved and H-intact flavocyt b_2. (Solid lines) calculations from Blandin-Capeillère et al.'s kinetic data (28,26). Here flavin data considered F_{SQ}/F_{red}, are thermodynamically significant and relates to the electron exchange involved in step 2. (Dotted lines) equilibrium data under identical conditions.

domains as observed in studies by Gervais et al. led us to think that they can move relative to each other. Would such a movement increase or decrease the probability of electron transfer? Is it because of such movements that one or other cytochrome globule can either accept electrons from the same flavodehydrogenase protomer or, alternatively, flavodehydrogenase before donating them to cyt c?

Indeed, if we now examine transfer to cyt c, the supposed natural acceptor, it has been possible (an exceptional case) to reach, at low ionic strength, conditions where the life time of the stable complex formed with flavocytochrome b_2 is long; so long as to permit a 1e-transfer with a 100% probability after each encounter. The electron tunneling rate within this complex has been determined to be $k=400\ s^{-1}$ at 5°C with activation energy of 3.3 kcal/mol (38). In a study reported elsewhere in this volume (37), it is shown that the interactions which stabilize the complex of cyt c to flavocytochrome b_2 concern two areas on the latter, one on globule 2, electrostatic in its nature with $Z_A Z_B = 3$–5 and the other on globule 1, much less effective and with a lower product of charge factor (three times).

The advantage of the gene fusion that joined the flavodehydrogenase with cyt b_2 is thus now understood. Indeed the isolated flavodehydrogenase moiety is neither able to react with cyt b_2 nor with cyt c, as recalled above, nor with their mixture. On the other hand, cyt c can react with the isolated cyt b_2 core but much more slowly than with flavocyt b_2, because the complex is less stable (39). Thus, in the whole system, the association of the flavodehydrogenase with its specific acceptor (that needs to be attached) provides the possibility of binding cyt c so that a regular electron flow, from bound lactate to flavin, flavin to heme b_2, heme b_2 to cyt c, is able to take place by means of 3 steps of equal rates, at least in H-flavocyt b_2, the more rapid of cyt c reductases.

References

1. Appleby, C.A. and Morton, R.K. (1954) *Nature* 173:749–752.
2. Morton, R.K. and Armstrong, J. (1961) In *Proc. of 5th Int Congress of Biochemistry, Moscow–V Intracellular Respiration*, pp. 213–224.
3. Morton, R.K., Armstrong, J., and Appleby, C.A. (1961) In *Hematin Enzymes*. New York: Pergamon Press, pp. 501–523.
4. Labeyrie, F. (1966) *Bull Soc Franc Physiol Veg* 12:17–28.
5. Hinkson, J.W. and Mahler, H.R. (1963) *Biochemistry* 2:209–216.
6. Ogura, Y. and Nakamura, T. (1966) *J Biochem* 60:77–86.
7. Labeyrie, F. and Baudras, A. (1972) *Eur J Biochem* 25:33–40.
8. Jacq, C. and Lederer, F. (1972) *Eur J Biochem* 25:41–48.
9. Gervais, M., Groudinsky, O., and Labeyrie, F. (1975) *10th Meeting FEBS* (Abst) 1248.
10. Naslin, L., Spyridakis, A., and Labeyrie, F. (1973) *Eur J Biochem* 34:268–283.
11. Gervais, M., Groudinsky, O., Risler, Y., and Labeyrie, F. (1977) *Biochem Biophys Res Comm* 77:1543–1551.
12. Gervais, M. and Tegoni, M. (1980) *Eur J Biochem* 111:357–367.
13. Labeyrie, F., Groudinsky, O., Jacquot-Armand, Y., and Naslin, L. (1966) *Biochim Biophys Acta* 128:492–503.
14. Keller, R., Groudinsky, O., and Wuthrich, K. (1973) *Biochim Biophys Acta* 328:233–238.
15. Guiard, B., Groudinsky, O., and Lederer, F. (1974) *Proc Natl Acad Sci* USA 71:2539–2543.
16. Guiard, B. and Lederer, F. (1979) *J Mol Biol* 135:639–650.
17. Mével-Ninio, M., Risler, Y., and Labeyrie, F. (1977) *Eur J Biochem* 73:131–140.
18. Gervais, M., Labeyrie, F., Risler, Y., and Vergnes, O. (1980) *Eur J Biochem* 111:17–31.
19. Gervais, M. and Risler, Y., unpublished data.
20. Baudras, A. (1973) In *Dynamic Aspect of Conformation Changes in Biological Macromolecules*. C. Sadron (ed.) Dordrecht, Holland: D. Reidel Publ. Co. pp. 181–205.
21. Prats, M. (1978) *Biochimie* 60:77–79.
22. Baudras, A. and Spyridakis, A. (1971) *Biochimie* 53:943–955.
23. Morton, R.K. and Sturtevant, J. (1964) *J Biol Chem* 239:1614–1624.
24. Iwatsubo, M. and Capeillère, C. (1967) *Biochim Biophys Acta* 146:349–366.
25. Lederer, F. (1974) *Eur J Biochem* 46:393–399.
26. Capeillère-Blandin, C., Barber, M.J., and Bray, R.C., this volume.
27. Pompon, D., Iwatsubo, M., and Lederer, F. (1980) *Eur J Biochm* 104:479–488.
28. Capeillère-Blandin, C., Bray, R.C., Iwatsubo, M., and Labeyrie, F. (1975) *Eur J Biochem* 54:549–566.
29. Iwatsubo, M., Mével-Ninio, M., and Labeyrie, F. (1977) *Biochemistry* 16:3558–3566.
30. Forestier, J.P. and Baudras, A. (1971) In *Flavins and Flavoproteins*. H. Kamin (ed.) Baltimore: University Park Press, pp. 599–604.
31. Capeillère-Blandin, C., Iwatsubo, M., Labeyrie, F., and Bray, R.C. (1976) In *Flavins and Flavoproteins*. T.P. Singer (ed.) Amsterdam: Elsevier Scientific Publ. Co. pp. 621–634.
32. Hiromi, K. and Sturtevant, J.M. (1966) In *Flavins and Flavoproteins*. E.C. Slater (ed.) New York: Elsevier, pp. 283–303.
33. Suzuki, H. and Ogura, Y. (1969) *J Biochem Tokyo* 67:277–289.
34. Pompon, D. (1980) *Eur J Biochem* 106:151–159.
35. Capeillère-Blandin, C. (1975) *Eur J Biochem* 56:91–101.
36. Iwatsubo, M., Baudras, A., di Franco, A., Capeillère, C., and Labeyrie, F., (1968) In *Flavins and Flavoproteins*. K. Yagi (ed.) Baltimore: Univ. Park Press, pp. 41–56.
37. Gervais, M., Thomas, M.A., Labeyrie, F., Favaudon, V., and Pochon, F., this volume.
38. Capeillère-Blandin, C., Iwatsubo, M., Testylier, G., and Labeyrie, F. (1980) In *Flavins and Flavoproteins*. K. Yagi and T. Yamano (eds.) Baltimore: University Park Press, pp. 617–630.
39. Capeillère-Blandin, C. and Albani, J. (1981) unpublished results.

Published 1982 by Elsevier North Holland, Inc.
Vincent Massey and Charles H. Williams, Editors
Flavins and Flavoproteins

CHAPTER 137

Localization of the Main Association Areas Between Flavocytochrome b_2 and Cytochrome c by Fluorescence Studies

M. Gervais,* M.A. Thomas,* F. Labeyrie,*
V. Favaudon,† and F. Pochon†

**Centre de Génétique Moléculaire du CNRS - 91190 Gif-sur-Yvette, France; †Fondation Curie, Institut de Radium, Section de Biologie - 91405 Orsay, France*

Among the components of the mitochondrial electron transfer chain, cytochrome c, which is easily prepared in large quantities, was a good candidate for studies on the structural modes of association between two electron carrier proteins. For a c-type cytochrome, it was suggested that this association is controlled by an interaction between some of its positively-charged lysine residues and complementary negatively-charged groups on the respective oxidoreductases (1–3). Microsomal cytochrome b_5 itself exhibits a cluster of negative charges located on the surface around the heme in the region of most accessibility to the solvent, in contrast with the positively-charged belt of c-type cytochromes. This situation led Salemme (4) to propose a hypothetical structure for an intramolecular electron transfer complex between cytochrome b_5 and cytochrome c. Flavocytochrome b_2, a cytochrome c reductase detected only in yeast, is able to form a stable complex with cytochrome c (very likely its physiological acceptor) in particular at low ionic strength (5). It has been established in our laboratory that each protomer of the tetrameric flavocytochrome b_2 has its polypeptide chain folded into three "globules" (6), one of which, n, the cytochrome b_2 core, is a cytochrome b_5-like domain (7) and the two others, ε and β, form together the flavodehydrogenase domain (6).

In this communication, we present information concerning the localization, among these three n, ε, β globules, of the main areas involved in the association with cytochrome c and the stability of the complex that each forms. Instead of ferricytochrome c, we have used a fluorescent derivative, Zn^{2+} cytochrome c.

Materials and Methods

Hansenula anomala flavocytochrome b_2 was prepared according to (8); proteolytic derivatives of flavocytochrome b_2 were prepared according to the methods developed in our laboratory (6, 9). These derivatives are: (1) the monomeric cytochrome core (n) and the tetrameric flavodehydrogenase (FDH_{V8}) which are the two functional moieties of this two-headed enzyme, (2) the monomeric α fragment which is formed by the covalent linkage of the cytochrome core (n) and the ε part of the flavodehydrogenase (6).

Zn-cyt.c was prepared by replacement of iron in Hansenula anomala ferricytochrome c by zinc as described by Vanderkooi et al. (10). Its concentration is measured using $\varepsilon_{418\ nm}=117\ mM^{-1}cm^{-1}$ (11).

The quaternary structure of Zn-cyt.c was investigated by several methods: (1) polyacrylamide (15%) sodium dodecyl sulfate gel electrophoresis patterns of Zn-cytochrome c and native cytochrome c show a single band with identical molecular weight ($\simeq$ 13000 daltons), (2) determination of cyt.c/flavocyt.b_2 and Zn cyt.c/flavocyt.b_2 molar ratios in complexes isolated by molecular sieving in the same conditions gives a mean value twice as high as Zn-cyt.c. (3) In gel filtration, the comparison of elution volumes of Zn cyt.c and cyt.c led to an estimation of molecular weight twice as high for the former, (4) ultracentrifugation equilibrium sedimentation patterns of Zn cyt.c at low ionic strength exhibit a monodisperse behavior with MW$\sim$26000. These results led us to the conclusion that Zn-cyt.c is a noncovalent dimer (possibilities of dimerization of cytochrome c itself have already been described.)

Fluorescence studies were carried out on a Farrand Mark 2 fluorimeter, at 10°C, using for Zn cyt.c: λ excitation: 545 nm, λ emission = 640 nm according to (10) in two different buffers: buffer A: 10 mM phosphate, 10 mM Dl sodium lactate pH 7.2; buffer B: 150 mM phosphate, 10 mM DL sodium lactate, pH 7.2.

Results

Binding of Zn-Cytochrome c to Flavocytochrome b_2

Addition of flavocytochrome b_2 to a solution of Zn cytochrome c results in a decrease of Zn cytochrome c fluorescence intensity according to a typical titration curve which reaches a plateau (Figure 1). Before analyzing such a curve, we have to note that (1) in spite of the light sensitivity of Zn-cytochrome c (10), the loss of fluorescence intensity resulting from successive illuminations is less than 2% of the initial level at the end of each experiment; (2) correction for screening and reabsorbing inner effects of Zn cytochrome c and the fluorescent quenchers in the region of both excitation and emission is less than 2%; (3) the fluorescence spectrum of Zn cytochrome c when excited at 545 nm remains unchanged after addition of any quencher.

These remarks lead us to consider that the quenching of fluorescence observed for each presented experiment significantly demonstrates the existence of actual complexes between Zn-cytochrome c and flavocytochrome b_2 or its derivatives.

The best computer fitting in the case of flavocyt.b_2-Zn cyt.c complex gives $K_d=10^{-7}$ M in buffer A (see the simulated curve, Figure 2). The stoichiometry (Figure 1 and Table 1) is one dimeric Zn cyt.c/protomer b_2. Such a value is different from that found by Vanderkooi et al. (12) for the complex horse heart Zn cyt.c-*S.cerevisiae* flavocyt.b_2 but is in accord with values independantly found by several authors for various cyt.c and *H.anomala* flavocyt.b_2 (6, 13, 14).

Binding of Zn Cytochrome c to Flavocytochrome b_2 Fragments

Flavocytochrome b_2 proteolytic fragments (n, α_c, α_T, FDH_{V8}) form complexes with cytochrome c, both at low ionic strength (Figures 1,2) and high ionic

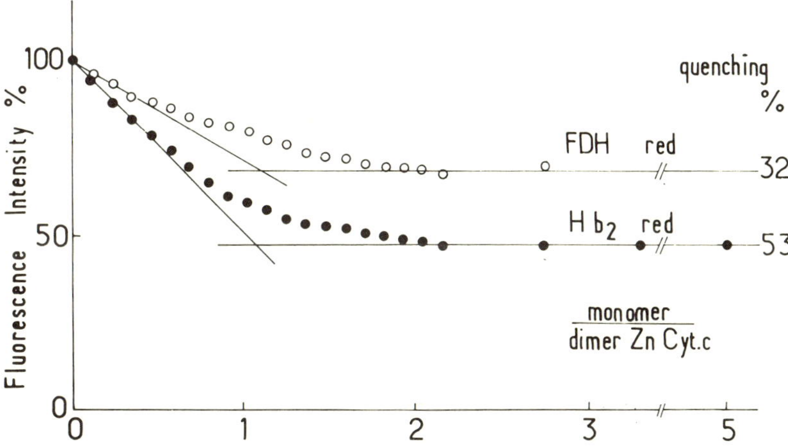

Figure 1. Fluorescence yield of Zn-cyt in presence of flavocytochrome b_2 (●) or its flavodehydrogenase derivative (○). (Zn cyt.c)=2 μM, buffer A.

strength. In buffer A (I=0.04), n excepted, all fragments lead to a titration curve very similar to that of the whole flavocytochrome b_2, while in buffer B (I=0.46) we find for each of them a saturation curve similar to that of n in buffer A. The main results of the experiments are reported in Table 1.

Discussion

These data show the existence of two areas involved on flavocyt.b_2 for its association with Zn-cyt.c. One area has been localized on each of the two functional domains of flavocyt.b_2. If the association with the flavodehydrogenase is essentially electrostatic ($Z_A Z_B = -3$), the one with cyt.b_2 core does not present such a character ($Z_A Z_B = -1$). This is in accord with other results from our laboratory (15). Moreover, α_c and α_T fragments (uncleaved assembly of n and ε), the flavodehydrogenase V_8 (assembly of ε and β) and flavocytochrome b_2 present similar characteristics of association with cytochrome c (K_d and stoichiometry). We conclude that, in the biglobular flavodehydrogenase domain, the globule ε is sufficient to assure cytochrome c binding. It seems that the dimeric character of Zn cytochrome c does not play any role in the interaction with flavocytochrome b_2. This association does not depend on the quaternary structure of flavocytochrome b_2 since the same results are found with monomeric α fragments and flavodehydrogenase or flavocytochrome b_2 which are tetrameric species.

Finally, let us consider a possible relationship between the formation of flavocytochrome b_2-cytochrome c complex and the electron transfer that takes place between these two proteins. We have found a different quenching level of Zn cytochrome c by oxidized and reduced flavocytochrome b_2 as noticed by Vanderkooi et al. (12). Such a difference has been explained by these authors as due to different spectral overlaps between Zn cytochrome c emission and oxidized and reduced flavocytochrome b_2 absorbances (12). In fact, we have

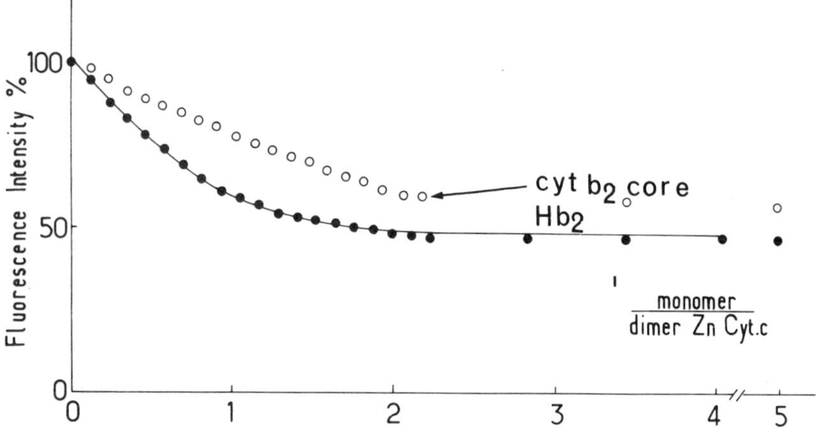

Figure 2. Fluorescence yield of Zn cyt.c in presence of flavocytochrome b_2 (●) or its cytochrome core (○) (Zn cyt.c)=2 μM, buffer A. The solid line corresponds to the simulated curve with $K_d = 10^{-7}$ M.

verified that, when sodium dithionite is used as a reductant, no difference in the quenching level could be observed for oxidized and reduced forms of flavocytochrome b_2. Actually, we only find some effect (9%) when lactate is the reducing agent. A control made with Zn cytochrome c alone in the presence of sodium lactate (DL or L+) leads us to conclude that the observed difference may be due to a direct lactate effect on Zn cytochrome c itself and not to the redox state of flavocytochrome b_2. Moreover, the existence of a similar quenching for the flavodehydrogenase (heme-free) and for the α_c fragment (heme carrying) indicates that the heme b_2-heme c energy transfer emphasized in (12) is probably not the only factor that provokes the quenching level observed.

Conclusion

In a recent work, it has been suggested that the same molecule of cytochrome c may interact with both subunits II and III of cytochrome oxidase (16). Our data concerning flavocytochrome b_2 (a cytochrome c reductase) and cytochrome c lead us to the same kind of conclusion. One interaction area has been localized on each of the two functional domains of flavocytochrome b_2, however the domain which does not seem to be able to transfer electrons to cytochrome c, the flavodehydrogenase domain, binds it with a high affinity, while the domain which transfers electrons to cytochrome c (cytochrome b_2 core) (14) is the domain that makes the minor contribution to the stabilization of the complex. Flavocytochrome b_2 was already proposed to result from a gene-fusion between ancestral genes coding for a cytochrome and a flavodehydrogenase. The above-evoked paradox might justify a selective advantage for such a gene-fusion.

Table 1. Fluorimetric study of complex formation between Zn-cyt.c and flavocytochrome b_2 or its Derivatives.

Species	Monomer/Dimer c at the equivalent point	% Quenching (plateau)	$K_d(\mu M)$ (10 mM phosphate)	$K_d(\mu M)$ (150 mM phosphate)
H	1.1 ± 0.1	53 ± 2	0.1	4
FDH V_8	1.1 ± 0.1	32 ± 2	0.1–0.2	3
α_C	1.0 ± 0.1	34 ± 2	0.1	4
α_T	1.0 ± 0.1	52 ± 2	0.15	3
η_T	[a]	[a]	$\simeq 1$	$\simeq 3$

[a] In spite of the absence of an actual plateau these curves were fitted with a quenching value of 52% which corresponds to the quenching observed for H and α_T.

H: *H.anomala* "intact" flavocty.b_2; T: trypsic; C: clostripaic.

References

1. Hettinger, T.P. and Harbury, H.A. (1965) *Biochem* 4:2585–2589.
2. Wada, K. and Okunuku, K. (1969) *J Biochem* 66:249–262.
3. Smith, L. and Minnaert, K. (1965) *Biochim Biophys Acta* 105:1–14.
4. Salemme, F.R. (1976) *J Mol Biol* 102:563–568.
5. Baudras, A. (1973) In *Dynamic Aspects of Conformation Changes in Biological Macromolecules.* Sadron, C. (ed.) Dordrecht-Holland: D. Reidel Publ. Co., pp. 181–205.
6. Gervais, M., Groudinsky, O., Risler, Y., and Labeyrie, F. (1977) *Biochem Biophys Res Comm* 77:1543–1551.
7. Guiard, B., Groudinsky, O., and Lederer, F. (1974) *Proc Nat Acad Sci USA* 71:2539–2543.
8. Labeyrie, F., Baudras, A., and Lederer, F. (1978) In *Methods in Enzymology.* Vol. 53. Fleischer, S. and Packer, L. (eds.) New York: Academic Press, pp. 238–256.
9. Gervais, M., Corazzin, S., Risler, Y., and Labeyrie, F., unpublished results.
10. Vanderkooi, J.M., Adar, F., and Erecinska, M. (1976) *Eur J Biochm* 64:381–387.
11. Thomas, M.A. and Favaudon, V., unpublished results.
12. Vanderkooi, J.M., Glatz, P., Casadei, J., and Woodrow, G.V. III, (1980) *Eur J Biochem* 110:189–196.
13. Prats, M. (1977) *Biochimie* 59:621–626.
14. Capeillère-Blandin, C., Iwatsubo, M., Testylier, G., and Labeyrie, F. (1980) In *Flavins and Flavoproteins.* Yagi, K. and Yamano, T. (eds.) Baltimore: University Park Press, pp. 617–620.
15. Capeillère-Blandin, C., personal communication.
16. Moreland, R.B. and Dockter, M. E. (1981) *Biochem Biophys Res Comm* 99:339–346.

Published 1982 by Elsevier North Holland, Inc.
Vincent Massey and Charles H. Williams, Editors
Flavins and Flavoproteins

CHAPTER 138

Differences in Electron Transfer Rates Among Prosthetic Groups Between Two Homologous Flavocytochrome b_2 (L-Lactate Cytochrome c Oxidoreductase) From Different Yeasts

C. Capeillère-Blandin, M.J. Barber,* and R.C. Bray*

*Centre de Génétique Moléculaire du CNRS 91190 Gif-sur-Yvette, France; *School of Molecular Sciences, University of Sussex, Brighton, BN19QJ United Kingdom*

Parallel and detailed EPR rapid-freezing and stopped-flow absorbance studies have been performed previously on the course of reduction of bakers' yeast flavocytochrome b_2 at 24°C (symbol S_x, a cleaved form of molar activity 65 s^{-1} at 5°C) (1). They allowed the characterization of the sequential electron exchange steps involved between the electron donors L-lactate and FMN, and the one-electron acceptor, the heme b_2. They showed the main limiting step of the reduction of flavin and heme to be at the level of the electron transfer between bound substrate and flavin; simulation studies afforded minimal estimations of the rate constants involved in the following intramolecular processes, i.e., the transfer of one electron from reduced flavin to heme (600 s^{-1}) and the dismutation exchange of one electron between two flavins (100 s^{-1}) at 24°C (2). Preliminary results from experiments performed at 5°C, supported that the proposed rate constants were not markedly underestimated (3).

The *Hansenula anomala* yeast contains a homologous flavocytochrome b_2. Nevertheless, marked differences are observed in their steady-state kinetic parameters; in particular the H-enzyme molar activity is 4 to 5-fold higher than that of the S_x-enzyme (4). In order to explain the kinetic differences and define the reaction steps rate-limiting in the molar activity, a detailed analysis of rapid reduction kinetics of the H-flavocytochrome b_2 (symbol H, intact form of molar activity 220 s^{-1} at 5°C), is presented in this work, along with a redox equilibrium study; kinetic data are compared with those from S_x enzyme reinvestigated for that purpose at 5°C.

Oxidation-Reduction Equilibrium Between Prosthetic Groups

The mid-point reduction potentials of FMN and heme b_2 were determined by two complementary methods: potentiometric titration by the substrate, L-lactate, with quantification of the reduced species [reduced heme, H_r and sum of semiquinone, F_{sq}, and hydroquinone, F_r, reduced forms of flavin, corresponding to $(1\text{-}F_0)$] performed by spectrophotometry at suitable wavelengths (Figure 1); and anaerobic titration of the enzyme by L-lactate, following the EPR signals of semiquinone and Fe^{3+} heme, H_o, species (Figure 2) (1). The titration confirms that the enzyme takes up 3 electrons per protomer, i.e., 1 on the heme and 2 on the flavin; no other electron acceptors are detectable in the enzyme.

Data from absorbance titration experiments carried out in the presence of dyes were interpreted using classical procedures to estimate the oxidation-reduction potentials and the number n of electrons donated (5); as illustrated in Figure 1, the $E_{m,7}$ values determined were 12 ($\pm$2) mV for heme (with $n=1.05\pm0.05$) and +5 ($\pm$2) mV for flavin (with $n=1.24\pm0.5$) at 10°C. The flavin titration curve ($1<n<2$) indicated that the semiquinone form is certainly present during the titration at the same time as the fully oxidized and reduced species. The correct estimation of the redox potential for the F_0/F_{sq} system can be calculated only using the measured content of F_{sq} in the course of similar titration followed by EPR detection as described in Figure 2. The amount of flavin radical reaches its maximal value of 38% when one mole of L-lactate per mole of protomer was added. When EPR and absorbance

Figure 1. Anaerobic potentiometric and absorbance titration of heme b_2 and FMN in H-flavocytochrome b_2. Conducted anaerobically by addition of L-lactate in the presence of mediator dyes (22 μM enzyme, 3 μM pyocyanine, 3 μM methylene blue in a 0.1 M phosphate buffer, pH 7.0 10°C). Absorbance titrations were performed at 557 nm for the heme b_2 isosbestic point] for FMN.

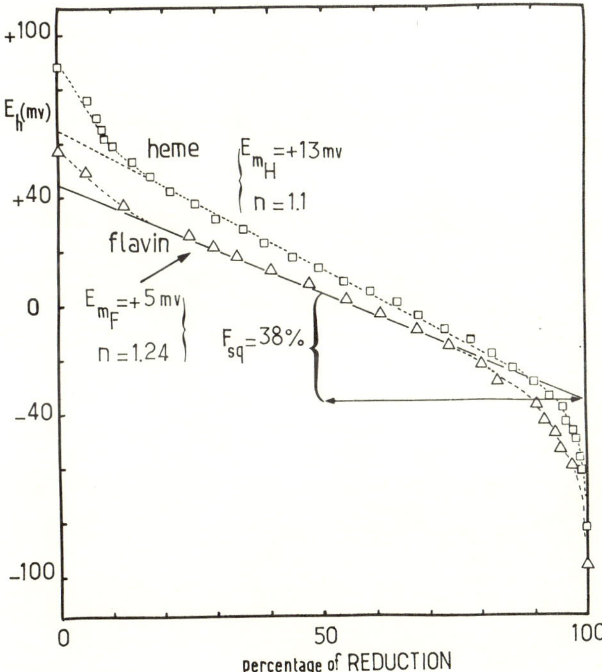

titrations data are pooled for comparison, it is clear that there is a good agreement for data concerning the Fe^{3+} heme in spite of problems of comparing measurements on liquid and frozen solutions. The best fit to the EPR data is obtained for the following set of $E_{m,7}$ parameters, assuming $E_{mHo/Hr} = +12$ mV; $+6$ (± 7) mV for F_o/F_{sq} and -4 (± 6) mV for F_{sq}/F_r and $+1$ (± 9) mV for F_o/F_r couple, at 5°C. The same fitting performed on our previous results on the S_x-enzyme, at 20°C (1) leads to different values -44 (± 8) mV for F_o/F_{sq}, -57 (± 9) mV for F_{sq}/F_r and -51 (± 16) mV for the F_o/F_r couple, with $+6$ mV for the H_o/H_r couple (1).

In conclusion, two differences are apparent between the titration results on H-enzyme and those on S_x-enzyme; (1) whereas the heme potential is unmodified, the potentials of flavin redox couples have shifted to more positive potentials for all of them, keeping constant the F_{sq} stability constant; (2) a rather smaller difference in potentials between the F_o/F_{sq} and H_o/H_r systems was found $+6$ mV instead of 50 mV possibly characteristic of the H-enzyme or due to a rather strong temperature effect (-3 mV/dg) affecting mainly the F_o/F_{sq} and H_o/H_r equilibrium reaction.

Rapid Time-Courses of Prosthetic Group Reduction

A clear understanding of the electron exchange between FMN and heme inside the enzyme molecule requires precise knowledge of the reduction time-course of each transient redox species. In Figure 3 is presented the pattern of concentration changes for the different redox species during the H-flavocytochrome b_2 reduction. This pattern does not change with enzyme

Figure 2. Anaerobic titration of heme b_2 and semiquinone by L-lactate. EPR measurements were carried out on a 90 μM flavocytochrome b_2 solution mixed with L-lactate at 5°C (0.1 M phosphate buffer pH 7.0) and frozen after a 20 s reaction time. Shaded areas are related to spectrophotometric determinations of H_o (lower) and F_o (upper) amounts in the course of similar titrations performed as illustrated in Figure 1.

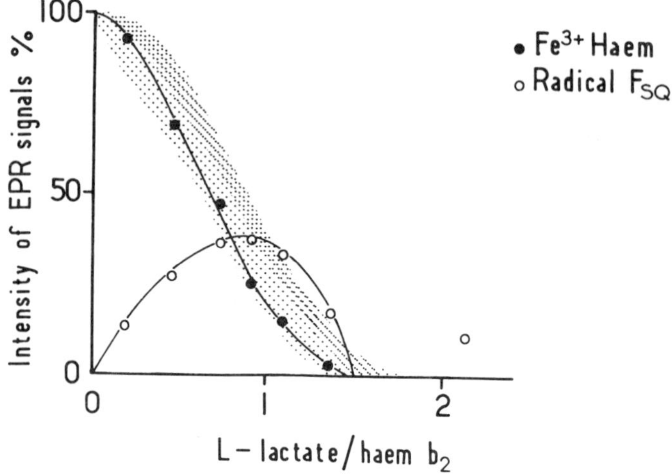

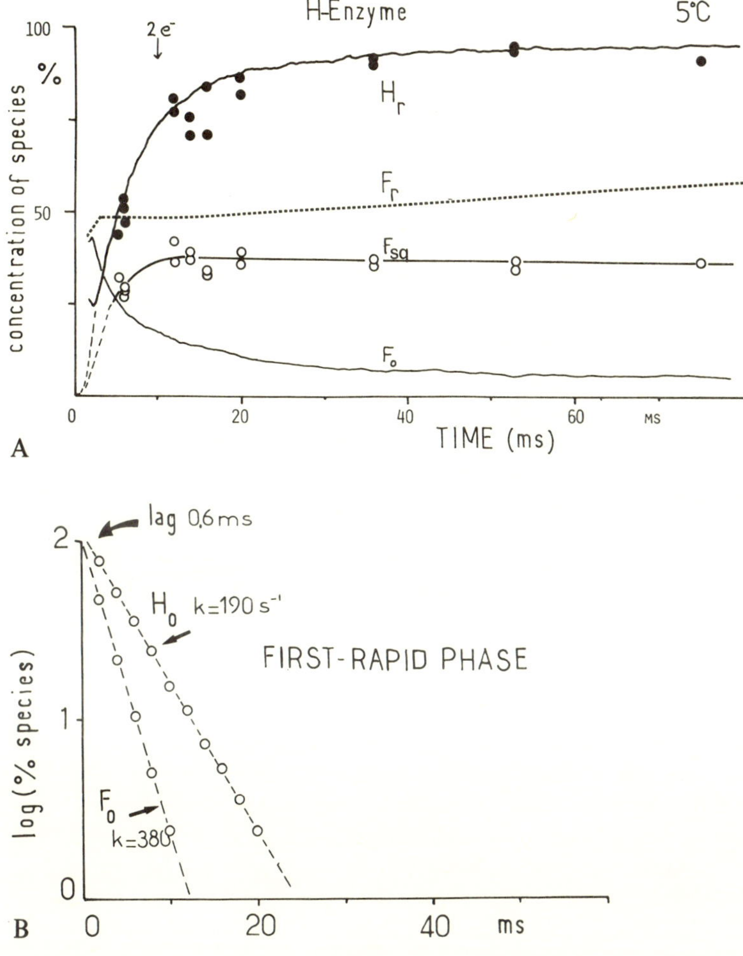

Figure 3. (A) Reaction pattern for the different redox species during H-flavocytochrome b_2 rapid reduction at 5°C. Along with stopped-flow absorbance recordings at two wavelengths, 557 nm for H_r and 438.3 nm for F_0 (10 μM flavocytochrome b_2, 10 mM L-lactate, 0.1 M phosphate buffer, pH 7.0), are given EPR rapid-freezing results for H_r (solid circles) and F_{sq} (circles) (94 μM flavocytochrome b_2, 10 mM L-lactate) and calculated curve for F_r (1-F_o-F_{sq}; dotted lines). (B) First-order plots of the stopped flow recordings for H_o and F_o. The heme line extrapolates to a time significantly different from zero: lag time: 0.6±0.2 ms.

concentrations over a large range; the EPR and absorbance results for H_r are in excellent agreement. Therefore the electron distribution very likely results from intramolecular electron exchange.

The main features of the time-courses are essentially the same (except when otherwise mentioned) as those previously observed by us for S_x-enzyme (1): (a) The reaction is biphasic (phases I (80%) and II) for F_o and H_r corresponding to the superposition of two exponential processes. A total of 2 electrons per protomer is distributed among H_r, F_{sq}, and F_r at 10 ms in the end of the rapid phase I, while the entry of the third electron per protomer occurs in the much slower phase II at a rate constant irrelevant to involvement in the turnover (Table 1). This result confirms that only rapid processes involved in phase I can participate in the overall catalytic reaction, as discussed in (1). (b) In contrast with the data on S_x-enzyme, there is a marked difference between the apparent reduction rate constants of F_o (k_{F_I}) and H_o (k_{H_I}) (Table 1); the time-courses observed for F_o and H_r are distinct and well separated on time scale with a significant lag time for heme reduction in the range of 1ms. The desynchronization (with $k_F > k_H$) and lag phase between flavin and heme reduction confirm the sequential electron transfer mechanism, i.e.,

$$\text{L-lact} \xrightarrow{k_F} \text{FMN} \xrightarrow{k_{F-H}} \text{Heme},$$

with k_{F-H} value not very high compared to k_F. (c) Moreover by considering the initial burst of F_r which precedes the appearance of F_{sq} and H_r at the very beginning of phase I, an estimation of the relative rate constant values for the rapid formation and consumption of F_r can be obtained; for H-enzyme 45% is observed instead of 25% for S_x-enzyme, this result is consistent with a faster formation process as observed in H-enzyme if the following step was unchanged.

Considering these data (Table 1) it appears that the main difference between H (intact) and S_x (cleaved) flavocytochrome b_2, lies at the level of the bound L-lactate to flavin electron transfer step which is no longer the rate-limiting step for the further electron transfers and becomes 7-fold faster in the H-enzyme. Consequently it allows for the first time an accurate evaluation of the rate constants involved in the intramolecular electron exchange between F_r and H_o of the same active site. Up to now minimal estimation of these parameters has been provided by simulation studies (1–3, 7, 8). Interpretation of our present results can first be attempted using an oversimplified scheme based on a two-step mechanism, with the following rate-constant values:

$$\begin{matrix} H_o & & H_o & & H_r \\ & \xrightarrow{380} & & \underset{110}{\overset{380}{\rightleftharpoons}} & \\ F_o & & F_r & & F_{sq} \end{matrix}$$

an equilibrium electron distribution between flavin and heme was assumed as supported by the very close redox potential difference between the 2 systems (see above). It leads to a reasonable fitting of the heme reduction which reaches its equilibrium level at the end of phase I. It also accounts for the lag time between F_o and H_r time-courses.

However, the simulated time-course of H_r is not biphasic, and it does not take into account the concentration of F_{sq} being always lower than that of H_r. Therefore we try to interpret in terms of elementary steps phase II which leads to the total reduction of the enzyme using previously published schemes (2, 8). Apparently preliminary data fail to give a reasonable simulation of the reaction pattern of the 4 redox species; due to the difficulty to explain, at the same time, the plateau of F_{sq} which extends over 200 ms before decaying with a $t_{1/2} = 1.3$ s and the continuous decrease of F_o. Nevertheless further developments of simulation studies are currently in progress.

Table 1. Comparison of Kinetic Data for H and S_X-Flavocytochrome b_2 at Saturating L-Lactate Concentration (5°C).

	Flavin, F_O		Heme, H_R		
	Phase I $k_{F_I}{}^a(s^{-1})$	Phase II $k_{F_{II}}{}^a(s^{-1})$	Phase I $k_{H_I}{}^a(s^{-1})$	Phase II $k_{H_{II}}{}^a(s^{-1})$	Lag-time (ms)
H-enzyme	380	25	200	13	.5–1
S_X-enzyme	42	1.2	36	1.4	.5–1.5

	Semiquinone, F_{SQ} Max (%)	Semiquinone, F_{SQ} $t_{1/2}$(ms)	Reduced flavin, F_R burst (%)	Molar activity TN(s^{-1})	Flavin reduction rate/ overall reaction rate $2k_{F_I}/TN$
H-enzyme	38	4	45	220	3.5
S_X-enzyme	40	20	25	65	1.3

[a]Apparent first-order rate constant.

If now the rates involved in the overall catalytic reaction of H-enzyme are compared with those of reductions of flavin and heme groups, it appears that the flow of 2 electrons via the flavin is more than sufficient to account for the overall molar activity ($2\ k_F > TN$). Data on S_x-enzyme show that the reduction rate of flavin essentially determines the rate of all the following sequence of steps (Table 1). In contrast, a sequence of two successive electron transfer steps which leads to the electron-pair splitting between prosthetic groups of the same active site, with formation of H_r and F_{sq}, would be mainly rate-limiting in the overall electron transfer catalysed by H-flavocytochrome b_2.

Methods

Same as described in reference 1.

ACKNOWLEDGMENTS
The authors are particularly indebted to Dr. F. Labeyrie in whose laboratory this work was carried out, for many stimulating discussions and helpful suggestions concerning the manuscript. One of us (C.C.B.) thanks the E.M.B.O.

References

1. Capeillère-Blandin, C., Bray, R.C., Iwatsubo, M., and Labeyrie, F. (1975) *Eur J Biochem* 54:549–566.
2. Capeillère-Blandin, C. (1975) *Eur J Biochem* 56:91–101.
3. Capeillère-Blandin, C., Iwatsubo, M., Labeyrie, F., and Bray, R.C. (1976) In *Flavins and Flavoproteins*. Singer, T.P. (ed.) Amsterdam: Elsevier, pp. 621–634.
4. Labeyrie, F., Baudras, A., and Lederer, F. (1978) *Methods in Enzymol* 53:238–256.
5. Clark, W.M. (1960) In *Oxidation-Reduction Potentials of Organic Systems*. Baltimore: Williams and Wilkins Company, pp. 182–203.
6. Bray, R.C., Lowe, D.J., Capeillère-Blandin, C., and Fielden, E.M. (1973) *Biochem Soc Trans* 1:1067–1072.
7. Pompon, D., Iwatsubo, M., and Lederer, F. (1980) *Eur J Biochem* 104:479–488.
8. Pompon, D. (1980) *Eur J Biochem* 106:151–159.

Published 1982 by Elsevier North Holland, Inc.
Vincent Massey and Charles H. Williams, Editors
Flavins and Flavoproteins

CHAPTER 139

Kinetics of Ligand Binding by the Flavin Moiety of *Chromatium Vinosum* Flavocytochrome c

M.A. Cusanovich* and T.E. Meyer**

**Department of Biochemistry, University of Arizona, Tucson, Arizona; **Department of Chemistry, University of California at San Diego, La Jolla, California*

Summary

The kinetics of adduct formation by the flavin moiety of *Chromatium vinosum* flavocytochrome c with sulfite, thiosulfate, glutathione (GSH), and cyanide have been investigated by following the loss of 480 nm absorbance. Adduct formation was found to be kinetically homogeneous with second-order rate constants determined over the pH range 4.5 to 10. Monovalent ligands (HSO_3^-, CN^-, and GS^-) reacted rapidly with flavocytochrome c; the rate constants were 10^4-10^6 $M^{-1}S^{-1}$. Divalent ligands ($S_2O_3^=$, $SO_3^=$) were reactive, but with much smaller rate constants. A protein linked ionization with pK 6.6, possibly due to a histidine residue, affects the thiosulfate and perhaps the sulfite adduct formation reactions. The flavin-sulfite adduct recolored at alkaline pH in a first-order reaction, which involves a second protein linked ionization with pK 8.4, consistent with participation of a cysteine residue. Based on the results reported, it may be concluded that these reactions are unique to the flavocytochromes c: (1) in the variety of reactive ligands, (2) in the relative rates of bisulfite and sulfite reactions, and (3) in alkaline decomposition of the sulfite adduct.

Introduction

Flavocytochromes c have been isolated from four species of purple and green phototrophic bacteria (1,2) which utilize thiosulfate as well as sulfide as reductants for photoautotrophic growth on carbon dioxide. Plasmid-encoded flavocytochromes c have also been isolated from a *Pseudomonas putida* in which they function as p-cresol dehydrogenases (3–5). The phototrophic bacterial flavocytochromes c catalyze oxidation of sulfide to sulfur in a reaction inhibited by cyanide but not by carbon monoxide (6,7). The cyanide presumably inhibits sulfide oxidation through formation of a flavin adduct, a reaction which was found to occur with thiosulfate, sulfite, and mercaptans in addition to cyanide (8). We have now studied the kinetics of adduct formation as a means of further defining the physiological role of thiosulfate and sulfite in the catalytic mechanism of flavocytochromes c from phototrophic bacteria.

Materials and Methods

Chromatium vinosum strain D flavocytochrome c was prepared using the procedure devised by Bartsch (9). The protein used in the experiments reported had a 280 nm/410 nm ratio of .54 and a 475 nm/525 nm ratio of 1.25, which is an indication of purity and of uncomplexed flavin respectively. Studies were performed over a range of pH in Tris-Glycine-Sodium Acetate-Potassium Phosphate (TGAP buffer), 25 mM in each component. Stock solutions of ligand and protein were prepared with this buffer. Kinetic studies were conducted in a Durrum-Gibson stopped-flow spectrophotometer with a mixing time of 3 msec. Experiments were performed in air at 20° and absorbance changes were monitored at 480 nm. Raw data were analyzed with the aid of a Data General Nova II computer.

Results and Discussion

For all reactions, addition of ligand resulted in a bleach of the flavin absorbance at 480 nm with concomitant formation of a broad charge transfer band in the near infrared centered near 670 nm as previously reported (8). The observed kinetics were identical whether observed at 480 nm or 670 nm. Further, in the presence of an excess of ligand, all reactions were accurately pseudo-first-order for at least three half-lives and yielded linear second-order plots (observed rate constants vs ligand concentration). Figure 1 presents the second-order rate constants determined as a function of pH for the ligands-$HSO_3^-/SO_3^=$; HCN/CN^-; $S_2O_3^=$; and GSH/GS^-. The solid lines given in Figure 1 are the calculated curves for a single ionization (n=1) at the pK values and limiting rates indicated in the figure caption. It must be pointed out that in some cases the pK values determined are estimates as *Chromatium* flavocytochrome c is unstable below pH 4.5 and above 10; hence the limiting rate constants were not well defined.

Figure 1A demonstrates that both $SO_3^=$ and HSO_3^- are reactive with the flavin moiety of flavocytochrome c. Sodium bisulfite has a rate constant approximately ten times that of sodium sulfite. The effect of pH on the observed second-order rate constant is consistent with a pK value of 6.7, which is in excellent agreement with that expected for $HSO_3^-/SO_3^=$ ionization (pK=6.75). However, the observation that bisulfite is more reactive than sulfite is contrary to the results of Muller and Massey (10), who reported that the rates of adduct formation between sulfite and synthetic flavin were maximal at pH values above 7. In view of these observations, it can be concluded that the protein moiety of flavocytochrome c must substantially alter the chemistry of the bound flavin relative to free flavins. The reaction of sulfite and bisulfite with 10 μM flavocytochrome c gave maximal flavin bleach even at dilutions as low as 30 μM sulfite/bisulfite, which indicates that the dissociation constant must be less than 20 μM. Direct titration at pH 5.5 gave a dissociation constant of 4 μM, which should be considered an upper estimate because of the following limitation. At pH values of 7 and above, sulfite adduct formation was followed by first-order recoloring of the flavin (observed at 480 nm). This recoloring reaction has a rate constant which increases with increasing pH and has a pK of 8.4 (Figure 1A), consistent with ionization of a

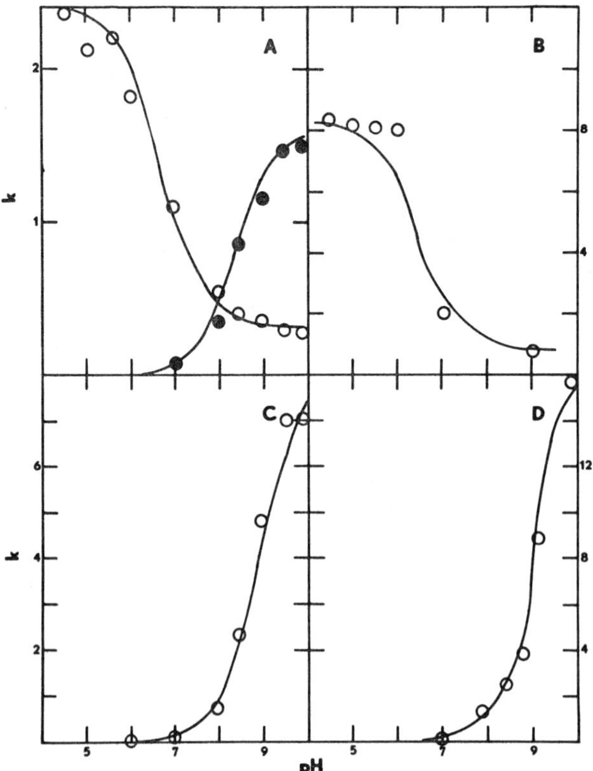

Figure 1. Effect of pH on adduct formation by *Chromatium vinosum* flavocytochrome c. (A) Reaction with sodium sulfite. (Open circles) adduct formation, $k \times 10^{-4}$ $M^{-1}s^{-1}$, solid line calculated with a pK of 6.7 with the limiting rate at low pH, $k_a = 2.4 \times 10^4$ $M^{-1}s^{-1}$ and the limiting rate at high pH, $k_b = 3 \times 10^3$ $M^{-1}s^{-1}$. (Filled circles) flavin recoloring, $k \times 10$ s^{-1}, solid line calculated with a pK of 8.4, $k_a = 0$, $k_b = 0.16$ s^{-1}. (B) Reaction with sodium thiosulfate. Solid line calculated for a pK of 6.6, $k_a = 8.2$ $M^{-1}s^{-1}$, $k_b = 0.8$ $M^{-1}s^{-1}$. (C) Reaction with sodium cyanide, $k \times 10^{-5}$ $M^{-1}s^{-1}$. Solid line calculated for a pK of 8.9, $k_a = 8 \times 10^5$ $M^{-1}s^{-1}$. (D) Reaction with reduced glutathione, $k \times 10^{-5}$ $M^{-1}s^{-1}$. Solid line calculated for a pK of 9.2, $k_a = 0$, $k_b = 1.8 \times 10^6$ $M^{-1}s^{-1}$.

cysteine (pK=8.3) (of which there are approximately six residues of undetermined oxidation state in the flavin subunit).

The reaction of sodium thiosulfate with flavocytochrome c was found to be pH dependent as shown in Figure 1B. The pH dependence can be fit with a pK of approximately 6.6, which might be due to ionization of a histidine residue (of which there are six in the flavin subunit). This ionization may also affect the sulfite/bisulfite kinetics in a manner similar to that of thiosulfate in light of the observation (10) that the effect of pH on the reaction of sulfite with free flavin is markedly different from what we observe with flavocytochrome c. At none of the pH values studied was thiosulfate adduct formation followed by recoloring as noted with sulfite on the time scale of the stopped flow apparatus. The reaction of thiosulfate with flavocytochrome c did not result in the same extent of bleach at each dilution, which allowed estimation of dissociation constants of 9 mM at pH 6 and 60 mM at pH 9.

The reactions of reduced glutathione (GSH) and of sodium cyanide with flavocytochrome c were found to be similar in that adduct formation was rapid and increased with increasing pH. As shown in Figures 1C and 1D, adduct formation could be described by pK values of approximately 8.9 and 9.2 for cyanide and glutathione respectively. The pK values are in reasonable agreement with those for the ionization HCN (pK=9.3) and the thiol of reduced glutathione (pK=8.7). For both the glutathione and cyanide, exact quantitation of the kinetic pK values is difficult due to the instability of flavocytochrome c above pH 10. Nevertheless, the results are consistent with CN^- and GS^- being the reactive species. No recoloring was observed when glutathione was the ligand. However, at pH 6.0 with cyanide as the ligand, a very slow recoloring was noted ($t_{1/2}$ on the order of minutes). This recoloring reaction appeared to be dependent on the HCN/CN^- concentration, but was not studied further. Under anaerobic conditions, adduct formation with glutathione was followed by a slow zero-order reduction of heme (approximately 10 μM min^{-1}). This result suggests that the flavocytochrome c glutathione adduct serves as an intermediate in the reduction of heme, presumably due to reaction with a second molecule of reduced glutathione. One possibility is that the products of the rate-limiting reaction step are oxidized glutathione and reduced flavin, which in turn rapidly reduces the heme and reforms an adduct with a third glutathione molecule. If this interpretation is correct, it suggests a mechanism by which sulfite and thiosulfate could be catalytically oxidized by flavocytochrome c through participation of a mercaptan cofactor analogous to the mechanism of action of Adenosine PhosphoSulfate (APS) reductase in which the flavin sulfite adduct serves as an intermediate in flavin reduction through participation of an adenosine monophosphate cofactor (11). The rate of reduction of flavocytochrome c heme by various mercaptans in the presence and absence of the adduct forming ligands is being investigated.

ACKNOWLEDGMENTS

This work was supported in part by the National Science Foundation, Grant PCM-7804349 and Public Health Service Center Development Award K04EY00013 to M.A.C. and National Institutes of Health Grant GM-18528 to T.E.M.

References

1. Bartsch, R.G., Meyer, T.E., and Robinson, A.B. (1968) *Structure and Function of Cytochromes.* Okunuki, K., Kamen, M.D., and Sekuzu, I. (eds.) Baltimore: Univ. Park Press, pp. 443–451.
2. Bartsch, R.G. (1978) *The Photosynthetic Bacteria.* Clayton, R.K. and Sistrom, W.R. (eds.) New York: Plenum Press, pp. 249–279.
3. Hopper, D.J. and Taylor, D.G. (1977) *Biochem* 167:155–162.
4. Keat, M.J. and Hopper, D.J. (1978) *Biochem J* 175:649–658.
5. Hopper, D.J. and Kemp, P.D. (1980) *J Bact* 142:21–26.
6. Yamanaka, T. and Kusai, A. (1976) *Flavins and Flavoproteins.* Singer, T.P. (ed.) Amsterdam: Elsevier, pp. 292–302.

7. Fukumori, Y. and Yamanaka, T. (1969) *J Biochem* 85:1405–1414.
8. Meyer, T.E. and Bartsch, R.G. (1976) *Flavins and Flavoproteins*. Singer, T.P. (ed.) Amsterdam: Elsevier, pp. 312–317.
9. Bartsch, R.G. (1971) *Methods in Enzymology* 23:344–363.
10. Muller, F. and Massey, V. (1969) *J Biol Chem* 244:4007–4016.
11. Bramlett, R. (1973) PhD Thesis, University of Georgia.

PART X D:

Other Complex Flavoproteins

Published 1982 by Elsevier North Holland, Inc.
Vincent Massey and Charles H. Williams, Editors
Flavins and Flavoproteins

CHAPTER 140

Flavoproteins in Sulfur Metabolism

Harry D. Peck, Jr. and Royce N. Bramlett

Department of Biochemistry, University of Georgia, Athens, Georgia

The biological reduction of sulfate to sulfide and the oxidation of reduced forms of inorganic compounds of sulfur to sulfate are extensive processes in our environment, and it has been estimated that fully 75% of the sulfur in the crust of the earth has been biologically cycled. Microorganisms play a major role in the gross cycling of sulfur compounds, but plants have the ability to reduce sulfate for the biosynthesis of amino acids and cofactors, and both plants and animals can oxidize reduced sulfur compounds to sulfate. The three predominant forms of sulfur in nature are sulfate, elemental sulfur, and sulfide. Sulfate, which is formed from the oxidation of sulfide or elemental sulfur, can result in the formation of H_2SO_4 with concomitant acidification of the environment (to pH 1.0). This in turn can result in injury to agricultural lands and, most spectacularly, the corrosion of stone. This aspect has been of particular concern where heavily industrialized areas pollute with H_2S which, on oxidation to H_2SO_4, has caused extensive damage to monuments of antiquity. The reduction of sulfate to sulfide and its impact on the environment is more obvious because of our olfactory sensitivity to H_2S. The environmental effects of sulfide are derived largely from the chemistry of H_2S and have been summarized by Postgate (1). These include the generation of anaerobic conditions by removing O_2, formation of alkaline environments by the spontaneous volatilization of H_2S, precipitation of heavy metals as metal sulfides, and removal of H_2 which contributes to the anaerobic corrosion of iron by sulfate-reducing bacteria. The economic manifestations of sulfate range from the purely aesthetic (few enjoy the odor of H_2S) to the formation of heavy metal ores.

An outline of the major types of microorganisms involved in the transformations of inorganic sulfur compounds is shown in Figure 1. The reduction of sulfate to sulfide can be divided physiologically into two types: biosynthetic sulfate reduction in which organisms reduce only enough sulfate to meet their nutritional requirements for reduced sulfur compounds, and respiratory sulfate reduction in which organisms utilize sulfate as their terminal electron acceptor and produce extensive amounts of sulfide. The former process is catalyzed by most aerobic and anaerobic bacteria and by plants; the latter process is restricted to the strictly anaerobic sulfate reducing bacteria, species of *Desulfovibrio* and *Desulfotomaculum*. Tetrathionate ($S_4O_6^{-2}$), trithionate ($S_3O_6^{-2}$), thiosulfate ($S_2O_3^{-2}$), and elemental sulfur (S^o) can also serve as terminal electron acceptors for the growth of microorganisms (2) but, except for S^o, these bacteria have not been extensively studied. Pfennig and Biebl (3) have

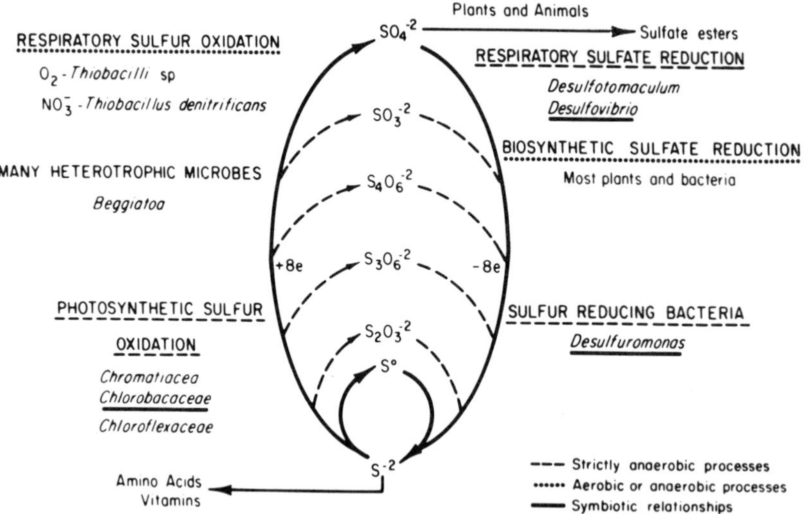

Figure 1. The microbial sulfur cycle.

recently described a new physiological type of strictly anaerobic bacteria, *Desulfuromonas acetoxidans*, which can utilize S° as its terminal electron acceptor for the complete oxidation of acetate, malate, and ethanol to CO_2 with the formation of sulfide. *D. acetoxidans* was first isolated from a consortium with a green sulfur bacterium, *Chlorobium sp.* which has the capability of oxidizing S^{-2} to S° in the light and fixing CO_2. This appears to represent a mini-sulfur cycle and suggests the possibility of other cyclic microbial associations involving $S_2O_3^{-2}$, $S_3O_6^{-2}$, $S_4O_6^{-2}$, and possibly SO_3^{-2}.

Two physiological types of microorganisms, the green and purple photosynthetic bacteria and the *Thiobacilli*, have been most extensively studied as regards the oxidation of reduced compounds of sulfur. Photosynthetic sulfur oxidation can involve $S_2O_3^{-2}$, $S_4O_6^{-2}$, $S_3O_6^{-2}$, S°, and S^{-2}; the oxidation of S^{-2} classically involves a two step process in which S^{-2} is first oxidized to S°, and subsequently, the S° to sulfate. In the case of the purple sulfur bacteria, the S° is deposited intracellularly, but the green sulfur bacteria deposit the sulfur extracellularly where it is available for reduction by other bacteria. Photosynthetic sulfur oxidation is an anaerobic process and under laboratory conditions can be linked to respiratory sulfate reductions and the sulfur cycle driven by light. In contrast, respiratory sulfur oxidation is an aerobic process except in the case of *T. denitrificans* where it is coupled to the reduction of nitrate to dinitrogen.

Two recent discoveries have reemphasized the importance of the reduction and oxidation of sulfur compounds in the biosphere. In marine environments which have a high sulfate concentration, usually about 28 mM, sulfate reduction rather than methane formation appears to be the terminal step in the anaerobic decomposition of complex organic compounds, and new species of microorganisms have been identified to account for these complex conversions

(4). The second discovery involves the finding of deep ocean thermal springs which support life in the dark (5). This ecosystem is not supported by photosynthesis but rather bacterial chemosynthesis driven by energy derived from the oxidation of reduced sulfur compounds originating from geothermal sources. These thiobacilli then serve as a food source for the exotic ecosystem.

An abbreviated representation of the enzymology of respiratory sulfur metabolism is presented in Figure 2. It should be noted that sulfite is a key intermediate in both sulfate reduction and the oxidation of reduced sulfur compounds and that similar, if not identical enzymes can be involved in both the reduction and oxidation of sulfur. Prior to its reduction or utilization in biological systems, sulfate must react with ATP to form adenylyl sulfate (APS) and inorganic pyrophosphate (6). APS can then be phosphorylated in the 3′ position to form 3′-phosphoadenylyl sulfate which is used for sulfate esterification reactions (7), reduced in the biosynthetic pathway of sulfate reduction by APS (or PAPS) sulfotransferase (8, 9), or reduced in the respiratory pathway to AMP and sulfite by the flavoprotein APS reductase (10). In the respiratory pathway, sulfite is next reduced to sulfide in a six-electron reduction by one of three nonheme iron siroheme bisulfite reductases which can be differentiated on the basis of their optical spectra (11–13). In the oxidation of sulfide and elemental sulfur to sulfite, the basic problem is the nature of the substrates and products. Flavocytochromes c have been isolated from *Chlorobium limicola* (14) and *Chromatium vinosum* (15) which exhibit sulfide-cytochrome reductase

Figure 2. Enzymology of the microbial sulfur cycle.

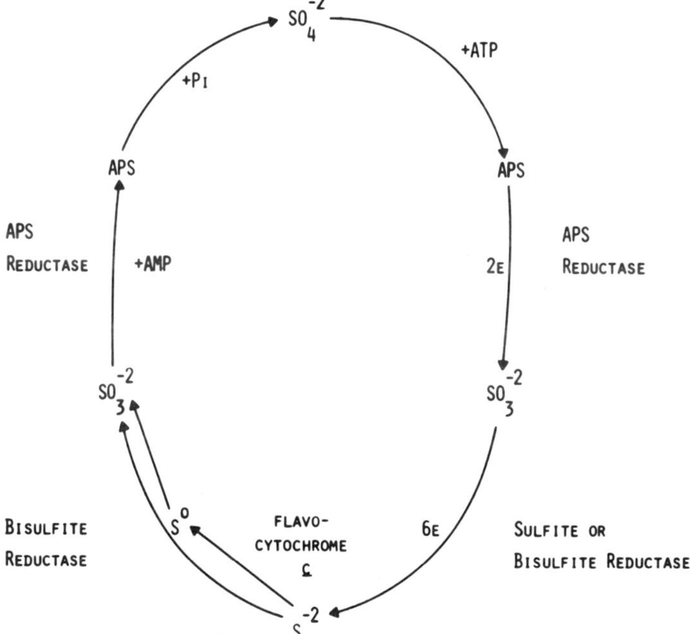

activity (16), but the product has not been described. In other systems, sulfide, elemental sulfur, and polysulfide are oxidized to sulfite but neither the enzymes nor substrates have been clearly defined (17–19).

Thiobacillus denitrificans, grown on thiosulfate and nitrate, contains a sulfite reductase which is similar to those found in the sulfate reducing bacteria (20) and, in the absence of a clearly defined physiological role for the reductase, it has been proposed that this sulfite reductase is responsible for the oxidation of S^{-2} to sulfite (21). Sulfite, once formed, is oxidized to sulfate by one of two mechanisms: a sulfite oxidase or APS reductase which, acting in concert with ADP sulfurylase, can generate ADP. APS reductase is the most extensively characterized flavoprotein exclusively involved in sulfur metabolism and its mechanism will be considered in detail.

APS reductases are found only in three physiological groups of bacteria: the sulfate-reducing bacteria, *Desulfovibrio* and *Desulfotomaculum* (22); in some but not all *Thiobacilli* (23); and in the purple and green sulfur photosynthetic bacteria (24). The properties of the reductases from different microbial types appear to be similar in that they all contain nonheme iron and one noncovalently-bound FAD except for some of the photosynthetic enzymes which contain, in addition, two c-type hemes per molecule. The reductase is soluble and occurs in amounts up to 3% of the total protein; purification is thus relatively easy except for the existence of multiple forms of the enzyme (25). The cofactor composition and some physical properties of APS reductase from *D. vulgaris* are shown in Table 1.

When the reductase is prepared in Tris-maleate buffer or in the presence of 10 mM AMP, it exists in a monomeric form with a molecular weight of about 220,000. Upon SDS electrophoresis, the enzyme exhibits subunits of 70,000 and 20,000 and, based on the molecular weight, it has been proposed that the monomeric enzyme contains three subunits of 70,000 and a single subunit of 20,000. When the enzyme is prepared with phosphate buffer or AMP removed by charcoal, the reductase exhibits at least three multimeric forms and, as shown in Table 1, the dimeric form has a molecular weight of 440,000. The monomeric form of the reductase contains 12 nonheme irons and 12 labile sulfides plus one FAD, and it has been proposed that the three large subunits contain the nonheme iron while the smaller subunit contains the FAD.

The purified enzyme catalyzes six different reactions. In the presence of reduced methyl viologen, APS is reduced to yield SO_3^{-2}, AMP and oxidized

Table 1. Structural Characteristics of Adenylyl Sulfate Reductase from *D. Vulgaris*.

Monomeric molecular weight (Tris-maleate buffer or 10 mM AMP)	218,500
$S_{20,w}$	9.8
Subunit molecular weights	72,000
	20,000
Dimeric molecular weight (phosphate buffer)	439,000
$S_{20,w}$	16.2
Moles iron/220,000 molecular weight	12.7
Moles sulfide/220,000 molecular weight	12.3
Moles FAD/220,000 molecular weight	1.1

methyl viologen (Reaction **1**). The natural electron donor has not yet been determined but this is the direction that is of physiological importance for respiratory sulfate reduction.

(**1**) $$APS+2MV^{+} \rightarrow AMP+SO_3^{-2}+2MV^{++}$$

The second reaction involves the oxidation of SO_2^{-2} in the presence of AMP and $Fe(CN)_6^{-3}$ to APS with the formation of $Fe(CN)_6^{-4}$ (Reaction **2**). In the

(**2**) $$SO_3^{-2}+AMP+2Fe(CN)_6^{-3} \rightarrow APS+2Fe(CN)_6^{-4}$$

oxidation, a high energy phosphosulfate bond is formed and, in the presence of Pi, ADP sulfurylase and adenylate kinase, can lead to the production of ATP. This is the physiologically important reaction in the photosynthetic bacteria and *Thiobacilli*; however, the natural electron acceptor has not been identified.

When APS reductase is free of AMP, the addition of sulfite partially bleaches the enzyme, and the difference spectrum is indicative of the reduction of FAD (Reaction **3**). This spectral change has been postulated to be due to the formation of a flavin-sulfite adduct (26).

(**3**) $$Enzyme+SO_3^{-2} \rightleftharpoons Enzyme\left[SO_3^{-2}\right]$$

The addition of AMP to the sulfite-reacted enzyme results in a further bleaching of the reductase, and the difference spectrum is indicative of the reduction of nonheme iron (Reaction **4**).

(**4**) $$Enzyme\left[SO_3^{-2}\right]+AMP \rightleftharpoons APS+Enzyme\ red$$

In addition to $Fe(CN)_6^{-3}$, the reductase can utilize cytochrome c and O_2 as electron acceptors (Reactions **5** and **6**), but the activities are much lower than those observed with $Fe(CN)_6^{-3}$.

(**5**) $$SO_3^{-2}+AMP+2\ Cyto\ c \rightarrow APS+2\ Cyto\ c\ red$$

As anaerobiosis and superoxide dismutase inhibit the reduction of cytochrome c (Table 2), it was concluded that the formation of APS observed in the presence of O_2 must result in the formation of superoxide.

(**6**) $$SO_3^{-2}+AMP+2O_2 \rightarrow APS+2O_2^{-}$$

The effects of a number of inhibitors on these different activities were investigated in order to establish the functions of FAD and nonheme iron. As shown in Table 3, pHMB was observed to selectively inhibit activity in the ferricyanide assay 55% while activity with cytochrome c and reduced methyl

Table 2. Effects of Anaerobiosis and Superoxide Dismutase on Cytochrome c Assay.

Additions	Inhibition %
Anaerobiosis	100
1×10^{-9}M superoxide dismutase	16.0
5×10^{-9}M superoxide dismutase	52.0
1×10^{-8}M superoxide dismutase	62.0
1×10^{-7}M superoxide dismutase	82.0

Table 3. Effects of pHMB on Various APS Reductase Assays.

Assay	Enzyme treatment	Specific activity (μmoles APS/min/mg)
Ferricyanide	None	3.73
	20 mM pHMB	1.68
Cytochrome c	None	.014
	20 mM pHMB	.014
Methyl viologen	None	1.33
	20 mM pHMB	1.17

viologen was only slightly inhibited, 1 and 12% respectively. Treatment of APS reductase with pHMB also induced a spectra change in the enzyme which, as shown in Figure 3, is suggestive of the removal of nonheme iron. The absorption peaks of the FAD are now clearly discernible and the difference spectrum of the two curves (insert in Figure 3) has a maximum at 390 nm, again suggesting destruction of the nonheme iron centers.

Low-temperature EPR measurements were undertaken in collaboration with Dr. D.V. DerVartanian in order to correlate changes in the oxidation-reduction state of the nonheme iron centers with the spectral and activity changes described here. The EPR spectrum of the isolated reductase free of AMP shows only an asymmetric signal at g=2.00, but no g=1.94 signal characteristic of reduced nonheme iron. In the presence of dithionite, the g=2.0 signal disappears and is replaced by a g=1.94 signal. The normalized signal ampli-

Figure 3. The effect of pHMB on the spectrum of APS reductase (a) spectrum-reductase as isolated; (b) spectrum-20 mM pHMP; (c) Insert—difference spectrum.

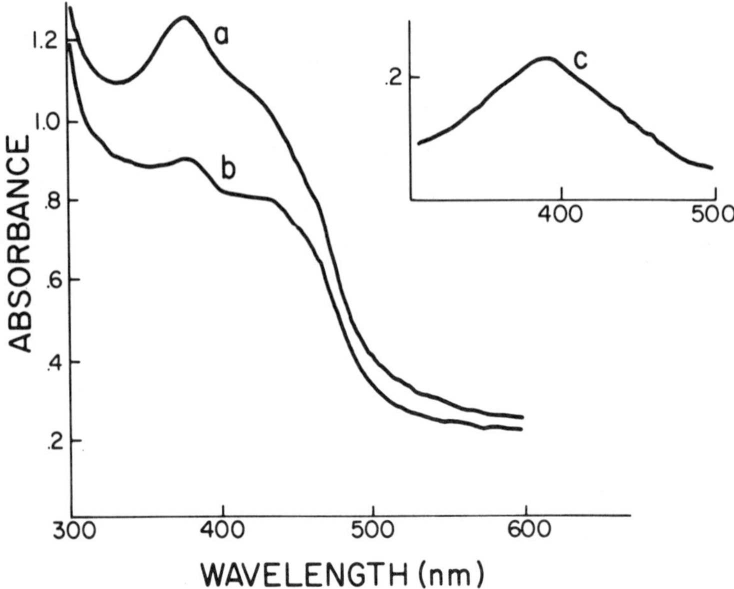

Table 4. EPR Studies on Adenylyl Sulfate Reductase.

Reactant(s)	Normalized signal amplitude at g = 1.94
Isolated enzyme	None
+ Sodium dithionite	139
+ Sulfite	19
+ AMP	None
+ Sulfite + AMP	107
+ AMP + Sulfite	113
pHMB + Sulfite + AMP	None
+ Sulfite + AMP + pHMB	101

tudes at g=1.94 after various additions to APS reductase free of AMP are shown in Table 4. With the addition of sulfite which reacts with FAD or AMP, very little g=1.94 signal was obtained. The small amount of signal obtained with sulfite may reflect the presence of residual AMP. In the presence of SO_3^{-2} plus AMP, a substantial g=1.94 signal is observed (80% of that obtained with dithionite) and the order of addition has little effect on the signal. pHMP-treated enzyme does not show a g=2.0 signal, and the addition of sulfite plus AMP, a g=1.94 signal, is not observed. The loss of ferricyanide activity and spectral changes caused by pHMB are prevented by the presence of AMP plus sulfite (27). Similarly, there is no loss of the g=1.94 signal when SO_3^{-2} and AMP are added prior to pHMB; however, there is a loss of the g=2.0 signal, suggesting that it is not catalytically important. Thus, the results obtained with EPR spectroscopy are consistent with spectral observations made with the reductase.

A general mechanism incorporating the data presented here is shown in Figure 4. Sulfite first reacts with the enzyme bound FAD to form a flavin-sulfite

Figure 4. Mechanism of APS reductase.

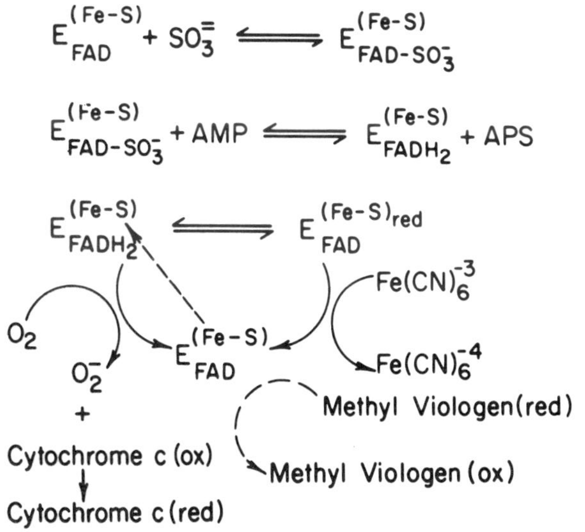

$$E^{(Fe-S)}_{FAD} + SO_3^{=} \rightleftharpoons E^{(Fe-S)}_{FAD-SO_3^-}$$

$$E^{(Fe-S)}_{FAD-SO_3^-} + AMP \rightleftharpoons E^{(Fe-S)}_{FADH_2} + APS$$

$$E^{(Fe-S)}_{FADH_2} \rightleftharpoons E^{(Fe-S)_{red}}_{FAD}$$

$$E^{(Fe-S)}_{FAD}$$

O_2 → O_2^- + Cytochrome c (ox) → Cytochrome c (red)

$Fe(CN)_6^{-3}$ → $Fe(CN)_6^{-4}$

Methyl Viologen(red) → Methyl Viologen (ox)

adduct; the nonheme iron can be either the oxidized or reduced state. The enzyme sulfite complex next reacts with a mononucleotide to yield the corresponding sulfur nucleotide. The reduced flavin can react directly with O_2 to yield O_2^- which can reduce cytochrome c. In the reverse direction, FAD is first reduced by reduced methyl viologen and then reacts with APS to form the flavin-sulfite adduct. The reduced FAD can also reduce the nonheme iron centers of the reductase which can be reoxidized by ferricyanide. It is interesting that only activity in the ferricyanide is dependent on the nonheme iron centers of APS reductase, and it will therefore be of considerable interest to establish the natural electron donor(s) and acceptor(s) in order to determine the role of flavin and nonheme iron of APS reductase in their biological activities.

ACKNOWLEDGMENTS
This work was partially supported by Grants GB-36006X and GB-1489 from the National Science Foundation and by a Grant from the Department of Energy, Contract number DEAS-09-79 ER-10499.

References

1. Postgate, J.R. (1979) *The Sulphate-Reducing Bacteria.* Cambridge University Press.
2. Tuttle, J.H. and Jannasch, H.W. (1973) *J Bacteriol* 732.
3. Pfennig, N. and Biebl, H. (1976) *Arch Microbiol* 110:3.
4. Abram, J.W. and Nedwell, D.B. (1978) *Arch Microbiol* 117:93.
5. Jannasch, H.W. and Wirsen, C.O. (1979) *BioScience* 29:592.
6. Robbins, P.W. and Lipmann, F. (1958) *J Biol Chem* 233:686.
7. Roy, A.B. and Trudinger, P.A. (1970) *The Biochemistry of Inorganic Compounds of Sulfur.* London and New York: Cambridge University Press.
8. Schmidt, A. (1972) *Arch Mikrobiol* 84:77.
9. Tsang, M.L.-S. and Schiff, J.A. (1976) *J Bacteriol* 125:923.
10. Peck, H.D., Jr., Deacon, J.E., and Davidson, J.T. (1965) *Biochim Biophys Acta* 96:429.
11. Trudinger, P.A. (1970) *J Bacteriol* 104:158.
12. Lee, J.P. and Peck, H.D., Jr. (1971) *Biochem Biophys Res Commun* 45:583.
13. Lee, J.P., Yi, C.S., LeGall, J., and Peck, H.D., Jr. (1973) *J Bacteriol* 115:453.
14. Meyer, T.E., Bartsch, R.G., Cusanovich, M.A., and Mathewson, J.H. (1968) *Biochim Biophys Acta* 153:854.
15. Bartsch, R.G. and Kamen, M.D. (1960) *J Biol Chem* 235:825.
16. Kusai, A. and Yamanaka, T. (1973) *Biochim Biophys Acta* 325:304.
17. Suzuki, I. (1965) *Biochim Biophys Acta* 110:97.
18. Adair, F.W. (1966) *J Bacteriol* 92:899.
19. Moriarty, D.J.W. and Nicholas, D.J.D. (1970) *Biochim Biophys Acta* 216:130.
20. Schedel, M., LeGall, J., and Baldensperger, J. (1975) *Arch Microbiol* 105:339.
21. Schedel, M. and Truper, H.G. (1974) *Biochim Biophys Acta* 568:454.
22. Skyring, G.W., Jones, H.E., and Goodchild, D. (1977) *Can J Microbiol* 23:1415.
23. Charles, A.M. and Suzuki, I. (1960) *Biochim Biophys Acta* 128:522.
24. Truper, H.G. and Peck, H.D., Jr. (1965) *Arch Mikrobiol* 73:125.
25. Bramlett, R.N. and Peck, H.D., Jr. (1975) *J Biol Chem* 250:2979.
26. Michaels, G.B., Davidson, J.T., and Peck, H.D., Jr. (1971) In *Flavins and Flavoproteins.* Third International Symposium. H. Kamin (ed.) University Park Press.
27. Bramlett, R.N. and Peck, H.D., Jr., unpublished data.

Published 1982 by Elsevier North Holland, Inc.
Vincent Massey and Charles H. Williams, Editors
Flavins and Flavoproteins

CHAPTER 141

Properties of Cobalt-Substituted D-Lactate Dehydrogenase from *Megasphera elsdenii*

Fraser F. Morpeth[1] and Vincent Massey

Department of Biological Chemistry, University of Michigan Ann Arbor, Michigan

Introduction

The pyridine nucleotide independent D-lactate dehydrogenase from *Megasphera elsdenii* is a zinc dependent flavoprotein. It has been shown to contain one essential zinc and one tightly bound FAD per 55,000 molecular weight subunit (1). The zinc may be selectively removed by extended dialysis against 0.1 M EDTA and activity restored by the addition of a stoichiometric amount of zinc or cobalt but not by any other divalent metal ion tested. In this communication, we report on the properties of the cobalt-substituted enzyme.

Results

Figure 1 shows that the spectral changes of the flavin are very similar on reconstitution of the zinc free enzyme either with zinc or cobalt. However, the cobalt enzyme has a new peak at 550 nm with a molar extinction coefficient of 180 M^{-1} cm^{-1}. This is within the range observed for tetrahedral cobalt (II) complexes (2).

D-lactate dehydrogenase is a fluorescent flavoprotein. Figure 2 shows that the cobalt-substituted enzyme is significantly less fluorescent than the native zinc enzyme. This suggests that the paramagnetic cobalt is very close to the flavin and is quenching its fluorescence.

The EPR spectrum of the oxidized cobalt enzyme is shown in Figure 3A and is characteristic of tetrahedral cobalt. On reduction by D-lactate or photochemically (3) there is a broadening of the cobalt signal (Figure 3B and C). The flavin semiquinone signal shown in Figure 3C cannot be saturated out by increasing the microwave power.

The catalytic properties of the cobalt enzyme differ from those of the native enzyme in several respects. When oxygen is the electron acceptor, there is substantial substrate inhibition by D-lactate at 25°C in potassium phosphate buffer, pH 8, ionic strength=0.23 μ. If potassium ferricyanide is the electron acceptor under the same conditions, the cobalt enzyme shows activation by both D-lactate and ferricyanide. At pH 8, the native zinc enzyme yields linear double reciprocal plots in both assays.

Kinetic coefficients determined from the linear parts of the curves are presented in Table 1 and are expressed using the nomenclature of Dalziel (4).

[1]Present address: School of Molecular Sciences, University of Sussex, Falmer, Brighton BN1 9Q5, England.

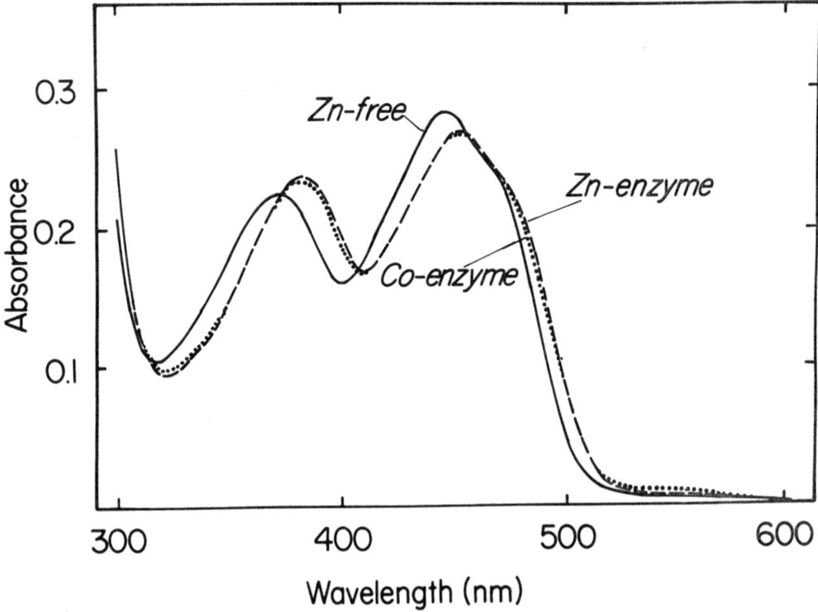

Figure 1. Absorption spectra of the zinc-depleted D-lactate dehydrogenase and the enzyme after titration with Zn^{2+} and Co^{2+}. The spectra were recorded in 0.1 M KPi pH 7 at 25°C. (—) zinc-free enzyme; (---) after addition of 1.08 equivalents of $ZnCl_2$; (···) after addition of 1.14 equivalents of $CoCl_2$ instead of $ZnCl_2$.

Figure 2. Fluorescence emission spectra (exciting at 455 nm) and fluorescence excitation spectra (emitting at 525 nm) of the zinc-depleted, zinc-reconstituted and cobalt-substituted D-lactate dehydrogenase. All spectra were recorded at 25°C 0.1 M KPi pH 7. The concentration of D-lactate dehydrogenase in all cases was 22 μM.

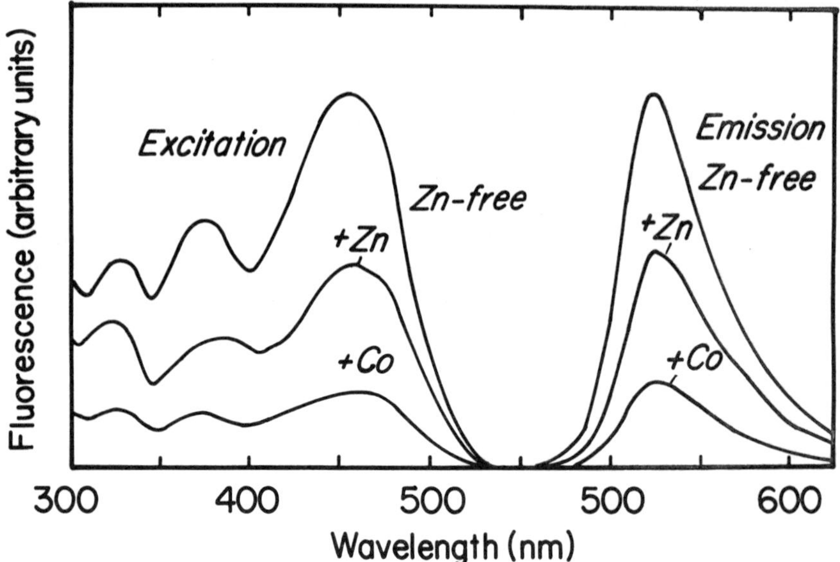

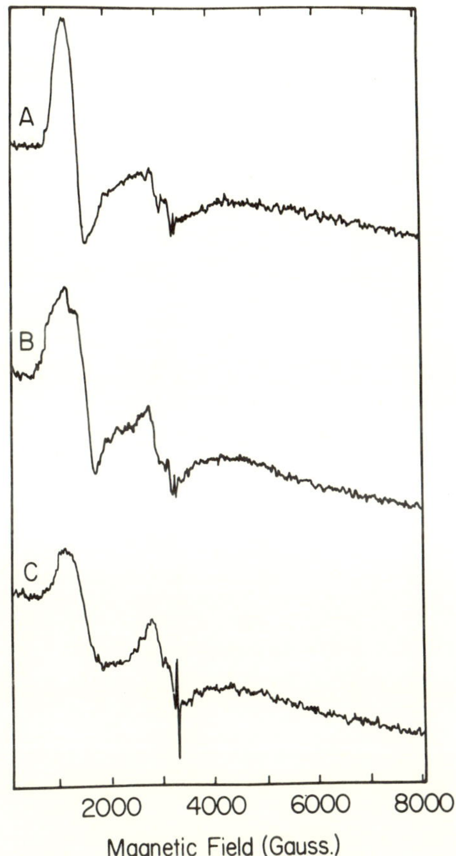

Figure 3. Low temperature EPR spectra of the cobalt substituted D-lactate dehydrogenase in 0.1 M KPi pH 7 at 19°K. (A) Oxidized Cobalt enzyme (0.09 mM). (B) Reduced with 50 mM D-lactate. (C) Cobalt enzyme (0.09 mM) plus 10 mM oxalate and 5 μM 5-deazflavin after irradiation for 30 seconds.

Table 1. Kinetic Coefficients Describing the Oxidation of D-Lactate by Oxygen and Potassium Ferricyanide Catalyzed by Native and Cobalt Substituted D-Lactate Dehydrogenase.

Electron acceptor oxygen	**ϕ_0(min) × 10^3**	**ϕ_1(mM min) × 10^2**	**ϕ_2(mM min) × 10^3**	**ϕ_{12}(mM^2 min)**	**ϕ_1/ϕ_0 mM**	**ϕ_2/ϕ_0 mM**
Native enzyme 25°C	4.5	2.9	1.3	—	6.4	0.278
Cobalt enzyme 25°C	5.7	0.178	1.8	—	—	0.315
Potassium ferricyanide						
Native enzyme 25°C	1.2	1.9	0.13	—	15.8	0.11
4°C	2.7	7.1	1.78	1.6	26.3	0.66
Cobalt enzyme 25°C	1.3	0.055	0.37	—	0.042	0.28
4°C	5.6	0.415	5.6	0.00326	2.4	1

The kinetic coefficients shown are those in the reciprocal rate equation

$$\frac{e}{v} = \phi_0 + \frac{\phi_1}{[S_1]} + \frac{\phi_2}{[S_2]} + \frac{\phi_{12}}{[S_1][S_2]}$$

Where e is the concentration of active sites based on the flavin concentration, S_1 is D-lactate and S_2 is potassium ferricyanide or oxygen. ϕ_1/ϕ_0 is the Michaelis constant for D-lactate and ϕ_2/ϕ_0 is the Michaelis constant for oxygen, potassium ferricyanide. All data were obtained in potassium phosphate buffer pH 8, ionic strength = 0.23 μ.

The activity to flavin ratio (AFR) of native enzyme as purified varies between 100 and 120 using the ferricyanide assay (1). All parameters with the native enzyme are corrected to an AFR of 130. The cobalt enzyme routinely has an AFR of 60–80 but for comparison all coefficients in Table 1 have been corrected to AFR 130 as well.

Inspection of Table 1 shows that ϕ_1 is significantly smaller for the cobalt enzyme. This is not inconsistent with the essential zinc/cobalt being involved in substrate binding and reduction.

ACKNOWLEDGMENTS
We would like to thank Dr. J.A. Fee for performing the EPR experiments. This work was supported by a Grant from the U.S. Public Health Service, GM-11106.

References

1. Olson, S.T. and Massey, V. (1979) *Biochemistry* 18:4714–4724.
2. Cotton, F.A. and Soderberg, R.H. (1962) *J Amer Chem Soc* 84:872.
3. Massey, V. and Hemmerich, P. (1978) *Biochemistry* 17:9–17.
4. Dalziel, K. (1957) *Acta Chem Scand* 11:1706–1723.

Published 1982 by Elsevier North Holland, Inc.
Vincent Massey and Charles H. Williams, Editors
Flavins and Flavoproteins

CHAPTER 142

Kinetic Mechanisms of Unactivated and Lipid-Activated Pyruvate Oxidase

Robert Blake, II[1] and Lowell P. Hager

Biochemistry Department, University of Illinois, Urbana, Illinois

Introduction

Pyruvate oxidase, a peripheral membrane flavoprotein isolated from *E.coli*, catalyzes the oxidative decarboxylation of pyruvate to form acetate and CO_2 (1). The pyruvate:ferricyanide reductase activity of the purified enzyme is stimulated 20 to 50-fold by a variety of amphiphiles (2,3). Initial velocity kinetic experiments (4) have indicated that the catalytic cycle of unactivated pyruvate oxidase can be conveniently represented by three sequential events: (a) decarboxylation of pyruvate to form a transient hydroxyethyl-thiamin pyrophosphate (RCHOH-TPP) intermediate, (b) reduction of enzyme-bound flavin by the RCHOH-TPP, and (c) oxidation of the $FADH_2$ by exogenous electron acceptors. These events constitute the catalytic cycle designated as Cycle I in Figure 1. Several of the steady-state kinetic patterns are altered when the enzyme is activated by an appropriate amphiphile (such as sodium dodecyl sulfate, SDS), and the events in Cycle II were added to explain the observations (4). According to the mechanism in Figure 1, two events can occur once the enzyme-bound FAD has been reduced by the RCHOH-TPP intermediate. The $FADH_2$ can either be oxidized by an electron acceptor (completing Cycle I) or the E^r species can decarboxylate another molecule of pyruvate to produce the $E^r \cdot$RCHOH-TPP complex (thus entering Cycle II). Unfortunately, due to the complexity of this kinetic mechanism, no firm conclusions could be reached from the steady state kinetic experiments about the identity of the step(s) which is(are) accelerated upon lipid activation of the enzyme. The rapid mixing kinetic experiments summarized below were performed to approach this question directly, as well as to verify the existence of Cycle II.

Results and Discussion

The mechanism of pyruvate oxidase is readily amenable to rapid mixing kinetic analysis. Cycle I can be conveniently divided into two partial reactions which can be studied individually in a stopped-flow spectrophotometer. One partial reaction results in reduction of the enzyme-bound flavin; the other partial reaction results in oxidation of the enzyme-bound flavin. The flavin

[1]Present Address: Department of Biological Chemistry, University of Michigan, Ann Arbor, Michigan.

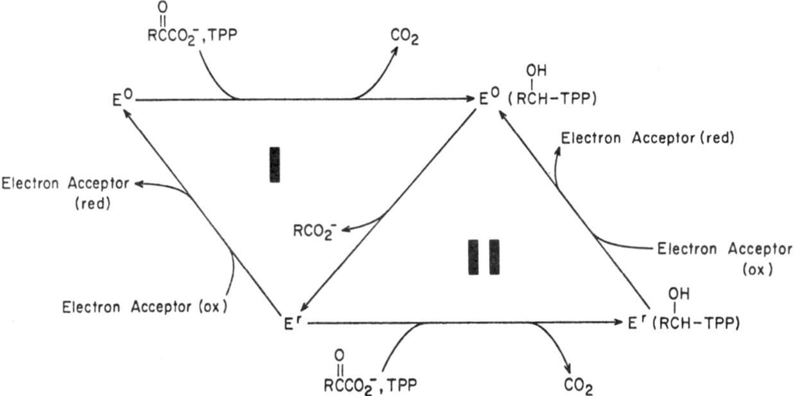

Figure 1. Kinetic mechanism proposed for the oxidative decarboxylation of α-keto acids catalyzed by pyruvate oxidase. The oxidized and reduced forms of the enzyme are represented by E^{o} and E^{r}, respectively. Thiamin pyrophosphate is abbreviated as TPP.

chromophore is an intrinsic spectrophotometric probe through which the transient oxidized and reduced states of the enzyme can be monitored with high sensitivity. The reductive partial reaction was studied by rapidly mixing the oxidized flavoprotein with pyruvate in the presence of TPP. The reduction of the flavin by pyruvate could be described mathematically as a single exponential function of time with apparent second-order rate constants of 40 $M^{-1}s^{-1}$ and 6500 $M^{-1}s^{-1}$ for unactivated and SDS-activated enzyme, respectively. The oxidative partial reaction was studied by rapidly mixing the reduced flavoprotein with an oxidized electron acceptor such as ferricyanide. The oxidation of the reduced flavin by ferricyanide could also be described as a single exponential function of time with apparent second-order rate constants of 1.5×10^{5} $M^{-1}s^{-1}$ and 8.3×10^{4} $M^{-1}s^{-1}$ for unactivated and SDS-activated enzyme, respectively.

A comparison of the above data with the turnover numbers of both forms of the enzyme under the experimental conditions of the steady state assay (2, 3) reveals that the rate-limiting step of the unactivated enzyme occurs in the reductive partial reaction, while the rate-limiting step of the SDS-activated enzyme is the oxidative partial reaction. Activation of the enzyme by SDS stimulates some event in the reductive partial reaction to such an extent that the reductive partial reaction is no longer rate-limiting.

The reductive partial reaction can be further subdivided into two steps: decarboxylation of the pyruvate to form the RCHOH-TPP intermediate and the subsequent two-electron reduction of the flavin by this intermediate. The contribution of the decarboxylation step to the overall rate of reduction measured in the stopped-flow spectrophotometer was determined by measuring the amount of CO_2 released from (1-^{14}C) pyruvic acid by SDS-activated enzyme under single turnover conditions using a rapid mixing chemical quench apparatus. Figure 2 shows the number of moles of CO_2 released per mole of flavin as a function of time; the ratio approaches 2 at longer time periods. The mechanism in Figure 1 correctly predicts that two equivalents of CO_2 should

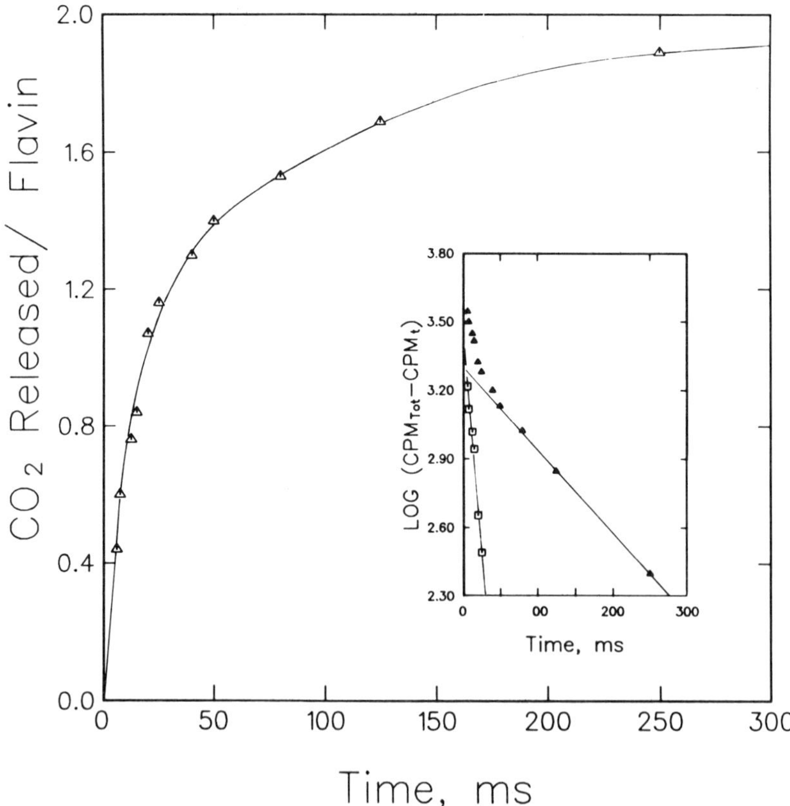

Figure 2. Stoichiometry and kinetics of the decarboxylation of pyruvate catalyzed by SDS-activated pyruvate oxidase. The release of CO_2 from (1-^{14}C) pyruvate was measured in the absence of an electron acceptor using a Precision Syringe-Ram system, Model VI-1001, built by Update Instruments, Inc. One drive syringe was filled with a solution containing 0.45 mg/ml oxidized pyruvate oxidase, 200 μM TPP, and 500 μM SDS. Another drive syringe was filled with a solution of (1-^{14}C) pyruvate (16 mM). Both solutions were buffered with 0.1 M sodium phosphate at pH 6.0 and contained 0.01 M $MgCl_2$. A third drive syringe was filled with 0.1 M NaOH (the quenching reagent). Activation of the ram forced the enzyme and the pyruvate solutions through a mixing chamber, along an aging hose, and into another mixing chamber into which the quenching reagent was also being pushed. The quenched enzyme reaction was then ejected from the apparatus. The radioactive CO_2 in the reaction mixture was released by acidification, trapped by hyamine hydroxide, and counted in a Beckman LS-230 Liquid Scintillation Counter.

be observed per flavin in the absence of an electron acceptor. The appearance of CO_2 can be described mathematically as the sum of two exponential functions of time, as shown by the nonlinear semilogarithmic plot in the inset of Figure 2. The rate constant for the second, slower process was obtained from the linear portion of the curved semilogarithmic plot, while the rate constant for the first, faster process was obtained by subtracting the contribution of the slow process from the primary data and then plotting the resulting difference in semilogarithmic form. The data in Figure 2 indicate that the first decarboxylation even catalyzed by the enzyme occurs faster than the second decarboxylation event. The redox state of the flavin appears to influence the rate of decarboxylation of pyruvate by the enzyme. Both the fast and the slow rate constants thus obtained were linear functions of the pyruvate concentration, as shown in Figure 3. Also shown in Figure 3 are the observed rate constants for the reduction of the enzyme-bound flavin obtained in the stopped-flow spectrophotometer under the same experimental conditions as those employed in the chemical quench experiments. Since both the first decarboxylation step and the overall reduction of the flavin by pyruvate display identical kinetic behavior, the slowest event in the reductive partial reaction must be the decarboxylation of pyruvate. From the above data and related observations, we conclude that it is some event in the α-keto acid

Figure 3. Comparison of the rates of decarboxylation of pyruvate with the rates of reduction of the flavin by pyruvate. (Triangles) the rate constants for the first decarboxylation event catalyzed by the enzyme. (Diamonds) the rate constants for the second decarboxylation event catalyzed by the enzyme. (Squares) the rate constants for the reduction of the flavin by pyruvate. The decarboxylation rate constants were obtained as described in Figure 2 and the text. Reduction of the flavin was monitored at 438 nm in a standard Gibson-Durrum stopped-flow spectrophotometer.

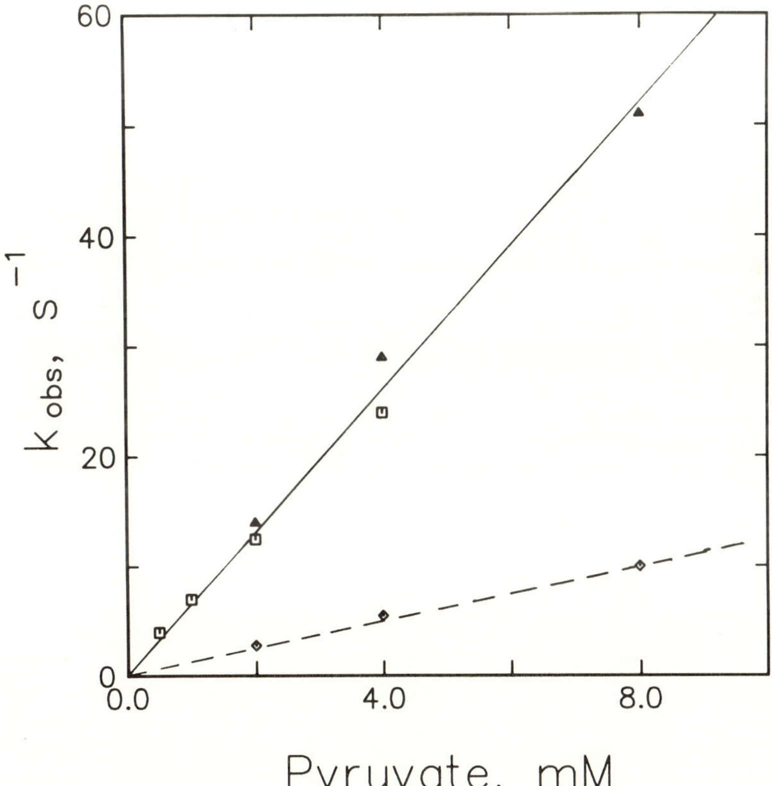

decarboxylation step which is greatly accelerated upon activation of the enzyme by amphiphiles.

ACKNOWLEDGMENTS
This research was supported by National Institutes of Health Grant GM 7768 and National Institutes of Health Predoctoral Trainee Grant GM 07283.

References

1. Hager, L.P. (1957) *J Biol Chem* 229:251–363.
2. Cunningham, C.C. and Hager, L.P. (1971) *J Biol Chem* 246:1575–1582.
3. Blake, R.C., II, Hager, L.P., and Gennis, R.B. (1978) *J Biol Chem* 253:1963–1971.
4. Blake, R.C., II (1977) PhD Thesis, The University of Illinois, Urbana, Illinois.

Published 1982 by Elsevier North Holland, Inc.
Vincent Massey and Charles H. Williams, Editors
Flavins and Flavoproteins

CHAPTER 143

Mechanism of *E.coli* Pyruvate Oxidase

Michael Mather,† Larry Schopfer,¶ Vincent Massey,¶ and Robert B. Gennis†

†*Department of Chemistry, University of Illinois, 505 S. Mathews Avenue, Urbana, Illinois;* ¶*Department of Biochemistry, University of Michigan, Ann Arbor, Michigan*

Introduction

Pyruvate oxidase is a flavoprotein electron transferase which couples the oxidation of pyruvate to the *E.coli* aerobic respiratory chain. The enzyme catalyzes the oxidative decarboxylation of pyruvate to acetate and CO_2 (1). Pyruvate oxidase is loosely attached to the inner side of the cytoplasmic membrane. Physical disruption of the cell dislodges the enzyme, allowing its purification as a water soluble protein (2). The purified enzyme consists of four identical subunits, each containing a tightly bound molecule of FAD. A second cofactor, Mg^{++}-thiamin pyrophosphate, is also required for catalytic activity. The natural primary electron acceptor has not yet been identified; *in vitro* activity is measured using artificial acceptors such as ferricyanide.

The addition of lipids to substrate-reduced pyruvate oxidase produces a dramatic activation of the enzyme, with a 20- to 30-fold increase in the turnover number and an order of magnitude decrease in the Michaelis constant for the substrate (3). A number of other changes in the enzyme have been observed upon reduction of the flavin by substrate, including enhanced affinity for lipids and detergents, exposure of a new region of the protein to endoproteases, and alteration of the solution properties of the protein toward those expected of an integral membrane protein (e.g., tendency to aggregate) (4–6).

Results and Discussion

The catalytic cycle of pyruvate oxidase is shown in Figure 1. The initial steps proceed according to the Breslow mechanism for thiamin pyrophosphate-requiring enzymes (7), leading to the formation of a nucleophile, intermediate IV, at the active site. Rapid kinetics experiments were performed in an attempt to determine which step(s) are accelerated upon activation of the enzyme. The rate of decarboxylation following the rapid mixing of substrate with enzyme was found to be equal to the minimum possible rate calculated from steady state kinetics data (see Table 1). Coupled with the absence of a carbon-13 isotope effect for CO_2 release (O'Leary, M.H., unpublished results), this result leads to the conclusion that the rate-determining process in the catalytic cycle of the unactivated enzyme precedes decarboxylation (step 3, Figure 1). Thus, the rate-determining process, which must be accelerated upon activation,

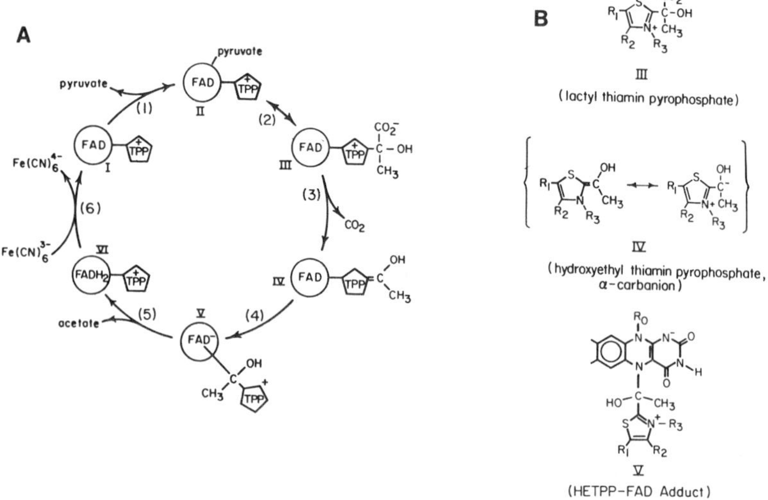

Figure 1. Catalytic cycle of pyruvate oxidase *in vitro*. (A) The catalytic cycle, with intermediate species identified by Roman numerals and reaction steps by Arabic numerals in parentheses. The circles represent a pyruvate oxidase subunit with a tightly bound flavin adenine dinucleotide (FAD) cofactor. The pentagons represent the thiamin pyrophosphate (TPP) cofactor, present at a saturating concentration. (B) More detailed structures of proposed intermediate species in the reaction cycle.

$R_1 = -C_2H_4 - P_2O_7^{3-}$

$R_2 = CH_3-$

$R_3 =$ NH$_2$, CH$_2$, N, N, CH$_3$

R_0 = adenine ribosyl group

occurs during steps 1 or 2 of the catalytic cycle as drawn in Figure 1. Table 1 summarizes the kinetic constants which have been determined.

One reasonable mechanism for the ensuing reduction of FAD is the addition of the nucleophilic intermediate IV to position N-5 of FAD to form intermediate V. Photoreduction and flavin replacement experiments provide some support for this mechanism: (1) Upon photoreduction in the presence of 5-deazaflavin, the anionic semiquinone of FAD is formed (Figure 2), in contrast to the neutral semiquinone observed with nicotinamide-nucleotide-dependent dehydrogenases. (2) Upon reconstitution of apopyruvate oxidase with 8-mercapto-FAD and 6-hydroxy-FAD, the benzoquinoid forms of these flavin derivatives are stabilized by the enzyme (Figure 2), again in contrast to the nicotinamide-dependent dehydrogenases. These same forms of the semiquinone and flavin derivatives are stabilized by the flavoprotein oxidases, for which a large body of evidence (reviewed in [8]) points to reduction via attack

Table 1. Summary of Rapid and Steady-State Kinetics Data.[a]

Activator	Rapid mixing experiments			Steady state parameters[b]	
	CO_2 Evolution[c]	Flavin reduction[d]	Flavin oxidation[d]	$k_{cat}/K_{0.5}$ (pyruvate)	$k_{cat}/K_{0.5}$ (ferricyanide)
None	69	67	150,000	70	7,500
Phosphatidylglycerol		>7,200	70,000	15,000	50,000
Sodium dodecyl sulfate		>6,400	83,000	13,000	60,000

[a]All results are expressed as second-order rate constants in units of $M^{-1}sec^{-1}$.

[b]These constants represent the minimum theoretical values for any second-order rate constant characterizing a portion of the reductive half (pyruvate) and the oxidative half (ferricyanide) of the catalytic cycle. k_{cat} is the catalytic center activity or turnover number per FAD. $K_{0.5}$ is the concentration of the indicated species at which half maximal activity is observed. Typical experimental procedures are described in (3).

[c]Chemical quench experiment. A solution containing the enzyme was rapidly mixed with a second solution containing (1-^{14}C)-pyruvate using a Precision Syringe-Ram System, Model VI-1001, (Update Instruments, Inc.). After a predetermined period of time, the reaction mixture was quenched by mixing with 0.2 N NaOH. The quenched reaction mixture was ejected into a Warburg flask, which was then sealed. 6 M H_2SO_4 was then injected into the mixture to liberate $^{14}CO_2$ produced by the enzymatic reaction. This $^{14}CO_2$ is trapped by phenethylamine which was placed in vial in the center well of the Warburg flask. After 40 mintues, the Warburg flask is opened and the trapping solution is removed for scintillation counting.

[d]Stopped-flow experiment. Experimental procedures are described in (9).

Figure 2. Absorption spectra of native and modified pyruvate oxidase. (Left hand panel) native enzyme in the oxidized, semiquinone, and fully reduced forms, all in 0.1 M sodium phosphate buffer, pH 7.0, 20°. (Right hand panel) spectra of pyruvate oxidase in which the FAD has been replaced by 8-mercapto FAD and 6-hydroxy FAD, both in 0.1 M sodium phosphate buffer, pH 6.0. Pyruvate oxidase containing modified flavins was prepared by incubating freshly prepared apoenzyme with an excess of the FAD analog for 40 min at 20°C and pH 6.0. The modified pyruvate oxidase was then separated from the excess FAD analog by gel filtration. Apopyruvate oxidase was prepared by a procedure similar to the modified method of Strittmatter (10).

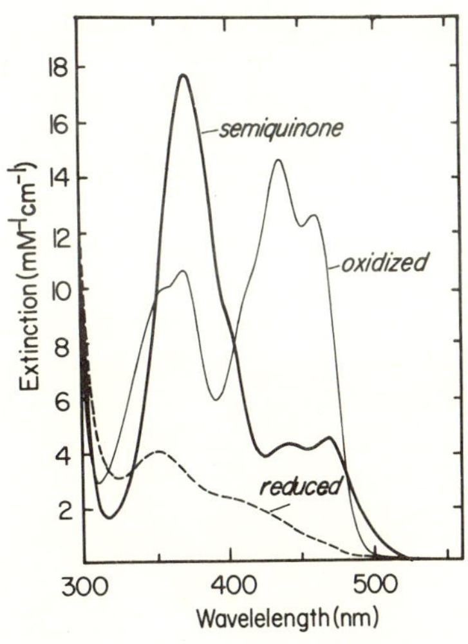

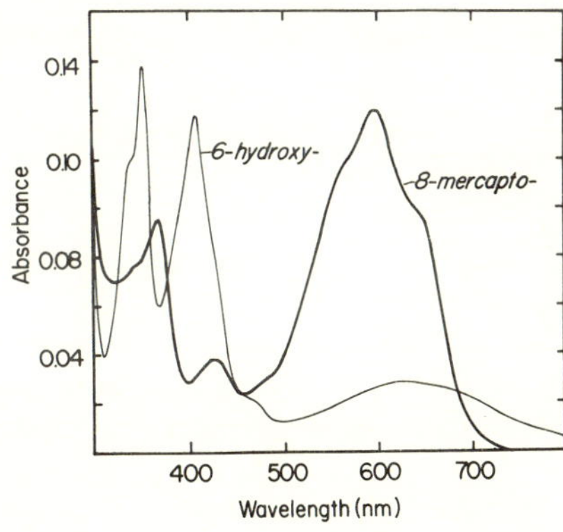

at position N-5. Perhaps class 4 flavoproteins (dehydrogenase/electron transferases) (8) should be divided into two subclasses, (a) the nicotinamide-linked dehydrogenases, which stabilize blue radicals, and (b) the nucleophile-forming dehydrogenases, such as pyruvate oxidase and succinate dehydrogenase, which stabilize red radicals.

ACKNOWLEDGMENTS
Thanks are due Mr. Michael Recny for assistance in developing a method for the preparation of apopyruvate oxidase. This work was supported by NIH Grant HL-16101 (to RBG).

References

1. Hager, L.P. (1957) *J Biol Chem* 229:251–263.
2. O'Brien, T.A., Schrock, H.L., Russell, P., Blake, R., and Gennis, R.B. (1976) *Biochim Biophys Acta* 452:13–29.
3. Blake, R., Hager, L.P., and Gennis, R.B. (1978) *J Biol Chem* 253:1963–1971.
4. Schrock, H.L. and Gennis, R.B. (1980) *Biochim Biophys Acta* 614:215–220.
5. Schrock, H.L. and Gennis, R.B. (1977) *J Biol Chem* 252:5990–5995.
6. Russell, P., Hager, L.P., and Gennis, R.B. (1977) *J Biol Chem* 252:7877–7882.
7. Hager, L.P. and Krampitz, L.O. (1963) *Fed Proc* 22:536.
8. Massey, V. and Hemmerich, P. (1980) *Biochenm Soc Trans* 8:246–257.
9. Blake, R. (1977) PhD Dissertation, University of Illinois, Urbana, Illinois.
10. Husain, M. and Massey, V. (1978) *Methods Enzymol* 53:429–437.

Author Index

Subject Index

N

O

P

Q

R

S

ABERYSTWYTH
LLYFRGELL
C.P.C.